Lecture Notes in Computer Science 4558

Commenced Publication in 1973
Founding and Former Series Editors:
Gerhard Goos, Juris Hartmanis, and Jan van Leeuwen

Michael J. Smith Gavriel Salvendy (Eds.)

Human Interface and the Management of Information

Interacting in Information Environments

Symposium on Human Interface 2007
Held as Part of HCI International 2007
Beijing, China, July 22-27, 2007
Proceedings, Part II

Springer

Volume Editors

Michael J. Smith
University of Wisconsin-Madison
Department of Industrial and Systems Engineering
2166 Engineering Centers Bldg., 1550 Engineering Drive, Madison, WI 53706, USA
E-mail: mjsmith@engr.wisc.edu

Gavriel Salvendy
Purdue University, Department of Industrial Engineering
Grissom Hall, 315 N. Grant St., West Lafayette, IN 47907-2023, USA
E-mail: salvendy@purdue.edu

Library of Congress Control Number: 2007930201

CR Subject Classification (1998): H.5.2-4, H.3-5, C.2, K.4, D.2, K.3

LNCS Sublibrary: SL 3 – Information Systems and Application, incl. Internet/Web and HCI

ISSN	0302-9743
ISBN-10	3-540-73353-1 Springer Berlin Heidelberg New York
ISBN-13	978-3-540-73353-9 Springer Berlin Heidelberg New York

Springer is a part of Springer Science+Business Media

springer.com

Printed in Germany

Typesetting: Camera-ready by author, data conversion by Scientific Publishing Services, Chennai, India
Printed on acid-free paper SPIN: 12083534 06/3180 5 4 3 2 1 0

Foreword

The 12th International Conference on Human-Computer Interaction, HCI International 2007, was held in Beijing, P.R. China, 22-27 July 2007, jointly with the Symposium on Human Interface (Japan) 2007, the 7th International Conference on Engineering Psychology and Cognitive Ergonomics, the 4th International Conference on Universal Access in Human-Computer Interaction, the 2nd International Conference on Virtual Reality, the 2nd International Conference on Usability and Internationalization, the 2nd International Conference on Online Communities and Social Computing, the 3rd International Conference on Augmented Cognition, and the 1st International Conference on Digital Human Modeling.

A total of 3403 individuals from academia, research institutes, industry and governmental agencies from 76 countries submitted contributions, and 1681 papers, judged to be of high scientific quality, were included in the program. These papers address the latest research and development efforts and highlight the human aspects of design and use of computing systems. The papers accepted for presentation thoroughly cover the entire field of Human-Computer Interaction, addressing major advances in knowledge and effective use of computers in a variety of application areas.

This volume, edited by Michael J. Smith and Gavriel Salvendy, contains papers in the thematic area of Human Interface and the Management of Information, addressing the following major topics:

- Communication and Collaboration
- Knowledge, Learning and Education
- Mobile Interaction
- Interacting with the World Wide Web and Electronic Services
- Business Management and Industrial Applications
- Environment, Transportation and Safety

The remaining volumes of the HCI International 2007 proceedings are:

- Volume 1, LNCS 4550, Interaction Design and Usability, edited by Julie A. Jacko
- Volume 2, LNCS 4551, Interaction Platforms and Techniques, edited by Julie A. Jacko
- Volume 3, LNCS 4552, HCI Intelligent Multimodal Interaction Environments, edited by Julie A. Jacko
- Volume 4, LNCS 4553, HCI Applications and Services, edited by Julie A. Jacko
- Volume 5, LNCS 4554, Coping with Diversity in Universal Access, edited by Constantine Stephanidis
- Volume 6, LNCS 4555, Universal Access to Ambient Interaction, edited by Constantine Stephanidis
- Volume 7, LNCS 4556, Universal Access to Applications and Services, edited by Constantine Stephanidis

- Volume 8, LNCS 4557, Methods, Techniques and Tools in Information Design, edited by Michael J. Smith and Gavriel Salvendy
- Volume 10, LNCS 4559, HCI and Culture, edited by Nuray Aykin
- Volume 11, LNCS 4560, Global and Local User Interfaces, edited by Nuray Aykin
- Volume 12, LNCS 4561, Digital Human Modeling, edited by Vincent G. Duffy
- Volume 13, LNAI 4562, Engineering Psychology and Cognitive Ergonomics, edited by Don Harris
- Volume 14, LNCS 4563, Virtual Reality, edited by Randall Shumaker
- Volume 15, LNCS 4564, Online Communities and Social Computing, edited by Douglas Schuler
- Volume 16, LNAI 4565, Foundations of Augmented Cognition 3rd Edition, edited by Dylan D. Schmorrow and Leah M. Reeves
- Volume 17, LNCS 4566, Ergonomics and Health Aspects of Work with Computers, edited by Marvin J. Dainoff

I would like to thank the Program Chairs and the members of the Program Boards of all Thematic Areas, listed below, for their contribution to the highest scientific quality and the overall success of the HCI International 2007 Conference.

Ergonomics and Health Aspects of Work with Computers

Program Chair: Marvin J. Dainoff

Arne Aaras, Norway
Pascale Carayon, USA
Barbara G.F. Cohen, USA
Wolfgang Friesdorf, Germany
Martin Helander, Singapore
Ben-Tzion Karsh, USA
Waldemar Karwowski, USA
Peter Kern, Germany
Danuta Koradecka, Poland
Kari Lindstrom, Finland
Holger Luczak, Germany
Aura C. Matias, Philippines
Kyung (Ken) Park, Korea
Michelle Robertson, USA
Steven L. Sauter, USA
Dominique L. Scapin, France
Michael J. Smith, USA
Naomi Swanson, USA
Peter Vink, The Netherlands
John Wilson, UK

Human Interface and the Management of Information

Program Chair: Michael J. Smith

Lajos Balint, Hungary
Gunilla Bradley, Sweden
Hans-Jörg Bullinger, Germany
Alan H.S. Chan, Hong Kong
Klaus-Peter Fähnrich, Germany
Michitaka Hirose, Japan
Yoshinori Horie, Japan
Robert Proctor, USA
Youngho Rhee, Korea
Anxo Cereijo Roibás, UK
Francois Sainfort, USA
Katsunori Shimohara, Japan
Tsutomu Tabe, Japan
Alvaro Taveira, USA

Human-Computer Interaction

Program Chair: Julie A. Jacko

Engineering Psychology and Cognitive Ergonomics

Program Chair: Don Harris

Universal Access in Human-Computer Interaction

Program Chair: Constantine Stephanidis

Jerzy Charytonowicz, Poland
Pier Luigi Emiliani, Italy
Michael Fairhurst, UK
Gerhard Fischer, USA
Jon Gunderson, USA
Andreas Holzinger, Austria
Arthur Karshmer, USA
Simeon Keates, USA
George Kouroupetroglou, Greece
Jonathan Lazar, USA
Seongil Lee, Korea
Ben Shneiderman, USA
Christian Stary, Austria
Hirotada Ueda, Japan
Jean Vanderdonckt, Belgium
Gregg Vanderheiden, USA
Gerhard Weber, Germany
Harald Weber, Germany
Toshiki Yamaoka, Japan
Mary Zajicek, UK
Panayiotis Zaphiris, UK

Virtual Reality

Program Chair: Randall Shumaker

Terry Allard, USA
Pat Banerjee, USA
Robert S. Kennedy, USA
Heidi Kroemker, Germany
Ben Lawson, USA
Ming Lin, USA
Bowen Loftin, USA
Holger Luczak, Germany
Annie Luciani, France
Gordon Mair, UK
Ulrich Neumann, USA
Albert "Skip" Rizzo, USA
Lawrence Rosenblum, USA
Dylan Schmorrow, USA
Kay Stanney, USA
Susumu Tachi, Japan
John Wilson, UK
Wei Zhang, P.R. China
Michael Zyda, USA

Usability and Internationalization

Program Chair: Nuray Aykin

Genevieve Bell, USA
Alan Chan, Hong Kong
Apala Lahiri Chavan, India
Jori Clarke, USA
Pierre-Henri Dejean, France
Susan Dray, USA
Paul Fu, USA
Emilie Gould, Canada
Sung H. Han, South Korea
Veikko Ikonen, Finland
Richard Ishida, UK
Esin Kiris, USA
Tobias Komischke, Germany
Masaaki Kurosu, Japan
James R. Lewis, USA
Rungtai Lin, Taiwan
Aaron Marcus, USA
Allen E. Milewski, USA
Patrick O'Sullivan, Ireland
Girish V. Prabhu, India
Kerstin Röse, Germany
Eunice Ratna Sari, Indonesia
Supriya Singh, Australia
Serengul Smith, UK
Denise Spacinsky, USA
Christian Sturm, Mexico
Adi B. Tedjasaputra, Singapore
Myung Hwan Yun, South Korea
Chen Zhao, P.R. China

Online Communities and Social Computing

Program Chair: Douglas Schuler

Chadia Abras, USA
Lecia Barker, USA
Amy Bruckman, USA
Peter van den Besselaar, The Netherlands
Peter Day, UK
Fiorella De Cindio, Italy
John Fung, P.R. China
Michael Gurstein, USA
Tom Horan, USA
Piet Kommers, The Netherlands
Jonathan Lazar, USA
Stefanie Lindstaedt, Austria
Diane Maloney-Krichmar, USA
Isaac Mao, P.R. China
Hideyuki Nakanishi, Japan
A. Ant Ozok, USA
Jennifer Preece, USA
Partha Pratim Sarker, Bangladesh
Gilson Schwartz, Brazil
Sergei Stafeev, Russia
F.F. Tusubira, Uganda
Cheng-Yen Wang, Taiwan

Augmented Cognition

Program Chair: Dylan D. Schmorrow

Kenneth Boff, USA
Joseph Cohn, USA
Blair Dickson, UK
Henry Girolamo, USA
Gerald Edelman, USA
Eric Horvitz, USA
Wilhelm Kincses, Germany
Amy Kruse, USA
Lee Kollmorgen, USA
Dennis McBride, USA
Jeffrey Morrison, USA
Denise Nicholson, USA
Dennis Proffitt, USA
Harry Shum, P.R. China
Kay Stanney, USA
Roy Stripling, USA
Michael Swetnam, USA
Robert Taylor, UK
John Wagner, USA

Digital Human Modeling

Program Chair: Vincent G. Duffy

Norm Badler, USA
Heiner Bubb, Germany
Don Chaffin, USA
Kathryn Cormican, Ireland
Andris Freivalds, USA
Ravindra Goonetilleke, Hong Kong
Anand Gramopadhye, USA
Sung H. Han, South Korea
Pheng Ann Heng, Hong Kong
Dewen Jin, P.R. China
Kang Li, USA
Zhizhong Li, P.R. China
Lizhuang Ma, P.R. China
Timo Maatta, Finland
J. Mark Porter, UK
Jim Potvin, Canada
Jean-Pierre Verriest, France
Zhaoqi Wang, P.R. China
Xiugan Yuan, P.R. China
Shao-Xiang Zhang, P.R. China
Xudong Zhang, USA

In addition to the members of the Program Boards above, I also wish to thank the following volunteer external reviewers: Kelly Hale, David Kobus, Amy Kruse, Cali Fidopiastis and Karl Van Orden from the USA, Mark Neerincx and Marc Grootjen from the Netherlands, Wilhelm Kincses from Germany, Ganesh Bhutkar and Mathura Prasad from India, Frederick Li from the UK, and Dimitris Grammenos, Angeliki Kastrinaki, Iosif Klironomos, Alexandros Mourouzis, and Stavroula Ntoa from Greece.

This conference could not have been possible without the continuous support and advise of the Conference Scientific Advisor, Gavriel Salvendy, as well as the dedicated work and outstanding efforts of the Communications Chair and Editor of HCI International News, Abbas Moallem, and of the members of the Organizational Board from P.R. China, Patrick Rau (Chair), Bo Chen, Xiaolan Fu, Zhibin Jiang, Congdong Li, Zhenjie Liu, Mowei Shen, Yuanchun Shi, Hui Su, Linyang Sun, Ming Po Tham, Ben Tsiang, Jian Wang, Guangyou Xu, Winnie Wanli Yang, Shuping Yi, Kan Zhang, and Wei Zho.

I would also like to thank for their contribution towards the organization of the HCI International 2007 Conference the members of the Human Computer Interaction Laboratory of ICS-FORTH, and in particular Margherita Antona, Maria Pitsoulaki, George Paparoulis, Maria Bouhli, Stavroula Ntoa and George Margetis.

Constantine Stephanidis
General Chair, HCI International 2007

HCI International 2009

The 13th International Conference on Human-Computer Interaction, HCI International 2009, will be held jointly with the affiliated Conferences in San Diego, California, USA, in the Town and Country Resort & Convention Center, 19-24 July 2009. It will cover a broad spectrum of themes related to Human Computer Interaction, including theoretical issues, methods, tools, processes and case studies in HCI design, as well as novel interaction techniques, interfaces and applications. The proceedings will be published by Springer. For more information, please visit the Conference website: http://www.hcii2009.org/

General Chair
Professor Constantine Stephanidis
ICS-FORTH and University of Crete
Heraklion, Crete, Greece
Email: program@hcii2009.org

Table of Contents

Part I: Communication and Collaboration

Part II: Knowledge, Learning and Education

Part III: Mobile Interaction

Part IV: Interacting with the World Wide Web and Electronic Services

Part VI: Environment, Transportation and Safety

Part I

Communication and Collaboration

The Karst Collaborative Workspace for Analyzing and Annotating Scientific Datasets

Linn Marks Collins[1], Diana E. Northup[2], Mark L.B. Martinez[1], Johannes Van Reenen[2], M. Alex Baker[1], Christy R. Crowley[2], James E. Powell[1], Brian Freels-Stendel[2], Susan K. Heckethorn[1], and Jong Chun Park[2]

[1] Los Alamos National Laboratory, Los Alamos, NM 87545 USA
{linn, mlbm, bakera, jepowell, heckethorn}@lanl.gov
[2] University of New Mexico, Albuquerque, NM 87131 USA
{dnorthup, jreenen, ccrowley, bfreels}@unm.edu, joumon@gmail.com

Abstract. Scientific fields of study such as astrobiology, nanotechnology, and cave and karst science involve the study of images and associated biological, physiochemical, and geological data. In order to ensure interdisciplinary analysis, it is important to make these kinds of datasets available for analysis and curation by the scientific community. The goal of this project is to design and develop an online workspace that enables scientists to collaboratively view, analyze, and annotate such datasets. The prototype contains scanning electron micrographs of karst and cave samples. The target users are the interdisciplinary community of scientists who study karst samples to learn more about critical biological and geological processes and the microbial communities often found in karst terrain. The prototype can inform the design and development of collaborative workspaces in other interdisciplinary fields.

1 Introduction

Collaboration tools have been integral to eScience projects since 1989, when the National Science Foundation in the US began using the phrase "collaboratory" to describe a virtual collaborative research center, or "a center without walls, in which the nation's researchers can perform their research without regard to geographical location." [1] As eScience has evolved, the nature of the collaborative work performed by scientists and their requirements for collaboration tools have evolved, as well. For example, recent reports by the National Science Foundation [2] and the National Science Board [3] describe three kinds of datasets: research, resource, and reference. The distinctions among the three relate to who publishes the datasets and who has access to them. Research datasets are available to the immediate research group, resource datasets are available to a more extended community of researchers, and reference datasets are curated, archived, and available to the scientific community as a whole. In order to ensure that the appropriate datasets can be curated and archived by scientists, it is necessary to provide them with user-friendly tools for performing these tasks.

M.J. Smith, G. Salvendy (Eds.): Human Interface, Part II, HCII 2007, LNCS 4558, pp. 3–12, 2007.

Karst science provides a particularly good context for designing, developing, and testing such tools. Karst terrain, which is characterized by caves, sinkholes, and underground drainage, occurs worldwide. As a consequence, karst science is conducted in many countries including China, Korea, Russia, Slovenia, Romania, France, Italy, Britain, and the United States. The karst community and its knowledge base are globally distributed, highly interdisciplinary, and fragmented, and consist of both professional scientists and amateur cavers. An online collaborative workspace that facilitates the timely analysis and annotation of karst datasets, as well as knowledge discovery among diverse individuals, is needed. Such a workspace will supplement other efforts that have been made to promote better communication within the karst community, such as the Karst Information Portal [4, 5].

Karst is important for many reasons. The drinking water supply of about 1.6 billion people depends upon the health of karst terrains and aquifers. [6, 7, 8] Karst hosts remarkable, but poorly understood, biodiversity and contains unique, rare, and endangered species. [9, 10, 11, 12] Karst terrain is characterized by spectacular geology and significant archaeological and paleontological resources of scientific, aesthetic, cultural, and economic value. [13]

Karst is the least understood and most vulnerable type of terrestrial landscape. [14, 15, 8] As more humans move into formerly unoccupied or lightly occupied karst lands, impacts on karst systems are intensifying, resulting in sinkhole collapses that sometimes swallow entire houses. Mitigation of geologic hazards in karst costs billions of dollars each year. Karst researchers need to collaboratively develop solutions to the environmental problems that threaten these terrains.

In addition, a growing interest in geomicrobiological studies in caves is producing a collection of images that shed light on a variety of research problems. Scientists from different disciplines such as microbiology and earth sciences focus on different features when they analyze the same set of scanning electron micrographs. Researchers in one discipline or part of the world who see features in their samples need to be able to communicate their interpretations to researchers in another discipline or part of the world who may be seeing the same features but interpreting them differently. A record of the interdisciplinary analysis of these samples is important to increased understanding of karst. Moreover, metadata that integrates multidisciplinary terminology is necessary to ensure that these datasets are accessible to the international, interdisciplinary community of karst scientists in the future.

Since more than 40% of the world's drinking water from groundwater is found in karst terrain, a collaborative woskspace that increases understanding of karst datasets is important in and of itself. Beyond this, however, the results of this work are applicable to other fields with a rich tradition of image datasets, such as astrobiology and nanotechnology, and to other problems where scanning electron microscopy is a critical tool, such as contamination of medical devices.

2 Design

The design and development of the Karst Collaborative Workspace was guided by user requirements. These requirements were determined by a group of karst scientists and communicated to the group of information scientists responsible for designing and developing the collaborative workspace. The design and development team included human-computer interaction researchers, developers, and professional librarians.

User requirements were determined in several steps. First, the general workflow, including the kinds of tools and technologies necessary to support the workflow, was developed. Second, a general social-technical model that captures and supports the workflow was developed. Third, the general social-technical model was refined into a specific social-technical model. The prototype was developed on the basis of this model. Fourth, specific scenarios describing the kind of functionality scientists would like built into the system in the future were developed.

2.1 General Workflow

The general workflow can be described as follows.

Image Capture and Storage. A researcher uses a scanning electron microscope to capture a high-resolution scanning electron micrograph image. He archives it for future reference by uploading the image to his institutional repository along with suitable metadata. He may include related digital objects such as audio files that contain discussions of distinguishing features of the image and additional files that capture information about the elemental composition of features within the image.

The image is automatically assigned a unique identifier in the repository. A lower-resolution version of the scanning electron micrograph is published in an RSS feed [16].

Collaborative Analysis and Annotation. The newly published image is stored in the image module of the collaborative workspace. Included in the feed are the image, metadata (including the unique identifier), and pointers to related digital objects.

Another researcher working in the collaborative workspace learns by reading the feed that a new image with features of possible interest is available to examine. Within the image module of the collaborative workspace, she can compare the new image to other images and annotate or tag it. The image might form the basis for further discussion in the researcher's weblog or a collaboratively authored paper.

Another researcher might bookmark, tag, and annotate a published paper that relates to the image in a social bookmarking tool.

All references to the image include its unique identifier. By virtue of being discussed, annotated and tagged in various tools, the image is now associated with a rich set of annotations.

A scientist with editing privileges in the collaborative workspace may determine that some of the annotations need to be archived. These annotations can be published as a feed and picked up for archiving in the repository housing the original archived image. This publishing cycle may be ongoing for the life of a specific research project or for many separate projects.

2.2 Social-Technical Model

After determining the general workflow, the design and development team mapped the workflow to the features available in several open-source tools.

The general social-technical model (Fig. 1) reflects several options in terms of user access, social software components, and institutional repository architectures.

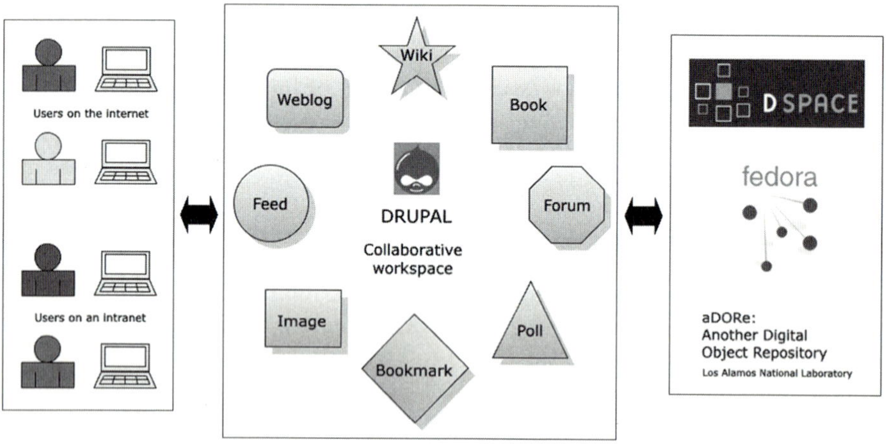

Fig. 1. The general social-technical model for the Karst Collaborative Workspace

The specific social-technical model (Fig. 2) reflects design decisions regarding user access, social software components, and institutional repository architectures. This model guided the implementation of the prototype.

Drupal is an open-source content management system that has been rated best-in-class by the IBM Internet Technologies group. [17, 18] DSpace is an open-source digital object repository platform developed by the Massachusetts Institute of Technology and Hewlett-Packard. [19] Both are relatively easy to install and configure, and are widely used around the world.

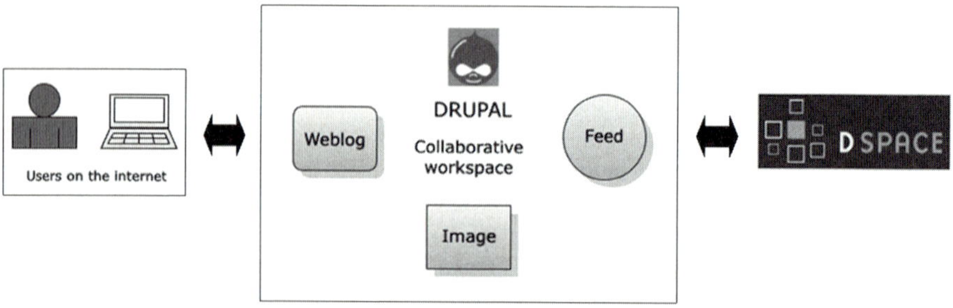

Fig. 2. The specific social-technical model for the Karst Collaborative Workspace

2.3 Future User Scenarios

In order to ensure that the system will be useful to scientists as eScience continues to evolve, user requirements for future versions were developed. These requirements take the form of eight user scenarios, two of which are included below. They include specific tasks of importance to scientists. When implemented, they will ensure that the Karst Collaborative Workspace evolves from an environment in which to discuss images to an environment in which scientists can ask more complex questions by drawing upon associated datasets.

Future Scenario 1. In this scenario, scientists will attempt to answer the following questions: Are the reticulated filaments found just in cave pool precipitates or also in other cave settings? Are they found just in limestone caves or in other types of caves? Are all reticulated filaments the same or are there variations? Have other groups found these morphologies? In order to answer these questions, scientists need to perform the following tasks:

1. View a variety of images with this morphotype.
2. Locate associated data:
- The width and length of objects in images
- The cave setting and type of cave
- The location of reticulated filaments (outside or inside of the speleothem)
- The mineralogical setting of the reticulated filaments
3. Create a new dataset consisting of the images and the associated data.
4. Run statistical tests on the data.
5. Access contact information for scientists who have added images with similar morphotypes to the repository.

Future Scenario 2. In this scenario, scientists will attempt to answer the following questions: Are ferromanganese deposits of different colors occurring in different geological/geochemical settings? Do the biological morphologies that occur with each color vary across colors? In order to answer these questions, scientists need to perform the following tasks:

1. View a Compass map of a cave with many ferromanganese deposits such as Lechuguilla Cave.
2. View a plot of where ferromanganese deposits occur and have been sampled.
3. View images of biological morphologies that have been sampled from these ferromanganese deposits.
4. Locate associated data:
- The color of the ferromanganese deposits from which the biological forms came
- Macroscopic images of the deposits in color
- The chemistry of the ferromanganese deposits and their substrate
- The mineralogical setting in which the ferromanganese deposits occur
- The geological setting, such as maps of where reef and backreef occur in the cave relative to the ferromanganese deposits

5. Create a new dataset consisting of the biological morphologies found in ferromanganese deposits and the associated mineralogical setting data.
6. Run statistical tests on the dataset.
7. Plot different geochemical trends on a map to give geographical information (both horizontal and vertical depth information).
8. Access contact information for scientists who have added images or data about ferromanganese deposits to the repository.

3 Development

The prototype was developed based on the social-technical model.

An instance of DSpace was installed at the University of New Mexico. A collection was created for Karst and Cave Studies. At this time, scientists have captured and stored 1076 scanning electron micrographs and 24 thumbnails of morphotypes. The home page of the collection (Fig. 3) includes thumbnail images of the 24 morphotypes.

Fig. 3. The University of New Mexico's DSpace / Karst and Cave Studies collection

An instance of Drupal was installed at the Los Alamos National Laboratory. At this time, user accounts have been created for the karst scientists working on the project, as well as the members of the design and development team.

Weblog, feed, and image functionality have been enabled and utilized. In order to supplement the image functionality, Gallery, an open-source image management system, has been embedded in the Drupal. [20] Additional functionality, such as forums, have been enabled but have not yet been utilized.

Users can view an array of thumbnail images. (Fig. 4) By clicking on the thumbnails, they can view larger versions of the images: either 640 x 480 or 1024 x 768. (Fig. 5) They can annotate the larger versions of the images by adding comments.

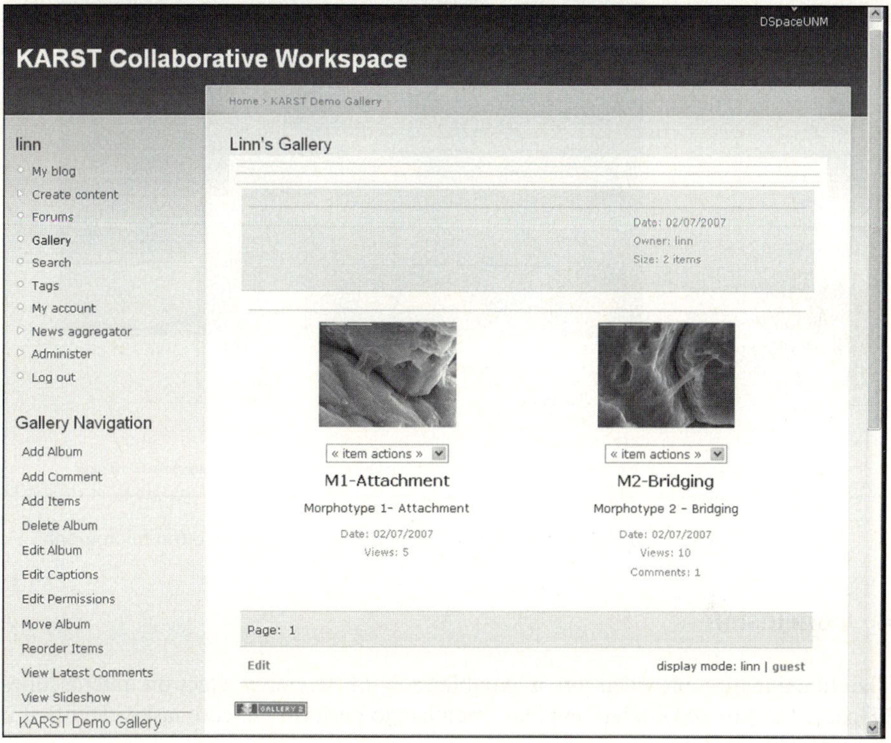

Fig. 4. The Karst Collaborative Workspace display of morphotype thumbnails

At this point in time, RSS feeds have been enabled and are being utilized. Several other mechanisms for exchanging digital objects between loosely coupled institutional repositories, such as DSpace, and social software tools, such as Drupal, are also being explored. This is an active area of research in the eScience and digital object repository communities [21] and the Karst Collaborative Workspace will utilize new mechanisms as they become available.

After the prototype has been tested, an instance of Drupal and Gallery will be installed at the University of New Mexico and made available to the international community of karst scientists.

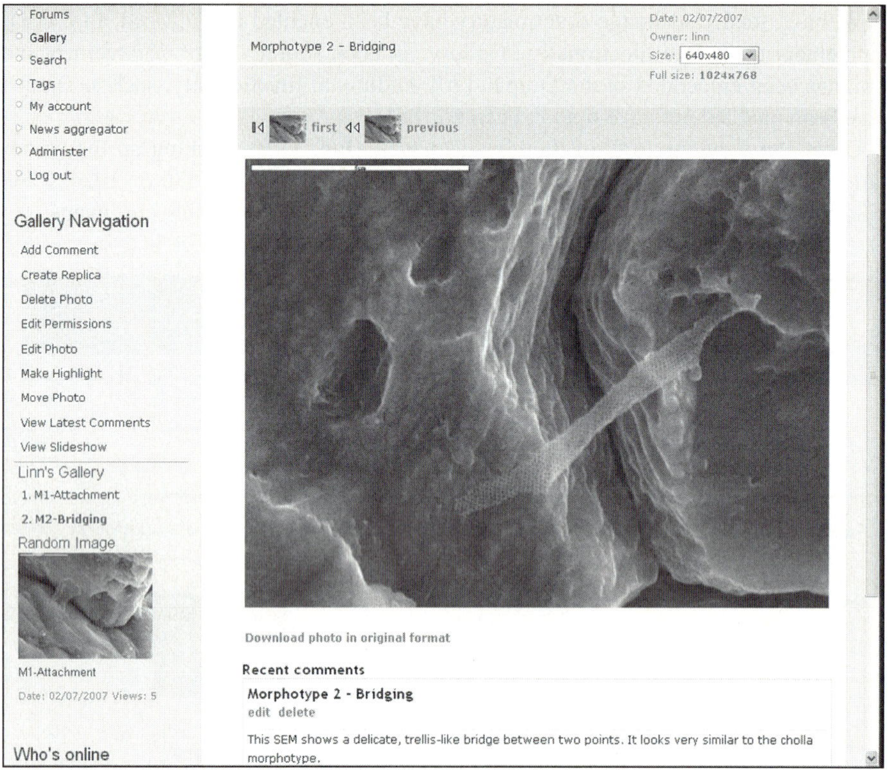

Fig. 5. The Karst Collaborative Workspace display of a scanning electron micrograph

4 Conclusion

The functionality provided by a combination of scanning electron microscopes, DSpace, Drupal, and Gallery enables scientists to perform the core tasks identified in the general workflow.

Image capture and storage are facilitated by scanning electron microsopes and DSpace. Scientists can upload scanning electron micrographs to an institutional repository along with suitable metadata. The scanning electron micrographs are automatically assigned a unique identifier in the repository. DSpace generates RSS feeds for new content and Drupal displays the feeds.

Collaborative analysis and annotation are facilitated by Drupal and Gallery. Scientists can compare new images to other images and annotate or tag them.

The Karst Collaborative Workspace thus makes it possible for scientists to collaboratively analyze and annotate scientific datasets consisting of scanning electron micrographs.

As eScience continues to evolve, two aspects of the system will change dramatically: the complexity of the tasks scientists can perform and the mechanisms for exchanging digital objects between repositories and social software tools.

The results of providing such a collaborative workspace for the international community of karst scientists will be a rich record of the collective interpretations of scanning electron micrographs, improved annotations and metadata for describing the morphological features found in the images, and a well-curated image dataset.

Acknowledgments. The authors wish to thank Chad Kieffer, Ketan Mane, Herbert Van de Sompel, and Zhiwu Xie for their contributions to this project.

References

1. Lederberg, J., Uncaphar, K.: Towards a National Collaboratory: Report of an Invitational Workshop at the Rockefeller University New York. National Science Foundation, Arlington, VA (1989)
2. National Science Foundation, NSF's Cyberinfrastructure Vision for 21st Century Discovery. Version 7.1 (2006) http://www.nsf.gov/od/oci/ci-v7.pdf
3. National Science Board. Long-lived Digital Data Collections: Enabling Research and Education in the 21st Century. Washington, D.C., National Science Foundation. (2005) http://www.nsf.gov/nsb
4. Karst Information Portal: http://www.lib.usf.edu/KIP/
5. Hose, L.D., Brinkmann, R., Northup, D.E.: The Global Karst Digital Portal: An Emerging Collaboratorium Will Enhance Information Exchange. In: Proceedings of the 17th National Cave and Karst Management Symposium, pp. 127–13 (October 31 to November 4, 2005) Albany, New York. The NCKMS Steering Committee (2006)
6. Ford, D.C., Williams, P.W.: Karst Geomorphology and Hydrology: Unwin Hyman, London (1989)
7. Smith, D.I.: The Nature of Karst Aquifers and Their Susceptibility to Pollution. In: Williams, P.W. (ed.) Karst Terrains: Environmental Change and Human Impact: Catena Verlag, Cremlingen-Destedt, Germany (1993)
8. Williams, P.W. (ed.): Karst Terrains: Environmental Change and Human Impact: Catena Verlag, Cremlingen-Destedt, Germany (1993)
9. Christman, M.C., Culver, D.C., Madden, M.K., White, D.: Patterns of Endemism of the Eastern North American Cave Fauna. Journal of Biogeography, 32, 1441–1452 (2005)
10. Culver, D.C., Deharveng, L., Gibert, J., Sasowsky, I.D. (eds.): Mapping Subterranean Biodiversity. Cartographie de la Biodiversité Souterraine: Karst Waters Institute Special Publication 6, Charles Town, WV (2001)
11. Culver, D.C., Master, L.L., Christman, M.C., Hobbs III, H.H.: Obligate Cave Fauna of the 48 Contiguous United States. Conservation Biology 14, 386–401 (2000)
12. Northup, D.E., Barns, S.M., Yu, L.E., Spilde, M.N., Schelble, R.T., Dano, K.E., Crossey, L.J., Connolly, C.A., Boston, P.J., Dahm, C.N.: Diverse Microbial Communities Inhabiting Ferromanganese Deposits in Lechuguilla and Spider Caves. Environmental Microbiology 5(11), 1071–1086 (2003)
13. LaMoreaux, P.: Foreword. In: Culver, D.C., White, W.B. (eds.) Encyclopedia of Caves, Elsevier Academic Press, Amsterdam pp. xvii-xviii (2005)
14. Drew, D., Hotzl, H. (eds.): Karst Hydrogeology and Human Activities: Impact, Consequences, and Implications: International Association of Hydrogeologists Contribution 20, Rotterdam, Netherlands (1999)

15. Veni, G., DuChene, H., Crawford, N.C., Groves, C.G., Huppert, G.N., Kastning, E.H., Olson, R., Wheeler, B.J.: Living with Karst: A Fragile Foundation. AGI Environmental Awareness Series 4. American Geological Institute, Alexandria, VA (2001)
16. Hammond, T., Hannay, T., Lund, B.: The Role of RSS in Science Publishing: Syndication and Annotation on the Web. D-Lib Magazine, vol. 10(12) (2004) http://www.dlib.org/dlib/december04/hammond/12hammond.html
17. Drupal: http://drupal.org
18. IBM Internet Technology Group: http://www-128.ibm.com/developerworks/ibm/library/i-osource1/
19. DSpace: http://www.dspace.org/
20. Gallery: http://gallery.menalto.com/
21. Open Archives Initiative Object Reuse and Exchange: http://www.openarchives.org/ ore/

Using Social Metadata in Email Triage: Lessons from the Field

Danyel Fisher[1], A.J. Brush[1], Bernie Hogan[2], Marc Smith[1], and Andy Jacobs[1]

[1] Microsoft Research, One Microsoft Way, Redmond, Washington. USA
[2] Department of Sociology, University of Toronto, 725 Spadina Avenue Toronto, ON M5S 2J4, Canada
{danyelf, ajbrush, masmith, andyj}@microsoft.com, bernie.hogan@utoronto.ca

Abstract. Email archives are full of social information, including how messages are addressed and frequency of contact between senders and receivers. To study the use of this rich metadata for email management, particularly email triage, we deployed SNARF, a prototype tool which uses social metadata to organize received email by correspondent, sort received email by past interactions, and filter email into multiple views. We discuss the lessons from a seven month deployment, including the value of organizing by personally addressed mail, the unexpected value of SNARF for email awareness, challenges with handling workflow, and ways to use social metadata in applications.

Keywords: Email overflow, email triage, social metadata, social sorting.

1 Introduction

The second time we meet a friend's dog it typically remembers us and so does not growl. Remembering how we have interacted with others in the past is basic to social interaction. Much like in-person interaction, email communication is rich with social information. How frequently have you previously corresponded with a particular sender? Was the message addressed directly to you or to a mailing list? Are you the only recipient? Unfortunately, email clients typically ignore this *social metadata* — the information about both a persons' history of interaction with their correspondents, and the ways that a message is addressed— treating all messages equally without regard to whether they come from a frequent correspondent or new friend.

During the task of email *triage,* people seek to separate relevant messages from a swath of less relevant messages. Neustaedter et al. [10] suggested that many social aspects of an email message are important to users when deciding what email to read during email triage including the sender and how the email was addressed to the recipient. That research, as well as others [15], also found that many users choose to employ a multi-pass strategy to read their email, skimming through to select ones that are particularly easy to handle or particularly important. Additional passes—to choose

M.J. Smith, G. Salvendy (Eds.): Human Interface, Part II, HCII 2007, LNCS 4558, pp. 13–22, 2007.

the next most important message—follow as time permits. Unfortunately, most email clients have limited support for the multi-pass practice, presenting email in order of receipt, rather than relevance.

While several research systems [7, 13] use social metadata in various ways, contemporary email clients do not provide users a way to directly view or access social metadata, nor do they utilize interaction histories. (This information is sometimes indirectly used to help fight spam.) Indeed, current email tools provide few social cues: besides instant messaging icons, email tools rarely give any feedback about our correspondents at all.

Although email facilitates communication and collaboration, email management is an intensely personal practice. To study the value of social metadata for email triage in a real world environment over time we extended and deployed a prototype application, SNARF, the Social Network and Relationship Finder [11]. At the core, SNARF implements a simple idea: present email organized by sender in lists that are ordered by the strength of relationship between the sender and user. Email from a colleague one corresponds with frequently will appear higher in the list than email from an infrequent correspondent. SNARF, then, uses the frequency of communication as an (imperfect) proxy for relationship strength.

SNARF employs social metadata to present users with alternative views of their email in three ways. It *filters* email by whether it has been read and how it is addressed; it *organizes* email by sender; and it *sorts* senders based on their history of interaction with the user. By allowing the user to focus on contextually-appropriate messages, and by bringing messages from frequent correspondents to the fore, we believe that SNARF allows users to more easily locate relevant messages, especially when pressed for time. SNARF supports the advantages of a multi-pass strategy without requiring users to repeatedly scan their entire inbox.

Over the course of seven months and two iterations of SNARF we gathered feedback through surveys, usage logs, and user suggestions from nearly a thousand people in our organization about their experience with SNARF.

The feedback (both positive and negative) that we collected from people using SNARF on their own email in everyday life taught us several lessons about the value of social metadata for email management. Most popular among SNARF users were the way to filter for messages explicitly addressed to themselves. Somewhat surprisingly, several SNARF users made unanticipated use of the tool to help with awareness of new mail. We also saw challenges for making use of social metadata around identity, particularly when people use multiple email addresses, and when people dramatically change their communication context, as when taking a new job.

In the rest of the paper we discuss related work (Section 2) and then describe the design of SNARF (Section 3) and our field deployment (Section 4). Section 5 presents our key findings from the field study and highlights how our experience with social metadata can help other designers and developers make informed decisions on how and when to incorporate social metadata in their applications (Section 6).

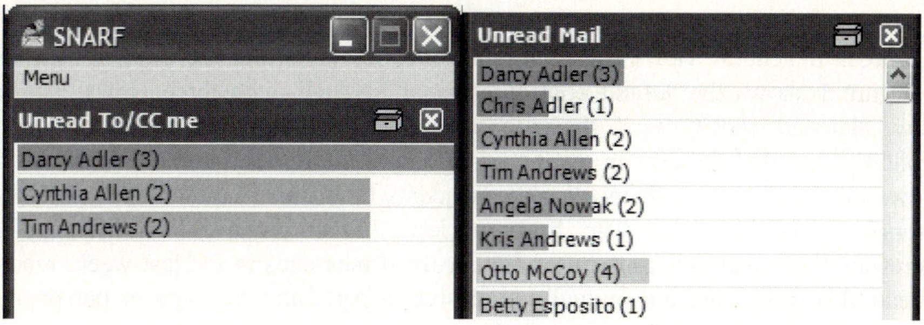

Fig. 1. Two views in the SNARF Interface: names are presented along with unread counts. Ordinarily, these two boxes would be arranged in a narrow vertical column.

2 Related Work

Assisting users in working with email has been addressed in many different ways in the HCI field from innovative interfaces [2, 7] to prioritizing systems [8]. IBM's ReMail [7] prototype includes a Correspondent Map, which organizes email by correspondent and organizational affiliation. Tyler and Tang [13] evaluates email histories to predict future availability and responsiveness. Nardi et al. [9] suggest that contact management takes a substantial amount of effort, and propose a socially-based interface that shows information oriented around people. In SNARF, we adapt this notion of orienting an interface around correspondents (rather than messages) for handling email.

3 Design of the SNARF Prototype

Takkinen and Shahmehri [12], followed by Venolia et al. [15] discuss multiple phases of handling email. SNARF is designed for email triage (or "busy mode"), a time when the user needs to locate particularly important messages and handle them quickly, and is to be used in addition to (rather than instead of) traditional email programs. The technical basis of SNARF's design was discussed in [11]. Here, we focus on the ways that SNARF uses social metadata to present email and the features needed to support day to day use in the field deployment.

3.1 Use of Social Metadata Within SNARF

SNARF uses information about past email behavior to display a user's email in three main ways.

Filtering Mail into Views. The main SNARF window is divided into several panes, each presenting one *view* showing a subset of available email correspondents[1]. Views filter mail based on how it was addressed: they separate mail that is addressed directly

[1] An email address can refer to a person, a mailing list, or an automated generator. For simplicity, we refer to all senders and receivers as correspondents.

to the user—mail that is more likely to invoke a new task—from mail that may not be as critical. In general, views are based on a collection of sorting and filtering settings.

Figure 1 shows the default configuration of SNARF, with three views. The top view, "Unread To/CC me," shows correspondents whose unread messages have explicitly included the user in the To or CC line. The middle view, "Unread Mail," shows unread messages from *all* correspondents (and thus may have been received through a mailing list). The bottom view, "This Week's Mail," shows all correspondents from whom the user has received messages in the last week, read or unread. The panes are not mutually exclusive; a particular message or person may appear in multiple panes if they match the criteria.

The original design for SNARF [11] showed only one view at a time; users would re-configure the display to match their current task. However, as the design evolved it became apparent that it is important to have multiple views so that *more* types of sorts could be shown. This reflects the difference between the relevance of a particular sender, and a sender's message. For example, Chris Adler may be very important to me, but messages from Chris sent only to me are often qualitatively different from messages sent by Chris to a mailing list we both subscribe to. The current display, which can accommodate three views of eight or so names without scrolling on a conventional screen, reflects this design.

Organizing by Correspondent. Messages are clustered together and linked to the sender of the message, in order to present a social perspective on the email. Each correspondent name is followed by the number of unread messages from them.

Sorting by Correspondence History. Each view presents names socially sorted: ordered by the degree of relation to the user that can be extracted from their interaction history. In the default view, names are ordered by the number of times that the user has sent mail to that person in the past year, although the time period is configurable. Frequent and recent correspondents bubble to the top, while infrequent correspondents sink to the bottom. In Figure 1, in the To/CC me view, the user has sent more email messages to *Darcy Adler* than to *Tim Andrews; Darcy* is therefore sorted higher on the display than *Tim.* The configuration options available on this display are discussed in more detail in [11].

Design for Field Deployment. In order to deploy SNARF, the prototype needed to be a tool that could support real day-to-day use. In particular, this required SNARF to respond to real-time updates and the ability to open mail and read it. SNARF therefore is a MAPI client, and connects itself automatically to the users' Outlook installation. The several views, like all aspects of SNARF, respond in real time to changes in email state. For example, if a message from a sender to the user is marked as read, the "Unread Messages" view is updated—it will no longer show that sender if no more unread messages remain.

When the user double-clicks on an entry showing just one message, SNARF opens the message in Outlook. If the correspondent is associated with more than one message—that is, there is more than one unread message from that person—then SNARF presents a message list in a second window.

4 Field Deployment

We gathered data about SNARF's use in the field in two phases. The first phase was a structured field study that took place over a six week period in July and August of 2005 where we conducted pre-surveys and post-surveys of SNARF users. Once the structured field study concluded, we shifted to a second phase where we released an updated version of SNARF and collected usage logs.

Phase 1: Field Study. The field study began in July 2005 when we sent out an invitation to use SNARF to people who had registered their interest at an internal demonstration of the system, as well as to two high-volume mailing lists of employees at our company that specifically target people with interest in advanced and speculative projects. In total, 574 people ran SNARF at least once within the first two weeks after it was made available. We stopped accepting new people into the study after two weeks, although they could still use SNARF. During the study we explicitly chose to make only one change to SNARF to address a performance issue when opening the message list window.

Data Collection. All users that ran SNARF in the first two weeks of the study were asked to complete an optional pre-survey related to their experiences with email. The survey followed the outlines of [10]; however, it added more detailed information about email habits within the user's current client. The survey included questions asking users how long they thought they spent triaging email, how much of their email they read, and their experience of stress associated with handling email.

In August 2005, at the end of the six-week study period, we sent out a post-survey. Users were asked to respond regardless of whether they had filled out the pre-survey or had continued to use SNARF after its initial installation. All users who reported that they used SNARF 'for a few days' or more were asked about their use in more detail on the post-survey. Users were asked to respond regardless of whether they had filled out the pre-survey or had continued to use SNARF after its initial installation. Participants that filled out the post-survey were entered into a lottery for gift certificates.

Study Experience. During the study we received extensive feedback from SNARF users. A total of 292 people completed the pre-survey, a response rate of 51% and 161 people filled out the post-survey (response rate of 28%), some of whom had not responded to the pre-survey. Post-survey respondents identified their job role as largely program managers (23%), developers (17%) and consultants (13%); sales (10%) and software testers (9%) made up the bulk of the remainder. The population was overwhelmingly male (92%).

Post-survey responses help us understand SNARF adoption. Of the respondents 4% of users reported that they could not get SNARF to work, 44% 'tried it once or twice', 29% tried it 'for a few days', 11% tried it 'for a few weeks' and 13% were 'still using SNARF'.

Phase 2: Ongoing SNARF Use. After the field study concluded, we released a series of updates to SNARF which addressed issues that the first version revealed. We promoted and released an improved SNARF several months later. 410 users tried the program at least once and provided log data during this second phase, which ran from late November to February of 2006. Usage data was logged and uploaded daily containing records of both any changes to the mail store and click data within the program itself. Of the 410 users, 25% used it on five or more days (the maximum was 103 days, almost seven days a week). The usage data provides a complementary picture to the qualitative survey data.

5 Lessons from the Field

In this section we describe the lessons we have learned from our seven month field deployment of SNARF, bringing together qualitative data gathered in the field study phrase and quantitative data from user logs as well as feedback provided directly to us by users. We classify our observations into a number of different aspects of using social history, including *social filtering*, *social awareness,* and *email workflow.* We examine places where SNARF has succeeded—and ways in which it has not.

5.1 Social Filtering

SNARF uses social filtering to give users multiple views of their email. We learned that people are particularly focused on whether a message is addressed to them and they took advantage of having multiple views. We also saw that the interface must be careful to support multiple overlapping views in a way that is understandable to the user.

Personal Mail Matters: *"[My favorite aspect of SNARF was:] Showing the email that was just to me or cc to me."* SNARF provides many different configurable views. Conceptually, we separate these views into two types. *Personal views* filter for messages that explicitly mention the user, such as "Unread To/CC me." In contrast, *aggregate views* show collections of messages are not addressed to the user, such as unread messages sent to lists ("Unread lists"). Personal views generally contain shorter lists of people; the latter type generally contains more people. On both the post survey and in the usage logs, we saw a preference for personal views

Multiple Views Were Used. SNARF provides multiple views of a person's email. On the survey, users "agreed" (mean = 1.17, N=69[2]) that "it is useful to filter mail into a number of panes." Of the 358 people that clicked once in any view, 225 of them clicked at least once in two or more views. This supports the idea that maintaining multiple perspectives on email is valuable.

Users found different views useful for different tasks. During the triage task, for example, users might concentrate on mail sent directly to them as they cleared out messages that directly needed their attention. List views were particularly useful for catching up on a high volume of mailing list messages. Skimming the top off the "Unread Mail" view allows a user to monitor interactions on mailing lists and discussion that involve their closet correspondents.

---

[2] On a -2 ("Strongly disagree") to +2 ("strongly agree") scale. This scale is used for all survey responses unless noted. The words "agree," "strongly agree", and similar in quotation marks refer to the median choice.

Challenges Showing Messages Multiple Places. Some users complained that they would see the same message in several different places, as the "Unread mail" would show messages that we also "To/CC Me". The experience of reading one message and having several parts of the interface change can be distracting or confusing. While SNARF does not fully handle this situation, any system that repeats information in multiple places must make this clear.

5.2 Organizing by Correspondent

SNARF supports the idea of organizing information socially. [10] suggested that users pressed for time feel a need to "cherry-pick" their mail, seeking out messages specifically from people they know are relevant. SNARF cherry-picks for them, first separating messages that are directly to them, then ordering sets of messages by how often the user and sender correspond with one another.

While we feared that users might complain that clustering by person would roll together irrelevant ("Lunch!") messages with relevant ("Need report soon"), users "agreed" (mean = 1.08, N=73) that listing "correspondents rather than individual messages is useful."

5.3 Social Sorting

SNARF shows that the world is divided into those people who I have previously interacted with and those who I have not. People who I have not sent mail to before are treated differently than people who I have. This is the concept behind white-lists, which are meant to hold off spam and bulk email; it can also be successful in handling email messages during triage. During those pressed times, emphasizing the top few messages from the most-contacted correspondents probably makes sense.

The goal of social ranking in SNARF is that the high-priority people a user interacts with often will float to the top of a list, while strangers would stay at the bottom. Survey respondents generally "agree" (mean = 0.95, N=58) that sorting correspondents socially is useful. One user commented, *"[My favorite aspect of SNARF was:] sorting mail by sender, first sender first."*

Organizational Information and Sorting. The default sorting in a SNARF view is based on the number of messages a user has sent to a correspondent. This method of sorting neglects to take into account important hierarchical information about organizational relationships. We have chosen not to incorporate organizational information into SNARF at this point. In practice, we have found that multiple views may handle issues of important email from an infrequent correspondent. As personal views are typically shorter (section 5.1.1), a personally addressed message from an important but infrequent correspondent is easily spotted.

5.4 Social Awareness: An Unexpected Gain

Perhaps the most surprising result of the deployment of the SNARF tool was the success of using SNARF as an awareness tool. The same minimalist display that addressed the triage task also meant that the application could be left active at a minimal cost of screen space. Post-survey respondents that told us they were "still using" SNARF at the time of the survey "agreed" that the tool was useful for keeping aware (mean = 0.78, N=18). In contrast, users who used SNARF just for "a few days" before stopping "disagreed" that SNARF was useful for keeping aware (mean = -0.43, N=40), a lower rating than they gave to "finding important email" or "triaging."

SNARF could have been helping handle the "flow" phase of email (to use [12] terminology): the time in which users keep up with incoming messages, is also a poorly-addressed feature in many current tools. For example, in many common mail tools, a small notification window appears in the corner of the screen for a short time when new mail arrives, and disappears shortly after. This is a "push" mechanism: if the user is to read the message, they must quickly evaluate the window, reading the sender and subject, decide if they wish to read it, and click, all within a few seconds. In contrast, SNARF shifts to a "pull" model. A user can glance at SNARF to see if new email they might be interested in has arrived and then decide whether to read the message now or later, without the sound or a distracting [3] visible change.

5.5 Email Management and Workflow

The email triage task, as explored by [10, 12, 15], focuses on deciding how to handle individual messages. The notion of a mailbox needing to be cleaned up or re-organized is left for a periodic "cleanup" sweep. In reality, users are accustomed to often being able to file or delete messages after they have been read, even if they often do not bother to do so.

Many "one-touch" strategies for dealing with email (such as the popular "Getting Things Done" [1]) combine *reading* mail with immediate *filing* or *processing* of messages. Many users seemed accustomed to reading a message, closing it, and then filing or deleting it. Once the message was read in SNARF, it was no longer able to be deleted, as it had disappeared from the interface. Users who wanted to delete messages would have to go to Outlook to do so. This meant users found that SNARF did not smoothly support the filing and processing phase of triage.

In addition, the inbox being a "habitat" [4] means that *already read* messages can be important for browsing or reviewing for tasks. Once a message had been read, it disappeared from the SNARF interface, making it poor for reviewing current tasks.

5.6 Social Metadata Challenges

Changes in Context. As we note above, social histories by necessity track a user's past. Yet sometimes, a user's context changes—by changing jobs, for example, or by switching projects—and their history can become less valuable as predictors of the relevance of content from new correspondents. Socially-sorted systems should consider ways to resolve this, either by reducing the importance of past interactions, by pre-populating information, or by incorporating organizational information.

Social History is More Than Email. SNARF treats the volume of past emails as a proxy for relationship strength. Obviously, this is a fairly limited metric—relationships and interactions with other people are far deeper, as anyone with a parent who dislikes email knows. SNARF's logs do not capture instant message conversations, telephone calls, or hallway interactions. It is clear that measuring

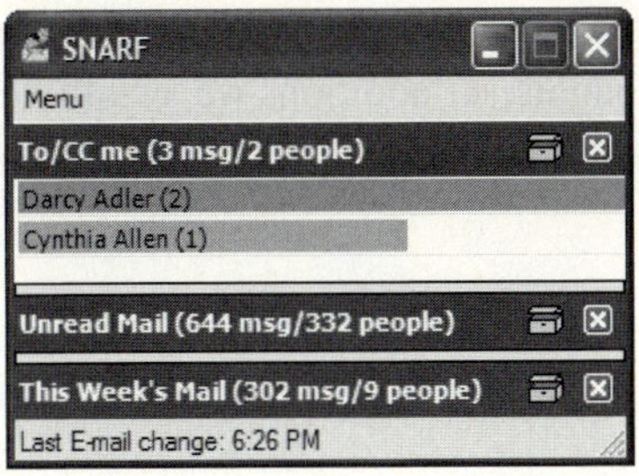

Fig. 2. SNARF collapsed in "Awareness Mode". Clicking any view title will open that view.

electronic communications may not, in the end, account for true social importance. Still, volume seems to be a comprehensible way of approximating interaction.

6 Concluding Remarks

SNARF shows one set of experiments into social sorting, as applied to email triage. We believe, however, that these concepts of applying social metadata to end-user tools to be broadly applicable. A number of users seemed particularly interested in using SNARF as a way to get a general overview of their email: who they have interacted with and how often, and to collect general statistics on their informational history. For example, one user commented *"[My favorite aspect of SNARF was:] being able to tell who the people are who send me the most mail—and having the evidence to persuade them to send me less!"* While much of this social metadata is collected in SNARF database, it is not yet reflected back to users in a general way.

While SNARF may not be for everyone, deploying an application that people could use on their own email on a day-to-day basis allowed us to collect a wealth of data. The deployment illustrated that personally-addressed mail matters, that social metadata can be helpful for staying aware of incoming mail, and that strangers are importantly different from people you know. We also saw challenges in accurately representing people's identities and in handling the lag in social history when people switch communication contexts. These lessons can be applied to future systems that incorporate social metadata.

One of the most popular features of SNARF was the personal views. We are pleased that aspects of this can be directly approximated in other mail clients today. Many email clients support "search folders," dynamic folders that list all email that match given criteria. A search folder of "unread mail, sent to me" allows people to adopt one use of social metadata into their current email experience. Of course there are some drawbacks: the view provided in the search folder would not combine multiple messages by one person, nor would it sort those people by their names.

Finally, we feel one of the strengths of SNARF was the *simplicity* of the underlying metrics. The notion of "how many emails have I sent to this person" is a comprehensible one and the transparency of the social sorting algorithms allowed users to understand the system. As we begin to take use social metadata in other places, as when auto-completing email addresses or ordering search results, emphasizing simple and straightforward metrics will prevent user confusion

Acknowledgements. We would like to thank Gina Venolia for valuable survey and user interface design thoughts; in addition, we would very much like to thank our users.

References

1. Allen, D.: Getting Things Done: The Art of Stress-Free Productivity. Penguin Putnam: New York (2001)
2. Bellotti, V., Ducheneaut, N., Howard, M., Smith, I.: Taking email to task: the design and evaluation of a task management centered email tool. In: Proceedings of CHI, pp. 345–352 (2003)
3. Cutrell, E., Czerwinski, M., Horvitz, E.: Notification, Disruption, and Memory: Effects of Messaging Interruptions on Memory and Performance. In: Proceedings of Interact 2001. Tokyo, Japan (2001)
4. Ducheneaut, N., Bellotti, V.: E-mail as Habitat: An Exploration of Embedded Personal Information Management. Communications of the ACM, pp. 30–38 (2001)
5. Fiore, A., LeeTiernan, S., Smith, M.: Observed Behavior and Perceived Value of Authors in Usenet Newsgroups: Bridging the Gap. In: Proceedings of CHI, pp. 323–330 (2002)
6. Fisher, D., Dourish, P.: Social and temporal structures in everyday collaboration. In: Proceedings of CHI, pp. 551–558 (2004)
7. Gruen, D., Rohall, S., Minassian, S., Kerr, B., Moody, P., Stachel, B., Wattenberg, E., Wilcox, E.: Lessons from the ReMail prototypes. In: Proceedings of CSCW, pp. 152–161 (2004)
8. Horvitz, E., Jacobs, A., Hovel, D.: Attention-Sensitive Alerting. In: Proceedings of UAI '99, pp. 305–313 (1999)
9. Nardi, B., Whittaker, S., Schwartz, H.: NetWORKers and their Activity in Intentional Networks. The Journal of Computer-Supported Cooperative Work 11, 205–242 (2002)
10. Neustaedter, C., Brush, A., Smith, M.: Beyond From and Received: Exploring the Dynamics of Email Triage. In: Proceedings of CHI 2005, pp. 1977–1980 (2005)
11. Neustaedter, C., Brush, A., Smith, M., Fisher, D.: The Social Network and Relationship Finder: Social Sorting for Email Triage. In: Proceedings of the, Conference on Email and Anti-Spam (CEAS) (2005)
12. Takkinen, J., Shahmehri, N.: Café: A Conceptual Model for Managing Information in Electronic Mail. In: Proceedings of the Thirty-First Hawaii International Conference on Software Systems (1999)
13. Tyler, J., Tang, J.C.: When Can I Expect an Email Response? A Study of Rhythms in Email Usage. In: Proceedings of the European Conference on Computer Supported Cooperative Work (ECSCW 2003), ACM Press, New York (2003)
14. Venolia, G.D., Neustaedter, C.: Understanding Sequence and Reply Relationships within Email Conversations: A Mixed-Model Visualization. In: Proceedings of CHI 2003, pp. 361–368 (2003)
15. Venolia, G.D., Dabbish, L., Cadiz, J.J., Gupta, A.: Supporting Email Workflow. Microsoft Technical Report TR-2001-88 (2001)

Fond Memory Management System by Using Information About Communities

Katsuya Hashimoto and Yoshio Nakatani

Ritsumeikan University
1-1, Noji-Higashi 1, Kusatsu, Shiga 525-0058, Japan
rs035025@se.ritsumei.ac.jp,
nakatani@is.ritsumei.ac.jp

Abstract. In this paper, we propose "yourStory" as an effective reminder management system to construct a life story from the viewpoint of communities to which the user has belonged. Our framework helps the user manage his/her personal history and tell his/her life stories to other users through sharing communities and episodes. This system provides four kinds of functions: building up a life story in accordance with community information, output of the "communityTree" and "lifeChart" data based on the life story, registration of an episode in relation to a community, and display of various kinds of data, including "communityTree," "lifeChart," photos and episodes.

Keywords: Fond memory, community, life story, episodic memory.

1 Introduction

The development and spread of information equipment, such as personal computers, digital cameras, digital camcorders, camera-equipped cell phones, make it possible to easily record and preserve fond memories as digital data. Nowadays, many people preserve their reminders, such as photos, videos, and diaries, on the computer.

The problem here is that people still waste time managing reminders. Although they realize that management of the folders is more convenient as compared with before [1][2], they often feel it troublesome to arrange, edit, and continuously manage the data. Additionally, they worry about the obsolescence of tools for managing the data and the destruction of data. Because of these concerns, people began to not only copy the data onto other disks, but also to print photos and create photo albums. Judging from these actions, we can say with fair certainty that people remain dissatisfied with computer software and reliability for preserving reminder data on a computer.

Our goal is to provide a computer system to support continuous and efficient management of reminder data on the computer and bring fond memories clearly into view on the screen. The characteristic of our approach is that we use communities to which people belong for classifying their fond memories. In this paper, we discuss a new framework to manage fond memories on the computer.

M.J. Smith, G. Salvendy (Eds.): Human Interface, Part II, HCII 2007, LNCS 4558, pp. 23–29, 2007.

2 Anxiety About Loss of Reminders

People can live on the faith of self-images that have not changed throughout the course of their life times. Many people write and manage diaries and take photographs as their reminders. They love seeing their diaries and photographs again and again. In psychological therapy, the "reminiscence method" was proposed and has been used widely in order to support the aged people recall their memories and to restore their vitality to live positive [3]. This method effectively uses reminders: the aged people tell their memories to one or more audiences, inspired by photographs and other reminders.

This can be seen more clearly in the event of great disasters. Every year in the world many people suffer from catastrophes. Among those victimized by catastrophes, the elderly require substantial time to recover while the younger recover earlier [3]. One reason for this is that the former have accumulated a wide range of experiences in daily life and the loss is very serious. For catastrophe victims to return to their pre-catastrophe lives and restart new lives, they need to rediscover their vividly fond memories and the meaning of their past lives [4].

In recent years, there has been a huge boom in blogging.. Many people publish diaries, photographs, and their feelings every day on the web. The authors of "blog"

Table 1. Examples of typical answers to "How do you feel about reminders you have been managing yourselves"

- Digital photography is high-resolution and high-quality, and it doesn't fade away as time goes by. But I feel anxious about hard disk crashes, product obsolescence of data, and the risk of virus infection.
- Human beings memorize a lot of things sequentially, and recall fond memories instantly triggered by time, scenery, a word, someone's voice, or anything else. From now on, if the computer cannot provide similar mechanism, there is no point in preserving reminder data in a computer.
- I would like to convert analog photos and analog videos into digital ones, but doing so is very troublesome because I have to spend much time and use much money to do that. I wish I could do that more readily and cheap.
- I like seeing photo albums. However, it is not interesting if the photos are merely being preserved on the computer, although I know that digital data is suitable for preserving photos.

want to tell his/her own life stories to their potential readers as self-disclosure [5]. However, this boom might be just a bubble. New media might be introduced to manage their diaries and pictures. If the stored remainders cannot be read on the new media, many people will lose their personal reminders forever. This might be a serious problem to those concerned.

We conducted our Questionnaire Survey of Managing Reminders in December 2005 and January 2006. The participants in this survey were 34 individuals (26 males and 8 females), aged twenty years and above. We have given examples of answer to a questionnaire item which is "How do you feel about reminders you have been managing for yourselves?" (Table 1). We surmise that most people feel anxious about the loss of reminders and feel dissatisfaction towards the halfway measures of present day systems. We do not want to overlook the fact that people wish to manage reminders more efficiently and more safely.

3 Architecture

In this paper, we propose a framework for expressing individual human life through using information about communities to which people had belonged (Figure 1). This framework will help people to manage their complicated reminder data, and will allow them to realize a prototype system, "yourStory." This system also facilitates communication with their colleagues and friends who use terminals on the network with this system.

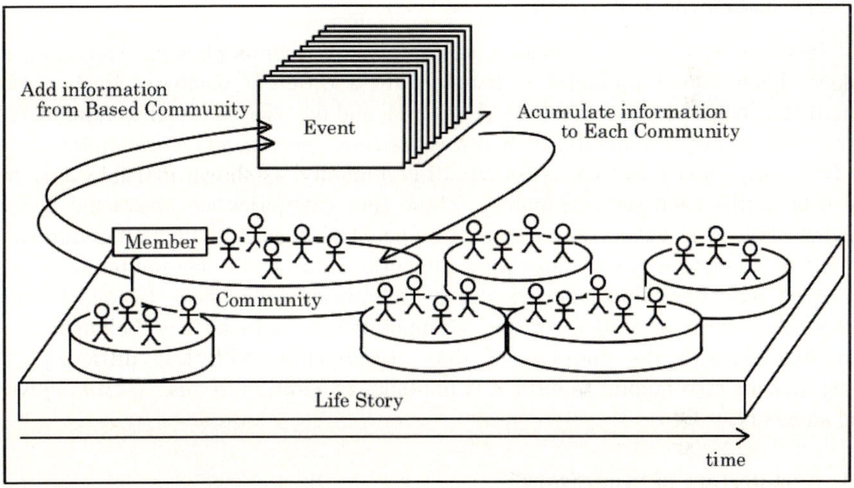

Fig. 1. "Attributed Assembly Model" of human life

3.1 Managing Fond Memories

In this framework, we define community as a certain group or field in which people could be themselves and spend meaningful time for a certain period. Generally, our self-images depend on the memory of such communities. People can easily describe their life stories from the viewpoint of communities through using this framework, and their narratives become more valuable than just a photo album.

3.2 Prototype System "yourStory"

In this section, we propose "yourStory," the reminder management system. Users of this system recall their past memories of communities to which they had belonged, and they provide some pieces of information about these communities by following system instructions.

In “yourStory”, attributes of each community consists of five elements:

- community’s name
- period of belonging to the community
- episodes concerning the community
- three typical photos of the community
- members of the community

Users can input information about episodes that concern each community. Each episode has four attributes:

- name of episode
- date of the episode
- picture diary of the episode
- major participants in the episode

Users can input digital reminders of an episode, such as photos, voice data, and videos. Each digital reminder is managed in a different database. Each episode inherits attributes of the community (Figure 1), and the user can refer to reminders by using attributes of community, such as *time*, *season*, *episode*, and *participants*.

The system provides four categories of community, as shown in Table 2, to help the user define each sub-community. These four categories are abstracted because they are thought to have a strong impact on peoples’ lives. Although all communities cannot be categorized into these four categories, almost all sub-communities were classified into these four categories in a preliminary experiment. Guided by these categories, user can remember the communities to which they belonged (sub-communities) and the duration of their participation with less difficulty. The categories are also helpful to retrieve a reminder in addition to *time*, *season*, *episode*, and *members*.

3.3 Architecture of “yourStory”

“yourStory” realizes four functions:

- Building up a life story according to community information that is input by the user.
- Output of the “communityTree” data and “lifeChart” data based on the life story.
- Registration of an episode in relation to a community.
- Display of a various kind of data, including “communityTree,” “lifeChart,” photos and episodes.

Figure 2 is an example of a “yourStory” screen image. If the user inputs information about communities for each category, “yourStory” shows the user reminders with a graphical interface such as this figure.

“communityTree,” shown on the left side of Figure 2, is a tree structure of communities registered in the system. In this field, the user can manipulate

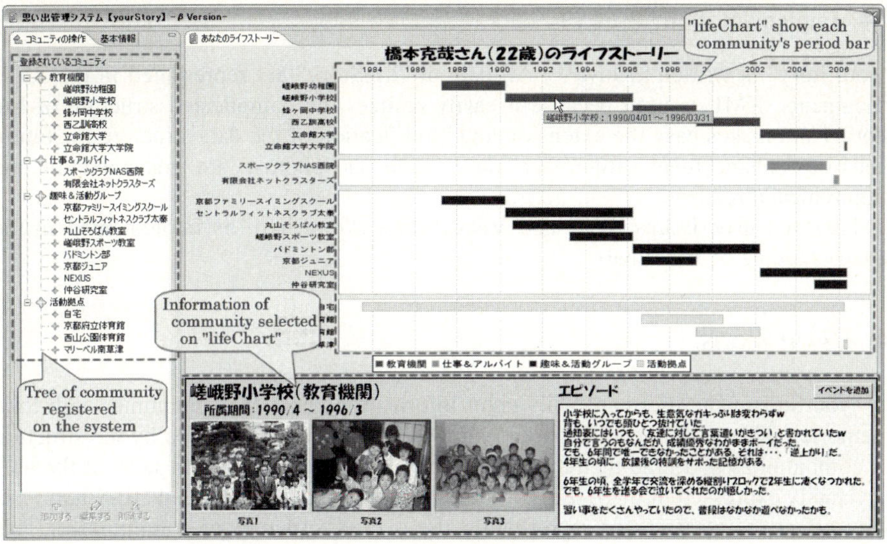

Fig. 2. Example of "yourStory" screen image

information of the communities: adding a new community to a category, editing a community name and structure, and eliminating a community from a category.

"lifeChart" on the right side of Figure 2 shows a period in each community defined in "communityTree." Each period is displayed as a bar whose color differs from category to category: *educational institution* is red, *work place* is green, *hobby* is blue, and *base region* is yellow. The user can expand and contract a period bar by selecting it with the mouse and dragging it to a certain period. Bars to be displayed in "lifeChart" can be limited by selecting communities in "communityTree."

Shown in the bottom of Figure 2 is information about a selected community. When the user selects a community on "lifeChart" or "communityTree," related photos and episode information are displayed in this field. The number of photos is limited to three. This is because the goal of this system is not to organize photos but to use photos in order to help the user describe his or her life story. Three photos for each episode are sufficient for describing an episode. This field helps the user talk to someone about memories that concern the community.

By composing the screen like this, digital photos can be managed by connecting them with communities. People usually organize photos by giving various kinds of information, such as name and date, to the photo files and folders. This makes it difficult to position photos in the order of a life story. An advantage of our system is that it enables the user to select and organize photos by visually mapping them with memories.

This system also enables the user to include a comment not only with each photo and each episode but also with a community. For example, the user can input a long-term episode such as "Those were the days" to a certain community.

3.4 Implementation

"yourStory" is written in Java™. All data of "lifeStory" are represented in the XML documents. XML is used because it easily realizes the complicated structure of life stories and makes easy the extension and modification of the data structure for future function enhancement. Reminder data, such as photos, are managed in the hierarchical directory.

The GUI uses EclipseSWT and VisualEditor. "lifeChart" is implemented in the library module of JfreeChart.

4 Discussions

In "yourStory," the user can easily enter information about the community with fond memories. There are not so many items for input and they are guided by the structure of communities and fond memories that were already input. The life story of the user is visually expressed through using information about community, and "lifeChart" is a very effective interface for representing the overall structure of communities. This framework helps each user to manage reminders and recall fond memories.

In comparison with other frameworks to express human life, the advantage of our framework is that it is effective not only for the management of personal memories but also for the promotion of communication with other people who use the same system. Our framework enables the user to compare his/her own memory with the memories of other people also involved in a similar community, and to select and talk about a memory that may attract the interest of other users based on the episodes registered in their systems. This kind of extended use of the system suggests the potentiality of this framework.

The number of digital data concerning fond memories is increasing. However, the essence of fond memories is not supported by the number of data, but by the content of the fond memory itself. Management of a significant amount of daily in-depth data, such as e-mail, cell-phone history, schedule, activity-log, and other such forms, causes low efficiency of reminder management and complicates system interface. The reminder management system should manage the necessary minimum number of photos.

5 Conclusions and Future Works

In this paper, we proposed "yourStory" as an effective reminder management system to construct a life story from the viewpoint of communities to which the user belonged. Our framework helps the user manage his/her personal history and talk about his/her life stories to other users through sharing communities and episodes.

Our future work includes implementation of a system with more effective functions to help people manage reminder data and recall their fond memories. We are planning to extend "yourStory" to implement a method of displaying two or more "lifeStory" at the same time. We think this method will arouse unexpected, appealing conversation among people. We also plan to conduct experiments to validate our framework.

This work was supported in part by the "High-Tech Research Center" Project for Private Universities: matching fund subsidy from MEXT, 2002-2006.

References

1. Yamashita, K., Nojima, H.: Portable Personal Memory for Communication 2001, Human Interface Symposium, pp. 503–506. (in Japanese) (2001)
2. Yamashita, K., Nojima, H.: An Arrangement Method of Digital Photographs Based on Importance of Photographs. The 19th Annual Meeting of the Japanese Cognitive Science Society, pp. 256–257. (in Japanese) (2002)
3. Butler, R.: The Life Review: An. Interpretation of Reminiscence in. the Aged. Psychiatry 26(1), 65–76 (1963)
4. Nakatani, Y.: Support System for Reconstructing Fond Memories Lost in Disasters. In: 11th International Conference on Human-Computer Interaction 2005 (CD-ROM) (2005)
5. Miura, A., Yamashita, K.: Why do people publish weblogs? An online survey of weblog authors in Japan, Human Perspectives in the Internet Society 2004, Wessex Institute of Technology, UK (2004)

Spatial Electronic Mnemonics: A Virtual Memory Interface

Yasushi Ikei[1], Hirofumi Ota[1], and Takuro Kayahara[2]

[1] Tokyo Metropolitan University, 6-6, Asahigaoka, Hino, Tokyo 191-0065, Japan
[2] Miyagi University, Gakuen 1, Daiwa, Kurokawa-gun, Miyagi 981-3298, Japan
ikei@computer.org, ota@krmgiks5.tmit.ac.jp, kayahara@myu.ac.jp

Abstract. This paper proposes a novel idea of the Spatial Electronic Mnemonics (SROM) that augments human memory by using electronically annotated and/or converted materials based on places, objects and people in the real world. The SROM provides effective recall cues for information to be memorized with visual materials such as digital images captured and modified for easy association. A basic function as well as structure of SROM and some plans to construct the SROM is presented. The initial implementation and a preliminary registration experiment of spatial virtual memory peg are demonstrated.

Keywords: Memory augmentation, Pegs, Images, Locations, Wearable computer.

1 Introduction

The amount of information that we come across is continuously growing by remarkable advance in information technology. However, we do not necessarily have sufficient method to effectively utilize the vast information in our daily activities. One of the powerful solutions would be information retrieval or mining by a computer, however semantic manipulation in the search for information has not successfully realized only to depend on a keyword. An effective information use in a truly creative task can not be achieved only by machine retrieval, but be performed virtually only by the human brain. A creative work to produce novel idea is only done through manipulating information in mind. So, more information available instantly will produce more ideas. Not all vast information perceived by the user are memorized and utilized naturally, however it will probably be possible to assist mental process of memorization of the user by wearable/ubiquitous computers. These new computing machines attached to the user or distributed in the environment provide a good base for a novel mnemonic technique for the effective use of large amount of information discarded so far due to overflow.

It is well known that mental images provide strong promoter in memorization [1]. One of the oldest methods to augment human memory is the method of loci that made use of visual image effectively. The history of mnemonics [2] supports this cognitive

M.J. Smith, G. Salvendy (Eds.): Human Interface, Part II, HCII 2007, LNCS 4558, pp. 30–37, 2007.

phenomenon. In general, however, the mnemonics itself is not necessarily easy to memorize. The user of mnemonics has to master the technique, which limited the number of users of large capacity mnemonics.

In this paper, a novel idea of electronic augmentation of human memory, the Spatial Electronic Mnemonics (SROM) is presented. The SROM is an augmented cognition interface that expands memory capability so that the user can effectively exploit the vast information obtained by him/her.

The SROM introduces edited and annotated photo/video images of a real space to form an external Virtual Memory Space (eVMS) where the user entirely subjectively places items to remember. Although the SROM inherits powerful characteristics of classic mnemonic devices, such as the method of loci, the peg system, etc., clearly new components to enhance the functions as well as specifically ease of learning or unconscious/enjoyable mastering are provided by the SROM.

2 Spatial Electronic Mnemonics

2.1 External Virtual Memory Space (eVMS)

The SROM provides the user with devices to effectively construct and make use of cognitive memory space based on electronic augmentation of real space and objects.

The cognitive memory space that refers to the real space is termed the External Virtual Memory Space (eVMS) in the present research. The eVMS resides in the user's mind. The addressing indices to the eVMS include places, objects, people and their electronic augmentations. The indices are generically called Virtual Memory Pegs (vMPegs).

The eVMS construction is performed with the following working hypothesis.

1. Basic principle: the item to memorize is associated to a vMPeg, and localized in the eVMS to effectively augment the recall probability of the item.
2. Create association: the item to memorize is associated with a vMPeg appropriately presented at the time the item occurred so that the item is located in the eVMS to increase recall probability effectively.
3. Recall promotion: implicit cued (spontaneous) recall is scheduled by making the eVMS dependent on a context so that the item is recalled when it is needed.
4. Deletion control: deletion of an item from the eVMS is promoted by interference or self decay based on increased manipulability of the eVMS.

2.2 Virtual Memory Peg (vMPeg)

A vMPeg is a tag to the location in the eVMS. Basic characteristics of the vMPeg are the same as usual memory pegs.

1. Framework for recall: a vMPeg provides a framework of memory addressing immediately recalled when needed. It is visual and concrete. The order of vMPegs is defined as an ordered set. Otherwise, an exhaustive scan in the set of vMPegs is possible for an unordered set.

2. High associability: a vMPeg is associable to other events and objects. It is visual, concrete, and easy to make a story.
3. Fast and easy to build: a vMPeg is created with effective assistance of wearable/ubiquitous computers faster and easier than conventional method (performed without computers) with high retention, large number of items, fast and accurate recall.

A vMPeg has three material types and three attributes for each type. The types are a *location* type, an *object* type and a *human character* type. The location type is based on places and their landmark where we can go or imagine standing before. The object type is created from any objects in the world. The human character type is based on any persons the user knows.

Three attributes are *familiar*, *real*, and *virtual*. A familiar type vMPeg is a real and familiar place, object, person easy to access for the user. A familiar vMPeg is episodically firm for the individual. A real type vMPeg is a real place, object, person not accessible or hard to access for the user. It includes famous places, persons, art works, or expensive products. A virtual type vMPeg is a place, object and avatar created by virtual reality system.

vMPegs are created from places, objects and people in the forms of annotated/animated photos and videos and other electronic devices/materials, augmented by IC tags, smartphones, and wearable/ubiquitous computers. A vMPeg is a fundamental element of the SROM and eVMS.

2.3 eVMS Operations of the SROM

The SROM system includes three basic functions to manipulate the eVMS through vMPegs.

1. Registration of vMPeg: to establish a vMPeg so that it works as a framework for addressing the eVMS by electronically augmenting places, objects and people. Fast acquisition, long term retention, and a large number of items are expected merits of vMPegs. The visual images provide efficient coding for memory in the operation.
2. Association of vMPeg: to fast establish relations to other events and objects of vMPegs with high retention. Optimized association for high recall ratio is explored during the extended process of association.
3. Recall Manipulation: to adaptively adjust vMPeg to control characteristics of functions of voluntary recall and deletion of memory.

3 Registration of Location-Type vMPegs

3.1 A Method to Register Location-Type vMPegs

As one method of vMPeg registration, we propose a building process for location-type vMPeg. The method involves three features: graphic digits, photograph of location, and graphics arrangement in the photograph by the user. The vMPeg is a series of places that the user selected to use as peg locations. A picture of a place or an object of interest at the place is captured by a camera. When taking a picture, the user observes a live video image of the place as well as an overlapped graphics that indicates a digit number shown in Fig. 1. The graphic digit is associated with the (background) scene of a place with the user's arbitrary arrangement inside a video frame. The sequence of places captured is traceable in a physically continuous path that helps the user recall each place in the sequence. The associated graphic digit strengthens the hookup between the places in addition to direct mapping of a digit to the place. The method was evaluated by a memorization experiment below.

Fig. 1. Graphic digits

3.2 Presentation Condition

The subjects memorized a series of locations in two conditions: SROM condition and no-SROM condition.

SROM condition

The image presented in the SROM condition is shown in Fig. 2(a). A graphics indicating the ordinal number of the place, in this case a swan for the second, is overlaid on the background image of the location. The subject arranged the swan in a cardboard box by positioning his body and rotating his head to orient the camera on the cap to the appropriate direction. The subject observes the image by a single-eye head-mounted display. The graphic digit moved and related to the place makes an impressive visual image for the observer with an episode of his intention to create the image. It is expected that these will produce effective recall cues for the location sequence.

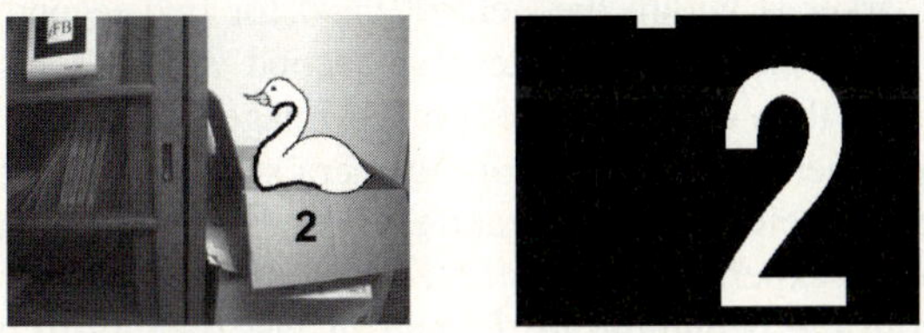

(a) (b)

Fig. 2. Presented images in (a)SROM condition, (b)no-SROM condition

no-SROM condition

In a usual condition to memorize many places, the images of the place are only recalled sequentially with its ordinal number in the mind. So, only the Arabic ordinal number assigned to each place is presented in the no-SROM condition. The displayed image is shown in Fig. 2(b).

In the both condition, the images in Fig. 2 are presented for ten seconds after the picture is taken. A rectangle progress indicator for the ten-second observation moved right at the top the display.

3.3 Experimental Setup

As shown in Fig. 3, a cap mounted small camera captures the user's view, and a single-eye head-mounted display (SV-6, MicroOptical Corp.) presents the overlaid video image or an Arabic digits image. In the experimental area, fifty AR markers (AR toolkit [3]) are put on a wall or objects. The display control is triggered when the camera detected one of the markers. A notebook PC in a backpack controls both sensing and displaying during the experiment. Subjects' responses are recorded as they performed pointing on a map displayed on the monitor screen of a desktop PC in addition to recording of their voices for think aloud protocol.

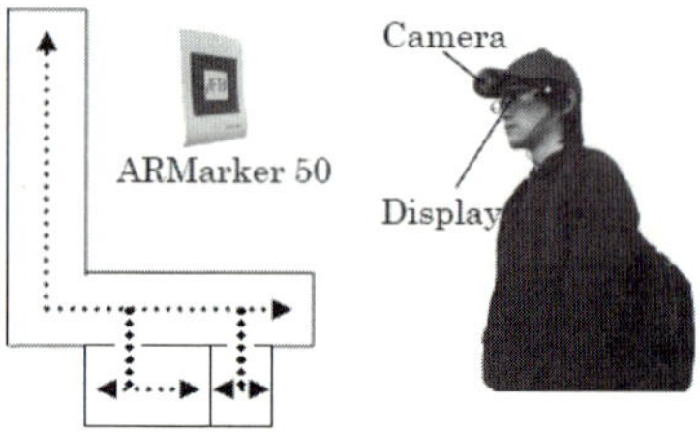

Fig. 3. Experimental workspace and wearable setup

3.4 Procedure

Within an about forty-meter path that had fifty AR markers, the subject selects twenty-five markers arbitrarily and memorize the place and its ordinal number starting from one end of the hallway. The system detects the user's selection of a marker (place) when the marker is within the field of view for two seconds and the system notifies the user it by a short sound. After five-second wait, the system captures the scene of the place and then displays a designated image for ten seconds. For SROM condition, the subject performs framing of a camera so that the graphic digit and the scene can be related to facilitate memorization. The subjects learned the graphic digits from 1 to 25 before the experiment.

The duration from the beginning to the end of last (twenty-fifth) presentation is about ten minutes. The subject was asked to recall all the places and point them on the map after two minutes rest with eyes closed. After reported the places at random order following the indication on the screen, the subject answered the places serially from the start. The answer was also video taped for protocol analysis. Each subject performed two conditions with separation of an hour rest. The start point and the goal point were exchanged between the two conditions to avoid learning effect. The recall procedures were performed three times: immediately after the experiment, two days later, and four months later. The subjects are six graduate and under graduate students with normal visual acuity and memory twenty-three years old on the average.

4 Results and Discussion

The average percentage correctly recalled and the average response time are shown in Figs. 4 and 5 (where error bar indicates standard error). Correct recall in the case of SROM two minutes after was 94.7 %, 89.3 % and 92.0 % for serial recall, random recall and the average, respectively. No-SROM condition exhibited statistically significantly lower correct recall ratio against the SROM condition. Significant difference was observed in the result obtained two days after the session. This indicates that SROM provided appropriate encoding for recall based on its three features of graphic digits, a photo of selected location, and the arrangement of a graphic digit in a background scene. The response time was longer in the SROM condition, which would be ascribed to transformation between the ordinal number to its graphic figure. The verbal report the subject showed during the response session supports this cause. It is expected that this extra time is reduced when the user is accustomed to the graphic figures. As compared to serial recall, random order recall seems more difficult for subjects. Although not statistically significant the correct recall was lower when recalled in a random order. The average response time was longer in the random recall than in the serial recall. The serial recall had a cue of places and it was performed after the random recall, which may account for the difference.

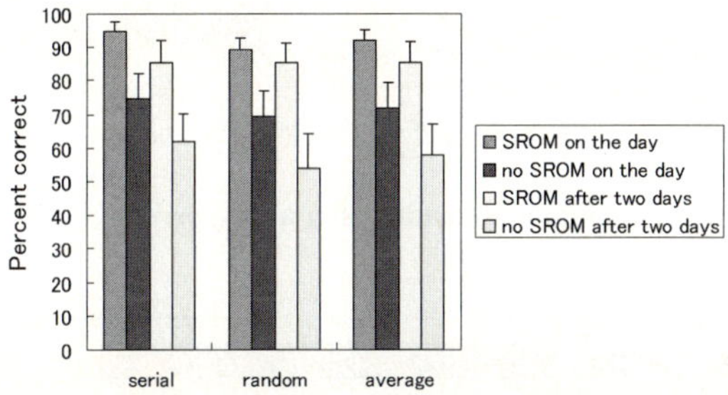

Fig. 4. Percentage correctly recalled

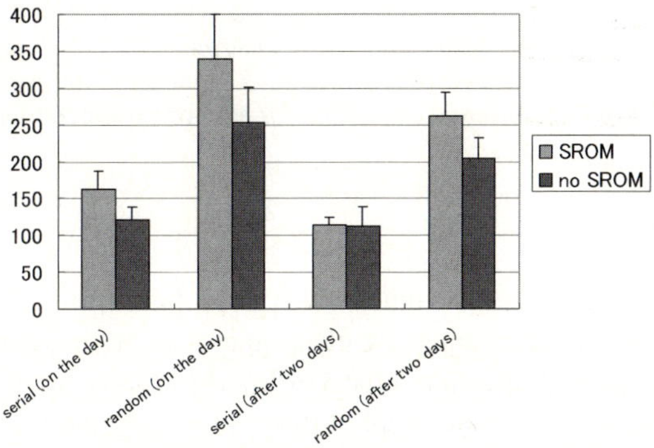

Fig. 5. Response time (in second)

Figure 6 shows the retention after a long period. SROM indicated a very high recall ratio of 62 % after four months while no-SROM was 15 % recall. It is considered that SROM could constructed effective cognitive organization for recall in view of short interval distances and less salient features of locations designated to include in the series to remember.

The result of subjective evaluation performed after the session is shown in Fig. 7. The question was "Was it easy to memorize places?" The scores differed remarkably between SROM and no-SROM conditions, 4.1 and 1.9, respectively. The subjects felt it much easier to memorize with SROM than no-SROM. They left such comments that memorability was increased even for the places with no distinctive features by creating characteristics on the fly relating graphic figures to the place, and that more increase in memorability is expected if more familiar places are used as a basis for vMPegs.

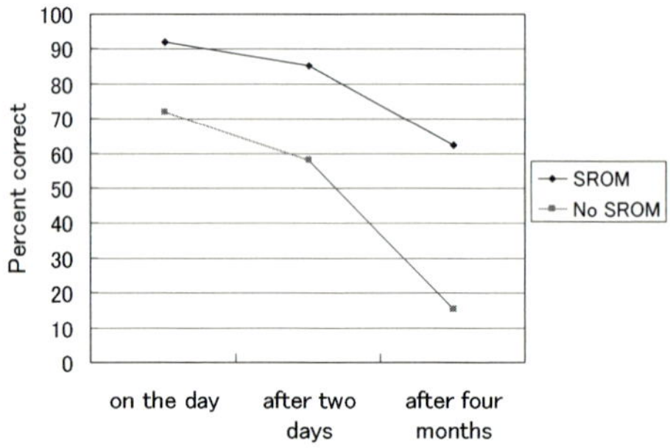

Fig. 6. Retention after a long period

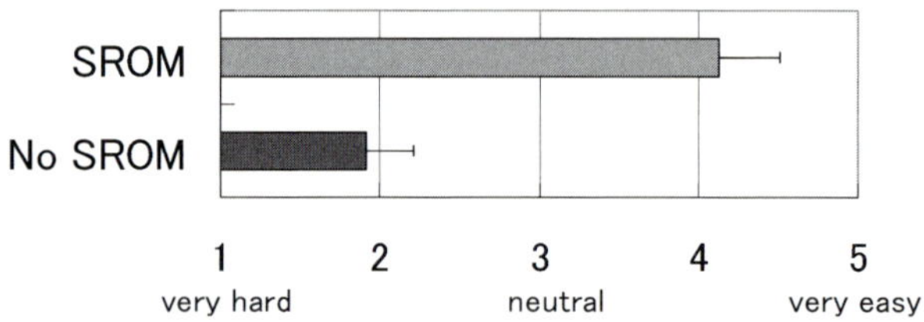

Fig. 7. Subjective evaluation for ease of memorization

5 Conclusions

A basic idea of spatial electronic mnemonics is proposed in which traditional techniques of mnemonics is augmented by electronic devices. In terms of a registration phase of SROM system, a location–type vMPeg is partially implemented and evaluated. A new technique to arrange graphic figures to relate them to a background image of location was evaluated experimentally. The result suggested this technique the user takes a photo of a place as overlapped with graphic figures significantly assisted memorization of locations as a basis of spatial mnemonics. Specifically a definite effectiveness was expected in subjective cognitive load of the user during memorization process.

References

1. Bellezza, F.S.: Mnemonic Devices: Classification, Characteristics, and Criteria. Review of Educational Research 51(2), 247–275 (1981)
2. Yates, F.A.: The Art of Memory. London: Routledge & Kegan Paul (1966)
3. http://www.hitl.washington.edu/artoolkit/

An Assistant Tool for Concealing Personal Information in Text

Tomoya Iwakura and Seishi Okamoto

Fujitsu Laboratories Ltd
Address: 1-1, Kamikodanaka 4-chome, Nakahara-ku, Kawasaki 211-8588, Japan
{iwakura.tomoya,seishi}@jp.fujitsu.com

Abstract. This paper presents an assistant tool for concealing personal information in text. Concealing personal information is one of the important roles for protecting privacy in disclosure of public documents, protection of accidental personal information leakages, and so on. However, concealing personal information is very time-consuming, because it is strongly depending on manpower. In order to alleviate tasks of concealing personal information, we have developed a graphical user interface (GUI) tool that has the following three characteristics: 1) Extracting candidates of personal information in text. 2) Presenting the candidates with colors indicating types of personal information. 3) Creating extraction rules for personal information from text including annotations of personal information. The experimental results on tasks of concealing person names in Japanese text showed that processing times of concealing personal names with candidates of person names were about 1.5 to 3.9 times faster than without candidates of person names.

1 Introduction

Protection of personal information is now one of the problems for disclosing documents, making ready personal information leakages, and so on. For example, Act on Access to Information Held by Administrative Organs [1] was effective fully in 1999 for providing for the right to request the disclosure of administrative documents. A problem of disclosing administrative documents is that documents have personal information. In order to disclose administrative documents, the Japanese low says that the protection of personal information in documents is obligation. Furthermore, the law related to protection of personal data applied to private company was fully executed in April of 2005 in Japan [2]. As the results, Japanese companies have to pay more attention than before.

One of the methods for protecting personal information is concealing personal information in documents, and it is necessary to disclose documents. However, concealing personal information is very time-consuming works, because the concealing process strongly depends on manpower. One of the crucial problems is finding personal information in text. Workers concealing personal information have to find personal information by reading all text in documents very carefully, because text written by natural language has no structures without processing texts.

M.J. Smith, G. Salvendy (Eds.): Human Interface, Part II, HCII 2007, LNCS 4558, pp. 38–46, 2007.

In this paper, we present an assistant GUI tool for concealing personal information in text. Our GUI tool has the following characteristics

- Extracting personal information: Our GUI tool extracts several types of proper nouns and numeric expression with their classes, such as person names, locations, organizations, and so on, as personal information candidates in text.
- Presenting candidates of personal information: Our GUI tool presents candidates of personal information with colors indicating their classes, and users can conceal the candidates by just clicking them.
- Creating extraction rules for personal information: Our GUI tool creates extraction rules from text including annotations of personal information to be concealed.

This paper is organized as follows. In section 2, we present examples of concealing personal information in texts. In section 3, we present our assistant GUI tool for concealing personal information. In section 4, we report the experimental results on tasks of concealing personal information in Japanese text. In section 5, we discuss what our approach for extracting personal information differs from the other approaches. Finally, we conclude in section 6.

2 Concealing Personal Information in Text

Text data may include information identifying person, such as person names, dates of birth, addresses, and so on. To protect privacy, we must recognize person identifiable information and transform them to different forms not identifying him or her.

For example, a court decision in Japan is not disclosed without concealing personal information [2]. In order to disclose such documents, we have to conceal personal information in documents. The following example shows concealing personal information. If underlined person names are anonymized or concealed, we can disclose documents.

- Original text: An accused man, Yamada, was released after agreeing to testify against the others. Mr. Yamada testified that …
- Text including concealed person names: An accused man, **X**, was released after agreeing to testify against the others. Mr. **X** said that …

From the above example, we see that even if person names are concealed, we can understand most part of information except for the person names. However, concealing personal information are very time-consuming, because the concealing process strongly depends on man power. Thus, we think that assistant tools for concealing personal information are required.

3 Our Assistant Tool for Concealing Personal Information

We present our assistant tool for concealing personal information in text. Our tool consists of a Named Entity (NE) extractor for extracting personal information, a GUI for concealing personal information, and a rule learner for personal information extraction.

3.1 Extracting Personal Information by Using Context Information

One of the problems for concealing personal information in text is that we have to correctly discriminate the meaning of words. For example, most Japanese location names are used as Japanese family names, and we have to use context information to discriminate meanings of words.

In order to correctly find personal information in text, we apply Named Entity (NE) extraction technologies. NE extraction aims to identify word chunks , such as proper nouns, numerical expressions, and classify the word chunks into their classes, such as persons, locations, organizations, dates, and so on, in text . Table 1 shows NE examples. In this paper, we extract NEs as candidates of personal information. We are now focusing on Japanese. We briefly describe our personal information extraction methods for Japanese.

Table 1. NE examples

Numeric Expression

NE class	DATE	MONEY	PERCENT	TIME
Example	May 5th	200 JPY	100 %	10 PM

Proper Noun

NE class	ARTIFACT	LOCATION	ORGANIZATION	PERSON
Example	Novel Prize in Chemistry	Japan	Fujitsu	Jorge White

NE representation in text: We use a method that classifies words into NE chunk labels by using context information for extracting NEs, because NEs consist of one word or more than one word. We used Start/End (SE) representation. SE representation uses the five tags, S, B, I, E and O, for representing word chunks becoming NEs [3]. S means that the current word is a chunk consisting of only one word. B means in the start of a chunk consisting of more than one word. E is the end of a chunk consisting of more than one word. I is the inside of a chunk consisting of more than two words. O is the outside of any chunk.

We used the five tags with NE class labels as NE chunk labels for representing NEs in text. For example, the person names and the outside of person names in

…, Mr. Michael W. White said to Mr. Brow…

are represented by using NE chunk labels as follows.
…, Mr./O Michael/B-PERSON W./I-PERSON White/E-PERSON said/O to/O Mr./O Brown/S-PERSON…
B-PERSON, I-PERSON and E-PERSON indicates the beginning, inside, and end of a person name, respectively. S-PERSON indicates a person name consisting of only one word.

Classifying word into NE classes: To classifying words into NE labels, we use information of surrounding two words as context in addition to current word information as context information. We use the following word information as features for extracting NEs.

- Words: Asian languages, such as Japanese and Chinese, have no word boundary. In order to segment words from Japanese sentences, we use a morphological analyzer.
- Part of speech (POS) tags: We use POS tags of words tagged by a morphological analyzer, which is used for segmenting words.
- Character types: We use several types of character types, such as 'Hiragana', 'Katakana', 'Chinese letter', 'capitalized alphabet', 'digit', 'sign', and so on, and these combination.
- Dictionaries: We use dictionaries for augmenting features, if available. The dictionaries include person title lists, address dictionaries. In this experiment, we used an NE dictionary created from news articles and web pages by using several NE extractors [4].

For example, to classify the word of "W." in the above sentence, word information of "Mr.", "Michael", "White" and "said" are used as context information to discriminate the NE class label of "W.".

Classifying character into NE classes: Japanese NEs include a part of a word becoming begin or end of an NE sometimes, because Japanese words have no explicit word boundary. For example, the "訪米(visit U.S.A)" in

田中使節団は訪米(Tanaka mission party visit U.S.A.)

dose not match with LOCATION "米(U.S.A)" because this sentence is tokenized by a morphological analyzer as

"田中(Tanaka) / 使節(mission) / 団(party) / は(particle) / 訪米(visited U.S.A)",
"/" indicates a word boundary.

To solve this problem, we apply a character-unit-chunking-based NE extraction algorithm [5]. We use character-unit-chunking based NE extraction after classifying words into NE classes. To classifying characters into NE labels, we use information of surrounding two characters as context in addition to current character information.

- Characters and words: We use words including characters within the current and surrounding two characters. Words are expressed with position identifiers to indicate where the character appears in words. We use B, I, E and S, which is the same as the SE representation
- POS tags: POS tags are annotated into words by a morphological analyzer. POS tags are expressed with position identifiers by SE representation, the same as characters' ones.
- Character types: We use Kanji, Hiragana, Katakana, digit, lowercase alphabet, uppercase alphabet, and the others.
- NE labels of words: We use NE labels of words tagged by a word-unit-chunking-based NE extractor with stacking [6]. NE labels of words are also expressed with position identifiers by SE representation.

- NE labels of preceding extraction results: We used NE labels of previous character. We classify characters into NE labels in the direction of the end to begin of a sentence.

Each character is classified into NE labels represented by IOB2 [5] representation. IOB2 representation uses the three tags, B, I, and O. B means in the start of a chunk. I is the inside of a chunk consisting of more than two words. O is the outside of any chunk. For example, the above example is represented with character-based IOB 2 representation as the following.

"田ORGANIZATION-B 中ORGANIZATION-I 使ORGANIZATION-I 節 団 ORGANIZATION-I は/O 訪O 米LOCATION-B

ORGANIZATION-B and ORGANIZTION-I indicates a beginning and inside of an organization. LOCATION-B indicates a beginning of a location, and O indicates outside of organizations and locations.

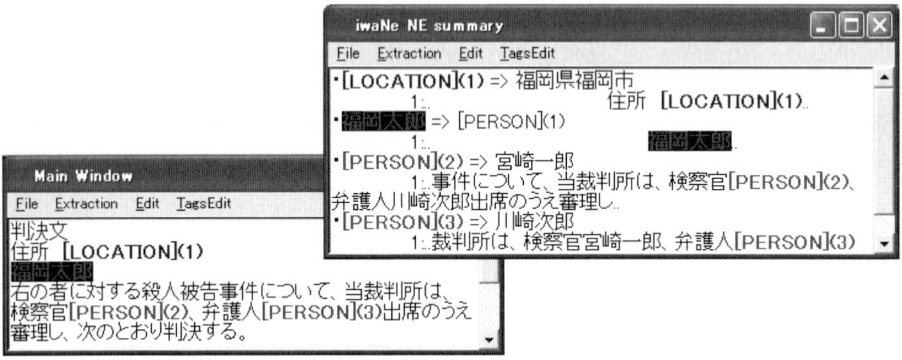

Fig. 1. A snap shot of our tool: The left window is presenting whole text. The right window is presenting a summary. Concealed person names and a location name are represented by their classes and numbers. The person name focused by a mouse pointer is with black back ground color.

3.2 Our GUI for Concealing Personal Information

One of the annoying procedures for concealing personal information is finding personal information in text. Since text has no structure differing from structured data like Relational Database, workers concealing personal information have to read whole text for finding personal information. In order to alleviate such finding procedures, our GUI tool presents candidates of personal information as follows.

- Presenting candidates: Our tool presents candidates of personal information with colors corresponding to types of personal information. By presenting the candidates of personal information with the colors of their corresponding classes, workers can find personal information by their colors. The left window in Fig 1 shows a snapshot of our tool. The characters in bold font indicate personal information. The black background place indicates that a mouse pointer is focusing on a candidate of personal information. Users of our tool can conceal the candidates by just clicking them. [LOCATION](1), [PERSON](2) and [PERSON](3) in Fig. 1 indicate concealed person names and a location name.

- Creating summary: Our tool creates summaries for extraction results of personal information by collecting candidates of personal information with their surrounding words. By presenting summaries of extracted candidates of personal information, workers concealing personal information can conceal personal information without reading whole text as far as personal information extracted correctly. The right window in Fig. 1 shows a snapshot of a summary created by our tool. User can conceal personal information in text by clicking candidates of personal information presented in the summary.

Our tool also provides users shortcut keys for modifying wrong extractions. Fig. 2 shows a snapshot for a modification of a wrong extraction. For example, if a user defines ‘key p with Alt-key’ as annotating person tags, the user can annotate person tags by selecting the places to be annotated with a mouse and pressing ‘key p with Alt-key’ Our tool also provides users short cut for deleting wrong extractions. Since our tool provides users shortcut keys for modifying and deleting wrong extractions, users can edit wrong extractions only few steps.

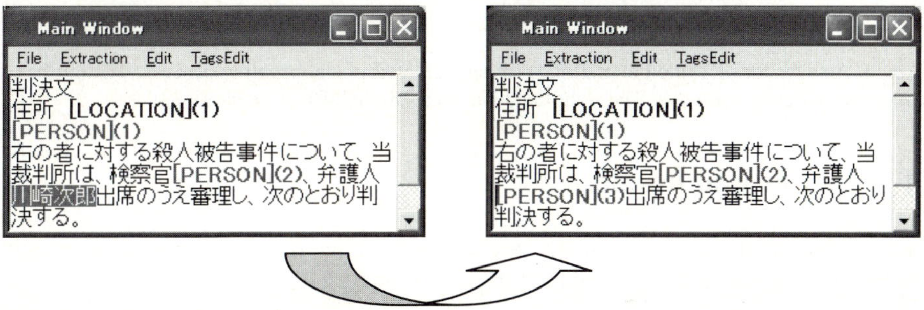

Fig. 2. A snapshot of modifying missed extraction. Users can annotate tags of personal information by selecting the parts of personal information and pressing shortcut key corresponding to personal information classes.

3.3 Creating Rules for Extracting Personal Information

There are two approaches for creating rules extracting personal information. First one is hand-crafted rule based approaches. The other is machine learning based approaches.

If we use hand-crafted rule based approaches for extracting personal information, we can revise extraction behaviors by adding new rules or modifying current rules. However, creation and modification of rules are very time consuming and it is necessary for rule developers to learn how to create rules. If we use machine learning based rule creation approaches, we can obtain extraction rules from text that personal information are concealed. However, it is difficult to control extraction behaviors.

The two approaches have different benefits and drawbacks. However, we apply a machine learning based approach, because machine learning based approaches can crate rules from new text including annotations of personal information created in routine work.

We use a boosting algorithm as the machine learning algorithm. Please refer to the paper [7] for the detailed explanation of boosting algorithms. Our rule learner creates rules extracting personal information from training data including annotations of personal information tagged with their class names. Fig 3 shows an example of a rule generation from data created in a concealing process.

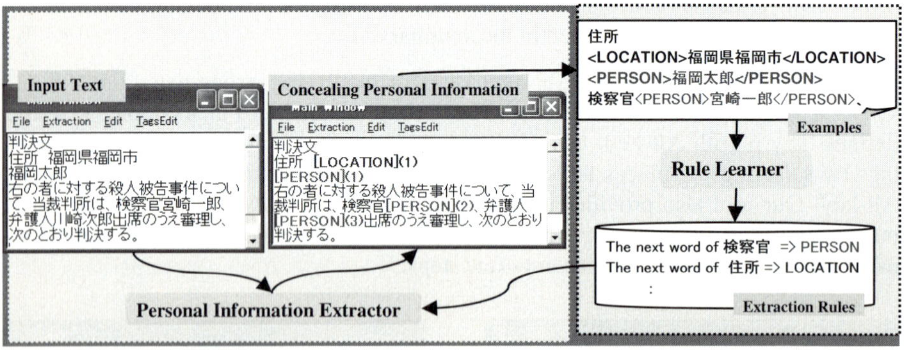

Fig. 3. A snapshot of a rule generation from data created in a concealing process: The left hand side is a concealing process. The right hand side is a sample of rule generation.

4 Experimental Results

We evaluated effectiveness of our tool by measuring times of concealing person names in Japanese text. We used the following data for our evaluation.

- A pseudo court decision: The pseudo court decision has 29 types of person names. The total number is 45. The size of the data is about 12KB. The person names appear in text all the here and there.
- A minute in a company: The minute has 25 types of person names. The total number is 76. The size of the data is about 12 KB. This minute is well organized compared to the pseudo court decision. For example, all the persons attending the meeting are presented at the beginning of the minute.

We conducted the evaluations as follows.

- Concealing person names: When we had candidates of person names presented by our tool, we used them at first for concealing person names. After that, we read whole text to find person names not concealed. If we did not have any candidates of person names, we concealed person names by reading whole text and using shortcut keys of our tool.
- Checking missed person names: We have done this phase for checking person name not concealed. We read whole text again.

We conducted all the tests using candidates before the tests without candidates for evaluating performance of our tool on fully unseen data. We used the rules extracting person names generated from training data including 7818 person names. The training data includes no text of data set for evaluation. We used MeCab, a morphological analyzer, for segmenting words from text[1]. We measured extraction accuracy of our GUI tool by using Recall, Precision, and F-measure, defined as follows.

Recall = NUM / (the number of correct person names)
Precision = NUM / (the number of person names extracted by our GUI tool)
F-measure = 2 * Recall * Precision / (Recall + Precision),

where NUM is the number of person names correctly extracted.

Table 2 shows the extraction accuracy of our tool. Our tool extracts person names more than 90 % F-measures.

Table 2. Our GUI tool performance on our evaluation data

Data Set	# of person names (# of types)	Precision	Recall	F-measure
A pseudo court decision	45 (29)	97.44	84.44	90.48
A minute in a company	76 (25)	98.65	96.05	97.33

Table 3 shows that the experimental results of concealing person names with two subjects for each task. The results showed that our tool enabled the subjects to conceal person names about 3.3, 3.9, 1.9 and 1.5 times faster than without using candidate of person names.

We think that the reasons are as follows. The first is that times for finding person names are greatly reduced by using candidates. The other is that times for concealing person names are also reduced because we can conceal them by just clicking them.

Table 3. Experimental results of concealing personal names

Subjects	Manual operation (a)	With Our tool (b)	Improvement((a)/(b))
A pseudo court decision			
A	938 seconds	243 seconds	3.9 times
B	706 seconds	218 seconds	3.3 times
A minute in a company			
C	486 seconds	245 seconds	1.9 times
D	578 seconds	384 seconds	1.5 times

5 Related Works

There are three approaches widely used for extracting personal information. The first approach is dictionaries based personal information extraction [8]. This approach has the advantage that users can control extraction behaviors by just adding new entries of

[1] http://mecab.sourceforge.net/

personal information into dictionaries. However, the approach can not discriminate meanings of words having meanings more than one. Furthermore, the approach tends to output many noises.

The second is hand-crafted rule based extraction [9]. This approach has the advantage that users can control extraction behaviors by creating and modifying rules. However, the creation of rules and the managements of rules require skills.

The third is machine-learning based extraction adopted by our tool. This approach crates extraction rules by machine learning algorithms. Although the approach has difficulties to control extraction behaviors, this approach has the advantage that users can crate new rules by creating data including annotations of personal information. Furthermore, this approach provides users to enhance extraction performance by incorporating data including concealed personal information created in daily work.

6 Conclusion

This paper has presented our GUI tool to help concealing personal information in text. Our GUI tool finds personal information in text by using a Japanese Named Entity extractor discriminating meanings of words. We have experimentally demonstrated that our GUI tool contributes to improve processing times of concealing personal information in text about 1.5 to 3.9 times.

We think that further improvements of concealing personal information will be realized by enhancing performances of personal information extractors. Future work should consider methods to combine approaches based on dictionaries, handcrafted rules, and machine learning algorithms, for further improvement.

References

[1] Act on Access to Information Held by Administrative Organs (Act No. 42 of 1999) web page. http://www.soumu.go.jp/english/gyoukan/060516_03.html

[2] Web page for disclosed court decision (in Japanese). http://www.courts.go.jp/saisinhanrei.html

[3] Uchimoto, K., Ma, Q., Murata, M., Ozaku, H., Utiyama, M., Isahara, H.: Named entity extraction based on a maximum entropy model and transformation rules. In: Proc. of the ACL 2000, pp. 326–335 (2000)

[4] Iwakura, T., Okamoto, S.: Improving Named entity extraction accuracy using unlabeled data and several extractors. CICLing (to appear 2007)

[5] Asahara, M., Matsumoto, Y.: Japanese named entity extraction with redundant morphological analysis. In: Proc. of HLT-NAACL 2003 (2003)

[6] Wolpert, D.H.: Stacked generalization. Neural Networks 5, 241–259 (1992)

[7] Schapire, R.E., Singer, Y.: BoosTexter: A boosting-based system for text categorization. Machine Learning 39(2/3), 135–168 (2000)

[8] Dehenken Web page (in Japanese): http://www.dehenken.co.jp/products/products-03/products-kansalib01.html

[9] Takemoto, Y., Fukushima, T., Yamada, H.: A Japanese Named Entity Extraction System Based on Building a Large-scale and High quality Dictionary and Pattern-matching Rules (In Japanese). IPSJ Journal 42(6), 158–159 (2001)

Effect of Providing a Web-Based Collaboration Medium for Remote Customer Troubleshooting Tasks

Chulwoo Kim[1], Pilsung Choe[1], Mark R. Lehto[1], and Jan Allebach[2]

[1] School of Industrial Engineering, Purdue University, West Lafayette, IN 47907, USA

[2] School of Electrical and Computer Engineering, Purdue University, West Lafayette, IN 47907, USA

{kim218, pchoe, lehto, allebach}@purdue.edu

Abstract. This study investigates the effect of providing a Web-based diagnostic tool as a collaboration medium on remote customer troubleshooting tasks with and without the assistance of a customer call center agent. The study tested three troubleshooting modes (Web tool alone, call center agent alone, and Web tool + call center agent). The hypothesis that the Web tool + call center agent mode would be faster than the other two modes was not supported. However, the results of the experiment showed that the Web-based self-help diagnostic tool can be a potentially cost-effective way of providing customer support. The performance using the Web tool alone was comparable to the call center agent.

Keywords: collaborative troubleshooting, communication mode, web-based diagnostic tool.

1 Introduction

Customer support operations are increasingly supported by communication technologies such as instant messaging, e-mail, etc [1]. However, the telephone is still the most popular form of customer support service. Customers make phone calls to a customer call center and work together with an agent to solve their problems.

This research focuses on the inefficiency originating from the interaction between the customer and the call center agent. Since the customer usually does not have the background information to troubleshoot his/her problem, the call center agent goes through the detailed troubleshooting procedures. It is a time consuming task. As a way of improving the efficiency of the interaction, this research considers adding a visual information channel with the assumption that adding a visual information channel, the Web-based diagnostic tool, to the audio channel, telephone, would shorten the interaction time.

1.1 Literature Review

Conversations during collaborative physical tasks involve the identification of objects, descriptions of procedures/actions, and confirmation of whether the actions are completed [2]. Therefore it is important to establish a *common ground*, which

M.J. Smith, G. Salvendy (Eds.): Human Interface, Part II, HCII 2007, LNCS 4558, pp. 47–53, 2007.

refers to mutual knowledge, understanding, and assumptions between the participants [3], for effective collaboration [4]. The collaborators establish the common ground through *grounding*, communicating what they do or do not understand over the course of a conversation [4]. Grounding becomes more critical for the joint activities between a novice user and an expert since the expert guides the novice user who significantly lacks common ground through a task.

This study assumes that providing a visual information channel would be effective for establishing common ground and improving communication between the two parties. The assumption that providing visual information channel with audio channel would help the collaboration for the collaborative physical tasks can be supported by media richness theory [5]. Richer media provided rapid feedback and multiple cues so that participants could reach a common interpretation by reducing information uncertainty and ambiguity. [6] indicates that the selection of communication mode needs to be based on the assessment of the richness of communicated information since the fit between the message and the communication mode influences the performance. A rich medium should be used for communicating difficult messages and a lean medium for simple messages.

1.2 Purpose of the Study

The literature provides evidence for the argument that adding a visual information channel to an audio channel would increase the performance of a collaborative task. Therefore, the research hypothesizes:

Providing a Web-based collaboration medium (Web-based diagnostic tool) for a physical collaborative task between the customer and the call center agent would reduce the task completion time.

The study also aims to check whether a Web-based diagnostic tool can be easily used by the customers without the assistance of the call center agent. Three troubleshooting modes are investigated in the context of a print quality troubleshooting process: Web tool alone, call center agent (CCA) alone, and Web tool + CCA. The Web tool is expected to reduce the interaction time between the customer and the call center agent by reducing time related to *grounding*.

2 Method

2.1 Web-Based Diagnostic Tool

The Web-based diagnostic tool consists of hyperlinked Web pages, which describe the necessary procedures to troubleshoot and repair print quality problems. The tool is composed of three layers as shown in Fig. 1. Layer 1 consists of printing test pages and then selecting a test page with a problem. Layer 2 consists of selecting a matching print quality issue. Layer 3 consists of performing troubleshooting procedures. The troubleshooting process is described in the detailed step-by-step procedures with explanatory images. The images of printer components and detailed explanation of procedures were provided as hyperlinks to help the understanding of the users assuming that the users have no previous knowledge about the printer. Several sets of

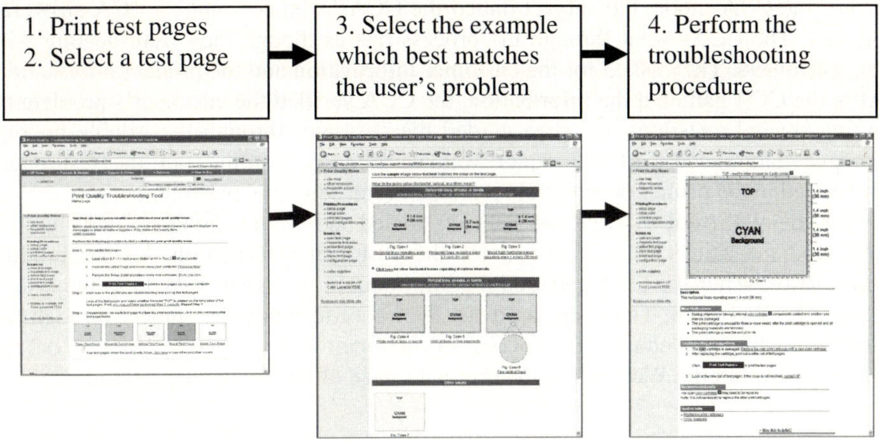

Fig. 1. Troubleshooting process of the web-based diagnostic tool

cognitive walkthrough, heuristic evaluation, and user testing were conducted during the development process to improve the usability of the website.

2.2 Experiment

Twelve students at Purdue University were paid to participate in the experiment. Real CCAs at a remote site participated in the experiment as a CCA. The subjects were composed of seven males and five females with an average age of 25.2 yrs. They had used the Internet more than two hours a day without previous experience with the printer used in the experiment.

A printer was placed right next to a speaker phone and a computer with a LCD monitor so that the subjects could work on the printer while looking at the website or talking with the CCA.

A 3x3 Latin square design was used. A total of nine tasks were generated by the full combination of three print quality issues and three troubleshooting modes (Web tool alone, CCA alone, and Web tool + CCA). The subjects were divided into three groups and each subject performed three scenarios of diagnosing and repairing a print quality problem.

Before starting the task, the subject was given a brief instruction on the purposes and the overall procedures of the experiment and a consent form to sign. After the initial briefing, the subject was given a defective printout with its original digital image on the computer monitor, and was asked to fix the problem using one of the three troubleshooting modes. After the task, a questionnaire was given to the subject to collect satisfaction measures on the troubleshooting mode. The subject repeated this procedure for the other two print quality issues.

For the Web tool alone mode, the subject only used the Web tool without the assistance of a CCA. Using the troubleshooting instructions on the Web tool, the subject fixed his/her problem, and verified whether the problem was fixed by printing out another set of test pages.

For the CCA mode, the subject dialed the CCAs' numbers until a CCA answered the call. The CCAs went through the process just as though they were dealing with real customers. They asked for the customer information and the product information. After the CCA gathered the information, the CCA verified the customer's problem by asking several questions and then guided the customer through the troubleshooting process. The success of the repair was verified by printing out the test pages embedded in the printer or the original image of the defective printout.

For the Web tool + CCA, the subjects dialed the CCAs' numbers until a CCA answered the call. The CCA asked for the customer information and the product information. After the CCA gathered the information, the agents asked the subjects to access the Web tool. The agent used the Web tool to guide the subject through the troubleshooting procedures. The success of the repair was verified by printing out the test pages within the Web tool or the original image of the defective printout.

3 Results

The tasks in the experiment were divided into two phases: diagnosis and repair. The diagnosis phase indicates the activities required until the subject or the CCA identifies or becomes confident about what the subject's issue is. The repair phase indicates the activities required until the subject fixes the printer by performing the troubleshooting procedures after identifying what the issue is. The time required to connect to one of the CCAs and to gather the customer and the product information is eliminated from the analysis.

Table 1 summarizes the diagnosis time, the repair time, and the total time for each troubleshooting mode. The diagnosis time for the Web tool is 46% faster than that of the CCA and is 53% faster than that of the Web tool + CCA. The repair time for the CCA is 25% faster than that of the Web tool and is 34% faster than that of the Web tool + CCA. The total time for the Web tool is 8% faster than that of the CCA and is 31% faster than that of the Web tool + CCA.

Table 1. Time to diagnose and repair a print quality problem (unit: sec.)

	Web tool		CCA		Web tool + CCA	
	Mean	S.D.	Mean	S.D.	Mean	S.D.
Diagnosis	271	139.6	503	295.4	582	388.9
Repair	585	295.4	432	156.1	650	279.4
Total time (diagnosis+ repair)	856	407.7	935	304.1	1232	458.7

Tables 2, 3, and 4 summarize the ANOVA results for the diagnosis time, the repair time, and the total time. The main interest of this study is the main effect of the troubleshooting mode. The main effect of the troubleshooting mode was significant for the diagnosis time ($F(2,16)=4.62$, $P=0.026$) and for the total time ($F(2,16)=6.04$, $P=0.011$) at $\alpha=0.05$. The results of the post-hoc analysis using Student-Newman-keuls test ($\alpha=0.05$) showed that the diagnosis time of the Web tool is significantly faster than those of the CCA and the Web tool + CCA. The total times of the Web

Table 2. ANOVA summary table for the diagnosis time

Source	d.f.	Mean Square	F-value	P-value
Troubleshooting mode	2	312510.0	4.62	**0.026***
Issue	2	103733.7	1.53	0.246
Troubleshooting mode x Issue	4	106483.6	1.57	0.229

* Significantly different at α =0.05

Table 3. ANOVA summary table for the repair time

Source	d.f.	Mean Square	F-value	P-value
Troubleshooting mode	2	151023.7	2.90	0.084
Issue	2	186933.0	3.59	0.052
Troubleshooting mode x Issue	4	33601.0	0.65	0.638

Table 4. ANOVA summary table for the total time

Source	d.f.	Mean Square	F-value	P-value
Troubleshooting mode	2	470211.1	6.04	**0.011***
Issue	2	410420.6	5.27	**0.018***
Troubleshooting mode x Issue	4	179638.0	2.31	0.103

tool and the CCA were significantly faster than that of the Web tool + CCA. There were no significant interaction effect between the troubleshooting mode and the issue.

Table 5 shows the average ratings of the subjective questionnaire measures for each troubleshooting mode. There are no significant differences among the three troubleshooting modes (F(2, 33)=1.90, P=0.166 for 1, F(2, 33)=2.54, P=0.094 for 2, F(2, 33)=0.64, P=0.531 for 3) other than the ratings for overall satisfaction (F(2, 33)=4.28, P=0.022). The results of the post-hoc analysis using Student-Newman-keuls test (α =0.05) showed that the overall satisfaction for the CCA is significantly higher than that for the Web tool. There were no significant differences between the Web tool and the Web tool + CCA and between the CCA and the Web tool + CCA.

Table 5. Average ratings of subjective measures

	Web tool		CCA		Web tool +CCA	
	Mean	S.D.	Mean	S.D.	Mean	S.D.
1) Easiness of identifying the issue	5.5	1.51	6.4	0.67	6.0	1.13
2) Easiness of performing the troubleshooting procedure	5.3	1.37	6.0	0.60	6.2	0.72
3) Easiness of performing the whole task	6.1	1.16	6.4	0.67	6.4	0.51
4) Overall satisfaction	4.9	1.31	6.3	0.65	5.6	1.44

4 Discussion

The hypothesis that adding a visual information channel to a voice channel would improve the interaction efficiency between the customer and the CCA was not supported. The Web tool + CCA mode was significantly slower than the other two troubleshooting modes. It can be explained in terms of the time required to specify the objects that the two parties were indicating. Especially, the subjects needed to pay attention to three objects during the Web tool + CCA mode: the printer, the Web tool, and the CCA. While listening to the CCA, the subject needed to look at/control both the Web tool and the printer. Although it might be helpful for the subjects to understand what the CCA was explaining, it took more time to confirm which object or which page the subject or the CCA was using. In addition, some subjects got confused when they did not locate the Web pages which the CCA was using.

It may be explained in terms of the amount of the information provided on the Web tool. The Web tool seems to have too much detail for the interaction between the customer and the CCA. This may suggest that a simpler version of the Web tool is needed for the collaboration between the customer and the CCA. Although the very specific details were helpful in guiding a novice user to perform the procedures, it took more time for the CCA and the subject to go through all the steps within the Web tool. When the CCA skipped some steps, customizing the steps according to the subject's situations, the subject was often confused and did not know which page the CCA was using. The Web tool used for the experiment seems more suitable when the customers try to solve their problems on their own without the assistance of a CCA.

The results of the experiment show a potential of the Web-based interactive step-by-step self-help tool in dealing with remote customer problems. The tasks used in the experiment were real troubleshooting tasks with some complexities. The troubleshooting procedures required many steps: performing paper setting, printing test pages, identification of print quality defects, matching print quality defects, replacing print cartridges, swapping print cartridges, etc. The tasks of performing the procedures took an average time of 16 minutes 48 seconds, and the longest one took 18 minutes 45 seconds. Even though the tasks were long and somewhat complicated, the diagnosis error rate was only 9.1% (1 out of 12), and the repair error rate was 0 %. The subjects were good at performing the procedures if the procedures are explained step-by-step in detail with the explanatory images. Especially, the Web tool was useful in diagnosing print quality issues. Given that the very nature of diagnosing print quality issues is very visual, the sample images of the print quality problems were very useful. The subjects were easily able to figure out their problems by comparing their images with the sample images on the Web tool. The diagnosis time for the Web tool was significantly faster than the other two modes.

References

1. Froehle, C.M., Roth, A.V.: New measurement scales for evaluating perceptions of the technology-mediated customer service experience. Journal of Operations Management 22, 1–21 (2004)
2. Kraut, R.E., Fussell, S.R., Siegel, J.: Visual information as a conversational resource in collaborative physical tasks. Human-Computer Interaction 18, 13–49 (2003)

3. Clark, H., Wilkes-Gibbs, D.: Referring as a collaborative process. Cognition 22, 1–39 (1986)
4. Clark, H., Brennan, S.E.: Grounding in communication. In: Resnick, L.B., Levine, R.M., Teasley, S.D. (eds.) Perspectives on socially shared cognition, pp. 127–149. American Psychological Association, Washington, DC (1991)
5. Daft, R.L., Lengel, R.H.: Information richness: a new approach to managerial behavior and organizational design. Research in Organizational Behavior 6, 191–233 (1984)
6. Keller, R.T.: Technology-information processing fit and the performance of R&D project groups: a test of contingency theory. Academy of Management Review 37(1), 167–179 (1994)

Natural Language Based Heavy Personal Assistant Architecture for Information Retrieval and Presentation

Algirdas Laukaitis, Olegas Vasilecas, and Vilnius Gediminas

Technical University, Sauletekio al. 11,
LT-10223 Vilnius-40, Lithuania
{algirdas.laukaitis,olegas}@fm.vtu.lt

Abstract. In this paper we present the progress of the natural language usage as the paradigm for information extraction and presentation in the enterprise environment. Distributed heavy personal assistant architecture and its implementation is presented as the solution to overcome difficulties related of the natural language use in the information systems development. A new methodology based on connectionist and symbol processing techniques for a knowledge worker to process his documents and utterance is suggested. Then we suggest the results from those processes to reuse for new documents classification and generation of small atomic applications. Finally the experiment is presented. We compare Microsoft EQ, IBM WebSphere Voice Server NLU toolbox and our solution for concepts identification accuracy.

Keywords: Information extraction, Natural language understanding, Ontology, Personal assistant, Enterprise Semantic web.

1 Introduction

Significant part of the artificial intelligent (AI) research has been dedicated for the natural language formalization. Nevertheless natural language processing (NLP) and understanding (NLU) paradigms are still struggling to find their way into information systems (IS) development beyond documents indexing techniques. In that sense, the history of natural language database interfaces (NLDBI) can be a good example of NLP use as a programming paradigm in one of the "*simplest*" case, i.e., when the program is one structured query language (SQL) sentence (see [1] for the field review).

There are several projects that have been lunched recently in the effort to use NL as the programming paradigm but no one have reached the level where the others not involved in the project can verify achieved results. Actually, except Microsoft English Query [10], we haven't found any other products for the investigation. In this paper we attribute several factors that prevented natural language programming to mature up to the industrial implementation level:

1. ***Over-simplicity*** from the user's point of view and over-complexity from the implementation point of view. The users of IS are not very much interested with the systems that supports only one sentence utterance and one sentence program (i.e. SQL

M.J. Smith, G. Salvendy (Eds.): Human Interface, Part II, HCII 2007, LNCS 4558, pp. 54–63, 2007.

sentence). In our system we suggest to build atomic applications with the natural language and then stack those atomic applications to implement business solution.

2. ***Distributiveness.*** When we are dealing with the natural language formalization we are faced with the overwhelming state space search problem. An idea suggested in this paper is that users are building their local knowledge bases via communication with the personal assistant. Then, natural language utterance is projected on the set of those local knowledge bases.

3. ***Integration*** of NPL into the whole life cycle of IS development. About 15 years ago, Kevin Ryan claimed that NLP is not mature enough to be used in software engineering [13]. Nevertheless, Internet has boosted NLP research and nowadays the natural language formalization it is not seen as an unachievable goal. In this paper we extend the methodology suggested by Laukaitis et. al. [8] to combine connectionist and symbolic processing techniques for NLP integration into requirements engineering, conceptual modeling and final reuse in the IS interfaces.

4. ***Open source and free*** of charge vs. close and commercial solutions. Presented solution would be impossible without the support from the open source community. In early stages of the project we tried the solutions from Microsoft and IBM (both solutions are compared at the final section of the paper) but they have been discarded due the problems of integrity with the other modules of the system. Those are the main modules we reused form academic community: GATE - the framework of the natural language processing [2], WordNet dictionary [11], self-organizing map SOM toolbox [15] , Galicia - formal concept analysis software [9], JMining - framework of the web applications [9]. Economical soundness of the architecture suggested in this paper can be proved only by use of free software as we suggest to replicate typical enterprise server architecture to each personal computer of the knowledge worker.

By keeping in mind all those factors we state the following problem: *How to create an environment for developing business applications by use natural language.*

The solutions of stated problem organize the rest of the paper as follows. First, we present the general framework of the model generation system from the IS documentation and engineers utterance. Then we describe personal assistant for the creation of local knowledge basis and advocate for NL use in distributed enterprise computing. Next we introduce JMining framework and language to which natural language statements are transformed. NLP module is described in details in section 4. Finally an experiment is presented where we compare several solutions.

2 General Framework of the Solution

Conceptual models offer an abstracted view on certain characteristics of the domain under consideration. They are used for different purposes, such as a communication instrument between users and developers, for managing and understanding the complexity within the application domain, etc. The presence of tools and methodology that supports coordination of textual documents and local knowledge bases is crucial for the successful IS development. Even more, we can say that the essence of modeling is the ability of modeler to classify the textual data and then to present it by some formal modeling language. Then artificial intelligence technologies that will attempt to automate IS modeling and development must follow that cognition process of human modeler.

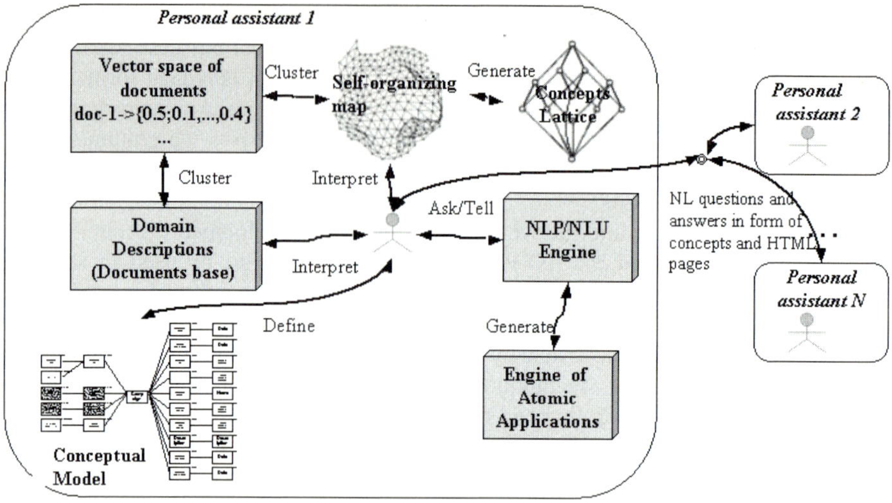

Fig. 1. Loopback of unstructured information flow

An overall process of automatically clustering the concepts and building knowledge basis is presented in Figure 1. First, the corpus is created form IS documents and utterance. Then, vector space of the corpus is created using natural language processing framework, domain ontology and WordNet ontology [11]. The self-organizing network [7] is build and used for cluster analysis. Next, with conceptual context and formal concept lattice [3] improvements are made in the understanding of clusters relationships. After the modeling stage the conceptual model self-organizing map and concept lattice are reused by natural language processing and understanding engine. The NLP/NLU engine combines map and lattice with such traditional NLP processing techniques like finite state traducer and gazetteer lookup.

Some philosophical arguments for the use of the connectionist paradigm in the context of information systems development can be found in the paper of Timo Honkela [5]. By the reference to the works of Von Foerster, he supposed that most information systems are developed as "trivial machines" to be predictable and controllable. All results in this paper can be interpreted as an implementation of these philosophical statements.

2.1 Heavy Personal Assistant

The idea behind heavy personal assistant architecture (HPAA) is to investigate the IS architecture that resembles society of humans in which natural language plays an important role. If the Figure 1 presents the general information flow processes in HPAA then the Figure 2 presents the technological implementation of the framework. As we can see, the central idea behind technological implementation of the heavy personal assistant is that traditional enterprise server architecture is replicated for each corporate personal computer to handle personal information flow and communication with other personal Web services in the Enterprise Semantic Web environment. We

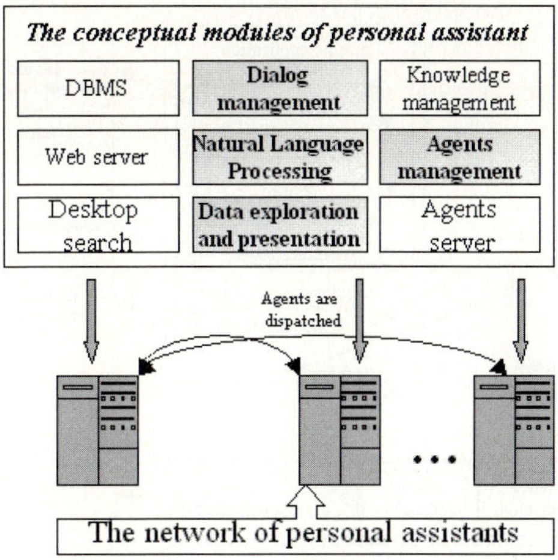

Fig. 2. Heavy personal assistant architecture

assume that all conceptual modules are installed on each personal computer involved in the information and knowledge management process. The grey shaded blocks represents the modules that have been originally build in current research project.

The white blocks represent software modules that we reused from open source community. All of them are free of charge (an important issuer if we are planning to install them on all working personal computers). The following example is the short description of the implementation we used in our framework (for more details we refer to [9]: 1. MySQL has been chosen as the database management systems. We store all annotated documents and instances of ontology classes in it. 2. We used the Apache Tomcat 5.5 Servlet/JSP container as the server of Web services. 3. Desktop search module represents software like Google desktop search. It's primary use in our architecture is to faster documents search for the agents that implements personal Web services. 4. For knowledge management we used *Protégé* application but problem with it is that it has difficulties for handling large OWL files.

3 JMining

As we mentioned above, our goal is the creation of the information systems developing environment by natural language. Previous section presented the framework for classification of the IS documents and utterance. By being able to identify classes and their semantics (i.e. relationships) we now present the JMining framework and language for the generation of business applications.

There were numerous attempts to use natural language as the programming paradigm. The most of them have concentrated on structural query language (SQL), however, some of them tried to lift a formal language representation complexity up to such languages like Java [12]. But neither of them have withstood more than few years. JMining language was created to overcome difficulties stressed by numerous projects in the field.

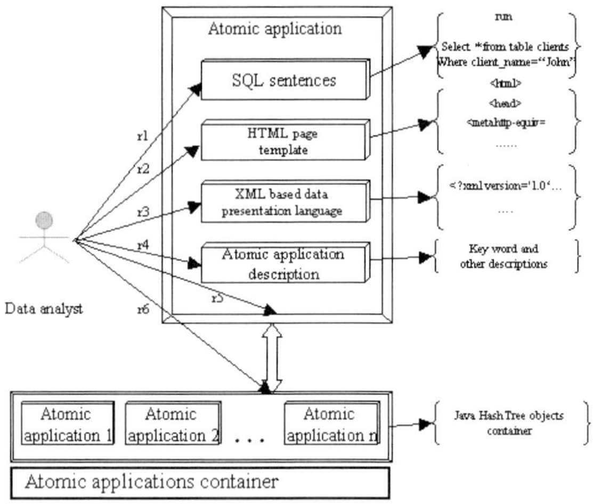

Fig. 3. Structures of the JMining framework

The central idea of the framework is an atomic application concept. An entire business solution is presented as a tuple of atomic applications liked by the URL references and sheared memory pool to store all parameter values entered by user. Figure 3 depicts the major structure of the atomic application. By its essence it is the small web application, which contains four components: {*database script, HTML page, data representation script (XML, XSL, etc.) and documentation page*}. No additional constrains are put on those components if they are accessed directly (i.e. manually corrected through Web Browser). If they are accessed using natural language interface, then all components must constrain their representation to the tuple calculus.

Each time the user enters NL sentence the system assumes that it is addressed to one of the six objects: container of the atomic applications, atomic application itself or one of the four objects that forms atomic application. Table 1 presents all state space variables that are used to control the system behavior. There are three futures of the JMining language that we attribute in the current research project:

1. *Expressiveness and complexity*. One of the reasons for the failure to lift NLDBI up to practical use is that most projects concentrated on the generation of one SQL sentence. However, practice shows that users are not interested in such interfaces and we solve this problem by addressing more objects than a single SQL sentence.

2. *Integrativeness*. No information systems development framework or programming language was developed to support IS development by means of NLP in all IS development life-cycle stages. In the previous section we introduced the modeling technique with the use of self-organizing maps. In the JMining language each atomic application has *documentation* object: the textual description of the atomic application. Each atomic application document has vector representation that can be measured by self-organizing map. Each time when the sentence is presented to the system it matches

self-organizing map fired neuron again atomic application vector. The closes match associates selected atomic application with the sentence.

3. *Distributiveness.* As we mentioned the state space produced by the natural language can be too huge to be matched by one ontology or conceptual model. In the JMining framework, if the personal assistant is not able to find the close match of the sentence in the local knowledge base then it tries to query other personal assistants. In the JMining framework remote personal assistant is using the same interface as the humans i.e. by using the Jakarta Commons HttpClient component it imitates Web Browser.

Table 1. The state space variables

Variable name	Example
agenda	Each item of the agenda is associated with some number: 0 - no agenda item selected, 1 - atomic application selected, 2 - get info from metadata storage, 3 - write script.
app. object	0 - no objects, 1-SQL, 2-HTML, 3-XML, 4-documentation
confidence	1 - if the object has been established, 0 -if not.
value_track	Tracks whether the system has obtained a value (no=0, yes=1).
number_of_ times	Tracks the number of times the dialog manager has asked for an attribute.

Primary the JMining has been developed to attribute IS development using only Web Browser on the "*thin*" computational device like mobile phone. And this means that all created atomic applications can be corrected by manually editing generated Jmining script using Web Browser.

4 Natural Language Processing

In this section we present NLP techniques because it is a core module in the HPA architecture. Figure 4 shows the main steps in the NLP processing. The goal of the engine is to produce vector representations and ontology annotation sets for each sentence that passes it. The vector space model (VSM) for documents transformation to the vectors is a well-known representation approach that transforms a document to a weight vector in automatic text clustering and classification. The method is based on the bag-of-words approach, which ignores the ordering of words within the sentence and uses basic occurrence information [14]. The NLP processes are depicted in Figure 4 and the main processes are described in details below.

GATE – general Architecture for Text Engineering - is a well-established infrastructure for customization and development of NLP components [2]. We briefly describe modules used in our research for building concepts vector spaces. The Unicode tokeniser splits the text into simple tokens and is used for the next steps of the natural language processing. The tagger is a modified version of the Brill tagger, which produces a part-of-speech tag as an annotation on each word or symbol. We used it to extract nouns and verbs and remove all other words from the dictionary. The gazetteer further reduces dimensionality of the documents corpus prior to classification. It uses the lists of named entities and annotates text with class labels such as cities, organizations, days of the week, etc. We replaced each named entity with the label of the class. Semantic tagger - provides finite state transduction over annotations based on regular expressions. It produced additional set of named entities and we replaced each named entity with the class label. Orthographic Coreference - the module adds identity relations between named entities found by the semantic tagger. Reduction of the state space dimensionality is achieved by replacing marked tokens with named entities class labels found by the semantic tagger. SUPPLE is a bottom-up parser that constructs syntax trees and logical forms for English sentences. We used it only to remove tokens not annotated by this module. All modules within the GATE produced annotations - pairs of nodes pointing to positions inside the document content, and a set of attribute-values, encoding linguistic information.

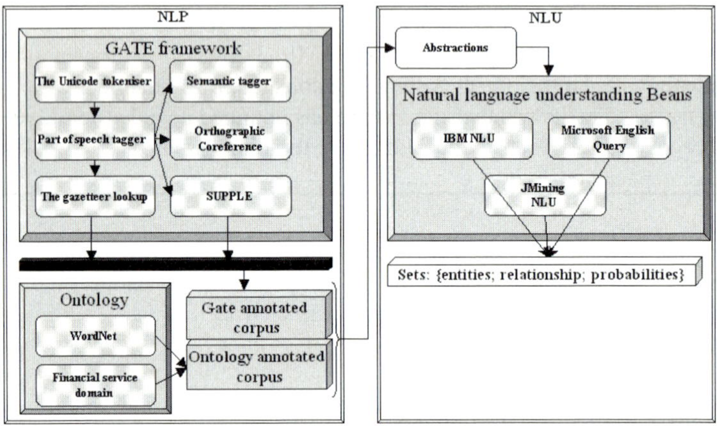

Fig. 4. Processes of the natural language processing and natural language understanding

Abstraction. The basic idea of the abstraction process is to replace the terms by more abstract concepts as defined in a given thesaurus, in order to capture similarities at various levels of generalization. For this purpose we used WordNet and annotated GATE corpus as the background knowledge base. WordNet consists of so-called synsets, together with a hypernym/hyponym hierarchy [4]. To modify the word vector representations, all nouns have been replaced by WordNet corresponding concept ('synset'). Some words have several semantic classes ('synsets') and in that case we used a disambiguation method provided by WordNet - the 'most common' meaning for a word in English was our choice. The words replaced by the GATE named entities annotation scheme were not included for the WordNet processing.

Vectors space. In our experiments we used vector space of the terms vectors weighted by *tfidf* (term frequency inverse document frequency)[14].

5 Experiment

In the previous sections we have shown how to build local knowledge bases for personal assistant and how to share those bases in the distributed enterprise environment of the personal assistants. In this section we provide description of two experiments we conducted to provide the answers for the following questions: 1. *What is rate of concept identification accuracy from user's utterance?* 2. *If the following hypothesis have any sound meaning? H0: If we measure the number of correctly answered questions of personal assistant to the questions brought by other personal assistants and then reword the knowledge worker to the extend of personal assistant performance, can we expect that the knowledge worker will be more interested in teaching it personal assistant.*

To answer the first question we conducted the following experiment. The Microsoft English Query [10] and IBM WebSphere Voice Server NLU toolbox V5.1 [6] have been chosen as benchmarks for the comparison with suggested solution in this paper. Microsoft product has been chosen because it's demo version is freely available and it comes as the most user friendly product available on the market (it seems that currently there is no other product left as several authors and companied declined our request to test their solutions). Comparison between Microsoft and IBM was interested because Microsoft utilizes rule based approach (symbolic processing) and IBM utilizes statistical approach (close to the connectionist paradigm). From the IBM presentation [6] it appear that the system is primarily intended to support database interfaces in the telecommunication market. It was a challenging task to test it on the more complex system e.g. a full Enterprise conceptual model for the financial market.

Table 2. Concept identification comparison between tree approaches and with two conceptual models

	Northwind database	CN=9	CN=50	CN=200	CN=400	CN=500
IBM NLU	30.82	36.82	17.26	14.82	11.15	8.22
Microsoft EQ	80.46	-	-	-	-	-
JMining	28.93	36.47	21.49	17.85	14.88	11.41

After the training process the following experiment has been conducted. A group consisting of 9 students has been instructed about two database models: 1) Northwind example comes with Microsoft EQ demo and IBM financial services conceptual model (see[9] for more details). They queried the system with about 20 questions and tried to identify the "*customer*" concept. For the Northwind example the second column in the table 2 shows the percentage rate of correct answers. No surprise that Microsoft scored the best as the model roles have been accurately written for this example. On the other hand for the second example where we interactively increased the set of concepts from 9 up to 500 there was no way to teach Microsoft approach due to limited resources. To answer the second question we made the following experiment with 3 students group. The 50 $ was the budget distributed proportionally with the knowledge contributed by the user working environment to the other users' personal assistants. The students where instructed about the IBM financial service model. At the first stage nor reword has been mentioned. As the consequence the personal assistants have shown very weak performance. As the reword has been introduced we noticed that the performance of the personal assistants increased about 50{\%}. Finally, we admit that the current experiment is not very sound but we conducted it anyway because question how to reword people for teaching their robots can be interesting for future research in the field of artificial intelligence.

4 Conclusion

Natural language based IS development can supplement traditional IS development approaches. In this paper we have shown how connectionist and symbol-processing paradigms can be combined within one single framework for information systems development. Distributed personal assistants architecture has been suggested to utilise this framework and create new working environment for the knowledge based professions. Provided evidences have shown that high quality documentation can be reused in all stages of the information systems development and increase company productivity.

References

1. Androutsopoulos, I., Ritchie, G.D., Thanisch, P.: Time, Tense and Aspect in Natural Language Database Interfaces. Natural Language Engineering 4, 229–276 (1998)
2. Cunningham, H.: GATE, a General Architecture for Text Engineering. Computers and the Humanities 36, 223–254 (2002)
3. Ganter, B., Wille, R.: Formal Concept Analysis: Mathematical Foundations. Springer, Berlin-Heidelberg (1999)
4. Hofmann, T.: Probabilistic latent semantic indexing. In: Research and Development in Information Retrieval, PP. 50–57 (1999)
5. Honkela, T.: Von Foerster meets Kohonen - Approaches to Artificial Intelligence, Cognitive Science and Information Systems Development. Kybernetes 34(1/2), 40–53 (2005)
6. IBM Voice Toolkit V5.1 for WebSphere Studio. Accessed (January 2007) http://www-306.ibm.com/software/
7. Kohonen, T.: Self-Organizing Maps. Springer-Verlag, Heidelberg (2001)
8. Laukaitis, A., Vasilecas, O.: Self-organizing map for conceptual modelling. In: Proceedings of the International Conference on Computer Systems and Technologies CompSysTech'06, pp. 141–147 (2006)
9. Laukaitis, A., Vasilecas, O., Berniunas, B.: JMining – information delivery web portal architecture and open source implementation. Information Systems Development. Advances in Theory, Practice and Education. Springer Science, pp. 199–206 (2005)
10. Microsoft corporation. SQL Server and English Query. Accessed (January 2007), http://msdn.microsoft.com/
11. Miller, G.A.: WordNet: A Dictionary Browser. In; Proc. 1st Int'l Conf. Information in Data, pp. 25–28 (1985)

12. Price, D., Riloff, E., Zachary, J., Harvey, B.: NaturalJava: A Natural Language Interface for Programming in Java. In: the Proceedings of the International Conference on Intelligent User Interfaces (2000)
13. Ryan, K.: The role of natural language in requirements engineering. In: Proceedings of IEEE International Symposium on Requirements Engineering, pp. 240–242. IEEE Computer Society Press, Washington (1993)
14. Salton, G.: Automatic Text Processing: The Transformation, Analysis and Retrieval of Information by Computer. Addison-Wesley, New York (1989)
15. Toolbox, S.O.M. Accessed (January 2007), http://www.cis.hut.fi/projects/somtoolbox/
16. Valtchev, P., Grosser, D., Roume, C., Rouane, H.M.: GALICIA: an open platform for lattices. In: de Moor, A., Ganter, B. (eds.) Using Conceptual Structures: Contributions to 11th Intl. Conference on Conceptual Structures, pp. 241–254 (2003)

"Seeing Is Not Believing But Interpreting", Inducing Trust Through Institutional Symbolism: A Conceptual Framework for Online Trust Building in a Web Mediated Information Environment

Emma Nuraihan Mior Ibrahim, Nor Laila Md Noor, and Shafie Mehad

Faculty of Information Technology and Quantitative Science,
University Technology MARA, 40450 Shah Alam, Selangor, Malaysia.
{emma, norlaila, shafie}@tmsk.uitm.edu.my

Abstract. Trust has become an essential prerequisite for customer relationship building. Implementing institutional mechanisms seems to be an obvious outcome in inducing trust, rather than the result of independently motivated decisions of the user's interpretation on the institutional design features to be trustworthy. Proponents of this idea argued by understanding the concept of a sign and its related properties, it is possible to improve the analysis and design of interface design elements within web environment to induce trust. Drawn from established theoretical work on institutional based trust and semiotic, the concept of Institutional Symbolism and its four underlying dimensions are proposed to facilitate thinking and perception as to how trust can be built through signs comprehension. Conceptualized in the context of web mediated information environment, the implications of these dimensions on web site design and future work is further discussed.

Keywords: Web Mediated Information Environment, Institutional based Trust, Institutional Symbolism, Semiotic.

1 Introduction

Research on trust has begun with reference to Aristotle concept of *ethos* in *On Rhetoric,* dated back 2000 years ago which discussed theoretical consideration on trust in interpersonal communication [1]. As the Internet becomes pervasive in our lives, the issues of trust have raised much interest among the scholars and practitioners from interpersonal communication to a buyer-seller interaction [7]. With the emergence of e-commerce mainly in B2B and B2C environment, trust becomes the key facilitator in establishing, building and maintaining consumer relationships and interaction. The development of virtual relations has given the trust issues a new edge. Consumers not only perform interpersonal or inter-organizational transaction within the electronic exchange model but unaware to many, consumers also involved in the knowledge transactions and exchanges within the information exchange mode [11]. This is due to the growing number of web sites that offer advices or disseminating information to

M.J. Smith, G. Salvendy (Eds.): Human Interface, Part II, HCII 2007, LNCS 4558, pp. 64–73, 2007.

consumer in sensitive fields such as religion, health, legal, politics, financial and so on. Hence we use the term *web mediated information environment* to collectively refer to the activities involving information exchange and dissemination among web sites to the users over the Internet, for example, users seeking legal advices (e.g: WorldLawDirect.com, legalserviceconnection.com) or obtain personal advice from free online advice services (e.g: FreeAdvice.com). One of the methods to engender trust within electronic environment is through trust building based on institutional structures or known as institutional based trust [26]. The operationalization on institutional trust focus on the underlying technology infrastructure, well known practices and control mechanisms that acts as a tool to ensure safety and the reliability of an e-commerce environment has been proven effective to monitor trustworthy action under transaction environment [25, 26].

However, research on institutional based trust has been limited to the traditional e-tailing environment, varies considerably in its dimensions and problems [16]. To the best of our knowledge very little attention has been paid to this aspect of the institutional based trust interfaces within the web mediated information environment as these sites introduce an additional element of trust beyond standards concerning information privacy and security. Furthermore, a lack of user understanding of effective mechanisms for self-protection on the Internet may be causing many people to ignore their use. Feelings of trust are not always rational. A decision to trust a piece of information does not necessarily imply what respondents described as trust; it may include strong intuitive, emotional component especially dealing with sensitive issues.

Following Social Learning Theory [2] suggests that people learn through association or modeling. Trusting behavior therefore comes to be associated with symbols that act as cues in environment. Meaning making in a scene is a product of both internal associations and the matching of internal and external cues [15]. To trust is therefore a problem of signs interpretation, considering what we know, what we see and what we interpret in something or someone, we can decide if we should trust or not. Of our interest, we further introduce the notion *institutional symbolism* and its underlying dimensions as the external cues of the institutional environment, a mechanism that signal trustworthy message in a form of signs on the web interface which will further discuss in details. Specifically the contributions of this paper are:

- Definition of concept on web mediated information environment institutional based trust and institutional symbolism.
- Identification of the four dimensions of institutional symbolism and its underlying values.
- To introduce a conceptual framework on institutional trust building based upon semiotics that allows a better understanding of the institutional phenomenon as on how users interpret the institutional signs of the interface.

2 Institutional Based Trust: A Theoretical Background

Drawing on the concepts and findings of the sociologists and economists, this approach to trust considers the development of trust between individuals and institutions in which they interact in their every day's life such as banking, law, government system and economy [5]. In the literature, this type of trust is also known as *system trust* [19, 23], *reputed credibility* [10], and is similar to the *transference process* described in [38] and *control trust* [38]. Lewicki et. al [28] describe institutional based trust as the trust that develops when individuals generalize their personal trust to a large organizations made up of individuals with whom they have low familiarity, low interdependence and low continuity of interaction. For example, we trust the government system to ensure the stability of a state or we place our trust in a juridical system to uphold the law enforcement and sanctions. The trust is not based on the sole property

Table 1. Summarizes institutional based trust dimensions

Dimension	Explanations	Sources
Situational normality	The beliefs that everything is "normal". This gives the trustor a sense of security, which positively supports his trusting decisions. An example in the e-business environment is that communication infrastructure is secure in the sense that authorization, integrity, non-repudiation and confidentiality are guaranteed	McKnight et. al. [19], McKnight and Chervany. [20]
Structural assurances	The belief of success is likely because of regulations, contracts, promises or guarantees are in place. It seeks to reassure the trustor that measures have been taken specifically to safeguard and reduce the risk of something going wrong.	McKnight et. al. [19], McKnight and Chervany. [20]
Facilitating Conditions	Shared standards, relationship values and common beliefs about behavior and goals. Examples: common processes for uniform product descriptions, the use of interoperable IT platforms and business management standards like EDI.	Pavlou, P.A. and D. Gefen [26]
Third party certification and Escrow services	Guaranty expected outcomes. The main tasks of these third party and guarantors are to create the appropriate conditions when there is a low level of familiarity in the environment.	Zucker [41]
Technology Trust	Ability of technological systems used to carry out interactions to support privacy, accuracy, authenticity or authorized parties and accountability of e-commerce transaction. Dimensions suggested are security services such as confidentiality mechanisms, authentication mechanisms and access control mechanisms.	Ratnasingam [31]
Internet Trust	Identified several dimensions of competence trust in the context of Internet, correctness, availability, reliability, security and survivability.	Schneider [35]

or state of the trustee but rather on the property of the system or institution within which the trust relation exists. Hence, the trust on the system property or institution is a reflection of the security one feels about a situation because of guarantees, safety nets, other structures [41] or institutionalized procedures [39]. Institutional based trust has been conceptualized and examines within e-commerce environment and regarded as a critical part of Internet transactions [20]. This is because the e-commerce brings together organizations with no familiarity or similarity [26] and thus this type of trust builds on shared structures, processes and routines to create stable transaction contexts. As most of the online users are not well knowledgeable or technically sound with the inner working systems, the users will place their trust as a belief in these institutional mechanisms that helps to engender trust in the electronic environment [20, 25]. Institutional based trust dimensions are mostly embedded in the socio-technical systems technology. It is seen as backdrop that envelopes and safeguards interaction in the online electronic environment [33].

As institutional based trust dimensions are largely based on socio-technical system and technology trust mechanisms, trust could also entails in non-technical mechanism that safeguard interaction on the web [33]. Our interest to the latter is to try to make it easy for human to reason about trust online by involving analysis of how people interpret or perceived the trustworthiness and credibility of information through the institutional signs and symbols they see online.

3 Trust and Website Design

When a potential customer visits a site for the first time, his or her initial perceptions of that web sites are based on features that are visible on the web site's main pages. This is the same in the case of offline world, consumer's exhibit attitudes and behaviors that are affected by intrinsic cues gathered from the physical environment in which they make a trust based decisions [32]. Self-perception theory posits that one's attitude towards another party is formed through interaction with that party and through circumstantial information [17] where trust usually involves an assessment of external information. These initial impressions drive the development of customer's relationship, with the web site's ability to persuade visitors to take the action that it wants them to take (e.g.: make purchase, information exchange), the first step to building trust. Several researchers have indeed considered the components and structure of an e-commerce website design that might induce or trigger trust in e-commerce [4, 6, 40, 42], factors such as branding, navigation, website designs, technology and security measures, third party seals and so on. Designing for trust in e-commerce is an increasing concern for the field of HCI [33]. Because of the considerable importance of trust in the web mediated information environment, the designers must create new social norms for professional institutional electronic environment by designing web interface that able to induce trustworthy feelings. However, the ability of the designers to promote trustworthy perceptions of institutional trust through the interface design of its own is unknown. Hence, institutional based trust features need to be understood from the structures of meaning producing events at both verbal and non-verbal levels [36]. As a new medium, it does not yet have the established standards or conventions needed to be accepted as trust communication tool, it lacks of semiotics.

3.1 Semiotics of Trust

Semiotic is the study of signs and meaning making that convey semantic (the meaning of signs), syntactic (the relationship between signs), pragmatics (the ways in which signs are used and interpreted) and their use as a communicative tool [14]. The web sites are complex interactive communication systems speaking two languages: the one of the "content" and the one of the "interface" [37]. Our interest is to the latter, which is referred to as a self-referential symbolic system consists of a set of "signs" (examples icons, symbols, words etc.) used to provide interaction to users [9]. In user interface design, semiotics can be used to convey particular aspects of a design to a user without requiring explicit labels or instructions [34] similarly to the concept of affordances described by Norman [22] as the perceived properties of a thing that determine how it can be used. Signs convey two meanings: a "content meaning" and also a "functional meaning". These signs should be understood by users in both ways: the user should be able to understand the message it's trying to convey, a structural characteristics facilitating both users' cognitive and affective apprehensions. And the expectations that lead to the behavior (trust, values, and beliefs) impose by the organization or institution [29, 30].

Semiotics in design serves as an experience link between user's feeling, interpretation and action in a situation. The nature of the entire user's interaction with the web site is mediated through the initial impression from the web pages they see. Symbols and signs act as cues in the environment, when they are associate with user's internal states or feelings, their physical presence can evoke the associated states and feelings to guide users towards a behavior for example the presence of trust cues may lead to increased perceptions of trust and reliability on the web. Several researchers have incorporated semiotics to the design of e-commerce websites, example a semiotic based methodology for web assessment [9], a semiotic model of software in which a metaphor affects the interpretation of semiotic features, measuring the effect of semiotic web design features on expectations of the performance criteria in an online web store explained in [32] and propose semiotic aspects of a web site to increase trust [14].

3.2 Trust as Cognitive and Affective Perceptions

Trust is deeply interwoven with risk, and both are based upon perception. Perception is the process by which individual select, filter, organize and interpret information to create a meaningful picture of the world [36].Cognitive psychology believes that perception is impacted by emotions, motivations, expectations, interplay of cognitive and affective dimensions. Conceptualizing individuals evaluative judgment as having cognitive (belief) and affective (emotion) based is a commonly used approach [7, 18]. Trust in the signs and symbols (social trust in strict sense) seem to consist of beliefs evaluation as the basis of reliance on the symbols properties and capabilities. Those beliefs can be well justified, warranted and based on reasons, as perceived the other's dependability, credibility and competence [6, 20]. This represents the "rational" part of trusting in symbols, a cognitive dimension when trust is based on well motivated, evidences and on good inferences when it constitute beliefs are well grounded. But those beliefs can also be not really warranted not based on evidence, more rather depending on the emotional security and bond users develop with the trustee, and those beliefs rely on faith, honesty, benevolence [6, 7, 18] that makes up the affective dimension.

As sign reading is a fundamental part of deciding whether to trust, it is then important to study the feeling of trust focusing on what happens into an individual mind when he/she trusted someone or something. Perception towards the signs envelope a broad area of human processes ranging from sensation to concept formation [36]. People encountering symbols, read these symbols through their own individual eyes and the symbols acquire meaning in the organization through recurring experiences. Importantly, only a connection between symbols and underlying values provides full understanding of both symbols and its trust warranting properties [27, 34].

4 Institutional Symbolism and Its Dimensions

We conceptualized institutional based trust as a form of symbolism, the system of representations and symbols through institutional trust inducing features on the web which we define as *institutional symbolism*. The term institutional symbolism refers to a visible, physical manifestation of the institutional characteristics, behavior and values. Institutional symbolism is the *trust marks*, signs that depicting and presenting connoted message of some *assurance*. Based on the work of Barber [3] says that trust implies expectations of the future, which can be broken down into three forms of expectations, technical competence, fulfill moral orders and fiduciary responsibility. On the basis of these expectations, we content that the trust marks represent the beliefs (values) and expectancies held by individual about the overall impersonal structures and situations construed as to network both cognitive and affective trust warranting properties. These beliefs implying that the institutional symbolisms carry its own disposition and attribution meaning, an institutional manifestation through textually or graphically presentation on the web site. We propose four dimensions of trust marks and its underlying values as an instantiation of institutional symbolism that serve as a starting point for measuring the effect of institutional trust inducing features on the web site. We contended, trust on this institutional symbolism act as a form of social trust where trust initiated from the social mechanism, behavior and values through the means of institutional symbolism representation.

Table 2. Framework of Institutional Symbolism Trust Inducing Features

Dimensions	Values	Measurements	Sources
Trust marks that reflect third party assurance or seals of approval.	A belief that it will perform a particular action, to monitor or to control that certain acts and behavior is warranted.	Trust marks that symbolized: 1. Protecting privacy 2. Providing security 3. Demonstrating consumer satisfaction 4. Providing reliability 5. Providing assurance or guarantee.	Hu et. al [12] Cheskin [4] Wang, Y.D and Emurian, H.H. [40]
Trust marks that reflect credibility of the web content	A belief that it has the ability and competency to carry out the obligations.	Trust marks that symbolized: 1. Competence *(knowledge, expertise and skill).* 2. Reliability *(accuracy, currency, coverage and believability).* 3. Predictability *(stability of information).*	Chopra, K., and Wallace, A.W. [5], Fogg, B., and Tseng, H [10] Corritore, et. al. [6]
Trust marks that evoke emotional assurance or security.	A belief that it will provide a sense of comfort that is reflective, thoughtful and careful.	Trust marks that symbolized: 1. Benevolence *(goodwill and objectivity)* 2. Honesty *(validity and, openness).* 3. Integrity *(fiduciary obligations).*	Chopra, K., and Wallace, A.W. [5] D. McAllister [7], Riegelsberger., J., et. al [33].
Trust marks that reflect trustworthy expectations derived from the message.	A belief that it signifies positive or prominent identities and values.	Trust marks that symbolized: 1. Reputation - *Offline reputation* (evaluation of all image aspects about an institution). - Perception of a *company's ability* to meet the expectation as intended (authority, certification reviews, references, ratings and awards). 2. Brand - *Brand Image* (perceived overall reputation) - *Brand Personality* (beliefs in the overall *competence, benevolence, integrity and predictability*).	Einwiller, S., and Will, M [8], Chopra, K., and Wallace, A.W. [5]

5 Conclusions and Future Work

We sought to heighten the awareness of the importance in designing information to meet what user's needs and in turn on how to better design sites to convey intended and desired meaning. Sophisticated web sites design does not always engender consumer trust in their privacy and security. Hence, more empirical supported guidelines together with new method in approaching the design issues are needed. Through semiotic analysis the study of signs and meaning making can help us to understand the properties of the institutional symbols on the web by addressing the intrinsic meaning present within the symbolic system and the perceived meaning decoded by a particular set of users (common shared meaning) [9].

The semiotic analysis will involved 5 levels, the first is to *analyze the nature* of web mediated information environment to look into the shared patterns of interaction, norms, values, expectancies and obligations among the users on trust issues, second the *pragmatics,* analyzing the user behavioural aspects on communication processes and shared assumptions in evaluating the sign, constituting the beliefs, intention and expectations within a specific culture, third the *semantics*, evaluating the meaning of trustworthy inducing trust features on the web, fourth the *syntatics,* the relationship among signs and the operations it may be subjected, an underlying rules associated with the sign properties and fifth, *empirics*, the establishement of guidelines with respect to designing trust inducing features on the web. In addition, we adopt Social Semiotic principles developed by Hodge and Kress [13] for further analyzing the effects of these signs and symbols on user's interpretation of trust.

Hence, communication and meaning making can be said, are linked through a belief or value in a shared understanding of signs that if a population of users interprets the design feature similarly, it can be used effectively in a system [32]. As such it is important to be sensitive as to how signs and symbols are used in the text and graphic to convey intended meaning and to induce trustworthy behavior to the users especially if it's addressing a specific culture. Our introduction to the concept of institutional symbolism introduce avenue for future research within institutional trust phenomenon. The framework presented will serve as the basis for future empirical studies that will be apply to the web mediated information environment. The operationalization of institutional symbolism trust inducing features will take place within the Islamic web based content environment, which introduce a critical process of meaning making and trust constructions on the level of users as it involves areas of sensitivity. A promising area to study trust that focus on a particular set of communicators, having its own cultural identities. The outcome will be benefits for designers design guidance for trust elements placement on the web, cultural embedded design features and effective security and privacy protecting behaviors for those who would want to involve in acquiring, communicating Islamic knowledge or information on the web.

References

1. Aristotle, On Rhetoric: A Theory of Civil Discourse. Trans. George A. Kennedy, New York: Oxford Up (1991)
2. Bandura, A.: Social Learning Theory. Prentice Hall, Englewood Cliffs, N.J (1977)
3. Barber, B.: The Logic and Limits of Trust, New Brunswick: Rutgers University Press (1983)
4. Cheskin Research Studio Archetype/ Sapient, Ecommerce Trust Study (1999), http://www.cheskin.com
5. Chopra, K., Wallace, A.W.: Trust in Electronic Commerce. In: Proceedings of the 36th Hawaii International Conference on System Sciences (HICSS'03). IEEE (2002)
6. Corritore, C.L., Wiedenbeck, S., Kracher, B.: The Elements of Online Trust. In: Proceedings of the Conference on Human Factors on Computing Systems (CHI'01), pp. 504–505. Seattle, USA (March 31-April 5, 2001)

7. McAllister, D.: Affect and Cognition based trust as foundations for interpersonal cooperation in organizations. Academy of Management Journal 381, 24–59 (1995)
8. Einwiller, S., Will, M.: The role of reputation to engender trust in electronic markets. In: Proceedings for the 5th International Conference on Corporate Reputation, Identity and Competitive, electronic publication, URL: http://www.communicationsmgt.org/modules/pub/view.php/communicationsmgt-11
9. French, T., Polovina, S., Adam, V.: Semiotics for E-Commerce: Shared Meanings and Generative Futures. In: Proceedings of BIT'99, (3–4 November, 1999), http://www.man-bus.mmu.ac.uk/confs/bit99/
10. Fogg, B., Tseng, H.: The Elements of Computer Credibility. In: Proceedings of CHI'99, ACM Press, New York (1999)
11. Forray, D.: The Economics of Knowledge. MIT Press, Cambridge (2004)
12. Hu, X., Lin, Z., Zhang, H.: Myth or reality: Effect of trust promoting seals in electronic markets, presented at WITS' 01, New Orleans, LA. USA (2001)
13. Hodge, R., Kress, G.: Social Semiotics. Cornell University Press, Ithaca, New York (1988)
14. Karvonen, K., Parkkinen, J.: Signs of Trust. In: Proceedings of the 9th International Conference on HCI, New Orleans, LA, USA (2001)
15. Kaplan, S., Kaplan, R.: Cognition and Environment: Functionin. In: An Uncertain World. Ann Arbor, MI: Ulrich's (1983)
16. Kim, D., Song, Y., Braynov, S., Rao, H.: A-B-to-C Trust model for on-line exchange. In: Proceedings of Americas Conference on Information Systems, pp.1–3. Boston, MA, USA (2001)
17. Kim, H.W., Koh, J., Xu., Y.: Trust Building in Internet Vendors, Comparison of New and Repeat Customers. In: Information Resources Management Association Innovations Through Information Technology, pp. 115–118. New Orleans, USA (May 23-26, 2004)
18. Lewis, J.D., Weigert, A.: Trust as a social reality. Social Forces 63, 967–985 (1985)
19. McKnight, D.H., Cummings, L.L., Chervany, N.L.: Initial Trust Formation in New Organizational Relationships. Academy of Management Review 23(3), 473–490 (1998)
20. McKnight, D.H., Chervany, N.L.: What trust means in e-commerce customer relationships: an interdisciplinary conceptual typology. International Journal of Electronic Commerce 6(2), 35–59 (2002)
21. Sutherland, P., Tan, F.B.: The Nature of Consumer Trust in B2C Electronic Commerce: A multidimensional Conceptualization. In: Proceedings of the International Conference of the Information Resources Management Association: Innovations Through Information Technology, pp. 611–614. New Orleans, USA (May 23-26, 2004)
22. Norman, D.: The psychology of everyday things. Basic Books, New York (1988)
23. Luhmann, N.: Trust and Power. John Wiley and Sons, Chichester, England (1979)
24. Pavlou, P.A.: Institutional Trust in Interorganizational Exchange Relationships: The Role of Electronic B2B Marketplaces. Journal of Strategic Information Systems, vol. 11(4) (2002)
25. Pavlou, P.A., Gefen, D.: Building Effective Online Marketplaces with Institution-Based Trust. In: International Conference on Information Systems. Barcelona, Spain (2002)
26. Pavlou, P.A., Tan, H.Y., Gefen, D.: The Transitional Role of Institutional Trust in Online Interorganizational Relationships. In: Proceedings of the 36th Hawaii International Conference on System Sciences (2003)
27. Pondy, L.R., Frost, P.J., Morgan, G., Dandridge, T.C: Organizational Symbolism. JAI Press, Greenwich, CT (1983)
28. Lewicki, R.J., McAllister, D.J., Bies, R.J.: Trust and distrust: New relationships and realities. Academy of Management Review 23, 438–458 (1997)

29. Rotter, J.B: Interpersonal trust, trustworthiness and gullibility. American Psychologist 35(1), 1–7 (1980)
30. Rotter, J.B.: Generalized expectancies for interpersonal trust. American Psychologist 26, 443–452 (1971)
31. Ratnasingam, P., Pavlou, P.A.: Technology trust in internet based interorganizational electronic commerce. Journal of Electronic Commerce in Organizations 1(1), 17–41 (2004)
32. Resnick, M.L., Montania, R.: Perceptions of Customer Service, Information Privacy and Product Quality From Semiotic Design Features in an Online Web Store. International Journal of Human Computer Interaction 16(2), 211–234 (2003)
33. Riegelsberger, J., Sasse, S.M., McCarthy, D.J: The mechanics of trust: A framework for research and design. International Journal Human-Computer Studies 62, 381–422 (2005)
34. Schein, E.H.: Organizational Culture. American Psychologist 45(2), 109–119 (1990)
35. Schneider, F.B. (ed.): Trust in Cyberspace. National Academy Press, Washington, D.C (1998)
36. Singh, N., Pereira, A.: The Culturally Customized Website, Customizing Website for the Global Marketplace. Elsevier Butterworth-Heinemann (2005)
37. Speroni, M., Bolchini, D., Paolini, P.: Interfaces: Do Users Understand Them? In: Trant, J., Bearman, D. (eds.) Museums and the Web: Proceedings, Toronto: Archives & Museum Informatics, published March 1 (2006), at http://www.archimuse.com/mw2006/papers/speroni/speroni.html
38. Stewart, K.J.: Trust transfer on the world wide web. Organization Science 14(1), 5–17 (2003)
39. Tan, Y., Thoen, W.: Towards a generic model of trust for e-commerce. International Journal of Electronic Commerce 5, 61–74 (2000)
40. Wang, Y.D, Emurian, H.H.: Inducing Consumer Trust Online: An Empirical Approach to Testing E-Commerce Interface Design Features. In: Proceedings of the International Conference of the Information Resources Management Association: Innovations Through Information Technology, pp. 41–44. New Orleans, USA (May 23-26, 2004)
41. Zucker, L.: Production of Trust: Institutional Sources of Economic Structure 1840-1920. Research in Organization Behavior 8(1), 53–111 (1986)
42. Yang, Y., Hu., Y., Chen, J.A: Web Trust Inducing Model for E-Commerce and Empirical Research. In: Proceedings of the Seventh International Conference on Electronic Commerce (ICEC' 2005), pp. 188–194. Xi'an, China, (August 15-17, 2005)

Exchanging Graphical Emails Among Elderly People and Kindergarten Children

Megumi Mitsumoto[1] and Sanae H. Wake[2]

[1] Doshisha Rohm Plaza Project,
[2] Department of Information and Media, Doshisha Women's College
Kyoto 610-0394, Japan
mmitsumoto@stu.kobe-u.ac.jp, swake@dwc.doshisha.ac.jp

Abstract. Email is a convenient tool, however, it is still difficult for elderly people and kindergarten children. If they can learn to use it, their ability to communicate may be remarkably increased. Therefore we developed two kinds of email software 'Gin-Mail' for elderly people and 'Ocha-Mail' for kindergarten children. They adopt a pen-based interface with a tablet PC. Users write email using Ink function, and software sends it as an attached image file. There are two modes in it, User mode and Maintenance mode. User mode offers only minimal functions to prevent misoperation. We devised each user interface in accordance with the user's characteristics. 'Ocha-Mail' and 'Gin-Mail' are accessible to kindergarten children and elderly people respectively. Our goal is to encourage interactions among different generations by through the use of email.

1 Introduction

In modern Japan, with the shift toward the nuclear family and the falling birth rate, communication between generations has become difficult. In the past, grandparents and grandchildren lived together, but now grandparents often live far away, and relationships with neighbors have become more distant as well, making interaction between generations all the more infrequent. If there is an easy communication means without regards to distance, the exchange between many generations will surely deepen. Though a system like the message board already exists, it is suitable only for a short 'sticky note' type message (Hutchinson, et al., 2002). This format is useful for encouraging basic communications, however it is not suitable for a more profound interaction which we hope to achieve. Email is easy to send and very convenient, therefore it is gradually becoming one of today's most important forms of communication for the youth and the middle-aged. However, the percentage of elderly people using email is still relatively low. The rate of using the internet (including email use) in 2005 exceeded 90% in 13-49 years old, but it stayed only 19.8% among people in their 70's and 7.2% among people over 80 (MIC, 2006). Likewise, children around 5 years old have a similarly low usage rate. We have not yet reached the point where everyone can use email easily, because of the difficulty in using the devices required to send email, like PCs and mobile phones. Since these

M.J. Smith, G. Salvendy (Eds.): Human Interface, Part II, HCII 2007, LNCS 4558, pp. 74–83, 2007.

devices have a lot of functions, their complexity makes it difficult for the very old or the very young to operate them easily. We believe it is possible that email can become a new form of communication, enabling people in all walks of life to engage in richer communication that transcends the boundaries of distance or generations, if all generations can use it easily.

Our goal is to enable those who have never used email before to use this communication tool with ease and to encourage interactions among all kinds of people, even across generations. Therefore we developed two kinds of email software, 'Gin-Mail' for elderly people and 'Ocha-Mail' for kindergarten children. We describe a summary of these software and an exchange experiment among three generations with these software in this paper.

2 Preparatory Investigations

First of all, we investigated characteristics of the target generation in order to create a communication tool suitable for each.

2.1 Characteristics of Elderly People

The greatest reason why elderly people do not use information instruments is the complexity of the devices. Because they are not familiar with a keyboard and a mouse, they are apt to give it a wide berth. Therefore we decided to use a tablet PC, because all of the operations can be carried out with a pen and there is no need for a keyboard or a mouse.

Before the development of email software, we performed an experiment to confirm whether reading and writing with the tablet PC were accepted by elderly people (Wake 2004). A summary of the results are as follows. Refer to bibliography (Wake 2004) for details.

2.1.1 Reading

As for the direction of the character, they liked vertical writing because they were accustomed to it. We compared handwriting with printing type. They preferred handwriting rather than printing type, because they could feel a human touch in handwriting.

2.1.2 Writing

We compared the pen input using the ink function with handwriting recognition. As for the pen input using the ink function, they all could write a character on the tablet PC without difficulty. On the other hand, the difficulty of the handwriting recognition was quite different depending on the person. The average recognition rate was 79.0%. When misrecognition occurr, handwriting recognition facility needs a second operation to correct it. It is very complicated for elderly people.

2.2 Characteristics of Kindergarten Children

Children are usually able to take care of a child who is smaller than them and to attempt communication with adults at five or six years old (Hayashi & Ishibashi,

2005). Their curiosity is strong and their desire to experiment with different ways of doing things is also large. When they begin to do one thing, they tend to give priority to their feeling over the surrounding situation (Akiba. & Shiraishi, 2006). They are poor at logical thinking compared with adults. Physically, a fine motion of their finger progresses significantly (Akiba. & Shiraishi, 2006). In terms of the intellectual aspect, they get interested in a sign or a character and their desire to learn becomes vigorous. Generally it is from this age they begin to learn reading and writing of characters. However, they haven't come to acquire literary words completely by that time (Akiba. & Shiraishi, 2006). Therefore an important means to express themselves concretely is to draw a picture. Although handwriting recognition is also considered an effective method of supporting their input, there are some problems in terms of practical usage (Read et al., 2002). As for operation, the ratio of children who play TV games has reached about 44% (Akiba & Shiraishi, 2006). So it is thought that many children are familiar with a game-like user interface.

Before the development of email software, we visited a kindergarten and observed small children in order to investigate their characteristics. As a result, it turned out that their curiosity was really strong and they have often applied an original rule into the action. The rule is like that they consider an empty box a washing machine and put slippers into it, or they decide not to approach a child playing with a toy telephone. If we can lead this habit of making a rule well, we might help them easily learn usage of software. It also turned out that they tend to like character items. So we decided to make an image character. We prepared for several kinds of plans and let them choose the favorite one and finally adopted the most popular plan. We describe the details of the adopted character in the following chapter.

3 Designing and Development of 'Gin-Mail' and 'Ocha-Mail'

We developed two kinds of email software which have based on the characteristics of each generation, 'Gin-Mail' for elderly people and 'Ocha-Mail' for kindergarten children. They adopt a pen-based interface with a tablet PC. We describe them separately as follows. Both work on the operation system Windows XP Tablet PC Edition which Microsoft developed. They can also work on Microsoft Windows XP Home Edition and Microsoft Windows XP Professional, if Microsoft Table PC Platform SDK is installed.

3.1 Gin-Mail

It is necessary to reduce the work load of the recognition of software so that elderly people may easily use Gin-Mail. Therefore, we divide use functions and maintenance functions, and provide as two modes, 'User-mode' and 'Maintenance-mode' respectively in order to simplify a usual operation. User-mode is what the target users themselves use, so we narrowed down the function offered in User-mode to the minimum, only for receiving, reading and writing of email. Maintenance-mode is the mode for a supporter to perform initial setting and email management.

3.1.1 User-Mode

When the PC is started, User-mode starts automatically and will begin to receive any new emails. Therefore even if the users don't know the method of starting software, they can use email. If the basic data are not set, Maintenance-mode starts to promote setting them. Once the basic data are set, User-mode will start automatically for the next time.

In User-mode, only three basic functions, i.e., receiving, reading and writing (sending is included) are offered. As for receiving and reading, it is possible to receive and display not only email from the same kind of software but also the usual text email from a general mailer. It is also possible to display the attached image file. As for writing, the Ink function of Tablet PC is used for email creation. Handwriting of a pen is recorded as it is, and it is not converted to ASCII. The contents are sent as an attached file of an image format.

There are three basic screens, [receiving email], [reading email], and [writing email]. These are changed with the tab on the lower part of the screen. Fig.1(a)-(b) shows example screenshots of Gin-Mail. We adopt a metaphor of the letter which elderly people are used to. A screen is likened with one letter paper. the direction of the character is vertical. We don't use scroll bar but perform a display in a letter paper unit.

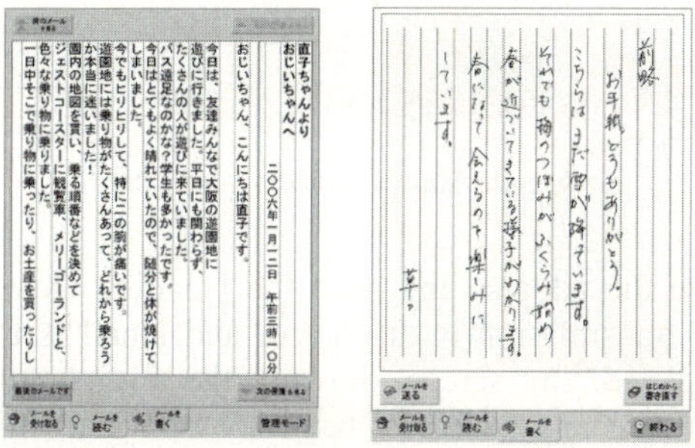

(a) (b)

Fig. 1. Screens of Gin-Mail : (a) [reading email], (b) [writing email]

[receiving email]

Email reception is performed at the time as starting Gin-Mail and pushing down a tab of [receiving email]. When the reception process is completed, the number of newly arrived email is displayed by the message. However, newly arrived email here does not mean email received by the reception process at that time but the unopened email in the preservation data. We decided to perform such an indication for the purpose of preventing unread email being forgotten.

[reading email]

The screen of [reading email] is displayed by pushing the tab of [reading email] under the screen. The oldest unopened email is displayed just after moving to this screen (Fig.1 (a)). In case there is other unopened email, the button to call email already read is not displayed. So users can read only an unread email until they finish reading all unread emails. When they read a long email, email can be changed into the opened state only if all the pages of email are displayed. These specifications are for the purpose of preventing unread email remaining unopened.

[writing email]

A blank Letter paper is displayed by pushing the [writing email] tab, and a handwriting input by the Ink function is enabled. Users can write one letter paper's worth of email with a pen (Fig.1 (b)). However the length is limited to a maximum of one page. Letter paper returns to a blank paper if the [rewrite from the beginning] button on this screen is pushed.

After finishing email, users can change to an address selection screen by pressing the [sending email] button. Fig.2 shows the address selection screen. Here, the addresses registered in the address book are listed by nickname.

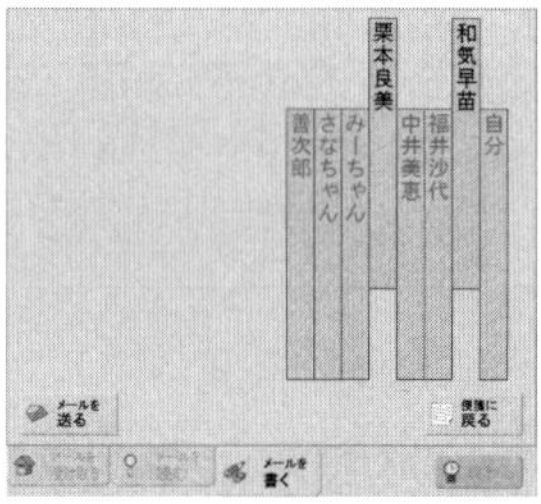

Fig. 2. Address selection screen

When users choose the address and push the [sending email] button, the screen showing the state of transmission is displayed, and email is transmitted. After the transmission is completed, the screen showing the completion of transmission is displayed. In the case that any addresses have not been chosen, the [sending email] button is nullified. Likewise, while choosing an address, every button is nullified excluding [sending email] and [returning to the letter paper], and other operations cannot be executed. Using this method, Gin-Mail has tried to suppress the danger that accidental screen transition and error create by giving restricting screen shifts in each state.

3.1.2 Maintenance-Mode

It is possible to establish initial settings, to delete email and to manage an address. Maintenance-mode rises automatically and promotes making settings in the functions that basic data are not yet set, such when we start for the first time. Once the basic data are set, we have to call this mode from the start menu of Windows by manual operation if it is necessary.

The users can check and read email in this mode. But unopened email will not become opened. Here, we read in order to delete. The users can delete or register the address by choosing a received email. It is not possible to write or to send in this mode.

Various settings are possible here. The users can set up what is required for communication, e.g. a server name, ID and password. The user can register and edit an address. The users can select some styles related to user interface, such as the direction of the character, fonts, colors or width of a ruled line.

3.2 Ocha-Mail

Children will not use if they do not like it at first glance. Therefore, the graphic design is very important. We made ‘Ochamin’ as an image character of Ocha-Mail (Fig.3). Ochamin is the most popular character among children mentioned above. We unified the screen design of Ocha-Mail with an atmosphere of Ochamin. In terms of function, we also provided 'User-mode' and 'Maintenance-mode' in Ocha-Mail by the same reasons as Gin-Mail.

Fig. 3. ‘Ochamin’, the image character of Ocha-Mail

3.2.1 User-Mode

User-mode of Ocha-Mail is basically the same as that of Gin-Mail. It starts automatically when the PC is started. It has three basic functions, receiving, reading and writing (sending is included). It is possible to receive and display email both from the same kind of software and from a general mailer. Also any attached image file can be displayed. The contents are recorded with the Ink function and sent as an attached file of an image format. However, the contents in Ocha-Mail are not a character but a picture. The children whom Ocha-Mail regards as the target users cannot necessarily master a character yet. Therefore we decided the most appropriate way for them to produce emails would be through drawing.

The greatest difference between Gin-Mail and Ocha-Mail is its user interface. The features of Ocha-Mail’s interface are simple, graphical and with sound effects. The parts arranged on the screen are minimized in order to prevent confusion. Not only a word but also a picture is used for explanation of parts, because it is easier for children to understand the meaning of a picture than a word. Moreover, in the case of buttons we wanted to encourage children to push, its design is changed when the pen is set on it. This makes it easy to attract children's interest. We also use sound effects to attract them.

There are four basic screens, [top], [receiving email], [reading email], and [writing email] in Ocha-Mail. Fig.4 shows the basic structure of screen transition.

[top]

[top] is the basic menu screen (Fig.4(a)). Users can move to other screen only from [top]. To move to each screen freely, the number of buttons on every screen should increase. Children's curiosity is strong, and if there is a button, they are likely to push

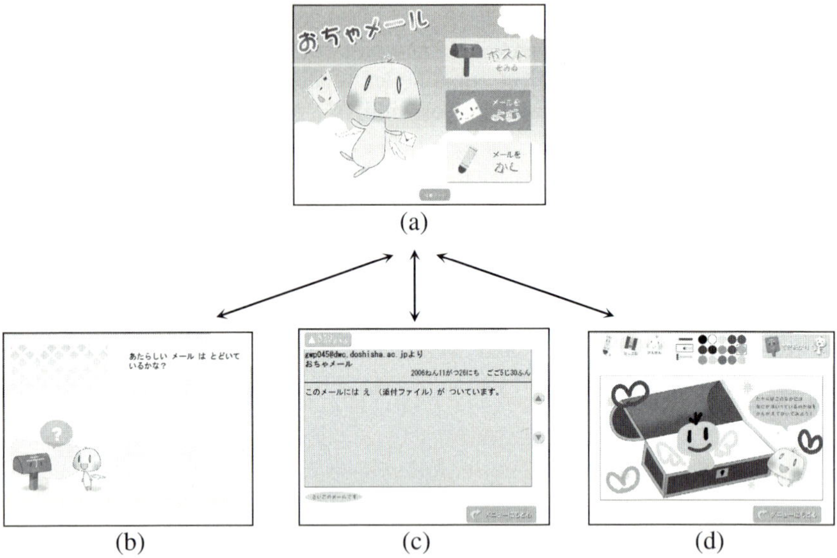

Fig. 4. Basic structure of screen transition (a) [top], (b) [receiving email], (c) [reading email], (d) [writing email]

it. Therefore we made [top] to prevent the unwilling screen transition by false operation while reading and writing. There are four buttons, [checking the mailbox], [reading email] and [writing email].

[receiving email]
If the [checking the mailbox] button on [TOP] is pushed, the screen changes to this (Fig.4(b)) and email is received and its result is displayed. According to the result, new button is displayed on the screen (Fig.5(a)-(b)). Exceptionally, it is possible to move from here to the other two screens. This is to promote reading and writing of email.

Fig. 5. (a) When a new mail arrived, (b) When no email arrived

[reading email]
If the [reading email] button is pushed, the screen changes to this (Fig.4(c)) and users can read email. Functionally this is the same system as that of Gin-Mail, but the direction of the character is horizontal and its interface is game-like.

[writing email]
If the [writing email button] is pushed, the screen changes to this (Fig.4(d)) and users can make the contents. There are buttons for email creation in the upper part, the canvas in the central part and the [returning to top] button in the lower part on this screen. The buttons in the upper part are classified into three groups. Three buttons on the left are for selecting the tool itself. There are three kinds of tools, the pen, the eraser and the letter paper. The buttons in the middle are for changing the attribute of each tool. There are three kinds of thickness and fifteen colors in the pen tool (Fig.6(a)). We decided these kinds and numbers based on the tools which children usually used. There are two kinds of erasers, one for deleting partially and another for deleting all (Fig.6(b)). There are ten kinds of letter papers (Fig.6(c)).

(a) (b) (c)

Fig. 6. Variations of Buttons depending on each tool (a) The pen tool, (b) The eraser tool, (c) The letter paper tool

The button on the right is for sending email. Sending process is the same as that of Gin-Mail. When the contents are blank, a warning message is displayed and email cannot be sent.

3.2.2 Maintenance-Mode

Maintenance-mode in Ocha-Mail is completely the same as that in Gin-Mail in respect of functions. Only the appearance is different. Its designs are unified with User-mode in Ocha-Mail.

4 Experiment on Exchanging Emails Among Three Generations

Now we are conducting an experiment on exchanging emails by using this software among three generations, i.e. elderly people, kindergarten children and college students. Each generation met for the first time in this experiment.

4.1 Acquisition of Software Usage

We made a brief manual for Gin-Mail. We made a movie and posters to explain instead of a manual for Ocha-Mail. We believe that users should be able to begin to use them immediately only after seeing these explanations. But this time we held a briefing to see concurrently how things were going. For elderly people, we explained how to use Gin-Mail together. They seemed to have understood rough operation quite quickly. For kindergarten children, we explained how to use it by playing a puppet show. They looked at it eagerly and tried to use Ocha-Mail by themselves soon after the explanation. Since it was the first experience for both groups to send email, they seemed to be at a loss about the work itself at first. But there were no signs that they were in much trouble with how to use software. Because of these results, we are sure that acquiring the necessary skills to use the software is not difficult.

4.2 Exchanging Among Different Generations

We are testing with six children, three elderly people and six students. We made some groups that consisted of one elderly person, one or two children and two students. Email is exchanged between participants in the group.

We also held a gathering to deepen friendship. Elderly people were looking forward to meeting the children and the students. We felt that elderly people were very motivated to interact with a different generation. Not a relative, children were able to enjoy themselves with elderly people.

Although the final result has not come out yet, the following are observed at present. The children tend to write email frequently. They are enjoying drawing pictures with a new tool. Many of them make contents as they please, but a few answer received email. The frequency that elderly people write email depends on their character. People who like postal letter are able to enjoy writing email with Gin-Mail. Many of elderly people are asking for an exchange with a different generation.

5 Conclusions

In this paper, we described the design philosophy and application of email software, Gin-Mail for elderly people and Ocha-Mail for kindergarten children. Before developing them, we investigated characteristics of the target generation. Both generations need a simple interface. Elderly people are familiar with a postal letter, whereas kindergarten children are familiar with a game machine. A picture is more suitable for children than a character to understand. We developed Gin-Mail and Ocha-Mail based on these results. Both programs have User-mode and Maintenance-mode. User-mode is what the target users themselves use, therefore its interface is quite simple and has only three functions. Maintenance-mode is the mode for a supporter to perform initial setting and email management.

We are now conducting an experiment on exchanging emails by using Ocha-Mail and Gin-Mail among three generations. All generations are enjoying exchanging emails. We believe it possible that email could become a new form of communication for every generation.

Acknowledgments. We acknowledge the generous financial support we received from Microsoft Corporation. We also received significant supported from Takarazuka Eden no Sono and Doshisha Kindergarten in this experiment. This work was done as a project of Doshisha Rohm Plaza Project.

References

1. Akiba, H., Shiraishi, E.: Five years old child.: The series of a child and childcare, ed. Osaka Child Care Research Institute, Kamogawa Publishing (in Japanese) (2006)
2. Hayashi, Y., Ishibashi, Y.: The perfect manual of training in a nursery and a kindergarten, Narumido Publishing (in Japanese) (2005)
3. Hutchinson, H., Plaisant, C., Druin, A.: Case Study: A Message Board as a Technology Probe for Family Communication and Coordination, Position Paper for CHI 2002 New Technologies for Families Workshop (2002), http://www.cs.umd.edu/hcil/interliving/chi02/hutchinson.htm
4. Ministry of Internal Affairs and Communications (MIC), Results of communication use trend investigations in 2005, p. 4 (in Japanese) (2006), http://www.soumu.go.jp/s-news/2006/pdf/060519_1_bt1.pdf
5. Read, J. C., MacFarlane, S., Casey, C.: Pens Behaving Badly - Usability of Pens and Graphics Tablets for Text Entry with Children, Poster at ACM UIST02 (2002), http://www.uclan.ac.uk/facs/destech/compute/staffprof/read_janet/documents/UISTlastagain.pdf
6. Wake, H.S.: Email software for aged persons -UI design based on an experiment using Tablet PC-, The Technical Report of the Institute of Electronics Information and Communication Engineers of Japan, WIT2003-59(2004-03), pp. 47–52 (in Japanese) (2004)

Shadow Arts-Communication: System Supporting Communicability for Encounter Among Remote Groups

Yoshiyuki Miwa[1], Shiroh Itai[1], Shoichi Hasegawa[2], and Daichi Sakurai[2]

[1] Faculty of Science and Engineering, Waseda University
[2] Graduate School of Science and Engineering, Waseda University
59-319, 3-4-1, Okubo, Shinjuku-ku, Tokyo, Japan
miwa@waseda.jp, hk-mark23@ruri.waseda.jp,
large.ground@gmail.com, itai@fuji.waseda.jp

Abstract. The present authors have developed a communication system WSCS (Waseda Shadow Communication System), with which your remote communicating partner (or you) can be positioned and appeared 3-dimensionally in your (or your partner's) real space, by exchanging body shadows each other. With this WSCS, we have reported previously that cooperative work and conversation can be achieved with your remote partner while taking the spatial distance ("Maai" in Japanese) between you and his shadow appeared in your space. For further development of the WSCS as Arts Communication System being supportive to a generation of co-creative "Encounter Ba", a creative expression technique has been investigated in the present study. With this technique, a relationship between you and your remote partner at the encounter occasion can be strengthened from the following two points of view. The first point is related to a supportive method to a self-organization of "Ba" at the encountering "Now, we are here", by enriching the co-existing feeling between remote participants. For this specific aim, the application software was developed with which remote participants can perform embodied interaction between remote groups using a virtual shadow ball. The second point relates to a method by which scene of the past time was reproduced across the time scale and the bodily expression of the concerned person can be trace-experienced. For this point, a co-experiencing type archive software was developed with which one can put himself at the past scene by processing of shadow images of recorded person(s), so that one can share experiences with concerned person(s) while entering the past scene across the time scale. After installing these software programs into WSCS, several communication experiments were conducted. It was found that (i), with the first software, a similar effect as an ice-breaking activities occurred at a workshop was recognized, and the software was found effective to develop a workshop which shares a mutual context among groups, and (ii), with the second program, since participants and the concerned person can be present at the same scene, changes in feeling/emotion of the concerned person can be conveyed directly via the embodiment.

1 Introduction

Majority of video communication systems which were previously proposed and developed was designed, aiming on performing cooperative work or conversation as

M.J. Smith, G. Salvendy (Eds.): Human Interface, Part II, HCII 2007, LNCS 4558, pp. 84–94, 2007.

if remote participants have met in the same space [1-6]. However, since these systems are relied on a common method of sending back and forth embodied actions that are expressed as a video image on screen or monitor on both remote parties, a co-sharing space, for example with ClearBoard [3] or HyperMirror [4], is confined within a space where actions of both parties are expressed – namely, space represented on screen or monitor. In other words, current systems [1-6] are designed for preferentially performing an isolation of an expression from the space where each participant exists and sharing/recognizing such isolated expression epistemologically. Hence, these systems do not principally provide the co-existing space for each participant to position each other's body subjectively [7-9]. Because of lucking of the co-existing space, these current systems are normally used on the assumption that both parties are known about what to be expressed, or its context is previously shared already. It is obvious that co-existing feeling, which was created when participants meet at the same space, enables that feelings can be conveyed, emotion can be understood, or experiences can be shared. Therefore, for establishing a co-sharing the context, a direct and simultaneous communication that is performed through the embodiment on the same stage becomes to play an important role [10 11]. What this statement implies is that it is difficult with current systems to integrate remote groups and to conduct a workshop simultaneously and co-creatively. Such remote groups consist of a plurality of people with different personal backgrounds and value-concepts; therefore they do not share the context preliminarily.

To solve the aforementioned problems, the present authors have previously developed the co-existing type communication system WSCS (Waseda Shadow Communication System), using body shadow as an agent of the existence [12-13]. When a plurality of WSCS is present at various remote places, each WSCS system is sending back and forth their information on position and movement of people in respective space to corresponded locations in remote space via a shadow. As a result, space of each system is accordingly changed to a space where a real person and a shadow sent from remote system are now co-existing. It is very important to point out here that, since a human has a capacity to sense an existing-feeling on a other's body shadow just as he does on real person in the same space, a co-existing feeling very similar to that taken place in a face-to-face case can be created by your body sensing the space where other person's shadow is with you as "space with activities" – which is "inseparable subject-object" space where subject and object are inter-related. and comes into your inside, and can be expressed by Japanese "Ba". Because of this unique "Ba" existence, the present authors have previously reported that, even people are separated in remote spaces, they can be ontologically positioned as if they were present in the same space, and they are able to make conversation while taking the spatial distance ("Maai" in Japanese) or behaving themselves in impromptu manner while taking mutual timing [14-15].

The present study has investigated methods to support the encounter "Ba" between remote groups who did not met before, and create an encountering "Ba" between present time person and past time person across the time scale at the past scene, utilizing unique characteristics of the WSCS based on novel communication concept. For creation of the encounter "Ba" between unknown remote groups, WSCS was developed to express interactions between the ball and groups at respective remote spaces simultaneously, by mutually expressing a ball and its movements on each space as virtual shadow image. Several encountering communication experiments were conducted. For the creation of the encountering "Ba" across the time scale, since

WSCS can record detailed information on actions and movements of people involved in the communication and can store these information in the actual field of the communication, a system was developed with which creative communication process can be trace-experienced across the time scale by reproducing the recorded/stored information on peoples' action and movements later on. Since this system, unlike the archive method [16] using a plurality of video images, can make you enable to participate directly the space where the concerned thing took place in the past time, one can encounter the person who existed in the past time and share his experience and express your feeling to him, while sensing the conditions and feeling of the past "Ba" via embodiment.

2 Virtual Shadow Ball, Being Supportive to Encountering "Ba"

For a co-existing space created by WSCS, the following items have been investigated which are necessitated to support embodied interaction like the ice-breaking activities, taken place at the encounter among groups.

1. easy to use without any special instructions
2. anyone, regardless of age and sex, can share to enjoy
3. embodiment can be unconsciously educed
4. impromptu way to play can be created naturally among groups
5. feeling of connection can be born by co-sharing experiences

As a result, a virtual shadow ball (which is applied to the WSCS) was developed as a tool which satisfies the above listed specifications. Fig. 1 depicts an image of the adapted shadow ball. The shadow ball in this development is a virtual ball, being projected by a projector, and possesses a unique character that it bounces not only when the partner's shadow touches the ball, but also even when your own shadow touches it. Using this uniqueness, one can play with a ball while positioning/transferring your remote partner into your current space. Hence, setting design specifications (as listed below) to realize the concept shown in Fig. 1, an application software for virtual shadow ball was developed.

(a) a collision was detected between the virtual shadow ball and all participants' shadow using the WSCS spaces, and the ball will bounce when it collides.

(b) movements of the ball should be shared by both spaces.

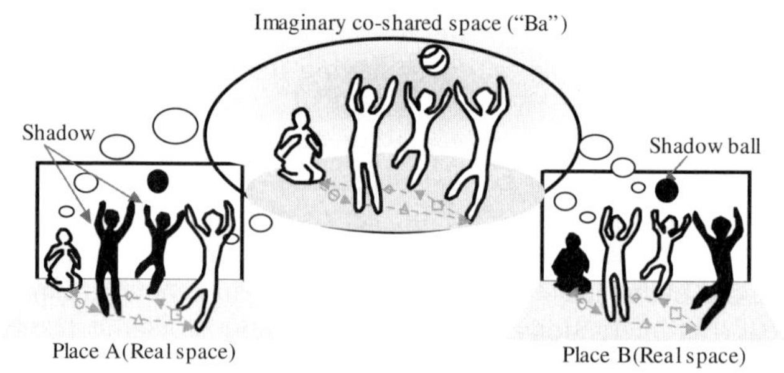

Fig. 1. Concept of Application of Virtual shadow ball

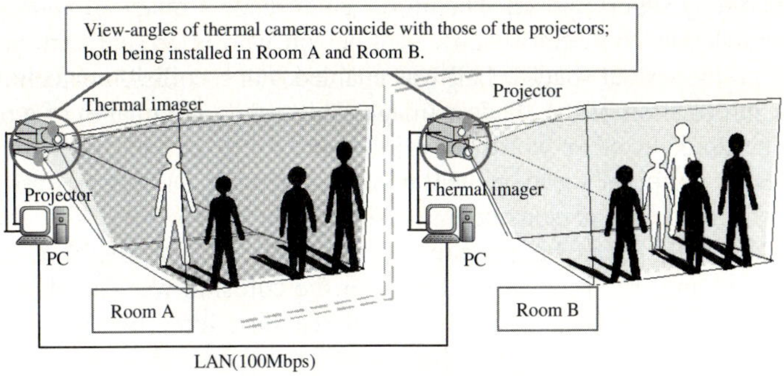

Fig. 2. System Structure of WSCS

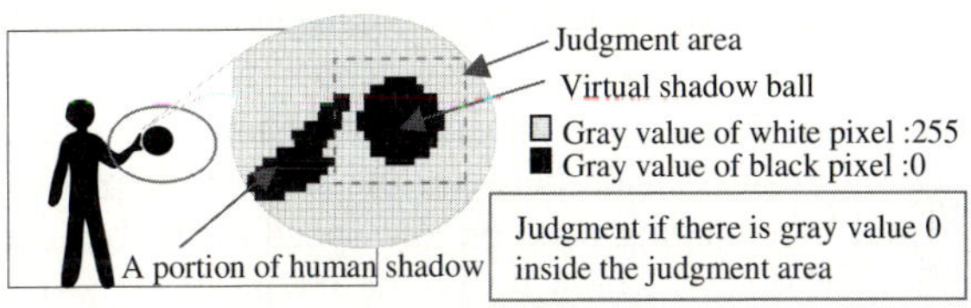

Fig. 3. Collision detection of human shadow and virtual shadow ball

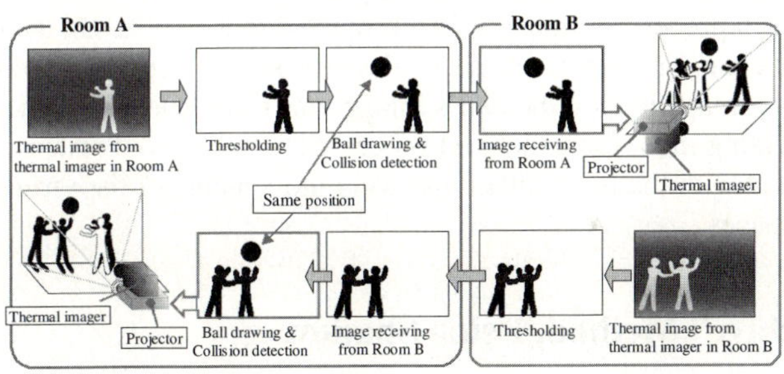

Fig. 4. Image processing for co-sharing the virtual shadow ball between remote spaces

Firstly we describe the way to realize the specification (a). As seen in Fig. 2, images of human generated by a thermal imager are sent back and forth in the WSCS, and the received image is projected by a projector. As a result, if one uses the shadow image (received shadow image) sent from a remote space, a collision detection between a shadow of a remote people and the virtual shadow ball was enabled. Moreover, a shadow of the other person who is present with you in the same space is created by a projecting light coming from the projector. At this moment, in order to adjust standing positions in the both spaces, angle of view of both the projector and thermal imager should be in correspondence with each other. Hence, when the shadow image (being created by the thermal imager) of a person who is present with you in your same space is projected by a projector which is in your own space, the projected shadow image and your own shadow being created by the projected light should be perfectly superimposed. Therefore, using shadow image of your own space, the collision detection between shadow of a person who is present with you in your own space and the virtual shadow ball was enabled. For a collision detection, the fact that a color information of the shadow image (8bit gray scale image) of a person is 0 (zero) can be used. In other words, in a judgment range around the virtual shadow ball, the presence of pixel's gray value 0 is judged (see Fig. 3). Hence, even when any part of the shadow of all persons being present in the image space is in contact with virtual shadow ball, it can be operative. The contact is considered as a perfectly elastic collision, and the speed of the ball after the collision was calculated provided that person's shadow is steady. Moreover, a descending force like gravity is always applied on the ball. The ball is furthermore programmed to perform a plane movement while a having a perfectly elastic collision at the boundary face of the screen of WSCS space. The initial position, speed and dimension of the ball can be arbitrarily determined. With regard to (b) specification, as seen in Fig. 4, the movement of the virtual shadow ball was controlled by collision detection and imaging the ball performed at one space (say, Room A). First of all, in Room A, a virtual shadow ball was drawn onto the created shadow image. Then, using this image, collision detection was made between the shadow of person in the Room A and the ball, followed by changing position and speed of the ball, accordingly. Such processed image is then sent to Room B. To the image received from Room B, the ball is imaged on the exact same position of the image which was previously sent to Room B. Later on, with this image, a collision detection between the image of person in Room B and the ball was made. Then the position and speed of the ball were again changed and such processed image is then projected to Room A. On the other hand, in Room B, after creating shadow image of the person, such image is sent to Room A and, the shadow image having the ball's image sent from Room A is then projected to Room B. Repeating the aforementioned processes between both spaces, body shadow can be interacted with the ball while co-sharing movements of the virtual shadow ball in respective spaces.

3 " Ba archive" Method, Using Shadow

In the co-experiencing type archive, the past scene was not objectively observed, rather it is required to present yourself at the same place when the concerned person was present. In this study, at the scene reproduced by the shadow, a new participant instead of the concerned person tries to get into its scene by changing the expression of the concerned person's shadow. Fig. 5 depicts an image of the co-experiencing type archive method. To achieve this task, it is necessary to record changes in positions of each person when recording the shadow of the person, and to add required processes to person's shadow who was arbitrarily selected upon reproducing the recorded scene. A reproducing software which can change the shadow's expression is installed to the WSCS as a position measuring system of each participant within the group. For developing such position measuring system, there were several specifications to be met. They include that (i) it should be able to measure positions of a plurality of person within a space of 3.6m x 2.7m which was used previously for WSCS space, (ii) there should not be any disturbances for the communication during measurements, and (iii) the system should be fit into the standard ceiling height (ie., 2.4m high) of Japanese houses. Measurements were preformed with two abreast installed cameras on the ceiling. Images obtained from these cameras were processed in a stereo system and further transformed into a real position coordinate system. During the process, with the method of acquiring the contour of the person a measurement error might be created due to changes of the light atmosphere upon the projecting the shadow, a sphere on which infrared LED are installed was placed on the top head portion as a marker to measure the position of the person. As a camera, CCD camera (Sony, XC-ST50CE) with a high spatial resolution and potential to detect the near infrared light was employed. A near infrared transmitting filter and fish-eye lens having a view angle of 160 degrees are installed to expand the measurable range and make it possible to image only the marker. The image obtained from two cameras is input to PC through the frame-grabber (Leutron, PicPort Color). The central coordinate of the marker within such obtained images is further transformed to the position coordinate in a real coordinate system to provide an existing position of the person.

Fig. 5. Concept of archive, using shadow

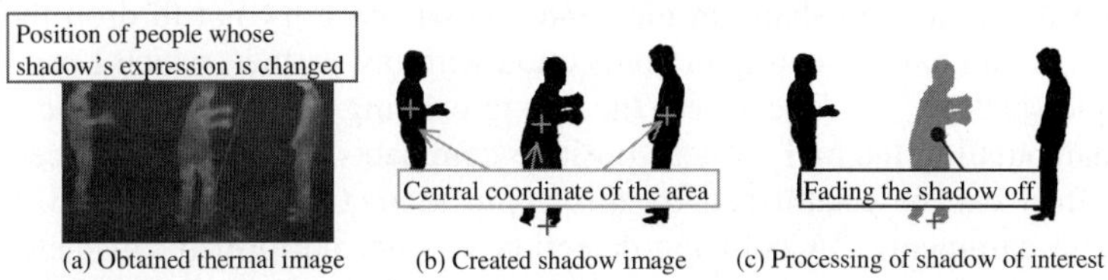

Fig. 6. Image processing to change shadow expression of a certain people

In the next step, an image processing software was developed to make it possible to operate a certain person's shadow in a thermal images, based on the position measuring system for the group which as previously developed. The main flow chart is shown in Fig. 6. The thermal images recorded in the video tape were reproduced by digital video cassette recorder (Sony, DSR-45), and input to processing PC through the frame-grabber. The measured position of a person was transformed in coordinate system, and the position of the person in the thermal images was identified (Fig. 6(a)). A range including the person was extracted by thresholding on this image and a portion which corresponds to the persons' shadow is appropriately selected from the range area. To each selected range, its central column coordinate (x-axis) was calculated (Fig. 6(b)). Then, image range which has the closest coordinate system of the person, who was selectively chosen to be replaced, was eliminated or its brightness was manipulated to fade the shadow off (Fig. 6(c)), followed by projecting it to the concerned scene. By the aforementioned processes, it is realized that a certain person can be arbitrarily selected from a plurality of recorded persons and his shadow can be eliminated or faded off to change its expression.

4 Communication Experiments

4.1 Communication of Encounter, Using Shadowy Ball

Two groups comprising of 7 to 8 persons (mixed male and female, age ranging from 10 to 75 years old) were requested to enter the two separate WSCS spaces for workshop (dimensions: 5.4m width, 3.6m depth, 2.7m height). Each participant for both spaces has not met before. They are asked to play freely with a virtual shadow ball, which has three different sizes of 30cm, 70cm, and 1m in diameter. These different sized-balls are properly exchanged for each experience. Typical example results are shown in Fig. 7. It was clearly observed that, a self expression was naturally established by taking the spatial distance ("Maai" in Japanese) between them and action timing by each participant, so that each participant gradually started to have a sense of being together at the same stage with the ever-changing the ball's position. Moreover, as seen in Fig. 8, by changing the size of the ball, participants started to create impromptu ball games such as playing catch, soccer, or rolling the ball back and forth. And it was found that, during such impromptu ball game, the context of "Ba" was co-shared unconsciously. Here are several comments from the participants; "Using the ball, I can enjoy to play with someone I've just met for the first time, and I feel the tension leaving away from my body.", "A group can unconsciously create and share an idea and mutual sense try not to drop the ball on the floor.", "I can have a feeling of connection with my partner, which I can not have only by the shadow.", or "I can feel the strong existing feeling toward the opponent who is manipulating the ball". These positive comments suggest that an ice-breaking activity effect can be established by this application. On the other hand, there are some useful comments; "A collision detection was not performed satisfactorily.", or "Not only the ball can bounce, there could be nice to have a function such as stopping the ball." These comments should be reflected to further study for development and modification of this system by considering the changes in balls' speed, and movement and dimension of the ball. All these concerns are related to the 3-dimentional spatial positions (in the depth direction).

4.2 "Ba" Communication, Across the Time Scale

As an archive experiments using the shadow, various actions of participants were recorded and reproduced. Such various activities included gymnastic play of a

Fig. 7. Communication sequences, using virtual shadow ball

Fig. 8. Various scenes, playing with the virtual shadow ball

plurality of people, a lesson of tennis form, and conversation and collaborative picture drawing done by many people. For the experiments, three cases were pre-set; (1) all shadows were reproduced just as they were before, (2) a shadow of one of the participants was completely eliminated, and (3) a shadow of one participant was faded off. These cases were randomly set. Furthermore, for case (1), the subject was indicated to enter the space freely as a new participant. For case (2), since one of recorded shadows was going to be completely eliminated, the subject was guided to where he should start and informed that he moved freely after that. And for the case (3), since one of the recorded shadows was going to be faded off, a subject was indicated to enter the same position of the faded-off shadow. Fig. 9 shows scenes which were recorded by shadows on the gymnastic play and were reproduced for the above cases (2) and (3), and features how the trace-experience taking place. Furthermore, setting 8 items from the points of view including the existing feeling of person which was reproduced by the shadow, trace-experience of the past scene, and co-experiencing, all participants were asked to evaluate their experiments with 7 steps in range of ±3. All evaluation results were tested to find a significant difference using the Wilcoxon signed-ranked tests. The results are shown in Fig. 10.

As a result, it was found that, for case (1) when recorded shadow was indicated as it was before, when participants were going to exchange their body arrangement, they appeared to have some difficulty to take the distance from the surrounding shadows. There was also that one participant felt as if he was isolated. For case (2) when one shadow was completely eliminated, the participant behaved himself while coordinating to surrounding shadow's movements or voices. More interestingly, even the participant had this experience for his first time, he moves to a similar position where the past person was and tried to participate the past scene. For this case (2), there was a comment indicating that the participant felt that he was standing there as if he was there in the past scene.

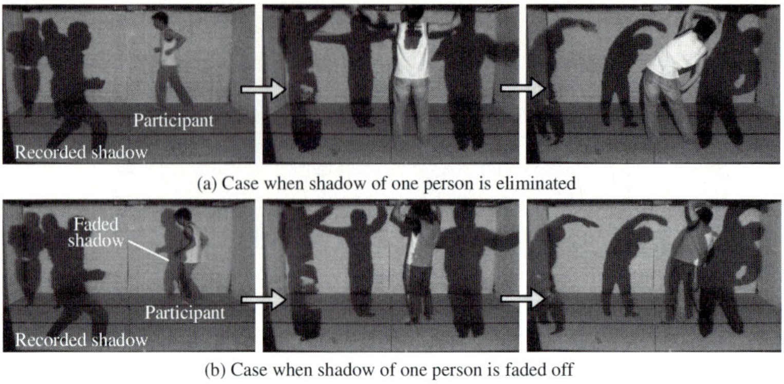

Fig. 9. Experiments of Ba archive" (a scene of gymnastics)

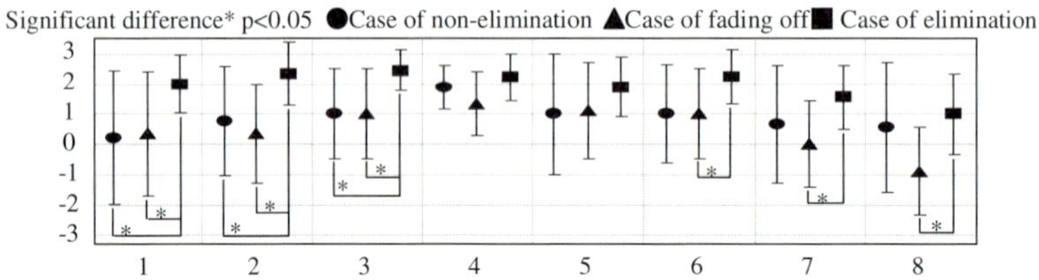

1. could take the spatial distance ("Maai" in Japanese) with surrounding shadow without any difficulties
2. felt as if I stand on the same stage as the shadow does
3. felt as if I am with the shadow representing person
4. sensed the livening up or down atmosphere and situation of actual field
5. felt as if I could a trace-experience the reproduced scene
6. felt as if I had co-experienced the reproduced scene with a person who was recorded along with conditions when the concerned things took place in past
7. could take timing naturally with surrounding people
8. felt tired while participating experiments (- sign for Yes, + sign for No)

Fig. 10. Results of questionnaire on experiments of Ba archive

For the case (3) when one shadow was faded off, although there were many participants who tried to coordinate themselves to position and action of the fading shadow, there were few who tried not to superimpose himself onto the fading shadow. Hence we had positive and negative comments to the case (3). According to the statistical tests, items 1, 2 and 3 showed the significant difference at a=5% level between case (1) and case (2), and items 1, 2, 3, 6, 7, and 8 exhibited the significant difference at a=5% level between case (2) and case (3). In both statistical tests, case (2) receives a higher evaluation. From the above, it can be concluded that the case when one shadow was eliminated provides the highest possibility for participant to have a co-experiencing the past scene. However, we also collected the following opinions; "When I made same mistakes which the fading shadow made previously, I felt that I can share the same feeling of the person", or "When I realized that I coordinated myself to surrounding, I felt as if I played gymnastics with them." Therefore, these comments suggested us that the fading technique might be also effective since the participant can share the same feeling of the past person.

Fig. 11. A scene of collaborative drawing among different groups (to draw a big tree by all participants; each of them has different role in impromptu manner)

5 Conclusions

Characteristic of virtual shadow ball application employed in this study is that peoples' shadow and ball's virtual shadow are simultaneously expressed between remote spaces, and these shadows are positioned at single integrated mutual existing space. Due to this characteristic, between body and body's shadow having an ontic inseparable relationship to the body, an implicit feeling of connection is established as if bodies are directly interacted through the ball. Therefore, at "Encounter Ba", it appears to be that the virtual shadow ball is playing a mediator's role between the space where existence is to be positioned for each other and the space where actions are expressed. This ball's role can be strongly related to the fact that one can sense a strong existence feeling on the person who is manipulating the ball. Such strong existence feeling can not be sensed by a communication using only the shadow. This strong existence feeling can be supported by and result from the fact that a co-share of the intention about the object (for example, not to drop he ball), was naturally created. Namely, because of the spontaneous expressions to make up other person's action continuously during the ball-play case, it can be considered that the alternant expressions" is taken place between participants, resulting in strengthening their co-existing feeling to make it possible for the feeling to be co-shared among groups. If so, a development of application with which a mutual complementary expression like the virtual shadow ball is naturally created possesses a promising indication for designing the arts communication which activates the encounter "Ba" among groups. Actually, when we conducted experiments on collaborative drawing among groups after groups played with the virtual shadow ball, a co-sharing feeling that they have drawn a better picture as a group effort was promoted among the groups. As seen in Fig. 11 as an example, each participant voluntarily decided each individual role and a picture was gradually completed as a group product, resulting in a well-organized picture.

On the other hand, in the archive of "Ba", since one can put his body at the place where the past thing took place even the time passed away, the context of the concerned place and feeling/emotion of the concerned person can be conveyed. These suggest that a co-experiencing with concerned person of the past time can be established. However, in order to grasp conditions and atmosphere of the scene when the concerned thing took place as close and actual as possible, extensive expression of not only human involved but also background as well as objects should be further investigated. As an Oriental saying "Ancestor plants a tree and descendant relaxes himself under the tree" indicates, a dream can become true for the encounter 'Ba" where descendant can meet the ancestor across the time scale, by shadows' function of body and involved objects.

Acknowledgment. This research was partially supported by Grant-in-Aid for Scientific Research (S), 18300044, from Japanese Ministry of Education, Science, Sports and Culture. This research also has been conducted under the project "Generation and Control Technology of Human-Entrained Embodied Media" and is supported by CREST of JST. We would like to express our sincere thanks to C. Ishibiki, visiting Lecturer of Wabot-House Laboratory, and our graduate students, K. Yoshiyama, Y. Kashiide for their assistants and collaborations for conducting delicate experiments.

References

1. Fish, R., Kraut, R., Chalfonte, B.: The videowindow system in informal communication. In: Proceedings of CSCW'90, pp. 1–12 (1990)
2. Tang, J.C., Minneman, S.L.: VideoWhiteboard: video shadows to support remote collaboration. In: Proceedings of the SIGCHI conference on Human factors in computing systems, Reaching through technology, pp. 315–322 (1991)
3. Ishii, H., Kobayashi, M., Grudin, J.: Integration of Inter-Personal Space and Shared Workspace: ClearBoard Design and Experiments. In: Proceedings of CSCW'92, pp. 33–42 (1992)
4. Morikawa, O., Maesako, T.: HyperMirror Toward Pleasant-to-use Video Mediated Communication System, CSCW'98, pp. 149–158 (1998)
5. Kuzuoka, H., Yamashita, J., Yamazaki, K., Yamazaki, A.: Agora: A remote collaboration system that enables mutual monitoring. In: Proceedings of CHI'99, pp. 190–191 (1999)
6. Apperley, M., McLeod, L., Masoodian, M. et al.: Use of Video Shadow for Small Group Interaction Awareness on a Large Interactive Display Surface. In: Proceedings of AUIC 2003, pp. 81–90 (2003)
7. Dreyfus, H.L.: On the Internet (Thinking in Action), Routledge (2001)
8. Kendon, A.: Conducting Interaction: patterns of behavior in focused encounters. Cambridge University Press, Cambridge (1990)
9. Miwa, Y.: Co-existing communication technology supported by integrating "Ba" (in Japanese). The Journal of the Institute of Electronics, Information and Communication Engineers 89(3), 218–225 (2006)
10. Shimizu, H.: The Thoughts of "Ba": Cereative Stage of Life (in Japanese), Universitey of Tokyo Press (2003)
11. Hall, E.T.: The Hidden Dimension, Garden City, NY (1966)
12. Miwa, Y., Ishibiki, C., Watanabe, T., Itai, S.: Design of Co-existing space by Shoji interface showing Shadow. Human-Computer Interaction Theory and Practice (HCII2003) 2, 508–512 (2003)
13. Miwa, Y., Ishibiki, C.: Shadow Communication: System for embodied interaction with remote partners. In: Proceedings of CSCW 2004, pp. 467–476 (2004)
14. Ishibiki, C., Watanabe, T., Miwa, Y.: Development of Co-existing Communication System using the Shadow as a Media Integrating "Ba" (in Japanese). Journal of Human Interface Society 7(3), 311–318 (2005)
15. Ishibiki, C., Miwa, Y.: Co-existing Communication between Groups using Body Shadows (in Japanese). Journal of Human Interface Society 7(4), 497–505 (2005)
16. Hirata, K., Harada, Y., Takada, T., Aoyagi, S., Shirai, Y., Yamashita, N., Yamato, J.: The t-Room-Toward the Future Phone. NTT Technical Review 4(12), 26–33 (2006)

Visual Feedback to Reduce the Negative Effects of Message Transfer Delay on Voice Chatting

Kazuyoshi Murata[1], Megumi Nakamura[2], Yu Shibuya[2], Itaru Kuramoto[2], and Yoshihiro Tsujino[2]

[1] Tokyo University of Agriculture and Technology, 2-24-16 Koganei-shi Nakacho, Tokyo. 1848588, Japan

[2] Kyoto Institute of Technology, Matsugasaki Goshokaidocho, Sakyo-ku, Kyoto, 6068588, Japan

kmurata@cc.tuat.ac.jp, {megumi, kuramoto}@hit.is.kit.ac.jp, {shibuya, tsujino}@kit.ac.jp

Abstract. Voice chat system is one of the most popular communication systems using computer networks. There is considerable delay to transfer the sender's voice to the receiver on usual voice chat system. The message transfer delay often causes unintentional interruptions and makes participants of the system feel confused. It is considered that such negative effects occur because participants can't know the status of their voice playback in remote site. In this paper, two types of remote status monitoring systems are introduced to reduce unintentional interruptions and confused feeling. Then, it is confirmed experimentally that these systems were effective to reduce these problems.

Keywords: Voice Chatting, Message Transfer Delay, Unintentional Interruption, Status Monitor.

1 Introduction

Voice chat system is one of the most popular systems for real-time communication using computer network. However, voice chat system still involves some problems to be solved. For example, message transfer delay is a problem one should not ignore. Message transfer delay on voice chatting often causes unintentional interruptions. These interruptions disrupt flow of the conversation and this disruption makes participants of voice chatting feel confused. It is considered that such problems occur because participants of voice chatting can't know the status of their voice playback in remote site. In this paper, two types of remote status monitoring systems are introduced to reduce the problem caused by delay, and, then, the effectiveness of these systems is confirmed experimentally.

There are many studies about the influence of delay on media communication. For example, Krauss and Bricker[1] analyzed effect of message transfer delay for audio-communication and they showed that delay caused disruptive influence on communication. Ruhleder and Jordan[2] focused on the impact of delay on distributed communication and they showed that delay disrupts the turn-taking system. Florian and Reichl [3] addressed influences of delay on communication interactivity using

M.J. Smith, G. Salvendy (Eds.): Human Interface, Part II, HCII 2007, LNCS 4558, pp. 95–101, 2007.

VoIP connection. Bouch et al.[4] focused on user's perceptions of QoS and presented minimum threshold of acceptable delay in various communication tools. There are many studies aimed to reduce transmission delay. In voice chat system, however, it is impossible to get rid of delay completely. The purpose of our research is not decreasing the length of delay, but reducing the influence of delay by improving user interface of voice chat system.

2 Impact of Delay on Voice Chat

Unintentional interruption is one of the most important problems caused by message transfer delay in voice chatting. Fig.1 illustrates an example of conversation in delayed situation. A1, A2, and B1 mean that participant A or participant B utters something. A1', A2' and B1' mean that participant A or participant B is hearing the utterance of the other participant. If there is not message transfer delay, the voice of participant A is transmitted to participant B's site right away and participant B can receive the voice immediately. If there is certain delay, however, each participant receives delayed utterance of other participant. Therefore the start timing and end timing of A1, A2, and B1 don't coincide with that of A1', A2', and B1'. In this example, participant A utters something first (A1), and waits participant B's response but there is no response. After a certain time, participant A starts talking about another topic (A2) because he/she assumes that participant B doesn't want to respond. Then participant A is interrupted by utterance of participant B (B1') suddenly and he/she is confused. On the other hand, in participant B's site, participant B receives delayed utterance (A1'). Participant B starts responding to participant A after a certain thinking time (B1), but he/she is interrupted by next utterance of A (A2') suddenly. Participant B is also confused because the topic of A2 is not related to A1' and B1.

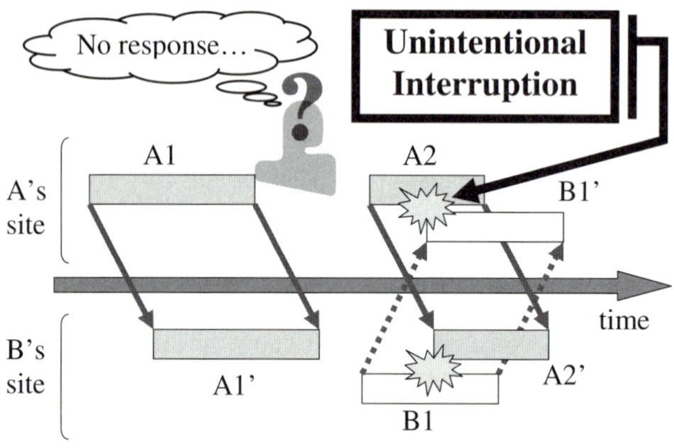

Fig. 1. Conversation in delayed situation

Thus, when there is certain delay in voice chatting, the sequence of conversation is shuffled and unintentional interruptions happen in both of A's site and B's site. These unintentional interruptions disrupt conversation and makes participants confused. Unintentional interruptions also happen even if there is little delay. However, these interruptions don't cause disruption of conversation because participants can recover from the disrupted situation easily. In contrast, if the participant utters something to recover in delayed situation, this utterance causes another interruption and this interruption makes the situation worse.

The reason for these problems is that participant A can't wait participant B's response on the assumption that participant B isn't willing to respond. Therefore, if participant A can recognize the length of waiting time caused by delay, it becomes easy for him/her to wait for participant B's response and unintentional interruptions will decrease.

3 Status Monitoring System

In this paper, Literal Indicator and Progress Bar Indicator are presented as the status monitoring systems for reducing the negative effects of message transfer delay on voice chatting. Each system has a simple message display window which displays the status message of voice playback in remote site. Users of these systems can understand that his/her partner of conversation is hearing his/her voice or not because he/she can recognize the start timing and end timing of voice playback in remote site easily. Fig.2 illustrates an example of display timing of the status message of voice playback in remote site. The display timing of the status message is delayed for the length of message transfer delay from actual start timing of the voice playback in remote site. That is, the length of the partner's thinking time in remote site (T_R) is equal to the length of it in local site (T_L). Consequently, the participant can distinguish between waiting time caused by message transfer delay and his/her partner's thinking time.

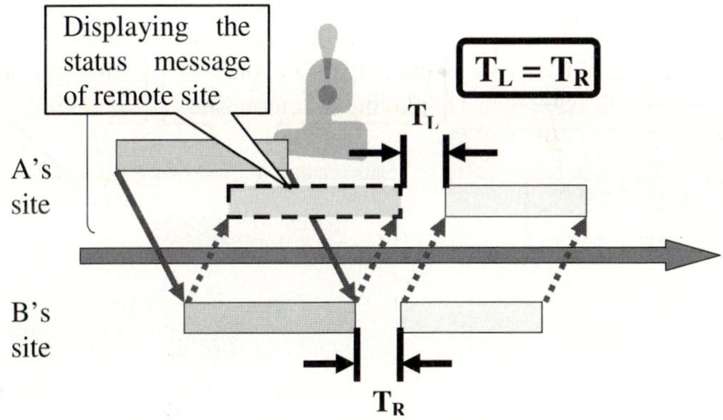

Fig. 2. Display timing of the status message of voice playback in remote site

There are some push-to-talk systems for voice chatting and these systems can prevent unintentional interruptions. However, these systems also prevent "intentional interruptions" in the same way. Intentional interruption is important for turn-taking system. Therefore, it should be avoided for smooth voice chatting to prevent the intentional interruption. Participants using our status monitoring systems can use intentional interruptions, if necessary.

3.1 Literal Indicator

Literal Indicator is illustrated in Fig.3(a). While a participant's voice is playing in remote site, Literal Indicator displays a message which indicates that his/her voice is playing in remote site. When the remote chat client starts playing the participant's voice, a start signal is sent from the remote chat client to the Literal Indicator. Then, Literal Indicator starts to display message when the signal reaches to it. In the same way, when the remote client stops playing participant's voice, a stop signal is sent from the remote client and Literal Indicator stops displaying the message. The participant using Literal Indicator can know if his/her partner in remote site is listening his/her voice or not.

3.2 Progress Bar Indicator

Progress Bar Indicator is illustrated in Fig.3(b). Progress Bar Indicator displays a white bar and a red bar, which indicates a process of playing participant's voice in local and remote site. When a participant using Progress Bar Indicator starts to talk, Progress Bar Indicator displays a white bar firstly. The white bar keeps extending while he/she speaks, which means how long the participant is speaking. Then, Progress Bar Indicator displays a red bar over the white bar when Progress Bar Indicator receives a start signal from remote site. The red bar keeps extending until Progress Bar Indicator receives a stop signal from remote site, which means how long the other participant is listening in remote site. The participant using Progress Bar Indicator can know the start timing and end timing of his/her voice playback and infer how long his/her voice will be played in remote site.

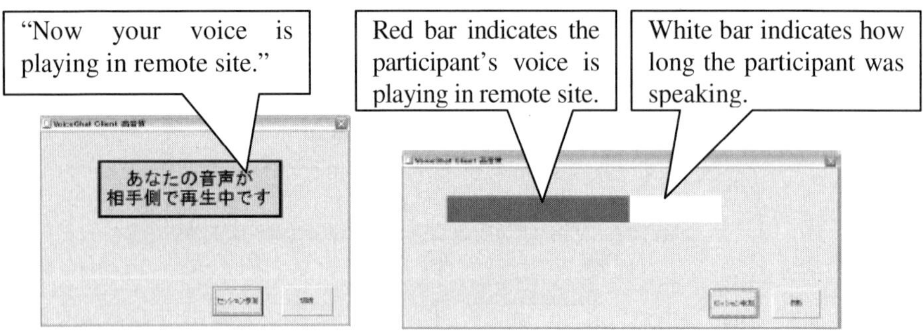

(a) Literal (b) Progress Bar

Fig. 3. Two types of status monitoring systems

4 Experiment

An experiment had been performed to confirm that presented status monitoring systems can decrease unintentional interruptions and reduce confused feeling of participants caused by these interruptions.

4.1 Experimental Setup

A simple server-client voice chat system was constructed for this experiment. This chat system had one chat server and two chat client. Each client was connected to the server through LAN and the server could generate some kinds of message transfer delay. This experiment was performed under three kinds of delay conditions with delay of 0.5[sec], 1.0[sec], and 1.5[sec]. There were 10 pairs of subjects who were recruited from our laboratory for this experiment. They had been known well each other and made daily face-to-face communication. Each subject of a pair was located to a soundproofing room and asked to make 10 minutes discussion with another subject. They had a list and there were 20 items in the list. Their task was to rank these items by their mutual consent. For example, there were 20 countries in the list and subjects were asked to discuss which country was good for next journey and rank the countries. Subjects were asked to perform this task by using Literal Indicator and Progress Bar Indicator. Moreover, they were asked to perform this task without these status monitoring systems as the controlled condition. Consequently, there were three kinds of delay conditions (0.5[sec], 1.0[sec], 1.5[sec]) and three kinds of status monitoring conditions (Literal Indicator, Progress Bar Indicator, Without status monitor).

In this experiment, the rate of unintentional interruptions was measured. For subjective evaluation, each subject was asked to complete a questionnaire which had a list of following questions after each discussion.

- Q1: You feel confusion about conversation in the discussion. (1: disagree ~ 5: agree)
- Q2: You couldn't utter when you wanted to. (1: disagree ~ 5: agree)
- Q3: Did you change conversational matter when you used the status monitoring system? (Free format)

4.2 Result and Discussion

The average rate of unintentional interruptions is illustrated in Fig.4. There was a significant main effect of status monitoring condition [$F(2,72)=15.272$, $p<.01$], and the rate of without status monitor condition was significantly higher than the rate of Literal Indicator and Progress Bar Indicator condition (Scheffe's test, $p<.01$). This result indicates that status monitoring systems could decrease unintentional interruptions as we expected. From the answers of Q3 ("Did you change conversational matter when you used the status monitoring system?"), many subjects addressed that they waited their partner's response until the displayed message disappeared. It is considered that this behavior resulted in decreasing unintentional interruptions.

Fig.5 illustrates the average score of Q1 ("You feel confusion about conversation in the discussion."). There were significant main effects of delay condition [$F(2,72)=7.097$, $p<.01$] and status monitoring condition [$F(2,72)=4.683$, $p<.05$]. This result indicates that subjects felt strong confusion about conversation when the length of delay was 1.5 [sec] but the status monitoring systems could reduce the confused feeling of participants. Especially, the score with Literal Indicator was significantly lower than the score without status monitor [Scheffe's test, $p<.05$].

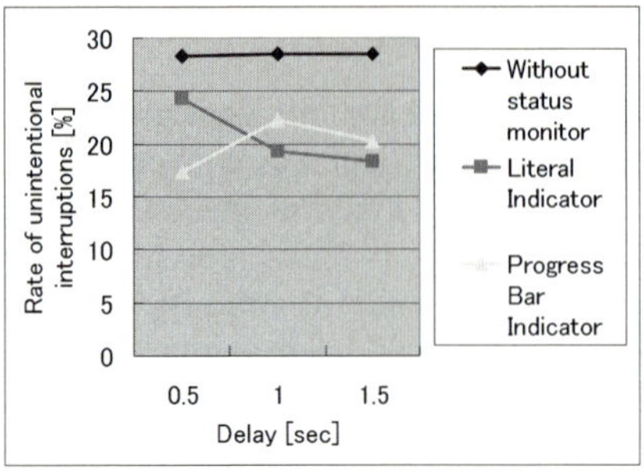

Fig. 4. The average rate of unintentional interruptions

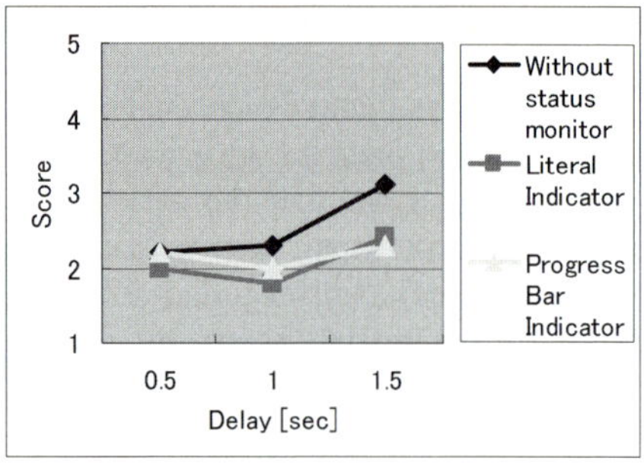

Fig. 5. The average score of Q1: You feel confusion about conversation in the discussion (1: disagree ~ 5: agree)

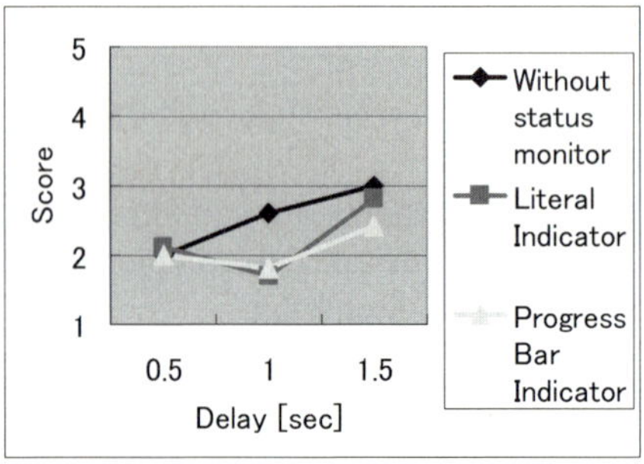

Fig. 6. The average score of Q2: You couldn't utter when you wanted to. (1: disagree ~ 5: agree)

Fig.6 illustrates the average score of Q2 ("You couldn't utter when you wanted to."). There were also significant main effects o f delay condition [F(2,72)=13.769, p<.01] and status monitoring condition [F(2,72)=4.871, p<.05], and the score with Progress Bar Indicator was significantly lower than the score without status monitor [Scheffe's test, p<.05]. It is important for voice chatting that participants can utter when they want. In this experiment, subjects without status monitoring system felt difficult to utter as the delay was increased. However, if they could use status monitoring systems, especially Progress Bar Indicator, they could utter when they wanted.

We considered that unintentional interruptions would increase with increasing delay before this experiment. However, the result showed that these interruptions didn't increase actually. On the other hand, the results of questionnaire showed that confusion of subjects increased with increasing delay as we expected. From these results, it is considered that the unintentional interruptions in long delay condition can confuse subjects strongly. Furthermore, it is more important to decrease these unintentional interruptions in long delay condition than in short delay condition. Literal Indicator and Progress Bar Indicator were effective to decrease unintentional interruptions in both of short delay condition and long delay condition, and, moreover, these could be effective for reducing the confusion about conversation in long delay condition.

It was not clear the difference between two types of status monitoring systems from the result of this experiment. However, many subjects indicate that it was easier for the Progress Bar Indicator to understand the process of their voice playback in remote site and to predict the timing of the response. This difference will be clarified by further experiment and investigation in the future.

5 Conclusion and Future Work

In this paper, we focused on unintentional interruptions caused by message transfer delay in voice chatting and confused feeling caused by the unintentional interruptions. To reduce the negative effects, we presented two types of status monitoring systems, Literal Indicator and Progress Bar Indicator. Then we had an experiment to confirm the effect of these systems. From the result of this experiment, we confirmed that these systems were effective to decrease unintentional interruptions and could reduce the confusion about conversation even if there were certain amount of delay. However, further investigation is needed for improving presented systems. For example, it is important for improvement of these systems to compare Literal Indicator and Progress Bar Indicator in detail. In addition, it is needed to evaluate other display styles instead of text or progress bar. Future work focuses on improving these systems and building them into practical applications.

References

1. Krauss, R., Bricker, P.: Effect of Transmission Delay and Access Delay on the Efficiency of Verbal Communication. The Journal of the Acoustical Society of America 41(2), 286–292 (1967)
2. Ruhleder, K., Jordan, B.: Co-Constructing Non-Mutual Realities: Delay-Generated Trouble in Distributed Interaction. Computer Supported Cooperative Work. 10(1), 113–118 (2001)
3. Florian, H., Reichl, P.: The Well-Tempered Conversation: Interactivity, Delay and Perceptual VoIP Quality. In: Proc. IEEE ICC'05, pp. 244–249 (2005)
4. Bouch, A., Sasse, M., DeMeer, H.: Of Packets and People: A User-centered Approach to Quality of Service. In: Proc. IWQoS'00, pp. 189–197 (2000)

Friendly Process of Human-Computer Interaction – A Prototype System in Nostalgic World

Seiko Myojin[1], Mie Nakatani[2], Hirokazu Kato[1], and Shogo Nishida[1]

[1] Graduate School of Engineering Science
Osaka University
1-3, Machikaneyama, Toyonaka, Osaka, 560-8531, Japan
seiko@nishilab.sys.es.osaka-u.ac.jp,
{kato, nishida}@sys.es.osaka-u.ac.jp
[2] Center for the Study of Communication-Design
Osaka University
1-1, Senri Expo Park, Suita, Osaka, 565-0826, Japan
mie@cscd.osaka-u.ac.jp

Abstract. In the interactive model considered herein, a user can be made to feel that the virtual world exists as tangibly as the real world does. Once "Friendly Process of Human-Computer Interaction" launches, a user can wait to see what happens before he or she takes action. Even if a user does nothing, the program still runs. A PC user is an audience. The computer program performs for its user audience. If the user is interested in the computer program's behavior, that user interacts with the PC as its audience. The purpose of our work was to examine the potentiality and significance of our interactive model. We integrated our concepts into our interactive design. In order to evaluate our concepts, we prepared three interactive movie system prototypes. The creation of our prototypes is well underway toward effectuating our concepts. After we improve our prototypes, we will then need to evaluate them.

Keywords: entertainment, emergent relationships, nostalgia, image recognition.

1 Introduction

Originally, the personal computer (hereinafter, "PC") was a business tool used for tasks such as document preparation or formula manipulation. A user commands the PC to manipulate a formula. A PC tells the user to push a button. This relationship was not an interactive one that elicited emotions or excitement, especially among those who had not worked with PCs before. In recent years, however, PC use has spread within the general public and now includes use as an entertainment tool. Everyone can enjoy telecasts or movies whenever and wherever they want. People can take pleasure in music by watching a live concert. Users can have fun chatting or playing Web camera games. Everywhere in contemporary society, you can find integrated circuit chips and sensors. People have no consciousness that they are being watched by computers. The systems that are not even regarded by users as computers

M.J. Smith, G. Salvendy (Eds.): Human Interface, Part II, HCII 2007, LNCS 4558, pp. 102–109, 2007.

have also increased. In the future, the relationship between users and computers will become even more varied.

We would like to propose herein an interactive model that is different from the traditional paradigm of a user's PC operations. In general, when a user does nothing, a PC will not start. If users want to avail themselves of that PC, they have to take necessary actions. Our concept offers a viewpoint differing from that of the traditional paradigm of a user operating a PC. With our design, characters from a virtual world will sometimes be able to suddenly tempt a real world user. However, this temptation does not beckon a user directly into a virtual world. A computer program performs for its user. If that user is interested in the computer program's behavior, the user can then interact with the computer as its audience. Even if no one is interacting with a PC, that PC still runs automatically. To the best of the authors' knowledge, no research has yet been carried out to discuss the potentiality and significance of such a relationship between a user and a PC. In the interactive model that we consider herein, a user can be made to feel that the virtual world exists, just like the real world. A user can wait to see what happens and what a PC thinks before that user takes action. Even if a user does nothing, the PC still starts. The PC user is an audience. A computer program performs for its user. If a user is interested in the computer program's behavior, that user interacts with the PC as its audience.

The purpose of our work described herein was to examine the potentiality and significance of our interactive model. We created original concepts and integrated them into our interactive design. In order to evaluate our concepts, we prepared three prototypical interactive movies, that is, three systems. The first system contains the minimum requirements necessary to bring our concepts to life as an interactive movie. The second system has the same function as the first system but an added feature is that, with the second system, a user can easily notice being recognized by that system. The third system uses a Web camera to recognize its user's motions. The creation of these prototype systems is well underway toward effectuating our concepts. In this paper, Chapter 2 discusses our concepts; Chapter 3 details the methods for realizing our concepts; Chapter 4 is an explanation of our prototype; Chapter 5 is devoted to a discussion about our prototype and related work; and, finally, Chapter 6 offers conclusions about this paper.

2 Our Concept

Our concept takes a viewpoint different from that of the traditional paradigm of a user's PC operation. We recognize a user before the user is able to notice that the PC recognizes him or her. A PC is also an invitation to the virtual world. A user faces the PC as an audience. A PC can prompt a user to think any of the following: "Maybe the PC recognizes me," "It is possible that the PC is responding to my actions," or "Perhaps I am a character in a story."

2.1 A User May Notice That He or She Is Recognized in the Virtual World

We consider the circumstances of using a PC. In general, a user realizes he or she can operate a number of characters and tools within a game before that user begins to play the game. However, in our work, a user does not know anything before entering into game play. The user has not even noticed that the game exists. For example, our scenarios look like circumstances from "The NeverEnding Story." Pastian, the protagonist, becomes aware that he is being recognized by the characters from the book that he is reading. We believe that if we can give PC users such awareness, we can then easily tempt them into the virtual world. Research on conventional entertainment systems shows the virtual world has increasingly grown in scale such that a user can be absorbed into it [1]. For example, CAVE [2] surrounds a user. It stimulates the user's five senses. We have tempted a user into the virtual world before he or she has noticed that even being recognized. The user can unconsciously open the door to the virtual world. As a result, that user can be absorbed into the virtual world.

2.2 A User May Sense What a PC Thinks

In general, there is always some button to push in response to a PC's command. After that, the PC reacts as the user would expect in response to that particular button. In this type of interaction, the PC is in control of its user. This type of interaction facilitates such programs as shooting games. We consider the interaction where a user becomes an observer. In daily life, such interactions regularly take place. Sometimes a person does not react to another person as expected. This interaction has interest. For example, your friends do not always react as you expect when you speak with them. People cannot necessarily reach clear understanding about each other. However, a person can certainly imagine what a friend thinks. Then many people are interested in that interaction. Now replace this interaction between the two friends with an interaction between a user and a PC. Although a PC does not always react to a user as expected, such unexpected responses can lead to new and interesting pathways.

2.3 Emergent Relationships

If a user can feel the virtual world to exist as surely as the real world does, we can more easily tempt him or her into the virtual world. To make this happen, the characters in the virtual world and the PC user must have some kind of relationship. For example, in "The NeverEnding Story," there was a queen awaiting a child's help. That child was Pastian. Pastian sensed this relationship between himself and that of the world of Fantargen. Pastian came to feel that the world of "Fantargen," the world found inside the book that the boy was reading, existed as surely as the real world did. We call this phenomenon "Emergent Relations," and we have incorporated this idea into our interface.

3 Our Methods

We took our original concepts and incorporated them into interactive designs. In order to evaluate our concepts, we prepared three prototypical interactive movies. The theme of these movies is nostalgia. For purposes of our discussion, "nostalgia" refers to a person remembering the good old days in a sentimental way. Nostalgia does not usually require that a person consciously summon up a nostalgic feeling. Rather, nostalgia is an emotion that spontaneously springs up in someone's heart as if a person suddenly crooned a favorite song without any reason. This is why nostalgia is a suitable theme to the application of our concepts and prototype production. Our movies could be something that provokes people into remembering their fond memories.

Our concept consists of three elements. First, *a user may notice that he or she is recognized in the virtual world.* Second, *a user might sense what a PC is thinking.* As a result of this second element, a user can wait to see what happens next, what their PC thinks, before that user takes any action. The third element is that of the *emergent relationship.* A user will feel that the virtual world exists as tangibly as the real world does. We have constructed our movies prototypes in order to effectuate the interface of these elements on a PC. We have done so for the following reasons: When a user does nothing, the PC can play our movie as if it were a screen saver playing. We can stimulate the user's awareness with our timing. We can also instigate interactions between a user and a population of characters and tools from within the virtual world. Furthermore, the virtual world characters and the user can develop a relationship.

We considered what movie features we should use to bring our concepts to life. We drew up a movie plan in which we project the characters' lines and narration both with sound and captions. Our movies are played from the perspective of the main character. Our movie regards a user as one of the characters within the movie. Even if a user does not interact with the movie characters, our stories would be designed to develop and finish automatically. In the characters' lines, our movies do not explicitly invite a user inside the virtual world by such exhortations as *Join with us* or *Write on it.* Rather, our movies express the characters' lines such that a user can somehow sense meaning from the context. For example, the user senses that he or she is a character within the movie by hearing or seeing dialogue such as *We were waiting for you to finish writing it.* By using a Web camera, our system activates when it recognizes a user in front of the screen. The challenge is to keep the user from noticing this. In the virtual world, multiplicities of characters are acting. By coordinating the virtual world timing, our system is able to recognize a user as one of the virtual world characters and the story develops in response to the user's appearance. The user does not necessarily notice that he or she is recognized by a Web camera. However, even if a user does notice the active Web camera, the user may still be interested in the characters' behavior and wish to interact with them. Nonetheless, the user does not yet know how to interact. The user needs to figure out how to interact. It is not difficult for users to learn how to interact. Our movie provides a number of clues.

Other points to note are as follows: In selecting the descriptions, the camera work, and the colors, we took note of what images might be remembered from scenery in the past. For example, we did not use realistic colors within the world of memories. Even if the picture is different from the true look of past scenery, it is still important to induce a nostalgic texture. For the keywords in the characters' lines, we set keywords that the average adult will not often use. In order to do this, we chose childlike expressions and youthful words. For our landscape, we used an ordinary school building and its adjacent park so that a user could easily relate to the scenery. We integrated sound effects to our benefit: the sounds of someone turning a page or footsteps, the noise of a crowd, a cicada's buzz, and other such sounds. We have attached importance to realistic sounds. The effectiveness of sound to induce a user's emotions and imagery is discussed in Reference [3], below. We considered the length of time of our movies. Our movies need enough time to make a user notice that he or she is being recognized by our system. They require sufficient time to interact. Our movies should appear the way a screen saver does, that is, disappearing while a user is wondering about the story. Our movies make a user understand a story in a short time. Therefore, when a user does not interact with our movie, that movie plays no longer than approximately five minutes. If a user does interact with our movie, we then play our movie for a maximum of about ten minutes.

4 The Prototype

4.1 Prototype Architecture

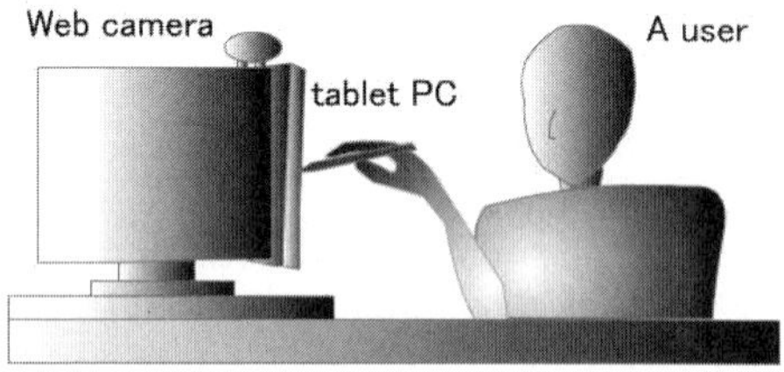

Fig. 1. Architecture of our prototype

Fig. 1 shows our prototype. We used Windows XP and Macromedia Flash (Action Script 2.0). In Graduation2, we also employed Microsoft Visual Studio.NET 2003 (Microsoft Development Environment 2003, .NET Framework 1.1, Visual C#.NET), and put "Vidgets[4]" to use as a library for image recognition. Our Web camera's specifications are as follows: The minimal-length subject distance is 20 cm; the video resolution is 640×480 with a CMOS sensor.

4.2 Movie Scenarios

Graduation1 (Fig. 2): The theme of this movie is a memory illusion about graduation. Students in Japan usually buy their school yearbook when they graduate from school. A school yearbook is those students' photograph collection. Students and teachers tend to write messages for their friends or students in these yearbooks. The protagonist of this movie visits his or her old school. The protagonist graduated from the school about ten years ago. He or she walks into classroom. After a while the protagonist meets past classmates. Then strange things happen there.

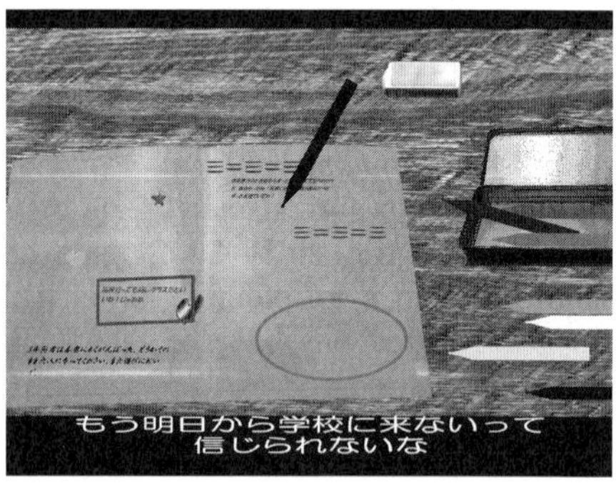

Fig. 2. Graduation1

Graduation2 (Fig. 3): Although the scenario of this movie is identical to that of Graduation1, this movie serves another function, which is to show random noise. The user is unable to view the movie clearly because of the random noise whenever the user's face gets too near to the screen. This is inspired by the idea of a dream state. If a person is having a pleasant dream, they try to sleep not to wake. However, their visibilities in the dreams become poor because they almost start to wake up.

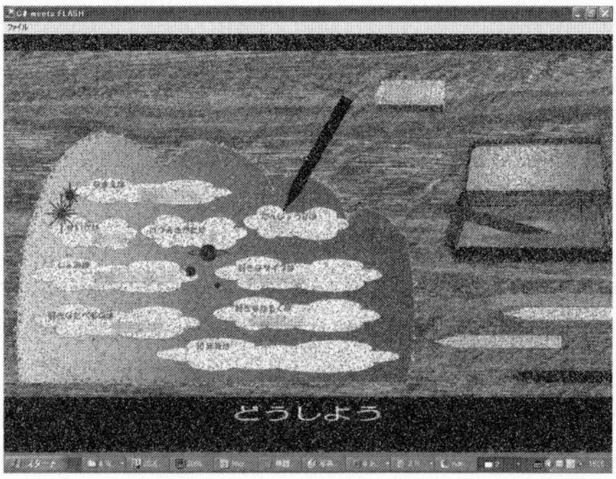

Fig. 3. Graduation2

Fig. 4. DARUMA-san 1

DARUMA-san1 (Fig. 4): This movie's theme is a memory illusion about "DARUMA-san ga Koronda." "DARUMA-san ga Koronda" is a kind of tag game, one that is the traditional play of Japanese children in a group. Children choose one child to be called "Oni." Oni has to catch other children. Oni turns around and faces a tree or wall. While Oni counts for ten seconds, the other children can scatter in all directions. After Oni has finished counting to ten, he or she says "DA RU MA - SA N GA KO RO N DA" and turns around, looking. Other children can draw closer to Oni provided that Oni does not see them. However, if a moving child is seen by Oni, Oni can then catch him or her. If Oni manages to catch all the children, Oni wins the game. In summary, Oni calls out "DA RU MA - SA N GA KO RO N DA" and then looks around to try and catch any moving children. If a child touches Oni without Oni's seeing their approach, then all the children get a free pass to scatter from Oni in any direction. Although there are other rules at play in various regions of Japan, this is the general method of "DARUMA-san ga Koronda".

4.3 Future Work

The following parameters sum up our prototypes: Each of our interactive movies includes three specific functions designed to effectuate our concepts. (1) We project the characters' lines and narration both through sound and caption; (2) Our movie incorporates the user as one of the characters within the movie; and (3) Even if a user does not interact with the other characters in our movie, the movie will nonetheless have to develop and finish automatically. The first movie system contains the minimum requirements to bring our concepts to life as an interactive movie. The second movie system serves the same function as the first system but, additionally, a user can easily observe being recognized by the system. Finally, the third movie system utilizes a Web camera to recognize the motions of its user. Although we have nearly completed our work, the systems still have some shortcomings. We will continue to improve our systems. We tested Graduation1 with one user. We observed that Graduation1 succeeded in getting the user to notice that she was a character within the movie. She could find almost how to interact. Once we have improved our prototypes, we will then need to evaluate our systems more thoroughly.

5 Related Works

Our concepts resemble Kohei Asano's "Shiawase wa soko ni aru[5]." Asano is in the Engineering Department at Tokyo Polytechnic University. In Asano's system, when a user waves a hand at a cat, the cat has some kind of reaction. A small camera incorporated into the display recognizes the motion of the user's hand and the system responds by playing a randomized video of cats. Cats are naturally capricious. Humans understand that a cat does not always react to a person's specific action. This is why, even if the system plays a randomized video of cats, a user still feels that he or she has interacted with a cat. Asano's system is predicated upon new content, rather than a new interactive model.

Although there are only a few journal articles that focus on inducing nostalgia, there are many that address works of art. For example, "Imagery Box"[5], "Interval"[5] "Kioku"[6]. Although the user may simply enjoy seeing such work, he or she is unable to feel becoming or being a character within the story. These projects are not interactive movies.

Toshio Iwai's "ElectroPlankton[7]" contains similarities to our project's concepts. In "ElectroPlankton," users do not know how to interact before they enter into the game. However, the characters in "ElectroPlankton" do not engage in any relationship with users. In that regard, our concepts differ from those of "ElectroPlankton."

6 Conclusions

In the interactive model that we have considered, a user can be made to feel that the virtual world exists as tangibly as the real world does. A user can wait to see what happens and what a PC thinks before he or she takes action. Even if a user does nothing, the PC still starts. A user is a PC's audience. A computer program performs for its user. If a user is interested in the computer program's behavior, then that user will interact with the PC and be its audience. The purpose of our work was to examine the potentiality and significance of our interactive model.

We have created original concepts and incorporated them within our interactive design. In order to evaluate our concepts, we prepared three prototypical interactive movie systems. The first system contains the minimum requirements to effectuate our concepts as an interactive movie. The second system serves the same function as the first system but also allows a user to easily notice that he or she is being recognized by the system. The third system utilizes a Web camera to recognize the motion of users. The creation of these prototypes is well underway toward effectuating our concepts. After we improve our prototypes, we will then need to evaluate them. In our evaluative experiment, we may have users download our software. Alternatively, we might choose a sparsely populated street, set up some screens there on the street, and then play our movie upon one of those screens. On the other screens, we would play a conventional movie. Afterward, we would conduct interviews with users to obtain those users' impressions.

References

1. Bowman, D.A., Raja, D., Lucas, J., Datey, A.: Exploring the Benefits of Immersion for Information Visualization, HCII2005 (2005)
2. Cruz-Neira, C., Sandin, D.J., DeFanti, T.A.: Surround-Screen Projection-Based Virtual Reality: The Design and Implementation of the CAVE. In: Proc.of SIGGRAPH'93, pp. 135–142 (1993)
3. Berman, A.D., Cruz-Neira, C., Berman, S.L., Tuch, L.: 9/11: Building A World of Survivor Memories, HCII2005 (2005)
4. Vidgets: http://qwik.jp/media2005/Vidgets.html (Tateo Igarashi uses Vidgets in his lecture Media Information Science) (in Japanese)
5. NHK: Digital Stadium (in Japanese), http://www.nhk.or.jp/digista/
6. TBS: DigiCon6+2 (in Japanese) (2006), http://www.tbs.co.jp/digicon/digicon/
7. Iwai, T.: ElectroPlankton, (in Japanese) http://www.watch.impress.co.jp/game/docs/20050408/ele.htm

Communication Environment for Sharing Fond Memories

Mie Nakatani[1], Seiko Myojin[2], Masumi Shimizu[2], Hirokazu Kato[2], and Shogo Nishida[2]

[1] Osaka University, 1-1, EXPO Park, Suita, Osaka, Japan
mie@cscd.osaka-u.ac.jp
[2] Osaka University, 1-3 Machikaneyama, Toyonaka, Osaka, Japan
{seiko, shimizu, kato, nishida}@nishilab.sys.es.osaka-u.ac.jp

Abstract. Fond memories provide not only warm feeling to those who possess them, but also bonds that tie people who share them together. Utilizing this property, we propose to develop an environment where people can share their feelings for fond memories, as well as empathize and communication with each other. We consider this environment like a park, where you drop in and take a rest on a bench, look at the fountain and feel comfortable atmosphere away from your daily life. Instead of benches or fountains, we provide some triggers for fond memories. We named this environment "Reminiscence Park". An example of the triggers for remembering in Reminiscence Park is the old popular song common to the users' younger days. We developed a prototype system modeling a music box. Opened by one or more people, the box starts to play a common song from their youth.

Keywords: fond memories, reminiscence, park, popular song, music box.

1 Introduction

Memories are kept in one's mind. Some of them are happy, some of them are too bitter to recall, and yet all of them are precious for the holder. What torments man who has lost his memory is reported that the loss of his identity rather than inconvenience of daily life[1].

Fond memories have different property from mere memory. Remembering "your old days" brings you feelings at the time of its occurrence and sense of reminiscent for the time around its occurrence at the same time with information about it. Contents of memories are easily transformed and edited. For example, a painful experience sometimes turns into a good memory after a long time, and even a childhood episode without remembrance becomes like a real memory after a number of mother's narrative. Fond memories are kept as private treasure but they are often used as topics of conversation. That topic may be about their common experience, like when you are looking a photo album with family or talking about your younger

M.J. Smith, G. Salvendy (Eds.): Human Interface, Part II, HCII 2007, LNCS 4558, pp. 110–117, 2007.

days at class reunion; also, it may be your own experience of your childhood, when the talk is with a person of first meet. Communication using memories have aspects to make people relive the emotion of the speaker or empathize reminiscent feelings of the speaker, more than to exchange information. When speaker and listener feel empathy successfully, they feel a bond with each other.

Our work is aimed to create communication environment that provide empathy to users by utilizing such function of memories mentioned above. This environment supports users increasing familiarity to each other via reminiscence, even if they know nothing or few about each other. The triggers for reminiscence are the popular songs common to the users' good old days, which can promote communication. For the devices to play songs, we propose music boxes.

2 Studies About Memory

Many studies about memories are conducted in the field of bioscience or social science. From the standpoint of information process, memory has three phases; memorization, retention and retrieval (remembering). Revealing a processing mechanism in each phase is the main purpose of these studies. There are also many studies about behavior of memory. Some studies focus on contents of retained information; episodic memory / semantic memory, and some focus on temporal factor of memory; short-term memory / long-term memory. Among these studies, we discuss joint remembering here, as it relates to a function of memory in the social interaction.

Joint remembering is the remembering by multiple people searching their memories through discussion. The studies in this field are broadly classified in two categories: the performance and the process of remembering[2]. The studies in the former category discuss the quantity and accuracy of remembering compared with individual remembering and they analyze the effect of interaction with others.

The studies in the latter category, which focus on the function of memory rather than mechanism, analyze conversation of people involved in joint remembering. One study reported that remembering and discussion common experience among group enhances group cohesiveness[3].

As one of the applications of the function of memory, we introduce a music therapy here, which mainly support old ages' mental rehabilitation. In this therapy, songs are used to elicit patient's active feelings, because of their close relationship with the situation or emotion of the time you had listened or sung a song[4].

"Fond memories" are tend to be thought as personal action (to memorize personal experience and to remember personally.) However, talking about your own memory to others to offer a fun topic or get empathy from others. It happens quite often in our daily life. If you find a similar experience to you in a parson, you would get fellow feeling to him. If you want to become close to a parson, you would like him to listen to your past and would like to know his past as well. We focuses on the function of memory that shortens the mental distance as mentioned above.

3 Reminiscence Park

3.1 Park as a Communication Environment

Joint remembering is a reminiscence of common experience among members belonging to a same group. Our goal is to create an environment which people can feel a bond with each other by sharing a sense for fond memory, even if they have just met for the first time.

Now we introduce the concept of "park." A park in a city is the place where various kinds of people come and go. In the park, there are trees, benches, ponds or fountains and ice cream or popcorn stands. You can have lunch, take your dog for a walk or read your favorite book in the park. You can spend your time in a park freely however you like. Yet it is public space, not like the office or home. It is a space where you interact with others.

Many people have special feelings or their own images to the word "park." They have memories such as playing with their families, wandering in thought, or walking with their girlfriends in the park. Thus, the park is a place full of memories and, at the same time, it is a public space. The park is a magic public space which offers people comfort. We aim to create "Reminiscence Park," a special park to express and exchange private memories with each other.

Reminiscence Park provides an environment that realizes interaction of fond memories, where you can reminisce your own memory and empathize with others memory. We don't support to transfer information. We propose a kind of communion of heart with heart.

In Reminiscence Park, there is a trigger to remind you of your old days, a cue to join in a reminiscence space of others and an air that allows you to join in there.

The possibility for the trigger includes songs, pictures, photos, smells and tastes. We utilized popular songs for the trigger, since popular songs are listened to by many people and they lead to many people's memory. We have developed a prototype system of Reminiscence Park as shown in the following section.

3.2 Interface for Reminiscence Park

In a real park, there are a lot of things that can tempt adults to stay or play. Such things like a fountain, a bench, a ball or a merry-go-round, are what you can enjoy regardless of your age. The interface in Reminiscence Park also needs to provide such enjoyment to draw passers-by around. We think it's desirable that the interface has the following features.

- An attractive feature that draws people into the park.
- A trigger that lets people reminisce and talk about it.

As described in previous section, we use old popular songs as the trigger because of its ability to bond many people's memory. So we need an appropriate player that meets the requirements. There are some possible instruments to realize the features. For examples, the following:

- Music Box:
  A music box is well-known equipment for playing songs and it can be intuitively understood the usage. Opening and closing the box is the switch for song to start

and stop. Furthermore the music box has an image of memory holder, because sometime the box is a gift from someone, sometime people save jewel or treasure in it.

- Ripple of water:

  When people look at a pond in a park, most people throw pebbles in a pond. The pebble causes circular wave pattern and it spread on the water surface. It's not known exactly why, but we all like to gaze at the course of ripples. If a song starts with a pebble splashing down and fades out with wave disappearing, it is a comfort and beautiful image. A pebble represents one of a memory of people.

  More than one pebble causes the same number of ripples, which cross each other land develop another wave. That is similar to the way someone's memory evoke others' memory.

In this paper, we adopt a music box as a player of old popular songs. However a music box is mostly small in size and sound, and is not able to attract passers-by's interest. Therefore, we add a resonance box to amplify sound and impact. Thus the interface in Reminiscence Park is music boxes and a resonance table. A resonance table means a very large resonance box. The all music boxes are set on the rim of the resonance table. Each box has respective theme and plays different old popular songs corresponding to the theme every time people open the box. People can open and close the box as many times as they want. When they meet their favorite song, they place it on the central area of the table with its lid open and the ripples of sound spread from the box and the song is played from the table (Fig.1). If two or more boxes are placed on the central area and ripples cross other ripples, songs fade away and another different old popular song in common floats out from the table. The song in common share fond memory that is common property among boxes on the central area, like the year of release, the singer and so on. People around there share time and space by listening to the old songs and looking at restless ripples. It becomes Reminiscence Park of the moment.

Fig. 1. Music boxes and Resonance table

This interface is enough large to catch to passers-by's eye and more than one person can enjoy it at the same time.

4 The Prototype System

4.1 System Architecture

The hardware architecture is shown in Fig.2. The music box is 20 cm x 15 cm x 15 cm in size. The table is approximately 1.4 meters wide, 1.1 meters depth and 0.9 meters height. In the central area of the table the frosted glass is embedded. This glass is used at once as a screen and as a special area to put the box on (apparently like a pond). On the rim of the table there are several music boxes.

Music Box:
A sensor, a transmitter, a radio and a marker are built in the music box. Information that the lid is opened or closed is sent to PC. This is the sign to start / stop playing a song. At the same time information of box position is recognized by an identification marker[5] attached to the bottom. When the box is not put on the central space of the table, the song is coming from the radio in the box. (Of course equipments in a box are concealed. Inside of the box looks like a jewel box.) When the box is put on the central space of the table, the song is played from the speaker of the table. The action that open or close the box is the switch for song to start or stop. It is so easy to handle that you can start and stop songs anytime you like.

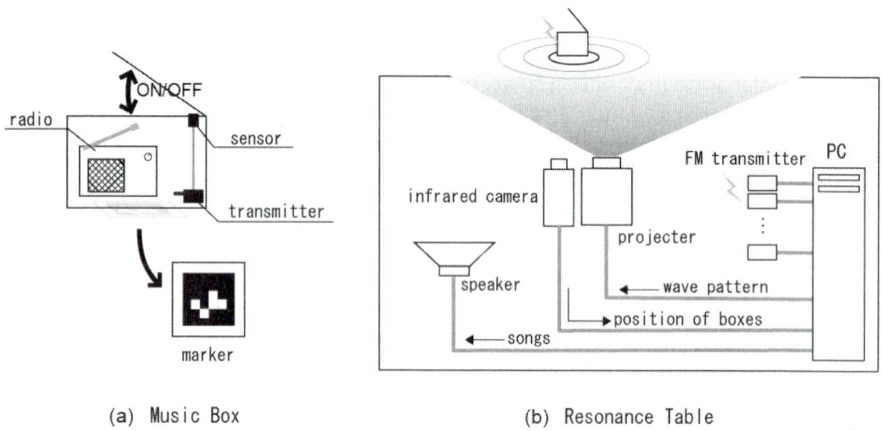

Fig. 2. Hardware architecture of the music box and the resonance table

Resonance Table:
An infrared camera, a projector, two or three speakers and a computer are built in the resonance table. The same numbers of the transmitters as the boxes are linked to the PC. The camera recognizes the position of the box. PC determines the destination of a song by the position. When the box is not on the frosted glass, a song is sent to the radio via a FM transmitter. When the box is on the frosted glass, a song is sent to the speaker and the projector projects ripple pattern around the box from below the table. At the same time, PC calculates wave pattern and changes a song when ripples crosses other ripples.

4.2 Functioning of the System

Here we explain about the music data and selection of songs. This time we prepare about 1300 songs and five boxes for the prototype system. The songs are divided into 5 categories, 80's hit, 90's hit, 00's hit, 60's-80's hit in English and title tune of TV programs and movies. Each category is assigned to one box and there are about 60~100 songs in each box. We make a database of all songs with properties; a release year, a singer name, a lyric writer, a composer, a program (drama, CM, cartoon, movie, etc.) that the song was used as a theme song for.

Each time of the box-open, a song is chosen randomly and played. A user can open and close the box as much as he or she likes. When the user meets a favorite song, he or she puts the open-box on the central area of the table and ripples of sound spread from the bottom of the box. The song comes from the side of the table, not from the box. If another user chooses the song and places the box on the table, new ripples spread and different songs is played from the table at the same time. When both ripples cross, former songs fade away and a different song that has the common property to former two songs is played from the table. For example of Fig.3, the first song is “Yesterday Once More” by Carpenters(70's) and the second song is “Love me do” by Beetles(60's). The property common to these two songs is covering because Carpenters cover Beetles’ “Ticket to Ride”. If there is no common property among songs, PC plays the song everyone knows like nursery rhyme, theme song of long-running show.

If more than three boxes are placed on the central area at the same time, the screen is filled with ripples and many songs from the speaker become only noisy. So when fourth box are placed on the central area at the same time, the screen is filled with white noise and all songs stop. Both the music box and the table play only one song at one time and don’t play two songs consecutively. If a user would like to enjoy more music, he or she close the box and open it again.

Fig. 4 shows the appearance of the resonance table.

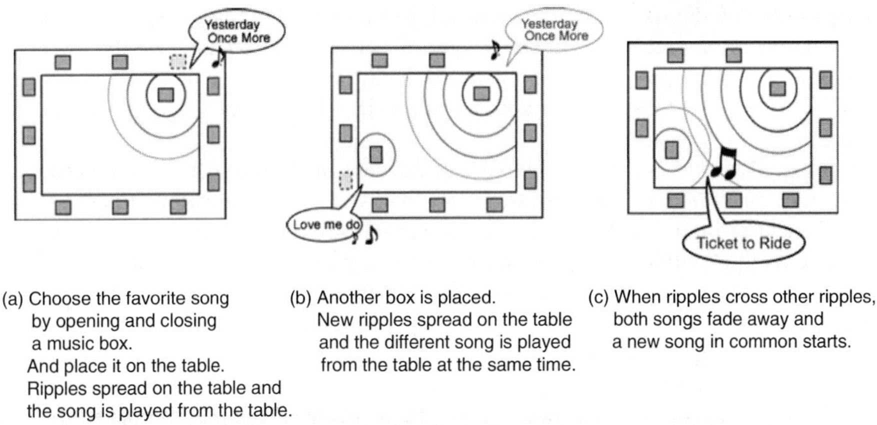

Fig. 3. Process of interaction with music boxes on the resonance table

Fig. 4. Image of the prototype system

5 Discussion

In the interface engineering field, studies dealing with fond memory have reported few. GrandChair[6] is aim to support storytelling about own experience by grandparents to grandchildren living in long distance. Storytellers tell their story to the child-looking agent on a screen and the story is recorded on videotape to send to their children later. This work is based on the recognition of the importance of reminiscence for the elderly and a listener for storytelling. GrandChair is intended to use with a teller alone, not with real listener. This means communication in GrandChair is almost one-way and not interactive one. (Agent reacts to storyteller, but do not talk about own story of agent.)

The concept of Omoide-engineering (fond memory engineering) advocated by Nojima is to recognize the importance of fond memory and establish communication based on this recognition. They have developed a portable photo memory to support photo-based communication[7]. The trigger of communication here are the memorial photos of users, which prepared by users. They are strictly connected with users memory, thus good for owner of a photo to recall, whereas not so good for others to share.

Compared with these studies above, our concept of Reminiscence Park is to elicit interactive and mutual communication, not one-way communication. Expressly our environment is intended the users even of non-close relationships. This prototype is designed as the "public" communication tool in a park. So anyone is able to touch it and enjoy in one's own manner without the preparation of one's private date. In some cases people enjoy alone, in other cases enjoy singing aloud with friends, in other cases listen to songs with passer-by.

In this paper we propose the concept of Reminiscence Park and show its realizability. We will improve our system based on the result of evaluating the prototype system. Though, there remain some technical tasks at the present stage. The first problem is the grouping of songs in boxes. This time old popular songs are classified according to the property of them, say, the ages like 60's, 70's, 80's and theme song of TV drama or cartoon film. We need more consideration about the grouping.

The second problem is the way to select the common song. When songs playing on the table don't share the common property, the system provides a nursery rhyme. But there may be a more pertinent solution on it. It is also the problem about intergenerational communication.

In addition, the number of boxes on the table is another subject of future work. If more than three boxes are put on this table, sounds becoming messed up and noisy. And that it will be too difficult to select the common song. We have to design the natural interface that sets a limit on the number of boxes. For example the frosted glass becomes dark when more than a certain number boxes are located on the grass.

6 Conclusion

We proposed a new concept of a communication environment named Reminiscence Park and showed an interface to realize this concept. The interface is composed of the music box and the resonance table. The music box is at once a player and the symbol of fond memory. The resonance table undertakes a role as an amplifier of sound and communication. People enjoy fond memory by listening old popular songs from the music box and by looking beautiful sound ripples on the resonance table. When the ripples cross other ripples on the table, the songs fade away and a different old popular song floats out from the table. It is an image that fond memories crowd in upon one's mind just like restless wimple in a pond. People share some nostalgic feeling with other people around. We developed a prototype system and it is now under testing. The result of this test will lead to develop another interface for Reminiscence Park.

References

1. Sacks, O.: Awakenings. Duckworth Pub, London (1973)
2. Takahashi, M.: Social aspect of memory —focusing on collaborative remembering —. In: Umemoto, T., Kawaguchi, J. (eds.) Today's Study for Cognition–toward the 21st century–, pp. 17–28. Baifukan co., ltd, Tokyo (in Japanese) (1999)
3. Kobayashi, T.: <urbanization and nostalgia> —the heart of people from amami in a city—. Technical report, Annals of human sciences (in Japanese) (1987)
4. Bando, H.: My Idea of Healing Music. Ongaku no tomosha corp., Tokyo (in Japanese) (2001)
5. Kato, H., Billinghurst, M.: Marker tracking and hmd calibration for a video-based augmented reality conferencing system. In: Proc. of 2nd Int. Workshop on Augmented Reality, pp. 85–94 (1999)
6. Smith, J.: GrandChair: Conversational Collection of Grandparents Stories. Master of science in media arts and sciences, Massachusetts Institute of Technology (June 2000)
7. Yamashita, I., Nojima, H.: Portable personal memory for communication. In Human Interface, Human Interface Society 2001, pp. 261–264 (in Japanese) (2001)

Experimental Comparison of Multimodal Meeting Browsers

Wilfried Post, Erwin Elling, Anita Cremers, and Wessel Kraaij

TNO Human Factors, Kampweg 5, 3769 DE Soesterberg, The Netherlands
{Wilfried.Post, Erwin.Elling, Anita.Cremers, Wessel.Kraaij}@tno.nl

Abstract. This paper describes an experimental comparison of three variants of a meeting browser. This browser incorporates innovative, multimodal technologies to enable storage and smart retrieval of captured meeting. Over a hundred subjects had to work in a design team in which they had to prepare and carry out a final meeting, while supported by one of the browser variants. In one condition, teams worked without such support. Measures on individual characteristics, the team, the process and outcome of the project, and the usability of the browsers were taken. The results indicate that a multimodal meeting browser can indeed improve meetings. Further analysis of the now available data will provide additional insight into how browsers can contribute to more efficient and satisfactory meetings, improved team performance and higher quality project outcomes.

1 Introduction

The research reported here is carried out as part of AMI, a European Integrated Project that aims at developing new multimodal technologies for supporting human interaction in the context of meetings [1]. These technologies enable storage, interpretation and retrieval of captured meeting interactions and will be used to develop a meeting browser that should make meetings more effective, efficient and pleasurable. The importance of improving meetings is evident when is realized that most professionals join (but not enjoy) at least one meeting a day [2]. Not so evident, however, is to establish that a meeting browser indeed improves meetings, and on which elements this depends. The objective of the presented study is to determine whether and how a multimodal meeting browser improves a meeting. Our ultimate aim is a more efficient and satisfactory project process and higher quality results.

2 Method

An experiment was set up to compare meetings with and without meeting browser support. For a better understanding of what form of support might work best, in total

M.J. Smith, G. Salvendy (Eds.): Human Interface, Part II, HCII 2007, LNCS 4558, pp. 118–127, 2007.

three variants of a multimodal meeting browser were part of the comparison. The setup is described next.

2.1 Stimulus Material

In [3] is pointed out that the success of a meeting is better determined from a series of meetings, such as in a project with a clear goal. Further, the success of a meeting, or a project, depends not only on the means used (e.g., a meeting browser), but also on (project or meeting) method, individual factors, team factors, type of task, organizational culture, environment, etc. These factors have been specified in the following experimental scenario. The subjects have to participate in a design project team, playing a specific role (project manager, industrial designer, user interface designer and marketing expert). They are told to take over a project carried out so far by a team that did not do well enough. The subjects have to use all the material used and produced by this previous team. They have to prepare and carry out a final design meeting in which they have to design a television remote control, according to specific requirements. Both preparation and execution of the meeting is carried out in meeting rooms in Soesterberg (Fig. 1) and Edinburgh, well instrumented research environments for four subjects, with individual workplaces (including a private computer), a shared workplace (including electronic presentation boards), and, depending on the experimental condition, a particular set of tools. The participants and their computer interactions are observed and recorded by means of video cameras, microphones, and screen videos.

Fig. 1. Research environment called "Team Cockpit". On the left, not shown, is a shared large presentation screen. The subjects are discussing a clay prototype remote control.

On preplanned points in time, subjects receive e-mails about the tasks to carry out (sent by a virtual head of the department), some hints (sent by a virtual coach), and a series of questionnaires and rating scales.

2.2 Subjects

Data from 27 project teams consisting of 4 participants were collected, of which 12 in Soesterberg and 15 in Edinburgh. Most of the 108 subjects were undergraduate students. All of them were paid €45,- for about 4 hours of work.

2.3 Independent Variable

The four experimental conditions are as follows.

1. In the basic condition, no computer support is provided, other than a folder structure in Microsoft Explorer, organized by project phase (Project Kick-off; Functional Design Phase and Conceptual Design Phase). In these folders, they can find documents, minutes, PowerPoint slides and audio/video recordings of three previous meetings.
2. In the second condition, all documents are still provided in a folder structure. For making use of the meeting registrations, however, a meeting browser is provided, which offers synchronization of the multimodal meeting recordings (i.e., synchronous browsing through PowerPoint slides, automatic speech transcripts or audio/video material).
3. The third condition is similar to the second one, except that the browser includes automatic generated abstracts - new technology that will be available in the near future. Fig. 2 shows a screenshot of the browser. In the upper middle part of the screen, one can browse through audio/video registration of a particular meeting within a project. In the lower left corner, one can browse through speaker segmentations: each participant has its own speaker activity log represented. In the lower right part, an automatic transcript is presented, which is browsable by scrolling and searchable by keyword search. In the upper left corner, the PowerPoint slide of a particular episode that is browsed is shown. All these parts are synchronized.
4. In the fourth and final condition, all material, including documents, are integrated from the perspective of a user in a task setting (i.e., carrying out role-specific work in a design project (see [4]). The Task Based Project Browser, as it is called, provides direct access to three different information sources via the tabs Meetings, Documents and Messages. In addition, it is possible to access these sources indirectly via three task-oriented tabs: *Project* (project details, people involved, different design phases), *Todo's* (see Fig. 3) and *Decisions*. All information items are hyperlinked to the original sources. This makes it possible to, for instance, immediately view a meeting clip in which a specific information item, such as an action item, is being discussed. The information provided in the Meetings tab is identical to the information in the meeting browser of the third condition.

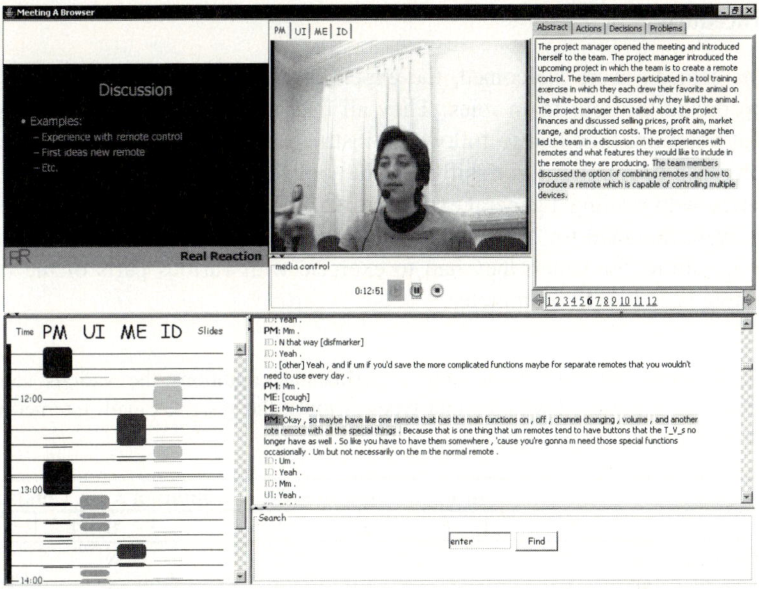

Fig. 2. Screenshot of the meeting browser used in condition 3. The browser used in condition 2 is similar, except for the automatic generated abstracts in the upper right corner.

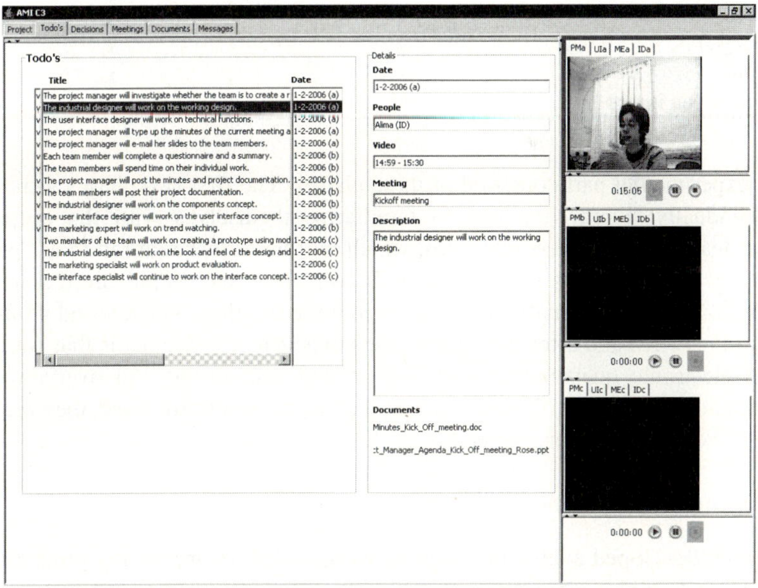

Fig. 3. Screenshot of the "Todo's" tab of the Task Based Project Browser used in condition 4

2.4 Procedure

After the subjects were welcomed, the experiment was explained. Next, they were asked to divide the four team roles. They all had to open their e-mail browser and from that point on, they had to follow the instructions sent to them. To prevent bias caused by lack of familiarity or skill with the tool set, the subjects were trained first. This started with reading a general explanation of the functionalities of the tool set the subjects were assigned to. Next, the subjects received e-mail instructions to answer specific questions for which they had to exercise with various parts of the tool set. The training lasted about 30 minutes.

Table 1. Overview of how in the four experimental conditions the various sources of information can be retrieved with the four types of tools: the system, Email, Internet browser, and Multimodal Meeting browser

Information	**Condition 1**				**Condition 2**				**Condition 3**				**Condition 4**			
	F	E	I	B	F	E	I	B	F	E	I	B	F	E	I	B
Minutes	x				x				x			x				x
Presentations	x				x			x	x			x				x
Emails/messages		x				x				x						x
Internet			x				x				x					x
Close up videos	x							x				x				x
Speaker activity log								x				x				x
Meeting transcripts								x				x				x
Abstracts	x				x				x			x				x
Actions/Todo's	x				x				x			x				x
Decisions	x				x				x			x				x
Problems/Open issues	x				x				x			x				x

The experimental part consisted of three phases. The first two phases were carried out individually. First, they had to familiarize themselves with the project, the previous team and their personal roles, which will be further referred to as "getting a gist". They had to do this in 15 minutes. Second, they had to prepare the upcoming meeting in 45 minutes. Finally, they had to join the meeting, which lasted 45 minutes. Before, in between and after the phases, the subjects received e-mails that were either links to electronic questionnaires or task related instructions. An overview of the questionnaires is found in Table 2. During the second and third phase, they received a warning five minutes before the end of it.

2.5 Dependent Variables

A specially developed evaluation instrument was used for measuring project process and outcome (see [5]). This instrument includes subjective workload rating scales and team questionnaires, and objective analysis of information transfer and project outcome. An overview is provided in Table 2.

Table 2. Overview of all the measurements taken during the experiment

When	What	How
Individual characteristics		
Before the experiment	Background, experience	Questionnaire
	Spatial orientation	Test
	Memory	Test
Team and project measures		
During the meeting	Behavior	Subjective observation
After the meeting	Dominance	7 pt scale
	Info processing	4 items
	Leadership	4 items
	Process satisfaction	3 items
	Cohesiveness	5 items
	Work pace	4 items
	Communication	4 items
	Supporting behavior	8 items
	Effectiveness	4 items
	Efficiency	7 items
	Outcome satisfaction	5 items
	Team satisfaction	2 items
	Project experience	3 questions
Analysis afterwards	Info transfer	# shared info
	Info outcome	# correctly applied info
	Quality product	Multiple Expert Assessment
Usability measures		
Before getting a gist	Mental effort	150 pt scale
During getting a gist	Use of browser	Logging
	Behavior	Subjective observation
After getting a gist	Browser usability	Questionnaire
	Mental effort	150 pt scale
During preparation	Use of browser	Logging
	Behavior	Subjective observation
After preparation	Browser usability	Questionnaire
	Mental effort	150 pt scale
During meeting	Use of browser	Logging
	Behavior	Subjective observation
After the meeting	Browser usability	Questionnaire
	Mental effort	150 pt scale
	Usability browser	Questionnaire
	Meeting room of the future	Questionnaire

3 Results

3.1 Individual Characteristics

In total, 108 subjects participated in the experiment. Eight teams of four subjects each were assigned to condition 1, seven to condition 2, seven to condition 3, and five to condition 4. Their mean age was 23 years old (s.d. 7 years), 56% were male and 44% female. Almost all participants were students (97%). Most of them were students of computer science or information science (50%), 16% were students of psychology, 10% of philosophy and the rest of varying or unspecified subjects.

Participants were asked a number of questions on their patterns of computer use. All participants use both the computer and the internet on a daily basis. Participants also received questions on their experience with meetings. Half of the participants participate in meetings on a weekly basis, one third on a monthly basis, one sixth never. More than half of the participants feel that most of the times objectives for their meetings are attained; one third feels that they are sometimes attained; one-sixth varies between the rest of the answers.

About 80% of the participants feel that either sometimes or most of the time, time for their meetings is well-spent. About half of the participants feel that most of the times they like to participate in meetings. The others either like it sometimes or always. Hardly anybody never likes to participate. When asked to characterize their typical meetings, about 80% answer that the meetings have an informal atmosphere, about 10% formal. Other qualifications used are either positive (good: 6%) or negative (tedious: 4%).

Finally, participants were asked about their experience with working in project teams. About 80% has either hardly any or average experience with working in project teams. The rest has either no experience or a lot of experience. About half of the participants have no experience at all in product or service development. The other half has hardly any or average experience.

3.2 Team and Project Measures

For the variables *Information processing, Leadership, Process satisfaction* and *Outcome satisfaction,* that were scored on a 7 point scale, notable results are found. *Information processing, Leadership* and *Process satisfaction* are in condition 3, significantly higher than in condition 1 (resp. $F(3,26)=3.621$; $p<5$; , $F(3,26)=3.015$, $p=0.5$); $F(3,26)=3.253$, $p=0.5$), and *Outcome satisfaction* is almost significantly higher ($F(3.26)=2.843$, $p=0.6$).

3.3 Usability

At three points in time during the scenario participants were asked fill out the browser usability questionnaire with questions about their opinion on the tool set that were available to them. Questions were asked referring to the usability aspects effectiveness, efficiency and satisfaction. For every aspect four questions were asked, in a random order. Participants answered the questions on a seven-point scale, varying from not applicable at all (1) to very much applicable (7).

Mean total scores per assessment and per condition were calculated. All mean scores varied between 3.45 and 4.67, indicating general moderate usability. There were no differences between scores for effectiveness, efficiency and satisfaction.

Significant effects of point of measurement and condition occurred. For all three aspects there was a significant effect of condition. For effectiveness, condition 1 differed significantly from both condition 2 and 3 ($F(3,23)=6.39$; $p<.01$). For efficiency, condition 1 differed significantly from condition 3 ($F(3,23)=3.68$; $p<.05$).

For satisfaction, condition 1 differed significantly from 3 (and almost from 4) ($F(3,23)=5.17$; $p<.01$).

For all three aspects a significant effect of point of measurement occurred. For effectiveness, measurement point 1 differed significantly different from measurement point 3 ($F(2,46)=6.39$; $p<.01$). For efficiency, measurement point 1 differed significantly different from measurement point 3 ($F(2,46)=4.37$; $p<.05$). For satisfaction, measurement point 1 differed significantly from both measurement points 2 and 3 ($F(2,46)=11.04$; $p<.001$). This can be considered a learning effect.

Table 3. Mean effectiveness, efficiency and satisfaction for condition 1 to 4

Effectiveness				Efficiency				Satisfaction			
1	2	3	4	1	2	3	4	1	2	3	4
3.45	4.26	4.62	4.19	3.67	4.07	4.40	4.17	3.74	4.18	4.67	4.47

Table 4. Mean effectiveness, efficiency and satisfaction after familiarization (1), preparation (2) and meeting (3)

Effectiveness			Efficiency			Satisfaction		
1	2	3	1	2	3	1	2	3
3.95	4.16	4.27	3.89	4.11	4.22	4.02	4.27	4.50

3.4 Afterwards

Afterwards, participants were asked three questions on their experiences working in the project, giving them five possible answers: never, hardly ever, sometimes, most of thc times, and always. Answers were generally positive. The mean score over all conditions on the question whether they felt the objectives for the project had been attained, was 3.82 (most of the times). There were no significant differences between the conditions. The mean score over all conditions on the question whether they felt the time for the project had been well-spent was 3.74 (most of the times). There were no significant differences between the conditions, although the difference between conditions 1 and 3 was almost significant ($F(3,23)=3.00$;$p=.054$). The mean score over all conditions on the question whether they had generally liked to participate in the meeting was 4.00 (most of the times). There was a significant difference between the conditions 2 and 3 $F(3,23)=4.11$;$p<.05$). The difference between conditions 2 and 4 was almost significant ($p=.057$).

3.5 Subjective Observations

Up to now we have only presented objective evidence, but some subjective material has been collected as well. Some personal impressions of the experimenter:

- In all conditions, there was a clear learning curve. Participants were mostly hesitant at the start. This gradually decreased and led to curiosity towards going straight to the information they were looking for.

- In all conditions, a small number of people suffered from information overload. They did not know where to start looking for, or lost track. It should be mentioned that the design task is quite complex.
- The browser variants were mainly used during *getting a gist* and *meeting preparation*, and not during a *meeting*.
- Whatever condition, it appeared that without a strong, goal-oriented, natural leader, the team process was slower and fewer ideas and information were exchanged.
- Only in the task based meeting browser condition, some groups seemed to have time left for monitoring meeting progress. Also, only in this condition, the norm solution was found: a remote control meeting all criteria, including the form of a banana.

4 Conclusions and Future Work

The most important result of this work is that we have found experimental evidence that a meeting browser indeed can improve a meeting. Especially meetings supported by the browser used in condition 3 have a significant higher perceived effectiveness, efficiency and satisfaction. Our expectation, however, that meetings supported with the Task Based Project Browser would be superior could not be shown. We provide some possible explanations for this result.

During experimentation, we were conscious of the limitations of the current experimental implementation of the Task Based Meeting Browser. Due to time and budget constraints, it contains less functionality and is less integrated than we originally had conceived (see [4] for a full description). Subjects regularly mentioned additional functionalities, alternative ways for organizing all information, other search possibilities - suggestions that were already part of our conceptual design. These are taken into account during further development.

Working on a design project is difficult, especially for our inexperienced and rather young subjects. This may have influenced the results in the sense that they were probably not fully aware of the requirements for meetings and functioning in a project team. They were experienced with using the computer, though, which suggests potential skills for using the tools and trust in the information offered. This could be a reason why they were quite apt at using the tools and were not really intimidated by the prospect of meeting in an instrumented meeting room.

A huge amount of data is currently waiting for analysis, such as the influence of individual characteristics, information flow analysis, using expert opinions about the resulting clay remote control prototype as outcome measure, and detailed analysis of tool usage using our logging data. We hope to see more beneficiary effect of the meeting browser, and expect to find some differences between the three variants. This would provide us explanatory material, and subsequently guidance for further development. Interesting is the effect we have found of experimental condition on *leadership*. If this effect is strongly related to the outcome of a project, this would indicate that a meeting support tool should focus more on, for example, "leadership support". Multimodal technologies can help recognize various types of social interactions, and can accordingly be applied to support social behavior itself.

Acknowledgments. This work was partly supported by the European Union 6th FWP IST Integrated Project AMI (Augmented Multi-party Interaction, FP6-506811, publication AMI-234). About half of the data collection has been carried out at the University of Edinburgh, thanks to Jean Carletta, Mike Lincoln and Melissa Kronenthal. Charley Penning, Erik Boertjes, Jonathan Kilgour, Mike Flynn and Stephan Lesch contributed substantially to development of the meeting browser variants.

References

1. AMI. Augmented Multi-party Interaction (AMI). Sixth Framework Programme, Priority 2, Information Society Technologies, Integrated project Annex 1 – Description of Work (2003)
2. Rogelberg, S.G., Leach, D.J., Warr, P.B., Burnfield, J.L.: Not another meeting! Are meeting time demands related to employee well-being? Journal of Applied Psychology (in press)
3. Post, W.M., Cremers, A,H.M., Blanson Henkemans, O.: A research environment for meeting behavior. In: Nijholt, A., Nishida, T. (eds.) Proceedings of Social Intelligence Design 2004, pp. 159–165 (2004)
4. Cremers, A.H.M., Groenewegen, P., Kuijper, I., Post, W.M.: The Project Browser: supporting information access for a project team. This volume (2007)
5. Post, W.M., Huis In't Veld, M.A.A., van den Boogaard, S.A.A.: Evaluation Meeting Support Tools. In: Personal and Ubiquitous Computing. (Accepted for publication)(2008)

How Participation at Different Hierarchical Levels Can Have an Impact on the Design and Implementation of Health Information Systems at the Grass Root Level – A Case Study from India*

Zubeeda Banu Quraishy

Department of Informatics, University of Oslo
Gaustadalleen 23, PO. Box. 080 Blindern, NO - 0316 Oslo, Norway
Mailing Address: G-3, M.I.G Flats, Fore-Shore Estate, Chennai-600 028, India
Tel.: 98 400 20114; Fax: + 91-44 -421 04 106
zubeedaq@ifi.uio.no

Abstract. To build the effective information systems Scandinavian tradition emphasizes and encourages the participatory design methods which involves both designers and users equally in building the systems. While the main intention of Scandinavian tradition is to refocus the dominant orientation of systems developers by involving users but one really wonders whether it can be applied in the context of developing countries particularly in a country like India where the structures are rigid, hierarchical and strongly bureaucratic in nature. In spite of the fact that health sector being a key social area in India's developmental administration it is highly fragmented and receives low priority. In India, while the general administration is mostly concentrated at the district level, health and family welfare administration has been centralized at state and central levels. This coupled with the excessive concentration of powers in the secretariat system of governance, has considerably centralized planning and monitoring. Keeping in view the above complexities and challenging conditions this paper tries to understand using case study as a method how participation at different levels affects the design and implementation of Health Information Systems in primary health care sector in India.

Keywords: Participatory Design, Health Information Systems Programme, Primary Health Centres, Users, Developing Countries.

1 Introduction

In developed countries where participation is welcomed and encouraged if not in all sectors at least in social sectors like education and health in developing countries in all contexts it is highly discouraged. In developing countries like India there was no tradition of participatory planning and decision making until December 1992 when

* The empirical material for the above case study was collected during my association with HISP project as Project Coordinator, HISP India from January 2001 to April 2005.

M.J. Smith, G. Salvendy (Eds.): Human Interface, Part II, HCII 2007, LNCS 4558, pp. 128–136, 2007.

72nd (Panchayats) and 73rd (Municipalities) constitutional amendment bills, were passed by both the houses of parliament and came in to force in 1993. The above legislations brought in to effect the much talked about decentralisation with a view towards promoting participation and improving the controlling function held by the lower levels of the political system and the communities as it was believed that decentralisation would bring government and its administration closer to the people. Also it was considered that the existence of local political arenas makes it easier for ordinary citizens to participate and exert influence. Thus the legitimacy of public bodies may be considerably boosted by the greater involvement and influence of ordinary people due to the greater 'proximity to politics' that decentralisation entails [2] as this tends to strengthen the state's capacity for implementation. Keeping this in view, the policies were framed in India to improve the involvement of both the decision-makers and the communities at the local level hoping that change occurred at the grass root level and in organisations will not only encourage users to participate in discussions and decisions, but will focus on the issues they identify as important.

Furthermore, it was believed that a decentralised bureaucracy will not only adapt more easily to local identities and norms, which may increase effectiveness but decision-makers who are rooted in the locality and have knowledge of conditions on the ground are better placed to formulate concrete policies for that area than functionaries from the centre [3].

Thus, in order to bring about the desired social change it was considered necessary to have the constitutional support and legislative measures but soon it became clear that they are not the sufficient condition to achieve that goal as social and political structures present in India are highly rigid, hierarchical and strongly bureaucratic in nature.

With this as a back ground the present paper using case study as a method tries to highlight how implementation of Health Information Systems Programme (HISP) with the main objective of strengthening information practices within the Primary Health Care (PHC) sector with the larger aim to improve processes concerning health care delivery for the rural community failed in getting stabilized and sustainable even after five years of implementation efforts as it was thwarted by many challenges mainly due to the lack of participation not only by users but by policy makers and other officials at various levels.

Health Information Systems Programme is an ongoing action research project related to the implementation of information systems for improved local control and use of information at district and sub-district levels in the health sector in Andhra Pradesh.

2 Situation Analysis

The Primary health care sector in Andhra Pradesh which is the main interface between the community and the health sector is fragmented and there is no cohesion between the various health departments for whom enormous amount of data is collected at the grass root level by the health worker. While the main duty of the health worker is to perform out reach services to the communities under her purview such as providing Ante Natal Care(ANC), performing institutional deliveries, giving immunization to infants and children, providing treatment on minor ailments, and motivating eligible couples for adopting spacing methods and sterlisations, conducting sputum tests of suspected TB patients, reporting and providing immediate care to patients on outbreak of epidemics etc but instead of performing outreach services the health worker is now burdened with achieving targets on performance indicators mainly sterlisations, immunizations, ANC, Post Natal Care and institutional deliveries etc. On 28th of every month all the health workers assemble in primary health centres to collate data collected on the above performance indicators into different registers. Generally the health workers enters the data after performing services into her field diary and then from her field diary the data is aggregated and entered in to the comprehensive register and then from the register the aggregated data is again entered in to different reports like CNA(Community Needs Assessment) report, school health and hostel reports, immunization, Eligible Couple report, Blindness control, Leprosy, Iodine deficiency, malaria and TB programme reports etc. It is quite tedious work and the health workers who are now used to this monotonous work cull out the figures from the registers and enter into reports to be sent to the various officers in charge for the different vertical programmes thus resulting in routine collection, collation, duplication and manipulation of data.

The data collected by the health worker at the cutting edge level is sent vertically upwards with little or no feed back and support from the top to the health worker in the field who needed the information most. In spite of the fact that community use, feedback and feed up is essential for a successful sustainable information system particularly in cases of sensitive issues such as Reproductive Tract Infection(RTI), Sexually Transmitted Infections(STI), HIV/AIDS cases and epidemics like Gastroenteritis(GE) and diarrhea it never happened in the health sector of Andhra Pradesh. It is at this local level that the information is available and where immediate action needs to be taken. On contrary, the use of information is minimally used at the top. Communities and service providers need to share relevant information on the local situation to develop and prioritize strategies together to improve the health situation. Sharing respectful dialogue and resources between communities and service providers can produce positive and lasting improvements by creating a shared vision, goal and objective through approaches that foster equity and shared responsibility.

The transition to an information system that is, an integral part of local governance and part of a social process for development is not only a question of changing of structures and technologies but it challenges deeply rooted values and perceptions with regard to society, technology and health care.

The quality of existing information management practices within the Primary Health Care sector in Andhra Pradesh is extremely poor as the data collected is of poor quality, redundant, aggregated at every level and by the time it reaches the top the situation in the peripheral areas is completely masked. The health workers are under tremendous pressure for data collection and transmission to the top where it is used minimally. Also, major portion of the time of health workers gets wasted in duplicating the data as they have to report almost the same data to various departments using approximately forty forms thus leading to poor quality and fragmentation of the data. The above work is done at the cost of neglecting of providing essential services to the community which is the prime duty of the grass root health worker.

Except for making the field staff attend routine camps to train them how to motivate eligible couples for sterilizations there were no attempts at all to equip them with additional technological skills, develop a culture of use of information, enhance their capacity in decision making at the PHC level. The structure is so bureaucratic and hierarchical that there is no way of participation of health workers even at the PHC level while participation in the decision making at the higher level is almost a dream. Decentralization which was so much talked about in India about not only promoting participation but also improving the controlling function held by the lower levels of the political systems making more ordinary people influence and involve in the political process thus strengthening the state's capacity for implementation remained only on paper with little or no effect. Thus, in this case too the health staff at the cutting edge level in all terms were marginalized.

3 Introducing HISP

It is in this context Health Information Systems Programme (HISP), a global initiative was introduced initially as a pilot study in Kuppam, a small geographical area consisting of 9 PHCs spread over 5 mandals(administrative revenue divisions) in Chittoor district in the state of Andhra Pradesh, India.

As the main objective of our project is to strengthen information practices within the Primary Health Care (PHC) sector with the larger aim to improve processes concerning health care delivery for the rural community initial months were spent in trying to understand the structure and functioning of the PHCs along with the complex and multi-level flows of health information from the rural community to the department in state head quarters through the intermediary layers of the PHCs and district offices.

The district Health Information System(DHIS) that has been developed by HISP includes software, methodologies for health information management, training programmes, health data standards and a data dictionary.

Based on the discussions with the grass root level health functionaries and the officials at various levels efforts were made to set up the necessary infrastructure by placing computers in all the 9 PHCs of Kuppam constituency for initiating change through DHIS (District Health Information System),an open source software for improving their routine health information processes since more than 60% of the time of health workers is spent on health information transactions (collection of data, collation of data in various forms and registers, and the transmission of data to upper levels of district and state). It was also considered that improving these processes will then directly have an impact on work, and the health staff can focus more time on providing care to the community as compared to filling up endless forms that ultimately are marginal to their work of providing health care to the community. So keeping in view the above perspective strategies were devised to provide training to the health workers and other staff in the PHC and at the District Medical and Health Office(DM&HO) with a view towards computerizing the 9 PHCs within Kuppam constituency and at the same time building capacity among health workers and other staff to work independently on the DHIS software, dealing with health information, and thus enabling a culture of "use of information.

3.1 A Brief Note About the Training Strategies

Two main types of strategies were identified for the training and they were 'On-site' and 'Off-site'.

On-site Training

In the above type of training a trainer was based in each of the PHCs selected for piloting of the project from morning till evening for a period of six months providing training to the health staff.

The trainers provided training beginning from explaining concepts of HIS, basics of computer to making them familiar with working on MS word, MS Excel and teaching of DHIS - customised software on HISP.

Initially both the trainers and the health staff had many problems on both sides - teaching as well as learning.

On teaching side the trainers faced the initial problem of gaining domain knowledge(health), explaining the concepts of HIS, advantages of HIS, familiarity of handling computer, time devoted by the staff in learning the concepts, and the software.

On learning side the heath staff faced the major problem of time and work load. Also in three of the PHCs among the nine in Kuppam the PHCs did not have the facility of keeping the computer(problem of infrastructure). As a result, the staff from the above three PHCs have to visit their neighbouring PHCs which in most cases were quite far from their own PHCs for getting training on DHIS software(problem of distances and remoteness). More over as there was only one computer there will be many people to be trained and in most of the cases many of the health staff will not get a chance to work practically on computer(shortage of resources).

Off-site Training

Once in a month a three day intensive training programme was organised in a computer training institute where HISP had entered into partnership for conducting training programmes for health staff.

In the above programmes the training was organised in such a way that some time is allocated for teaching in class about explaining the HIS concepts, various components about the DHIS software, importance of maintaining regular practice of doing data entry, importance of maintaining data quality, data analysis and how data can be converted into information and then how it can be used for making crucial decisions for instance, about relevant drug requirements, identifying reasons for epidemics or common outbreak of diseases in their wards or regions. Also during ITPs(Intensive Training Programmes) the health staff were trained to self monitor their work, the work of their sub centres and also make comparisons between the Sub centres in their PHC and also between different PHCs pertaining to any one or a cluster of performance indicators.

At the end of the ITP a feed back session was organised where the participants express their opinion about the training methodology, training content and their requirements for further training.

In spite of the hectic schedule and pressures from the higher authorities to achieve the irrational targets set on performance indicators majority of the health workers showed an interest in learning computers and the DHIS software and within seven months of time since the project was initiated most of the health workers quickly grasped the knowledge of using computers and DHIS software for doing data entry and doing simple analysis to monitor both individual and as well institutional performance in relation to the targets and indicators set by the higher authorities.

Many health staff in all PHCs started to use the application by regularly entering and updating the routine health monthly data in to the two forms of DHIS designed by the programmers in consultation with the users of the technology. The grass root health workers including the other health staff and the medical officer based at the PHC soon realised the benefits of using the application which not only reduced their manual work, duplication of collecting and collating data but helped them to analyse both individual and institutional performance both across time, indicators and institutions.

At the end of the eighth month after the project was initiated it was evaluated by the higher officials and once the authorities at the district and at the state were convinced of the use of computers, use of DHIS software and the quick technical skills gained by the health staff orders were issued from Commissioner of Family Welfare Department which mainly controls all the activities of the PHCs in the state for making transition to using reports generated from DHIS software in the place of using manual reports.

4 Challenges

While getting a Government Order from Commissioner of Family Welfare Department (CFWD) to institutionalise HISP was a great achievement for the implementation team but soon the team members realised that they have to face many problems in trying to stabilise the project due to non participation of officials at various levels. In the case of institutionalisation of HISP it was expected that all health staff including the health workers based at the cutting edge level concerned with HMIS will use the above application by doing data entry, analyse the data, generate reports and make use of the information using the DHIS software but in actual practice the senior staff within PHCs and authorities based at the District Medical and Health Office did not participate in the use of application mainly in generating of reports thus putting the whole process of implementation of HIS at stake. The problem was that the officials at DM & HO refused to learn the computers and as well the DHIS software giving vague excuses such as work pressure, non availability of time and even expressing scepticism about the application itself thus insisting from the health staff from PHCs to submit manual reports along with a copy of reports generated from the DHIS application.

Both the health workers and other staff at the PHC felt that by submitting both the manual reports as well as the reports generated from computer using the DHIS application as demanded by the authorities at the district level that they are made to work in parallel on both the systems rather than on one as promised by the implementation team and the officials at the district level thus increasing their burden of work. This resulted in health workers losing interest in using the above application.

Another challenge that posed serious threat to the sustainability to the project itself was again the generation of reports by health workers using DHIS. Little did the health staff realised at that time that generating reports using DHIS software would bring to the fore the stark realities of manipulation of data in the PHCs that has been going on for years. In Andhra Pradesh the health sector has selected 16 performance indicators such as Sterlisations, immunisations, institutional deliveries, number of Eligible Couples being motivated for undergoing button hole tubectomy(BHT), number of pregnant women registered for ANC before twelve weeks etc are some of the important performance indicators against which the PHCs were given targets to accomplish which again filters down to the field staff. Without taking into consideration the ground realities such as size, type and composition of the population and other duties of the health workers the policy makers based at the state head quarters fix the irrational targets for the health workers based at the field to achieve them. In one case a health worker responsible for a Population Project(PP)unit in Kuppam constituency was asked to complete 11 cases of sterlisations per month. To achieve that target the PP Unit should have a population of 11000 population but the above unit had only 5000 population and while many times the health worker has highlighted the mismatch between the size of the population and the targets fixed on sterlisations and pleaded with the authorities to fix the rational targets it has not been taken into consideration and in turn the authorities kept pressurising the health worker to achieve the targets. These targets have been unrealistic in the sense that in majority of the cases they are beyond the stage of being fulfilled. Instead of questioning the irrationality of targets and demanding for rational targets from higher authorities the practice for years has been to manipulate the data particularly at the time of collating and reporting in the formats at the PHC level. This is done by the health workers, health supervisors and medical officers to escape the wrath of the higher authorities and to get good grading for their institutions. The participation of the users in the process of implementation of HIS remained fine till the point of data entry but little did they realised that manipulation of data cannot be done while generating reports.(The DHIS generated monthly progress reports of PHC,PPU,UFWC,UHC, Family Welfare Performance report, Family welfare performance institution-wise report and MTP[Medically Terminated Pregnancy report] and many more required by the DM&HOs office).After completing data entry in DHIS and when reports were generated both the supervisor and medical officer were shocked that figures shown in the reports were far below the targets given by the authorities. Since the reports were printed using the application they could not change the figures. And there was no option to generate empty formats of the reports in the DHIS software and so the opportunity of filling the figures manually was not present. Soon the staff at the PHC realised that computerised reports designed with their participation did not allow them to manipulate the data which they had been used to.

Within short span of time the numbers of PHCs generating reports using DHIS software dwindled. Initially they gave excuses but soon they came with the request of developing the option in the software to manipulate the figurers. Since improving quality of data is one of our main objective the team discussed the above issue with the medical officers in getting their targets fixed rationally using the information about their geographical area, population and other criteria but it is of no use as the concept of decentralised information which is much spoken is still far behind in practice.

While many of the staff in the PHC justify the manipulation of data to the unreasonable targets that are fixed by the policy makers without taking into consideration the actual situation in the field that exists others mutely accept the fact and to save themselves from the wrath of being punished and humiliated do not want to use the systems. Most of them looked for loopholes in the systems where the figures can be changed automatically while the reports are printed to be submitted to the district officials but it was futile. The state proclaims several things but rarely reaches down to the base level of society[4] .

At that time, implementation team continuously motivated the users to complete the data entry in the DHIS software and generate the reports using the above application and then using the information demand for rational targets to be fixed from the higher authorities. But the users were in dilemma. Since the users at the higher level that is, the policy makers based at the state head quarters did not participate in the implementation of HIS the staff at the field level reverted back to manual systems of collecting, collating and making manual reports. Thus implementing change is more difficult and contrary to the notion that in developing countries anything which comes free particularly technology will be easily accepted and can be implemented is a fallacy. Thus non participation of officials at various levels and to some extent by the users based ay the grass root level has put all the efforts of implementation team and as well the project itself at stake.

5 Analysis

Though much is spoken about decentralization in India still everything remains centralized and even the basic right to information which is necessary for citizens to participate in governance, especially at local levels is denied by those in authority as according to them information is power. The bureaucracy in India still keeps up the colonial culture of secrecy, distance and mystification [5].

According to Ostrom's study (1990) in order to bring about sustainable reforms, a process of change must take place in the local community. Central government can facilitate such a development in several ways, but it cannot accomplish the process by itself and this can only be done by local actors on the ground. To a great extent it is a matter of mobilizing and empowering the local community. Hence, local democratisation is an essential part of the 'reform package'.

But generally what happens is that due to the basic weakness that prevails in centralisation decision-making is inflexible to varying local demands and conditions. In addition, central actors are severely restrained when it comes to monitoring the actual implementation of programmes out in the field. Involving local actors with an independent responsibility for accomplishing certain activities may mitigate these drawbacks. This is a fundamental argument for decentralisation. The problem, however, is that existing local decision-makers often have their own policy agenda and are therefore likely to divert programmes for their own ends, to the detriment of the community at large [6]. To counteract this tendency, it is necessary to inspire the stakeholders – the intended beneficiaries – to become involved in the process .

So, in a country like India where hierarchy and bureaucratic structures are strong it is important first to bring about a change in the organization and hierarchical structures. During the process of any implementation it is imperative to give importance to both technical and organizational issues since organisational issues play a crucial role in making any implementation a success.

References

1. Bhattacharyya, D.: Limits of Legal Radicalism: Land-Reforms and the Left Front in West Bengal. Calcutta Historical Journal 16(1), 57–100 (1994)
2. Ostrom, E.: Governing the Commons: The Evolution of Institutions for Collective Action. Cambridge University Press, Cambridge (1990)
3. Hadenius, A.: Decentralisation and Democratic Governance.Experiences from India,Bolivia and South Africa. ed. Almqvist & Wiksell, International, Stockholm (2003)
4. Mander, H., Joshi, A.S. The Movement for Right to Information in India: People's Power for the Control of Corruption, Commonwealth Human Rights Initiative (reprinted 2001) (1999)
5. Manor, J.: The Political Economy of Decentralisation, The World Bank, Washington D.C (1997)

Media Sharing and Collaboration Within Mobile Community: Self Expression and Socialization

Younghoo Rhee, Kiran Pal Sagoo, Jayoun Lee, Juyoun Lee, Doekwon Kim, and Youngwan Seo

Samsung Electronics, SW Lab, Mobile Software Team, Interaction Part
Metan-3Dong, Gyeonggi, Suwon, 82-31-277-7326
{yh.rhee,kp.sagoo,jayoun.lee,florial.lee,doekwon.kim, sywpro}@samsung.com

Abstract. The importance of social relation has never been as vital as of now, in an age of sound bites, aging population and busy work schedule, contemplating social behavior has become a rarity. The 'Structuration' theory [2] elaborates the fact that how much social relation's are important in human life. Here we re-define basic concept and category (types) of mobile community. A realization of mobile communities through unified UI for mobile devices is believed to provide users efficient communication ways while satisfying properties of various communities. Such unified UI led to drive an adaptive and intuitive user interfaces that depict real-time interaction involving multiple parties, contents and activities. Many different scenarios emerge with the proposed UI to give a deeper thought and understanding about the role of mobile community in everyday life and help promoting user's communication experience.

Keywords: Mobile Community, Visualizing Social Interaction, Community Architecture and concept.

1 Introduction

What is fun? The definition could differ, however it generally stands for the activities that are enjoyable or amusing. After extensive survey and interview with college students who were in their early twenty's, the authors concluded that "fun" experience is related to many factors which range from achievement, challenge, exploration, learning, storytelling, fantasy, immersion, violence, humor, physical activity, thrill, aesthetic, self expression, and social interaction. The authors constructed an affinity diagram that reflects four fun domains using fifteen fun factors, which are as follows: fun related to interaction between people; accomplishment from challenge or learning; indirect experience or nostalgia; cognitive or affective feeling (Figure 1).

In this paper, "fun" is more related to new experiences through proactive activities like self expression and meeting new people rather than playing game or shopping alone. Such experience is similar to features of a social network. A social network is

M.J. Smith, G. Salvendy (Eds.): Human Interface, Part II, HCII 2007, LNCS 4558, pp. 137–146, 2007.

described as a structure made of individual or social connections. It indicates the ways in which they are connected through various social familiarities ranging from casual acquaintance to close familial bonds. The term was first coined in 1954 by J.A. Barnes [3]. Our concept related to social network includes not only relations with existing relations such as friends and colleagues but utilizes new ties between people through casual meeting. A social relation generally depicts dynamic interactions across human communities in belonging society and implies personal participation in that society. Such definition, recently, expands to online community in internet era and is embodied in cyber space. Blogs, online café, and mini homepages such as myspace, second life, and cyworld are representatives of community in virtual space. They generally provide a history of communication in community and relationships among members who have same interests. Recently services such as - chatting or online meeting/dating are one of main drivers to mobile business. For example, ImaHima-IM service in Japan, Time2Flirt in England, mobile community service Upoc in United States recently release a free SMS Flirt service. However the aforementioned services are generally served using central server or though web technology such as IMS / RSS / DCD, regardless of user's context. And looking through mobile perspective none of them serves personalized or intelligent services matching user's context to construct community. Considering the limitation of the existing community services, the community in the present study expands to conceptual definition: a group of people who has same goal. In turn, friends or a family relation itself is a goal, the reasons of existence has nothing to do with gaining some benefits. On the other hand, a group who has same interests has different reasons to exist: their goal may be achieving fun /entertainment. In addition, there are group of people for wish to efficiently and expect better service due to the supply and demand policy.

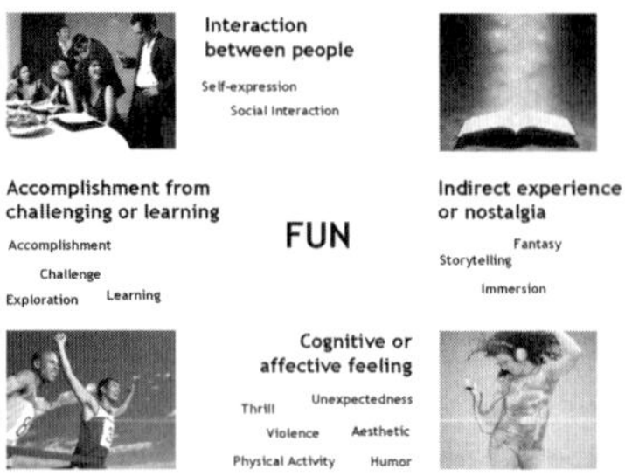

Fig. 1. Fun factors and affinity diagram

The relations and entertainment communities have more sociable features as compared to the group of people for service or works. On the other hands, service or entertainment group people may be considered as more dynamic and possess flexible features than works and relations communities: once goal within their communities is fulfilled, the community disappears – it is temporary. If then, what features are important to construct a community if it is real matter to human life? This question is a ground point for our present study. "Seamless communication" environment between community members is defined as main purpose for the question. How do we define seamless communication? It simply implies the implicit and intuitive interaction methods to effectively communicate between members. Detail functional explanation of different types of communities will be provided in concept development section. Mobile community in the present study has three distinguished features compared to other online or offline community: goal, context, and temporal. A communication is likely to happen when people have problem given context and they wants to solve it, that setup a personal goal. Such a problem very rely on the individual context including location, emotion, and activity. However, such that goal or context is not static, but very temporal since the individual problem always changes given their dynamic contextual feature.

Goal, Context, and Time are the distinguished features that construct a mobile community and their scope not only covers human-to-human relations but also device-to-device and device-to-human. Briefly, the community in the present study realized through an individual perspective. Namely, individual plays each role across four types of communities. A particular role is temporarily accepted due to the individual's goal or exists permanently through genetic or social relations. The present study believe that designing community concept by utilizing tree structure has a limitation since such that user interface and interaction has been developed owing to new technology development such as SMS , MMS, mobile TV that makes phone menu / structure disordered. It is generally believed that poor information architecture and low usability on device makes people hard to find a specific function and makes user frustrated. Furthermore, recent trends including sharing and collaboration may not be integrated well with previous functions on mobile device. One of contributions in the present study is developing new mobile community UI that accompanies user's needs related to simple and intuitive features. Based on the relations and socio-fun communities, the present study developed four scenarios. The purpose of scenarios in the present study is to present concept proof and verify feasibility of the conceptual User Interface regarding user's interaction. The reason that the present study initially concentrated on such communities is people spend their life for fun, family and future: current trends reports that people spend most of resources for their family, fun, future. Such mega trends make us to facilitate communication within community by inventing new mobile community platform and applications.

2 Concept Development

2.1 Personalization and Socialization

In today's moment of time the life of an individual involves a set of interactions, these interactions are part of the individual's daily life. An individual possess its own personal identity, likes or dislikes and personality. Since the human evolution the interactions of the users have also evolved into something more in terms of communicating his/her opinion. Thus the concept of 'personalization' has emerged in which the user may change the way things work or function reflecting owners' personality. In this day and age of technological advancement various technologies assist user in a very personalized way. Various facilities are available to the user which allows users to document their daily life; daily calendar, events, pictures, music, video and share or consume the same in various social context.

Also today an individual is part of far larger population than before and its involvement in the number of communities is determined by its participation, wherein an individual has to perform many roles, as a friend, a family member at home, a colleague at office or its relationship with external entities. These roles or personas are now not limited to the basic role and interaction but are also being advanced with the technology. Nowadays a user can remain connected with various communities, which could be part of his workplace, friend circle, family and common interest etc all this could be achieved with the help of single mobile device. Mobile *'Life Diary'* [1] is such an application developed for caching the life of the user on the Mobile device. The embedded application automatically collect, arrange and store users moments during his day, for example: - call logs, payments, schedules, entertainment etc and by analysis of which user personal preferences could be evaluated and could be acted upon. The *'Life Diary'* user interface, which includes layout, interaction, and intuitiveness, is designed cautiously so as to enhance users experience and social interaction in a tangible way.

2.2 Mobile Community: The Conceptual Model

The above mentioned background and motivational keywords led us to the proposal and formation of a holistic conceptual model of Mobile community as shown in the diagram below. The social interaction of the individual is divided into four quadrants keeping user/individual/peer at the center. The social interaction is divided into four quadrants and are termed as 'Relations', 'Office', 'Service' and 'Entertainment' and each quadrant has an associative property of being efficient, social, static or dynamic.

The terminology of the different type of social interactions is derived from three main keyword's i.e. 'Goal' 'Content' and 'User'. These keywords also encouraged to develop conceptual User Interfaces (UI), which are described in detail design section. The keyword 'Goal' drives the interaction between various entities in life of the 'User' for example the goal of the individual could be to build a relationship with other community members, within family members, office workers, with entertainment services and service providers. Focusing on the first quadrant we realize that the social interaction is mainly based on the relationship of user with other members of the community. The members could include friends, relatives etc. The first quadrant possesses properties 'Static' and 'Social'. The reason behind this would be as the relationship between individuals needs to be strong and steady so as to improve the social ties between individuals. Also the property 'Social' refers to a common shared goal of the community and the way the people behave and interact in groups. The second social interaction quadrant is 'Colleagues/Co-workers'. This quadrant plays important part in user's life as recently people spend a great amount of time at workplace, forming communities at workplace is imperative nowadays. This quadrant possesses one different property from previous quadrant i.e. 'Effectiveness', while keeping the property 'Static' same. The requirements of community within the workplace differs from that of 'Relations' as the workplace /office environment demands more effective and clear communication between colleagues for example a user could be member of technical club where the interaction could be professional, comprehensive and demanding, this requires more efficient dealings within group members.

Life diary main view Social network view Detail view (SMS)

Fig. 2. Life Diary

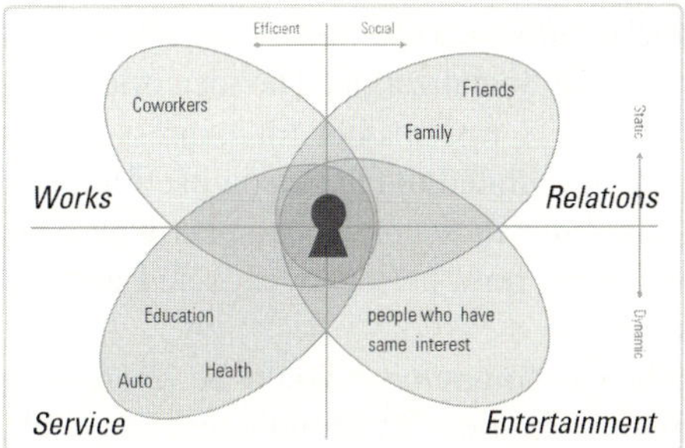

Fig. 3. Mobile Community Conceptual Model

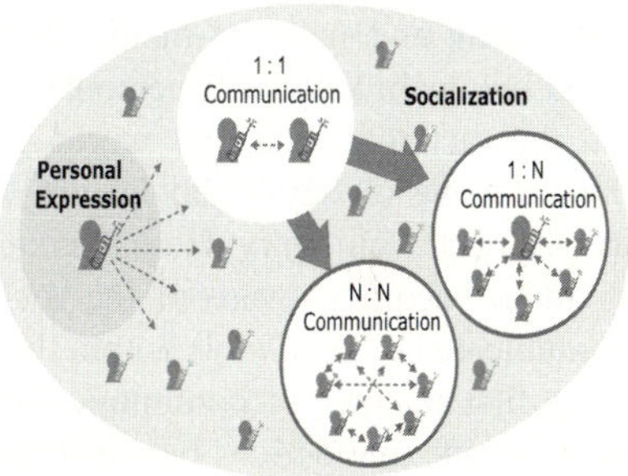

Fig. 4. Next generation communication

The last two quadrants of social interaction enjoy 'Dynamic' property by virtue of the activities within the community and its external environment. The 'Service' quadrant implies the interaction of user with varied services provided by service provider and the communities formed around those services. The property 'Dynamic' in this quadrant stands for the wide-variety of applications provided by various service providers for example on-line education where the user registers for some course and becomes a part of the specific educational community. Also, health services are important part of daily lives, which can range from personal medical prescription services to identifying diseases etc, the options are unlimited. The user in such cases could form communities around a specific topic like disease prevention and collaborate and contribute to the same. The 'Effectiveness' property in this quadrant plays significant role as the information gathering and data analysis in 'Services' interaction should be accurate and to the point. The last and most socially interactive quadrant is 'Entertainment'. The entertainment sector has seen a major growth since last few years due to he explosion of various Multimedia devices/services, it promises plenty of opportunity for social interaction among users. What make this quadrant 'Dynamic' is the prompt formations of communities based on some common interest or trend and disappearance of the community after the event. For example the introduction of a new PC game initially would attract lot users worldwide; the users with same interest would form communities/chat rooms etc for the same. Introduction of another game would leave the previous trend at the back but same people could form new communities based on new trend. The property 'Social' for this quadrant serves the same role as for Relations quadrant i.e. shared goal and topic of same interest permit dynamic social interactions among individuals and communities.

2.3 Next Generation Communication

As the networking technologies have evolved through time so as the ways of an individual to get networked with other users which expand human to human communication experience/scope. After we defined the concept and categories of mobile community, idea grouping was conducted given human to human communication scope. The Fig2 below shows vibrant structure of mobile community which can be broadly classified into three type's of communication mode: one-to-one, one-to-many and many-to-many. These three communication modes are developed around two main keywords personal expression and Socialization. The idea originated from the fact that in a real world the individual is responsible for expressing his *personal* views on day-to day events, which are not limited to the person itself but to a part of larger community. Here the concept of *socialization* comes into play where the individual becomes aware of the society and learn how he/she is expected act and communicate in many ways within the community.

The first manner of communication within the community would be *one-to-one* which include interactions between two individual entities. The communication could be a direct face-to-face or through a mediating device such as mobile phone. Such devices have strong influence on the social behavior and consequences of the communicating parties. Such that communication is elaborated with two main domains: self expression and location based service. Ideas regarding self expression included expressing moods, communication with avatar, message with history, and status sharing. The second manner of interaction includes communication from individual to several other individuals which themselves could be a community. This type of interaction is stated as *one-to-many*. The mobile device being the go-between the parties here as well facilitates the individual goal of corresponding with larger audience. Many use cases can be result from such interaction for example: push to talk, 1:N payment, network game, rating, voting, location based messaging service etc are all involved in one to many communication type at idea brainstorming process. The last manner of interaction i.e. *many-to-many* includes both of the above mentioned communication way's. It consists of interactions between larger group of individuals or communities and can span many demographics especially size, culture, networks etc. Many varied models of correspondence can arise from this dynamic interaction. Multimedia sharing, booking, schedule sharing, collaborated storytelling, network game, large screen, orchestra phone, etc are generated.

The above explained community dynamics helped us to realize three main 'activities' within the mobile community: *Sharing*, *connecting* and *collaborating*. These basic activities formed the basis of our concept generation process and helped us in creating strong functional use cases and intuitive user interfaces.

3 Detail Design

The three community activities introduced above i.e. *Sharing*, *connecting* and *collaborating* forms the topic of discussion in this section. The basic activity of *'Sharing'* starts of with sharing something with other person, and then you have it, use it or occupy it. The concept is not just limited to physical objects but is also extended to user experience with someone; in that case both the entities involved share the same experience because you are together at the same time. In the digital world this can be translated to communication activity, which is executed within a circle of friends, office colleagues etc. There exist many technologies in today's world to share objects and an experience, for example Location based services (LBS) allows us to share our position in a city or a locality to my friend or family. User may well be sharing his status during a particular event such as in a meeting; 'Connected Planet' a project carried out at Philips Design Research illustrates such a concept. A user might also share media objects like mp3, Video, Photo files etc to full fill his desire for *'Sharing'* within the community.

The term connecting *'Connecting'* may also be referred to as *'involvement'*, or as two things joined together. During this activity there a subtle relationship is established between the two participating entities. The connection could be one way or mutual, but in both cases both are benefited. The concept revolves within community members which could be friends, family etc as above. Some of the examples could be narrated for this activity as putting out personal experience about an event with the community members. As per the factual means of communicating, it can be relayed with the help of a mobile device with specialized application for the same. The method and device involved for this activity may diverge and could be user specific.

The final activity *'Collaboration'* forms the important aspect of the community. The group effort or teamwork in the current time and in the pas has played a vital role for the survival of mankind. In short it could be described as the act of working together to produce a piece of work with a common goal. The activity may be achieved within the same period of time and duration or a long period of time. And the result of collaboration can be shared and owned by the all the people involved in the *'Collaboration'* activity. An existent example could be a birthday card signed and decorated by family members for their elder, thereby every member of the community (family) collaborate to reach a certain goal i.e. to forward their well wishes to a particular member of the family. The tool utilized could be paper card or an e-card, sent through post or directly on the mobile phone, irrespective of the communication the cooperation of the members result in well defined social activity.

3.1 Mobile Community: User Interface Design

The focus of our user interface design and scenario development has been concentrated around three important factors; *people*, *contents* and *goal*. The Figure 5 demonstrates user interface designs that are centered on these factors.

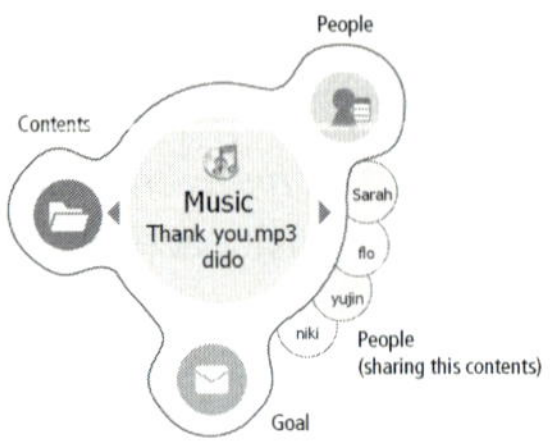

Fig. 5. Mobile Community User Interface Wireframe

During our research on the topic of mobile community and especially 'socialization' we realized that any activity/event has a common *goal* to achieve, which is initiated by the *people* involved and the goal could be or in most of the cases involves some *content* sharing, creation etc. In the above example a 'voting' application in executed from the suite of various mobile community applications followed by selection of *people* which are involved and finally the *contents* of the voting are typed/entered. In the above right diagram the focal factor can be modified effectively to reflect the user's preference for any of the three factors and is not fixed. The second diagram shows media content as the initiation point which could then be shared with the other users of interest.

Fig. 6. Mobile community scenarios

3.2 Mobile Community: Scenario Development

The above figures depict four different scenarios wherein both the concept of 'personal expression' and 'socialization' are utilized to achieve the final common goal. The scenarios fall into the category of 'Relations' and 'Entertainment' quadrant from the Figure 1 and possess the properties of being static and dynamic to social. In those community scenarios, a total of seven features / functions are proposed as one application: voting, video collage, rolling paper, schedule sharing, mobile DJ, personal broadcasting, emotional message. Such seven applications share and reuse user's communication information. In turn, a log regarding communication utilized in voting application is reused and shared in predefined any other application such as schedule sharing or video collage. The interaction feasibility and concept is justified through scenarios and concept UI is willing to be elaborated with additional features including call, SMS, contact.

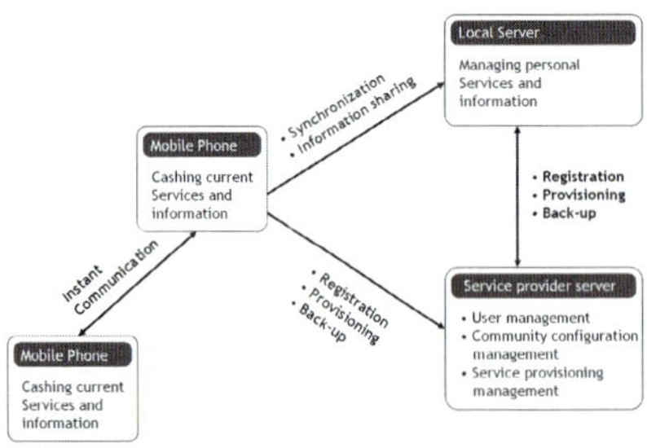

Fig. 7. Mobile community platform architecture

4 Technology

A technology vision for mobile community has three key components: mobile service, network and hardware, and service software. Mobile phone is not only device for voice communication, but point of access for seamless and pervasive service. Such services include tribalism, ad-hoc social network, and goal oriented interaction. A network & hardware aspect has technological business model and roadmap which enables to connect heterogeneous devices (phone to PC or IPTV). A software as a service means a service platform that implements future user scenario. For such technology vision, a service middleware platform exists between previous mobile device service and network layer which reduces the dependency of hardware and network and enables to connect and reuse services. Figure 8 explains the role of mobile devices, local server, and server of Network service provider.

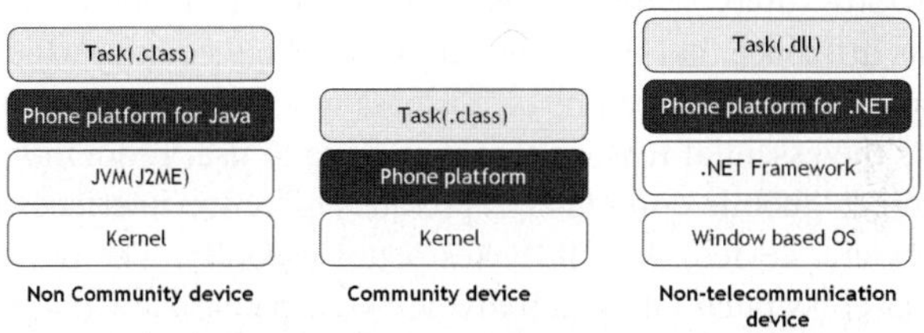

Fig. 8. Mobile Community application architecture

Figure 8 explains the communication ways across community, non community, and non telecommunication devices. The present study utilizes JVM (J2ME) to execute community service between non community devices such as previous versions of standard phone, which have java phone platform. It executes java class file type task. On the other hand, community service is possibly operated by net phone platform at non telecommunication network device including PC / PDA.

5 Conclusion and Future Work

5.1 Virtual Anthropology/Being

In previous chapter, we introduced two projects "life diary" and "community". If a life diary reflects personal identity and expression by automatically generated diary in phone, on the other hand, mobile community deals with the group identity or social interaction- the ways individuals effectively communicate to conserve and support their communities. Next step (future work) of diary and community is to propose new virtual anthropology given web 2.0 trends. The term refers to nothing but the increasing popularity of the virtual identity of the human beings. One of good examples is 'Second life' of Linden Labs. Companies and individuals around the world have discovered this powerful representation for expressing themselves and create business in and around the same. The 'Second life' can represent the personality as the real being and can get involved in the social interactions as well.

Even if we are ongoing to develop a concept for virtual anthropology, but it may reflect personal and social aspects, which get individuals have personal portal – individuals may have not only their history and private information as personal presence, but social interaction through sharing or collaboration with someone who has same goal as well. In addition, identity/profile provided from diary and community and common sensing computing capability on web intelligently collect information on the web which provides users benefits. By constructing virtual space, users may interact or meet with not only friends and family but people who has same goal – virtual serendipity.

5.2 Discussion: Improving Ease of Use and User Community Experience

How could we improve user experience (UX)? – The improvement of UX is not always adjusted by new technology or features like mobile TV service. For example, recent hot technology - wibro, dmb, wimax- enabling seamless mobility driven use case (scenarios) seems affluence, but those are very technology centric, which become being exclusion by end users and limitations of business model. Because such that stories did not provide users the essential reasons that they have to use. From those lessons, at the mean time, a project "mobile community" provides effective interfaces between human and computer (mobile device), which makes better usability. The present study names community UI for conceptual UI, which divide two separate view point: readability and execution. Such introducing revolutionary community UI is believed to give user better interaction experience than previous UI for communication. In mobile community, humans become the social entities and mobile device is utilized as digital technology that drives new value to end users and business model within social context. Figure 9 explains how new value is created (adopted from Beyond UI, Kim Jinwoo)

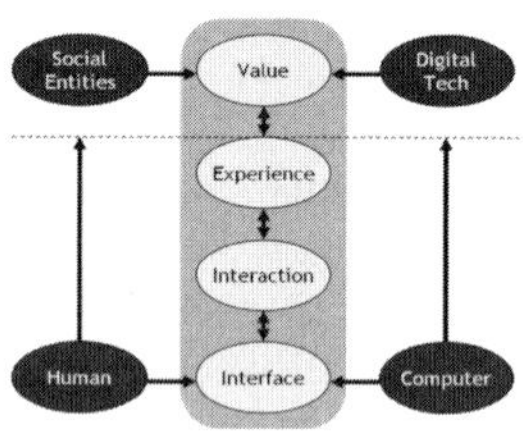

Fig. 9. User experience and value creation in mobile community

References

1. Rhee, Y., Kim, J., Chung, A.: Your phone automatically caches your life. ACM, Interaction (July-August 2006)
2. Giddens, A.: New rules of sociological method: a positive critique of interpretive sociologies, 2nd edn. Polity Press, Cambridge, UK (1993)
3. Barnes, J.: Class and Committees in a Norwegian Island Parish. Human Relations 7, 39–58 (1954)

Adaptive Information Providing System for R&D Meeting Environments

Sang Keun Rhee, Jihye Lee, and Myon-Woong Park

Intelligence & Interaction Research Center,
Korea Institute of Science and Technology,
39-1 Hawolgok-dong, Seongbuk-gu, Seoul, Korea
{greyrhee,bluesea,myon}@kist.re.kr

Abstract. As extensive amount information is required within a R&D environment, an automated information system could save a considerable amount of time and efforts. There are currently many researches to overcome the information overload phenomenon, including information systems, recommender systems and adaptive hypermedia, and this paper presents an adaptive information providing system utilising some characteristics of these technologies to support R&D and collaborative meeting environments. The system described in this paper aims to present the potential synergy between recommender systems and adaptive hypermedia systems.

Keywords: Adaptive Hypermedia, Context Awareness, Information System, Personalisation, Recommender System.

1 Introduction

Within a R&D environment, a project is generally performed as a team-based collaborative work, and researchers and developers spend a significant amount of time and effort on searching, collecting and organising the necessary information. Not only the published papers but also informal documents created by individual researchers within R&D projects are valuable resources, and an automated system for managing, utilising, and recommending such resources can be beneficial to the R&D process. Moreover, as there usually are many meetings during a project, and the participants often require some additional information during a meeting, such an information system can also help the meeting processes.

Currently, there are various technologies developed to overcome the information overload phenomenon, including search engines and algorithms, recommender systems, adaptive hypermedia systems, and so on. These technologies play different roles depending on the applications, from decision making support to electronic learning aids. To support a R&D process, we propose an automated information system utilising some characteristics of these technologies,

This paper presents an adaptive information providing system that helps organising various resources and provides automated recommendation of necessary data with context-awareness and personalisation, within a R&D and meeting environment.

M.J. Smith, G. Salvendy (Eds.): Human Interface, Part II, HCII 2007, LNCS 4558, pp. 147–155, 2007.

First, the design principles of our system are discussed in four aspects – system role, resource management, context & preference, and interface & visualisation. Secondly, some related works are presented in this field including recommender systems, adaptive hypermedia, and R&D support systems. Thirdly, the structure of our system is described with examples, and finally, it is concluded with discussions and some suggestions.

2 Design Principles

As mentioned in the previous section, the main role of our system is helping researchers and developers by providing the necessary information in the current context. However, there could be different methods and services that can achieve this purpose from various aspects, hence the four design principles of our system are discussed in this section.

2.1 System Role

The system is to be used in a R&D environment, both in individual work processes and in meetings. The main role of this system is recommending necessary information based on the current context and individual user's characteristics, and the recommendation is to be generated automatically without explicit user queries. Regarding the environment, this system is to be used within a single project, in which the participants create or collect data and the resources are shared only among the team members. Furthermore, it provides some additional information to project managers including schedule, organisations, etc. Finally, this system can be implemented in a tangible meeting room for intelligent information services, which has been the fundamental motivation of this project within the Intelligent Responsive Space (IRS) project at KIST.

2.2 Resource Management

There might be various different kinds of information necessary in a R&D environment. A researcher may be looking for specific technical documents or academic papers in a certain area, a project manager may seek a person with certain skills and backgrounds, a person may search for a certain project within his/her field, and so on. Therefore, it is necessary to clarify what is the information this system would provide to support a R&D environment.

Though it might be often underestimated, the materials created or collected during a project – including informal or ill-structured documents – are in fact valuable information and it is not rare that people need some of those materials but have a hard time in finding them. Moreover, it becomes even harder when it comes to finding other person's material. Therefore, in our system design, the term *resource* is defined as the material created or collected by each participants of a certain project.

Instead of managing the resources as documents, the data objects are managed as units of *fragments*. A fragment is defined to be a reusable unit data object, and it can be either extracted from document resources or independently created. The basic unit data object is called an *atomic fragment*, and a *compound fragment* is a collection of these atomic fragments and/or other compound fragments. Each fragment has a specific set of metadata describing its characteristics. The main purpose of managing the resources as fragments is to provide adaptive documents – in other words, re-structured and re-organised documents that only contain the contents suitable for the user's requirement.

2.3 Context and Preference

To generate a recommendation without an explicit user query, it is necessary to trigger an automated query generation process based on context information. In a R&D environment, the environmental information is the overall project information, including the milestones and tasks, participants and their roles, meeting information – both past meetings and future meeting plans – and so on. Instead of providing the same recommendation to every participant, it is also determined to design personal profiles reflecting individual characteristics and preferences so that the recommendation is both suitable to the current environmental status and personalised. More details on the context parameters can be found in Section 4.1 and also in [1].

2.4 Visualisation and Interface

Many recommender systems present recommendations as an ordered list [2][3]. Although this approach provides some benefits, it might be not the best technique for providing internal resources. Furthermore, during a meeting, it is not advisable to provide a long list of documents since the participants cannot spend too much time on reading the recommended data while others are in an active discussion. Therefore, it is decided to provide information in a single re-composed document form, whose contents is a collection of various fragments. Such document is to be generated each time the query results are given, and the contents are regarded as more important than the overall style since it is assumed that the users can understand the contents well enough even though the document is not in a perfectly well-structured form. Also, the main purpose of the generated document is to provide people with necessary information on the fly, not for personal study or authoring a formal document.

Compared to recommending a list of original documents, this approach can save some time for finding and understanding the required information. Furthermore, it is particularly useful for managing and re-using informal material created by individuals, and it also shows the benefit when the system is used within a meeting, where a user cannot spend a long time in reading his own material.

3 Related Works

Since our system shares some common characteristics with other works, it is worth mentioning some related works in this field before describing the details of our system.

3.1 Recommender Systems

In 1997, Resnick & Varian defined recommender systems as computational systems in which 'people provide recommendations as inputs, which the system then aggregates and directs to appropriate recipients' [4]. Since then, the recommender systems have been advanced with various technologies [2], and applied in many different areas. In general, a recommender system 'offers recommendations to users based on their preferences' [3], and current recommender systems generally provide a list of recommended objects to support decision making.

Amazon or eBay is an example of successful implementation of a recommender system in e-commerce. In R&D related domains, the Foxtrot system [5] is a good example which analyse the user's topics of interest and recommends academic papers in those topics.

3.2 Adaptive Hypermedia Systems

Adaptive hypermedia systems connects hypermedia and user modelling to adapt to the needs of users [6][7]. The two typical adaptation types are adaptive presentation and adaptive navigation support [7], and our system applied the former approach. Adaptive hypermedia systems have a strong connection with semantic web technologies and ontology, and are mostly advanced in e-learning area – SCARCE [8] and WHURLE [9] are good example applications.

In e-learning, adaptive hypermedia systems select, organise and present the contents based on user's background knowledge, goals, preference, etc., but usually the contents and their structures are clear and well-defined. In addition to the various existing issues in adaptive hypermedia, one great challenge of our system design was utilising the resources that are created by different individuals, without specific format.

IBM [10] and Nokia [11] developed this technology concentrating on device-based adaptive presentation, and UBiquoto [12] added user preference, location and time information to provide travel information.

3.3 Systems Supporting Collaborative Work Environment

Apart from information systems, there are active researches in Human-Computer Interaction area to support collaborative work environments. Some of them also have information services as well as intelligent and interactive interfaces.

Access Grid [13] is a well known technology to support video conferencing, and has been implemented in various forms of creative and academic collaborations around the world. MIT project Oxygen [14] is another successful example of collaborative work support system,

4 The System

The main functionality of our system is providing adaptive information to researchers and developers, and it shares some common features with recommender systems and with adaptive hypermedia systems. The prototype system presented in this paper provides the necessary data resource during a R&D process and in meetings without explicit queries from users, by searching and selecting the relevant data based on the user profiles and other context information. The query results are then re-structured and organised to form a single document, yet it lacks navigation support unlike other adaptive web information systems (AWIS) [6][7]. This system has slight emphasis on its usage within meetings, but it can also be applied for general R&D processes. The system overview is shown in Fig. 1.

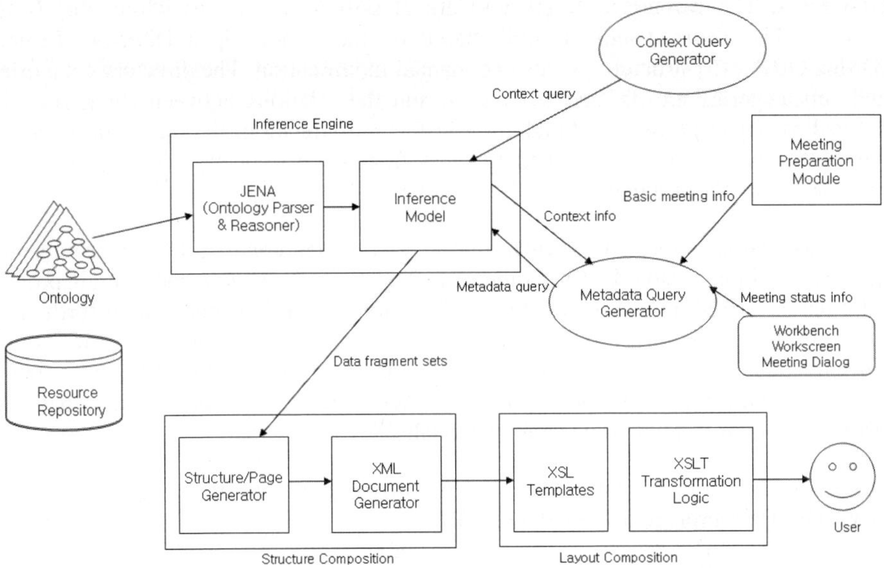

Fig. 1. The system structure

4.1 The Information Model for Adaptive Composition

To organise the data objects and provide the right information by context-awareness, it is firstly required to model the various elements and properties of a R&D environment. Considering the design principles of our system, the Information Model for Adaptive Composition (IMAC) [1] was suggested which defines the fundamental structure and base schema of the data resources, people, and environment within the R&D domain (Fig. 2 shows the IMAC overview). Our system utilised the IMAC in part, and each domain and its elements are structured and represented in ontological form.

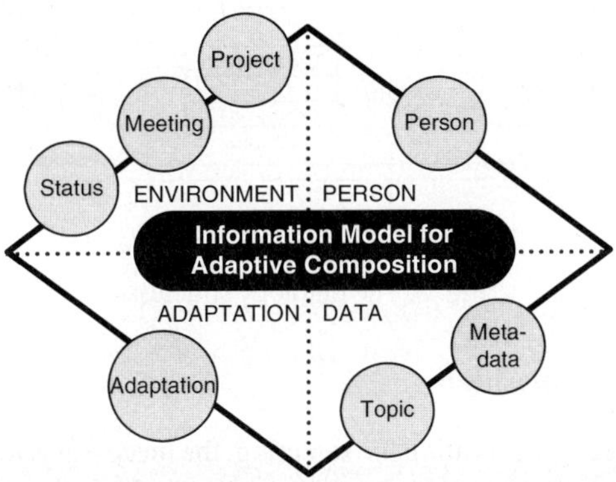

Fig. 2. The IMAC overview. More details of the IMAC can be found in [1]

Resources. The resources (fragments) are classified and stored within the *Topic Domain*. The *Topic domain* is built based on the DMoz Open Directory Project (DMoz ODP) [15] structure, with some manual modification. The directory categories and subcategories are defined as classes, and the relations between categories are defined as object properties. In addition to this classification, the metadata based on the *Metadata Domain* is assigned to each fragment to specify its properties and relations with other fragments.

Contexts. To generate the correct query at a given context, the environmental information is also stored in the ontology. The *Project Domain* the overall project information including task hierarchies, related topics, participants, and so forth. The *Meeting Domain* analyses the meeting information and their agenda, and finally the current situation is computed utilizing the *Status Domain*. In addition to these environmental information, personal tasks, roles, preferences, etc. for individuals are stored in *Person Domain* to generate personalised results.

Ontology. The domain and elements of the IMAC is represented as an ontology in OWL DL using Protégé 3.1.1[16]. (Fig. 3.)

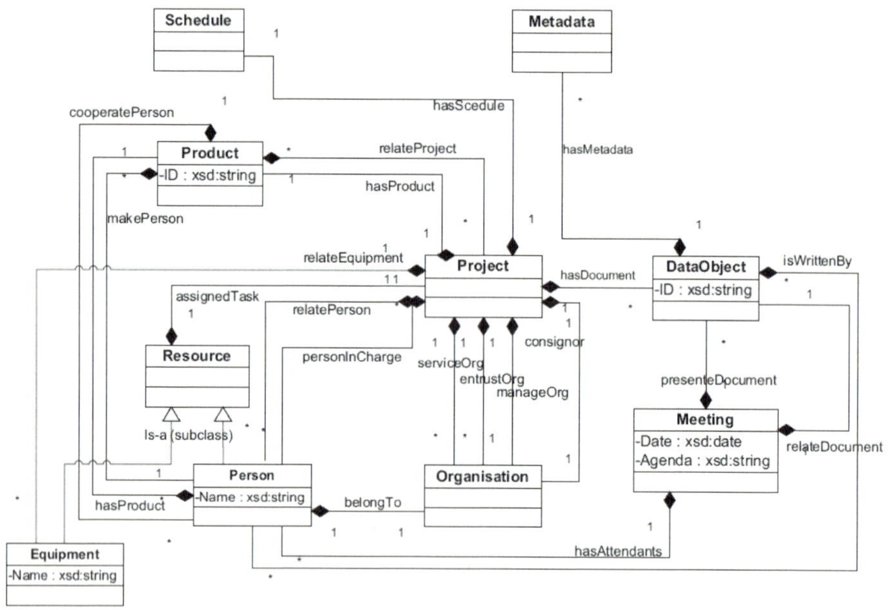

Fig. 3. The ontology structure

4.2 Selection and Composition

The first step for query generation is selecting the necessary topics for the current context. The system sees the topic structure as a labelled directed graph, with each topic as a node and each relation between topics as an edge. Each edge is labelled with one of the four relation types – is-a(super topic), narrow(subtopic), related, and symbolic. For each relation type, a numeric value is assigned representing the *distance* between two topics, and the value is set according to several tests.

To find the topic(s) required within the current context, the project, tasks and related topic information is extracted from the *Project* and *Person* domains. In addition, personal task, role, and preferences are analysed for personalization, and the meeting agenda and related tasks from the *Meeting* domain are also included in case this system is used within a meeting. Based on this context information, an ordered list of the related topics is generated by navigating the topic graph.

Once the topics are selected, the system then gathers the appropriate fragments within the selected topics. Concerning the user's knowledge and language level, task, the relation value (distance) between the selected topics, and the relations between each fragments, the final target fragments are selected and these are structured into an XML document format. This XML document is then parsed to generate a HTML page via XSL/XSLT. Fig. 4 presents an example result page, which may seem to be an ordinary single document but is actually a structured set of around 40 fragments.

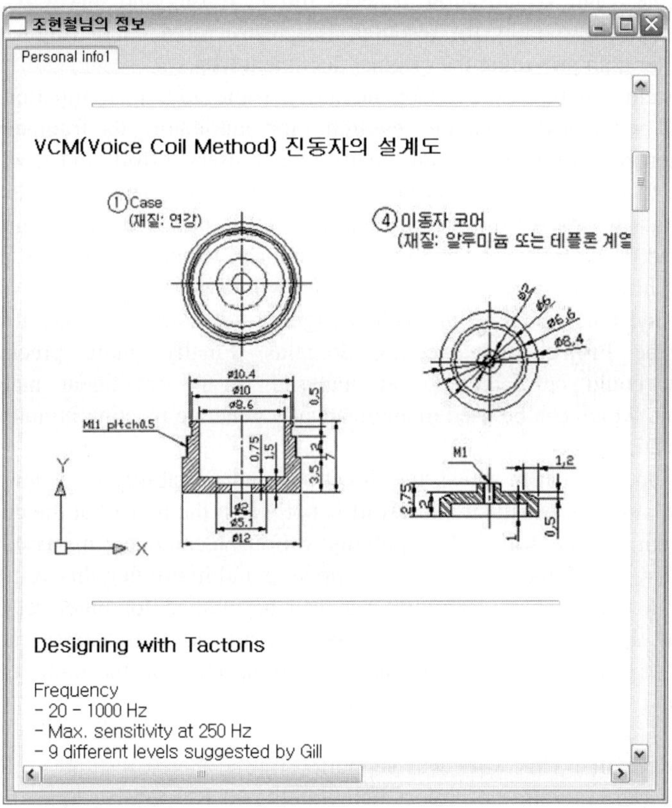

Fig. 3. An example of a generated document. In this example, titles, a picture, and texts are all separate fragments, and the horizontal lines indicate groups of fragments with related contents.

5 Discussion

The system presented in this paper is a prototype that was designed and developed to actively provide personalised information in an R&D environment. The main objective of this paper is to present and discuss the possibility of utilising the methods and techniques used in recommender systems and adaptive hypermedia systems. The adaptive composition method used here enables users to receive the core contents that can be read and understood reasonably quickly but there are still several issues and much work is left to be done in the future.

Currently, the system requires too much users' effort. The users, including the project manager and/or system administrator, need to set various context information and user profiles manually. Also, when updating a new data fragment, its metadata needs to be assigned manually. In addition, at the early stage of a project, this system can hardly be of any help since there are few resources to provide, apart from initialisation of context information and preferences. This problem, known as the cold-start problem, appears to be much severe than other recommender systems. Regarding the result visualisation, it needs further testing and survey to answer the question whether providing the resource as a restructured set of fragments form is actually better than providing the original document lists.

This system can be improved by including some additional functionalities, and those would be the next step in our research. First, automating the fragments and their metadata generation process would greatly reduce users' effort, and it would enable the system to crawling and gathering external resources. The access level to each fragment can also be set for security and privacy issues. Second, automatic profile update based on both implicit and explicit feedback would improve the precision of the result. Third, this system can be combined with a generic project management or scheduling system, which would help assigning some of the context information including the Project and Meeting domains. Finally, more precise meeting information could be gathered and analysed in the intelligent meeting room environment, which can be used to automatically generate meeting minutes and better recommendations.

To sum up, this paper shows the directions and possibility of a new method in information systems, but the users' effort is not worth the reward at the current stage. This problem can be solved by applying various techniques discussed above. In addition, there are various other ways of applying and improving this system design – developing a more generic ontology that can be applied for other related domain applications, applying more individual characteristics such as learning style [17] or other cognitive properties [18] for personalisation, adapting the multi-agent system structure, and so on.

Acknowledgments. This work was supported by the Korea Institute of Science and Technology (KIST) with the project IRS (Intelligent Responsive Space), which is a part of TSI (Tangible Space Initiative) project.

References

1. Rhee, S.K., et al.: Information Modelling for Adaptive Composition in Collaborative Work Environment. Enformatika 14, 151–156 (2006)
2. Burke, R.: Hybrid Recommender Systems: Survey and Experiments. User Modeling and User-Adapted Interaction 12, 331–370 (2002)
3. Terveen, L.G., Hill, W.: Beyond Recommender Systems: Helping People Help Each Other. In: Carroll, J. (ed.) HCI in the New Millennium, Addison Wesley, New York (2001)
4. Resnick, P., Varian, H.R.: Recommender systems. Communications of the ACM 40(3), 56–58 (1997)
5. Middleton, S.E., Shadbolt, N.R., De Roure, D.C.: Ontological User Profiling in Recommender Systems. ACM Transactions on Information Systems 22(1), 54–58 (2004)
6. Brusilovsky, P.: Methods and Techniques of Adaptive Hypermedia. User Modeling and User Modeling and User-Adapted Interaction 6(2-3), 87–110 (1996)
7. Brusilovsky, P.: Adaptive Hypermedia, User Modeling and User-Adapted Interaction 11, 87–110. Springer, Netherlands (2001)
8. Iksal, S., Garlatti, S.: Adaptive Web Information Systems: Architecture and Methodology for Reusing Content. In: Proc. 1st International Workshop on Engineering the Adaptive Web, Eindhoven, the Netherlands (2004)
9. Moore, A., Brailsford, T.J., Stewart, C.D.: Personally tailored teaching in WHURLE using conditional transclusion. In: Proceedings of the 12th ACM conference on Hypertext and Hypermedia, 163–164 (2001)
10. Britton, K.H.: Transcoding: Extending e-business to new environments, IBM Systems Journal, 40(1) (2001)
11. Metso, M., et al.: Mobile Multimedia Services - Content Adaptation. In: 3rd International Conference on Information, Communications and Signal Processing, Singapore (2001)
12. Cena, F., et al.: Integrating Heterogeneous Adaptation Techniques to Build a Flexible and Usable Mobile Tourist Guide. AI Communications 19(4), 369–384 (2006)
13. Access Grid Project Homepage, http://www.accessgrid.org/
14. MIT Project Oxygen Homepage, http://www.oxygen.lcs.mit.edu/
15. DMoz Open Directory Project, http://dmoz.org/
16. The Protégé Ontology Editor and Knowledge Acquisition System, http://protege.stanford.edu/
17. Grigoriadou, M., Papanikolaou, K., Kornilakis, H., Magoulas, G.: INSPIRE: An Intelligent System for Personalized Instruction in a Remote Environment. In: Proceedings of 3rd Workshop on Adaptive Hypertext and Hypermedia, Sonthofen, Germany, 13–24 (2001)
18. Tarpin-Bernard, F., Habieb-Mammer, H.: Modeling Elementary Cognitive Abilities for Adaptive Hypermedia Presentation. User Modeling and User-Adapted Interaction 15(5), 459–495 (2005)

Disseminating and Sharing Information Through Time-Aware Public Displays

Marcela D. Rodríguez[1], Angel G. Andrade[1], Maria Luisa González[1], and Alberto L. Morán[2]

[1] Ingeniería en Computación, Facultad de Ingeniería, UABC, Mexicali, B.C., México
[2] Ciencias Computacionales, Facultad de Ciencias, UABC, Ensenada, B.C., México
University of Baja California (UABC), Mexicali, México
{marcerod,angel_andrade,alberto_moran}@uabc.mx, mlglez@yaqui.mxl.uabc.mx

Abstract. Organizations use several methods to communicate formal and informal information. Currently, message boards or corkboards are used as a means for sharing information, coordinating activities and communicating with others. Using our organization as a case study, we observed that there are many message boards around the physical environment and most of them are covered with documents announcing upcoming meetings, seminars, conferences and other events. Besides that, some of the documents were out-of-date which makes it more difficult for users to find information that is relevant for their activities. To support the dissemination of information in organizations, we are proposing a Time-aware Public Display System that enables organization personnel to become aware of relevant information in a timely fashion. We used the Perspective Wall visualization to enable the categorization of the information into topics and dates. We evaluated the system design which shows that users considered appropriate and useful the presentation of information on the Time-aware Public Display.

Keywords: Information presentation, Workplace Study, Software Agents.

1 Introduction

Organizations use several methods to communicate formal and informal information. These methods can be electronic means, such as e-mails and web pages, and others that depend on the users' location in order for them to get the information. For instance, corkboards are located throughout the physical environment in order to advertise important information for the users' activities. Taking our organization as a case study, a campus of the Autonomous University of Baja California (UABC), we observed that there are many physical corkboards around the campus and most of them are covered with documents announcing upcoming meetings, seminars, conferences and other events. Besides that, some of the documents were out-of-date which makes it more difficult for users to find relevant information for their activities or role (such as students, full professors and research professors). We noticed that publishers on the corkboards use several techniques for attracting the attention of the

M.J. Smith, G. Salvendy (Eds.): Human Interface, Part II, HCII 2007, LNCS 4558, pp. 156–165, 2007.

target public. For instance, information considered as urgent and which is used for coordinating activities is presented in documents with a highlighted message. Based on the above observations we identified some issues that should be investigated. For this, we carried out a case study based on the Rapid Contextual Design technique [2]. Based on the results of the case study, we identified the functional requirements of our system and propose a Time-aware Public Display System to support the dissemination of information in organizations by allowing organization personnel to be aware of the information addressed to them. The public display-based system uses the Perspective Wall [1] visualization technique to attract the attention of the target users and enables them to easily locate information that may be relevant for their activities.

Before presenting our solution, we first present in section 2 other related projects, section 3 describes the workplace study we conducted to get a deeper understanding of the sharing and dissemination of information in our organization; section 4 explains the functional requirements of our system; section 5 describes the functionality and architecture of the proposed system; section 6 presents some of the results of the evaluation of the system design; and finally, section 7 presents conclusions and future work.

2 Related Work

Other projects have also proposed the use of electronic public displays for sharing information. Notification Collage [4], IM-Here [5] and MessyBoard [6] use public displays for presenting documents that were arbitrarily posted on the public display. As the public display can become saturated with information, it may be difficult for users to be aware of important information. The difference between these projects and The Plasma Poster Network, is that this last system was created not only with the aim of keeping users informed, but to encourage users to initiate conversations. For this, it is recommended that one must "create attractive, inviting interfaces where content changes regularly" [7]. They proposed a plasma display for publishing dynamic multimedia content, which changes every 60 seconds, and enables users to pause the presentation to see the information of his interest. However, even though this technique could be attractive for readers, users must spend time locating information that may be important for them, and when they walk in front of the display they may not be aware of information of their interest.

3 Case Study

With the purpose of getting a good and solid understanding of the dissemination of information in organizations by using message boards, we used qualitative methodologies, such as observations and informal interviews. Thus, we got a preliminary understanding of the use of message boards by observing the different boards located throughout our University Campus. We observed that message boards or corkboards are mainly located in the university laboratories. Figure 1 shows a corkboard at the entrance on the Computer Laboratory. We realized that different kinds of documents are posted on corkboards located in places where persons usually walk (such as the laboratory entrances and halls) and on the windows of the managers' offices, which very often have to be visited by the organization personnel. We noticed that the same documents can be published in several corkboards; a corkboard contains different types of documents (i.e., for coordinating academic activities, advertising the selling of computers) with information addressed to students and/or professors; and finally, we noticed that some of the published documents were out-of-date. The above preliminary observations may indicate that organization personnel are not opportunistically aware of the information relevant for their activities since for locating and identifying relevant documents they must spend time everyday in order to be aware of new information.

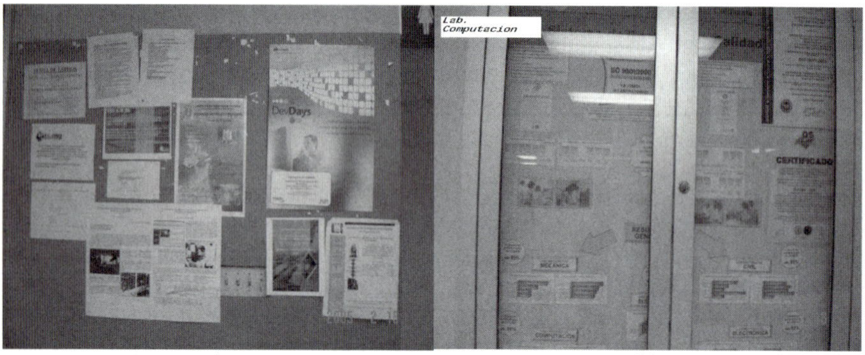

a) b)

Fig. 1. a) Public message board and b) Private message board located in the hall of the Computer Laboratory

Based on the above observations we identified some issues that should be investigated through our case study. For this, we based our case study on the Rapid Contextual Design technique [2]. We interviewed the organization personnel that frequently publish information (Career Managers and Laboratory Managers that have to coordinate many activities) and those to whom the information is addressed (such as part-time and full-time professors). Figures 2 and 3 show some of the diagrams illustrating the results we obtained from the interviews. The information we collected from our interviewees was categorized by using post-it notes of different colors. Similar answers and comments (which were written on yellow post-it notes) given by the interviewees were categorize by topics (blue post-it notes) that gave us an idea of their practices for publishing or consulting information, and then we grouped the related issues in high level topics (pink post-it notes) that enabled us to identify the requirements of a Public Display-based System for disseminating information.

3.1 Publishing Practices

To obtain information regarding publishing practices, we interviewed four (4) Career Managers and four (4) Laboratory Managers who manage the information on the corkboards. We asked them questions related to the priorities for publishing, type of information they usually publish, document organization, and information maintenance.

Our findings were grouped into the following categories:

The information is published in several places. Our interviewees agreed that they publish different types of documents (call for meetings, courses for professors, conferences, or news related to academic events). Based on the urgency or importance of the document they decide where to publish the document. Thus, the most relevant information is published in several places around campus, and in some places from which this information can not be removed by anyone, such as the windows of the Managers' offices.

There are priorities for publishing information. We found that for publishing information, message boards can be classified as public, semi-public and private. Any person (professors, managers, students, and others who may not be personnel of the organization such as sellers) can publish or remove any document from the public message boards. For publishing on the semi-public message boards, users are required to ask for permission from the Managers. These boards contain exclusively academic information and information used for coordinating the personnel activities. And finally, only the Managers can publish information on the private message boards. The private message boards, as shown in figure 1b), have information that due to its relevance, should remain for a period of time. For instance, it may contain information related to the rules for using laboratory equipment.

There are priorities for removing documents from the message boards. We found that only managers have authorization for removing documents from private and semi-public boards. From public boards, any person can remove a document, i.e. when there is no space for posting a document, the publisher removes an out-of-date document, or posts it on top of an expired document. As public boards can become saturated with documents, they are cleaned at the end of each course period (semester).

Different types of information are published by managers. They must make information available that is relevant for the organization personnel, such as academic events; and information relevant for coordinating their activities, such as upcoming meetings. Thus, message boards are used by managers for communicating with the organization personnel.

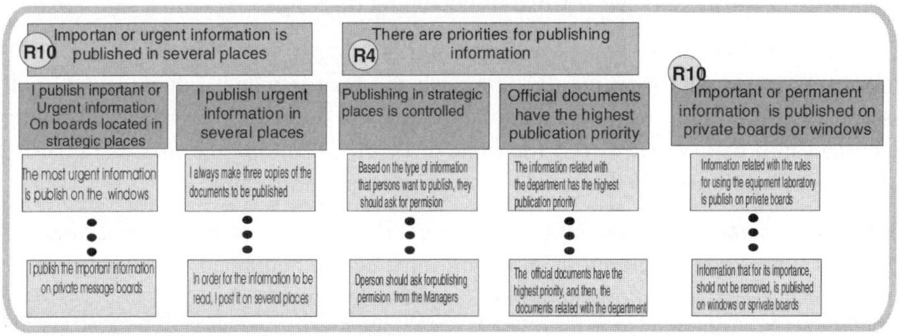

Fig. 2. Findings from the interviews for finding out the publishing practices

3.2 Consulting Practices

We also interviewed eight (8) professors. We asked them about when and where they look for information, what usually attracts their attention, and about the relevance of the information they find on the message boards. We ask them to tell us about the last time they consulted a corkboard, and the last time they found information of interest to them. This stage of the case study enabled us to mainly identify the different types of information that the subjects considered more important for their activities and define the categories in which the information can be classified (see figure 3). We also found out how our informants locate and keep the important information. These findings are described next.

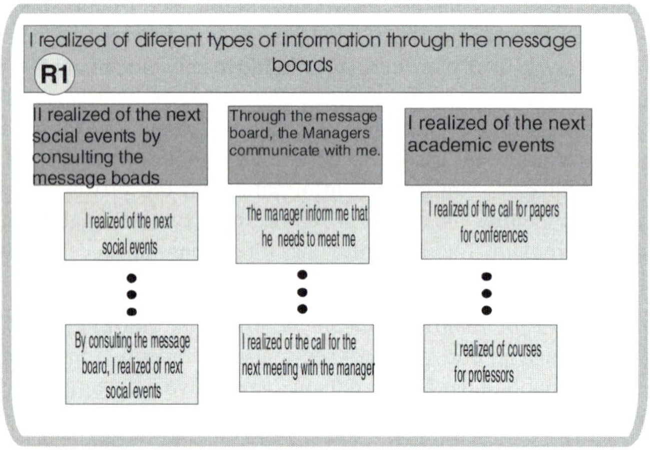

Fig. 3. Findings from the interviews for finding out the consulting practices

Message boards are used to communicate different types of information. The interviewees stated that they usually get information about upcoming academic events such as conferences and courses, social and cultural events from the public message boards. They also stated that these boards are used by managers for communicating with them, i.e. professors are notified of upcoming meetings.

Sometimes users did not view important messages in a timely fashion. The informants stated that they usually approach the boards when a document visually attracts their attention, or sometimes they explicitly approach the board for finding important information. However, for some of them, consulting the message boards is not a constant activity and they tend to miss important information.

Users keep the important information with them. Some of the interviewees said that they make note of the most important information contained in a document, for instance dates of an event or meeting, URL of the document.

4 Functional Requirements

From the results of our case study, we identified the following requirements of a public-display based system that enables organization personnel to be aware of the information addressed to them.

The public display system must be time-aware. The system must automatically remove out-of-date documents and be aware of upcoming deadlines to publish reminders. Thus, the system will enable users to not get distracted by the documents that are not relevant anymore.

Public displays should be located in places that are the most visited by the users. Our findings indicated that information should be pervasive to reach the users. To do this, several public displays should be located throughout the physical environment and in strategic places in the organization to keep personnel informed.

The system should provide mechanisms for authorizing users to publish or modifying information. The public display-based system should provide mechanisms for deciding who has permission for publishing or modifying the content of each of the public displays available in the environment.

The system should facilitate the publishing of different types of documents from the users' computer. The public display system should facilitate the publication of different types of information which may be contained in documents of different formats, such as images, text documents, or web files.

The information should be classified in categories. We identified four main categories in which documents can be easily located by users: Academic Events (i.e. Conferences, Courses for professors), Notifications (i.e. notifications of upcoming meeting), Social and Cultural Events, and News (news related to the University Campus).

5 Time-Aware Public Display System

Our solution uses the Perspective Wall visualization technique in order to enable users to easily find the information relevant for their activities. This technique enables the presentation of information which is classified in topics, and it also emphasizes the information that is currently important. This technique has the advantage that it requires the peripheral attention from users. Thus, users may decide to approach the public display to consult a document. In the following sections we explain the design of the user interface based on this visualization technique and then we present the design of the system architecture.

5.1 Design of the User Interface

The public display is divided into three perspectives: left, center and right as illustrated in Figure 4. The center and right perspectives have vertical axes which define three categories (Academic Events, Notifications, and Social and Cultural Events) and horizontal axes that indicate the time in which the information is relevant.

For instance, a Call for papers for a Conference will always appear in the first row (Academic Events) and the column of the current week (July 9-15). This document will also appear in the first column of the right perspective with a notification attached to it. This notification will indicate to the users that there is a deadline on July 30-31. This type of document will appear on the row of the current week until the deadline is reached. The left perspective of the Wall presents dynamic content of recent news of the organization. Finally, the left perspective also enables users to visualize more information about any of the documents they select from the center or right perspectives.

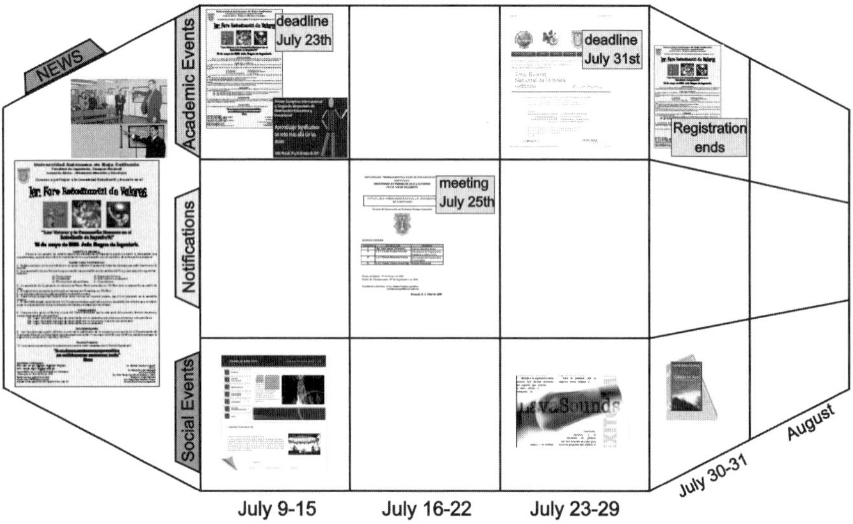

Fig. 4. Time-aware public display

5.2 Architecture of the System

Figure 5 shows that the system consists of several nodes executing components. Some of the system's components were identified as software agents that may act as proxy to a device, or carry out a functionality that must be hidden to the user. These agents autonomously decide how to act based on the information they perceived [3]. The system's nodes and components are the following:

Display Server Node. This is a node containing the components that publish, remove or modify the information presented on the Public Display. It contains a Web Server which enables any authorized user to publish information from any Client with a Web Browser.

Web Server Node. It contains the components for enabling the user to upload a document to the public-display with are the following:

- *Submission Form* which is a web form that the users can invoke from any *Client* with a Web browser. The Submitted Form requests the document to be published and other information related to the document (meta-information) such as the category, the deadline, the URL of the web page in which there is more information related to the document, and the identity of the person who posts the document.

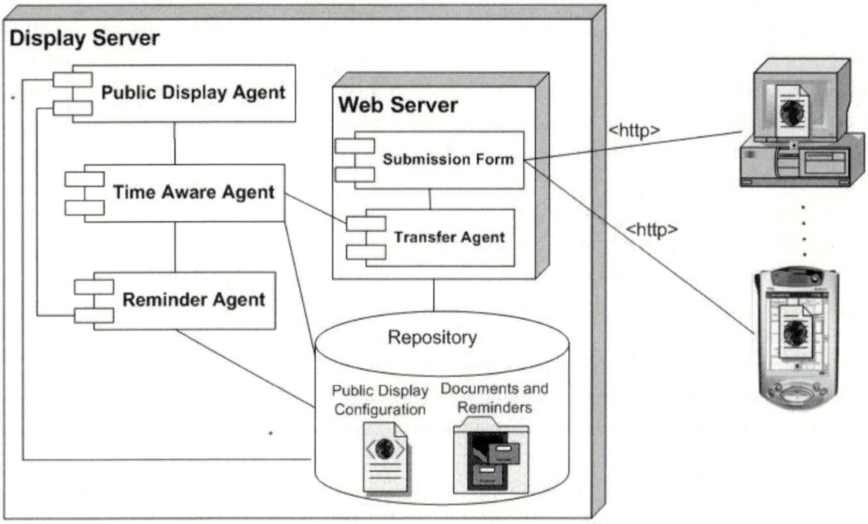

Fig. 5. Architecture of the Time-aware Public Display System

- *Transfer Agent.* This sends the document to the Repository. It also gets the meta-information introduced by the user, and transforms it into an XML format which is delivered to the Time-aware Agent and added to the Public Display Configuration of the Repository component.

Time-aware Agent. It analyzes the document's meta-information in order to identify where to publish the new document (category and date). Then, the Time-aware Agent notifies the Public Display Agent that a new document has to be posted. The Time-aware Agent will be checking when a new document has expired to notify it to the Public Display Agent, and will be aware of documents deadlines in order to ask the Reminder Agent to create a reminder.

Reminder Agent. This agent creates reminders about deadlines. When a new reminder is created, the Reminder Agent saves it in the Repository component and modifies the Public Display Configuration. Then, the Reminder Agent informs the Public Display Agent that there is a new reminder to be published.

Public Display Agent. This agent acts as a proxy to the public display by publishing or removing information from the public display. The Public Display Agent is informed by the Time-aware Agent or the Reminder Agent if there is a new document or reminder to be published which have to be retrieved from the Repository component by the Public Display Agent.

6 Evaluation of the System Design

We have already evaluated the system design with the aim of knowing the users' perception of ease of use and utility of our technological solution. For participating in the evaluation we selected 20 professors. The evaluation consisted in a 2 hours session in which: i) we first found out how and when they consult the organization's message boards, and what type of information they mainly consult; ii) we presented two scenarios of use to illustrate the functionality of the Time-aware Public Display System; iii) finally, we asked questions to know the opinion of the participants with respect to the utility of the system and the facility to locate information relevant for their activities.

All the participants (20) said that they approach a message board at least once a week. Three (3) participants stated they approach a message board looking for information that may be relevant for them, thirteen (13) participants stated that they approach a message board only when a document attracts their attention and four (4) participants mentioned the above two reasons. The information most consulted was the documents posted by Managers and the Principal of the Engineering School (such as notifications of upcoming meetings), following by information about Conferences and Symposiums, and the information less consulted was related with Cultural and Social Events.

The evaluation results of the design of the Time-aware Public Display System showed that the participants considered useful the system. One of the participants commented that: "*as the documents are not posted in all message boards of the organization, sometimes we do not read important information,... the system will enable us to notice when there is important information*"; other participant stated: "*the system will enable us to organize information, and therefore it will also facilitate us to locate important information*". Finally, all the participants considered appropriate to visualize the information categorized (by topics and dates), and that the reminders will enable them to be opportunistically aware of important dates.

7 Conclusions and Future Work

In this paper we propose to support the dissemination and sharing of information in organizations through a Time-aware Public Display System by enabling the organization personnel to be opportunistically aware of the information addressed to them. The evaluation results showed that currently the message boards are mainly used for communicated information from Managers to the organization personnel. Thus, the public display-based system will enable this communication carries out efficiently by enabling the organization personnel to be aware of important information opportunistically. An important system's characteristic is the Perspective Wall [1] visualization technique to attract the attention of users, which requires the peripheral attention from users in order for them to realize that there is new information that may be important for them.

We are planning to incorporate mechanisms that enable users to transfer the information published, to their personal information space. For this, we are envisioning that users can transfer the relevant information they found on the Public Display by using their mobile phones or Personal Digital Assistants.

Acknowledgements

We thank the UABC professors that participated in the case study and the evaluation session. This work was funded by UABC and CONACYT under contract J51582-R.

References

[1] Mackinlay, J.D., Robertson, G.G., Card, S.K.: The Perspective Wall and Context Smootly Integrated. ACM Conference on Human Factors in Computing System (CHI), pp. 174–179 (1999)

[2] Holtzblatt, K., Burns-Wendell, J., Wood, S.: Rapid Contextual Design (Ed), p. 316. Elsevier, New York (2005)

[3] Rodriguez, M.D., Favela, J., Preciado, A., Vizcaino, A.: Agent-based ambient intelligence for healthcare. In: AI Communications, vol. 18(3), pp. 201–216. IOS Press, Amsterdam (2005)

[4] Greenberg, S., Rounding, M.: The Notification Collage. ACM Conference on Human Factors in Computing Systems (CHI), pp. 514–521 (2001)

[5] Huang, E.M., Russel, D.M., Sue, A.E.: IM Here: Public Instant Messaging on Large, Shared Displays for Workgroup Interactions. ACM Conference on Human Factors in Computing Systems (CHI), pp. 279–286 (2004)

[6] Fass, A., Forlizzi, J., Paush, R.: MessyDesk and MessyBoard: two Designs Inspired by the Goal of Improving Human Memory. DIS, pp. 303–311 (2002)

[7] Churchill, E.F., Nelson, L., Denoue, L., Girgensohn, A.: The Plasma Poster Network: Posting Multimedia Content in Public Places. NTERACT, Ninth IFIP TC13 International Conference on Human-Computer Interaction (2003)

A Tactile Emotional Interface for Instant Messenger Chat

Heesook Shin[1], Junyoung Lee[1], Junseok Park[1], Youngjae Kim[2], Hyunjoo Oh[2], and Taehwa Lee[2]

[1] Smart Interface Research Team, Post PC Research Group, Electronics and Telecommunications Research Institute, 161 Gajeong-dong, Yuseong-gu, Daejeon, Korea
[2] Digital Media Lab, Information and Communications University, 517-10 Dogok-dong, Gangnam-gu, Seoul, Korea
{hsshin8,dozob,parkjs}@etri.re.kr, {smart,hoh,thlee}@icu.ac.kr

Abstract. Instant Messaging (IM) is becoming as the dominant means of communication over the Internet. The popularity of IM has led to significant developments in the sophistication of the software. There have been a number of attempts to deliver emotional content through IM. The sense of touch has been a focus in a number of these research projects [1], [2], but there have been no implementations of tactile IM that have successfully bridged the gap to public, widely deployed use. In this paper, we describe a tactile interface designed to convey emotional content and focused on intuitive input and sensorial output methods to be applied in a public IM client. The initial stage of this research involved a fundamental investigation into IM usage patterns. We distributed a questionnaire to gain insight into user's requirements, and received completed results from 79 people in their twenties (32 male and 47 female). Regarding how the tactile stimulus should be integrated into the environment, 50% felt a mouse (26%), a glove (14%), or a keyboard (10%) based display would be appropriate and afford easy physical contact during IM. Furthermore, they expressed a desire for tools to enable them to easily author their own tactile sensations. From the data gathered in this survey we developed a design concept for the physical interface. And we selected a set of emoticons most used in IM and defined both tactile and visual information to help emotional expression for each emoticon. We named these enhanced emoticons TCONs (TouchCON, Touch Emoticon). We defined 6 TCONs and 3 intuitive tangible input methods. We played our TCONs on a physical interface device, called the TCON Display, during public IM. The TCON Display consisted of three components, two devices shaped like cartoon hands, the other like a pair of lips. These devices included various output and input systems: vibrating motors, pin actuators, Heat coil, pressure sensors, buttons and LEDs. We also implemented a TCON Editor to allow users to create TCONs and for testing our research. The TCON was applied to a public IM system, NateOn, just like graphical emoticons. We tested the TCONs and tactile emotional interface system with 12 users test in their twenties (6 male and 6 female) and received a positive response.

Keywords: Instant Messaging (IM), Emotional Interface, Sensorial Interface, Intuitive Input, Tactile Output.

M.J. Smith, G. Salvendy (Eds.): Human Interface, Part II, HCII 2007, LNCS 4558, pp. 166–175, 2007.

1 Introduction

According to a report by the Radicati Group (www.radicati.com), the number of instant messaging (IM) subscribers is increasing 20% per year in contrast with a decrease in the number of e-mail users. Internationally, MSN, AOL and Yahoo messenger are used regularly (conScore Media Media Metrix, 2005.7.), but in the case of the Korean domestic market, NateOn is the most popular IM Client, followed by MSN (koreanclick 2005.10.). In particular, users in their twenties prefer NateOn messenger while MSN is favoured by older users.

We conducted a questionnaire survey of users in their twenties to gain information on the usage patterns and the users requirements for new IM features.. The increase in IM use makes improving the functionality of messenger S/W important. The first IM client, offered by AOL, only supported a text based interface. More recently many IM systems support multimedia contents such as moving icons and avatars, background music, image sharing, and Flash emoticons. It is becoming increasingly common for them to support voice and video chatting, and to use sophisticated voice tone TTS (Text To Speech) technologies. Other, more advanced features such as 'Availabot' which provides more accurate monitoring of the user's availability are also becoming available [3]. In addition, there are several research projects using tactile expressions for emotional communication [1], [4], [5]. Although we can get visual and auditory information with IM easily, interfaces for the tactile expression are still under research. However, a number of authors have suggested that tactile information is valuable for interpersonal communication [1], [2] and is especially important in remote Computer-Mediated-Communication (CMC).

In this research, we focus on tactile expression for an emotional interface in IM and support this with limited visual information shown by variously colored LEDs. We aim to support emotional expression with an intuitive tangible input method.

2 Related Works

Many authors have investigated emotional communication between remote users. *Rivera.* 'Rivera' [6] introduced Emotional Icons to CMC in 1996. In current IM use, emotional icons (in the form of text emoticons like ':)' and ';(') are both commonplace and arguably indispensable.

inTouch. Another system that uses tactile information is 'inTouch' [1] first described in 1997. It introduced the method of applying haptic feedback to interpersonal communication providing a physical haptic link between users separated by distance.

Rovers. 'Rovers' [4], [5] suggested augmenting the communication of text messages with haptic effects and hapticons. The research described a HIM (haptic instant messaging) open framework supporting small chatting communities to explore the design and to use hapticons and haptic IO devices. But it was not available for a public IM system and the IO devices had to be actively controlled by one hand. This made it difficult to be used in the course of a normal text chat.

ExMS. 'ExMS' [7] proposed an avatar based messaging system for asynchronous communication. In this research, graphical avatars express feelings, reactions and moods which relate to the IM. This make the IM chat more similar to face to face communication.

SenToy. 'SenToy' [8] described an interface based on using a person's physical movements to control a physical doll. We think that IM also has an entertainment factor, because most of its regular users view it as an extension of their personal conversions with their friends (koreanclick 2005.10.).

eMuu. 'eMuu' [9] study introduced an embodied character for the expression of emotion. The physical embodiment of character displayed visual and auditory information. However, in our study we included tactile information in emotional expression and therefore the physical embodiment of the TCON Display differs significantly from the eMuu's one.

comTouch. 'comTouch' [10] was designed to enhance the expressive capacity of a communication held between two people on a mobile phone, by the addition of a direct pressure-to-vibration mapping between devices. Squeezing one device led to a vibration on the other. No special vibration patterns were used to express specific meanings, comTouch simply offered another channel to communicate on a mobile phone.

Others. 'Lover's Cup' [11], 'LumiTouch' [12] and even a car, named 'Pod' (Toyota 2002) expressed emotions using color in order to support the intimate relationships between remote users (except 'Pod'). In particular, Ståhl[13]'s study defined different colors, shapes and animations for expressing emotion. When applied to mobile phones, it was shown that an emotional gesture interface fits the area of a circle [14]. Ståhl's research was adapted for our study of the presentation of visual information representing emotion. Many more references attesting to the important of emotion in communication exist in the basic research on human emotion [15], and also the area of affective computing [16] and tangible and sensorial interfaces [17], [18], [19].

3 Tactile Emotional Interface

3.1 Survey of Emoticons in Public IM

User Questionnaire Survey. First of all, we performed a basic investigation into the usage pattern of IM and user's requirement for novel functions. We achieved this through a questionnaire survey of 79 people in their twenties (32 males and 47 females). We briefly summarize the results as follows.

•57% of respondees used the graphical emoticons for the emphasizing their emotion in IM chat.
•Second, in males tending to use graphical emoticons for the representation of mainly positive emotions (48%). On the other hand, women more often expressed negative emotions using graphical Emotions (41%). Moreover, males suggested they would express 'negative (25%)' and 'surprise (25%)' emotions using a tactile physical interface, while woman favoured a winsome (39%)' emotion.
•65% suggested that tactile or touch words were associated to sensing information through the fingertips.

•Fourth, the respondees suggested a glove (14%), a mouse pad (13%), a kind of mouse (13%), a kind of keyboard (9%) and the back of a chair (9%) might be suitable shapes for touch or tactile devices. We concluded that users preferred that the shape of the apparatus by easily available for touch during chatting.
•Fifth, respondees requested that they retain the ability to easily switch the tactile devices on and off. They also indicated that they would like the ability to create and edit the tactile patterns that can be transmitted, but that most of them would not in fact take advantage of this feature. From the questionnaire, we also extracted a short list of the kind of emotions which can be expressed emphatically using tactile interface.

Based on these results, we decided to manufacture the physical interface, called the TCON Display, in the form of keyboard wrist pad (a small device intended to be place directly in front of the keyboard and designed to support the wrists while tying) for display touch stimuli. These took the form of two hand shaped devices. To present visual information, we also designed a separate TCON Display which can be attached to the frame of a computer monitor. This took the form of a pair of lips. To ease user acceptance and provide an intuitive interface we built our devices with a cartoon-like, intentionally 'fun' design ethic. In our study, we separated the graphical emoticons into 5 emotion categories based on the Ekman's research [20] and the special features of IM. We then selected the most used graphical emoticons within each of the emotion categories.

3.2 TCON Definition

We defined the TCONs by mapping the visual and tactile expression for each selected graphical emoticon, as illustrated in figure 1.

Emotion Category	Emotion	TCON						
		Visual Expression	Tactile Expression					
			Vibrating Motor				~~Heat~~ Others	
			Voltage (min:0.1 ~ max:2.8)	Duration (ms)	Delay (ms)	Number of Times	Pin Actuator On/Off	Peltier Display ON/OFF
Positive	Grin		quick, light and regular vibration					
			1.9	100	100	5	off	off
Negative	Cry		slow-moving vibration, hight -> weak -> mid -> weak stress					
			2.6 ->0.3 ->1.9->0.3	500->700 ->900->1100	0	1	off	off
	Anger		one of strong vibration + pin operation					
			2.6	200	0	1	on	off
Surprise	Surprise		one vibration of thick and mid-stress					
			1.9->0.3	800->200	0	Heat	off	off
Winsome	Kiss		being strong gradually and continuously + Peltier display operation					
			0.1->0.8 ->1.9->2.8	800->800 ->800->1500	0	1	off	on
State of Body	Sleepy		slow and regular vibration					
			0.3->1.9 ->0.3->1.9	1500	0	2	off	off

Fig. 1. Definition of TCON

In the initial study, we defined the tactile cues based on our subjective experience and informed by a survey of the available literature [20], [21]. We then conducted an informal user test involving 10 people in their twenties. The goal of this test was to determine whether the tactile cues were suitably expressive and easy to learn [22], [23]. The visual icons were based on the Ståhl's research [13].

We also defined input patterns for our emotional interface system. As our system supports the creation of input patterns using the TCON Editor, we did evaluate these in this study. We took just comments during a user test and a public exhibition. From those comments, we plan to define suitable input patterns for IM. Table 1 shows examples of input patterns using our TCON Display.

Table 1. Examples of input pattern

Input device	Action (sensor)	Mapped TCON
Shape of palms	Several tapping (pressure sensor)	Grin
	One big slapping (pressure sensor)	Anger
Shape of a pair of lips	Touching on the lips (cds)	Kiss

3.3 TCON Display

The TCON Display comes in the form of a pair of devices in the shape of palms and a pair of lips. They include several visual and tactile actuators for expressing TCONs. The each palm has three vibrating motors, four pin actuators and one heat coil. There is also one power button and two control buttons to enable management of the device. They include pressure sensors which act as input devices. The lips were equipped with 8 LEDs and CDS photocells. The LEDs present the visual patterns of TCONs for 7 color (red, green, blue, purple, white, yellow and cyan) and photocells serve as an additional input device. For the input example the 'Kiss' TCON is generated and sent to the remote user by the action of touching on lips.

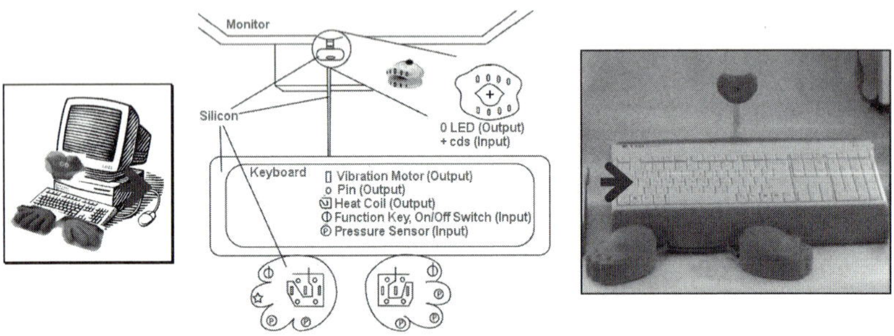

Fig. 2. TCON Display: concept, design and implementation

3.4 TCON Editor

The TCON Editor supports creating, editing and playing of TCONs. It is a software application which provides a simple GUI to support this process. It is implemented on the .Net Framework with C#. The TCON Editor allows users to make tactile/visual patterns and then test them immediately. All information relating to the patterns is saved in an XML based format. Therefore, users can also make new TCONs using a standard XML editor. Also users can define the input pattern using the TCON Editor. We designed the TCON Editor to for creating and playing TCONs easily.

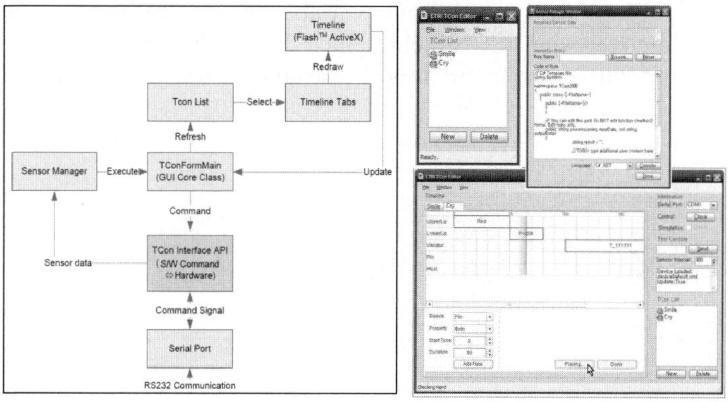

Fig. 3. Aarchitecture of TCON Editor and screen shot

3.5 Emotional Interface System Based on TCON

The system architecture of our tactile emotional interface system is shown in figure 1. Essentially, the TCON Editor creates TCONs, which are then registered in the TCON Library. The TCON I/O Module is a plugin to an IM client, and it sends and receives TCONs to and from users. When the TCON I/O Module receives a TCON, it finds the pattern information in the TCON Library and activates the TCON Display through the Device Control DLL API. We made implemented a version of the system which supports sending and receiving entire TCON patterns thus supporting the independent development of new TCONS. We implemented this system as a plugin to NateOn, a popular IM client in Korea.

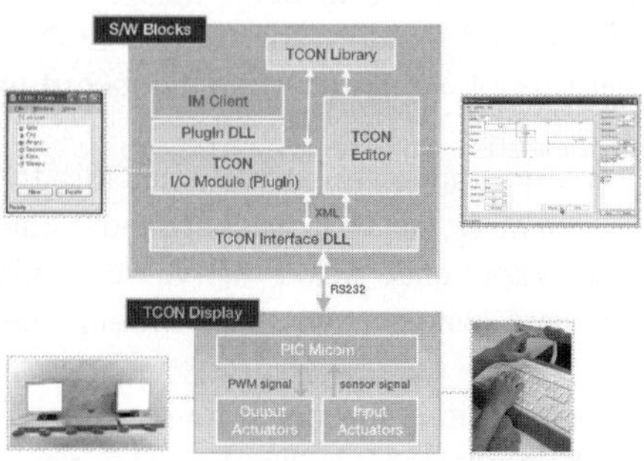

Fig. 4. System architecture of our tactile emotional interface

4 Exhibition and Evaluation

We evaluated our TCONs with 12 people in their twenties (male 6, female 6). We first played participants TCONs then asked them to select one emoticon (from a list of 6) which best matched this. We used no training nor provided explicit information about the tactile and visual cues. In the evaluation, participants correctly identified emoticons an average of 4 times from 6 trials. Second we asked participants to express how effectively the TCONs represented the emoticons. Average results were 1 on a scale from -2 to 2. The figure 5 shows the above two evaluation results.

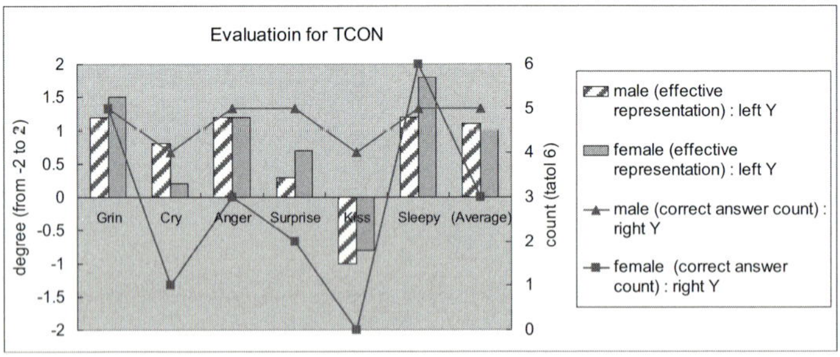

Fig. 5. Evaluation Result for TCON

The two lines indicate first evaluation result (referenced on the right Y axis.) and the bars indicate second result (referenced on the left Y axis.). The results indicate that our 'kiss' TCON does not accurately represent the kiss emotion, but 'sleepy' and 'grin' were highly effective. We also note considerable differences between females and males: females tended to answer with the correct TCON less frequently and feel that the TCON patterns were less effective than males.

Thirdly, we gave a decision making problem to 6 pairs of users, and left them to chat freely for 10 minutes. We observed that all but one couple used the TCONs, and that they tended to do so when they needed to express their emotion actively. Furthermore, the amount of TCON usage increased gradually during the chat session. Although the TCON interface is more difficult than text emoticons because users have to interact with a separate window for using TCONs, the mean number of TCONs used in each chat was 14, while the mean number of text emoticons was only slightly greater at 17.

Finally, we conducted a questionnaire survey after this free usage of the TCON system. We asked how enjoyable the chat was, and how emotional, how fresh and new and how absorbed the experience was. Finally, we inquired how easy it was to feel the tactile stimulus and whether we had chosen suitable patterns for the expression of emotion.

Participants gave the most positive responses for fresh and new chat and enjoyable chat (-2 is the most negative grade and 2 is the most positive grade.). We also observe that males were more satisfied with the experience than women.

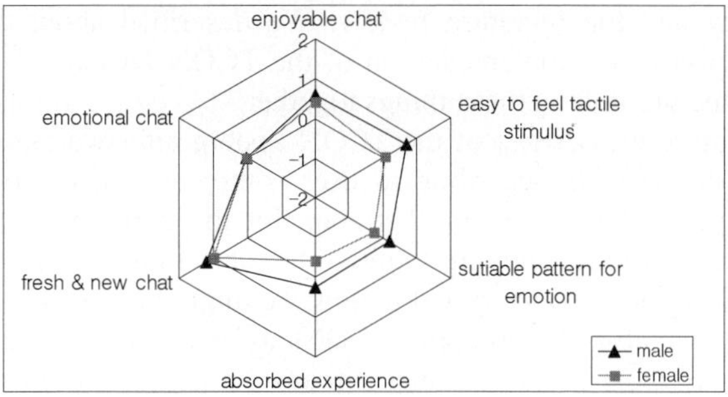

Fig. 6. Questionare Result

Our system was exhibited in the Next Generation Computing Exhibition held in November 2006 in Seoul, Korea. From the exhibition, we recieved many comments from general users and conducted simple interviews with people who used the TCON system.

Most people commented it was interesting system. We observed a significant difference in the responses according to the age of the users. Teenage and those in their twenties really liked our system, while those over the age of forty reacted negatively. When users took the time to creat their own TCONs using the TCON Editor, they seemed especially satisfied and pleased with the system.

It seems unlikely that the same tactile pattern can convey the same emotional information to all people. We expect that there are individual differences in presenting and accepting of human emotion through this kind of channel. Consequently, our system validates the idea that the tactile and color patterns can be combined cohesively to support the expression of emotion in a more active personal manner. We believe that intuitive and sensorial communication based on tactile and abstract color information can be a powerful emotional interface to support remote users as they chat on IM.

Fig. 7. Exhibition in the Next Generation Computing Exhibition

5 Conclusion and Progressing Works

In this paper, we proposed a tactile emotional interface using TCONs delivered over an IM system. We surveyed and analyzed useful emotional expressions in IM based on a user survey and the literature review. We described about the definition of TCONs and discussed the implementation of the TCON Display, TCON Editor and the system architecture linking these things together.

We plan a long term user test of the TCON system after we optimize the TCON Display, refine the TCONs according to user comments we received and make a simpler UI for the TCON Editor. Furthermore, for the purpose of immersion in the emotional interface, we are studying intuitive tangible input methods using physical body movement. We are also interested in how we might be able to evaluate the input methods. Finally, we also plan to apply TCONs to mobile devices, and in particular, mobile phones. As most mobile phones already have a vibration motor and color display, we can apply our TCON emotional interface more easily.

Acknowledgments. This work is supported by MIC (Ministry of Information and Communication, Republic of Korea) and ETRI (Electronics and Telecommunication Research Institute) through the "Development of Smart Haptic Interface" project. We extend our thanks to all members of the smart interface research team. We also would like to acknowledge the contribution of ICU Media Lab students for work on the initial user survey and implementation of the TCON Display.

References

1. Brave, S., Dahley, A.: inTouch: A Medium for Haptic Interpersonal Communication. ACM, Computer-Human Interaction, pp. 363–364 (1997)
2. Robles, G.: The Importance of the Sense of Touch in Virtual and Real Environments. IEEE, Haptic User Interfaces for Multimedia systems (2006)
3. Availabot, http://schulzeandwebb.com/2006/availabot/
4. Rovers, A.F., Van Essen, H.A.: HIM: A Framework for Haptic Instant Messaging. ACM, Computer-Human Interaction, pp. 1313–1316 (2004)
5. Rovers, A.F., Van Essen, H.A.: Design and evaluation of Hapticons for enriched Instant Messaging. Eurohaptics (2004)
6. Rivera, K., Cooke, N.J., Bauhs, J.A.: The Effects of Emotional Icons On Remote Communication. ACM, Computer-Human Interaction, pp. 99–100 (1996)
7. Persson, P.: ExMS: an Animated and Avatar-based Messaging System for Expressive Peer Communication. GROUP (2003)
8. Paiva, A., Costa, M., Chaves, R., Piedade, M., Mourão, D., Sobral, D., Höök, K., Andersson, G., Bullock, A.: SenToy: an affective sympathetic interface. International Journal of Human-Computer Studies (2003)
9. Bartneck, C.: eMuu - an embodied emotional character for the ambient intelligent home. Ph.D. thesis, Eindhoven University of Technology, Eindhoven (2002)
10. Chang, A.: ComTouch: A Vibrotactile Mobile Communication Device. M.D. thesis, Media Arts and Sciences, MIT (2002)
11. Chung, H., Lee, C.J., Selker, T.: Lover's cup: drinking interfaces as new communication channels. ACM, Computer-Human Interaction, pp. 375–380 (2006)
12. Chang, A., Resner, B., Koerner, B., Wang, X., Ishii, H.: LumiTouch: An Emotional Communication Device. ACM, Computer-Human Interaction, pp. 313–314 (2001)
13. Ståhl, A., Höök, K., Sundström, P.: A Foundation for Emotional Expressivity. International Journal of Business and Design (2006)

14. Sundström, P., Ståhl, A., Höök, K.: eMoto-Affectively Involving both Body and Mind. ACM, Computer-Human Interaction (2005)
15. Ekman, P.: Facial expression of emotion: New findings, new questions. Psychological Science 3, 34–38 (1992)
16. Picard, R.W.: Affective Computing. MIT Press, Cambridge (1997)
17. Ishii, H., Ullmer, B.: Tangible Bits: Towards Seamless Interfaces between People, Bits and Atoms. ACM, Computer-Human Interaction, pp. 234–241 (1997)
18. Bonanni, L., Lieberman, J., Vaucelle, C., Zuckerman, O.: PlayPals: Tangible Interfaces for Remote Communication and Play. ACM, Computer-Human Interaction, pp. 574–579 (2006)
19. Chang, A., Ishii, H.: Sensorial Interfaces. DIS, pp. 55–59 (2006)
20. Kenji, S., Shuji, H.: FeelLight: A Communication Device for Distant Nonverbal Exchange. In: Proceedings of the 2004 ACM SIGMM workshop on Effective Telepresence, pp. 40–44 (2004)
21. Deepa, M.: vSmileys: Imaging Emotions through Vibration Patterns. Feelings & Games'05 (2005)
22. Jung, C., Paek, J., Hahn, M.: Emphasizing Emotional Expression through Vibration Patterns in Online Environments. HCI Korea 1, 411–416 (2007)
23. Shin, H., Lee, J., Park, J., Oh, H., Kim, Y., Lee, T.: Emotional Interface on the Internet Messenger. HCI Korea 1, 875–880 (2007)

A Theoretical Framework of Co-purposing in Systems Design

Modestos Stavrakis, Nikos Viorres, Panayiotis Koutsabasis, and John Darzentas

University of the Aegean,
Department of Product and Systems Design Engineering,
GR 84100, Hermoupolis, Greece
{modestos,nviorres,kgp,idarz}@aegean.gr

Abstract. The paper outlines a theoretical framework for the conceptualisation and design of collaborative systems for design organisations. The intent is to examine and support the idea of 'co-purposing' as an influential design metaphor for the analysis of design teams and the development of design support collaborative systems in contexts where systems' designers collaborate with their clients. By offering associations to the work activities and the social interaction of designers in actual design workspaces, we propose that the metaphor of 'co-purposing' can stand as an alternative perspective in the analysis and design of potentially new realisations of computer supported cooperative work systems for supporting teamwork.

Keywords: co-purposing, collaborative systems, e-collaboration, postmodern systems design.

1 Introduction

In the light of postmodern organisation theory, discourses that have historically shaped the status of organisation analysis, namely romanticism and modernism, is believed that they have lost their cogency [10]. According to the philosophical shift of the late 20th century (late modernists and postmodernists positions), the essence of theoretical frameworks is not based on their foundational concepts but on their intelligibility which provides the grounds for their usefulness. In this sense, postmodern organisations, are seen as flat hierarchies, based on decentralised decision making processes with greater capacity and tolerance for ambiguity, permeable boundaries, capable of renewal, self-organising constitutions with self-integrating coordination mechanisms [6]. The viability of these often geographically, culturally and organisationally dispersed teams is well depended on the use of information and communication technologies, but not simply driven by them in an instrumental sense [14]. The reliance on perspectivistic viewpoints of analysis based on insulated ontological observations by external (systems) designers frequently resulted on the fragmental support of collaboration processes and usually assisted to the development of systems that weaken communication couplings among teams members. A richer, human to human interaction within a socio-cultural environment, mediated by

M.J. Smith, G. Salvendy (Eds.): Human Interface, Part II, HCII 2007, LNCS 4558, pp. 176–184, 2007.

technology is needed and it must be approached with the partnership of organisation members and information and collaboration systems designers in mind.

This paper analyses the current situation in systems design tradition for collaboration, from a theoretical perspective of discursive practice[1] related to the interfacing of both modern and postmodern practices and theoretical viewpoints. It focuses to pose forward the importance of an integrated systems design methodology where systems designers and their clients (possibly other designers in different contexts) collaborate.

It attempts to contribute an understanding in the following key issues:

a) In the second section we provide the theoretical basis of our arguments. We analyse earlier definitions of 'communication', 'coordination', 'cooperation' and we consider their strengths and weaknesses through the prism of our research investigation.

b) Section three outlines the design framework that provides research directions to system designers for the future development of information technologies for the support of design organisations. Based on the metaphor of 'co-purposing' this framework provides the aspects of design considerations that system designers can possibly elaborate for the conceptualization, design and evaluation of collaborative technologies.

e) Finally, we conclude by discussing the findings of this work. These indicate that the proposed framework of 'co-purposing' offers an alternative perspective for supporting the conceptual design of collaborative systems.

2 Communication, Coordination, Cooperation and Collaboration

Communication, coordination, cooperation and collaboration are notions widely used in literature of information systems development and communication studies with a variety of meanings often contradicting or unjustified. This section provides a short review of current literature and outlines the concepts in an appropriate way for the needs of this work.

2.1 Communication

Communication in information systems development is believed to provide the basis for interaction. Collaboration and social interaction is considered to occur through communication between human agents while technological advances in telecommunications, the internet and information systems development appear to have profound effects in mediating communication and collaboration. Previous approaches in defining communication are considered as "the command and control paradigm" or "simple linear models", where communication is thought possible when there is exchange of information and or data between a sender and a receiver via a communication medium [15]. Formally, this happens when the sender is informed that the receiver has received the message. According to Mumby [16] "such a

[1] The term ***discourse*** is used here in Foucault's sense of a system of possibilities for the creation of knowledge [9].

representational model is unable to conceive of the possibility that communication is anything other than an empty conduit for effectively communication an already existing set of conditions…The act of communication and the world about which one is communication remain firmly separated" [16].

On the other hand the interpetivist approach to communication (or pragmatic approach) place subject and object in a productive dialectical tension. Communication in this collaboration paradigm entails the mutual understanding of the meaning of messages that have been exchanged among participants during the course of a communicative event. These models recognize the importance of context and social negotiation of meaning, and have been established in explaining the complexities of communication. This point of view is related to the early Habermasian [11] argument that any attempt at communication or action presupposes a horizon of meanings shared with others.

Although the interpretivist approach to communication is focusing in the analysis of the participatory construction of a meaningful world through communicative practices, the critical project and the theorists of Frankfurt school [13] are characterised by an articulation of hermeneutical suspicion. In his late writings Habermas argues that the acknowledgement of the possibility of success lies within communicative failures through the implicit recognition of shared criteria constitutive of our communicative practices. By following a pragmatic tradition, he speaks of "universal and unconstrained consensus"; the possibility of an ideal speech situation of collective comprehension [12]. This is a kind of mutual pre-understanding of the criteria for successful understanding, shared by human beings even in contexts of misunderstanding, describe a social constructionist view of the world but criticise the interpretivist inability "to explore issues of power and ideology and the processes through which certain realities are privileged over others" [12].

By contrast, postmodern thinkers disagree with those who adopt ideals of community based on an ideal form of communication, and support the idea that any successful communication contains the alternative possibility of its own failure. This possibility of failure is the condition of any communicative event – makes communication possible. This failure of communication embedded in communication calls for the recognition that we might not fully share criteria for success, or might make a mistake about what is shared. Contrary to Habermas who supports the idea that mutual understanding is implied in day to day language and actions, Derrida argues that this ideal is impossible [7]. The misunderstanding is integral in communication and action; participants in a communicative event are in a state of continuous *aporia* between successful and not successful communications. This idea of a continuous flux in meaning creation is based on "speech acts" that take place in relation to what does not 'belong' to 'speech acts". Every context of communication is enfolded by another context, so that the context is never stable, without a definitive fixing of its meaning.

In the case of communication in design teams the characteristics of the context are in a continuous negotiation and are thought to be provided by the actors themselves (beings), the situation governing the status of the actors' behaviour (social) including the medium/vehicle of communication (technology), the cultural environment, language etc. For the purposes of this paper we will follow a discussion among the four paradigms mentioned above and begin the analysis by the employment of the three ontologically observed characteristics of a system of interaction [Fig. 1.] within the communication discourse that are more frequently used in information and collaboration systems' design: being (participants, actors or agents), social environment (community, organisation or team) and communication medium (technology: communication and collaboration systems, language).

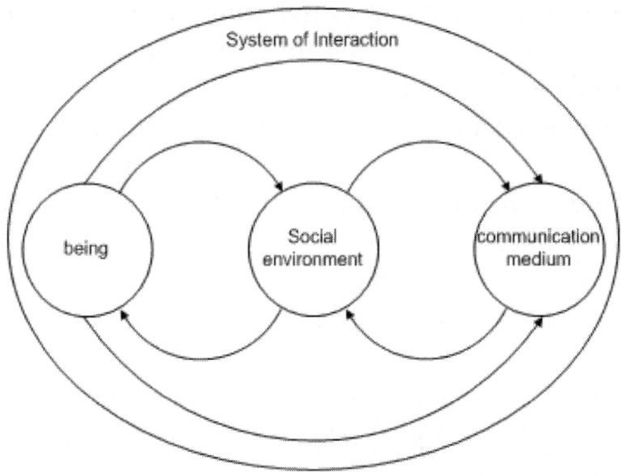

Fig. 1. System of interaction in the communication discourse

2.2 Coordination

Coordination in group work, is considered as a regulation process; the normal and routine flow of interaction where individuals are following their scripted roles. These scripts are thought as parts of product or process design documents and are coded in written rules, plans and schedules or in tacitly assumed traditions and norms [4]. The differentiation processes in organisation contexts, i.e. division of the work between specialists, groups, departments, are coupled by coordination processes to adjust the activities of individuals and groups to common goals. Andriessen's [2] deconstructive approach in defining coordination involves the possibility that participants, who work together under the framework of a common goal, do not necessarily share the same ideas and personal goals. He argues that *"...this common goal does not exclude that the individuals involved may also have different goals and/or conflicts. If however there is no commonality whatsoever, coordination is not needed."*

Often, the commonality of goals imply interdependency of activities; a notion that is considered central to coordination. This degree of task interdependence refers to the degree to which participants in a design context, rely on each other to carry out their individual responsibilities successfully. This degree of interdependence regulates the degree of coordination; the higher the task interdependence the higher the need for coordination [2].

Coordination is considered a central concept in cooperation/collaboration studies and often is used to define them. In this sense Thompson [17] supports the idea that task interdependence is one variable that affects team performance in terms of coordination. Type of interdependence is another. According to his work three types of interdependence can be identified:

1. *Pooled* interdependence (discrete roles, no interdependence of tasks, no common goal, externally coordinated): in this type, each participant provides a discrete contribution to the design team by collating or pooling its obtained experience (information and knowledge). No need for direct individual interaction is needed. A typical example of this is a group of individual labourers who produce work and are coordinated by a supervisor responsible for adjusting the contribution of each members output to the task of the group.
2. *Sequential* interdependence (discrete roles, interdepended, no common goal, externally coordinated): in this type, the product or labour work of a participant is highly dependent upon the output of another. A classic example of this type is the work on an assembly line.
3. *Mutual or reciprocal* interdependence (discrete and specific roles, interdependence, common goal, internal or external coordination): in this type, participants set critical contingencies for each other that have to be determined before taking action. Participants have different but specific roles and interact in a meaningful way for a common goal, e.g. a surgical team. This is often described as a *collaborative* type of interaction [3].

Team or group interdependence (roles are unspecified-emergent, interdependence, common goal, self-coordinated) defined as a fourth type of interdependence by Van de Ven [18] in order to include the possibility where participants with similar or not discrete roles, jointly cooperate in order to accomplish tasks. In this case the organisation or team self-organises and autonomously defines its course of action. This is a typical example of a design team trying to complete a design goal or project and often is described as a *cooperative* type of interaction [3].

2.3 Cooperation and Collaboration

The 'collaboration' versus 'cooperation' debate is more complex compared to the distinction that Bair [3] provided in terms of coordination. Many researchers use these terms interchangeably and one can without doubt identify that there is an obvious disagreement amongst the authors themselves [8][2]. For the purposes of this work and in acknowledgement of the distinctions that others in the field have made, we provide definitions for the terms *cooperation* and *collaboration* that take into account previous approaches and also expand the meaning of both terms in an attempt to involve the paradigm of postmodern thinking that has been analysed in the previous section of communication analysis.

Therefore, we describe both terms within the limits of the three ontologically observed characteristics of the system of interaction in communication discourse outlined earlier. These can include:

1. Type of participants' roles: specific, discrete, emergent
2. Type of symbiosis: synergy, interaction, relations, interdependence, association
3. Group characteristics: size, proximity of members, virtuality
4. Division of labour
5. Type of teleological purpose (goal, aim, end)

6. Type of management/organisation (coordinated, autonomous, self-organised, indeterminate)
7. Type of environment (passive, competitive)

Cooperation is considered as a means to an end that involves gains and losses on the part of each participant (being a designer or an organization). This type of symbiosis is defined in terms of a continuous interaction and interdependence, within a contingent environment. This behaviour can sometimes encourage a competitive environment, and parties need not necessarily to carry a relationship beyond the accomplishment of the task at hand. The goal is considered static and is imposed by a manager/coordinator. In cooperation there is not a strong sense of team integrity beyond the goal but participants are aware of the interdependence that emerges because of the common goal that is imposed externally.

Collaboration is considered in terms of autonomy and self-organisation. Similarly to cooperation this type of symbiosis is described in terms of continuous interaction and interdependence, within a contingent environment. It is important that encouragement of competitive environment is considered necessary but its competitiveness is continually redefined in terms of the team coherence. There is not an external goal to be accomplished but all participants are focusing in carrying on interacting with other participants in order to secure their own participation in the network and as a result the autonomy of the team. Moreover the structural couplings that are produced during the course of the interaction provide the ground for the emergence of temporal goals that need to be accomplished. In this context a strong sense of team integrity and situation awareness is produced.

3 Analysis of Theoretical Framework

Our theoretical standpoint originates from the four characterisations (conservative, pragmatic, critical and radical) that Coyne [5] introduced for thinking about information technology in the postmodern era. Similar discourses have also been identified by Mumby [16] and Anderson et al. [1] in communication studies. These discourses [Fig. 2.] are described in terms of the aforementioned system of interaction (being, social environment, medium of communication) within the communication discourse [Fig. 1.].

Conservative is based on ontological representations and suggests the uncovering of original meanings that are placed in a real world. It is found on the Cartesian legacy of the separation of subject (researcher) from object (inquiry/knowledge). In this sense, there are 1) technological characteristics (data, information, knowledge, communication etc) that are quantifiable entities and can be measured and manipulated by technological means, 2) human activities, behaviour and needs that are determinable qualities while 3) social engagement remains a matter of causal explanations. For the purposes of this framework, the conservative discourse is useful for re-establishing the objects of our concern (e.g being, social environment, communication medium). The usefulness of conservative discourse is seen only within an iterative/recursive process of exploitation of the framework (see pragmatic).

Pragmatic discourse is not exclusive to the first but sees this separation of object and subject emerging from a continuous productive dialectical tension. It focuses in describing how the ontological concepts that are found in the conservative mode, interactively work together and tries to identify their characteristics within the system of interaction of communication. The observation of these characteristics (participants' roles, type of symbiosis, group characteristics, division of labour, purpose, etc.) provides feedback to both the conservative mode and the subsequent critical mode of analysis.

Critical is interested primarily in analysing the ways that observers in the pragmatic mode, co-construct meaningful contexts of communicative practices and focuses in exercising scepticism about the outcome. This is a reflexive mode of "suspicion" for the arguments of the previous modes of analysis. This position adopts a critical posture towards the explanation of issues of power, ideology, and marginalisation of ideas found in the conservative and pragmatic discourses by the observers. This mode of analysis focuses in identifying the way that the metaphor which is under conceptualisation in this framework, is to be treated with caution by the information systems designers who formalise it in design scenarios.

Radical discourse unsettles any claims to the importance of the proposed metaphor by showing that any ideas developed about the centrality of the metaphor might turn out to demonstrate the opposite. This "discourse of vulnerability" [16] represents the crisis of the validity of the framework itself by questioning all of its previous presuppositions. It provides systems designers with the idea that the "emancipation" that is offered by this very framework is elusive. It treats it as a text in the Derridian sense, where an endless play of signs takes place [7]. This framework is revealed and concealed through the incessant play of difference and contradiction.

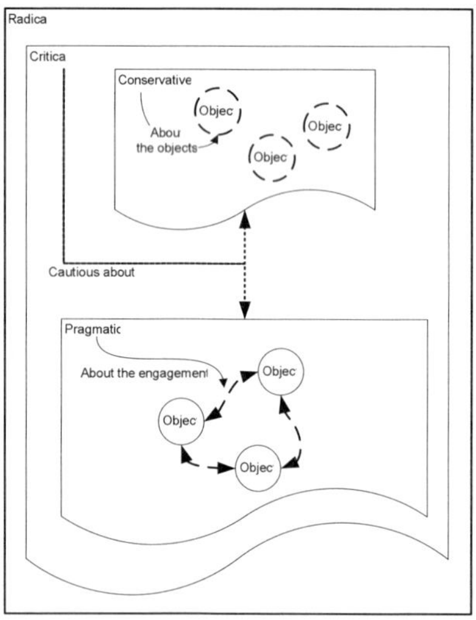

Fig. 2. The interaction of the four theoretical characterisations that Coyne [5] introduced in information technology theory

3.1 Co-purposing

At this point we define this interplay of ontological observations for knowledge acquisition, practice, critical apprehension and representational deconstruction, with the notion of "co-purposing" and we locate it among the communal practices, of communication, coordination, cooperation and collaboration that have been outlined previously and take place in all levels of systems design. According to the argument that this paper develops, 'co-purposing', as a metaphorical system of reference, is thought to offer mobility of the design context, by supporting the exchange of contextual awareness of work activities among the design participants and the information and collaboration systems designers, and thus provide the possibility of the design of a flexible setting for collaborative work. 'Co-purposing' itself is often mediated by information and communication technology tools and can be related to an organisational context of human activity within a socio-cultural environment.

The notion of 'co-purposing' is not considered as a static belief of instrumental communication or data manipulation or even a rigid systemic construct, but is seen as a continuous, social interaction and reflexive production of meanings among designers of various levels, including systems designers and their clients. These design participants continually strive to purposefully collaborate by constructing and deconstructing (differentiate and defer) meaningful structures of systems designs and design methodologies that they produce throughout their interaction.

4 Conclusion

This paper presented an empirical approach of analysing design organisations and it is though to operate by its continuous redefinition. The language which is produced by the use of its empirical process does not equate with an increasingly accurate correspondence of reality. Instead, it represents a process of professional self-justification. According to this postmodern view, research and design analysis proceeds on the basis of discourses which are already shared within particular systems design community. The evidence that is produced by the use of the aforementioned framework is interpreted and justified only within a restricted domain of linguistic constancy and it is possible to provide an inspiring methodological supplement for the design of information systems.

Acknowledgments. This work has been partially supported by the EUREKA F-JEWEL project.

References

1. Anderson, J.A., Baym, G.: Philosophies and Philosophic Issues in Communication, 1995-2004. Journal of Communication (Blackwell Synergy) 54(4), 589–615 (2004)
2. Andriessen, E.: Working with Groupware. In: Understanding and Evaluating Collaboration Technology, Springer Verlag, London (2002)

3. Bair, J.H.: Supporting cooperative work with computers: addressing meetingmania. In: Thirty-Fourth IEEE Computer Society International Conference: Intellectual Leverage, Digest of Papers. IEEE (1989)
4. Bardram, J.: Designing for the Dynamics of Cooperative Work Activities. In: Proceedings of the 1998 ACM conference on Computer supported cooperative work, pp. 89–98. ACM, Seattle, Washington, United States (1998)
5. Coyne, R.: Designing Information Technology in the Postmodern Age. MIT Press, Cambridge, Massachusetts (1995)
6. Daft, R.L., Lewin, A.Y.: Where Are the Theories for the'New'Organizational Forms? Organizational Science, vol. 4(4) (1993)
7. Derrida, J.: Writing and Difference, Routledge London (1978)
8. Dillenbourg, P.M., Baker, A., Blaye, A., O'Malley, C.: The evolution of research on collaborative learning. In: Learning in Humans and Machine: Towards an interdisciplinary learning science, pp. 189–211. Elsevier, North-Holland (1996)
9. Foucault, M.: Power/knowledge: Selected interviews and other writings, pp. 1972–1977 (Gordon, C., Marshall, L., Mepham, J., Soper, K., Trans.) New York, Pantheon (1980)
10. Gergen, K.J.: Organization Theory in the Postmodern Era. Rethinking Organization (Sage), pp. 207–226 (1992)
11. Habermas, J.: Communication and the Evolution of Society. Beacon Press, Boston (1979)
12. Habermas, J.: The theory of communicative action. Beacon Press, Boston (1987)
13. Horkheimer, M., Adorno, T.: Dialectic ofenlightenment (Cumming, J., Trans.) Continuum, New York (1988)
14. Leonardi, P.: Technology and Work in the Virtual. Encyclopedia of information science and technology (Idea Group), pp. 2753–2756 (2005)
15. McQuail, D.: Communication [Book]. Longman, London (1975)
16. Mumby, D.K.: Modernism, postmodernism, and communication studies: A rereading of an ongoing debate. Communication Theory (Blackwell Synergy) 7(1), 1–27 (1997)
17. Thompson, J.D.: Organisations in action. McGraw-Hill, New York (1967)
18. Van de Ven, A.H., Delbecq, A.L., Koenig, R.: Determinants of coordination modes within organizations. American Sociological Review 41(2), 322–338 (1976)

Toward Adaptive Interaction – The Effect of Ambient Sounds in an Ultra-Realistic Communication System

Noriko Suzuki, Ichiro Umata, Tatsuya Kitamura,
Hiroshi Ando, and Naomi Inoue

National Institute of Information and Communication Technology/
ATR Cognitive Information Science Laboratories
2-2-2 Hikaridai, Seika-cho, Soraku-gun
Kyoto, 619-0288, Japan
{noriko,umata,kitamura,h-ando,na-inoue}@nict.go.jp

Abstract. Our research focuses on *ambient sounds* as a key to implementing an ambient atmosphere of people, that is, the feeling of another's proximity or presence. Such knowledge is sought for constructing an ultra-realistic communication system with shared reality. This study examines whether people's impressions and behaviors are affected by (a) the existence or (b) non-existence of ambient sounds. A pilot study is conducted by using the ambient sounds of a newspaper's pages being turned and a glass being placed on a table in an adjacent room. No difference in the subjects' surface behavior between the two ambient-sound conditions is observed explicitly. However, the preliminary results suggest that participants feel an ambient atmosphere of people in the next room when the ambient sounds are output. They also suggest that the participants become nervous as they feel the presence of other people. This study demonstrates the importance of investigating the fundamentals of an ambient atmosphere in interaction via an ultra-realistic communication system.

Keywords: Ultra-realistic communication system, Ambient atmosphere, Ambient sound, behavioral evaluation, psychological evaluation.

1 Introduction

Many people have had the experience of visiting the haunted house or fun house at an amusement park. The Haunted Mansion at Disneyland is a typical example of a walk-through ghost house. In these haunted houses, why do people feel so scared of ghosts? Why do people feel that ghosts actually exist around them? Lighting, sounds and objects produce the perceived presence of ghosts in a haunted house. Sounds in particular seem to play an important role in facilitating the perceived presence of ghosts.

The authors have joined the Ultra-realistic Communication System Group of the National Institute of Information and Communication of Japan (NiCT). The purpose of the project is to construct a system for embodied interaction with remote partners by improving the reality of interaction through multi-modality techniques, i.e. using

M.J. Smith, G. Salvendy (Eds.): Human Interface, Part II, HCII 2007, LNCS 4558, pp. 185–194, 2007.

3-D video, 3-D audio, olfaction and the tactile sense. As a way to improve the reality of interaction, our research focuses on the effect of ambient atmospheres or existences of people, especially the ambient sounds that people make.

Most conventional systems for supporting communication with remote partners have been developed to provide the physical presence of people [1-3]. Some of these studies have utilized both verbal cues like speech and non-verbal cues like gestures to enhance the presence of remote partners [2, 3]. On the other hand, in face-to-face interaction, people might consciously or subconsciously perceive an ambient atmosphere of partners by listening directly to ambient sounds such as the noises that partners make as a result of interaction between a human body and objects or noises from a human body itself. Ambient sounds seem to play an important role in creating a shared reality among participants. However, little research has focused on the ambient sounds made by people.

This paper examines whether ambient sounds affect both the psychological impressions and behaviors of people, with the aim of demonstrating the effect of ambient sounds on creating a shared reality. A pilot study was conducted with a between-participants design under two different sound conditions: (a) existence and (b) non-existence of ambient sounds. There are many kinds of ambient sounds that people make in everyday life, e.g. footsteps, the rustle of clothing, and breathing. In this pilot study, we examined two types of ambient sounds: a newspaper's pages being turned and a glass being placed on a table. We divided the participants into two groups according to the two sound conditions. They were asked to wait in an experimental room with or without ambient sounds from the next room. Participants answered a personality test and a post-questionnaire about their impressions of the experimental room's atmosphere before and after the experiment, respectively. In addition, their surface behavior was observed during the waiting period. The preliminary results were analyzed, and three factors were examined: (i) the difference in behaviors between the two sounds conditions, (ii) the difference in psychological impressions of the atmosphere between the two conditions, and (iii) the correlation between psychological impressions and the personalities of participants.

2 Related Works

Acoustics

In the research field of acoustics, the focus of studies on ambient sounds has mainly been their induced effect of spatial attention in visually impaired people [4, 5]. These research efforts have studied the relationship between the reflected sound localization and the sense of obstacles held by such people; it was found that they seem to recognize the type of obstacle or their distance from the obstacle by using the reflected sounds that they make themselves. Moreover, by using both acoustics and virtual reality techniques, a spatial navigation system has been developed for the visually impaired. These conventional studies thus examined the relationship between the reflected sounds from an obstacle and the existence of an obstacle. However, little attention has been given to the relationship between ambient sounds that people make and the presence of people.

Ambient interface

In the research field of ambient interfaces, most conventional studies have tried to adapt the sound presence of people to other media, e.g. the lighting or color of an object including a lamp or a wall. Some studies have regarded ambient sounds as a subtle cue for calling attention to the physical presence of others, such as a hamster or people [6-8]. In these studies, the sound presence of creatures was modified to other kinds of ambient sounds, such as the natural soundscape or the sound of material. The level or the intensity of ambient sounds has been controlled according to the change in the level of the original sound source or the number of people. However, little attention has been given to the direct utilization of the original sound presence of people.

3 Pilot Study

Our research aim is to construct an ultra-realistic communication system featuring a shared reality. As a method for establishing a shared reality among people at multiple points, we focus on the ambient atmosphere arising from communication partners. In face-to-face communication, people can consciously or subconsciously perceive the ambient atmosphere of communication partners by listening to the ambient sounds that people make as a result of interaction between the human body and objects or from the human body itself. We believe that ambient sounds can play an important role in facilitating a shared reality among participants through an ultra-realistic communication system. As a first step, we conducted a pilot study to examine the effect of ambient sounds on the psychological impressions and behavior of people when they hear such sounds from an adjacent room.

Types of ambient sounds

People make many kinds of ambient sounds, such as footsteps, the rustle of clothing, and breathing. In this pilot study, we selected two kinds of ambient sounds that a human makes while sitting in a chair in the course of everyday life: turning the pages of a newspaper and placing a glass on the table.

Recording of ambient sounds

Figure 1 shows the recording environment for the above two ambient sounds. These sounds were recorded in an anechoic room via a microphone (NEUMANN U87A) and a microphone pre-amplifier (AMEK System 9098) to a DV recorder (SONY DSR-45) (sampling rate 44.1 kHz, 16 bit, monaural). The sound spectrogram of these ambient sounds, especially turning the newspaper's pages, spread out over a wide frequency bandwidth. Ambient sounds are output for five minutes, and they include 15 instances of page turning and 12 instances of placing the glass on the table.

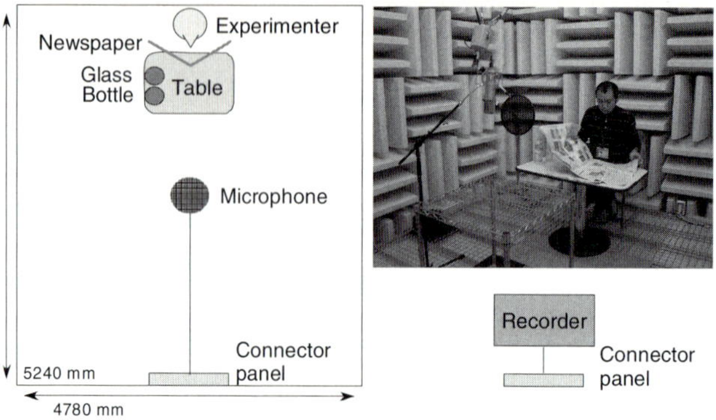

Fig. 1. Recording environment for ambient sounds: wiring diagram and appearance of the anechoic room

Experimental environment

Figure 2 shows the experimental environment for the pilot study. A soundproof room was used as the experimental room. This room was divided into two areas by partitions: a waiting room for the participants and an area used for playing ambient sounds from the speakers. Ambient sounds were played from two speaker sets: high-frequency sounds from JBL Studio Monitor 4410 and low-frequency sounds from BAG END 115 Cabinet via power amplifiers AMCROM D75A and AMCROM D-150A Series II, respectively. The sound source was DV tape and the player was the SONY DSR-45 DV player/recorder. The door of the experimental room was open, but it was difficult for the participants to see the status of the next room across the partitions.

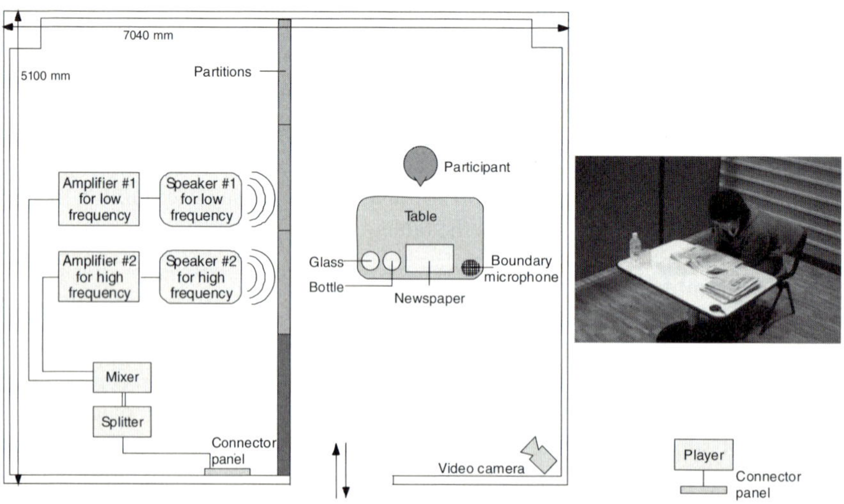

Fig. 2. Experimental environment for the pilot study: wiring diagram and appearance of the experiment

Participants
The participants in the pilot study were 22 people aged from 21 to 36 years old. They were told simply that they would be participating in a study on human tendencies in an ultra-realistic communication system. They were divided into two groups according to the conditions described below.

Procedure
The pilot study was conducted with the two ambient-sound conditions in a between-participants design. On arriving at the laboratory, each participant was asked to answer the Big Five Personality Test. Participants were brought to the experimental room and instructed to sit down and wait in a chair until the experiment was prepared. Some newspapers, a bottle of water and a glass were put on the table in front of each participant. The experimenter explained why these objects were placed there when the participants asked about them. The participants waited for five minutes. After that, they were asked to answer a post-experimental questionnaire about their impressions of the waiting time. At the end of the experiment, the experimenters also interviewed them. In addition, their behavior was observed during the waiting period with a video camera and a boundary microphone.

Conditions
Two conditions were used for ambient sounds, and the participants were divided into two groups based on these conditions.
EAS: Existence of ambient sounds from two speakers in the next room.
NAS: Non-existence of ambient sounds from the next room.

Measures
(i) Personality evaluation: In a Big Five Personality Test conducted before the experiment, the participants were asked to answer 70 questions about their personality via a computer. The score of each basic personality dimension was calculated from the results of the test. The basic dimensions of personality in a Big Five Personality Test are as follows: E (Extroversion), A (Agreeableness), C (Conscientiousness), N (Emotional stability/Neuroticism), and O (Openness for experience/Intelligence).

(ii) Psychological evaluation: In a psychological evaluation conducted after the experiment, the participants were asked to answer a questionnaire on the following evaluation items. All of these items were evaluated based on a 7-point scale by the participants.

1. Nervousness: Degree of nervous feeling held by the participant during waiting.
2. Space of the experimental room: Degree of feeling the space in the experimental room.
3. Oppressiveness: Degree of oppressive feeling held by participant in the experimental room.
4. Presence of other people: Degree of feeling the presence of other people in the next room held by the participant.
5. Existence of sound: Degree of feeling the existence of sound from the next room held by the participant.
6. Room layout: Degree of belief by the participant that there is another door for other people beyond the partition in the experimental setting.

(iii) Behavior observation: While they waited in the experimental room for five minutes, the participants' behavior was observed by the experimenters and recorded through a video camera and a boundary microphone. In this paper, the experimenter simply analyzed the participant's surface behavior during the first 1 minute according to the following three criteria:

Synchrony:

(a) Whether the participant read the newspaper while listening to the ambient sounds of someone turning newspaper pages.
(b) Whether the participant picked up the glass while listening to the ambient sounds of someone placing a glass on a table.

Awareness:

(c) Whether the participant looked around the experimental room when she/he becomes aware of the ambient sounds from the next room.

Hypothesis
In this pilot study, we assumed the following hypotheses to examine the effects of participants' impressions and behavior in response to the existence of ambient sounds, even when the ambient sounds do not include human speech but simply noises resulting from the interaction between a human body and an object, i.e. a newspaper or a glass.

Psychological evaluation:

- Participants may be aware of ambient sounds when they are assigned to either the EAS group/NAS group.
- Participants may feel the presence of people in the next room when they hear ambient sounds from the next room under either the EAS condition/NAS condition.

Human behavior:

- Participants may change their behavior, such as lifting their face or gazing around the experimental room, when they are aware of ambient sounds under either the EAS condition/NAS condition.
- Participants may pick up a newspaper or a glass when they hear ambient sounds from the next room under either the EAS condition/NAS condition.

4 Results

Behavior observation
Figure 3 shows behavior observation results for both EAS and NAS conditions. NP means the participants who picked up a newspaper and read it. GR means the participants who gazed around the experimental room. WT means the participants who simply waited without gazing around the room while sitting in the chair. No significant difference in people's surface behavior between the two conditions was observed explicitly. The NP participants were 6 for the EAS group and 4 for the NAS group, and the GR participants were 4 for the EAS group and 5 for the NAS group. In

post-experimental interviews, participants explained the reasons for their behavior. Under both conditions, the NP participants answered that they picked up the newspaper just because they found it on the table. In the NAS group, the GR participants answered that they gazed around because they were curious about the structure of the experimental room or the installation of the video camera and microphone. In addition, the GR participants in the EAS group answered that they were also curious about the sounds from the next room. Consequently, these results did not verify our hypotheses of human behavior directly.

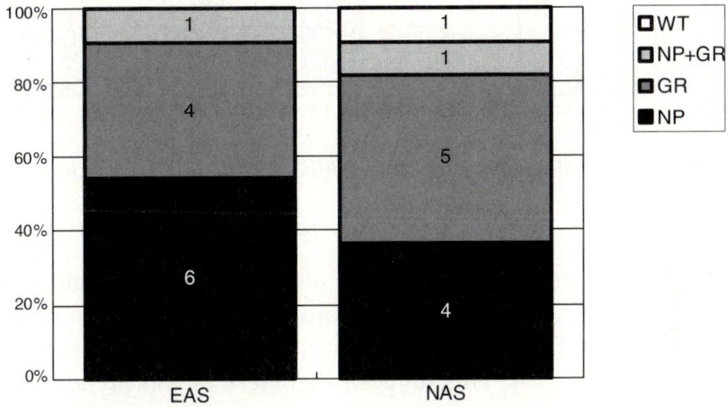

Fig. 3. Behavior observation results under two conditions: EAS and NAS

Psychological evaluation

Analysis 1: Comparisons between EAS and NAS groups

Figure 4 shows the results of comparing the psychological evaluations of the EAS and NAS groups. There were significant differences in both evaluation items 4 (Presence of other people) ($p = .0001$) and 5 (Existence of sounds) ($p < .0001$) by using the Mann-Whitney U test. On the other hand, there were no significant differences in 1 (Nervousness) ($p = .9171$), 2 (Space of experimental room) ($p = .1222$), 3 (Oppressiveness of room) ($p = .8869$), and 6 (Room layout) ($p = .3326$). In the post-questionnaire, participants who scored 5 points or more on question D were asked how many people were in the next room. All of these participants answered just one. This result shows that the participants under the EAS condition were aware of ambient sound from the next room and felt the presence of other people there. These results support our hypothesis.

Analysis 2: Correlations among items

EAS group: There is a positive correlation between 4 (Presence of other people) and 5 (Existence of sounds from the next room) ($\rho = .780$, $p = .0031$). This result supports our hypothesis.

NAS group: There is a positive correlation between 1 (Nervousness) and 5 (Existence of sounds from the next room) ($\rho = .788$, $p = .0025$). This result suggests that the participants in the NAS group may be sensitive to sounds around them, including noise from the air conditioner during the waiting period.

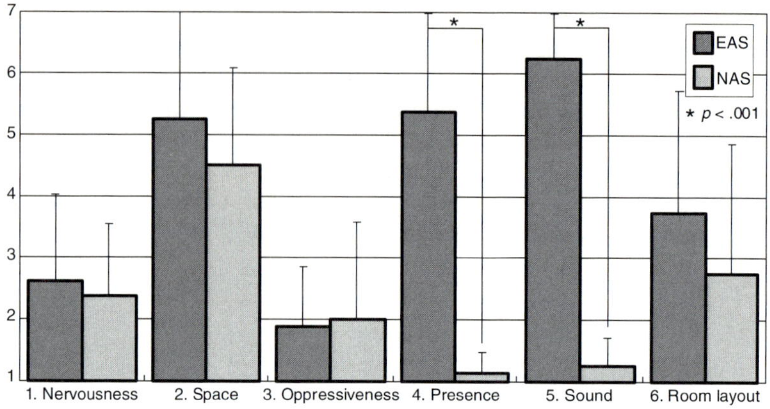

Fig. 4. Psychological impressions between EAS and NAS groups

Analysis 3: Factor analysis

EAS group: Factors were extracted by the unweighted least squares method, and the promax rotation was adopted. Two factors were extracted by considering the decay of the eigenvalues (Table 1). The inter-factor correlation between two factors is weak, with a value of only .084. This suggests that these factors are independent.

NAS group: No factor was extracted.

These results suggest that the participants in the EAS group became more nervous as they felt the presence of other people in the next room by listening to the ambient sounds. In addition, the oppressive feeling of the participants in the EAS group is associated with the room layout of the experimental room. These results supported our hypothesis.

Table 1. Factors in EAS group loading after promax rotation

	Factor I	Factor II
4. Presence	0.847	0.076
5. Sound	0.803	-0.196
2. Space	0.585	0.365
1. Nervousness	0.568	-0.166
6. Room layout	-0.004	0.813
3. Oppressivenss	0.122	-0.725

Correlation between personality of participants and psychological evaluation

EAS group: There is a positive correlation between C (Conscientiousness) and 3 (Oppressiveness) (ρ = .643, p = .0309). In addition, there is a negative correlation between N (Emotional stability/Neuroticism) and 1 (Nervousness) (ρ = -.665, p = .0233).

NAS group: *There was no correlation.*

We found two correlations between personality of participants and psychological evaluation in the EAS group but no correlation in the NAS group. These results suggest that ambient sounds provide an interpersonal effect, since the Big Five Personality Test includes interpersonal factors.

5 Discussion

The above results partly support our hypotheses. No difference in participants' surface behavior between the two ambient-sound groups, EAS and NAS, was observed explicitly. On the other hand, the participants in the EAS group were aware of ambient sounds from the next room and felt the presence of other people there, even when ambient sounds did not include human speech but simply the sounds of turning the pages of a newspaper or placing a glass on a table. In addition, the above results suggest that the participants in the EAS group felt nervous as they felt the presence of other people in the next room.

The main reason for this is the difficulty of selecting the appropriate surface behavior for behavior observation. In this pilot study, we selected two kinds of surface behaviors, synchrony and awareness. For synchrony behavior, we did not induce such behavior from the participants, since we did not succeed in making the participants recognize the correct sound source. We set newspapers, a glass and a bottle of water on the table, since we expected that the participants would synchronize their behavior to the contents of ambient sounds. Almost none of the participants in the EAS group recognized the contents of the ambient sounds of turning a newspaper's pages and placing a glass on the table, although they were aware of the existence of sounds. In the post-questionnaire, they answered that these sounds came from turning A4-sized papers over or touching a plastic bag, not from a newspaper. They also assumed the sounds of operating equipment, i.e. a computer mouse, rather than from a glass. This also involves the technical problem of recording and playing ambient sounds with a wide frequency range. In addition, people are apt to pick up a newspaper to "kill time," regardless of the existence or non-existence of ambient sounds. For awareness behavior, we did not identify the participants' behavior of gazing around the room as a result of awareness. The participants in both EAS and NAS groups gazed around the experimental room. In the post-experiment interview, the participants in the NAS group explained that they did this because they were interested in the structure of the room and equipment such as the video camera and microphone. On the other hand, the participants in the EAS group answered that they gazed around because the sounds from the next room had caught their attention.

These preliminary results have theoretical and practical implications. The key theoretical implication is that ambient sounds have the possibility of inducing the presence of other people; this is even true for sounds that do not include human speech or any noise from the nose/throat but that instead consist of noises people make as a result of interaction between the human body and an object. This could lead to examination of other kinds of ambient, implicit factors as fundamental determinants of feeling the presence of other people.

As a practical implication, this pilot study shows a new direction in a design concept for a system that supports communication with remote partners, i.e. an ultra-realistic communication system. Moreover, it was found that even ambient sounds offer the possibility of facilitating a shared reality among participants by inducing the presence of other people.

6 Conclusions

In this paper, we focused on the effects of the ambient sounds of a newspaper's pages being turned and a glass being placed on a table. These effects were investigated as a fundamental aspect of facilitating a shared reality among participants through an ultra-realistic communication system. We examined how ambient sounds affect psychological impressions and behaviors of people through a pilot study. We analyzed the experimental data through behavior observation, psychological evaluation, and a personality test. This pilot study was conducted with a between-participants design, by dividing the participants into two groups: EAS (existence of ambient sounds) and NAS (non-existence of ambient sounds). From the pilot study, we obtained the following results:

- Participants in the EAS group were aware of ambient sounds from the next room and felt the presence of other people there.
- Participants in the EAS group became nervous as they felt the presence of other people in the next room.
- No difference in the participants' surface behavior between the two groups was observed explicitly.

We confirmed that ambient sounds affect human impressions of the presence of other people, even though the ambient sounds consist of only noises resulting from an interaction between the human body and objects, i.e. a newspaper or a glass.

This work represents a first step toward examining the relationship between ambient sounds and the shared reality among participants achieved through an ultra-realistic communication system. Perhaps there are many other ambient and implicit factors that can provide more effective fundamentals. We will investigate such candidates in our future research.

References

1. Tang, J.C., Minneman, S.: Video Whiteboard: video shadows to support remote collaboration. In: Proceedings of CHI91, pp. 315–322 (1991)
2. Paulos, E., Canny, J.: Social tele-embodiment: understanding presence. Autonomous Robots 11(1), 87–95 (2001)
3. Miwa, Y., Ishibiki, C.: Shadow communication: system for embodied interaction with remote partners. In: Proceedings of CSCW 2004, pp. 467–476 (2004)
4. Ifukube, T.: From sensory substitute technology to virtual reality research. Artificial Life. Robotics 4(2), 145–150 (1998)
5. Ohuchi, M., Iwaya, Y., Suzuki, Y., Munekata, T.: Cognitive-map formation of blind persons in a virtual sound environment. In: Proceedings of ICAD 2006 (CD-ROM) (2006)
6. Pedersen, E.R., Sokoler, T.: AROMA: abstract representation of presence supporting mutual awareness. In: Proceedings of CHI97, pp. 51–58 (1997)
7. Ishii, H., Wisneski, C., Brave, S., Dahley, A., Gorbet, M., Ullmer, B., Yarin, P.: AmbientROOM: integrating ambient media with architectural space. In: Proceedings of CHI98, pp. 173–174 (1998)
8. Hallnas, K., Redstrom, J.: Slow technology – designing for reflection. Personal and Ubiquitous Computing 5, 201–212 (2001)

Evaluation of the Participant-Support Method for Information Acquisition in the "Multiplex Risk Communicator"

Hiroshi Yajima*, Tomohiro Watanabe, and Ryoichi Sasaki*

School of Engineering, Tokyo Denki University
2-2 Kandanishiki-cho Chiyoda-ku Tokyo
* RISTEX of the Japan Science and Technology Agency
yajima@im.dendai.ac.jp, sasaki@im.dendai.ac.jp

Abstract. There has been a diversification of the social risks to information-based society leading to complex social issues, and risk communication is necessary in order to solve the complicated social problems that arise concerning stakeholders with various levels of knowledge and differing standards regarding risks. In this paper, we introduce a portal system that assists participants to achieve an optimal combination of countermeasures. The characteristics of the proposed method are: first, to classify the stage when participants acquire information; second, to enable smooth transitions during the information acquisition stage; and third, to support information acquisition by offering information portals.

Keywords: information exploration, information search, information retrieval, risk communication.

1 Introduction

With the growth and development of information technology there has been a diversification of the social risks to information-based society leading to complex social issues. Both businesses and individuals have been made victims of, for example, the leaking of personal information by computer viruses, or breaches of authors' copyrights due to unauthorized duplication. These social issues involve a complex assortment of risks regarding security, loss of privacy, erosion of user-friendliness, and management. In order to solve these issues, it is essential to consider how to reduce these various risks, and how to resolve the opposing interests of decision-makers, participants and stakeholders.

Solving these complex social issues does not merely require an analysis of the risks involved. An exchange of information and opinions—information-sharing and communication—among people either directly or indirectly related to the risks is also important so as to prevent misunderstandings or insufficient understanding among participants. A risk-communication process is therefore essential to build a consensus among participants with differing standpoints and perspectives, such as decision-makers, stakeholders and specialists [1].

M.J. Smith, G. Salvendy (Eds.): Human Interface, Part II, HCII 2007, LNCS 4558, pp. 195–203, 2007.

However, there many are cases when it is difficult for participants with sparse specialist knowledge to grasp the structure of the risks, trade-offs among various risks, and solution space. This means that risk communication cannot proceed smoothly. Summarizing these problematic points from the perspective of participants, the following three points may be stated: (1) it is not known what kind of perspectives are required for information gathering, (2) it is not known what the techniques are for revealing the kind of information necessary to understand the risks, and (3) it is not known how the obtained information stands when regarded from different perspectives.

In order to solve these problems, it is necessary to consider systems that support the determination of the combination of risk countermeasures and individual desires, and enable participants to gather essential information smoothly based on incomplete initial information. We are therefore currently developing a Multiplex Risk Communicator (MRC) as a risk-communication support tool in order to resolve the issues of social risk and social consensus building in an information-based society [1].

In this paper, we present a support system for participants to smoothly obtain the information needed to identify and evaluate their optimal combination of countermeasures using the portal system of the MRC's participant-support component. The characteristic features of the proposed system include the pattern-based breakdown of the steps in information gathering, the efficient gathering of information that participants require by means of a smooth migration between each of the information gathering methods, and the facilitation of planning in order to support the identification and evaluation of an individual's optimal combination of risk countermeasures. The utility of the system was validated experimentally.

2 The Risk-Communication Problem

2.1 Characteristics of Risk Communication

Risk communication is an interactive process whereby decision makers, participants, specialists and facilitators exchange opinions and information regarding risks, and communicate their intentions mutually. The objective in this paper is to present an interactive process for planning solutions using the MRC [1] [2][3]as a means of risk communication.

2.2 Outline of the Multiplex Risk Communicator

Until now, risk has been thought of as a single entity, but in fact there is a multiplicity of risks, and in order to establish risk countermeasures it is also necessary to obtain a consensus among the interested parties. The MRC is currently being developed as a tool for supporting the acquisition of an optimal combination of countermeasures while considering the various risks against this background, and planning communication between multiple participants.

The MRC supports the following four basic processes: (1) analyzing the risks according to a consideration of the tradeoffs among multiple risks, (2) selecting an optimal combination of countermeasures from multiple possible combinations, (3) identifying both the information needed for each of the participants' desired combination of countermeasures, and corresponding information gathering methods, and (4) building a consensus among the participants, decision makers, interested parties and specialists. The MRC specialist input/output component supports the basic processes numbered (1) and (2), and the MRC participant-support component supports those numbered (3) and (4). This paper deals with the MRC participant-support component and is focused on process (3).

2.3 Outline of Risk Communication in the MRC Participant-Support Component

The MRC participant-support component conducts risk communication over the following three steps [4][5].

1st risk communication
Information is gathered in order for each participant to clarify their opinion.
2nd risk communication
Mutual understanding among each of the participants is developed.
3rd risk communication
A consensus is established among all of the participants.

This paper deals with the portal-based information-gathering support system for the 1st risk communication occurring in the MRC's participant-support component.

2.4 Problems Regarding Participants with Insufficient Information

Many participants have only sparse specialist knowledge. This makes it difficult for them to comprehend the structure of the risks in their desired solution, the tradeoffs among the risks, and what kind of position the solution occupies in the space of possible solutions. It is thus also difficult for them to compare and evaluate their desired solution with respect to an optimal solution recommended by specialists.

In such a context, participants will find it difficult to understand and interpret the issues upon which the specialists based their optimal solution. The causes may be summarized as the following three problems.

Problem 1 It is common for participants with only sparse specialist knowledge to know neither what kinds of perspective are appropriate for information gathering, nor the structure formed by the risk space.

Problem 2 It is common for participants with only sparse specialist knowledge to be unable to conduct information gathering activities fulfilling the information gathering aims and objectives.

Problem 3 Participants with only sparse specialist knowledge cannot grasp the position of the information obtained when regarded from alternative perspectives and how it relates to peripheral information.

3 Basic Principles of Problem Solving

3.1 Aims

Aim 1: To provide multiple structured perspectives constituting the basic points needed to conduct information gathering in information space.

Aim 2: By categorizing the state transitions of the participants' information gathering process, to present the information gathering objectives and to offer information appropriate to the state so as to support comparatively specific information gathering (convergent activity) rather than abstract information gathering, and also to support comparatively abstract information gathering (divergent activity) rather than specific information gathering.

Aim 3: To provide support in order to grasp an understanding of what kind of position the information obtained occupies when regarded from alternative perspectives, and how it relates to peripheral information.

3.2 Basic Principles

Basic principle 1: Provide a portal for the structured information constituting the basic points of information gathering in order to fulfill Aim 1.

Basic principle 2: In order to fulfill Aim 2, participants' information gathering is categorized into three steps, and information appropriate to each information gathering state is offered. Smooth transitions between different information gathering methods are also supported.

It is common for participants with sparse specialist knowledge not to have a clear knowledge of the aims and objectives of the information gathering needed to clarify their own opinions. Information gathering for this kind of participant must involve the smoothly facilitation of three information gathering activities: making clear what kind of aims and objectives are appropriate (exploration activity), making clear what kind of structure is represented by obtained information according to a given perspective and what kind of peripheral information exists (search activity), and investigating obtained information through further detailed information (retrieval activity).

Basic principle 3: In order for fulfill Aim 3, visualizations are presented showing the position of the obtained information according to multiple differing perspectives within the structure of risk information, and showing peripheral information. The changing of perspectives is made possible.

The positional relationships within the information obtained from the multiple structured information portals, and the peripheral information relating to alternative perspectives are therefore represented visually. The promotion of convergent and divergent information gathering activities is also supported, by facilitating the changing of perspectives.

3.3 Definition of the Information Gathering Steps (Basic principle 1)

The following three information gathering steps are defined. The content of the information presented at each step is also shown.

1. Information exploration: objective-setting step
2. Information search: search step, for information related to the objectives
3. Information retrieval: examination step

Information is presented as appropriate, according to each of the three information gathering steps mentioned above.

3.4 Overview of the Information Portals (Basic principles 2 & 3)

The information portals described below which gather relevant information in a single place, are used as tools to support participants' understanding of information sources from various perspectives.

a. Portals effective for grasping risk structure

- Risk-analysis structure portal (Fault Tree Analysis (FTA), Event Tree Analysis (ETA), *etc.*)

b. Portals effective for grasping the tradeoffs among risks and the solution space

- Solution space structure portal
- Solution space search process portal (archive history)

c. Portals effective for studying risk

- Document structure portal (countermeasures, assumed risks, *etc.*)

The information portals are not used in isolation, but rather the following advantages may be achieved by combining a number of information portals and shifting between them.

1. An understanding of the risks is helped by visually establishing the relationships within the information obtained from a variety of perspectives, based on a grasp of the risk structure, the tradeoffs among multiple risks, and the solution space.
2. Convergence and divergence of information gathering activities may be freely conducted according to participants' concerns and interests, by using multiple information portals, and without being limited to a single information gathering step.

By using multiple structured information portals, participants may deepen their understanding of information obtained for the purpose of grasping the positional relationships of the obtained information within the structure of a portal, and the positional relationships of the obtained information according to different perspectives (portals). Thus, by using the information portals and conducting each information gathering step, participants may grasp an understanding of the related information they require, the risk structure, the solution space, and the tradeoffs among multiple risks.

4 The Specific Support Procedure Occurring in the 1st Risk Communication

The risk communication in the MRC's participant-support component is conducted using the participants' desired solutions and compromise constraint conditions obtained in the 1st risk communication. In this 1st risk communication, desired solutions and compromise constraint conditions that reflect the participants' opinions are established by gathering information for participants to clarify their own opinions. The support process for participants' information gathering in the 1st risk communication is proposed.[4].

According to the support process, participants make use of information portals in order to clarify their own opinions, and make state transitions through the steps involved in gathering information appropriate for the information-gathering aims and objectives.

The following steps are conducted during the exploration phase: (1) presenting what kind of information can be obtained from each portal facilitates the gathering of information involved in the objective setting steps, and (2) presenting keywords for relevant portals facilitates the gathering of information involved in the objective setting steps.

Following the exploration phase, the input for the search phase is based on the portals used for exploration. The gathering of information related to the objectives is thus conducted in the search phase by presenting structured information from the portals, and information related to the area under focus as required using overlays.

Following the search phase, the input for the retrieval phase is generated by selecting refining keys and determining the area under focus. By presenting detailed information related to the refining keys and area under focus during the retrieval phase, information is gathered for the steps of the investigation. Participants conduct information gathering activities in the three phases described above, while repeating the state transitions of each information gathering step based on their own queries. Also, while refining the information required, participants evaluate their desired solutions and validate their applicability. As a result, the desired solution and compromise constraint conditions constituting each participant's own opinions are established.

5 Empirical Validation

5.1 Experimental Objectives

The experimental objectives were to validate whether or not participants could establish the following points (in their own opinions) by conducting the 1st risk communication in the MRC's participant-support component portal system.

(1) Determining the objective function and items in the constraint conditions [2] [3]

Objective function:

Minimum total cost of the "sum of the cost of the countermeasures and the risk of personal information leaks"

Constraint conditions:

(a) cost of countermeasures (Japanese Yen)
(b) privacy burden for employees (%)
(c) user-friendliness burden for employees (%)
(d) personal information leak rate (% = cases per year)

(2) Election of the set of countermeasures [2][3]
8 countermeasures were elected for the set

1. Could the desired solution be established?
2. Could the compromise constraint conditions be established?

5.2 Preconditions

The issue of personal information leaks was treated as an object of the risk communication. The subjects were 10 male university students in their early twenties, and were assigned roles as managers or employees.

5.2.1 Establishing the Assumed Target Enterprise

The target business was assumed to be a service provider company with 1,000,000 personal information retained.

5.2.2 Preconditions of the Risk-Analysis Background

The risk analysis procedure regarding personal information leaks was conducted as follows .

5.2.3 Establishing the information portals

The information portals comprise a set of four portals: the countermeasures portal, the assumed-risk portal, the scatter-plot portal, and the FTA portal.

5.3 Experimental Results

Managers' query: Given the security revolution, what cost of countermeasures should be accommodated in response to risks? Managers followed an information gathering process in order to resolve their query. It reveals that after going through the information gathering process, managers ultimately selected the solution endorsed by specialists as their desired solution. Also, they selected a leak rate of at most 20%, and a cost of at most 80 million yen for countermeasures as constraint conditions.

An example information search shows that by using the scatter-plot portal, participants were able to investigate what kind of structure the whole solution space takes when defined according to two constraint-condition axes, where the optimal and desired solutions exist in solution space, and what kind of solutions exist in the surrounding area of solution space.

Example overlay of information portals shows that by using a superposition of multiple information portals, participants were able to grasp an understanding of what other sets of countermeasures exist, what kind of detailed information is available regarding other sets of countermeasures, and where in solution space the sets of countermeasures incorporating the desired countermeasures exist.

6 Discussion of Experimental Results

Among the various results of the information gathering process, managers mainly centered their information gathering activities on the scatter-plot portal. They developed an understanding of the risk structure through gathering a quantitative perspective on the solution space by searching it using the scatter-plot portal. In addition, by way of auxiliary information gathering, they used the FTA portal to search for information from a qualitative perspective regarding the causes of risks and positional relationships among the countermeasures. They also used the countermeasures portal and assumed-risk portal to retrieve information about the risks and countermeasures, making transitions among each of the portals to solve problems 1, 2 and 3.

It was possible to conduct smooth transitions during information gathering according to the process described above using basic principles 1, 2 and 3. Managers were able to resolve their initial queries, and establish their desired solution and compromise constraint conditions.

7 Conclusion

In this paper we proposed a support process for smooth information gathering aimed at allowing participants with sparse specialist knowledge to grasp the risk structure, the tradeoffs among multiple risks, and the solution space itself in the context of the multiplex risk communicator.

A characteristic of the proposed method is the pattern-based categorization of participants' information gathering process. In addition, the smooth transitions among each of the information gathering steps were planned in order to make the gathering of information that participants regard as essential, smoother and more efficient.

The utility of the proposed method was validated empirically by conducting experiments using the issue of personal information leaks as the object of an example risk communication.

Acknowledgements

We received much advice regarding the participant-support component of the multiplex risk communicator (MRC) from members of the multiplex risk communicator working group, to whom we are very grateful, and we are also very grateful to the people who took part in our experiments.

This research is supported in the activity of security and safety working group in applied security forum, and is studied in the research on "Challenges in identifying vulnerabilities hidden in our highly sophisticated information society and exploring solutions" of the Research Institute of Science and Technology for Society in Japan Science and Technology Agent. We thank Dr. Murayama and Dr. Yoshiura for their useful comments and advices

References

1. Sasaki, R., Ishii, S., Hidaka, Y., Yajima, H., Yoshiura, H., Murayama, Y.: Development Concept for and Trial Application of a Multiple Risk Communicator. In: 5th IFIP Conference e-Commerce, e-Business, and e-Government (I3E (2005)
2. Sasaki, R.: Sturdy of Relationship between security and ptrotekution of personal information SITE2003-14, pp. 1–6 (2003)
3. Sasaki, R.: Compatibility and Conflict between security and protekution of personal information. Cyber Security Management 6, 20–24 (2003)
4. Yajima, H., et al.: Support System for Decision Maker in Multiple Risk Communication. CHI, vol. 7 (2005)
5. Yajima, H., Matsumoto, S., Sasaki, R.: Proposal of Risk Communication Supporting Method in Multiplex Risk Communicator. In: Proceeding of 11th IEEE International Conference on Emerging Technologies and Factory Automation, pp. 255–260 (2006)

Resolving Assumptions in Art-Technology Collaboration as a Means of Extending Shared Understanding

Yun Zhang, Alastair Weakly, and Ernest Edmonds

Creativity and Cognition Studio, University of Technology, Sydney, Australia
yunzhang@it.uts.edu.au,
alastair@weakly.org.uk,
ernest@ernestedmonds.com

Abstract. This paper extends the knowledge and understanding of art-technology collaboration. It reports upon a close empirical study of how computer programmers interacted with a digital artist to develop a computer-based interactive artwork. Analysis of the data collected showed that the joint uncovering and resolving of assumptions made by each party led to increased shared understanding. The contribution of this paper is to provide a better understanding of creative collaboration, particularly focusing on how developing the artefact increased the understanding between the artist and technologists.

Keywords: Art-technology collaboration, computer-based interactive artwork, case study.

1 Introduction

In art-technology collaboration, as in other multidisciplinary projects, it can be very hard to develop shared understanding (Candy & Edmonds 2002) (Zhang & Candy 2006a) (Zhang & Candy 2006b). In such projects, the two parties typically have only limited experience of one another's domain. In addition, the artist typically starts the project with an ill-defined vision of what they would like to create. The translation of this vision into programmatic code by the technologists is a very important activity. Problems may arise because it may be very difficult for the programmer to grasp or relate to the artist's vision. Similarly, the artist typically does not understand the intricacies of the code involved.

Shared understanding comes about where two people's spheres of experience overlap to some degree. In terms of the nature of the shared understanding, there is a very important concept; the 'boundary object' (Star 1989). Boundary objects are defined as concepts or ideas that exist in this shared space and which have the same meaning to each party (Wenger 1998). Fischer et al (2005) pointed out that boundary objects allow different knowledge systems to interact by providing a shared reference

M.J. Smith, G. Salvendy (Eds.): Human Interface, Part II, HCII 2007, LNCS 4558, pp. 204–211, 2007.

that is meaningful within both systems and they serve to communicate and coordinate the perspectives of various constituencies communicate.

Following Fischer, this research aims to examine the process where the shared understanding between artists and technologists accumulates during the collaboration and also understand further how boundary objects shared between artists and technologists affect the shared understanding in between. In order to achieve this, we conducted an in-depth case study of an art-technology project named the GEO narrative landscapes project.

In the next section, a brief background of the GEO project and how the study was carried out are provided.

2 Study Contexts

2.1 Background of the GEO Project

The GEO narrative landscapes project is the result of an art-technology collaboration led by the artist Chris Bowman. Phase one of the project was carried out from June to December in 2006 within a multi-disciplinary research team comprising an artist, two technologists, one graphic designer and one observer. All participants were co-located in the same city. The artist was the designer of the digital artefact and also the leader of the team. The designer mainly worked on the video sequences. The two technologists were in charge of different parts of the technical functions of the project: one was mainly working with the software application MAX/MSP (Cycling74 2001), the other was mainly working with the programming language Java (Java). The observer, who is also the first author of the paper, remained on site for five months observation. During the five months observation period, there were eighteen group meetings that were observed. The artist, the technologists and the observer attended 95% of the meetings and the designer attended 35% of the meetings.

The basic collaboration mode for this project was via group meetings. The duration of the group meetings was between thirty minutes and one hour based on on-site-demand and the frequency of the meetings was once a week. In addition to the weekly meeting, the remaining time was spent working individually or in person to person communication by email or telephone. The meeting location was in the Creativity and Cognition Studio (CCS) at the University of Technology, Sydney where a power Macintosh G5, a projector, and interactive floor pads were installed (see Fig. 1).

In this paper, we will give an overview of the system in practical terms, without going too deeply into its artistic aims. The system is an immersive environment where the user can explore a remote location seen through the eyes of other people. It is designed for installation in a museum location. The user interacts with the system by moving around in front of a wall-sized rear projection screen. Beneath the carpet, a six-by-six grid of pressure sensors detects where the users are standing. Displayed on the screen is video and text material relating to the remote location. Initially the user is presented with a 360 degree panorama of the site. As they move around the space, the panorama rotates and tilts up and down. As the interaction continues, the system enters another, deeper, mode of interaction where the user can choose from a number of locations around the panorama by stopping its rotation at the appropriate point. When a location is selected, the display transports to that location, where the user is presented with material specific to a particular person's experience of that location. This material is again video, text and sound.

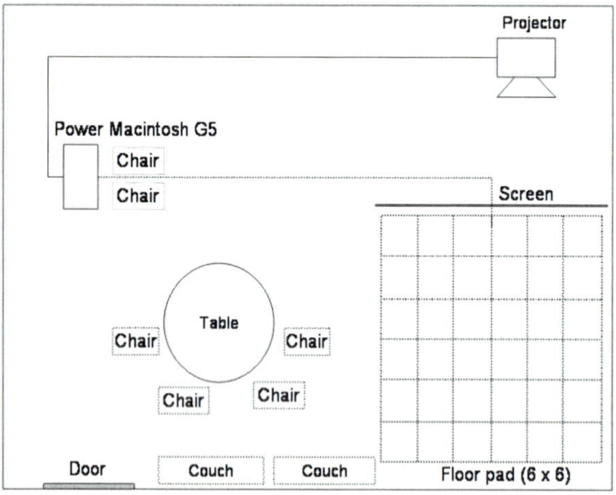

Fig. 1. Equipment layout of the Creativity and Cognition Studio

Within the GEO system, the user can explore the remote space, at several levels of interaction, and explore different people's views of particular aspects of the site. Three views of the GEO interface in a test laboratory are shown in Fig. 2. The final outcome of this work will be exhibited in April 2007 at Beta-space in the Sydney Powerhouse Museum. Beta-space is an exhibition space where the general public can engage with art outcomes of leading research in art and technology (Costello et al. 2005).

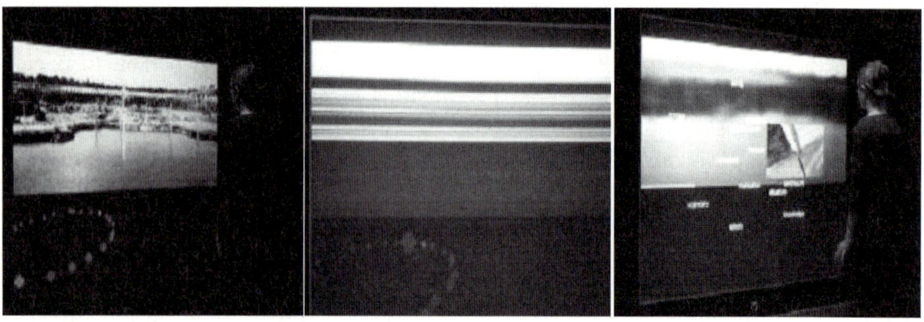

Fig. 2. Three views of the GEO artwork

During the project, team meetings were video-recorded and field notes were taken by the observer over a period of five-month collaboration. More detail about the data collection methods in this study could be found in the following section.

2.2 Study Approaches

The main data collection approach adopted in this research is a combination of non-participant observation and semi-structured interviews (Yin 2002) (Valerie 1998) (Denzin & Lincoln 1998). Non- participant observation is a method where researchers 'stand to one side' and view the experience or the environment (Slack & Rowley 2001). During the observation of this project, the observer used video facilities to record the meetings between participants over the five-month period from July 2006 to December 2006. In addition, field notes were taken for each meeting by the observer. As non-participant observation cannot provide cognitive information, such as attitudes, beliefs, motivation or perceptions (Slack & Rowley 2001), we overcame this by having the observer ask questions and conducting informal conversations with participants subsequent to the video observation. At the end of the five-month observation, a semi-structured interview was also conducted with each participant. In addition, emails between participants relating to the project were also included as part of the data collected in order to arrive at as detailed a picture of the collaboration as possible.

Within these collected data, we analyzed the video data of these observed meetings with the assistance of the analysis software application INTERACT (Mangold 2005). From the data, we have found that the protagonists clarified and improved understanding between one another using verbal communications, gestures and sketches drawn on paper or the whiteboard. In this paper, we specifically concentrate on how they used prototypes of the system under development to build up shared understanding between each other. In the next section, the detailed findings with specific examples extracted from the data are provided.

3 Findings

From the analysis of the empirical data within the GEO project, we found that the understanding between the artist and technologists improved when the technologists showed a working prototype to the artist and occasionally, these prototypes enlightened the artist's creative ideas. Furthermore, we also found that the technologists, both deliberately and non-deliberately, made assumptions about what the artist required. Acting on these assumptions, they produced prototype systems which they then demonstrated to the artist. The prototype systems did not always exactly embody what the artist had in mind. It was often in uncovering shortcomings in the prototypes that new understandings arose. At such times, assumptions made by the technologists were revealed. Similarly, the artist's assumptions about what could be achieved programmatically were exposed on review of the prototypes developed. In order to illustrate this process more closely, the following are some examples extracted from the data project where technologists built prototype systems based on the assumptions of the artist' requirement.

Example 1

The GEO system is controlled by a person standing on the pressure pads in front of the screen. During the development of the project, one of the technologists raised the issue of how the system should behave if there were two people present in the room.

In a number of discussions, he suggested a range of alternative strategies to the artist. These were that the system should find some average position between the two (or more) people, and use that as the position of an imaginary person in the room, alternatively it was suggested that just the person nearest to some point in the room should control the system - other inputs would be ignored. The technologist suggested that, given the time limits imposed on the project, it would be too difficult to track a single user's interaction (or that of many users) with the system as other people came into or left the room.

The artist was reluctant to commit himself to either course without further investigation or thought and for this reason the technologist decided that the person nearest the entrance to the room should control the piece. In the museum setting where the piece was to be shown, this would be the person furthest from the screen and to the right. The rationale for this, on the part of the technologist, was that this would mean that the system would start to respond to a person as soon as they walked into the room, hopefully encouraging them to come in further.

However, in an evaluation carried out in the test laboratory, this choice of corner proved less than ideal. Some audience members were sitting at the back of the room and their feet activated floor pads there, preventing the person actually interacting with the system from being able to control it properly. From a technical point of view, then, a number of assumptions had been made in arriving at the architecture of the system under development. At this point in the project, a number of the decisions taken had to be reversed and the system architecture was redeveloped. Therefore, the assumptions made by the technologists and embedded in the prototypes systems did not achieve the way the artist requires and the development had to be redone.

Example 2
The GEO system presents to the user data (video, text and so on) which is location-specific. At the same time that the video was captured, a GPS track log was recorded about the spatial position of the camera person. The intention was that a visualisation of the track in space would be presented to the user at the same time as the video material to help them orientate themselves within the virtual space. In the storyboards, which are proposals the artist made specifically for how the system should behave in response to specific actions, the visualisation of the GPS data was designed to be achieved in three steps: reading the co-ordinates from the GPS data, scaling them appropriately and then drawing a line in a 3D space from each co-ordinate to the next.

As an experiment, and unprompted by the artist, one of the technologists rewrote the visualisation module so that it displayed points on the track as spheres in a three dimensional space. As the user's location in the virtual space passed over an individual sphere, its diameter was increased. The technologist here was assuming that, because this added additional functionality to the system, the artist would be pleased with this new representation. It would, of course, have been difficult to give an indication of the current location using just a single line representation.

At the next project meeting, the artist was shown the new representation at a relatively early stage of development: there were a few problems with the implementation still to be solved. The artist asked for a new representation, which was a development of the spheres display and not a return to the single line. This is an example of the technologist's assumption being accepted into the artist's view of the system and then developed further by him.

Example 3
In the early stages of the GEO project, the system was envisaged by the technologists as one that reacted to user input. The approach taken to the design was an attempt to create a system that exhibited certain behaviours. The aim was to develop a general system in which data about any site could be displayed, and that this display would be controlled by the behaviours built in to the system. The artist wanted a high degree of control over what material would be displayed by the system as well as over how this would be shown. It was at this point that it became clear that there had been an assumption made on the technological side that had influenced the work to date. One of the technologists described it as the artist wanting to be much more like a movie director than he had previously imagined. That is, the artist wanted to be in much closer control of the moment-by-moment audience experience.

Within each of the three examples above, an assumption was made by the technologist about what the artist either required or would like. In example 1 the technologist's assumption was deliberate. He knew that a decision had to be made on a particular point and assumed that his choice would be acceptable. In testing it turned out that the choice of orientation was not ideal for the test laboratory, but at the same time it became clear to all how the system was behaving. In uncovering a shortcoming in the system, the technologist's assumption was exposed, understood and open for further discussion. In example 2, aspects of the modification to the system, which the technologist assumed the artist would approve of, were adopted for use in the rest of the project. Here the technologist assumed largely correctly that he understood the artist's aims for the system and acted upon these assumptions. It was immediately clear to the artist, on sight of the prototype, what had been done and how it needed to be adjusted to bring it into line with the artistic requirements. In the third example, the artist did not realise that his storyboards would necessitate a change in the implementation of the system. Both parties, here, had made assumptions about the other's domain. Looking together at the storyboards helped them to uncover the assumptions and gave them something to refer to in subsequent discussions.

We suggest that during art-technology collaboration, the external representation of the artefact serves as a very important mediating tool to bridge the understanding between artists and technologists. By this exposing and clarification of assumptions, the shared understanding between artists and technologists was increased. We suggest that uncovering the assumptions caused the boundary zone of shared understanding to grow. In the next section, a mode built based on this finding is presented.

4 Model

As described in the previous section, the examples suggest that joint study of an artefact, either a software prototype or in one case a set of storyboards, helped the collaborators to discover and clarify assumptions they had made about one another's domains. Furthermore, in this study, we found that the knowledge, ideas and understanding that went into the creation of the programmatic artefacts come from the boundary object between artists and technologists combined with the assumed objects and programming knowledge of the technologist. Therefore, a model was developed, where we defined a space "assumed space", which extend the existing model of a boundary space containing boundary objects (Arias & Fischer 2000; Fischer et al. 2005)(Fig.4).

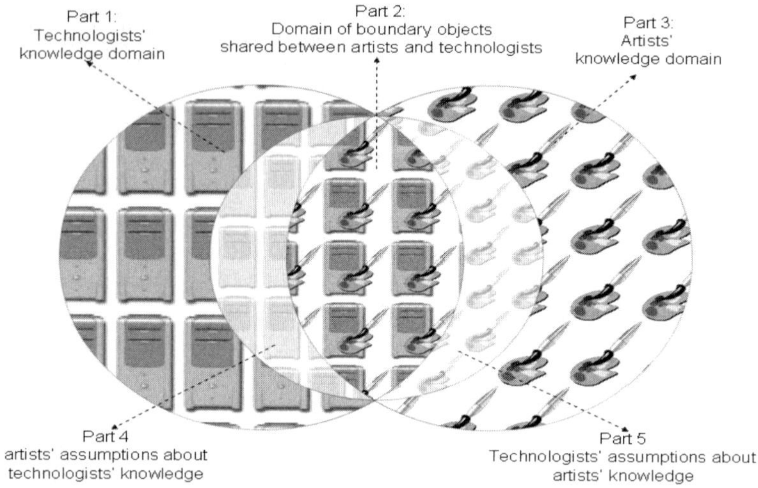

Fig. 3. Model of "assumed space" between artists and technologists

As we see from Fig. 3, boundary objects (part 2 in Fig 4.) allow both artists' and technologists' knowledge systems to interact with each other by providing a shared reference that is meaningful within both artists' and technologists' knowledge system. In contrast to the boundary objects, the assumed objects (part 4 and part 5 in Fig 4) are artist's and technologists' assumptions about one another's domains. They initially lie outside the boundary zone of shared understanding. In the GEO project the technologists occasionally built computer prototypes based on these assumed objects (Part 5 in Fig. 4). Similarly, artists occasionally expressed their artistic plans or requirement based on assumptions about the technical domain (Part 4 in Fig. 4)'.

When these assumptions were exposed both parties' understanding increased. Part of the assumed space therefore became included within the shared space. We suggest that this is one mechanism by which shared understanding can grow. Knowingly or unknowingly, one party makes an assumption about the other's domain. Some time later, perhaps when studying a prototype in which the assumption is embodied, the assumption and its shortcomings, if any, are revealed. Both parties then have a new, shared boundary object to which to refer in the future.

5 Conclusions and Future Work

In this paper, we presented a study based on an observation of a recent art-technology collaboration project. The outcome of this study is a model which extended the boundary object concept to cover part of the assumed space. The contribution of this study is that it has extended our knowledge of how shared understanding was developed during the art-technology collaboration. In the future, the model will be testified and validated further by applying more studies in the similar research contexts.

Acknowledgements. The authors wish to thank Chris Bowman, Shigeki Amitani and Andrew van der Westhuyzen for the generous support through the project. The research work was partly conducted within the Australasian CRC for Interaction Design, which is established and supported under the Australian Government's Research Centres Programme.

References

1. Arias, E., Fischer, G.: Boundary Objects: Their Role in Articulating the Task at Hand and Making Information Relevant to It', ICC'2000, pp. 567–574. ICSC Academic Press, Wetaskiwin, Canada (2000)
2. Candy, L., Edmonds, E.A.: Explorations in Art and Technology. Springer-Verlag, London (2002)
3. Costello, B., Muller, L., Amitani, S., Edmonds, E.: Understanding the experience of interactive art: Iamascope in Beta_space. In: Proceedings of the second Australasian conference on Interactive entertainment, Sydney, Australia, pp. 49–56 (2005)
4. Cycling74 2001, Max 4: Getting Started. viewed (October 2006) <http://www.synthesisters.com/download/Max4GettingStarted.pdf>
5. Denzin, N., Lincoln, Y.: Entering the field of qualitative research. In: Denzin, N., Lincoln, Y. (eds.) Strategies For Qualitative Inquiry, pp. 1–34. Sage, London (1998)
6. Fischer, G., Giaccardia, E., Edena, H., Sugimotob, M., Yea, Y.: Beyond binary choices: Integrating individual and social creativity. International Journal of Human-Computer Studies. Special Issue on Computer Support for Creativity 63, 482–512 (2005)
7. Java (2006) <http://www.apple.com/macosx/features/java/>
8. Mangold, INTERACT user guide, Mangold International GmbH, Arnstorf, Germany (2005)
9. Slack, F., Rowley, J.: Observation: Perspectives on Research Methodologies for leisure Managers. Management Research News. 24(1/2), 35–42 (2001)
10. Star, S.L.: The Structure of Ill-Structured Solutions: Boundary Objects and Heterogeneous Distributed Problem Solving. Volumgan Kaufmann Publishers Inc, San Mateo, CA (1989)
11. Valerie, J.J.: Stretching Exercises for Qualitative Researchers. Sage Publications, Thousand Oaks, CA (1998)
12. Wenger, E.: Communities of Practice— Learning, Meaning, and Identity. Cambridge University Press, Cambridge, U.K (1998)
13. Yin, R.K.: Case Study Research: Design and Methods, 3rd edn. Sage Publications, Thousand Oaks, Calif (2002)
14. Zhang, Y., Candy, L.: Investigating collaboration in art and technology. CoDesign: International Journal of CoCreation in Design and the Arts. 2(4), 239–248 (2006a)
15. Zhang, Y., Candy, L.: Investigating Interdisciplinary Collaboration: Case Studies in Art and Technology. In: Proceedings of the third international conference on Qualitative Research in IT & IT in Qualitative Research, Griffith University Press, Institute for Integrated and Intelligent Systems, Griffith University, Brisbane, Australia, pp. 173–183 (2006b)

A Communicative Behaviour Analysis of Art-Technology Collaboration

Yun Zhang and Linda Candy

Creativity and Cognition Studios, University of Technology, Sydney, Australia
yunzhang@it.uts.edu.au,
linda@lindacandy.com

Abstract. This paper presents an approach to investigating interdisciplinary collaboration between an artist and a technologist based on case study methods. The aim of the research is to understand how artists and technologists communicate with each other during a collaborative process. The paper begins with a brief account of the art-technology context, and goes on to describe how the data was collected and how the analysis framework was developed specifically for this context. At the end of this paper, we discuss the preliminary findings which illustrate the characteristics of participants' communication behaviours in art-technology collaboration.

Keywords: Art-technology collaboration, case studies, protocol analysis, INTERACT.

1 Background

In order to support, understand and investigate art-technology collaboration, making artists-in-residency programs available is one of the ways to provide a better environment for artists to develop their creative ideas (Harris 1999). Among those artists-in-residency programs, the COSTART project (COmputer SySTems for Creative Work: An Investigation of ARt and Technology Collaboration), was one of the few which not only provided a conducive environment for artists to explore their creative ideas, but also moved a further step forward by establishing a research environment for social scientists to study how the process of collaboration took place between artists and technologists by collecting empirical data during the residencies (Candy & Edmonds 2002a). Previous research based on the COSTART project revealed that the forms of collaboration varied significantly with different types of work and people. It also claimed that the quality and ease of communication was likely to influence the effectiveness of the collaboration(Candy & Edmonds 2002b) (Edmonds et al. 2005).

Following the earlier research, the authors of this paper have recently conducted a series of case studies based on the data collected in COSTART by applying the ethnographic analysis approach (Zhang & Candy 2006a) (Zhang & Candy 2006b). One of the findings from these studies is that the mediation of computers may help artists and technologists overcome the difficulties of lacking of common language. As

M.J. Smith, G. Salvendy (Eds.): Human Interface, Part II, HCII 2007, LNCS 4558, pp. 212–221, 2007.

the research continues, we are investigating this issue further specifically focused on exploring communication behaviours between artists and technologists. In order to achieve this, we conducted protocol analysis of three COSTART cases. In the next section, a brief background description of each case is provided.

2 Three Cases

As mentioned previously, the protocol data analysed in this paper comes from the COSTART project, which carried out research into creativity between artists and technologists, bringing experts together in Human Computer Interaction, Creative Media and Digital Art Practice. A full account of the COSTART project may be found in Candy and Edmonds (2002a) and Edmonds et al (2005). There were many types of data sources, which includes field notes, residency diaries, audio and video recordings of collaboration. Each case from the COSTART project involved three kinds of participants: artists who submitted a residency proposal, technologists who had the expertise artists needed and observers who collected data during the residency by direct observation over a period of five days. Preparatory work was begun in advance of the actual period in residence in order to identify required tools and expertise necessary to support each project (Candy & Edmonds 2002b).

The data used for each case in this research comprises selections of recorded conversations between artists and technologists that took place over three days in each residency. The data represents sequences of conversations between artists and technologist that took place over three days. The brief background information of each case is provided below.

In case A, the proposal by the artist A_1 was made to develop a 3D computer model based upon an existing physical model. During the residency, the goal was to model a physical object by using a software- application called 3D Studio Max. The artist A_1 in this case had a great deal of experience in the digital art research field but she was not familiar with the 3D Studio Max software. The technologist T_1 had advanced expertise in the 3D Studio Max software but he had very limited knowledge about art.

In case B, the artist A_2's objective was to create a composition of percussion and orchestral samples which could be executed in real-time by several participants across the Internet. A_2 came from a music background and he also had some knowledge about computers and software. The technologist T_2 in this case was a fully-trained computer programmer and he also worked as a designer for years.

In case C, the artist A_3's objective was to develop a drawing machine which could be used in performance to create a visual image. A_3 was a digital artist who was major in interactive art and he also had a great deal of experience in artist-residency collaboration. The technologist T_3 came from computing science background and was familiar with many applications in Mac and PC. However, T_3 did not have much practical experience in some specific software program of Mac, such as Max MSP (Cycling74 2001) and Director (Adobe). In the next section, the analysis method is presented and analysis process is illustrated.

3 Methods

For the data analysis, we applied the protocol analysis method (Ericsson & Crutcher 1991), which has been broadly applied in cognitive science research (Suwa, Purcell & Gero 1998) and collaboration research (Cross, Christiaans & Dorst 1996). Protocol analysis makes use of categories of fundamental knowledge such as concepts, attributes, values, tasks and relationships. A coding scheme is a group of these categories, which reflect researchers' interests focusing on particular aspects relevant to researchers' questions (Suwa, Purcell & Gero 1998). The coding scheme applied in this research contains a group of communication behaviours identified in a previous study by the authors in a similar art-technology collaboration context (Zhang & Candy 2006c) (Table 1).

Table 1. Communication behaviour code scheme

Code	*Abbr.*	*Definition*
Evaluation	EVA	Analyzing an object according to a criterion. It can be either positive or negative.
Agreement	AGR	Agree or disagree with others.
Justification	JUS	Arguing or explaining the rationale of a certain choice.
Enquiry	ENQ	Ask questions, started with 'when', 'where', 'how', 'what' etc.
Solution generation	SO-GE	Suggestions for problem solving and specific applicable plans.
Recall	REC	Retrieve knowledge about previous experience
Inform	INF	Hand out information with respect to the nature of an object or criterion; illustrate a fact; explain some knowledge new to others
Fragment	FRA	Sentences do not make very much sense

The process of protocol analysis in this research was assisted by software INTERACT (Mangold 2005), which is a computer based video protocol analysis package. The following figure presents a snapshot of analysis carried out in this software (Fig. 1). In this figure, part 3 is the area where the video data is played, part 1 shows the segmentation of the video data with precise starting and ending time, part 2 shows the detail of the coding for each segmentation, part 4 shows the annotation of the coding segment where transcriptions or research thoughts about this particular segment are recorded, part 6 shows the coding scheme.

The analysis process in each case may be described in three steps. First, the duration of each coded data segment under each node in Table 1 was calculated. Second, distributions of the duration of each node in the previous step were made. Third, those distributions were further compared within two different communication environment settings, which have been observed in the cases: the first one is named 'non-assisted environment', where conversations are conducted when participants talked with each other without using any medium, such as computers, physical objects etc; the second condition is named 'computer-assisted environment', where conversation are conducted when participants talk with each other without using any medium, such as computers, physical objects etc.

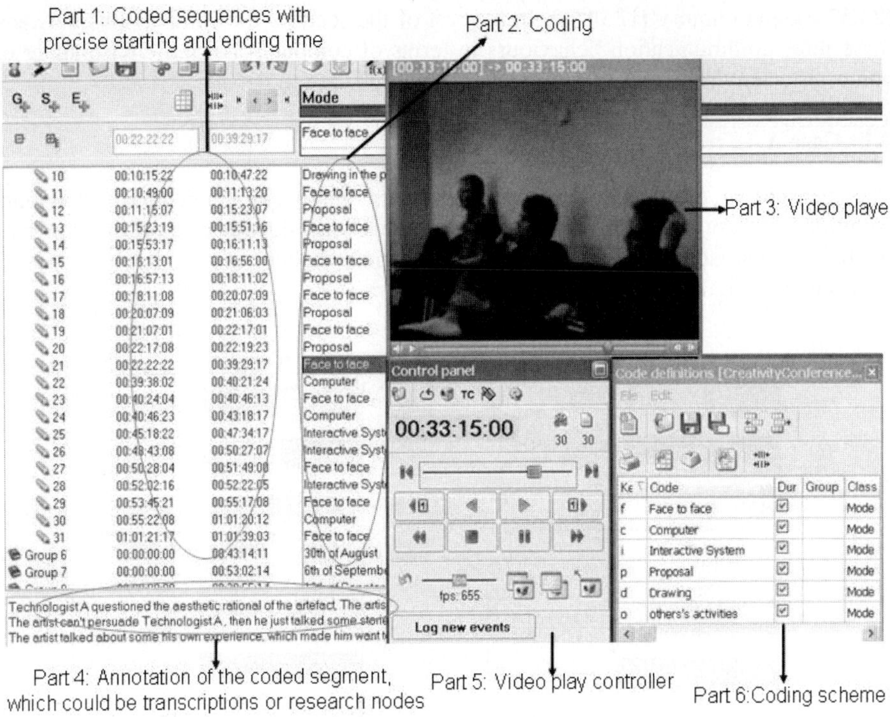

Fig. 1. Snap shot of analysis carried out in INTERACT

4 Analysis Results

In this section, the statistical data related with different kinds of communication behaviours, which were listed as nodes in Table 1, were presented. The following figure (Fig.2) shows the overall distribution of communication behaviours within three cases.

From Fig.2, we can see that within three cases, participants mainly spent most of the time on three communication behaviours: inform (37.63%), solution-generation

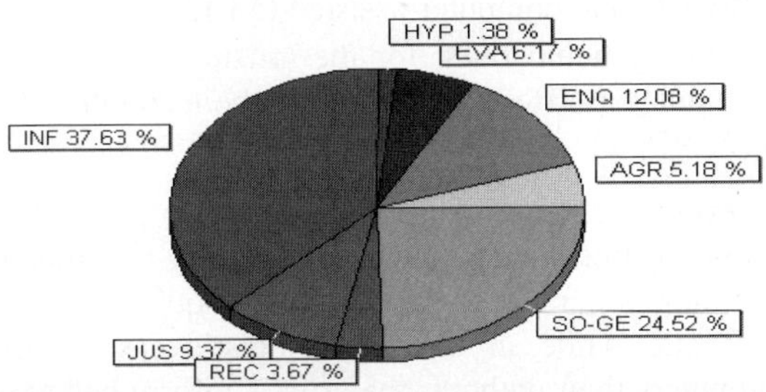

Fig. 2. Distribution of communication behaviours within three cases

(24.52%), and enquiry (12.08%). In the rest of the section, the comparisons between these three communication behaviours in terms of computer assisted environment or non-assisted environment are presented within each case in detail.

4.1 Case A

The distribution of the duration of participants' three communication behaviours (solution-generation, inform and enquiry) in case A was illustrated as follows in terms of the comparisons between non-assisted environment and computer assisted environment (Fig. 3).

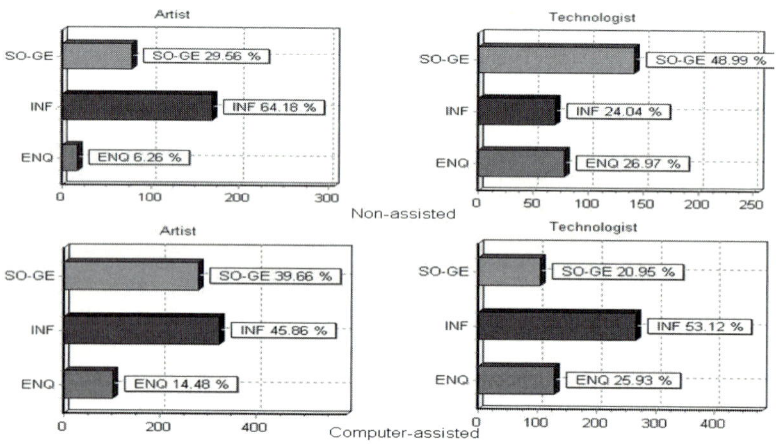

Fig. 3. Duration of participants' communication behaviours during non-assisted environment and computer assisted environment in case A

From Fig 3, we can see that

- Participants' solution-generation behaviours: for the artist, it increased from non-assisted (29.55%) to computer-assisted (39.66%); for the technologist, it decreased from non-assisted (48.99%) to computer-assisted (20.95%).
- Participants' inform-behaviours: for the artist, it decreased from non-assisted (64.18%) to computer assisted (45.86%); for the technologist, it increased from non-assisted (24.04%) to computer-assisted (53.12%).
- Participants' enquiry behaviours: for the artist, it increased from non-assisted (6.26%) to computer-assisted (14.48%); for the technologist, it decreased from non-assisted (26.97%) to computer assisted (25.93%).

In summary, in terms of changes from non-assisted to assisted, the artist's and the technologist's communication behaviours were affected by computers in a different way: the artist's solution generation behaviour and enquiry behaviour increased with computers than without, while in the technologist's case, these two behaviours decreased with computers than without; the artist's inform behaviour decreased with computers than without, while the technologist's inform behaviour increased with computers than without.

4.2 Case B

The distribution of the duration of participants' three communication behaviours (solution-generation, inform and enquiry) in case A was illustrated as follows in terms of the comparisons between non-assisted environment and computer assisted environment (Fig. 4).

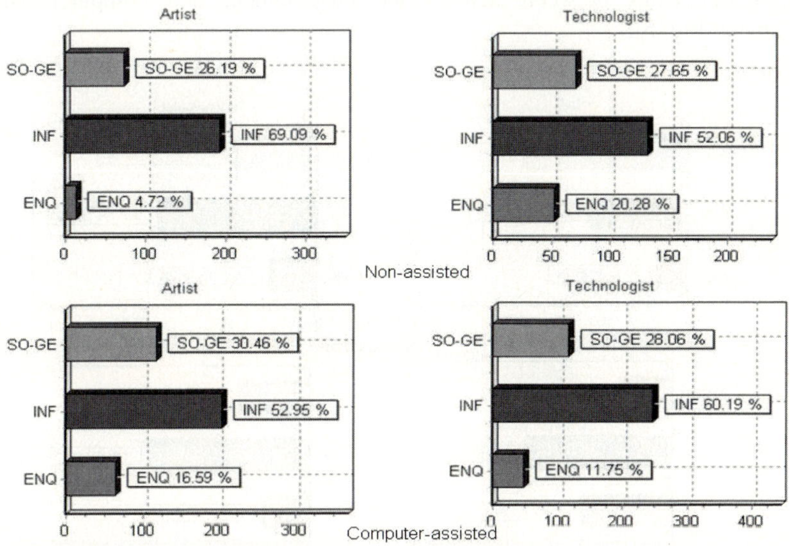

Fig. 4. Duration of participants' communication behaviours during non-assisted environment and computer assisted environment in case B

From Fig 4, we can see that

- Participants' solution-generation behaviours: for both the artist and the technologist, it increased from non-assisted (A: 26.19%; T: 27.65%) to computer-assisted (A: 30.46%; T: 28.06%).
- Participants' inform behaviours: for the artist, it decreased from non-assisted (69.09%) to computer assisted (52.95%); for the technologist, it increased from non-assisted (52.06%) to computer-assisted (60.19%).
- Participants' enquiry behaviours: for the artist, it increased from non-assisted (4.72%) to computer-assisted (16.59%); for the technologist, it decreased from non-assisted (20.28%) to computer assisted (11.75%).

In summary, in terms of changes from non-assisted to assisted, the artist's and the technologist's inform and enquiry communication behaviours were affected by computers in a different way: the artist's enquiry behaviour increased with computer than without and the technologist's enquiry behaviour decreased with computer than without; the artist's inform behaviour decreased with computers than without and the technologist's inform behaviour increased with computers than without. Moreover, the artist's and the technologist's solution-generation behaviours were affected by computers in a similar way: both of participants' solution-generation behaviours increased with computers than without.

4.3 Case C

The distribution of the duration of participants' three communication behaviours (solution-generation, inform and enquiry) in case A was illustrated as follows in terms of the comparisons between non-assisted environment and computer assisted environment (Fig. 5).

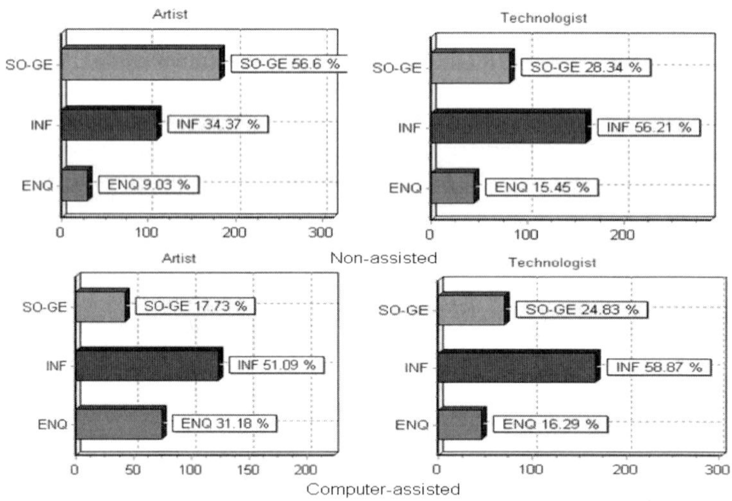

Fig. 5. Duration of participants' communication behaviours during non-assisted communication and computer assisted communication in case C

From Fig 5, we can see that

- Participants' solution-generation behaviours: for both the artist and the technologist, it decreased from non-assisted (A: 56.6%; T: 28.34%) to computer-assisted (A: 17.73%; T: 24.83%).
- Participants' inform behaviours: for both the artist and the technologist, it increased from non-assisted (A: 34.37%; T: 56.21%) to computer assisted (A: 51.09%; T: 58.87%).
- Participants' enquiry behaviours: for both the artist and the technologist, it increased from non-assisted (A: 9.03%; T: 15.45%) to computer-assisted (A: 31.18%; T: 16.29%).

In summary, in terms of changes from non-assisted to assisted, the artist's and the technologist's communication behaviours were affected by computers in a similar way: both the artist's and the technologist' inform behaviour and enquiry behaviour increased with computers than without; both the artist's and the technologist' solution-generation behaviour decreased with computers than without.

5 Comparison of the Case Study Results

In this section, comparisons between the results from the three cases described in the previous section are carried out and the implications discussed.

First, for the solution-generation behaviour, comparing the statistical analyses of three cases outlined in Section 4, two out of three technologists exhibited the solution-generation behaviour more often with computers than without, whilst two out of three artists had the solution-generation behaviour more often without computers than with. As the solution-generation behaviour was defined in Table 1 previously, as "suggestions for problem solving and specific applicable plans", this finding could indicate that computers might help the technologists to provide more solutions during the collaboration with the artists. However, computers do not appear to offer a similar effect for the artist's solution-generation behaviours. It would seem that the artists can produce more solutions without computers than with computers. The reasons for this differentiation may possibly lie in the degree of ease or familiarity with the medium for each person and also in the role each plays in the collaborative situation.

Second, for the inform behaviour, it was observed that the information type the technologists were mainly concerned with was related to technological knowledge, such as functions of software applications. On the other hand, the artists were mainly concerned with artistic knowledge, such as artistic concepts and aesthetics. Comparing the statistical results illustrated in the previous section, the inform behaviour of the technologists in all three cases increased from non-assisted to assisted and two out of three artists' inform behaviours decreased from non-assisted to assisted. This suggests that the artists used inform behaviours to illustrate their artistic knowledge more often in the non-assisted environment than the computer-assisted environment and on the contrary, technologists used inform behaviours more often in computer-assisted environment than non-assisted environment. The reasons for this differentiation may also lie in the artists' and technologists' familiarity with the medium, and in their preference for the way ideas are communicated.

Third, for the enquiry behaviour, comparing the statistical analyses of three cases listed in Section 4, the enquiry behaviour of all the artists in three cases increased from non-assisted to computer-assisted while two out of three technologists had less enquiry behaviours when they were with computers than without. In other words, the artists asked more questions when they were under computer-assisted environment than non-assisted environment. This finding may suggest that working in front of computers could raise more questions for artists to ask and the more questions the artists asked, the better the artists and the technologists could understand each other. For example, the artists could understand better the potential of the technology for their artistic goals and the technologists could become more aware of what the artistic requirements were and what were the artists' concerns.

In summary, based on these three cases, computers affect the artists' and the technologists' solution-generation behaviour, inform behaviour and enquiry behaviour differently. For the behaviours of solution-generation and inform, computers can increase these behaviours of the technologists and decrease these behaviours of the artist. On the contrary, for enquiry behaviour, computers can increase this behaviour of the artists and decrease this behaviour of the technologists. From these findings, we can see that during computer assisted environment, the artists asked more questions, generated less solutions, informed less their own knowledge than they did in non-assisted environment. In comparisons, during computer-assisted environment, the technologists asked fewer questions, generated more solutions and informed more their own knowledge. These findings reveal some features of the role of the computers during the communication between artists and technologists. Computers, as a mediation tool, could help artists asked more questions and help technologists to tell the artists more about the technology itself and also it could help technologists to produce more solutions to the problems through the discussion with the artist.

6 Conclusions

The contribution of this paper is two-fold. First, it describes an innovative approach to analyzing qualitative data that may be adapted by future research in other domains. Secondly, the communication behaviour patterns of the artists and technologists are a novel way to analyse the nature of the interaction during a creative collaborative process. These findings illustrate: firstly, how the artists' and the technologists' communication behaviors were distributed, and secondly, in terms of the impact of the role of computer mediation, what are the differences and similarities between artists and technologists' three behaviours they mainly spent time on (solution-generation, inform, enquiry). This research study is one of the first to provide a communicative behaviour analysis of the nature of collaboration between artists and technologists for developing interactive art. The ultimate significance of the research is to help us to understand the impact of computers and their software as a mediation tools in art-technology collaboration. Based on these understandings, in the future, strategies could be developed to optimize the computer-assisted collaboration process in order to facilitate the success of collaboration between artist and technologists and also, computers could be used intentionally during the collaboration in order to overcome the disadvantages inherent in collaborative processes between people of different backgrounds.

References

1. Adobe, viewed (October 2006) http://www.adobe.com/products/director
2. Candy, L., Edmonds, E.A.: Explorations in Art and Technology. Springer-Verlag, London (2002a)
3. Candy, L., Edmonds, E.A.: Modeling Co-Creativity in Art and Technology. In: Hewett, T., Kavanagh, T. (eds.) Proceedings of the Fourth International Conference on Creativity and Cognition, pp. 134–141. ACM press, New York (2002b)
4. Cross, N., Christiaans, H., Dorst, K.: Introduction: The Delft Protocols Workshop. In: Cross, N., Christiaans, H., Dorst, K. (eds.) Analysing design activity, pp. 1–14. John Wiley & Sons, Baffins Lane, Chichester (1996)

5. Cycling74 2001, Max 4: Getting Started. viewed (October 2006) http://www.synthesisters.com/download/Max4GettingStarted.pdf
6. Edmonds, E.A., Candy, L., Fell, M., Pauletto, S., Weakley, A.: The Studio as Laboratory: Combining Creative Practice and Digital Technology Research. International Journal of Man.-Machine Studies, Special Issue on Creativity and Computational Support 63(4), 452–481 (2005)
7. Ericsson, K.A., Crutcher, R.J.: Introspection and verbal reports on cognitive processes - two approaches to the study of thought processes: A response to Howe. New. Ideas in Psychology 9(1), 57–71 (1991)
8. Harris, C.: Art and innovation: the Xerox PARC artist-in-residence program. MIT Press, Cambridge (1999)
9. Mangold, INTERACT user guide, Mangold International GmbH, Arnstorf, Germany (2005)
10. Suwa, M., Purcell, T., Gero, J.: Macroscopic analysis of design processes based on a scheme for coding designers cognitive actions. Design Studies 19(4), 455–483 (1998)
11. Zhang, Y., Candy, L.: Investigating collaboration in art and technology. CoDesign: International Journal of CoCreation in Design and the Arts. 2(4), 239–248 (2006a)
12. Zhang, Y., Candy, L.: Investigating Interdisciplinary Collaboration: Case Studies in Art and Technology. In: Proceedings of the third international conference on Qualitative Research in IT & IT in Qualitative Research, Griffith University Press, Institute for Integrated and Intelligent Systems, Griffith University, Brisbane, Australia, pp. 173–183 (2006b)
13. Zhang, Y., Candy, L.: A Study of Interdisciplinary Collaboration in Art and Technology. In: Edmonds, E., Muller, L., Turnbull, D. (eds.) Proceedings of CCS/ACID Symposium: Engage: Interaction, Art & Audience Experience, Creativity and Cognition Studio Press, Creativity and Cognition Studio, University of Technology, Sydney, Australia, pp. 282–290 (2006c)

The Vision of Ubiquitous Media Services: How Close Are We?

Maria Åkesson and Carina Ihlström Eriksson

Halmstad University, P.O. Box 823,
301 18 Halmstad, Sweden
{maria.akesson,carina.ihlstrom_eriksson}@ide.hh.se

Abstract. The high penetration of mobile devices implies that we are closing up to the vision of ubiquitous media environments. This paper reports from a broad survey about barriers, use patterns, and motivations for using mobile media services. The aim is to contribute to the understanding of mobile service adoption process by studying users in different phases of the adoption process. The findings imply that mobile services are still not adopted and ubiquitously ingrained in use patterns. There is still a long way to go.

Keywords: ubiquitous media services, mobile service adoption.

1 Introduction

The visions of ubiquitous information environments encompass anytime, anywhere access to mobile services in mass-scale through a multitude of devices [1; 2; 3] The penetration of mobile devices such as mobile phones in everyday use was in 2006 as high as 110% in both Italy and Sweden [4]. During recent years, new and improved technology such as high-speed wireless broadband, hotspots and, high capacity mobile devices has brought us closer to the vision. Given this development, hopes for a new arena for media services have grown among service providers.

However, in spite of the high penetration and technological maturity the adoption of mobile services have not reached the levels hoped for [5; 6]. A growing body of research into adoption of mobile service has sough to understand the reasons for the slow adoption rate of mobile services [7; 8; 9; 10; 11]. Among the explanations of this phenomena are usability factors, technological factors and factors related to the business models. However, these explanations are not enough if we want to understand more about the process of adopting mobile services in ubiquitous information environments. Adoption process is in literature described as the relative speed by which an innovation is adopted by members of a social system [12]. To understand the process of mobile service adoption in ubiquitous information environments we need to uncover the influencing factors and barriers of adoption in relation to adoption level of anytime anywhere use of services.

In this article we aim at contributing to the understanding of mobile service adoption process by studying users in different phases of the adoption process. The aim is to identify use patterns in time and space, motivating factors as well as to

M.J. Smith, G. Salvendy (Eds.): Human Interface, Part II, HCII 2007, LNCS 4558, pp. 222–232, 2007.

identify barriers that influence the adoption. The intention is to give a reality snapshot from April 2006 to provide an idea of how close to the vision of a ubiquitous information environment we are. With this in mind, we have conducted a broad survey among users of mobile media services offered by media organizations, i.e. a ubiquitous media environment [13]. This context is very suitable since media services are easily accessed through for example portals, there is a fairly large range of services. Further, it meets an everyday need and thereby has a wide potential audience compared to services satisfying specific user needs.

2 Related Theory

Commonly, studies on how mobile services are taken into use and ingrained in peoples everyday use have applied frameworks based on diffusion and innovation theory [12], theory of technology acceptance [14; 15], and the unified theory of acceptance and use of technology [15].

The theory of diffusion and adoption by Rogers [12] describe communication processes and features of users and technological innovations to explain the process of how innovations are taken in use and spread. In this body of theory, users are categorized on the basis of how inclined they are to adopt new innovations. These groups are *innovators*, *early adopters*, *early majority*, *late majority*, and *laggards*. Innovators are the first to adopt a new innovation followed by the others. An individual's decision to adopt or reject an innovation is a process consisting of five steps *knowledge, persuasion, decision, implementation,* and *confirmation.* The characteristics of the innovation that influence this process are *relative advantage*, *compatibility*, *complexity*, *trialability*, and *observability* in relation to other products or services. Theory of technology acceptance [14; 15] explains that user's intention to use a new system is determined by *perceived usefulness* and *perceived ease of use*, in turn influencing actual use. The unified theory of acceptance and use of technology [15] posits that four key factors are determinants of intention to use and actual use. These are *performance expectancy*, *effort expectancy*, *social influence*, and *facilitating conditions*. The impact of these is mediated by gender, age, experience, and voluntariness of use.

Many of the existing studies of adoption of mobile devices and services have applied these theoretical frameworks. Studies have been addressing drivers for adoption and intentions to adopt mobile services [16], factors influencing adoption [17; 18], adoption patterns [8; 19; 20], and attitudes towards using mobile services [21]. However, there are still more issues to explore as the diffusion of new services is increasing and since there are a lot of hopes that has not been fulfilled, mobile services are still marginally adopted [5; 6]. In this study, we are addressing the barriers for adoption of mobile media services among people that have tried but rejected further use of mobile services and we are approaching users that regularly use mobile services to investigate their motivations for use and use patterns in time and space. Our argument here is that understanding adoption patterns require an understanding of the adoption process in relation to adoption level. Malhotra and Segars [19] addressed the adoption of wireless web services clustering respondents according to the categories of adopters according to Rogers's framework [12] and comparing them. They found that customer segmentation and together with an evolutionary approach are plausible strategies for reaching full customer adoption of new mobile services. However, we argue, that to understand the adoption process in ubiquitous media environments we need to address the ideas of anytime and anywhere usage in the analysis.

Sarker and Wells [10] took an approach grounded in users actual practice and designed a framework for studying key issues related to mobile device use and adoption, including aspects of mobility. This framework is built as an input-process-output model. The inputs are *user characteristics*, *communication/task characteristics*, *technology characteristics*, *modality of mobility*, and *surrounding context*. The process consists of *exploration and experimentation* as one sub process and *assessment of experience* as another. Output refers to actual adoption behaviors, such as continuity of use over time. As this framework is based on study of motivations and circumstances surrounding individual's adoption and use of mobile devices, some of the issues described in the framework are related to communicational tasks such as voice communication, sms, e-mail etc.

In this study we are investigating adoption and use of mobile media services with content such as sports results or weather forecasts. Therefore, the content needs to be taken into consideration along with the factors described above such as technology characteristics and attitudes to technology to understand user's adoption decisions. In this study we have clustered the respondents according to adoption outcome as described by Sarker and Wells [10] to analyze barriers among non-users as well as use patterns and motivations among users. We have included the anywhere and anytime perspective when describing use patterns, and content value aspects to describe motivation for use.

3 Method

The results in this paper build on a broad online survey concerning electronic news and mobile service use and motivations. Data was collected with a questionnaire at the web sites of three Swedish newspapers Aftonbladet, Göteborgs-Posten, and Sundsvalls Tidning during April 2006. These news organizations were all partners in the DigiNews project. Aftonbladet is a tabloid with the most visited news site in Sweden, Göteborgs-Posten is a local morning paper covering Göteborg (the second largest city in Sweden) and its surroundings, and Sundsvalls Tidning is a local morning paper in northern Sweden. All of these newspapers offer mobile media services to their audience.

As argued by [22], web samples can be regarded as representative as traditionally collected samples because of the heterogeneity of the online population. Admittedly, there are problems with controlling for example duplicate answers from respondents. In this case we blocked duplicate submissions from the same IP number in order to reduce that risk.

The questionnaire was divided in four parts concerning background data, business models for electronic news, preferences for future electronic news and, finally, use of mobile media services. The fourth part about mobile media services was only presented to the respondents that had answered that they use mobile services in part one. In total, 127 questions were asked and 4339 respondents answered the questionnaire. The respondents that had given an age under 15, those who did not complete or answered the questions included in this study contradictorily were excluded form the data set. The dataset contains 3626 respondents of whom 1388 (38.3%) are mobile services users and 2238 (61.7%) do not use mobile services. The user group was used to analyze use patterns in time and space and the non-users to analyze adoption barriers.

In the first part of the questionnaire the respondents that do not use mobile services regularly answered a multiple choice question about why not, to identify adoption barriers. The fourth part of the questionnaire concerning use patterns and motivations for use of mobile media services contained 41 questions. These questions were about motivation and what types of media services the respondents regularly use.

The respondents were clustered into four categories of users based on their response to adoption output (quantity of services that they regularly use). This allowed us to analyze the use patterns and motivations in relation to different levels of adoption. Use patterns are compared between the user groups on different adoption levels whereas the barriers for not using mobile services regularly could not be compared between the four groups as they were only answered by those who are non-users. By calculating the means of motivating factors we could compare scores of between the three different cluster groups of users.

4 Findings

In the ***cluster analysis***, questions about adoption outcome were used as input. The respondents were asked to state if they have ever tried using any kind of mobile service and in that case what services (see Table 1).

Table 1. Mobile services that the respondents have experienced

Service	No	Proportion
Download ring tones	1430	39.4%
News services	1185	32.3%
Information services (phone numbers, maps etc.)	1120	30.9%
Bank services	688	19.0%
Time tables (busses, trains etc)	621	17.1%
Sports results	605	16.7%
Download music	537	14.8%
Traffic information	512	14.1%
Download music videos	231	6.4%
Ordering services (flowers, tickets etc)	178	4.9%
Payment of parking fee	174	4.8%

As shown by Table 1, the services that have attracted most users to try mobile services are downloading ring tones followed by news services and information services. Services that have not been so attractive are ordering services and parking fee payment. However, there might be other reasons than lack of interests that explains this. The later services are not as widely offered as the services higher on the list, which might be one additional explanation. The cluster analysis based on continuity of use resulted in four groups with different levels of adoption; *Non-adopters*, *Low level*, *Medium level*, and *High level adopters* (see Table 2).

Table 2. Adoption level and individual characteristics

Cluster groups	No serv-ices	No of respon-dents	Gender W%	Gender M%	Av. age	Income level	Edu. level	Mobile phone penetr.
Non adopters 61,7%	0	2238	48.1	51.9	39	Low	Low	95.1%
Low 15,1%	1-2	546	35.5	64.5	37	Medium	Medium	99.6%
Medium 17,2%	3-5	625	28.0	72.0	36	Medium	High	100%
High 6,0%	6 +	217	18.9	81.1	33	High	High	100%
All	--	3626	38.9	61.1	37	Medium	Medium	97.5%

As can be seen in Table 2, the sizes of the different group differs, the by far largest group is the non-adopters. The low and medium level adopters are 15.1% respectively 17.5% of the total sample and the group with a high adoption level is 6.0% of the total sample. The individual characteristics are different between the groups. The gender distribution is more imbalanced the higher adoption level. Men are overrepresented in higher levels of adoption compared to the overall distribution between sexes in the entire sample. The average age is higher the lower adoption level. Further levels of income and education are higher the higher adoption level. The penetration of mobile phones is high in all cluster groups, however slightly lower among non-users.

To investigate ***adoption barriers***, the respondents in the non-adopter group answered a multiple choice question concerning the reasons why they do not use mobile services. In Table 3, these reasons are presented in frequency order.

Table 3. Percentage of non-users giving reasons for not using mobile services regularly

Reasons for not using mobile services	Proportion
It is too complicated to use mobile services	41.1%
I am not interested in trying out new technology	23.6%
Mobile services are too expensive	22.0%
I have no interest in mobile services	21.1%
My mobile phone does not have the capabilities required	16.9%
The available services are not interesting	13.9%
Navigating services in a mobile phone is too difficult	13.5%
Accessing services is too time consuming	12.9%
I am not acquainted with the selection of mobile services	4.0%

As shown in Table 3, among non-users 16.9% state that they are not in possession of a mobile phone with the capabilities required to use mobile media services, in turn this means that 83.1% are. The strongest barrier is that mobile services are perceived as being too complicated to use. Lack of interest in new technology and/or services are also important barriers. Noticeable is that services are perceived as being too expensive to use is as strong barrier as lack of interest. Interestingly, lack of knowledge about mobile services that are available is the least important barrier for the non-adopters.

To compare ***use patterns*** among mobile service users, the respondents were asked about how often they use mobile services in general (see Table 4).

Table 4. Use pattern regarding frequency of use per cluster group

Frequency of use	Low (546)	%	Medium (625)	%	High (217)	%
Almost never	14	2.6	2	0.3	0	0
Seldom	417	76.4	313	50.1	31	14.3
Once a week	75	13.7	159	25.4	62	28.5
Several times per week	17	3.1	83	13.3	49	22.6
Daily	9	1.6	28	4.5	16	7.4
Several times per day	14	2.6	40	6.4	59	27.2
Total	546	100	625	100	217	100

As can be seen in Table 4, use of mobile services in general is higher the higher adoption level. A vast majority of the Low level adopters use mobile services seldom. About half of the medium level adopters use services seldom and a quarter every week. Weekly or daily use is the pattern among high level adopters. The respondents were also asked at what time of day they in general use mobile services. As shown in Table 5, most users in all three cluster group use mobile services independently of time, e. g anytime. In general, very few respondents have the user pattern connected to a specific time of day.

Table 5. Use pattern concerning time of day per cluster group

Most common time of day	Low (546)	%	Medium (625)	%	High (217)	%
In the morning	10	2.0	3	0.5	6	2.8
At lunchtime	22	4.5	20	3.3	7	3.3
In the afternoon	34	6.9	29	4.8	7	3.3
In the evening	66	13.4	73	12.1	8	3.7
During the entire day	190	38.7	266	44.1	80	37.4
24/7	169	34.4	212	35.2	106	49.5
Missing	55	-	22	-	3	-
Total	546	100	625	100	217	100

On the topic of use context the results show that the use pattern differs between low level adopters compared to medium and high level adopters. Low level adopters use mobile services at home to larger extent than medium and high level adopters (see Table 6). On travel and on free time outside home are common contexts among all three cluster groups.

The respondents were also asked about their use patterns in relation to more specific categories of media services. They were asked what type of mobile media services they use, what time of day and how often they use each type of service (see Table 8).

Table 6. Use pattern concerning context where mobile services are used per cluster group

Use context	Low (546)	%	Medium (625)	%	High (217)	%
At home	137	27.7	111	18.3	32	15.0
On travel	123	24.8	154	25.4	61	28.6
Commuting to work	35	7.1	83	13.7	33	15.5
At work	66	13.3	80	13.2	26	12.2
Out of home on free time	134	27.1	178	29.4	61	28.6
Missing	51	-	19	-	4	-
Total	546	100	625	100	217	100

Table 7. Proportion of users within adoption level that regularly use mobile media service, and use pattern regarding how often and where the service is used

Type of service	%	Low (546) Use pattern	%	Medium (625) Use pattern	%	High (217) Use pattern
Sport services	30	Seldom At home	49	Seldom or a few times per week At home or on free time outside	67	Daily or several times per day At home or outside the home on free time
Weather	48	Seldom A home or on travel	67	Weekly or more seldom At home or on travel	80	Weekly or more seldom At home or on travel
News	59	Seldom At home or on free time outside home	79	Several times per day or weekly At home or on free time outside	87	Several ties per day All places
Stock market	18	Seldom At home	27	Seldom At home	45	Seldom At home or on free time outside home
TV-tableau	34	Seldom or weekly At home	48	Seldom or weekly At home	62	Several times per week At home
Dating	10	Seldom At home	10	Seldom At home	17	Seldom At home
Wine tip	12	Seldom At home	19	Seldom At home	31	Seldom or weekly At home
Down loading services	51	Seldom At home	69	Seldom At home	74	Seldom or weekly At home or on free time outside home
Mobile internet services	54	Seldom or weekly Home or outside home on free time and on travel	76	Several times per day or week Home or outside home on free time and on travel	80	Several times per day or week Home or outside home on free time and on travel

As can be seen in Table 7, the most commonly used type of service in all three cluster group is news services followed by mobile internet and downloading services. The types of services that are used by the least respondents are dating, wine tips and stock market services. It is interesting that many services in all three groups are used in the home. Use in the home is more common for the low adopter group while the high adopter group use more services on travel or outside the home. Further, many services among all three cluster groups are used seldom. This is of course related to type of service, news, weather and stock market information are of interest daily whereas wine tips or dating might not be so interesting for daily use. Still, many of the services that can be expected to be of daily interest are used seldom, such as weather and stock market services.

To investigate ***motivation factors***, the respondents were asked about what use experiences that motivates them to use mobile services. They graded the suggested experiences on a 7-grade Likert scale. The results are reported in Table 8.

Table 8. Experiences that motivates use per cluster group

	Low		Medium		High	
Experience	Mean	Std. dev	Mean	Std. dev	Mean	Std. dev
To be entertained	2.68	1.83	3.07	1.90	3.69	1.94
Be the first to know	3.24	2.04	3.86	2.02	4.59	1.97
Get surprised	2.87	1.88	3.30	1.87	3.68	1.97
Connecting with people	5.09	2.04	4.96	1.98	5.04	1.83
Learn new things	4.29	2.03	4.66	1.91	5.00	1.76

The motivators are similar between the three cluster groups. "Connecting with other people" is the most important experience for all three cluster groups followed by "learning new things". Surprisingly, "to be entertained" and "get surprised" scored the lowest in all cluster groups. It is noteworthy that the experiences generally are scored higher the higher adoption level. The respondents were also asked how they assess mobile services on a 7-grade Likert scale (see Table 9).

Table 9. Assessments of services per cluster group

	Low		Medium		High	
Services are the most valuable to me when	Mean	Std. dev	Mean	Std. dev	Mean	Std. dev
They are locally anchored	4.37	1.70	4.37	1.58	4.39	1.56
I am situated in specific locations	3.93	1.67	4.08	1.70	4.37	1.73
Content is locally adapted	3.97	1.81	4.19	1.73	4.43	1.75
Content is adapted to where I am	4.20	1.71	4.31	1.56	4.50	1.60
Content is available at the right time	5.53	1.51	5.89	1.29	6.20	1.02
Content is adapted to my personal interests	4.98	1.60	5.09	1.55	5.63	1.34
Content is adapted to time of day	3.83	1.80	3.94	1.83	4.06	1.89
I can use the same service wherever I am	4.37	1.75	4.54	1.73	4.73	1.70
I am traveling	4.65	1.95	4.97	1.85	5.26	1.67
They support my private needs	4.93	1.62	5.28	1.44	5.20	1.59
They help make my every day life easier	4.96	1.67	5.27	1.46	5.39	1.50
I can use the service in my work	4.05	2.10	4.73	1.99	5.32	1.79

The pattern is similar between the three cluster groups also for assessment and again the higher adoption level the higher score in general. All three cluster group find "availability at the right time" to be very important. "Support for everyday life" and "private needs" as well as "adaptation to personal interests" are considered valuable as well. "When I am traveling" and "wherever I am" score in the mid range. "Local adaptation" and "place" as well as "time of day" score the lowest. The high level adopters value "use for working purposes" higher than "support for private needs", while the other two groups value "support for private needs" higher.

5 Discussion and Conclusion

In this article, we have studied the interesting phenomenon of how users adopt mobile services, in order to learn about how close to the vision of ubiquitous media services we are. We have done so by identifying adoption barriers among those who have decided to reject services and use patterns and motivation factors among those who have adopted mobile services. The later were analyzed according to adoption level in terms of adoption outcome, based on continuity of use [10].

This study confirms that the adoption level of mobile services in general is low in spite of the high penetration of mobile devices in Sweden. Considering the high penetration it could be expected that the interest in mobile services would be higher. The maturity among users to ubiquitously use mobile services anytime and anywhere is low. The average user is still uninterested or in a trial and error phase. This implies that market communication of the contents and values of mobile services is very important to accelerate the adoption of mobile services as opposed to communicating technological innovation as in traditional adoption theory [12]. The results from this study on barriers, use patterns and motivation offer an explanation to this.

The barriers identified in this study are the opinion that mobile services are too complicated to use, lack of interest in new technology and new services as well as the cost level. These barriers are there even though the device as such is adopted for use as the majority of the non-users are in possession of devices with the capacity required. ***The use patterns*** among the three levels of adoption differs in that high level adopters use services more often, 24/7 and in physical and social mobility contexts. Interestingly, the home is very common, not only among low level and medium level adopters, but also fairly common among high level adopters. Based on this, it is plausible to say that mobile services compete with the home computer and TV as medium for services. The relative advantage often communicated for ubiquitous media services is the anytime and anywhere accessibility of mobile media services. A response to this is not reflected in the use patterns identified in this study. ***The motivation*** for using mobile was in this study among other things found to be the experience of connecting to other people and to learn new things, rather than to be entertained or surprised. Assessment of services are according to this study based on support for private needs, and for high level adopters also for work related needs rather than adaptations to context switches such as time and/or place.

From these findings we can conclude that the adoption rate has not changed, it is still as slow as reported in previous research [5; 6] in spite of service providers efforts to improve services and increased service offerings. This study implies that as long as the ubiquity of mobile services is ingrained in use patterns the adoption rate is slowed down. While mobile services offered in mobile phones or other handheld devices suffer from the limitations of small screens, in-established business models etc. they cannot compete with services offered by for example the home or work computer, nor the TV set.

To conclude, the findings of this research contributes to the understanding of mobile service adoption in ubiquitous media environments by shedding light on how close to this vision we are. We can conclude that we are not there yet, and it seems to be a long journey ahead.

References

1. Abowd, G.D., Mynatt, E.D.: Charting Past, Present, and future research in ubiquitous computing ACM Transactions on Computer-Human Interaction vol. 7(1), pp. 29–58 (2000)
2. Lyytinen, K., Yoo, Y.: Research Commentary: The Next Wave of Nomadic Computing. Information Systems Research 13(4), 377–388 (2002)
3. Weinberger, G.: The new millennium: wireless technologies for a truly mobile society. In: Solid-State Circuits Conference, Digest of Technical Papers, IEEE International, USA, pp. 20–24 (2000)
4. 30 Countries Passed 100% Mobile Phone Penetration in Q1. Retrieved January 30th (2007), Available at: http://www.telecommagazine.com/newsglobe/article.asp?HH_ID=AR_2148
5. Carlsson, C., Carlsson, J., Hyvönen, K., Puhakainen, J., Walden, P.: Adoption of Mobile Devices/Services – Searching for Answers with the UTAUT. In: Proceedings of the 39th Hawaii International Conference on System Sciences, Hawaii. (CD-ROM) Computer Society Press (2006)
6. Constantinou, I.D., Damsgaard, J., Knutsen, L.A.: Beware of Dane-geld: Even if Paid, M-Service Adoption Can be Slow. In: Proceedings of European Conference on Information Systems (2005)
7. Baldi, S., Pyu-Puy Thaung, H.: The Entertaining Way to M-Commerce: Japan's Approach to the Mobile Internet - A Model for Europe? Electronic Markets, vol. 12(1) (2002)
8. Carlsson, C., Hyvönen, K., Repo, P., Walden, P.: Asynchronous Adoption Patterns of Mobile Services. In: Proceedings of the 38th Hawaii International Conference on System Sciences (2005)
9. Samtani, A, Leow, T.T., Lim, H.M., Goh, P.G.J.: Overcoming Barriers to the successful Adoption of Mobile Commerce in Singapore. International Journal of Mobile Communications, vol. 1 (2003)
10. Sarker, S., Wells, JD.: Understanding mobile handheld device use and adoption. Communications of the ACM, vol. 46(12) (2003)
11. Vrechopoulos, A., Constantiou, I., Sideris, G.: The critical role of consumer behaviour research in mobile commerce. International Journal of Mobile Communications, vol. 1(3) (2003)
12. Rogers, E.M.: Diffusion of innovations. The Free Press, New York (1995)
13. Åkesson, M., Ihlström, C.: Towards an Ubiquitous Media Environment - adding the e-newspaper channel. In: Proceedings of TAGA, Vancouver (2006)
14. Davis, F.D., Bagozzi, R.P., Warshaw, P.R.: User acceptance of computer technology: A comparison of two theoretical models. Management Science 35(8), 982–1003 (1989)
15. Venkatesh, V., Morris, M.G., Davis, G.B., Davis, F.D.: User acceptance of information technology: Toward a unified view. MIS Quarterly 27(3), 425–478 (2003)
16. Anckar, B., DÍncau, D.: Value-Added Services in Mobile Commerce: An Analytical Framework and Empirical Findings from a National Consumer Survey. In: Proceedings of the 35th Hawaii International Conference on System Sciences, Hawaii, IEEE Computer Society Press, Washington (2002)
17. Khalifa, M., Cheng, S.K.N.: Adoption of Mobile Commerce: Role of Exposure. In: Proceedings of the 35th Hawaii International Conference on System Sciences, Hawaii, IEEE Computer Society Press, Washington (2002)
18. Hong, S.-J., Tam, K.Y., Kim, J.: Mobile data service fuels the desire for uniqueness. Communications of the ACM, vol. 49(10) (2006)

19. Malhotra, A., Segars, A.H.: Investigating Wireless Web Adoption Patterns in the U.S. Communications of the ACM 48(10), 105–110 (2005)
20. Knutsen, L.A., Constantinou, I.D., Damsgaard, J.: Acceptance and Perceptions of Advanced Mobile Services: Alterations during a Field Study. In: Proceedings of International Conference on Mobile Business, pp. 326–332 (2005)
21. Knutsen, A.: M-Service Expectancies and Attitudes: Linkages and Effects of First Impressions. In: Proceedings of the 38th Hawaii International Conference on System Sciences (2005)
22. Buchanan, T., Smith, J.L.: Using the Internet for psychological research: Personality testing on the World-Wide Web. British Journal of Psychology 90, 125–144 (1999)

Part II

Knowledge, Learning and Education

A Framework for Text Processing and Supporting Access to Collections of Digitized Historical Newspapers

Robert B. Allen, Andrea Japzon, Palakorn Achananuparp, and Ki Jung Lee

College of Information Science and Technology
Drexel University Philadelphia, PA
{rba,acj26,pa442,kl324}@drexel.edu

Abstract. Large quantities of historical newspapers are being digitized and OCRd. We describe a framework for processing the OCRd text to identify articles and extract metadata for them. We describe the article schema and provide examples of features that facilitate automatic indexing of them. For this processing, we employ lexical semantics, structural models, and community content. Furthermore, we describe visualization and summarization techniques that can be used to present the extracted events.

1 Introduction

Worldwide, there are many projects to digitize historical newspapers. One of the largest of these is the National Digital Newspaper Program (NDNP) of the U.S. National Endowment for the Humanities (NEH) and the Library of Congress (LC) that proposes to digitize several million pages of newspapers which were microfilmed as part of the United States Newspaper Program (USNP). We describe a framework for indexing and providing access to that material. We employ an iterative process of developing modular models and improving the corpora based on the materials in the NDNP collections (Figure 1). We have received extensive files of the Washington Times (1900-1902) from NEH/LC and we are using them to test our framework and tools.

We are focusing at first on improving access to the content and then on categorizing articles in the collection. Typically, we will first apply knowledge in which we have high confidence and then move to inference techniques that employ probabilistic inference. Text processing will help to categorize and index the articles and will be used to develop rich semantic representations. Moreover, the metadata and indexing derived from text processing can provide input for improved user interfaces. Both improved interfaces and representations of the historical events reported in the newspapers will organize, and facilitate access to this large news archive.

2 OCR Processing

2.1 Processing the OCR's Text

The words in the newspaper images have been extracted with OCR. The quality of the OCR is variable. In some places, the quality is high because the original text and the

M.J. Smith, G. Salvendy (Eds.): Human Interface, Part II, HCII 2007, LNCS 4558, pp. 235–244, 2007.

NDNP microfilms are also high quality. However, in other places, where the print is faded or smudged, the quality is marginal. The OCR'd text is delivered in METS ALTO XML (http://www.loc.gov/ndnp/techspecs.html). LC itself has been digitizing the Washington Times and we obtained the XML files for it. We process the METS ALTO OCR files to extract the article metadata. A gradual convergence with inferences across several levels is needed so the processing needs to be modular and adaptive. Figure1 shows the extracted and lightly-processed text along with page coordinates and fonts. Figure 2 shows a longer passage.

```
HEAD(ID11): HP: 1176.0 OS: 1806.0 VP: 4056.0 WC:   1.0 HD:   1.0 ST: ID3 T: 20  CO:  Numerous
HEAD(ID11): HP: 1902.0 OS: 2256.0 VP: 4047.0 WC:   1.0 HD:   1.0 ST: ID11 T: 21  CO: Other
HEAD(ID11): HEAD(ID11): HEAD(ID11):
begin TEXTLINE
HEAD(ID11): HP: 1176.0 OS: 1308.0 VP: 4242.0 WC:   1.0 HD:   1.0 ST: ID11 T: 22  CO: of
HEAD(ID11): HP: 1389.0 OS: 1602.0 VP: 4236.0 WC:   1.0 HD:   1.0 ST: ID11 T: 23  CO: the
HEAD(ID11): HP: 2004.0 OS: 2274.0 VP: 4233.0 WC:   1.0 HD:   1.0 ST: ID3 T: 24  CO:  hers
HEAD(ID11): HP: 2367.0 OS: 2676.0 VP: 4221.0 WC:   1.0 HD:   1.0 ST: ID11 T: 25  CO: Slain
##(accept current)
HEAD(ID11): HP: 2811.0 OS: 3387.0 VP: 4209.0 WC:   1.0 HD:   1.0 ST: ID11 T: 26  CO: Captured
HEAD(ID11): ##(skip current)
begin TEXTLINE
HEAD(ID11): HP: 1173.0 OS: 1311.0 VP: 4434.0 WC:   1.0 HD:   1.0 ST: ID11 T: 27  CO: or
HEAD(ID11): HP: 1383.0 OS: 1860.0 VP: 4413.0 WC:   1.0 HD:   1.0 ST: ID11 T: 28  CO: Injured
```

Fig. 1. Example of extracted OCR for the METS ALTO. The text on the right is news about the Boer War the word "Boers" is incorrectly identified as "hers".

> The petitioners say that living [I]n the [] near the mill is a burden and they assert [Chat] the smoke is not only a menace to the [h]ealth of the neighborhood but that clothing bedding etc are injured by the smoke and that the [I]nteriors their houses are practical[] ruined [] [I]t.

Fig. 2. A longer, relatively clean, fragment of processed OCR of text from the Washington Times (1902). Errors are indicated with brackets.

2.2 Article Segmentations

The basic unit of newspapers is the article. The METS ALTO standard used by awardees includes segmentation by text block, which generally means by paragraph rather than by article. Article segmentation for some historical newspapers is done manually. Therefore, we are developing an article segmentation utility to provide this function. For that tool, we will use a pixel analysis of the original TIF files as well as zones identified by word-coordinates from the OCR files and we will use knowledge about the layout for each newspaper (such as the layout of the masthead, see next section). Gatos et al. [14] have reported considerable success segmenting articles in one newspaper but we believe this is a difficult task for most historical newspapers. Beyond the clues from the image, we will apply semantic-based segmentation (e.g., Hearst, 1994). We have developed an interface for human identification and tagging of articles (Figure 3). We have tagged a test set of articles which can serve as the ground-truth for tuning the automated article-tagging process.

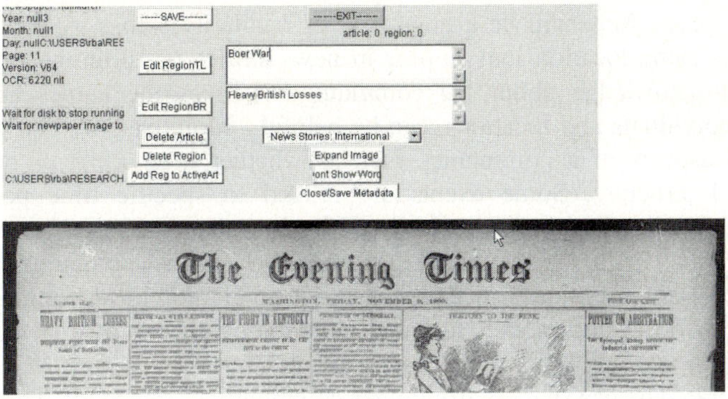

Fig. 3. Interface for extracting ground truth about articles from newspapers. The upper section is a control panel. The user can mark out articles on the newspaper image. In addition, the word positions identified by the OCR can be presented. Note that the article and word boundaries are not shown, but well reproduced, in this image.

2.3 Newspaper Structure

Some basic information about the structure is helpful for effectively tagging and processing the articles. Indeed, this is already used by some OCR extraction companies. Both the content and layout of individual news articles are highly structured. There is often redundancy within a single newspaper (across time) and across different newspapers. We can use the structure and redundancy to improve the accuracy of the text processing. We are developing hierarchical descriptions from sections, regions, and pages. The newspaper structure goes beyond the definitions of ALTO structure. This would include the types of material we can expect to find on specific pages and in specific sections. The newspaper model will characterize the structure and style of the material typically found in the newspapers. Some newspapers have continuation of articles. Some have advertisements on the front page. Initially, it will be tuned for the Washington Times. The parameters include continuation, column width. Moving to higher level content such as where advertisements are likely to appear and the appearance of features are expected.

2.4 Layered Content and Community Models

Newspaper content is very different from other types of material; typically, it is full of named-entities, that is the names of Organizations, People, Places, Roles, Dates, and Holidays. Moreover, there is a distinctive discourse or style [21]. Further, the collection of historical newspapers is distinctive because we have the complete history of news events rather than processing them as they are streamed. Indeed, in many cases (at least for national and international stories), we will be able to compare the way that the articles are presented across newspapers.

For local news, there will be a great deal of regularity across stories. The same person or the same location may appear in news stories on different topics. Thus, incorporating knowledge about the community, such as the names and roles of prominent individuals and locations, can be helpful. Moreover, we can characterize higher level aspects of a community such as whether it is urban or rural and its governmental structure. Some events are related to specific time periods. This includes both predictable cyclic events (e.g., elections, harvests, snow storms) as well as unique events (e.g., a World's Fair). Even more broadly, the community constraints will reflect the traditions and distinctive activities of individual communities. For instance, we will develop models for two specific localities --- a rural model for the files from Utah and an urban model based on files for Washington, DC. The model will be elaborated as newspapers from other communities are added. The model will also include events scripts; in some cases, these will be generic (e.g., what types of events are likely to be reported following a flood) and in other cases, the event instances will be elaborated in detail (e.g., a specific flood).

2.5 Extracting and Using Named Entities

As suggested above, named-entities are very helpful for processing news articles. Entity-lists form a sort of lexicon employed for the community constraint model. There are many sources of ontologies and of tools for developing ontologies [10] [30], but none of these is entirely appropriate for the application here in terms of the time-period covered. However, other types of material such as directories, gazetteers [19] and databases such as those available at Ancestory.com would be useful and we may explore research agreements to use them.

For processing historical U.S. newspaper articles, it will be essential to have systematic information about the relevant time periods. As an initial step, we are developing structured descriptions of prominent U.S. politicians with governmental structure both past and present. This extends the Federal Register's US Government Manual and should have independent value. Of course, the named-entities in the articles need to be matched to named-entities in entity lists and lists expanded. However, Crane and Jones [15] have described the difficulties in named-entity extraction from historical materials. For instance, people are referred to in many different ways.

For processing the article text, we expect to use Gate, a standard named-entity extraction tool [16]. While it is unclear how OCR degradation as seen in Figure 1 affects POS tagging, we expect to be able to repair the OCR sufficiently so that POS tagging performance will not be significantly degraded.

2.6 Article Metadata

Articles will be categorized by genre (e.g., editorials, reviews, advertisements) and by topic (international news, sports). This extends the earlier work of Allen and Schalow [7]. The International Press Telecommunications Council (http://www.iptc.org/). Both types of categories can be useful as metadata. Automatic genre categorization is based on the structural constraints. For instance, in some newspapers, the editorials appear on the center page. Figure 4 illustrates a possible descriptive metadata schema based on the Dublin Core. This is human generated and therefore costly in terms of time but thorough in description.

Element	**Scheme**	**Example of content**
DC Title		Tuan in Open Rebellion : Boxer Leader and General Tung-fuhsiang take the Field
DC.Subject	LCSH	China--History--Boxer Rebellion, 1899-1901
DC.Subject	NITF	Unrest, conflicts and war--civil unrest--rebellions
DC.Subject	Personal Names	Tuan, Tung-fuhsiang, Bazaure, Chwang, Yu Hsien
DC.Subject	Topic	International
DC.Type	NITF-newscode – genre	Actuality
DC.Coverage.Geographic Location.Historic		
DC.Coverage.Geographic Location.Current		Gansu Cheng - China; Kansu
DC.Coverage.Date		November 15, 1900
DC.Identifer.Serial	Standard No, LCCN	sn82-16310
DC.Identifer.Issue		1654
DC.Identifer.Page		01
DC.Identifer.Article		
DC.Relation	Use this to connect related articles	

Fig. 4. Qualified Dublin Core descriptions proposed for articles

After news and advertisements, notices are the next largest group of articles in the Washington Times. Figure 5 lists the article headings for a variety of notices. Notices are the most varied in terms of topic and structural presentation. Personal notices are distinguished from corporate or social entities through the font and formatting of the headings. Notices contain a variety of topics from death announcements to sales of coal.

Notices		
	Death	Personals
	Special Notices	Birds, Dogs, Etc.
	Commissioner's Notices	Recovery from Illness
	Society announcements	Educational
	Crime report	Foreign Mails
	Collaterals Forfeited	Help Wanted

Fig. 5. Subdivisions of Notices

The primary indication of a unique article of the Washington Times is the shape of an article given by its borders and visual cues. Advertisements are typically the easiest to detect with dark solid lines for borders or borders that are laced asterisk styled characters (e.g., Figure 6). Further, advertisements are the only type of article that contains graphic images as part of an article. Political cartoons are the only other form of images in the Washington Times.

The first line in the heading of a news article is in all capital letters and the largest font size of all titles/sub-titles in the heading. If there is a sub-title to the article, then a thin line of about a half inch in length is drawn. After this line, the subtitle appears and it is bold faced but not in all capital letters. If there is an additional subtitle, then another thin line will appear before the next subtitle which will be smaller in font size than the first subtitle. Articles that are not from DC will begin with the location of city and sometimes the state, followed by the date and a dash for example, “BALTIMORE, Md., Nov. 15--”.

This consistent way of formatting news articles is helpful in discriminating articles that are not news. There are many medical cures advertised that try to pass for news articles but lack the formatting particular to news articles.. Also, as the Washington Times fills the entire column, many very small articles are included in its pages. Typically, these are a one-to-three sentence excerpts from other newspapers; these excerpts are consistently formatted as well.

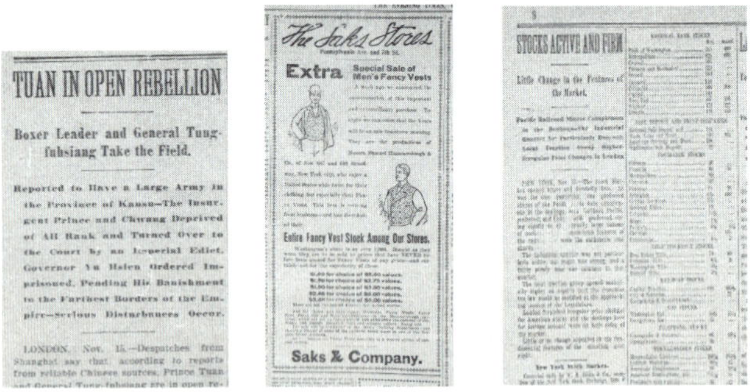

Fig. 6. Three examples of newspaper structures that can facilitate automated processing. In the article on left, note the distinctive layout and fonts. In the advertisement, center, note the dark border. For the stock reports on the right, note the distinctive tabular structure.

There is usually a large number of clues in each article relevant to category assignment. We will develop predictors of such clues for each category. For topic categorization, two approaches are commonly used: (a) the distribution of words that appear in the text [20]; and (b) the presence of specific named-entities.

2.7 Refining the Constraints and Named-Entity Lists

We are exploring the use of back-propagation neural networks learning about and adapting to the constraints [28] but refining the constraints (models) and entities will necessarily be a joint human-machine activity. The members of the project staff will make revisions from their own knowledge and we will encourage librarians and community groups to enter local knowledge. Thus, we will provide interfaces for people to work with the constraints and named-entities. Furthermore, because this may be a hybrid human-machine update process, we will track the sources of all revisions (i.e., data provenance) in case the learning drifts and needs to be reset and restarted. Because the extent of community participation may vary greatly, the text processing tools we will also need the option of tools that can operate with minimal human intervention.

2.8 From Historical Facts to Histories

Once we will have extracted entities, we can use them in many ways. For instance, users of the system could browse the people mentioned in the news for a given year.

These brief descriptions for individuals can link to specific news articles. We would also seek to provide summaries of major news topics and even to synthesize responses to open-ended queries about the news events in a given period. These are increasingly ambitious text processing tasks and they depend increasingly on statistical inferences.

Clustering News Stories and Topic Threads. News articles often report an evolving news story so that later articles expand upon earlier ones. A person viewing any one of these stories, may want to know about the other ones. To identify similar articles, a cluster analysis can be conducted. There has been a great deal of work on topic detection and Tracking (TDT) [1][17]. In TDT, nascent news topics are detected and tracked as they are received. In this project, we are looking for topics in an historical collection rather in a news stream. Thus, we can examine the entire corpus when trying to identify a topic thread. In fact, we could identify the topic first and work backwards to find the first story.

Multi-Document Summarization of Articles. One benefit of identifying topics sets of articles as described in the previous section is that we can pick articles as summaries from that set. Some articles will provide more background and could even provide overviews. For instance, an article appearing at the end of a sequence of articles on a given topic might summarize the earlier articles. However, none of the articles may be an adequate summary. Thus, we might be able to analyze the contents in more detail and to summarize them. For example, someone interested in the history of Washington, DC might ask the question "Describe the results of the 1902 commission to redesign Washington" and a summary could be provided.

Multi-document summarization combines evidence from several documents. The simplest method of summarization is extracting sentences; this is often accomplished by picking sentences with salient terms [29]. However, such extractive summaries are often unsatisfactory, and it may be helpful to consider a wider range of features such as the rhetorical structure of articles --- that is, the discourse purpose of different sections [26]. Techniques developed for dynamic collections (i.e., streaming news sources) may also be adapted. For instance, "NewsBlaster" is a news summarization tool [23]. As with TDT, news stories are processed in NewsBlaster as they are received, and we could likewise adapt this technique to fixed news collections.

Even more broadly we may want to understand the historical context of news articles. Consider a user who has identified an ancestor from a rural county and found news stories mentioning that ancestor that referred to unfamiliar events in the county's history. An interface may be developed to present summaries of those events to provide context about that ancestor. Finally, we will explore interfaces for coordinating with other historical resources.

2.9 Interfaces and End-User Access

NDNP will greatly benefit from improved end-user access to its content. The current generation of interfaces for digitized historical newspapers depends largely on either browsing by date or searching the OCR'd text. Many other types of navigation and search are possible. A graphical view may be the most effective way to present large amounts of information. Specifically, timelines which display sets of temporal events would be the central organizing metaphor. This could be sets of articles [31] or it could be attributes of Entities such as events in a person's life [2][4][5]. Figure 7 shows a novel timeline developed and applied to Brooklyn history. Significant events of Chinese immigration in Brooklyn are shown to the right of the center line. Broader historical context is shown to the left of the center line. Both (manually generated) text summaries of an article on Chinese immigration in Brooklyn and related news articles are shown in pop-up windows.

While simple timelines show sets of events, we'd also like to show dependencies among events. Events may be shown as causal relationships and the causal relationships may be woven into narratives. In previous work, we have explored graphical approaches showing causal relationships and narratives suitable for presentation in a Web browser. Specifically, a model of causation between events in scientific explanations was proposed by Allen, et al. [4]. In Allen and Acheson [6], narrative sequences of events were identified for a children's story. We will explore extending these techniques to large and complex descriptions of historical events as drawn from the newspapers.

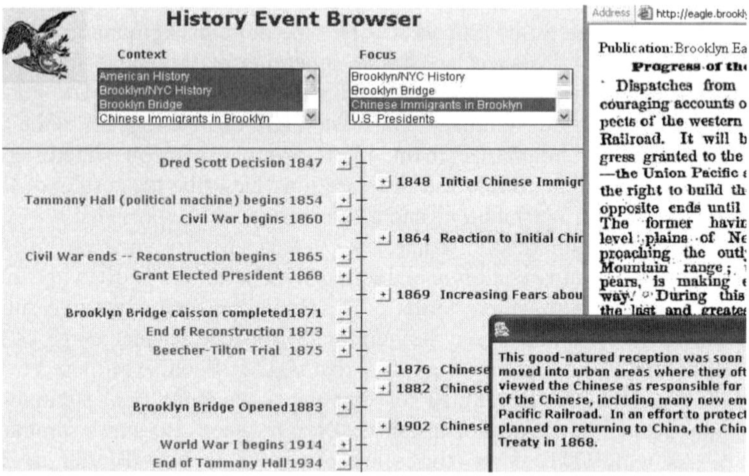

Fig. 7. A timeline browser for the Brooklyn Eagle (from Allen, 2005a). Focus events are shown right of center while sets of contextual events are shown on the left side.

3 Conclusion

We have outlined a framework for processing the digitized historical newspapers such as those from the NDNP. The first step is article segmentation based on image characteristics and the OCR'd text. Articles will be categorized based on their text, their position in the newspapers, and relationship to other articles. This text processing will facilitate access to the articles and eventually to the extraction of events.

Acknowledgments. This work is funded in part by NSF grants #0541637 for "Interacting with Threaded Event Scenarios".

References

1. Allan, J.: Introduction to topic detection and tracking. In: Allan, J. (ed.) Topic Detection and Tracking: Event-Based Information Organization, pp. 1–16. Kluwer, Dordrecht (2002)
2. Allen, R.B.: Timelines as information system interfaces. In: Proceedings of International Symposium on Digital Libraries, pp. 175–180 (1995)
3. Allen, R.B.: Developing a query interface for an event gazetteer. In: Proceedings of IEEE/ACM Joint Conference on Digital Libraries, pp.72–73 (2004)
4. Allen, R.B.: A focus-context timeline for historical newspapers. In: Proceedings of IEEE/ACM Joint Conference on Digital Libraries, pp. 260–261 (2005a)
5. Allen, R.B.: Using information visualization to support access of archival records. Journal of Archival Organization 3, 37–49 (2005b)
6. Allen, R.B., Acheson, J.: Browsing structured multimedia stories. In: Proceedings of ACM Digital Libraries Conference, pp. 11–18 (2000)
7. Allen, R.B., Schalow, J.: Metadata and data structures for the Historical Newspaper Digital Library Project. In: Proceedings of ACM Conference on Information and Knowledge Management, pp. 147–153 (1999)
8. Allen, R.B., Wu, Y.J., Jun, L.: Interactive Causal Schematics for Qualitative Scientific Explanations. In: ICADL 2005. LNCS, vol. 3815, pp. 411–415. Springer, Heidelberg (2005)
9. Beitzel, S.M., Jensen, E.C., Grossman, D.A.: Retrieving OCR text: A Survey of Current Approaches. Symposium on Document Image Understanding Technologies (SDUIT) Greenbelt, MD (2003)
10. Bontcheva, K., Maynard, D., Cunningham, H., Saggion, H.: Using Human Language Technology for automatic annotation and indexing of digital library content. In: Proceedings of the 6th European Conference on Research and Advanced Technology for Digital Libraries, pp. 613–625 (2002)
11. Caplan, P., Barnett, B., Bishoff, L., Borgman, C., Hamma, K., Lynch, C.: Report of the workshop on Opportunities for Research on the Creation, Management, Preservation and Use of Digital Content (2003) http://www.imls.gov/pubs/pdf/digitalopp.pdf
12. Cohen, W.W.: Infrastructure components for large-scale information extraction systems. Conference on Innovative Applications of Artificial Intelligence (2003)
13. Cox, R.J.: Documenting Localities: A Practical Model for American Archivists and Manuscript Curators, Scarecrow Press (2001)
14. Gatos, B., Mantzaris, S.L., Chandrinos, K.V., Tsigris, A., Perantonis, S.J.: Integrated Algorithms for Newspaper Page Decomposition and Article Tracking. In: Proceedings of the Fifth International Conference on Document Analysis and Recognition, p. 559 (1999)
15. Crane, G., Jones, A.: The challenge of Virginia Banks: an evaluation of named entity analysis in a 19th-Century newspaper collection. In: Proceedings of IEEE/ACM JCDL (2006)
16. Cunningham, H., Bontcheva, K., Tablan, V., Ursu, C., Dimitrov, M.: Developing language processing components with Gate (user's guide), Technical report, University of Sheffield, U.K (2002), http://www.gate.ac.uk/
17. Fiscus, J.G, Doddington, G.R.: Topic detection and tracking evaluation overview. In: Allan, J. (ed.) Topic Detection and Tracking: Event-Based Information Organization, pp. 17–31. Kluwer, Dordrecht (2002)
18. Hearst, M.: Multi-paragraph segmentation of expository text. In: Proceedings of Association for Computational Linguistics, pp. 9–14 (1994)

19. Hill, L.: Core elements of digital gazetteers: Placenames, categories, and footprints. In: Proceedings of the European Conference on Digital Libraries, pp. 280–290 (2000)
20. Hovy, E., Lin, C-Y.: Automated text summarization in SUMMARIST. In: Mani, I., Maybury, M. (eds.) Advances in Automated Text Summarization, MIT Press, Cambridge (1999)
21. Liddy, E.D, McVearry, K.A., Paik, W., Yu, E., McKenna, M.: Development, implementation and testing of a discourse model for newspaper texts, Human Language Technology (1993)
22. Mantzaris, S.L., Gates, B., Gouraros, N., Tzavelis, P.: Integrated Search Tools for Newspaper Digital Libraries
23. McKeown, K.R., Barzilay, R., Evans, D., Hatzivassiloglou, V., Klavans, J.L., Nenkova, A., Sable, C., Schiffman, B., Sigelman, S.: Tracking and summarizing news on a daily basis with Columbia's Newsblaster, Human Language Technology (2002)
24. Murray, R.: Towards a metadata standard for digitized historical newspapers. In: Proceedings of IEEE/ACM JCDL, pp. 330–331 (2005)
25. Petras, V., Larson, R.R., Buckland, M.: Time Period Directories: A Metadata Infrastructure for Placing Events in Temporal and Geographic Context. In: Proceedings of IEEE/ACM JCDL (2006)
26. Radev, D.R.: A common theory of information fusion from multiple text sources step one: Cross-document structure, ACL SIGDIAL (2000) http://www.sigdial.org/workshops/workshop1/proceedings/radev.pdf
27. Riloff, E., Lehnert, W.: Information extraction as a basis for high-precision text classification. ACM Transactions on Information Systems 12, 296–333 (1994)
28. Rumelhart, D.E., McClellan, J. (eds.): Parallel Distributed Processing, vol. 2. MIT Press, Cambridge, MA (1986)
29. Salton, G., Allan, J., Buckley, C., Singhal, A.: Automatic analysis, theme generation, and summarization of machine-readable texts. Science 264, 1421–1426 (1994)
30. Smith, D.A: Detecting events with date and place information in unstructured text. In: Proceedings of ACM/IEEE Conference Digital libraries, pp. 191–196 (2002)
31. Swan, R., Allan, J.: Automatic generation of overview timelines. In: Proceedings of ACM SIGIR, pp. 49–56 (2000)

A New Method for Teachers and Students to Record Daily Progress in a Class

Akinobu Ando[1] and Kazunari Morimoto[2]

[1] Miyagi University of Education
andy@staff.miyakyo-u.ac.jp
[2] Kyoto Institute of Technology
morix@kit.ac.jp

Abstract. The purpose of this study is to help to improve teachers' instructions by using information and communication technology. We developed a new method with the LAPP (Linux, Apache, PostgreSQL, and PHP) system. We adopted a mobile-phone as a terminal device. Using this system, a teacher could record easily and quickly students' learning situations also with taking pictures and evaluation data. A teacher could understand how well a particular student and whole students during a lecture – numerical evaluation values, average and the standard deviations of classroom average. We used this system actually for improving a lecture of programming and checking students' wood works. In both using, our system could work without any trouble. Understanding trends in students' learning, we found that this system could help teachers' instructions for students and reduce teachers' load.

Keywords: mobile-phone, interaction, learning situation, taking picture, improving a lecture.

1 Introduction

This paper is about a using information and communication technology to record various data for a teacher and students as an effective tool. We adopted a mobile-phone as the means. This is a really small device, which is able to connect to the Internet and has a camera function without the wire. We developed a network server system for a mobile-phone. This system was used in several classes as a trial. As a result, it was concluded that this system and method could help teachers to record data and to improve their class. In addition to, we found that we had to improve user-interfaces of this system.

2 Background and Purpose

In the Japanese educational field, a teacher has to record students' evaluation data [1]. In the past, teachers usually do it using a pen and papers from the old days. The amount of this data is huge and a teacher can not keep remembering. Thus, a teacher

M.J. Smith, G. Salvendy (Eds.): Human Interface, Part II, HCII 2007, LNCS 4558, pp. 245–251, 2007.

could not just record, but their lecture and use of data would not improve either. We think that a teacher has to utilize evaluation data for improving his/her way of lecture. In many cases, teachers record evaluations data on paper. After recording, they input huge amount of evaluation data to a PC for calculation. It is obvious that this operation is hard work and troublesome. And some teachers often make a mistake. To solve these problems, we are in much need of a new tool for evaluations. As a similar antecedent study, a handy terminal - called "bar code reader" - is used.

The way this antecedent study works was to collect evaluation data using a handy-terminal, bar-code and PC efficiently [2][3]. However there were three problems. First, it was so particular that the terminal lacked versatility. Second, it was troublesome to cooperate with a PC. To do it, a user must prepare a special software and hardware drive. Finally, the bar-code dirties or tears easily [4].

In this study, our purpose is to develop a versatile evaluation-system for teachers to use it efficiently and easily. A mobile-phone was selected as its means.

3 Essential of Evaluation Data and Validity of a Device

In this chapter, we will describe essential evaluation data for a teacher and the valid device to collect them. Table 1 shows evaluation data, which a teacher collected during a lecture.

Table 1. Evaluation data

Kinds of Information	Description
Information (numerical calculations)	Criterion of evaluation.
Information (characteristics)	To describe a reason or appearance of numerical information.
Information (illustrations)	Such as pictures or drawings.
Information (audio or sound)	Such as students' speech
Information (visuals)	Such as students' work or movement.

Other various information of evaluation may be needed by a teacher. In many cases, these five kinds of information are sufficient. We will not feature audio (sound) information and moving information because of size of data.

Next, we will consider the device to input information. According to a situation which a teacher will use the device in a class, we make a specification of the device (Table 2)

There are small laptop-PCs or PDAs or mobile-Phones (smart phone) which apply to this specification. Mobile-phones are a popular device these days. The number of contractor of mobile-phones is 78% of the Japanese population [5]. A mobile-phone is one of the information and communication terminal devices because of the high-rate of contractors, its small size and the ability to connect to Internet [6][7].

Table 2. A specification of the device

- Easy to prepare
- Easy to use and handy
- Ability of treating much information (numerical information, described information, illustrated information and so on)
- Easy to collect
- Easy to save, record and manage
- Easy to understand students' records.
- Good PC compatibility

4 Outline of the System

4.1 Circumstance of Development and Execution

We developed the system under the circumstance shown as table 3.

Table 3. The spec we used

• CPU	AMD Athlon1600+
• Memory	512MByte
• HDD	40GByte
• OS	FedoraCore1
• Database	PostgreSQL7.4.3
• CGI	PHP5
• Httpd	Apache2.5

This is a common system as a network-server using LAPP: Linux, Apache, PostgreSQL and PHP. This system can be used by mobile-phones which has a web browser function for example 2G and 3G models. The database tables of this system were defined as followed.

1. Teachers table, 2. Students table, 3. Subjects table, 4. Unit table, 5. View points table, 6. Evaluation value table, and 7. Image table.

4.2 Function of This System and Executing Screen Capture

This system has mainly three functions. First, inputting numeric value. Second inputting comments. Finally, taking pictures. A teacher selects one numeric value and a target student when he/she records evaluation information. Then comment information is written at a text box form. When a teacher wants to look for students' data, the average of all data is shown as a radar-chart. Also, the standard deviations of classroom average and their average can be shown (Figure 1).

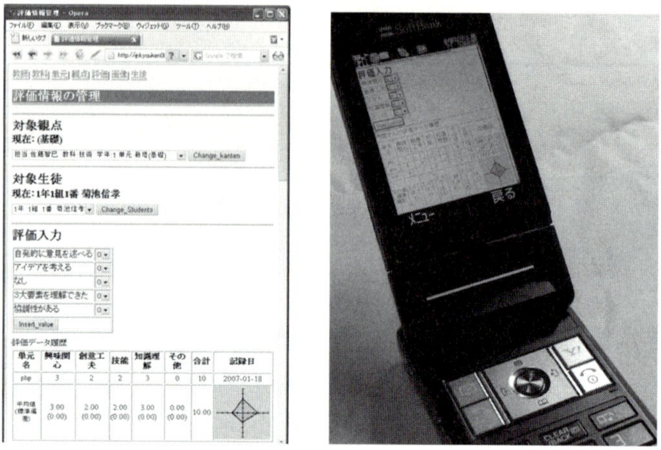

Fig. 1. Understanding trend in students' learning with a radar-chart on a PC screen (left) and on a mobile-phone (right)

Because this radar-chart is shown together on the screen for inputting data, a teacher can input students' data with understanding statistically about them - i.e. what are the students' strong or weak points.

All of these data are recorded in a database-server. Whenever a teacher needs data, he/she can download them as CSV format. Thus a teacher can deal with data using a spread-sheet - i.e. MS-Excel.

5 Demonstration in a Class

We tested our system in two situations to see if it works normally without trouble.

One is a situation where a teacher uses it in an actual class. Another is a situation where a teacher checks a student's work. This demonstration was used in university classes. The numbers of participant were 13 persons and we used it for a month (for a total of four times). In this class, an instructor teaches programming grammar and the students are writing a program every time. This system was used to record for a teacher how students write program (Figure 2). This was recorded on four view-points. First point is "Interest", second is "Idea", third is "Skill" and fourth is "Knowledge". Each view-point was based on a scale from 1 to 5. "1" means "worst", "3" means "normal" and "5" means "very good". "Interest" is how hard a student works. "Idea" is how far a student contrives his/her own works. "Skill" is how realizes a student's intention. "Knowledge" is how well a student remembers what a teacher taught - whether a student remembers programming grammar.

It was trouble and hard work for a teacher to keep remembering how students wrote a program code a few weeks ago. Using this system, a teacher can understand trends in students learning. If a teacher notices that students keep maintain a low level of "knowledge", a teacher will give homework to help students understand programming grammar or, if a teacher notices students maintain a low level of "skill", a teacher will teach know-how about programming. In this class, many students had

Fig. 2. Recording with a mobile-phone

knowledge about programming, but they could not write a program code using their own knowledge at the beginning of the first and second classes. The teacher could understand this deficit vaguely. However it was impossible for him/her to understand it in detail without our system. Concretely, students could not write a program code for reading all lines in a file with using the "while" command. Because of this, at the beginning of the third class, a teacher prepared some sample program codes and changed the method of teaching by indicating some similar algorithms. Then, at the beginning of the third and fourth classes, students got higher "skill" levels – i.e. using the "for" and the "while" statements.

Next, a teacher checked students' works using our system and a mobile-phone with a camera function, as shown in Figure 3 and 4.

Fig. 3. Checking students' works

Until now, a teacher used a pen and paper to record students' situations or conditions and evaluation data. After writing, the teacher had to re-type data into a PC in order to tally up data on a PC. Moreover, it was difficult for a teacher to remember situations in detail because he/she had only numerical values or only characteristic data. In this trial, total twenty persons used our system. They tried taking "smooth snap" – a method of taking pictures with a digital camera and uploading them to the server – pictures of students' works with numerical evaluation data. The steps were as follows: First, a teacher snapped pictures of students' works. Second, teachers accessed the system server using a web-browser on a mobile-phone, Third, he/she input data in forms on web-pages. Finally, he/she uploaded pictures via the Internet which had already been taken.

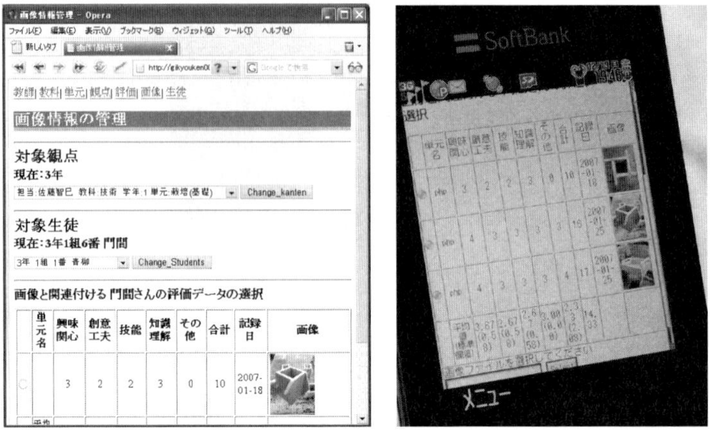

Fig. 4. A student's evaluation data and a picture shown in a web-browser on a PC screen (left) and on a mobile-phone (right)

This system was used by five teachers and fifteen students who were studying to be a teacher in a particular class which they performed some instruction as a teacher role. We had an interview with total twenty subjects, where they were asked about usefulness of the system to understand about trends in their students learning and convenience for their improving classes. All of the subjects answered that they thought it was meaningful to record both pictures and numerical data together, as this would allow them compare more effectively current conditions to past conditions. 90% of the subjects answered that they could upload pictures easily with numeric evaluation data. 35% of the subjects answered that they wanted to know a students' name when they looked at a file name. At the time, file names for pictures were a student's serial number. In the newest version the file name contains the student's name with a time stamp. Furthermore, 18% of the subjects answered that it was at times, a little hard to select a target student from pull-down menus on web pages.

6 Discussion

These trials show that such usage can help expand the potential of a mobile-phone. It is small, wireless, contains a digital camera, and can connect the Internet – it is the most suitable tool for this purpose. It is really hard for a teacher to carry a laptop-PC with single-handed, and a PDA is a minority device in Japan. Our system used some "form-parts" in web pages to input data. These kinds of "form-parts" have poor usability. It was not convenient when a teacher wanted to change or find a target student freely and smoothly from a lot of students' lists. In many cases, a teacher does not check a student in turn. If we could use a PC which has a wide screen and high performance, we could design a new user-interface using Java Script. However, many of the mobile-phones don't have performance to work Java Script without trouble. As long as the technology to run complex processes on the mobile-phone was not developed, it is better for us to use only versatile method – i.e. traditional HTML.

7 Conclusion

In this study, we developed the system for a teacher to record students' data and to help students learning using "form parts" in web pages on a mobile-phone web browser. In Japan, Most of the mobile-phones have "web browser" with limited function. Our system was coded with using simple statement. So our system can be almost worked on mobile-phones. In addition, the communication channel will be able to be encrypted by SSL (Secure Socket Layer). Using this system, a teacher could consider and arrange his/her way of teaching with understanding trend in students' learning, which maybe miss them if he/she had used old method. And we have to consider a new user interface to expedite to change and find a target student and to use smoothly a camera function.

References

1. Ministry of Education, Culture, Sports, Science and Technology (2000) http://www.mext.go.jp/b_menu/shingi/12/kyouiku/toushin/001211.htm
2. TAMURA, T., KOMURO, K.: Development of the Method of Recording Evaluation which Utilized the Bar Code, while Teaching between Besks. The. bulletin of Ibaraki University 54, 147–156 (2005)
3. TAMURA, T., KOMURO, K.: Design of the Evaluation Record System which Utilized a "Handy-terminal" as an Information Terminal. Japan Journal of Education Technology 29(4), 617–625 (2006)
4. ANDO, A., ABIKO, H., SATO, T.: Development of the "Fish-eye" User Interface on a Mobile-phone, Journal of the Japanese Society of Technology Education Tohoku Branch 24th, pp. 29–30 (2006)
5. Telecommunications Carriers Association (2007) http://www.tca.or.jp/
6. ANDO, A., ABIKO, H., YAMADA, T.: A statistical Evaluation of Participants' Awareness of Using e-mail with a Cellular-phone during a University Class, International Association of Societies of Design Research, International Design Congress, CD-ROM (2005)
7. ANDO, A., ABIKO, H., KINEFUTI, M.: Enhancing Interaction in School Hours by Using a Cellular –phone, International Ergonomics Association, vol.3, XVth Triennial Congress, CD-ROM (2003)

Evaluation Method of e-Learning Materials by α-Wave and β-Wave of EEG

Michiko Anse and Tsutomu Tabe

Graduate School of Science and Engineering, Aoyama Gakuin University
5-10-1, Fuchinobe, Sagamihara, Kanagawa, 229-8558, Japan
anse@ise.aoyama.ac.jp, tabe@ise.aoyama.ac.jp

Abstract. In e-Learning study, the materials are the only environment. Therefore, the influence which these have on a students' emotion has big influence on progress of study. The purpose of this study, from a standpoint of making easy-to-follow e-Learning materials, is to construct the foundation for a methodology to evaluate the materials from emotional aspect. At first we re-corded the basic EEG during mental calculation. Second, we tested two different levels of materials which had 3 important learning points. We recorded subjects' EEG, then analyzed and compared with the basic EEGs. The result suggested us that at the learning point, where the α-wave power was greater than the basic one and the β-wave was less than it , the content was not able to understand.

1 Introduction

Instructional Design (ID) that is a guideline to prepare materials defines two-step evaluation of e-learning materials. The first is the formative evaluation for im-proving materials in the development phase and the second is the overall evaluation for evaluating the effect of education in a comprehensive way after completing the development phase.

The purpose of this study is to select user-friendly materials that with higher learning effect among e-learning materials already developed and to lay the foundation for a method for evaluating materials from a viewpoint of affective aspect considering storyline. Kirkpatrick's four-level model is the most widely used and popular model for the evaluation of training and learning. The four levels of Kirkpatrick's evaluation model essentially measure: 1) reaction of student - what they thought and felt about the training, 2) learning - the resulting increase in knowledge or capability, 3) behavior - extent of behavior and capability improvement and implementation/application and 4) results - the effects on the business or environment resulting from the trainee's performance. They are calculated from the results of test, questionnaire and interview (Japan Society for Educational Technology ed., 2000). If classified roughly, 2) Learning can be evaluated by the results of tests, and 1) Reaction of student, 3) Behavior and 4) Results may be hardly evaluated from the viewpoint of affective aspect. Conventionally, information based on the subjective of learners has been used for evaluating the affective aspect. However, it is nonobjective because the criteria are un-clear.

M.J. Smith, G. Salvendy (Eds.): Human Interface, Part II, HCII 2007, LNCS 4558, pp. 252–259, 2007.

Therefore, researchers began to use biological information. Biological information is operant responding of learners. Biological information provides objective and quantitative data (Nakayama, Shimizu, et al., 2003). Meanwhile, the questionnaire and interview survey hardly evaluate materials considering change over time because they are performed after learning. Therefore, they are not appropriate for performing overall evaluations on materials completed including story line. In that respect, EEG that is one of the biological information is excel-lent in time resolution.

The purpose of higher education targeted in this study is to learn basic knowledge first and to understand and apply knowledge in a comprehensive way. Affective effect is highly requested in the higher education. Therefore, a method for evaluate materials from a viewpoint of affective aspect is important. As just mentioned, the affective aspect can be evaluated through an objective perspective by using EEG among biological information. Additionally, storyline can be also evaluated because EEG has better time resolution.

Previous works suggest that learners dedicate themselves to learning in comfortable atmosphere (learners learn deliberately in a relaxed and comfortable environment, make advances in their learning and are hardly distracted) when learners advance in knowledge at the essential points of learning (basic knowledge is out-lined and learners are made to apply knowledge). At the time, alpha wave is ob-served significantly (Murakami, 2003).

In this study, two materials with different difficulty levels were used for evaluating the affective aspect. EEG was measured over time for each material. The relationship between EEG change and final achievement of learning target was made clear in order to evaluate materials from a viewpoint of affective aspect. Each material had 3-graded learning targets. Essential points of learning are indicated at the end of each grade.

The evaluating method is as follows: 1) A reference status for evaluating the affective aspect (atmosphere in which learners learn deliberately in a relaxed and comfortable manner, make advances in their learning and are hardly distracted) is created. EEG is measured to determine the basic EEG. 2) Subjects are made to learn with e-learning materials. EEG at that time is measured. 3) Subjects take an exam for evaluating achievement of learning target. Verbal reports about materials are collected. 4) EEG at each phase of learning is calculated from the EEG during learning. Calculated EEG is compared to the basic EEG (matching level). 5) An evaluation method is obtained from the relationship among the matching level, examination and matching level.

2 Method

Basic EEG value is determined from the EEG when learners dedicate themselves to learning in order to evaluate the affective aspect during learning based on EEG. Then, EEG during learning by using materials with essential learning point known is measured. The EEG value calculated from EEG at the essential point of learning is cross-checked with the basic EEG value in order to evaluate the affective aspect of learning. Subjects are made a verbal report and take an examination in order to support the result. The procedure is shown below:

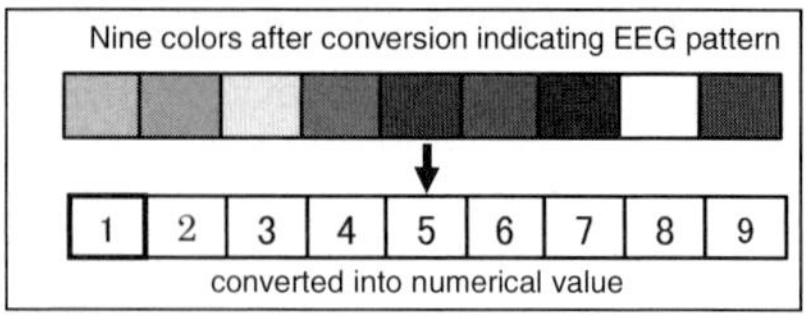

Fig. 1. Conversion of EEG pattern to numerical values

STEP1: The EEG when learners dedicate themselves to learning in comfortable atmosphere is recorded to obtain the basic EEG value. Topography of alpha and beta wave in the occipital region is prepared from the EEG. The pattern is converted into a numerical value with a digitizing program (Fig. 1). "Basic EEG value" is obtained from the average of processing results from 5 subjects.

STEP2: EEG of multiple subjects, who learn by two materials (different difficulty levels), is measured. After completing learning, verbal reports are recorded in order to capture subjective impression of the lesson, and subjects are made to take an examination in order to explore their level of understanding. Topography of alpha and beta wave in the occipital region is prepared for three essential points of learning. The pattern is converted into a numerical value with a digitizing program.

STEP3: The matching level of EEGs of the three essential points of each material is calculated on the basis of the basic EEG value. The denominator of the matching level is the basic EEG value obtained in STEP1. The numerator of the matching level is the EEG value during learning obtained in STEP2. The matching level is expressed in percentage.

STEP4: The relationship among the matching level and scores of the verbal report and examination is discussed.

3 Experimental Result

Subjects include master course students and senior students in this department. Five subjects do mental calculation in order to obtain the basic EEG value. E-learning materials during learning include two materials (referred to as material P and Q) with different difficulty levels from problems of production scheduling that is a part of "production management technology" in which more than 80% of subjects par-ticipate. Each experimental group consists of five subjects. The prepare of the mate-rials made material P and Q so that they included three grades respectively. As a re-sult of analyzing each material, subjects reach the essential point of each grade at 3:20, 5:51 and 16:20, and 5:30, 10:30 and 21:10 after starting learning (Fig. 2). The EEG at the point where the amount of beta wave becomes maximum within 30 seconds of the essential point of each grade is considered to be the data of the essential point. After that, the essential point is called "evaluation point" because the essential point is considered to be a point for evaluating the material.

After completing the experiment, subjects take an examination in order to explore their level of understanding each material. Additionally, a verbal report is obtained in order to (1) understand the subjective impression of the whole lesson and each

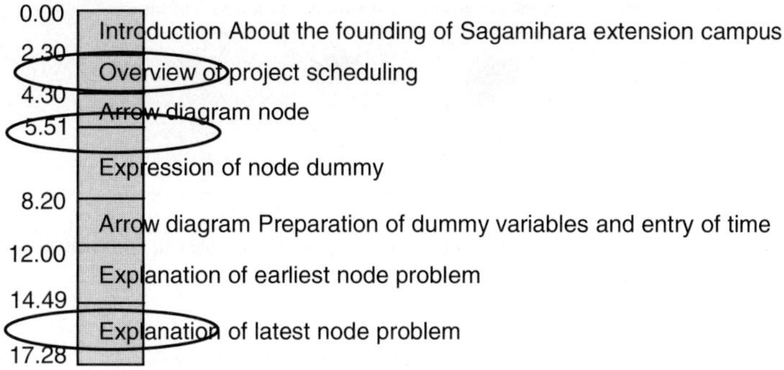

Fig. 2. Configuration of material P and three points

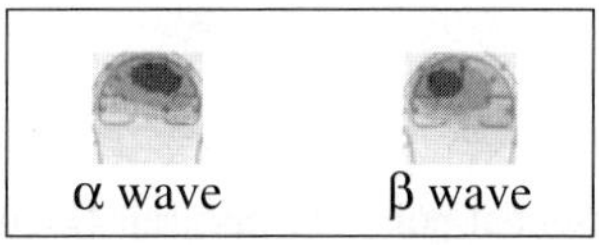

Fig. 3. Topographies of occipital region during mental arithmetic (subject a, maximum potential: alpha wave: 1.6μV, beta wave: 1.1μV)

Table 1. Basic EEG value

Subject	a	b	c	d	e	Average
a wave	1470	1265	1259	1536	1207	1347
b wavc	1441	1607	1532	1500	1676	1551

chapter, (2) to confirm that subjects consider the evaluation points to be essential points, and determine evaluation points based on a series of EEG data.

3.1 Basic EEG Value

Subjects do mental arithmetic (integer addition: 1+2+3, 3+3+6, 6+4+10, 10+5+15…) keeping their eyes open for two minutes. Based on the verbal report of subjects, they are dedicating themselves to learning in comfortable atmosphere when the sum is approximately 100. Therefore, the EEG of their occipital region (Fig.3) is FFT-analyzed at this point for two seconds to obtain a topography of wave and wave (Fig. 3). The topography is converted into a numerical value with a digitizing program (table 1).

3.2 EEG Value During Learning with e-Learning Materials

Figs. 4 and 5 show topographies of alpha and beta wave in occipital region for three essential points in material P (basic problems) and Q (applied problems) respectively.

As shown in Fig.4, alpha wave is dominant over beta wave at each evaluation point of subject a for the material P. As shown in Fig.5, beta wave is dominant over alpha wave at each evaluation point of subject a for the material Q.

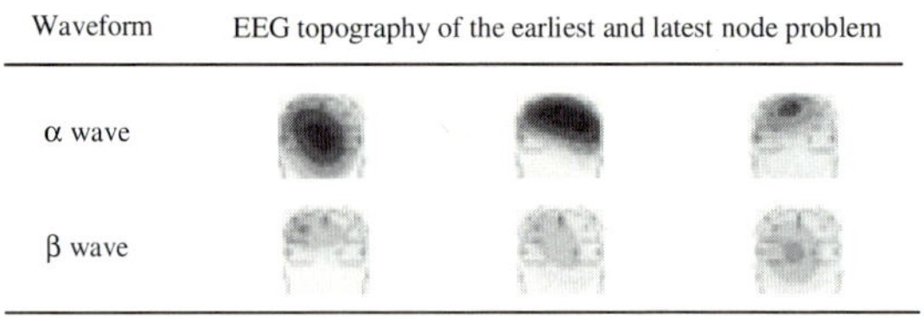

Fig. 4. Topographies at three essential points of material P (subject a, maximum potential: alpha wave: 1.6μV, beta wave: 1.1μV)

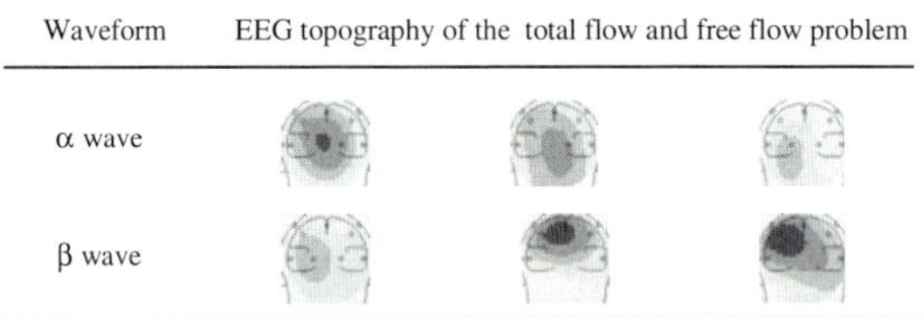

Fig. 5. Topographies at three essential points of material Q (subject a, maximum potential: alpha wave: 1.6μV, beta wave: 1.1μV)

Potential of each point of topography (maximum potential is indicated as level 9) of brain alpha and beta wave in the occipital region obtained by this method is converted into values shown in Fig. 1. The summation is calculated to obtain the EEG value at the evaluation point.

3.3 Cross-Check Between the Basic EEG Values and Numerical Values Converted from the EEG During Learning with Materials

The matching level is defined in order to indicate the learning status of an evaluation point by using the "basic EEG value" established as a criterion of agreeable advancement of learning. The matching level is defined by the following equation:

$$\text{Matching level} = \text{EEG value at evaluation point}/\text{basic EEG value} \times 100 \qquad (1)$$

Figure 6 shows the matching level of material P and Q at evaluation points. In both materials, the matching level of alpha wave decreases and beta wave increases as learning advances. However, the matching level of beta wave exceeds 100 for material Q.

As shown in table 2, one-way analysis of variance showed that the matching levels of alpha wave were different between evaluation points for both material P and Q (significant level: 5%). Additionally, as shown in tables 3, the matching levels of beta wave were different between evaluation points for both materials (significant level: 1%).

The relationship between the matching levels of alpha and beta wave for material P and Q, and evaluation points of materials will now be discussed. X axis is alpha wave matching level. Y axis is beta wave matching level.

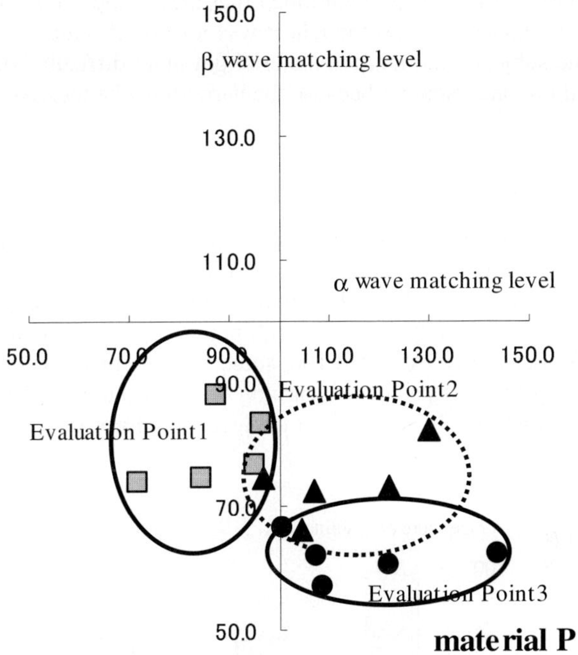

Fig. 6. Matching level of evaluation point 1, 2, 3 of material P and Q

Table 2. ANOVA on the difference of α and β wave power at evaluation point for material P

wave form	Factor	s.s	d.f	d.f	F_0
α wave	Between evaluation points	2525.93	2	1262.97	6.56*
	Within evaluation point	2311.80	12	192.65	
β wave	Between evaluation points	811.37	2	405.68	14.99**
	Within evaluation point	324.69	12	27.06	

Table 3. ANOVA on the difference of α and β wave power at evaluation point for material Q

wave form	Variation factor	s.s	d.f	d.f	F_0
α wave	Between evaluation points	564.60	2	282.30	6.08*
	Within evaluation point	557.19	12	46.43	
β wave	Between evaluation points	2129.95	2	1064.97	18.92**
	Within evaluation point	675.29	12	56.27	

3.4 Results of Examination and Verbal Report by Material

The average score of examination for the material P was 97.2. For the material Q, the average scores were 78.9 and 44.4 for the middle and final point respectively. These values show that the general cognitive assessment of the material Q is higher than Q.

In verbal reports, some subject mentioned that they could focus on the evaluation point 3(last) of the material P and thought it was not so difficult. With regard to the material Q, some subjects said that the material got more difficult from the middle of it and they could not understand it because the narrator spoke too fast.

4 Discussion

The relationship among evaluation points, and alpha and beta wave shows that (1) the matching level of beta wave gets higher and that of alpha wave gets lower at evaluation points in latter part. The brain is activated, alpha wave is suppressed and beta wave increases as learning advances. (2) with regard to the evaluation point 3 of the material Q where beta wave exceeded 100, some subjects mentioned in the verbal reports after learning that (a) pacing of learning was too intense to understand it. (b) The test score was 44. So, it seemed that subjects could not understand the material because it was too difficult.

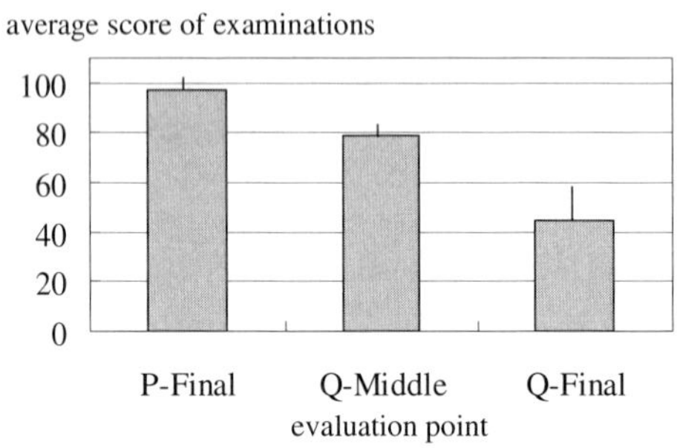

Fig. 7. Test score by material(by evaluation point

The results shown above suggest that combination of alpha and beta wave provide us with information for making a decision of the state of learning where learners can advance the study agreeably and judging the difficulty level of materials. The method may be used for evaluating cognitive aspect. Therefore, also for affective aspect, it is supposed that materials with which alpha wave decreases and matching level of beta wave exceeds 100% (the second quadrant in Fig. 7) after combining alpha and beta wave are not appropriate.

5 Summary

As a result by advancing this research, it was confirmed that that the matching level of alpha wave tended to fall and that of beta wave tended to rise when the brain was activated. Additionally, the accuracy rate and verbal reports showed that learners could not understand the material enough at the points of materials in which the matching level of the beta wave exceeded 100 because the narrator spoke too fast or the difficulty level of problems was too high. When evaluating a material by using biological information, if there were multiple essential points in the material, it was suggested that all evaluation points could be evaluated in a comprehensive manner.

After this, the state of understanding during learning will be clarified by using the simultanagnosia chart, and EEG changes generated before and after "understanding" and the state within the brain will be made clear through experiments and analyses.

References

1. Nakayama, M., Shimizu, Y.: Research. Trend on Educa tional Evaluation of Learning Behaviors with Biological Information. Jpn. J. Educ. Technol 24, 15–21 (2000)
2. Okuma, T., M.D.: Clinical Electroencephalography, 5th edn. Igaku Shoin Ltd, Tokyo (1999)
3. Iwata, M., Ohigashi, Y., Kato, M., Tanabe, H., Yamadori, A.: Brain and mind. To-kyo, Kyoritsu (2002)
4. Bloom, B.S., Hastings, J.T., Madaus, G.F.: Handbook on formative and summative evaluation of student learning. McGrraw-Hill, New York (1971)
5. Lee, W.W., Owens, D.L.: Multimedia-based instructional design: com-puter-based training, web-based training, distance broadcast training, Lee, W.W. Owens, D.L., Jossey-Bass/Pfeiffer (2000)

Webcasting Made Interactive: Persistent Chat for Text Dialogue During and About Learning Events

Ronald Baecker[1,2], David Fono[1], Lillian Blume[1], Christopher Collins[1,2], and Delia Couto[1,2]

[1] Department of Computer Science
[2] Knowledge Media Design Institute,
Univ. of Toronto, 40 St. George St. #7228, Toronto Ontario M5S2E4 Canada
rmb@kmdi.toronto.edu, fono@dgp.toronto.edu,
lilblume@kmdi.toronto.edu, ccollins@cs.utoronto.ca,
delia@kmdi.toronto.edu

Abstract. This paper presents a "persistent chat" extension to the ePresence Interactive Media webcasting infrastructure to support real-time commenting on and discussing of issues that arise during a learning event, followed by ongoing asynchronous dialogue about these issues while viewing the archives after the event. We report encouraging results of a field study of use of the system by students and a teaching assistant in a computer science class on communication skills, which encouraged students to review, think critically about, and improve their public speaking abilities.

Keywords: webcasting, streaming media, eLearning, digital media, digital video, persistent chat, asynchronous communications, public speaking.

1 Introduction

Because geographically distributed teachers and learners are usually not available concurrently, asynchronous tools such as listservs, threaded discussions, wikis, organizational memories, and courseware management systems are the dominant vehicles for eLearning. Yet asynchronous tools rarely establish the immediacy, interactivity, and shared purpose that result from face-to-face interactions. We have been developing a scalable Internet technology infrastructure that enables effective remote attendance at learning events, both concurrently and retrospectively, with maximum engagement, interactivity, and support for community.

Our approach uses a technology known as *webcasting* — the Internet broadcasting of streaming media to be viewed via a Web browser on a personal computer. Today's streaming media engines do significant buffering in order to provide smooth Internet media delivery to potentially very large numbers of viewers, resulting in 15-25 second delays between when events happen and when they are viewed. Consequently, while webcasting is scalable to large numbers of participants, it is typically a one-way non-interactive broadcast medium.

M.J. Smith, G. Salvendy (Eds.): Human Interface, Part II, HCII 2007, LNCS 4558, pp. 260–268, 2007.

Webcasts are also typically ephemeral media that result in no permanent record, or archived recordings that only can be played from the beginning, or that offer very weak methods for finding particular sections of interest. Our approach differs in being significantly interactive, and producing structured, embeddable, navigable, searchable, and taggable archives.

This paper presents an experimental "persistent chat" extension of our ePresence Interactive Media webcasting infrastructure [1], [2], [3] to support real-time text annotation and discussion of issues arising during a learning event, followed by continuing asynchronous dialogue about these issues while viewing the archives[1]. Annotations and asynchronous threaded discussions are bi-directionally linked to particular moments in the webcast. We also report results of a field study of use of this environment in a computer science class dealing with communication skills, in which the technology was used to help students improve their public speaking skills.

2 Background and Previous Work

We can enable asynchronous dialogue to follow and respond to issues raised in real-time discussion by integrating "persistence" features into synchronous, ephemeral media such as chat. *Persistence* refers to the availability of old conversational content. Most commercially available chat or instant messaging applications incorporate very limited forms of persistence. The traditional chat interface contains a history pane that displays a chronologically sorted list of recent messages, which are lost when the user logs out. Some chat tools allow users to save transcripts of chat sessions as local text files, which can then be browsed and searched. Halverson (2004) found that this feature facilitates recovery of useful information from old conversations, although it can sometimes be difficult to find the desired information amongst a large collection of transcripts with limited metadata [5].

A number of research projects have experimented with greater degrees of persistence. Erickson, et al. (1999) developed Babble, a chat application that stores all conversations that take place within the system, and makes them available to all participants [6]. The authors found that this "conversation as a single document" approach supported group awareness, and helped foster an ongoing narrative of the group as the persistent conversation continuously evolved. Ribak, et al. (2002) developed ReachOut, a peer support tool that features fully persistent conversations with limited lifespans [7]. In this case, persistence was found to generate additional ideas and dialogue which may not have otherwise emerged, since users were able to observe previous discussion before deciding to contribute their own thoughts. Robbins-Sponaas and Nolan (2005) have made similar observations about MOOs, which have many chat-like properties [8]. In particular, they noted that persistence allows a blend of synchronous and asynchronous interaction, a combination that makes for a dynamic collaborative environment.

[1] A companion paper in this conference — Webcasting Made Interactive: Integrating Real-time Videoconferencing in Distributed Learning Spaces [4] — discusses enhancing real-time video and voice webcast interactivity. For more current information on ePresence, also see http://epresence.tv.

These findings indicate that persistence in chat can provide a promising environment for learning and for encouraging reflection and participation. However, these benefits are contingent on the ability of users to read and keep track of lengthy chat histories, as well as interact over them. Several researchers (e.g., [5-6]) report that this is difficult to do using existing tools.

While there has been some research into strictly text-based persistent chat, no integration with video has been explored. There have been numerous investigations into text-based discussion over live or archived video, but these focus on either synchronous or asynchronous discussion as disparate modalities. Pea (2006) and Stevens (forthcoming) provide good examples of annotation over archived videos [9], [10]. Bargeron, et al. (2001) and Fishman (forthcoming) provide good examples of asynchronous discussion over archived video [11], [12]. White, et al. (2000) and Baecker, Moore, and Zijdemans (2003) provide good examples of synchronous, non-persistent chat during live webcasts [13], [14]. In contrast to these systems, our aim has been to develop a persistent chat system that supports a seamless blend of synchronous and asynchronous discussion over live and archived videos.

According to a survey by Hobbs (2006), video is frequently used in schools, but primarily as a means for content delivery [15]. Teachers reported low levels of actual use of video technology such as camcorders. They gave only a few examples of more student-engaging methods of using video "to create or analyze information" (p. 45) through student video production, or "to document student performance" (p. 46).

Such instructional strategies can be beneficial to a student's active learning process and help improve academic performance. A common example involves the use of video to encourage dialogue between student and teacher regarding the student's filmed performance. In particular, Jambor and Weekes (1995) describes how allowing physical education students to direct the reviewing of their performance videos, while teachers act as facilitators, helps them take an active role in discussion and better comprehend their motor skills [16]. Traditional uses of video, especially those involving content delivery, typically lack such student feedback and engagement.

Student-created video projects have also shown to be beneficial in student learning, improving the retention of foreign language vocabulary of secondary school German students [17]. Video projects were also found to be successful in "encouraging [nursing] students to promote pattern recognition of characteristic features of common illnesses, to develop teamwork strategies, and to practice their presentation skills in a safe environment among their peers" (p. 558) [18]. Students also expressed satisfaction from producing their videos, felt that it added to the learning process, and reported great value in being able to later review what they did.

3 A Persistent Chat System for ePresence

ePresence has always had a chat subsystem that allows dialogue over a webcast. Typical uses of chat are for exchange of ideas and commentaries about the topic, social interchange, administrative discussions, and questions and comments about webcast technology and experience [19]. A persistent chat system would allow discussions about issues raised during the lecture to continue asynchronously into the future, discussions that could also include individuals who encountered the material by viewing the archives.

Backtalk [20], [21] is a persistent chat system that preserves the traditional chat interface while adding capabilities for preserving content, engaging in threaded discussions, tagging messages via resizing and colouring, displaying overviews of message history, filtering the messages according to criteria such as author or tag, and allowing temporal and contextual (i.e., in terms of threads) displays of sequences of messages.

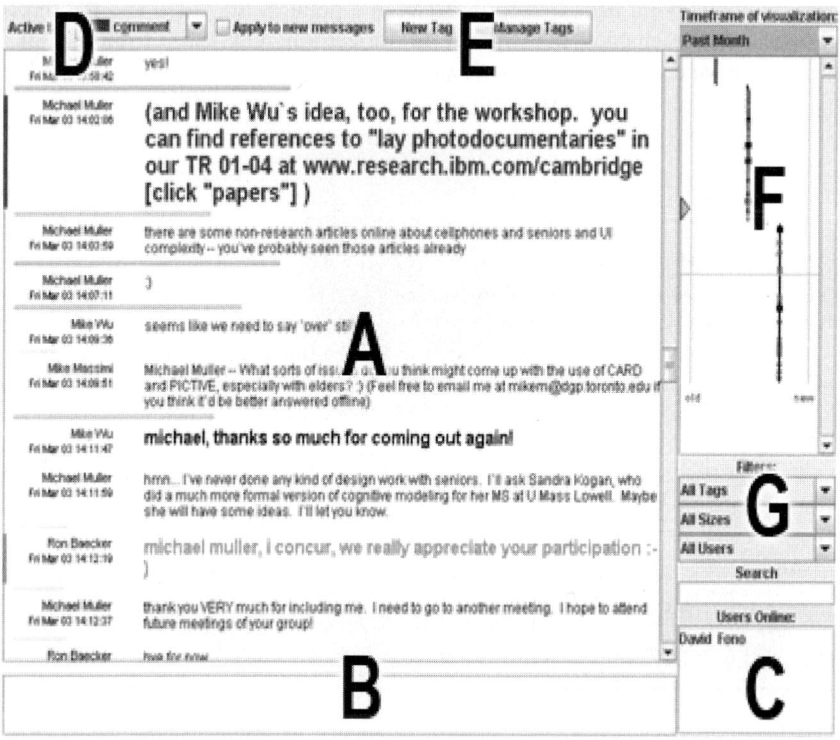

Fig. 1. The BackTalk user interface

A typical Backtalk screen is portrayed in Figure 1: A is the chat history, B the message entry window, C the list of concurrent users, D and E the tagging controls, F the message visualization, and G the filtering controls. Tagging controls allow messages to be distinguished and made more or less salient through the use of colour and size. Messages are displayed chronologically; filtering can be used to reduce those displayed to a manageable subset of the entire message history, based on colour, size, or sender. Threaded replies can be left on any message in the chat history, and reply chains are visualized using the tree structure popularized by newsgroup and e-mail clients. The visualization provides an overview of a large subset of the chat history, including temporal data, tagging data, and reply status for each message. The visualization can also be used to quickly browse a lengthy history by clicking and dragging within the pane, which rapidly scrolls the chat history window.

Figure 2 shows a specially tailored version of Backtalk integrated with an ePresence video archive of a student presentation with a persistent chat dialogue about the presentation. We see a student giving a speech, and a dialogue about that speech between a teaching assistant (TA) and the student. Dialogues are threaded discussions in which seed messages may be annotations made by the TA during the talk or added later by the TA in reviewing the archive. Bi-directional links, represented by timeline markers under the video frame, and video icons to the right of discussion items, point back and forth between a point in the video and the threaded discussion, which enables the student to reflect on and learn from his or her "performance" and the TA's comments on it.

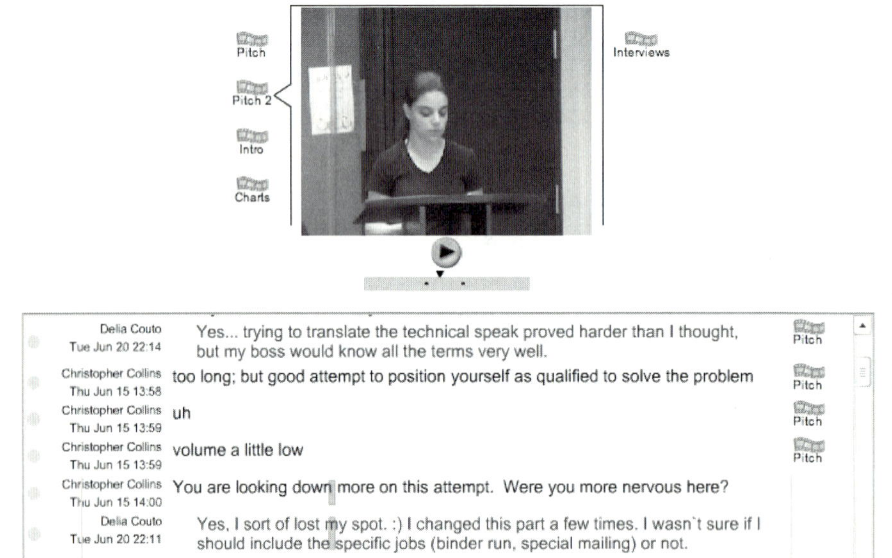

Fig. 2. BackTalk integrated with ePresence, as applied to instruction on public speaking. Christopher is the TA, making comments on a speech by student Delia. Delia reviews her performance and Christopher's comments, and responds to some of them.

4 A Field Trial of ePresence BackTalk Applied to the Public Speaking Component of a Communication Skills Course

In the summer and fall of 2006, we deployed this experimental system for use in sections of a 2nd year Computer Science class dealing with writing, speaking, and interpersonal communications. The class meets one evening per week for 3 hours. Six times during the term, each student makes a 30 second to 5 minute oral presentation. Students typically improve significantly in skill and poise over the semester, but it is hard for them to appreciate how far they have come and to reflect critically upon their strengths and weaknesses based only on their memories of their speeches and written critiques from a teaching assistant (TA).

We therefore used ePresence BackTalk to capture four of their talks over a period of nine weeks, and allow the 15 students to review their performances. The TA annotated each presentation in real-time as it was given. The TA and each individual student could then review that individual's presentations, and discuss over the semester what the student did well, what could be improved, and how the student's performance evolved over the term. Since students were generally nervous about their speaking abilities, we did not use the full power of the technology, which would have allowed class-wide discussion of all presentations by all students. At the end of the course, students filled out a questionnaire detailing their experiences with BackTalk, and participated in a group interview on this subject.

A total of 710 messages were posted over the course of the study. Of these, 418 were notes written by the TA while presentations were being recorded, 171 messages were from the TA written outside of class, and 121 messages were from students. For each ongoing discussion between a student and the TA, the mean number of messages posted by the TA was 11.4, and the mean number posted by the student was 8.1.

Each discussion typically followed a set pattern. After a new recording was uploaded along with the TA's in-class notes, the students would log in and respond to the TA's feedback, occasionally making unrelated critiques of their own. The TA often responded in turn, but only rarely did students perpetuate a thread of discussion after that point. Discussion then generally ceased until the next presentation was recorded and uploaded, at which point the process would repeat. All discussion was carried out asynchronously, without any synchronous interaction between the TA and students.

Of those messages posted after the initial recording, 265 were linked to a specific point in a video, and 27 messages were linked to a specific video but no specific point. No messages unrelated to a specific presentation were posted. 88 messages were posted as a direct response using the reply feature, compared to 204 that were not.

The questionnaire that students filled out yielded mixed metrics of satisfaction. Students agreed that watching the recordings and reading the feedback from the TA were helpful in developing their presentation skills. Students also agreed that they could communicate effectively with the TA, and they felt comfortable using the system. They indicated that they would recommend that the system continue to be used in the course.

During the interview, the majority of students were enthusiastic about the system's potential in these regards. Many students expressed feelings such as "this method of receiving feedback was much, much better than other ways." They appreciated the ability to easily view moments in their presentations that the TA referenced in his feedback, and the ability to verify the TA's claims about their performance.

However, students used the system primarily as a mechanism for personal consumption and reflection, rather than dialogue. As such, they were not as enthusiastic about the ability to respond to the TA's feedback, and there was only a mild agreement that this aspect was helpful.

Students did have qualms with various flaws in the implementation, resulting in a high rating for perceived difficulty of using the system, as well as a low rating for overall enjoyment. The most oft-cited problem was the students' inability to determine when the TA had responded to their comments, which forced them to continually login to the system in order to check. Many suggested a system for e-mail notification of updates to the discussion. We had noted the need for such a mechanism in our design goals for BackTalk, but failed to fully implement a solution. The other major problem encountered by students was general clutter in the interface caused by the number of features that used pop-up visual elements. While the use of pop-ups was not fundamentally problematic, students became confused when numerous dialogs, bubbles, and other components simultaneously occupied the same display space.

When asked about the possibility of applying this technology towards other purposes in other domains, students had numerous ideas, most of which reiterated the theme of providing feedback on video-recorded performances in alternative contexts (mainly sports). Additional ideas included discussion or Q&A over lecture recordings, as well as discussion by researchers over videos recorded as part of ethnographic studies. These suggestions are evocative of several existing research prototypes that enable collaboration over video and that are reviewed in a forthcoming edited volume [22].

The TA[2] found through using the system he was able to provide more detailed feedback to the students than was possible when he evaluated short public speaking assignments using the 'pen and paper' method. He found it easy to provide specific and detailed feedback — often on things that would otherwise go by so quickly that one may not have time to take sufficient notes. Using BackTalk, he was able to review the presentations after the class and add additional comments or clarify comments he had made in real-time as the webcast was recorded. Additionally, he found the archives and discussions with students to be useful aids for recognizing student improvement on the particular aspects of presentation skills that had been noted as problem areas in previous assignments.

For the final exam, students were asked to review the videos and the dialogue with the TA in order to find specific places where they demonstrated confidence and other places where they seemed less confident. They were asked what steps they would take to improve the speaking deficits they identified. The exam answers showed that with a video record of their speeches and feedback linked to specific points on the video, the students gained much more self-awareness than with any other speech evaluation method used by the instructor[3] in 15 years of teaching public speaking.

5 Summary, Discussion, and Conclusions

We have presented a persistent chat extension to an interactive webcasting system that allows dialogue among students and members of the teaching staff to occur during a live webcast of a learning event, and then be continued while viewing the webcast archives after the event. The paper also reports on a use of this technology in an undergraduate computer science course on communication skills.

Results from a questionnaire and a group interview suggest that the ability to review their presentations is viewed by the students as highly beneficial, both in terms

[2] Christopher Collins.
[3] Lillian Blume.

of specific understandings about technique as well as growing self-confidence in their skills. The TA feels that the technology enables him to give more thoughtful and more constructive feedback to the students. The instructor believes that this system and procedures for its use enable her to teach public speaking significantly more effectively than any other method she has used in 15 years of teaching.

Yet this particular study did not realize the full potential of the technology in two respects. Because only the TA had access to an ePresence system during the class presentations, there was no synchronous interaction. In other words, the TA's writing of messages functioned as personal bookmarking and notetaking, and not as a real-time dialogue with other members of the class. Also, asynchronous dialogue after the class consisted of 15 pairs of individual dialogues between the TA and 15 individual students, and not a more general discussion among all class members. The former limitation is removed in a version of BackTalk integrated with another experimental version of ePresence that is discussed in the companion paper [4].

Acknowledgements. We are grateful to Peter Wolf, Chief Architect of ePresence Interactive Media, and Kelly Rankin, Business Manager. This work was supported in part by the Natural Sciences and Engineering Research Council of Canada through a scholarship to David Fono and through the Network for Effective Collaboration Technologies through Advanced Research (NECTAR) Research Network Grant.

References

1. Baecker, R.M.: A Principled Design for Scalable Internet Visual Communications with Rich Media, Interactivity, and Structured Archives. In: Proc. CASCON 2003, pp. 83–96 (2003)
2. Baecker, R.M., Wolf, P., Rankin, K.: The ePresence Interactive Webcasting System: Technology Overview and Current Research Issues. In: Proc. Elearn 2004, pp. 2532–2537 (2004)
3. Rankin, K., Baecker, R.M., Wolf, P.: ePresence; An Open Source Interactive Webcasting and Archiving System for eLearning. In: Proc. Elearn 2004, pp. 2888–2893 (2004)
4. Baecker, R., Birnholtz, Causey, R., Laughton, S., Rankin, K., Mak, C., Weir, A., Wolf, P.: Webcasting Made Interactive: Integrating Real-time Videoconferencing in Distributed Learning Spaces. In: Proc. HCI International 2007, July 2007 (to appear)(2007)
5. Halverson, C.A.: The value of persistence: a study of the creation, ordering and use of conversation archives by a knowledge worker. In: Proc. HICSS 2004, pp. 108–117 (2004)
6. Erickson, T., Smith, D.N., Kellogg, W.A., Laff, M., Richards, J.T., Bradner, E.: Socially translucent conversations: social proxies, persistent conversation, and the design of "Babble". In: Proc. CHI 1999, pp. 72–79 (1999)
7. Ribak, A., Jacovi, M., Soroka, V.: Ask before you search: peer support and community building with ReachOut. In: Proc. CSCW 2002, pp. 126–135 (2002)
8. Robbins-Sponaas, R.J., Nolan, J.: MOOs: polysynchronous collaborative virtual environments. Workplace Internet-Based Communication: Industry and Academic Perspective, Idea Group, pp. 130–155 (2005)
9. Pea, R.D.: Video-as-data and digital video manipulation techniques for transforming learning sciences research, education and other cultural practices. In: Weiss, J., Nolan, J., Trifonas, P. (eds.) Int'l Handbook of Virtual Learning Environments, pp. 1321–1393. Kluwer, Boston (2006)

10. Stevens, R.: Capturing ideas in digital things: A new twist on the old problem of inert knowledge. In: Goldman, R., Pea, R., Barron, B., Derry, S., (eds.) Video Research in the Learning Sciences, Lawrence Erlbaum Associates (in press) (To appear)
11. Bargeron, D., Gupta, A., Grudin, J., Sanocki, E., Li, F.: Asynchronous Collaboration Around Multimedia and Its Application to On-Demand Training. In: Proc. HICSS, pp. 4042–4050. IEEE Press, New York (2001)
12. Fishman, B.: Linking online video and curriculum to leverage community knowledge. In: Goldman, R., Pea, R., Barron, B., Derry, S., (eds.) Video Research in the Learning Sciences, Lawrence Erlbaum Associates (in press) (To appear)
13. White, S.A., Gupta, A., Grudin, J., Chesley, H., Kimberly, G., Sanocki, E.: Evolving use of a system for education at a distance. In: Proc. HICSS, pp. 1–10. IEEE Press, New York (2000)
14. Baecker, R.M., Moore, G., Zijdemans, A.: Reinventing the Lecture: Webcasting Made Interactive. In: Proc. HCI International 2003, June 2003, vol. 1, pp. 896–900. Lawrence Erlbaum Associates, Mahwah (2003)
15. Hobbs, R.: Non-optimal uses of video in the classroom. Learning, Media and Technology 31(1), 35–50 (2006)
16. Jambor, E., Weekes, E.M.: Videotape feedback: Make it more effective. Journal of Physical Education, Recreation & Dance 66(2), 48–50 (1995)
17. Sildus, T.I.: The effect of a student video project on vocabulary retention of first-year secondary school german students. Foreign Language Annals 39(1), 54–70 (2006)
18. Epstein, C.D., et al.: Lights! Camera! Action!: Video Projects in the Classroom. Journal of Nursing Education 42(12), 558–561 (2003)
19. Zijdemans, A., Moore, G., Baecker, R.M., Keating, D.P.: ePresence Interactive Media and Webforum 2001: An Accidental Case Study on the Use of Webcasting as a VLE for Early Child Development. In: Weiss, J., Nolan, J., Trifonas, P. (eds.) International Handbook of Virtual Learning Environments, pp. 1395–1428. Kluwer, Dordrecht (2006)
20. Fono, D.: Structuring and Supporting Persistent Chat Conversations, M.Sc. Thesis, Department of Computer Science, University of Toronto (October 2006)
21. Fono, D., Baecker, R.M.: Structuring and Supporting Persistent Chat Conversations. In: Proc. CSCW, pp. 455–458 (November 2006)
22. Goldman, R., Pea, R., Barron, B., Derry, S. (eds.): Video Research in the Learning Sciences, Laurence Erlbaum Associates (in press)

Webcasting Made Interactive: Integrating Real-Time Videoconferencing in Distributed Learning Spaces

Ronald Baecker[1,2], Jeremy Birnholtz[2], Rhys Causey[1,2], Simone Laughton[3], Kelly Rankin[2], Clarissa Mak[2], Alison Weir[3], and Peter Wolf[2]

[1] Department of Computer Science
[2] Knowledge Media Design Institute, Univ. of Toronto, 40 St. George Street, Toronto, ON M5S2E4, Canada
[3] Univ. Toronto Mississauga, 3359 Mississauga Rd. N. Mississauga, ON L5L 1C6 Canada
rmb@kmdi.toronto.edu, jbirnhol@gmail.com, r.causey@gmail.com, slaughto@utm.utoronto.ca, kelly@kmdi.toronto.edu, clarissa.mak@utoronto.ca, aweir@utm.utoronto.ca, wolfpet@kmdi.toronto.edu

Abstract. This paper presents an extension to the ePresence Interactive Media webcasting infrastructure to support real-time voice and video conferencing for a few attendees while concurrently streaming an event to many others. We emphasize the tools for enhancing awareness of the remote attendees in the lecture hall or seminar room, and for facilitating communication between remote attendees and the lecturer. We present preliminary results of a field study of the use of this environment for a multi-campus university class.

Keywords: webcasting, streaming media, eLearning, digital media, digital video, videoconferencing, synchronous communications, awareness.

1 Introduction

Synchronous technologies such as video can link learners and instructors at the "same time, anytime and anywhere", allowing them to learn from each other [1]. Video learning systems provide enriched interaction through words, voice tone, environmental and nonverbal cues, and immediate feedback [2]. These technologies are used in academe and business [1], [3], [4], but questions of technical implementation and pedagogical methods and effectiveness remain. We have been developing and studying an Internet streaming media infrastructure that enables remote attendance at learning events such as lectures, both concurrently and retrospectively, with maximum engagement, interactivity, and support for community learning.

Audio/web conferencing and videoconferencing are often used for real-time Internet communication and collaboration. Audio/web conferencing (e.g., www.webex.com) allows the real-time multipoint transmission of voice and slides, but lacks the media richness, sense of presence, and ability to engage participants afforded by dynamic visual media. Internet desktop video conferencing (e.g.,

M.J. Smith, G. Salvendy (Eds.): Human Interface, Part II, HCII 2007, LNCS 4558, pp. 269–278, 2007.

www.microsoft.com/windows/netmeeting/) supports real-time multipoint audio and video communications as well as shared workspaces, but still does not provide reliable video performance at reasonable cost, and is not scalable to large numbers.

Our approach is based on a third technology known as *webcasting* — the Internet broadcasting of streaming media to be viewed via a Web browser on almost any suitably Internet-connected personal computer. Webcasting is scalable to large numbers of participants, but, due to buffering requirements, is typically a one-way broadcast medium with only limited provision for interaction via time-delayed text messages. Yet one could imagine a webcasting system that would allow significantly greater interactivity just in those cases where it is needed.

This paper presents an extension [5], [6][1] of our ePresence Interactive Media infrastructure [8-11] that supports real-time voice and video conferencing for a *few* attendees while streaming an event to *many* others[2]. We also report preliminary results of a field study of using the environment in a multi-campus university class.

2 Background and Previous Work

To allow the scalability and low barriers to access that characterize webcasting, while also supporting real-time interaction between speaker and audience, we have begun to combine webcasting and conferencing in ePresence [5], [6] (Figure 1).

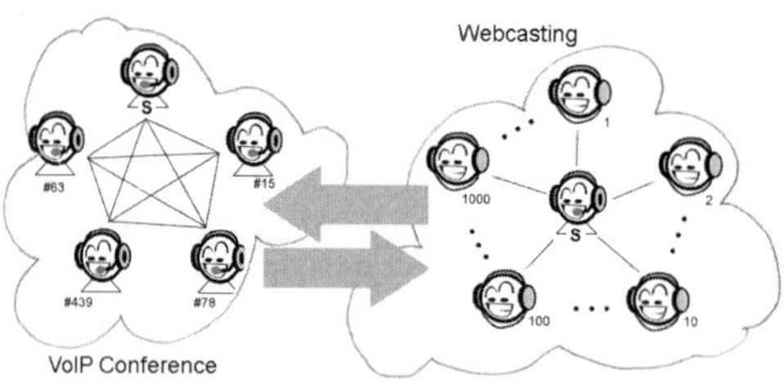

Fig. 1. Bridging webcasting for the many with conferencing for the few

To make such technology effective, at least the speaker and ideally also others in the presentation room must be aware of the remote audience. For example, the speaker may be interested in knowing answers to the following questions: How many are watching? Who is watching? Are they paying attention? Do the remote viewers seem to comprehend what is being said? Do any of them have questions?

The value of a pioneering system for distributed presentations was demonstrated in [13]. The design and evaluation of a large wall display that presents outlines, still pictures, and video streams of up to 50 remote participants, and supports audience interaction via text, is discussed in [14]. The Virtual Auditorium employed video conferencing occupying real estate on a huge wall display to link a speaker with as

---

[1] Design of this system was grounded in part on the results of an observational study of classrooms [7].

[2] A companion paper [12] discusses enhancing interactivity via text chat both live over a webcast and asynchronously over archives of a webcast.

many as 25 remote participants [15]. Methods for providing the speaker with other forms of remote audience feedback were also investigated [16].

3 Conferencing and Awareness Extensions to ePresence

We made two assumptions: 1) The new conferencing/awareness and existing webcasting capabilities should not be interdependent; thus a passive observer can watch the webcast from a PC that does not support the new features. 2) The system should be scalable and customizable based on available bandwidth and user needs.

3.1 Usage Scenario: Lecture Environment

In a distributed classroom, our approach allows a few individuals or remote viewing groups to engage actively in a videoconference discussion with the speaker at any time, while tens or hundreds of individuals and groups can watch the webcast and participate by submitting text questions. The remote audience may include passive observers who want to simply watch the webcast, and interested students who cannot attend the lecture in person, e.g., because they are in a remote area, ill, or traveling. We therefore have four distinct stakeholders: the speaker, the local audience, the remote audience and, to a lesser extent, a webcast moderator[3] who configures various aspects of the system (e.g., which remote audience members are permitted to speak).

Figure 2 depicts the layout of a small lecture hall we have configured for webcasting. There are two projection screens. The one in front shows the speaker's PowerPoint slides. The second displays a large "in-room awareness display" (explained below) containing visual representations of remote participants, and also their text questions, comments, and contributions to discussion. Thus the speaker and local audience members can be aware of the remote audience, which facilitates interaction.

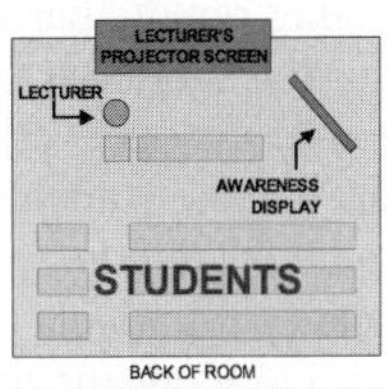

Fig. 2. Lecture hall organized for a webcast

Remote participants have the option of using slow-frame-rate video representations or still photos of themselves. Video conveys more information to the speaker and likely allows for improved engagement, but may be seen as too intrusive by some participants who wish to watch passively or even multitask without the knowledge of the presenter. Remote participants can watch the webcast and actively participate via voice interaction (using VoIP) or text chat.

[3] Our use of a moderator parallels the use of a "technology facilitator" to aid distributed teams working together with desktop conferencing and application sharing [17].

3.2 The In-Room Awareness Display

This display (Figure 3) is intended to give the speaker the ability to quickly assess the composition of the remote audience and their level of engagement. There is also a persistent chat window (implemented with BackTalk [12], [18]) that displays text questions, comments, and contributions to discussion. Text is not intended as the primary means of communication, but instead as a back channel [19] to augment voice conversation and also help the speaker gauge audience comprehension. Participants can color-tag or increase the size of text they contribute on a message-by-message basis; with this, participants could, for instance, emphasize certain contributions, or distinguish with pre-defined tags questions to the speaker, comments, and within-audience chat that the speaker need not necessarily see.

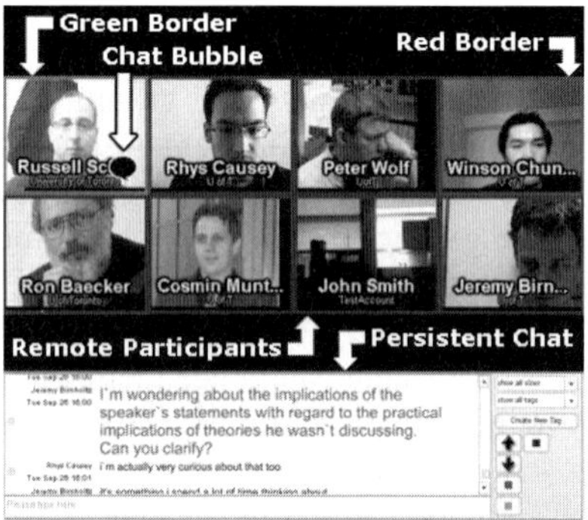

Fig. 3. The in-room awareness display

Video representations give the speaker the ability to visually monitor attentiveness and facial expressions, and see whether remote participants are engaged in the lecture. While our system does not allow for the subtleties of gaze awareness or simulated eye contact, it does provide information to the speaker that can be useful in tailoring the lecture to audience needs. While all remote students may not choose to use webcams, we conjecture that many of those who elect to let the presenter see them will likely be more engaged. Our lecturer interview participants (see section 4) suggested that these are often the ones on whom they tend to focus the most.

The awareness display also simplifies voice interaction between remote participants and the speaker. When remote participants interact by voice, they are transferred from the one-way, slightly time-delayed webcast video and audio feed to a "voice conversation," which allows for two- or multi-way voice interaction with a video feed with a nearly-imperceptible delay. This voice conversation then becomes part of the webcast that is streamed to remote participants not engaged in the voice conversation.

On the awareness display, colors and icons identify the states of remote audience members to make it easy for the speaker and local audience to determine who wants to speak and who is speaking. Remote students who wish to speak indicate this by clicking an icon to "raise their hand." When this icon is clicked, their representation on the awareness display has a red border. Once given permission by the moderator to enter the voice conversation, the participant's representation has a green border.

Because there can be multiple simultaneous participants in the voice conversation, we additionally distinguish between those who simply have permission to speak and those who are actively speaking. When the system detects that a remote participant is speaking, a chat bubble icon appears over his representation. This allows the speaker and local audience to identify which remote student(s) are speaking.

3.3 The Presenter Interface

The presenter can also use the ePresence presentation tool to easily send material to the remote audience. This tool also makes it easy for the presenter to display and send Powerpoint slides, videos, fully-navigable websites, and live desktop demonstrations.

3.4 The Remote Viewing Interface

Our primary goal for the remote viewing interface is to increase the ability of remote participants to engage and interact naturally in real-time. We allow participants to see each other through slow-scan, low-resolution video or digital images, and give them the option of interacting with text or voice. The remote viewing interface is shown in Figure 4. As with traditional webcasts, the participant receives the video feed (Fig. 4a) in sync with presentation material, such as slides or video (Fig 4e).

Questions, comments and discussion can be sent to other remote participants and the in-room display using the BackTalk persistent chat tool (Fig. 4f). Fono and Baecker [18] report that such a system, by incorporating features such as threading, annotation, and filtering, can create an environment for rich, collaborative, and ongoing conversation while maintaining the intuitiveness of more typical chat clients.

Remote participants are also shown their own chosen representation—the video image from their webcam or uploaded still image—so they see how the speaker and others will see them (Fig 4b). When remote participants are granted permission to enter the voice conference, their status bar, which is typically gray (Fig. 4b) and says "watching," turns green and says "on air." This would be the case whether they raise their hand or are "called on" by the speaker. It is also possible, in small classes or seminars where it may be desirable for all students to speak, or where certain students may have more to contribute than others, for the moderator or speaker to designate certain remote students as "VIPs" who can enter into and out of the voice conversation without having to "raise their hand."

For awareness of others watching remotely, there is a full list of remote participants (Fig 4d), with small visual representations. The system automatically detects and displays a list of the remote participants who are currently speaking. The first participant in the list is given a larger representation above the full user list (Fig. 4d) as the "Active Speaker" (Fig. 4c). Showing only a single speaker at a time works well because speech from remote participants typically only overlaps very briefly, and only during turn transfer. Just as a group will place focus on a person who is speaking in a traditional lecture, and a video camera will focus in on the current speaker in a traditional webcast, remote speakers are given a similar focus in our system, enabling remote participants to be aware of current speaker's identity, whether local or remote.

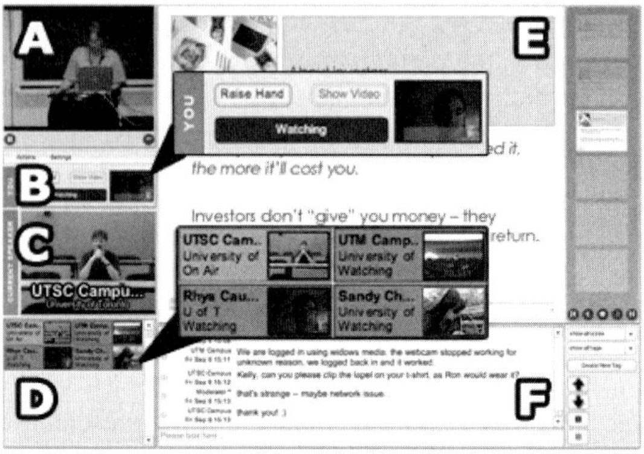

Fig. 4. The remote viewing interface: (a) webcast video; (b) user controls; (c) active speaker; (d) remote participant list; (e) presentation content; (f) chat

3.5 The Moderator Interface

As speakers have many styles for engaging students, we provided flexibility in the moderator interface. Moderators can control voice permissions, tweak automatic speech detection settings (for current speaker identification), and remove certain remote clients from being displayed on the in-room awareness display (e.g., to hide those connected locally). The moderator can also control frame rates and capture resolutions of individual remote participants' webcams. This enables the moderator to optimize the system based on available bandwidth (or even on the basis of speaker interest in certain remote participants), to divide bandwidth unequally between users (e.g., to give a larger share to students who may have more to contribute), or even to prevent certain users from using their webcam if this is necessary for any reason.

4 A Field Trial in a Multi-campus Class

In fall 2006, we deployed this experimental system in a 4th year Computer Science evening course dealing with creating new software ventures. The course was offered to students on two campuses of the same university, located 25 kilometers apart. Approximately 60 students were enrolled at the "main" campus (where the instructor was based), and 15 at the "satellite" campus. The class met weekly for 3 hours, and consisted primarily of lectures by the instructor and weekly guest speakers.

These presentations were, for the most part, delivered in a lecture hall on the main campus, though there were occasional presentations to the entire class by a teaching assistant at the satellite campus, and two of the guest speakers delivered their presentations from California and Korea, respectively. Students had the option of attending the course at whichever campus was most convenient for them, or, when this was not possible, to sign in and "attend" from any Internet-accessible location.

At the main campus, the room was configured as shown in Figure 2. There were two video cameras at the main campus, and two staff members who operated the cameras, moderated the discussion and selected the camera shot for the webcast.

Students at the satellite campus sat in a smaller room where the webcast presentation (using the remote participant interface described above) was displayed on a large screen at the front of the room. At the satellite campus, there was a single camera, and one staff member was responsible for setting up the technology and operating the camera. To provide for VoIP interaction, there was also a wireless, handheld microphone at the satellite campus. At this campus, students indicated by raising their hand that they wished to speak, and then the microphone was brought to them.

A comprehensive evaluation of student and speaker experience is underway. Due to space constraints, we discuss here only the initial reactions of students and presenters.

4.1 Evaluation Methods

Four questionnaires were administered to all students at periodic intervals. The first of these gathered baseline demographic data and student attitudes toward and experience with technology, using an instrument described in [20]. The remainder assessed student experience in the course and with ePresence.

Interviews lasting 30-60 minutes were conducted with 7 students, with 3 students interviewed multiple times during the term. Interviews were also conducted with 5 of the guest speakers, and the teaching assistant at the satellite campus. All interviews were fully transcribed and preliminarily coded using qualitative coding software.

Field observations were conducted at both campuses. Three independent observers conducted eleven 1-3 hour observation sessions and recorded detailed field notes that were later typed and expanded. Four observation sessions were conducted at the satellite campus, seven at the main campus. One observer visited both campuses.

4.2 Student Experience

Preliminary analysis of questionnaire data indicates that there were some significant differences in technology-related attitudes between this group and those in the Caruso & Kvavik study. A higher percentage of the students in this class have higher self-reported skills ratings, and prefer a higher level of information technology within their courses compared with those surveyed in the prior study. This may be partly due to many students being computer science majors. There are also some differences in the student profiles between the two campuses. For example, the students at the "main" campus were on average two years older and had substantially more work experience in fields related to information technology than those at the "satellite" campus.

Students at both campuses were satisfied equally with the smoothness of the performance of ePresence. Evidence that students at both campuses learned from the course is that they, on average, indicated agreement or strong agreement with the statement "I learned a lot from taking this course," (M_{Main} = 5.5, $M_{Satellite}$= 5.6, on a 7-point Likert scale). Students at the "satellite" campus indicated that it was slightly more difficult to participate (M= 3.57 out of a possible 7), than those at the "main" campus (3.09), but this difference was not statistically significant.

4.3 Presenter Experience

Presenters' experience with the system was also generally positive in that it allowed them to deliver their presentations and interact with remote students, but they also indicated some concerns about being able to get adequate information about remote audience members, and about interacting naturally.

Most speakers indicated that they periodically looked at the awareness display, but that it was good mostly for very high-level information about their audience. They reported being able to see the approximate number of students and whether or not they were walking around the room. They could not see the facial expressions of individual students or assess their level of comprehension.[4]

Another issue one presenter noticed was that the camera location was at the back of the room, while the awareness display was near the front. This meant that giving students at the satellite campus the impression that one was looking at them necessarily meant not looking at their images. This presenter indicated that, once he figured that out, he made conscious effort to look at the camera periodically, and to make comments to try to engage the students:

> *And then I would try to say something that made them realize that I wasn't just randomly looking straight into the camera so that would indicate that I was intentionally looking into the camera to look at them (GS1).*

Our observational data suggest, however, that while some speakers made a special effort to engage students at the satellite campus and those logged in remotely, not all of them did so. Nonetheless, most speakers did feel that the technology did help them to reach a larger audience than they would otherwise have been able to, and that it did not detract from their interaction with and ability to engage the local audience.

4.4 Technical Issues

As might be expected, there were numerous technical issues with this experimental system deployed temporarily (and physically set up anew each week) in an environment designed primarily for face-to-face lectures. In some cases, problems were unavoidable. Network traffic and quality of service are unpredictable, and small "hiccups" can have a large effect if they interrupt critical moments in VoIP dialogue. There were several occasions where the satellite campus was switched out of the "conference mode" and into the "webcast" to afford smoother transmission. This had the unfortunate side-effect of causing a delay when students at the satellite campus wished to participate, and both students and speakers indicated in interviews that they found this distracting and an impediment to natural interaction.

Some problems could be overcome with a more permanent facility designed for the effective capture and transmission of audio and video. Examples of these problems are the awkwardness of passing a single microphone around the room for questioners, and the difficulty of seeing adequate detail in visual images of remote participants.

[4] This was likely due in part to the low-resolution video the awareness interface currently supports, which was designed mostly for individually located remote users, with one camera per person, rather than one camera for a small group of students as was done here.

Solutions for these problems include arrays of fixed-location microphones and higher-resolution cameras, possibly with automated control capabilities.

5 Summary and Conclusions

We have presented the design of a novel combination of webcasting, conferencing, and awareness technologies. Our prototype implementation of the concept worked sufficiently well for both students and presenters to convince us that the concept is sound. Further positive evidence comes from a current use in a research seminar where some participants and speakers are remote from the seminar room.

We are currently re-architecting and re-implementing the system for better performance and better scalability for both wecbast and conference participants. Research goals include more automation, including remote camera control, better remote audio capture, and less need for intervention by the moderator. A future version should allow the moderator to select the location where remote students will appear on the awareness display (to capitalize on presenters' spatial memory). We are also analyzing data from this study in more detail. A future in-depth study of the pedagogical impacts of the use of ePresence involving both instructors and learners may help to identify innovations — pedagogical as well as technical — that will better support teaching and learning.

Acknowledgements. This work was supported in part by the Natural Sciences and Engineering Research Council of Canada through a grant to the Network for Effective Collaboration Technologies through Advanced Research (NECTAR).

References

1. Patcha, A., Scales, G.: Development of an internet based distance learning program at Virginia Tech. In: Poster, 6th Conf. Information Technology Education, Newark, NJ (2005)
2. Kupritz, V.W., McDaniel, S.K.: Pedagogical reflections from an instructional technology workshop. Educational Technology & Society 2(4), 119–121 (1999)
3. Heeler, P., Hardy, C.: A preliminary report on the use of video technology in online courses. Journal of Computing Sciences in Colleges 20(4), 127–133 (2005)
4. Mizukoshi, T., Kim, Lee, J.Y.: Instructional technology in Asia: focus on Japan and Korea. Educational Technology Research and Development 48(1), 101–112 (2000)
5. Baecker, R.M., Baran, M., Birnholtz, J., Chan, C., Laszlo, J., Rankin, K., Schick, R., Wolf, P.: Enhancing Interactivity in Webcasts using VoIP. In: Proc. CHI 2006, pp. 235–238 (2006)
6. Causey, R., Birnholtz, J.B., Baecker, R.M.: Increasing Awareness of Remote Audiences in Webcasts. In: Proc. CSCW2006, Conference Supplement, pp. 59–60 (November 2006)
7. Birnholtz, J.P.: Back to School: Design Principles for Improving Webcast Interactivity From Face-to-Face Classroom Observation. In: Proc. DIS 06, pp. 311–320 (2006)
8. Baecker, R.M.: A Principled Design for Scalable Internet Visual Communications with Rich Media, Interactivity, and Structured Archives. In: Proc. CASCON 2003, pp. 83–96 (2003)

9. Baecker, R.M., Moore, G., Zijdemans, A.: Reinventing the Lecture: Webcasting Made Interactive. In: Proc. HCI Int'l 2003, vol. 1, pp. 896–900. Lawrence Erlbaum Associates, Mahwah (2003)
10. Baecker, R.M., Wolf, P., Rankin, K.: The ePresence Interactive Webcasting System: Technology Overview and Current Research Issues. In: Proc. Elearn 2004, pp. 2532–2537 (2004)
11. Rankin, K., Baecker, R.M., Wolf, P.: ePresence; An Open Source Interactive Webcasting and Archiving System for eLearning. In: Proc. Elearn 2004, pp. 2888–2893 (2004)
12. Baecker, R.M., Fono, D., Blume, L., Collins, C., Couto, D.: Webcasting Made Interactive: Persistent Chat for Text Dialogue During and About Learning Events. In: Proc. HCI International 2007 (to appear) (July 2007)
13. Isaacs, E., Morris, T., Rodriguez, T.K.: A forum for supporting interactive presentations to distributed audiences. In: Proc. CSCW 1994, pp. 405–416 (1994)
14. Jancke, G., Grudin, J., Gupta, A.: Presenting to Local and Remote Audiences: Design and Use of the TELEP System. In: Proc. CHI, pp. 384–391 (2000)
15. Chen, M.: Design of a virtual auditorium. In: Proc. ACM Multimedia 2001, pp. 19–28 (2001)
16. Chen, M.: Visualizing the pulse of a classroom. In: Proc. ACM Multimedia 2003, pp. 555–561 (2003)
17. Mark, G., Grudin, J., Poltrock, S.E.: Meeting at the Desktop: An Empirical Study of Virtually Collocated Teams. In: Proc. ECSCW 1999, pp. 159–178 (1999)
18. Fono, D., Baecker, R.M.: Structuring and Supporting Persistent Chat Conversations. In: Proc. CSCW 2006, pp. 455–458 (November 2006)
19. McCarthy, J.F., Boyd, D.: Digital backchannels in shared physical spaces: experiences at an academic conference. In: Proc. CHI 2005, pp. 1641–1644 (2005)
20. Caruso, J.B., Kvavik, R.B.: ECAR study of students and information technology, 2005: Convenience, connection, control, and learning. Retrieved February 13, 2007 (2005) from http://www.educause.edu/LibraryDetailPage/666?ID=ERS0506

A Quantitative Approach for the Design of Academic Curricula

Carlos Castro[1], Broderick Crawford[1,2], and Eric Monfroy[1,3]

[1] Universidad Técnica Federico Santa María, Valparaíso, Chile
[2] Pontificia Universidad Católica de Valparaíso, Chile
[3] LINA, Université de Nantes, France
Carlos.Castro@inf.utfsm.cl,
Broderick.Crawford@ucv.cl,
Eric.Monfroy@univ-nantes.fr

Abstract. We present experimental results of solving mathematical models, using very efficient complete and incomplete techniques, for designing balanced academic curricula. Solutions to these models give curricula with academic load for each period as similar as possible. Based on this work, a software tool has been developed to help administratives as well as students in the planning of academic curricula.

1 Introduction

When designing an academic curriculum a lot of factors are taken into account. Based on the objectives defined for the career under consideration, experts have to propose the courses covering the fundamentals in each domain. Roughly speaking, each domain is represented by a set of courses and some precedence relationships are also imposed among them. Indeed, the academic load of each course, representing the amount of effort required to successfully follow the course, is expressed in some way (generally called credit). Some explicit restrictions can also be imposed when developing a curriculum. For example, number of academic periods, maximum/minimum number of courses per period, and maximum/minumum academic load per period. Once this information is available, the assignment of courses to each academic period is carried out using a trail and error approach until an adequate curriculum is reached.

Assuming that a balanced academic load favors academic habits and facilitates the success of students, we are interested in designing balanced academic curricula. Considering that the academic load of each period is given by the sum of the credits of each course taken in a period, the problem of designing balanced academic curricula consists in assigning courses to periods in such a way that the academic load of each period will be balanced, i.e., as similar as possible. In order to obtain such a balanced curriculum we think that a quantitative approach, instead of the traditional trail and error approach, can be very useful.

M.J. Smith, G. Salvendy (Eds.): Human Interface, Part II, HCII 2007, LNCS 4558, pp. 279–288, 2007.

Although the mathematical models representing these situations are hard to solve (in fact, they are Combinatorial Optimization problems [6]), the development of powerful complete as well as incomplete methods in the last years allows to tackle real-life instances. Complete techniques aim to find the global optimal solution, however, when problems become too hard, these techniques cannot find the optimal solution in a reasonable time. Hopefully, incomplete techniques have been used to deal with these hard problems obtaining, in general, good local optimal solutions very quickly. Currently, a lot of research is done on the development of hybrid methods integrating complete and incomplete techniques aiming to solve hard problems very efficiently.

This paper is organized as follows: section 2 describes the problem we are interested in. Section 3 presents an integer programming model for the balanced academic curriculum problem. In order to better explain this model, we present, in section 4, a very simple example. In section 5, we present the solution to this example, we present experimental results when solving three real cases, and we describe some applications of the model. Finally, in section 6, we conclude the paper and give some perspectives for further works.

2 The Balanced Academic Curriculum Problem

In [2], we introduce the Balanced Academic Curriculum Problem (BACP), we propose a mathematical model, and we solve some real-life instances. The problem was then studied by others authors to evaluate the performances of some solving techniques using different models [3]. BACP was included as a new benchmark (prob030) in the library CSPLIB, available at www.csplib.org, a set of Constraint Satisfaction Problems [4]. As a general framework we consider administrative as well as academic regulations of the Technical University Federico Santa María:

- Academic Curriculum: An academic curriculum is defined by a set of courses and a set of precedence relationships among them.
- Number of periods: Courses must be assigned within a maximum number of academic periods.
- Academic load: Each course has associated a number of credits or units that represent the academic effort required to successfully follow it.
- Prerequisites: Some courses can have other courses as prerequisites.
- Minimum academic load: A minimum amount of academic credits per period is required to consider a student as full time.
- Maximum academic load: A maximum amount of academic credits per period is allowed in order to avoid overload.
- Minimum number of courses: A minimum number of courses per period is required to consider a student as full time.
- Maximum number of courses: A maximum number of courses per period is allowed in order to avoid overload.

In this work, we concentrate on three particular instances of the balanced academic curriculum problem: the three Informatics careers offered by the Technical University Federico Santa María.

3 Integer Programming Model

In this section, we present an integer programming model for the balanced academic curriculum problem. Classical examples and very good explanations on the subject of modeling in integer programming can be found in [8].

- Parameters

Let

- m: Number of courses
- n: Number of academic periods
- α_i: Number of credits of course i; for all i = 1, …, m
- β: Minimum academic load allowed per period
- γ: Maximum academic load allowed per period
- δ: Minimum amount of courses per period
- ε: Maximum amount of courses per period

- Decision variables

Let

- x_{ij} = 1 if course i is assigned to period j; 0 otherwise for all i = 1 , …, m; for all j = 1, …, n.
- c_j: academic load of period j; for all j = 1, …, n.
- c: maximum academic load for all periods

- Objective function

$$\text{Min } c = \text{Max}\,\{c_1, \ldots, c_n\}$$

- Constraints

- The academic load of period j is defined by:

$$c_j = \sum_{i=1}^{m} \alpha_i \times x_{ij} \quad ; \forall\, j = 1, \ldots, n$$

- All courses i must be assigned to some period j:

$$\sum_{j=1}^{n} x_{ij} = 1; \forall\, i = 1, \ldots, m$$

- Course b has course a as prerequisite:

$$x_{bj} \leq \sum_{r=1}^{j-1} x_{ar} ; \forall j = 2, \ldots, n$$

- The maximum academic load is defined by:

$$c = \mathrm{Max}\,\{c_1, \ldots, c_n\}$$

This can be represented by the following set of linear constraints:

$$c_j \leq c ; \forall j = 1, \ldots, n$$

- The academic load of period j must be greater than or equal to the minimum required:

$$c_j \geq \beta ; \forall j = 1, \ldots, n$$

- The academic load of period j must be less than or equal to the maximum allowed:

$$c_j \leq \gamma ; \forall j = 1, \ldots, n$$

- The number of courses of period j must be greater than or equal to the minimum allowed:

$$\sum_{i=1}^{m} x_{ij} \geq \delta ; \forall j = 1, \ldots, n$$

- The number of courses of period j must be less than or equal to the maximum allowed:

$$\sum_{i=1}^{m} x_{ij} \leq \varepsilon ; \forall j = 1, \ldots, n$$

4 Example

In order to better explain the general model developed in section 3, we present a reduced version taken from an informatics career offered by the Technical University Federico Santa María. Table 1 presents 18 courses assigned to 4 academic periods, each course identified by its code, the number of credits and the precedence relationships.

As parameters, we consider a maximum of 16 credits per period, a minimum of 3 credits per period, a maximum of 6 courses per period, and a minimum of 1 course per period.

Of course, constraints concerning minimum and maximum number of courses and credits per period are satisfied by the curriculum under consideration, so this assignment represents a possible (but maybe not optimal) solution to the model.

Using this solution students have an academic load of 13, 16, 16, and 10 credits per period. The maximum load per period is 16 credits and the maximum difference between loads per period is 6. We guess that this unbalance load does not help students to successfully follow the courses.

Table 1. A reduced academic curriculum

Academic period	Code	Credits	Req1	Req2
01	DEW100	1		
01	FIS100	3		
01	HCW310	1		
01	MAT190	4		
01	MAT192	4		
	Total	13		
02	FIS101	5	FIS100	MAT192
02	IWI131	3		
02	MAT191	4	MAT190	
02	MAT193	4	MAT190	MAT192
	Total	16		
03	FIS102	5	FIS101	MAT193
03	HW1	1		
03	IEI134	3	IWI131	
03	IEI141	3	IWI131	
03	MAT194	4	MAT191	MAT193
	Total	16		
04	DEW 0	2	DEW101	
04	HCW311	2	HCW310	
04	IEI132	3	IEI134	
04	IEI133	3	IEI134	
	Total	10		
	Total	55		

This solution can be represented by the values for the decision variables as presented in Table 2.

Table 2. Current assignment to variables x_{ij} for the reduced example

i/j	1	2	3	4	5	6	7	8	9	10	11	12	13	14	15	16	17	18
1	1	1	1	1	1	0	0	0	0	0	0	0	0	0	0	0	0	0
2	0	0	0	0	0	1	1	1	1	0	0	0	0	0	0	0	0	0
3	0	0	0	0	0	0	0	0	0	1	1	1	1	1	0	0	0	0
4	0	0	0	0	0	0	0	0	0	0	0	0	0	0	1	1	1	1

We now explain how constraints are satisfied by this solution.

– All courses i must be assigned to some period j:

$$\sum_{j=1}^{n} x_{ij} = 1 \ \forall i = 1,\ldots,m$$

This is verified by the sum of the elements in each column of the assignment matrix presented in Table 3.

Table 3. Current solution assigning each course to some period

i/j	1	2	3	4	5	6	7	8	9	10	11	12	13	14	15	16	17	18
1	1	1	1	1	1	0	0	0	0	0	0	0	0	0	0	0	0	0
2	0	0	0	0	0	1	1	1	1	0	0	0	0	0	0	0	0	0
3	0	0	0	0	0	0	0	0	0	1	1	1	1	1	0	0	0	0
4	0	0	0	0	0	0	0	0	0	0	0	0	0	0	1	1	1	1
Sum	1	1	1	1	1	1	1	1	1	1	1	1	1	1	1	1	1	1

– Number of courses of period j greater than or equal to the minimum allowed:

$$\sum_{i=1}^{m} x_{ij} \geq \delta \ j = 1,\ldots,n$$

This is verified by the sum of the elements in each row of the assignment matrix presented in Table 4.

Table 4. Current solution assigning a feasible number of courses per period

i/j	1	2	3	4	5	6	7	8	9	10	11	12	13	14	15	16	17	18	Σ	δ	ε
1	1	1	1	1	1	0	0	0	0	0	0	0	0	0	0	0	0	0	4	6	1
2	0	0	0	0	0	1	1	1	1	0	0	0	0	0	0	0	0	0	4	6	1
3	0	0	0	0	0	0	0	0	0	1	1	1	1	1	0	0	0	0	6	6	1
4	0	0	0	0	0	0	0	0	0	0	0	0	0	0	1	1	1	1	4	6	1

- The academic load of period j must be greater than or equal to the minimum required:

$$c_j \geq \beta \ \forall j = 1,\ldots,4$$

- The total number of credits per period is obtained by the multiplication of the assignment matrix and the vector containing the number of credits of each course:

$$\begin{pmatrix} 111110000000000000 \\ 000001111000000000 \\ 000000000111110000 \\ 000000000000001111 \end{pmatrix} \begin{pmatrix} 1 \\ 3 \\ 1 \\ 4 \\ 4 \\ 5 \\ 3 \\ 4 \\ 4 \\ 5 \\ 1 \\ 3 \\ 3 \\ 4 \\ 2 \\ 2 \\ 3 \\ 3 \end{pmatrix} = \begin{pmatrix} 13 \\ 16 \\ 16 \\ 10 \end{pmatrix}$$

These sums also verify the minimum academic load per period allowed $\beta = 3$.

- The maximum academic load is defined by:

$$c = \text{Max}\{c_1, c_2, c_3, c_4\}$$

This can be represented by the following linear constraints:

$$c_j \leq c \; \forall \; j = 1, \ldots, 4$$

Table 5 presents academic loads per period and the value c satisfying this constraint.

Table 5. Academic load per period (c_i)

I	c_i	$c = Max(c_1, c_2, c_3, c_4)$
1	13	16
2	16	16
3	16	16
4	10	16

– Course b has course a as prerequisite:

$$x_{bj} \leq \sum_{r=1}^{j-1} x_{ar} \; \forall \, j = 2, \ldots, n$$

For example, considering the precedence between MAT191 and MAT190:

$x_{82} < x_{41}$

$x_{83} < x_{41} + x_{42}$

$x_{83} < x_{41} + x_{42} + x_{43}$

5 Experimental Results and Applications

Solving the reduced version of a balanced academic curriculum problem, presented in section 4, we obtain the optimal solution presented in table 6. This assignment of courses to periods gives now 14, 14, 14, and 13 credits per period. The maximum load per period is 14 credits and the maximum difference between loads per period is 1. That is what we are looking for, the most balanced academic curriculum.

As real-life instances we first considered the three informatics careers at the UTFSM, involving 8, 10, and 12 academic periods, and 36, 42, and 62 courses, respectively. This means that the mathematical model representing each situation involves, 36 x 8, 42 x 10, and 62 x 12 binary variables, respectively. We have solved all these instances of BACP using complete as well as incomplete techniques. We have used lp_solve to apply Operations Research techniques and Oz to apply Constraint Programming techniques. lp_solve is a software toll for solving integer and linear programming models that is available free of charge at tech.groups.yahoo.com/group/lp_solve. Details about Operations Research techniques for solving integer programming models can be found in [6]. Oz [7] is a software implementing Constraint Programming techniques that is available for free at www.mozart-oz.org. Details about Constraint Programming techniques can be found in [1]. Although both complete techniques used have found the optimal solution for all instances very efficiently, we have also implemented a Genetic Algorithm [5] to apply the incomplete approach. We have obtained very good solutions in a very short time, but still not the optimal solutions obtained by the complete techniques.

Considering the successful results obtained dealing with these instances, we have tried to validate the real applicability of this approach. On one hand, BACP has already been used by the UTFSM to improve several academic curricula. On the other hand, a user-friendly interface has been developed to help students to plan their curricula when they have not approved some courses. Indeed, we have extended the basic model presented in this paper to consider several careers at the same time in order to assign common courses to the same academic period. Due to the higher complexity of this problem, we had to develop hybrid algorithms integrating complete and incomplete techniques because none of the previous ones was able to solve some problems alone. This more complex problem has been very interesting in terms of economic resources for the UTFSM because solutions allow to reduce the number of common courses dictated, assigning some of them in the same academic period, and so reducing the number of lecturers.

Table 6. Optimal solution for the reduced academic curriculum

Academic period	Code	Credits	Req1	Req2
01	FIS100	3		
01	MAT190	4		
01	MAT192	4		
01	IWI131	3		
	Total	14		
02	DEW100	1		
02	FIS101	5	FIS100	MAT192
02	MAT191	4	MAT190	
02	MAT193	4	MAT190	MAT192
	Total	14		
03	IEI134	3	IWI131	
03	IEI141	3	IWI131	
03	MAT194	4	MAT191	MAT193
03	DEW 0	2	DEW101	
03	HW1	1		
03	HCW310	1		
	Total	14		
04	FIS102	5	FIS101	MAT193
04	IEI132	3	IEI134	
04	IEI133	3	IEI134	
04	HCW311	2	HCW310	
	Total	13		
	Total	55		

6 Conclusions

We have presented a quantitative approach for designing balanced academic curricula. Using solving techniques implemented in software available for free we have been able to solve very efficiently all instances we have considered in this work. We strongly believe that this kind of quantitative approach should be used in order to support several decision problems that arise in the educational field. Many fruitfull experiences have already been carried out for solving, for example, timetabling problems using a similar approach.

Acknowledgments. The first author has been partially supported by the Chilean National Science Fund through the project FONDECYT 1070268. The third author has been partially supported by the Chilean National Science Fund through the project FONDECYT 1060373. All authors have also been partially supported by the project INRIA-CONICYT VANANAA II.

References

1. Apt, K.R.: Principles of Constraint Programming. Cambridge Univ. Press, Cambridge (2003)
2. Castro, C., Manzano, S.: Variable and Value Ordering When Solving Balanced Academic Curriculum Problems. In: Proceedings of The Sixth Workshop of the Working Group on Constraints, ERCIM 2001, Prague, Czech Republic (June 2001)
3. Hnich, B., Kiziltan, Z., Miguel, I., Walsh, T.: Hybrid Modelling for Robust Solving. Annals of Operations Research 130(1-4), 19–39 (2004)
4. Mackworth, A.K.: Constraint Satisfaction. In: Shapiro, S.C. (ed.) Encyclopedia of Artificial Intelligence, 2nd edn. vol. 1, Addison-Wesley, New York (1992)
5. Michalewicz, Z., Fogel, D.: How to Solve It: Modern Heuristics. Springer, Heidelberg (1999)
6. Nemhauser, G.L., Wolsey, L.A.: Integer and Combinatorial Optimization. In: Wiley-Interscience Series in Discrete Mathematics, Wiley, New York (1999)
7. Smolka, G.: The Oz Programming Model. In: van Leeuwen, J. (ed.) Computer Science Today. LNCS, vol. 1000, pp. 324–343. Springer, Heidelberg (1995)
8. Williams, H.: Model Building in Mathematical Programming, 4th edn. J. Wiley and Sons, New York (1999)

Privacy and the Public Educator

Melissa Dark and Clewin McPherson

Department of Computer and Information Technology, Purdue Univerisity
401 North Grant Street, West Lafayette, IN 47906-2021
{dark, mcpherso}@purdue.edu

Abstract. As of 2003, 99% of K-12 schools in the United States use the Internet. With the increased use of databases and other technologies to manage data in K-12 institutions, an inherent threat to the privacy and confidentiality of that information has also increased. Over the last decade, with the increased use of these technologies, there has also been an increase in the number of privacy-related violations that have occurred, both in industry and in the K-12 environment. There are a plethora of security technologies that can be used to improve privacy; however, organizations such as K-12 schools often cannot afford to hire IT staff versed in the state of the art. Furthermore, technology alone will not address security and privacy problems; policy is an essential ingredient for any organization. This study is a gap analysis investigating privacy practices of public educators in the Midwestern portion of the United States. The significance of the work is that we cannot improve the practices unless we understand deficiencies in current privacy practices and perceptions.

Keywords: Privacy practices, privacy perceptions, privacy risk, public education.

1 Introduction

In 1999, the United States participated in the International Math and Science to determine their ranking in the mathematics and science achievement when compared to the world. The results of the study suggested that new technologies needed to be introduced into the educational system to increase the ranking. As a result, by the end of 1999 the ratio of students to computers in K-12 schools was 6 to 1 [1]. A study conducted in 1995 observed the trends in educational technology and concluded that 1) computers are pervasive in schools, granting almost every child access to a computer, 2) networking is being increasingly used, and 3) educational technology applications have grown significantly and so have their delivery systems [2].

These increasing technologies have not only affected the way teachers deliver their content and students do their research, but also the way in which administrators manage school business. School corporations have become a virtual warehouse of personal information due to the expansive task of managing student, staff and administrative records. As of 2003, 99% of K-12 schools use the Internet [3]; that number has most likely increased since that report.

M.J. Smith, G. Salvendy (Eds.): Human Interface, Part II, HCII 2007, LNCS 4558, pp. 289–298, 2007.

With the increased use of databases and other technologies to manage data in K-12 institutions, an inherent threat to the privacy and confidentiality of that information has also increased. Privacy can be defined in many ways, but can generally be summed up as an individual's ability to control what information is revealed about oneself and who has access to that information [4]. Over the last decade, with the increased use of these technologies, there has also been an increase in the number of privacy related violations that have occurred, both in industry and K-12.

1.1 Privacy Regulations

The increased reliance and utilization of information technology in society has created a need for regulation of the use and abuse of these systems. As technology has increased, the number and scope of the regulations that have been enacted to protect the privacy of information stored and used by these systems has also increased (as illustrated in Figure 1). While not all of the enacted regulations affect K-12 organizations, it is important to understand the general protections that have been placed around information. There are currently over 20 established privacy and privacy-related regulations that affect our daily lives. FERPA, COPPA and HIPAA directly affect the way K-12 organizations are managed. In addition, there are 17 bills that are pending in congress, three of which may affect K-12 institutions [5].

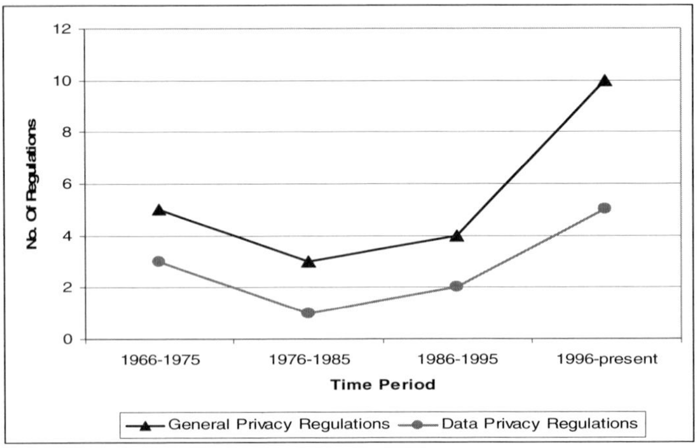

Fig. 1. Privacy related regulations over the last 40 years

Family Educational Rights and Privacy Act of 1974 (FERPA). FERPA was enacted in 1974 and is applicable to all schools that receive funding from the U.S. Department of Education. The regulation protects the privacy of student educational records, which includes any information about a student that is handwritten, in print or stored in electronic format. Educational records include much more than test scores or class standing. They can include health information, description of physical appearance, family economic circumstances, criminal history, ethnic background, political and religious affiliations, and psychological test results. The information can be a fact such as birth date or a Social Security number, and it can include teachers' opinions of your child [6]. In 1994, several components of FERPA were amended by the Improving America's Schools Act in order to increase the privacy assurances provided for students and families [8].

Children's Online Privacy Protection Act of 1998 (COPPA). Passed in response to the increased online marketing targeting children, COPPA protects the privacy of children (defined as anyone under the age of 13) by requesting consent from a child's parent or legal guardian to collect or use any personal information [9]. While typically the law is focused on commercial websites, it is also applicable to School and District websites that may require the students to provide personal information to access or contribute to the website.

Health Insurance Portability and Accountability Act of 1996 (HIPAA). In 1996 the US Department of Health and Human Services developed a series of rules governing health information. The goal of the rules was to standardize the communication of health information between health care providers and to protect the privacy and security of personally identifiable health information. The information that is protected under HIPAA is any Individually Identifiable Health Information that relates to or includes 1) the individual's past, present or future physical or mental health or condition, 2) the provision of health care to the individual, or 3) the past, present, or future payment for the provision of health care to the individual [10].

It should be noted that the Privacy Rule excludes employment records that a covered entity maintains in its capacity as an employer and education records subject to, or defined in, the FERPA. As a result, while student medical information is covered under FERPA and not HIPAA, any health care information maintained for the purposes of staff and faculty of the school corporation is covered under HIPAA.

Technology is critical to the daily operation of K-12 organizations. These organizations are regulated, much in the same manner as corporations in industry. Unlike these corporations, most K-12 institutions do not have the resources (both financial and technical) to effectively comply with these regulations, while staying abreast of technological advances. It is important to understand exactly where K-12 organizations are in their implementation of privacy protecting principles and technologies in order to design programs and system to enable them not only to be effective in their compliance, but also to take full advantage of the resources and technologies available to them.

2 Current State of K12 Privacy

In March, 2006, an assessment tool, which was designed to assess the state of K12 institutions with regards to various aspects of privacy, was given to a number of technology directors and administrators employed in K-12. The tool was intended to identify the problem space of privacy in the context of each school environment and assess the existence of some suggested privacy preserving strategies and techniques.

2.1 Instrument

The questionnaire was administered online and distributed to 36 schools in the Midwest United States. The survey consisted of 62 questions across five sections – privacy basics, malware, phishing, responsibility and privacy policies. Each section was assigned a proportionate weight based on the degree of importance to the K12 community in understanding relevant privacy issues. The importance was assigned based on conversations with subject matter experts. Privacy basics, responsibility and privacy policies were weighted equally (25%) as these issues were pivotal to understanding the K12 privacy space. Each question in each of the sections was also assigned a weighting based on how effective the question would be in evaluating the privacy practices of a school corporation and a similar survey provided by Thomas Peltier in his book *Information Security Risk Analysis*, as part of the risk assessment process. For example, in the Privacy basics section, the question "*Is there anyone that is responsible for privacy related issues in the school corporation?*" was weighted higher than "*Are areas containing sensitive information properly secured?*" due to the responsibilities of someone who is responsible for privacy issues in implementing and maintaining privacy preserving strategies and most likely including the securing of sensitive information.

The survey takers were asked to rate each question on a four point Likert scale. The responses were "Yes", "Being Implemented", "Being Discussed" or "No". A "Yes" response would signify total agreement with the question asked. "Being Implemented" means that the corporation is currently implementing systems or processes that would lead to a "Yes" response for that question. "Being Discussed" indicates that the corporation has identified that the practice addressed in the question could be implemented to solve a privacy issue. A "No" response signifies complete disagreement with the stated question. Each response is assigned a numerical weighting, (in a positively worded question, an answer of "Yes" would elicit a score of 4, "Being Implemented" would be scored 3, "Being Discussed would score 2 and "No" would elicit a score of 1). A relative score for the question is then tabulated based on the weighting that was assigned to the question (the takers of the survey did not have any knowledge that the survey was weighted) as shown by Formula 1 below.

$$Q_{1\ rel\_weight} = Sum\ (Q_{1\ score} * Q_{1\ weight}) \tag{1}$$

The relative percentile score for the section is also tabulated as shown by Formula 2.

$$Section\_Score = Sum(Q_{1\ rel\_weight} : Q_{7\ rel\_weight}) \tag{2}$$

2.2 Population and Sample

The survey was completed by a sample of 26 persons. Ten of those persons were Superintendents, Principals or Assistant Principals, while the other 16 were Technology Directors. The email was sent to the Superintendents and they either completed it themselves or assigned it to the technology director to complete.

3 Results

The results of the survey are reported in the modules according to how the survey was divided. Figure 2 shows that the Privacy Policies section of the assessment yielded the lowest results from the respondents, while the malware section was significantly higher than the other areas. The maximum score attainable in any section was 100%. Respondents scored over 80% in only one category, and under 50% in two categories.

3.1 Module I – Privacy Basics

The goal of this module was to determine the level of importance and attention that privacy and privacy related issues received in their organizations. Table 1 shows the results of the first module. Overall, the respondents scored just under 60%, which indicates that there is still work to be done in this area of privacy within the various school corporations.

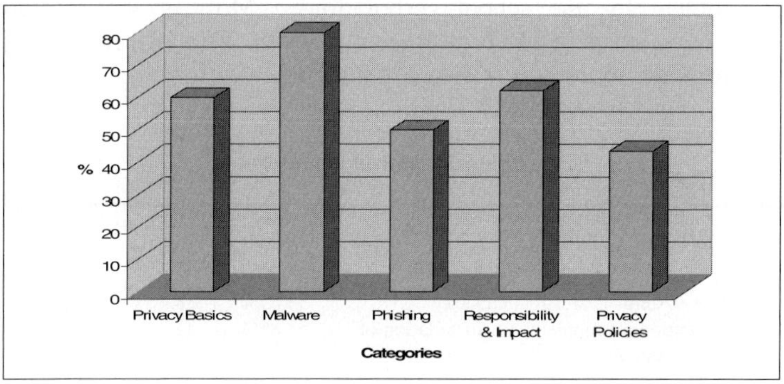

Fig. 2. Survey results by module

Table 1. The privacy basics results

Question	Yes	Being Implemented	Being Discussed	No
1. Does your school corporation separate security and privacy?	26.92%	3.85%	11.54%	57.69%
2. Is there anyone responsible for privacy related issues in the school corporation?	46.15%	0.00%	19.23%	34.62%
3. Is privacy a line item in the budget?	7.69%	0.00%	3.85%	88.46%
4. Is an annual report on privacy compliance issued to management?	0.00%	3.85%	3.85%	92.31%
5. Are areas containing sensitive information properly secured?	61.54%	23.08%	0.00%	15.38%
6. Is confidential information properly secured?	69.23%	15.38%	3.85%	11.54%
7. Are passwords and accounts being shared?	23.08%	3.85%	7.69%	65.38%

Table 1 shows that over 50% of the respondents did not separate privacy and security in their organization and that there is no annual report on compliance with privacy regulations. In most of the schools, there was not a specific person responsible for privacy related issues. This is significant because of the unique privacy specific threats that exist, in addition to fact that privacy preservation in school corporations is mandated by FERPA. Practically none of the schools had even discussed placing privacy and privacy issues as an item in their budget. It is noted that while a specific focus is not placed directly on privacy, some of the issues are addressed in the organizations, such as the protection of confidential information and the lack of sharing of user accounts and passwords.

3.2 Module II – Malware

The goal of this module was to determine the level of readiness for malware incidents and protection against possible incidents to which these organizations are susceptible. Overall, this module had the highest score of all the modules. Table 2 shows the overall results for this module and indicates that the malware issues were not a significant problem for the school corporations. Most of the institutions had procedures or technology in place to prevent malware from affecting their systems and could identify the sources of malware in their organization.

Table 2. The malware results

	Yes	Sometimes	Rarely	No
1. Do your computer systems block banner ads and "pop ups"?	80.77%	19.23%	0.00%	0.00%
2. Do your computer systems perform at capacity?	53.85%	38.46%	3.85%	3.85%
3. Do you have prevention mechanisms for installation of free and shareware on systems?	53.85%	23.08%	7.69%	15.38%
4. Are administers, teachers and staff prevented from downloading and installing software?	38.46%	34.62%	11.54%	15.38%
5. Is software downloaded from underground warez sites installed on your computer systems?	12.00%	20.00%	24.00%	44.00%
6. Do your browser settings remain constant and unchanged?	65.38%	15.38%	7.69%	11.54%
7. Additional toolbars on your browser are disallowed?	50.00%	26.92%	11.54%	11.54%
8. You organization has mechanisms to ensure that spyware does not negatively affect privacy.	69.23%	23.08%	7.69%	0.00%
9. Viruses have never negatively impacted your corporation.	42.31%	15.38%	30.77%	11.54%
10. Can you identify the major sources of malware in your corporation?	48.00%	28.00%	4.00%	20.00%
11. Are there any technological solutions in place to mitigate malware on your systems?	56.00%	36.00%	0.00%	8.00%
12. Are your employees aware of the dangers of opening attachments in emails?	73.08%	26.92%	0.00%	0.00%

3.3 Module III - Phishing

While the school corporations were generally aware of malware and had sufficient mechanisms in place to preserve the privacy and confidentiality of information from malware, they were not prepared for phishing attacks, as indicated in Table 3.

3.4 Module IV - Responsibility

The responsibility of preserving privacy was the second highest rated area based on the survey (see Table 4). This means that the school corporations recognize the

Table 3. The phishing results

	Yes	Being Implemented	Being Discussed	No
1. Is there any awareness program about phishing attempts?	32.00%	12.00%	12.00%	44.00%
2. Do you think your employees are able to detect phishing attempts?	28.00%	12.00%	12.00%	48.00%
3. Have phishing attacks caused privacy or security problems in the past year?	12.00%	4.00%	8.00%	76.00%
4. Do you block executables?	80.77%	3.85%	3.85%	11.54%
5. Do you block HTML transmitted through email?	29.17%	0.00%	12.50%	58.33%
6. Are you a subscriber to any of the popular phishing scam corporations (ebay, chase, etc)?	15.38%	0.00%	0.00%	84.62%

Table 4. The responsibility results

	Yes	Being Implemented	Being Discussed	No
1. Do you encourage your staff and students to read the privacy statements on websites before registering on it?	73.08%	0.00%	7.69%	19.23%
2. Do you encourage your staff and students to share their personal information only after reviewing privacy statements?	53.85%	3.85%	7.69%	34.62%
3. Are employees trained to be cognizant of privacy matters?	52.00%	16.00%	8.00%	24.00%
4. Is access to sensitive/confidential information by contractors monitored?	73.08%	0.00%	3.85%	23.08%
5. Do employees receive training on privacy relative to their experience and responsibilities?	38.46%	15.38%	7.69%	38.46%
6. Are employees receiving both positive and negative feedback regarding privacy on their performance reviews?	11.54%	0.00%	7.69%	80.77%
7. Are administrators given additional privacy specific training?	7.69%	0.00%	15.38%	76.92%
8. Is there a regular privacy awareness program in place (newsletter, etc)?	0.00%	3.85%	11.54%	84.62%
9. Are audit logs in place for systems containing sensitive information?	38.46%	3.85%	0.00%	57.69%
10. Are violations to privacy tracked?	23.08%	19.23%	11.54%	46.15%
11. Are procedures in place for the proper disposal of confidential information?	48.00%	16.00%	4.00%	32.00%
12. Are unsecured and temporary accounts restricted from sensitive information and disabled in a timely fashion?	80.00%	8.00%	0.00%	12.00%
13. Have employees been trained on proper password management?	56.00%	12.00%	16.00%	16.00%
14. Are permissions being properly set (only those who need the information have access to it)?	92.31%	7.69%	0.00%	0.00%
15. Are users of all network resources required to change the initial default password?	61.54%	7.69%	3.85%	26.92%
16. Are Access Control Lists maintained on a regular basis?	65.38%	11.54%	0.00%	23.08%

importance of privacy in their daily operations and the problem may lie in them not having the resources available to sufficiently deal with privacy risk.

3.5 Module V – Privacy Policies

Privacy policies rated the lowest based on the survey. While all the respondents indicated that they had websites, not all had privacy policies on their websites. In addition, the ones that did have privacy policies had very poorly written policies, in terms of content. In addition, there were no policies or documents that indicated what information was collected by the organization in general or how that information is stored, who has access to it and how it is maintained.

Table 5. The privacy policies results

	Yes	Being Implemented	Being Discussed	No
1. Do you have a website?	100.00%	0.00%	0.00%	0.00%
If the answer to Q1 is yes continue below… If the answer is no, go to Q17				
2. Does your website target group include children under 13?	69.23%	0.00%	0.00%	30.77%
3. Does your website collect email addresses?	19.23%	0.00%	3.85%	76.92%
4. Does your website collect personal identifying information other than email address?	3.85%	0.00%	3.85%	92.31%
5. Is there a privacy policy for your website?	41.67%	12.50%	12.50%	33.33%
If the answer to Q5 is yes continue below… If the answer is no, go to Q16				
6. Does the Privacy Policy mention anything about what specific personal information is collected?	7.69%	0.00%	7.69%	84.62%
7. Does the Privacy Policy mention how the website may use personal information it collects for internal purposes?	11.54%	0.00%	7.69%	80.77%
8. Does the Privacy Policy mention anything about the website's use of personal information collected to send communications to visitors?	12.00%	0.00%	8.00%	80.00%
9. Does the Privacy Policy mention anything about the website's right to disclose personal information to third parties?	8.00%	0.00%	4.00%	88.00%
10. Does the Privacy Policy say whether the website allows users to review, modify and delete some of the personal information collected about them?	4.00%	0.00%	12.00%	84.00%
11. Does the Privacy Policy say what measures the website takes to provide security?	8.00%	0.00%	12.00%	80.00%
12. Does the Privacy Policy say what measures are taken to provide security for personal information during the transmission from the consumer to the website?	4.00%	0.00%	12.00%	84.00%
13. Does the Privacy Policy say what measures are taken to secure personal information after it has been collected?	4.00%	0.00%	12.00%	84.00%
14. Does the Privacy Policy say whether the website places cookies or not?	3.85%	0.00%	11.54%	84.62%

Table 5. (*Continued*)

15. Does the Privacy Policy say whether third parties may place cookies or collect personal information on the website?	3.85%	0.00%	11.54%	84.62%
No privacy policy for the website				
16. Does your corporation have any plans to develop a privacy policy for your website? Answer and then go to Q17	0.00%	12.50%	12.50%	75.00%
No website in the corporation				
17. Do you have a document that describes the type of information collected in your corporation?	26.92%	0.00%	15.38%	57.69%
18. Do you have a policy that documents the privacy practices of your corporation?	34.62%	0.00%	15.38%	50.00%
19. Do information providers have any options when disclosing information?	24.00%	0.00%	16.00%	60.00%
20. Do you have a policy that states the right to disclose personal information to third parties?	30.77%	0.00%	11.54%	57.69%
21. Do you have a policy that states how sensitive information collected is stored and secured?	25.00%	0.00%	12.50%	62.50%

Fewer than half of the respondents (all of whom had a website) had a privacy policy for the website. This is probably not as severe as expected since less than 20% collect email addresses and less than 5% collect personally identifiable information. The section was important, however, since many of the school corporations were considering implementing online student management systems, where parents can access information about their children.

4 Discussion and Implication

One of the most significant things revealed through the results of this study is the discrepancies within responses between the administrators of the school and the technology directors. This indicates that there needs to be a series of conversations between Technology Administrators and Superintendents, where the topic is privacy and privacy related issues. An important precursor to this conversation is the other main observation that can be drawn from this study – the need for privacy education for all those involved in the administration and operation of school corporations. This is significant since 3 of the five areas yielded very poor results (under 60%). The conversation between these education leaders cannot be effective unless and until both parties can use the same vocabulary, understand the important of privacy in their organizations, understand their role in privacy preservation and identify the threats and vulnerabilities that their organizations face.

5 Conclusion

Privacy is important to K12 organizations. Not only is it important because of civil responsibility, but it is also important because it is mandated by law. K12 organizations are currently unprepared to deal with the threats to privacy and in most cases, do not understand their role in protecting the privacy of their constituents. It is important that K12 personnel be educated about privacy within the constructs of their organizations and that management have a common vocabulary and understanding of these issues.

Acknowledgments. This material is based upon work supported by the National Science Foundation under Grant No. 0430274. Any opinions, findings, and conclusions or recommendations expressed in this material are those of the author(s) and do not necessarily reflect the views of the National Science Foundation.

References

1. Smerdon, B., Cronen, S., Lanahan, L., Anderson, J., Iannotti, N., Angeles, J.: Teachers' tools for the 21st century: A report on teachers' use of technology (NCES Publication No. 2000-102). Washington: U.S. Department of Education, Office of Educational Research and Improvement, National Center for Education Statistics. Retrieved October 15, 2006 (2000) from http://nces.ed.gov/pubs2000/2000102A.pdf
2. Ely, D., Plotnik, E.: Trends in Educational Technology 1995. ERIC Digest, ERIC Clearinghouse on Information and Technology, Syracuse New York (1996) ED398861
3. U.S Department of Education, National Center for Educational Statistics. Internet Access in Public Schools and Classrooms: 1994 – 2002, NCES 2004-011. Washington, D.C (2003)
4. Brandeis, L., Warren, S.: The Right to Privacy. Harvard Law. Review 4, 193 (1890)
5. EPIC, EPIC Bill Track: Tracking Privacy, Speech, and Cyber-Liberties Bills in the 109th Congress. Electronic Privacy Information Center (EPIC). Retrieved October 22, 2006 (2005) from http://www.epic.org/privacy/bill_track.html
6. Privacy Rights Clearinghouse, Privacy in Education: A Guide for Parents and Adult-Age Students. Privacy Rights Clearinghouse. Retrieved October 22, 2006 (2006) from http://www.privacyrights.org/fs/fs29-education.htm#3
7. U.S Department of Education. Family Educational Rights and Privacy Act (FERPA). Ed.gov. Retrieved on 16th October 2006 (2005) from http://www.ed.gov/policy/gen/guid/fpco/ferpa/index.html
8. U.S Department of Education. Legislative History of Major FERPA Provisions. Ed.gov. Retrieved on 16th October, 2006 (2004) from http://www.ed.gov/policy/gen/guid/fpco/ferpa/leg-history.html
9. Federal Trade Commission. Children's Online Privacy Protection Act of 1998. Retrieved on 17th October, 2006 (1998) from http://www.ftc.gov/ogc/coppa1.htm
10. Privacy Rights Clearinghouse. HIPAA Basics: Medical Privacy in the Electronic Age. Privacy Rights Clearinghouse. Retrieved October 22, 2006 (2006) from http://www.privacyrights.org/fs/fs8a-hipaa.htm

HCI for m-Learning in Image Processing by Handhelds

Danco Davcev, Marjan Arsic, Dalibor Ilievski, and Andrea Kulakov

Sts. Cyril and Methodius University, Faculty of Electrical Engineering
Karpos 2, bb, 1000 Skopje, Macedonia
etfdav@etf.ukim.edu.mk, amarjan@on.net.mk, i_dalibor@yahoo.com

Abstract. The objective of this paper is to present a part of m-learning process developed at our University at the Faculty of Electrical Engineering in the field of image processing. The basic courses in this field are on the Faculty Web. The multimedia illustration of the basic methods in image processing is realized both on Desktop PC and on handheld (PDA) devices equipped with cameras and could be used individually by each student. The students can take photos with the cameras and interactively learn about the results of the image processing algorithms. For efficient use of the handheld devices we developed a suitable HCI. According to the surveys with 20 students at the last year of study, their experience with our specially developed tools for m-learning is very positive.

Keywords: Mobile learning, image processing, HCI, Gabor 2D, Pocket PC.

1 Introduction

Pocket PCs have become very popular in the last few years due to the increase of availability of different wireless services [1][2][3][4].

Mobile learning (m-learning) is defined as the ability of using handheld devices to access learning information like electronic books and other materials anytime and anywhere. M-learning has received a lot of attention these days as the next wave of learning [5][6][7].

The possibility of the handheld devices, like handhelds' mobility and touch-screen interactivity made them ideal for laboratory, classroom and terrain use. For this purpose an application which enables students to analyze images using the Gabor 2D transformation was developed. The students can also view the results of the transformation of images captured from camera, or from existing previously saved bitmap images.

In this paper, we will present our Human Computer Interface (HCI) for m-learning in Image Processing by Handhelds by illustrating one of the standard methods in this field. The multimedia illustration is available on handheld devices such as PDA devices and could be used individually by each student. It is also available on Desktop PC. The related work is presented in Section 2. The Section 3 gives the design and implementation of our application. Section 4 gives some examples of images captured from camera and from BMP file, which are used for transformation. In Section 5 the

M.J. Smith, G. Salvendy (Eds.): Human Interface, Part II, HCII 2007, LNCS 4558, pp. 299–308, 2007.

evaluation study is presented while the results of the evaluation are presented in Section 6. Section 7 concludes the paper.

2 Related Work

The theory-informed design, implementation and evaluation of a handheld learning device is described in [8], which is intended to support 9-11 aged children to capture everyday events such as images, notes and sounds, to relate them to web-based learning resources, to organize these into a visual knowledge map, and to share them with other learners and teachers. Our work is intended for the students in faculties who learned the subjects in signal processing.

The experience of designing a collaborative learning activity for a traditional historical/cultural museum is described in [9]. Based on a "Mystery in the Museum" story, a group of students collaborated to solve a mystery inside the museum, interacting through mobile devices. The application described in this paper is a puzzle game, and can be used for educational purposes, but with some limitations. Unlike this, our application is intended for students which can interactively use the interface by capturing images of landscapes using the Pocket PC camera.

In [10] some special issues in processing and streaming lecture videos were addressed. New approach for real-time content analysis and adaptive transmission of these videos over wireless networks is proposed. The effectiveness and scalability of the e-learning system was demonstrated. Experimental results of adaptive transmission of instructional videos over 802.11b wireless networks to mobile devices were also given.

The rationale for teaching human-computer interaction, primarily to upper-level undergraduates in computer science and in information sciences and technology is described in [11]. The structural schema and a browser is developed for some case studies. The case-based activities employed in the courses together with the experiences of instructors and students are also given in the paper.

An implementation of the concept of object replication approach for integrating synchronous and asynchronous distributed work using time-line based user interfaces, called Classroom Bridge can be found in [12]. It supports activity awareness by facilitating planning and goal revision in collaborative, project-based middle school science.

Results from a study of an automated capture and access system – e-Class are presented in [13]. The e-Class was designed to capture the materials presented in college lectures for later review by students. In [14] the benefits and challenges of using handheld computers to support learners in creating concept maps (a type of visual outline) were explored. The Pocket PiCoMap, as a learner-centered concept mapping tool for handheld Pocket PCs is presented.

The product named as e - Class was designed to capture the materials presented in college lectures for later review by students. Another comparison of the traditional (classroom) learning and e-learning can be found in [15].

In [16] a project for integration of the handheld machines into the classroom was documented. The design of three learner-centered handheld tools used as a part of a nine-month classroom study involving some eighth grade students is given in [17]. A review of related work identifies some of the challenges of building educational software within the constraints of handheld screens. Two broad design guidelines are presented. The first focuses on decomposing the learning activity and the second focuses on methods for implementing scaffolds.

Like all of these papers, our paper also describes HCI, where the handheld devices are used for image processing.

3 Design and Implementation of Our Application

In this paper we will show the design and implementation of a HCI for support the education process in the field of image processing. The multimedia illustration of the basic methods in image processing (useful for students in faculties which study the signal processing courses) is realized on handheld (PDA) devices equipped with cameras and could be used individually by each student. The students can take photos with the cameras and interactively learn about the results of the image processing algorithms.

During the design process several problems had to be solved; one of them being the method of taking photos with the Pocket PC camera and analyzing individual pixels, while another one was the method of dealing with previously saved images. In the second case, it was necessary to adjust the dimensions of the image to the display of the Pocket PC. The results from the image processing are saved in BMP format and are later displayed by coding HTML pages including the BMP images and Java Scripts for navigating the images obtained from the transformation.
For the image processing application, the Gabor filter method is used and the image is convoluted with a lattice of possibly overlapping banks of Gabor filters [18] at different orientations and frequencies.

The goal of the 2D Gabor transformation [19][20] is to represent a digital image $f(x, y)$, where x and y represent spatial coordinates, either exactly or in some optimal sense (e.g., minimizing the mean squared error between the reconstructed image and the original image) by projecting it onto a set of 2D Gabor elementary functions. For a finite extent image $f(x, y), x = 0,1..., X-1; y = 0,1,..., Y-1$ partitioned into $K \times L$ non-overlapping lattices of size $M \times N$ where it is assumed that $X = KM$ and $Y = LN$, the approximated or reconstructed image $\hat{f}(x, y)$ can be written as

$$\hat{f}(x, y) = \sum_{m=0}^{K-1} \sum_{n=0}^{L-1} \sum_{r=0}^{M-1} \sum_{s=0}^{N-1} a_{mnrs} G_{mnrs}(x, y). \tag{1}$$

In this representation, there are $KM \times LN$ coefficients a_{mnrs}'s that have a symmetry property, i.e., the real part has an even symmetry, and the imaginary part has an odd symmetry. Thus, only half of the coefficients will be sufficient for the image representation. Each Gabor elementary function G_{mnrs}'s in (1) constitutes a sine and cosine wave modulated by a Gaussian window, i.e.

$$G_{mnrs}(x,y) = w(h_1(x) - mM, h_2(y) - nN) \cdot \exp\left(2\pi j\left[\frac{h_1(r)h_1(x)}{M} + \frac{h_2(s)h_2(y)}{N}\right]\right) \quad (2)$$

where $h_1(z) := z - \frac{M-1}{2}$, and $h_2(z) := z - \frac{N-1}{2}$ are introduced to center the Gaussian window at the center of the lattices; parameters M and N define the spatial distance between the centers of the elementary functions, which are usually called the Gabor logons, and $w(x, y)$ is the Gaussian window function of the form

$$w(x,y) = \left(\sqrt{2}\alpha\right)^{\frac{1}{2}}\left(\sqrt{2}\beta\right)^{\frac{1}{2}} \cdot \exp\left[-\pi\left(\alpha^2 x^2 + \beta^2 y^2\right)\right]. \quad (3)$$

The parameters α and β in $w(x, y)$ define the scaling of the Gaussian in the spatial domain along x and y coordinates, respectively. The values of these parameters are typically chosen to be the same, in which case, we get

$$w(x,y) = \left(\sqrt{2}\alpha\right) \cdot \exp\left[-\pi\alpha^2\left(x^2 + y^2\right)\right]. \quad (4)$$

From (4) and (2), it can clearly be seen that the Gabor elementary functions are parameterized for an invariant Gaussian window, which is positioned in fully overlapping Cartesian lattice location $\{x_m, y_m\} = \{mM, nN\}$, where (m,n) is the index for the lattice. The complex exponential that modulates this overlapping Gaussian is parameterized for the Cartesian lattice of 2D spatial frequencies: $\{u_r, v_s\} = \{r/M, s/N\}$ for integer increments of (r,s) spanning their range inside each lattice $[-(M-1)/2, (M-1)/2]$, $[-(N-1)/2, (N-1)/2]$, respectively.

For fixed values of M and N, the choice of the Gaussian scale α determines the amount of effective overlap of the Gabor elementary functions across the neighboring lattice locations. It also determines the required support size (number of pixels) of each Gabor elementary function so that the truncation of the Gaussian tail produces negligible error. Therefore, if the truncation of the Gaussian is done outside each lattice the value of α should be chosen so that most of the energy of the Gaussian window lies inside the lattice. It must also be pointed out that since the image is partitioned into lattices and the Gabor elementary functions for each lattice are all

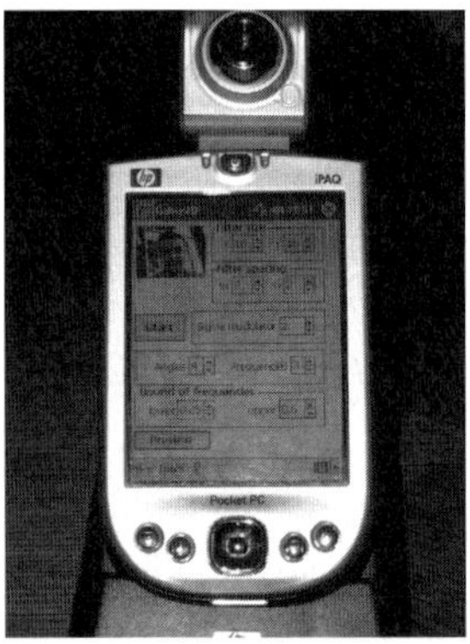

Fig. 1. Preview of the Gabor 2D application on Pocket PC device

centered, only translations that are equal to multiples of the lattice dimensions will result in a simple shifting relation of the transform coefficients.

The corresponding algorithm is implemented in the Gabor 2D application for Pocket PC, which was developed using Microsoft eMbedded Visual C++ 4.0 [21] and Microsoft Pocket PC 2003 SDK [22][23]. The preview of the above application is shown on Fig. 1.

4 Experimental Results

The interface of the Gabor 2D application for Pocket PC is shown on Fig. 2. Here the Gabor 2D parameters, like filter size, filter spacing, sigma modulator, number of angles and frequencies, lower and upper bound of frequencies can be set. The maximum values of the parameters are the following: filter size up to 32x32; filter spacing up to 8x8; sigma modulator up to 10; up to 9 different angles; up to 9 different frequencies; lower and upper bound of frequency up to 10. Greater values are not recommended, because the time needed for image processing can be too long.

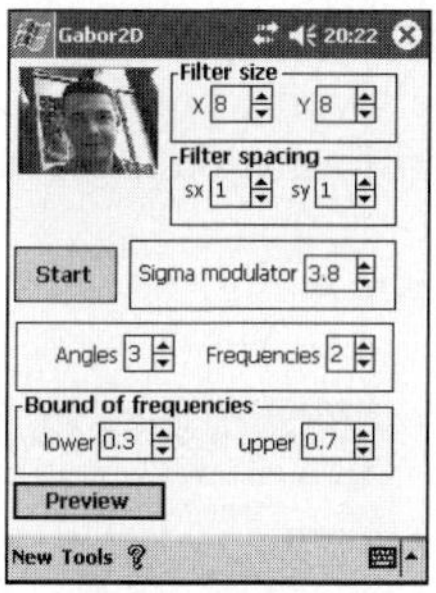

Fig. 2. Gabor 2D application for Pocket PC

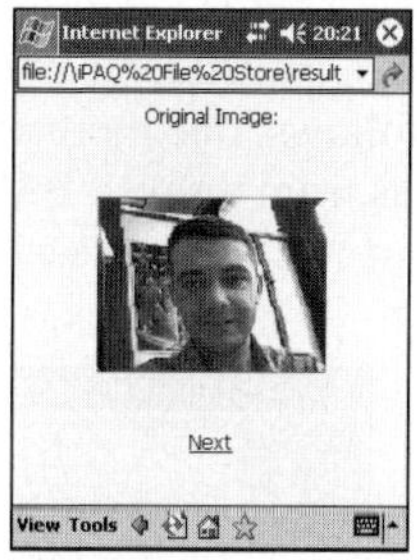

Fig. 3. Source image from Pocket PC camera

The image used for the transformation can be captured from the Pocket PC camera or an existing BMP file can be used. These options are available on the Tools menu. The original image and the results of the transformation, together with the filters can be viewed later with a standard web browser. One example image captured from the Pocket PC camera is shown on Fig. 3.

The parameters used for the transformation of this image are the following: 8x8 fitter; 1x1 spacing of the filter; sigma modulator with value 3.8; three angles 0°, 60° and 120°; two frequencies; lower bound of the frequency equal to 0.3 and upper bound of the frequency equal to 0.7. After taking photos with the Pocket PCs, these parameters can be selected by students. In that way, the students can experiment the properties and different outputs of the Gabor 2D filter method.

After applying the Gabor 2D transformation on the image shown on Fig. 3. , we get real and imaginary results from the 8x8 filters for all angles and frequencies. Some of the resultant images are displayed on Fig. 4.

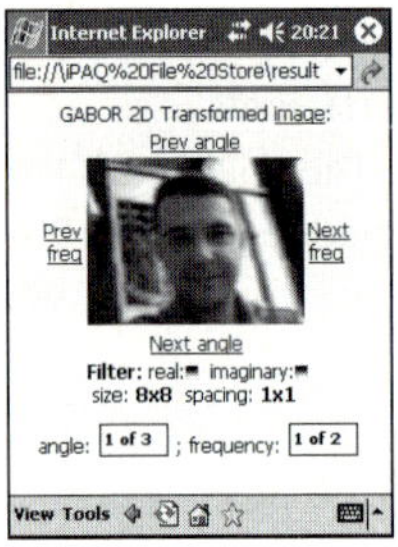

Angle of 0° and lowest frequency

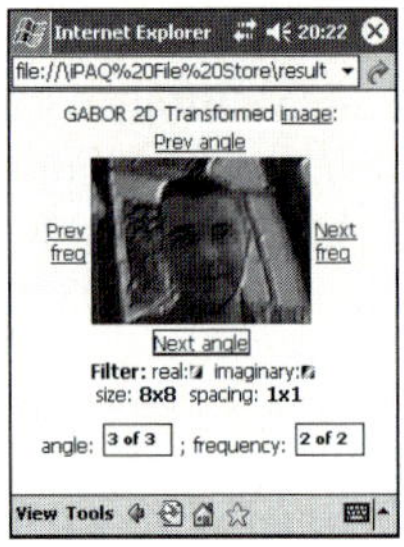

Angle of 120° and highest frequency

Fig. 4. Transformations of the image on Fig. 3

The obtained results can be viewed on appropriate links on the results web pages using the links for choosing angles and frequencies. Also, we can make comparisons with the original and transformed image.

Beside the resultant images, the real and the imaginary part of the filter, the size and the spacing of the filter are also displayed, as well as the selected angle and frequency.

In some cases, it is useful for the professor to give the same picture (as an example for analysis) to the students in order to emphasize some specific points.

In what follows, one more example of Gabor 2D transformation is given, where the used image is from a saved BMP file.

The parameters for processing of this image are shown on Fig. 5.

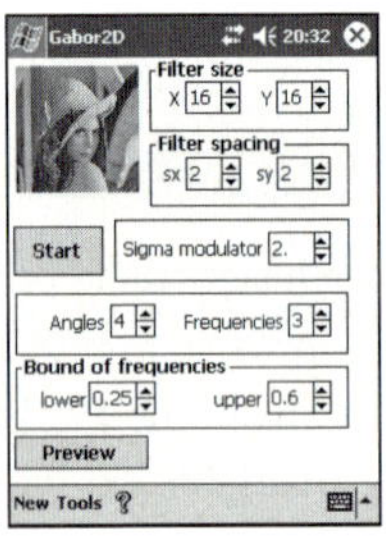

Fig. 5. Gabor 2D application with parameters for the second image

Fig. 6. Source image from BMP file

The image ready for transformation is shown on Fig. 6.

The parameters used for the transformation of this image are the following: 16x16 fitter; 2x2 spacing of the filter; sigma modulator with value 2; four angles 0°, 45°, 90° and 135°; three frequencies; lower bound of the frequency equal to 0.25 and upper bound of the frequency equal to 0.6. Some of the resultant images for these parameters are shown on Fig. 7.

From both transformations we can see that the transformed images look better and more similar as the original for lower frequencies, but for higher frequencies they look darker and less detailed.

Angle of 90° and lowest frequency   Angle of 45° and highest frequency

Fig. 7. Transformations of the image on Fig. 6

5 Evaluation Study

The evaluation was done by questionnaire. Twenty randomly chosen examinees were surveyed anonymously. All of them were students at the 4th year of study at the Faculty of Electrical Engineering at the University "St. Cyril and Methodius" in Skopje – Macedonia, at the Department of Computer Engineering.

For the purposes of the evaluation, the examinees tested the Gabor2D application on Pocket PC devices in the Laboratory of the Faculty for 45 minutes.

5.1 Evaluation Criteria

The criteria by which the evaluation was performed for this particular study are as follows:

- **Familiarity with the educational problem**
  1. Have the students been going on lectures of the signal processing subjects? (possible answers: all, some, none)
  2. Did they pass the exams of those subjects? (possible answers: all, some, none)
- **Functionality of the HCI**
  3. Are they satisfied with the arrangement (design of elements) of the pictures, spins etc. on the screen?
  4. How interactive is the HCI (Could they easily select the parameters of the algorithms on screen)?
- **Usability of HCI**
  5. For which degree, curriculum, group of students etc. this HCI is suitable (appropriate)?
- **Visibility**
  6. Were the results of the processing algorithms clearly displayed?
  7. How much the process of visualization helped students during the studying of their exams?
  8. Do the students like the HCI from the point of view of the quality of visualization?

The questions from 3 to 7 were answered by ranking from 1 to 5, where 1 is lowest, and 5 is highest rank.

6 Results of Evaluation

Based of the survey for the familiarity with the educational problem, the obtained results for both questions are displayed on the Table 1 and 2 respectively as follows:

Table 1. Results of the Question 1

	Number of students	Percent
All subjects	9	45 %
Some of the subjects	11	55%
No one	0	0%

Table 2. Results of the Question 2

	Number of students	Percent
All subjects	7	35%
Some of the subjects	11	55%
No one	2	10%

The functionality of the HCI was evaluated with ranks between 1 and 5 and average results of 4.53 for Question 3 and 4.85 for Question 4 were obtained. It can be seen that the students evaluated the above criteria for functionality with high rank. However, most of the students prefer tapping on the Pocket PC screen using the stylus than typing on the keyboard.

Concerning the usability of this HCI, we provided some comments from a group of students in different fields of study at the Faculty of Electrical Engineering. The basic survey (presented in section 6 in this paper) was provided for the Computer Science group of students. It was also tested with a group of students from Electronics and Telecommunications Department (ETD) as well as a group of students from Control Systems Department (CSD). For the students from ETD the interface was very convenient, while for the students from CSD some additional help about the parameters used for Gabor 2D transformation is welcome.

One of the key evaluation criteria is the visibility, where the students answered on 3 questions.

For the first question the average rank was 4.8. Their explanations were that the application has clear interface and excellent preview of the results.

About the second question, the results were that the interface was not over-laden and they were easily got on the results of the transformation. However, this question is with low rank 3.38 because some of the students had prepared the exams theoretically and more time was needed for preparing. If they would have the application, the theoretical results would be obtained visually for shorter time.

The third question was ranked as shown on the Table 3.

Table 3. Results of the Question 8

	Number of students	Percent
Yes	19	95%
No	0	0%
I don't know	1	5%

As can be seen on Table 3, the HCI from the point of view of the quality of visualization was high ranked.

The explanations of the students were that the application has nice visual presentation of the theory.

General conclusion of the results of the survey is that the program is useful for the exams of signal processing area. For most of the questions the students gave high rank, between 4 and 5.

7 Conclusion

According to the surveys with 20 students at the last year of study, their experience with our specially developed HCI for m-learning is very positive. The students are very interested to be able to interactively understand the theoretical concepts about image processing presented by the professor, during their experiments with different images captured from the Pocket PC cameras. In that way, they can become familiar with the impact that different parameters can have on the image being processed. These examples could be very useful for explaining how Gabor 2D transform is used as pre-processing in image classifiers.

References

1. Pierce, J.S., Mahaney, H.: Opportunistic Annexing for Handheld Devices: Opportunities and Challenges. Georgia Institute of Technology (2004)
2. Fallon, M.A.C.: Handheld Devices: Toward a More Mobile Campus, Syllabus (November 2002) http://www.campus-echnology.com/article.asp?id=6896
3. Laberge, S., Vujosevic.: Building PDA Databases for Wireless and Mobile Development. Wiley Publishing, Inc, Indianapolis, Indiana (2003)
4. Roschelle, J., Pea, R.: A Walk on the WILD Side: How Wireless Handhelds May Change Computer-Supported Collaborative Learning. Int'l J. Cognition and Tecnology 1, 145–168 (2002)
5. Aggarwal, Anil.: Web-based learning and teaching technologies - opportunities and challenges. Idea Group Publishing, Hershey, Pa (2000)
6. Finn, M., Vandenham, N.: The Handheld Classroom: Educational Implications of Mobile Computing, Australian Journal of Emerging Technologies and Society, 2(1) (2004)

7. Kinshuk, Goh, T.: Mobile Adaptation with Multiple Representation Approach as Educational Pedagogy. In: Uhr, W., Esswein, W., Schoop, E., (eds.) Wirtschaftsinformatik, Dresden, Germany (2003)
8. Sharples, M., Corlett, D., Westmancott, O.: The Design and Implementation of a Mobile Learning Resource. In: Personal and Ubiquitous Computing, 6, 220–234. Springer, London (2002)
9. Cabrera1, J.S., Frutos1, H.M., et al.: Mystery in the Museum: Collaborative Learning Activities using Handheld Devices. MobileHCI'05, Salzburg, Austria (September 19-22, 2005)
10. Liu, T., Choudary, C.: Realtime Content Analysis and Adaptive Transmission of Lecture Videos for Mobile Applications. MM'04, New York, USA (October 10-16, 2004)
11. Carroll, J.M., Rosson, M.B.: A Case Library for Teaching Usability Engineering: Design Rationale, Development, and Classroom Experience. ACM Journal on Educational Resources in Computing, Article 3, vol. 5(1), (March 2005)
12. Ganoe, C.H., Somervell, J.P., et al.: Classroom BRIDGE: using collaborative public and desktop timelines to support activity awareness. UIST '03, Vancouver, BC, Canada
13. Brotherton, J.A., Abowd, G.D.: Lessons Learned From eClass:Assessing Automated Capture and Access in the Classroom. ACM Transactions on Computer-Human Interaction 11(2), 121–155 (2004)
14. Luchini, K., Quintana, C., Soloway, E.: Pocket PiCoMap: A Case Study in Designing and Assessing a Handheld Concept Mapping Tool for Learners. CHI 2003, pp. 5–10, Ft. Lauderdale, Florida, USA (April 2003)
15. Zhang, D., Zhao, J.L., et al.: Can e-Learning Replace Classroom Learning?. Communications of the ACM, vol. 47(5) (May 2004)
16. Jipping, M.J, Krikke, J., et al.: Using Handheld Computers in the Classroom: Laboratories and Collaboration on Handheld Machines, SIGCSE 2001 2101 Charlotte, NC, USA (2001)
17. Luchini, K., Quintana, C.: Soloway, Design Guidelines for Learner-Centered Handheld Tools, CHI 2004, Vienna, Austria (April 24–29, 2004)
18. Anderson, D.: Image Object Search Combining Colour With Gabor Wavelet Shape Descriptors; B.Sc., University of Victoria (1997)
19. Ibrahim, A., Azimi-Sadjadi, M.R.: A Fast Learning Algorithm for Gabor Transformation. IEEE Transactions on Image Processing 5(1), 171–175 (1996)
20. Teuner, A., Hosticks, B.J.: Adaptive Gabor transformation for image processing. IEEE Trans. Image Processing 2(1), 112–117 (1993)
21. Microsoft eMbedded Visual C++ 4.0 (Microsoft Corporation 2004) http://msdn.microsoft.com/mobility/othertech/eVisualc/default.aspx
22. Pocket PC Downloads – Power Toys 2004 (Microsoft Corporation, 2004) http://www.microsoft.com/technet/prodtechnol/wce/downloads/ppctoys.mspx#EFAA
23. Windows Mobile Based-Pocket PC (Microsoft Corporation, 2004) http://www.microsoft.com/windowsmobile/pocketpc/default.mspx

OntoGen: Semi-automatic Ontology Editor

Blaz Fortuna, Marko Grobelnik, and Dunja Mladenic

Department of Knowledge Technologies, Institute Jozef Stefan,
Jamova 39, 1000 Ljubljana, Slovenia
{blaz.fortuna, marko.grobelnik, dunja.mladenic}@ijs.si

Abstract. In this paper we present a semi-automatic ontology editor as implemented in a new version of OntoGen system. The system integrates machine learning and text mining algorithms into an efficient user interface lowering the entry barrier for users who are not professional ontology engineers. The main features of the systems include unsupervised and supervised methods for concept suggestion and concept naming, as well as ontology and concept visualization. The system was tested in extensive user trails and in several real-world scenarios with very positive results.

Keywords: Ontology Editor, Ontology Learning, Text Mining, Machine Learning.

1 Introduction

The rapid growth of documents, web pages and other types of textual content pose a great challenge to modern content management systems. Ontologies offer an efficient way to reduce the amount of information overload by encoding the structure of a specific domain and offering easier access to the information for the users. However, all major ontology editors (such as Protégé[3], OntoStudio[4], ...) are fully manual and offer little support to the users for structuring domains.

OntoGen [1, 2] is a semi-automatic and data-driven ontology editor focusing on editing of topic ontologies (a set of topics connected with different types of relations). The system combines text-mining techniques with an efficient user interface to reduce both: the time spent and complexity for the user. In this way it bridges the gap between complex ontology editing tools and the domain experts who are constructing the ontology and do not necessarily have the skills of ontology engineering. The two main characteristics of the system are the following.

- **Semi-Automatic** – The system is an interactive tool that aids the user during the ontology construction process. It suggests: concepts, relations between the concepts, names for the concepts, automatically assigns instances to the concepts, visualizes instances within a concept and provides a good overview of the ontology to the user through concept browsing and various kind of visualization. At the same time the user is always in full control of the systems actions and can fully adjust all the properties of the ontology by accepting or rejecting the system's suggestions or manually adjusting them. This lets the user to establish a trust

M.J. Smith, G. Salvendy (Eds.): Human Interface, Part II, HCII 2007, LNCS 4558, pp. 309–318, 2007.

towards the system in a way that he has a full control over all the modifications to the edited ontology.

- **Data-Driven** – Most of the aid provided by the system is based on the underlying data provided by the user typically at the beginning of the ontology construction. The data reflects the structure of the domain for which the user is building ontology. The data is provided as a document corpus where ontological instances are either the documents themselves or name-entities occurring in the documents. The system supports automatic extraction of instances (used for learning concepts) and co-occurrences of instances (used for learning relations between the concepts) from the data.

The rest of the paper is structured as follows. In Section 2 we provide a general overview of OntoGen system, following with detailed description of implementation of the major features in Section 3. We conclude the paper with description of user trails (Section 4).

2 Overview of OntoGen System

Major features of the system generally serve one or both of the two major design goals of OntoGen: (1) visualization and exploration of existing concepts from the ontology and (2) addition of new concepts or modification of existing concept trough simple straightforward procedures aided by machine learning, text mining.

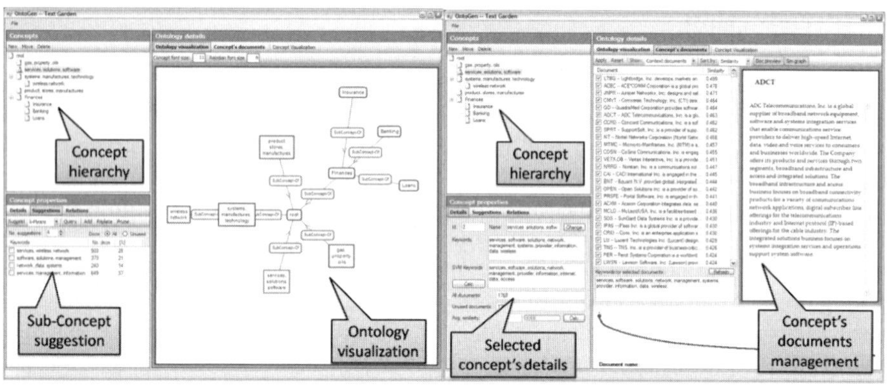

Fig. 1. *Left*: the user is getting suggestions for the sub-concepts of the selected concept (left bottom part); the ontology is visualized as a concept hierarchy in textual mode (left upper part) and in graphical mode (right central part). *Right*: the user is investigating the selected concept by browsing through the concept statistics (bottom left), the concept's documents including their content (upper right) and the graph of similarities of the documents towards the concept (bottom right).

The main window of the system provides multiple views on the ontology. A tree-view on the ontology, as it is intuitive for most users, presents a natural way to represent a concept hierarchy. This view is exposed as a standard Windows control used to show folder structure and as a visualization offering a one-glance view of the whole ontology. Each concept from the ontology is further exposed by the set of the most informative keywords for the target concept being automatically extracted using text-mining techniques [1] and through a topic-map of the documents belonging to the concept (more details in Section 3). Figure 1 shows the standard layout of the system.

Concept suggestions play a central part in the system. We provide unsupervised and supervised methods for generating suggestions. **Unsupervised** learning methods automatically generate a list of sub-concepts for a currently selected concept by using k-means clustering and latent semantic indexing (LSI) techniques [1, 8] to generate a list of possible sub-concepts. **Supervised** learning methods on the other hand require the user to have a rough idea about a new topic – this is identified through a query returning the documents. The system automatically identifies the documents that correspond to the topic and the selection can be further refined by the user-computer interaction through an active learning loop [1, 10] using a machine learning technique for semi-automatic acquisition of the user knowledge. The system also provides means of defining concept by selecting areas in the corresponding topic-map.

3 Implementation of Main Features

In this section we describe the main features of OntoGen including concept hierarchy visualization, management of a concept providing the concept details and suggestions for the concept naming, formation of a new concept based on unsupervised and supervised machine learning methods, concept visualization. We also provide a brief description of some additional features that enable collaborative editing of ontologies and user profiling.

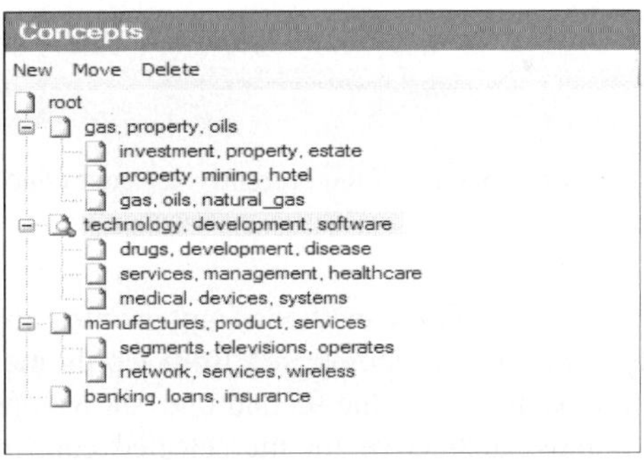

Fig. 2. Tree-view visualization. The user can reposition an existing concept to a different position in the hierarchy or delete an existing concept from the ontology. When moving a concept, the program pops up a dialog window asking the user to select the destination concept.

3.1 Concept Hierarchy Visualization

The part of the main window which is always visible to the user is the tree-view visualization of the concept hierarchy (the upper left in Figure 1 "Concept hierarchy", isolated in Figure 2). It offers a quick overview of all the concepts with their position in the concept hierarchy, direct access to commands for manipulating the hierarchy.

Ontology is also visualized in real time on the right side of the main window ("Ontology visualization" in Figure 1). The root concept is displayed in the centre of the visualization; sub-concepts of the root concept are displayed in the circle around the root concept, and so on. The user can select a concept by clicking on it on the visualization. The currently selected concept is drawn in red color, other concepts are drawn in blue color and the relations are drawn with green color.

3.2 Concept Details and Management

Detail and properties of the selected concept are presented in the bottom left part of the main window ("Selected concept's details" in Figure 1 and left side of Figure 3). In this part the user can change the name of the selected concept, check the concept's main keywords and see the number of instances in the concept (*All documents*), number of instances not in any of the sub-concepts (*Unused documents*) and average similarity of the documents within the concept (*Avg. Similarity*).

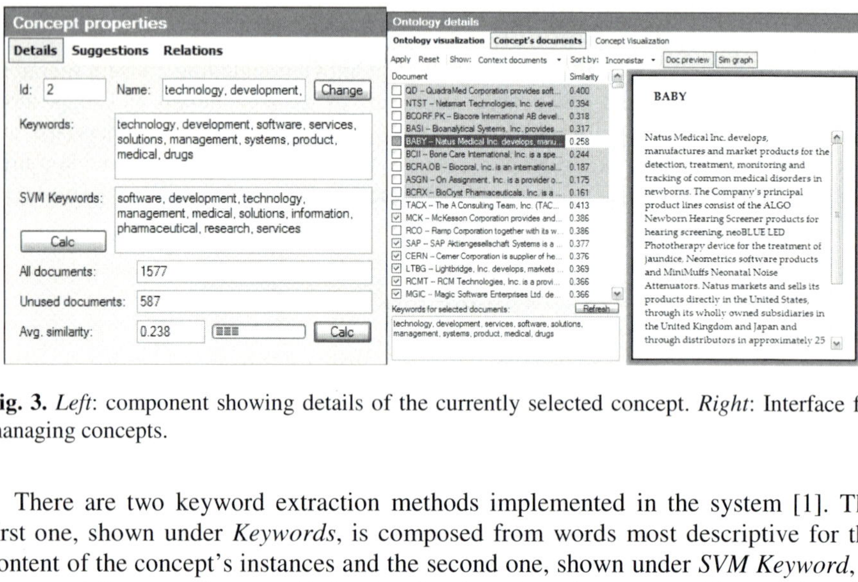

Fig. 3. *Left*: component showing details of the currently selected concept. *Right*: Interface for managing concepts.

There are two keyword extraction methods implemented in the system [1]. The first one, shown under *Keywords*, is composed from words most descriptive for the content of the concept's instances and the second one, shown under *SVM Keyword*, is composed from words most distinctive for the selected concept with regards to its sibling concepts in the hierarchy. Since SVM keywords are expensive to extract they are computed only on user's request (with button *"Calc"*).

After a new concept is added to the ontology, either with unsupervised or supervised methods, the system automatically assigns instances to it. Since this assignment is often not perfect the user can manually reassign instances to and from the concept through the interface shown in right part of Figure 3. Interface is composed of the list of instances where the instances from the concept are checked. Assignment of instances can be simply changed by checking or un-checking them. The content of the selected instance is displayed on the right side. OntoGen also has functionality for detecting specific inconsistencies inside ontology. When an instance does not belong to the selected concept but is in one of its sub-concepts, then it has highlighted using red background (for example check Figure 3).

3.3 Concept Suggestions

One of the main parts of the system is suggestions of possible new concept to add to the ontology. There are two different approaches implemented for concept learning. In the *unsupervised* approach the system provides suggestions for possible sub-concepts of the selected concept. In the *supervised* approach the user has an initial idea of what a sub-concept should be about and enters it into the system as a query.

3.3.1 Unsupervised

Unsupervised suggestions are based on clustering methods from text-mining [1] and clusters of instances from selected concept are treated as sub-concept suggestions. The user can supervise the number of clusters the clustering methods should return and on which instances the clustering should be performed (*All* for all instances in the concept and *Unused* for instances that are currently not in any of the concept's sub-concepts). An example of suggestions is given in Figure 4. User can accept and add to the ontology any of the provided suggestions by simply checking it and clicking *Add* button.

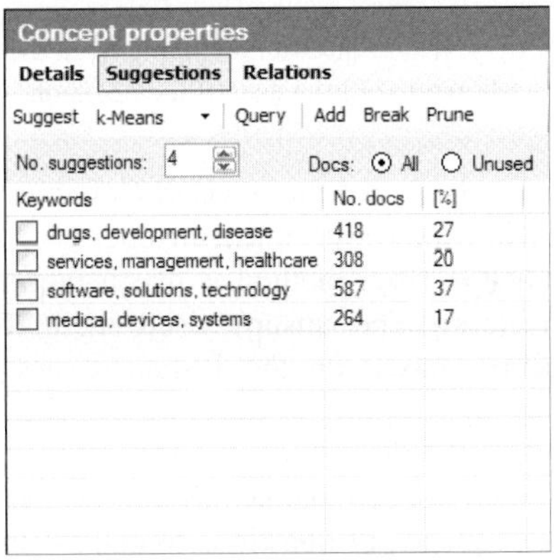

Fig. 4. The clustering algorithm prepares suggestions and the program displays them in the list (see lower half). Each suggested sub-concept is described by the extracted main keywords (using the centroid method) and the size. A longer list of keywords is displayed when the user moves with mouse over the suggestion.

3.3.2 Supervised

A new feature in OntoGen is a supervised method for adding concepts. It is based on SVM active learning method [2] and is only applied to the instances from the selected concept. The user supervision is provided first by a query or a set of keywords describing the concept that the user has in mind and followed by a sequence of questions (see figure 5).

On each step the system asks if a particular instance belongs to the concept and the user can select *Yes* or *No*. The questions are chosen from the instances on the border between being relevant to the query or not and are therefore most informative to the system. The system refines the suggested concept after each reply from the user and the user can decide when to stop the process based on how satisfied he is with the suggestions. After the concept is constructed it is added to the ontology as a sub-concept of the selected concept.

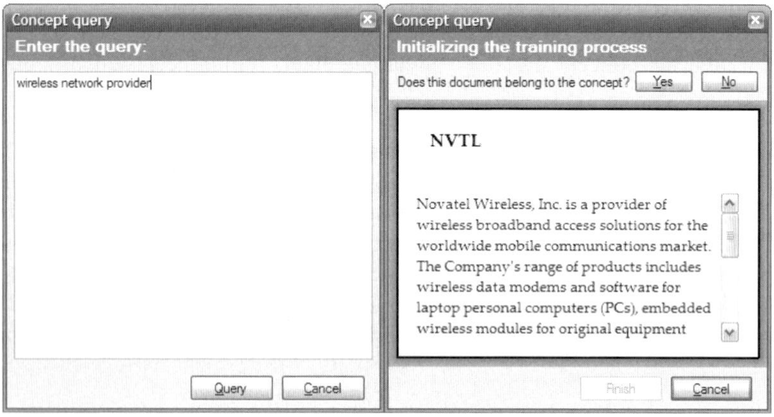

Fig. 5. *Left*: user entering a query. *Right*: question example from the system for the given query. Instance and its description is shown to aid the user at the decision.

3.3.3 Example of Unsupervised and Supervised Concept Suggestion

There is a fundamental difference between the unsupervised and supervised methods. The main advantage of unsupervised method is that it requires very little input from the user (number of clusters). The unsupervised method provide well balanced suggestions for sub-concepts based on the instances and can be very useful for exploring the data.

The supervised method on the other hand requires more input. The user has to first figure out what should the sub-concept be, he has to describe the sub-concept trough a query and go through the sequence of questions to clarify the query. This is intended for the cases where the user has a clear idea of the sub-concept he wants to add to the ontology but the unsupervised methods do not discover it. Idea for a query can come from concept visualization presented in the next section.

3.4 Concept Visualization

Instances of selected concept can be visualized using text-document visualization techniques described in [6]. The instances are presented as points on a map in such a way that each instance located close to similar instance and far from less-similar instances. The similarity between instances is calculated based on their textual description using *cosine similarity* (well know similarity measure from text-mining and information retrieval). The density of instances in specific part of the map is presented by the background texture. Most common keywords are shown for each area of the map and when the user moves the mouse around the map an extended list of the most common keywords is shown for the area around the mouse. The user can also zoom-in to see specific areas in details.

The visualization is tightly integrated into the system. The user can add new sub-concepts to the ontology by simply selecting a group of instances from the map. This is particularly useful since visualization often represents the groupings of similar instances into clusters. There is an example of visualization in Figure 6.

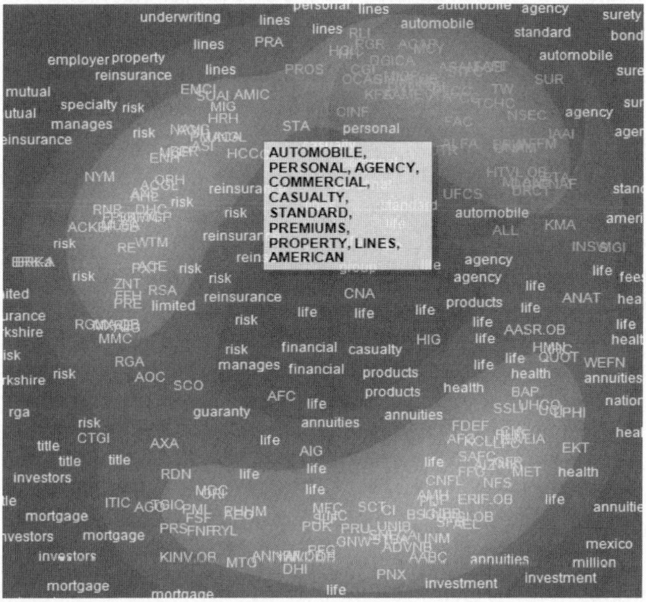

Fig. 6. The visualization is showing a map where each instance is a description of one insurance company. We can see that instances nicely group into three distinct clusters: reassurance, home and car insurance and life insurance. The user selected the instances from home and car insurance using mouse (the pink instances) to add them as a sub-concept to the ontology.

3.5 Collaborative Editing and User Profiling

The system also offers collaborative editing of ontologies where the user can use topics and relations which were previously constructed by other users. The system supports the user by suggesting similar topics/relations from the collection of ontologies.

User profiling is also used to tune the human ontology interaction based on the previous work of the users. This is done by recording previous choices that the user made when constructing ontologies and using them as an extra input to provide a personalized view on the underlying data and the ontology constructed so far trough "personalized word weighting schemas" (instead of using predefined schemas such as TFIDF [7]).

4 Evaluation

The first prototype of the system [1] was already successfully applied in case studies of several commercial projects from the domain of business, legislation and digital libraries. The users participating in the case studies were domain experts with limited experience in ontology construction. The feedback we got from the user was positive and we have used it to further improve the user interface. For example, simple method for moving concepts in the hierarchy (Section 3.1) is based on the feedback from users. In particular, the system enabled the users to model ontologies which would be significantly more difficult and expensive to model otherwise.

4.1 User Trail Setup

The latest version of the system [2] was extensively tested trough user trails at Faculty of Arts and Sciences in Rijeka [11]. The general goal of user trials was to gain as much useful information as it could be done in order to objectively assess the current stage of development of the OntoGen and its functionalities [9].

For that purpose two different groups of users were chosen to provide the useful data from two different points of view; one group consisted of 48 Computer sciences students and other from 43 Psychology and Pedagogy sciences students. That kind of user selection enabled the gathering of data from subjects who, generally, have good computer skills, only one group has more than average informational knowledge and understands the technical background of software development, and other group has more than common knowledge about cognitive processes of mental data structuring and organizing.

An average length of user trail session took 90 minutes and consisted of three phases. The first phase was used to introduce the users to the purpose of the task and to explain the procedures which will follow in the second and third phases of the trail. Main data collection is second phase of user trials and it had three main parts – before, during and after the demonstration of OntoGen system. In these parts we recorded several measure points for data collection:

1. Filling in the Initial questionnaire
2. In the second part the students tested the ontology construction system OntoGen. After demonstration of the OntoGen system by the facilitator, they were asked to fulfill the following two tasks:
   - Construct ontology which captures the areas covered by the companies in the given collection of descriptions of 7177 companies.
   - Construct two different ontologies on top of the same collection of 5000 news articles from Reuters news agency. The first ontology should group instances based on geographical properties, and the second based on topic of news articles.
3. In the third part, through the End questionnaire, they were asked about their experiences with the tested system. After the trial students also had a possibility to discuss their experiences with other participants involved in trials.

Conclusion is the third phase of user trial in which it is necessary to thank all the users for participating in user trial. Also, at this moment, it is very important to discus, if necessary, any question or thought aroused from the tasks during the trials.

4.2 User Trail Results

The general impression of OntoGen tool is marked "mostly good" to "very good" on scale from "not good at all", "sufficient", "mostly good", "very good" to "excellent". The great majority of users think that the tool is useful, but there is some resentment in content presentation of the main areas; the overall mark would be "mostly good" for Ontology visualisation, Concept tree and Concept Properties. The similar situation occurred in assessment of graphical interface; overall impression is "mostly good", but there were objections expressed concerning the attractiveness. Users repeatedly pointed out that there should be more colours.

Qualitative analysis revealed that one of the main advantages of OntoGen is managing large data bases with easiness. According to users, the tool is efficient, saves time and effort and gives a lot of space for user intervention. Disadvantages, as perceived, mostly concern unattractive look, abstract conception and occasional slowness and need to learn how to use it first.

Users also think that there is a clear distinction between two ways of ontology creation (unsupervised and supervised). Here the users only draw attention to the improvement of the graphical layout. Lack of detailed instructions and need for all the options in one place were pointed out as the main difficulties in generating concepts.

Visualization of concepts proved to be a great help in choosing sub-concepts for majority of users. Most of the users found the current state of visualization appropriate, but in need of more attractive graphical solutions and even bigger window or perhaps larger monitors. Furthermore, users think that with adding of concepts, picture becomes cluttered which makes it hard to inspect.

When it comes to concept documents management, users think that it is quite good manageable. There are no significant differences in assessing that part of OntoGen; general mark for all the operations included in document management within the concepts is "very good".

5 Conclusions

In this paper we presented an extension of ontology editors based on extensive use of machine learning and text mining methods which aid the user in the major steps of ontology construction process. The inherent danger of such approach is than an introduction of new text mining methods could make the whole system hard to comprehend and use by the domain experts, which in general do not have any knowledge of text mining. However, our user trails showed that tool OntoGen managed to avoid such missteps and that it can be used by people without background from machine learning.

As part of future work we plan to use the comments of the user trails to improve the user interface and user interactions. We also plan to extend the system with support for user collaboration when editing larger ontologies and to support for reuse of ontologies or parts of ontologies produced by other users.

Acknowledgments

This work was supported by the Slovenian Research Agency and the IST Programme of the European Community under SEKT Semantically Enabled Knowledge Technologies (IST-1-506826-IP), NeOn Lifecycle Support for Networked Ontologies (IST-4-027595-IP) and PASCAL Network of Excellence (IST-2002-506778).

References

1. Fortuna, B., Mladenić, D., Grobelnik, M.: Semi-automatic construction of topic ontologies. In: Ackermann, M., Berendt, B., Grobelnik, M., Hotho, A., Mladenič, D., Semeraro, G., Spiliopoulou, M., Stumme, G., Svátek, V., van Someren, M. (eds.) Semantics, Web and Mining. LNCS (LNAI), vol. 4289, pp. 121–131. Springer, Heidelberg (2006)
2. Fortuna, B., Grobelnik, M., Mladenic, D.: Semi-automatic Data-driven Ontology Construction System. In: Proceedings of the 9th International multi-conference Information Society IS-2006, Ljubljana, Slovenia (2006)
3. More information on Protégé (Stanford University) and the download of the latest version is available from http://protege.stanford.edu
4. Information about OntoStudio (Ontoprise GmbH) is available through http://www.ontoprise.de/content/e1171/e1249/index_eng.html
5. Fortuna, B., Grobelnik, M., Mladenic, D.: Visualization of Text Document Corpus. Informatica 29, 497–502 (2005)
6. Fortuna, B., Grobelnik, M., Mladenic, D.: Background Knowledge for Ontology Construction. In: Proceedings of the 15th International World Wide Web Conference WWW 2006, Edinburgh, Scotland (May 23-26, 2006)
7. Manning, C.D., Schutze, H.: Foundations of statistical Natural Language Processing, MIT Press.
8. Mladenic, D., Grobelnik, M.: Evaluation of ontology generation for case studies, Technical Report IJS-DP, J. Stefan Institute, Ljubljana (January 2007)
9. Tong, S., Koller, D.: Support Vector Machine Active Learning with Applications to Text Classification. In: Proceedings of 17th International Conference on Machine Learning
10. Ilijasic Misic, I., Taksic, V., Kovacic, B., Mohoric, T.: Design of a user study and evaluation of software tool OntoGen V 2.0, Technical Report FAS Rijeka (December 2006)

User Interface for Knowledge Sharing Using Knowledge Gardening Metaphor

Afdallyna F. Harun and Nor Laila Md. Noor

Faculty of Information Technology and Quantitative Sciences, Universiti Teknologi MARA, 40200 Shah Alam, Malaysia
afdallyna.f.harun@gmail.com, norlaila@tmsk.uitm.edu.my

Abstract. Knowledge sharing is presumed to be better facilitated through knowledge gardening metaphor. Nonetheless, there is a research gap in the metaphor application, where there is yet a formal elicitation undertaken towards it. This paper intends to answer three questions: How does knowledge gardening metaphor supports knowledge sharing within a user interface? What properties should the overall user interface have to support user-oriented environment? How can the knowledge gardening metaphor be elicited? Henceforth, a description on the user interface design incorporating knowledge gardening metaphor is discussed. Steps undertaken in a metaphor elicitation are also provided.

Keywords: Knowledge Sharing, User Interface Design, Metaphor, Metaphor Elicitation.

1 Introduction

To enable knowledge sharing activities, proper infrastructures and connectivity must be provided in order to empower and sustain a knowledge culture and its resulting initiatives. Related human behaviours must also be considered as prior studies have consistently find that knowledge sharing is positively related to having status similarity (Cohen and Zhou, 1991) and a history of prior relationship (Krackhardt, 1992). However, there is yet a system that provides all the features that good 'gardeners' of academic knowledge ecologies would want to have (Pór and Molloy, 2000).

Knowledge gardening is defined by having in place a specific activity to support, a database, allowing metadata vocabulary definition and search on shared items as well as monitor changes made. This approach is applicable to any digital artifacts created and found in the course of daily work (Udell, 2004). This has motivated our work of testing the applicability of the knowledge gardening metaphor in an academic digital sharing environment.

M.J. Smith, G. Salvendy (Eds.): Human Interface, Part II, HCII 2007, LNCS 4558, pp. 319–327, 2007.

2 Conception of Knowledge Gardening Metaphor in an Academic Web Space

The conception of knowledge gardening metaphor in our study was driven by two research questions; *1. How does knowledge gardening metaphor supports knowledge sharing within a user interface?* and 2. *What properties should the overall user interface have to support user-oriented environment?*. We developed a user interface prototype to support knowledge sharing activities by identifying five knowledge gardening processes, i.e. knowledge seeding, knowledge landscaping, knowledge transplanting, knowledge harvesting and knowledge weeding.

2.1 Knowledge Seeding

Knowledge seeding is the introduction of article(s) into a knowledge repository with the intention of sharing and distributing its benefits. It can take place in two ways; users contributing articles of their possession or submit articles found through external search.

Seeding Own Articles
Users are provided with a form-fill interface where one is required to upload the article source file and provide mandatory information like article's title, author names, abstract and a list of keywords. These fields are made mandatory as they can be unique in context and can easily help with distinguishing one article to another. These mandatory details when submitted would allow users to be more well-informed about the article effortlessly. The information can also infuse users for further reading and help users make quick decisions on retrieval decisions.

Keyword listing plays an imperative factor for the success of knowledge gardening (Udell, 2005) as it helps in the clustering of relevant papers during article search.

Seeding External Articles
It is proposed that users are not confined to articles available within the system's repository alone. Users should be able to seek articles in other search engine (our study used Google search engine) by attaching it within the system. Hence, users would be able to contribute articles found externally which indirectly encourage them to not selfishly hold good materials to themselves. It may seem that users are not able to retrieve external materials for personal keeping and that they are forced to share when using this system. However, some duress is needed in order to ensure continuous sharing.

2.2 Knowledge Landscaping

Knowledge landscaping is about adding and editing supplementary information regarding the article in order to attract a wider audience. It comes in the form of article rating, recommendation and commentaries as well as the editing of supplementary information.

Rating, Recommendation and Commentary Features

Many academic websites provide rating capabilities but most do not provide description on what a score truly represents. This creates ambiguity as one user's perception of a score differs from another user's. This in turn generates a false impression on the overall rating the article has received. In our study, we have explicitly describe what a score represents, allowing users to choose a specific value that best represents their opinions.

The recommendation feature acts as a venue for advice in influencing why and to whom the article would be worthy and desirable. Commentaries feature on the other hand consist of statements that express a personal opinion when interpreting a particular article. Recommendation and commentary features are provided as sharing and debating knowledge is far more powerful than simply uploading new article in a database. It is a way to test ideas, collect evidence, and mediate discussions electronically (Brown, 1999).

Reservations on the reliability of recommendations and comments are expected. In that matter, Neus (2001) mentioned that users will eventually know whose recommendations and comments they should take heed. They will be able to identify who has the right mixture of being knowledgeable and accessible in the continuing dialogue in due time.

Editing Features

Knowledge landscaping also involves the editing of supplementary information through the edit function. This feature allows users to make constant revision to improve the accuracy and relevance of the information in regards to the article.

2.3 Knowledge Transplanting

Knowledge transplanting is defined as reallocating an article from one location to a more fitting location. It is done in order for the article to better-reach targeted readers/retrievers. This metaphor simply involves users reallocating an article into another subject area.

2.4 Knowledge Harvesting

Knowledge harvesting is an activity that (i) captures the expertise of article authors and knowledge contributors (through recommendations and commentaries), (ii) prevent knowledge loss and (iii) ensure knowledge continuity. In the context of our study, users can harvest knowledge via knowledge repository or via external search engine.

Knowledge Harvesting via Knowledge Repository

Users can seek articles through keyword search, article title search or author search. Users are also able to refine their search query by selecting publication year, subject areas and paper types (i.e.: seminal, current or archive papers). These options are convenient for users who have different information seeking style (Ingwersen, 1992).

Users can choose article of choice from the results list where each selection will provide users with the article's general information, its ratings, recommendations and commentaries. This allow users to weigh the contribution potential of the article in regards to their academic activities, hence supporting a better search and retrieval process.

Users are able to retrieve articles by (i) adding to book collection, (ii) berrypick article or (iii) save article via adobe interface. The bookcart feature plays an important role during the retrieval process. It contains four folders which stores short cut links for each retrievals depending on retrieval type, as seen in Table 1. It is used to track article retrievals as well as track articles' 'landscapes' easily.

Table 1. Bookcart's Folder Description

Folders	Description
Book Collection	Store user's favourite articles from the knowledge repository.
KCG Berrypick	– Temporarily store articles retrieved from knowledge repository within a 3 day time frame. – Useful for users who wants to defer reading. – Suitable for information foragers who constantly modify their search strategies.
Google Berrypick	Works the same as KGC Berrypicks, but concerns with materials found via Google search engine.
KGC Retrievals	Store links of articles retrieved from the knowledge repository and saved in user's own hard disc.

For each article link in the bookcart, users can be briefly reminded of the article in terms of authors, keywords and one third of the abstract by hovering on the article's title. Example can be seen in Figure 1.

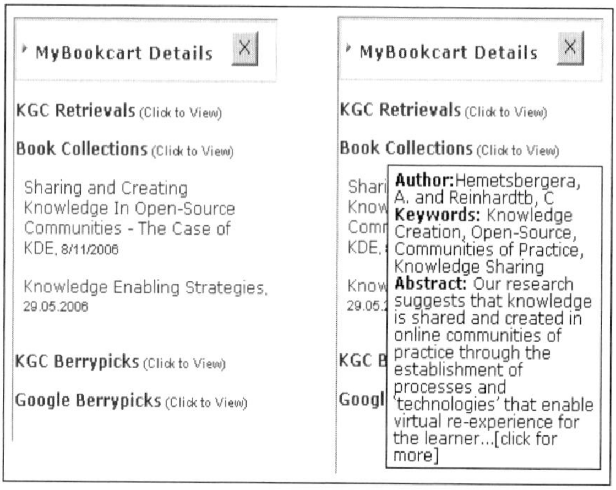

Fig. 1. Brief Reminder Provided By Hovering on the Article's Title

Knowledge Harvesting via External Channel

By providing an external channel, users are able to harvest more materials. Even though search and retrieval is done within Google, knowledge harvesting can still be experienced as the process involved is similar to knowledge harvesting via knowledge repository. However, users are not able to directly retrieve the articles derived from Google search. They can only berrypick them.

2.5 Knowledge Weeding

Knowledge weeding is defined as the suppression of untrustworthy and irrelevant articles. This is done to ensure better and reliable knowledge delegation and sharing. In the context of our study, it involves users duly request knowledge removal. However, it is the Super Gardener who will make the final decision as they are in the better position of evaluating the rationale of the request and relevance of the article to academic trends. The privilege for material deletion is not provided to all as material deletion affects site content and navigational structures. It also greatly affects article retrievers who constantly track the article for 'landscape' updates.

2.6 Human Behaviour Support

The conception of knowledge gardening metaphor has also incorporated human behaviours. This section shall briefly describe how human behaviours are incorporated through the user interface design.

Feedback through Recommendation and Commentaries
In a knowledge sharing organization, appropriate feedback is very critical. Respond of others through recommendation and commentaries contributes to self-worth development and better sense of self-competence (Covington and Beery, 1976). Social ties are also created by responding to each others' posting (Wasko and Faraj, 2005).

Seminal, Current and Archived Paper Categorization
Paper categorization acts as information scent. Users require proximal cues to identify desirable and relevant information sources and to navigate towards them (Pirolli, 2001). Allowing users to recognize a pattern of categorization would allow him to go "beyond the information given" by using probability and prediction (Bruner, 1973). This help increase the amount of relevant information acquired by user. If a user can attend to more information per unit time, then the user's information processing capacity is increased, amplifying cognitive capital.

Article Logging
Users are provided with a sense of awareness by informing who is around and what actions are done by them (Roseman, 1996) through article logs. Article logging also supports the need for chronological awareness (Chen and Gaines, 1996), where users need to be aware when a resource exists and changed. Modification must be tracked so that contributors of change can be credited and allow selective reversing of changes.

3 Experimental Set Up

The experimental set up was driven by the following research question, *How can knowledge gardening metaphor be elicited?.*

3.1 Participants

Three lecturers from the Human Computer Interaction Special Interest Group of the Universiti Teknologi Mara were selected as participants. As they are regarded as usability specialist, 2 to 5 participants would be sufficient to conduct inspection evaluation (Koyani et. al, 2006).

3.2 Usability Test

The usability test mainly involves users performing task scenarios via a prototype which are closely related to the knowledge gardening metaphor. Question asking protocol was also carried out to get the participants to react to the interface, to understand their mental model of the system and the tasks, and to identify where they have usage trouble.

After completing the task scenarios, users were given post-task questionnaire which was designed base on Nielsen's (2003) definition of usability. Questions were grouped into four constructs; learnability, efficiency, memorability, and satisfaction. Questions acquiring participants' perceived social and intellectual capital gained were also included.

3.3 Metaphor Test

Metaphor test involves assessing users' mental models on the tested concept as well as assess the metaphor structure. This comes in the form of card sorting exercise and post-task questionnaire respectively.

Before card sorting exercise, users are briefed on the knowledge gardening metaphor. Participants are then asked to allocate 33 index cards in one of the knowledge gardening metaphor categories or under the unknown category (if they do not know to which category the card belongs to). In our study, the index cards contained only screen captures, but no description to avoid influencing participant's judgment during card sorting.

In order to resolve ambiguity, we classified knowledge gardening metaphor into 5 categories; knowledge seeding, knowledge transplanting, knowledge landscaping, knowledge harvesting and knowledge weeding. The categories are displayed with posters, labeled with specific gardening names and description. The allocation of the cards under each category reflects their understanding of the metaphor.

Card Scoring

In this research, the scores are given based on the following descriptions, seen in Table 2.

Table 2. Card Scoring

Characteristics	Score
User allocate required object correctly, where i(th) object ≡ j(th) object	1
User did not allocate required object, where i(th) object = Ø, j(th) object = 1	0
User allocate incorrect object, where j(th) object = Ø, i(th) object = 1	-1

The i(th) object represents user's allocation while j(th) object represents the required object within a category set. On a spreadsheet, card numbers were listed sequentially along the x (horizontal) and y (vertical) axes. The scores as seen in Table 2 were written into the intersection accordingly. In order to assess how well participants understood the metaphor, a simple statistical method of calculating the percentage of correct allocation was done.

Post-Task Questionnaire

Four structure constructs were used to gather participants' perception on the metaphor structure. They are amount, applicability, representation and suitability.

4 Findings

We have found that the conceptualization of an academic web space user interface using knowledge gardening metaphor is able to support knowledge sharing activities. This was evident from participants being able to perform all required task scenarios and associating them to the metaphor without much difficulty. This was made possible through the categorization of knowledge gardening metaphor into knowledge seeding, knowledge landscaping, knowledge transplanting, knowledge harvesting and knowledge weeding. The categorization helps identify tools and other functionalities required during user interface design. This in turns allow better knowledge sharing facilitation within an academic web space.

Other than identifying tasks requirements and conditions, the knowledge gardening metaphor has also help identify the need to support human behaviours. The analysis on the usability test conducted showed that this was well accorded for through out the user interface design. Four usability constructs used for assessment showed that the user interface is efficient to use as the information is organized intuitively and appropriately placed. Good navigation layout and simplified task steps have helped participants performed required tasks efficiently as well. Memorability factor was supported mainly through the bookcart feature as it enables users to jog their memory of what have been retrieved and where it is located. It is also found that the user interface provides relevant knowledge sharing functions, is pleasant and the content well organized. This makes it easy and intuitive to perform tasks, enabling participants to enjoy using the interface, increasing the satisfaction factor. However, more can be done to improve the learnability aspect. Novices who have little experience in using knowledge sharing system may experience some difficulties.

From the findings gathered it is suggested that the user interface design is improved by providing:

a. Buttons and links that is easily understood on the first time basis,
b. More logical and consistent content layout, and
c. More colour schemes in differentiating functions and sub-functions.

The capital findings in terms of social and intellectual also adds to the evidence that human behaviours are well supported. Both social and intellectual capital plays very important role in knowledge sharing. This is accorded for by having the system taking note on users' every form of contribution, paper categorization, as well as providing ratings, commentaries and recommendation features.

Finally, the experiments carried out have shown that metaphor elicitation can be carried out through a series of card sorting exercise and responding to post-task questionnaire. Through the card sorting exercise, it is gathered that the metaphor was well understood, with harvesting being the least and weeding the most understood. As card sorting reflects users' mental model, it was found that each metaphor component is relatively easy to understand. In terms of metaphor structure however, there is still a need for improvement. The amount of metaphor structure is found to be poor. This was reflected during card sorting exercise where participants had difficulties in allocating some of the index cards correctly. Hence, improvements need to be done to increase the 'feel' of the metaphor. Nonetheless, it was also found that knowledge gardening metaphor would be useful and applicable in various areas. This is so as the underlying concept was elementary in the sharing and maintaining of knowledge. Moreover, the underlying concept of knowledge gardening metaphor was also found to be suitable for knowledge sharing. Finally, the metaphor was well represented in the user interface. It plays on intuitiveness by relating closely to the common gardening activities.

5 Conclusion

This section concludes the paper by describing the specific outcomes of the study and importance. This research has formally studied the capacity that knowledge gardening metaphor embodies to support knowledge sharing within an academic web space. The outcome of this research presents a testimony that knowledge gardening metaphor is adeptly capable to support collective knowledge sharing within an academic institution. The capability of the metaphor to support social and intellectual human behaviours has also shown that the metaphor is not just about being task oriented. It is capable to provide better knowledge sharing experience, hence encouraging future use.

This study is significant as currently the application of knowledge gardening metaphor is lacking a systematic approach in assessing its structure applicability. Hence, this paper has underlined steps that can be taken to elicit a metaphor once it is fully conceived in a user interface design. It has proven that responding to post-task questionnaire after card sorting exercise provides a more comprehensive and validated insight to user's mental model on the tested concept. This in its own provides empirical evidence that metaphor elicitation needs to be tested in terms of user's understanding, structure applicability and usability. Researchers, designers or developers of knowledge gardening metaphor should apply the approach taken to construct better and more effective knowledge sharing system that meet users needs.

References

1. Brown, J.S.: Sustaining the Ecology of Knowledge. Leader to Leader 12, 31–36 (1999)
2. Bruner, J.: Beyond The Information Given: Studies in Psychology of Knowing. (1973). In: Arglin, J. M., (ed.) New York: W. W. Norton & Co In Malhotra, Y., Information Ecology and Knowledge Management: Toward Knowledge Ecology for Hyperturbulent Organizational Environments. Proceedings of the Encyclopedia of Life Support Systems 2002. UK: Oxford (2001)
3. Chen, L.L.-J., Gaines, B.R.: Methodological issues in studying and supporting awareness on the World Wide Web. In: Maurer, H., (ed.) Proceedings of WebNet96. Charlottesville, VA, Association for the Advancement of Computing in Education (1996)
4. Cohen, B.P., Zhou, X.: Status Processes in Enduring Work Groups. American Sociological Review Vol. 56, pp. 179–188 (1991). In: Wasko. M., Faraj, S. (eds.) Why Should I Share? Examining Social Capital And Knowledge Contributio In Electronic Networks Of Practice. MIS Quarterly, vol. 29, pp. 35–57 (2005)
5. Covington, M.V., Beery, R.G.: Self-worth and school learning. New York: Holt, Rinehart, and Winston (1976)
6. Ingwersen, P.: Information retrieval interaction. Taylor Graham London 1992. In: Bystrom, K. and Jarvelin, K. (eds.) Task Complexity Affects Information Seeking and Use. Information Processing and Management, vol. 31, pp. 191–213 (1995)
7. Koyani, S.J., Bailey, R.W., Nall, J.R.: Research-Based Web Design and Usability Guidelines. US Department of Health and Human Service (2006)
8. Krackhardt, D.: The Strength of Strong Ties: The Importance of Philos in Organizations. (1992). In: Nohria, N., Eccles, R., (eds.) Organizations and Networks: Structure, Form, and Action. Harvard Business School Press. Boston (1992) In: Wasko. M., Faraj, S., (eds.) Why Should I Share? Examining Social Capital And Knowledge Contribution In Electronic Networks Of Practice. MIS Quarterly, vol. 29, pp. 35–57 (2005)
9. Neus, A.: Managing Information Quality in Virtual Communities of Practice: Lessons learned from a decade's experience with exploding Internet communication. In: Proceedings from the IQ 2001: The 6th International Conference on Information Quality at MIT (2001)
10. Nielsen, J.: Usbaility 101: Introduction to Usability. (2003) Retrieved from Alertbox Website: http://www.useit.com/alertbox/20030825.html
11. Pirolli, P.: Exploring and Finding Information. In: Toward a Multidisciplinary Science of Human-Computer Interaction. Working Draft, Morgan Kaufmann Publishers, San Francisco (2001)
12. Pór, G., Molloy, J.: Nurturing Systemic Wisdom through Knowledge Ecology. Pegasus Communication, vol. 11 (2000)
13. Roseman, M., Greenberg, S.: TeamRooms: Network Places For Collaboration. University of Calgary: Department of Computer Science (1996)
14. Udell, J.: Collaborative Knowledge Gardening. (2004) Retrieved from InfoWorld Website: http://www.infoworld.com/article/04/08/20/34OPstrategic_1.html
15. Wasko, M., Faraj, S.: It is What One Does: Why People Participate and Help Others in Electronic Communities of Practice. Journal of Strategic Information Systems 9, 155–173 (2000)

A Study of an Effective Rehearsal Method in e-Learning

Toshiaki Horinouchi, Shinsuke Wakita,
Michiko Anse, and Tsutomu Tabe

Graduate School of Science and Engineering, Aoyama Gakuin University
5-10-1, Fuchinobe, Sagamihara, Kanagawa, 229-8558, Japan
c5606112@cc.apyama.ac.jp, c5606138@cc.aoyama.ac.jp,
{anse, tabe}@ise.aoyama.ac.jp

Abstract. When information is memorized, the brain rehearses it. We can preserve information over a long period of time if this rehearsal is effectively performed in studying. This research aims 1.To clarify the effect of the rehearsal in e-learning, 2. To clarify the effective method for summary-making using a prototype system that studies structure of a stock (equity in a company). For this purpose, we 1.Compare the difference among three rehearsal methods, 2. (From result of 1) compare the results of an experiment of "Summarize" rehearsal method. As a result we find that the "Summarize" rehearsal is effective, especially when used for memorization constraints imposed in the number and types words used

Keywords: e-learning, rehearsal, Cognitive psychology, Dispersion-analysis, summarize.

1 Introduction

E-learning with IT technology is used as an educational tool for school education, company training, and life study with no constraints on time or place [3] [4]. E-learning has yet to satisfy various demands, however, and its forms for introduction are still being development. The prohibitive costs of introducing e-learning have dissuaded educators from assuming a leadership role in disseminating e-learning, and there are still only few teaching materials available. Lee [2] advocate a method of creating a teaching material called Instructional design (ID) from the point of system. KOGO[9] argue that e-learning system based on learning theory can be used as a superior professor system. Action analysis study is a framework for study aiming at "Body." Situation study theory is a framework for study aiming at "Mind." Cognitive psychology, on the other hand, provides a framework for study aiming at "head." The volume of memory is a kind of index in the learning. There is no clear effective method for maintaining memory with cognitive psychology in e-learning. If we can find a method by which the memory can be effectively maintained in study, then we can promote study in e-learning as one of the new methods for making teaching materials.

This research focuses on "elaborative rehearsal" [5] in cognitive psychology. The aim is to confirm the effectiveness of the rehearsal function in the e-learning teaching material through experiments to clarify the volume of memorized information. We

M.J. Smith, G. Salvendy (Eds.): Human Interface, Part II, HCII 2007, LNCS 4558, pp. 328–336, 2007.

perform these experiments by constructing a prototype study support system for teaching the "structure of the stock," then we clarify an effective method for making summaries. As to be described, this research begins by demonstrating an outline of the "Rehearsal" function and the structure of the prototype learning systems that include the rehearsal function. Next, an experiment is conducted to show the effectiveness of rehearsal by comparing four prototype learning systems: a "non-rehearsal" system (a system without rehearsal), A "Questions" system (subjects solve questions), a "Summary" system (subjects make summaries), and a "questions + summary" system with rehearsal (subjects solve questions and make summaries). Lastly, an experiment is conducted to clarify a summary method by comparing four different learning systems: a "non-indication" system (a system without indication), a "limited number" system (a system with limited numbers of summary words), a "limited words" (a system with limitations of the types of summary words), and "limited numbers and types of words" (a system with limitations of the types and numbers of summary words), which the reasons for the learning system are changed based on the results of a pre-experiment.

2 An Effective Method to Maintain the Memory

2.1 What Is "Rehearsal"?

The structures of human memory are geared to "take in," "maintain," and "remember" information [1]. The mind depends on its long-term memory to fend off oblivion. Several models of maintaining memory have been developed. Fig. 1 illustrates the "multistore model" proposed by Waugh & Norman [1]. This model proposes that the mind has two stores to retain the memory, the short-term store (STS) and long-term store (LTS). The memory kept in STS is called primary memory, while that kept in LTS is called secondary memory. Primary memory can only be maintained temporarily, for 20 seconds at most [8]. Secondary memory can he maintained forever, or until the mind slips into oblivion [8]. The memory of which we are most aware is of course the latter, secondary memory. Information taken in is initially kept in primary memory. The information will be either forgotten or kept in the LST with rehearsal. The model also defines two types of memory rehearsals that significantly enhance the likelihood that information will remain in LTL: first, elaborative rehearsal; second, maintenance rehearsal to reinforce retention in STS [7] [10]. If rehearsal is effective, we can maintain the memory for a long time [6]. Elaborative rehearsal can be further broken down into three types: "Questions" by reproduction, "Summary" by renewal, and "Questions and Summary" by reproduction and renewal. In this research we confirm the effectiveness of rehearsal as a means of elaborative rehearsal, and confirm "Summary" alone in an experiment for clarifying the summary method.

2.2 Prototype System

Our purpose, in making prototype systems with rehearsal, is to demonstrate the effectiveness of rehearsal methods and to clarify a summary method. Fig. 2 shows the structure and concept of the prototype system.

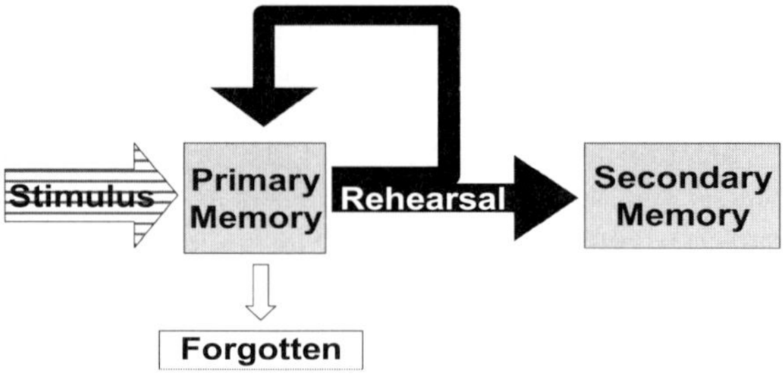

Fig. 1. Multistore model by Waugh & Norman (1965) [1]

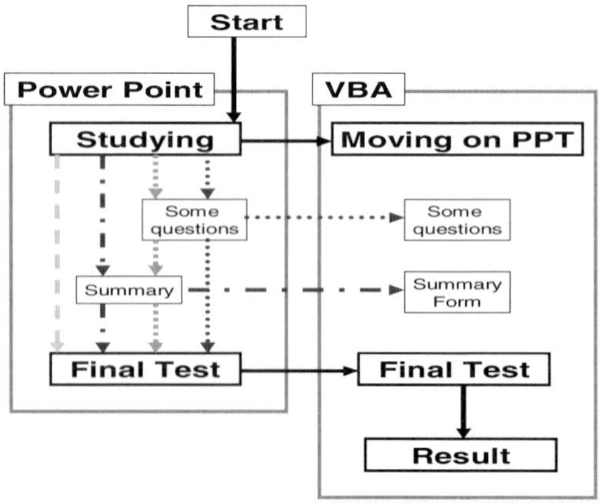

Fig. 2. Concept of the prototype system

Fig. 3. Example of Power Point Content

The prototype system is made by Microsoft Power Point (PPT) and Visual Basic for Application (VBA). The learning content is to "understand the structure of the stock." Fig. 3 shows an example of the content. This prototype system mainly consists of a study activity using PPT. It disposes of movement between slides, or shows the form of "summary" rehearsal by VBA. It is a prototype system that enables the learner to move between all of the slides with mouse clicks in order to freely inspect all of the learning material while studying. The learning content consists of five units, and a rehearsal work task is placed at the end of each unit.

3 A Measurement Experiment of the Effective Rehearsal Method

3.1 A Method of the Experiment

To measure the effectiveness of rehearsal, we ask the subjects study the prototype system and afterwards we measure the amounts of information retained in memory. The experiment is carried out in three steps: 1. Learning by prototype; 2. Questionnaire for the learner; and 3. Final Test. In the first step, the subjects asked to split up into groups of three and to study one of four prototype systems: "non-rehearsal," which has no rehearsal, "question-solving" to solve questions," "summary-making," and "question-solving and summary-making with rehearsal." In the second step, the subjects answer a questionnaire written up to ascertain whether the learners themselves noted any enhancement in the learning effect through rehearsal. In the third step, the subjects are given a final test the next day to confirm whether the learning content is memorized and to assess the amount memorized based on the rate of correct answers. The Final Test requires the subjects to answer words and phrases that come out in the prototype system in a newly generated form. It consists of 28 questions. Fig. 4 shows an example of the Final Test.

1. We publish a stock as the evidence that you (1) some money with (2). This money is money without (3) and calls the person who took out money a (4).

Fig. 4. Sample question from the Final Test

3.2 A Result of the Experiment

The bar chart in Fig. 5 plots the correct answer rate on the final test in the experiment conducted to measure the effectiveness of rehearsal.

The average of correct answer rates for the rehearsal types are 50% for "non-rehearsal," 64% for "question-solving," 68% for "summary-making," and 71% for "question-solving and summary-making." Table 1, a decentralized analysis table of these correct answer rates, confirms significant differences in the effectiveness of the rehearsal types (P<5%). The multiple comparisons shown in Table 2 validate the difference between the rehearsal types. Significant differences are found in all inter-group comparisons (P<1%), except for the comparison between "Summary" and

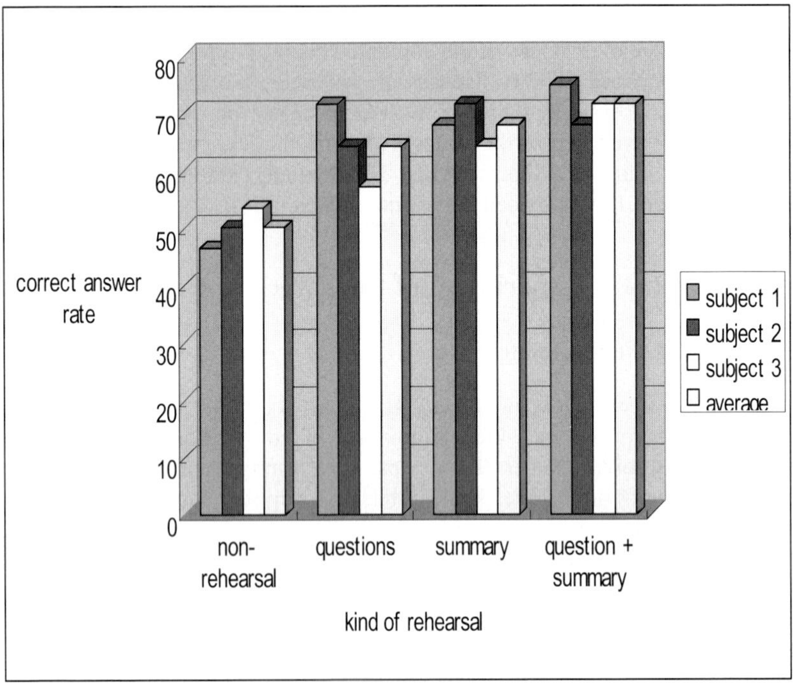

Fig. 5. Correct answer rates of rehearsal types

"Small Problem + Summary." These results demonstrate a difference between the prototype systems with and without rehearsal and show that the summary-making rehearsal can maintain the memory more effectively than the question-solving rehearsal.

To clarify the effects of the rehearsal, the questions of the final test are separated to four patterns: "questions without rehearsal," "questions with rehearsal for question-solving," "questions with rehearsal for summary-making," and "questions with rehearsal as test in the prototype system", and compare these four patterns. Table 2 shows the results. The average of the correct answer rate rises as the level of rehearsal rises. However, no difference was seen in the correct answer rates between the problems exhibited when studying with the "Question" prototype system. The correct answer rate was question > attainment level test > summary, regardless of the rehearsal type. This is thought to be attributable to the difficulty of the final test.

The effects of rehearsal in this experiment are thought to be higher when the steps are carried out in an order such as "non-rehearsal < question < summary <= question

Table 1. Dispersion-analysis

	S	f	V	F_0
m	4270.951	3	1423.65	3.29107*
e	14275.11	33	432.5791	
T	18546.06	36		

Table 2. Multiple comparison

	question	summary	question + summary
non-rehearsal	12.14947 **	20.25794 **	21.30952 **
question		8.108466 **	9.160053 **
summary			1.051587

** P<1%
* p<5%

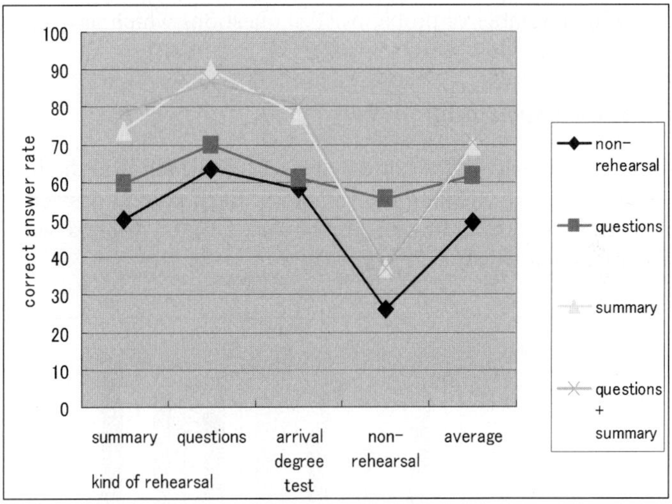

Fig. 6. Relation between correct answer rate and rehearsal

+ summary." According to the questionnaire given after the learning, all of the subjects felt that rehearsal was useful for learning and effective for enhancing memory retention.

4 A Clarification Experiment of a Summary Method

4.1 A Method of the Experiment

The pre-experiment described in chapter 3 demonstrates that "Summary" is effective as a rehearsal. Next, we examine the method used for summary-making. The experiment is carried out in the following steps: 1. A questionnaire before learning; 2. Learning by prototype systems with "summary" rehearsal; 3. A questionnaire after learning; 4. the Final Test.

1. The questionnaire before learning checks how much knowledge the subjects have to begin with.
2. The learning by prototype systems with "summary" rehearsal finds the differences between four types of "summarize-only" rehearsal: A. "non-indication," with no

indication; B. "limited number," with a limited number of summary words; C. "limited words," with limited types of summary words; and D. "limited number and words." The experiment is con-ducted with a prototype system that builds on these four summary methods. The summary frequency in this prototype system is corrected three times, a new group of four subjects study the prototype systems and four resulting patterns are compared.

3. A questionnaire after learning measures how subjects felt about the "summarize" activity.
4. The Final Test measures the amount of learning content retained in memory. To clarify the differences in the amounts retained, the test is organized into 40 questions expressing a number of problems. The questions which have rehearsal in the proto-type system are Nos. 1-10, 15-19, 25-34, and 38-40.

4.2 A Result of the Experiment

Figure 6 shows the bar chart of the correct answer rate which is in the final test of the experiment which to clarify making the summary. Each correct answer rate be-came that "non-indication" is 60%, "limited number" is 58.1%, "limited word" is 63.1% and "limited number and word" is 70.6%.

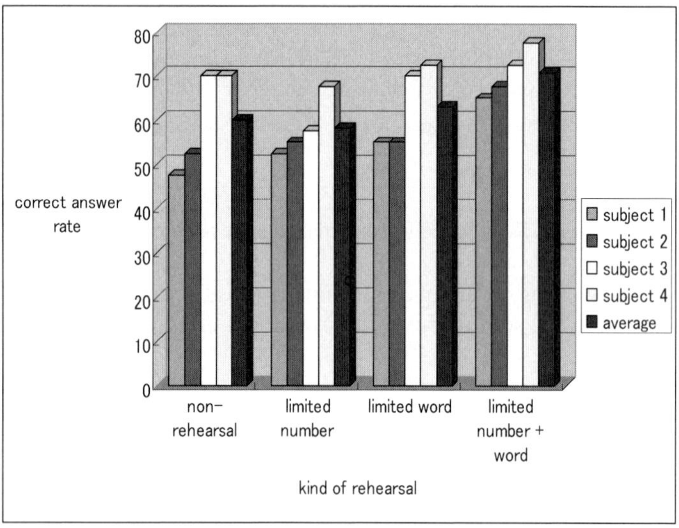

Fig. 7. Correct answer rate each summary rehearsal

Fig. 8 shows the correct answer rates from the problem to be rehearsed in the confirmation test conducted to measure the differences in the amounts of memory retention between the different summary methods. Table 3 shows the results of a decentralized analysis of these correct answer rates. Our results confirm that there are no significant differences (P<5%) among any components of the entire prototype system. Table 4 shows the results of multiple comparisons to validate whether there are differences between rehearsal types. Significant differences were found in the following comparisons in Table 4: the comparison between "limited number" and "limited word," the comparison between "non-indication" and "limited number and word," and the comparison between "non-indication" and "limited number and word" (P<%5). Thus, our results indicate that it is most effective to summarize specifying "limited numbers and types of words." Four-fifths (80%) of the subjects answered that "limited words" or " limited numbers and types of words" were easier to perform as methods than "non-indication," "limited number," "limited types of words," and "limited numbers and types of words."

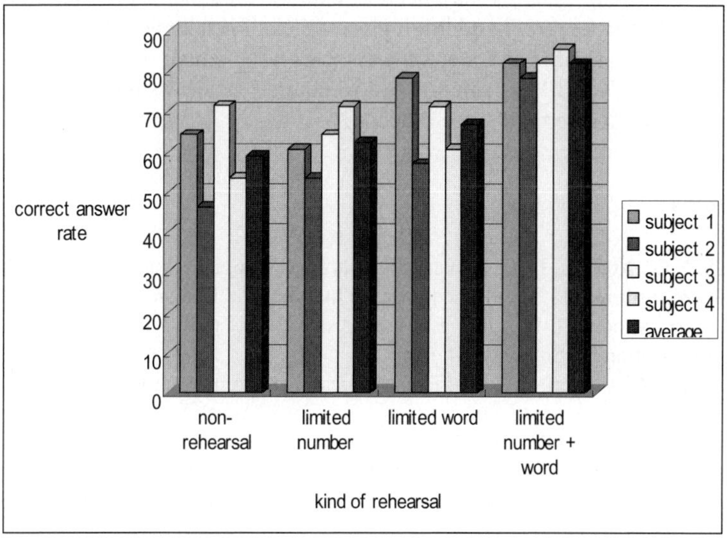

Fig. 8. Correct answer rate for summary rehearsals

Table 3. Dispersion-analysis 2

	S	f	V	F_0
m	1354.432	3	451.4773	6.4113*
e	845.0255	12	70.41879	
T	2199.458	15		

Table 4. Multiple comparison 2

	Limited word	Limited number	Limited word + number
non-rehearsal	8.022357	3.558071	23.20093 **
limited word		4.464286	15.17857
limited number			19.64286 *

** P<1%
* p<5%

5 Conclusion

Two experiments were conducted in this research. The first experiment confirmed the effectiveness of rehearsal in e-learning. The second experiment clarified an effective summary-making method for learning. In e-learning, summary-making is effective for learning content that slips out of long-term memory, and content can be effectively memorized by specifying key words with set numbers of characters to summarize. The results of the study itself and the level of satisfaction of the learners seem to coincide. Two important problems remain to be investigated and solved henceforth. First, some e-learning systems may need to exclude students, as the learning venues may include enterprises, etc. in addition places sites of academic training. Second, the subjects who were asked to make summaries later commented that "To make summaries was surprising" and "I worried about whether it was possible to summarize it." These types of resistance need to be reduced.

References

1. Waugh, N.C., Norman, D.A.: Primary memory. psychological Renew 72, 89–104 (1965)
2. Lee, W.W., Owens, D.L.: Multimedia-Based Instructional Design
3. Marginson, Simon,: E-learning: Expanding the Bottle to Fit the Guide. International Higher Education 30, 5–8 (2003)
4. Huffaker, D.A., Calvert, S.L.: Journal of Educational Computing Research, vol. 29(3), pp. 325–334 (2003)
5. Dark, V.J., Loftus, G.R.: The Role of Rehearsal in Long-Term Memory Performance. Journal of Verbal Learning and Verbal Behavior 15(4), 479–490 (1976)
6. Meunier, G.F.: Overt Rehearsal and Long-Term Retention. Journal of Experimental Psychology 102(5), 913–914 (1974)
7. Dark, V.J., Loftus, G.R: The role of rehearsal in long-term memory performance. Journal of Verbal Learning and Verbal Behavior 15, 479–490 (1976)
8. Human Memory: The Processing of Information: GR Loftus, EF Loftus – (1976)
9. Kogo, C.: Foundations of E-Learning: Behavioral, Cognitive, Situated Learning Theories and Their Integration
10. Meunier, G.F.: Rehearsal of Individual Items in Short-Term Memory. Journal of Experimental Psychology 95(2), 465–467 (1972)

Function Interfaces Assessment of Online Game Websites in Great China Area Using Content Analysis

Chi-I Hsu[1], Chaochang Chiu[2], Chich Hua Su[3], Ping-Ju Wu[4], and Yuan Jie Yang[5]

[1] Dept. of Information Management, Kainan University, Taiwan, ROC
[2,3,4] Dept. of Information Management, Yuan Ze University, Taiwan, ROC
[5] Dept. of Information Management, National Central University, Taiwan, ROC
[1]imchsu@mail.knu.edu.tw, [2]imchiu@saturn.yzu.edu.tw,
[3]m93221007@ms2.knu.edu.tw, [4]ceci.wu@gmail.com,
[5]944403017@cc.ncu.edu.tw

Abstract. Online game websites can not only provide players with a hyperspace for enjoying game playing, but also with a service platform for diverse community interactions, information sharing, and online transactions. Based on the four virtual spaces proposed in Angehrn's (1997) ICDT model: virtual information space (VIS), virtual communication space (VCS), virtual distribution space (VDS) and virtual transaction space (VTS), this research developed a content architecture for analyzing the function interface of online game websites. 26 online game websites are chosen as sample sites. This research found that most of the websites investigated provide good service functions in both VIS, but require further improvement of their VTS functions. Certain suggestions are proposed to the online game websites in Great China area.

1 Introduction

Due to the dramatic development of the Internet, surfing the Internet has become part of humans' everyday life. Through the Internet, users are engaging in more and more online activities such as shopping and entertaining games, not just simply for searching information and using email services. Online game is a newly fast-growing entertaining activity with the development of the Internet; it not only provides a channel for online game players to relax, but also brings them great satisfaction and achievement through accomplishing those tasks designed in online games. Besides, online game offers complete interactive functions, such as online chatting and virtual objects exchange. Compared to the traditional single-machine game, online game comes with much more interactivity and simultaneously satisfies four online demands: relation, transaction, interest and fantasy (Hagel et al., 1997).

The revenue model of online games has been mainly from hourly playing fee, membership subscription, and game software's selling. The hourly playing fee can be charged by purchasing a game points card or monthly game pass sold by the game company. However, due to the high competitive environment and low entry threshold, online game industry is facing with the "market saturation" at present. In order to

M.J. Smith, G. Salvendy (Eds.): Human Interface, Part II, HCII 2007, LNCS 4558, pp. 337–346, 2007.

attract more players, online game practitioners have converted the revenue model based on selling game software and membership subscription into selling virtual items related to the game itself, and this change has brought them more profits than those gained from the traditional model. This is mainly because of the uniqueness and variety of virtual items that make game role-playing more vivid and attractive. Thus, in addition to understanding and satisfying customers' needs, providing perfect services and adopting a suitable revenue model become key factors to a successful business.

This research investigates the contents of major online game websites of the Great China area, including Mainland China, Hong Kong and Taiwan. Our study is to analyze the content of services provided on the target websites and their revenue models. Online game websites can not only provide players with a hyperspace for enjoying game playing, but also with a service platform for diverse community interactions, information sharing, and online transactions. Based on this feature, this study adopts the four virtual spaces proposed in Angehrn's (1997) ICDT model: virtual information space (VIS), virtual communication space (VCS), virtual distribution space (VDS) and virtual transaction space (VTS), as the basis for content architecture of analyzing the function interface of online game websites, and practical suggestions will be proposed in the final section.

2 Literature Review

2.1 Online Game

Online games are one sort of entertainment-oriented and Internet-based information technology. Typically they are multiplayer games for users to fantasize and entertain themselves. Online players can enjoy user-friendly interfaces and multimedia effects via graphical operations. In addition, users are allowed to play a variety of fantasy roles and interact with one another in a virtual world. Several factors may contribute to user intention to play online games. To investigate the reasons why people play online games, Hsu and Lu (2004) conducted a study applying an extended technology acceptance model (TAM) with social influences and flow experience as two constructs to predict users' acceptance of online games. The results from their study indicated that social norms, attitude, and flow experience explain about 80% of game playing. A few of researchers have conducted studies investigating the impact of human behaviors on playing computer games. For example, Chou and Tsai (2004) tried to explore gender differences among adolescents playing computer games and found out that although both genders agreed that playing computer games has bettered the quality of their friendships, there was still a slight gender difference with regard to the predictors of Taiwan high school students' game playing enjoyments. In Gao's (2005) paper, a set of guidelines which help online game marketers understand how users form their trust in online games as well as a theoretical overview of the factors influencing user trust are provided. Gao pointed out that because an online game player is both a consumer of web-based entertainment and a computer user, factors related to the web environment and to the games have to be examined in order to identify what could possibly affect user trust in online games. Three groups of factors were discussed: perception about the company, perception about the online gaming site, and perception about an individual game. Griffiths, Davies and Chappell (2003) conducted an online questionnaire survey to examine various factors of online computer game players. The results showed that the time a player spent each week playing computer games would be prolonged if the age of the player was younger. Chen, Huang and Lei (2005) analyzed a 1356-million-packet trace from a sizeable MMORPG (Massive Multiplayer Online Role Playing Games) and pointed out that the arrival of game sessions within 1 hour can be modeled by a Poisson model, while the duration of game sessions is heavy-tailed. Hsu and Liang (2004) conducted a web questionnaire to investigate the domestic players' needs for personalized services in a MMORPG and the result showed that personalization in MMORPG should reflect on real-life experience, which means that the experience of interaction in the real world should be shifted to a virtual MMORPG world.

2.2 Content Analysis

A number of scholars have given content analysis different definitions and explanations based on their own perspectives. Weber (1990) defined "content analysis" as a research method to be used for the procedure of accurate deducing process in a subject. Krippendorff (1980) proposed three different categories for the reliability design of content analysis: stability, reproducibility and accuracy. Stability refers to the consistency in the analysis process; reproducibility refers to the similarity of the reproducing process by different coders in different environments and places; accuracy refers to the degree whether the coding process fits the definite standard function or expected result. No matter which one the reliability refers to, they are all used to evaluate the agreement degree between observers and coders to deal with data individually. The reliability can be expressed by Formula 1 and Formula 2. Kassarjian (1977) suggested the reliability coefficient be higher than 85%, which means the coding result can be accepted.

$$\text{Intercoder Agreement} = \frac{2M}{N_1 + N_2} \quad \text{......} \text{(Formula 1)}$$

where M = the number of coding decisions upon which the two coders agree

N_1 = number of coding decisions made by coder 1

N_2 = number of coding decisions made by coder 2

$$\text{Reliability} = \frac{n \times \text{Mean of Intercoder Agreement}}{1 + [(1+n) \times \text{Mean of Intercoder Agreement}]} \quad \text{......} \text{(Formula 2)}$$

n = the number of coders

3 Content Architecture of Online Game Websites

The four virtual commercial spaces in the ICDT model (Angehrn, 1997) are utilized in this research for establishing a content analysis structure of online game services. ICDT presents four different virtual spaces generated with the development of the Internet, as well as analyzes and classifies the commerce-based Internet strategies through systematic research methods. The four virtual spaces include: Virtual Information Space (VIS), Virtual Communication Space (VCS), Virtual Distribution Space (VDS), and Virtual Transaction Space (VTS). By dividing four different virtual spaces, the idea of using the Internet as an extension from the traditional market is emphasized, and the ICDT model is developed and used to analyze how traditional products and services can be redesigned through the Internet, and to point out that new competitive strategies for enterprises can be created via the information exchange and communication, and distribution of various products and services in the cyberspace. Figure 1 clearly illustrates the classification of the ICDT model and the main function of each space, as explained below:

1. VIS includes new Internet channels provided for an agent to display the related information to products or services offered. The main function of this space is to establish new channels for product sales and services by Internet marketing, mass advertising, and low cost.
2. Instant information exchange can be achieved through the interactive mode of VCS; furthermore, the access of products and services can be improved to enhance the communication relationship between up and down-stream companies.
3. To satisfy the delivery for a variety of products and services, VDS classifies services into three categories: online delivery (e.g., e-books), delivery without real substances, and assistance service. The main function is to reduce the cost and improve the quality and innovation of product and service distribution.

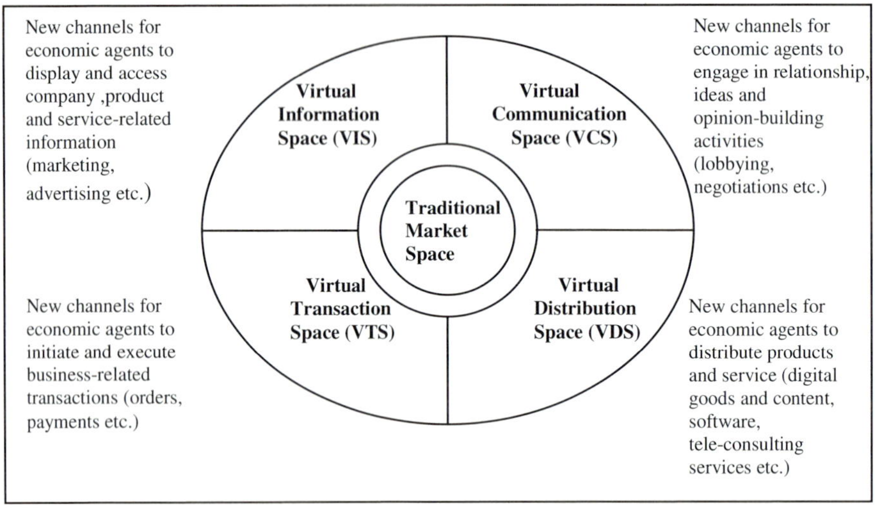

Fig. 1. The four virtual spaces of ICDT model (Angehrn ,1997)

4. VTS includes all the transaction activities carried out online, such as placing orders, issuing invoices, and making payments, etc. The purpose of simplifying the transaction process can be accomplished through online transaction services and hence to increase competitive advantages.

The four virtual spaces of the ICDT model are used to check the service functions offered by selected online game websites. As shown in Table 1, an architecture is proposed for the detailed structure of each stage included in the content analysis of this research.

Virtual Information Space (VIS) mainly provides related information to products or displays service information; therefore, those functions related to products and services offered should be provided in this space. At the same time, the two major online transaction procedures proposed by Liang et al. (1998) – search and comparison are

Table 1. The Architecture of Online Game Websites

Virtual Space	Category		
VIS	Authority Info	News User Contract Make-up Mechanism Game Rules Regular Maintenance Schedule Advertisement Column	
	Game Info	Game Guide Beginner's Guide Payment Explanations	
	Downloads	Support Tools Desktop Pictures Animations	
VCS	Account	Create/Manage Account e-Paper	
	Forums	Community Vote Forum	
	Support	FAQ Technical Support	
VTS	Purchase Determination	Product	Introduction, searching by key words, product details
	Transactions	Order	Order online, purchasing places
		Security	Emblems issued by other organization, standard security emblems
		Payment	Credit card, phone calling payment, ISP, postal/inter bank transfer, e-pocket, others
		Revenue Models	Pre-paid card, equipment exchange, membership fee, game software
	Post-Purchase Interaction	Customer Service	E-mail account, service line, online service
VDS	e-Delivery		
	Point-of-Sale Pickup		

adopted in this research to ensure the concept of demand and construct the variables required in VIS, which can be divided into three main sections: authority information, game information and downloads sharing. In Virtual Communication Space (VCS), the concept of examination from online transaction procedures proposed by Liang et al. (1998) is used to construct the required variables. Examination is defined as to examine the purchased products. Thus, based on this concept, the functions of VCS can be classified as membership accounts, discussion forums and supporting offers. For Virtual Transaction Space (VTS), this research divides online transaction activities into three main stages from a purchaser's viewpoint: purchase determination, purchase consumption and post-purchase interaction (Kalakota, 1996). Following these three stages, the item sorting and variable design are carried out, which include product purchase determination; order, security, payment and revenue models for purchase consumption; customer service for post-purchase interaction. Finally, for Virtual Distribution Space (VDS), the *delivery* from online transaction procedures proposed by Liang et al. (1998) and *receipt of product* from online transaction behavior proposed by Kalakota (1996) are referred to for considering the fact that most products ordered from online game websites are directly delivered to the game players through the Internet. As a result, the functions should be equipped with both e-delivery and point-of-sale pickup option from conventional shopping, with the expectation of achieving a more accurate analysis.

4 Function Interfaces Assessment

26 online game websites are chosen as sample sites mainly based on the reports of Top 10 of the Online Game Downloads in Mainland China (iResearch, 2006), Online Game Rankings of Yahoo! Hong Kong website (http://hk.games.yahoo.com/pc /index.html), and Hot Game Bulletin Board of Bahamut (http://www.gamer.com.tw/), during the period of August 7th to September 12th, 2006. Based on the constructing categories and coding principals of Content Analysis, three investigators analyzed the contents of these sample sites independently from Oct. 8th to Oct. 26th, 2006.

In order to examine whether the variable definitions and coding principals are clear, and the coders understand the procedures and principals related to the analysis, it is necessary to predict the reliability before starting coding. If the prediction reliability is low, then communication and review must be carried out until the reliability meets the acceptance level to initiate the formal coding process. To predict the reliability, this research selects 20% from online game websites in different areas as samples. The intercoder agreement degrees are 0.859, 0.805 and 0.751, respectively, and the prediction reliability is 0.925. Based on the content coding done by three investigators, the reliability reaches the standard level (>0.85), which means that the coding quality of the research meets the requirement of objectivity.

Based on the classification method publicized in DFC (2006), this research classifies online games as Casual Games, Children's Entertainment, Sports/Fighting/Racing, First-Person Shooters/Action, Strategy/Simulation Games, and Massively Multiplayer/ Persistent Worlds (MMOGs). According to the classifying result shown in Table 2 below, the most popular online game websites in Great China area are of MMOGs category, and are mainly designed by Korean manufactures.

The main function of VIS is to help an enterprise establish new channels for product sales and customer services through Internet marketing with extensive advertisement and low cost. The coding statistic results of the three parts of VIS are shown in Table 3. Through VCS, instant information exchange can be made by the interactive mode on the Internet, meanwhile the acquisition of products and services can be improved to enhance the up- and downstream communication relationship. The coding statistic results of VCS are shown in Table 4. VTS is used for online transactions, including order, invoices and payments, etc. Through the new online transaction services, business owners can simplify the operation process to promote their competitive advantages. The coding statistic results of VTS are shown in Table 5. The main function of VDS is to help e-retailing businesses reduce cost and improve product quality, and further enhance distribution efficiency and innovation. The coding statistic results of VDS are shown in Table 6.

The provided service rate of these four virtual spaces is also shown in Table 7. The "provided service rate" is defined as the percentage of actual service number to all provided service number in these four virtual spaces. For example, the total provided service number of VIS is 201 from 26 online game websites, and the average service number of VIS from each website is 12, hence the provided service rate is 201/12x26 = 64.4%.

Table 2. The Classification of Online Game Websites

	MMOGs	Casual Games	Action	Sports	Racing
Mainland China	4	1	0	1	1
Hong Kong	4	1	1	0	0
Taiwan	11	0	1	0	0

Table 3. Variable Statistics of VIS

Category		Percentage	Mainland China	Hong Kong	Taiwan	Average
VIS	Authority Info	News	100%	100%	100%	100%
		User Contract	87.5%	100%	100%	95.8%
		Make-up Mechanism	12.5%	0%	16.7%	9.73%
		Game Rules	12.5%	50%	50%	37.5%
		Regular Maintenance Schedule	12.5%	50%	66.7%	43.1%
		Advertisement Column	50%	50%	25%	41.7%
	Game Info	Game Guide	100%	100%	100%	100%
		Beginner's Guide	87.5%	100%	91.7%	93.1%
		Payment Explanations	25%	50%	41.7%	38.9%
	Downloads	Support Tools	100%	100%	100%	100%
		Desktop Pictures	75%	66.7%	91.7%	77.8%
		Animations	62.5%	83.3%	83.3%	76.4%

Table 4. Variable Statistics of VCS

Category		Percentage	Mainland China	Hong Kong	Taiwan	Average
VCS	Account	Create/Manage Account	100%	100%	91.7%	97.2%
	Forums	e-Paper	12.5%	16.7%	33.3%	20.8%
		Community	100%	50%	91.7%	80.6%
		Vote Forum	62.5%	16.7%	50%	43.1%
	Support	FAQ	100%	100%	100%	100%
		Technical Support	37.5%	0%	41.7%	26.4%

Table 5. Variable statistics of VTS

category			percentage	Mainland China	Hong Kong	Taiwan	Average
VTS	Purchase Determination	Product	Introduction	100%	83.3%	100%	94.4%
			Searching by key words	25%	0%	33.3%	19.4%
			Product details	0%	16.7%	66.7%	27.8%
			Best sellers	50%	0%	25%	25%
	Transactions	Order	Order online	100%	33.3%	100%	77.8%
			purchasing places	62.5%	83.3%	83.3%	76.4%
			License vendor	25%	16.7%	16.7%	19.5%
		Security	Emblems issued by other organization	0%	0%	0%	0%
			Standard security emblems	0%	33.3%	66.7%	33.3%
		Payment	Credit card	100%	33.3%	91.7%	75%
			Phone bill charge	75%	0%	75%	50%
			ISP	12.5%	0%	50%	20.8%
			Postal/inter bank transfer	62.5%	16.7%	66.7%	48.6%
			e-pocket	0%	0%	25%	8.3%
			Others	100%	66.7%	75%	80.6%
		Revenue Models	Pre-paid card	87.5%	66.7%	58.3%	70.8%
			Equipment exchange	87.5%	50%	75%	70.8%
			Membership fee	0%	0%	16.7%	5.6%
			Game software	12.5%	66.%	66.7%	48.6%
	Post-Purchase Interaction	Customer Service	E-mail account	87.5%	66.%	83.3%	79.2%
			Service line	100%	66.%	91.7%	86.1%
			Online service	25%	66.7%	66.7%	52.8%

Table 6. Variable Statistics of VDS

Category	Percentage	Mainland China	Hong Kong	Taiwan	Average
VDS	e-Delivery	100%	50%	100%	83.3%
	Point-of-Sale Pickup	25%	83.3%	75%	61.1%

Table 7. Average provided service percentage of four virtual spaces

	VIS	VCS	VTS	VDS	Average
Mainland China	60.4%	68.8%	50.6%	66.7%	61.6%
Hong Kong	70.8%	47.2%	34.8%	62.5%	53.8%
Taiwan	66.7%	68.1%	60.6%	87.5%	70.7%
Average	66.0%	61.4%	48.7%	72.2%	62.0%

5 Conclusions and Suggestions

This research uses content analysis to investigate the contents of major online game websites of the Great China area, including Mainland China, Hong Kong and Taiwan, and constructs a research model based on ICDT to analyze the current service functions provided in those online game websites. Based on the research results, we propose the following suggestions:

1. Online game websites in Great China area should improve their internal security system. The lack of a complete security platform will easily make customers feel insecure during the online transaction process. Only with a provision of complete security policies can effectively convince customers to purchase online with assurance.
2. The online game websites in Hong Kong and China should improve the detailed introduction and explanations of products. Since most products sold are game equipments to be delivered directly to customers through the Internet, an absence of detailed explanations cannot make the players have full trust on the products. Also the lack of probation period and return policy can cause the players' insecure feeling towards such websites.
3. The provision of a complete make-up mechanism is missing in the majority of online game websites in Great China area. As games are played online through the Internet, when a server problem hinders players' connection to the website, the game website must clearly set up a mechanism and policy to make up the players' loss. For example, a function that can keep all complete records of the game users' history (including playing time and transactions) is suggested to be added to the web design. The establishment of a complete make-up system is necessary to win customers' trust.

References

1. Angehrn, A.: Designing Mature Internet Business Strategies: The ICDT Model. European Management Journal 15(4), 361–369 (1997)
2. Chou, C., Tsai, M.J.: Gender differences in Taiwan high school students computer game playing. Computers in Human Behavior, pp. 812–824 (2007)
3. Chen, K.T., Huang, P., Lei, C.L.: Game traffic analysis: An MMORPG perspective. Computer Networks, pp. 3002–3023 (2006)
4. DFC Intelligence, Online Games: Game Genres and The Online Gamer, Market Research Report (2006)
5. Gao, Y.: Factors influencing user trust in online games. The. Electronic Library 23(5), 533–538 (2005)
6. Griffiths, M.D., Davies, N.O., Chappell, D.: Online computer gaming: a comparison of adolescent and adult gamers. Journal of Adolescence, pp. 87–96 (2004)
7. Hagel, J., Armstrong, A.G.: Net Gain: Expanding Markets Through Virtual Communities, Mckinesy and Company (1997)
8. Hsu, C.L., Lu, H.P.: Why do people play on-line games? An extended TAM with social influences and flow experience. Information & Management 41(13), 853–868 (2004)
9. Hsu, C.C., Liang, C.C.: A Study of Design of Personalized Service in Massively Multiplayer Online Roleplaying Game. Journal of Educational Media and Library Sciences 42(2), 221–242 (2004)
10. iResearch, Top 10 of the Online Game Download in Mainland China (2006) Available: http://www.iresearch.com.cn/html/online_game/detail_views_id_33830.html

11. Krippendorff, K.: Reliability in Content Analysis Some Common Misconceptions and Recommendations. Human Communication Research 30(3), 411–433 (2004)
12. Kalakota, R., Whinston, A.B.: Frontiers of Electronic Commerce, Readings. Addison-Wesley, MA (1996)
13. Kassarjian, H.H: Content analysis in consumer research. Journal of Consumer Research 4(1), 8–18 (1977)
14. Liang, T.P., Huang, J.S.: An empirical study on consumer acceptance of products in electronic markets: a transaction cost model. Decision Support Systems, pp. 29–43 (1998)
15. Weber, R.P.: Badic Content Analysis, Susan McElroy (1990)

Impact of Interactive Learning on Knowledge Retention

Mohamed Ibrahim and Osama Al-Shara

ADWC, Higher Colleges of Technology
PO Box 41012, Abu Dhabi, UAE
Mohamed.ibrahim@hct.ac.ae,
Osama.Alshara@hct.ac.ae

Abstract. There is amble evidence that supports true interactivity, both in the interface and in the presentation methodology, will further enhance learning and knowledge retention among students. The real challenge is how to implement it, measure and evaluate this outcome. This paper presents findings that look into the efficacy of the combination of virtual environments and interactive teaching techniques in enhancing learning and increasing the knowledge retention level of students.It is now widely accepted that the conventional way of lecturing students has certain limitations. E-learning and the use of ICT, interactive tools in the education field has proven to be effective in assisting the teaching and learning experience. It has been observed that the one way communication method of lecturing with limited or no opportunity for feedback from students and audiences in general has a major impact on their attention span and the retention level. There are issues related to the establishment and implementation of interactive learning tools as an aid for teaching and especially in continuous evaluation of students' achievements. We point out that it is not only the 'teaching' tasks that are impacted by the use of ICT but also current methods of evaluation including written examinations which have proven to be time consuming and inaccurate reflection of students' achievements. The introduction of interactive learning and evaluation has several advantages and this paper will give a convincing support to the fact that the advantages are more than the fact that it makes *the teaching and evaluation process easy, simple and rapid.* Educators should not fail to adapt and respond to the needs of 21st century bust and multi-tasking students for innovative learning methods, i.e. interactive learning. The role of education institutions and educators in the future will become demand–driven; to satisfy technically sophisticated student clients. The paper recommends that we should not ignore the use of interactive technology used as it should be, as a tool, to enhance sound pedagogical methods, however, we should also be able to measure its contribution in enhancing teaching and learning. It is pertinent to synthesize a framework that encompasses ways to compliment existing methods of teaching with interactive learning.

Keywords: interactivity, simulation, elearning, pedagogy.

M.J. Smith, G. Salvendy (Eds.): Human Interface, Part II, HCII 2007, LNCS 4558, pp. 347–355, 2007.

1 Introduction

Teaching and learning in the post e-learning era has become more interesting and challenging. Educators are continuously challenged to deliver new and innovative ways to educate, and deliver higher quality content to achieve better outcomes. Keeping students motivated and interested in learning is a tremendous challenge for educators particularly when there are so many alternative methods to teach and learn. The digital era introduced an effective means to communicate, teach, and provide learning opportunities to all. Educators and students are at the forefront of this challenge, always looking for innovative ways to ensure deeper learning and greater retention of knowledge, and with ever shrinking budgets, at a lower cost.

There is an ongoing debate on the effects of ICT based teaching, i.e. e-learning on students' achievement. This paper looks into this issue focusing on knowledge retention and whether the students in the e-learning environment in general were as successful academically as students in more traditional classroom settings. The finding presented in the paper are based on the authors' experience in using innovative teaching methods i.e. Learning by Doing (LBD)[1] environment on the internet as a means of communication with students and utilising interactive learning tools. We observed that the students learn more and knowledge in greater if ICT based learning tools enable more interactivity rather than moving paper based content into an electronic format.

This paper takes an introductory look at interactive learning within the context of e-learning and how the digital technology can make an immediate impact on improving learning initiatives. It is intended for learning and technology decision makers and begins by describing the theory behind this new method of teaching and learning in section two, there is an ongoing debate on the effects of ICT based teaching, i.e. e-learning on students' achievement. This paper looks into this issue focusing on knowledge retention and whether the students in the e-learning environment in general were as successful academically as students in more traditional classroom settings. The finding presented in the paper are based on the authors' experience in using innovative teaching methods i.e. Learning by Doing (LBD) environment on the internet as a means of communication with students and utilising interactive learning tools. We observed that the students learn more and knowledge in greater if ICT based learning tools enable more interactivity rather than moving paper based content into an electronic format.

---

[1] **Learning by Doing Environment (LBD) on the internet:** The LBD environment and the program design afforded a new paradigm for a successful online synchronous, real-time distance learning experience, conducive to the practice of identified skills and lasting knowledge a synchronous, online e-classroom available over the Internet and designed specifically for the experiential "learning-by-doing" pedagogy—uses a real-time, interactive HTML formatted text, image, and animation messaging system. The e-classroom, consisting of a main room and four breakout rooms for small e-group experiential e-activities and e-discussions, was password protected, monitored, and archived. Most students logged into the e-classroom from their homes. Text and image based lecture materials were posted to the e-classroom in real-time, and the "learning-by-doing" e-group activities offered, in this medium, facilitated learning through practice and discussions (Lobel, et al., 2002).

2 Theoretical Underpinnings

There is currently extensive literature on technologically delivered, mediated, monitored and valuated education systems. This overwhelming evidence supports this paper's proposition. More than a decade ago, over 30 studies have found that interactive technologies reduce learning time requirements by an average of 50 percent (Miller 1990).

A number of studies in recent years have demonstrated that information and communication technologies (ICT) has positive effects on the students' achievements, improvement of learning processes, development of educational skills, concentration, communication, met cognition, motivation and creativity (Monteith, 1998; Becta, 2001; Leask and Pachler, 1999, Barton, 1998; Hotle, 1998). However, some researches on the technology based learning indicate that there is no significant statistical difference in instructional effectiveness and exam scores (D'Alessandro et al,1993; Smeaton and Keogh 1999). In these studies, both student grades and test scores as measures of student achievement has been used (Hoyt, 1999). There are also many studies on the teaching with ICT in which methodologies are discussed (Driscoll, 1997; Shaw et al., 1997; Berge, 1999; Watson and Rossett, 1999; Price, 1996; Pool, Blanchard, and Hale, 19995; Whalen and Wright, 1999; Sherry, 1996).

With ICT based education, students interact with pre-produced sources for course replacing much of the traditional interaction between students and instructors. Emotional involvement, easy access to information, and a friendly chat contribute to learning pleasure and student motivation—which increases learning and knowledge retention. Teaching effectiveness is "demonstrated by students' learning of what has been taught" (Simonson, Schlosser and Hansen, 1998; Hoyt 1999). The core proposition of the paper is to further enhance the learning approach that uses ICT in order to "support", "enhance", and "extend" both the teaching and learning processes with serious interactive methods that grasp the students' imagination and transform the learning process into a natural extension of the students' daily activities.

Teaching and learning can be expanded in the sense that active learning ingredients can be and should included in the education content delivered to the students through the use of interactive learning objects, simulation applications, and virtual platforms that enable the students to practise and do experiments, solve problems as often as necessary to ensure abstract concepts are understood. Of course this is more important in the engineering, science and medical education environments where conducting experiment or performing surgical practice can be very expensive or in some cases unrealistic.

Technology is already used within instruction in the form of computer simulations, and at home as a means of interaction and self study. Considering the various aspects of teaching and learning with the help and the richness of available educational technology today, educators should utilize interactive systems based learning methodologies. This can be extended to the use of Learning Object which will use visual imagery and audio to vivify the learning and achieve interactivity in the true sense of the word—beyond mere mouse-clicks and multiplechoice questions. The penetration of technology in the education sector has reached a level that challenges the current teaching methods. Educators should be asking how to utilize these innovations, how to engage technology savvy learners? It is no longer acceptable that the 'theory' happens in the learning product and the 'application' happens only in real life? There has to be new ways of merging the two, and this is the power on interactive learning. For example, a simulation enables the learner to practice in a failsafe environment before going into the real world with the added advantage of enabling the educators to monitor and evaluate the learner's progress. As in many other technical sectors, definitions are vogue in the online learning industry today. In summary the description given above closely associates with the constructivist view of learning, which emphasizes the need for individuals to create their own models of knowledge. Therefore simulations can be viewed as cognitive tools.

3 Significance

E-learning educators are increasingly challenged to develop courses which promote interaction and reflective thinking in a virtual learning environment. The potential of the interactive learning approach in supporting knowledge retention that will equip student with the necessary skills for a long term job opportunities as well as life long learning.

Interactive learning is one of many innovative solutions that present itself as imaginative and creative use of technology in the education sector. We consider as a fundamental paradigm shift in the opportunities provided by new these technologies such as the Internet and the World Wide Web, CD-ROM data bases, multimedia presentations, and other instructional uses of computers which require considerable reflection and debate as to whether and under which conditions; they will enhance the quality of learning and teaching. To what extent do the newer interactive technologies improve the quality of the overall learning of students studying completely online? In particular, we must accept that it is important to establish whether the technology can function as a tool for improved learning rather than simply operate as a medium for the delivery of content.

We highlight the significance and the need to revisit the use of simulation approaches and techniques in teaching and learning. Many researchers found a sufficient evidence to support this approach as a successful mechanism to enhance the teaching of for example medical or engineering related topics. They argue that some of the most interesting current interactive applications are simulations. In general, such programs provide a platform for highly interactive learning according to Lander (1999).

It is now widely recognized that the effectiveness of online learning tasks depends on the way they generate interaction. Interactivity is a necessary and fundamental mechanism for knowledge acquisition and may be the key to successful online learning. (Barker, 1994; Mesher, 1999). Lander highlights that the fuller uses of new information technologies cannot be realized without a paradigm change toward learner-centered, interactive, outcomes-oriented instruction. Such a shift requires new role definitions for both faculty (teachers) and students (learners). "It may also be noted that, through interaction, online learning tasks can be used to generate the 'variation' in experience identified by Bowden & Morton (1998) as critical in many forms of learning" (Lander, 1999)

It is important that we recognize the importance of developing new models for students to equip them for the dynamic environment in which they will need to be life-long learners in order to survive and thrive. The issue, however, is to address the appropriate use of the technology.

4 Changing Teaching Methods

Interactive learning involves more than a simple content resource; it will enhance the teaching and learning model. The students will be able interact with subject matter experts and each other, attend live classes, or become a member of a discussion group focused on an immediate learning need. The education sector has recognized this and using this method will involve accepting the need for teaching methods. The existing technology in today's education institutions enable the delivery of educational content, its storage, and repeated delivery, broadcasted, etc. as well as on demand access.

Because the technology utilized other advanced feature within the ICT arena, education content delivered using digital media, for example, can be indexed and easily updated, learners can discover relevant and current content for the tasks at hand or they can reuse at a later stage. Learning retention can be increased because auditory and visual learning preferences are easily accommodated through the use of interactive learning tools based on digital media presentations. The most effective learning outcomes will fulfill the requirements of both students and the educators who can easily track, monitor and evaluate the students' progress.

One example that demonstrated this capability is StoBLs™ , a methodology developed by Tata Interactive Systems (TIS)[2], global e-learning pioneer, this methodology uses stories as the instructional approach to deliver learning effectively. The StoBL object utilizes visual imagery and audio to bring learning content to life. The students become independent and this interactivity platform helps them make decisions that control the movement of the story, as it were. This teaching and learning methodology does combine many aspects of outstanding pedagogical principles, the following is a summary that in a nutshell provides the profound shift in thinking on how to deliver educational content in a way the enhances retention (www.tatinteracive.com).

> *Contextual:* A story with a context conveys emotions, triggers individual and group memories, and provides intuition and insights to events; this in turn enables appreciation and application of the learning points.
>
> Realistic: Stories are based on reality, or are converted to reality, making them, and therefore the learning derived from them, credible—and learners, receptive.
>
> *Unusual:* A story must be memorable—with an "a-ha" moment—promoting retention of the associated learning.
>
> Natural: The events in a story appear natural, not contrived, enabling easier learner acceptance.
>
> *Concrete:* Stories deal with specific people, things, and events, which learners can relate to.

---

[2] www.tatainteractive.com

Human: There is a human angle in stories, even if they are not about human beings all the time, ensuring learner empathy and receptivity.
Easily Accepted: People lower their defenses to absorb a story instead of questioning it. The curiosity and readiness to be entertained translates into high acceptability and retention of the underlying concepts.
Discovery-oriented: Adults love to learn when they discover the learning themselves—which a story-based approach facilitates

The concept of simulations and interactive learning as an instructional technique has been in existence for a long time, especially in the form of games, models, and role-playing. What is not immediately obvious is how this method of teaching and learning enhance knowledge retentions? The US Air Force is one of the oldest and most prominent users of simulations for training pilots and flight crews, and they have been fairly successful in this. The use of simulations for learning has emerged from the need to provide hands-on practice. Besides, simulations promote higher order thinking skills, such as decision making, analytical reasoning, and problem solving[3].

What is of greater relevance in the education field is the need for high retention of small modules of e-learning information provided "just in time" in support of on-the-job activity is attributed its perceived higher relevance and "digestibility", as compared to lengthier, traditional, "just-in-case" training. The Research Institute of America found that, over time, the retention of knowledge from classroom lectures dissipates, with only 15% of such knowledge being retained three weeks after the course (Forbes 2000)

Students in this environment generally absorb the same instructional material in less time. By enabling them to navigate through material at their own pace, this minimizes the time that knowledgeable learners must spend in such training activity. And, for all learners, graphical presentation of material leads to faster comprehension. These two factors help to explain the results of several studies showing significant time saved per course[4], i.e. students requiring from 40% to 60% less time than the same material delivered in a traditional classroom setting or “reduced time to achieve given instructional objectives (30%) — or increased student skills and knowledge (30%) — depending on whether achievement or time was held constant."[5]

5 Interactive Learning Experience

Students interact with educational content similar to how they play games within a Web page whether the content can be located on a computer or a portable device. The students expects a quick and easy way of accessing and viewing content and so is the

[3] A Military organization found an increase in ability of personnel to correctly diagnose and repair aircraft systems the "first time" by more than 80%. · ROI with Blended E-Learning, White Paper Report, http://www.mentergy.com/blended/roi.html, 2001.

[4] *Return on Investment and Multimedia Training*, Brandon-Hall, 1995.

[5] *SCORM, Version 1.1*, Advanced Distributed Learning Initiative, quoting J.D. Fletcher (2001), "Evidence for Learning from Technology-Assisted Instruction"; H.F. O'Neil Jr. and R. Perez, "Technology Applications in Education: A Learning View", Hillsdale, NJ, Lawrence Erlbaum Associates.

educator who can in 'real time' or at a later time monitor what the students are doing or have done. It is this instant-on or always-on interaction that makes the whole learning and teaching experience more enjoyable, measurable and revisable where necessary.

Supplementing conventional teaching with interactive e-learning can lead to savings in time, effort, and money. This is quickly emerging as an effective and efficient learning aid in the education industry which is the main reason for many educational institutions turning to this as a primary means of improving their students' learning experience. The benefits may include:

- The access to visual imagery, which facilitates retention. Most people tend to remember 70 percent to 80 percent more of what they see than what they read. Live and on-demand scenarios bring the learning event closer to the participant.
- It enables learning institution to broaden it's by providing participants with greater and more flexible access to the educational content and other material, i.e. assessment and evaluation documentations.
- Instructional designers, implementers of education content delivery systems, can provide their expertise in a consistent way, reducing instructor and content variation and inaccuracy.

We need to keep in mind that interactive learning and simulation techniques used in teaching and learning are complimentary tools rather than replacement to all other teaching tools. The main advantage is the possibility that students can learn on their pace, concentrate on specific areas, etc. The baseball analogy demonstrates this point succinctly; consider a baseball player. He will practice his pitching separately. He will practice his batting separately. And he will practice his catching separately. Even within this, he will practice different aspects separately. The reason for this is not difficult to see. A baseball player needs to be good in all aspects of the game to be successful. Therefore, he needs to concentrate on all aspects separately. On match day, all these come together. "Practice in parts, play together"

In life, there are many concepts one learns and applies together. However, will it not help if one can practice the concepts separately? Will it not help if a learner can apply different concepts deeply so that the learner can understand the nuances of that concept? Interactive learning will allow learners to focus on specific areas of their learning teaks as often as necessary.

Flexibility and accessibility of the interactive learning environment offers learners a risk-free setting, enabling them to make mistakes and experiment with new ideas without risk, and allowing for reflection and review. We mentioned that interactive simulations make learning more interesting and engaging, resulting in increased retention. However, an area many neglect is the adaptability for complex concepts; Training on complex topics or processes is an area traditionally accepted to be challenging and difficult to handle, this has now managed in a successful manner with the help of interactive learning.

6 Conclusions and Recommendations

Increasing student retention and achievement is the focus and core function of all who are involved in the teaching and learning business. Research indicates increasing interactions during class between faculty and students and among students increases their achievement in the class and retention. Interacting with faculty, and interacting with students, are both obvious ways students acquire knowledge. In fact, a student's interaction style is another dimension of that student's learning style. Interactive learning can occur both inside and outside the class making learning a part of the students' daily activities. The learner's ability to apply in real life what he/she has learnt is probably the best indicator of effective learning and this can be achieved systematically when interactive learning tools are given to learners.

At a different level, we have accepted the notion that "The half-life of knowledge is decreasing, It's a matter of maybe three years in most professions and is down to months in the information technology field. The time pressures are only accelerating."[6] This will reinforced our proposition to let the learner have access to interactive learning tools that can be updated easily as speed of knowledge progression increase.

Based on the literature reviewed and our own experience over the last few years, we came to the conclusion that the impact of interactive learning has been underestimated and educators should revisit this valuable method of teaching and learning. There are areas that require further studies and deeper research in teaching methodologies and the inherent value of knowledge retention. We highlighted the need to reevaluate the fundamental notion that supports the validity of knowledge retention itself in a 'wired' and technologically advanced world. Should educators limit themselves to providing the necessary tools to facilitate learning and allow the learner to explore, gather, and apply the necessary knowledge when and where necessary? And transforming evaluation and assessment tools into a routine application, practicing, rehearsing and interacting constantly, not dissimilar to what pilot and athletes do before flights and races respectively.

References

1. Barker, P.: Designing interactive learning. In: de Jong, T., Sarti, L. (eds.) Design and Production of Multimedia and Simulation-based Learning Material, Kluwer Academic, Dordrecht (1994)
2. Barton, R.: IT in which way now? London: Routledge.Practical work. Assessing and increasing the value added. In: Wellington, J. (ed.) Practical work in school. Science (1998)
3. Berge, Z.L.: Interaction in post-secondary web-based learning. Educational Technology 39(1), 5–11 (1999)
4. British Educational Communications and Technology agency (Becta) The importance of activities in the teaching of science: a constructivist perspective on teaching science (24/1/2001) http://www.owu.edu/~mggrote/pp/construct/f-construct.html
5. D'Alessandro, M.P., Galvin, J.R., Erkonen, W.E., Albanese, M.A., Michaelsen, V.E., Huntley, J.S., McBurney, R.M., Easley, G.: The instructional effectiveness of radiology multimedia textbook (hyperlung) versus a standard lecture. Investigative Radiology 28(7), 643–648 (1993)

[6] Andrew Sadler, vice president of strategy and alliances for IBM Corp.'s Mindspan Solutions. As mentioned in Schultz, J. Effective, Inexpensive E-Learning Expands, 06/25/02 Washington Technology.

6. Driscoll, M.: Defining Internet-based and web-based training. Performance Improvement, vol. 36(4), pp. 5–9
7. Hotle, F.: Technology and the science class: going beyond the walls of the disk Drive. In: Cannings, T. R., Finkel, L. (eds.) The technology age classroom. Wilsonville: Franklin, Breddle and Associates (1993)
8. Hoyt, J.: Does the Delivery Method Matter? Comparing Technologically Delivered Distance Education with On-Campus Instruction (1999)
9. Lander, D.: Online learning; ways to make tasks interactive. Ultibase, Royal Melbourne Institute of Technology (1999)
10. Leask, M., Pachler, N.: Learning to teach using ICT in the secondary school. London: Routledge (1999)
11. Mesher, D.: Designing interactivities for Internet learning. Syllabus, vol. 12(7) (1999)
12. Miller, R.L: Learning Benefits of Interactive Technologies, Multimedia and Videodisc Monitor, pp. 14 (February 1990)
13. Monteith, M.: Information technology for learning enhancement. Exeter: Sweets and Zeitlinger Publishers (1998)
14. Pool, T.S., Blanchard, S.M., Hale, S.A.: From over the Internet: Users discuss a new direction for learning. Tech. Trends 40(1), 24–28 (1995)
15. Price, R.V.: A model for the on-line college-level guided study course. Tech. Trends 41(4), 39–43 (1996)
16. Shaw, R.E., Effken, J.A., Fajen, B.R., Garrett, S.R., Morris, A.: An ecological approach to the on-line assessment of problem-solving paths: Principles and applications. Instructional Science 25, 151–166 (1997)
17. Sherry, L.: Supporting a networked community of learners. Tech Trends (October 28-32, 1996)
18. Simonson, M., Schlosser, C., Hanson, D.: Theory and distance education: A new discussion. In: Gillespie, K.H. (ed.) The Impact of Technology on Faculty Development, Life and Work, pp. 60–75, San Francisco: Jossey Bass (1998)
19. Smeaton, A.F., Keogh, G.: An analysis of the use of virtual delivery of undergraduate lectures. Computers & Education 32(1), 83–94 (1999)
20. Watson, J.B., Rossett, A.: Guiding the independent learner in web-based training. Educational Technology 39(3), 27–36 (1999)
21. Whalen, T., Wright, D.: Methodology for cost-benefit analysis of web-based tele-learning: Case study of the bell online institute. The American Journal of Distance Education 13(1), 24–44

Design and Development of Computer-Based Discussion Support Tool for Science and Technology Communication Exercise

Kyoko Ito[1], Eriko Mizuno[2], and Shogo Nishida[2]

[1] Senri Expo Park 1-1, Suita, 5650826 Osaka, Japan
[2] 1-3, Machikaneyama-cho, Toyonaka, 5608531 Osaka, Japan
{ito, emizuno,nishida}@sys.es.osaka-u.ac.jp

Abstract. Science and technology communication is very important for a lot of people, engineers, scientist, and students. In order to study science and technology communication method, some education programs have started in some universities of Japan. In this study, to support education programs for science and technology communication, a support tool using computer is designed and developed. The target of support is group discussion of the exercise, and the developed software tool was used in the exercises in Osaka University.

Keywords: Science and technology communication, education, liberal arts of graduate schools, discussion, support tool.

1 Introduction

Recently, communications concerning the science and technology have been paying to attention in Japan for a social problem over the science and technology [1]. Some universities began the lectures concerning the science and technology communicator training program or the science and technology communication in Japan. The purposes of these classes are various. Osaka University is developing some education programs by which postgraduates becoming specialists in the future apply attainments of the science and technology communication. The postgraduates have time for group discussion. It is not easy for the postgraduates, who have different specialties and gather from all universities, to discuss a problem concerning science and technology, different from their own specialty.

The purpose of this study is to develop a discussion support software in graduate school education program of science and technology communication. The software supports the discussion of the students with different specialties on a topic concerning the science and technology.

The discussion support with a computer has been researched so far [2][3]. In this study, it aims to develop a new, effective tool from the viewpoint of the science and technology communication and the education.

M.J. Smith, G. Salvendy (Eds.): Human Interface, Part II, HCII 2007, LNCS 4558, pp. 356–363, 2007.

2 A Method Supporting Science and Technology Communication Exercise

2.1 Outline of Science and Technology Communication Exercise

First, the class to learn the method of the science and technology communication at Osaka University is concretely considered.

A trial program was conducted as a trial of "Science and technology communication exercise" in August, 2005 before starting an official educational subject for the graduate school in fiscal year 2006 by Osaka University. The trial program was conducted as an intensive program of five days, and bovine spongiform encephalopathy (BSE) problem was set as an example of the problem concerning the science and technology. The theme of "conditions for restart of importing beef from the United States of America" was considered by participants from two or more viewpoints through five days. Total of 17 students participated, and they are postgraduate who belongs to each research course of science, engineering, engineering science, literature, law, and human science. A rough flow of the exercise simulates one of participatory technology assessment, "Consensus conference"[4]. The students discuss on the theme three times during the exercise as follows.

1. To write up 3 points of controversy (Key-question) on "conditions for restart of importing beef from the United States of America" (Discuss it in the group of the research course mixture of 5-6 people).
2. To answer to Key-questions presented in the discussion of (1) (Discuss it in the group in each research course of 2-6 people).
3. To discuss about "conditions for restart of importing beef from the United States of America" (Discuss it by all students).

2.2 For Supporting the Science and Technology Communication Exercise

This study aims at a proposal of a new method of communications for the science and technology communication exercise, and the support of the discussion concerning the science and technology topic or problems is considered. The discussions about above-mentioned (1) - (3) have respectively the following purposes.

1. The problems are enumerated for the given topic, and they are brought together in three Key-question.
2. Appropriate answers are made for the Key-questions.
3. The ideas are bring brought to "conditions for restart of importing beef from the United States of America" together.

In this study, the discussion about (1) is paid attention to because from the viewpoint of science and technology communication whose important process is to recognize the topic from the participant's own position and bring the ideas of all participants together. And, the supporting method is considered in this study.

3 Analysis of Science and Technology Communication Exercise

3.1 Procedure of Analysis

For the discussion about "Key-question" making about the science and technology communication exercise, the content of the discussion is analyzed to search the problem of face-to-face communication, and to explore an appropriate supporting method. In the analysis, the videotape in which the exercise was recorded is used. The analysis targets two groups (G1, G2). The time that had been allocated in the discussion was four hours in total of 11:00-16:00 (On the way, the rest includes one hour at lunch). The instruction concerning the method for conducting discussions was not especially given.

The procedure of the analysis is shown below.

1. A rough content in the discussion is understood by browsing the videotape.
2. All remarks in the discussion are made a text.
3. The remarks are classified.

3.2 Results of Analysis

1. Outline of Discussion. The outlines of the discussion of each group were brought together by browsing the videotape.The number of discussion participants was six in both G1 and G2. Figure 1 shows the appearance of the room of beginning to discuss in G1 and G2. 1A-1E and 2A-2E show the discussion participants of G1 and G2 respectively.

Each participant used one personal computer (PC) in G1 and G2, and the PCs were connected on the Internet. The PC screen was projected by using the projector in G1 and G2, and the both discussions proceeded referring to the projected contents. When beginning to discuss, G1 pooled the idea to brainstorm by using the tag paper and the thick paper. On the other hand, one participant easily took the memo as a typist using PC when beginning to discuss, and recorded the content of the discussion in G2. G1 required about five hours and G2 about four hours to make three Key-questions.

2. Analysis of Content of Discussion. The contents of the remarks were made a text based on the videotape, and the number of remarks and the number of remark

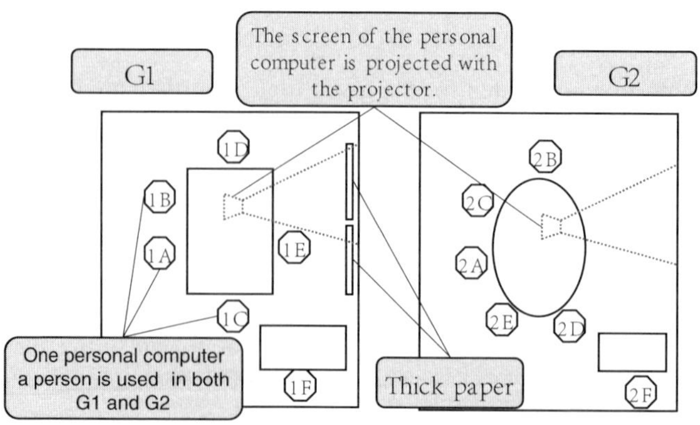

Fig. 1. Alignment in G1 and G2 in the discussion

characters of discussion participants were brought together. The number of remarks shows how many times to have made remarks while the discussion participants were discussing it. The number of characters shows how many characters in the text and the characters bring Chinese character and hiragana together.

3. Classification of Remark. The remarks were classified by using the text of the remarks. The classification was done from the viewpoints of role of remark in the discussion. It was referred to DAMSL (Dialog Act Markup in Several Layers) [5][6][7] for classifying the role of utterances from the viewpoint of communication. Based on the outline of the exercise, 11 kinds of category were selected based on DAMSL, form the viewpoint of science and technology communication exercise.

Table 1 shows the classification. In this study, the classification item when the remark is classified is named "Tag".

Table 1. Tags for the classification of the remarks

Statement
Question
Proposal
Information supplement
Accept
Reservation
Rebuttal
Consent
Answer
Addition of answer
Others

4. Summary of Results. Based on the discussions of results of G1 and G2, the features of the discussions of this science and technology communication exercise are summed up as follows.

1. The images to the key word are different depending on the participant of the discussion.
2. The remark degrees are different in each participant.
3. The frequency of the role of remarks is different depending on the topic in the discussion.
4. It is difficult for all members to always understand the flow of the discussion.

4 Design and Development of Software Supporting Science and Technology Communication Exercise

4.1 Target

In this study, the following viewpoints are considered aiming at the support of the discussion in the science and technology communication exercise.

1. The theme of the discussion is about the science and technology, and society.
2. The target is education to aim at science and technology communication
3. The class is an exercise liberal arts education for the postgraduate.

As to (1), the support for the perception gap to the point under discussion is important so that the students who have different expertise may discuss some problems in the contemporary society including various points under discussion. As to (2), to improve the ability of the science and technology communication, the student's participation in the discussion equally and positively is important. As to (3), an efficient discussion process is important because of the restriction of time and human recourses as the class.

4.2 Outline of Supporting Tool

In this study, it aims at developing of a software tool that supports the science and technology communication exercise, and the outline of the made software is considered in this section.

First of all, it is considered to show the features of the discussion to the participants as a whole. The important point is not to present all the contents of the discussion as text but to present the features of the discussion efficiently. It is considered to extract necessary information for the remark under the discussion to the participants. Because of keeping from obstructing the flow of the discussion, the third people other than the participants should input the feature of the remark. And, the input contents should be appropriately presented in the form that the participants use easily.

Therefore, the outline of the proposed supporting tool is shown in Figure 2. First of all, "Keyboarder" is prepared excluding the discussion participant. The keyboarder inputs necessary contents listening the content of the discussion. The keyboarder uses a PC, and the results of the input are projected by using the projector in an appropriate form. The participants discuss it referring the presented screen.

Fig. 2. Outline of the proposed software tool

4.3 Presented Information

Based on summary of the features of the discussion about the science and technology communication exercise, the information presented to support the discussion is considered.

In this study, it proposes the following information presentation.

1. *Keyword*: The keywords that closely relate to contents of the discussion are properly presented. When discussion participant's background knowledge is different, the support to keeping from their perception gap can be expected by presenting the keywords.
2. *Ratio of the number of remarks of each discussion participant*: To offer the equal opportunity of remark and to impress the participation in the discussion to the participant, each participant's remark ratio is presented.
3. *Classification of remark*: The results of classifying the remark are presented by using the tag in Table 1. It is expected for the discussion participant to grasp the features of the discussion.
4. *Flow of discussion*: In what flow the remark of which role appears is presented. It is expected for the discussion participant to grasp the flow of the discussion.

To present the above information, the keyboarder input the following contents.

A) Role of the individual remark
B) Keywords under discussion

From the viewpoint of workload of keyboarder's input, the utilization of role tags was narrowed down to 6 kinds of tags; statement, question, answer, rebuttal, accept and other.

The proposed information presentation screen is shown in Figure 3.

Figure 2 is divided into three parts, left, upper right, and lower right. The left side presents the flow of the discussion, upper right the classification ratio of the remark and the remark of each participant, and lower right a memo column by the participants.

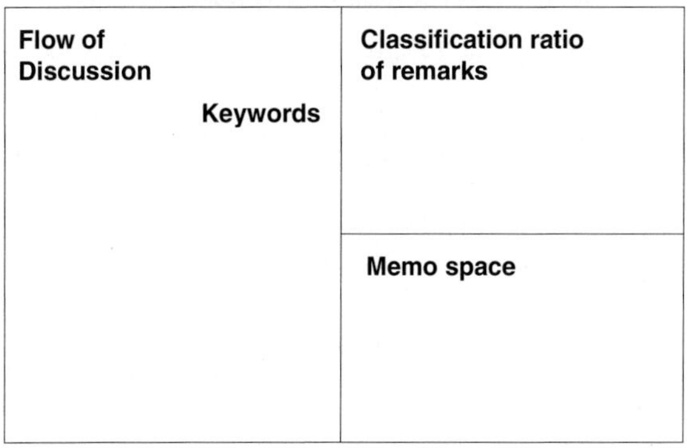

Fig. 3. Information presentation Layout

4.4 Development of the Designed Software

Based on the design, a software tool for the discussion of science and technology communication exercise has been developed.

5 Utilization of the Developed Software

The developed software was utilized during the class of "Science and technology communication: Theory and practice " in September, 2006 as a liberal arts of postgraduate in Osaka University. The theme of the discussion was about "Disposal of the high-level radioactive nuclear waste". The participants were 15 postgraduates and the members of groups A were 8 students, and groups B 7 students. Figure 4 and figure 5 show respectively the appearance of the discussion using the developed software and the screen presented for the students.

Fig. 4. The appearance of the discussion

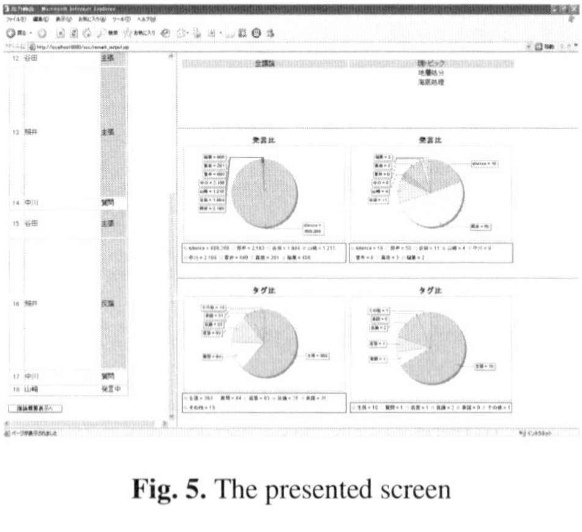

Fig. 5. The presented screen

The following points were shown from the results of the observation by examinees and the questionnaire for the discussion participants.

- It is difficult to keep seeing the screen presented while discussing it.
- It is difficult to understand a lot of information in the presented screen.
- It is possible to look back on an individual amount of the remark by presenting the ratio of the remark of the participants.

It is considered to improve the developed software based on the above-mentioned points.

6 Conclusion

In this study, at first, the problems of science and technology communication exercise in the university were analyzed for the development of a discussion supporting tool for science and technology communication education. The extracted problems were considered, and a discussion support tool for science and technology communication exercise was designed and developed. The developed supporting tool was utilized in the exercise of postgraduate liberal arts. In the future, the development of an more effective discussion supporting tool for the science and technology communication will be improved.

References

1. The Ministry of Education, Culture, Sports, Science and Technology(eds.): White Paper on Science and Technology in 2004 - Science and technology and society of the future (2004)
2. Conklin, J., Begeman, M.L.: gIBIS: A Hypertext Tool for Exploratory Policy Discussion. In: Proc. ACM Conference on CSCW, pp. 140–152 (1988)
3. Horita, M., Enoto, T., Iwahashi, N.: A pluralist approach to visualization of policy discourse: argumentation support as socio-technical systems. In: Proc. Research of science and technology for society, vol. 25–37 (2003)
4. Kobayashi, T.: Who considers about science and technology? In: A trial of "Consensus conference", The University of Nagoya Press, Nagoya (2004)
5. Shriberg, E., Dhillon, R., Bhagat, S., Ang, J., Carvey, H.: The ICSI Meeting Recorder Dialog Act (MRDA) Corpus. In: Proc. HLT-NAACL Conference (2004)
6. Switchboard, S.W.B.D.-D.A.M.S.L.: Shallow-Discourse-Function Annotation Coders Manual, Draft 13 (February 2006) http://www.dolorado.edu/ling/jurafsky/manual.august1.html
7. Core, M.G., Allen, J.F.: Draft of DAMSL: Dialog Act Markup in Several Layers (February 2006) http://www.cs.rochester.edu/research/cisd/resources/damsl/RevisedManual/RivisedManual.html

Context Aware Human Computer Interaction for Ubiquitous Learning*

Chulho Jeong and Eunseok Lee

School of Information and Communication Engineering Sungkyunkwan University
300 Chunchun Jangahn Suwon, 400-746, Korea
chjeong@ece.skku.ac.kr, eslee@ece.skku.ac.kr

Abstract. In the past year, the majority of conventional learning schemes have been transformed offline, gradually adapting E-Learning and M-Learning through the evolution of IT technology. In order to achieve effective learning, existing static schemes of learning must be transformed in all domains, to deliver true personalized learning depending on various user characteristics. Offering a conceptual background that supports this environment is the foundation of ubiquitous learning. Ubiquitous learning exists in the physical space of everyday life. It is possible to compose learning components existing in the activity space of an intelligent network. Adaptive learning in such ubiquitous environments is a significant learning step. In a general sense, our aim regard Ubiquitous computing is pursuit in the same light with Human Computer Interaction is pursuit by related works. However, in order to achieve sufficient improvement, ubiquitous learning requires 1) learning system has to provide learner centered learning contents what learner wants, 2) and through consideration of learner's learning device and learning environment, it has to offer learning contents. In this paper, a system supporting the dynamic configuration of user optimized learning objects, and the transmission of learning content irrespective of the learner device used, is designed and implemented. In addition, through the development of an effective learning model, the effect on learning is maximized.

1 Introduction

Recently, IT technology and internet improvement bring to various change in our society. These growths are made close to human and computer, so called name of the "Ubiquitous Computing", and is developed that our society enable to what is computing through anytime, anywhere and each device. There are features in Ubiquitous Computing as follows. Firstly, real space of computer is connected by communication net between wire and wireless. Secondly, these environment are simulated, it is provided user-oriented information and service. Finally, ubiquitous environment is

* This work was supported in parts by Ubiquitous Autonomic Computing and Network Project, 21th Century Frontier R&D Program, MIC, Korea, ITRC IITA- 2006-(C1090-0603-0046), Grant No. R01-2006-000-10954-0, and Basic Research Program of the Korea Science & Engineering Foundation. Corresponding author: Eunseok Lee.

M.J. Smith, G. Salvendy (Eds.): Human Interface, Part II, HCII 2007, LNCS 4558, pp. 364–373, 2007.

composed with computer resources. In a general sense, our aim regard Ubiquitous computing is pursuit in the same light with Human Computer Interaction is pursuit by related works.

This recently changes of technical circumstances have led a lot of changes in educational domain. The existing studying system is going to Ubiquitous Learning through E-Learning and M-Learning. Ubiquitous Learning which surpassed physical and cyber space after Ubiquitous environment applied to education makes that people can learn into customized courses in their life regardless of time and place. It will be able to be accomplished by using the computing environment for ubiquitous learning, smooth interaction between learners, and offering of learner centered learning contents. To achieve this, 1) learning system has to provide learner centered learning contents what learner wants, 2) and through consideration of learner's learning device and learning environment, it has to offer learning contents.

Existing Learning Systems based on Ubiquitous environment have used the method to construct various learning contents in advance or the method using the exclusive software developed for the learner's device. At the same time, the research, relevant to learning system based on HCI technology, is focused on the management system for instructor not learner. When we define the user of learning system as learner and instructor, it is important to research the system focused on instructor. But, if we see that the number of learner is much more than that of instructor, we can understand the research focused on learner is more important.

In this paper, we described following three techniques for learning based on HCI.

1) 'Context Aware Learning Contents Delivery Engine' providing learning contents focused on each learner. 2) "HCI-based Dynamic Interface for U-Learning' to support the smooth interaction between learner and learning system and to provide learner with optimized use of the system. 3) Lastly, the method to effectively evaluate the two techniques stated, based on HCI technology.

In this paper, we present concrete methodologies for designing and implementing mentioned above. In the 'Context Aware Learning Contents Delivery Engine' where provides learner-centered learning context, in order to provide composition of learning for learners properly, it's required to consider the Contexts as follows: Learners' learning taste information and learners' situation information. Learners' learning taste information is basic learning profiles, including basic information for mastering learners, learners' capability for considering learners' learning level, and History Data that learner received currently. Learners' situation information is information which decides alterative learning situation along with learners' position; it contains learners' position and Device Control actions. This paper consists of the following. "Related Work" presents related U-Learning research in the 3rd chapter, a description for the proposed system is presented, the 4th chapter presents an experiment, and evaluation of the proposed system. Finally, in the 5th chapter, this research is concluded, with a note regarding future work.

2 Related Works

Presently, at home and abroad, research on E-Learning and M-Learning using mobile devices is being widely pursued, and greater than 100 universities in Korea confirm the possibility of constructing and managing a Wireless LAN. However, as mentioned, the three core technologies stated above are indispensable for efficient U-Learning and no solution has been achieved that uses all of these technologies.

- Dynamic reconfiguration of learning objects is a technology offering exact, personalized, and well-organized learning contents to learners in various learning environments. However, well-organized research is not being conducted, although much research and standardization has been performed to compose a learning object, such as SCORM [2].

- The technology to inter-transmit learning content among heterogeneous devices and networks is indispensable in satisfying ubiquitous environments. Although much research, has been undertaken, it is limited in application to U-Learning.

- Though existing E/M-Learning emphasizes self centered leading learning and learner centered educational concepts is a fixed idea. The thought that the teacher is the informer and learner is the consumer, is a common thought, and the system for quality of resources for learning is at an early level. Furthermore, devices for education are demanded, and flexibility is insufficient, because learning content is fixed in devices.

Table 1 presents related research and associated problems.

Table 1. Related research and associated problems

Organization	Technique	Research and Problem
Dalhousie Univ. (Canada)	Composition of Learning Objects	• Proposal of information architecture which can reuse learning objects. • the learning object should be revised for composition.
Virginia Univ.(USA)	Event based Middleware (DsWare)	• supports integration of real time service of data. • provides various real time events.
Stanford Univ. (USA)	Collaboration Service Technique	• Absence of collaboration technique which provides real-time service and understands real time attributes among agents.
OMG(USA)	The technique of real time Channel processing of Middleware	• Absence of technique and analysis, creating a real time event channel mechanism.
IBM(USA)	Self-Configuring	• Self-Recognize and auto configuration • problem of learning the content delivery technique

In this paper, the above problems are overcome, and a ubiquitous learning system is proposed for a learner centered ubiquitous learning system appropriate for new ubiquitous environments.

3 Proposed System

The learner centered ubiquitous learning system proposed in this paper, is applied to the multi-aspect modeling technique of learners and the dynamic reconfiguration of learning object.

3.1 Context Aware Learning Contents Delivery Engine and HCI-Based Dynamic Interface

Context Aware Learning Contents Delivery Engine and HCI-based Dynamic Interface technique consists of:

1) A technique to correctly understand and analyze various learning situations of learners. 2) A technique to dynamically reconfigure different learning objects in a distributed environment.

These techniques are based on a modeling technique to perfectly analyze learner's location, device, learning, and environment information, in order to formulate optimized learning contents specifically for the learner's situation. In addition, these techniques consist of integration, permitting distributed learning objects on the Internet to be dynamically reconfigured. The ability to provide optimized learning content for each learner is based on research relating to location, device, learning and environment information. The overall architecture is as follows.

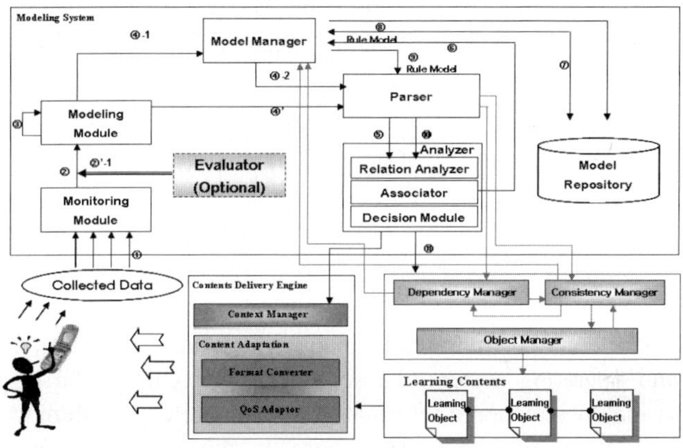

Fig. 1. Overall architecture

The modeling system monitors the learner's location, learning device, learning information, and environment information, modeling this data for effective use. The modeling system initially collects information relating to the learner's location, through a ubiquitous sensor network. The Learner's device information includes learning device type, CPU of device, memory of device, and display size of device. The learning environment includes network status, type of protocol, location of gateway and proxy server, and executing external system. Therefore, this information is analyzed and refined in the Model Analyzer.

The Modeling Module models the specification of leaner's device architecture, operational resources, behavior of learner, and environmental information in terms of network bandwidth. The modeling data generated in this module are separated using the three aspect models.

-Resource Model
The Resource Model represents platform independent elements which have an effect on system operation. It consists of static and dynamic aspects of resources, as in the following example.

Table 2. Example of resource model

	Pre-defined elem ents	Collecting element i n runtime	Real collected data
Static	Device (Type)	Device(Type)	MobileDevide(PDA)
	Os(Type)	Os(Type)	PPC(Pocket PC)
	Display (Size, Resolution)	Display (Size, Resolution)	320*240
Dynamic	Port(Number)	Port(Number)	2000
	CPU(type,Total_s ize)	CPU(Type, Current_ Usage)	(Intel Xscale, 640Mh z)
	Memory (Total_size)	Memory (Current_Usage)	(512M, 140M)
	Power (Total Size)	Power(MaxUpTime, remainUpTime, Battery Level)	(240mins 144mins, 60%)

Static aspects include resource information that have less dynamism, such as device type, and operating system. These data do not change during runtime. The dynamic aspect includes resource information such as CPU and Memory usage that is able to have an effect on the system, because these change continuously. The Deployment diagram is used to model these data.

-Behavior Model
The Behavior Model represents the dynamic behavior of learner, to validate that the learner's preference. The sequence diagram and activity diagram are used to model this dynamic behavior. The activity and sequence diagram is one of the UML diagrams modeling the dynamic aspects, and representing scenarios.

- Environment Model
The Environment Model represents platform independent elements which have an effect on system operation. The model includes network bandwidth, available protocol, and location address of accessible device.

The Architecture of the Model Analyzer and execution mechanism is as follows.

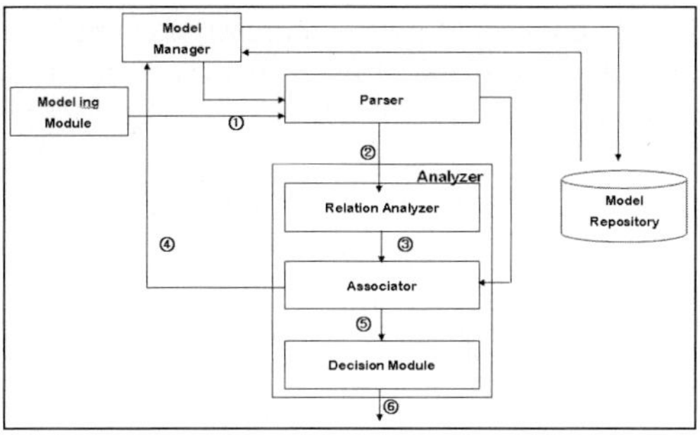

Fig. 2. Architecture of model analyzer

1. Transport the XML description, which includes the modeling data, to the Parser
2. Transmit parsed modeling type, element information, and modeled data to the Relation Analyzer.
3. The Relation Analyzer analyzes the relationship between parsed modeling data and each element and mapping rule. This information is transmitted to the Associator
4. The Relation analysis is completed, and the Associator request association model is transmitted to the Model Manager.
5. Associator constructs model information and the set is received from the Model Manager, generating the mapping rule.
6. The Decision Module decides on the reconfiguration type, information or learning object, and action information for reconfiguration.

An example of the Rule Model using an Associator is presented below.

- Rule 1
If(Bandwidth<=MaximUM)Associate(ResourceModel.CPU);
else(Bandwidth>Minimum)Associate(EnvirtModel.Usersatisfication);
....
- Rule 2
If(currentUsageCPU>80%)TransformReformatinge(GCIF);
If(currentUsageCPU>30%)TransformResolution(HIGH);

The dynamic learning object reconfiguration system is a module provided to learners to reconfigure learning content, according to learning object, learner's location, learner's device information, learning information, and environment information gained by the learning object in the modeling system.

Our system is composed to the Modeling System, Adaptive Interface, Object Manager, reconfiguration Manager, and Content Delivery Engine, which dynamically creates learning content. The dynamic learning objects should be composed to SCORM standards, in order to compose learning content to the learning object. Also, learning object is consisted assets. Asset is an atomic element for learning, as video clip, image, voice, and etc.

The Reconfiguration Manager is a module which uses each individual learning object, and composes of learning content, fitted to the various contexts of the learner. The learning object is created from SCORM standardization, and specification regarding the information exists. The necessary learning objects for the learner are detected, and the learning object is connected by the learning content based on this specification. Conceptual architecture of SCORM based learning objects and Content package is as follows.

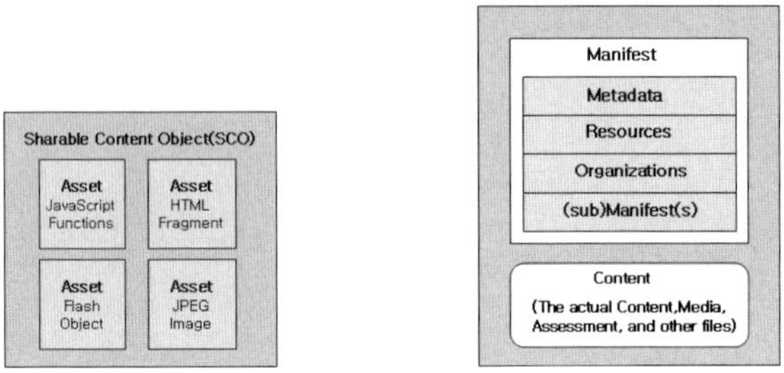

Fig. 3. Conceptual architecture of SCO and Content Package

Contents Delivery Engine consists of Context Manager and Content Adaptation. Content Adaptation consists of the Format Converter and QoS Adaptor. The Format Converter is a module transforming the format of learning content according to software type like that of the user's browser and media viewer. QoS Adaptor is a module that adapts the QoS(Quality of Service) of asset to environment modeling data in the network environment. As mentioned above, asset is an atomic element of learning contents. This table presents monitoring information and a formula used to decide the QoS of the asset.

Table 3. Information collected from Monitor module

Type	Monitoring Elements
Video	Video X Size, Video Y Size, Chrominance, Luminance, Frame rate
Audio	Audio Channels, Audio Codex, Audio Frequency
Platform	Port, CPU rate, CPU rate when Max Bandwidth, CPU rate when min Bandwidth
Network	Max Bandwidth, Available Bandwidth

'HCI based on Dynamic Interface for U-Learning' supports smooth interaction between learners and learning system and the use of optimum learning system for learners, and provides smooth and optimum learning system environment for learners under considering learners and learning system. Context consists of learners' learning machine information and learners' situation information here. Learners' learning machine information contains the Display Size of learning machine and usable Applications, and information about learners' positions information and present using application is included in learners' situation information. Learners' learning machine information is needful information in order to provide suitable form of learning context for learners' learning machine. However, learners' situation information is information for controlling alterative application information along with changing learners' positions. We can consider the information as indispensable information that the learners use learning system smoothly and it provides efficient learning opportunity sufficiently, and it can be seen that it is approaching the goal that HCI is about to pursuing. We use the same context to make possibility of dynamic interface implementation for learners. For this, we combine Decorator pattern of GOF Design Pattern and the function of dynamic reconfiguration in executing time of component-based application. This figure shows the composition of Decorator pattern and reconfiguration

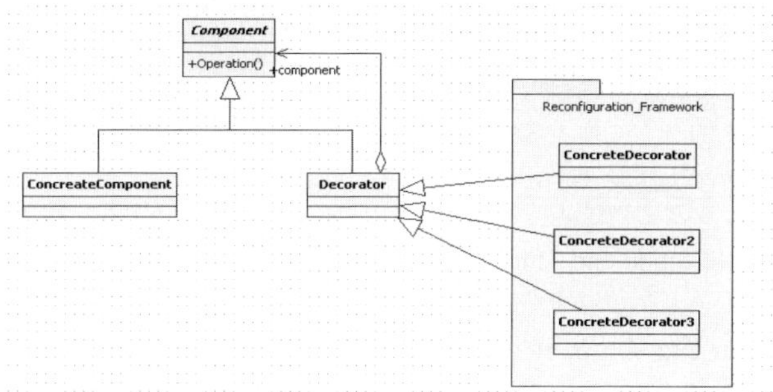

Fig. 4. Dynamic Interface architecture based on Decorator Pattern

4 Evaluation

The proposed designed and developed system can provide learning content to learners without awareness of the learner's characteristics. In this section, experiments where the system provides a more effective service for each three different type of device are achieved. The degree of satisfaction for receiving video data from local education broadcasting is evaluated. The experimental environment is as follows

1) Three learners are provided learning content from Education Server. 2) First learner's learning device is PDA using 802.11a in a wireless network environment. 3) Second learner's learning device is Desktop PC in a wired network environment. 4) Third learner's learning device is a Notebook computer using 802.11a/b/g.

When each of the three learners have accessed the education server, the latency rate for PDA user and packet loss rate for Notebook user are evaluated. In addition, in comparison with Desktop Users, the identified degree of satisfaction for PDA and Notebook users is much lower for E-Learning.

Fig.5 presents PDA user's latency rate and packet loss rate.

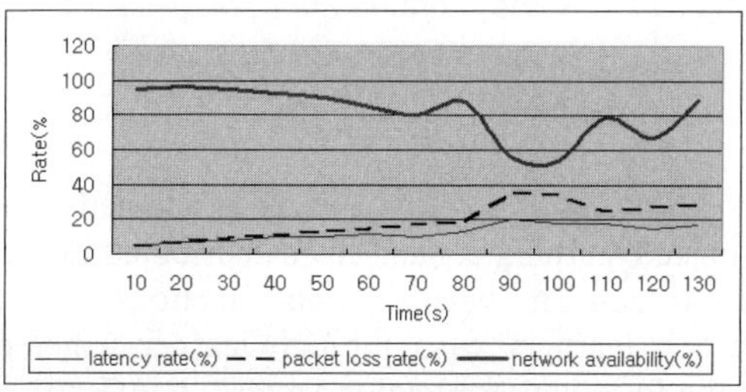

Fig. 5. First Learner's latency rate and packet loss rate

In fig.4, PDA user's network availability is unstable. However, the PDA user who received optimized-content demonstrated stable network availability.

Fig.5,6 presents desktop and notebook user's latency rate and packet loss rate.

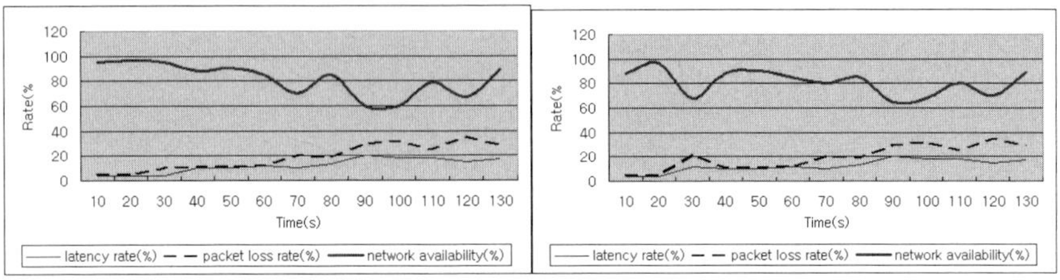

Fig. 6. Second and Third Learner's latency rate and packet loss rate

As presented in Fig 4,5,6, it can be confirmed that the same learning content is used, even though different learning devices are used, and different environments exist. Table 3 presents adapted video data for each device.

Table 4. Adapted video data for each device

	Video Stream Size(kbps)	Buffering Time
PDA(802.11a, 320*240, 64Mbyte RAM, X scale-400 Mhz) – Wireless	128Kbps	15sec
NoteBook (802.11a/b/g, 1024*768, 256Mb, Pentium M 1ghz) - Wireless	192kbps	12sec
Desktop Pc (100M bps, 1024*768, 512Mb, Pentium 2.0Ghz)	500kbps	5sec

Table 5. Evaluation Results

Effectiveness	Efficiency	Satisfaction
Effectiveness is evaluated about the ability of successful execution of given test by users.	Efficiency is about the use of physical and mental resources which users consume to achieve specific objectives.	Satisfaction expresses the degree of user's satisfaction with offered tool, application and service.
Achieved Learning : 100%	Attempt Num : 50 Achieved ratio : Mobile : 87% Desktop : 100%	Mobile : 85% Desktop : 83%

Though the proposed system, learning content is adapted to various devices, and learner received optimized learning content from distributed Education servers.

Finally, in order to confirm whether two functions as stated above deliver the learning data which learners wish through learner-centered methodology, we evaluate it based on the following evaluation factors. 1) Effectiveness: in the relevant definition of usability of ISO 9241, Effectiveness is evaluated about the ability of successful execution of given test by users. 2) Efficiency: Efficiency is about the use of physical and mental resources which users consume to achieve specific objectives. 3) Satisfaction: Satisfaction expresses the degree of user's satisfaction with offered tool, application and service. 4) User Feedback: it's a sincere discussion about evaluation of service after user receives offered service. This table shows evaluation results. (Candidate Num : 100, Mobile:80%, Desktop: 20%)

5 Conclusion

In this paper, a learner modeling mechanism for a ubiquitous learning environment is proposed, and the dynamic reconfiguration mechanism of the learning object is presented. The contribution of this paper is as follows.

1. Correct judgment of learning situation, making use of a variable model
2. Reuse of distributed learning objects
3. Delivery of learning content without learning device awareness
4. Adaptation of learning content by considering the learning environment
5. Reflection of systematic learning model for ubiquitous learning environment

Through this merit, learning technology suited to a Ubiquitous environment is developed and, using this, learner centered learning can be effectively delivered. However, although this effective learning model is applied, the system is still in the evaluation stage of development.

References

[1] LEUNG, C.-H., CHAN, Y.Y.: Mobile Learning: A New Paradigm in Electronic Learning. In: Proceedings of the The 3rd IEEE International Conference on Advanced Learning Technologies (ICALT'03)

[2] ADL. Sharable Content Object Reference Model version 1,2, Advanced Distributed Learning (2002) http://www.adlnet.org/ADLDOCS/Document/SCORM_1.2_RunTimeEnv.pdf, http://www.adlnet.org/ADLDOCS/Document/SCORM_1.2_CAM.pdf

[3] Sommers, K., Gesler, J., Bostick, J.: Little Guys Make a Big Splash: PDA Projects at Virginia Commonwealth University. In: Proceedings of the 29th annual ACM SIGUCCS conference on User services(SIGUCCS'01) (2001)

[4] Baloian, N., Pino, J., Hoppe, H.: A Teaching/Learning Approach to CSCL. In: Proceedings of the 33rd Hawaii International Conference on System Sciences, Hawaii, IEEE Computer Society Press, New York (January 4-7, 2000) [CDROM, 10 pages]

[5] Campus Alberta Repository of Educational Objects (CAREO). Repository in a Box. (2002) Retrieved January 22, 2003, from http://careo.netera.ca/

[6] Dunn, R., Dunn, K.: Teaching Students Through Their Individual Learning Styles: A Practical Approach. Prentice-Hall, Virginia (1978)

[7] Kellar, M., MacKay, B., Zhang, R., Watters, C., Kaufman, D., Borwein, J.: Architecture to Support Dynamic Composition of Math Lesson Plans. In: Proceedings of the 2003 ACM Symposium on Applied Computing, Melbourne, Florida, pp. 569–574. ACM, New York (March 9-12, 2003)

Involving Users in OPAC Interface Design: Perspective from a UK Study

Elahe Kani-Zabihi and Gheorghita Ghinea

School of Information Systems Computing and Mathematics, Brunel University, Uxbridge, UB8 3PH, UK
{Elahe.Kani, George.Ghinea}@brunel.ac.uk

Abstract. The purpose of this study was to determine user suggestions for a typical OPAC (Online Public Library Catalogue) application's functionality and features. An experiment was undertaken to find out the type of interactions features that users prefer to have in an OPAC. The study revealed that regardless of users' Information Technology (IT) backgrounds, their functionality expectations of OPACs are the same. However, based on users' previous experiences with OPACs, their requirements with respect to specific features may change. Users should be involved early in the OPAC development cycle process in order to ensure usable and functional interface.

1 Introduction

Electronic library catalogues over the Internet have evolved into applications called OPACs (Online Public Access Catalogues). Although there have been many investments undertaken by academic and research libraries targeting online resources, with academic libraries in the United Kingdom spending around £25 million on electronic resources [4], there are, however, still issues outstanding with respect to the usability of these systems [3]. Generally these systems have a lack of understanding of their users and have ignored users' perspectives. Hence, in designing and implementing OPACs users' abilities and experiences have to be taken into account, which forms the focus of the work reported in this paper.

2 Background

There is a great deal of research which targets users' interactions with OPACs' interfaces. However most existing user studies have mainly focused on OPACs' search engine's capabilities [1], [5], as well as on identifying user's retrieval methods [6], [8], so that users are able to find the information they need with minimal effort. Whilst such research disregards the fact that environmental conditions affect people's behaviour more than their personal goals [2], an even more important omission, in our opinion is that they implicitly assume that a good interface design has been executed in the first place. Nonetheless good interface design should ensure that OPACs'

M.J. Smith, G. Salvendy (Eds.): Human Interface, Part II, HCII 2007, LNCS 4558, pp. 374–383, 2007.

interfaces are friendly enough to deliver user satisfaction with the quality of their results first time round, irrespective of environmental conditions and any potential inhibiting influences these might exert.

A close look at how users interact with OPACs helps to understand users' needs and consequently to design these systems in a way appropriate to satisfy its users. We can go one step further by actually involving users in the design stage. Research shows that involving users in the development stage has resulted in delivering usable systems. Moreover, user involvement has resulted in user satisfaction at the end [7]. There are different approaches to this. In our study we are particularly interested in introducing elements of user-centred design and participatory design.

User-centred design methods focus on users through planning, design and development of a product since only users can provide "sufficient knowledge and understanding of the context of use". In a survey of user-centred design practice, conducted by Vredenburg et al. [12], results showed that there was improvement in product usability and that user-centred design methods are extensively used in industry. Moreover participatory methods allow all participants (users, designers, and developers) to contribute equally and influence OPAC design. Therefore, users as well as designers and developers have an important impact in the product outcome [9].

However, in involving users, irrespective of whether one employs a user-centred design method or participatory method, it is important to have users with variety of IT backgrounds, as their use of OPACs is very much dependent on their previous knowledge and experiences in IT. This was shown in a study measuring online (database) searching experiences, investigated by Hsieh-Yee [10] involving two groups of users. The first was made up of 32 professional online searchers (who had at least one year of search experience or had taken courses on online searching), whilst the second comprised 30 novice searchers (who had little or no search experience and had not taken any courses on online searching). This particular study investigated how people use their search methods to search an online database and also whether their search experience and subject knowledge had an effect on their searching methods. The conclusion of this study showed that in fact both the subject expertise and search experience had an effect on the search method employed by users.

From a related but a different perspective, a study carried by Fields et al. [11] observed librarians while helping students who did not have sufficient skills to find the information they need using the catalogue. Two librarians were videotaped and were asked to think out loud whilst they were interacting with the students. Here again, the conclusion of their study is that expert users are more successful at finding information than non-experts, thus supporting our view that OPAC design has to take into account user IT skills. The purpose of the study described in the present paper is therefore to inclusively involve users with a wide spectrum of IT abilities, by soliciting their opinions on what OPAC functionality should comprise, so that their expectations are satisfied.

3 Experiment

In order to understand users' expectations and preferences we conducted an experiment observing their behaviour and interactions with OPACs. Moreover, as one of the focii of this experiment was to find out how users search for materials and what sort of difficulties they may experience, the results obtained have potentially a more general applicability than just strictly in the realm of OPACs.

3.1 Participants

In our study we were particularly interested in how the level of IT sophistication of users impacted the quality of suggestions for OPACs. Accordingly, 54 users (31 female and 23 male), aged between 19 and 49 years old, with a variety of IT knowledge and academic backgrounds participated in our experiment. They were grouped into three different categories (depending on his/her IT specification): novice, intermediate and advanced – with an equal number of users in each category [16].

3.2 Procedure

The experiment involved users interacting with OPACs from three UK universities (UniA, UniB, and UniC), all with differentiating of distinctive functionality features, users mastery of which was examined through the four tasks that they had to undertake. After each user had attempted these tasks, s/he was asked to complete a questionnaire soliciting her/his views on OPAC features and functionality.

The UniA OPAC updated its features in 2004-05, with users being now able to see more details about an item in the OPAC, such as the abstract of a book. However, in introducing new interfaces, tasks such as placing a hold on an item have arguably became more complex. To examine this, task one of our study targets exactly this facility of the UniA OPAC. As anecdotal evidence pointed to the possibility that the UniA OPAC help service is poor, in the second task of our study we asked users to find an inter-library loan service (not provided by the OPAC), and thus explore whether the support available is sufficient to guide users in this way.

As opposed to UniA, UniB allows users to search for books using the ISBN number from the OPAC home page, via a drop down menu from the search bar. Moreover the OPAC displays all books nearby the shelf of the searched item. Users are therefore able to see other books which might also be helpful to them, much in the same way as they would pick a book from its shelf in a physical library. Task three was thus allocated to UniB to find out whether users are able to use these services.

Finally, task four concerned the UniC OPAC. Services provided by this OPAC were relatively different to those used in UniA and UniB, with users having access to a rich functionality array. For instance, here users were able to save their searches; view their previous searches; add an item to 'Basket' for future use; email the content of an item to their email account; see the details of recently searched item; search for items using different category such as DVD, Books, music and etc; and finally users were able to clear 'Basket' and 'Previous Searches' lists by clicking on 'End session' button on the screen. Task four, as shall be shown below, explored whether users would use the whole functionality spectrum at their disposal. Specifically, the tasks were:

Task one: Search the library for a book, which contains material on *Distributed Systems and Java*. Once you have located such a book, place a 'hold' on it.

Task two: Find out whether you can order an article through the inter-library loan service.

Task three: Search for the book with ISBN number: 0471346098. When you find it, look for at least four other books within the same subject area (the Java programming language) on the same shelf as this book.

Task four:

Step 1 - Search for the book with title *XML in a nutshell*; view full details of the book; add the book to the 'Basket' so you can view it again; email the details of the book to yourself.

Step 2 - Search for *The Godfather* DVD; find out who stared in the movie; add it to the 'Basket'.

Step 3 - See the list of items you have added to 'Basket'.

Step 4 - See the list of searchers you have done so far.

Step 5 - Go to the result list of the search for the film The Godfather.

Step 6 - Clear all your searches to start a new search with one click.

All users were observed by the same single person, who recorded their action and their achievement. The observer also made sure each user understood the terminology used in the questionnaire and the purpose of each task given to them. Users were also asked to think out loud while interacting with OPACs.

To find out users' opinions about each task using the related OPAC, we asked them to fill in a questionnaire (Table 1) after accomplishing each task. This contained a set of five statements on which the user expressed their opinion on a five-point scale (strongly agree, agree, natural, disagree, and strongly disagree). Each statement describes whether the user is happy with the way the task is delivered by the OPAC or if the user would prefer to do the task using a physical library instead.

Table 1. OPACs' Questionnaire

It was easy to do the task.
It was easy to read information provided in the OPAC.
There were sufficient help available.
I prefer to use this OPAC instead of a physical library for this task.
I will use this OPAC again to do this task.
It is easy to be familiar with this OPAC.

The last part of the experiment involved users replying to a questionnaire soliciting their views on OPACs in general (Table 2). The questionnaire solicited users' opinions on the features that they would prefer the OPACs to contain.

Table 2. Users' preferences

From the three OPACs systems (UniA, UniB and UniC) that you have used to do different tasks, which one of them was more helpful and friendly to you?...........
Specify your reason here: ....
In your opinion, state here the main requirements that an OPAC should satisfy (stating with the most important one):.................................

Data collected was analysed with the Statistical Package for Social Sciences (SPSS), version 11.5. The t-test and Analysis of Variance (ANOVA) were applied to identify differences, and to test potential significant differences between the three user types. The significance value used for this study was $p < 0.05$.

4 Results

4.1 Users' Interaction with OPACs

As mentioned above, participants were asked to use three OPACs while their interactions were observed. Almost 76% of users ($n = 41$) did not accomplish tasks correctly - and this does not exclude experienced users. Moreover, the observer noted that users often blamed their own lack of knowledge of IT when they did not achieve a task and not the application. This section describes how users achieved each task using the three OPACs.

4.1.1 UniA OPAC

The UniA OPAC updated its interfaces in the 2004-2005 academic year. The new version has more functions available and displays more details about a selected book. However, users who had used the catalogue in previous years commented that they prefer the old version, as it was less complicated, and tasks such as placing a hold on a book had previously been much more straightforward.

An ANOVA on our study data shows that all three different types of users said they will use UniA OPAC again ($p = 0.012$, $F = 4.801$) even though 31.48% of users ($n = 17$), mostly novice, could not accomplish Task 1. Fig. 1 shows the average credit given by users in their responses to the questionnaire (Table 1).

Apart from Task1-Q3 (Table 1), there was general agreement ($df = 53$, $p = 0.000$) among all users as regards their ranking opinion of each task. Although all three types of users indicated it was easy to do this task, they did prefer to use a physical library for this task, especially advanced and intermediate users. This may point towards the fact that the OPAC did not satisfy users' needs and their expectations were not fulfilled. Surprisingly, all users surveyed in our study indicated that they found it easy to familiarise with the OPAC (Task1-Q6). Hence, although users were confident to

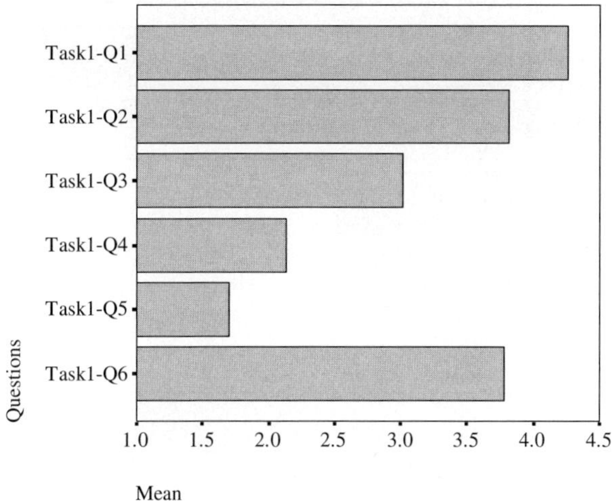

Fig. 1. Users' responses to Task 1

use the OPAC, they were not happy with the way this task could be accomplished in the OPAC.

Users, especially intermediate ones, were not in favour of using this OPAC for the task. 72.22% of novice users (n = 13) struggled to find the 'Place Hold' link. These users suggested that it could have been better if this option was available at the result page and this link should be placed in a more obvious place on the page. The observer noted that 18.52% (n = 10) of users gave up using the Help function after their first attempt. The rest did not notice the link.

4.1.2 UniB OPAC

As opposed to the UniA OPAC, users were much more comfortable with using the UniB OPAC. Fig. 2 shows the average scores given by users in their responses to the questionnaire (Table 1). Apart from Q3 and Q4 (Table 1), there was general agreement (df = 53, p = 0.000) among all users as regards their ranking opinion of each task. It seems that users did not all agree whether the help provided was sufficient and if they would prefer to use the OPAC in future. However, all users agreed that it was easy to use the OPAC and to read the information provided. Our results show that Intermediate users seemed to like this OPAC more than users from the other two categories.

An option to search for books by ISBN number was available under a drop down menu list on the first page of searching. Only six users (two from each category) could not find this option and hence did not find the book at the first attempt. 88.88% (n = 48) of users managed to find the book on the same shelf number, by clicking on the right link. The rest, however, either tried searching by the shelf number or subject name.

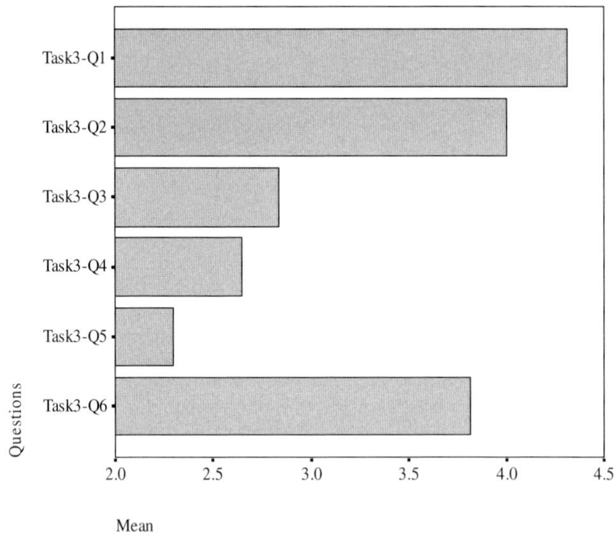

Fig. 2. Users' responses to Task 3

4.1.3 UniC OPAC

The result from the UniC OPAC shows that users were also happy with the interfaces of this catalogue. Fig. 3 shows the average credit given by users in their responses to the questionnaire (Table 1).

Here, we have a similar results profile to those obtained in Task 1 and Task 3. Thus, apart from Q3 (Table 1) there was general agreement (df = 53, p = 0.000) among all users as regards their ranking opinion of each task.

More advanced users were in favour of UniC OPAC than the previous one. Although all users from three categories have said tasks done on this OPAC were easy, and it was easy to be familiar with the OPAC not all users accomplished their tasks correctly.

35.19% of users (n = 19), mainly from the novice category, could not find the 'Add to basket' link. These users were clicking on 'Basket' button instead, which was at the top of the page. One typical comment given on this function was that the terminology 'Basket' or 'Add to basket' is used for purchasing items on e-shopping websites rather than OPACs. Therefore, different phrases should be used for these functions.

90.74% of users (n = 49) did not find the DVD search option under 'Subset' menu. These users searched for the DVD in much the same way as they searched for a book. It was not obvious to users that the 'Subset' button contains the list of different materials available in the OPAC e.g. DVD, Books, Journals. Some users looked for a DVD search option under the drop down menu on the first page. Only four advanced users and one intermediate user managed to find this option.

Our result indicates that the majority of users managed to view the content of the 'Basket'. Users, who did not find the option, were from all three categories. However,

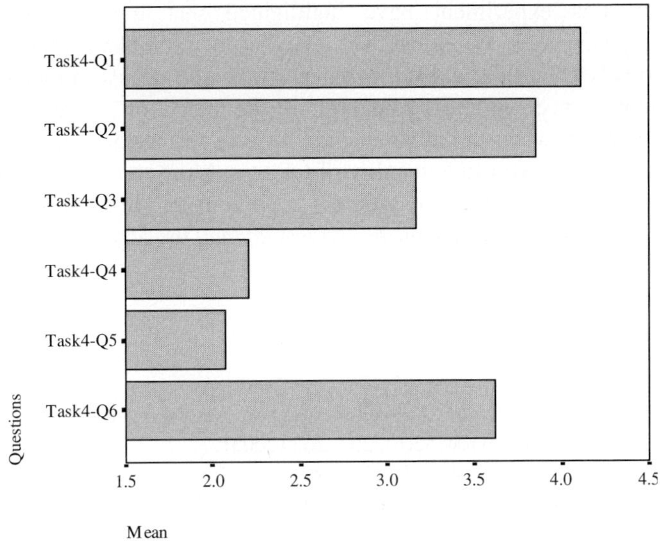

Fig. 3. Users' responses to Task 4

35.19% of users (n = 19) did not click on the 'Previous Search' button which, was in the same style as 'Basket' button, this may indicate that users were not familiar with the phrase used and therefore could not make a connection between the labelled button and the task. A similar observation applies to the 'Results List' button.

Another difficult task for users was when they were required to clear their previous searches using one click. Only 29.63% (n = 16) of users employed the 'End Session' button, which was designed for this purpose. These users were from all three categories. In concluding we remarked that users were generally more confident in using the UniC OPAC than the other two. Nonetheless, users commented that UniC OPAC had introduced new features, some of which were unnecessary and rather confusing. Therefore the number of people achieving Task 4 satisfactorily was low.

5 Conclusion

In this paper we have reported the results of a study which sought to explore the gap between users' self-reported OPAC requirements and the functionality and usability of such systems in practice. We believe that the integration of user-defined requirements of OPACs, along with other Human-Computer Interaction considerations, offer a better understanding of user perceptions and expectations with respect to OPACs and ultimately result in the design of truly user-centred OPACs. Accordingly, in our work, as opposed to previous studies ([13], [14], and [15]) in the area, which have mainly focused on information retrieval and usability issues in OPACs, we have elicited users' opinions regarding features that OPACs should have from users with a wider spectrum of IT skills, ranging from novice to advanced users.

Results from this experiment have highlighted that users often struggle in accomplishing their tasks. We noted that novice users often blamed their own lack of knowledge of IT when they did not achieve a task and not the application. Novice users were more concerned about readability of the interfaces and receiving feedback from the application, whenever an interaction is done. Intermediate users often did not take advantage of the extra functionality of OPACs. Moreover both intermediate and advanced users highlighted the insufficient support from Help services in all three OPACs. Most of the advanced users suggested that the inter-library loan facility should be available in all OPACs.

References

[1] Babu, R.B., Tamizhchelvan, M.: An investigation into the features of OPACs in Tamil Nadu (India). Library Review 52(6), 257–267 (2003)

[2] Blandford, A., Stelmaszewska, H.: Shooting The Information Rapids. In: Vanderdonckt, Blandford, Derycke. (eds.) Ihm-HCI 2001, pp. 51–54 (2001)

[3] Bordeianu, S., Carter, C.E., Dennis, N.K.: Delivering electronic resources with Web OPACs and other Web-based tools: needs of reference librarians. Reference Service Review 28(2), 111–118 (2000)

[4] Creaser, C., Hamblin, Y.,Davies, J.E.: An assessment of potential efficiency gains through online content use. Program: electronic library and information systems 40(2), 178–189 (2006)

[5] Fox, E.A., France, R.K., Sahle, E., Daoud, A., Cline, B.E.: Development Of A Modern OPAC: From Revtolc To Marian. In: Korfhage, R., Rasmussen, E., Willett, P. (eds.) ACM Press, pp. 248–259 (June 27-July 01, 1993)

[6] Jansen, B.J., Pooch, U.: A review of web searching studies and a framework for future research. Journal of the American Society for Information Science and Technology 3(52), 235–246 (2000)

[7] Kujala, S.: User involvement: a review of the benefits and challenges. Behaviour & Information Technology 22(1), 1–16 (2003)

[8] Tenopir, C., Hitchcock, B., Pillow, S.A.: Use and Users of Electronic Library Resources: An Overview and Analysis of Recent Research Studies. Council on Library and Information Resources (2003)

[9] Gulliksen, J., Boivie, I., Göransson, B.: Usability professionals-current practices and future development. Interacting with Computers 18(4), 568–600 (2006)

[10] Hsieh-Yee, I.: Effects of Search Experience and Subject Knowledge on the Search Tactics of Novice and Experienced Searchers. Journal of the American Society for Information Science 3(44), 161–174 (1993)

[11] Fields, B., Keith, S., Blandford, A.: Designing for Expert Information Finding Strategies. In: Fincher, S., Markopoulos, P. (eds.) People and Computers XVIII - Design for Life, pp. 89–102. Springer, Heidelberg (2004)

[12] Vredenburg, K., Mao, J., Smith, P.W., Carey, T.: A survey of User Centred Design in practice, pp. 471–478 (2002)

[13] Kan, M., Poo, D.C.C.: Detecting and supporting known item queries in online public access catalogues. In: Proceedings of the 5th ACM/IEEE-CS joint conference on Digital libraries, pp. 91–99. ACM Press, New York, USA (2005)

[14] Agosti, M., Masotti, M.: Design of an OPAC database to permit different subject searching accesses in a multi-disciplines universities library catalogue database. In: Belkin, N., Ingwersen, P., Pejtersen, A.M. (eds.) Proceedings of the 15th annual international ACM SIGIR conference on Research and development in information retrieval, pp. 245–255 (June 21-24, 1992)

[15] Mcmullen, S.: Usability testing in a library Web site redesign project. Reference Service Review 29(1), 7–22 (2001)

[16] Kani-Z, E., Ghinea, G., Sherry, C.: Digital Libraries: What users want? Online Information Review 30(4), 395–412 (2006)

The Effectiveness of Educational Technology: A Preliminary Study of Learners from Small and Large Power Distance Cultures

Elizabeth Koh and John Lim

School of Computing
National University of Singapore
3 Science Drive 2 Singapore 117543
Singapore
{elizabethkoh, jlim}@nus.edu.sg

Abstract. The cultural background of learners has been highlighted as crucial in determining the effectiveness of educational technology. This paper focuses on the influence of power distance in determining the effectiveness of educational technology. Utilizing a multiple case study, we examined the perception of learners from small and large power distance societies in terms of satisfaction with learning, self-efficacy with educational technology and perceived learning. Our findings show that the availability of educational technology enhances the learning outcomes of both cultures. The study suggests the notion that learning outcomes differ for learners from small and large power distance cultures.

Keywords: Educational Technology, Culture, Power Distance, Effectiveness.

1 Introduction

Many educational institutes and corporate organizations are utilizing educational technology as a tool for learning and training. The market for online higher education is forecast to reach US$69 billion by 2015 [1]. Similarly, worldwide corporate educational technology revenues are estimated to hit US$23.7 billion in 2006, up almost fourfold from US$6.6 billion in 2002 [2]. Correspondingly, cross-cultural collaborations among organizations and universities are growing. Online education also facilitates the expansion of distance education towards other countries. In addition, there is an increasing diversification of student and organizational populations. However, educational technology for one culture might not be as effective in another [3]. The cultural background of learners has been highlighted as crucial in determining the effectiveness of educational technology [4, 5].

Whereas culture can be examined from many perspectives, seminal research on culture by Hofstede [6] is considered pertinent to the examination of learning outcomes [7]. Hofstede's [6] taxonomy includes five cultural dimensions - power

M.J. Smith, G. Salvendy (Eds.): Human Interface, Part II, HCII 2007, LNCS 4558, pp. 384–393, 2007.

distance, individualism/collectivism, masculinity/femininity, uncertainty avoidance and time-orientation. This paper focuses on the influence of power distance in determining the effectiveness of educational technology. Hofstede found that cultures differ in the level of power distance. Large power distance cultures tended to have a high regard for authority while small power distance cultures preferred to see others as equals. Research has found that power distance has a strong influence on learning outcomes [7, 8]. Students from large power distance cultures perceived a greater gap between themselves and instructors while students from small power distance societies regarded instructors as their peers. This has implications for the satisfaction, self-efficacy, perceived learning and academic achievement of learners [9]. Hence the effectiveness of educational technology would hinge on the degree of power distance.

The objectives of the paper are, to discover how educational technology affects learning outcomes, and to explore the role power distance plays in influencing the effectiveness of educational technology. The paper begins with a presentation of the research framework followed by the research methodology. It then focuses on the findings of the two case studies. Subsequently, the role of power distance is discussed before providing several suggestions for future research and practice in the conclusion.

2 Research Framework

We propose that power distance moderates the relationship between the availability of educational technology and learning outcomes. Figure 1 depicts the proposed relationships [9]. Furthermore, we stipulate that the effect of the availability of educational technology on learning outcomes, consisting of satisfaction with learning, self-efficacy with educational technology, perceived learning and academic achievement, will be greater on small power distance learners than on large power distance learners. The definitions of the constructs are presented in Table 1.

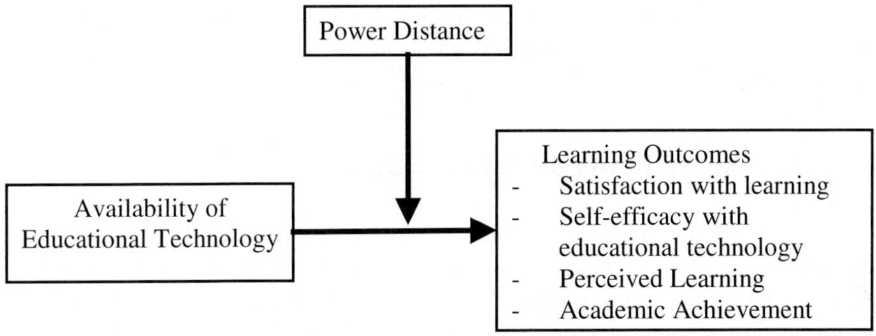

Fig. 1. Research Framework [9]

Table 1. Definitions of Constructs (adapted from [9])

Power Distance	The extent to which the less powerful people in society respond to inequality in power and consider it as normal [6]
Satisfaction with learning	The degree to which a learner feels a positive association with his or her own educational experience [10]
Self-efficacy with educational technology	The degree to which learners feel confident of learning from educational technology [5]
Perceived Learning	Changes in the learner's perceptions of skill and knowledge levels before and after the learning experience [11]
Academic Achievement	Actual cognitive development of learners; this concept is related to task performance and typically measured by project or course grades [5]

3 Research Methodology

Utilizing a multiple case study approach, we examined the perception of learners from small and large power distance societies. Data was collected from interviews, observation and documentation. We employed multiple techniques to analyze data such as data displays and pattern matching [12]. Our research framework served as a guide to the analysis. We investigated the perception of students from both cultures in terms of satisfaction with learning, self-efficacy with educational technology and perceived learning. Academic achievement was not captured owning to the lack of access to student's grades.

There were two cases in total – the first consists of students from the smaller power distance culture while the second comprises students from the larger power distance culture. The unit of analysis is at the individual level. Informants in Case A were selected from the student exchange program in a large university in Singapore (host University). Informants in Case B were chosen from the student population in the same university. Within case analysis and cross case analysis were performed.

4 Case A: Small Power Distance Learners

The small power distance learners hailed from Europe – 3 were from Finland and 2 from Germany. They were pursuing a master's degree that took 4.5 to 6 years to complete. They had been studying in the host University for 3 months on the exchange program when they were interviewed. Table 2 records the profile of the learners.

Table 2. Profile of Small Power Distance Learners

Informants	Year of Study	Age	Gender	Major, Faculty
S1	5	26	M	Computing, Computing
S2	5	25	M	Mechanical, Engineering
S3	5	24	M	Process, Engineering
S4	4	24	F	Southeast Asian Studies, Arts
S5	4	27	F	Southeast Asian Studies, Arts

4.1 Satisfaction

For small power distance learners, their satisfaction with educational technology is affected by the *ease of communication*, the *ease of searching for information*, and the *flexibility of studying anytime*. For instance, one student (S2) responded, "*It's easier with educational technology. You can share your knowledge, get materials, [even] extra materials and help for [the] difficulties [you encounter]. You don't have to read through books [to search for answers] for a little [problem]. You can ask things that are hard to find.*"

As noted in the literature, small power distance learners relish in using educational technology to connect with others [7]. Sharing information with each other is part and parcel of communicating with others and is satisfying to small power distance learners.

Another reason for their satisfaction is the ease of searching for information. They liked the integration of course information in the host University's virtual learning environment and the ability to search for course material online. The flexibility of studying anywhere and anytime was highlighted as another reason for their satisfaction. Students saved time by watching webcast lectures and sending email at their own convenience.

Thus, small power distance learners were clearly satisfied with the usage of educational technology. However, a possible reason for their enthusiasm was their limited experience in using educational technologies. These students did not have access to a similar virtual learning environment in their home university where learning materials, information about modules, assessments and interactive features were integrated into an information system.

4.2 Self-efficacy

Students were interviewed on their confidence in using educational technology. Most students expressed high self-efficacy. For example, in response to a question on how confident she would be to navigate a new website, S4 answered, "*I can do it on my own. Sometimes I do not really find every hidden detail or something but, I[will] click on this and that... it [may] take me 2 hours, but at least I find out where they are. I don't mind spending some time.*"

An exception was S3. S3 admitted that he "*can't figure out how to use IT*" and was "*not so good*" at using it. This lower self-efficacy can be attributed to his lack of experience with educational technology in his home university that did not make much use of IT for instruction. However, S1 and S2 were also from the same home university and they were quite confident in using IT. A more likely reason could be due to the nature of his academic discipline. He was studying process engineering which would involve lesser use of educational technology as compared to a computing course. The IT and business disciplines generally make use of educational technology more than other fields [13]. Nevertheless, the data seems to suggest that the self-efficacy of small power distance learners was high with the availability of educational technology.

4.3 Perceived Learning

All informants agreed that educational technology helps in their learning. From the data, we identify 3 main causes. First, educational technology shortens the time needed to search for educational material which directs energy toward learning. It makes information much easier to find such as a portal storing all the course materials or a digital library for e-books and e-journals. The convenience afforded by these educational tools allows students to spend less time and energy gathering the materials and devote their time and energy to learn from these materials.

Second, educational technology enhances the knowledge of learners through social learning. Students noted the use of the electronic forums and email as a medium to learn from other learners and the instructors. "*It's a good way to share things. If somebody doesn't know something it's easy to ask classmates [in the forum]."(S3).*

Third, educational technology was perceived to increase learning when it was designed and utilized well. Educational technology could be developed by organizations or by individual instructors. S2 remarked that educational software helps in his learning if the software was designed appropriately by software companies. Students also found that instructors who made effective use of educational technology in their teaching facilitated their learning.

Although educational technology was seen to enhance learning, students also cautioned the replacement of traditional face to face modes of teaching with educational technology. "*If I [attend] lectures I can learn more. I cannot concentrate [as well] with webcasts. But maybe someone [else] can." (S1).* Students found it harder to concentrate on educational technology as they could be easily distracted by other leisure activities online. They seemed to prefer educational technology to supplement and not replace traditional face to face teaching.

In sum, the availability of educational technology on small power distance learners raised their satisfaction, self-efficacy and perceived learning. The paper now proceeds to describe the second case of large power distance learners.

5 Case B: Large Power Distance Learners

The large power distance learners originated from Singapore, Malaysia and China. All of them were studying full-time in the host University. Most were undergraduates in 3 or 4 year Bachelor degree programs. A total of 9 students were interviewed. The profiles of the informants are shown in Table 3.

Table 3. Profile of Large Power Distance Learners

Informants	Year of Study	Age	Gender	Major, Faculty
L1	4	24	M	English Language, Arts
L2	4	24	M	Lifescience, Science
L3	4	22	F	Civil, Engineering
L4	3	23	M	Finance, Business
L5	3	30	F	Bioengineering, Engineering
L6	1	19	F	Statistics, Science
L7	1	19	F	Arts
L8	1	19	F	Chemistry, Science
L9	1	21	M	Mechanical, Engineering

5.1 Satisfaction

Large power distance learners were satisfied with educational technology due to the *ease of communication and knowledge sharing*. Many highlighted the specific educational tool such as the virtual community feature in the virtual learning environment that enabled them to share knowledge with other learners.

Ya a lot of times I'm satisfied. Sometimes when we are doing something together, we need to share our information. Sometimes we cannot find a lot of time to meet together, so in this way [using educational technology] it is easier for us to share our information,[and] update one another. (L5).

In general, large power distance learners had positive attitudes towards educational technology and were satisfied with the availability of educational technology.

5.2 Self-efficacy

Many students believed that they could handle educational technology relatively well. They had no difficulty in navigating through the virtual learning environment of the host University. One even went so far as to claim that the system was *"idiot-proof" (L9).* Many had been using IT in their earlier years of education and felt quite capable of using it. Hence, the data reveals that the *past experience* of students in using IT affects their degree of self-efficacy with the availability of educational technology.

Although they seemed to be confident in using educational technology, students were reluctant to admit they were IT experts but only acknowledged that they were average users – 7 out of 9 students claimed they were average users. This is so even though almost all informants accessed educational technology typically once a day. Their self-efficacy towards using educational technology is less than their self-efficacy towards other educational means such as textbooks and their instructors.

5.3 Perceived Learning

Perceived learning seemed to be higher with the availability of educational technology than without. First, students highlighted how educational technology *supplements face to face learning* methods. Educational technology is "*a way to let you learn and teach you about the lesson. [It is] a way I can learn [by] myself besides [through the] lecture. It is helpful.*" *(L6).* Students placed the spotlight on the additional knowledge they can gain from educational technology.

Second, students were able to *share knowledge* and learn from others over the online medium. Essentially, this is social learning at work. Educational technology gives more opportunities for students to communicate with each other. L2 spoke of how he contributed to the learning of his classmates, "*I will read through [the discussion forum] to see the topics that actually make sense. When I feel that [the posting] is a good contribution then I'll post something. I usually post when I have something substantive. Sometimes I will reply to other people's discussions.*" Hence, the availability of educational technology enhances the learning outcome of large power distance learners.

6 Role of Power Distance

6.1 Satisfaction

The analysis of small power distance learners showed that they were clearly satisfied with the availability of educational technology. They were effusive of their praise for educational technology. In contrast, large power distance learners were milder in their affection for educational technology. It seems that these students took for granted the ease of searching for information and the flexibility of studying anytime via educational technology.

Moreover, we believe that large power distance learners were not as satisfied with the sharing of information as small power distance learners. Some large power distance learners did not appreciate the ability to use the online discussion forum to interact and participate in the online class. "*I usually don't go [into the forum] unless it's required. You know some modules they require forum participation. Too much time is required...*" *(L2).* L2 found it time-consuming to participate in the online forum and would only post when it was a requirement of the course. It seems then that large power distance learners do not see a need to learn and share with others in the online discussion forum whereas learners from small power distance cultures enjoyed using educational technology to connect with others.

A possible reason for the difference in satisfaction of large and small power distance learners is the degree of instructor support. A large power distance user commented that the reason she does not actively use the forum is the lack of feedback from instructors. "*I use it [the online forum] occasionally only because not many people use it to post questions online. And there are no answers from the lecturer or tutors. So I use it only occasionally."(L8).*

Indeed, this is reflective of learners from the large power distance society which views the teacher as the knowledge source. Learners in large power distance societies focus their attention on absorbing knowledge from the instructor. Hence, in online forums where the instructor has minimum participation, students do not believe they can gain much knowledge from the social interaction and are less inclined to participate. From these analyses, we suggest that the effect of the availability of educational technology on satisfaction will be greater on small power distance learners than large power distance learners.

6.2 Self-efficacy

Four out of the five informants from the small power distance culture regarded themselves as having a high level of self-efficacy with educational technology. This is a higher number compared to large power distance learners. Two out of nine large power distance learners felt they were able to use educational technology very well while the rest preferred to admit that they were average users. This difference is detected despite the fact that all of them use educational technology almost daily.

The above observation is conceivably due to large power distance learners' higher dependency on the instructor. They are then less comfortable in independent learning with educational technology and have less self-efficacy than low power distance learners. Students see instructors as knowledge fountains and are extremely reliant on their teachers [6]. "*I usually would rather go straight to the lecturer or tutor. They should know what they are talking about."(L2).* Large power distance learners prefer to go directly to the instructor rather than search for the answer themselves via educational technology.

Small power distance learners on the other hand, have a low dependency on instructors and are familiar with an active mode of learning. "*If there is a problem with my computer I won't go to someone and ask them for help but I will study how to do it and do it myself."(S2).* Small power distance learners seem more confident on learning on their own. Educational technology facilitates this independence and their self-efficacy of learning with it will increase. Thus, the availability of educational technology will lead to greater self-efficacy for small power distance learners compared to large power distance learners.

6.3 Perceived Learning

In terms of perceived learning, the data suggests that learners from both small and large power distance cultures can learn equally well with educational technology. There does not seem to be any difference in perceived learning among the cultures. There were students from both cultures who gained knowledge and skills from educational technology.

Some researchers observe that students from high power distance cultures are constrained by the public nature of the discussions [7]. However, we find that students were not inhibited by the online medium but were more willing to participate using educational technology than in the face to face classrooms. "*I prefer online over face to face as I will self censor when it's face to face. Online I will not restrict myself - things I thought are stupid, I'll be able to say them online. I feel more relaxed. I feel embarrassed to ask people in the lecture or tutorial too. Also, there is no chance for interaction in the lecture or tutorial." (L7).* As noted by other researchers [14], educational technology enables large power distance learners to communicate and gain knowledge more effectively.

In sum, we find that the availability of educational technology has a similar effect on the perceived learning of small and large power distance learners. Students from both cultures are equally comfortable in learning with educational technology.

7 Conclusion

This case study has served to illuminate antecedents for our research framework. It seems that the ability to share knowledge and interact with other learners and instructors is a critical part of educational technology for learners from both large and small power distance cultures. We believe that these two antecedents deserve further study. Moreover, these preliminary findings serve to advance our understanding of the role of power distance in influencing the effectiveness of educational technology. This research framework could be further tested using a quantitative approach. Our next step involves a sample survey.

An important message of this study is that power distance is a key moderating influence on the link between the availability of educational technology and learning outcomes. Therefore, educators, instructional designers and researchers need to consider its influence in teaching and learning situations. Specifically, further research should examine issues related to the design and structure of educational technology to enhance the satisfaction and self-efficacy of large power distance learners.

Educators can help to increase the satisfaction and self-efficacy of large power distance learners by actively participating and giving feedback using online media such as electronic discussion forums. Larger power distance learners see their instructors as knowledge sources and have a high respect for this figure of authority. Online discussions by the instructor will then enhance the knowledge gained for students. We caution that instructors should encourage the participation of all learners in the electronic forum by challenging the students rather than spoon-feeding them with answers. Next, the design of educational technology must enhance the ease of knowledge sharing. As this has been highlighted by both cultures, it should be made easily available in educational technology.

References

1. Research and Markets.: Global E-learning Opportunity for U.S. Higher Education(2006) Accessed (September 17, 2006) from http://www.researchandmarkets.com/reports/c42197
2. eMarketer.com.: E-Learning Gains Momentum (2003) Accessed March 29, 2006 from http://www.emarketer.com/Article.aspx?1002352
3. Morse, K.: Does one size fit all? Exploring asynchronous learning in a multicultural environment. Journal of Asynchronous Learning Networks 7, 37–55 (2003)
4. Leidner, D.E., Jarvenpaa, S.L.: The Use of Information Technology to enhance management school education: a theoretical view. MIS Quarterly 19, 256–291 (1995)

5. Chang, K.T., Lim, J.: The Role of Information Technology in Learning: a Meta-Analysis. In: Carbonara, D.D. (ed.) Technology Literacy Applications in Learning Environments, pp. 14–36. Idea Group Publishing, Hershey (2005)
6. Hofstede, G.H.: Culture's Consequences: International Differences in Work-Related Values. Sage Publications, Beverly Hills (1980)
7. Wang, M.: Designing online courses that effectively engage learners from diverse cultural backgrounds. British Journal of Educational Technology (2006) Electronic document found at http://www.blackwell-synergy.com/, doi: 10.1111/j.1467-8535.2006.00626.x
8. Paulus, T.M., Bichelmeyer, B., Malopinsky, L., Pereira, M., Rastogi, P.: Power distance and group dynamics of an international project team: a case study. Teaching in Higher Education 10, 43–55 (2005)
9. Koh, E., Lim, J.: The Role of Power Distance in Determining the Effectiveness of Educational Technology. In: Khosrow-Pour, M. (ed.) Proceedings of the Eighteenth Information Resources Management Association International Conference, forthcoming (2007)
10. Shin, N.: Transactional Presence as a Critical Predictor of Success in Distance Learning. Distance Education 24, 69–86 (2003)
11. Alavi, M., Marakas, G.M., Yoo, Y.: A Comparative Study of Distributed Learning Environments on Learning Outcomes. Information Systems Research. 13, 404–415 (2002)
12. Yin, R.K.: Case Study Research, Design and Methods, 3rd edn. Sage Publications, Thousand Oaks (2003)
13. OECD.: E-learning in Tertiary Education (2005) Accessed (October 15, 2006) from www.oecd.org/publications/Policybriefs
14. Bauer, C., Chin, K.L., Chang, V.: Web-Based Learning: Aspects of Cultural Differences. In: Hansen, H.R., Bichler, M., Mahrer, H. (eds.) Proceedings of the Eighth European Conference on Information Systems, pp. 1396–1402 (2000)

Human-Mediated Visual Ontology Alignment

Monika Lanzenberger[1] and Jennifer Sampson[2]

[1] Institute of Software Technology & Interactive Systems, Vienna Univ. of Technology, Austria
lanzenberger@ifs.tuwien.ac.at
[2] Norwegian University of Science and Technology, Trondheim, Norway
sampsonj@idi.ntnu.no

Abstract. We develop a multiple-view tool called AlViz, which aims at supporting the ontology alignment process visually. Combing views on several levels of abstraction, the tool tries to make the 'relatedness' between entities accessible. Based on a literature study we identified relevant phases emerging in ontology alignment. We extended a general alignment framework in order to reflect the adoption of visualization techniques. This framework builds the background for our user study. We evaluate visual ontology alignment with AlViz in three stages: (1) Participative software development, (2) usability evaluation, and (3) utility study. The evaluation methods proved viable even though our study design is challenging.

Keywords: Ontology Alignment, Information Visualization, Evaluation.

1 Introduction

Ontologies represent a shared meaning of domain knowledge. Being enabler for many applications such as data integration, agent communication, peer-to-peer systems, e-commerce, and semantic web services, ontologies provide content represented in a form that is easily machine-processable. With an increase in the development of ontologies comes the need for tools and techniques for solving heterogeneity problems between different ontologies. Therefore we need ontology alignment, which helps us to bring different knowledge representations into mutual agreement. But this is largely a human-mediated process, although there are existing tools which can help with identifying differences among ontologies. User interaction is still essential in order to control, approve, and optimize the alignment results. However, evaluation of human-mediated components in ontology alignment is an open question. To that end we describe the tasks involved when doing visual ontology alignment. This is our starting point when we develop a user study in order to investigate the utility of a visual ontology alignment tool called AlViz [1].

Based on a literature study we identified relevant phases emerging in ontology alignment. We extended this phases in order to reflect the adoption of visualization techniques and then described the relevant steps of our framework for visual ontology alignment. This framework builds the background for our user study. The user study enables us to find answers to the following questions: Do the users find missing alignments? Do the users identify weak alignments? Do the users understand the relatedness of both ontologies?

M.J. Smith, G. Salvendy (Eds.): Human Interface, Part II, HCII 2007, LNCS 4558, pp. 394–403, 2007.

With this paper our contribution is twofold. First we describe a core framework listing the relevant phases in visual ontology alignment. Second we develop a user study based on this framework in order to evaluate visual ontology alignment tools.

In the next section we describe the tool AlViz. Thereafter, we explain the extended alignment process and give a practical example of alignment assessment. In section 5 we discuss the evaluation methodology and the user study in more detail.

2 AlViz - A Tool for Visual Ontology Alignment

Visual metaphors encode large, multi-dimensional data sets effectively and support human interpretation and understanding. Information visualization (InfoVis) uses such visual metaphors in order to communicate the relevant information in an intuitive way and to foster new insights into the underlying processes [2] and patterns. InfoVis aims at making complex data accessible for interactive investigation by the user. Graphical primitives such as point, line, area, or volume are utilized to encode information. These objects are characterized by position in space, size, connections & enclosures, shape, orientation, and visual cues like color and texture, with temporal changes, and viewpoint transformations [3].

We develop a prototype for visual ontology alignment called AlViz [1]. It is implemented as a multiple-view plug-in for protégé-owl [4] in order to visually support alignment of two ontologies by making the type of similarity between entities explicit. The tool consists of different views coupled by the linking and brushing technique. J-Trees [5] are one type of the views applied in AlViz. Such trees consist of a root node, expanded or collapsed branch nodes and leaf nodes displaying the hierarchical structure by indentation. They support the access and manipulation of instances and single items within classes quite effectively and are well established within the Protégé community. However, J-Trees have several shortcomings regarding the representation of large or complex ontologies because they become cluttered and do not provide adequate overview functionality. To overcome this problem we integrate another visualization type: small world graphs [6]. In a second view, such graphs help the user to examine the structure of the ontology more intuitively. This method uses clusters to group the nodes of a graph showing a specific level of detail. The nodes represent the entities (concepts or instances) connected to each other according to the selected relations, also called object properties, such as `IsA`, `IsPart`, `IsMember`, `locatedIn`, `IsDistinct`. So, each source ontology is visualized as a clustered graph, where the edges represent the selected relations among entities. A combination of object properties or relations is possible and interesting as well. In section 4 we discuss an example involving different types of relations. The Foam algorithm [7] is used to get preliminary alignment results which are visualized in AlViz.

Figure 1 depicts AlViz showing the alignment of two ontologies about tourism with a high degree of detail and one specific entity marked with a blue shadow in all four views. In particular, the concept `Ereignis` (event) is highlighted in both ontologies. Clustering the nodes would emphasize the overall structure of the graph. The colors of the nodes and the dots next to the entities' names represent the degree of similarity or type of association. We distinguish six categories of association between entities:

'equal', 'syntactically equal', 'similar', 'broader than', 'narrower than', and 'different'. Based on the results of the mapping algorithm, we generate the input file for AlViz.

By default an entity of one ontology equal to an entity in the other ontology is colored red; a syntactically equal entity is colored orange; an entity broader than an entity in the other ontology appears blue; a narrower entity is colored violet, a similar entity is colored green; and finally, an entity different from all entities in the other ontology is colored yellow. In the graphs the clusters of nodes inherit the colors of the underlying nodes in accordance to the selected comparison strategy. A variety of interaction functionality is implemented in AlViz in order to conveniently support the exploration and assessment tasks [1].

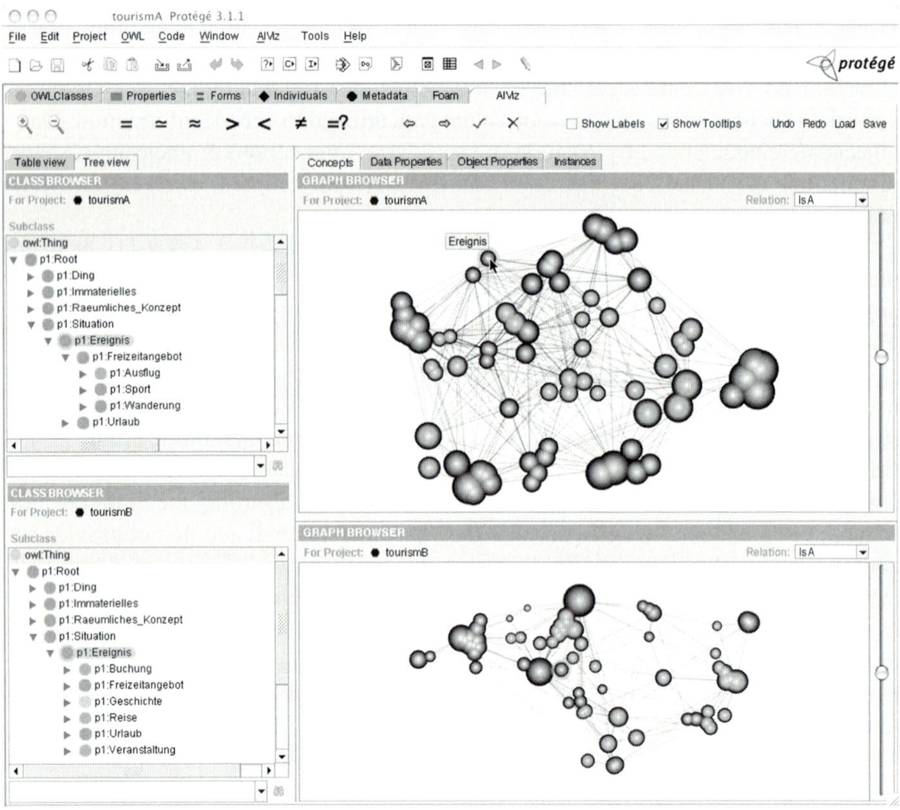

Fig. 1. AlViz [1]: the four views of the tool visualize two ontologies named tourismA and tourismB. The nodes of the graphs and dots next to the list entries represent the similarity of the ontologies by color. The size of the nodes results from the number of clustered concepts. The graphs show the `IsA` relationship among the concepts. Green indicates similar concepts available in both ontologies, whereas red nodes represent equal concepts. The sliders to the right adjust the level of clustering. One concept (called 'Ereignis') is highlighted in all views and shown with the tooltip text in tourismA.

For the calculation of the alignment values we have extended the general alignment process of Foam [8] to include transformation, evaluation and visualization of the output results. We claim that it is necessary to transform the output from lists of entity pairs into a format that can be understood and evaluated by domain experts. Furthermore, a lot of information regarding how entities are related between ontologies is hidden when just examining lists of entity pairs. The extended alignment process is described in more detail in the next section.

The development of AlViz is an ongoing process and first results of our evaluation indicate that the tool is suitable for ontologies up to 1000 entities. With the small world graph approach it shares its good functionality of making semi-structured data accessible on different levels of details. This feature helps the user when aligning ontologies visually.

3 Ontology Alignment

The basic elements of ontologies are concepts, properties, and instances. They describe (by necessary conditions) or define (by necessary and sufficient conditions) a knowledge domain for a certain purpose. Ontology alignment means to analyze these elements or entities in order to identify overlapping parts in two ontologies.

According to Ehrig [8] entities are the same, if their features are the same or similar enough. In ontologies features represent a certain meaning of an entity. Beneath we list some examples of such features. If an alignment algorithm determines low similarity among entities of two ontologies, this may not give evidence for an alignment. But if high similarity is found we may have strong evidence. In order to estimate the similarity we need to consider the features of the neighborhood because entities may be the same if their neighborhood is the same or similar enough. That means entities are also defined by their position in the ontological structure. Structure similarity is expressed through the similarity of the other entities in the structure.

In a first round of the alignment algorithm only basic comparison methods such as string similarity are applied or even manually pre-defined alignments are used. In further rounds already computed pairs are the basis to measure structural similarity. However, not every type of similarity is of equal importance. Therefore, we need to judge and emphasize the relevant similarities. The alignment process is executed iteratively and it stops either after a fixed number of iterations, a specific time, or the number of newly identified alignments is below a certain threshold defined beforehand. Like [8] we used ten iterations which is efficient and appropriate in practice. Thereafter, in the second part human interaction is involved. Here, the results are visualized and presented to the user. She or he investigates and perhaps corrects the alignments in an iterative manner as well. This may be followed by an automatic re-calculation of the similarities going back to the first part. In this case the manually changed alignments are additional input for the algorithm such as the pre-defined entity pairs. The user decides whether or how often this re-calculation is done. In the following we describe such a general alignment process (based on the work of [8]). We extended this process to include the human-mediated alignment steps.

1. Alignment algorithm
   (a) Feature engineering and compatibility: Determining the input ontologies for alignment. When a choice of ontologies can be made, the user may assess the ontologies for compatibility with respect to the scenario and complexity of the ontologies, and also validate the ontologies prior to alignment. Additionally, in this step we extract characteristics of both ontologies, i.e., the features of their ontological entities. Such features are for instance ontology language primitives (e.g., range and domain definitions for properties), identifiers (e.g., unified resource identifiers - URIs), derived features exploiting the inheritance structure, and external features (e.g., additional text files describing the instance data). So, when measuring similarity this features will allow for comparison of entities.
   (b) Search step selection: Two initial entities from the two ontologies are chosen for comparison. Every entity pair may be compared or only an efficient subset may be chosen using heuristics to lower the number of mappings [8]. We can reduce the number of candidates by comparing entities of the same type only.
   (c) Similarity assessment: Based on the features of the selected candidate pairs we do the similarity computation. As stated in [8] comparing two ontology entities goes far beyond the syntax level. It considers their relation to the real world entities they are representing (their meaning), as well as their purpose in the real world (their usage). Therefore, a semiotic view on ontologies is used and the similarity measures are classified into three layers: data-, ontology-, and context layer. On the first layer data types such as integers, strings etc. are compared by operations like relative distance and edit distance. Second, on the ontology layer we measure the similarity of ontological objects such as comparing the domains and ranges of two properties. Finally, we consider the context: for example two books may be similar if their authors have many coauthored publications. For all three layers we take domain knowledge into account, e.g. domain specific vocabularies like Dublin Core in the bibliographic domain.
   (d) Similarity aggregation: The individual measures from the previous step are input to the similarity aggregation. They are weighted and combined using a sigmoid function. This emphasizes high individual similarities and de-emphasizes low individual similarities.
   (e) Determination: The aggregated numbers, a threshold and an interpretation strategy are used to generate the actual alignments. Borderline alignments are marked as questionable. The threshold is used to determine whether similarity is high enough to align the candidate pair. Everything above is aligned, lower values indicate different entities.
2. Human-mediated alignment
   (a) Visualization: A visualization tool gets the input from the algorithm and renders the preliminary alignments. Each ontology is shown as a detail and overview visualization capable of displaying a high number of items.
   (b) Clustering and navigation: The user changes the level of detail according to her or his needs. For validation the user selects individual items which are connected in the different views by linking and brushing.

(c) Competency questions: To check whether specific alignments make sense the user collects competency questions which should be answered by the aligned ontologies. Moreover, results of earlier competency questions are reviewed again.
(d) Validation: Appropriate changes may be made to the alignments based on the visualization analysis. The user may verify questionable alignments, either by accepting or rejecting the suggestions. Moreover, all alignments can be subject to changes if the user decides to assign other types of alignments.
(e) Ontology update: Ontology alignment may involve changes to one or both source ontologies. In order to get better targets for the similarity assessment the user can create additional entities. Before such ontology updates are done the user needs to consider the usage of the source ontologies carefully.
(f) Set similarity threshold: The algorithm's threshold used for identifying alignments is a crucial value. Sometimes it is necessary to adapt the threshold in order to get better alignment results. If the user finds many missing alignments and the preliminary results derived from the algorithm show a high amount of different entities this could indicate that the threshold is too high.

These phases are not applied in a linear process but rather the alignment results are re-fined when users iteratively work through the various thematic or hierarchical subareas of the ontologies on different levels of detail. The next section gives an example of the human-mediated alignment process.

4 Interpretation of Alignments

To ease the understanding of the process only parts of the two source ontologies are visualized. AlViz (compare Fig. 1) looks a bit different because it arranges the two ontologies horizontally and provides interaction functionality. However, to describe an example of a verification task practically, it is appropriate to reduce the complexity of the user interface. So, besides the data, only the clustering sliders are depicted. Figure 2(a) shows a subgraph of the first ontology called tourismA.

Sometimes it is not necessary to see the whole ontology - clustered or not. Moreover, too many visual objects not relevant for a specific tasks may impede exploration. Therefore, the user can focus on a certain entity, so that the graph only represents the related entities, emphasizing the context of this specific entity. This view includes all sub-entities (transitive relation) and directly related entities (non-transitive relation), supplemented with all relations and entities among them within a beforehand defined number of hops (relations). The second source ontology (tourismB) is visualized in Fig. 2(b).

In this particular example, the labeled nodes represent concepts, the edges represent three different types of relations, and the entity 'Urlaub' (vacation) is focussed. In tourismA the depicted relations are: `IsA`, `HasPeriod` (hatReisedauer), `GoesTo-Region` (hatZiel), `UsesVehicle` (hatReisemittel). The `IsA` paths are shorter than the other because we gave these edges a higher weight. In order to distinguish different types of relations such as functional, transitive, or non-transitive we apply different weights, which can be modified by the user according to the exploration needs.

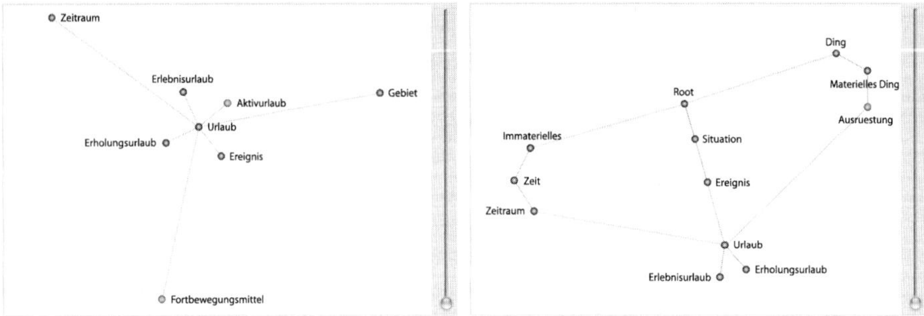

(a) Subgraph of ontology tourismA (b) Subgraph of ontology tourismB

Fig. 2. Small world graph visualizations of two ontologies in the tourism domain: the focus of the graph is on the entity 'Urlaub' (vacation) showing all related concepts for both ontologies. Labeling is activated, the colors represent the type of alignment based on the results of the alignment algorithm: orange means, there is at least one syntactically equal entity in the other ontology; green indicates similarity with at least one entity of the other ontology based on the neighborhood. In this view the nodes are not clustered meaning each node of the graph represents one entity.

In a simpler version of graph visualization the user may select one type of relation only. However, showing all relations of the central entity at once gives an impression of its context. On the right hand side the subgraph of tourismB consists of the relations: `IsA`, `HasAPeriod` (hatEineDauer), `EquipmentNeeded` (manBenoetigtAusruestung), `HasAGoal` (hatEinZiel). Whereas tourismA has three subclasses of the central concept, tourismB has only two.

Although the entities `Erholungsurlaub` (recreation) and `Erlebnisurlaub` (adventure) are identified as syntactically equal in both ontologies, the third called `Aktivurlaub` (active vacation) in tourismA with no directly corresponding entity in tourismB, was found to be similar to another entity by the alignment algorithm. By highlighting this concept we perceive the associated concept in the other ontology. It is `Erlebnisurlaub`. Based on the similar object properties which are inherited from the superclass (`Urlaub`), on the one hand, and a smaller distance (in terms of characters to be changed) between `Aktivurlaub` and `Erlebnisurlaub` than between `Aktivurlaub` and `Erholungsurlaub`, on the other hand, the alignment algorithm calculates the association which says `Aktivurlaub` and `Erlebnisurlaub` are similar. Similar concepts are colored green, whereas syntactically equal concepts are of orange color.

Obviously, in this subgraph the algorithm found many syntactically equal concepts but just a few similar concepts. For the moment we ignore the other green nodes and investigate the concept of `Aktivurlaub` only. The graphs show that the neighborhood of `Aktivurlaub` in tourismA and of `Erlebnisurlaub` in tourismB are similar which strengthens the claim that both concepts are closely associated. But `Erlebnisurlaub` in tourismA is also related to `Erlebnisurlaub` in tourismB: both concepts are syntactically equal. Now the user needs to decide whether changes of one or both ontologies are required and confirm or reject the alignment of type 'similar' between `Aktivurlaub` in tourismA and `Erlebnisurlaub` in tourismB.

Although our application example is static, the process of exploration is a highly interactive process. In terms of our alignment framework this example involves steps of clustering/navigation and validation. In practice the exploration of alignments is more complex as described in the previous section. However, to give an idea how interpretation of ontology alignments could be done, this simplified example is complex enough.

5 Evaluation Methodology and Study Design

The evaluation of AlViz should on the one hand yield information on its quality. But the proposed methods of evaluation should on the other hand give indications on the question which specific features of the visualization supports the users in the ontology alignment process. The use of diverse evaluation methods enables different views on the technology under investigation. We will use heuristic evaluation, thinking aloud, observation, software logs, interviews, focus groups, and benchmarks.

We evaluate visual ontology alignment with AlViz in three stages: (1) Participative software development, (2) usability evaluation, and (3) utility study.

The first stage is a formative evaluation done together with two ontology engineers who are involved in the development of the software. In general, the development of new methods and their implementation is a time-consuming task and involves different users having different types of models of understanding. Norman [9] pointed out that the mental models programmers have about software systems can differ fundamentally from the models the users have. Making these different models explicit can help to solve problems in software development projects at an early stage. In the AlViz project potential users (two ontology engineers) are integrated into the development process to a high degree. This enabled us to get informal feedback about usability and utility in an early stage.

Second, usability evaluation is done in order to identify problems and enhance the user interface of the software. We use Heuristic Evaluation [10]. It is well known usability evaluation method. A small group of evaluators (usually experts in the field of usability) checks the conformity of a given software with 10 usability principles. These principles are of a very general nature and therefore called "heuristics".

In the third stage we adopt a bundle of methods capable of evaluating the utility of visual ontology alignment software. Two ontology engineers (domain experts) and ten computer science students (semi-experts who attended two ontology engineering classes with excellent grades) do the evaluation. First, about one week before we conduct the evaluation, the subjects receive an introduction to the software tool using two simple ontologies to get a first impression. They get information about the functionality of the visualization tool, the data, and the alignment results derived from the algorithm. This is done with all subjects at the same time and takes around 30 minutes.

For the evaluation itself we include log functionality in AlViz in order to monitor the user activities. Only simple activities are logged: time stamps, view numbers, entity names, clustering levels, and validation actions. The advantage is that software logging does not interrupt the user's activities because it is not visible for the user. But it has shortcomings and therefore we need additional methods. For the individual evaluation these are: thinking-aloud, observation, and interviews.

One part of the utility evaluation is individual testing of AlViz for 60 minutes. Subject's overall goal in ontology alignment is bringing the source ontologies into mutual agreement and getting a maximum of correct alignments. Additionally, subjects are asked to report competency questions which arise in the process of aligning ontologies. A tutor and an observer attend the evaluation of each subject. The task of the tutor is to explain the evaluation procedure and to help if severe problems occur, which would make further evaluation impossible. The observer takes notes about important actions of the user, such as change of alignments or threshold, movements to other tabs, usage of linking and brushing and labeling, selection of views, execution of the alignment algorithm, etc. In addition, the subject is asked to describe her or his thoughts while using the tool. This method is called thinking-aloud (cf. [11]). The subject's voice is recorded and this gives more details about the intention of the subject. And finally, subjects report their competency questions in written form. This is necessary because subjects need to see and review their competency questions while doing the evaluation.

Immediately after the individual evaluation the observer changes the role and interviews the subject for around 10 minutes. The questions are: Did the subject have enough time to align the ontologies? Did she or he think the tool is appropriate for ontology alignment? How confident does the subject feel about her or his performance? Did the subject thinks she or he understood the knowledge represented by both source ontologies and the alignments?

After completing all individual evaluation procedures, the second part of the utility evaluation is done with the group of all subjects using the focus group method. Focus groups [12] often give reasons and additional subjective opinions. This also ensure correct interpretation of the results of other the evaluation methods. One major strength of this method lies in the interaction of participants. This can lead to valuable information on the discussed subject none of the participants would have stated in individual interviews. Once again a moderator and an observer are present. The moderator monitors the time and topic constraints and ensures equal chances of participation for all subjects. In a focus group setting the subjects discuss the strengths and shortcomings of the alignment tool. Afterwards, they get a list of all competency questions identified previously. They are asked to discuss and classify the questions in respect to their significance for representing the aligned ontologies. The following categories are given by the moderator: Crucial- , relevant- and further competency questions. Then the moderator asks each subject what helped and what hindered the finding of their competency questions during the individual evaluation. Finally, the subjects get a short questionnaire to give feedback about the evaluation. In the analysis of the evaluation we use the crucial competency questions as benchmarks for evaluating the alignment results of the subjects.

6 Conclusion

The tool implementation and the user study is an ongoing process. So far we have collected the data of 12 subjects and are currently analyzing the data. The evaluation methods proved viable even though our study design is challenging.

In particular, the competency questions reflect the users' understanding of the knowledge domains of both ontologies and the alignments. They allow for adequate handling of an open knowledge space. The benchmark evaluation of the alignment process results with the outcomes of the focus groups (rated competency questions) will give information on the quality of different problem-solving strategies. A combined interpretation of all the results will enable for a profound assessment of AlViz and furthermore provide knowledge on the ontology alignment process in general.

Acknowledgments. We are grateful to Markus Rester for his interesting inputs to our evaluation discussion. Furthermore, we thank Slaven Banovic for his help with the implementation of the software.

References

1. Lanzenberger, M., Sampson, J.: Alviz - a tool for visual ontology alignment. In: Society, I.C.S. (ed.) Proceedings of the IV06, 10th International Conference on Information Visualization, London, UK (July 2006)
2. Tufte, E.: The Visual Display of Quantitative Information. 2 edn. Graphics Press (2001)
3. Card, S., Mackinlay, J., Shneiderman, B. (eds.): Readings in information visualization. Morgan Kauffman, San Francisco (1999)
4. Stanford Medical Informatics: protégé-owl. Online `http://protege.stanford.edu/overview/protege-owl.html`
5. Sun Microsystems: Class J-Tree. Online `http://java.sun.com/j2se/1.5.0/ docs/api/javax/swing/JTree.html`
6. Van Ham, F., van Wijk, J.: Interactive visualization of small world graphs. In: Proceedings of the IEEE Symposium on Information Visualization 2004, pp. 199–206, CSPress, (2004)
7. Ehrig, M., Sure, Y.: Ontology mapping - an integrated approach. In: Bussler, C., Davis, J., Fensel, D., Studer, R. (eds.) Proceedings of the First European Semantic Web Symposium (2004)
8. Ehrig, M.: Ontology Alignment - Bridging the Semantic Gap. Springer, New York (2007)
9. Norman, D.: Cognitive engineering. In: Norman, D., Draper, S. (eds.) User Centered System Design, Lawrence Erlbaum, Hillsdale, N.J (1986)
10. Nielsen, J.: 2. In: Heuristic Evaluation, pp. 25–62. John Wiley & Sons, New York (1994)
11. Boren, T.M., Ramey, J.: Thinking aloud: Reconciling theory and practice. IEEE Transactions on Professional Communication 43(3), 261–278 (2000)
12. Kuniavsky, M.: User Experience: A Practitioner's Guide for User Research. Morgan Kaufmann, San Francisco (2003)

Towards Guidelines on Educational Podcasting Quality: Problems Arising from a Real World Experience

Marco Lazzari and Alberto Betella

University of Bergamo,
Faculty of Educational Sciences,
Piazzale Sant'Agostino, 2 – I-24129 Bergamo Italy
marco.lazzari@unibg.it

Abstract. This paper presents an experience of educational podcasting set up at the University of Bergamo (Italy), and derives from that experience some remarks upon the quality of podcasting services, in order to promote the definition of guidelines on podcasting quality. We discuss three main attributes of a podcasting environment: quality of the production environment (recording and editing), quality of the product (content and communication style), quality of the distribution environment (paratext and management).

Keywords: podcast, educational podcasting, quality of podcasting, distance education, distance learning, mobile learning, collaborative learning, computer science education, experimentation, open source, freeware/shareware.

1 Introduction

This paper presents an experience of educational podcasting set up at the university of Bergamo (Italy), and derives from that experience some remarks upon the quality of the podcasting services, in order to promote the definition of guidelines on podcasting quality.

A short description of our experience shall allow us to introduce some of the most critical issues that should be taken into account by podcasters.

Our podcasting service (Pluriversiradio, http://www.pluriversiradio.it), was implemented by developing Podcast Generator, an open source library for building and managing podcasting services. Besides the University of Bergamo, our software library is currently used by at least five hundred podcasting services in Italy and abroad.

Pluriversiradio has been used during the second semester of the academic year 2005-2006 for courses on multimedia communication and human computer interaction at the University of Bergamo; after that first experiments, other colleagues began recording podcasts for their students.

The development and management of an educational podcasting service led us to face the problem of the quality of the podcasts and of the service itself. On the ground of our experience we try to highlight some guidelines that could be used for developing quality models for (educational) podcasting services.

M.J. Smith, G. Salvendy (Eds.): Human Interface, Part II, HCII 2007, LNCS 4558, pp. 404–412, 2007.

2 The Use of Podcasting in Education

The increasing popularity of mobile media players has made podcasting an interesting and attractive way for distributing educational materials: the use of podcasting is spreading throughout the education sector and several universities have set up podcasting services (among the first ones Michigan, Duke, and Stanford), but the literature is still poor of examples, case studies and evaluations [11, 16, 20].

Nevertheless, despite the current lack of quantitative data about the effects of podcasting on education, many educators consider podcasting as an exciting learning paradigm of impressive pedagogical potential [2, 4, 7]: limited technical skills and efforts are enough to produce materials, such as course lectures, interviews, workshops reports, which can be used to meet individual's learning or teaching needs [10, 17]; students can be involved in producing, editing and delivering podcasts by themselves as a part of their assignments, developing in such a way collaborative and social networking activities and stimulating more effective learning processes [3, 5, 8, 9, 13]. Moreover, learning through listening is greatly appreciated by those students whose learning style is mainly auditory ([14]), while visual learners benefit from seeing videos from which they can catch teacher's expressions and body language.

On the ground of these positive feelings, we started in 2005 a project aimed at developing a podcasting service to integrate the elearning facilities of the University of Bergamo.

Since 1999 the University of Bergamo has enabled its students to access learning facilities through several elearning services: an elearning platform for asynchronous activities, tools for synchronous lessons, web procedures for distributing materials and managing all the bureaucratic aspects of the academic life.

Such efforts are appreciated by the students for two main reasons:

1. first of all, as rather usual in Italy, there is a high number of part-time students, who are very keen on any distance learning facility, which enables them not to commute to the university. Moreover, students declare to appreciate the availability of online tests and assignments, the ability to interact with lecturers, tutors and other students via online discussion facilities and to actively participate by posting questions and replying to discussion threads, as well as reading or downloading files [12].
2. on the other hand, the number of the students enrolled at the University of Bergamo doubled over the last five years, and this enormous increase induced some structural problems, which could be more easily faced by means of the elearning services: for instance, the Linguistic Laboratory has been virtually extended by developing on the elearning platform a large set of lessons incorporating the digital version of the audio and video materials already available at the laboratory (several hundreds of lessons for 24 different courses on English, French, German, Russian and Spanish, as well as Italian for foreign students, are currently available on the elearning site of the Faculty of Foreign Languages and Literatures).

As a natural evolution of our work on audio files, we decided to try the podcasting as an educational tool. Therefore, we started a project with the aim of developing an open source, free and sharable software library to manage a podcasting service, which could be freely downloaded and easily installed by other potential podcasters; and the release of an open podcasting environment, freely accessible by any Internet user.

As a result, in 2005 we designed and implemented Podcast Generator (PG), an open source library for building and managing podcasting services (http://podcastgen.sourceforge.net); using PG we built Pluriversiradio, a podcasting service freely accessible via web (http://www.pluriversiradio.it) or through a feed aggregator (http://www.pluriversiradio.it/feed.xml).

3 Podcast Generator and Pluriversiradio

Podcast Generator (PG) is a free PHP script released under GPL, that lets users upload media files via a web form and automatically creates rss 2.0 w3c-compliant podcast feeds, which are fully compatible with Juice and iTunes. Moreover, Podcast Generator acts as a content management system, able to manage a web site that hosts one or more podcast channels: when we say podcast channels we mean thematic subsets of the documents of a podcasting service, that is, for instance, all the recordings on a specific subject, or related to a single course.

Podcast Generator comprises three main modules:

1. a database to store multimedia files and the catalog of such recordings;
2. a procedure for loading podcasts into the database;
3. a CMS (Content Management System) for the management of a web site for the distribution of the multimedia materials (our recordings can be downloaded either through podcast catchers or via web).

Since the first beta release of PG, the database has been available in csv version, to be easily installed on any web platform, and subsequently developed in MySql; it can hold and host 21 different multimedia formats and manage for each file a record with several fields: title, author, short description (text), long description (text), image, categories, iTunes keywords, metadescriptions (based on the Dublin Core Metadata Element Set) to facilitate future "intelligent" search agents.

The loader provides a user friendly form to load the audio / video podcast and all the data which describe it. The loader validates all these data, load them into the database, load the podcast into the multimedia archive, and updates the RSS feed of the podcasting service.

The CMS module manages a dynamic web site that allows web surfers to access all the podcasts, either searching through the whole set of podcasts or by looking at the subsets based on the categories declared at loading time.

Using the first beta release of PG, in March 2006 we created Pluriversiradio, a web site to collect podcasts and to provide web surfers with the proper feeds for getting files from their podcast aggregators. This content of this site is strictly related to courses, seminars and conferences held at the University of Bergamo.

Despite born as a small size experimental project, Pluriversiradio was subsequently hosted by the Interdepartmental Center for elearning of our university; besides the University of Bergamo, our software library is currently used by at least 500 podcasting services in Italy and abroad (for instance, the university of Pisa and Bari in Italy, of Delaware in the U.S.A., of Aarhus in Denmark).

The quick diffusion of PG is due to its efficiency, ease of use and installation: the open source nature of the program allowed the cooperation of a set of kind users, who translated the installation procedure, originally written in English and Italian, into Chinese, Dutch, French, Japanese, and Portuguese; it will be soon available in Catalan, Danish, Galician, and Spanish. Moreover, the Dutch foundation Kennisnet Ict op School used PG for publishing a screencast series about podcasting with open source and promoted the use of PG in schools: for this purpose they have also developed a video podcast the explain the installation procedure of PG (http://files.ictopschool.net/podos/mp4/09.mp4).

Pluriversiradio has been used during the second semester of the academic year 2005-2006 for two undergraduate courses on multimedia communication and human computer interaction at the Faculty of Arts and Philosophy (currently Faculty of Educational Sciences) of the University of Bergamo; after that first experiment, other colleagues began recording podcasts for their students, whilst some lessons from Pluriversiradio have been used by colleagues of other universities (for instance, the university of Verona) or stored into their open archives (for instance, university of Perugia): this is possible because our recordings are freely accessible and distributed under a Creative Commons By-Nc-Nd license.

The second of the multimedia communication courses which used Pluriversiradio (Laboratory of multimedia communication) aimed at introducing students to the foundations of the human computer interaction, to basic elements of graphic interfaces and to the management of audio files and the use of podcasting.

Podcasting was used first of all by the lecturer for recording syntheses of the theoretical lessons and for distributing them to the students. As soon as the students became rather skilled with audio files management (using the free software Audacity for recording and editing sounds), we started an experiment with the full time students (eighteen students were involved): each of them had to create his/her own podcasts. The experiment was based on three assignments: two re-elaborations of pre-existing interviews and the development of a lesson about one of the themes of the theoretical course not yet dealt by the lecturer. For this third task students were given some constraints (duration, sampling rate, bit rate), but were free to choose their favorite format: some of them arranged a formal lesson, others (a group of three students) simulated a moderated debate, others a desperate phone call between friends the night before the exam. (more details on the experiment and its evaluation are provided in [15]).

In order to understand whether the use of podcasting affected learning, we exploited exam results for evaluating student performance, by comparing exam grades over the past three academic years, using overall grade average. The goal was to determine first of all if there were significant differences in student performance this year, and subsequently if there were significant differences between the students involved in the experimentation and the other ones.

At first sight we found that the results of the current year were better than the previous ones. Nevertheless, after a first round of data processing [1], we demonstrated that this increase is not necessarily explained by the innovation of the podcasting.

Therefore, we separated the grades of full time students, who took part to the experiment, and part time students: these figures showed that full time students usually perform better than part time ones, but also that whilst the latter provided this year results similar to those of the past years, the former have seen a considerable increase. In this case data processing confirmed that there is a statistically significant increase of the average grades.

These results seem to support two conflicting ideas:

1. the grades of the part time students say that the use of podcasting for distributing course materials did not affect the results: this might be interpreted as a failure of the podcasting itself, or simply suggest that part time students ignored those integrative documents;
2. on the other hand, the grades of the full time students suggest that their involvement in creating podcasted lessons enhanced their learning experience in a very effective manner; by figures and observation we can say that podcasting design, recording, and editing spurred the development of reflective learning skills, stimulated students to go deep into the questions they had to face, fostered collaborative behaviors. Similar considerations arose from the analysis of the student satisfaction surveys, which are part of the institutional audit process of the University of Bergamo, and from the qualitative analysis, based on the grounded theory [18], which was conducted on transcriptions of colloquia with students and on their open answers to the student satisfaction questionnaire. The results emphasize the impact of the creative use of podcasting on the perceived quality of the course, but also on the ability of the students to assess their own understanding of the topics of the course and to deepen their competence beyond the walls of the classroom.

4 The Quality of Podcasting

The development and management of an educational podcasting service led us to face the problem of the quality of the podcasts and of the service itself. Unfortunately, the literature provides little help on the subject: web pages, scholarly papers and books about podcasting quality are very few and they are mainly concerned with recommendations about the quality of the audio (or video) files and the relevance of the bandwidth [6], but a really applicable quality model does not exist.

We do believe that a quality model for podcasting services should address multiple issues and we wish also to underline that the aforementioned requirements, even if they are in some way “user-centered”, are rather general and do not take into account that different podcasts could have different audiences.

Therefore, the aim of this paper is to foster a discussion about the quality of the podcasting services: we are far to define our own quality model for podcasters, but on the ground of our experience we try to highlight some guidelines that could be used for developing quality models for (educational) podcasting services.

Our proposal is based on our experience of podcasters, on a survey of podcasting services over the Internet, on discussions with colleagues and other podcasters, on interviews and colloquia with our students aimed to understand the current use of the service to get a clear picture of users' needs, and also on feedbacks that we have got from Internet surfers, who visited our site and downloaded our podcasts directly from the site or by subscribing to the podcast service (we logged some thousands downloads for each of our podcasts).

We believe that the first step towards the development of a podcast of quality should be done by identifying three main components [19]: primary users, primary tasks, context of use. Therefore, we have to answer to some fundamentals questions: who are the main users and which are their characteristics? What are their main tasks? What is their computing / reproducing environment? Note that the context is not only defined by the technical facilities, but also by the fruition modality: listening to a podcast at home through the PC is different than listening to it via a mobile device while commuting from home to the university. Moreover, listening to an entertainment podcast is rather different than listening to an educational one.

With reference to the fundamental question about users, in our experience we have to consider that our typical audience is mainly composed by students belonging to three different categories:

1. central users: students who regularly attend the lessons and use the podcasting as a source of summaries or deepening;
2. proximal users: students who sporadically attend the lessons, and therefore exploit the podcasts for integrating their knowledge;
3. distal users: students who rarely attend the lessons, who exploit the podcasts as a source of distance learning.

Each category could benefit from different types of podcasts from the point of view of the contents and their organization. Therefore, metrics for evaluating the quality of an educational podcasting should be weighted on the main final user of the service.

Moreover, the academic environment suggests different ways to manage the podcasting, that can be used for recording lectures, but also workshops, conferences, discussions, interviews, and also for training students, who can produce podcasts by themselves as a part of their assignments. With reference to the lectures, some colleagues interviewed on this topic said that they prefer to record the whole classroom lesson, while others do not like such intrusive way to record their talks and prefer to set up syntheses of the lessons, which can be more effective and less annoying for the listeners.

On the ground of such experience we wish to address the attention to some attributes that should be taken into account for developing high quality podcasting services - we have identified three main attributes, each of them has two main subcategories, each subcategory has its own attributes that should be taken into account:

1. Quality of the production environment: recording and editing.
2. Quality of the product: content and communication style.
3. Quality of the distribution environment: paratext and management.

For the sake of simplicity, in the following we shall deal with audio podcasts, but the same guidelines could be considered as a starting point for video podcasts as well.

With reference to the quality of the production environment, web sites on podcasting are rich of suggestions about the quality of recording, based on the type of microphones, recording environments, set-up phase, sampling rate and bit rate. A quality model for podcasting environments should provide a metric to evaluate all of these attributes, but we do believe that they should be weighted according to the type of podcasting service: whilst high sampling rate and bit rate, for instance, are very important for an entertainment podcasting that broadcasts music, the relevance of these attributes for educational podcasting, according to our experience, to literature and to colleagues, is rather low and this type of podcasting does not require very effective devices (we use 32kbps and 11,025/22,050 kHz).

When we say "quality of recording" we address the activities required to produce the file to be podcasted, that is filtering and mixing voices and choosing the proper file formats and compression levels. For our experiment we simply used mp3 files, but discovered after the first lectures that a musical background is considered by students an agreeable way to fill gaps between sentences or parts of the lesson.

The quality of the product is, of course, a key attribute: the content is required to be complete, credible, sound; relevance, coherence, completeness, length (time: we usually limit our recordings to 15 minutes), dimension (memory) are the attributes to be evaluated.

But podcasters must take into account the importance of the quality of the communication style: format, style, enunciation, evidence are the keywords to be considered. Educators must remember, as stated above, that the choice of the right format of the podcast is not optional: complete lectures can be perfect for distance learners, but could be rather annoying; nevertheless, they can be a good chance to catch extemporaneous speech that could happen to arise during the lecture. The podcasts produced by our student show that a non-academic format can be useful as well; and that the involvement of students in producing podcasts can be a way to improve learning processes.

With reference to the quality of the distribution environment, we do believe that podcasters should avoid delivering all their materials to each user independently by user's needs. Unfortunately, a survey of some dozens of professional podcasters showed that this is not the case on the majority of them: even pioneer podcasting services on iTunes U provide many different thematic feeds, but they appear too general (Faculty lectures: audio; Faculty lectures: video) to be really user centered.

We suggest that a podcasting service should at least provide multiple feeds for multiple channels. This is part of the attention that should be paid to the paratext, together with the compliancy with *de iure* or *de facto* standards (w3c, iTunes, Juice), and the meta-tagging for allowing current or future search engines to properly index the podcasts. At the present time we are working to design and develop an ontology of podcast objects, in order to allow enriching each podcast with meta-descriptions or tags suitable to be shared with software agents that could extend the poor access facilities currently provided by the aggregators; each podcast could be properly described and therefore would be suitable to be matched by communication agents instructed on the ground of detailed user's requirements. This would require more sophisticated agents than common feed aggregators, such as plugins of the current media players, able to build playlists (for instance all the podcasts of a specific author on a specific theme) on the ground of semantic descriptions based on shared ontologies; otherwise, this matching could be performed on the ground of preferences expressed by the user during a registration phase on the web site of the podcasting service – this solution would be more convenient for those services which apply access restrictions or fees. Both solutions aim at reducing the number of "useless" downloads and saving bandwidth; in some sense, the imperative is that users should be able to download exactly *quod erat downloadandum*, that is *what they needed* – therefore we call this approach *quodcasting*.

Finally, the quality of the management is related to server's functionality, availability and bandwidth, time to answer, and bug fixing – in this case the management of a podcasting service does not significantly differ from that of a conventional web site. An important issue is that of the storage platform: podcasters might decide to use their own storage area or upload the podcasts to a public database; they can accept registered users only or allow anonymous users to download files; they can set up a pure podcast service accessible via feed or provide their users with a mirror web site for direct download: each of these options should be chosen according to the management politics of the podcasting service.

5 Conclusions

We have presented an open source software library (Podcast Generator) that we developed and used to set up an academic podcasting service (Pluriversiradio). We have presented an experience of educational podcasting that involved students of human computer interaction both as podcasting users and producers. From our experience, literature, surveys, discussions with colleagues and experts, and feedback from users we have derived some guidelines that we propose as a starting point to develop quality models for podcasting services and portals.

Acknowledgments. Many thanks to Sara Piazzalunga, Daniela Iovino, Claudia Roffeni Tiraferri, Federica Baroni, Andrea Redolfi, Paola Bolognini, Stefania Fumagalli, Massimo Campana, Sophie Cenati, Francesca Bianchi, Emanuela Buizza, Caterina Ceroni, Cristiana Colombi, Miriam Ignoto, Andrea Mazza, Mattia Ferrari, Barbara Satta, Marco Mingone, Elisabetta Zuccotti, Francesco Pirelli and Giovanni Fiorentini for their contribution to the first steps of Pluriversiradio.

References

1. Bohrnstedt, G.W., Knoke, D.: Statistics for Social Data Analysis. F.E. Peacock (1994)
2. Brittain, S., Glowacki, P., Van Ittersum, J., Johnson, L.: Podcasting Lectures. EDUCAUSE Quarterly 3, 24–31 (2006)
3. Bruner, J.: Acts of Meaning. Harvard University Press (1990)
4. Cambell, G.: There's Something in the Air: Podcasting in Education. EDUCAUSE Review, 40(6) (2005)
5. Card, D., Polin, L., Parra, J., Rhoads, J.B., Sartori, T.: Can You Hear Me Now: the Return of Voice to Distance Learning. In: Proceedings of the 5th IASTED International Conference on Web-based Education, Puerto Vallarta, Mexico, 2006, ACTA Press (2006)
6. Caylor, J.: 10 Podcasting Quality Tips (2005) http://www.productiondisc.com/onedollar/2005/11/10-podcasting-quality-tips.html
7. Cebeci, Z., Tekdal, M.: Using Podcasts as Audio Learning Objects. Interdisciplinary Journal of Knowledge and Learning Objects, vol. 2 (2006)

8. Constant, D., Sproull, L., Kiesler, S.: The Kindness of Strangers: The Usefulness of Electronic Weak Ties for Technical Advice. Organization Science, vol. 7 (1996)
9. Crawford, C., Smith, R.A., Smith, M.S.: Podcasting in the Learning Environment: from Podcasts for the Learning Community, towards the Integration of Podcasts within the Elementary Learning Environment. In: Proceedings of Society for Information Technology and Teacher Education International Conference 2006 (Chesapeake, VA, 2006), AACE (2006)
10. Fok, A.W.P.: Peonto - Integration of Multiple Ontologies for Personalized Learning. In: Proceedings of the 5th IASTED International Conference on Web-based Education (Puerto Vallarta, Mexico, 2006) ACTA Press (2006)
11. Frydenberg, M.: Principles and Pedagogy: the Two P's of Podcasting in the Information Technology Classroom. In: Proceedings of the 23rd Annual Conference for Information Systems Educators ISECON 2006 (Dallas, TX 2006)
12. Gnudi, A., Lorenzi, A., Malvisi, L.: An Evaluation of an Online ICT Course: Student Perception, Support Issues and Lecturer Involvement. In: Proceedings of the 3rd EDEN Research Workshop (Oldenburg, Germany 2004)
13. Greer, J., McCalla, G., Collins, J., Kumar, V., Meagher, P., Vassileva, J.: Supporting Peer Help and Collaboration in Distributed Workplace Environments. International Journal of Artificial Intelligence in Education, vol. 9 (1998)
14. Honey, P., Mumford, A.: The Manual of Learning Styles. Peter Honey (1992)
15. Lazzari, M.: Podcasting in the Classroom: Involving Students in Creating Podcasted Lessons. In: Proceedings of HCI Educators 2007 - Creativity3: Experiencing to Educate and Design (Aveiro, Portugal 2007)
16. Malan, D.: Podcasting Computer Science E-1. In: Proceedings of the 38th ACM Technical Symposium on Computer Science Education SIGCSE '07 (Covington, KY 2007)
17. Ractham, P., Zhang, X.: Podcasting in Academia: a new Knowledge Management Paradigm within Academic Settings. In: Proceedings of the 2006 ACM SIGMIS CPR Conference on Computer Personnel Research (Claremont, CA 2006)
18. Strauss, A., Corbin, J.: Basics of Qualitative Research: Techniques and Procedures for Developing Grounded Theory. Sage, Thousand Oaks (1998)
19. Theofanos, M., Stanton, B., Bevan, N.: A Practical Guide to the CIF: Usability Measurements. Interactions, vol. XIII(6) (2006)
20. Wolff, T.: Podcasting made Simple. In: Proceedings of the 34th ACM SIGUCCS Conference on User Services SIGUCCS '06 (Edmonton, Canada 2006)

Context Modeling and Inference System for Heterogeneous Context Aware Service

Seungkeun Lee

INRIA Rhône-Alpes
Montbonnot Saint-Martin, France
Seung-Keun.Lee@inrialpes.fr

Abstract. Context can be utilized as an effective source of information for supporting the user system interface under the ubiquitous environments. The context awareness function forms the basis of the studies regarding the ubiquitous computing environment for creating numerous smart spaces, and a context awareness service is usually based on the context awareness system or middleware. However, conventional context modeling approaches based on ontology bear several problems. First, a context awareness service must share the context information with other context awareness services in the designing phase. Second problem is a context uncertainty, which can arise in the process of deducing the data acquired from the sensor into the context information based on the dynamic modification of the context ontology, must be resolved. This study proposes the context management system based on the dynamic context ontology management method, which involves the hierarchical context management method using common context and the paper proposes a method for resolving the context uncertainty problem of interpreting the data acquired from the sensors as two or more types of context information.

Keywords: Context Aware, Context Modeling, Ubiquitous Computing.

1 Introduction

The context awareness function forms the basis of the studies regarding the ubiquitous computing environment for creating numerous smart spaces, and there are several studies under way in universities and research institutes[1,2,3,4]. A context awareness service is usually based on the context awareness system or middleware. A con-text awareness system has a structure for generating the context information using the data acquired from the sensors and forwarding the data to the context awareness service. The context awareness services provide specific services to the user according to the context information received from the context awareness system. Therefore, the context awareness service must be able to clearly understand the significance of the context information conveyed from the context awareness system. There have been various studies on the techniques of context modeling, and the ontology modeling approach is receiving much attention, thanks to its capability to enable significance exchange among systems.

M.J. Smith, G. Salvendy (Eds.): Human Interface, Part II, HCII 2007, LNCS 4558, pp. 413–422, 2007.

However, conventional context modeling approaches based on ontology bear several problems. First, a context awareness service must share the context ontology with the context awareness system in the designing phase[5]. In turn, when a context awareness service is dynamically added to or deleted from the system, the new context awareness service cannot share the context ontology with the system under the circumstances where other context awareness services are not affected. In addition, the problem of context uncertainty, which can arise in the process of deducing the data acquired from the sensor into the context information based on the dynamic modification of the context ontology, must be resolved[6,7,8].

This study proposes the context awareness middleware based on the hierarchical con-text ontology management method, which involves designing of the common ontology hierarchy for the context ontology to be dynamically shared between the context awareness service and the middleware. The context awareness service dynamically defines the context ontology that it needs using the common ontology, which is forwarded to the middleware. The middleware integrates the context ontology received from the context awareness service with the domain ontology that is managed by the middleware for management. The context awareness middleware does not transfer all the deduced context information to all context awareness services, and only the context information that the context awareness services take interests is forwarded. Furthermore, the paper proposes a method for resolving the context uncertainty problem of interpreting the data acquired from the sensors as two or more types of context information. The context awareness middleware is designed using the proposed hierarchical context ontology management approach. The designed middleware uses the data acquired from the sensors to deduce the context information, which is delivered to the requesting context awareness service using the event delivery method. Compared to the existing studies, the designed context awareness system allows dynamic management of the context ontology according to service addition and deletion, and also has the advantage of resolving the context uncertainty issue.

2 Hierarchical Context Management Model

A context awareness service must be able to exchange context information based on identical understanding of the content among the user, devices and services. This chapter presents a method for dynamically modifying the context information required by the context awareness services operated from the context awareness middleware, as well as the hierarchical context ontology management and context uncertainty resolution methods in order to deliver the changes in the middleware so that the corresponding context information can be received from the middleware.

2.1 Hierarchical Context Ontology Structure

The context information managed by the middleware is constructed into the domain context information that is delivered to all services in the middleware and the individual context information defined for each service. All of the context information is defined by ontology. If a service is newly allocated in the middleware, the middleware must be able to integrate the existing domain context information and the individual context information separately defined in the service for operation. Here, the correlation between the two sets of context information must be guaranteed, for which purpose the common context information layer is located between the two context information layers. The common context information defines the basic elements required for the context awareness application as the Person, CompEntity, Location, Environment and Activity, and the domain context information and the individual context information are designed by inheriting the common context information. By providing correlation among sets of context information inherited from an identical parent class, the information is used to integrate the two sets of context information in the middleware. Fig 1 displays the hierarchical context ontology designed in this paper. Every type of ontology is inherited from the context class for creation. The common ontology comprises Person, CompEntity, Location and Environment and Activity Class inherited from the highest Context class, as well as Service, Device and Network inherited from CompEntity. Context may include multiple attributes to describe the corresponding situation, as well as other situations as properties. For example, in home network ontology, if there is a Room context inherited from the Location class and a Temperature context inherited from the Environment class, the Temperature context can be used as the property of the Room context.

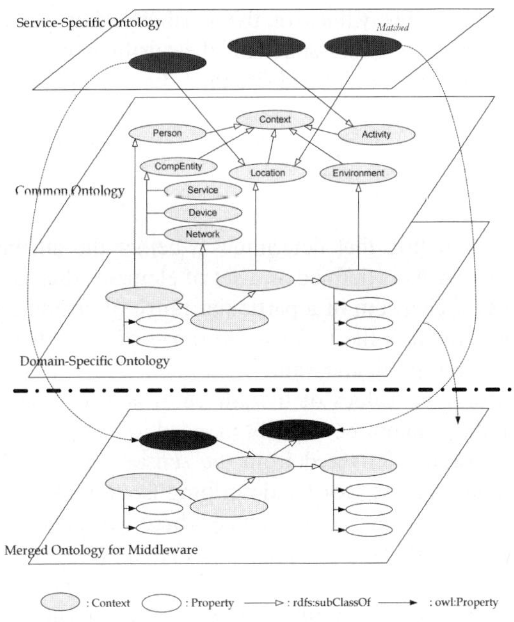

Fig. 1. Hierarchical Context Management bundle

The domain ontology author inherits the common ontology to define the context in-formation that is identically analyzed in all services of the middleware. Each context awareness service developer inherits the common ontology and defines the individual ontology that can be separately interpreted in each service. Here, the context awareness system assesses the correlation of the context ontology between the two ontology layers using the common ontology in the middle layer and manages it.

2.2 Uncertainty Problem Management

A context is determined according to whether various attributes collected from the sensors at a specific time are included in a particular realm. For example, if the user's body temperature received from the temperature sensor is from 36.021°C to 36.521°C, the user can be regarded as being in the "healthy state", and in the "ill state" otherwise. Hence, the system must have a method of analyzing the situation based on the data values received from the sensors. Eq 1 expresses the set of attributes delivered at a particular time*(t)* from the sensors linked to the middleware designed in this paper.

$$S(t) = \sum_{i=1}^{n} A_i(t) \tag{1}$$

In Eq 1, $A_i(t)$ indicates the attribute value at a particular time t. Context attributes are used in describing the elements that can be utilized in deducing the situation. Context attributes are linked to virtual or physical sensors. A context space is defined with the range of permissible values of the attributes that can induce a particular situation. A context space is expressed with the attribute values acquired from the sensors that correspond to the predefined situation. Eq 2 expresses the context space.

$$C_i = \sum_{i=1}^{n} S(t) \quad \mathrm{S}(t) = \{A|P(A)\} \tag{2}$$

P is the Predicate function that determines whether the attribute value from the sensor is permissible, and Ai is defined as a set of elements that satisfy the function P. Context(*C*) is expressed as a sum of a particular range of attribute values(*A*) received from the sensors. A_i, which is the permissible range of the values received from the sensors, comprises a set of particular values.

Although the permissible values of the attributes and the context information are linked through C(*t*), they cannot be directly mapped to a specific context information through the attribute values delivered from the sensors. This is due to the fact that there can be overlapping regions for C_A describing Context A and C_B for Context B. The following Fig 2 displays the circumstance where two contexts overlap. Since S_A is included in C_A in Fig 2, it can be mapped to Context A. Since S_B is not included in C_A or C_B, it cannot be mapped to any context. On the other hand, S_B is located in the overlapping region of C_A and C_B. In this case, since the attribute values recognized by the sensors are located in more than one context space, a filtering process is necessary to determine which context space it belongs to.

If the set of attribute values received from the sensors is included in various context spaces as in Eq 3, they are compared with the inclusive context spaces to deduce an adequate context. In order to do so, this paper defines a context space

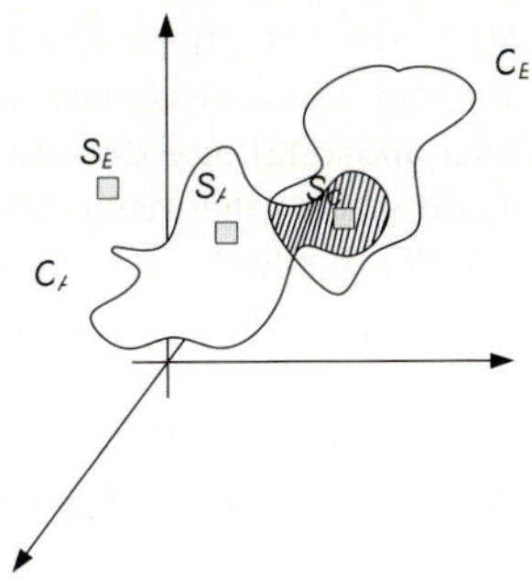

Fig. 2. The Overlap of Context Information

variation function to determine the difference among the attribute values and the context space. Using the space variation function, the variation values are mapped to the small context space. The context space variation function is defined as in Eq 3.

$$ContextSpaceDerivation = \sqrt{\sum (A_i - A_r)}$$

$$A_i \text{ : Attributes from Sensors, } A_r = \frac{MaxValue(A)+MinValue(A)}{2} \quad (3)$$

3 System Design

The context awareness middleware deduces the context information using the data received from the sensors and delivers the information to the desired context awareness service. In the development process of the context awareness service, the context awareness middleware collects, manages, combines the context information and for-wards it to the context awareness services.

In order to deduce the context creator for creating the basic context and the basic context into complex contexts, the context awareness middleware is constructed into the ontology deduction engine, the deduction engine interface for interaction, the context awareness service, the event broker for delivering events in the supply/register format, and the translator for analyzing the commands received from the context author and the context awareness service. The author defines the context ontology required for the domain using the common context ontology defined in Chapter 3. The defined ontology is constructed with OWL files, which are sent to the translator. The translator provides a method for registering the context information types that the context awareness service needs. The context ontology defined by the context author and the rules for deducing the complex context are delivered to the context deduction engine through the deduction engine interface, and the context space information is stored in the database of the basic context generator. The basic context generator creates an appropriate basic context using the data collected from the sensors as well as the context space information stored in the database and the context conflict processor. The basic context information is delivered to the event deliverer through the deduction engine interface, converted into a fact for complex context deduction, and forwarded to the deduction engine. The basic context delivered to the deduction engine goes through the consistency test and the complex context deduction process using the rules and ontology defined by the context author. The generated complex context is again sent to the event deliverer through the deduction engine interface and finally to the context awareness service that requires the corresponding context information. Fig 3 is the system configuration of the overall context manager.

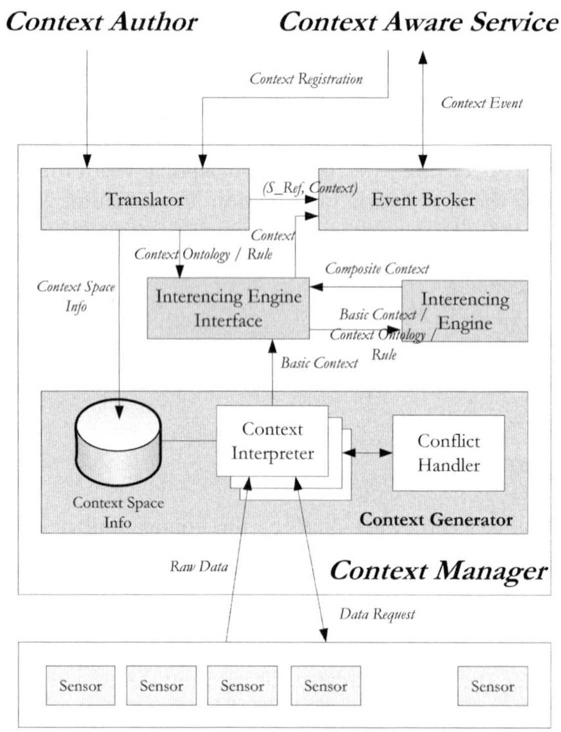

Fig. 3. An Overview of Context Awareness Middleware

The context awareness middleware was designed as a bundle on the OSGi framework foundation. Therefore, the middleware inherits the BundleActivator interface of the OSGi framework and uses the Jena.jar file for ontology deduction. The translator receives the domain context information and rules from the ontology author, and the readFile() method that allows reception of individual context information in a file format from the context awareness service, internally parses the OWL file, which is delivered to the ontology deduction interface and the translator. The basic context translator compares the information acquired from the sensors and the context space information to generate the basic context. If there is a conflict among the context spaces, an appropriate context space is sought using resolveConflict() of the conflict processor.

The context ontology information input by the middleware manager in the OWL file format and the basic context information generated by the context generator are transferred to Jena through the deduction engine interface. The deduction engine interface induces the complex context deduction using Jena, which is the ontology deduction engine. Here, insertContext() and removeContext() are for adding and deleting the basic context created by the basic context generator to and from the deduction engine in the Fact form. The complex context deduction rules defined by the middleware manager are managed using insertRule() and removeRule(). The deduction engine interface is constructed into six methods for adding and removing facts, rules and ontology. Each element receives an ID when being added to the deduction engine, and the ID is used to manage the corresponding element.

The basic context generator receives the data from the sensors and compares them with the context space information to deduce the appropriate basic context. The context information is given a context ID to be managed by the middleware, converted into a Fact through the deduction engine interface, and retransmitted to the deduction engine. The deduction engine checks consistency of the context information and generates complex context through ontology deduction and user rule-based deduction using the facts and the rules. The generated context information is delivered to the event broker, which forwards the information to the context awareness service that has registered the context information.

4 Experiment

This chapter explains the process of conducting tests by implementing a smart home network to evaluate the functions and performance of the context awareness middleware proposed in this paper. The server used for the home network service was IBM eServcer X206, 2.8GHz, 512MB RAM, and it was operated with Windows Server 2003 using the OSGi framework Knopflerfish 1.3.3 and HP Semantic web toolkit Jena2[9,10]. Fig 4 displays the home network environment implemented for the test.

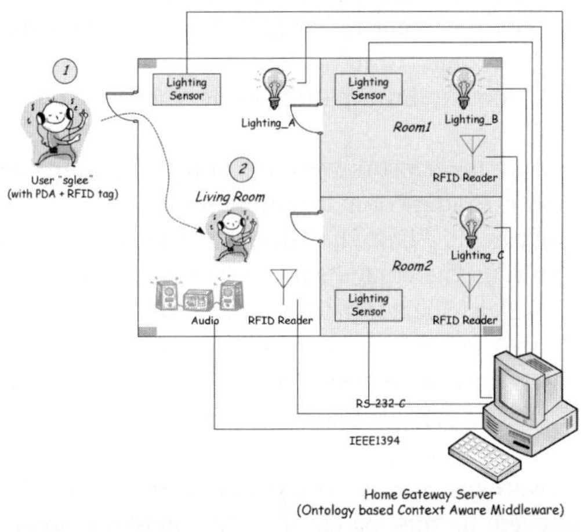

Fig. 4. Prototype of Smart HomeNetwork

The test scenario involved creating context information that is generated when a user comes into the house, and the information was delivered to the context awareness service. The light control service is a context awareness service that automatically turns on the light nearest to the user if the lighting is too dim. The light control service uses the information acquired by the illumination intensity sensor to recognize the brightness of each location. In order to assess the level of illumination in the house, light sensors are installed in Room1, Room2 and LivingRoom.

The sensor in each location detects the light intensity and expresses it in 10 bits, creating a value between 0 and 1023. The context is deduced to be "dim" if the value is below 512, and "bright" otherwise. The light control service defines the independent contexts of the "user location dim" and the "light must be turned on" as in [List 2].

List 2. The Rule for Context Inference

```
(?LightSensor locateIn ?place), (?LightSensorValue
bigger 512) -> (?place lightLevel "Bright")

(?LightSensor locateIn ?place), (?LightSensorValue
smaller 512) -> (?place lightLevel "Dim")

(?user locatedIn ?place), (?place lightLevel "Dim") ->
(?place needed "lighting")
```

When the user "sglee" moves from location 1 to 2 in Fig 4, the RFID reader 3 reads the user ID from the RFID tag attached to the user. The RFID reader delivers the data to the RFID management service, which parses the corresponding data to forward the user ID value to the middleware. Then the basic context generator creates the information ("sglee" locatedIn "LivingRoom"). The generated basic context information is delivered to the ontology deduction engine, as well as the light control context service.

The illumination intensity in each room and the living room is regularly checked using the sensors, and the values are delivered to the middleware. According to List 2, the middleware generates the context "bright" for a sensor value acquired in each location greater than 512 and "dim" otherwise. Fig 5 describes the process of deducing the complex context information using the data obtained from the light sensors.

The values acquired by the sensors were ("Room1" lightingValue 284), ("Room2" lightingValue 653) and ("LivingRoom" lightingValue 327), which created the basic contexts ("Room1" lightLevel "Dim"), ("Room2" lightLevel "Bright") and ("LivingRoom" lightLevel "Dim"). The contexts are delivered to the deduction engine, which used the knowledge ("sglee" locatedIn "LivingRoom"), (?user locatedIn ?place) and (?place lightLevel "Dim") -> (?place needed "lighting") to deduce the complex context information ("LivingRoomg" needed "lighting"). The deduced complex context is delivered to the light control service that has registered the context through the event broker.

Since the context awareness service can dynamically register the context ontology, the middleware proposed in this paper has the ontology integration overhead in addition to the time required for loading the ontology to the memory. As a result of measuring the overhead, the integration time was not a significant burden compared to the loading time as indicated in Fig 5. Moreover, the amount of increase in the integration time was not substantial compared to the amount of increase of the domain context ontology value and the developed context ontology value. Therefore, due to the time required for integrating the context information, it can be suggested that the middleware proposed in this study is not adequate for the real-time service platform, but useful for general application services.

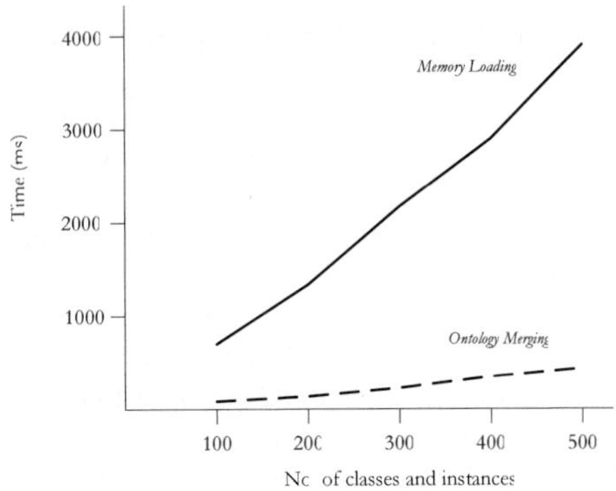

Fig. 5. Overhead of Merging Context Ontology

5 Conclusion

This paper proposed an ontology-based context awareness service management model with hierarchical context management. This hierarchical context management provides the resolution method for context uncertainty. And This study proposes the context management system based on the dynamic context ontology management method, which involves the hierarchical context management method using common context and the paper proposes a method for resolving the context uncertainty problem of interpreting the data acquired from the sensors as two or more types of context information. The context awareness service also can be registered to system dynamically. The designed middleware was implemented under the OSGi framework environment to connect with physical devices such as sensors and appliances. Finally, the home network context awareness service was implemented to validate the feasibility of the middleware designed and proposed in this paper. Functions and performance of the middleware were evaluated through the test, and it was confirmed that the overhead of the proposed hierarchical context ontology management model makes the middleware unfit for the hard real-time ubiquitous computing environment. However, it was determined that the model can be applied to the general ubiquitous computing environment. It can be concluded that the ontology-based context awareness middleware proposed in this paper can be applied in the service gateway for various ubiquitous environments such as the home network, telematics and smart office for providing context awareness services.

Acknowledgement. This work was supported by the Korea Research Foundation Grant funded by the Korean Government(MOEHRD)" (KRF-2006- D00163).

References

1. Lee, S., Lee, J.: Dynamic Context Aware System for Ubiquitous Computing Environment. In: Shi, Z.-Z., Sadananda, R. (eds.) PRIMA 2006. LNCS (LNAI), vol. 4088, Springer, Heidelberg (2006)
2. Loke, S.: Context Aware Artifacts: Two Development Approaches. IEEE Pervasive Computing (2006)
3. Hung, N.Q., Lee, S.Y., Hung, L.X.: A Middleware Framework for Context Acquisition in Ubiquitous Computing Systems. In: Proceedings of the Second International Conference on Computer Applications (2004)
4. Gu, T., Pung, H.K., Zhang, D.Q.: An Ontology-based Context Model in Intelligent Environments. In: Proceedings of Communication Networks and Distributed Systems Modeling and Simulation Conference, pp. 270–275 (2004)
5. Euzenat, J., Pierson, J., Ramparany, F.: A contex information manager for pervasive computing environments. In: Proc. 2nd ECAI workshop on contexts and ontologies (C&O), Riva del Garda (IT), pp. 25–29 (2006)
6. Truong, B., Lee, Y., Lee, S.: Modeling Uncertainty in Context-Aware Computing. In: Proceeding of the 4th Annual ACIS International Conference on Computer and Information Science (2005)
7. Padovitz, A., Loke, S., Zaslavsky, A.: On Uncertainty in Context-Aware Computing: Appealing to High-Level and Same-Level Context for Low-Level Context Verification. In: Proceeding of the 1st International Workshop on Ubiquitous Computing (2004)
8. Gu, T., Pung, H., Zhang, D.: A Bayesian approach for dealing with uncertain contexts. In: The Proceeding of the Second International Conference on Pervasive Computing (2004)
9. Open Services Gateway Initiative. http://www.osgi.org
10. Knopflerfish. http://www.knopflerfish.org

Skill Transfer from Expert to Novice – Instruction Manuals Made by Means of Groupware

Chung-Yong Liu[1,2,*] and Yasufumi Kume[1]

[1] Department of Mechanical Engineering, School of Science and Engineering, Kinki University, Japan
[2] Industrial Technology Research Institute, Taiwan
liu@itri-tokyo.jp

Abstract. Modern manufacturing industry changes dramatically following the fast progress of mechanical improvement and informatics development. Young people focus on learning new techniques in order to catch up the fast progress. Traditional industrial skills are missing step by step that will be a problem for modern industrial manufactures. The technological generation gap also causes damage in manufacturing industry. Elder experts who accumulated many traditional industrial skills faced the problems of retirement and strict competition. Young novices receive modern industry technique trainings, but still require traditional skill to maintain the industrial manufacture. The skill transmission from elder expert to young novice is known as an important course. In Japan, industry pays a lot of concern at the decline in mastery of skills technology. Based on the importance and difficulties of skill transmissions, an idea model was searched and established in this study. We choice screw manufacture as a representative industry in this study because screw manufacture required traditional skills and modern techniques. A manual introducing traditional skill and integrating computer science could transmit technologies smoothly from elder expert to young novice. Computer science here is used to link generation gap, to introduce traditional skill delicately in order to keep all the detail tricks and lively in order to be accepted by novice. The transmission efficiency and efficacy both will be considered.

1 Introduction

20th century is entering into information age step by step. Mechanization industry is followed this changes. During these changes, novel mechanized techniques in the production line are especially important in manufacture industry. In Japan, industry moves abroad seeking cheap manpower after the establishments of mechanized techniques. This phenomenon causes some problems during manufactory skill conservation in Japanese manufactory industry, such as skill transfer gap from expert to novice and traditional skill gradual loss. There are some problems make the skill

* Presentor: Chung-Yong Liu, Japan business group, International business center, Industrial Technology Research Institute, Taiwan.

M.J. Smith, G. Salvendy (Eds.): Human Interface, Part II, HCII 2007, LNCS 4558, pp. 423–429, 2007.

transfer more difficult, named as the aging of society, young people transfer their job from time to time and the main company and its manufactory field are distantly separated. According to this concern, an organization processes should be produced in order to make the skill conservation smooth including the success transfer from information to knowledge, from concept to language, from hand operation to manual. Based on this proposal, it might be possible to decrease the gap as well as industry risk through a series of organization processes, including integration the various skills into knowledge, making obscure concept into language, combining the individual skills into systematic procedure and consulting the above into protocol after modification by expert and novice several times.

Based on the above considerations, some strategies are claimed to solve the skill passage problems in production line in manufactory field. Here shows the strategies in the screw manufacturing industry to establish a model to transmit expert skills to young beginners in the production line. The efficacy and deposition of conversion from beginning introduction to the ending termination through operating guider is discussed in this paper.

Traditional operation guides were written by expert alone to list the manufacture technique procedures (personal technology formula). In this kind of operation manual, individual's cognitive difference causes the manual difficult to carry by beginners or other workers. Here present a technology conversion method to transmit personal technology formula into a organized technology transduction. All personal technologies will be guided into SECI model through socialization, externalization, combination and internalization.

2 From Knowledge to Manufacture Process

Knowledge can be grouped depending on its creature characters, one is tacit knowledge which is hard to express into language and the other is explicit knowledge which can transmit to others easily through formula. Figure 1 shows four transmission procedures. 1. Socialization: Integration the tacit knowledge from individuals through common experience. In this step, the knowledge conversion is from tacit knowledge to tacit knowledge (Shown in figure1). 2. Externalization: transmitting the tacit knowledge into explicit knowledge by using charts, tables and figures. In this step, the knowledge conversion is from tacit knowledge to explicit knowledge. 3. Combination: linking the individual explicit knowledge together into organized set. In this step, the knowledge conversion is from explicit knowledge to explicit knowledge. 4. Internalization: solidify the organized knowledge set into manufacture production line through small experimental production group. In this step, knowledge conversion is from explicit knowledge to tacit knowledge. The integration of those 4 steps here is so shortly called SECI model using their first letter. Three of the four types of knowledge conversion, named socialization, combination and internalization, have been discussed from various perspectives in organizational theory. For example, socialization is connected with the

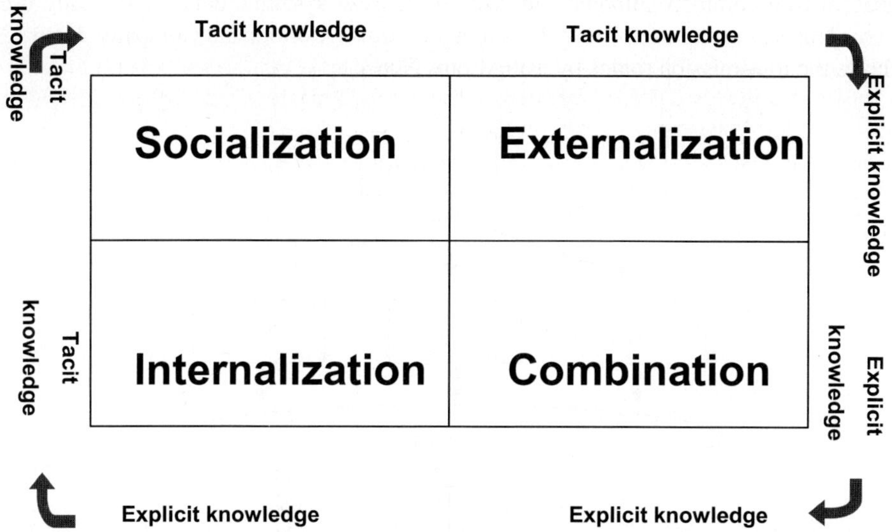

Fig. 1. Four models of skill conservation from tacit knowledge to explicit knowledge. The four steps are socialization → externalization → combination → internalization. The four steps will be modified by experts and novices through several circles. After modification, the mature protocol will be gained in order to conserve the expert's skills in manufactory fields.

theories of group processes and organizational culture,; combination has its roots in information processing; and internalization is closely related to organizational learning. However, in this three step model, externalization has been somewhat neglected. In our theory, we add externalization to help knowledge conversion from tacit knowledge to explicit knowledge smoothly.

3 Protocol Outline Analysis

Protocol analysis is a psychological research method that elicits verbal reports from research participants. Protocol analysis is used to study thinking in cognitive psychology, cognitive science and behavior analysis. It has found further application in the design of surveys and interviews, usability testing and educational psychology. In this SECI model, the protocol is firstly recorded through language individually. Individual protocol will analyzed their similarities and difference to express into an organized protocol.

4 Software Grouping

Grouping software means a computer interface can integration various data, thus to establish a data bank. This computer systematic data bank supports the users of similar goals with a common computer interface to analyze systemic data. This study use Lotus Notes as grouping software belonging to Lotus Development Company. Figure 2 shows the transmission routes by using Lotus Notes.

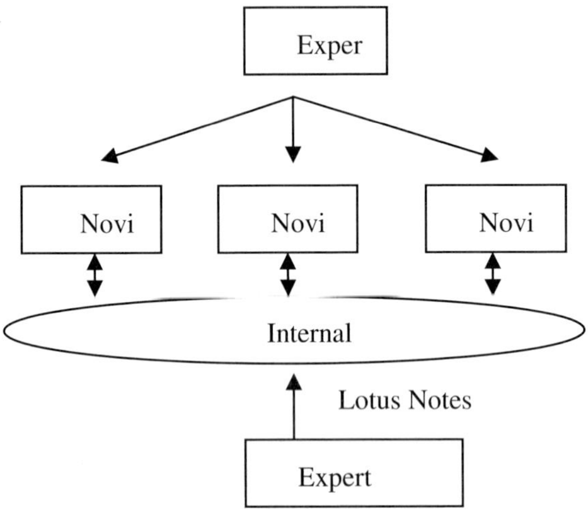

Fig. 2. Knowledge conversion is shown here by using Lotus Notes

5 Technology Transmission Using SECI Model

Figure 3 shows to transmit technology through SECI model. Four procedures are used as mentioned above – socialization, externalization, combination and internalization. All individual expert's experience and skills are integrated and externalized into common protocol guideline. Real practices and experiments will be carried under this common protocol guideline. The technology transmission procedures will be listed below:

Investigation of expert's technology
↓
Technology analysis and integration into knowledge
↓
Protocol guideline production
↓
Actually operation using protocol guideline

5.1 Inveatigation of Experts' Technologies

In this step, Notes software is used to investigate the expert's technology during manufacture production line and their technologies will transmit into Notes formulae. Recording by speech or video during the experts presenting their skill both can be used in those investigations.

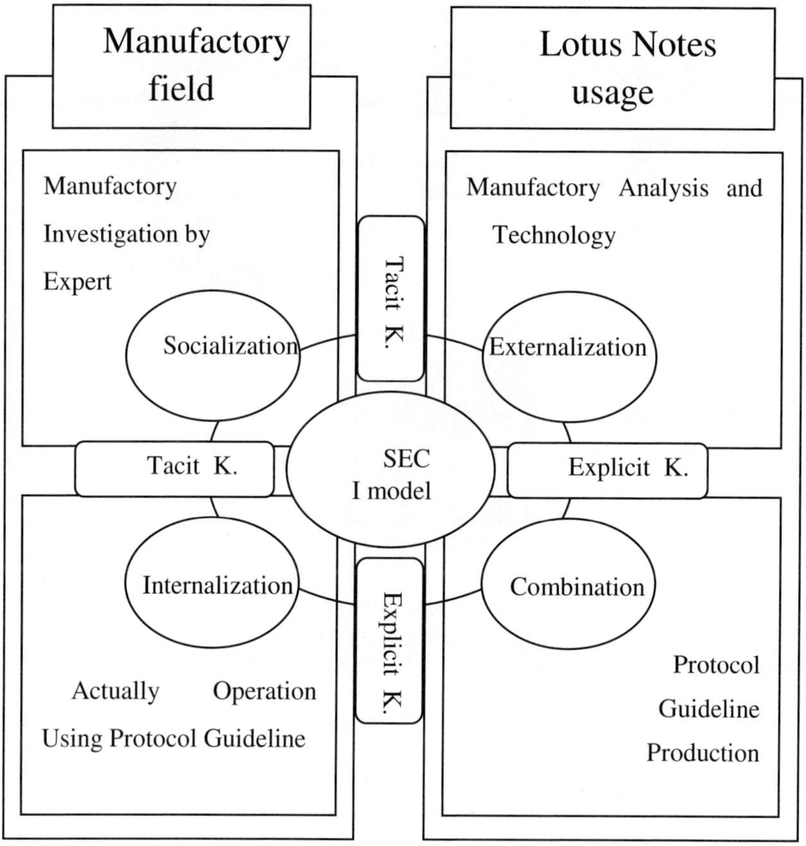

Fig. 3. Technique conservation uses SECI model. Technique knowledge will be conserved through tacit K. (knowledge) to explicit K. (knowledge) by SECI four steps round and round.

5.2 Investigation Analyses and Transmission from Technology into Knowledge

Transmitting technology into knowledge using words, lectures, or pictures is an important procedure of externalization. Sympathized knowledge about consumers' wants may become explicit conceptual knowledge about a new-product concept through socialization and externalization. Such conceptual knowledge becomes a guideline for creating systemic knowledge through combination. A new-product concept steers the combination phase, in which newly developed and existing component technologies re combined to duilf a prototype. Systemic knowledge turns into operational knowledge for mass production of the product through internalization. Experience-based operational knowledge often triggers a new cycle of knowledge creation. The users' tacit operational knowledge about a product is often socialized, thereby initiating improvement of an existing product or development of an innovation.

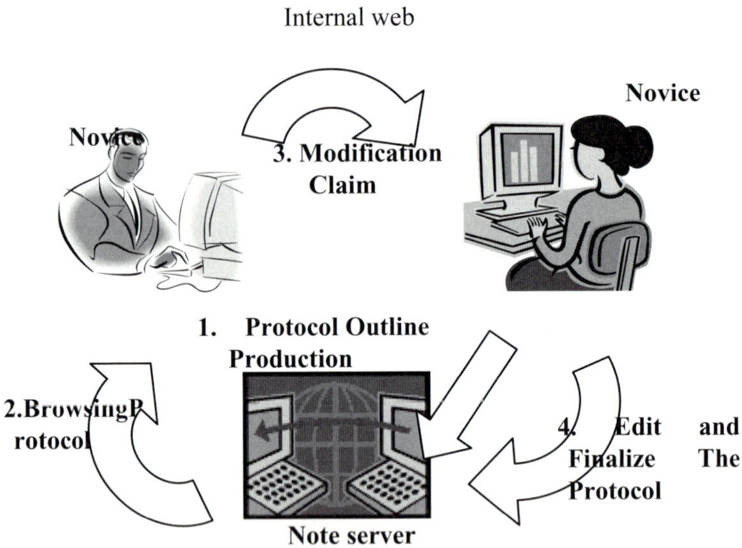

Fig. 4. The procedure of modification and editing the protocol outline divides into four steps. In the 1st step, the expert will write a brief manufactory protocol by using Lotus Note. A server is established based on this purpose. In the 2nd step, the novice will browse the brief protocol by using the internal web. The novice will go through reading and claims some questions in the 3rd step. The expert will edit the protocol in the 4th step. After 1st circle, the edited protocol will pass to novice in the 2nd step of the 2nd run. The circle will repeat till the novice can understand the protocol, then operating themselves. After actual operation, novices will claim new questions and expert will edit the protocol till the mature and final protocol done.

Discussion is especially important during these processes. Communication through email might be a good way since web is very popular and the email can be kept and used in the later steps.

5.3 Protocol Guideline Production

After externalization, the techniques are recorded into digital formula. The protocol guideline will be produced and modified as followed:

Step 1: As shown in figure 4, not only experts but also beginners must participate the integration and production of protocol guideline. The editor will collect the original information and discussion from experts and beginners as a basis to list the data into Lotus Notes software that can be integrated later.

Step 2: The beginners must browsing on all the information on Lotus Notes to help editor to improve the data bank of Lotus Notes.

Step 3: The editor will collect the original information from beginners and checked by experts. Beginners will claim the modification when carries their works based on the protocol recommend by expert.

Step 4: Repetition from step 2 to step 3 is necessary till no claimed modifications.

Step 5: Experts will aid the beginners to make products through protocol guideline. During this process, the beginners will modulate the protocol and then expert will correct those modulations.

Step 6: Experts will point out the improvements in protocol guideline after checking the beginner's production and review the protocol guideline. Then, the beginners will modify the protocol guideline again.

5.4 Protocol Guideline Modification Through Operations

In the process to produce protocol guideline, the beginners become successors. The externalized knowledge will internalized as self knowledge to those successors. Now the technology passage is successfully done. There is one case of screw manufactory in Japan, the expert cost 151 sec in SECI circle and the beginners cost 1421 sec to produce a final protocol guideline.

6 Inspection and Limitation

Some limitation will occur when using SECI to produce final protocol guideline. Discussed below:

1. The limitation of technology externalization

Some techniques are difficult to express in language, especially some skill by using figures or body movements. Those difficulties cause limitation.

2. The limitation of Transmission limitation

Experts are familiar with their techniques but not expressions. Some gaps will be occurred during transmissions.

3. The limitation of un-transmissible techniques

Some techniques is characterized as difficult to transmit. It is very difficult to transmit some screw productions no matter using a lot of time and transmission procedures in the screw manufacture industry. The real operation will be necessary in those cases. In those cases, real operations will help the technique transmissions.

7 Conclusion

This study introduces a SECI model to transmit technology into knowledge step by step. All the technologies, possible externalized, can be integrated into knowledge to be transmitted from generations to generations, from experts to beginners. The training of successors of traditional skills is very important in aging and informatical society. This study will help the transmission of technologies easier and more efficacy.

References

1. Ikujiro Nonaka David Teece: Managing industrial knowledge, SAGE Publications, pp. 13–122 (2001)
2. Fujimori, Y.: Lotus Note 4.6Guidebook, Ohmsha, Ltd, pp. 20–398 (1998)

Development of a Skill Acquisition Support System Using Expert's Eye Movement

Takashi Nagamatsu, Yohei Kaieda, Junzo Kamahara, and Hiroyuki Shimada

Faculty of maritime sciences, Kobe University
5-1-1 Fukae-Minami, Higashi-nada, Kobe, Japan
nagamatu@kobe-u.ac.jp, 058w103n@stu.kobe-u.ac.jp,
{kamahara, shimada}@maritime.kobe-u.ac.jp

Abstract. The number of experts in various fields such as manufactures, traditional arts/crafts decreases because of their retirement in Japan. In the field of nuclear power generation, it also becomes problem especially in maintenance. The goal of this study was to develop an advanced instruction video system to support skill acquisition. We proposed to use eye movement as one of the expert's skill information, and developed the system for supporting recording skill (scene video and the point of regard of an expert), making teaching materials in XML format and learning skills with the special user interface. Evaluation will be necessary by experiment to confirm the effectiveness of the training method.

Keywords: Eye tracking, Skill, Training, Expert, XML.

1 Introduction

The work presented here is of a novel support system for skill acquisition.

Nowadays, the number of experts in various fields such as manufactures, traditional arts and traditional crafts decreases because of their retirement in Japan. Therefore it becomes difficult to transfer expert's skills including various experiences, know-how, motor skills etc. to the next generation. We call the problem a "skill inheritance" problem.

In the field of nuclear power generation, it also becomes difficult to employ young talented persons because the number of youths learning nuclear engineering decreases as well as the number of construction of a new plant. In particular to keep the health of aging old plants and complicated new plants, skilled maintenance workers are strongly needed.

One solution of skill inheritance problems is automation of skill to do the same task as an expert. Some skills were analyzed and transferred to machines, but others are difficult to automate. Another solution is to support the non-expert's skill acquisition process effectively, so we focus on a self-training system that supports skill acquisition.

When one wants to learn a skill by oneself, he/she usually watches an instruction manual or video. An ordinary instruction video shows scenes of performing a task by an expert with characters and/or graphic information on the image. However, very little

M.J. Smith, G. Salvendy (Eds.): Human Interface, Part II, HCII 2007, LNCS 4558, pp. 430–439, 2007.

attention was given to the expert's eye movement, although eye movement reflects one's attention. Sadasivan et.al studied scanpath-based feedforward training for visual inspection task [1]. They used expert's eye movement for training, but they applied to only visual search task and their system was constructed as a virtual reality environment instead of real environment.

The goal of this study was to develop an advanced instruction video system to support skill acquisition.

This paper suggests the use of eye movement as one of the skill information, the design of the skill data representation and the design of whole the system from recording skill to imparting skills.

2 System Design

2.1 How to Support Skill Transfer

In fields of transferring skill to an apprentice (non-expert) from a master (expert) such as apprentice system and on-the-job training, the way to transfer skill is usually the following strategy.

- An expert lets a non-expert observe expert's skilled performance.
- The non-expert imitates it.
- The expert observes the non-expert's behavior and gives advice.

Experts suffer from this process because this process is highly stressful for a long time to complete. Although we adopted this training strategy, we were required to make this process shorten or easy to complete. Therefore we have proposed an advanced instruction video system to computerize this training strategy.

The proposed system provides a self-training environment to the user. The user is expected to watch the expert's performing video and imitate the skilled task. The user checks the expert's performance carefully and practices repeatedly in order to master skills. While the user practices the user can see beneficial information such as advices and know-how on the user interface interactively.

Before the user uses the system, the teaching materials must be prepared. Therefore we designed the skill transferring process that is divided into three steps: 1) recording expert's skill, 2) making teaching materials and 3) self-training. Three subsystems were constructed to realize these three steps.

2.2 Eye Tracking

We proposed the use of eye movement as expert's skill. There are several kinds of eye tracking methodologies such as electro-oculography (EOG), scleral contact lens/search coil, limbus tracking and video-based tracking. The video-based tracker is commonly used these days and consists of relatively inexpensive cameras and computers for image processing. Therefore we adopted the video-based eye tracking methodology and constructed the video-based eye tracker as a part of the system.

2.3 System Diagram

As Polanyi said, "We know more than we can tell", tacit knowledge is important to skill transfer [2]. In our system we treats the scene image and the point of regard (POR) as tacit knowledge. On the other hands, explicit knowledge is the text that may be derived from the interview of an expert. Both of the data are important in recording the expert's skill, so we designed the system that treats both of data.

This system we designed consists of three subsystems: the subsystem for recording skill, the subsystem for making teaching materials and the subsystem for learning skills.

The subsystem for recording skill consists of a head mounted device and two computers. The head mounted device has two cameras and gets the scene image and the eye image. The eye-PC calculates the pupil center position from the eye image, and the position data are transmitted to the record-PC. The record-PC calculates the POR data by transforming the position of the center of the pupil into view camera coordinate system using calibration data. Then the POR data and the scene video are saved separately. The diagram is shown in Fig.1.

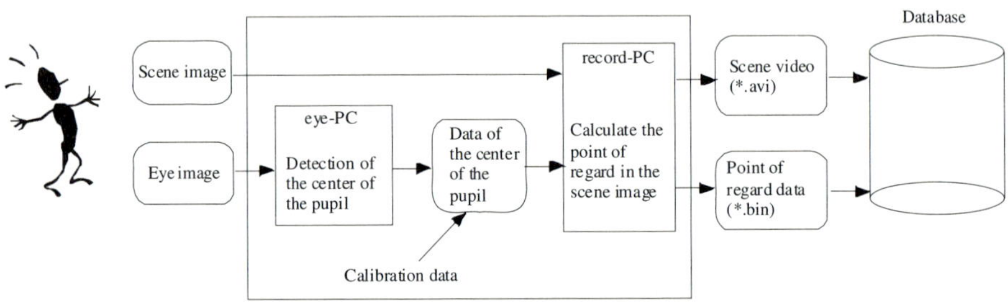

Fig. 1. Subsystem for recording skill

The subsystem for making teaching materials consists of the authoring-PC. The authoring-PC reads the scene video file and the POR data file. An expert is interviewed to get the explicit knowledge. After inputting the explicit knowledge data, the skill file can be saved. The data of an expert's skill (skill data) are saved as a skill file in our original XML format. The file includes a scene video file and the POR file and the text data (advices, failure examples, know-how etc.). The skill file format is described in the next section in detail. The diagram is shown in Fig.2.

The subsystem for learning skills consists of the learn-PC. In learn-PC, the image and data of an expert's skill are provided to a user with a user interface program. The learn-PC reads the data, and the data are displayed being synchronized with the scene video. In addition this subsystem has the following function. The user interface program can change the playback rate. If a user plays it slowly when the user wants to confirm the scene of a difficult task, the detailed information that was not displayed in normal speed is displayed. On the other hand, a user who already mastered an

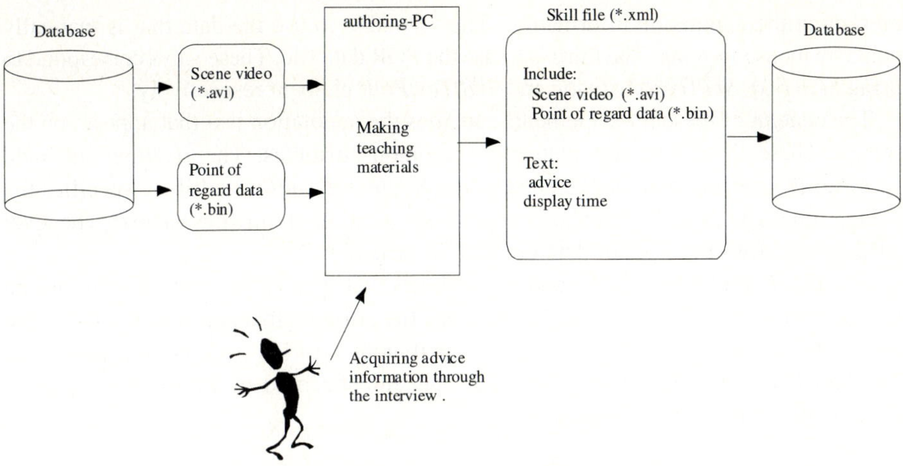

Fig. 2. Subsystem for making teaching materials

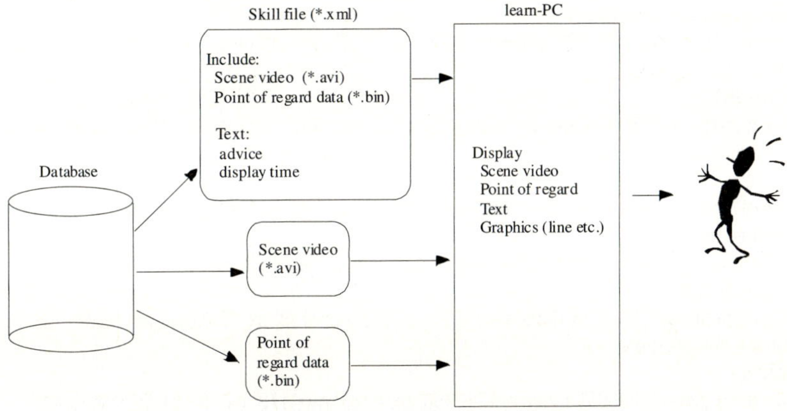

Fig. 3. Subsystem for learning skills

individual task can get essence/outline information by putting up playback rate. The diagram is shown in Fig.3.

2.4 Skill File Format

The skill file is written in XML and several tags are defined. The Fig.4 shows an example of the skill data file.

The *PORFile* element's content (Sample.bin) indicates the POR data file. The *VideoFile* element's content indicates the scene video file (out.avi).

There are three ways that are used to specify the position of drawing a line for annotation information. The first is to describe directly the x and y coordinates of presenting the annotation information. The second is to use the data that is manually made by mouse moving. The third is to use the POR data file. These ways correspond to using *SkillText*, *SkillTextDynamic* and *SkillTextPOR* element respectively.

The content of these three elements specifies the annotation text that appears on the screen. These three elements also have common attributes. The *starttime* attribute specifies the starting time of presenting the text and the *endtime* attribute specifies the ending time. The *cx* and *cy* attributes specify the location of drawing the text. The level attribute specifies the level of detail of skill information.

The *SkillText* element has *gx* and *gy* attributes that specify the starting point of line for annotation text. The *SkillTextDynamic* has file attribute that specifies the file name instead of *gx* and *gy*, and the file has the continuous position data moving with time. The *SkillTextPOR* element uses the POR data file specified by *PORFile* element to draw a line from the point of regard.

```
<?xml version="1.0" encoding="UTF-8" ?>
<a>
<PORFile>Sample.bin</PORFile>
<VideoFile>out.avi</VideoFile>
<SkillTextPOR starttime="4508" endtime="8055" cx="145" cy="-20" level="3">
Check whether packing is torn
</SkillTextPOR>
<SkillText starttime="4508" endtime="12063" gx="182" gy="187" cx="217" cy="271" level="3">
Fit the holes and set it
</SkillText>
<SkillTextDynamic starttime="15071" endtime="42165" cx="15" cy="271" level="3"
file="flange1.bin">
Install the bolt, plain washer and spring washer and tighten them by hand
</SkillTextDynamic>
<SkillText starttime="15071" endtime="42165" gx="138" gy="138" cx="68" cy="-26" level="1">
Tighten the bolt securely
</SkillText>
<SkillText starttime="220003" endtime="233956" gx="180" gy="172" cx="0" cy="271" level="3">
If you can't tighten them any further by hand, tighten the bolts with a wrench
</SkillText>
<SkillText starttime="220087" endtime="258722" gx="170" gy="17" cx="185" cy="-23" level="3">
Next is the one opposite
</SkillText>
</a>
```

Fig. 4. Skill data file

3 System Implementation

The prototype system was developed on windows platform using JAVA, JMF (Java Media Framework API), SAX (Simple API for XML) etc.

3.1 Subsystem for Recording Skill

The expert wears a headgear with two cameras (a scene camera and an eye camera) for recording skill as shown in Fig. 5. The scene camera that is attached to the front of expert's head captures the scene of performing the skilled task and the eye camera with IR-LED captures the expert's eye. The eye-PC processes the eye image to detect the center of the pupil. Fig.6 shows the detected center of the pupil.

The center of the pupil is transformed to the point of regard by record-PC. Then the POR data and the scene video are saved. Fig.7 is a user interface of the record-PC. The interface has three tabs. If the user changes the tabs the user can see the data of the center of the pupil, the POR data or control buttons (e.g. save button). Fig.8 shows a recording scene that a worker assembles flanges with this subsystem.

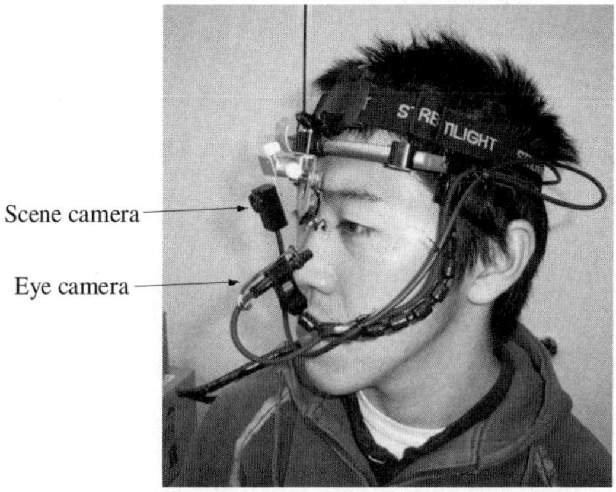

Fig. 5. Expert wearing the headgear with eye camera and scene camera

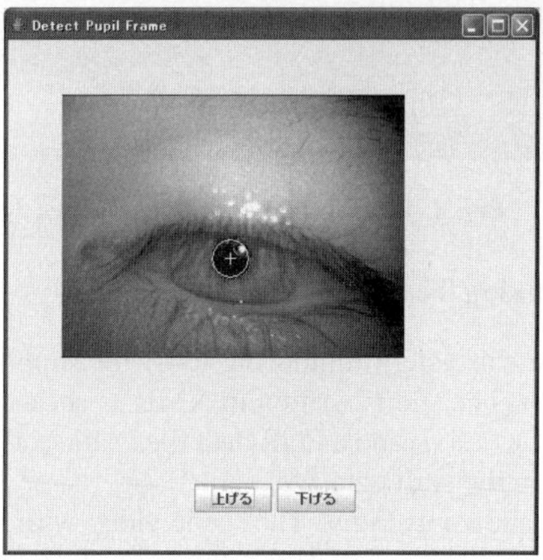

Fig. 6. Eye image and the detected center of the pupil

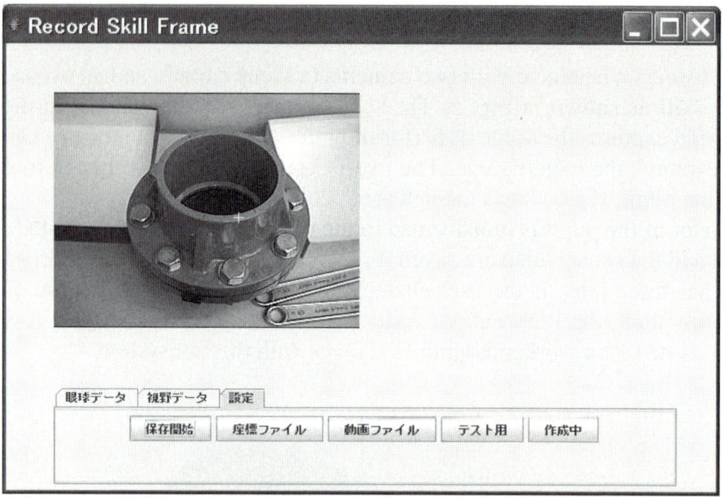

Fig. 7. The user interface of the subsystem for recording skill

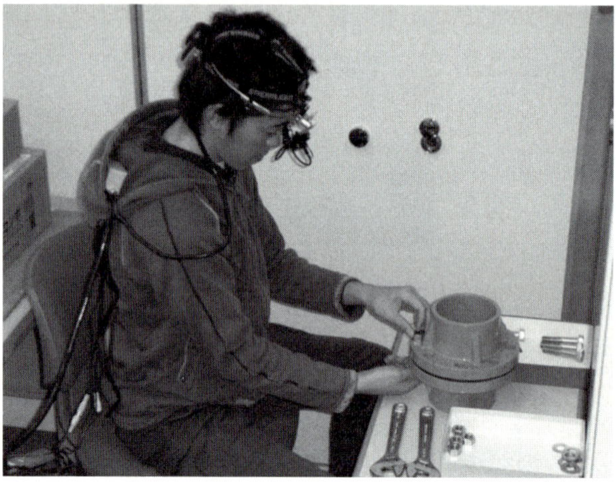

Fig. 8. Worker who assembles flanges

3.2 Subsystem for Making Teaching Materials

We have developed the subsystem (authoring tool) for supporting to make teaching materials (skill data file). The file is written in XML as shown in Fig.4. This data file includes a scene video file (*.avi) and a POR data file (*.bin). The explicit knowledge is described as XML text in the skill data file.

The user interface is shown as Fig.9. The scene video with the speed slider and position slider is displayed on the top left of the window. The text input area that helps to generate the skill data is on the bottom left and the generated skill data in xml format is displayed on the right of the window.

The user interviews an expert to get the explicit knowledge that includes when and what advices, know-how etc. must be given to a non-expert, and inputs the explicit knowledge data to this subsystem. After that the skill file can be saved in XML format. Thus the user can easily make the skill data file with this subsystem.

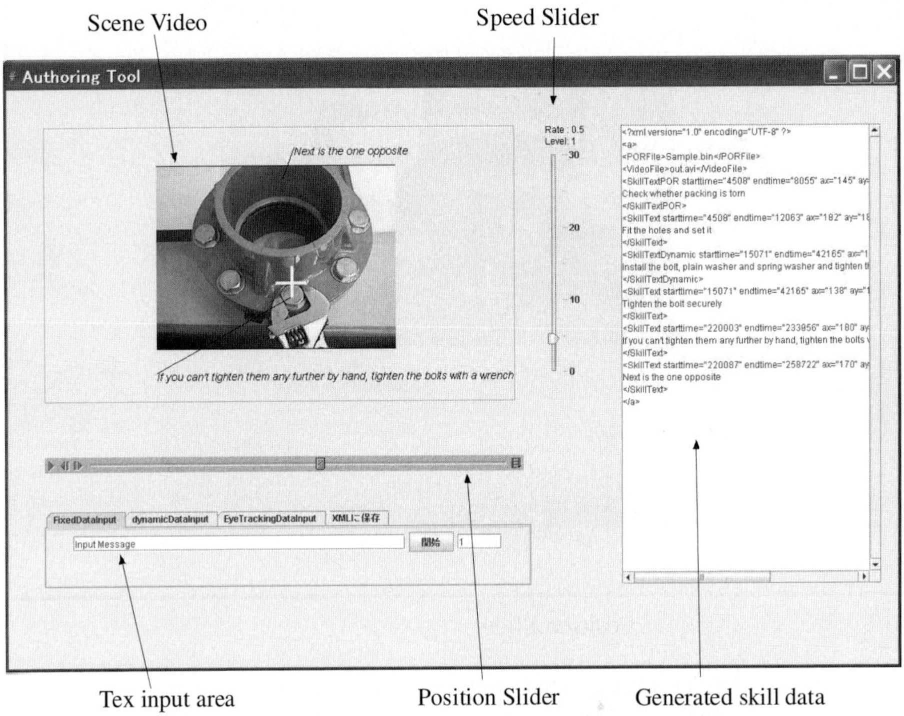

Fig. 9. The user interface of the subsystem for making teaching materials

3.3 Subsystem for Learning Skills

Fig. 10 shows the interface of subsystem for learning skills. The video is played on the center of the window. The expert's POR is displayed in the shape of cross. The explicit knowledge is displayed as annotations with line whose starting point is specified by the fixed position described in *SkillText* element, the continuous position data file described in *SkillTextDynamic* element or the POR file described in *SkillTextPOR* element.

The user is expected to watch the expert's performing video with this user interface, check the expert's performance carefully and practice repeatedly in order to master skills. While the user practices the user can see beneficial information such as advices and know-how on the user interface interactively.

The playback speed is used to change information presented on the video. For example, when a user watches the scene where the expert's behavior is quick or the task is very difficult, if the user plays the video at a slow speed, the detailed instruction

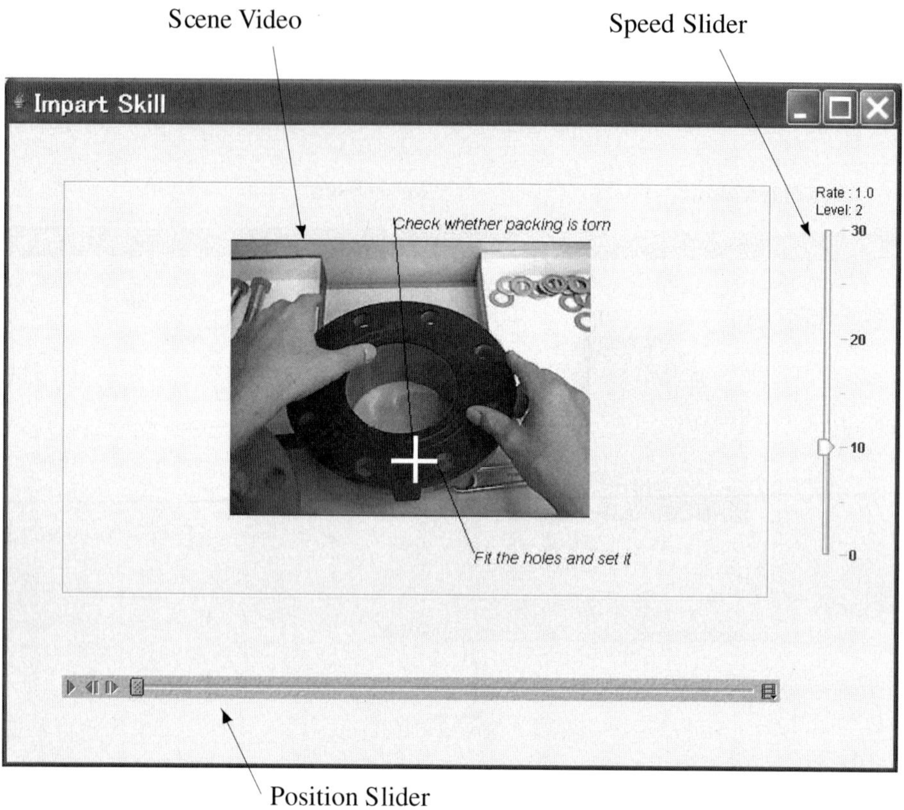

Fig. 10. The user interface of the subsystem for learning skill

information is presented. On the other hand, a user who already mastered such a detail way can get only outlined information of the whole task by putting up a playback speed.

4 Discussion

As Sadasivan et.al examined the training using expert's eye movement but they apply to only visual search task. Therefore the training method using eye movement requires further study on other tasks such as assembling and maintenance that are our targets.

This paper describes development of the novel advanced instruction video system. The system includes three functions recording skill, making teaching materials and user interface for self-training. Compared to the ordinary instruction video, the developed system has an ability of presenting video, texts and POR flexibly because those data are saved separately. Furthermore the system displays the information depending on the playback speed.

By using this developed system recording skill and making teaching materials can be achieved. However evaluation will be necessary by experiment to confirm the effectiveness of the training method.

By the way some skills needs continuous use of hands, so the user can not move the sliders to check the video and advice information. In that case some input methods without hands are necessary for example foot switches or eye input. Hence the applications to particular tasks require further study about user interface.

5 Conclusion

The goal of this study was to develop an advanced instruction video system to support skill acquisition. We proposed to use eye movement as one of the skill information, and developed the system that supports the recording skill (scene video and the point of regard of the expert), making teaching materials in XML format and learning skills with the special user interface.

References

1. Sadasivan, S., Greenstein, J.S., Gramopadhye, A.K., Duchowski, A.T.: Use of eye movements as feedforward training for a synthetic aircraft inspection task. In: Proc. Of CHI 2005, pp. 141–149 (2005)
2. Polanyi, M.: The Tacit Dimension. Garden City, New York: Doubleday & Co (1966)

MOCET: A MObile Collaborative Examination Tool

Sergio F. Ochoa[1], Andrés Neyem[1], Gabriel Bravo[1], and Emilio Ormeño[2]

[1] Computer Science Department, Universidad de Chile
Blanco Encalada 2120, Santiago, Chile
{sochoa,aneyem,gabravo}@dcc.uchile.cl
[2] Institute of Informatics, Universidad Nacional de San Juan
Av. I de la Roza and Meglioli, Rivadavia, San Juan, Argentina
eormeno@iifo.unsj.edu.ar

Abstract. This paper presents a mobile collaborative application to support the Collaborative Examination Technique (CET). This software tool, named MOCET, keeps the metaphor of the student's paper notebook used to support CET processes. In addition, it provides solutions to overcome the two main limitations identified on CET: (a) the inefficiency to support the building and sharing knowledge processes carried out by students during the pre and post-test activities, and (b) the high cost in time spent by instructors to carry out the examination process. Although MOCET is able to run on several computing devices, it has been specially designed to be used on Tablet PCs and PDAs and interact with the users using a pen device. This tool replaces the student's paper notebook during exams and supports students and instructors in every phase of the examination process. MOCET has been tested in two computer science courses and interesting results were obtained.

Keywords: Student Examination Process, Mobile Collaborative Tool, On-Line Tests.

1 Introduction

According to Hay, the objective behind the realization of an examination can be (a) to measure the real level of the students' knowledge, and (b) to measure the students' capability to relate and integrate the acquired knowledge [5]. Meanwhile, Ebel and Frisbie argue the objective of examinations is to measure the students' achievements, to motivate and direct their process of learning [3]. These are the most accepted interpretations of the examinations main goal. However, Ochoa et al. stated examinations can also be used for learning, especially if collaborative learning activities can be integrated into them. In order to demonstrate such hypothesis an examination technique was designed and tested: *CET (Collaborative Examination Technique)* [9]. CET involves three stages composing a collaborative examination process: *pre-test*, *test* and *post-test* (see Figure 1). Each phase is carried out in a consecutive lecture time space, for example, a lecture. The pre-test main goal is to help students to understand and assimilate, in an early stage, the knowledge that is going to be considered in the test phase. During a pre-test, students follow a

M.J. Smith, G. Salvendy (Eds.): Human Interface, Part II, HCII 2007, LNCS 4558, pp. 440–449, 2007.

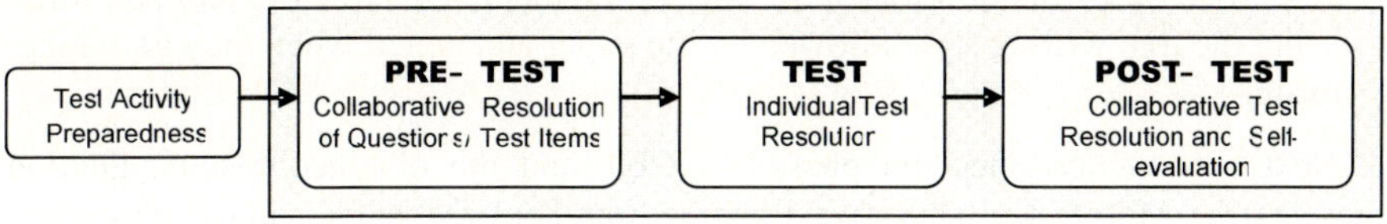

Fig. 1. CET Phases

collaborative process similar to Jigsaw [1] in order to find consensual responses to questions/test items provided by the instructor.

In CET the test phase involves the individual resolution of an exam, similar to a traditional examination processes conducted by using paper and pencil. Then, the post-test intends to reach two main goals: (1) students find the right answers to the test questions/items through a conducted discussion, and (2) students can identify the right points and the mistakes in their answers.

This technique has been recurrently applied in computer science and engineering courses at University of Chile. The obtained results have shown students increase their knowledge during these collaborative examination activities and they feel these techniques are fair evaluating the acquired knowledge [9]. However, CET has shown two main limitations that jeopardize its applicability: (a) the inefficiency to support the building and sharing knowledge processes carry out by students during pre and post-test, and (b) the high cost in time spent by instructors to carry out the examination process. These limitations are a consequence of the manual methods used by instructors to conduct each CET phase and the poor mechanisms used to record and share the knowledge among the people involved in the examination process. Next, these two limitations are briefly explained.

- *Inefficiency to support the building and sharing knowledge processes.* The knowledge building and sharing processes carry out by students doing pre-test and post-test activities are currently supported by hard copy and verbal information. This type of support limits the students' capability to represent, keep and distribute the knowledge among their partners. Typically, the people memory and motivation play an active role when they have to share the knowledge with the partners. In addition, the knowledge building and evolution process require co-located teammates provided the information synchronization is slow and difficult to do when the knowledge is recorded in a paper and in the students' mind. However, students feel comfortable with the free writing style allowed by the paper and pencil, and they think these are the best tools they have to support the building and sharing knowledge processes.
- *High cost in time for instructors.* In addition, the effort required by instructors to plan, execute and analyze each examination phase is high. If instructors want to get detailed information about the examination process execution in order to adjust it or improve it, they have to record the information in video and then analyze it in a manual manner. On the other hand, the rating process of a student's exam requires considering his/her contributions during the test and post-test phases. Such information is recorded in paper and it should be manually processed in order to get the final student's score. For that reason, the process becomes time-consuming

and prone to instructor's mistakes. Similar to students, instructors feel comfortable using the free writing style allowed by the paper and pencil when they are rating an exam.

Next section describes the phases of CET and the obtained results. Section 3 presents the Mobile Collaborative Examination Tool (MOCET) and it describes its functionality. Section 4 shows the obtained results and section 5 presents the conclusion and the further work.

2 Collaborative Examination Technique

Collaborative Examination Technique (CET) introduces collaborative learning activities in the pre and post-test phases in order to promote discussion and negotiation among students about the response to questions pre-established by the instructor. Each phase is carried out during consecutive lectures (Figure 1) and all of them contribute to the exam final score for a student.

2.1 Pre-test Phase

The pre-test first activity is the creation of a list of questions with complexity similar to those that will be included into the exam. These questions should be related to the knowledge considered for the test (exam) and they have to be open to promote the discussion and negotiation among students.

Once the instructor created the pre-test questions, she defines the work groups, trying to obtain heterogeneous groups, in regard to levels of knowledge. For this purpose, she can base her decisions on previous evaluations or the students' history. If that information is not available, groups should be formed at random. Each group should have four or five students [6]. The set of questions handed by the instructor should be the same length as the number of group members. All the groups must be the same size. Once the groups are formed, the instructor numbers the groups and hands over the questions. Then, each member of the group is assigned a question, and his responsibility is to become an expert on that subject (individual responsibility).

In the solution discussion in pairs or expert teams, each individual leaves his group and looks for another student who was assigned the same question (it is recommended that it be done in pairs, but it can be done in groups of three, and maximum four). These students have to interact and negotiate their viewpoints, with the objective of solving the assigned problem. They can aid themselves with any didactic material (books, notes, etc.) while doing this activity. The instructor decides the length of this activity and sees that pre-established times are globally fulfilled. This revision in pairs can be repeated more than once, until all solutions have reached all groups.

During the revisions in pairs, the students with the same assigned questions group again in pairs, but different pairs than in the previous activity. Then, the solution that was found in the previous phase is revised, and the presentation of this solution to the group is posed. Finally, during the information presentation, the students return to their original groups, where each member informs his fellow group members about the solution to the problem that had been assigned to him.

The positive interdependence present in this phase is of the "reward" type since one of the analyzed questions will appear in the test. The instructor must guarantee that this will happen. Consequently, if each member of the group does his part of the work well, everybody wins, because they have solved one of the test questions beforehand.

2.2 Test Phase

In that phase, each student answers his test individually. The idea is that part of the knowledge built or reviewed in the pre-test should be related with the questions or problems included in the exam during the test phase. Typically six questions (or problems) are included in the exams, and the students have 90 minutes to respond it.

2.3 Post-test Phase

During the class after the test, the students receive a photocopy of their exams. Then, groups of four or five students are formed to solve the test questions using discussion and negotiation. Finally, each student grades his/her own exam.

Students use the solution outline constructed by the group for that purpose. If the grade a student assigns to an exam item differs in no more than 0.3 points (i.e., plus or minus 5% of the grade range) from the grade assigned by the corrector (instructor/assistant), the final grade for that question will be equal to the corrector's grade, plus a 0.5 point bonus. If a student gets a bonus, it is because he/she understood what is right and wrong in his/her exam responses. This bonus shall not be assigned to those grades that required a justification but do not have one. Nor will it be assigned to those answers that agree with the grade, but have a wrong or insufficient justification.

During the post-test a member of the group must keep track of the time, because at the end of the activity all items must have been corrected. The positive interdependence stems from the fact that good solutions to the questions, constructed with the aid of everybody, help everyone be more objective in his own evaluation.

2.4 CET Results

This technique was tested with various Computer Science courses in three different universities (Table 1): Universidad de Chile (UCH), Pontificia Universidad Católica de Chile (PUC) and Universidad Nacional de San Juan, Argentina (UNSJ) [9]. These courses correspond to college and graduate levels.

For some of these courses, the proposed technique was evaluated more than once (several semesters). During the experimentation process four exams were taken in each course: two traditional exams and two exams taken using the proposed technique. All were done in the same order and following the same patterns.

At the end of each course, an opinion poll was conducted, to get the students' opinions. The obtained results shown that CET is able to reduce the impact of the following typical problems:

1. *Ambiguous Writings:* It is possible to find more than one meaning for the same exam question. This can lead to the appearance of differences between the student's and the examiner's interpretation.
2. *Anxiety:* The students show anxiety before and after the exams. The pre-test anxiety is due to the student's uncertainty about the knowledge that will be evaluated. On the other hand, the post-test anxiety is due to the fact that examination results take too much time to be corrected, so many students are anxious to know their results. This anxiety can interfere with the student's learning process [2].
3. *Lack of uniformity in correction criteria:* Usually, courses have more than one Teaching Assistant (TA) in charge of examining tests. Even though they receive a correction outline, the same correction criteria are not typically used. In many cases, an important part of the final grade depends on the corrector's criterion.
4. *Sensation of injustice:* Many students do not dare to ask the instructor or the correctors about dubious correction of their exams. It is also possible that the corrector does not give a convincing answer to a student's protest. More so, there might be more than one correct answer for one exam item, which many times are not considered by the corrector. This leads to the student feeling a sensation of injustice in relation to the instructor and the correctors, and his work.

On the other hand, the limitations explained in section 1 were identified: (a) the inefficiency to support the building and sharing knowledge processes carry out by students during pre and post-test, and (b) the high cost in time spent by instructors to carry out the examination process. Because these limitations are a consequence of the manual methods used by instructors to conduct each CET phase and the poor mechanisms used to record and share the knowledge among the people involved in the examination process, the authors developed a software tool that support students, assistants and instructors in such process. The tool was named MOCET (Mobile Collaborative Examination Tool), and it is presented in the next section.

3 The Mobile Collaborative Examination Tool

MOCET was designed to overcome the previously mentioned CET limitations, but keeping the advantages of such technique. The tool systematizes the examination process and supports students/instructors during every phase of CET (see Figure 2). It provides four main functions for instructors: (1) examination activity preparedness, (2) monitoring of pre-test, test and post-test activities, (3) rating support for every exam and (4) statistical information provision of the examination process. The tool also provides three main functions for students corresponding to (1) pre-test, (2) test and (3) post-test activities.

Services for both students and instructors use the asynchronous data sharing support, the persistency and the synchronization mechanisms provided by the PASIR platform [7, 8]. Shared objects, such as images, text documents or exams, represent basic pieces of information that can be shared on-demand among members of a Mobile Ad-hoc NETwork (MANET) session. Such objects are distributed among the mobile devices used by students, which compose an ad-hoc distributed shared memory systems. Each object is shared with a set of information layers (XML files) linked to it. Each layer represents a voice or textual comment from a student to the knowledge represented in the shared object. The instructor is the only role able to act as a hub of shared objects. In fact, this capability should be used to support the test and post-test phases.

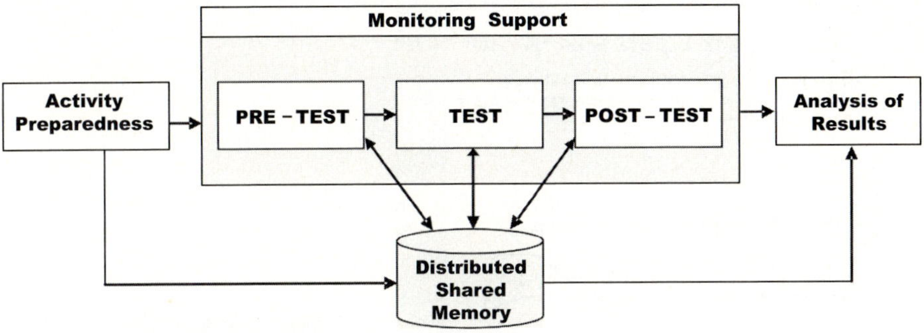

Fig. 2. Basic Architecture of MOCET

Provided that students and instructors feel comfortable using free writing style, MOCET was conceived to be used on Tablet PCs and PDAs (see Figure 3 and 4) with a pen device. Therefore, particularly relevant was the support included in the tool for using hand-written information and multimedia resources (images, audio/video records, text and strokes) during each phase of the examination process (*pre-test*, *test* and *post-test)*.

The system design assumed each student will use his/her personal computing device to do the exams. In such cases the tool provides more flexibility and capability to create/evolve/share the information in several scenarios (traveling, visiting or wandering) [4]. However, students can also use mobile computing devices provided by the education institution they belong, just to do the exams. In this case, students have to export the knowledge built and shared in each phase of the process to a remote personal computer in order to use such knowledge as study material or backup.

In both cases, students are able to evolve the shared knowledge (shared objects with their XML information layers) in an asynchronous way, and synchronize on-demand such knowledge with one or more partners just pushing a button in the MOCET user interface. Provided this mechanism works in one-hop and multi-hop networks, the synchronization process can be done in a large range of physical scenarios (e.g. on the street, a bank, a library or a coffee shop). This becomes flexible the process of sharing information.

MOCET provides different functionalities for instructor/assistant and students. For instructors and assistants the tool allow them to (1) design the pre-test, test and post-test, (2) group students, (3) deliver documents (e.g. exams) among the students and receive their responses, (4) correct the exams and determine if a student gets the bonus or not, and (5) monitor the students' activities.

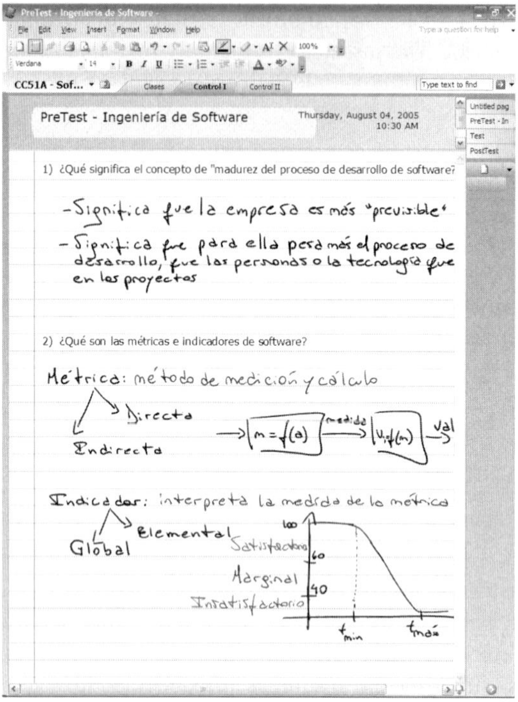

Fig. 3. MOCET version for Tablet PC

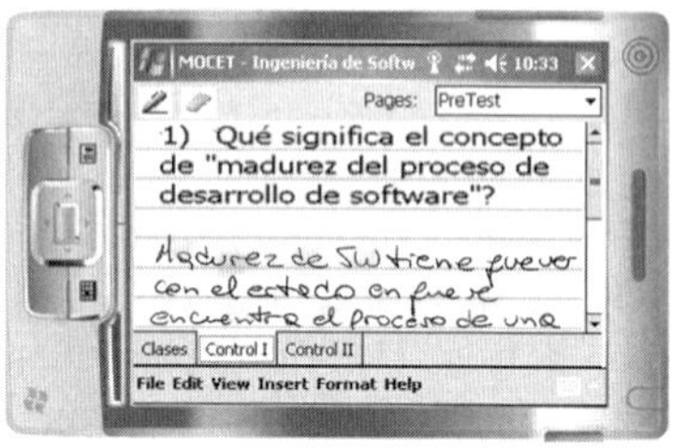

Fig. 4. MOCET version for PDA

On the other hand, MOCET allows students to: (1) share knowledge and the questions included in the pre and post-test, through a student mobile workspace, (2) receive, respond and return an exam to the instructor, and (3) receive the instructor correction. Next section presents the experimental results obtained applying MOCET in computer science courses.

4 MOCET Obtained Results

This tool was used once in two different computer science courses, corresponding to the sixth term and ninth term respectively, in the computing engineering carrier at University of Chile. Each experiment involved 10 students (6 using Tablet PCs and 4 using PDAs) and 2 instructors (one using a Tablet PC and the other using a PDA). These computing devices were lent by the Computer Science Department to students and instructors for ten days in order to execute this experiment. Students and instructors had previous experience using CET without technological support.

The pre-test and test were based on 6 questions. The obtained results were similar in both experiences. The later analysis shown that about 80% of the students learned during the examination process, especially during the post-test. This percentage is similar to the CET historical values. After the exercise an opinion poll was conducted, to get the students' and instructors' opinions. Most of the students (70%) said that MOCET is easier to use than CET; particularly for the sharing information. People using Tablet PCs said that hand-writing functionalities allowed them to operate in a similar way to use a paper and pencil. However, people using PDAs found hand-writing functionalities worse than paper and pencil, but still usable. In addition, approximately the 80% of the students considered the examination and rating process was fair, which is a value similar to CET historical records. If we consider these students know how to use CET, but do not MOCET, we could expect these percentages improve in the future.

Fig. 5. Experimental application of MOCET

On the other hand, the instructors considered the tool reduced, at least to the half, the effort required to apply CET. All of them considered the examinations and rating processes were fair. Furthermore, the knowledge sharing services and the hand-writing support were also the most relevant features for instructors.

The number of testing instances of MOCET is still low because this is a recent developed tool. Although, it is not possible to get strong evidence of the tool contributions, the preliminary results are encouraging. We envision this tool could also be used to support traditional examination process in a broad range of instructional scenarios: from middle-school courses to postgraduate programs. Currently, authors are designing a massive experimentation process using MOCET in computer science and engineering courses at the Engineering School of the University of Chile.

6 Conclusions and Future Work

Collaborative Examination Technique (CET) has shown good results in engineering education as examination and learning activities. However, this technique has also presented two main limitations: (a) sharing information among students is inefficient and (b) an important effort should be invested in order to carry out the examination process. Clearly, these two limitations jeopardize the applicability of such examination techniques. This paper presented a mobile collaborative examination tool (MOCET) that helps overcome these limitations. Particularly, the tool implements a variation of CET and it has been formally used in collaborative examination activities in courses of computer science and engineering undergraduate carriers. Although the obtained results are still not enough, it is expected this solution produces the following benefits to students and instructors/assistants:

Benefits for Students. Students using Tablet PCs and PDAs could execute part of the pre and post-test in a distributed and flexible way (traveling in a subway, waiting the bus or just walking), taking advantage of the mobility of these devices, their capabilities to form ad-hoc networks and to share information in an easy way. Provided the shared information (strokes, images, text, audio) is stored in XML format, the knowledge distribution among participants can be easily done using XML file synchronizations. It means the effort a user has to spend to be synchronized with the instructor and his/her partners is just push a button on the users interface (low cost). Because the tool supports for hand-writing, students are able to complete an exam in similar way to the traditional exams (writing a paper), avoiding being victims of the change of paradigm. Once completed each phase of the examination process, students and instructors can get immediate evaluations and download the results in their mobile devices for a later analysis/study. In addition, we expect students can get the benefits reported in [9]: (a) an the increment of their knowledge because of these collaborative examination activities and (b) a feeling this technique is fair to evaluate the acquired knowledge by them.

Benefits for Instructors/Assistants. Instructors/assistants use mobile devices (Tablet PC or PDA) to collect, review and correct the students' outcomes in every phase of the examination process. The correction process and the feedback delivery can be done by the instructor in several places (traveling in a bus or staying in a waiting room) and circumstances (talking in the sidewalk or in a coffee shop). Instructors can use such devices during preparedness of the activity and during the analysis of the obtained results. Because the process phases are automated by MOCET, the effort the instructor has to spend in order to apply the examination technique is low. In addition, the settings defined for an examination process can be reused of mixed with other ones. It allows instructors to share these setting with the partners and to get new ones, reducing thus the examination preparedness effort. We expect these advantages can be reach by students and instructors/assistants in a broad range of instructional scenarios: from middle-school courses to postgraduate programs.

Provided the obtained results are encouraging, we envision this proposal can provide a transversal solution able to support examination on several educational institutions (from middle-school to universities) and adhering to several instructional strategies, such as: Computer-Supported Collaborative Learning (CSCL), Computer-Integrated Classroom (CIC) and Active Learning, among others. In the short term, we have planned to apply MOCET in other engineering courses in Chile.

Acknowledgements. This work was partially supported by Fondecyt (Chile), grant N°: 11060467 and by MECESUP (Chile) Project N° UCH0109.

References

[1] Aronson, E., Patnoe, S.: The Jigsaw Classroom: Building Cooperation in the Classroom. Longman, New York (1997)

[2] Center for the Improvement of Instruction. Teaching at the University of Wisconsin-Madison: A Sourcebook. Chapter 16: Exams & Quizzes. Madison, WI: University of Wisconsin at Milwaukee (1990)

[3] Ebel, R., Frisbie, D.: Essential of Education Measurement, 4th edn. Prentice-Hall, Englewood Cliffs, NJ (1986)

[4] Guerrero, L., Ochoa, S., Pino, J., Collazos, C.: Selecting Devices to Support Mobile Collaboration. In: Group Decision and Negotiation, vol. 15(3), pp. 243–271. Springer, Netherlands (2006)

[5] Hay, I.: Examinations: Achievement Tests. Journal of Geography in Higher Education 20(1), 137–142 (1996)

[6] Kagan, S.: Cooperative Learning. San Juan Capistrano, CA. Kagan Cooperative Learning, CA, USA (1992)

[7] Neyem, A., Ochoa, S., Pino, J., Guerrero, L.: Sharing Information Resources in Mobile Ad-hoc Networks. In: Fukś, H., Lukosch, S., Salgado, A.C. (eds.) CRIWG 2005. LNCS, vol. 3706, pp. 351–358. Springer, Heidelberg (2005)

[8] Neyem, A., Ochoa, S., Pino, J.: A Strategy to Share Documents in MANETs using Mobile Devices. Accepted in ICACT 2006. Phoenix Park, Korea (February 2006)

[9] Ochoa, S., Guerrero, L., Pino, J., Collazos, C., Fuller, D.: Improving Learning by Collaborative Testing. The Journal of Student Centered Learning (JSCL). Forum Press. USA. 1(3), pp. 123–135 (2003)

Requirement Analysis of the Portal Site Serving Distributed Climate Data for e-Science

Taezoon Park[1] and Lan Zhao[2]

[1] School of Industrial Engineering, Purdue University
315 N. Grant St. West Lafayette, IN 47907, USA
[2] Rosen Center for Advanced Computing, Purdue University
151 S. Grant St. West Lafayette, IN 47907, USA
{tzpark, lanzhao}@purdue.edu

Abstract. The growth of grid computing facilitates the discovery of scientifically meaningful findings from the result of computational simulations by providing enormous amount of computing power and huge volume of storage area. As the computer technology advances, the relative portion of the time required for preparing and analyzing the result of the calculation increases because the complexity of the problem to be solved by grid computing also increases. In order to provide transparent and flexible interface to various groups of users who have different objectives, multi-layer design is applied for developing data portal for climate data. This paper presents three-layer model with four user groups which represents student, teacher, researcher, and administrator. The functionalities for job submission, post-processing, data management, content management are distributed into three different layers of complexity which is categorized as basic, intermediate, and advanced.

Keywords: grid computing; portal; distributed data system; user requirements.

1 Introduction

Grid computing is one of the most prevailing ways to discover scientific finding when solving computationally intensive problems, such as numerical simulations in biology, pharmaceuticals, and the physical sciences. Research in those area is becoming increasingly computation based, with problems of interest requiring peta-flops of computational power that consume weeks, months, or years of computer time. Grid technology enables us to handle complex problems which were nearly impossible to solve in one cluster or supercomputer. Also, efficient using of existing computational resources is possible by applying it.

As the complexity of the problem increases, however, managing large amount of data and finding out a meaningful result from enormous sets of result is getting harder and harder. Therefore, it is expected that the usability would be critical factor when enhancing the total throughput of the problem solving in e-Science. Building a portal service which enables researchers to submit jobs, retrieve results and interpret implications is suggested as an alternative way other than conventional terminal based

M.J. Smith, G. Salvendy (Eds.): Human Interface, Part II, HCII 2007, LNCS 4558, pp. 450–456, 2007.

approach. However, functions, data, and information should be well organized and properly offered in order to provide uniformed and seamless service. Especially, problems in computational science usually deals with huge amount of data, which is, in most case, very complex format. In addition, analyzing those complex data requires quite amount of cognitive processing, and easily exceed the limit of normal processing range without proper assistance of supporting software. Since there exists variety of data and functions required and the expertise of the user is diversified widely, arranging the contents and functions of e-Science portal is both challenging and essential for providing effective and efficient services.

This study focuses on building a portal service for observation, model building and simulation of the global climate. As a single access point to various datasets, we have are developing the Purdue Environmental Observatory Data Portal that allows researchers in the field and students in environmental engineering to easily access the data without having to learn the underlying software technology. For the first step, the necessary contents and functions required in each step of experiment are collected from in-depth interview with researchers, and organized by group task analysis. Functional groups found are data browsing/searching, job submission, monitoring, post processing and visualization. Functions which are necessary for a general portal service are also included. Among the functions found and refined above, grid operation functions are divided into three groups according to the persona which represents student, teacher, researcher, and administrator each. By the expertise of each user group, each functional group is divided into three levels; basic, intermediate, and advanced. It is found that automated sequence and guided workflow are important for the novice, whereas more user-control is anticipated for intermediate and advanced users.

2 Backgrounds

2.1 High Performance Computing and Grid Technology

Grid technologies differ from other major technologies like distributed computing, by providing a highly coordinated resource sharing and problem solving. The coordination of resources is quite complicated, and requires agreements of various stakeholders who have different interests. Developing grid enabled tools is difficult because of diverse system components, unpredictable interactions, rapidly evolving environment, widespread scalability, and interwoven software environments [8]. Moreover, a lot of supplementary features have been added to ensure the securities of the system, one of the primary issues of grid computing, which decrease the usability of the system as a result [1, 2]. The grid systems inherently have complexities not only because the grid technology itself is complicated, but also the computational problem itself is difficult.

The purpose of constructing portal service for grid computing is relieving the burden of unnecessary work which could be automated behind the scene. In other words, portal enables to hide distributed computational resources and to make unified access of the resources regardless of its physical locations [3].

2.2 Universal Usability

The concept of universal usability is first introduced by Ben Schneiderman [11] in order to meet the requirements from various types of users with different knowledge and skills. During the system development, one of the most common pitfalls is considering very small range of users which might be involved in the development process. For the successful development and deployment, however, developers should consider the designing for a large group of users, all with different levels of knowledge of computer and the task itself. Multi-layered design has been proven to provide a feasible solution as a flexible and differentiated interface design to meet various demand of the user [6, 12]. Multi-layered design suggests splitting the interfaces into several layers which different level of complexity. Each layer has different features according to the expertise and role of the user. By adding more layers the flexibility of the software system to support various users increase, and the user can perform more difficult tasks. It also facilitates learning by providing appropriate level of complexity that the user can handle without difficulties. For the user groups, three categories of user are suggested like novice or first time user, knowledgeable intermittent users, and expert frequent users. Novice user is assumed to have little knowledge about the concept of the interface and functions implemented with it, whereas expert user has plenty of knowledge about the interface concept and the functionality. For the expert group, the information, the dialogs and feedback should be minimal and fast, for example, by providing shortcut commands.

Proper number of design layers needs to be determined by a systematical way. Since the complexity of the system increases when adding a layer, which also increase the development cost, the designer should consider what would be the appropriate number of layers. Moreover, too many layers with similar complexities do not contribute to the increase of usability. For the multi-layered design, at least two layers are necessary; one is novice knowledge layer and the other is full knowledge layer. Novice knowledge layer contains minimal complexities, functions, and information, which sits on the starting point of learning curve, whereas full knowledge layer includes all the complicated features of the system to meet the needs and skills of an expert user, which is at the end of the learning curve. However, it is not very efficient to handle the requirements if the system has only two layers. Appropriate number of layers and the complexity of novice knowledge layer should be adjusted according to the expertise of target user group.

3 Case Study: Purdue Environmental Observatory Data Portal

In order to understand the climate systems on Earth, computational models have been developed, which are based on the combinations laws of physics that govern the evolution of the climate system. The climate system is modeled through complex mathematical equations that describe the temporal evolution of climate. The huge sized and complicated simulation model can be solved with the help of the advances in high performance computing. Also, the observation data, such as, satellite image, Doppler radar data, and weather station data, are collected and stored for the researcher to compare them with simulation result, and to use them as an input to the other simulation models.

3.1 System Overview

The Purdue Environmental Observatory Data Portal, which is under development and in a partial operation, provides a single access point to multiple resources for different groups such as, student, educator, and researcher. Resource includes climate dataset, computational resource, storage space, and reading materials.

The datasets serviced through the portal includes the Laboratory of Applications of Remote Sensing (LARS) image data collections [7], Purdue Terrestrial Observatory (PTO) image data set [9], National Weather Service (NWS) Doppler radar dataset [10], and Community Climate Simulation Model (CCSM) outputs. The LARS data image includes multi-spectral and hyper spectral image data collected from 1972 to 2004 mostly in the state of Indiana. PTO dataset is a satellite image currently captured by GOES-12, which covers the whole globe. NWS Doppler radar data is collected from the 159 weather surveillance radar sites across the State and a few selected overseas locations. The level II Doppler radar data comprises reflectivity, radial velocity, and spectrum width, which is processed and provides secondary information as the level III format. CCSM dataset is provided by Climate Dynamics Lab in the Department of Earth and Atmospheric Science at Purdue University, which simulates global climate change. CCSM result is expected to be used as a boundary condition for finer simulation for the regional area [13].

As a gateway to the computational climate simulations on TeraGrid, the portal provides function sets that could facilitate running simulations and analyzing the results. Major function implemented in the portal sites are preparation for running jobs, job submission, checking submitted job status, storing and retrieving result, post proccssing simulation result, and data searching and browsing. Contents serviced through data portal include research papers, presentations, course materials, and tutorials.

3.2 Underlying Technologies Applied on the Development

The Purdue Environmental Observatory Data Portal is developed and deployed using the Gridsphere Portal Framework – an open source web portlet container which allows users to develop and deploy JSR-168-compliant portlets onto a standard portal container like Apache Tomcat. It consists of customized Storage Broker (SRB) portlets that are web applications providing dynamic content from SRB. The current implementation of data portal supports collection browsing, metadata display, metadata query, file download, and preview for HTTP URL SRB objects.

Datasets are stored in Storage Resource Broker developed by San Diego Super Computing center, which supports distributed data storage and access for heterogeneous datasets. Also, the Thematic Realtime Environmental Distributed Data Services (THREDDS) serves the Doppler radar data using Open-source Project for a Network Data Access Protocol (OPeNDAP), which is prevailing protocol for climate data [4]. For the workflow engine, JOpera is employed to serve construction and execution of dynamic data-driven workflow. JOpera is a workflow composition tool that allows users to define their custom workflows using nodes or programs, which are the individual execution units [13, 5].

3.3 Tasks Required for Data Portal

When conducting a computational climatic research through the portal, the first step is preparing for the job submission. Job submission usually requires dividing jobs into atomic unit, preparing data, finding computational and/or storage resources, and staging in the scripts. When the job is submitted, it is necessary to monitor the status of jobs. If some problem happens, fixing and restarting functionality make it possible to avoid unnecessary resubmission because crash of a single job can happen anytime just by the defect of the system itself. After completing jobs, the system sends a notification message to the user. In order to find scientifically meaningful result, it is necessary to provide post-processing functionalities, for example, basic manipulation of the data, visualization, and simple statistical test. Among the huge simulation result, meaningful output needs to be stored and retrieved later for reference. Since the data usually has a huge volume, organizing and maintaining the data in a manner which facilitates search and browse is crucial, which could be achieved by organizing the structure and content of metadata.

Table 1. Tasks categorized by the user groups

User Category	Student	Teacher	Researcher	Administrator
Tasks	basic job submission post processing data storage and retrieval read contents learning module	basic job submission post processing data storage and retrieval create and maintain contents learning module	job submission post processing data storage and retrieval read and publish paper	maintain job submission, post processing, and data storage/retrieval privileged access on resources manage contents site management

3.4 Task Distributed on Layers

The user group of data portal is divided into four groups by the primary objective and role of the users; student, teacher, researcher, and administrator. Student includes high school, undergraduate and graduate students, and teachers are accordingly matched. The role of graduate student and professors in the university overlap with the role of researcher, but those positions focus on learning and teaching rather than research activities. Usually, student has more focus on reading the contents, like course material, research papers, whereas teacher and researcher generate and organize reading materials and research papers.

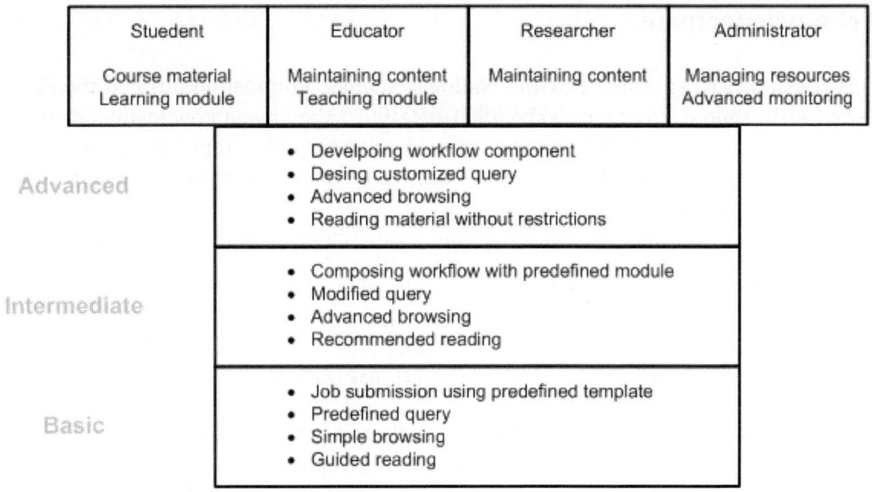

Fig. 1. Mushroom model for climate data portal

For the developing system, the number of layers is determined as three, and the contents are organized according to the role of users. Since there exist common functionalities, such as job submission, data search/browse and reading the material, regardless of the roles in the user group, the features of those functions are distributed into each layer according to the level of complexity in Fig 1. In the basic layer, job submission is allowed only by using predetermined template, and the query is also predefined for novice user. In the intermediate layer, the users can compose the workflow by combining pre-designed components. Also, intermediate user can modify query at his/her own conveniences using functions in intermediate layer. In the advanced layer, developing workflow component is supported, and advanced query can also be composed by the expert user. The content management modules and learning modules are specialized according to the requirement of each user group. Especially for the administrator group, advanced monitoring and detailed logging function is supported.

4 Conclusions

In this paper, the overall structure of data portal is suggested through the design process of Purdue Environmental Observatory Data Portal. From the requirement analysis, four user groups consisting of student, teacher, researcher, and administrator are found, and the tasks required in each group are itemized. Applying principles of multi-layer design gave three-layered structure that has common functionality of job submission, data manipulation, and content management. Three layers have different level of complexity according to the expertise of target user group.

Acknowledgement

This research was sponsored by the National Science Foundation through the Purdue TeraGrid Resource Provider award #SCI-0503992. Views and conclusions contained in this report are the authors' and should not be interpreted as representing the official opinion or policies, either expressed or implied, of the Government, or any person or agency connected with them.

References

1. Butt, A.R., Adabala, S., Kapadia, N.H., Figueiredo, R.J., Fortes, J.A.B.: Grid-computing portals and security issues. Journal of Parallel and Distributed Computing 63(10), 1006–1014 (2003)
2. Bubak, M., Funika, W., Bals, B., Wismuller, R.: A Concept of Grid Application Monitoring. In: Wyrzykowski, R., Dongarra, J., Paprzycki, M., Waśniewski, J. (eds.) PPAM 2001. LNCS, vol. 2328, pp. 307–314. Springer, Heidelberg (2002)
3. Choong, T., Sheong, C.S.: Enhancing the collaborative use of Grid Computing Applications with a User-Oriented Portal Dashboard. In: Proc. of International Conference on Scientific & Engineering Computation (2004)
4. Domenico, B., Caron, J., Davis, E., Kambic, R., Nativi, S.: Thematic Real-time Environmental Distributed Data Services (THREDDS): Incorporating Interactive Analysis Tools into NSDL. Journal of Digital Information, vol. 2(4) (2002)
5. Kalyanam, R., Zhao, L., Park, T., Goasguen, S.: A Web Service Enabled Workflow System for Climate Modeling Data Processing in TeraGrid. Teragrid, Indianapolis, IN (2006)
6. Kang, H., Plaisant, C., Shneiderman, B.: New Approaches to help User Get Started with Visual Interfaces: Multi-Layered Interfaces and Integrated Initial Guidance. In: Proceedings of the National conferences on Digital Government Research (2003)
7. Laboratory for Applications of Remote Sensing, http://www.lars.purdue.edu
8. Pancake, C.M.: Usability Issues in Developing Tools for the Grid - And How Visual Representations Can Help, Parallel Processing Letters, vol. 13(2) (2003)
9. Purdue Terrestriall Observatory, http://www.itap.purdue.edu/pt o /
10. Purdue Level II Radar Data, http://roskilde.eas.purdue.edu/ level2/index.html
11. Shneiderman, B.: Universal usability. Communications of the ACM 43(5), 84–91 (2000)
12. Shneiderman, B.: Promoting universal usability with multi-layer interface design. In: Proc.of 2003 Conference on Universal Usability, Vancouver, British Columbia, Canada, pp. 1–8. ACM Press, New York (2003)
13. Zhao, L., Park, T., Kalyanam, R., Lee, W., Goasguen, S.: Purdue Multidisciplinary Data Management Framework Using SRB. In: Proceedings of SRB Workshop, San Diego, CA (2006)

Efficient Creation of Multi Media eLearning Modules

Hans-Martin Pohl, Patrycja Tulinska, and Jan-Torsten Milde

Fulda University of Applied Science
HCI Research Center
Heinrich-von-Bibra-Platz 3, 36037 Fulda, Germany
hans-martin.pohl@verw.hs-fulda.de, tulinska@aol.com,
jan-torsten.milde@informatik.hs-fulda.de

Abstract. ECampus is a project spanning all departments at the University of Fulda. It has been started to create a uniform learning environment at the university. The objective is to research and develop a user-friendly easy-to-use editor to generate SCORM 2004 conform eLearning modules. This editor is based on Open Source software and new technologies such as XSL transformations and the Google web toolkit. Some eLearning modules can be developed with the system immediately. These modules are now being used during the lessons with great success.

Keywords: SCORM 2004, XSLT, transformation, creation of content, eLearning, modules, lesson, user friendly, style sheet, LOM.

1 Introduction

Creating eLearning units in SCORM 2004 (see [8]) format is a complex task. SCORM, as a technical standard, combines a number of existing eLearning specifications and formats. A SCORM compliant eLearning unit may consist of arbitrary files containing the actual content of the unit. In most cases, the unit's textual content is stored in (X)HTML files, which can be displayed with a standard web browser. Multi media files, e.g. images, flash animations, audio and video files, are stored separately in their proprietary formats.

In addition to the content files, a SCORM 2004 module contains a set of XML files describing the structure of the unit ([12]). Metadata is added in order to facilitate the management of the unit within the Learning Management System (LMS). Sequencing information is also part of these external files, including a difficult to understand specification of conditional constraints for the sequence progression.

For the average author it is almost impossible to generate such a file structure by hand. A small number of editors for SCORM files exist (e.g. Reload, *http://www.reload.ac.uk/scormplayer.html*). Even with these systems, the user still has to have a good understanding of the SCORM file structure. This technical barrier is much too high for most tutors. As a result, standard compliant eLearning content is not produced.

M.J. Smith, G. Salvendy (Eds.): Human Interface, Part II, HCII 2007, LNCS 4558, pp. 457–465, 2007.

2 State of the Art

While there are a lot of different Learning Management Systems at present, the use is limited to but a few in Germany. Hereby ‘moodle’ (see *http://www.moodle.org*) and ‘Ilias’ (see *http://www.ilias.de/ios/*) play the greatest role.

In contrast to the LMS there are only a limited number of editors for creating an eLearning module. HTML editors are widely-used (HyperText Markup Language). But this doesn’t provide adequate functionality. Furthermore eLearning modules generated by an editor can mostly only be used within a proprietary LMS.

Additionally a lot of effort is required to learn how to use these editors. Such a system fails entirely if there is to be consistent development and presentation of tests and learning success.

The advantage of these systems is the functionality for communication between users or between tutor and student. In our case this can be neglected because this functionality is captured by the LMS ‘System2Teach’ (see also *http://www.system2teach.de*) which is a proprietary development of the University of Applied Sciences Fulda. This paper focuses on easy and comfortable development of eLearning modules based on textual content by using the eCampus framework.

3 The eCampus Project

This research therefore focuses on the development of a user-friendly system for writing eLearning units, allowing authors to concentrate on the content and the didactic concept of the unit, instead of worrying about the underlying technology.

It is part of the eCampus project at the University of Applied Sciences Fulda. In this interdisciplinary project, content is being created for the faculties N*utritional Sciences*, F*ood and Consumer Sciences*, A*pplied Computer Science*, N*ursing & Health Care* as well as F*ood Technology*. Within the first year of the project nine eLearning modules were produced. All modules address prominent introductory

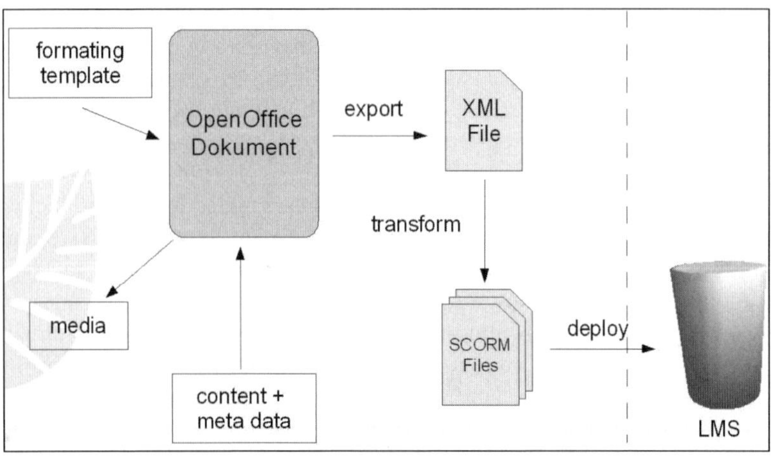

Fig. 1. Scheme of eCampus Framework

courses of the particular program of study. This was only possible by using the eCampus framework which is described below.

The central target of the project is the implementation of a process that decreases the complexity of the process of learning module production. The central objectives of the learning module production are an increase in the interactivity of the course, as well as an increase in the quality and quantity of self learning.

A secondary target of the project is the development of an LMS that is SCORM 2004 compliant. Here we would like to integrate the ADL reference implementation into our LMS System2Teach.

4 Overview eCampus Framework

The learning module production takes place in two phases (see Figure 1):

- the *structural* annotation of the textual content
- the *automatic* generation of the eLearning unit

In phase one, most of the (textual) content is created in a standard word processor (Open Office, [10]). Here the text is structurally annotated using a predefined formatting template. The text is stored and automatically converted into a simple XML file, which provides the basis for subsequent transformation steps. This approach works very well for static content and is used in a number of systems (e.g. eLAIX (see *http://www.boldt-media.de/*)).

In the second phase, the file prepared in phase one is transformed into a SCORM 2004 compliant eLearning module automatically. For this, the document is unpacked and analyzed. To get by on cascaded XSL transformations (EXtensible Stylesheet

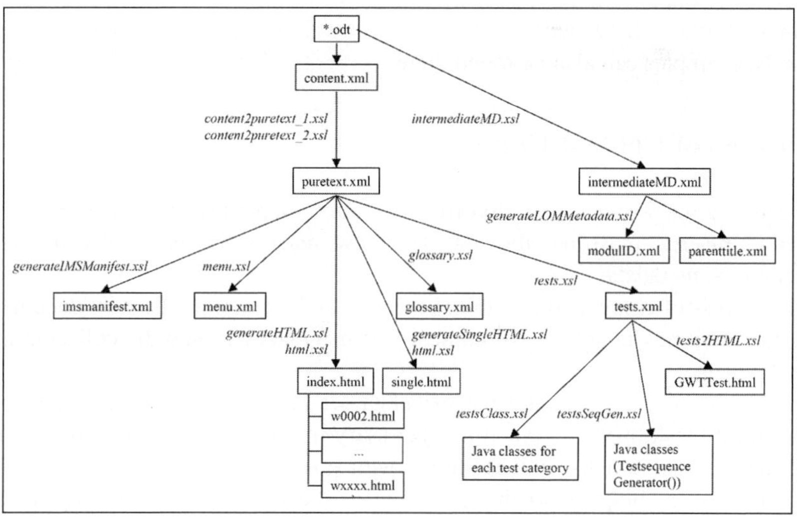

Fig. 2. Transformation Tree of eCampus Framework

Language) each element is identified and transformed with specific instructions (see Figure 2). Finally all contents are packed into an eLearning module and multiplexed with the necessary auxiliary files.

The available metadata is separated and attached to the module in a LOM (Learning Objects Metadata) compliant form.

The complete process is controlled by the Apache ant build tool and uses saxon8 (see *http://saxon.sourceforge.net/*) as a transformation engine ([4]).

5 Structure and Hierarchy

ECampus *modules* consist of *sessions* that are divided into *sections*. This hierarchy roughly corresponds to a *course*, that is organized as a sequence of *lectures*, each of them addressing a set of *topics*.

At a low level the s*ection* is implemented. The section is the least self contained element within an eLearning module. Sections can be arbitrarily nested. Therefore theoretically more than three levels are possible. The development of new sessions is possible by combining different existing sections. New modules arise by combining different sessions. Thus the hierarchical structure contributes to re-use of learning objects.

6 Installation of the eCampus Framework

By running the installation file the framework is installed on the machine. Hereby the necessary programs and styles are copied onto a local disk drive. Furthermore the java development kit is installed if not yet available. Additionally the environment variables are adapted and the path variable enhanced. After installation all parts of the eCampus framework needed are in a user definable folder. Similarly the user guide and some examples can also be found there.

7 Process of Content Compilation

After starting the writer application from the open source office suite ‚Open Office', executing macros must be allowed. The new macros are especially needed for collecting the metadata.

The process of content compilation begins with the creation of a new document by using the eCampus template. All content and metadata can now be collected in this new document.

The menu entry for creating the metadata for the whole module is first chosen. Then a form mask opens in which the necessary metadata is requested. If a choice is to be made possible a list box is shown with possible entries. Fixed data is either superimposed or compiled in the document in the background. This top hierarchy level entry is allowed only once per module.

After this the first session can be generated by choosing the appropriate menu entry. After filling in the metadata form it can create the sections. Once the metadata is acquired the author is able to start writing the content. The author can add further sections to structure the lesson. For more substantial contexts further sessions are also possible. The pure content can be written into the document directly or can be integrated by copy and paste from existing documents.

All common media types can be used for content. Beside text these can be images, pictures, tables, lists, definitions, links, glossary entries or arbitrary references to external media files.

The most important task is the subsequent annotation task. Hereby the content is annotated using predefined style templates. Depending on the specific element the transformation can identify text, pictures, headlines, etc. There are therefore more than 20 styles available. These styles are stored and linked to the content, so the transformation recognizes each specific element and can transform it.

By saving the document the process of content collection is completed. The document is now ready for transformation.

8 Process of Transformation

The complete process is controlled by the Apache ant build tool and uses saxon8 (see *http://saxon.sourceforge.net/*) as a transformation engine ([4]) supported by a graphical user interface (GUI).

After starting transformation the written Open Office document is unpacked. After that the file 'content.xml' which contains the content is accessible and can now be transformed (see Figure 3).

The first step of the transformation is to transform this content.xml into a new file called 'puretext.xml'. This new file is in an intermediate format. Thereby the annotated elements are assigned an explicit format template. Furthermore all elements without annotation are deleted. The result of this transformation is a universal intermediate format without direct correlation to the producer application.

The transformation is controlled by the XSLT stylesheets, content2pure_1.xsl' and, content2pure.xsl'.

The advantage of this intermediate format is that the resulting transformations are totally independent of the content collecting process or application. As a result a content collecting process using other applications is also possible in the future, e.g. Microsoft Word. Only the transformation into the intermediate format has to be adapted.

This file in intermediate format can be transformed into the output format (here HTML) by applying the XSLT stylesheets ,generateHTML.xsl' and ,html.xsl'

For this purpose each element from ,puretext.xml' is analyzed and the corresponding transformation instruction is searched. If the correct instruction is found, it is applied to the specific element. The transformation result is stored into the HTML file.

To print out the content easily a further version is generated. In contrast to the first form all learning objects, sessions and sections, are stored within one page. This page can easily be printed by using the print functionality of the browser.

```
<text:p text:style-name="EC_5f_Page" />
<text:p text:style-name="EC_5f_Text">Food law, as with all law, is
  enacted by states. Today the world consists of about 200
  independent states. This leads to more than 200 food law systems,
  which follow different trade theories.</text:p>
<text:p text:style-name="EC_5f_Text">
  <text:span text:style-name="T1">Centuries ago, the absolutistic
    states created a mercantile system. They cared only for the
    advantages of domestic production and used deep government
    interventions and control over the economy. Mercantilism is not
    the leading economic theory anymore. The liberalisation during
    the 19</text:span>
  <text:span text:style-name="T2">th</text:span>
  <text:span text:style-name="T1">century brought more and more
    freedom to private ventures. The states restricted themselves
    to setting general rules (e.g. contract rules) and controlling the
    conditions for foreign products (e.g. via customs and standards
    for imported goods). In times of protectionism, these rules
    were strict to give maximum room and better possibilities for
    the state owned products. Socialistic state law brought about
    similar results. At present, socialism and open protectionism is
    reduced to only a handful of states. Neoliberalism and the free
    trade of goods has become the leading economic theory in all
    continents. Economists believe in the power of worldwide
    liberalised markets.</text:span>
</text:p>
```

Fig. 3. Abstract of the unpacked *.odt

An overall glossary is generated by applying the stylesheet, glossary.xsl'. Hereby all elements which are annotated as glossary entries are transformed into a separate glossary file.

To use the learning module without an LMS there must be extra navigation. In this case all headlines from sections and sessions are transformed into the file 'menu.xml' supported by 'menu.xsl'

The metadata which is stored within the ,content.xml' is read and transformed into a further intermediate format, intermediateMD.xml'. From this the , generateLOMMetadta.xsl' generates the ,modulID.xml' and parenttitle.xml' which conforms to the LOM standard. Subsequently the complete content and all metadata are transformed into XML files.

9 Process of Test Transformation

Creating the static content of the eLearning module has become relatively simple. A much harder problem is the XML-based generation of dynamic tests. Here online forms have to be created, JavaScript code is needed for data pre-processing and data submission to the learning management system. The tests should work offline and online and the student should not be able to manipulate the results or to cheat by looking at the source code of the online test. Finally, the tests should run on as many platforms as possible.

In order to meet these requirements, we chose to use the GWT (Google Web Toolkit, see [5]), a new free software technology provided by Google for creating AJAX-based online applications. It centres around a powerful Java to JavaScript cross compiler. Using this approach the programmer, instead of coding in JavaScript directly, implements the online application in Java, compiles it and embeds the dynamic components into XHTML host pages. The GWT automatically generates

JavaScript code for the four most prominent web browsers. The generated code is obfuscated by the compiler, thus hiding the internal structure of the online application.

With this technology we have implemented a widget library framework for online tests. The widgets provided are flexible, configurable and easy to use UI components for multiple choice tests, fill in forms and free form questionnaires. Special forms have been designed for tests, where a set of given answers has to be ordered by the student dynamically. Multimedia elements, e.g. images or video sequences, may be integrated into all forms. Additional widgets for online help, user authentication and user guidance have been developed and can be combined with the test forms. The widgets are placed in XHTML host pages, where anchor elements define their positions (an example is shown in figure 3). Cascading style sheets are used to specify the visual appearance and the layout of the widgets. As a result the user interface design of the online tests is not fixed and can be adapted to user specific requirements.

In order to create SCORM 2004 compliant online tests, we need to be able to submit the results to the server. The ADL reference implementation of SCORM 2004 provides a simple JavaScript API for data persistence. When loaded, the eLearning unit registers itself with the LMS and initializes the API. In our framework, we used the JavaScript native interface (JSNI) of the GWT to connect to this API. This interface makes it possible to call arbitrary external JavaScript code from the generated application. The form data can be serialized into JSON (see [1]) and XML (see [11]). We also provide functions to encrypt and decrypt the form data.

10 A Preparatory Module in Microbiology

As an example we would like to give a short overview of one of the modules created using the eCampus approach. Figure 4 shows the module as it is displayed in the ADL reference implementation of SCORM 2004.

For the faculty of Nutritional Sciences a preparatory module for a biochemical practical course was developed. Up to now, a two week course had to be taken by the students introducing them to the biochemical laboratory of the faculty. The course has now been reduced to a single week. Beforehand the students receive a copy of the new module. It explains the laboratory setup and the basic working methods in microbiology. The module's content is based on an existing written handout. The handout was annotated with Open Office. Images were taken and integrated into the text. In addition multi media material was produced. A number of video sequences show the central steps of the working methods (e.g. how to produce a sterile culture medium). Students are requested to conduct a number of statistical analyses. Online simulations have been developed, making it possible, to gain some experience in virtual experiments. In order to control the students learning progress, a set of 100 questions have been annotated and integrated into the module. These questions are automatically transformed into online quizzes.

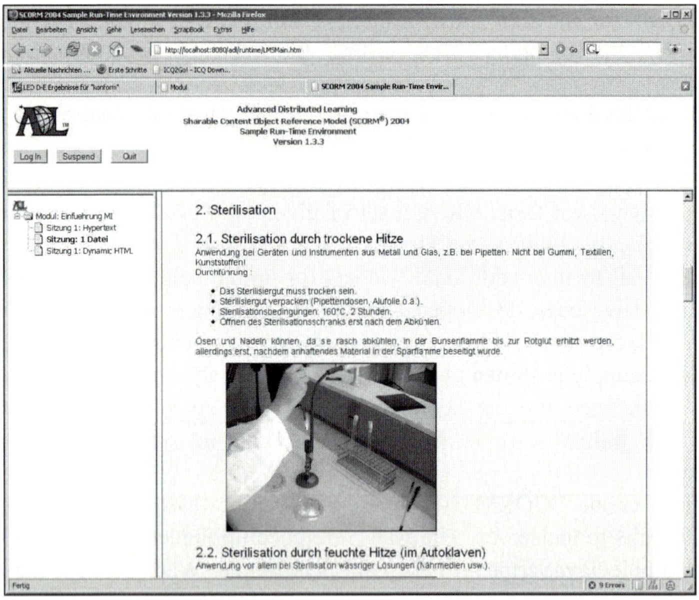

Fig. 4. Example of an eLearning Module within the ADL Reference Server Implementation

11 Advantages and Disadvantages of the eCampus Framework

The advantage of the eCampus framework is the easy production of SCORM 2004 conform eLearning modules.

Most of the learning material is drafted and written with the support of a word processor. With the eCampus framework there is no need of change the working environment. Even for beginners producing eLearning modules, using the templates to annotate the content is simple. All users with experience with the framework confirm this.

A further advantage is the independent intermediated format. Alternative applications for collecting content can be used in the future.

In contrast to most other editors the most important advantage is the conformity to the standard SCORM 2004. Through consequent implementation the eLearning modules can be used within arbitrary LMS which can administrate SCORM 2004 conform eLearning modules. Therefore a high re-use is possible. Also the reorganisation or editing with any standard conform editor is feasible. Especially adding constraints or information about sequencing is possible.

As well as small problems with the robustness of the framework which is due to the early prototype state of the software, the fundamental problem is the lacking route back from the module to the editor. The generated modules can not be transform back to the original document. Necessary editing of the content requires transformation once again.

Small corrections within a section are not problematic. Changes of the structure of the module are not possible because the storage of the state and results of tests can not be kept consistent.

With the eCampus project a new kind of teaching can be offered at Fulda University. ELearning modules can be created efficiently even by non-experts.

Already existing resources can be integrated into the curriculum and will be used in online, offline and blended learning situations. The use of SCORM 2004 compliant eLearning modules allows more complex student to LMS interactions, thus facilitating the design of electronically supported courses of a higher didactic complexity ([9], [7]).

References

1. Crockford, D.: RFC 4627. The application/json Media Type for JavaScript Object Notation (JSON). Online, accessed (17-11-2006) http: //www.json.org/
2. Eisenberg, J.D.: OASIS OpenDocument Essentials. LuLu.com (2004)
3. Hodgins, W., Duval, E., et al.: Draft Standard for Learning Object Metadata. Report, Learning Technology Standards Committee. IEEE, Washington (2002)
4. Holzner, S.: Ant - The Definitive Guide. O'Reilly Media (2005)
5. Google Inc. GWT: the Google Web Toolkit. Online, accessed (17-11-2006) http://code.google.com/webtoolkit/
6. Kay, M.: XSLT 2.0 programmer's reference, Indianapolis, IN. [u.a.] Wiley, Chichester (2004)
7. Koper, R.: Modeling units of study from a pedagogical perspective (2001) http://eml.ou.nl/introduction/articles.html
8. Advanced Distributed Learning. SCORM 2004. Online, accessed (17-11-2006) http://www.adlnet.gov/scorm/index.cfm
9. Schulmeister, R.: Grundlagen hypermedialer Lernsysteme: Theorie - Didaktik - Design. München [u.a.]: Oldenbourg Verlag (2002)
10. OpenOffice.org team. OpenOffice. Online, accessed (17-11-2006) http://www.openoffice.org/
11. W3C. XML. Online, accessed (17-11-2006) http://www.w3.org/MarkUp/
12. Wilde, E., Lowe, D.: XPath, XLink,XPointer, and XML. Addison-Wesley Professional, London (2002)

Simulation-Based Automated Intelligent Tutoring

Barbara Sorensen[1] and Sowmya Ramachandran[2]

[1] Air Force Research Laboratory, Mesa, AZ USA
Barbara.sorensen@mesa.afmc.af.mil
[2] Stottler Henke Associates, Inc., San Mateo, CA USA
Sowmya@stottlerhenke.com

Abstract. Simulation-based training has traditionally relied on live simulations for the specification of training scenarios. However, both the logistics and cost of live simulation constrain how often that form of training can be accessible. As the healthcare community begins to invest significantly in simulation-based training, it is running against the limitation of requiring expensive hands-on instructor-led facilitation to make it effective. New technological advances applied to simulation-based training provide automated coaching and feedback which reduce the need for instructor facilitation, thus making simulation-based training cost-effective and feasible. The design and development of SimCore (**Sim**ulate, **Co**ach, **Re**view), a simulation based framework that adds intelligent performance assessment and coaching facilities to training simulations, is discussed. Through the use of SimCore, simulation developers are able to convert their simulations into automated, intelligent tutors. SimCore is a unique application that includes a scenario authoring tool tailored to the medical domain that can be used to rapidly customize existing scenarios or to create new ones.

Keywords: Simulation-based training, Intelligent Tutoring, Scenario Authoring Tool.

1 Introduction

Simulation-based training can significantly benefit learners by providing them the opportunity to practice and learn from applying their skills and knowledge to realistic training scenarios [1,2,3]. Traditionally, simulation-based training has meant live simulations. The logistics and cost of such simulation constrain how often they can be presented. Typically such simulations deliver one or two scenarios. With computer-based simulations, it is possible to deliver cost-effective training with the added benefit that trainees can play them anywhere, anytime. In order to support such self-paced learning, simulations must be accompanied by performance assessment and feedback. As the healthcare community starts to invest significantly in simulation-based training [4], it is running against the limitation of requiring hands-on instructor-led facilitation to make it effective. This is prohibitively expensive. Automated coaching and feedback reduces the need for instructor facilitation, thus making simulation-based training cost-effective and feasible.

M.J. Smith, G. Salvendy (Eds.): Human Interface, Part II, HCII 2007, LNCS 4558, pp. 466–474, 2007.

According to Kolb [5], experiential learning is a cyclic process where a student works with a concrete situation, reflects on his experience, creates abstractions of the knowledge gained from the experience, and finally tries his new knowledge on other related situations. Simulations provide the concrete experience. Without reflection and abstraction, the experiential learning cycle would be incomplete. Typically instructors work with students to help complete the learning cycle; however, this makes the cost of simulation-based training considerably higher. Additionally, providing a personalized learning experience requires one-on-one instruction which can increase the cost substantially. Techniques from the fields of Artificial Intelligence and Intelligent Tutoring systems can be applied to automate one-on-one personalized instruction including automated performance assessment, coaching and review. Incorporation of these techniques will close the loop on the experiential learning cycle while freeing simulations from the requirement for dedicated instructors.

We are currently developing technologies for applying the concept of Intelligent Tutoring Systems (ITS) to healthcare simulations. ITSs are software tutors that are designed to provide one-on-one tutoring, much like humans [6]. As a first step in this direction, we are developing SimCore (**Sim**ulate, **Co**ach, **Re**view), a tool that adds intelligent performance assessment and coaching facilities to training simulations. Using SimCore, simulation developers can convert their simulations into automated, intelligent tutors. Freed from the need for hands-on instructor-led support, simulation systems can be used to deliver not one but a variety of scenarios, giving students the opportunity to learn by applying their newly acquired skills under varying conditions. SimCore also allows for complex free-play simulations. It can be used to actively guide a student towards performing an action or a procedure, even as the simulation allows them the freedom to explore. Authoring tools are typically needed to make ITs a viable option [7]. SimCore includes a scenario authoring tool that can be used to rapidly customize existing scenarios or to create new ones, thus, precluding the additional time and expense of acquiring a compatible authoring tool.

2 Architecture

SimCore employs a client-server architecture with two server-side components (Figure 1): the authoring tool and the runtime server. The authoring tool enables the specification of training scenarios in abstract, high-level terms. The runtime server drives the simulation, initiates events, responds to and evaluates student actions, and provides after-action reports. The system has been designed to interface with third party simulators with minimal customization of the runtime engine or the authoring tool. A default Flash-based simulator is included with the SimCore package.

The runtime server connects to a simulator via a simulator Application Programming Interface (API). This simulator API must be customized to work with each specific Simulator. For the included simulator, communication between the runtime server and the simulator is performed using Extensible Markup Language (XML). In general, the simulator is responsible for maintaining the user interface for one or multiple players. The runtime server is responsible for maintaining the state of the simulated world.

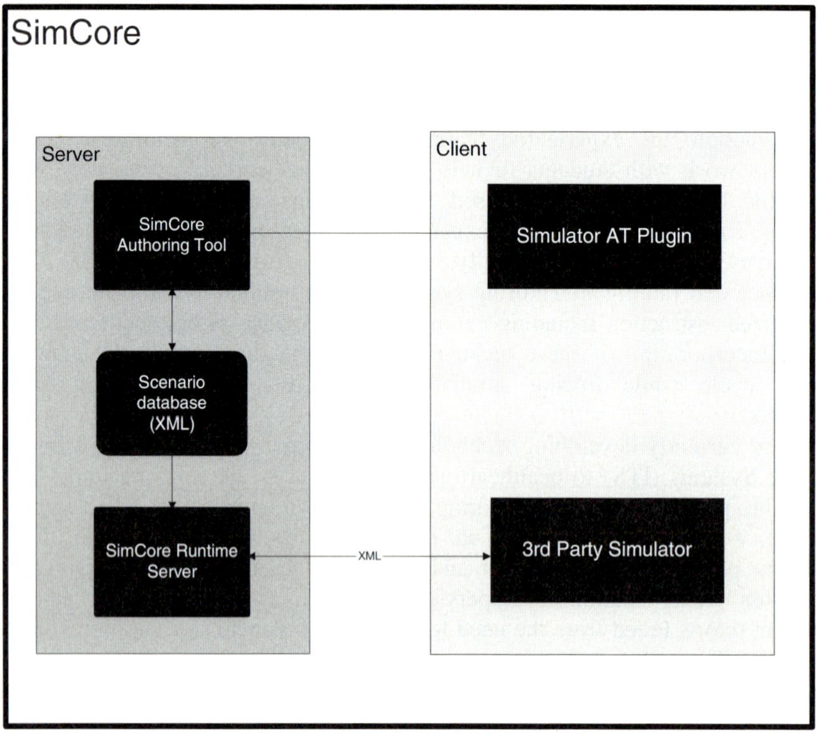

Fig. 1. SimCore client-server architecture with two server-side components

In a typical case, with the SimCore Runtime Server running, the player would run the Simulator. The Simulator would have a list of scenarios from which the player can choose. The list of scenarios is generated by the Runtime Engine. After selecting a scenario, the player uses the Simulator to start the simulation. The Simulator notifies the Runtime Engine when the simulation begins, and the Simulation Engine sends packages of information that contain the current location of the player and other locations the player can move to, all of the props in the location, and all of the actions that can be performed on those props or on the location itself.

The Simulator receives this package of information and is responsible for organizing these options for the player. The Simulator is responsible for displaying a representation of the current location; whether text based, 2D graphical or 3D graphical. In general, the player has two options: perform an action on a prop (or location) or move to a new location. Once the player chooses one of these options, the Simulator sends a message to the Simulation Engine containing the option chosen.

Upon receiving a move message, the Simulation Engine moves the player to the new location. It calculates the props and the actions that can be performed on these props and packages them up into a new message to be sent to the Simulator. When this message is sent to the Simulator, the Simulator acts just as when the game started: it displays a representation of the new location, and sets up a mechanism for the player to access all of the possible actions for the new location.

Performing an action on a prop (or at a location) is the core mechanism of the simulation. The action is tied to the item or location and, performing an action potentially changes the state of the simulation world.

In the Runtime Engine, there are a series of events that are triggered by actions or other events. These events do things like change the simulation state, set up timers (for triggering other events,) end the simulation, etc.

Eventually, an event or a performed action will cause the simulation to end. At this point, the Runtime Engine evaluates the player's performance and sends a message to the Simulator that contains the score data. The Simulator displays this data and the simulation ends.

3 Scenario Model

SimCore is based on a generalized, domain-independent scenario model. In the runtime server, the actions, props, etc. are represented by a set of Java Objects. These building blocks are created by the authoring tool, stored in a series of XML files, and read in by the runtime server. A scenario is composed of locations, props, NPCs (non-playing characters), and events. Locations represent the various different logical places in a scenario. For example, a side of the street which is the scene of an accident can be a location. An ambulance may be a second location. Props are scenario objects in locations and can have actions performed on them. The car involved in an accident is an example of a prop. NPCs are virtual human characters in the scenario. The person hurt in the accident, as well as his companions in the car would be modeled as NPCs. Props and NPCs are associated with attributes that represent their states at any given time. For an NPC, for instance, the modeled attributes may include his/her vital signs, age, medications, allergies. The scenario model includes actions that be performed on the NPCs and Props.

Each action has a unique command name, a textual representation to display to players, a series of possible textual replies upon performing the action, a series of sub-areas on the target where the action can be performed, a set of preconditions as to when this action can be performed, and a set of effects that are executed after the action is performed. For example, "Check Blood Pressure" may be an action with the precondition that the blood pressure monitor should be available for performing this action. The action definition would also include rules specifying the simulators response when the player performs this action. A scenario also includes events. Sometimes it is necessary to change the scenario state when the player does not initiate an action. Events are the way to accomplish this. Events are a set of effects that will be performed if a set of preconditions are met. They also contain a trigger set that, if equal to true, will cause this event's preconditions to no longer be checked and the event will be made inactive. An example is a box (prop) that is supposed to explode in 30 seconds if it has not been opened. An event is created that has a precondition that checks to see if 30 seconds has passed. If so, run the effect that causes the box to explode. The inactive section of the event checks to see if the box has been opened. If so, then the event's preconditions will no longer be checked, and the box won't explode because of this event. Authoring a scenario entails instantiating the scenario model. The Runtime Server executes this model to deliver a training scenario.

4 Automated Performance Assessment and Coaching

Built specifically to support human interaction with computers, an automated performance assessment component was developed and included in SimCore. This performance assessment component also guides students towards an optimal solution path. Each scenario can be associated with a solution template using the SimCore authoring tool. A solution template is a generalized procedure or protocol that is required to complete the scenario successfully. For the purpose of understanding the solution template, a generalized procedure, is one that can include unordered sets of actions and conditional actions. This solution template is not visible to the player; it is used behind-the-scenes to assess student performance. The solution template defines the actions the student should perform. In the development of the scenarios, the scenario author can also specify specific error rules that represent actions the student should not perform. This capability allows for more in depth scoring and feedback to the player.

During each scenario, the simulator software sends notification messages to the SimCore engine that describe each student action and reports the values of each simulation state variable. SimCore evaluates each student action by comparing it with the solution template and the error rules contained within the scenario definition. This information is passed back to the simulation which then shows the appropriate feedback. In addition, this information is included in the after-action review. The solution template is also used by SimCore to provide just-in-time hints which are associated with the solution template during authoring.

5 Authoring Tool

The authoring tool is a very important component of the SimCore framework. It enables the rapid development and modification of training scenarios. The authoring tool has multiple, tailored, visual editors for defining the scenario model described earlier. These editors include domain specific “wizards” which guide the author in creating the building blocks of the scenario. In addition, there is a section in the authoring tool for specifying the solution templates and error rules. Hints and feedback to help coach can also be specified using these editors. Figure 2 below shows the Props editor where scenario author can create Props, define their attributes, and the actions that can be performed on them.

To easily understand and participate in the building of a scenario, it can be broken down into the following parts: (1) designing the physical world, (2) populating the world, (3) adding events to the world, (4) linking the scenario world to the simulator Graphical User Interface (GUI), and (5) keeping track of what the student is doing.

Designing the physical world entails laying out a series of locations for a world map. Where can the players go? Are there locations that are unreachable at the start of the scenario? After establishing the locations, items or Props are added to the scenario. Props are inanimate objects such as cars, trees, houses or anything that the players can interact with in the scenario. If the player does not need to interact with an object, it does not need to be defined as a Prop. In addition, there are special Props called Tools that can interact with Characters in the scenario.

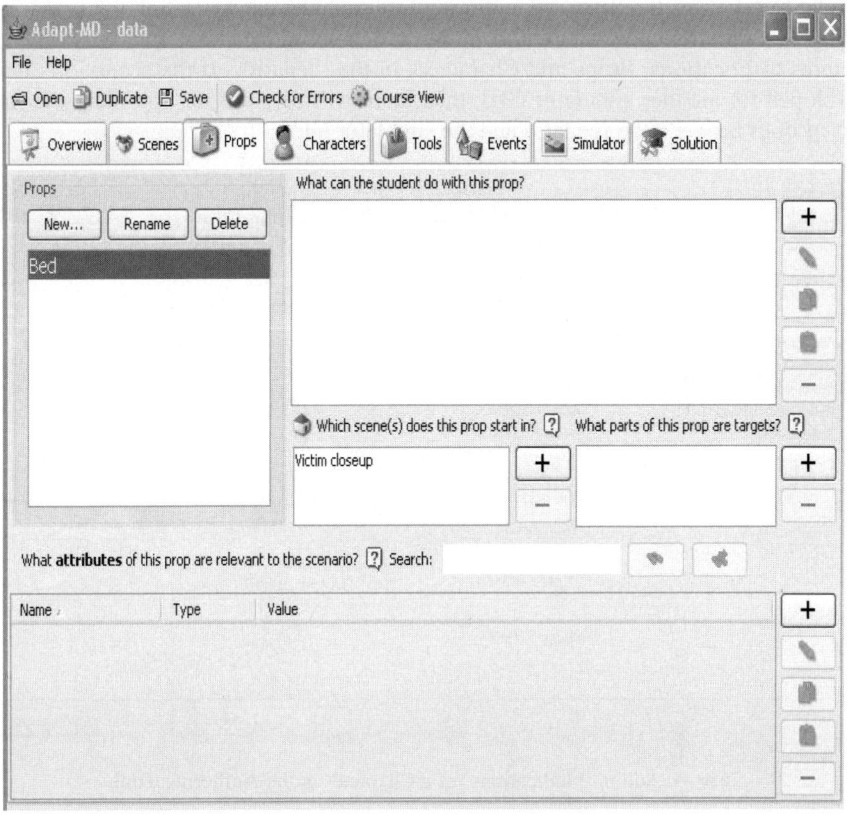

Fig. 2. SimCore Authoring Tool

Following the physical layout of the scenario build, the Characters are added to the scenario. Characters represent victims, bystanders and other people that are not players in the scenario. Locations, Props, Tools, and Characters can all have Actions associated with them. Actions either cause something to happen or report on the state of something in the scenario. For example, "check pulse" would be an Action that reports the state of a Character's pulse, whereas "bandage arm" would be an Action that causes something to happen.

Characters and Props have a series of attributes that allow the simulator to keep track of the state of the object during the simulation. The authoring tool includes domain specific wizards that help the author populate a number of default attributes (Figure 3). The types of data collected from the wizards were based on feedback from experienced EMS trainers.

Events are series of Actions that are not tied to Locations, Props, Tools or Characters. Events trigger on timers and other world state changes and execute their Actions without a player's direct influence. In addition, an event can end up triggering another event, allowing the author to create event chains for specific triggers.

Eventually, it is necessary to link the scenario to a simulator of some sort. The standard version of the SimCore authoring tool includes a way to link the scenario to a Flash based simulator. This allows the author to add pictures and connect these pictures to Locations, Props and Characters in the scenario. If the scenario is being developed for another simulator GUI, then there will be a tab for providing the same sort of links between the scenario and the simulator GUI.

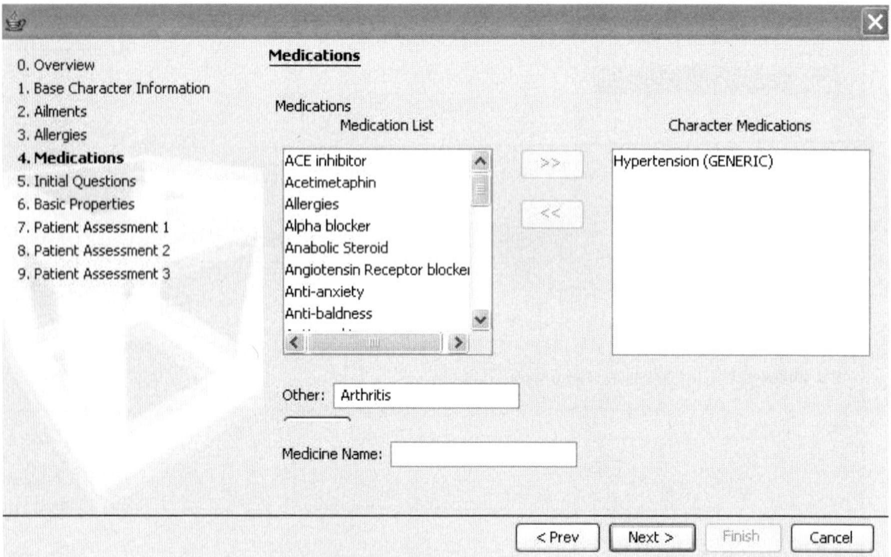

Fig. 3. Adding Medications for a Character in the Authoring Tool

Finally, there is a section in the authoring tool for outlining the performance assessment criteria expressed as a specification of the correct steps the student player is supposed to perform in the scenario. A solution template is a tree made of nested nodes which can be simple task nodes or group task nodes. Each simple task node recognizes a single student action. It specifies:

- An action pattern that can be matched by certain actions,
- An optional simulation state condition that specifies a constraint on the values of one or more simulation state variables that must be satisfied in order for the task node's action to be appropriate at a given time,
- zero or more principles that are demonstrated when the student carries out a matching action when the simulation state condition is satisfied,
- optional text that is displayed as a hint to the student, and
- optional text explaining the justification for the action. This text is shown during after-action review to explain the significance of a particular action.

Each group task node contains one or more simple task nodes and/or lower level group task nodes. It has the following information associated with it:

- An optional description that serves mostly as annotation
- A condition field whose intent changes depending on whether the node has been marked mandatory or conditional. Mandatory nodes have an associated pre-condition field that specifies the condition that must be satisfied before the actions in the group can be performed. Nodes marked Conditional have a condition field that states when the node is valid.
- Zero or more learning objectives that are demonstrated when the student carries out the actions that are recognized by the simple and (sub) group task nodes in this group.
- Justification that explains the reason for the task.

Each group task node can be configured so that it is satisfied when the student carries out actions that match (or satisfy) all of its children nodes, or just one of the children nodes.

In addition, there is a mechanism to incorporate expected errors as specified by the author. Expert instructors are often aware of the typical mistakes made by novice learners. With error rules, SimCore provides scenario authors with a way to recognize these types of mistakes and provide specific feedback to the student. For example, an instructor may know from experience that novice paramedics typically forget to wear gloves before administering fluids. An error rule can be used to recognize this and provide feedback on the importance of wearing gloves to avoid possible infections.

Associated with each error rule is text feedback that will be sent to the simulator when an error rule is matched by a student action. Similarly, authors can specify explanations with error rules that are sent to the simulator as a part of the after-action review. This is intended as a detailed contextual explanation of the error with enough information to help the student avoid this error in the future.

6 Evaluation

The SimCore framework was developed in three spirals. The software at the end of each spiral was demonstrated to experienced EMS trainers. Each time, the expert suggestions have strongly influenced the design for the next spiral and for the final system. We are currently making the product available to beta testers and we will be collecting feedback from these users.

7 Related Work

A few intelligent tutoring systems have been developed for healthcare domains [8,9]. In most such systems, the tutoring is tightly coupled with the simulation or the problem-solving interface. SimCore is based on a domain-independent scenario model and can thus be used for a broad range of domains. It is designed to be plugged into third-party simulators with minimal effort. Murray [7] presents an in-depth discussion of authoring tools for Intelligent Tutoring Systems. The trade-off between power and usability is a central dimension that characterizes authoring tools. Our objective is to develop a tool that is usable by healthcare training experts who have little programming experience. Hence our first version of SimCore trades power for authoring simplicity. Future versions of the system will make the framework increasingly powerful while involving users in the development process to ensure that usability does not get sacrificed.

8 Future Work

SimCore beta is now available. SimCore is re-usable framework for developing low-cost, customizable simulations and training games for training EMS and other healthcare professionals. Using SimCore, subject matter experts and course developers can rapidly create simulations to test students' ability to apply their skills and decision-making capability in realistic scenarios. SimCore provides the facility to incorporate automated performance assessment and feedback, enabling organizations to create rapidly powerful, skill-application oriented self-paced training. A Flash simulator that is included as a part of the SimCore package makes the system usable by organizations that do not have a third-party simulator at their disposal.

In the near future, our primary objective to collect feedback from beta users on the system, understand how well it meets their needs, and gaps that must be addressed by future versions. Efficiency related improvements are also in the near-term agenda.

Our long-term vision for the product includes improved usability, enhanced student modeling, adaptive coaching, and reflective after-action review with Socratic dialogs.

References

1. Schank, R.: What We Learn When We Learn by Doing, Technical Report no. 60, Institute of Learning Sciences, Illinois (1995)
2. Johnson, W.B., Norton, J.E.: Modeling student performance in diagnostic tasks: A decade of evolution. In: Regian, J.W, Shute, V.J. (eds.) Cognitive Approaches to Automated Instruction, Lawrence Erlbaum Associates, Mahwah (1992)
3. Madni, A.M., Madni, C.C., Sorensen, H.B., Garcia, S.K.: Intelligent agents for individual and team training applications. In: IEEE Proceedings of the Man, Systems and Cybernetics Conference (October 2005)
4. Sorensen, H.B., Babbitt, Bettina,: Training medical professionals using interactive simulation. In: Research on Instructional Competencies, AACE, World Conference on Educational Multimedia, Hypermedia and Telecommunications (June 2003)
5. Kolb, D.A.: Experiential Learning. Prentice-Hall, Englewood Cliffs, NJ (1984)
6. Forbus, K.D., Feltovich, P.J.: Smart Machines in Education. AAAI Press, California (2001)
7. Murray, T., Blessing, S., Ainsworth, S.: Authoring Tools for Advanced Technology Learning Environments: Toward Cost-Effective Adaptive, Interactive, and Intelligent Educational Software. Springer, Heidelberg (2003)
8. Shaw, E., Ganeshan, R., Johnson, W.L., Millar, D.: Building a Case for Agent-Assisted Learning as a Catalyst for Curriculum Reform in Medical Education. In: Proceedings of the Int'l Conf. on Artificial Intelligence in Education (July 1999)
9. Kizakevich, P.N., McCartney, M.L., Nissman, D.B., Starko, K., Ty Smith, N.: Virtual Medical Trainer: Patient Assessment and Trauma Care Simulator. In: Westwood, J.D., Hoffman, H.M., Stredney, D., Weghorst, S.J. (eds.) Medicine Meets Virtual Reality - Art, Science, Technology: Healthcare (R)evolution, pp. 309–315. IOS Press and Ohmsha, Amsterdam (1998)

Virtual Pop-Up Book Based on Augmented Reality

Nobuko Taketa, Kenichi Hayashi, Hirokazu Kato, and Shogo Noshida

Graduate School of Engineering Science, Osaka University
1-3 Machikaneyama-cho, Toyonaka, Osaka 563-8531 Japan
{taketa, hayashi}@nishilab.sys.es.osaka-u.ac.jp,
{kato, nishida}@sys.es.osaka-u.ac.jp

Abstract. In this paper, we introduce a virtual pop-up book system using Augmented Reality technology. This system displays 3D virtual objects on the real book based on pose and position estimation of the camera. Although many marker-based methods have been proposed, a picture book with markers looks unattractive. Our system does not use any markers. We describe four advantages for the virtual pop-up book. Firstly, 3D rendering helps readers understand scenes. Secondly, characters look lively by using motions. Thirdly, author of the picture book can use the new representation, which mixes 2D and 3D rendering. Lastly, it can express time changes using animation.

Keywords: pop-up book, Augmented Reality, texture tracking.

1 Introduction

Picture storybook is a widely used tool for providing amusement and educational effects to children. There are many kinds of picture books. For example, fairy tails all over the world, scientific stories, lyric prose, nonfiction, arithmetic picture books, fancywork and cooking books and so on. Many kinds of books are proposed to provide exciting impressions, such as gimmick books, pop-out books, perforated books, hologram books, sound books, and so on. As just described, sometimes changes when, wide variety of material is implemented and the structure of book itself [1][2]. Under this situation, we would like to propose a new framework.

Recently, characters and scene are showed in 3D computer graphic in movie theaters or theme park attractions. This style attracts the gallery. Like this, by adding 3D computer graphics into 2D scenes, users feel the scene alive and they become involved in the story. We apply this style to picture storybooks. We call this style of book "Virtual Pop-up book".

In case of movies, it is easy to display 3D objects on the screen because the screen is fixed. However, in our case, it is difficult because user can move the book. It is necessary to adjust 3D objects to the picture book constantly. This problem is proper for geometric compliance in Augmented Reality (AR) [3]. AR technology supplements the real world with virtual object that appear to coexist in the same space as the real world. We apply these methods, which were proposed by us before, to this problem.

M.J. Smith, G. Salvendy (Eds.): Human Interface, Part II, HCII 2007, LNCS 4558, pp. 475–484, 2007.

The structure of this paper is as follows. Section 2 provides several related works. Section 3 describes concepts of virtual pop-up book. Section 4 describes system overview. Section 5 describes the proposed methods. A facture of virtual pop-up book is presented in section 6 and conclusions are drawn in section 7.

2 Related Works

A virtual pop-up book using AR was introduced by Rekimoto's work [4]. They picked up examples of AR system using 2D matrix code. Other examples of the virtual pop-up books are "Magic book" [5] proposed by Kato, "Little Red" [6] proposed by Saso and "Vivid Encyclopedia" [7] proposed by Shibata. Magic Book used a book as a transitional interface among the real environment, augmented reality environments and virtual environments. A user of the system can fly in/out the virtual environment. The story of Little Red forks at certain points, in accordance to the user's actions. Vivid Encyclopedia is MR pictorial book of insects. The system displays various 3D CG insects. A user of the system can transfer the insects by his hand and scale them with an interaction device. Magic Book and Little Red need particular square markers for estimation of the camera position. However, the markers disturb the pictures and makes users feel uncomfortable. Vivid Encyclopedia does not use square markers, but the pages have to be consisted of black background and picture of insects. Black background helps for estimating the camera position. We aim for the realization of a virtual pop-up book that does not use these specific expressions. In particular, our system estimates the camera position using natural features of the pictures.

3 Concepts of Virtual Pop-Up Book

We call a picture book using AR technology "Virtual Pop-up book". A user can see 3D CG objects on the pictures of the book. This system is very user-intuitive. User reads the book turning pages as with popular pictorial books. Therefore, everyone can use the system without many instructions.

We will give four points of advantage with our system compared to existing picture books.

- 3D virtual objects help readers understand scenes.
- Characters look lively by using motions.
- New means of expressions is possible by mixing 2D and 3D rendering.
- It can express time changes using animation.

We will show these points in detail in the following.

3.1 3D Virtual Objects

Our system uses 3D virtual objects. Those help the user to understand a scene that is difficult to express with 2D illustration. For example, Fig.1 is a scene of a girl going to a red roof house rendered by 3D CG. User can peek into the house by moving the illustration.

Fig. 1. Viewing a scene from different angles

3.2 Characters Look Lively

It is important to draw characters lively in the book to attract readers. Painters normally express character's motions such as jumping and running through 2D drawing. However, it is difficult to express little motions such as a character being annoyed of waiting or a sleeping persons breathing. Using CG animation, such little motions can be expressed easily.

3.3 Mix 2D and 3D Rendering

Illustrations of original picture books are 2D. However, with our proposed book, 2D illustration, 3D animation and mixed 2D and 3D rendering is possible, providing multiple presentations. Especially, mixing 2D and 3D rendering is the most significant feature in our new system. Our system calculates the position of the illustration. Therefore, the 3D objects can be displayed at the proper position just as the author had designed. In Fig .1, a house, which has red roof, is rendered in 3D and the other objects are 2D illustration. The 3D house is rendered where the 2D house originally existed without any strangeness. In Fig.2, 2D painted flowerpot is overlapped with 3D rendering flower. As can be seen, there is no strangeness in mixing 2D and 3D rendering. This can make readers pay attention to the 3D object.

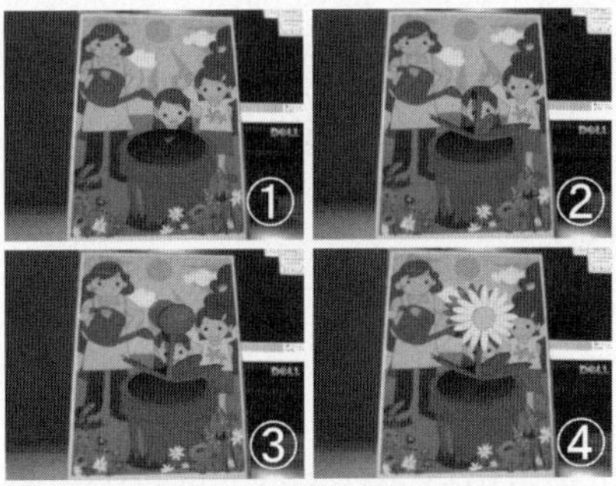

Fig. 2. Examples of time change

3.4 Express Time Changes

In original picture books, it is difficult for readers to convey passage of time and sounds using pictures. Instead, readers understand them by the text [8]. However, our book can convey time passage by an animation on an illustration. Fig.2 shows a flower growing. The scene expresses time change.

4 System Overview

Our system has two kinds of display types. One is PC monitor (Fig. 3). The other is Head Mounted Display (Fig. 4). The PC monitor type is better suited for uses by children who feel HMD as a load or demonstrations where many users see the content at the same time. On the other hand, HMD type offers immersive sense to the user.

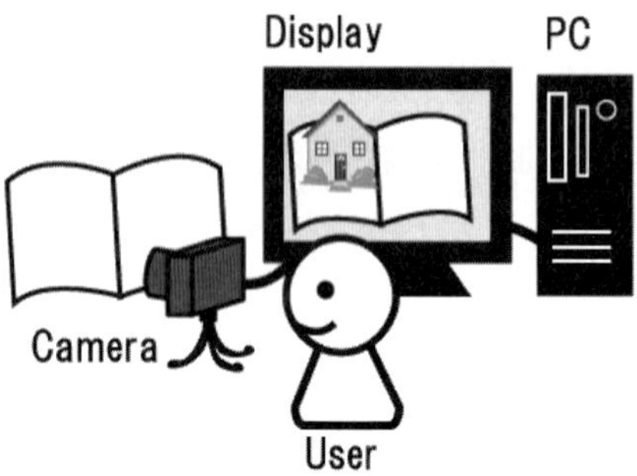

Fig. 3. PC monitor-based display system

Fig. 4. HMD-based display system

5 Methods

We use ARToolKit [9] for obtaining image from the camera and rendering CG objects. It is necessary to get camera pose and position to the book for rendering 3D CG objects at the proper position. This is a common problem in AR research field. Many kinds of vision-based methods have been proposed so far. Our system needs marker-less registration method. We take on the method that uses natural features. Fig.4 is our system's process flow. We introduce the three parts of the flow in which are, "Page determination", "Initial pose estimation" and "Tracking".

5.1 Page Determination

Flow of page determination is described in the following. (Fig. 6)

1. Template of each page is prepared. The templates are cut out from input image by using image-editing software.
2. Input image is reduced to 1/16 by averaging four pixels twice.
3. Searching the templates in 1/16 input image by using active search [10].
4. Adapting a page that has the highest similarity template.

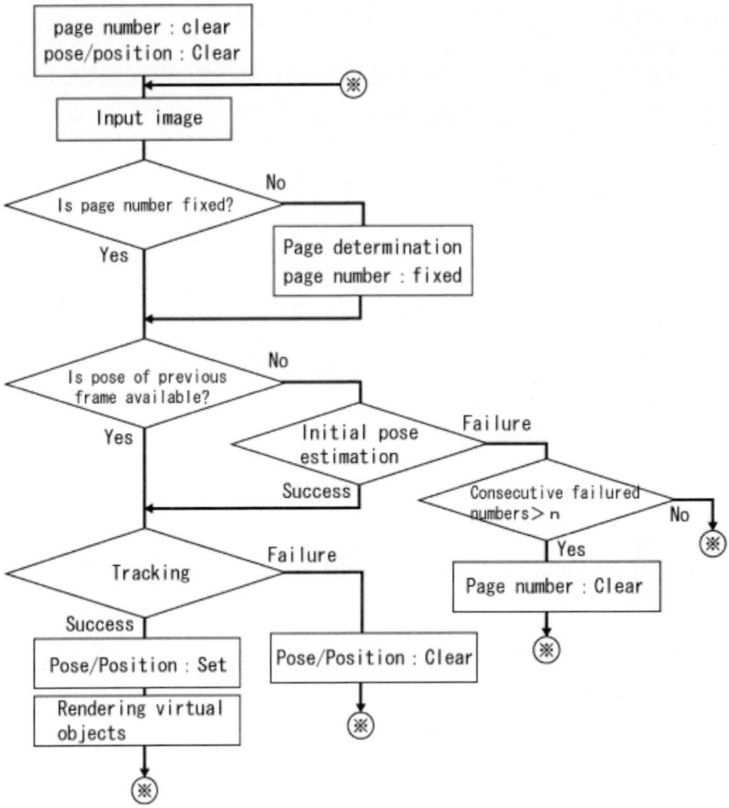

Fig. 5. Flow of processing

5.2 Initial Pose Estimation

If we can get a camera pose that means the relationship between the world coordinates and the image coordinates, 3D objects are displayed in proper position. It is necessary for the pose estimation to search for the feature points those world coordinates are already known from the input image. Detected feature points are on a 2D planar surface, because this system treats book illustrations. Under this condition, four pairs of world coordinate and image coordinate of feature points lead the camera pose.

Pose estimation consists of two phases, which are initial pose estimation and tracking. In tracking phase, tracking history can reduce the search area of feature points. The feature points do not move so much from the positions of the previous frame. However, in initial pose estimation phase, the search area is the entire input image. Therefore, it takes very long time for searching. To deal with this problem, advance active search method [10][11] is adapted. The flow of the initial pose estimation is described as follows.

1. Making templates of feature points. (Off-line processing)
2. Searching some feature points from the input image.
3. Calculating camera position by using image coordinates of detected feature points.

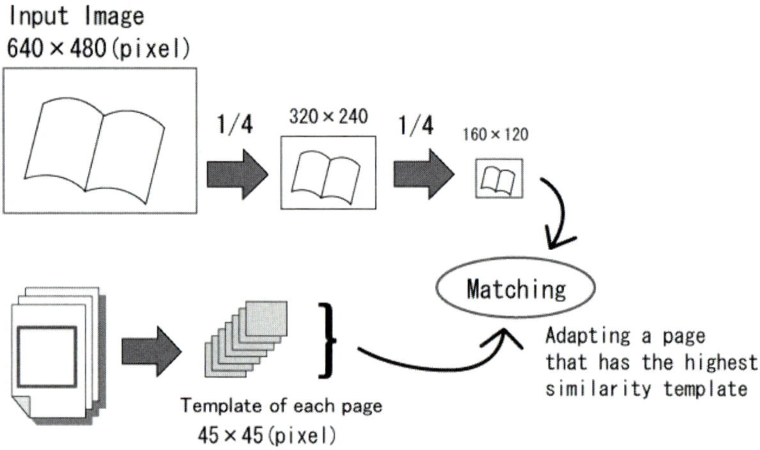

Fig. 6. Page determination method

In the template-making, the templates of the area that matches to the following two conditions are generated.

- The area with various kinds of colors.
- The area where active search frequency decreases.

Then, four templates are selected from all templates so that those templates widely spread on the target illustration. The fifth template or later is selected based on order of registration. Fig.7 is an example of the selected templates.

This system applies circularizing templates, normalizing brightness of RGB and multiplexing histogram for increasing matching accuracy.

We made a prototype system and tested for pose estimation. The test environment is as follows. The input image size is 640x480. The template size is 25x25. 5 templates are used per an illustration. The objects for estimation are multiple color illustrations. We found from the testing result that our method works even when the object is declined (z axis (camera axis):360°, x and y axis: -45°-45°) and brought close to the camera or turned from the camera (20cm~60cm). Processing speed is 5 fps. This speed is acceptable, because once camera pose is acquired, the process switches with tracking.

5.3 Tracking

Once getting the camera pose from initial estimation, the process switches with tracking.

We employ the texture tracking [12] for the tracking method. This method does not use special markers in the iterative tracking phase. Several natural feature points are extracted from the tracked object by off-line image analysis. While tracking the

Fig. 7. Example of selected templates

object, some of these feature points are selected for template matching, and the object pose and position are calculated. Flow of this method is as follows.

1. Forecasting the image coordinate of the feature points.
2. Making templates based on the image coordinate of the feature points.
3. Template matching using the generated templates.

This process finds the pairs of world coordinates and image coordinates. With over 3 pairs, the camera pose can be acquired.

The feature points are registered as many as possible with respect to each resolution (Fig. 8).

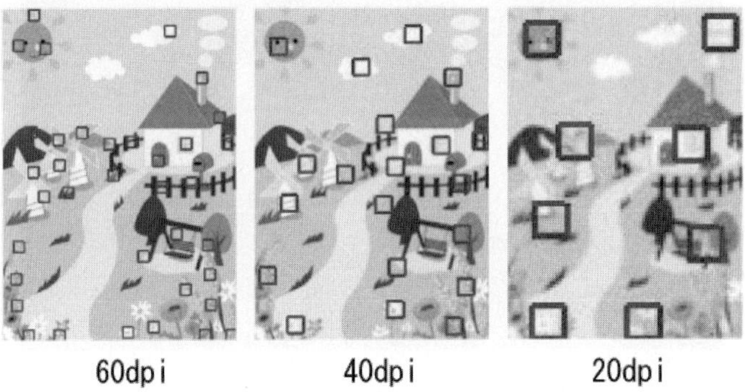

Fig. 8. Example of registered templates

We made a prototype system and tested its performance. The environment is as follows. Input image size is 640×480. Template size is 24×24. As a result, it worked excellently in real time (30fps).

6 A Facture of Virtual Pop-Up Book

We present a facture of virtual pop-up book. The display type of this facture is monitor-based system. A PC and a camera is implemented this system. The PC uses a 3.2GHz Pentium 4 processor, 1GB memory and the OS is Fedora Core 3. The camera is Dragonfly manufactured by "point gray". Its resolution is 640x480 pixels.

Fig. 9. Book illustrations

The book is consisted of seven scenes. A scene has a text part (right) and an illustration part (left). The size of the book is A4. 3D virtual objects are displayed on an illustration. Fig.9 shows the book illustration. Fig.10 shows the 3D virtual objects.

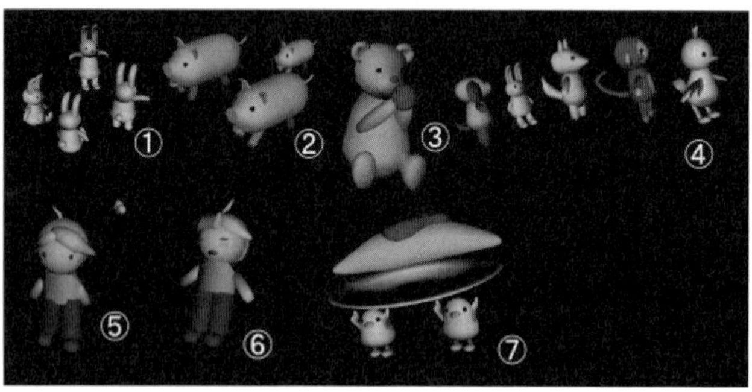

Fig. 10. 3D objects and page number

Fig. 11 is a picture of the book in use. Right hand side pictures of this figure are pictures through the camera. There are no discomfort between 3DCG character and 2D illustration. Processing speed of the system is 30fps, except the initial estimation process. Users are not much concerned about this, because the initial process runs once when the user change the pages.

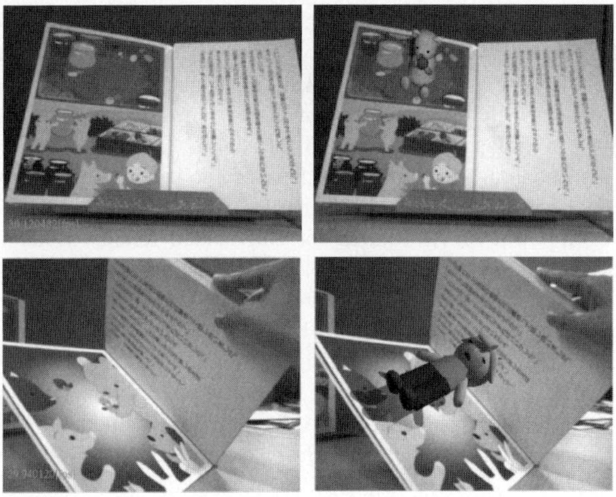

Fig. 11. Displays 3D virtual objects on the real book (Left: Before, Right: After)

We demonstrated this system in MIRU 2005 [13]. It seemed that the users amuse oneself with the system moving the camera in different directions. Users mentioned that combining 3D object and 2D illustration is interesting, and it will give pleasure to children. On the other hand, some users indicated that the letters of the book was not easy to read

7 Conclusion

In this paper, we proposed a virtual pop-up book and described its characteristics. Then, we outlined the methods. Finally, we showed a facture, and described the demonstration.

For the feature works, we will make the system more interactive and discuss about more robust and high-speed methods.

References

1. Nihon kodomo no hon kennkyukai.: Kodomo to ehon no gakkou, PORUPU-SHUPPAN Publishing,(in Japanese) pp. 9–17 (1988)
2. Nakagawa, M.: Ehon ha art –Hirakareta ehonron wo mezasite-, KYOIKU-SHUPPAN Publishing (in Japanese) pp. 80–83 (1991)
3. Yokoya, N.: Mixed Reality: Merging real and virtual World – 1 What is Mixed Reality, Systems. Control and Information (in Japanese) 49, 489–494 (2005)
4. Rekimoto, J.: Matrix: A Realtime Object Identification and Registration Method for Augmented Reality. In: Proc. Of Asia Pacific Computer Human Interaction (APCHI '98), pp. 63–68 (1998)
5. Billinghurst, M., Kato, H., Poupyrev, I.: The MagicBook: A Transitional AR Interface. Computers and Graphics, pp. 745–753 (2001)

6. Issa Saso, T., Iguchi, K., Inakage, M.: Little Red: Storytelling in Mixed Reality, SIGGRAPH 2003 (2003)
7. Shibata, F., Yoshida, Y., Furuno, K., Sakai, T., Kiguchi, K., Kimura, A., Tamura, H.: Vivid Encyclopedia: MR Pictorial Book of Insects. In: The VR Society of Japan Annual Conference the 9th (in Japanese) pp. 611–612 (2004)
8. Matsumoto, T.: Ehonron-Atarashii geijutu hyougen no kanousei wo motomete. IWASAKI bibliopole (in Japanese) 65(67), 123–125 (1982)
9. Kato, H.: Development of ARToolKit for Augmented Reality system, PRMU 2001(in Japanese) vol. 232, pp. 79–86 (2002)
10. Murase, H., Vinod, V.V.: Fast Visual Search using focused color matching-active search. System and Computers in Japan 31(9), 81–88 (2000)
11. Taketa, N., Kato, H., Nishida, S.: Fast Visual Search Considering Interval and Order of Matching. In: MIRU 2006 Proceedings (in Japanese), pp. 55–160 (2006)
12. Kato, H., Shiosaki, N., Tachibana, K.: A Registration Method for Augmented Reality based on Matching Templates Generated from Texture Image. TVRSJ (in Japanese) 7(2), 119–128 (2002)
13. Taketa, N., Kato, H., Nishida, S.: Virtual Pop-Up Book using Texture Tracking. In: MIRU 2005 Proceedings (in Japanese), pp. 1616–1617 (2005)

A Mobile Environment for Chinese Language Learning

Chang-Chih Tseng[1], Chun-Hung Lu[2], and Wen-Lian Hsu[2]

[1] Institute of Information Systems and Applications, National Tsing-Hua University, Taiwan
[2] Institute of Information Science, Academia Sinica, Taiwan
{jasper, enrico, hsu}@iis.sinica.edu.tw

Abstract. Due to the diversity of mobile device and wireless network's progress, mobile learning has become a possible and convenient way to support learning activity, whether it was in or out of classroom. There have been some researches showed that using such devices in the course can improve learners' performance and arouse their motivation, cases like bird watching, plant observing, museum navigation and tourist guidance. Chinese language was considered difficult to learn all the time, because of its complicated shape, different pronunciations and multiple meanings. In this paper, we propose a system that works on PDA for foreigners to learn Chinese. The design concept derived from geographic information sharing, people will use PDA to explore the city they live, and gradually find virtual messages around them, thus enhance their Chinese communication skill.

Keywords: Mobile learning, Chinese learning, GIS, Geotagging.

1 Introduction

Owing to the diversity of mobile devices and wireless network's progress, mobile learning has become a possible and convenient method to support teaching activities, whether it was in or out of classroom. Some researches have demonstrated that using such devices in the courses can improve learners' performance and arouse their motivation, cases like bird watching, plant observing, museum navigation and tourist guidance [2, 3, 4, 5, 6, 8]. According to our requirements, a mobile environment can make communication between teacher, learner, system and user's surroundings.

Follows are those researches on mobile assisted language learning, Duke University adopts ipod as their assistant teaching devices in Spanish lessons [1]; BBC provides students with java games and mp3 files on their GCSE website to help them learning English, French, Germany and Spanish [8]; Griffith University in AU uses SMS to advance students' understanding of Spanish vocabularies [7]; National Century University in Taiwan assists teachers and learners with the teaching and practice of English courses via e-dictionary [9]; Tokushima University in Japan combines PDA and PHS to form a teaching environment in real lives [10].

Chinese language was considered difficult to learn all the time, because of its complicated shape, different pronunciations and multiple meanings. In this paper, we propose a system called GeoLearn that works on PDA for foreigners to learn Chinese.

M.J. Smith, G. Salvendy (Eds.): Human Interface, Part II, HCII 2007, LNCS 4558, pp. 485–489, 2007.

The design concept derives from geographic information sharing, and combines with task-based learning thus enhance foreign learner's Chinese communication skill.

2 Background

2.1 GIS and Geotagging

Under the rise of web2.0, it has become more easily now that internet users can choose or edit the contents which they really interest to browse, and share them to others effectively. In recent years, several famous companies like Google and Yahoo were proposed their web services based on geographic information, to be known as Google Map and Yahoo Map. People who use the services can post their own messages onto the web-based map according to the longitude and latitude; the subjects inside may be some useful information about a living area, or someone's travel note. Furthermore, people are gradually being familiar with the various combinations of mobile devices and GPS, one of the most well-known usages is satellite navigation. Accompany with specific teaching strategy in this research, we implemented GPS into the system to carry out mobile learning.

2.2 Wikipedia

Wiki is a type of website which encourages people to edit its content jointly, and maybe the most famous one is Wikipedia. Till now, more than 210 language versions of Wikipedia were established. The amount of English vocabularies in it has exceeded one million, and Chinese vocabularies have accumulated to about sixty thousands in the meantime. Considering the rich content of Wikipedia, we employed it as the supplement database to built-in textbook vocabulary, and try to provide learners more complete learning experience.

3 Pedagogic Design

The theory of situated learning holds that learning itself is the outcome of mutual interactions between learner and his environment, not just about acquiring knowledge [11]. Besides realizing its interpretation, since language can be seem as a social tool, it's necessary to realize how to use it properly toward different people. Language learners in traditional classroom may not have sufficient opportunities to taste these subtle differences because of some external restrictions, while they can experience deeper cultural implications in regular life through chatting with native speakers surrounding them.

Another theory that sometimes be mentioned with previous one is task-based learning. It means that learners will be assigned several specially designated tasks correlate to certain learning objectives [12]. While in the process of finishing these tasks, the learners' capabilities will also grow up. Once applying the concept of task-based learning into second language acquisition, learners are setting free from dull practice in traditional classroom that mainly stresses on the repeat of words and sentences. Under the authentic conversation, learners need to pay more attention to the others' wording, and adjust their expression to communicate with them, therefore achieve the learning effect.

4 System Architecture

There are three major components in GeoLearn, called learning map, textbook lexicon and common dialogue query.

4.1 Learning Map

The interface of learning map presents a mini scale map which is relative to real world. Through the connection of GPS, learner will find himself as an icon on the map that represents his real location in the city. Besides, there are also many virtual tags can be seem and clicked to open on this map in different forms, each of them stands for a specific category. Learner may filter these tags by selecting the category in pull-down menu, or publish a new one.

4.2 Textbook Lexicon

The textbook lexicon helps learner to review or look up vocabularies he has learned in previous classes. It allows learner entering an English word or the Chinese phonetic alphabet called Hanyu Pinyin, then retrieve its interpretation and example sentence. If

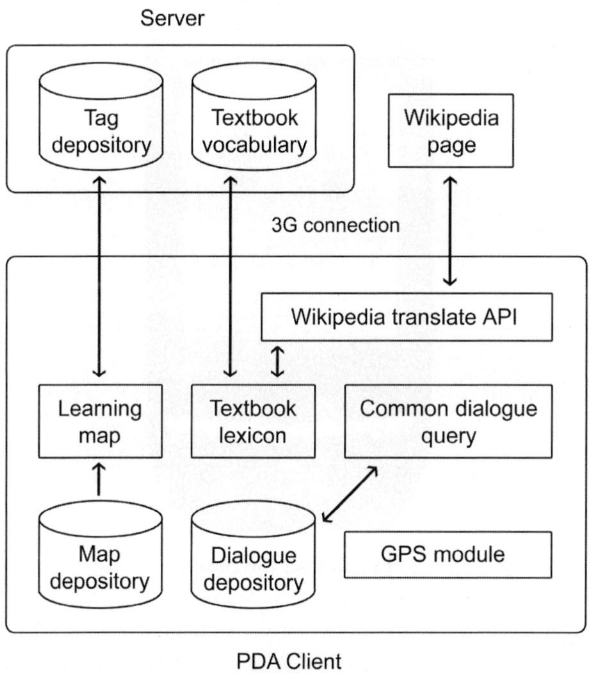

Fig. 1. System Architecture

there was no match result in our lexicon, the system will use wikipedia translate api to get relevant definition from wikipedia.

4.3 Common Dialogue Query

According to different occasions, the common dialogue query provides learner a convenient way to learn practical sentence with Hanyu Pinyin that fulfills his need at the moment.

5 Learning Activity

The main purpose of GeoLearn is to encourage learners exploring the city they live, and take down some interesting things in their regular life to share with others. Thus learners' Chinese communication skill may become mature, and also experience more social impact in the meantime.

The teacher will give a lesson first then arranges a script including a set of tasks, which learners can solve them by using vocabularies they acquired in the classroom. Among the progress of this script, all participators have to chat with local inhabitants in order to ask for helps or information to complete their tasks. Besides, they also need to draw supports from various tags which were posted by the teacher in advance, so finally they can reach the ending of this script.

Fig. 2. User Interfaces

6 Conclusion

There have been many cases for educators to inspect since the idea of mobile learning been proposed, and we mainly focus on language learning in this paper. Comparing to study only the literal meanings and pronunciations of the language, maybe the cultural features inside were more attractive and worth to note. Thus we employ the concept of message sharing and hope learners can keep their minds open to deal with local inhabitants more frequently. It is our thought that language learning should be an active social participation, not only occurs in the classroom. Afterwards, we will estimate the system's usability, user satisfaction, and learning outcomes. Then expand the scope of it to more other languages like Japanese and Korean.

References

1. Duke University ipod First Year Experience Final evaluation Report (2005)
2. Chen, Y.S., Kao, T.C., Sheu, J.P.: A mobile learning system for scaffolding bird watching learning. Journal of Computer Assisted Learning 19(3), 347 (2003)
3. Chen, Y.S., Kao, T.C., Yu, G.J., Sheu, J.P.: A mobile butterfly-watching learning system for supporting independent learning. In: Proceedings of the The 2nd IEEE International Workshop on Wireless and Mobile Technologies in Education (WMTE'04) (2004)
4. Chinnery, G.M.: EMERGING TECHNOLOGIES: Going to the MALL: Mobile Assisted Language Learning. Language Learning & Technology 10(1), 9–16 (2006)
5. Godwin-Jones, R.: Emerging Technologies: Messaging, Gaming, Peer-to-Peer Sharing Language Learning Strategies & Tools for the Millennial Generation. Language, Learning & Technology, vol. 9 (2005)
6. Hawkey, R.: Learning with Digital Technologies in Museums, Science Centres and Galleries. FUTURELAB SERIES, King's college, London (2004)
7. Horstmanshof, L.: Using SMS as a way of providing connection and community for first year students. ASCILITE Conference 2004, Perth, Western Australia (2004)
8. Laura Naismith, P.L., Vavoula, G., Sharples, M.: Literature Review in Mobile Technologies and Learning. FUTURELAB SERIES, University of Birmingham (2004)
9. Liang, J.K., Liu, T.C., Wang, H.Y., Chan, T.W.: Integrating wireless technology in pocket electronic dictionary to enhance language learning. In: Proceedings of the 5th IEEE International Conference on Advanced Learning Technologies (ICALT'05) (2005)
10. Paredes, R.G.J., Ogata, H., Saito, N.A., Yin, C., Yno, Y., Oishi, Y., Ueda, T.: LOCH: supporting informal language learning outside the classroom with handhelds. In: Proceedings of the 2005 IEEE International Workshop on Wireless and Mobile Technologies in Education (WMTE'05) (2005)
11. Lave, J., Wenger, E.: Situated Learning: Legitimate Peripheral Participation. Cambridge University Press, Cambridge, England (1991)
12. Willis, J.: A framework for task-based learning. Longman, London (1996)

Game Player Modeling Using D-FSMs

Tae Bok Yoon, Dong Moon Kim, Kyo Hyeon Park, Jee Hyong Lee*, and Kwan-Ho You

School of Information & Communication Engineering
Sungkyunkwan University, Suwon, Korea
{tbyoon, skyscrape, megagame}@skku.edu,
{jhlee, khyou}@ece.skku.ac.kr

Abstract. Recently, various ways are being explored for enhancing the fun of computer games and lengthening the life cycle of them. Some games, add realistic graphic effect and excellent acoustic effect, and make the tendencies of game players reflected. This paper suggests the method to collect and analyze the action patterns of game players. The game players' patterns are modeled using FSM (Finite State Machine). The result obtained by analyzing the data on game players is used for creating NPCs (Non-Player Characters) which show new action patterns by altering the FSM defined previously. This characters are adaptable NPCs which is learnable the action patterns of game players. The proposal method can be applied to create characters which play the role of partners with game players or the role of enemies against game players.

1 Introduction

User Modeling is the user cognitive process in which the data on users occurred under a specific system environment are collected and analyzed to achieve the objective of the system more effectively [1]. User modeling technique is being exploited diversely to analyze the profile information of users in the many fields such as games, educations, ubiquitous technologies and web mining [2].

Particularly in the game environment, the services based on analyzing gaming data of players, collected in the process of game activity, and utilizing the user modeling technique is being researched actively. In the game 'Black & White by Peter Molyneux', a partner NPC named Creature, has a different appearance in accordance with the game operation of the game player, and in the game Sims, an NPC also has a different appearance according to the player's game operation[3]. However, until now, in the field of games, user modeling techniques which have been used in console games and package games, rely on specific features of individual games.

This paper suggests D-FSM (Dynamic Finite State Machine) method which can create new NPCs behavior patterns to which the game players` tendencies are reflected. This method alters the FSMs for NPCs which were set with static action patterns in the initial phase of games, by adding information on the initial FSMs based on the data on game players which are collected while of gaming. In this paper, for

* Corresponding author.

M.J. Smith, G. Salvendy (Eds.): Human Interface, Part II, HCII 2007, LNCS 4558, pp. 490–499, 2007.

analysis of the collected data, decision tree method which is a kind of machine learning methods is used; the initial FSMs are altered with the analysis result. Along with this, script languages which set action patterns are defined so that D-FSM can be used flexibly for games in various categories D-FSM suggested in this paper can be used for following cases:

First, the suggested D-FSM can be used as a tool for designing NPCs in the initial phase of game production. Ordinarily game developers spend a lot of time and efforts to create NPCs in the initial phase of game production, in creating NPCs` behavior patterns. In designing computer games, if real game players` information is used, it will be much easier to create NPCs and thus the developing time and effort will be reduced. Moreover, it is possible to create more realistic NPCs of which behaviors are similar to human players.

Second, the suggested D-FSM can be used for creating partner NPCs in games. There are cases where NPCs share the same objectives with game players and cooperate with them. Since the D-FSM can build NPCs from human players' play record, NPCs which has abilities similar to game players and play games can easily be created.

Third, the suggested D-FSM can be used for creating hostile NPCs in games. Since the action patterns of monster NPCs in most games are also static, human players easily penetrate the patterns. This fact can be a factor which dwindles the life cycle of the games. The model dynamically generated from human players' play record can be used for creating hostile NPCs. Then, the NPCs will have a similar the abilities and action patterns adaptive to game players.

In Section 2, the related research is introduced and in Section 3, the player modeling using D-FSM and its application to NPCs is described. The experiment result is examined in Section 4. Finally, Section 5 concludes this paper and suggests future works.

2 NPC(Non-player Characters) and FSM(Finite State Machine)

2.1 Decision Tree in Games

Decision tree method, which is a technique widely used for data classification in data mining illustrates the patterns in the data in a tree structure by analyzing the given data. The reason why it is frequently used is that its' output is in a simple formal and can be easily understood by users. It has been applied to 'Black & White' and has a big favor from users. In the game, the player can give positive or negative feedbacks by caressing or beating the creature. A decision tree is generated from the feedbacks. Sometimes a creature checks if given objects will release hunger using the decision tree and sets the degree of aggressiveness. For example, if the player treats a creature with violence rather than praise more often, the creature gets violent more and more.2.1 FSMs for NPCs

2.2 FSM (Finite State Machine)

The state machine or finite state machine, which is one of widely used software designing patterns, is a machine which has finite numbered states [4]. An FSM is defined with the finite collection of states and transitions, between states. In applications, each state of an FSM may be attached with a specific action. NPCs conduct the actions attached to the present event state as their tasks. When a certain occurs, a state transits into another state will follow. FSMs are most widely used technique in games as it has a simple structure. FSMs are sometimes used mixed with other artificial intelligence techniques. Figure 1 illustrates an FSM.

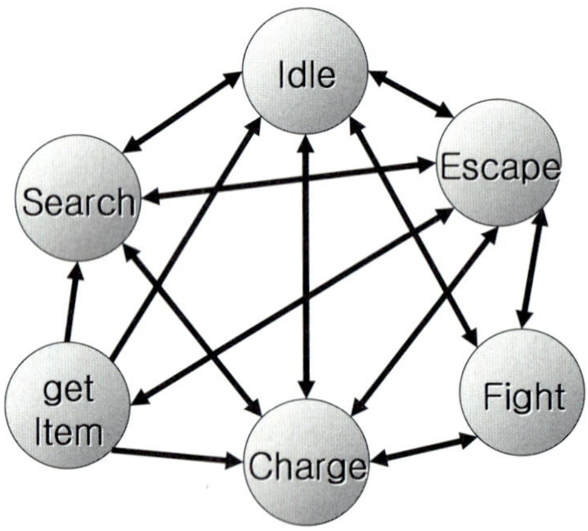

Fig. 1. 6 states and transition relation

Here it has 6 states and each state can be transited to the other states. For example, if the NPC in a state of "Idle" is attacked by the player, its' state will change into a state of "Fight" and if the NPC is injured, its' state will change into a state of "Escape".

2.3 Elementary NPCs and Improved NPCs

NPCs are agents that respond to game players using various elements. They can be merchants in stores, guides showing ways, teammates with game players, monsters, environmental elements such as the climate and geographic varieties and the tool items which game players use. The behavior patterns of these NPCs are directly set in source code or script, so never change during game playing. NPCs created by an improved method, usually show various responses according to the users` tendencies.

Table 1 and 2 illustrate the functions of NPCs in usual games and the examples of games to which elementary and improved NPCs are applied.

For elementary NPCs, the action patterns are defined in source codes directly in most cases. This causes the disadvantage that improving NPCs is difficult and that NPCs have static action patterns. Adaptable NPCs adjust their ability and play with

Table 1. Elementary NPCs

Description	· Game players` abilities are not reflected to by NPCs. · NPCs show the same response to the same situation. · NPCs` game patterns can be grasped. · NPCs show the same static action patterns according to situations.
Applied Examples	Space Invaders, Pac-Man , Donkey Kong and the like

Table 2. Improved NPCs

	Ability-Adaptable NPCs	Behavior-Adaptable NPCs
Description	· NPCs` abilities vary according to game players' abilities. (hitting accuracy rate, speed, energy and the like) · NPCs control the level of game difficulty dynamically.	· NPCs` action patterns vary according to the tendencies and abilities of game players. · NPCs are dynamically developed in games. · NPCs contribute to enhancing fun by raising the realistic feeling of games.
Applied Examples	FIFA, Call of Duty, Medal of Honor and the like	Black & White, The Sims and the like

the new ability in the next games. Ability-adaptable NPCs change only numeric parameters, such as the level of difficult, hit ratio, power, etc. and do not change the behavior patterns. On the contrary, behavior-adaptable NPCs show dynamic response to the game players by analyzing the game player's gaming data. Behavior-adaptable NPCs change behavior patterns or even take the creative and more developed actions which were not designed in the initial phase of game production. The D-FSM suggested in this paper can be utilized as a fundamental technique for providing behavior- adaptable NPCs through adjusting the initially defined FSM of NPCs using the gaming data of game players.

3 Dynamic-FSM Method Using Player's Gaming Data

This paper proposes the D-FSM which is the method to alter transition rules of FSMs defined in the initial phase using the gaming data of players collected in games. To create NPCs which can adapt to game players, the gaming data of players should be collected first. The modeling on game players are conducted by analyzing the collected data. The model of game players will be created in a form of FSMs. The FSMs will be applied to NPCs to change the behavior patterns of NPCs. The diagram in Figure 2 illustrates the D-FSM process.

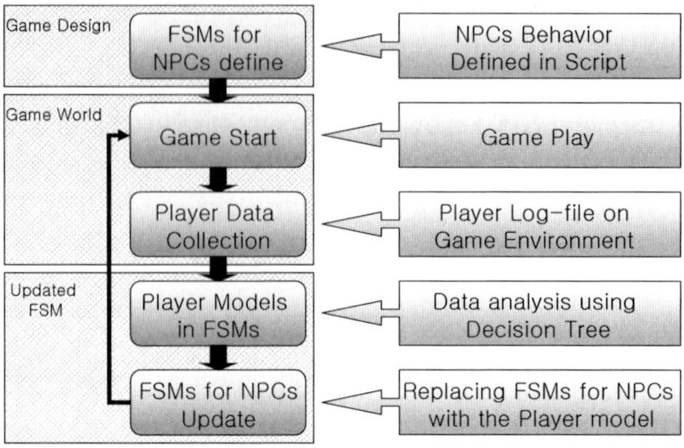

Fig. 2. Work-flow for D-FSM

3.1 FSMs for NPCs

Figure 1 shows the state transition of NPCs which is generally used in FPS (First Person Shooter) or MMORPG (Massively Multi-player Online Role Playing Game).

Table 3. State transition rules

S_{from}	PH	NH	HD	D	SD	PD	S_{to}
Search	H	H	0	O	1	1	Search
Search	M	H	0	I	-1	0	Charge
Search	M	M	1	I	1	-1	Idle
Search	H	L	1	I	0	0	Escape
Charge	H	H	1	I	0	-1	Charge
Charge	H	M	1	I	1	-1	Idle
Charge	L	L	1	I	1	-1	Fight
Charge	L	M	0	O	1	0	Search
⋮	⋮	⋮	⋮	⋮	⋮	⋮	⋮
getItem	M	M	0	O	0	0	Idle
getItem	M	H	0	O	1	1	Search

NPCs take the actions defined in its' current state. The transitions between states are usually defined with situation variables.

The representative situation variables are the body strength of PC (Player Character) and NPC, and the difference of body strength, attack power, moving speed and distance between PC and NPC. As additional variables, the items which characters own can be considered. Various other abilities of characters and the existence of comrade characters in the vicinity also may be counted.

After the states for NPCs are defined and the variables which are used for state transitions are identified, the state transition rules as shown in Table 4 should be generated.

Table 4. Variables used for defining transitions and their description

Variable	Attribute values	Description on Attribute values
PC Health Point (**PH**)	Low(L), Mid(M), High(H)	0~40=L, 41~70=M, 71~100=H
NPC Health Point (**NH**)	Low(L), Mid(M), High(H)	0~40=L, 41~70=M, 71~100=H
Health Difference of PC and NPC (**HD**)	0 ,1	PH < NH = 0, PH > NH = 1,
Distance of PS and NPC (**D**)	Near(N), Inside(I), Outside(O)	N : attack possible distance. I : inside of view sight. O : outside of view sight.
Speed Difference of PC and NPC (**SD**)	-1,0,1	If PC is faster than NPC, 1. If NPC is faster than NPC, -1. If the speed is the same, 0.
Power Difference of PC and NPC (**PD**)	-1,0,1	If PC is stronger than NPC, 1. If NPC is stronger than NPC, -1. If the power is the same, 0.

The transitions in Table 3 are defined using the values of 6 variables shown in Table 4. For example if the NPC is currently in the state of “Search”, the health of PC and NPC are 80 and 39 respective, PC is in the view sight and the moving speed and the attack power of PC and NPC are the same, the next state of NPC will be “Escape”, which is what the fourth rule says. Defining state transitions by scripts is one of most frequently used techniques because it is easy in the maintenance and expansion of NPCs.

3.2 Modeling Players with FSMs

This section will describe how to model players' palying patterns with FSMs which will be applied to NPCs. Players' playing data can be modeled in various ways. The reason for modeling with FSMs is to make the model easily adapted for updating NPCs' FSMs. If players' models are applied to NPCs, an NPC will play intelligently or in a way similar to how players' play in the game.

In order to model players' playing patterns with FSM, it is assumed that players also have a hidden FSM and choose actions based on the FSM. However, it is not easy to infer what states the player has it is also assume that the player's FSM has the same states as NPCs'. The current state of the player is inferred from observations and some heuristics.

NPCs in FPS games usually have six states shown as in Figure 1. The description of states and heuristics for inferring the player's current state are as follows:

Idle – In this state is that the game player takes no action. He may wait for the comrade or the enemy, or takes a rest on a specific location. If situation variables do not change for a certain time interval (5~10 sec.), the player is regarded as in “Idle” state.

Charge - This is the state to approach toward the enemies. If the player is in enemy’s view and approaching to the enemy, the player is regarded as in “Charge” state.

Fight - This is the state to combat with enemies which are in attackable distance. If an enemy is within attackable distance and attack keys are inputted, the player is regarded as in “Fight” state.

Escape - This is the state to retreat from enemies when game players feel disadvantageous in combat or when game players were in low body strength condition. If the player is in enemy’s view and the distance between them is increasing, the player is regarded as in “Escape” state.

getItem - This is the state where game players search tool items near from him to recover the body strength. If the player gets near to items and the player’s current state is “Escape”. The player is regarded as in “getItem” state.

Search - This is the state to move around to search enemies. So, if the player is out of enemy’s sight and moves around, the player is regarded as in “Search” state.

When a game player’s action and the values of situation variables are consistent with the heuristic condition of a state, the game player is regarded as in that state. For example, if the enemy is out of the player's sight and the player is moving around, it is concluded that the player is currently "Search" state. If the enemy gets into the player's sight and the player gets closer to the enemy, the player is regarded charging the enemy, i.e., in "Charge" state.

The next step is acquiring the player's transition rules between states. For acquiring transition rules, the values of situation variables are recorded whenever the player's state changes. That is, the values is collected in a form of (S_{from}, PH, NH, HD, D, DS, PD, S_{to}). The collected data is pre-processed to remove duplications and decision tree method is applied to find out transition rules in the data.

The result of decision tree learning is expressed in a tree structure which can be described in a table format like tables. The rules newly created will replace the NPC’s transition rules. That is, it is used to make NPCs intelligent or human-like. The newly created NPC rules are not static rules. These rules can be created from the game player’s gaming data whenever new rules need.

4 Experiments

The experiment was conducted using the Half-Life by Valve Software[7] game and Jeffrey’s HPB Bot[8], where a PC and NPC combat were used under the method suggested in this paper. This game initially sets the states and transition rules of the NPC by reading the scripts and collects the player’s gaming data while the game is going on. The player can move forward, backward, left and right, and attacks the NPC.

The PC and the NPC initially have the same moving speed and power. The power given to the PC and the NPC in the initial phase is 100. If the values of situation variables do not change for more than 5 seconds, it is defined that the player is in “Idle”.

During game play the values of the body strength of the PC and the NPC, the difference of the body strengths, the distance between the PC and the NPC, and the difference of the moving speeds and the attack powers are collected, whenever the state of the game player changes.

Fig. 3. DeathMatch Class Game in MOD of Half-life and Game Screenshot

The game recognizes the state of the PC on the basis of the heuristics previously described. After one round of the game, the NPC learns the collected data using decision tree method and updates its' transition rules. Figure 4 illustrates the initial transition rules in Table 3. Those are represented in a tree format. Figure 5 represents the newly learned transition rules from the player's gaming data. It is noted that the newly learned transition rules are quite different from the initial one. It has more complex rules for "Charge" than the original but simpler ones for "Search". The NPC takes actions more intelligently than with the initial ones.

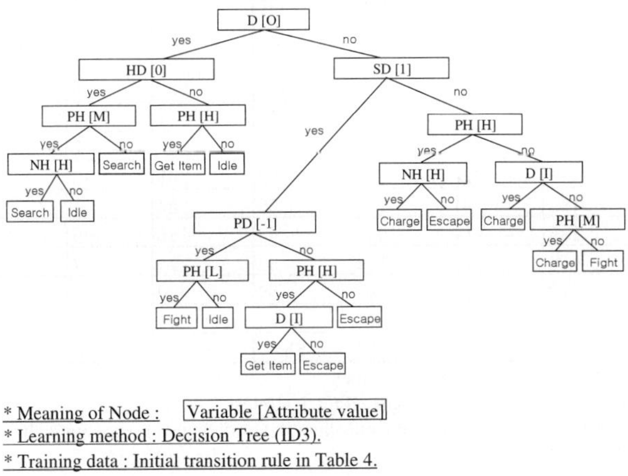

Fig. 4. Initial transition rules in a tree format

In experiment, 15 players play the game with the initial rules for NPC's, and then play again the game with the new rules learned from his previous game play data. We measure how long each player play the game and many times each player die during play. We also ask a few questions to the players: the degree of difficulty and the degree of enjoyment. Table 5 shows the result. The players answer that the game with learned rules are more interesting and difficult to win. The average playing time is increased from 217.3 to 251.0.

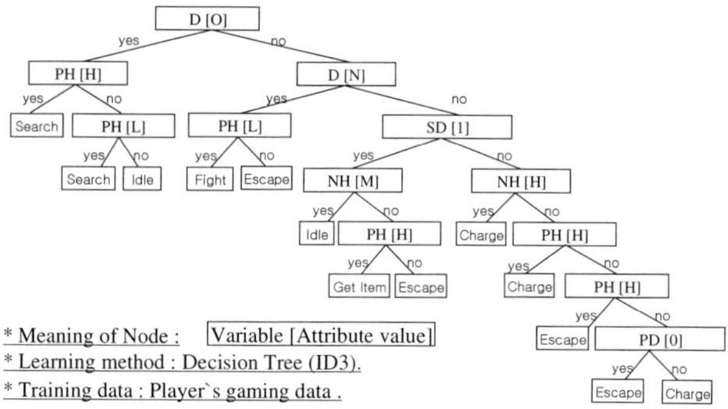

Fig. 5. A newly learned transition rules from the player's data

Table 5. Experiment result using D-FSM

Players	NPC of initial transition rules				NPC of learned transition rules			
	Game Data		Satisfaction (0~100)		Game Data		Satisfaction (0~100)	
	Game progress time(sec.)	Death number of times	Degree of difficulty	Degree of enjoyment	Game progress time(sec.)	Death number of times	Degree of difficulty	Degree of enjoyment
Player 1	184	5	55	65	272	7	80	85
Player 2	130	5	60	70	172	5	90	90
Player 3	205	4	65	60	402	4	100	100
Player 4	271	9	55	45	221	5	70	98
Player 5	328	7	60	50	196	6	60	95
Player 6	253	8	70	60	265	9	80	95
Player 7	266	7	50	70	287	6	90	85
Player 8	204	6	40	45	131	2	70	90
Player 9	200	5	50	55	261	6	40	80
Player 10	133	2	60	50	205	3	80	85
Player 11	182	4	70	65	184	7	90	90
Player 12	234	5	55	50	290	5	70	70
Player 13	178	6	65	30	221	7	80	60
Player 14	278	5	40	50	324	6	80	80
Player 15	213	4	30	70	334	7	90	85
Ave.	217.3	5.5	55.0	55.7	251.0	5.7	78.0	85.9

* Mission of this game is killing NPCs 10 times.
* Game progress time: It is cost time to kill NPCs 10 times.
* Death number of times: It is number of times that player dies during mission achievement.
* Degree of difficulty: If it is near to 0, easy. If it is near to 100, difficulty.
* Degree of enjoyment: If it is near to 0, boring. If it is near to 100, interesting.

5 Concluding Remarks

NPCs provided in games are usually have static action patterns. These static actions of NPCs are one of factors which degrades the fun of games and makes the life-cycle of games short. This study is on an approach which can provide creative and diversified NPCs by updating FSMs through collecting and analyzing the gaming data of players. This method can be used in other the process of game development, such as designing initial FSMs for NPCs, creating partner NPCs of players which cooperate with players and creating hostile NPCs in games. To generate a more compact rule set pruning method which reduces the number of rules will be applied as a future work.

References

[1] Brajnik, G., Guida, G., Tasso, C.: User modeling in expert man-machine interfaces: a case study in intelligent information retrieval. IEEE Trans. SMC 20(1), 166–185 (1990)

[2] Cha, H.J., Kim, Y.S., Lee, J.H., Yoon, T.B.: An Adaptive Learning System with Learning Style Diagnosis based on Interface Behaviors. In: International Conference on E-learning and Games, Edutainment (2006)

[3] Rabin, S.: AI Game Programming Wisdom 2, Charles River Media (2002)

[4] Loura, M.D.: Game Programming Gems, Charles River Media (2000)

[5] Laird, J.E.: Using a Computer Game to Develop Advanced AI. Computer IEEE Journal 34(7), 70–75 (2001)

[6] Spronck, P., Kuyper, I.S., Postma, E.: Online Adaptation of Game Opponent AI in Simulation and in Practice. In: Proceedings of the 4th international Conference on Intelligent Games and Simulation, GAME-ON, pp. 93–100 (2003)

[7] Half-Life Game site: http://www.half-life.com

[8] PHB Bot site for Half-Life game: http://hpb-bot.bots-united.com

Part III

Mobile Interaction

Readability of Character Size for Car Navigation Systems

Kazuhiro Fujikake[1], Satoshi Hasegawa[2], Masako Omori[3], Hiroki Takada[4], and Masaru Miyao[5]

[1] Graduate School of Information Science, Nagoya University, Japan
fujikake@nagoya-u.jp
[2] Department of Information Culture, Nagoya Bunri University, Japan
hasegawa@nagoya-bunri.ac.jp
[3] Faculty of Home Economics, Kobe Women's University, Japan
masako@suma.kobe-wu.ac.jp
[4] Gifu University of Medical Science, Japan
takada@u-gifu-ms.ac.jp
[5] Information Technology Center, Nagoya University, Chikusa-ku Nagoya, Japan
miyao@itc.nagoya-u.ac.jp

Abstract. An experiment was conducted on the ease with which a car navigation system could be read by elderly, middle-aged and young people. 127 subjects (19-76 years of age) evaluated ease of reading of the displayed characters. The experiment evaluated characters displayed on a liquid crystal display (LCD) similar to those used in car navigation systems, and evaluated readability in 9 stages. Five character sizes for traffic terms and two languages (Japanese and English) were used in the experiment. The experiment was conducted with the LCD in 2 positions: directly in front of the subject and diagonal to the left (about 30 degrees). The results showed that the directly frontward position was significantly more readable than the position of diagonal to the left.

Keywords: Car navigation system, Character size, Readability.

1 Introduction

Use of liquid crystal displays (LCDs) in communication terminal equipment has spread rapidly, and standards for them have been established. While the sizes of LCD screens have been increased, power-saving features have also been added by reducing their size and weight. Consequently, LDCs are now used in car navigation systems. Few reports, however, have been made on the readability of characters displayed on car navigation system screens in terms of the size of the character font used or with a focus on elderly people.

In this study, the appropriate size of character fonts used in car navigation system was examined on people of all ages, including the elderly.

M.J. Smith, G. Salvendy (Eds.): Human Interface, Part II, HCII 2007, LNCS 4558, pp. 503–509, 2007.

2 Experiment

2.1 Experiment Description

This experiment was conducted indoors and the subjects evaluated the readability of characters displayed on LCD screen. A 9-point scale was used for the evaluation: Very easy to read (9 points), Easy to read (7), Neutral (5), Difficult to read (3), Very difficult to read (1). The subjects were told that the readability was evaluated by the legibility of the displayed characters, not by the understandability of the content. They were also instructed to tell the experimenter when the characters were too small to read.

Five character sizes for traffic terms (character height: 2 mm, 4 mm, 6 mm, 8 mm, and 10 mm) and two languages (Japanese and English) were used in the experiment. The fonts used in this experiment were Japanese MS P Gothic font in boldface type (e.g. in Japanese "Yakan-Koutu-Dome") and English Arial font (e.g. traffic jam: TRAFFIC JAM).

Photo. Experiment Scene

2.2 Display Device

A personal computer monitor (ADTEC TL700A) was used as an LCD with white characters on a black background; the ratio of luminance contrast (i.e., differences in luminance between characters and background) was 17. The illuminance was 404 lx in the vertical direction and 283 lx in the horizontal direction.

To make the LCD the same size as a car navigation system's screen, the LCD screen was covered with a black paper with an 8-cm height/14-cm width square cutout. The characters displayed within the square cutout were evaluated as if displayed on a car navigation system's screen.

The LCD was set in 2 positions: directly in front of the subject and diagonal to the left (about 30 degrees). In both positions the LCD was set at a distance of 70 cm from the subjects' eyes. In addition, the subjects were told to move their head only when they looked at the screen in the left diagonal direction.

2.3 Subjects

127 subjects (range 19-76 years old) with normal vision were divided into 3 groups: 72 young people (range 19-44 years old), 30 middle-aged people (range 45-64 years old), and 25 elderly people (range 65-76 years old).

The average vision at a distance of 70 cm was 1.2 for the young, 0.9 for the middle-aged, and 0.7 for the elderly. In addition, subjects who used eyeglasses or contact lenses in riding a bicycle on an everyday basis used them also in the experiment.

3 Results

The readability was evaluated for each language, LCD position, character height, and age group. Figure 1 shows the average values and standard deviations for readability of Japanese characters when the LCD was in the frontal position, for each font size and age group. Figure 2 shows the average values and standard deviations for readability of Japanese characters, with LCD in the diagonal position, for each font size and age group. Figure 3 shows the average values and standard deviations for readability of English characters, with LCD in the frontal position, for each font size and age group. Figure 4 shows the average values and standard deviations for readability of English characters, with LCD in the diagonal position, for each font size and age group.

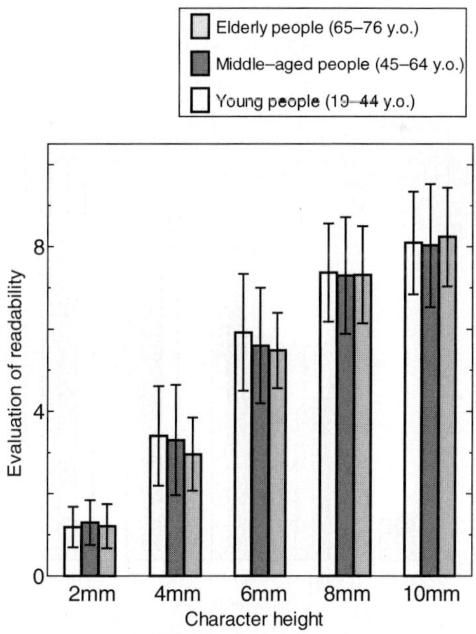

Fig. 1. Readability of Japanese characters (front view)

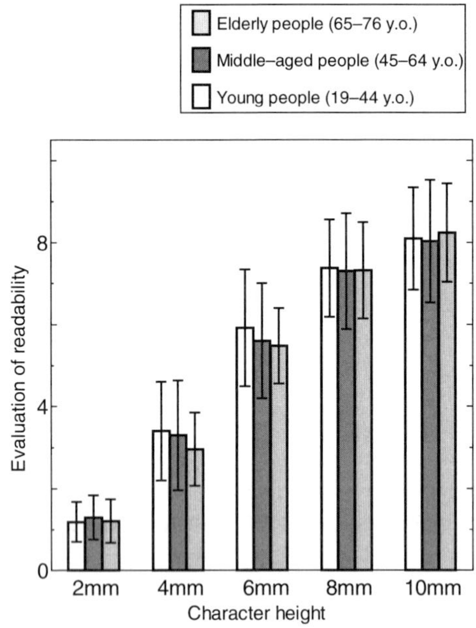

Fig. 2. Readability of Japanese characters (diagonal view)

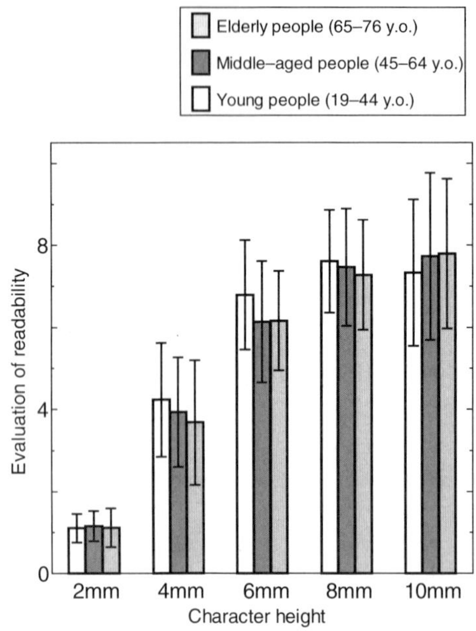

Fig. 3. Readability of English characters (front view)

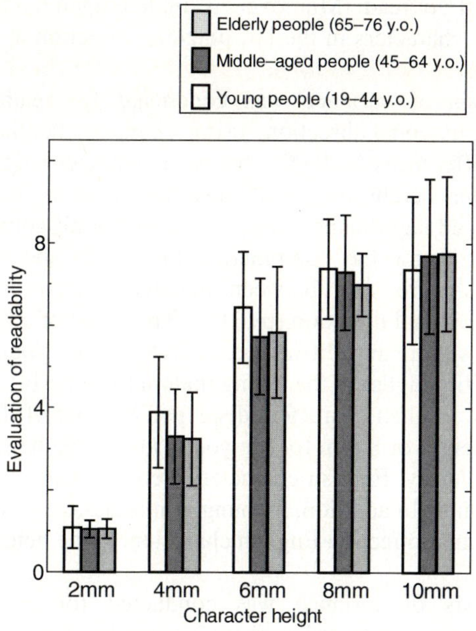

Fig. 4. Readability of English characters (diagonal view)

The overall comparison in readability between the frontal and the left diagonal positions of the LCD (paired t-test) revealed significant differences ($p<0.001$) and it was demonstrated that readability is higher when the LCD is in the frontal position than in the left diagonal position. Readability was then evaluated according to character height (paired t-test), both in the frontal and left diagonal positions, in the two languages. Significant differences were observed ($p<0.001$) in both languages, and it was demonstrated that readability is higher in the frontal position than in the left diagonal position.

As Figure 1 shows, Japanese characters displayed in front of the subjects were the most readable at the height of 10 mm in all age groups and the less readable at the height of 2 mm in all age groups. In addition, 2 young, 3 middle-aged, and 6 elderly subjects reported that Japanese characters in the frontal direction at the height of 2 mm were too small to read.

A two-way analysis of variance was conducted for readability of Japanese characters in the frontal direction, using character heights and age groups. It revealed significant differences ($p<0.001$) between character heights, with readability increasing as character height increased. The result of a follow-up analysis of character height showed significant differences between all combinations of heights ($p<0.001$ in all). In the readability of characters in the frontal direction, however, little difference was seen between the age groups.

As Figure 2 shows, Japanese characters displayed in the left diagonal were the most readable at the height of 10 mm in all age groups and the least readable at the height of 2 mm in all age groups. In addition, 3 young, 4 middle-aged, and 9 elderly subjects reported that Japanese characters in the left diagonal direction at the height of

2 mm were too small to read. Moreover, 1 middle-aged and 1 elderly subjects reported that Japanese characters in the left diagonal direction at the height of 4 mm were too small to read.

A two-way analysis of variance was conducted for readability of Japanese characters in the left diagonal direction, using character height and age group. It revealed significant differences ($p<0.001$) between character heights, with readability increasing as character height increased. The result of a follow-up analysis of character heights showed significant differences between all combinations of heights ($p<0.001$ in all), except for the combination of 8 mm and 9 mm. In addition, significant differences were seen between the age groups in the readability of characters in the left diagonal direction ($p<0.01$). The result of a follow-up analysis of age groups showed significant differences between the young and the elderly ($p<0.05$), with readability higher in the young than in the elderly.

As Figure 3 shows, English characters displayed in front of the subjects were the most readable at the height of 8 mm for the young and at the height of 10 mm for the middle-aged and the elderly. English characters were the least readable at the height of 2 mm in all age groups. In addition, 3 young, 4 middle-aged, and 7 elderly subjects reported that in the frontal direction English characters at the height of 2 mm were too small to read.

A two-way analysis of variance was conducted for readability of English characters in the frontal direction, using character heights and age groups. It revealed significant differences ($p<0.001$) between character heights, with readability increasing as character height increased. The result of a follow-up analysis of character heights showed significant differences between all combinations of heights ($p<0.001$ in all), except for the combination of 8 mm and 9 mm. In the readability of characters in the frontal direction, however, little difference was seen between the age groups.

As Figure 4 shows, English characters displayed in the left diagonal direction were the most readable at the height of 8 mm for the young and at the height of 10 mm for the middle-aged and the elderly. English characters were the least readable at the height of 2 mm in all age groups. In addition, 5 young, 9 middle-aged, and 7 elderly subjects reported that in the left diagonal direction English characters at the height of 2 mm were too small to read.

Finally, a two-way analysis of variance was conducted for readability of English characters in the left diagonal direction, using character heights and age groups. It showed significant differences ($p<0.001$) between character heights, with readability increased as the character height increased. The result of a follow-up analysis of character heights showed significant differences between all combinations of heights ($p<0.001$ in all) except for the combination of 8 mm and 9 mm.

4 Conclusion

Using a self-rated evaluation, display positions and character font sizes used in car navigation system were examined for readability by people of all ages, including the elderly.

It was found that readability decreased as the font size decreased in all age groups, and in both frontal and left diagonal directions. In the readability of Japanese characters, significant differences in were found between the age groups when the display was in the left diagonal direction but not in the frontal direction. Moreover, since Japanese characters displayed in the left diagonal direction were less readable than those in the frontal direction, smaller Japanese characters in the diagonal direction were less readable for the elderly.

On the other hand, English characters were more readable than Japanese ones, one reason for which may be that English characters are simpler than Japanese ones.

References

1. International organization for standardization, Ergonomic requirements for office work with visual display terminals (VDTs) (1992) ISO 9241-3
2. International organization for standardization, Road vehicles - Ergonomic aspects of transport information and control systems - Specifications and compliance procedures for in-vehicle visual presentation (2003) ISO 15008
3. Hasegawa, S., Matsunuma, S., Omori, M., Miyao, M.: Aging effects on the visibility of graphic text on mobile phones. Gerontechnology 4(4), 200–208 (2006)
4. Hasegawa, S., Sato, K., Matsunuma, S., Miyao, M., Okamoto, K.: Multilingual disaster information system: Information delivery using graphic text for mobile phones. AI & Society 19(3), 265–278 (2005)
5. Omori, M., Watanabe, T., Takai, J., Takada, H., Miyao, M.: Visibility and characteristics of the mobile phones for elderly people. Behavior & Information Technology 21(5), 313–316 (2002)

Readability of Characters on Liquid Crystal Displays in Mobile Phones

Satoshi Hasegawa[1], Masako Omori[2], Kazuhiro Fujikake[3], and Masaru Miyao[4]

[1] Department of Information Culture, Nagoya Bunri University, Inazawa Aichi, Japan
hasegawa@nagoya-bunri.ac.jp
[2] Faculty of Home Economics, Kobe Women's University, Suma-ku Kobe, Japan
masako@suma.kobe-wu.ac.jp
[3] Graduate School of Information Science, Nagoya University, Chikusa-ku Nagoya, Japan
fujikake@nagoya-u.jp
[4] Information Technology Center, Nagoya University, Chikusa-ku Nagoya, Japan
miyao@itc.nagoya-u.ac.jp

Abstract. Subject performance in reading characters on mobile phone liquid crystal displays was researched with the use of: (1) sentences without meaning, made from randomly ordered alphanumeric characters having 3 different sizes of characters and 4 different contrasts (n=78; age=39.9±17.6), (2) Japanese characters of 5 different sizes and 2 types of font (n=98; age=44.5±18.5), and (3) Japanese characters vertically enlarged 1-4 times (n=120; age=46.9±18.6). Readability was higher with higher contrast displays, and with Gothic than with Mincho font in Japanese. Readability of Japanese characters improved when they were vertically enlarged to approximately twice the width.

1 Introduction

The use of character information on mobile phones (MPs) has expanded to include business information, social information, disaster information [1], and other types of information in the current ubiquitous computing age. MPs are one of the most important types of character terminal and an effective information delivery tool. Universal designs allow the transmission of such important information to all people including foreigners [1], elderly [2] and those with disabilities. There have been many studies on the readability of characters on cathode ray tubes (CRT) or liquid crystal displays (LCD) of computers, and recommended character sizes have been standardized in terms of video display terminal (VDT) work. However, MPs have far smaller screens than personal computers (PCs) and differ markedly from PCs in being hand-held. Reading behavior may be expected to different from that with PCs. The purpose of this study was to research the reading performance of users who read characters on LCDs in MPs, and to examine the relation of character size, display brightness and contrast to readability. Graphic characters in the Portable Network

M.J. Smith, G. Salvendy (Eds.): Human Interface, Part II, HCII 2007, LNCS 4558, pp. 510–517, 2007.

Graphics (PNG) format that had readability consistent with that of ordinary binary fonts in MPs [1] were used in this study. The use of graphic characters allows experimental display of characters of various sizes, shapes and contrasts that are beyond the normal text display functions of mobile phones.

2 Methods

2.1 Method of Experiment 1

Sentences without meaning, made from randomly ordered alphanumeric characters according to ISO 9241-3 Amendment 1 [3], in which 240 characters included 15% space as in normal English sentences, were displayed on the LCDs of MPs using

Table 1. Luminosity and contrast of the graphic text displayed in experiment 1. Luminosity were values (cd/m^2) measured (Minolta LS-110) in the display backgrounds.

Code	Character	Back ground	Displayed on Type A		Displayed on Type B	
			luminosity	contrast	luminosity	contrast
W	white	black	9.1±0.5	11.9	3.2±0.0	53.8
X	black	white	107.5±2.5	11.9	171.4±1.2	53.8
Y	gray	white	105.1±2.2	2.5	171.0±1.8	4.4
Z	black	gray	47.0±2.0	5.2	38.2±0.8	12.2

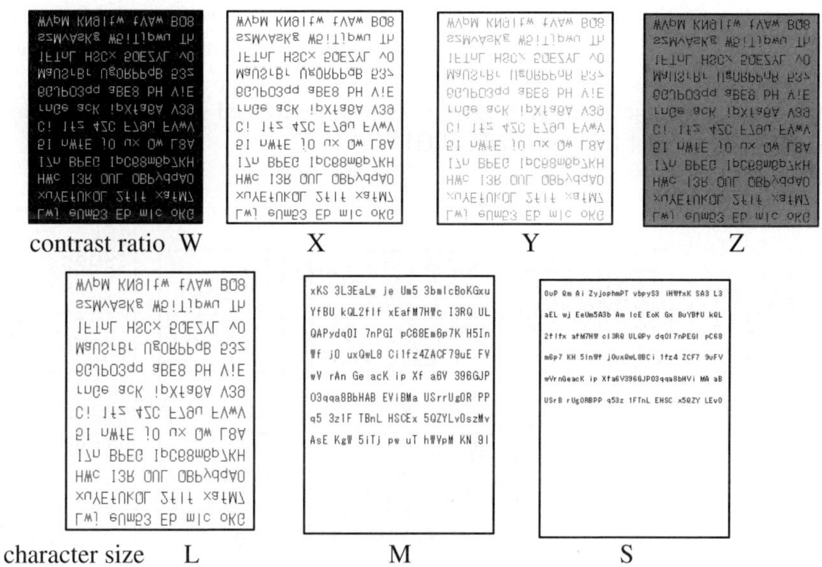

Fig. 1. Samples of graphic characters for experiment 1

graphic characters. Subjects (n = 78, age = 39.9±17.1, 19-76) counted the number of times a target letter appeared by searching the strings as if they were reading the sentences carefully and as fast as possible. Searching times, viewing distances, target character count, and subjective evaluations of readability were recorded. Two different types of MPs, Type A (Kyocera WX310K: 240x320 dot 2.4 inch color LCD) and B (Kyocera W43K: 240x320 dot 2.6 inch color LCD), and three different character sizes, L (~2.5 mm height), M (~2.0 mm) and S (~1.0 mm), were examined with four different contrasts consisting of W (negative), X (positive), Y (gray characters with white background) and Z (black characters with gray background), as shown in Table 1 and Fig. 1.

2.2 Method of Experiment 2

Japanese sentences with two different fonts (Gothic and Mincho) and five character sizes (G1-5 and M1-5) were prepared as shown in Fig. 2. G1 and M1 have the smallest characters (8 pt, 2.8 mm) in this experiment. G2 and M2 were 1.25 times size 1 (10 pt, 3.5 mm), and size G3 and M3 were 1.5 times size 1 (12 pt, 4.2 mm). G4 and M4 were vertically enlarged 1.5 times size 1. G5 and M5 were vertically enlarged 2 times size 1. The MP used in this experiment (Sharp SH53) has an LCD of 240x320 dots (2.4 inch, color CG Silicon). Subjects (n=98; age=44.5±18.5) read aloud these sentences. The reading time, viewing distance, number of errors and subjective evaluation were recorded.

2.3 Method of Experiment 3

Graphic texts with the height of size G1 (Fig. 2), and 1.25-, 1.5-, 2-, and 2.5-times the height of G1 without changes in width were evaluated by the same subjects as in experiment 2.

Other Japanese subjects (n=120; age=46.9±18.6) performed with oblong characters of G1, and 2-, 3-, and 4-fold the height of G1.

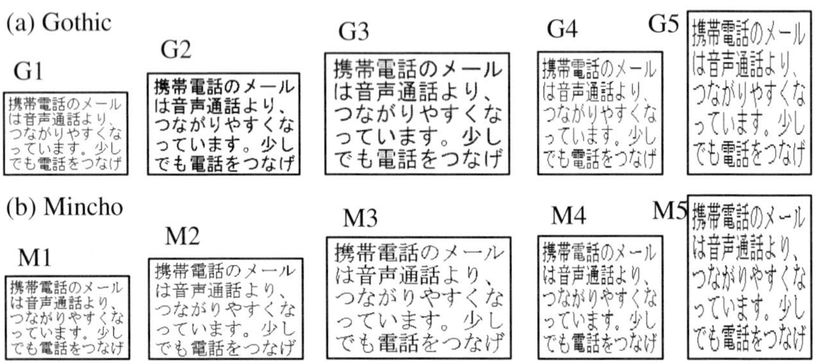

Fig. 2. Graphic characters used in the experiment 2

3 Results

Results of experiment 1 are shown in Fig. 3(a)-(e). A two-way ANOVA with factors of MP types (Type A and Type B of character size L) and contrast (W, X, Y and Z: shown in Fig. 1 and Table 1) was conducted for each parameter. Both MP type and contrast showed significant differences ($p<0.0001$) only in subjective evaluation without significant interaction. Readability was higher in type B than in type A. There were no significant

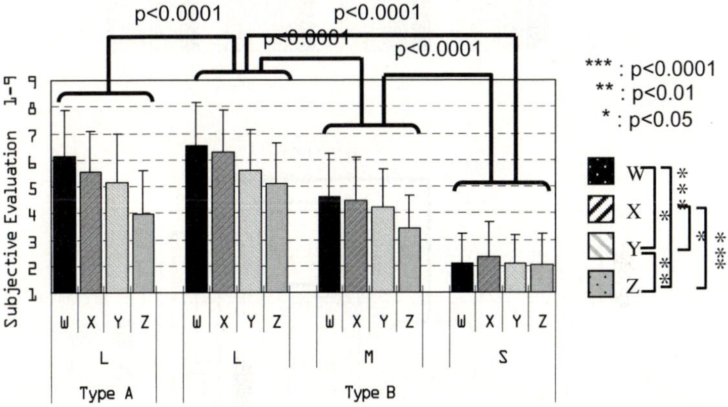

(a) Subjective evaluation

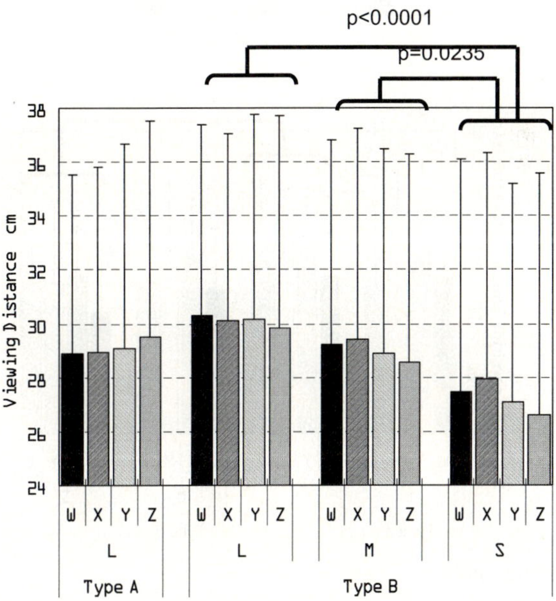

(b) Viewing distance

Fig. 3. Results of experiment 1

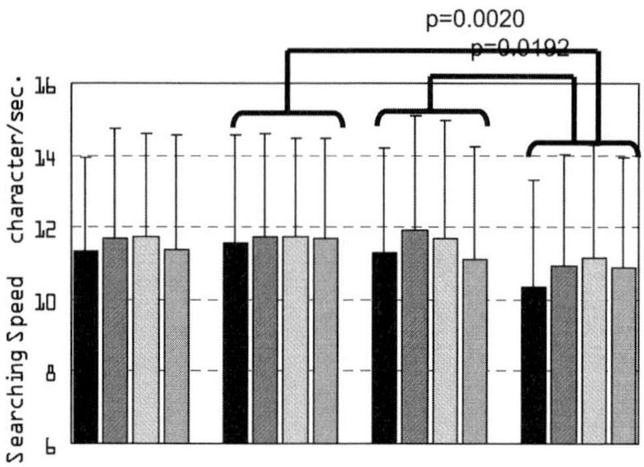

(c) Searching speed

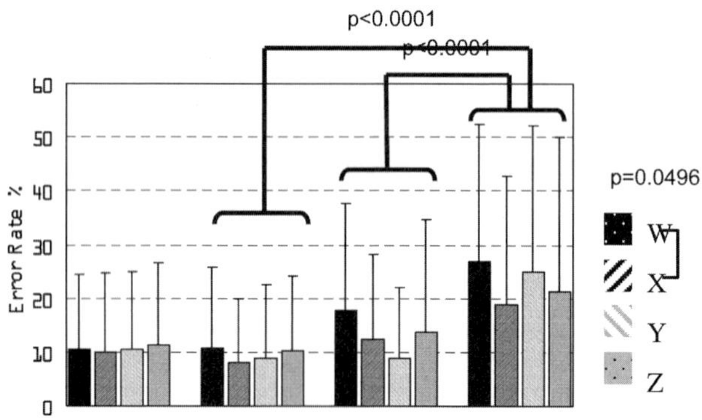

(d) Error rate

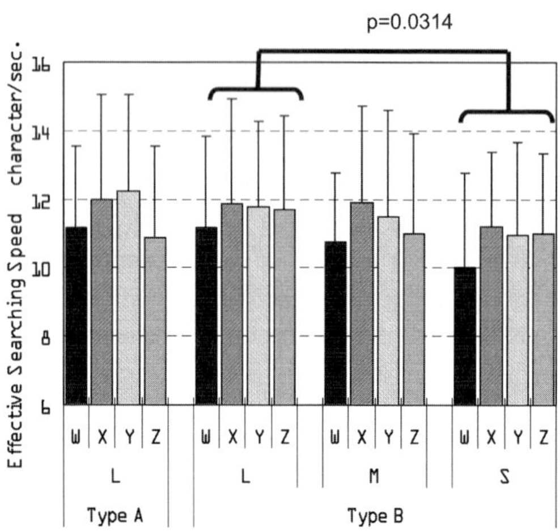

(e) Effective searching speed

Fig. 3. (*continued*)

differences between W and X; however, W-Y (p<0.0001), W-Z (p<0.0001), X-Y (p=0.0382), X-Z (p<0.0001), and Z-Y (p=0.0001) were significant (Fig. 3(a)).

Two-way ANOVA with factors of character size (L, M, and S of Type B) and contrast (W, X, Y and Z) was also conducted for each parameter. The differences were statistically significant in both character size and contrast (p<0.0001). The interaction was also significant (p=0.0009) in subjective evaluation. L-M, L-S and M-S were all significant (p<0.0001). In the contrast, W-X had no significant difference. W-Y (p=0.0111), W-Z (p<0.0001), X-Y (p=0.0366), X-Z (p<0.0001), and Z-Y (p=0.0066) were significant.

Significant differences in viewing distance (Fig. 3(b)) and searching speed (Fig. 3(c)) were seen only in the character size, especially in L-S and M-S. Error rate of target count (Fig. 3(d)) showed significance with contrast (p=0.0347) as well as

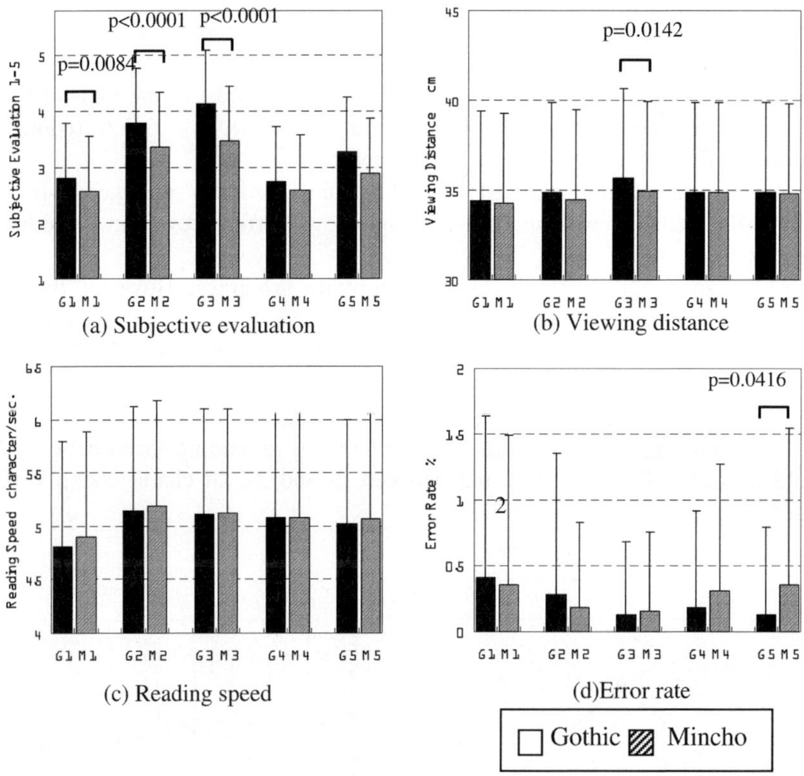

Fig. 4. Results of experiment 2

char-acter size (p<0.0001). Effective searching speed, calculated from data of correct count number, (Fig. 3(e)) showed significance with character size (p=0.0314).

Results of experiment 2 are shown in Fig. 4. Experiments using Japanese sentences were performed with an illumination of 624.3±7.4 lx on the desk and 294.0±7.9 lx on the screen of MPs. The luminosity and contrast for MP used in this experiment were: luminosity Black, 5.8±0.2 (cd/m2); White, 205.8±0.8 (cd/m2); and contrast, 35.7.

Significant differences were observed by two-way ANOVAs in subjective evaluation among sizes 1, 2, and 3 ($p<0.0001$) and between the Gothic (G) and Mincho (M) fonts ($p<0.0001$) (Fig. 4(a)). Reading speed ($p=0.0235$) and error rate ($p=0.0304$) as objective parameters significantly differed only among size 1, 2, and 3 (Fig. 4(b)-(d)).

Two-way ANOVA with factors of character vertical length (size 1, 4, and 5) and fonts (G and M) showed significance in both vertical length ($p<0.0001$) and fonts ($p=0.0001$) with no interactive effect in subjective evaluation only. P-values shown in Fig. 4 are the results of paired t-tests between the two fonts, G and M.

Results of experiment 3 are shown in Fig. 5. Subjective evaluation and reading speed are shown in Fig. 5(a) and (b), respectively. The conditions of the experiment com-paring characters vertically enlarged 1-, 1.25-, 1.5-, 2-, and 2.5-fold were the same as those in experiment 2.

Another experiment comparing characters vertically enlarged 1-, 2-, 3- and 4-fold was performed with an illumination of 447.3 ± 15.2 lx on the desk and 396.0 ± 16.8 lx on the MP screen, and the following luminosity and contrast of the LCD of the same MP types: luminosity Black, 7.4±0.9 (cd/m2); White, 197.6±1.6 (cd/m2); and contrast, 26.7.

Subjective evaluation showed that characters vertically enlarged 2-fold were most easily readable, with marked and significant differences from the other sizes (Fig. 5(a)).

Reading speed did not significantly differ by the group test. However, the paired t-test among characters vertically enlarged 1-, 1.25-, 1.5-, 2-, and 2.5-fold showed significant differences in reading speed between 1-fold and 1.5-fold characters ($p=0.0039$), between 1-fold and 2-fold characters ($p=0.0214$), and between 1-fold and 2.5-fold characters ($p=0.0429$). Comparison among characters vertically enlarged 1-, 2-, 3-, and 4-fold showed significant differences in reading speed between 1- and 2-fold characters ($p = 0.0009$) and between 2- and 3-fold characters ($p = 0.0040$) (Fig. 5(b)).

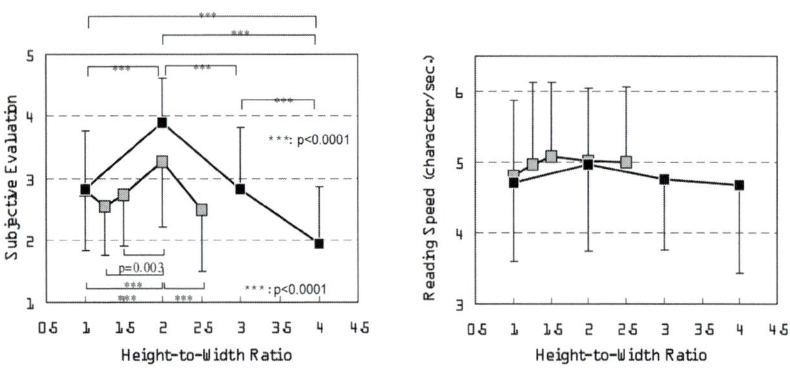

Fig. 5. Results of experiment 3

Error rate was also the lowest for characters vertically enlarged 2-fold. Thus, objective parameters also showed an improvement in readability through an approximately 2-fold vertical character enlargement. Visual distance negligibly changed with the vertical enlargement.

4 Conclusion

Readability of characters on LCDs in MPs deteriorates remarkably as the characters become smaller. The trade-off between convenience of compactness and readability of characters on the MP display, including its resolution, brightness, contrast and character size, should be examined. The LCDs of MPs may require a higher resolution and contrast than those of PCs because MPs are used in the hand at a nearer distance and/or outside of the room. Readability was higher with higher contrast displays. For Japanese characters, readability was higher for the Gothic font, which has a constant stroke width, than for the Mincho. Readability of Japanese characters improved when they were vertically enlarged to approximately twice the width.

In recent years, functions of MPs have spread to not only character email applications but also web access, television transmission, and animation images. The comparison among different contrast conditions on a display in this study is also useful in evaluating the accessibility of color contents in MPs. Various character sizes, vertical length and contrasts, including negative display mode, should be selectable by users in order to make mobile information available to all people with various visual functions to realize completely barrier-free human-computer interfaces for a fair ubiquitous society.

References

[1] Hasegawa, S., Sato, K., Matsunuma, S., Miyao, M., Okamoto, K.: Multilingual disaster information system: Information delivery using graphic text for mobile phones. AI & Society 19(3), 265–278 (2005)

[2] Hasegawa, S., Matsunuma, S., Omori, M., Miyao, M.: Aging effects on the visibility of graphic text on mobile phones. Gerontechnology 4(4), 200–208 (2006)

[3] ISO 9241-3. Amendment 1. (2000). ISO 9241-3. Ergonomic requirements for office work with visual display terminals (VDTs) –Part 3: Visual display requirements. International Standard (1992)

Study on Public User Interface

Atsushi Hiyama, Kotaro Hashimoto, Tomohiro Tanikawa, and Michitaka Hirose

Department of Mechano-Informatics,
Graduate School of Information Science and Technology,
The University of Tokyo,
7-3-1, Hongo, Bunkyo-ku, Tokyo, Japan
{atsushi, kotaro, tani, hirose}@cyber.t.u-tokyo.ac.jp

Abstract. In this paper, we describe life-log content mash-up application called "Sharelog" by using widespread consumer mobile device. By using consumer mobile device as a interface for ubiquitous and mobile computing application, the system developers are enabled to focus on the content side of an application. We have demonstrated our application "Sharelog" at the museum exhibition and realized the capability of consumer mobile device as an interface for ubiquitous computing in public space. We would like to call this kind of mobile devices as public user interface.

Keywords: mobile device, ubiquitous computing, human interface, mash-ups.

1 Introduction

The number of personal mobile devices such as cellular phones, music players, game machines and digital cameras are dramatically increasing according to the price-reduction and downsizing of computer devices. It might be easy to develop a system which can be accessed by the general public if we can design a system which is promised on the widespread use of such mobile devices. This attempt is quite different from existing ubiquitous and mobile computing developments that are focused on designing mobile devices and system infrastructure. Considering above mentioned background, we discuss on the characteristics and issues in designing information systems based on ready-prevalent mobile devices as user interface. We may call this kind of interface as public user interface.

1.1 Growing Popularity of Consumer Mobile Devices

Cellular phones are the most common mobile device. In June 2006, contracted quantities of cellular phones are 93,000,000 in Japan. The Japanese population is about 120,000,000. Therefore, almost every Japanese people carry a cellular phone.

Mobile music players for mp3 format music files are rapidly spread worldwide in recent years by the release of Apple Inc. iPod (Figure 1). 184,600,000 mobile music players are manufactured in 2006 and it was 36% increase from 2005. In Japan, 6,000,000 mobile music players are sold and it was 328.6% increase from previous year.

M.J. Smith, G. Salvendy (Eds.): Human Interface, Part II, HCII 2007, LNCS 4558, pp. 518–525, 2007.

Fig. 1. Apple iPod [1]

Nintendo DS is a mobile game machine released in December 2004 (Figure 2). It has double screen and touch panel interface. It also has wireless communication interface and speech recognition functions. Various kinds of game software using these characteristics are already produced.

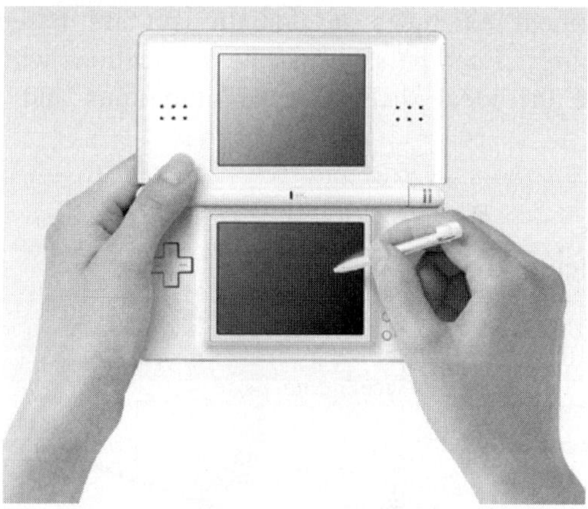

Fig. 2. Nintendo DS [2]

Electronic money is one of the rapidly widespreding mobile devices. Felica is an IC card system developed by sony. Felica system is based on noncontact interface and it is a de-fact standard of electric money in Japan.

1.2 System Designing in Web2.0

Web2.0 applications were realized by utilizing networked PCs as a new foundation of information systems. Picture sharing web site Flickr[3] or movie sharing web site

YouTube[4] are representatives of Web2.0 web applications. Developers of these web sites are only providing a place where general public users can share their contents and they do not make application contents by their selves. This type of system developments have new methodology that developers also do not need to deal with fundamental system architecture development.

2 New System Design Methodology in Ubiquitous Computing

Existing system development in ubiquitous and mobile computing is mainly focused on constructing system architecture and not much pay attention to mass user participation in general public. On the contrary, recent system development researches in this field are beginning to consider mass participation of general public as an implementation method such as Place Engine[5] or The ESP Game[6][7].

Conventional ubiquitous and mobile computing researches are focused on two objectives. One is to cover wider range of physical space where we can provide information services and the other is to personalize those information services in efficient way[8][9]. In addition to these two grouping axis, we bring up another grouping axis of degree of mass participation of general public. Resent research examples can be mapped in this three dimensional space as shown in figure 3.

System development examples in domain (A) are based on new system designing methodology. I can be said that the applications which will be mapped in domain (B) are the ideal realization of ubiquitous and mobile computing application.

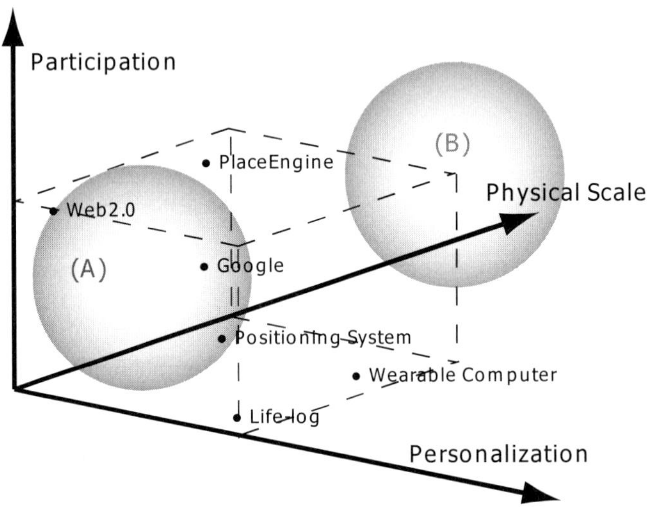

Fig. 3. Classification of System Developments in Ubiquitous Computing Field

3 Content Mash-Up Application

3.1 Electronic Money Card "Suica" as Ubiquitous Mobile Device

Requirements for the mobile device in our application are,

1. device must be widely used
2. device can carry an individual usage history
3. device can obtain position related data
4. device can be used as a input/output interface in the system

By considering above requirements, we chose electronic money card "Suica", which is mainly used as an IC card type train ticket in Japan (Figure 4).

Since 14,000,000 "Suica" cards are used in Japan (1), and recent 20 usage history (date, used station name, remains, usage) are recorded in a card (2)(3). As for a data input/output interface we can use uncontact interface (4). Therefore, we chose "Suica" card.

Nevertheless, most widely used mobile device is cellular phone, but current cellular phones are not able to use without any inhibition in requirements (3), (4).

Fig. 4. "Suica" card

3.2 System Configuration

Figure 5. illustrates the system configuration. The system is consists of client PC that collects the usage history data from "Suica" card and server PC that handles the data management and visualization. For the data input from "Suica" card, we adopted the same interaction style when we use usually use "Suica" card at the automatic ticket wicket at the station.

3.3 Interaction Design

It is important to demonstrate the capability of proposed system that can deal with mass participants in general public. Thus, we primarily designed the interface to make

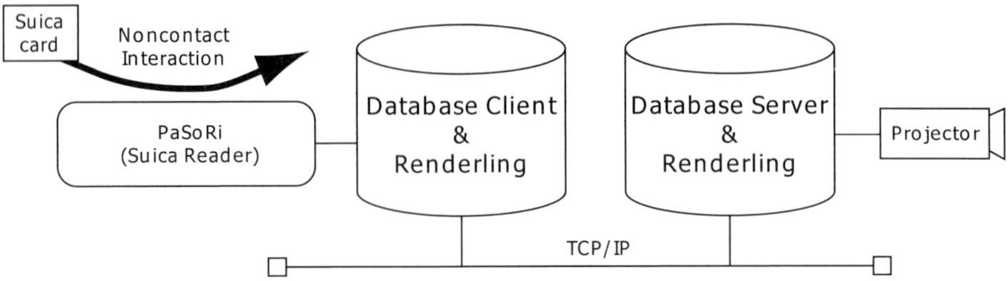

Fig. 5. System Configuration

visitors to voluntarily participate in interaction with developed system. For the implementation we took care of following two goals.

1. make participants enjoy by the interaction
2. afford participants how to make an interaction

Most of "Suica" card users do not know that their usage histories were recorded inside the memory. Therefore, visualizing these recorded data may enjoy the participants in a sense that they can remind their past experiences. As mentioned above, we design the package similar to actual automatic ticket wicket as possible (Figure 6).

Fig. 6. Interaction Style using "Suica" card

Developed application is named "Sharelog". In "Sharelog", if participants have their own "Suica" card can interact with the system and the system will display their used station records on the map from the oldest data. Traveling pass expected from used station records displayed in animation on the map is accumulated on the display so that each participant can compare and share their experiences in city of Tokyo (Figure 7). Each participant record is kept in anonymity.

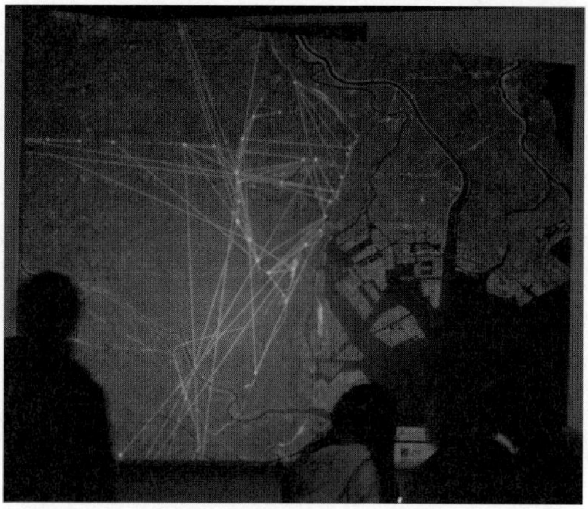

Fig. 7. Visualized Traveling Pass data from Participants' "Suica" cards

4 Experiment Results

For the experiment we exhibited our "Sharelog" at the exhibition of media art at Tokyo Metropolitan Museum of Photography. At the exhibition, a docent is staffed for giving commentary to visitors. Then general visitors can also make an interaction with "Sharelog", if they carry a "Suica" card. We made an evaluation about "Sharelog" by observing participants behavior and asking for their impressions.

4.1 Museum Installation

The exhibition period was 10 days and we had 3,201 participants for "Sharelog". Total 60,355 recorded used station data was input to the system. Figure 8. shows the changes in total visitors, expected users and participants of "Sharelog". The number of expected users is the estimated visitors who carry "Suica" card. From statistics, 48.5% of people have "Suica" card in Tokyo. We calculated the number of expected users from this the number of visitor multiplied by this ratio.

4.2 User Studies

As an evaluation, we observed following reactions from the participants. Many participants had an experience of reminding their past experiences after a time, but they could not remind at the moment they saw displayed their train boarding records. Also, we observed many times that from reminded experience memory encourages communication to speak her/his personal memories among accompanying person or a docent. Accumulating all the traveling pass of participants and coloring ongoing participant's traveling pass in different color seemed to have an effect of encouraging communication. In many cases, communication starts by referring her/his memory about the place they went and then she/he refers to places where other participants went. In addition, the system projects circles on the map as representatives of stations and projected circles grow larger according to the number of used station records from participants. This presentation had an effect to notice that the particular stations

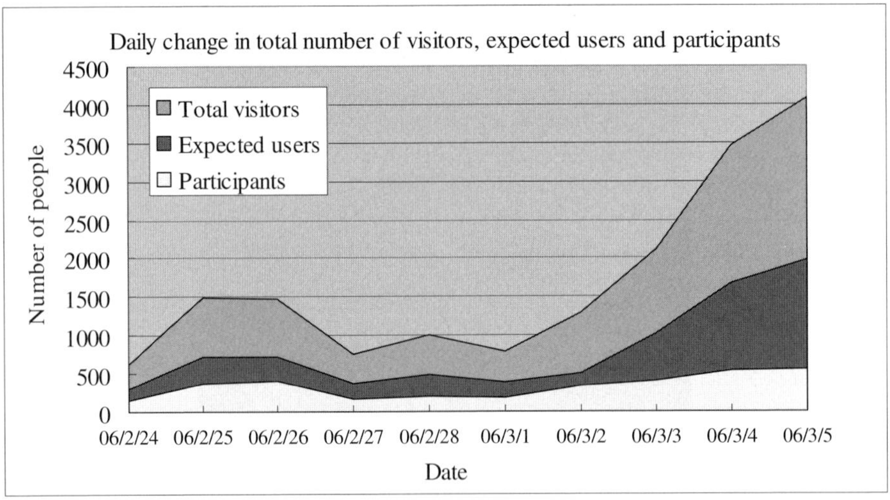

Fig. 8. Daily change in total number of Visitors, Expected Users and Participants

Fig. 9. Highlighted Stations (circle parts) in the Map of Tokyo

were used heavily in Tokyo (Figure 9). Many participants mentioned to this point. Some participants felt uncomfortable feeling in visualizing their personal traveling data. Especially, participants who did not explained the system characteristics well before they make interaction to "Sharelog" had this tendency.

5 Conclusion

Utilizing widespread consumer mobile devices as system platform enabled to develop an application that can be experienced by general public with ease. There are many advantages for system developers by using these kinds of mobile devices as an interface for ubiquitous computing application, since developers are enabled to focus on their work on application content side. For our future work, we would like to examine a way to adopt cellular phones equipped with GPS and camera as a public user interface for our new application.

References

1. http://www.apple.com/ipod/ipod.html
2. http://www.nintendo.co.jp/ds/
3. http://www.flickr.com/
4. http://www.youtube.com/
5. Sony Computer Science Laboratory. http://www.placeengine.com/
6. http://www.espgame.org/
7. von Ahn, L., Dabbish, L.: Labeling images with a computer game. In: Proceedings of the SIGCHI conference on Human factors in computing systems, pp. 319–326 (2004)
8. Sakamura, K., Koshizuka, N.: Ubiquitous computing technologies for ubiquitous learning. In: IEEE International Workshop on Wireless and Mobile Technologies in Education (WMTE'05), pp. 11–20 (2005)
9. Kidd, C.D., Orr, R.J., Abowd, G.D., Atkeson, C.G., Essa, I.A., MacIntyre, B., Mynatt, E., Starner, T.E., Newstetter, W.: The aware home: A living laboratory for ubiquitous computing research. In: Streitz, N.A., Hartkopf, V. (eds.) CoBuild 1999. LNCS, vol. 1670, Springer, Heidelberg (1999)

A Navigation System Using Ultrasonic Directional Speaker with Rotating Base

Kentaro Ishii[1], Yukiko Yamamoto[1], Michita Imai[2], and Kazuhiro Nakadai[3]

[1] Graduate School of Science and Technology, Keio University
[2] Faculty of Science and Technology, Keio University
3-14-1 Hiyoshi, Kohoku-ku, Yokohama 223-8522, Japan
[3] HONDA Research Institute Japan Co., Ltd.
8-1 Honcho, Wako-shi, Saitama 351-0114, Japan
{kenta,yukiko,michita}@ayu.ics.keio.ac.jp,
nakadai@jp.honda-ri.com

Abstract. This paper proposes new method for object reference which enables a person to find surrounding objects and develops a navigation system named CoCo. CoCo employs an ultrasonic directional speaker with a rotating base. Ultrasonic wave transmitted by an ultrasonic directional speaker is converted to audible sound when it reflects on floor, wall, or object. Based on this property, CoCo can emit audible sound from arbitrary place using a rotating base. When CoCo navigates a person to an object, CoCo transmits ultrasonic wave so as to hear audible sound continuously from the line between the ultrasonic directional speaker and the object. We designed and implemented an object management system as an application using 3D position sensors and a navigation system CoCo. This paper shows how CoCo navigates a person to an object using position data from an object management system.

Keywords: object reference, ultrasonic directional speaker, ultrasonic wave, position sensor.

1 Introduction

Recent sensor technology has realized small and light position sensors which can be attached to surrounding objects [1], [2]. Using those sensors, we can manage the positions of surrounding objects. In such a situation, we can find an object which we want even if a person forgot where he/she put away the object or another person moves the object.

In general, position data acquired from position sensor are coordinates on each sensor system. Those who are not specialists in sensor system can not directly realize a real world position from corresponding coordinates. It is essential that a system which refers real world objects has an interface which enables a person to realize a real world position directly.

M.J. Smith, G. Salvendy (Eds.): Human Interface, Part II, HCII 2007, LNCS 4558, pp. 526–535, 2007.

There are previous works for object reference system by intuitive method instead of position coordinates. [3] generates a natural language sentence based on a priori rule and displays the sentence. [4] develops a robot which points to an object and speaks to a person. [5] throws light on an object. [6] attaches a speaker and an infrared receiver. When a person with an infrared transmitter approaches to a receiver attached object, an infrared receiver receives an infrared signal and a speaker emits sound.

In case of natural language display, a person has to find a real world area where is corresponding the displayed sentence. In addition, since this system employs RFID tags with landmark sub-stations rather than position sensor, this system cannot recognize actual position. In case of pointing by a robot and light projection, if some objects occlude the line between the referred object and the robot or light source, it is difficult to detect the referred object. In case of attaching speaker, the speaker doesn't emit sound unless a person with an infrared transmitter approaches to an infrared receiver.

In this research, we propose a navigation system named CoCo. CoCo employs an ultrasonic directional speaker [7] with a rotating base. Ultrasonic wave transmitted by an ultrasonic directional speaker is converted to audible sound when it reflects on floor, wall, or object. Based on this property, CoCo can emit audible sound from arbitrary place using a rotating base. When CoCo navigates a person to an object, CoCo transmits ultrasonic wave so as to hear audible sound continuously from the line between the ultrasonic directional speaker and the object. Using this system, a person can directly find a real world area since audible sound is emitted from real world. CoCo takes advantage in case of occlusion since a person can hear audible sound continuously from the line between the ultrasonic directional speaker and the object rather than one point reference such as pointing or spotlighting. In addition, since our system knows the position of the referred object, CoCo can emit sound before a person approaches to the referred object.

We designed and implemented an object management system as an application using 3D position sensors [1] and a navigation system CoCo. This paper shows how CoCo navigates a person to an object using position data from an object management system.

The rest of this paper is organized as follows. Section 2 describes conditions and definitions. In section 3, this paper proposes new method for object reference and describes the details of a navigation system CoCo. Section 4 describes an object management system using position sensors and CoCo. Section 5 discusses the advantages and limitations of CoCo. Section 6 concludes this paper with our future plans.

2 Conditions and Definitions

2.1 Conditions

We assume that following conditions in the environment in which our system is used.

- Absolute object position can be acquired from position sensor
- There is space to install our system where the speaker look on the room
- There is no loud sound in the room

The first condition comes from the object reference method. Our object reference method is realized by changing the direction of the speaker in its tracks. The system has to know where to turn. It is different from the landmark reference method as in [3] or the exploratory method as in [6].

The second condition comes from the property of ultrasonic wave. Ultrasonic wave goes straight through in the air. In order to make sound from the object, ultrasonic wave must get through to the object although our method can be robust over occlusion to some extent.

The third condition comes from using sound for object reference. A person can interpolate the sound for split-second mishearing, since the person can hear audible sound continuously from the line between the ultrasonic directional speaker and the object. In the situation that continuous loud sound exist in the room, however, a person can not hear the sound from the speaker.

2.2 Definitions

This section describes definitions of terms used in the rest of this paper.

System user is used for the person who uses CoCo for object reference or the object management system to find surrounding objects. We also use simply *user* for this meaning.

System employer is used for the person who sets up CoCo or the object management system.

Reference line is used for the line between the ultrasonic directional speaker and the object. CoCo transmits ultrasonic wave onto this line.

3 CoCo

3.1 Hardware Configuration

We developed CoCo from an ultrasonic directional speaker [7] and a rotating base with 2 DOFs. Fig. 1 shows CoCo appearance. Each rotating unit has a DC motor controlled by a DC motor controller with an encoder. White rotating base rotates by 2 rotating units and change the direction of the ultrasonic directional speaker to arbitrary direction. The ultrasonic directional speaker produces an ultrasonic carrier wave modulated by an audible sound. Ultrasonic wave transmitted by the ultrasonic directional speaker goes straight through in the air. Ultrasonic wave is converted to audible sound when it reflects on floor, wall, or object. Based on this property, CoCo can emit audible sound from arbitrary place. Since human can not hear ultrasonic wave, human feel as if floor, wall, or object makes a sound.

3.2 Software Configuration

Fig. 2 illustrates system configuration of CoCo. CoCo receives object position and rotates its base with transmitting ultrasonic wave. CoCo consists of 3 software modules: command interpreter, motor control module, and sound control module. The rest of this section describes the details of each soft module.

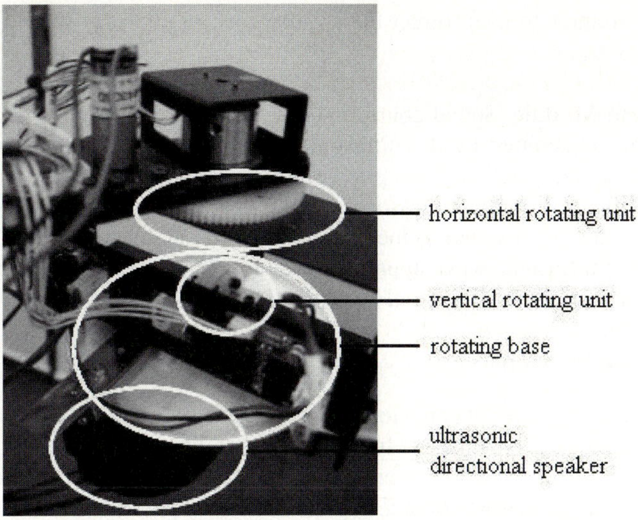

Fig. 1. CoCo appearance

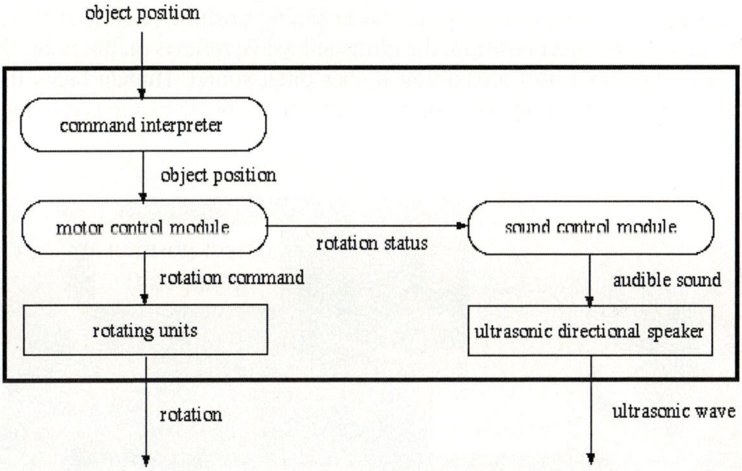

Fig. 2. System Configuration of CoCo

Command Interpreter. Command interpreter receives operation command with object position as input. When command interpreter receives operation command, it stores operation command in its internal queue. If no command is in operation, the stored command is immediately executed. Otherwise, command interpreter picks up the head command in queue and starts operation after previous operation is finished.

Motor Control Module. Motor control module handles exact motion of the motors of rotation units. Motor control module turns rotating base so that the reflecting point of ultrasonic wave moves on reference line from its feet to object position. It calculates goal encoder counter so as to direct the speaker toward object position. After that, it linearly moves motors.

Sound Control Module. Sound control module manages the original sound which is modulated and transmitted by the ultrasonic directional speaker. It receives rotation status of rotating units, and changes pitch of the original sound depending on rotation status. If the reflecting point of ultrasonic wave is near the speaker's feet, sound control module passes low pitch sound to the ultrasonic directional speaker. The more the reflecting point of ultrasonic wave approaches to object position, the higher pitch sound is passed to the ultrasonic directional speaker.

3.3 Navigation by Sound

This section describes overall behavior of CoCo. CoCo set motor to base position when the previous operation is finished. In this case, the ultrasonic directional speaker is pointing its feet. After receiving reference position, the ultrasonic directional speaker starts to turn toward reference object. Fig. 3 shows the appearance at that moment. In Fig. 3 the ultrasonic directional speaker is pointing its feet as the start point of the reflection.

White line in Fig. 3 means reference line, which is reflection trajectory of ultrasonic wave. The reflection point of ultrasonic wave changes position from start point to goal point. On the way to object position, the ultrasonic wave reflects on the floor, the chair, the desk, and the cup with transmitting higher pitch sound. Human hears the sound from the direction of the reflecting point and feels as if sound source is moving toward the object.

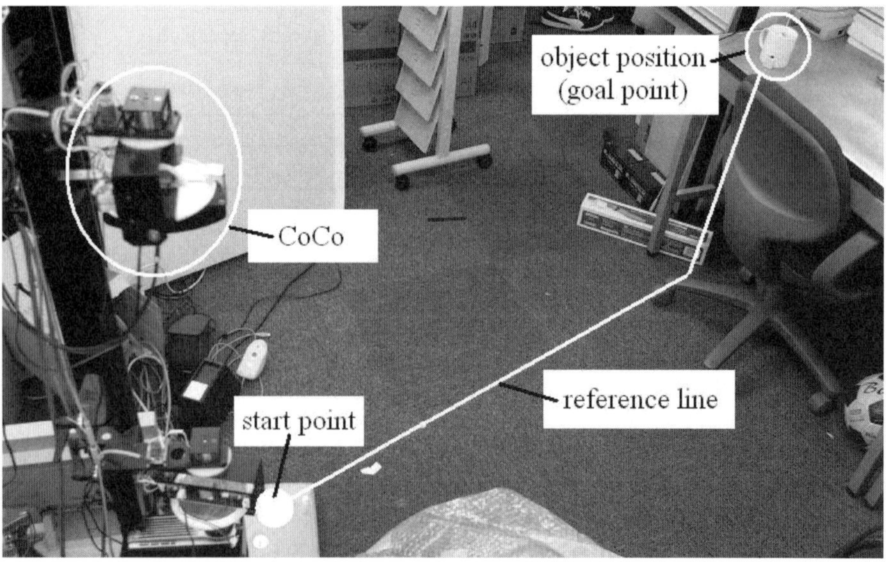

Fig. 3. Reflection Trajectory of Ultrasonic wave

At the point under the chair, the ultrasonic goes in the back of desk. In that instant, audible sound does not get to human ear. However, since human can interpolate the sound for split-second, human does not miss hearing sound movement in this situation. On the same score, CoCo can also navigate human toward the object position in case of occlusion. Human may hear as if sound source jumps other place, if some object including human gets into the way of the ultrasonic wave. However, human can recognize the direction toward the object.

4 Object Management Using CoCo

We designed and implemented an object management system as an application using 3D position sensors and a navigation system CoCo. This section describes configuration and behavior of the object management system.

4.1 System Configuration

Fig. 4 illustrates overall system configuration of the object management system. 3D position sensors obtain 3D positions of surrounding objects in background. Position data are stored in the corresponding tuple in object DB. Each tuple in object DB also has property information stored in advance such as object name, object color, object owner, object image, and so on.

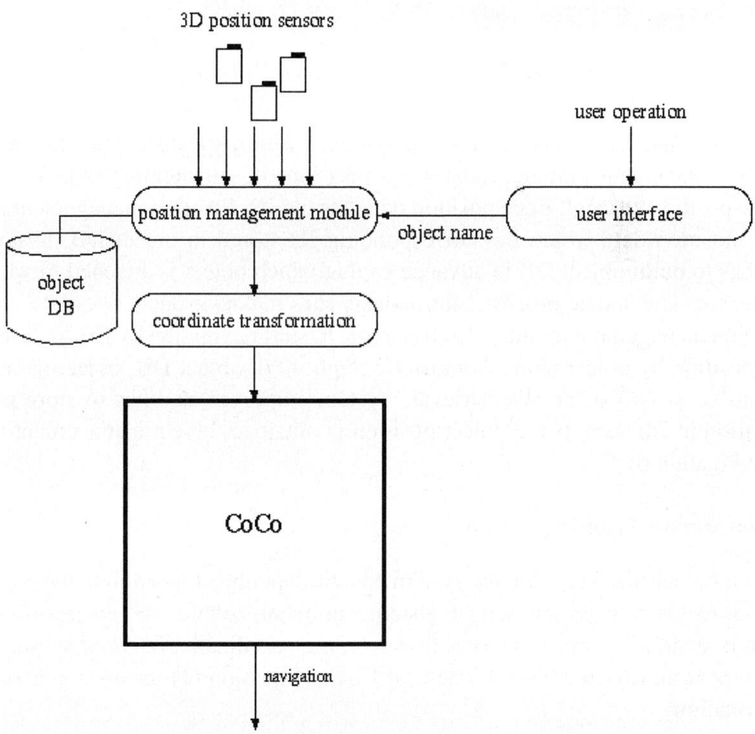

Fig. 4. System Configuration of the Object Management System

When system user want to find an object, system user operates the object management system via user interface. Position management module starts an operation when it receives reference object name from user interface. It loads object position from object DB and sends 3D position to CoCo after coordinate transformation.

In the following section, we show the details of each part of the object management system.

4.2 Position Management

We employ 3D ultrasonic tagging system [1] for position sensor. In this system, small tags emit ultrasonic wave and many ultrasonic receivers embedded in environment receive ultrasonic wave. We attached small tags to surrounding objects and installed ultrasonic receivers fixed to the ceiling. Fig. 5 shows sensor attached cup and installed ultrasonic receivers.

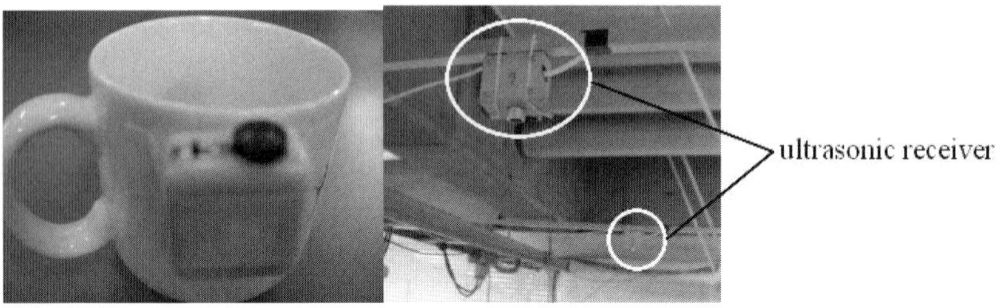

Fig. 5. 3D Ultrasonic Tagging System

Position management module has 2 functions: position update and position retrieval. Position management module updates 3D positions of surrounding objects acquired from 3D position sensors. Each position data has tag ID. Position management module updates position data which has corresponding ID stored in object DB. System employer has to build object DB in advance so that which object is attached to which position sensor. The update process continuously runs in background.

Position management module also has a role to start navigation by CoCo. It retrieves object position by object name from user operation. In object DB, object name is expected to be provided for this retrieval. System employer also has to store property information in advance. If the object position is obtained, object management module starts navigation by CoCo.

4.3 Coordinate Transformation

Before CoCo starts navigation, the system has to adapt object position to the coordinate system of CoCo. We assume sensor system can obtain absolute object position as described in section 2.1. In most cases, however, the coordinate system of sensor system is different from the coordinate system of CoCo. Equation (1) means a general space transformation.

$$\begin{pmatrix} x_C \\ y_C \\ z_C \\ 1 \end{pmatrix} = \begin{pmatrix} & \mathbf{R} & & \mathbf{t} \\ & & & \\ 0 & 0 & 0 & 1 \end{pmatrix} \begin{pmatrix} x_s \\ y_s \\ z_s \\ 1 \end{pmatrix}. \quad (1)$$

Where (x_C, y_C, z_C) is the coordinates of CoCo, (x_s, y_s, z_s) is the coordinates of sensor sytem, **R** is 3 dimension rotation matrix, **t** is 3 dimension translation vector. System employer has to find rotation matrix and translation vector in his/her own environment.

4.4 User Interface

The object management system has a user interface with which system user can operate the object management system from remote host. The object management system has a web server, and system user can access the user interface as a web page. When user operates the web page, a CGI script runs to start navigation.

Fig. 6 shows appearance of the user interface. User can select an object in 2 ways: by name or by image. The page contains a drop-down list in its upper part. The drop-down list has object names. When user clicks the button beside the drop-down list, the system starts navigation to the selecting object in the drop-down list. The page also contains images of objects in its bottom part. If system employer prepared object images, the page shows object images. When user clicks an image or a button below the images, the system starts navigation to the selecting object.

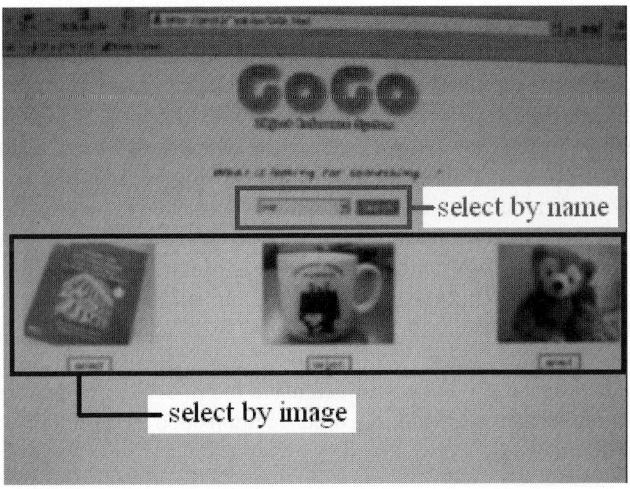

Fig. 6. User Interface of the Object Management System

4.5 Error Handling

The object management system has an extra function. Although user selects an object, there is a possibility that no corresponding position data is found in object DB.

Someone may bring the object outside the room, or the position sensor on the object may malfunction unfortunately. In those cases, the object management system informs user that there is no corresponding position data.

The object management system turns the rotating base toward the computer which user is accessing from. And then, the object management system says "the object is not found" with the ultrasonic directional speaker. The position of the computer is acquired from 3D position sensor in our system. Fig. 7 is a picture in this situation.

Fig. 7. A Extra Function to Inform Result

5 Discussion

We use sound for object reference. This choice provides many advantages. First of all, it makes this method an intuitive method. System user can recognize accurate position instantaneously by hearing sound even if he/she does not look at the object. Another advantage appears in fault situation especially in case of occlusion. Human can interpolate sound movement, and this property makes our method robust.

There are some limitations for our method. As described in section 2.1, we assume there is no continuous loud sound in the room. We can not use our method in a bustling environment such as an exhibition. We also have a limitation of space range. Although the ultrasonic wave goes straight through in the air, ultrasonic wave attenuates over distance. In addition, the longer distance ultrasonic wave goes, the more possibility of occlusion we have to consider.

We believe that our method is not an exclusive method. We believe that our method gets more powerful effect by combining other method such as spotlighting, pointing, or language generation.

6 Conclusion

We proposed new object reference method and developed an object management system using an ultrasonic directional speaker. In further research, we think we have to make an evaluation for this method, although our method make powerful effect for lost object searching.

References

1. Nishida, Y., Aizawa, H., Hori, T., Hoffman, N.H., Kanade, T., Kakikura, M.: 3D Ultrasonic Tagging System for Observing Human Activity. In: Proc. of IEEE/RSJ Intl. Conf. on Intelligent Robots and Systems, pp. 785–791 (2003)
2. Vicon.: Vicon motion capture system. http://www.vicon.com
3. Yap, K-K., Srinivasan, V., Motani, M.: MAX: Human-Centric Search of the Physical World. In: Proc. of The 3rd ACM Conf. on Embedded Networked Sensor Systems, pp. 166–179 (2005)
4. Sugiyama, O., Kanda, T., Imai, M., Ishiguro, H., Hagita, N.: Three-Layered Draw-Attention Model for Humanoid Robots with Gestures and Verbal Cues. In: Proc. of IEEE/RSJ Intl. Conf. on Intelligent Robots and Systems, pp. 2140–2145 (2005)
5. Nakada, T., Kanai, H., Kunifuji, S.: A Support System for Finding Lost Objects using Spotlight. In: Proc of the 7th Intl. Conf. on Human Computer Interaction with Mobile Devices and Services (2005)
6. Shinnishi, M., Iga, S., Higuchi, H., Yasumura, M.: Hide and Seek: Physical Real Artifacts which Responds to the User. In: Proc. of World Multiconference on Systemics, Cybernetics and Informatics, vol. 4, pp. 84–88 (1999)
7. Nakadai, K., Tsujino, H.: Towards New Human-Humanoid Communication: Listening During Speaking by Using Ultrasonic Directional Speaker. In: Proc. of IEEE Intl. Conf. on Robotics and Automation, pp. 1483–1488 (2005)

Mobile Social Networking Based on Mobile Internet and Ubiquitous Web Services

Yung Bok Kim

Department of Computer Engineering, Sejong University,
KunJa-Dong, Kwang-Jin-Ku, Seoul, Korea 143-747
yungbkim@sejong.ac.kr

Abstract. For online social networking service in the ubiquitous computing and networking environment, a ubiquitous Web service accessible with a mobile user interface was studied using single-character domain names among multi-lingual domain names, especially for mobile social networking. As convenient user interface for accessing a Web service for a social community, instead of long URL strings, the multi-lingual single-character domain names as indexing keys to social information in a ubiquitous Web service are convenient mobile interfaces to search social information as well as to register information and knowledge for mobile social networking service. We studied the convenience and performance of accessibility to ubiquitous Web and studied resource utilization for a mobile social networking service.

1 Introduction

A **social network** is a social structure made of nodes, which are generally individuals or organizations, indicating the ways in which they are connected through various social familiarities ranging from casual acquaintance to close familial bonds. The term was first coined in 1954 by J. A. Barnes (in: *Class and Committees in a Norwegian Island Parish*, "Human Relations"). The maximum size of social networks tends to be around 150 people and the average size is around 124 (Hill and Dunbar, 2002) [1]. Social networking also refers to a category of Internet applications to helping connect friends, business partners, or other individuals together using a variety of tools. These applications, known as online social networks are becoming increasingly popular [2], and the mobile social networking services are emerging on the basis of the evolution of mobile communication technology and multimedia mobile devices.

Researchers have examined social information flows for strong and weak ties in Internet communications and online spaces, how online socializing impacts people's psychological health, and how online socializing affects face-to-face interactions in communities [3]. We studied the convenience of ubiquitously accessing a Web service for various social communities, e.g. an alumni association, relatives, various offline associations, clubs, a companion group, and even a lecture group. Researchers have begun to examine the user's finding and refinding behaviors as well as the limitations of existing search technologies [4]. We should extend this concept of

M.J. Smith, G. Salvendy (Eds.): Human Interface, Part II, HCII 2007, LNCS 4558, pp. 536–545, 2007.

finding and refinding behaviors with the HCI concept to access to information for lifetime social networking service, especially with the mobile Internet and ubiquitous Web service.

There are many Internet portals providing various information, e.g. Web page, music, image, video, Web site, dictionary, encyclopedia, blog, mini-homepage, community café, and UCC (User Created Contents), etc. However, the behavior for finding and refinding the right information is difficult or confusing even for skilled users with a PC in the wired Internet environment. Considering mobile users in the mobile Internet and ubiquitous Web service, user interfaces (UIs) for mobile devices play a more crucial role in facilitating a pleasant and rich user experience with the above various contents; and the provision of appropriate mobile UIs becomes increasingly important [5]. Therefore even special mobile phones for mobile Internet, i.e. Google phone or Yahoo phone by Samsung, are scheduled to be commercially available.

Social network services such as Orkut and Friendster have accumulated data representing vast Webs of personal connections; however, their networks include numerous connections irrelevant to people's specific activities and lack relevant connections [6]. Search engines using over several tens of billion URLs cannot and should not index every page on the Web [7, 8]. Moreover, Web browsing on small-screen mobile devices will continue to be a major constraint for mobile users [9], and we need to consider new requirements using mobile devices [10]. Therefore we need to study more convenient mobile user interfaces and simplified Web services appropriate in the mobile Internet and ubiquitous Web sites, especially for the unified and lifetime service until the aged, beyond short term applications mainly for a specific younger generation.

In the following sections, we will discuss multilingual domain names for online social networking; and we will introduce the single-character multilingual domain names instead of long URL strings in order for convenient access to mobile Internet and ubiquitous Web service, especially for mobile social networking service. Information access in ubiquitous Web service will be discussed with consideration of services provided by commercial portals. Then, we will discuss the convenience metrics and performance metrics in ubiquitous Web access, and resource utilization for online social networking services including mobile social networking service. The empirical results will be discussed on the basis of implementation of a ubiquitous Web site for mobile Internet and ubiquitous Web service, accessible with multilingual single-character domain-names as well as capable of Text to Speech (TTS) functionality prior to speech UI (user interface) for elderly people in the Aged Society. Finally we will conclude our study with a consideration of future research.

2 Domain Names for Online Social Networking Service

The usefulness of the multilingual domain names for ubiquitous Web service in online social networking was studied, especially for the convenience of single-character multilingual domain names for mobile social networking service. For fast and convenient service, the speed of typing long and complex URLs will be one of the dominating factors in terms of performance for mobile phone users. In a previous work, the performance and UI (User Interface) issues in ubiquitous information network were studied by Kim et al. [11].

The user interface to type in the domain name for access to Web service, in wired Internet as well as in mobile Internet, should be as simple as possible. Single-character multilingual domain names satisfy a convenient user interface for the mobile Internet, including wired Internet. The scheme for multilingual domain names has been standardized worldwide by IETF (Internet Engineering Task Force) and has been approved by ICANN (Internet Corporation for Assigned Names and Numbers). The plug-in program for converting from Unicode or extended ASCII (e.g. Window codepage 949, etc.) to ASCII code had been the i-Nav program by VeriSign Inc., and the auto-conversion functionality for the standardized multilingual domain name services has been embedded in the Web browsers as a built-in functionality, e.g. from the version of MS IE7.0 and Mozilla Firefox, Opera, etc.

For information retrieval and access in online social networking services, the performance of the ubiquitous Web service with mobile Internet is important for providing QoS (Quality of Service) with cost-effective and inexpensive system solutions. We studied the requirement for performance and its metrics in the ubiquitous computing and networking environment. We introduce a social networking service, based on single-character multilingual domain names, and we show also the useful results of the implementation of the ubiquitous Web site for online social networking, using single-character multilingual domain names.

For consistency of information as well as for convenient user interface in online social networking service, we need a unified and ubiquitous Web DB for wired Internet and mobile Internet. Fast and convenient access of information as well as notification is required for online social community services including mobile social networking service. Information or advertisements in the ubiquitous Web site must be accessible with single-character multilingual domain names instead of long URL strings for convenient Web information access. However, even though convenient access and search is required for community information in many information portal sites using commercial search engines, the writing of community information or advertisements of Web sites or special notices for online social network is difficult or not possible, especially in terms of speed and convenience for the elderly during entire life of the service.

There are many different kinds of portals based on the wired or mobile Internet, and various community sites are proliferating as online social networking service, e.g. MySpace, Orkut, Friendster and CyWorld etc. The complexity for consistent information access from the portals or community sites has been increasing, and the inconvenience of the user interface for information access becomes serious even for the skilled users. Emerging mobile portals mainly used by the younger generation are more complicated for the elderly, in terms of UI for HCI. We suggest a ubiquitous Web DB, which is based on **'hand-board'** with meaning of online white**board** service using **hand**-sized information in PC or **hand**held phones, accessible with simple single-character multilingual domain names as a convenient mobile UI.

For the ubiquitous information portal for social networking service, the unified information service is indispensable in online social networking. The ubiquitous Web server should have the capability of showing unified contents, i.e. the HTML contents for the wired Internet as well as the mobile contents for many different kinds of mobile Internet and various contents using different mobile markup languages, e.g. WML (Wireless Markup Language) and mHTML (mobile Hypertext Markup Language). Easy typing of the URL is also important especially for mobile Internet. The single-character multilingual domain names, which could become the indexing key character for information retrieval from a ubiquitous Web DB and for information registration into a ubiquitous Web DB, were considered a convenient mobile user interface even for the elderly in the lifetime of the online social networking service.

3 Convenience and Performance in Ubiquitous Web Access

For typing the information in fast and convenient way, the user's typing speed is one of the important performance factors for information retrieval in online social networking service, especially with the mobile phone. Depending upon the keypad stroke number, the level of convenience of user interface could be estimated as a *convenience metric*. Considering the *convenience metric*, i.e. defined as the keypad stroke number, we could understand the usefulness of the single character domain names instead of long URL-string for access to unified and ubiquitous Web service, especially in the mobile Internet environment. For a ubiquitous information network, even the input of characters becomes important for retrieval from the unified Web DB as well as for registration into the unified Web DB, especially with keypads in the mobile phone. To access the unified Web DB ubiquitously, the user interface for information retrieval should be as convenient as possible for typing in the domain names or URLs for searching the right community Web site and accessing social information in the ubiquitous Web site.

We can access social information in the unified and ubiquitous Web site with multilingual single-character domain names as root nodes in information tree for online social networking. With multilingual single-character .net domains, we can access the required information, especially the unified and consistent information in a ubiquitous information network for the social community. If we looked at the user interface for handheld phones for a mobile information service, and even for the URL typing interface for information access in the wired Internet. A multilingual single-character is very convenient and useful because it is like a root node in the tree of the information-access hierarchy in the Web DB to generate any multilingual domain names or words for information access in China, Japan, Korea, and other countries using multi-lingual domain names.

Many different kinds of portals and Web sites may be a source of confusion, even though they may provide us with valuable information. Moreover, even though the single character for Web information access is very convenient, from commercial search portals in service we cannot retrieve any meaningful information even with huge number of results after searching with single character. We need a unified and ubiquitous Web information portal, based on the 'hand-board' we introduced, for ubiquitous information access in the ubiquitous networking environment, with convenient accessibility to the ubiquitous Web service for the mobile social networking service.

Let's look at mobile user interface in the mobile Internet for the mobile social networking service. The development and application environment are very different from the existing wired Internet environment, i.e. mainly based on MS Explorer as an Internet browser. Most of the Web pages in existence today are designed for desktop PCs, and viewing them on mobile Web browsers is extremely difficult. Chen et al. [12] have studied the adaptation of Web pages for small-screen devices. The searched Web information (e.g. mini-homepages, blogs, club, café, etc.) is very difficult to browse with a mobile phone, thus we need to consider the ubiquitous Web service with a convenient user interface and unified (for PC and mobile devices) online bulletin board, i.e. the so called 'hand-board' for text-based information adequate for fast TTS (text to speech) application.

We need to consider the performance of accessibility to the ubiquitous Web site as well as the performance of the mobile user interface for the ubiquitous Web service. We used a single Web server for information access as a unified service for simplicity of management and for cost-effectiveness. This method gives effectiveness and efficiency for the access of information and utilization of resources, in terms of the bandwidth for communication and the size of disk storage to build a ubiquitous Web site for the mobile social networking service.

We studied the important performance metric, e.g. delay, from the perspectives of normal users as well as the elderly to prevent the Digital Divide gap. We studied the performance metric, e.g. delay, not only with the time in the network and server, but also with the spent time by the user and the input time with the keypads for URLs or information for notification with the Web site for mobile social community services.

As *performance metrics* for mobile user, we assume that the random variables, the round-trip response time for user's single interaction in a session, from user to the contents in DB through wired/mobile Internet before next interaction with mobile phone is ***R***. That is composed of the preparation time for any user in online social networking to get mobile phones in his hand is ***U***. The time spent by the user with mobile phone to do appropriate action for service is ***D***. The aggregate time to the Web server after the mobile device through wired/mobile Internet for mobile service is ***S***. The time depending upon mobile contents is ***C***.

For the polling service, we may order the dominating factors in the overall performance at the user's perspective as follows. In general, the relationship between the delay metrics for the mobile Internet with the mobile phone could be ***U>D>S>C***. Here, we need to decrease the major times ***U*** and ***D*** for the polling service, as well as the network and server time ***S*** (or access time to contents in DB, ***C***). We need to decrease the times ***U*** and ***D*** for the polling service in the mobile Internet for mobile social networking service.

To be time-deterministic for the application with QoS (Quality of Service), e.g. searching time in one minute, the mobile user interface of the device should be convenient, the server should be efficient and have high performance for the dedicated application if possible, and finally the contents should be simple with a simplified and efficient format, considering the *convenience metric* we discussed. The user's preparation time for information access, or ubiquity metric, ***U*** will decrease depending upon the proliferation of the ubiquitous computing and networking environment. The user's preparation time for information access could be shortened depending upon the proliferation of ubiquitous application devices, e.g. mobile Internet devices. The average keypad-press number with the mobile phone is related to the time spent according to the mobile user interface. The aggregate time is related to the mentioned *convenience metric* and *performance metric*, which can be improved with a simple user interface and a text-based portal based on so-called 'hand-board' service, especially for the mobile social networking service.

4 Resource Utilization for Mobile Social Networking Service

We used a single Web server for the simplicity of management and the cost-effectiveness of implementation of text-based Web information portal for the mobile social networking service. We considered the effectiveness and efficiency for the management of Web information and the utilization of resources for social networking, in terms of the bandwidth for communication and the size of the Web DB for the mobile social networking service. Most commercial portals are providing rich-media contents for the community, e.g. UCC (user created contents), mini-homepages, blogs, clubs, and cafés; therefore the resource requirement for Web services are continuously increasing with more advanced computer systems as well as with higher network bandwidth. The investment of resources for multimedia Web services is quite great in comparison to our text-based approach; this is self-evident even without further analysis. Moreover the finding and refinding of the right information in most commercial portals is very difficult and is becoming increasingly complicated with mobile devices in the mobile Internet environment.

We analyzed the resource consumption based on commercially available services, e.g. SMS or e-mail service, mini homepage, blog, UCC, **'hand-board'** services related to the social networking service, as follows. With the push-style service (sometimes it becomes **spam-style** service, we use the subscript ***s***) of information notification based on SMS (short message service) or e-mail service, we considered the disk size of the mail/message server for n users. We considered the information access for social networking services based on **polling-scheme** (we use the subscript ***p***). We considered the consumed bandwidth proportional to the number of packets for transmission of each mail/message. In the case of the suggested method based on the polling services, e.g. mini homepage, blog, UCC, **'hand-board'** services, r was used as the right number of requesting/polling users, and $r \leq n$, sometimes the r is negligible. We considered the time spent by many mail/message users browsing through and deleting the mails/messages. We also considered the consumed time for information management for online social networking with the polling scheme. We primarily considered the server-side resource, ignoring the client-side resource for the cost-effective server, and we found the order of complexities of consumed resources, R_s and R_p (*s* means *pushing* scheme like *spam*, *p* means *polling* scheme) on the basis of three random variables (i.e. disk size, bandwidth for telecommunication and time consumption mentioned above), for *pushing* and *polling* scheme respectively. In general, we can assume $n \gg r$, and we could derive the relations as

$$n \gg r, O(R_s) = O(n^3) \gg O(R_p) = O(r^2) \quad (1)$$

Therefore we need a rather efficient way for resource utilization with the cost-effective Web server using the polling rather than the push-style method in the online social networking service.

We have analyzed further to compare the polling schemes more in detail, e.g. mini homepage, blog, UCC, **'hand-board'** services in terms of the consumption of the resources, i.e. telecommunication bandwidth and disk space. As a scheme for a '1 to many' service (i.e. one information provider notifies social information to many participants or customers) the above polling scheme seems to satisfy effective resource utilization, however the levels of efficiency and effectiveness are very different from service to service. Moreover the resource consumption is very different from scheme to scheme. We don't need to analyze each scheme in detail; instead we differentiate the above services in two schemes as follows. One scheme is based on multimedia contents, i.e. mini homepage, blog, UCC services, and another scheme is using only text-based contents with limited size, i.e. our proposed **'hand-board'** service. We chose the **'hand-board'** service because of the efficiency and effectiveness of resource utilization and TTS functionality for the mobile social networking service with hand-held mobile devices.

5 Implementation of a Ubiquitous Web Site

The implemented system is based on wired or mobile Internet with many multilingual single-character domains for fast access to a mobile social networking service in which the required information or advertisement can be registered in any time and any place using wired or mobile Internet with multilingual single-character domain names.

We measured the time to type-in the 'ktrip.net' as a simple example to show the critical time in terms of performance. We measured also the aggregate time spent by the Web site and network, and we observed the time to read the right content after a click of the title in the displayed information list on the screen of mobile phone. We observed that the typing time for full domain names or long URL-string with mobile phone was very slow in terms of performance. Therefore we need single-character domain names instead of long URL-string for mobile Internet UI with mobile phones, as we discussed; and the typing time is faster than 3 seconds, in most cases with the multilingual single-character is typed in 1 second.

In terms of packet cost, one packet, i.e. 512 Bytes, costs 0.5 Cents (U.S.), the minimization of delivered packet number from ubiquitous Web DB to the mobile phone is important for cost-effective service with the ubiquitous Web portal. We considered both the cost-effective packet size and the number of packets for delivery in our implementation of ubiquitous Web portal for the mobile social networking service; and moreover the TTS (text to speech) functionality in service was easy to be implemented because of the tiny size and text-based 'hand-board' contents.

We did some experimental research with students as a mobile social network for a lecture group, as follows. The number of students was around 700 across six semesters, i.e. around 117 students per semester composed of 4 months as a similar number 124 of averaged social network size [1], and they used frequently this notification bulletin board so-called 'hand-board' using the site 'ktrip.net'. The cumulative number of clicks was around 37,000 times, and that means the average click number in one semester by one student was 52.9 (i.e. around 13 clicks per month by each student), thus every student clicked around 3 times per week.

Therefore, for the special application where the usage frequency in our social network for the lecture group is around 3 times per week, this suggested method is useful. As an example of Web services for online/offline social networking service, special application for information registration/retrieval from/into a ubiquitous Web DB can be done with the notification bulletin board, the so-called 'hand-board' for alumni association in social networks; here the size of association members may be from ten people to a couple of hundred people in online/offline social networking.

This polling can be considered similarly to the sampling theorem, the polling frequency $f_{poll} \geq 2 * f_{notify}$, where f_{poll} is the frequency of polling the ubiquitous Web site for mobile social networking service, and f_{notify} is the frequency of notification in 'hand-board' by the association for the mobile social networking service. Usually many alumni associations are held within the time interval between one month and one year; that means the information polling frequency by each member is enough to be notified with the frequency between once every two weeks and once every six months.

The speed of on-line registration of the advertisement/notification as well as the speed of access of special information with the ubiquitous Web DB is fast enough for real applications. Moreover, the effectiveness of the Web DB usage for the mobile social networking service is anticipated if we consider the applications for various communities, mini-homepages, club, and blog, UCC services based on rich media, as far as the consumed disk storage and cost for operation and administration related to mobile social networking are concerned.

We implemented also the Text to Speech (TTS) functionality, because the simple and text-based information in 'hand-board' service was easy to implement the TTS functionality based on speech synthesis. The conversion time of 1 Kbytes text-based information was around 1 second; and in most cases, the information size for mobile social networking service was enough at around 1 Kbytes with our experience. This TTS functionality will be very helpful for the elderly; moreover, the telematics service for auto drivers will be applicable with convenient user interface for listening to the contents in 'hand-board' with transformed audio contents.

Commercial portals cannot give meaningful information after searching with multilingual single-character or alphabet, even though it is very convenient to search with single-character as searching index key for ubiquitous Web DB access in the mobile social networking service, especially in the mobile Internet using handheld phones with small screen. We are considering the premium Web information service instead of various Web information on the basis of using text-based Web DB for mobile social networking service; therefore the listed information, e.g. meaningful information searched based on a multilingual single-character domain name, is not so much as the commercial search portal services provide with meaningless information.

We are also considering the location-based services in mobile social networking services using mobile phones, then the listed information related to that specific location will be decreased tremendously for mobile social networking service in the ubiquitous networking environment with mobile Internet.

The convenience and usefulness of multilingual single-character domain names will be remarkable for accessibility to the ubiquitous Web site for the mobile social community service. With any mobile devices even for the elderly in Silver Society, without Digital Divide in the forthcoming Knowledge Society, the lifetime social community service will be feasible in the ubiquitous computing and networking environment.

6 Conclusions

A ubiquitous Web site accessible with multilingual single-character domain names instead of long URL-string for convenient information access has been studied for mobile social networking service. The convenience of multilingual single-character domain-names and the tiny 'hand-board' for mobile Internet service was discussed with the empirical results based on the implementation in the mobile Internet as well as in the wired Internet. The performance for information access using multilingual single-character domain names with mobile phone in the mobile Internet was also discussed, considering the keypad-press number with the mobile phone as a *convenience metric*, especially for the mobile social networking service.

The results of the implementation show the usefulness and convenience of multilingual single-character domain names for accessibility to the ubiquitous Web site in mobile social networking service. Additionally, the Text to Speech (TTS) functionality based on the text based contents in the 'hand-board' service for the mobile social community was implemented in the unified site 'ktrip.net' to increase convenience in the ubiquitous Web site for the elderly people in the Aged Society. With an expanding ubiquitous network environment, the delay time related to the user interface will become more critical; thus we need a more efficient user interface for mobile social networking services, especially using mobile Internet with mobile devices.

For further research, the reliability of the ubiquitous Web service in mobile Internet will be studied for the mobile social networking service.

References

1. Hill, R., Dunbar, R.: Social Network Size in Humans. Human Nature 14(1), 53–72 (2002)
2. Keyword: Social Network, Web Site: Wikipedia: http://wikipedia.com
3. Churchill, E.F., Halverson, C.A.: Social Network and Social Networking, pp. 14–19. IEEE Internet Computing, Los Alamitos (2005)
4. Capra, I.R.G., Perez-Quinones, M.A.: Using Web Search Engines to Find and Refind Information, pp. 36–42. IEEE Computer, Los Alamitos (2005)
5. Subramanya, S.R., Yi, B.K.: User Interfaces for Mobile Content, pp. 85–87. IEEE Computer, Los Alamitos (2006)
6. Konomi, S., Inoue, S., Kobayashi, T., Tsuchida, M., Kitsuregwa, M.: Supporting Colocated Interactions using RFID and Social Network Displays, pp. 48–56. IEEE Pervasive Computing, Los Alamitos (2006)
7. Hawking, D.: Web Search Engines: Part1, pp. 86–88. IEEE Computer, Los Alamitos (2006)
8. Hawking, D.: Web Search Engines: Part2, pp. 88–90. IEEE Computer, Los Alamitos (2006)

9. Hua, Z., Lu, H.: Web Browsing on Small-Screen Devices: A Multiclient Collaborative Approach, IEEE Pervasive Computing, Los Alamitos, pp. 78–84 (April-June, 2006)
10. Seyff, N., Grunbacher, P., Maiden, N.: Take Your Mobile Device Out from behind the Requirements Desk, IEEE Software, pp. 16–18 (July/August, 2006)
11. Kim, Y.B., Kwak, M., Cho, D.: Performance Analysis for User Interface in Real-Time Ubiquitous Information Network. In: Masoodian, M., Jones, S., Rogers, B. (eds.) APCHI 2004. LNCS, vol. 3101, pp. 171–179. Springer, Heidelberg (2004)
12. Chen, Y., Xie, X., Ma, W.Y., Zhang, H.J.: Adapting Web Pages for Small-Screen Devices, IEEE Internet Computing, pp. 50–56 (January/February, 2005)

Ubiquitous Hands: Context-Aware Wearable Gloves with a RF Interaction Model

Jong Gon Kim, Byung Geun Kim, and Seongil Lee

[1] Sungkyunkwan University, Department of Industrial Engineering
Suwon, 440-749, Korea
gon0325@gmail.com, billkim@limeorange.com, silee@skku.edu

Abstract. In this paper, we describe the development of gloves that can be used in a ubiquitous computing environment. The ubiHand gloves were developed to access information devices in various wireless environments, including mobile computing, games, and in-vehicle telematic systems. The gloves are equipped with chording keyboard mechanism for flexible input and control of the wireless devices, an embedded RFID reader and a set of accelerator sensors for gathering information from the users' various hand gestures. The system configuration and keymaps for ubiquitous information access as well as the interface for input and control for the gloves are presented.

1 Introduction

For years, many research efforts have been made in designing ubiquitous technologies for smart environments. To make access and retrieve information efficiently, users need a convenient access tools and interfaces to give the users primary role to manage the information by themselves. This includes how portable their computing tools are to carry anywhere and anytime, and how suitable their computing tools and interfaces are for navigation and information presentation as well as data input.

Like many other emerging technologies, ubiquitous computing is characterized by being different from traditional computing methods in the physical appearance and the contexts in which it is used. This paper described our efforts of developing the ubiHand, a wearable ubiquitous device that can provide mobile users with accessibility and usability to various computing environments. Gloves equipped with a wireless communication module can be a good interactive device to many embedded computers since the hands and fingers are the most natural and dominantly used parts of the body. Many simple gestures of the hands reveal so many different contexts which the users deal with. Gloves equipped with a RFID reader also facilitates applications that are triggered by handling tagged physical objects, enabling the system be applied to context-aware mobile computing. Using hands, humans can naturally develop unique and effortless strategies for interacting with ubiquitous environment. In other words, hands do not have to hold and manipulate interfacing devices, but hands can be the interfacing devices themselves for ubiquitous environments to interpret.

M.J. Smith, G. Salvendy (Eds.): Human Interface, Part II, HCII 2007, LNCS 4558, pp. 546–554, 2007.

2 Background

Several types of glove-based devices recognizing hand gestures or contact gestures directly have been widely proposed as input devices to computers. These devices are well suited for use in a mobile environment because the gloves can be worn instead of just being used to hold a device, are lightweight and easy to store and carry. It is, however, difficult to recognize enough separate gestures to allow useful text input. Some glove-based input devices, though, have capabilities to make decent text input in addition to their intended functions of gesture recognition and space navigation. Pinch Gloves [3, 15] are glove-based input devices designed for use in virtual environments, mainly for 3D navigation, and N-fingers [7] is a finger-based interaction technique for wearable computers also utilizing finger pinching. Pinching is basically the motion of making a contact between the tip of thumb and a fingertip of the same hand. It uses lightweight gloves with conductive cloth on each fingertip that sense when two or more fingers are touching.

For use of many small portable electronic products, chord keyboards have also been proposed as input devices [5, 10, 11, 14]. A chord keyboard is a keyboard that takes simultaneous multiple key pressings at a time to form a character in the same way that a chord is made on a piano. In chord keyboards, the user presses multiple key combinations, mainly two-letter combinations, to enter an input instead of using one key for each character. Pressing combinations of keys in this way is called chording. Since chord keyboards require only a small number of keys, they do not need large space, nor the many keys of regular keyboards such as the QWERTY keyboard. For example, the Handkey Twiddler is a one-handed chord keyboard with only 12 keys for fingertips and a ring of control keys under the thumb, and the Microwriter with only 6 keys [14].

Rosenberg and Slater [12, 13] proposed a glove-based chording input device called the chording glove to combine the portability of a contact glove with the benefits of a chord keyboard. In their chording glove, the keys of a chord keyboard were mounted on the fingers of a glove and the characters were associated with all the chords, following a *keymap*. Their design of the keymap for the chording glove for Roman-alphabetic text input left room for improvement according to the experimental results. Pratt [11] also designed a device-independent input device and language code called "thumbcode" using chording gloves targeting for PDAs.

All the previous works on glove-based input devices were intended for usage with general computers, not targeted for mobile computing environments. The current works investigate the utilities, functionality, and usability of the chording gloves as a smart device for ubiquitous information access to mobile devices.

Recently, there has been research on wearable systems for detecting objects with RFID tags [4, 6, 8]. These systems mainly tried to design and implement a system for accessing contextual information using the wearable computer. The wearable computer came with a variety of forms, but the most common forms were gloves with a RFID tag.

3 System Configuration

The design of the ubiHand chording gloves and its controller began with the arrangement of the sensors on the gloves which function as control keys. The interface of the system is based on the human-friendly nature of finger movements, particularly finger flexion. All the operations by the chording gloves consist of either a series of finger tapping on a hard surface with ground levels or finger contacts between the fingers with silicon inks. The ubiHand glove also contains a RFID reader so that it can sense what touches and where it is by reading RF tags providing such information.

3.1 General Structure

The ubiHand chording gloves developed in the study which use finger tapping to make various types of input is shown in Fig. 1. An input can be made by contacts between the keys on the fingertips and either the thumb or hard surface. Chording is possible by making simultaneous tapping with multiple fingertips to the thumb or hard surfaces. Four keys are placed on the fingertips and phalanges of the index and middle on the palm side of leather gloves. The keys are made of conductible silicon ink applied to the fingertips with the rectangle of 1.5 cm by 1.3 cm. The keys become "pressed" once the voltage through the silicon ink rises above 3.5 V with contact with the thumb or any hard surfaces. The voltage outputs of chord gloves are connected to an embedded controller that translates chord information into its corresponding character. The corresponding finger force to generate 3.5 V is set to 3.50 N to avoid unwanted activations with light contact between the fingers and driving wheel. The glove with the RF reader and controller on weighs approximately 55.0 g and the right-hand glove with only keys weighs 28.5 g.

3.2 Control and Communication

The 16F874 microprocessor is used for controlling and analyzing the signals from the chording gloves. The controller also works for functions of converting the signals to text codes and sending them to PC or PDA through a Bluetooth 1.1 (aircode) wireless adapter. The controller works with a 9V battery.

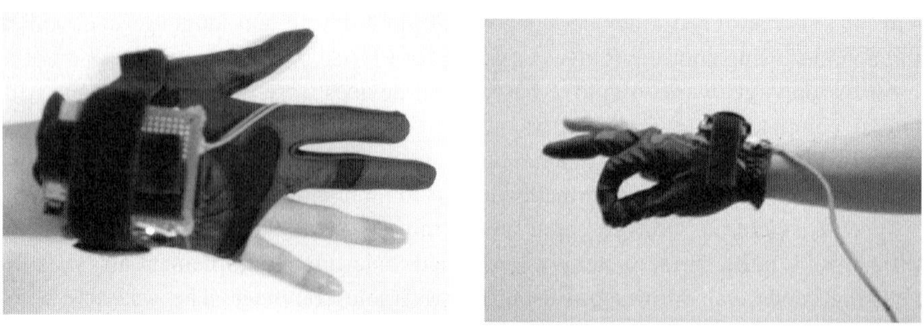

Fig. 1. The UbiHand Glove. The thumb acts as a ground to conduct electricity to the keys on the fingers made of silicon ink. It also contains an RFID reader so that it can sense and record what it touches or where it is.

The voltage outputs of chording gloves are sent to an embedded system that translates chord information into its corresponding code, character, or number. The chord glove-based input device can be used in connection with computers or PDAs directly as a portable keyboard. The voltage outputs of chording gloves are sent to an embedded system that translates chord information into its corresponding code, number, or signals. The controller transmits signals from the gloves to the PDA, mobile phones, and other telematic systems with a Bluetooth module to control them.

One in the pair of the ubiHand chording gloves are also equipped with a RFID reader.

4 Ubiquitous Access

4.1 Access to In-Vehicle Transport Telematics

Multimedia and telematic systems have become important parts to automobiles. A substantial amount of effort has been invested in transport telematics and in-vehicle multimedia systems over the last decade. The increase in use of smart electronic products such as navigation systems, wireless internet, mobile phones, and other audio/video devices in automobiles provide mobile information accessibility and convenient and pleasant driving conditions. Automobile infotainment systems for today's luxury cars provide noise-compensation and surround sound technology, communication systems with telematics and hands-free digital phone, navigation systems are equipped with DVD-based programs including maps and precise audio/visual directions [1, 2, 9]. Control or manipulation of each device, however, requires human attention and inevitably distracts driver's attention from driving tasks, resulting in degrades in driving performance and potential danger of accidents. Emerging concept for central control of these multimedia devices suggests for needs of interface design that can provide straightforward, effortless, and logical activation and manipulation without driver's performance degrade in the main task of driving.

The traditional display screen-keyboard interaction model is inappropriate for this ubiquitous information access capability in in-vehicle multimedia interface, and new interaction paradigms need to be investigated in multimodalities using speech or haptic input and output.

The ubiHand chording gloves we developed can provide drivers with safe, all-in-one telematics control functions with an integrated and wireless wearable multimedia interface. Gloves equipped with a WPAN-based wireless communication module can be a good controller to multimedia systems inside automobiles since the hands and fingers are the most dominantly used parts of the body. Therefore, drivers can naturally develop unique and effortless strategies for interfacing with multimedia and telematics in the car without moving their hands away from the driving wheel. In addition, the drivers do not have to move their eye gazes to control the devices while driving, which serves as a very important factor for safe driving. Control input can be made by simple tapping or chorded tapping of the fingers on the hard surface of driving wheel. The i-PAQ PDA with a Bluetooth module is working as a server for other in-vehicle information devices that are wired and as an access point for the wirelessly connected devices in our system (Fig. 2).

In the in-vehicle telematics mode, the glove for the left-hand basically works as a device selection keypad. The thumb, together with the thumb of the right hand, works for the power on/off of the entire unit. That means the system can be turned on or off only with the "thumb + thumb" chord tapping. The index finger selects the mobile phone, while the middle finger selects the navigator on the PDA. Once a device is selected, the interface changes to the control mode for the selected device until the other device is selected with the left-hand fingertip tapping. With the thumb excluded, a total of 6 devices can be selected from single finger tapping and two-finger chord tapping combinations. Pilot experiments showed that users could easily memorize and activate devices with chording combinations of adjacent fingers such as "index + middle" fingers, but had hard time to memorize and activate the functions assigned to the chords with "the other hand" chording.

Fig. 2. Use of chording gloves in an in-vehicle infomatics and telematics environment

4.2 Access to Wireless Internet

Users can access to information from the wireless internet either by the explorer program installed on the PDA or by the wireless internet services provided on the mobile phone's carrier company inside the car. Since the ubiHand chording gloves can control the PDA's operating system (Windows CE) as well as application programs, users can select internet navigation browser program and then type in the URL addresses in the text mode to access the information on the internet.

The other way to access information on the wireless internet is to use the wireless internet services provided by mobile phone's communication services carriers. The mobile phones equipped with a Bluetooth module can be controlled by the chording gloves, and a corresponding service function buttons can be activated by chording. In this way, text messages (SMS) or emails transmitted to the internet services on mobile phones can be accessed by the chording gloves. The accessed internet information, however, is not displayed on the LCD on the back of the left hand, but presented only on the mobile phone's display.

4.3 Access to Home Network

The ubiHand gloves can also be used for wireless access to and control of home network systems. A PC with an access point connected to an iPAQ serves as a server for the home network. The gloves can work in the same way with the in-vehicle system. The specific keymaps and wireless network protocols were not yet finalized in this area using the chording gloves. User can always carry a PDA and the gloves as a set, and select the appropriate mode and then activate the corresponding keymap. Future studies on situation-awareness can help the system be more efficiently used in many different environments.

5 RF Interactions for Context-Aware Computing

5.1 Experiments in a Classroom Domain

We performed an experiment in a classroom domain how useful the ubiHand gloves can be for context aware wearable devices. Since the ubiHand gloves are equipped with a RFID reader, it can sense the RFID tags in the surrounding environment, and hence can change the status of modes automatically based upon the context it infers. We set up a classroom with many relevant objects equipped with an RFID tag in a laboratory. A student and a teacher who is wearing the ubiHand on one hand simulated class acts such as lecturing and discussion on a whiteboard. The experimental set-up is described in Fig. 3. The concept of a wearable RFID-tag reader or wireless wearables for detecting objects have well been explored in other studies [####].

We also developed a matrix-based context-aware model for the ubiHand gloves application in the same domain. The matrix can describe the interactions between the RFID tags of the objects and also of the users and RFID readers of the ubiHand gloves. The model can extract data on context of each users by analyzing the RFID interactions on the matrix, and can change mode of operation of the gloves and computers. This process of context-aware model is described in Fig. 4, from the primary context data gathering by the ubiHands, to matrix-based operation of fusion context extraction, and the final context inference. The final context contains the data on why the user did such an action and how it is done at a certain time within the classroom.

Such objects as the whiteboard, student's chairs, podium, chalks, pencils were tagged with a small RFID. The teacher and the student were also attached with a RFID tag. The final context results were categorized into 5 contexts: student's attendance, teacher's attendance, lecturing, discussion on the whiteboard, end of class. The ubiHand changes its mode according to the context inferred. For example, the teacher's ubiHand glove changes its mode to a presentation pointing device when its context aware system recognizes it in a lecturing state, and to a communication or other controller mode when the class is over. The RFID tagging data was logged into a main computer every second for 12 minutes of experiment. At the end, the accuracy of correct context inference was measured at around 69.7%, with the analysis of video frame data.

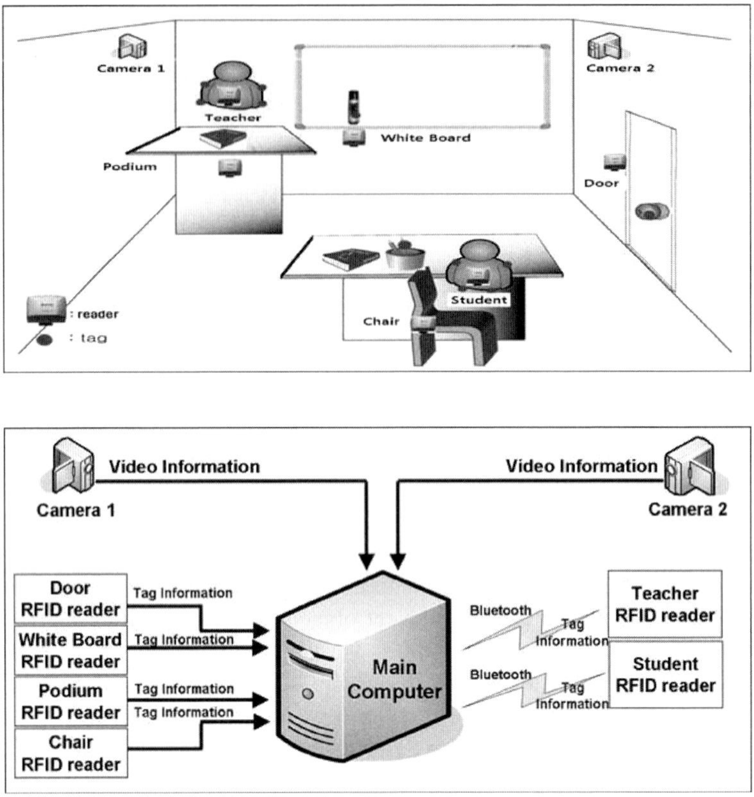

Fig. 3. Configuration of the ubiHand gloves experiments. A teacher and a student wearing the ubiHand gloves simulated class acts while touching and being near the classroom objects with a RFID tag. All the data along with the video frames are sent to a main computer to log the RFID tagging at every second.

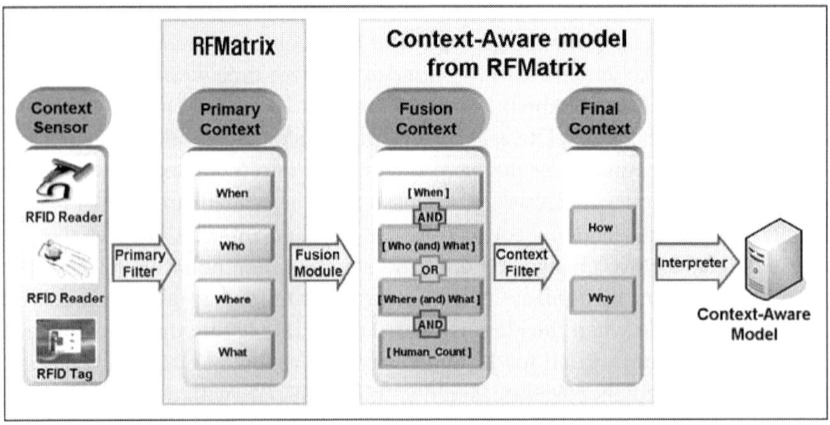

Fig. 4. A concept diagram of the context-aware model developed for the ubiHand application in the classroom domain

6 Conclusion

Our goal was to make a wearable controller for mobile computers capable of sensing the users and their current state, exploiting context information to significantly reduce demands on human attention. We came up with a glove-based context-aware computing system with a wireless communication module and a RFID reader. The ubiHand glove was supplemented with a RF interaction matrix to infer context data for it to adjust its function mode to its user's situation and states. Even though the accuracy of the RF interaction matrix-based context-aware system in a classroom test domain, it still shows possibility to be used as a ubiquitous controller for a context-aware mobile computing system.

We first tested the ubiHand gloves in various environments such as vehicles and homes, and then developed an appropriate context-aware computing model for the wearable controller that can be used in environments with many objects tagged with RFIDs. We developed a model that infer the context out of the information from multiple RF tags to define the tag and reader relationship based on a matrix that define the primary context (When, Who, Where, What) into Fusion context (Primary context with operators of AND and OR) to infer the final context of Why and How, using the ubiHand gloves. By combining a context-aware computing model with a wearable controller, we have been able to design and implement a user-friendly ubiquitous input device and controller. Even though the RF interaction matrix was not accurate enough to support the user's context in highly reliable fashion, it was still a powerful and useful system.

Acknowledgments

This research was supported by the MIC(Ministry of Information and Communication), Korea, under the ITRC(Information Technology Research Center) support program supervised by the IITA(Institute of Information Technology Assessment), IITA-2006-(C1090-0603-0046).

References

1. Akesson, K., Nilsson, A.: Designing Leisure Applications for the Mundane Car-Commute. Personal and Ubiquitous Computing 6, 176–187 (2002)
2. Alpern, M., Minardo, K.: Developing a Car Gesture Interface for Use as a Secondary Task. In: Proceeding of CHI 2003, Ft. Lauderdale, Florida, USA, pp. 932–933. ACM Press, New York (2003)
3. Bowman, D.A., Wingrave, C.A., Campbell, J.M., Ly, V.Q.: Using Pinch Gloves for both Natural and Abstract Interaction Techniques in Virtual Environments. In: Human-Computer Interaction; Proceedings of the HCI International 2001, pp. 629–633. Lawrence Erlbaum Associates, Mahwah (2001)
4. Fishkin, K.P., Philipose, M., Rea, A.: Hands-On RFID: Wireless Wearables for Detecting Use of Objects. In: Proceedings of the 9th IEEE International Symposium on Wearable Computers, Osaka, Japan (2005)

5. Fukumoto, M., Tonomura, Y.: Body Coupled Finge Ring: Wireless Wearable Keyboard. In: Proceedings of CHI '97, Atlanta GA, March 1997, pp. 22–27. ACM Press, New York (1997)
6. Krause, A., Smailagic, A., Siewiorek, D.P.: Context-Aware Mobile Computing: Learning Context-Dependent Personal Preferences from a Wearable Sensor Array. IEEE Trans. On Mobile Computing 5(2), 113–127 (2006)
7. Lehikoinen, J., Roykkee, M.: N-fingers: a finger-based interaction technique for wearable computers. Interacting with Computers 13, 601–625 (2001)
8. Patterson, D.J., Fox, D., Kautz, H., Philipose, M.: Fine-Grained Activity Recognition by Aggregating Abstract Object Usage. In: Proceedings of the 9th IEEE International Symposium on Wearable Computers, Osaka, Japan (2005)
9. Pieraccini, R., et al.: Multimodal Conversational Systems for Automobiles. Communications of the ACM 47(1), 47–49 (2004)
10. Porosnak, K.M.: Keys and Keyboards. In: Helander, M. (ed.) Handbook of Human-Computer Interaction, pp. 475–494. Elsevier, New York (1988)
11. Pratt, V.: Thumbcode: A Device-Independent Digital Sign Language. In: Proceedings of the 13th Annual IEEE Symposium on Logic in Computer Science, Brunswick, NJ (1998)
12. Rosenberg, R.: Computing without Mice and Keyboards: Text and Graphic Input Devices for Mobile Computing. Doctoral Dissertation. University College London (1998)
13. Rosenberg, R., Slater, M.: The Chording Glove: A Glove-Based Text Input Device. IEEE Trans. On Systems, Man, and Cybernetics-Part C: Applications and Reviews 29(2), 186–191 (1999)
14. Sturman, D., Zeltzer, D.: A Survey of Glove-based Input. IEEE Computer Graphics & Applications (Jan 30-39, 1994)

Power Saving Medium Access for Beacon-Enabled IEEE 802.15.4 LR-WPANs*

Joongheon Kim[1] and Wonjun Lee[2,**]

[1] Digital Media Research Lab., LG Electronics, Seoul, Korea
[2] Department of Computer Science and Engineering, Korea University, Seoul, Korea
wlee@korea.ac.kr

Abstract. To construct constraint-free human-computer interaction (HCI) environments, the wireless communication is one of the most essential research issues. Among the numerous wireless communication technologies, IEEE 802.15.4 low-rate wireless personal area networks (LR-WPANs) is the most suitable for ubiquitous and pervasive HCI environments, such as home networking applications, monitoring systems, and so on. In LR-WPANs, the power-efficiency is one of the most attractive issues. Therefore, this paper proposes a power saving medium access scheme for star topology beacon-enabled LR-WPANs. To achieve power efficiency, we turn off communication modules when the time slots are not used for data communications. On the other hand, we turn on the modules. The proposed scheme predict the time for data communications in a given superframe structure. Our simulation result shows that our algorithm achieves desired properties.

1 Introduction

Currently, many research efforts on IEEE 802.15-compliant wireless personal area networks (WPANs) have become one of the most active research topics in wireless networking technologies [1] by their advantages in human computer interaction (HCI) applications [2]. Among the IEEE 802.15-operated WPANs, An IEEE 802.15.4 is developed to support low-rate, low-speed, and low-power WPANs. Under this basic design rationale, IEEE 802.15.4 low-rate WPANs (LR-WPANs) is generally used for power-limited embedded system applications such as wireless sensor networks (WSNs). Sensor nodes (SNs), the main components of WSNs, are deployed over the networked sensing fields, and perform specific tasks with the data processing, event sensing, and data communicating [3]. Due to their limited power source, power consumption has been concerned as the most critical factor when designing WSNs protocols. Facing these challenges and research issues, several approaches to prolong the lifetime of the sensor networks with power-efficiency have been investigated in the aspects of IEEE 802.15.4 LR-WPANs MAC [2].

* This work was supported by ITRC Project supervised by ITTA-2005 (C1090-0501-0019).
** Corresponding author.

M.J. Smith, G. Salvendy (Eds.): Human Interface, Part II, HCII 2007, LNCS 4558, pp. 555–562, 2007.

There are two separated kinds of power management concepts in IEEE 802.15.4 LR-WPANs MAC, named 'macro power management' and 'micro power management'. The 'macro power management' [4]-[6] is a method for the power control by controlling packet transmission. That is, if network devices reduce the number of transmission of packets, the network devices can save transmission power. On the other hand, the `micro power management' [7] is the power management within one packet. Within the frame structure of a packet, time slots are existed. If some time slots are not used in a certain time, we can turn off a communication module for power saving. This concept is called as 'micro power management'. Previously, the research for macro power management was done frequently [4]-[6]. However, to achieve more power saving in power-limited IEEE 802.15.4 applications, we need to consider more strict power saving methods such as micro power management schemes. To predict the unused time for data communications, the MAC need to use a time division multiple access (TDMA) method. In a carrier sense multiple access with collision avoidance (CSMA/CA) method, we can not know the time of data communications by its random access nature. To use a TDMA method in IEEE 802.15.4 MAC, we must use beacon enabled star topology networks. In the network architecture, we can predict the unused time of data communications. In the time we turn o® a communication module to reduce power consumption.

The remainder of this paper is organized as follows. In Section 2, the basic feature of IEEE 802.15.4 LR-WPANs network topology is introduced. Section 3 describes our proposed power-saving medium access for beacon-enabled IEEE 802.15.4 TDMA MAC. In Section 4, performance of the proposed scheme is evaluated. Section 5 concludes this paper and presents the directions of our future work.

2 IEEE 802.15.4 Topology Overview

The IEEE 802.15.4 LR-WPANs MAC provides application support for the creation of two network topologies, i.e., star topology and peer-to-peer topology. Several application scenarios, mostly in home networking applications, drove the vision for star networks, while industrial and commercial applications drive the vision for peer-to-peer networks. The peer-to-peer topology enables the creation of the more exotic ad-hoc like natures, self-organizing wireless and mobile networks.

2.1 Peer-to-Peer Topology Networks

The peer-to-peer topology network allows any full-function devices (FFDs) to communicate with any other FFD within its range, and to have messages relayed to FFDs outside its range by using multi-hop routing methods. This topology enables the formation of more complex networks, including ad-hoc like operating, self-organizing, and self-healing natures. IEEE 802.15.4 MAC does not specify the details of these networks. It only defines the MAC functionalities. Fig. 1(a) shows a typical peer-to-peer topology networks.

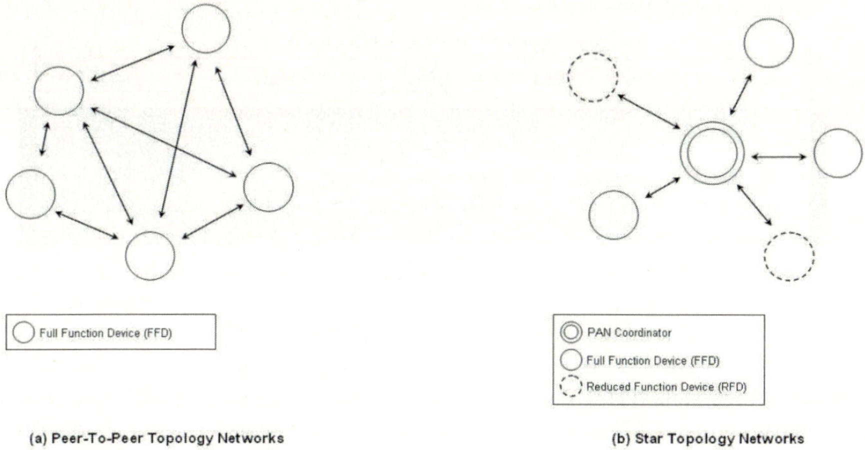

Fig. 1. Two Types of IEEE 802.15.4 Topologies: (a) Peer-To-Peer Topology Networks (b) Star Topology Networks

2.2 Star Topology Networks

In the star topology networks, communication is controlled by a unique PAN coordinator that operates as a network master or a cluster head of clustering-based networks, sending beacon messages for network device synchronization and maintaining association management. In this star topology networks, the network devices communicate only with the PAN coordinator. Fig. 1(b) shows the typical star topology networks. Any full-function devices (FFDs) may establish their own network by becoming a PAN coordinator. Star topology networks operate independently from their neighboring networks. The PAN coordinator of a new star topology network in formation must choose a network identifier that is not used by any other network. After this procedure is performed, the PAN coordinator can start sending beacon messages and permitting association to network devices requesting it. Network devices wishing for joining a star topology network must listen for beacon messages sent by the PAN coordinator. After a beacon message is received, the network device can send an association request back to the PAN coordinator, which in turn makes the decision to allow the association.

In star topology networks, an optional superframe structure is used. The superframe is managed by the PAN coordinator and is bounded by beacon messages sent by the PAN coordinator. Each beacon message contains information that will help network devices synchronize to the network. A superframe is divided in 16 time slots. The fundamental figure of superframe is shown in Fig. 2. The time slots are divided into two groups, i.e., contention access period (CAP) and contention free period (CFP). The CAP and CFP are operated by a carrier sense multiple access with collision avoidance (CSMA/CA) method and a time division multiple access (TDMA) method, respectively. In the CFP, to guarantee robust communications, the concept of

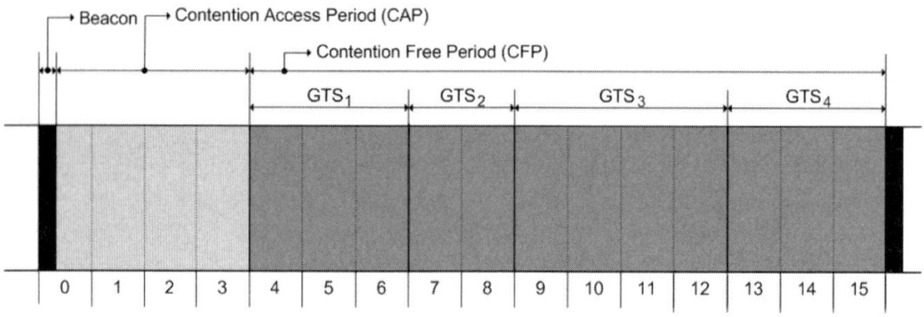

Fig. 2. IEEE 802.15.4 Superframe Structure

guaranteed time slots (GTSs) is used. If some network devices request more communication robustness, PAN coordinators give GTS to the network devices. Furthermore, if the network devices need high level robustness, the GTS has more time slots than the other GTSs. For example, in Fig. 2, among the 16 time slots, the front 4 time slots are allocated for CSMA/CA medium access. The other time slots are allocated for TDMA medium access to support more robustness. Also, 4 network devices acquires GTSs and the network devices take 3 time slots, 2 time slots, 4 time slots, and 3 time slots.

As shown in Fig. 1(b), a star topology network is similar to the clustering-based network architecture. By using this network architecture, several advantages in power-efficiency can be obtained, i.e., (1) improving spatial reuse and (2) mitigating the MAC-level medium contention. Therefore, we consider this star topology network architecture in this paper.

3 Power-Saving Medium Access for Beacon-Enable IEEE 802.15.4 MAC

This section describes the design rationale of the proposed power-saving medium access algorithm and the detailed algorithm description.

3.1 Design Rationale

As mentioned in Section 2, we use a beacon-enabled IEEE 802.15.4 superframe structure in this paper. The IEEE 802.15.4 superframe is bounded by two beacon messages. This beacon-enabled superframe structure uses time division medium access. Therefore, the network devices in the topology can be given time slots for data communication with a PAN coordinator. Under this network system, the network devices can recognize the time slots to communicate. Within the reserved time for communicating, the network device turns on its communications modules. The other times mean that the network devices do not communicate each other. Therefore, the communication modules of the network devices can be turned off their communication modules.

Table 1. Notations for describing the power saving medium access scheme

Notation	Description
ND_k	The network device named k
N_{slot}	The number of time slots of IEEE 802.15.4, i.e., 16
GTS_i	The number of allocated time slots of ND_i
$T_{active}(ND_k)$	The active time of the transmitter of the ND_k
T_{beacon}	The times for a beacon message
T_{SF}	The times for a superframe

3.2 Algorithm Description

The proposed power saving medium access scheme in this paper turns off the power of the transmitter of the network device for saving power consumption when CAP and GTS times are for the other network devices. By turning off the transmission power, the power consumption of the network device can be achieved drastically as verified in [8]. The allocated activation time of the network devices is calculated by following Eq. (1). All notations used in all equations in this paper are presented in Table 1.

$$T_{active}\left(ND_k\right) = \left[0, T_{beacon}\right] + \left[\frac{T_{SF}}{N_{slot}}\left(N_{slot} - \sum_{i=1}^{n-k+1} GTS_i\right), \frac{T_{SF}}{N_{slot}}\left(N_{slot} - \sum_{i=1}^{n-k} GTS_i\right)\right] \quad (1)$$

In Eq. (1), $T_{active}(ND_k)$ means the active time of the transmitter of the network device named k. T_{beacon} and T_{SF} mean the times for a beacon message and a superframe. GTS_i means the number of allocated time slots of ND_i. In other hands, if a network device does not have GTSs, the network device acquires its channel at CAP. Therefore, the active time of the transmitter can be formulated as follows.

$$T_{active}(ND_k) = \left[0, \frac{T_{SF}}{N_{slot}}\left(N_{slot} - \sum_{i=1}^{n} GTS_i\right)\right] \quad (2)$$

The calculated time is the one for CAP. The pseudo code for the proposed power-saving medium access is described in Algorithm 1, which is in Appendix.

4 Simulation Results

To verify the performance enhancement by the proposed power saving medium access method, we show the simulation results in this section. It shows the performance enhancement of the proposed scheme when the slot number of CAP is reduced.

In Fig. 3, the CAPs have 1, 3, 5, 7, and 9 time slots. If the CAP has more time slots, the IEEE 802.15.4 MAC has longer CSMA/CA interval. Generally, a TDMA based MAC has less collision probability than a CSMA/CA MAC. The reason of the phenomena is that TDMA is based on the contention free concept while CSMA/CA is based on the contention based concept. Under this basic property, the longer CAP

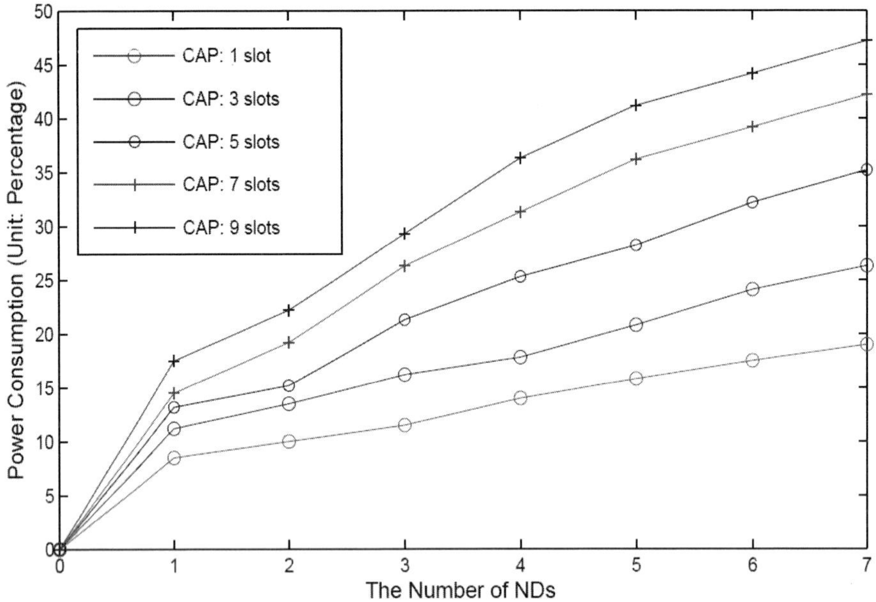

Fig. 3. Simulation Results

period makes higher collision probability. Furthermore, as shown in Fig. 3, the proposed scheme in this paper, a power saving medium access scheme, has more efficiency in a TDMA MAC, the longer CAP has less efficiency. Fig. 3 shows this situation clearly. When the number of time slots of CAP is 1, the other 15 time slots are a TDMA-based MAC. In this case, we can take the most energy efficiency than the others.

5 Conclusions and Future Work

This paper proposes a power saving medium access scheme for beacon-enabled IEEE 802.15.4 low-rate wireless personal area networks (LR-WPANs). In the proposed power saving scheme, we can have power efficiency by taking a inner superframe power management concept. In previous power management schemes, they do not consider the power management within a superframe structure, just try to reduce the useless transmission of data packets. In this paper, to obtain more power efficiency in each IEEE 802.15.4 mobile terminal, we consider the inner superframe power management. In a Superframe structure, we turn on communication modules when they are required, and visa versa. By turning off communication modules when they are not use, we can drastically reduce power consumption of each IEEE 802.15.4 embedded device. Simulation result shows that the proposed power saving medium access scheme can have the desired power efficient properties.

References

1. IEEE LAN/MAN Standards Committee, Part 15.4: Wireless Medium Access Control (MAC) and Physical Layer (PHY) Specifications for Low-Rate Wireless Personal Area Networks (LR-WPANs) IEEE Standard 802.15.4-2003 (2003)
2. Callaway, E., Gorday, P., Hester, L., Gutierrez, J.A., Naeve, M., Heile, B., Bahl, V.: Home Networking with IEEE 802.15.4: a Developing Standard for Low-Rate Wireless Personal Area Networks. IEEE Communications Magazine 40(8), 70–77 (2002)
3. Akyildiz, I.F., Su, W.L., Sankarasubramaniam, Y., Cayirci, E.: Wireless Sensor Networks: A Survey. In: Computer Networks, vol. 38(4), pp. 393–422. Elsevier, North-Holland (2002)
4. Anastasi, G., Conti, M., Francesco, M.D., Passarella, A.: An Adaptive and Low-Latency Power Management Protocol for Wireless Sensor Networks. In: Proceedings of the ACM International Workshop on Mobility Management and Wireless Access (MOBIWAC), Terromolinos, Spain (2006)
5. Dam, T., Langendoen, K.: An Adaptive Energy-Efficient MAC Protocol for Wireless Sensor Networks. In: Proceedings of the 1st ACM International Conference on Embedded Networked Sensor Systems (SenSys), Los Angeles, CA, USA (November 5-7, 2003)
6. Hohlt, B., Doherty, L., Brewer, E.: Flexible Power Scheduling for Sensor Networks. In: Proceedings of the 3rd IEEE International Symposium on Information Processing in Sensor Networks (IPSN), Berkeley, CA, USA (April 26-27 2004)
7. Guo, Z., Yao, R., Zhu, W., Wang, X., Ren, Y.: Intra-Superframe Power Management for IEEE 802.15.3 WPAN. IEEE Communications Letters 9(3), 228–230 (2005)
8. Min, R., Bhardwaj, M., Cho, S.-H., Sinha, A., Shih, E., Wang, A., Chandrakasan, A.: An Architecture for a Power-Aware Distributed Microsensor Node. In: Proceedings of the IEEE Workshop on Signal Processing Systems (SiPS) (October 2000)

A Appendix: The Pseudo Code

In this appendix section, we show the pseudo code of the proposed power saving medium access scheme for beacon-enabled IEEE 802.15.4 LR-WPANs in Algorithm 1. As shown in the pseudo code, a network configuration procedure is operated at first. After the procedure, the network device checked whether the network device is a PAN coordinator or not. If the network device is a PAN coordinator, it broadcast a message (`Broadcasting_Message();`) to configure neighboring network devices. By judging under the received replies, the PAN coordinator can configure its neighboring network devices. To receive replies, the PAN coordinator executes a function named `Receive_Replies();`. In addition, within the replies, the neighboring network devices contain information: (1) the request of GTS (2) the number of time slots for the GTS. Under the acquired information of all neighboring network devices, the PAN coordinator makes a decision for the active and inactive time of the communication modules of each network devices. After the decision, the PAN coordinator broadcast the information. On the other hand, network devices make a decision for whether a GTS time requires or not. If the GTS time is required, each network devices also make a decision the number of time slots for the GTS time. After the decision, each network device transmits the information toward the PAN coordinator to help its prediction for active time of each network devices.

Algorithm 1 Power Saving Medium Access Scheme

```
loop
  if Network Construction == TRUE then
    if The Network Device == PAN Coordinator then
      //A PAN coordinator (PNC) broadcasts a message for network configuration.
      Broadcasting_Message();
      //A PNC waits for the replies of all neighboring network devices.
      Receive_Replies();
      if a PNC receives replies from all neighboring network devices then
        //A PNC receives replies from neighbor network devices for the decision
        of GTS allocation and the amount of time slots for the GTS.
        //PNC makes decisions for all network devices
        if The network device requests a GTS then
          //Calculate GTS for the network device
          Calc: Eq. (1)
        else
          //Calculate a CAP time
          Calc: Eq. (2)
        end if
      end if
      //A PNC broadcasts the result of calculation
      Broadcasting_Message();
    else
      //The Network Device != PAN Coordinator (PNC)
      //A Network device (ND) waits for the broadcasted data from PNC.
      Receive_Data();
      //An ND transmits their request to the PNC.
      //The below pseudo code is for the calculation of the information to transmit
      toward the PNC.
      if a GTS time is required then
        REQ_GTS = 1;
        //Calculate the number of time slots (α) for the GTS
        NUM_TIMESLOT_GTS = α;
      else
        //A GTS time is not required.
        REQ_GTS = 0;
      end if
      //An ND transmits the calculated information.
      Broadcasting_Message(REQ_GTS, NUM_TIMESLOT_GTS);
      //An ND waits for the broadcasted data from PNC.
      //In the data, there are the predicted active times are contained.
      //An ND must turn on its communication module in the predicted time.
      Receive_Data();
    end if
  end if
end loop
```

Dynamic Cell Phone UI Generation for Mobile Agents

Gu Su Kim, Hyun-jin Cho, and Young Ik Eom

School of Information and Communication Engineering, Sungkyunkwan Univ.,
Chunchun-dong 300 Jangan-gu, Suwon, Gyeonggi-do, Korea
{gusukim, hjcho, yieom}@ece.skku.ac.kr

Abstract. A MA(Mobile Agent) is a computer program which autonomously migrates within the network and works on behalf of a user. With the progress of ubiquitous computing environments, MA paradigm becomes more useful paradigm because MA paradigm makes it possible for various user-centric services, such as follow-me service, private secretary service, and etc. Also, in ubiquitous environments, the cell phone has very important role, but the cell phone cannot support the MA technology because J2ME(Java2 Micro Edition) CLDC(Connected Limited Device Configuration), which is the java platform for cell phone, does not support the essential facilities for the mobility of the mobile agents such as reflection, dynamic class loading, and object serialization due to security problem of the cell phone. So, in this paper, we propose the scheme that the MAP(Mobile Agent Platform) dynamically generates the UI(User Interface) according to the user's device. Also, we show the experimental results of our proposed scheme. By using our proposed scheme, the user gets the service of MA by the cell phone.

Keywords: UI generation, cell phone, mobile agent, mobile agent platform.

1 Introduction

In ubiquitous computing environments, various computing devices will be embedded in our environments, and applications will run on the devices such as PDAs, celluar phone, touch panel, and so on [1]. In this ubiquitous computing environment, portable devices play an important role. PDAs and cell phones enable mobile users to handle information wherever they are. Therefore, the exact amount of information that the user needs should be presented.

The mobile agent platform is the execution environment for the mobile agent. Many mobile agent platforms have been implemented with Java technologies. Due to platform independence of the Java technologies, the mobile agents can be implemented on various devices. Among the three platforms of Java, J2ME(Java 2 Micro Edition) is the Java platform for consumer and embedded devices. J2ME configurations are divided into CLDC(Connected Limited Device Configuration) and CDC(Connected Device Configuration) according to the resources of device. CLDC is designed for devices with intermittent network connections, slow processor, and limited memory, such as cell phones, two-way pagers, and smart phones. CDC is designed for devices that have more memory, faster processors, and greater network bandwidth [4, 5].

M.J. Smith, G. Salvendy (Eds.): Human Interface, Part II, HCII 2007, LNCS 4558, pp. 563–571, 2007.

J2ME CLDC doesn't support the essential facilities for the mobility of the mobile agents such as reflection, dynamic class loading, and object serialization. So, the existing lightweight mobile agent platforms have been implemented on J2ME CDC platform. But, now, it is necessary to consider ubiquitous devices with J2ME CLDC.

In this paper, we'll describe our lightweight mobile agent platform, called *sMAP*(small Mobile Agent Platform), for J2ME CDC environments and *CUIP*(Cell phone User Interface Platform) for J2ME CLDC/MIDP environments, and describe the cooperation between *sMAP* and *CUIP*.

The rest of the paper is organized as follows. In Section 2, we describe related work on the J2ME and existing lightweight mobile agent platforms. Section 3 describes our two platforms, *sMAP* and *CUIP* and mobile agent migration scheme using the SOAP message. Section 4 shows our experimental results. Finally, Section 5 concludes with a summary.

2 Related Works

Java is the most popular development language for the mobile agent platform. Java supports three kinds of platform: J2EE(Java2 Enterprise Edition), J2SE (Java2 Standard Edition), and J2ME(Java2 Micro Edition). The lightweight mobile agent platform is implemented with J2ME environments. Dynamic class loading, serialization, and reflection are essential facilities for developing the mobile agent platform with Java. But these technologies are not supported in J2ME CLDC. Reflection is not supported on CLDC/MIDP environment to protect inflow of malignancy code like computer viruses to mobile-phone. A Java virtual machine supporting the CLDC must have a built-in class loader that cannot be overridden or replaced by the user. This is mainly for security reasons. Object serialization is not supported on CDLC/MIDP environment for the same reason as reflection. Also, network program in J2ME CLDC environment can be implemented by using javax.microedition.io package and this program can not create server sockets, but can create only client sockets [6, 7]. For these reasons, it is very hard to implement the mobile agent platform in J2ME CLDC environment.

For above reasons, script based mobile agent is used on cell phone instead of using the Java technology. Flipcast [8], which is developed at Toshiba Corp. in Japan, can be built in communication devices such as home robots, appliances, and cell phones. Flipcast consists of platform and scripts. Flipcast platform is the base software which runs on each device and the script consists of operations list to run on cell phone and acts as a mobile agent.

3 UI Generation Scheme for Cell Phone

In this Section, we describe the architecture of *sMAP*(small Mobile Agent Platform), which is a kind of LMAP(Lightweight Mobile Agent Platform), executable on J2ME CDC environments such as PDA, and *CUIP*(Cell phone UI Platform) executable on J2ME CLDC environments such as cell phone.

3.1 *CUIP*(Cell Phone UI Platform)

It is very hard to develop the mobile agent platform with Java standard technology under J2ME CLDC/MIDP environment because the object serialization, reflection, and custom class loading are not supported in the CLDC/MIDP environment.

In order to control the MA and MAP by cell phone, we also developed the *CUIP*. Our *CUIP* is implemented with Wireless Internet Platform for Interoperability(WIPI), which is wireless internet platform standard at Korea and is based on J2ME CLDC(Connected Limited Device Configuration). Figure 1 shows the software architecture at the cell phone side, *CUIP*. This platform consists of Communication Manager(CoM), Information Manager(IM), UI Generator(UIG), and UI Panel(UIP).

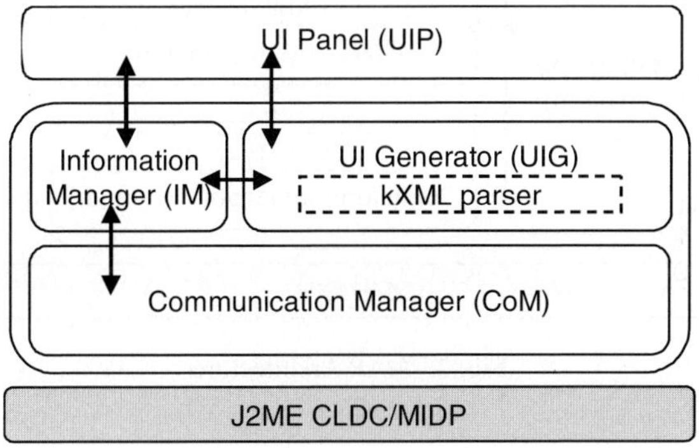

Fig. 1. *CUIP* Architecture

In the Figure 1, CoM is in charge of the communication with MAP or MA. CoM receives the xml-based status message from MAP or MA through HTTP protocol, and passes this message to IM. In order to generate the UI suitable for cell phone, IM passes the status message to UIG. *CUIP* communicates with *sMAP* through SOAP. We use kXML[9] as XML parser for parsing the SOAP message. UIG composes the SOAP message for the *sMAP*, and interprets the SOAP message received from *sMAP* by kXML parser and generates UI suitable for cell phone and shows the UI to the user by UIP module. IM module gets the input of the user through UIP, and converts the information in UIP to xml-based control message, which is the command message to the MAP or MA.

3.2 *sMAP*(Small Mobile Agent Platform)

We developed *sMAP*, which is a kind of LMAP, executable on J2ME CDC environment. The *sMAP* is a lightweight mobile agent platform for J2ME CDC environment which is available on PDA, smart phone, and so on. Figure 2 shows the *sMAP* architecture that is composed of the following 6 components:

- MA(Mobile Agent)
- AMS(Agent Management System)
- SC(Status Converter)

- CI(Command Interpreter)
- CM(Content Manager)
- CoM(Communication Manager)

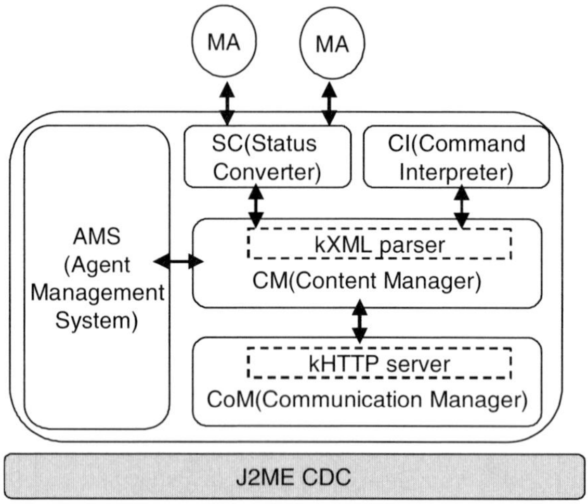

Fig. 2. *sMAP* Architecture

In the Figure 2, AMS(Agent Management System) manages the life cycle of MAs such as mobile agent creation, clone, migration, termination, and so on. CoM(Communcation Manager) has micro-HTTP server, kHTTP server, and is in charge of the communication with other MAP, MA or *CUIP*. In order to provide the interface for web and *CUIP*, CoM has the micro-HTTP server, kHTTP[10]. CM(Content Manager) interprets or generates the xml-based message through kXML parser. In this architecture, SC module is key component for converting the internal status to xml-based status message for *CUIP* or HTML message for Web user. This module investigates the member fields of the MA class through JAVA reflection and converts their values to xml-based status message. CI(Command Interpreter) module interprets the command message received from the cell-phone user or Web user, and executes the command in the message. The *sMAP* provides two user interfaces. One is the web interface and another is the interface for *CUIP*. CoM generates HTML pages for management of *sMAP* such as agent creation, *sMAP* status monitoring, and so on.

The *sMAP* uses the HTTP/SOAP for communication with other *sMAP*s. Also, the *sMAP* transfers a mobile agent to other *sMAP* through SOAP. When a mobile agent requests migration, the *sMAP* stops the execution of the mobile agent and converts the status of the mobile agent to byte stream by object serialization. Source *sMAP* sends MigrationRequest message, including the query about the need of mobile agent's class files, to target *sMAP*. If target *sMAP* accepts the migration request of the mobile agent and requires the class flies of the mobile agent, source *sMAP* generates the SOAP migration message including the mobile agent status and its class files. The *sMAP* compresses the class files by Zip and encodes compressed image by base64. The agent's serialized byte stream is also compressed by Zip and encoded by base64. The encoded class files and agent status are included in SOAP migration message and transmitted to the target *sMAP*. The target *sMAP* extracts encoded class files from the SOAP migration message, decodes and decompresses it, and registers class files into the its class loader. After registering class files to its class loader, the target *sMAP* decodes and decompresses the mobile agent's status, and restores the mobile agent from the byte stream format by unserialization. Finally, target *sMAP* registers the restored mobile agent into its AMS and restarts the mobile agent. Figure 3 and 4 show the algorithms of the mobile agent migration process.

The *sMAP* manages MAFPT(Mobile Agent Forward Pointer Table) for tracing the current location of the mobile agent. The entry of MAFPT has the following format:

$$<MA\_ID, Pre\_MAP, Next\_MAP> \quad (1)$$

The *MA_ID* field is the ID of the mobile agent and *Pre_MAP* field is the IP address of the mobile agent's previously visited platform and *Next_MAP* field is IP address of the mobile agent's next visiting platform. The entry of MAFPT is added to the table when the mobile agent enters into the current platform. At this time, AMS sets up *MA_ID* and *Pre_MAP* fields in the entry. When the mobile agent leaves the platform, AMS sets up the *Next_MAP* field in the entry.

```
migrateAgent(MobileAgent ma, MAP target)
{
   char *pClass, *pMa;
   byte *pdata;

   ma.stop();
   Response rsp = SendMigrationReqeust(target, ma.name);
   if(rsp.result == OK) {
      if(rsp.bclassfile == true) {
         pdata = GetClassFiles(ma);
         pdata = CompressByZip(pdata);
         pClass = EncodeBase64(pdata);
      }

      pdata = Serialization(ma);
      data = CompressByZip(pdata);
      pMa = EncodeBase64(pdata);
      SOAPMessage msg = MakeSOAPMessage(MIGRATION, pClass, pMa);
      SOAP_Delivery(target, msg);
   }
}
```

Fig. 3. The algorithm of migration at source *sMAP*

In order to get the current status of the mobile agent, the GETCURRENTSTATUS message is delivered to the *Next_MAP* host of the corresponding MAFPT entry. The *sMAP* received GETCURRENTSTATUS message checks whether the mobile agent exists in the platform or not. If the mobile agent doesn't exist in the platform, the

```
immigrateAgent(char *pClass, char *pMa) {
    byte *pdata;

    pdata = Base64Decode(pClass);
    pdata = UncompressByZip(pdata);
    RegisterToClassLoader(pdata);

    pdata = Base64Decode(pMa);
    pdata = UncompressByZip(pdata);
    MobileAgent ma = Unserialization(pdata);
    ResiterToAMS(ma);
    ma.restart();
}
```

Fig. 4. The algorithm of immigration at target *sMAP*

sMAP delivers this SOAP message to the next *sMAP* in the entry of its MAFPT. This SOAP message is delivered to the *Next_MAP* host of the corresponding entry of its MAFPT. This process is repeated until the mobile agent is found.

4 Experiment

In this Section, we describe the cooperation procedure between *sMAP* and *CUIP*. Existing mobile agent platforms can not be operated on the cell phone because of the limitation of J2ME CLDC environments. However, the cooperation of *CUIP* and *sMAP* enables us to get the service of the mobile agent on the cell phone with J2ME CLDC environments.

The user executes *CUIP* on the cell phone with J2ME CLDC/MIDP environment. The user logins the remote *sMAP* through *CUIP* interface and gets the right to use the *sMAP*. The cell phone user can send commands such as agent creation, agent tracing, and agent current status extraction to the *sMAP*. All commands are delivered through SOAP/HTTP. Table 1 describes these commands.

4.1 Mobile Agent Creation

When *CUIP* receives the command message from *sMAP*, *UIG* in *CUIP* generates the UI for the command message and shows the UI through UIP. The cell phone user fills up arguments of the command and returns the command message to MAP. Figure 5 shows the example of message to control MAP and the snapshot of the cell phone.

4.2 Tracing Agent and Extracting Its Current Status

When MA completes its execution or wants to receive the command from the user, MA generates its xml-based status message through SC module in MAP, and sends this message to *CUIP*. UIG in *CUIP* parses the xml-based status message and dynamically generates the UI suitable for cell phone. Figure 6 shows the example of the xml-based status message and the snapshot of the cell phone.

Table 1. Commands

Command	Description
GETAGENTLIST	Gets the list of mobile agents instantiated by the user
CREATEAGENT	Creates a mobile agent
GETCURRENTAGENTSTATE	Gets the current state of the mobile agent
GETAGENTRESULT	Gets the result of the mobile agent
EXITAGENT	Stops the mobile agent
GETAGENTIMAGE	Brings the mobile agent status image to the cell phone
RESOTREAGENT	Restores the mobile agent from the cell phone on another *sMAP*

```
<env:Envelope>
<env:Header>
      <content_class>MAP command</content_class>
</env:Header>
<env:Body>
      <cuip:command>Agent Creation</cuip:command>
      <cuip:agument type=input_string>Agent name</cuip:argument>
      <cuip:agument type=input_string>Migrate To</cuip:argument>
</env:Body>
</env:Envelop>
```

Fig. 5. The example of xml-based command massage and the snapshot of *CUIP*

4.3 User Mobility Support Through *CUIP*

The mobile agent can be used to support the user mobility service. But it is very hard for the mobile agent to trace and service the user when he takes a long travel or his destination is not determined a priori. In order to support these situations with the

```
<env:Envelope>
<env:Header>
      <content_class>MA status</content_class>
</env:Header>
<env:Body>
      <cuip:agentname type=output_string>MP3Search</cuip:agentname>
      <cuip:user type=output_string>GS Kim</cuip:user>
      <cuip:message name="Results" type=inputoutput_string>
               Total 2Files, 18.3MB, Tears_in_Heaven.mp3, Let_it_be.mp3
       </cuip:message>
      <cuip:command type=select>
            <option>Exit</option>
            <option>Go to another</option>
      </cuip:command>
</env:Body>
</env:Envelop>
```

Fig. 6. The example of xml-based status message and the snapshot of cell phone

mobile agent, many complicated infrastructures such as sensing the location of the user and identifying the user are required.

Before the user starts his journey, his *CUIP* brings the mobile agent status information and serializes the mobile agent, from *sMAP* to the mobile phone. The *CUIP* stores the mobile agent status information into its RMS. When the user arrives at the destination, the *CUIP* searches the *sMAP* near the user and restores the mobile agent to the *sMAP* found. This process is the same as the normal mobile agent migration process. The restored mobile agent at the *sMAP* now can service the user.

5 Conclusion

The mobile agent is very useful technology in distributed or ubiquitous environments. But, in order to apply the mobile agent concept to the ubiquitous environments, the mobile agent platform should be lightweight and should be designed with the consideration on the characteristics of the ubiquitous devices such as limited resources. We developed the lightweight mobile agent platform, called *sMAP*, executable on J2ME CDC environments and *CUIP* executable on J2ME CLDC/MIDP environments. Our proposed scheme dynamically generates the UI for MAP or MA according to the devices such as cell phone, which cannot support the MA technology, or devices only supporting Web UI. By using our UI generation scheme, the user can control the MAP or MA, and get services of MA through the device at which MAP is not installed.

Acknowledgments

This research was supported by the MIC(Ministry of Information and Communication), Korea, under the ITRC(Information Technology Research Center) support program supervised by the IITA(Institute of Information Technology Assessment), IITA-2006-(C1090-0603-0046).

References

1. Satoh, I.: SpatialAgents: integrating user mobility and program mobility in ubiquitous computing. Wireless Communication and Mobile Computing, vol. 3(3) (2003)
2. Aneiba, A., Rees, S.J.: Mobile Agents Technology and Mobility. In: Proc. of the 5th Annual Postgraduate Symposium on the Convergence of Telecommunications, Networking and Broadcasting, pp. 14–20 (2004)
3. Cho, K., Hayashi, H., Hattori, M., Ohsuga, A.: picoPlangent: An Intelligent Mobile Agent System for Ubiquitous Computing. In: Barley, M.W., Kasabov, N. (eds.) PRIMA 2004. LNCS (LNAI), vol. 3371, pp. 43–56. Springer, Heidelberg (2005)
4. Ledoux, T., Bouraqadi-Saadani, N.: Adaptability in Mobile Agent Systems using Reflection. Middleware 2000 Workshop on Reflective Middleware (RM2000), New York, USA (April 2000)
5. Mahmoud, Q.H.: Understanding Network Class Loaders. Developer Technical Articles & Tips (October 2004)

6. Mobile Information Device Profile(MIDP), http://java.sun.com/products/midp
7. Mahmoud, Q.H.: J2ME APIs: Which APIs come from the J2SE Platform? Sun Developer Network (January 2001)
8. Ueno, K., Kawamura, T., Hasegawa, T., Ohsuga, A., Doi, M.: Cooperation between Robots and Ubiquitous Devices with Network Script Flipcast. In: Proc. of Network Robot Systems integrated with environments (IROS 2004 Workshop) (2004)
9. kXML, http://kxml.objectweb.org/index.html
10. kHTTP, http://khttp.objectweb.org/index.html

TCP NJ+ for Wireless HCI

Jungrae Kim[1], Jahwan Koo[2], and Hyunseung Choo[1,*]

[1] School of Information and Communication Engineering,
Sungkyunkwan University, Suwon, Korea
{witjung, choo}@ece.skku.ac.kr
[2] School of Information and Communication Engineering,
Sungkyunkwan University, Suwon, Korea
jhkoo@songgang.skku.ac.kr

Abstract. TCP application in wireless HCI environments communicates to each other frequently through wireless networks. Therefore, it is important that significant TCP performance is guaranteed in wireless network. TCP New Jersey, known as the best existing scheme in terms of goodput in wireless networks, improves wireless TCP performance using the available bandwidth estimation at the sender and the congestion warning at intermediate routers. Although TCP New Jersey achieves 17% and 85% improvements in goodput over TCP Westwood and TCP Reno, respectively, we further improve TCP New Jersey by exploring enhanced available bandwidth estimation and suitable error recovery mechanism. Hence, we propose TCP NJ+. It outperforms other TCP variants by 19% to 104% in terms of goodput even when the network is in bi-directional congestion. Hence TCP NJ+ can provide the best services to user in wireless HCI environments.

Keywords: TCP, wireless HCI, Congestion Control, TCP New Jersey.

1 Introduction

Transport control protocol (TCP) is connection-oriented transport layer protocol which ensures the reliable data transmission [1]. However, TCP in wireless environments such as cellular networks, wirelessLANs, mobile computing, and wireless HCI experience severe performance degradation. The main reason is that the congestion control mechanism [2] decreases congestion window *cwnd* when the packet loss is detected because the sender assumes that the cause of all packet losses is congestion. Because packet losses is occurred due to wireless link error as well as the congestion, the congestion control in standard TCP is main reason that performance is decreased. Hence it need to intelligent congestion control mechanism which reflects the characteristic of wireless environments [3].

Human-computer interaction (HCI) is a discipline concerned with the design, evaluation and implementation of interactive computing systems for human use and

---

* Corresponding author.

M.J. Smith, G. Salvendy (Eds.): Human Interface, Part II, HCII 2007, LNCS 4558, pp. 572–581, 2007.

with the study of major phenomena surrounding them. Wireless HCI means HCI in wireless environments [4]. It can interact between users (mobile nodes) and mobile nodes through the wireless networks. For improving quality of services, it needs the high performance transport protocol in wireless networks because applications have to response user demands quickly.

TCP Jersey and New Jersey in wireless environments are based on the integration of the sender-side available bandwidth estimation (ABE) algorithm and the receiver-side packet loss differentiation scheme, thus, resulting in higher throughput than any other TCP variants [6], [7]. However, TCP New Jersey may suffer the degradation of throughput depending on background traffic to bring unexpected available bandwidth estimation. In addition, it cannot increase the reduced sending rates due to the packet loss effectively according to the cause of packet loss which contains both the packet loss by bit-error-rate (BER) and by network congestion. Hence we propose TCP NJ+ for improving the quality of service of TCP applications that modifies available bandwidth estimation and error recovery mechanism to ensure the high throughput. As a result of simulation using NS-2 [11], TCP NJ+ shows 3% and 19% improvement under 1% and 5% wireless link error rate, respectively over TCP New Jersey.

The remainder of the paper is organized as follows. Section 2 reviews the related work. Section 3 describes in detail the improved mechanisms of TCP NJ+. Section 4 presents performance evaluation via NS-2 network simulator under various network conditions. Section 5 offers some concluding remarks.

2 Related Work

In this section, we describe wireless HCI, TCP Reno [5] which is conventional TCP in wired networks, TCP Westwood [8], and TCP New Jersey that are existing schemes in wireless networks.

2.1 Communication in Wireless HCI

HCI is the study of interaction between people (users) and computers. Wireless HCI Wireless HCI means HCI in wireless environments. It can interact between users (mobile nodes) and mobile nodes through the wireless networks. That is, the purpose of study for wireless HCI is to provide the best quality service to the user processing user demands in the wireless environments. Fig. 1 shows wireless HCI scenario.

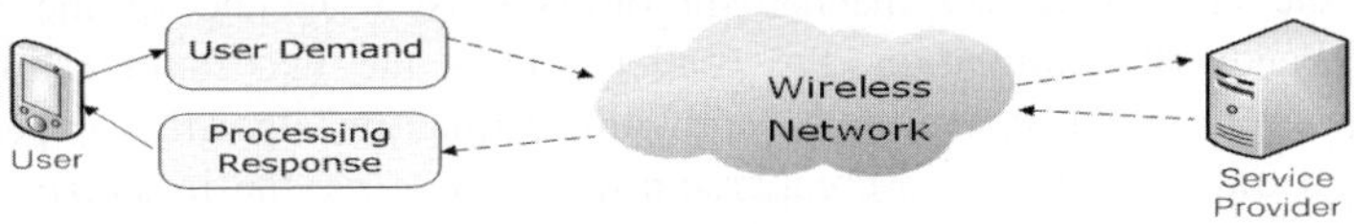

Fig. 1. Wireless HCI communication scenario

The TCP application sends user demands to the server through the wireless networks. Also the server sends services to the user through the wireless networks. If the user uses TCP application, it employs TCP/IP protocol and it can communicate each other frequently according to the user. Hence it needs high performance transport protocol for reacting to user demands quickly.

2.2 TCP Variants

TCP Reno is the regular TCP. It guarantees the reliable data transmission in wired networks. The multiple packet losses which occur in wireless networks force the TCP Reno to invoke a slow down in the recovery of the dropped transmission rates.

TCP Westwood is wireless TCP using end-to-end proactive congestion control. TCP Westwood estimates the current network bandwidth at the sender side. The sender estimates network bandwidth by exploiting the rate and pattern of returning ACK through the reverse links. Because it does not distinguish the cause of packet loss, it has a problem that decreases the throughput in high BER wireless networks such as wireless HCI environments.

2.3 TCP Jersey and New Jersey

TCP Jersey is developed to improve performance including congestion warning (CW) that to differentiate the packet loss by network congestion between the packet loss by wireless link error and available bandwidth estimation (ABE). In TCP Jersey, when the ACK packet arrives at the sender, the ABE algorithm computes the current available bandwidth based on the time interval of ACK packets. And the optimized window *ownd* is computed based on this estimation. However, TCP New Jersey employs different ABE algorithm comparing TCP Jersey. TCP New Jersey computes current available bandwidth based on the time interval of data segments at the receiver.

CW modifies explicit congestion notification (ECN) [9]. It is set to a threshold with fewer parameter settings at the intermediate router that shall mark all the packets when the average queue length exceeds the threshold. The sender who receives the marked packets decreases its *cwnd* based on ABE, otherwise it maintains the current *cwnd*.

3 TCP NJ+ in Wireless HCI

We propose TCP NJ+ which enhances the available bandwidth estimation and modifies the error recovery mechanism of TCP New Jersey. In the enhanced available bandwidth estimation, the sender selects the maximum estimation to overcome the problem of available bandwidth estimation algorithm depending on background traffic direction in comparison of two factors, the interarrival time of ACK packets at the sender and the interarrival time of data segments at the receiver. And also TCP NJ+ employs the suitable error recovery mechanism which increases the reduced *cwnd* more effectively when the sender detects the packet loss or RTO expiration.

3.1 Enhanced Available Bandwidth Estimation

Both TCP Jersey and TCP New Jersey estimate the current available bandwidth based on Eq. (1)

$$R_n = \frac{R_{n-1} \times RTT + L_n}{(t_n - t_{n-1}) + RTT} \tag{1}$$

Here R_n is the estimated bandwidth when ACK packet n arrives at time t_n at the sender, t_{n-1} is the previous ACK packet arrival time at the sender, L_n is the size of data packet n, and RTT is the round trip time at time t_n in TCP Jersey. Meanwhile the data segment arrival time at the receiver is found using the timestamps [10] option in the TCP header instead of using the ACK packet arrival time at the sender. Hence t_n and t_{n-1} in Eq. (1) are the data segment arrival time of the n th data packet and its previous data packet arrival at the receiver, respectively, in TCP New Jersey.

Because the ABE on TCP New Jersey is computed by the packet sending time for the receiver, it can explore precise estimation, if the network condition is degraded by background traffic in reverse links. On the other hand, it cannot estimate the accurate available bandwidth, if the network condition is deteriorated from the background traffic in forward links. Accordingly, both TCP Jersey and TCP New Jersey suffer from the problem that the ABE depends on the background traffic direction that brings the degradation of performance.

TCP NJ+, therefore, estimates the available bandwidth comparing two estimations at the sender. First, it estimates the available bandwidth by Eq. (2):

$$R_{s_n} = \frac{RTT \times R_{s_{n-1}} + L_n}{(t_{s_n} - t_{s_{n-1}}) + RTT} \tag{2}$$

where R_{s_n} is the estimated bandwidth when ACK n arrives at time t_{s_n} at the sender, $t_{s_{n-1}}$ is the previous ACK arrival time at the sender, L_n is the size of data segment n, and RTT is the round trip time. Second, it calculates the available bandwidth by Eq. (3):

$$R_{r_n} = \frac{RTT \times R_{r_{n-1}} + L_n}{(t_{r_n} - t_{r_{n-1}}) + RTT} \tag{3}$$

where R_{r_n} is the achieved bandwidth by using timestamps option when the data segment n arrives at time t_{r_n} at the receiver, $t_{r_{n-1}}$ is the earlier data segment arrival time at the receiver. Finally, it chooses the higher value comparing the R_{s_n} with the

```
(ACK packet received by sender)
if (timestamp)
   bwe_sender_n = (rtt * double(bwe_sender_n-1) + seg_size)
                  / (st_n – st_n-1 + rtt);
   bwe_receiver_n = (rtt * double(bwe_receiver_n-1) + seg_size)
                  / (rt_n – rt_n-1 + rtt);
   if (bwe_sender_n > bwe_receiver_n )
      current_bwe = bwe_sender_n;
   current_bwe = bwe_receiver_n;
   end if
end if
```

Fig. 2. Enhanced available bandwidth estimation

R_{r_n} that ensures proper *cwnd*, consequently. TCP NJ+ resolves the problem that the ABE in TCP New Jersey depends on the background traffic direction. It achieves higher throughput in wireless HCI where may have various background traffic regardless of the direction of background traffic. Fig. 2 represents the ABE algorithm of TCP NJ+.

3.2 Suitable Error Recovery for Wireless HCI

An optimized window (*ownd*) in TCP New Jersey is calculated by Eq. (4) shown below.

$$ownd = \frac{R_n \times RTT}{segment\_size} \tag{4}$$

Where R_n is the value of the ABE. When the RTO is expired by the n th packet, if packet loss is caused by network congestion, the sender sets the *cwnd* to one and the *ssthresh* to $ownd_n$. Otherwise, if packet loss is caused by BER, it adjusts the *cwnd* and *ssthresh* to $ownd_n$. Since RTO is caused by BER instead of network congestion and the network condition becomes poor incidently, we can utilize the rest of bandwidth due to *cwnd* sets to one. Thus, it is inappropriate to set the *cwnd* to $ownd_n$ which is estimated by the minimum estimation when the network condition is degraded temporarily.

For this reason, RTO due to network congestion, TCP NJ+ sets the *cwnd* to one and the *ssthresh* to $ownd_n$ as in the TCP New Jersey behavior. And for RTO due to the BER, TCP NJ+ sets the *ssthresh* to $ownd_n$ and the *cwnd* to the value according to the algorithm given in Fig. 3 which is to employ $ownd_{n-1}$ as well as $ownd_n$. In TCP NJ+, $ownd_{n-1}$ value is higher than $ownd_n$ because it is computed when the network condition is in a good state. TCP NJ+ achieves higher *ownd*, average of $ownd_{n-1}$ and $ownd_n$, and results in higher *cwnd* than TCP New Jersey, because the average of $ownd_{n-1}$ and $ownd_n$ are higher than $ownd_n$ at all times. Hence the throughput is guaranteed to decrease the recovery time of the dropped *cwnd* in order to avoid the network congestion and to utilize link bandwidth effectively in TCP NJ+.

```
when (packet loss is dectected)
if (RTO expired)
        if (Congestion Warning)  /* if RTO due to congestion */
                cwnd = 1;
                ssthresh = ownd_n;
        else                      /* if RTO due to BER */
                cwnd = (ownd_n + ownd_n-1) / 2;
                ssthresh = ownd_n;
        end if
end if

if (3 dupack received by sender)
        if (Congestion Warning)  /* if packet loss due to congestion */
                ssthresh = ownd_n;
                if (ssthresh < cwnd)
                        cwnd = ownd_n;
                end if
        else                      /* if packet loss due to BER */
                ssthresh = ownd_n
                cwnd = cwnd + 1;
        end if
end if
```

Fig. 3. Suitable error recovery

When the TCP New Jersey receives 3-dupack, the error recovery operates as follows. First, if the packet loss is caused by the network congestion, *ssthresh* is set to $ownd_n$, then if *cwnd* is lower than *ssthresh*, it is set to the current *cwnd*. However, if the *cwnd* is higher than the *ssthresh*, it is set to $ownd_n$. Second, if the cause of packet loss is the BER, TCP New Jersey keeps the current *ssthresh* and *cwnd*. Therefore, when the sender receives the 3rd duplicated ACK, it does not increase the *cwnd*. This means that the relative loss of *cwnd* is increased according to growing packet loss.

TCP NJ+ overcomes that problem. When TCP NJ+ receives 3-dupack, the suitable error recovery is performed as follows. First, if packet loss is caused by network congestion, TCP NJ+ operates the same as TCP New Jersey. Second, when packet loss caused by BER occurs, *ssthresh* is set to $ownd_n$ and *cwnd* is increased by one maximum segment size (MSS). The reason why the *cwnd* is increased by one MSS is to compensate the lost *cwnd* by the 3-dupack. However, the packet loss caused by the BER may utilize the remaining bandwidth. Hence the adjustment of the *cwnd* effectively is to invoke the improvement in performance. In TCP NJ+, the suitable recovery, which achieves the higher *cwnd*, ensures the significant performance improvement comparing to other wireless TCP schemes in wireless environments.

4 Performance Evaluation

We experiment the goodput of TCP NJ+ using the NS-2. One is a simple network topology and the other is a more realistic network topology with background traffic, as shown in Fig. 4 and Fig. 6, respectively. Various simulation parameters we use are presented in Table 1 [12].

Table 1. Simulation parameters

Bandwidth	Wired : 100MB / Wireless : 2MB
Packet size	762byte
Propagation Delay	Wired : 10 ~ 20ms / Wireless : 1ms
Queue size	20 ~ 200 packet

4.1 Performance in Simple Environment

Goodput is the effective amount of data delivered through the network. It is a direct indicator of network performance. We evaluate the goodput of TCP NJ+, New Jersey, Westwood, and Reno on various wireless link error rates using the simple topology shown in Fig. 4.

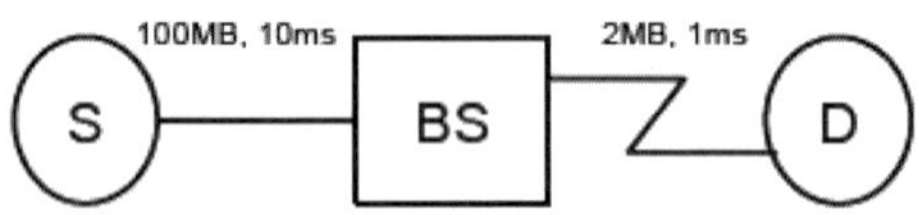

Fig. 4. Simple network topology

The source (Node S) connects to Node BS via a 100MB wired link with 10ms propagation delay. Node BS is linked to the destination (Node D) via a 2MB wireless link with 1ms propagation delay. The queue size of the wired link is set to 150 and the wireless link queue size is set to 20 respectively. The goodput result is shown in Fig. 5. TCP NJ+ shows a higher goodput performance, when wireless link error rate increases. Especially, in a 5% wireless link error rate, TCP NJ+ outperforms TCP New Jersey by 19% and TCP Westwood by 54%.

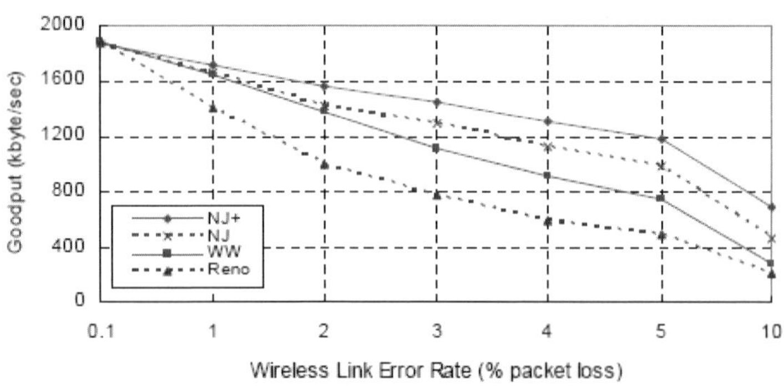

Fig. 5. Simple environment simulation result

4.2 Performance with Cross-Traffic Network

In TCP NJ+, ABE guarantees considerable performance regardless of the background traffic direction. As illustrated in Fig. 6, we measure the goodput of TCP NJ+, TCP New Jersey, TCP Westwood, and TCP Reno on various wireless link error rates under bi-directional background traffic. The source (Node S) connects to Node R1 via a 100MB wired link with 10ms propagation delay. R1 is linked to Node BS via a 100MB wired link with 20ms propagation delay. The asymmetric wireless link from BS to the destination (Node D) is represented by the differing bandwidth on the downlink (2MB) and uplink (1MB) with 1ms propagation delay. The cross-traffic flow is FTP background traffic via a 100MB wired link with 10ms propagation delay. The queue size of the wired link is set to 200 and the wireless link queue size is set to 20.

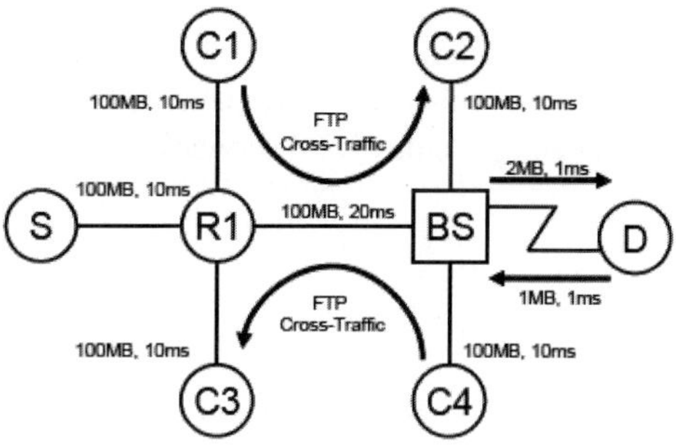

Fig. 6. Realistic topology with background traffic

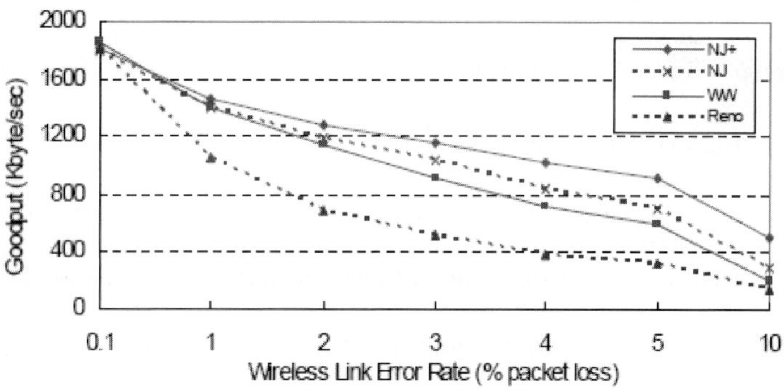

Fig. 7. Realistic environment simulation result

The goodput result of the FTP bi-directional background traffic is presented in Fig. 7. TCP NJ+ outperforms New Jersey by 31% and Westwood by 56% in a 5% wireless link error rate. The simulation result shows that TCP NJ+ achieves higher goodput than New Jersey regardless of the background traffic pattern when the wireless link error rate increases.

5 Conclusion

We propose TCP NJ+, to improve the performance of TCP New Jersey. Three enhanced mechanisms are proposed in TCP NJ+. First, the enhanced ABE guarantees higher throughput regardless of the background traffic direction because it always ensures the optimal *cwnd*. Second, when packet loss or RTO due to BER occurs, the suitable error recovery inflates by reducing the *cwnd* more quickly.

Results from the simulations demonstrate that TCP NJ+ improves the performance even when wireless link error rates increase. Particularly, TCP NJ+ outperforms New Jersey by 19% and Westwood by 54% in a 5% wireless link error rate with no background traffic. Under a 5% wireless link error rate with bi-directional background traffic, TCP NJ+ outperforms New Jersey by 27% and Westwood by 52%. In conclusion, TCP NJ+ with the enhanced ABE and suitable recovery are robust to wireless HCI environments and provides good services to users which use mobile TCP application.

Acknowledgments

This research was supported by the MIC(Ministry of Information and Communication), Korea, under the ITRC(Information Technology Research Center) support program supervised by the IITA(Institute of Information Technology Assessment), IITA-2006-(C1090-0603-0046). Corresponding author: H. Choo.

References

1. Postel, J.: Transmission control protocol. RFC 793 (1981)
2. Allman, M., Paxson, V., Stevens, W.: TCP Congestion control. RFC 2581 (1999)
3. Tian, Y., Xu, K., Ansari, N.: TCP in wireless environments: problems and solutions. IEEE Radio Communications 43, 27–32 (2005)
4. Steve, L.: Understanding mobile Human-Computer Interactions. Butterworth-Heinemann (2005)
5. Stevens, W.: TCP slow start, congestion avoidance, fast retransmit and fast recovery algorithms. RFC 2001 (1997)
6. Xu, K., Tian, Y., Ansari, N.: TCP-Jersey for wireless IP communications. IEEE Journal of Selected Areas in Communications 22, 747–756 (2004)
7. Xu, K., Tian, Y., Ansari, N.: Improving TCP performance in integrated wireless communications networks. Computer Networks 47, 219–237 (2005)
8. Casetti, C., Gerla, M., Mascolo, S., Sanadidi, M.Y., Wang, R.: TCP Westwood: bandwidth estimation for enhanced transport over wireless links. ACM/IEEE MobiCom, pp. 287–297 (2001)

9. Floyd, S.: TCP and explicit congestion notification. ACM Computer Communications Review 24, 10–23 (1994)
10. Jacobson, V., Braden, R., Borman, D.: TCP extensions for high performance. RFC 1323 (1992)
11. UCB/LBNL/VINT Network Simulator [Online]. Avilable: http://www.isi.edu/nsnam/ns
12. Cen, S., Cosman, P.C., Voelker, G.M.: End-to-end differentiation of congestion and wireless losses. IEEE/ACM Transactions on Networking 11, 703–717 (2003)

Use of Chinese Short Messages

Dafei Ma[1], Fumiko Ichikawa[2], Ying Liu[3], and Li Jiang[1]

[1] NO.10 (Jia), Datun Road, Chaoyang District, 100101 Beijing, China
[2] Arco Tower, Shimomeguro 1-8-1 Meguro-ku okia, TOKYO 153-0064, Tokyo, Japan
[3] Nokia House 1, No.11, He Ping Li Dong Jie, 100013 Beijing, China
motorfly5317@yahoo.com.cn,
{fumiko.ichikawa, ying.y.liu}@nokia.com, jiangli@psych.ac.cn

Abstract. Short text message (SMS) as a key communication means in China received a lot of attention in research community. 114 subjects attended the study, sharing totally 10843 SMS they sent and received daily. We divided the SMS into two categories (instrumental and expressive), analyzed vocabularies, functions and effects of demographic factors and SMS categories on SMS lengths and found: 1) Top 400 Chinese characters occupied 85% and top 388 words occupied 73% in SMS. Punctuations appeared frequently (18%), while Smiley appeared very little (less than 0.1%). 2) People sent both instrumental and expressive messages regardless of their age. Female users tended to send longer SMS. Retired people sent longest SMS, followed by working people and students. People exchanged SMS with close friends and families. Expressive SMS have more words than instrumental SMS. 3) People over 40s exchange more SMS with children than with friends.

Keywords: SMS, China, instrumental, expressive.

1 Introduction

Short messaging service (SMS) has seen a phenomenal growth in China after its proper release in 2000. In the same year, the total number of SMS sent by Chinese mobile users already reached up to 1 billion [1] and during the year of 2006, the amount reached 420 billion [2]. Since the penetration of mobile phone in China is still low and keeping increasing in near future, the SMS sent be Chinese users will keep growing too [1].The primary reason for using SMS appears the same as in other countries: It is relatively cheap, convenient, and conspicuous [9, 10, 11].

Compared with communication behaviors of 137 million Chinese internet users [3], SMS was the very first technology enabling personal online communication in such a wide scale in China. The phenomena aroused many researches and discussions in terms of its impact to the society.

Just like many countries have reported, it is teenagers and young adults who drive the usage of short messages. In 2002, a local magazine Xin Zhou Kan sampled residents of large cities including Beijing, Shanghai and Guangzhou, and asked about their SMS use [4]. The results showed that two age groups, 15 to 19 year olds and 20 to 25 year olds had highest ratio of SMS users (83.3% of 15 to 19 year-old mentioned

M.J. Smith, G. Salvendy (Eds.): Human Interface, Part II, HCII 2007, LNCS 4558, pp. 582–591, 2007.

they use SMS, while 89.0% of 20 to 25 year-old). In China, there is another factor which is the education. They found out that the higher the education of the user, the higher SMS usage rate. 31.1% of people with 9 years of compulsory education mentioned they use SMS, while it was 79.8% for the university students. Additionally, 49.3% of the non-SMS users mentioned that it is difficult to use. The research also investigated purposes of using SMS and the results indicated SMS are mainly used to greet people, chat, coordinate things e.g. daily work, and share contents e.g. jokes [4]. Lai and others studied college students in Shanghai, and reported their primary usage of SMS being making daily contacts, having a small chat, or showing care between acquaintances [5]. They also pointed out that SMS is preferred mode of communication as compared to other communication tools because SMS enables flexibility in terms of time of sending, time of reading, and whether to respond [5].

Another aspect which had received a lot of attention was the linguistic properties of text messages. In 2005, a collection of SMS poetry was published, followed by another collection of SMS novel [6]. On academic side researchers analyzed rhetoric, grammar and characteristics of SMS languages. In particular, seasonal greetings such as messages sent during Chinese New Year period had been analyzed by Zhang [7]. Messages during this festive time are peculiar, because of its content and the quantity. Fairly large amount of people would visit online to select short stories, jokes and poems developed by semi-professional SMS writers. Many sent messages in bulk as the year changes, and it was reported that over 10 billion SMS were sent on that specific moment [8]. Zhang classified these messages six categories by content and four by its format [7]. However, the study focused on this festive content, and had not extended to the content of everyday nature.

SMS is not always beneficial. Liu noted that although the peer-to-peer nature of SMS allows individuals to protect their privacy and enable personal discussions and free communications, he reminded that message can contain spam, and false content that would be harmful for the society [9]. Indeed, behind the impressive growth figure of SMS, spam messages have steadily increased and created social problems. Spam contents vary from harmless advertisement on flight tickets to false lottery winnings and in some rare cases, prostitution, but nevertheless, were out of control of individual's will to receive them.

There are a lot researches on SMS behavior in other countries. Ling mentioned SMS had found a niche in our communication needs, categories of the SMS can give us a clear picture of this niche [10]. Ling listed the most common themes used in SMS messages which are coordination, grooming, answers, questions, information, commands or requests, personal news and diverse other categories [10]. Grinter and Eldridge's research focused on teenagers and found their message categories are: communication coordination, planning activities, chatting and other (e.g., jokes, reminders, greetings etc) [11]. Those categories mentioned above give us some information about what people are talking about through SMS, however, they are somewhat not in the same level and some of them are overlapped.

The objectives of our study include 1) to identify the most frequently used words among Chinese mobile users; 2) explore characteristics of SMS Chinese users are sending and try to answer questions like: who are sending the longest messages and what are the contexts or occasions to send them; and 3) what are the key purposes for using SMS in China.

2 Method

We used questionnaire to collect SMS and user profiles. Users can choose to use paper and pencil or computer to record their SMS and share with us. We instructed users to record SMS without any change but totally according to their original form. Users can decide freely SMS they would like to share with us, if they think some are too private to share, that is also ok. User profiles collected in the study include demographic information including age, gender, education background, occupation, location, and income and phone relevant information including phone model(s), payment for phone bill, operator, using frequency of applications, and main input method for Chinese SMS.

We don't use other efficient ways e.g. SMS gateway and forwarding SMS to collect SMS because of either higher cost or not willing to limit the study to certain group of users.

2.1 Participants

114 participants attended the study, sharing with us totally 10,843 SMS. Before recruiting participants, we pre-defined 5 users groups and size of each age group, which is according to the ratio of SMS users in mainland China. We used stratified and snowballing sampling method to recruit users. At the same time, occupations (students, working people and retired) were balanced in user groups. Table 1 shows a whole picture of the 114 participants.

Table 1. Demographic information of participants

	Occupation		
Age	Students	Working people	Retired
16~24(41)	31	10	0
25~34(40)	3	37	0
35~44(20)	0	20	0
45~(13)	0	4	9
total	34	71	9

2.2 Procedure

Firstly, we designed the questionnaires for profile and SMS collection. Secondly, we decided to adopt snowballing sample method, which can ensure the sample size, to recruit participants. Thirdly, we collected the data according to our plan. In the phrase, we at first contacted 10 coordinators who were typical representatives in the 5 age groups and asked them to find more peers who would like to participate. The coordinators were also responsible for keeping contact with participants and ensuring the SMS recording and sharing. Totally, we found about 130 participants and finally 114 of them followed the study to the end. Numbers of SMS provided by each participant ranges from 50 to 200 items. A benefit of the snowballing method is that when we lose contacts with some participants, it is easy to reload with new ones. The data collection process lasted about two months. Finally, we transferred all SMS and user profiles to computer and analyzed them. In the analysis phrase, we focused on three topics: 1) vocabularies; 2) functions of SMS and 3) effects of factors like gender, age and functions on SMS length.

3 Findings and Discussions

In below, we presented our research results in four parts: 1) vocabularies; 2) categories; 3) effects of gender, age and category on SMS length; and 4) other results. In 1), we presented vocabulary results including the often used characters, words and punctuations in Chinese SMS. In 2), we described a method and result of categorizing Chinese SMS according to their functions. In 3), we analyzed the effects of gender, age and SMS category on SMS length and presented the results. In 4), we presented analysis result on relationships of SMS interlocutors across age groups. We also discussed the results after presenting them in each part.

3.1 Vocabularies

Table 2 and Table 3 listed the most often used Chinese characters and 2-character words in Chinese SMS. Further analysis indicated that the top 400 Chinese characters occupied 85% of the whole characters used in Chinese SMS and the top 388 words (including one-character words and two-character words) occupied 73% in Chinese SMS. The trend is very similar as in English [11].

Table 2. Top 20 frequently used characters

Rank	Word	Meaning	Frequency	Rank	Word	Meaning	Frequency
1	我	I	5.5%	11	想	think	0.9%
2	你	You	5.1%	12	就	then	0.9%
3	了	(past tense)	3.6%	13	给	give	0.8%
4	的	of	2.6%	14	去	to go	0.8%
5	不	no	1.6%	15	没	didn't	0.7%
6	在	at	1.3%	16	呢	question ending	0.7%
7	吗	question ending	1.2%	17	说	to say	0.6%
8	吧	suggestion ending	1.0%	18	啊	endin g	0.6%
9	好	good	0.9%	19	也	also	0.6%
10	是	be	0.9%	20	一	one	0.6%

Table 3. Top 16 frequently used 2-character words

Rank	Word	Meaning	Rank	Word	Meaning
23	什么	what	54	明天	tomorrow
32	现在	now	57	电话	call
37	知道	to know	62	你们	you
40	我们	we	63	可以	can
41	没有	don't have	70	就是	is
43	怎么	how	73	自己	myself
44	不是	be not	77	时候	when
47	今天	today	79	回来	to come back

Punctuations appear frequently (18%) in Chinese SMS and table 4 listed the most often used punctuations including comma, period, question mark, and exclamation mark. One thing to note is that Chinese users often mix using punctuations of half-width and full-width. In Table 4, the first comma, question mark and exclamation mark are in the form of full-width while the second ones are in half-width format. Surprisingly, English period is one of the most often used punctuation too. It might be because some phone platforms support only English period. Some users replace any punctuation with space in Chinese SMS, which is more convenient since space is always available and can be input directly in any input mode.

Table 4 also presented results about Smiley. Proportion of smiley was very small, less than 0.1% of entire corpus. Both vertical and horizontal Smiley were found, however, the result showed that in Chinese SMS, people primarily used vertical ones.

Table 4. Main punctuations and identifiable smiley

Rank	Word	Percentage	Rank	Word	Percentage
2	，	5%	320	:)	0.03%
3	。	4%	680	:-)	0.01%
7	,	2%	973	:(	0.01%
8	？	2%	2980	:-(	0.00%
9	.	2%	4509	:>	0.00%
11	?	1%	6973	;)	0.00%
13	！	1%	8740	:<	0.00%
18	!	1%	10666	;-)	0.00%
158	、	0	10658	^_^	0.00%

3.2 Categories

Based on the categorization of SMS in past researches [10, 11], we divided the SMS into seven categories according to their functions, purposes and moods. The seven categories are coordination, information, soft response, command and request, soft inquiry, personal information and emotional messages. Three researcher worked on the categorization of all collected SMS and everyone needed to categorize all SMS into the seven categories to make sure the categorization was valid. To keep coherence among researchers, researchers randomly selected some SMS to make trials. Everyone exercised on the SMS until the categorization coefficient reached 0.758 at the third round. During the categorization process, if two or all researchers put a SMS to a same category, we believed the SMS belonged to the category. And if the three researchers had different idea on a message, they would discuss on it until they agreed.

Based on natures of each category, we fatherly abstracted two general categories: instrumental and expressive SMS. The former five categories belong to instrumental SMS and the later two belong to expressive SMS in terms of their natures.

Table 5. Definition, example and proportion of 7 kinds of SMS

Category	Definition	Example	Percentage
1 Instrumental			
11 Coordination	Mutual planning of meetings & events.	*Laoguai, we will have a football match at 6 Am next morning at the playground. Join us on time!*	23.96%
12 Information	Ask or inform something that concerning both ends.	*Please tell me LaoLi's telephone number.*	25.10%
13 Soft response	Response to others' SMS.	*Ok. I think you might be busy, so I'll not disturb you any more.*	1.85%
14 Command and request	To ask for help directly.	*Buy me some bread!*	6.53%
15 Soft inquiry	Similar to Requests, but more descriptive and polite, and typically lengthy, describing situation and asking for some advice or help.	*To get an internship chance during vacation is really difficult. Could you give me a hand?*	0.28%
2 Expressive			
21 Personal information	To bring updates and in some cases gossips.	*Do you know XiaoLi has married?*	1.76%
22 Emotional messages	Chatting about personal or work lives, a way to show care to family or close friends.	*Forget the unhappiness, I wish you happy.*	40.53%

Table 5 listed all the categories with definition, example and proportions in the whole corpus. Emotional short messages occupied the highest proportion, followed by coordination and information. The other four categories are small in proportion and together they occupy about 10 percent. From the results, we can presume the key functions of SMS in China: 1) to show care to family and close friends and share personal and emotional issues; 2) to coordinate daily events and 3) to exchange and share information. When compared with SMS collected by Richard Ling [10], we can see both studies reported high proportion of coordination, but since the methodologies of SMS collection and classification are different, the proportion results are somehow different.

3.3 Effects of Gender, Age Group and Category on SMS Length

We also explored the effects of gender, age group and category on SMS length. Table 6 listed the average length of SMS of different user groups. Table 7 presented the MANOVA test results. The test results showed that:

1) Female users write SMS with more words than male users. As shown in table 6 and table 7, SMS written by female users are 1.26 words longer than those by male users, and the difference is significant according to ANOVA test (sig. =0.01). The result is compatible to our common sense, which is that female users are more sensitive and patient and inclined to write longer SMS.

2) SMS sent by retired people have the most words, followed by working people, and SMS by students have the least words. By MANOVA test, main effect of age is significant (sig. =0.00). Post-hoc test revealed that differences between any two groups are significant, which means retired people write SMS with the most words, followed by working people and the last is student group. The reason might be because the retired people have more urgent and strong needs to communicate with others on the fly. And since working people need to coordinate more daily events e.g. working issues than students, they might send more instrumental SMS than expressive ones, which will decrease their SMS length on average.

3) Expressive SMS are longer than instrumental ones. Averagely, expressive SMS have 16.92 words, significantly longer than instrumental ones as indicated by MANOVA test. The difference likely lies in the different functions of the two kinds of SMS. Instrumental messages are to coordinate daily issues and share information and a clear and concise style of writing would be applied. On the other hand, since

Table 6. SMS length by gender, age group and category

	Variables	Average (SMS length)	S.D.
Gender	Male	14.82	0.46
	Female	16.08	0.20
Age Group	Students	14.28	0.17
	Working people	15.17	0.16
	Retired	17.00	1.08
Category	Instrumental	13.98	0.37
	Expressive	16.92	0.34

Table 7. Statistical test results

Source	Sum of Squares	df	Mean square	F	Sig.
gender	795.52	1	795.52	6.22	0.01
age	2,696.63	4	674.16	5.27	0.00
category	4,323.55	1	4323.55	33.79	0.00
gender * age	6,165.91	4	1541.48	12.05	0.00
gender * category	303.32	1	303.32	2.37	0.12
age * category	4,077.25	4	1019.31	7.97	0.00
gender * age * category	2,365.75	4	591.44	4.62	0.00
Error	1,371,513.54	10,720	127.94		
Total	3,794,732.00	10,740			
Corrected Total	1,417,788.78	10,739			

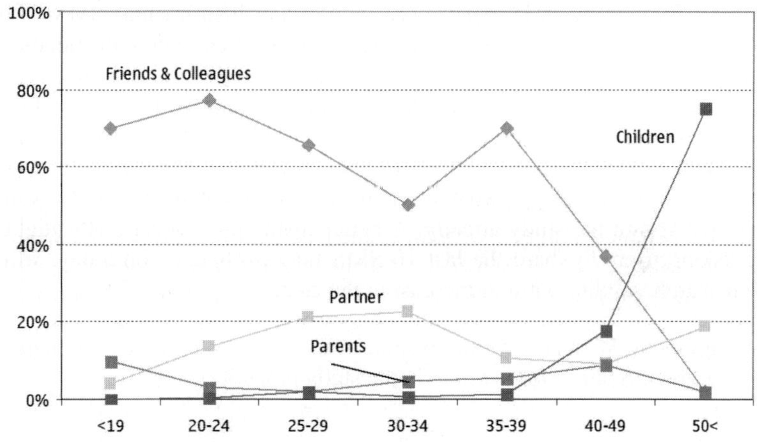

Fig. 1. Relationship between interlocutors via SMS by age

expressive messages are to share emotional and personal issues, there would be more affection expressions involved in them, which would increase the SMS length. In the study, we also found that some people even tell a small story in SMS.

Interactions are also found as shown in Table 7. But they are hard to explain in the framework of the study. More controlled study might be needed to address them.

3.4 Other Results

Earlier we assumed that expressive messages are only for youth. However, the outcome showed that the expressive nature is found across generation.

We also looked at the relationship between interlocutors and Figure 1 showed the results. For users over 50's, children (most likely in 20s) became the primary SMS counterpart. It might be because SMS is a new technology introduced by some children to parents and parents don't necessarily to use SMS to communicate with their counterparts.

4 Conclusions and Future Works

In the study, we have collected 10843 Chinese SMS from 114 local users. We divided the SMS into 2 categories and 7 sub-categories according to their functions. Finally, we analyzed vocabularies, functions of Chinese SMS and the relationships between some demographic factors and characteristics of SMS. The results indicated: 1) Top 400 Chinese characters occupied 85% and top 388 words occupied 73% in SMS. Punctuations appeared frequently (18%), while Smiley appeared very little (less than 0.1%). 2) People sent both instrumental and expressive messages regardless of their age. Female users tended to send longer SMS. Retired people sent longest SMS, followed by working people and students. People exchange SMS with close friends and families. Expressive SMS have more words than instrumental SMS. 3) People over 40s exchange SMS exchanges more SMS with children than with friends.

In future work, we should conduct a field study of SMS users to understand their motivations and habits of using SMS. We believe if we can go to users and talk with them in their living or working contexts, we can have a better understanding of the data and results got in the study. Method of collecting SMS should be improved in the study. Since users can decide what to share and what not in the study, some bias might be involved in the study already. A better method to collect SMS might be for example asking users to share the last 10 SMS they exchanged on a day, which can improve research validity without increasing the cost.

Acknowledgments. We thank Nokia to sponsor the study. We thank Li Yongjuan from China Academy of Science for her help in the study process.

References

1. Zhou, Y.: General Introduction of Chinese SMS Phenomena (in Chinese) (2005) http://news.xinhuanet.com/newmedia/2005-05/12/content_2949398.htm
2. Statistics of Chinese telecommunications (in Chinese) (2007) http://www.mii.gov.cn/art/2007/01/22/art_27_28238.html
3. Annual Chinese Internet survey 2007 by CNNIC (in Chinese) (2007) http://www.cnnic.net.cn/uploadfiles/doc/2007/1/22/212245.doc
4. Did You Send SMS Today (2002) http://culture.163.com/partner/weekly/editor/ 020809/020809_64441.html
5. Lai, Y., Wang, Y., Wang, S.: Investigation of SMS Behavior of Shanghai University Students. Journalists, vol. 2 (2004)

6. Ning, B.: The First Novel about SMS Sells 5000 Yuan in China (2005) http://news.xinhuanet.com/mobile/2005-11/18/content_3799342.htm
7. Zhang, H.: Discussion of the Spring Festival SMS, Transaction of BeiHua University (Social Science), vol. 2 (2005)
8. Jin, J.: Analysis of 10 billion SMS in 2005 Spring Festival: Information Supported Culture, Shanghai Information, vol. 3 (2005)
9. Liu, J.: How SMS Change Our Life. Chinese Media Technology, vol. 2 (2005)
10. Ling, R.: The socio-linguistics of SMS: An analysis of SMS use by a random sample of Norwegians. In: Mobile communications: Renegotiation of the social sphere, pp. 335–349. Springer, London (2005)
11. Rebecca, E.G., Margery, E.: Wan2tlk?: Everyday Text Messaging. In: Proceedings of ACM SIGCHI Conference on Human Factors in Computing Systems, Florida, USA, pp. 441–448 (2003)

Multilingual Disaster Information for Mobile Phones in Japan

Masaru Miyao[1,4], Kumi Sato[2], Satoshi Hasegawa[3,4], Kazuhiro Fujikake[4], Shozo Tanaka[4], and Kohei Okamoto[5]

[1] Information Technology Center, Nagoya University, Chikusa-ku Nagoya, Japan
mmiyao@med.nagoya-u.ac.jp
[2] Graduate School of International Development, Nagoya University, Nagoya, Japan
avenues@sato.email.ne.jp
[3] Department of Information Culture, Nagoya Bunri University, Inazawa Aichi, Japan
hasegawa@nagoya-bunri.ac.jp
[4] Graduate School of Information Science, Nagoya University, Nagoya, Japan
s.tanaka@tc.kawai-juku.ac.jp
[5] Graduate School of Environmental Studies, Nagoya University, Nagoya, Japan
miyao@itc.nagoya-u.ac.jp

Abstract. A multilingual disaster information system (MLDI) has been developed to overcome the language barrier during times of natural disaster. MLDI is a web-based system that includes templates in nine languages so that translated texts can be made available immediately. Mobile phone e-mail with graphic text is a useful tool for delivering multilingual disaster information. We assessed the performance of the translation system and the multilingual graphical characters, and found that they are sufficiently useable. We discussed the usage of mobile phones for multilingual disaster information delivery system including Early Earthquake Warning system. Multilingual mobile information and measures is useful for a safe and barrier-free society.

1 Introduction

There are more than 2 million foreign residents from more than 180 countries in Japan. Disaster information or other important information must be made available to all people in a country, including foreign residents and travelers who cannot understand the native language of the country. However, it is not easy to translate various information into various languages rapidly and accurately in times of emergency. We introduce a multilingual disaster information system (MLDI) that translates a range of information on natural disasters, including earthquakes, and tsunami (tidal waves), quickly and accurately into four languages. And we are now developing graphical character short message services (SMS) for Chinese and Korean characters.

2 Template Translation for Multilingual Disaster Information

Although automatic translation systems have been studied and developed, complete translation of natural languages into other languages is difficult. In the case of disaster information, correct translation is essential. Fortunately, most disaster-related

M.J. Smith, G. Salvendy (Eds.): Human Interface, Part II, HCII 2007, LNCS 4558, pp. 592–599, 2007.

information can be expressed in short sentences with fixed form (Fig. 1). Template translation is one possible means to produce multilingual information quickly and accurately.

Disaster information needs to be sent to all disaster victims including foreign residents, and to be made available to travelers in a language they understand. Such information should immediately reach victims no matter where they are. Reports from residents in disaster spots about traffic conditions or other local information for daily

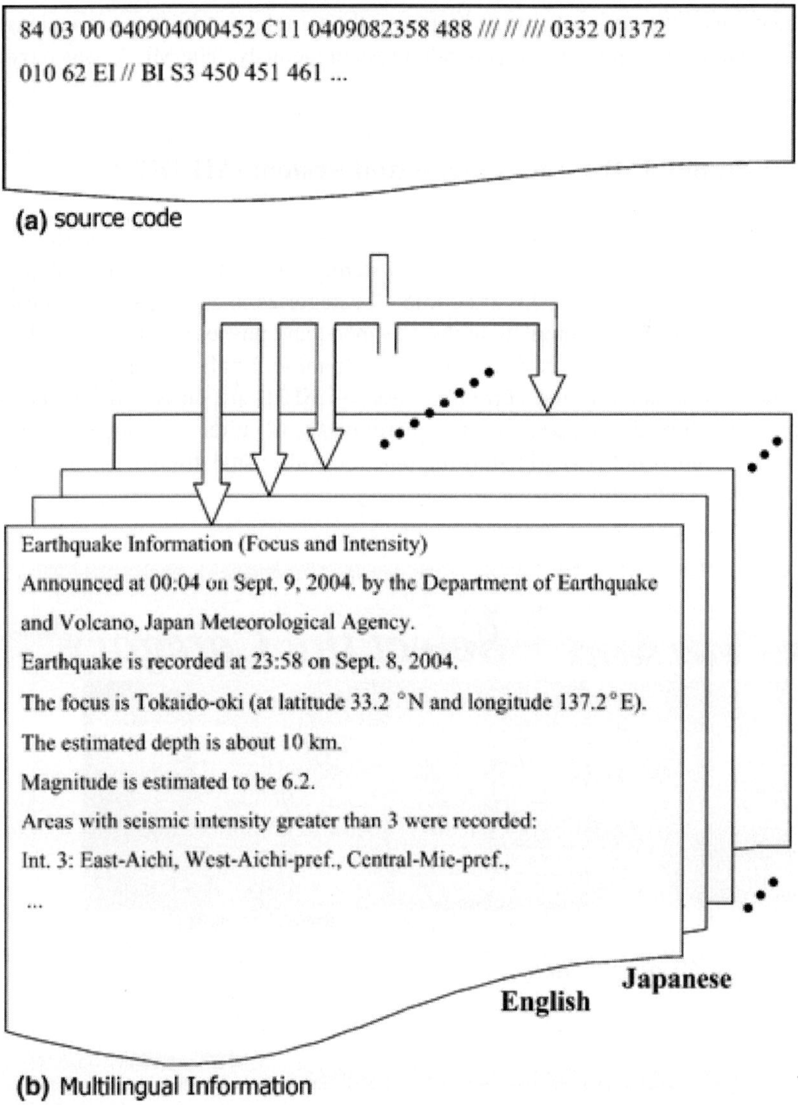

Fig. 1. Samples of graphic characters for experiment 1

life is also required so that victims can get survival goods or move on to appropriate shelter. Moreover, the true meaning of provided information is, in some cases, difficult to decipher by foreigners who have insufficient knowledge about the culture.

A template translation method would seem to be useful to provide appropriate information to foreigners who need additional explanations or special information.

The promising approach is to use multilingual templates prepared previously for both original and translated messages. Templates should be prepared in sufficient number and with sufficient contents in several languages including the original language. Once the user fills in the blanks in the template of his/her native language, multilingual information will be generated instantaneously. The MLDI was developed under this concept.

3 Multilingual Disaster Information System (MLDI)

The MLDI has been developed by the Multilingual Disaster Information System Consortium. The MLDI is a web-based system. This system has more than 1,100 different template sentences classified into 40 categories in 7 groups. Once a user fills in blanks in the Japanese template in the edit mode and pushes the translation button, translated sentences appear in five languages, Japanese, English, Korean, Chinese and Portuguese. Templates in the current version of MLDI are classified into the seven groups: (1) before the occurrence of an earthquake, (2) after earthquake occurrence, (3) damage and relief, (4) life information, (5) traffic and lifelines, (6) safety and consultation, and (7) full information.

Fig. 2. Template Translation System for Earthquake Information

4 How to Use MLDI

The categories of the templates can be selected by pushing a button (Fig. 2). The contents of the templates are limited to earthquake and living information for victims. A user can easily choose the appropriate template from the list displayed after the category is selected. The language used to edit the template is also limited to only Japanese in this prototype version. The blanks in the Japanese text can be filled with the numbers, Roman letters or Japanese characters representing the name of places, dates, times, telephone numbers, or other information. In this system, the edited Japanese templates are translated into complete texts in four languages, English, Korean, Chinese and Portuguese, just by pushing the Translate button.

Translated texts can be downloaded as texts coded in Unicode (UTF-8) so that the user can combine them or lay them out suitably using MS-Word or other editors

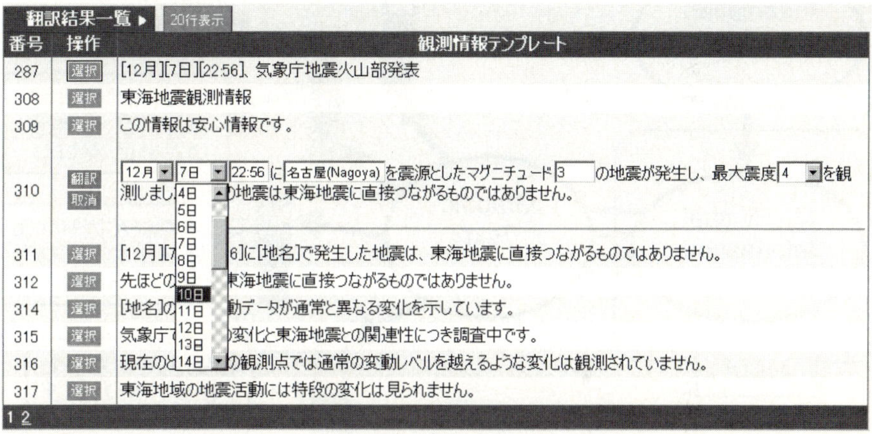

Fig. 3. Fill with the blanks in the Japanese text

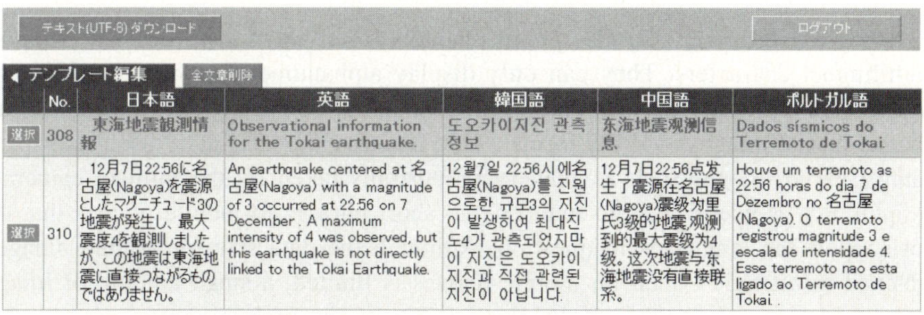

Fig. 4. Translated texts can be downloaded as texts coded in Unicode (UTF-8). Japanese, English, Korean, Chinese and Portuguese sentences are shown. Blanks in the template are filled with the name of a location in Japanese and Roman letters.

supporting Unicode. This translated matter can be printed and distributed or put up as notices. The translations can also be sent by e-mail. Sending e-mails to mobile phones using graphical character e-mail on mobile phones is a particularly useful method to deliver disaster information.

5 Information Delivery

Although there are many possible methods to deliver multilingual disaster information, mobile phone (MP) e-mail is superior to methods such as distributing printed matter or e-mail sent via personal computers, in terms of its quickness and independence of the location or supply of electric power (Fig. 5).

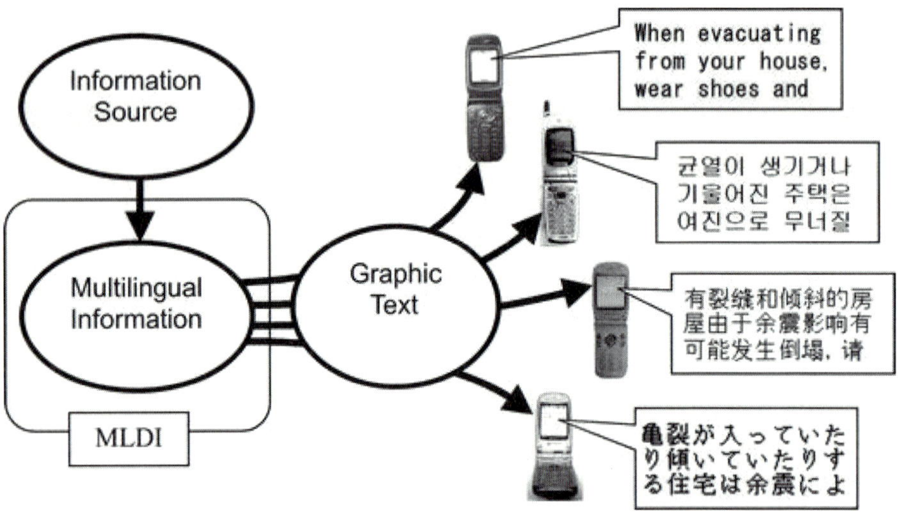

Fig. 5. Multilingual information delivery using graphic text for mobile phones

However, ordinary mobile phone models used in Japan do not support multilingual characters. They can only display alphanumeric characters, Japanese and several symbols. On the other hand, almost all of these mobile phones can display digital images in JPEG format. This function was originally used to display natural photo images such as scenery or portraits taken by the built-in cameras. Graphic text is useful to send multilingual short phrases to mobile phones that do not support the languages in question. Although the data size for graphic images attached to a single e-mail on mobile phones is limited, a small amount of image data is suitable for display on the small LCDs of the mobile phones, and is also suitable for the short phrases of multilingual disaster information generated by MLDI as shown in Fig. 6.

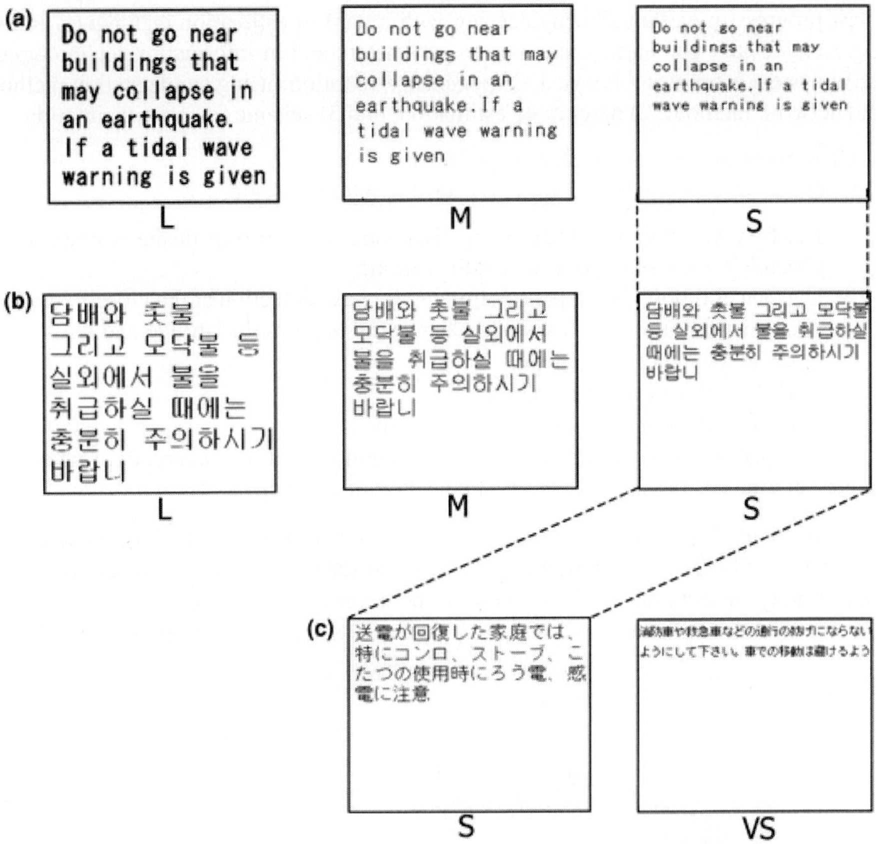

Fig. 6. Examples of graphic text. Short sentences are presented in the JPEG format. We checked the character size and font resolution.

6 Discussion on Disaster Information Delivery

Graphical textis as visible as text in the built-in font on LCDs in mobile phones. Short messages like the disaster information generated by MLDI can be sent by this method. However, the disaster information needed in an emergency varies. Urgent messages such as warnings for tsunami or tidal waves must reach people within a few minutes after an earthquake in some cases. On the other hand, living information such as the places of the shelters or the traffic conditions can be made available any time. Disaster information should be delivered to people who need it, at an appropriate time in an appropriate format, and in an appropriate language. We hope some media will be developed soon as a strong tool to deliver disaster information quickly and reliably in a calamity situation. Diversified media should be used to deal with various situations in times of calamity.

The Japanese Meteorological Agency (JMA) started delivering Earthquake Early Warning (EEW) since February, 2005. This system has been conducted involving the latest progress in earthquake engineering with regard to estimation techniques on the hypocenter, fault and earthquake motion parameters and in information technologies. EEW system estimates 3 items: 1) hypocenter estimation (using single station method and network method), 2) magnitude estimation, and 3) seismic intensity estimation.

Dissemination Criteria of EEW is as follows:
Over 3.5 for magnitude or 3 for JMA seismic intensity scale.

1. The first warning: The result using 2seconds waveform or onsite warning over 100cm/s/s (without hypocenter information)
2. The cancel report: There is no following P phase detection at the closest station.
3. The updated warnings: Threshold Change 0.2 degrees for latitude and longitude,
   - 20km for hypocentral depth,
   - +0.5, -1.0 for event magnitude
   - +0.5, -1.0 for maximum seismic intensity.
4. The final warning: when the estimated magnitude value converged, or when the pre-set time has passed after the first detection of the P phase.

EEW should be simple stereotyped information. MLDI is a web-based system that includes templates in five languages so that translated texts can be made available immediately. We think EEW is easy to be translated into foreign languages as templates. Mobile phone e-mail with graphic text is a useful tool for delivering EEW.

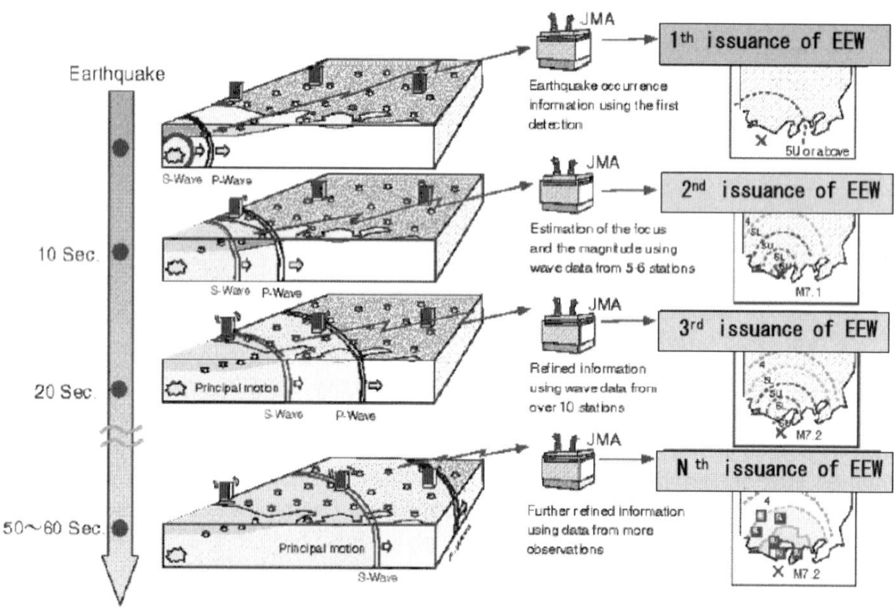

Fig. 7. Conceptual image of seismic wave propagation and Earthquake Early Warning Mitsuyuki HOSHIBA (2007) Earthquake Early Warning, -Information before strong ground motion - , Japan Meteorological Agency January 12, 2007.

7 Conclusion

A template translation technique and graphical character e-mail on mobile phones were used in this system. These techniques are essential to our MLDI system. Multilingual information will be helpful in creating a barrier-free society, including many foreign residents. We assessed the performance of the translation system and the multilingual graphical characters, and found that they are sufficiently useable. We discussed the usage of mobile phones for multilingual disaster information delivery system including Early Earthquake Warning system.

References

1. Great Hanshin-Awaji Earthquake Memorial Research Institute, Great Hanshin-Awaji Earthquake Research Paper (2004) http://www.hanshin-awaji.or.jp/kyoukun/eng/
2. Niigata Prefecture Earthquake information (2004) http://www.pref.niigata.jp/sougouseisaku/kokusai/english/
3. Hasegawa, S., Sato, K., Matsunuma, S., Miyao, M., Okamoto, K.: Multilingual disaster information delivery using graphic text for mobile phones. AI & Society, vol. 19(3), pp. 265–278. Springer, Heidelberg (2005)
4. Miyao, M., Okamoto, K.: Multilingual Disaster Information System Consortium (Representatives) (2004), http://tagengo.seis.nagoya-u.ac.jp/
5. Mitsuyuki HOSHIBA Earthquake Early Warning, -Information before strong ground motion -, Japan Meteorological Agency (January 12, 2007) https://www.prime-intl.co.jp/geoss/materials/PDF/Earthquake/2day/No,4_Dr,Hoshiba_geoss.pdf

Visibility Evaluation of Characters on Liquid Crystal Displays in Mobile Phones, and Visual Function

Masako Omori[1], Satoshi Hasegawa[2], Kazuhiro Fujikake[3], and Masaru Miyao[4]

[1] Faculty of Home Economics, Kobe Women's University, Suma-ku Kobe, Japan
masako@suma.kobe-wu.ac.jp
[2] Department of Information Culture, Nagoya Bunri University, Inazawa Aichi, Japan
[3] Graduate School of Information Science, Nagoya University , Chikusa-ku Nagoya, Japan
[4] Information Technology Center, Nagoya University , Chikusa-ku Nagoya, Japan

Abstract. The purpose of this study was to examine the relationships between the visibility of mobile phone displays and aging effects on subjects' visual functions, based on measurements of vision, including near visual acuity and cataract cloudiness. In the relation between Character size, CC, and NVA, visibility was lower with smaller character size "S". In the relation between Contrast, CC, and NVA, the organoleptic evaluations showed significant differences in contrast. The subjects' attached a high value to the negative display mode, especially when the subject's CC was more than 200. Such considerations are particularly important for elderly users who have marked deterioration of visual function.

Keywords: mobile phone, cataract cloudiness, near visual acuity, character size, contrast, visibility, elderly people.

1 Introduction

Today, mobile phones (MPs) are essential as IT devices for not only young people but also for most people aged 40 to 60 years. There are few studies on the visibility of MPs, however, that include elderly subjects [1][2]. Today's MPs can send and receive not only sound but also letters and pictures, and the functions of MPs will continue to progress. However, the size of the display is restricted. When we think about the universal design of MPs, it is important to consider the properties of MPs and users' visual functions. The purpose of this study was to examine the relationships between the visibility of mobile phone displays and aging effects on subjects' visual functions, based on measurements of vision, including near visual acuity and cataract cloudiness.

2 Subjects and Methods

The subjects were 78 people aged 19 to 76 years (age = 39.9±17.1), with normal or corrected-to-normal vision. We used text without meaning, made from randomly ordered alphanumeric characters according to ISO 9241-3 Amendment 1 [3], in which 15% spaces were included as in normal English sentences. This text was displayed on the liquid crystal displays (LCDs) of MPs using graphic characters. Subjects counted

M.J. Smith, G. Salvendy (Eds.): Human Interface, Part II, HCII 2007, LNCS 4558, pp. 600–607, 2007.

the number of times a target character appeared by searching the strings as if they were reading the text carefully and as fast as possible. The questioner recorded reading times, viewing distances, target character count, and subjective evaluations of visibility. Two different types of MPs and three different character sizes, large (~2.5 mm height), medium (~2.0 mm) and small (~1.0 mm), were examined with four different contrasts of negative, positive, gray characters with white background and black characters with gray background. Luminance was measured (Minolta LS-110), and is shown in Table 1 and Fig. 1.

The subjects' visual functions of cataract cloudiness (CC) and near visual acuity for a 50 cm distant target (NVA) were measured. Subjects' CC was indicated on a 256 gradation and classified into four groups, (-), (+), (2+), and (3+), where (-) means 0-99, (+) 100-149, (2+) 150-199, and (3+) 200-256. NVA was classified as either good (1.0-1.2), fair (0.5-0.7), or weak (0.1-0.3). In a two-way ANOVA, subjects' reading performance as measured by reading time (RT), character count in reading (Count), visual distance (VD) and organoleptic evaluations (OE) of visibility as the dependent

Table 1. Character size and Mean Luminance (cd/m^2) in dark and light areas on the text in this experiment

Character size	Characters per line	Lines per page	Vertical length of characters (mm)	Width length of character (mm)
L	20	12	2.5	1.3
M	30	8	2.0	1.0
S	40	6	1.0	0.5

Code	Character color	Back ground color	Displayed in Type A		Displayed in Type B	
			Mean luminance	Contrast ratio	Mean luminance	Contrast ratio
W	white	black	9.1±0.5	11.9	3.2±0.0	53.8
X	black	white	107.5±2.5	11.9	171.4±1.2	53.8
Y	gray	white	105.1±2.2	2.5	171.0±1.8	4.4
Z	black	gray	47.0±2.0	5.2	38.2±0.8	12.2

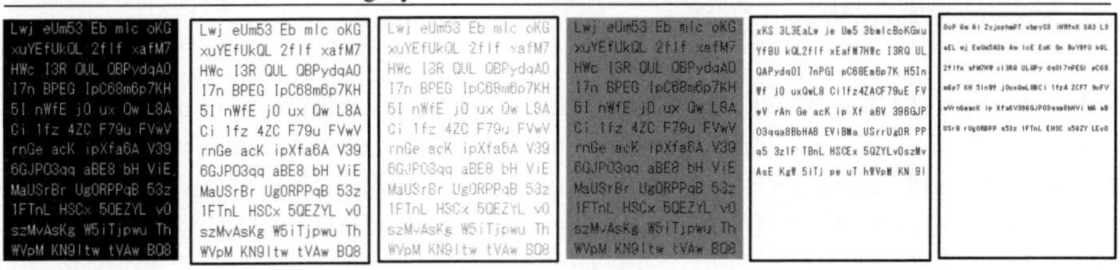

size L W X Y Z size M size S

Fig. 1. Example of graphic text prepared for the experiment variables. For ANOVAs, two independent variables were taken from among the 4 variables. One variable was CC or NVA, the other was taken from Character size and Contrast. Eventually, 16 ANOVAs were calculated.

3 Results

3.1 Result of Visual Functions

The results of this study indicated that there are close relationships between age, 30 cm NVA and CC. Figure 2(a) shows 30 cm binocular near vision by age group. Geometric means of the NVA with correction were 1.1 for 20- to 44-year-old subjects, 0.4 for 45- to 64-year-olds and 0.5 for 65- to 76-year-olds. Due to presbyopia, middle-aged and elderly subjects had weaker near eyesight than the younger subjects. The NVA in these subjects deteriorates after 45 years of age.
Using 256 levels of CC, figure 2(b) shows the CC by age. The mean cloudiness values were 60 for those 20 to 44 years old, 115 for those 45 to 64 years old and 157 for those 65 to 79 years old. Similarly for NVA, CC was severer in middle-aged and elderly subject groups than in the younger group.

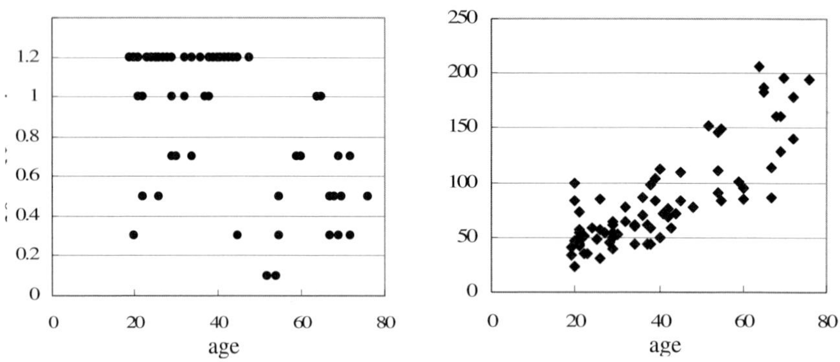

(a) Binocular near vision (30 cm) by age (b) Cataract cloudiness by age

Fig. 2. Age dependency of visual function of CC and NV

3.2 Result of Two-Way ANOVA

3.2.1 Cataract Cloudiness and Properties of MPs: Character Size and Contrast

We performed a two-way ANOVA on whether subject performance was influenced by the CC and character size. The result of RT (Fig. 3(a)) showed significant differences in both CC ($P<0.001$) and Character size ($p<0.01$). VD (Fig. 3(b)), Count (Fig. 3(c)) and OE (Fig. 3.(d)) all showed significant differences in both CC ($P<0.001$) and Character size ($p<0.001$).

Next, we examined the effect of the CC and Contrast on RT, VD, Count and OE. CC was found to have significant differences in RT (Fig. 4(a)), VD (Fig. 4(b)) and Count (Fig. 4(c)). Contrast showed no significant differences. However, OE showed significant differences in both CC (P<0.001) and Contrast (p<0.001) (Fig. 4(d)).

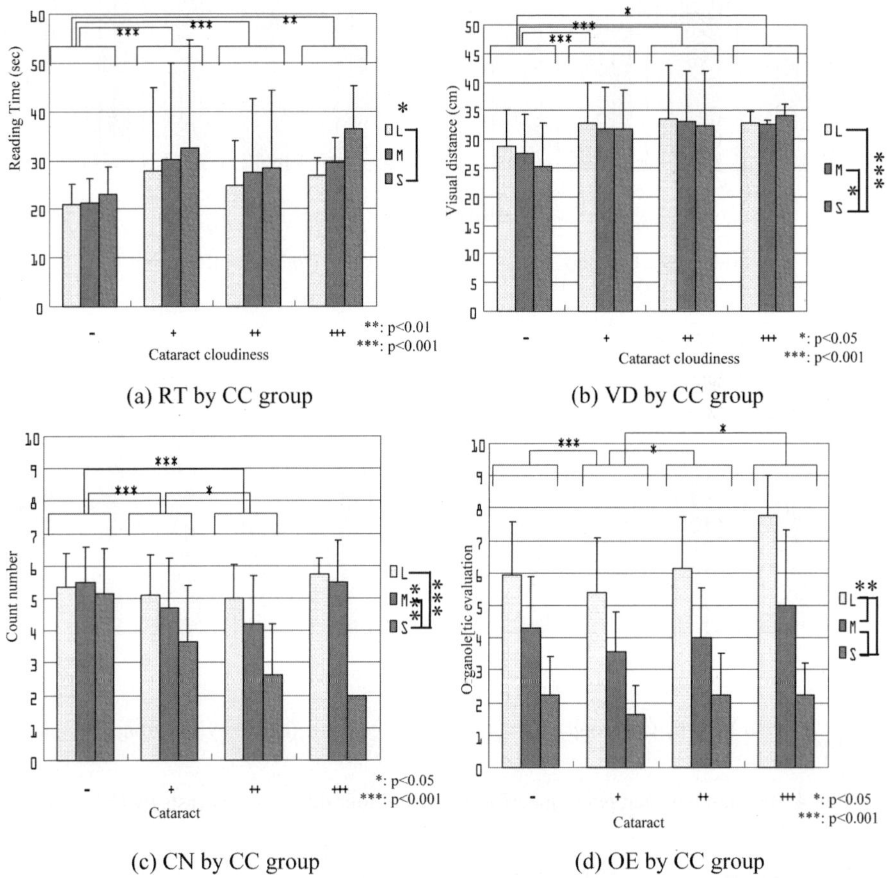

Fig. 3. Relation between reading performance and three different character sizes by CC group

3.2.2 30 cm NVA and Properties of MPs: Character Size and Contrast

Similarly, we examined the effect of the NVA and character size on RT, VD, Count and OE. RT showed significant differences in both NVA (P<0.001) and Character size (p<0.01), while VD and Count showed significant differences in both NVA (P<0.001) and Character size (p<0.001). OE showed significant differences in Character size (p<0.001). However, NVA showed no significant differences. In the relation between CC and NVA, RT, VD and Count showed significant differences in NVA. However, they showed no significant differences in contrast. OE showed significant differences in contrast.

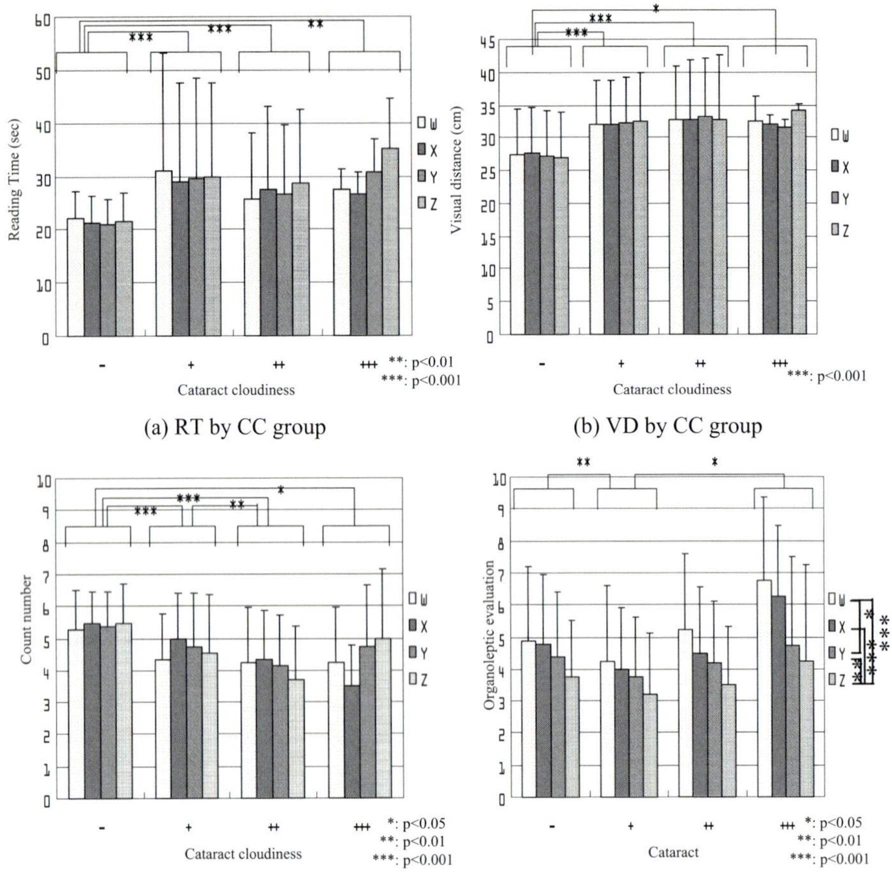

(a) RT by CC group

(b) VD by CC group

(c) CN by CC group

(d) OE by CC group

Fig. 4. Relation between reading performance and four different contrasts by CC group

4 Discussion

In this study, we measured visual function and analyzed the visibility of MPs. Deterioration of near visual acuity in healthy people is accelerated after 45 years of age (Weale 1975)[4]. We found a similar tendency in near vision in this experiment. Similar to presbyopia, cataract cloudiness gradually becomes severer with age after the middle years.

In two-way ANOVA, significant differences were found in RT, VD, Count, and OE for CC and character size. It was found that RT was prolonged and VD became shorter, and that Count decreased as character size became smaller. In the relationship between character size and CC, RT was prolonged and Count decreased significantly as CC exceeded 100 in the present subjects. The influence was large particularly in the severe group with CC of 200 or more. Thus, it was shown that cataract cloudiness severity and small character size strongly influenced reading performance.

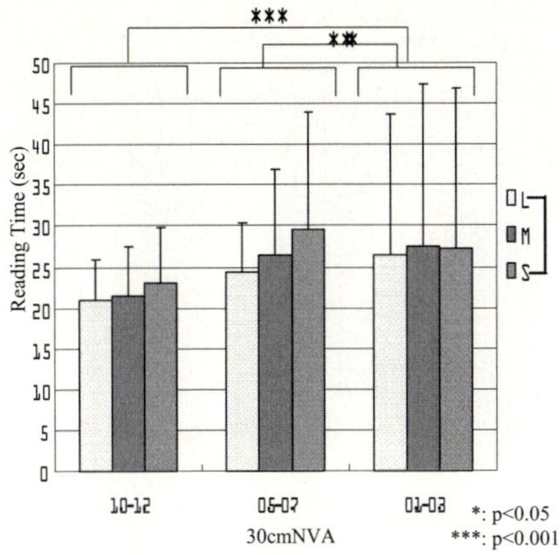

(a) RT by 30 cm NVA groups

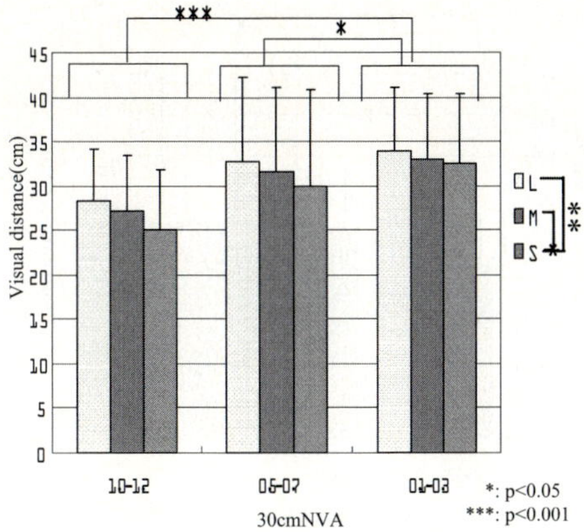

(b) VD by 30 cm NVA groups

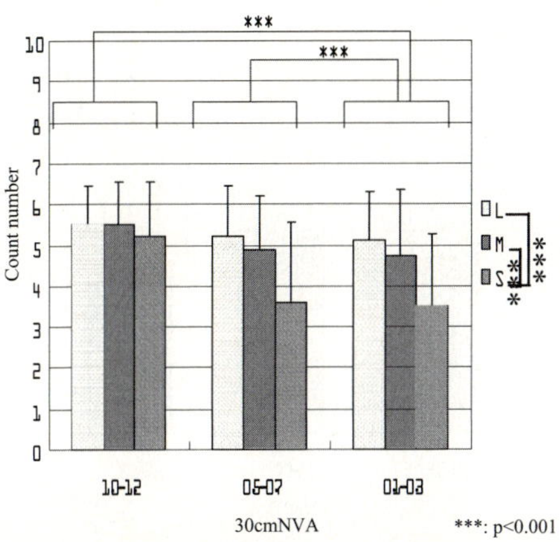

(c) CN by 30 cm NVA groups

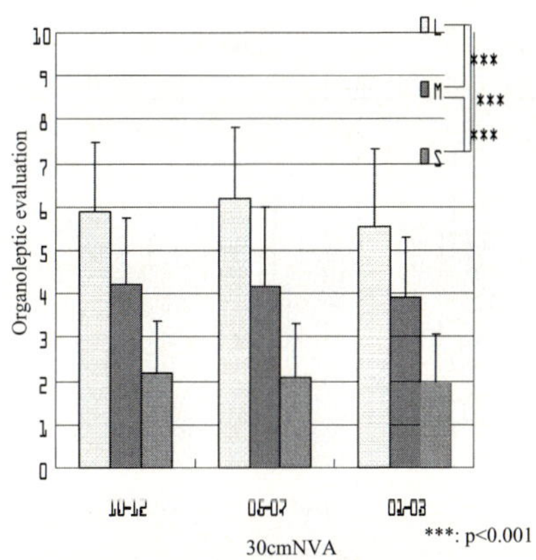

(d) OE by 30 cm NVA groups

Fig. 5. Relation between reading performance and three different character sizes by 30 cm NVA group

In the relationship between NVA and character size, a similarly tendency was found with the results of CC. In the low visual acuity group (0.1-0.3), RT was prolonged, VD was shorter, and Count decreased as character size became small. Thus, it is suggested that small characters and low visual function make text more difficult to see starting in middle age.

In OE and the relationship between visual function and character size, evaluations with character size "S" and "M" were below the average. However, character size "L" was the highest in the each group for CC and NVA. Character size "L" was high in the severe CC group, especially those with 200 or more. This suggests that character size needs to be "L" or larger.

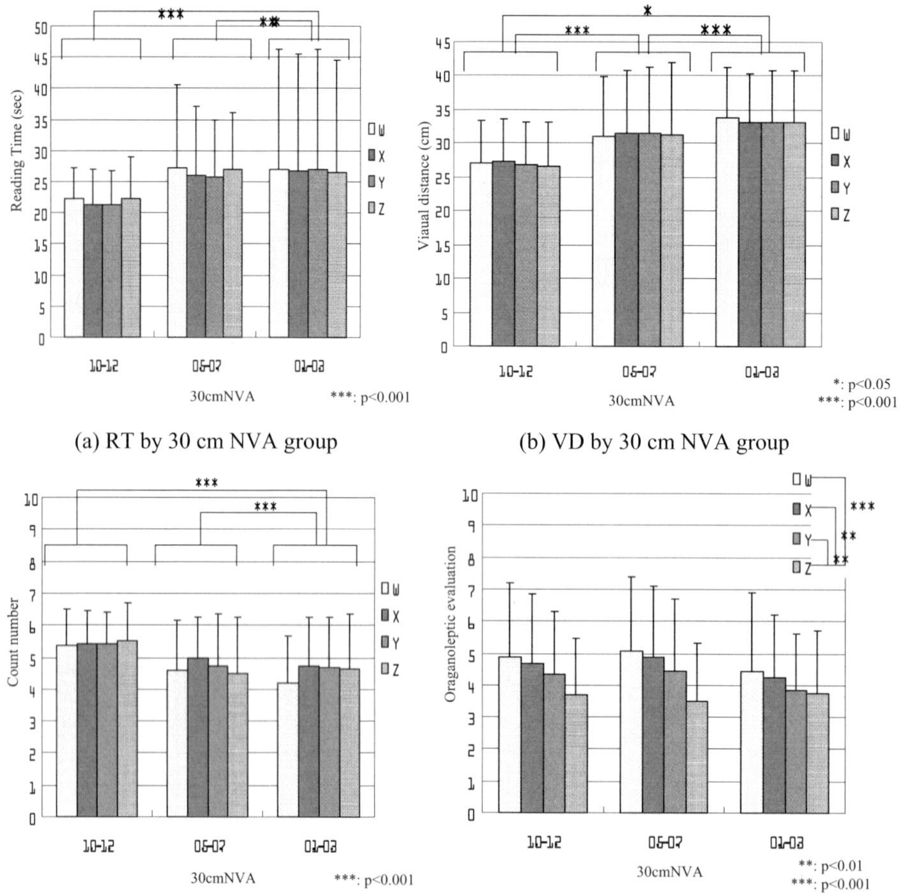

(a) RT by 30 cm NVA group

(b) VD by 30 cm NVA group

(c) CN by 30 cm NVA group

(d) OE by 30cm NVA group

Fig. 6. Relation between reading performance and four different contrasts by 30 cm NVA group

ISO 9241-3 (1992) [5] recommendations for character height on VDTs are that "the minimum character height shall be 16 minutes of arc and the maximum character height shall be 24 minutes of arc for tasks in which readability is important. Character heights of 20 to 22 minutes of arc are preferred for reading tasks." This is a preferred character height 3 mm (21 minutes of arc for a 50 cm visual distance) or more. Elderly people have low visual function, and their vision should be properly corrected for comfort. For this, vertical length larger than 3 mm is preferred [6]. In this experiment, character size "L" (2.5 x 1.3 mm) was not large enough, because it was smaller than the preferred size (3 x 3 mm). This result suggests that the visibility of character size "L" is lower as the visual function of MP users declines.

With respect to the influences of visual function and contrast on visibility, visual functions showed significant differences for RT, VD, and Count. However, Contrast showed significant differences only with OE. In the CC and NVA, the highest value was with the Negative display. The value was especially high in the group with severe CC of 200 or more. This suggests that negative displays have a beneficial effect for MP users increase with high CC.

5 Conclusion

In the relation between Character size, CC, and NVA, visibility was lower with smaller character size, "S". It was demonstrated that visibility is influenced by the deterioration of visual function. In the relation between Contrast, CC, and NVA, the organoleptic evaluations showed significant differences in contrast. The subjects' attached a high value to the negative display mode, especially when the subject's CC was more than 200. Such considerations are particularly important for elderly users who have marked deterioration of visual function. It is desirable that display settings such as character sizes and display mode can be easily adjusted, so that elderly users can adjust the displays to meet their individual needs.

References

1. Omori, M., Watanabe, T., Takai, J., Takada, H., Miyao, M.: Readability and characteristics of the mobile phones for elderly people. Behaviour & Information Technology 21, 313–316 (2002)
2. Hasegawa, S., Sato, K., Matsunuma, S., Miyao, M., Okamoto, K.: Multilingual disaster information system: Information delivery using graphic text for mobile phones. AI & Society 19(3), 265–278 (2005)
3. ISO 9241-3. Amendment 1 (2000). ISO 9241-3. Ergonomic requirements for office work with visual display terminals (VDTs) –Part 3: Visual display requirements. International Standard (1992)
4. Weale, RA.: Senile changes in visual acuity. Transactions of the Ophthalmological Societies of the United Kingdom 95(1), 36–38 (1975)
5. International organization for standardization, Ergonomic requirements for office work with visual display terminals (VDTs) (1992) ISO 9241-3
6. Miyao, M., Omori, M., Watanabe, T.: Usability and Characteristics of Mobile Phones for Elderly People. In: Proc. IEA 2003, pp. 560–563 (2003)

Integrated Multi-view Compensation for Real Sense Video Interfaces

Jongbin Park[1] and Byeungwoo Jeon[2]

School of Information and Communication Engineering
Sungkyunkwan University, Korea
{[1]feynmanc, [2]bjeon}@ece.skku.ac.kr

Abstract. Multi-view video is a new multimedia service which provides immersion and realism using the multiple view channels. However, as the number of cameras increases, the enormous data volume generated by them inevitably calls for better video compression and processing algorithms. However, the multiple images taken at any certain time instants always have inconsistency among the views in their intensity, color, sharpness, and so on. It not only degrades the compression efficiency greatly but also causes subjective quality loss such as dizziness and unnaturalness in random view access. To solve this problem, in this paper, we propose a new multi-view compensation method which uses the Kalman filter. It estimates the illumination discrepancy parameters autonomously both in encoder and decoder. Therefore, our method needs not transmit any additional data for illumination discrepancy information.

Keywords: Multi-view video coding, Illumination compensation, Inter-view prediction.

1 Introduction

Multi-view video is a ultimate human-oriented interface that will enhance the efficiency of "human-computer" or "human-computer-human" interaction by providing almost real sensation of "being present." Compared to the traditional stereoscopic video, the multi-view technology which employs N cameras provides much more accurate acquisition and more natural rendition of real world which is an essential part of most convenient human computer interaction [1]-[4]. In addition, it turns out to be a less expensive solution than the active 3D sensing system [1]-[6]. For these reasons, the multi-view problems have been recently received intense attention from the research community both in video processing and human-computer interaction fields [1]. However, as the number of cameras increases, the enormous data volume generated by them inevitably calls for better video compression and processing algorithms.

Most of the known approaches for multi-view coding are a direct extension of the conventional single view compression method into multiple camera settings [3],[7]. The multiple images taken at any certain time instants always have inconsistency among the views in their intensity, color, sharpness, and so on. It not only degrades the compression efficiency greatly but also causes subjective quality loss such as

M.J. Smith, G. Salvendy (Eds.): Human Interface, Part II, HCII 2007, LNCS 4558, pp. 608–617, 2007.

dizziness and unnaturalness especially in random view access [8],[9]. In this context, various researches are reported on using pre-processing to compensate the differences in intensity and/or color among different views [10]-[12].

The most simplest, but, yet an effective approach is to compensate the discrepancy based on the linear model, $Y = a \cdot X + b$, where X and Y respectively refer to any two different view images; and a and b are the compensating parameters of the discrepancy. The parameters, a and b, are estimated by minimizing the difference energy between X and Y [10]. To simply the process, one may further assume that the mean and the variance of X and Y should be the same [11].

It is noteworthy that the previous approach has at least two important problems. Firstly, due to occlusion or newly-appearing objects, there is a non-trivial possibility of erroneous pre-processing caused by wrong propagation of local intensity difference or statistically local anomaly into global compensation. Secondly, the compensation requires floating-point operation. Since a and b are, in general, floating-point numbers, the compensated pixel value is also a floating-point number, thus causing accuracy loss in forcing it to be an integer. When the pixel value is out of a pre-defined dynamic range, another inaccuracy comes from clipping [10]-[12]. In summary, once the multi-view images are compensated in their intensity or color, then it is impossible to have the bit-to-bit lossless original images back.

To compensate the illumination change in multi-view video coding, some researches were carried out [15]-[17]. The weighted prediction (WP) is a new video coding technique of H.264/AVC international video coding standard, known effective for encoding of fading video [13],[14]. The extended weighted prediction method is proposed for multi-view video coding [15]. Also developed is the illumination compensation method [16] which uses the mean removed sum of difference (MR-SAD) instead of sum of absolute difference (SAD) metric for the disparity estimation. The inter-view balanced disparity estimation (IBDE) method [17] based on the linear illumination difference model is also proposed.

It is worthwhile to note that the methods reviewed above all have to transmit additional data for illumination parameters, and this causes rate increment [15]. In this paper, we propose a new multi-view illumination change estimation and compensation method based on a Kalman filter which is known the optimum in filtering out the Gaussian random noises.

2 Inter View Prediction Method Using Kalman Predictor

In this section, we will show a model of illumination discrepancy in multiple cameras, and then propose a new multi-view compensation method which uses the Kalman filter which estimates the illumination discrepancy parameters autonomously in encoder and decoder.

2.1 Single Camera Response Model

Scenes of real-world objects are acquired as image sequences by image sensors such as charge-coupled device (CCD) or complementary metal-oxide-semiconductor

(CMOS). In the acquisition process, the images are corrupted by various noises such as fixed pattern noise, shot noise, white balancing noise, thermal noise, etc [19],[20]. Without including the shot or thermal noises in the model, the acquired response of single camera response is modeled by equation (1).

$$g(i,j)=\alpha\cdot f(i,j)+\beta \quad (i\in N_{rows}, j\in N_{cols}), \tag{1}$$

where $f(i,j)$ and $g(i,j)$ are respectively noiseless ideal real-world and acquired images. N_{rows} and N_{cols} are the number of rows and columns of the acquired image. α and β are the combined scaling factor and intensity offset [20]. To calculate the parameters (α , β) in the single camera response model in equation (1), some optimization methods are available [19],[20].

2.2 Illumination Discrepancy Model in Multiple Cameras

We can derive a model of illumination discrepancy in multiple cameras starting from equation (1). Equation (2) shows camera response model of current and reference images. The current and reference images refer to images obtained at the same time instance for the same scene, but different only from in their camera views.

$$\begin{aligned} g_{cur}(i,j)&=a_1\cdot f_{cur}(i,j)+b_1 \quad (i\in N_{rows}, j\in N_{cols}) \\ g_{ref}(i,j)&=a_2\cdot f_{ref}(i,j)+b_2 \quad (i\in N_{rows}, j\in N_{cols}), \end{aligned} \tag{2}$$

where $f_{cur}(i,j)$, $f_{ref}(i,j)$ are ideal noiseless current and reference view images. $g_{cur}(i,j)$, $g_{ref}(i,j)$ are respectively corresponding actual acquired images from different cameras. Now, we can derive equation (3) explaining relations between the current and reference view images.

$$\begin{aligned} g_{cur}(i,j)&=a_1\cdot f_{cur}(i,j)+b_1=a_1\cdot\left(f_{ref}(i-dy,j-dx)+N_{occ}+N_{app}\right)+b_1 \\ &=a_1\cdot f_{ref}(i-dy,j-dx)+a_1\cdot(N_{occ}+N_{app})+b_1 \\ &=a_1'\cdot f_{ref}(i-dy,j-dx)+b_1' \quad (i\in N_{rows}, j\in N_{cols}) \quad , \end{aligned} \tag{3}$$

where dx , dy are x- and y-components of the disparity vector. N_{occ} , N_{app} are noises caused by occlusion and newly appearing area.

2.3 Parameter Estimation Using Minimum Mean Squared Error Optimization

Equation (4) is derived from equations (2) and (3).

$$g_{cur}(i,j)=a\cdot g_{ref}(i-dy,j-dx)+b \quad (i\in N_{rows}, j\in N_{cols}), \tag{4}$$

where a and b are the compensating parameters of the discrepancy. a is a scaling factor and b is offset of illumination. They are estimated by minimizing the difference energy between $g_{cur}(i, j)$ and $g_{ref}(i-dy, j-dx)$ [10]. Equation (5) shows how to estimate the parameters under the least squared error criterion [10].

$$a = \frac{\text{cov}\{\mathbf{g}_{cur}(i,j), \mathbf{g}_{ref}(i-dx, j-dy)\}}{\text{var}\{\mathbf{g}_{ref}(i-dx, j-dy)\}}, \quad b = \mu_{cur} - a \cdot \mu_{ref}$$
$$\{\mu_{cur} = mean(\mathbf{g}_{cur}(i,j)), \quad \mu_{ref} = mean(\mathbf{g}_{ref}(i-dx, j-dy))\} \tag{5}$$

However, this method has to find the corresponding points between the current and the reference view images. It is a disadvantage.

2.4 Simple Parameter Estimation Under Wide Sense Stationary Assumption

It is possible to simplify the parameter estimation method under the wide sense stationary (WSS) assumption. Let's denote random variables by capitalizing corresponding variable names: for example, $f_{ref}(i, j) \rightarrow F_{ref}$, $f_{cur}(i, j) \rightarrow F_{cur}$, $g_{ref}(i, j) \rightarrow G_{ref}$, $g_{cur}(i, j) \rightarrow G_{cur}$. We assume that the means and variances of F_{cur} and F_{ref} should be the same. Under this stationary assumption, we can derive equation (6) from equation (4).

$$G_{cur} = a \cdot G_{ref} + b \tag{6}$$

Equation (7) shows that the new reference view (G^*_{ref}) from the current (G_{cur}) and the reference views (G_{ref}).

$$G^*_{ref} = a \cdot G_{ref} + b \tag{7}$$

The mean and standard deviation of new reference view (G^*_{ref}) in equation (7) are the same mean and the standard deviation of the current view (G_{cur}). Equation (8) is the derived compensating parameter results [11],[17].

$$a = \sqrt{Var(G_{cur}) / Var(G_{ref})}, \quad b = E(G_{cur}) - a \cdot E(G_{ref}) \tag{8}$$

2.5 Illumination Discrepancy Prediction Using Kalman Filter

Statistical difference between views depends both on camera inner parameters and outer environments. Fig.1 shows differences between view0 and view1 in terms of mean and standard deviation in two multi-view video sequences ("Race1", "Golf2"). In the "Race1" sequence, view0 and view1 are similar in their statistics. The small deviation shown in Fig.1 is due to subtle differences in camera parameters. In general, we can assume that the views at the same time instance mostly depend on camera inner parameters. However, in "Golf2" sequence, there is large deviation for the duration of frames 0 to 48. Visual inspection of the sequence reveals that a new object appears only in the view0 during the interval. This explains that the deviation among views also depends on outer camera environments such as occlusion or new objects appearing only in subset of views. We can assume that acquired image statistics from the same view strongly depends on outer camera environments. However, the assumptions we have made have uncertainty. We note that this uncertainty can be modeled by Gaussian random noise and can be processed by the Kalman filter [21] which is able to reconstruct noise-removed optimal output by repeatedly adjusting the discrepancy between actual observed output values and those predicted by the model.

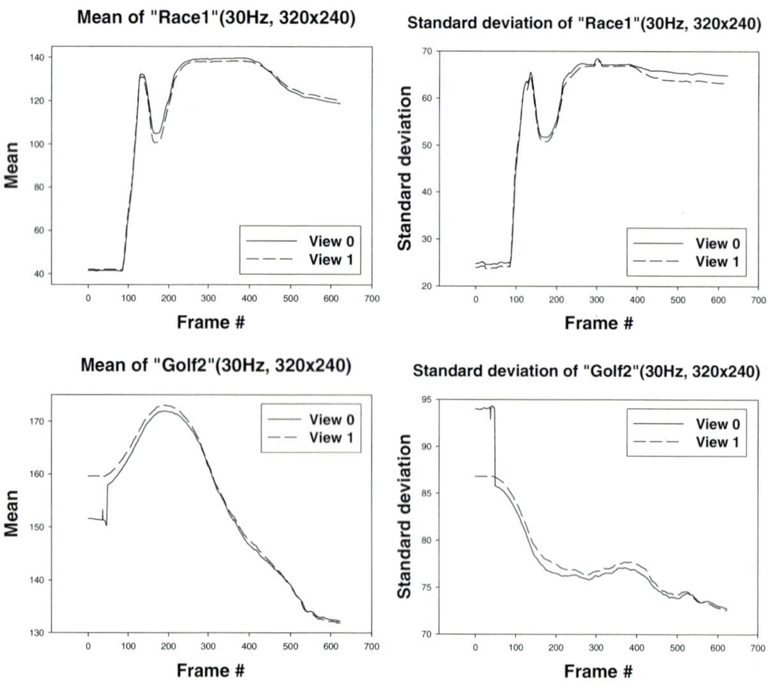

Fig. 1. Mean and standard deviation of frames of "Race1" and "Golf2" sequences

Equation (9) and (10) show the Kalman filter model, used in this paper, in a state space representation. It can be recursively updated by the system and measurement models shown in Table 1. Note that the Kalman filter is a recursive filer which does not require storing past measurements in the process of computing present estimates [21]. This is a clear advantage of no additional memory.

$$\mathbf{x}_t = \boldsymbol{\varphi}_{t-1} \cdot \mathbf{x}_{t-1} + \mathbf{w}_{t-1} = \begin{bmatrix} \boldsymbol{\mu}(t) \\ d\boldsymbol{\mu}(t) \end{bmatrix} = \begin{bmatrix} 1 & 0 \\ 0 & 1 \end{bmatrix} \cdot \begin{bmatrix} \boldsymbol{\mu}(t-1) \\ d\boldsymbol{\mu}(t-1) \end{bmatrix} + \begin{bmatrix} \mathbf{w}_{\mu}(t-1) \\ \mathbf{w}_{d\mu}(t-1) \end{bmatrix}, \mathbf{w}_t \sim N(\mathbf{0}, \mathbf{Q}_t), \quad (9)$$

where $\mathbf{x}_t$ represents the tth state, and $\boldsymbol{\mu}(t)$ denotes the mean of input image. $d\boldsymbol{\mu}(t)$ is the mean difference between the current and the reference views. $\mathbf{w}_{\mu}(t)$ and $\mathbf{w}_{d\mu}(t)$ are uncertainty factors for the system model. We assume that the uncertainty follows Gaussian random noise having zero mean and covariance matrix $\mathbf{Q}_t$.

$$\mathbf{z}_t = \mathbf{H}_t \cdot \mathbf{x}_t + \mathbf{v}_t = \begin{bmatrix} 1 & 0 \\ 0 & 1 \end{bmatrix} \cdot \begin{bmatrix} \boldsymbol{\mu}(t) \\ d\boldsymbol{\mu}(t) \end{bmatrix} + \begin{bmatrix} \mathbf{v}_{\mu}(t) \\ \mathbf{v}_{d\mu}(t) \end{bmatrix}, \mathbf{v}_t \sim N(\mathbf{0}, \mathbf{R}_t), \tag{10}$$

where $\mathbf{z}_t$ is the tth measurement data. $\mathbf{v}_{\mu}(t)$ and $\mathbf{v}_{d\mu}(t)$ are uncertainty factors for the measurement model. We also assume that the uncertainty in the measurement model is due to Gaussian random noise having zero mean and covariance matrix $\mathbf{R}_t$.

Table 1. Discrete Kalman filter [21]

* System model	$\mathbf{x}_k = \boldsymbol{\varphi}_{k-1} \cdot \mathbf{x}_{k-1} + \mathbf{w}_{k-1}, \quad \mathbf{w}_k \sim N(\mathbf{0}, \mathbf{Q}_k)$
* Measurement model	$\mathbf{z}_k = \mathbf{H}_k \cdot \mathbf{x}_k + \mathbf{v}_k, \quad \mathbf{v}_k \sim N(\mathbf{0}, \mathbf{R}_k)$
* Initial conditions	$E[\mathbf{x}(0)] = \overline{\mathbf{x}}_0, \quad E[(\mathbf{x}(0) - \overline{\mathbf{x}}_0) \cdot (\mathbf{x}(0) - \overline{\mathbf{x}}_0)^T] = \mathbf{P}_0$
* Other assumptions	$E[\mathbf{w}_k \cdot \mathbf{v}_j^T] = 0, \quad \text{for all } j, k$
* State prediction	$\overline{\mathbf{x}}_k(-) = \boldsymbol{\varphi}_{k-1} \cdot \overline{\mathbf{x}}_{k-1}(+)$
* Prediction error covariance	$\mathbf{P}_k(-) = \boldsymbol{\varphi}_{k-1} \cdot \mathbf{P}_{k-1}(+) \cdot \boldsymbol{\varphi}_{k-1}^T + \mathbf{Q}_{k-1}$
* State update	$\overline{\mathbf{x}}_k(+) = \overline{\mathbf{x}}_k(-) + \mathbf{K}_k \cdot (\mathbf{z}_k - \mathbf{H}_k \cdot \overline{\mathbf{x}}_k(-))$
* Error covariance update	$\mathbf{P}_k(+) = (\mathbf{I} - \mathbf{K}_k \cdot \mathbf{H}_k) \cdot \mathbf{P}_k(-)$
* Kalman gain matrix	$\mathbf{K}_k = \mathbf{P}_k(-) \cdot \mathbf{H}_k^T \cdot (\mathbf{H}_k \cdot \mathbf{P}_k(-) \cdot \mathbf{H}_k^T + \mathbf{R}_k)^{-1}$

2.6 Proposed Illumination-Compensated Inter-view Prediction

Fig.2 shows the proposed encoder and decoder for illumination compensated inter view prediction. Both of the encoder and decoder have illumination parameter estimator using Kalman filter. Since the Kalman filter estimates illumination parameter from reconstructed frames, there is no need to transmit the illumination parameter.

Fig. 3 illustrates how the illumination-compensated reference view is synthesized in the proposed method. The Kalman filter estimates the parameter deciding the intensity

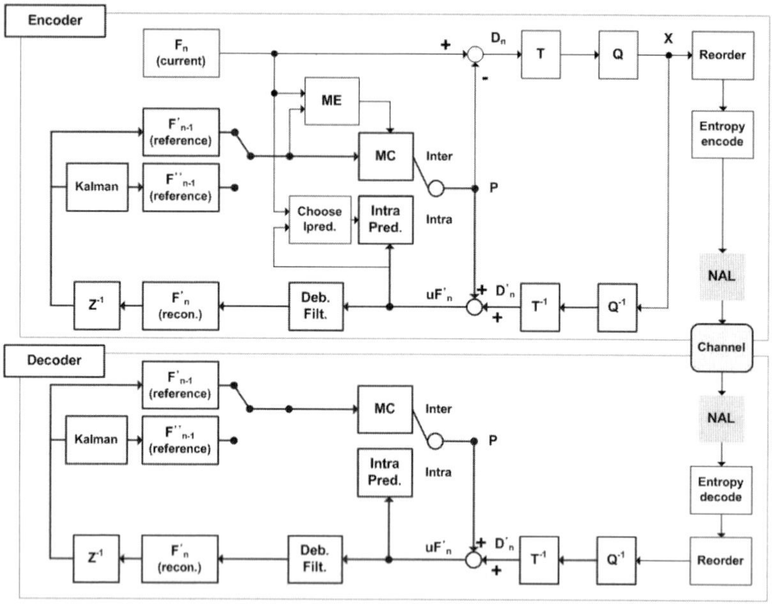

Fig. 2. Proposed encoder and decoder for illumination-compensated inter view prediction

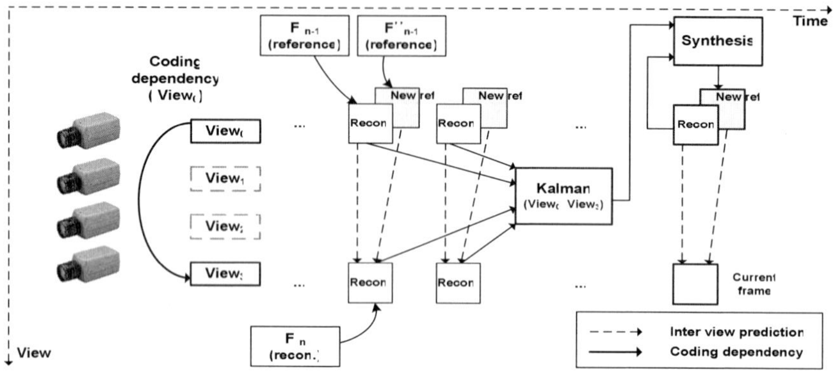

Fig. 3. Illumination-compensated reference frame synthesis using Kalman filter

discrepancies between views using the reconstructed frames. The illumination-compensated view is generated using the estimated parameter and the reference view image. This same process is performed both in encoder and decoder.

3 Experiment Results

To compare the proposed scheme with the conventional one, block based disparity estimation and MVC coding method are used. We use the test sequence of KDDI "Race1"(30Hz, 320x240) and "Golf2"(30Hz, 320x240).

3.1 Interview Prediction Performance

Fig. 4 shows disparity compensation result (exhaustive search, 16x16 block, search range [-24,24]). The graph denoted by "Ref. v0" is the squared disparity-compensated errors in terms of PSNR without illumination compensation. The graph indicated by "Ref. v0 Kalman_v0" is the same errors obtained after illumination compensation. In the experiment with the "Race1" and "Golf2" sequences, the proposed illumination compensation estimation scheme have shown better PSNR by up to 0.1~1.5dB. The lower disparity-compensated error means better coding performance.

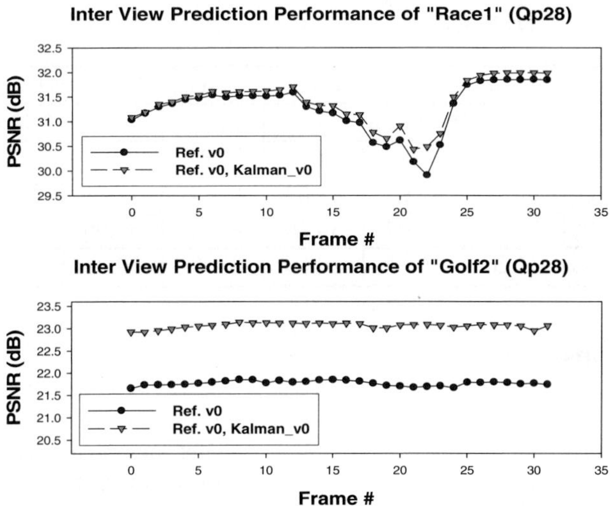

Fig. 4. Disparity compensation result (exhaustive search, 16x16 block, search range [-24, 24])

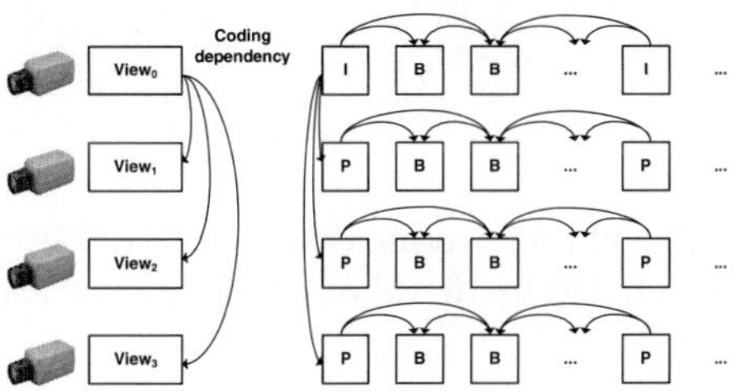

Encoding frames of view	128
I-frame period	12
The # of reference frames	12
Search range	+/- 32
Symbol mode	CABAC

Fig. 5. Prediction structure and encoding condition

3.2 Video Coding Performance

We perform the encoding experiment using JMVM2.1 reference software [18]. Fig. 5 shows encoding dependency and condition for this experiment. View1, 2, and 3 use the view0 for reference. As shown in Table 2 and 3, the proposed scheme obtains PSNR gain up to 0.03 dB.

Table 2. KDDI "Race1" (30Hz, 320x240) result

Qp	Without Illumination Compensation				With Illumination Compensation			
	Bit-rate [kbps]	PSNR [dB]			Bit-rate [kbps]	PSNR [dB]		
		Y	Cb	Cr		Y	Cb	Cr
20	367.50	41.58	43.51	44.26	364.15	41.56	43.55	44.26
28	129.36	36.93	39.48	41.15	127.77	36.94	39.50	41.11
36	51.57	32.47	36.95	38.52	50.85	32.50	36.96	38.63

Table 3. KDDI "Golf2" (30Hz, 320x240) result

Qp	Without Illumination Compensation				With Illumination Compensation			
	Bit-rate [kbps]	PSNR [dB]			Bit-rate [kbps]	PSNR [dB]		
		Y	Cb	Cr		Y	Cb	Cr
20	243.65	41.81	43.02	43.03	241.72	41.81	43.05	42.99
28	91.20	36.94	39.02	39.19	89.86	36.96	39.09	39.14
36	39.36	32.20	36.34	36.69	38.86	32.21	36.45	36.64

4 Conclusions

In this paper, we proposed a new multi-view compensation method which uses the Kalman filter. By letting the encoder and decoder autonomously estimate the illumination discrepancy parameters using the Kalman filter, the proposed scheme does not have any additional illumination discrepancy information to be transmitted for the illumination compensation. Our simulation result showed that the proposed scheme has better PSNR by up to 0.03 dB in the multi-view video coding (MVC) codec.

Ackwnowledgements

This work was supported by the Korea Science and Engineering Foundation (KOSEF) through the National Research Lab. Program funded by the Ministry of Science and Technology (No.M10600000286-06J0000-28610).

References

1. Smolic, A., Kauff, P.: Interactive 3D Video Representation and Coding Technologies. In: Proceedings of the IEEE, Special Issue on Advances in Video Coding and Delivery, vol. 93(1), pp. 98–110 (January 2005)
2. Requirements on Multi-view Video Coding, vol. 4, Poznan, Poland (July 2005) ISO/IEC JTC1/SC29/WG11 N7282
3. Survey of Algorithms used for Multi-view Video Coding, Hong Kong (January 2005) ISO/IEC JTC1/SC29/WG11 N6909

4. Matusik, W., Pfister, H.: 3D TV: A Scalable System for Real-Time Acquisition, Transmission, and Autostereoscopic Display of Dynamic Scenes. ACM Transactions on Graphics (TOG) SIGGRAPH 23, 814–824 (2004)
5. Brown, M., Majumder, A., Yang, R.: Camera-Based Calibration Techniques for Seamless Multiprojector Displays. IEEE Transactions on Visualization and Computer Graphics 11, 193–206 (2005)
6. Shum, H.-Y., He, S.B., Chan, S.-C.: Survey of lmage-Based Representations and Compression Techniques. IEEE Trans. CSVT 13(11), 1020–1037 (2003)
7. Ohm, J.R.: Encoding and reconstruction of multiview video objects: Looking at data compression in the context of the MPEG-4 multimedia standard. IEEE Signal Processing Mag. 16, 47–54 (1999)
8. Lim, J., Ngan, K., Yang, W., Sohn, K.: A Multiview Sequence CODEC with View Scalability. Signal Processing: Image Communication 19(3), 239–256 (2004)
9. Guo, X., Lu, Y., Gao, W., Huang, Q.: Viewpoint switching in multiview video streaming. Circuits and Systems, 2005. ISCAS 2005. In: IEEE International Symposium 4, 3471–3474 (2005)
10. Jeon, Y., Jeon, B.: Intensity Compensation for Efficient Stereo Image Compression. Journal of the Institute of Electronics Engineers of Korea, 101–112 (2005) 2005-42SP-2-11
11. Mancini, A.: Disparity Estimation and Intermediate View Reconstruction for Novel Applications in Stereoscopic Video, Master Thesis, McGill University (February 1998)
12. Kamikura, K., Watanabe, H., Kobayashi, N., Ichinose, S., Yasuda, H.: Video Coding Using Global Motion and Brightness-Variation Compensation with Complexity Reduction. Electronics and Communications in Japan 86(2), 80–93 (2002)
13. Boyce, J.: Weighted prediction in the H.264/MPEG4 AVC video coding standard. IEEE International Symposium on Circuits and Systems(ISCAS), pp. 789–792 (May 2004)
14. Yin, P., Tourapis, M.A., Boyce, J.: Localized weighted prediction for video coding. In: IEEE International Symposium on Circuits and Systems(ISCAS) 5, 4365–4368 (2005)
15. Illumination compensation for multi-view video coding, Montreux, Switzerland (April 2006) ISO/IEC JTC1/SC29/WG11 m13150
16. Joint Video Team (JVT) of ISO/IEC MPEG & ITU-T VCEG JVT-U052, CE11 : Illumination Compensation, 21st Meeting: Hangzhou, China (October 20-27, 2006)
17. Joint Video Team (JVT) of ISO/IEC MPEG & ITU-T VCEG JVT-T114, Results on CE2 using IBDE for multi-view video coding, 20th Meeting: Klagenfurt, Austria (July 15-21, 2006)
18. Muller, K.: Multi-View Coding Software, http://iphome.hhi.de/mueller/MVC_SW.htm
19. Healey, G., Kondepudy, R.: Radiometric CCD camera calibration and noise estimation. IEEE, Pattern Analysis and Machine Intelligence(PAMI) 16(3), 267–276 (1994)
20. Tsin, Y., Ramesh, V., Kanade, T.: Statistical calibration of CCD Imaging Process. In: IEEE International Conference on Computer Vision (ICCV'01) (July 2001)
21. Gelb, A.: Applied Optimal Estimation. MIT Press, Cambridge (1974)

Analysis of Naturalistic Driving Behavior While Approaching an Intersection and Implications for Route Guidance Presentation

Toshihisa Sato and Motoyuki Akamatsu

Institute for Human Science and Biomedical Engineering,
National Institute of Advanced Industrial Science and Technology (AIST)
Central-6 bidg., 1-1-1 Higashi, Tsukuba, Ibaraki 305-8566, Japan
toshihisa-sato@aist.go.jp,
akamatsu-m@aist.go.jp

Abstract. This study focuses on an analysis of the naturalistic driving behavior before making a right turn at an intersection. We conducted experiments on a public road and measured driver behavior, vehicle state, and headway and rear distances. The results suggest that the positions of the front and rear vehicles and the vehicle velocity have an influence on the onset location of covering the brake pedal. Structural equation modeling (SEM) was applied to estimate these relationships quantitatively. The results imply that the SEM with latent variables can represent the hypotheses obtained from the analysis of the measured data. We propose a detection method of unusual driver behavior by predicting the driver's preparatory behavior using the SEM, and possible new application of in-vehicle navigation systems is discussed.

Keywords: Mobile interactions, Route guidance information, Driver behavior, Naturalistic driving study, Structural equation modeling.

1 Introduction

In-vehicle navigation and communication systems are now popular all over the world. Drivers use route guidance information via the in-vehicle navigation systems while driving from the origin to the destination. The route guidance instruction helps drivers to choose and maintain the routes with lower mental workload. The auditory information on the route guidance is provided for drivers three times while approaching a target intersection. For example, "Please turn to the right at the intersection 700m ahead!", "Please turn to the right at the intersection 300m ahead!", and "Please turn to the right soon". These presentation timing is usually constant in any road traffic conditions or changes in correspondence to the vehicle velocity [1]. However, the presentation timing is too early or too late if driver's usual onset timing of the preparation for making a turn is later or earlier. It is essential to develop a human-centered system design, i.e. to develop a presentation timing adapted to the typical driver's preparatory behavior based on the investigation on natural driving behavior under real road traffic environments.

M.J. Smith, G. Salvendy (Eds.): Human Interface, Part II, HCII 2007, LNCS 4558, pp. 618–627, 2007.

In this study, we focused on the presentation timing of the voice instruction just prior to making the turn. This information can trigger a change in driving behavior from straight mode to preparation mode while approaching a target intersection.

1.1 Objectives and Approaches of This Study

The objectives of this study are as follows:

(a) Analyze naturalistic driving behavior before making a right turn at a specific intersection.
(b) Investigate the influence of traffic conditions around the target intersection on the onset location of the driver's preparatory behavior.
(c) Construct a driver model for describing the relationships between the driver behavior and the traffic conditions and for predicting the location where a driver begins preparation for making a right turn.
(d) Propose new application of in-vehicle navigation systems based on the behavior analysis and the driver model construction.

We conducted experiments using an instrumented vehicle on a public road in order to measure the natural driving behavior before making a right turn at an intersection. We focused on leading and following vehicles as the traffic conditions in the vicinity of the target intersection. In this study, the leading vehicle is defined as a forward vehicle that travels straight toward the target intersection. We did not analyze the traffic situations in which the forward vehicle turned to the right, as did the driver's own vehicle. The following vehicle is defined as a vehicle that follows the driver's vehicle in the same traffic lane, regardless of whether or not the following vehicle turns to the right at the turning point.

We used a structural equation modeling technique for investigating the impact of the traffic conditions on the onset location of the driver's preparatory behavior. The structural equation modeling is a modeling methodology depicting relationships among multiple variables hypothesized by theory and empirical research.

2 Experiment

2.1 Participants

Four drivers (three males and one female) participated in the experiments. The average age of the participants was 34.8 years (age range: 22 - 52 years). The average driving experience was 16.3 years (experience range: 3 - 33 years). All participants drove a passenger vehicle almost every day in their daily lives. They are not professional drivers, i.e. truck or taxi drivers.

2.2 Experimental Vehicle

Passenger vehicles with various sensors and a recorder system were developed to detect the vehicle driving state, including driving velocity, acceleration and geographical position, and to measure the driver's behavior, including accelerating, braking, and steering operations [2]. The relative distance and relative velocity to the leading and following vehicles were recorded with laser radar units that were fixed within the front and rear bumpers. The position of the driver's right foot (covering the accelerator or brake pedal without pressing) was measured by laser sensors fitted above the pedal surfaces. The turn signal activation was detected by adding encoders to the lever.

In summary, the following data sets were collected by the instrumented vehicles.

- Vehicle velocity, vehicle acceleration, vehicle angle, vehicle angular velocity
- Geographical position by a GPS sensor
- Relative distance and relative velocity to a leading or following vehicle
- Position of the right foot (covering the accelerator or brake pedal)
- Stroke of the accelerator and brake pedals
- Steering wheel angle, turn signal
- Visual images by CCD cameras (front view, rear view, right and left traffic lines, driver's face, driver's glancing view)

The data was recorded on a laptop computer and mobile hard disks via a driving recorder system. The driving recorder system was fixed inside the trunk of the experimental vehicle to encourage naturalistic driving behavior of the participants.

2.3 Experimental Design

We carried out repeated experiments using the four experimental vehicles on a public road in Tsukuba. The selected driving route was a 30-minute trip (total mileage: about 15 km) that included several left and right turns. The four participants started driving on the identical route per 10 minutes in order to measure each participant's driving behavior under similar traffic situation.

The total trips were 40 trials (over a period of about 8 weeks), and each participant made the recorded drives once a day on weekdays. Practice drives were made before the measurement trials so that the participants could drive from the origin to the destination without using a map. The participants were instructed to drive in their typical manner.

2.4 Target Intersection and Variables

Figure 1 shows an image of the analyzed intersection on the experiment route. The intersection has a designated lane for making a right turn. There is a long straight road about 2 km as far as the target intersection, which has two traffic lanes.

Driver preparatory behavior before making the right turn corresponds to activating the turn signal and deceleration. We focused on the foot movement to cover the brake pedal as the decelerating operations.

The behavioral events were quantified by using the remaining distance to the center of the target intersection (the stop line for turning to the right) at the onset of each preparatory maneuver. We also analyzed the vehicle velocity when the turn signal was activated and the brake pedal was covered by the driver's right foot, as well as the relative distances and speeds toward the leading and following vehicles before the driver entered the designated right-turn lane.

Fig. 1. Overview of target intersection

Within the total of 40 experimental trials for each participant, we did not use the data from the traffic conditions in which a lead vehicle also turned to the right at the target intersection or a vehicle had stopped at the stop line, and the data when there was a red light at the intersection. Valid data sets under appropriate conditions were about 30 trips for each participant.

3 Experimental Results

There were variations in the onset locations of covering the brake pedal and activating the turn signal over all of the experimental trials. Especially, large variations were found in the remaining distance to the center of the intersection when covering the brake pedal. It is considered that the onset location has an increasing or decreasing tendency when drivers are familiar with the experimental route. The results obtained from the long-term observation suggest that the adaptation to a driving route does not influence the onset location of the driver's preparatory behavior.

3.1 Relationship Between Vehicle Velocity and Driver's Preparatory Behavior

Figure 2 presents the results of the remaining distance to the center of the target intersection at the onset of each preparatory maneuver, based on the vehicle velocity when covering the brake pedal. The onset locations of covering the brake pedal are positively correlated with the driving speeds. When the driving velocity while approaching the target intersection is higher, the remaining distance when covering the brake pedal is longer. The slower driving speed leads to the closer onset location to the target intersection.

The onset locations of the turn signal activation while approaching the intersection at slower driving speeds tended to be closer to the center of the intersection than those at faster speeds. However, the decreasing tendencies were not remarkable compared to the results for covering the brake pedal. The drivers tended to activate the turn signal at a remaining distance of between 80m and 100m to the center of the target intersection, independent of the vehicle velocity.

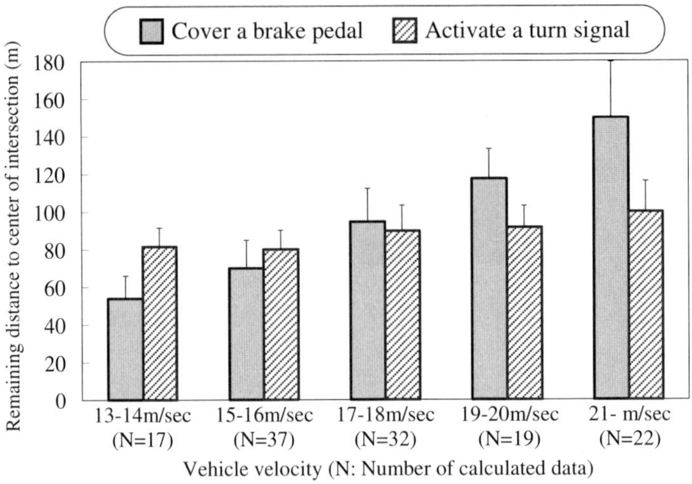

Fig. 2. Results of the onset location of each behavioral event based on the vehicle velocity when covering the brake pedal. The remaining distance has been calculated within all participants based on a classification of the driving speed. We adopted the driving speeds when the driver moved their right foot to cover the brake pedal before making a right turn, because the vehicle velocity at the onset of covering the brake pedal was almost equal to that at the onset of activating the turn signal.

3.2 Relationship Between Relative Distance to the Lead Vehicle and Driver's Preparatory Behavior

Figure 3 presents the results of the onset location of each behavioral event based on the relative distance to the leading vehicle, including the absence of a leading vehicle. The remaining distances to the center of the target intersection when covering the brake pedal were shorter compared to no leading vehicles. The shorter headway distance caused a closer onset location to the center of the intersection, indicating a linear relation between the decelerating point and relative distance to a lead vehicle.

The remaining distances when the turn signal was activated while approaching the target intersection with a leading vehicle were almost equal to those without the lead vehicle. The onset location of the turn signal operation was independent of the existence of and relative distance to a leading vehicle.

These tendencies are similar to the relationship between the vehicle velocity and the driver's preparatory behavior before making the right turn.

There were almost no differences in the relative speed between the driver's vehicle and the lead vehicle. The results imply that the drivers approach the turning point in a car-following condition.

3.3 Relationship Between Relative Distance to the Following Vehicle and Driver's Preparatory Behavior

Figure 4 presents the results of the remaining distance to the center of the intersection when each behavioral event occurs based on the relative distance to the following

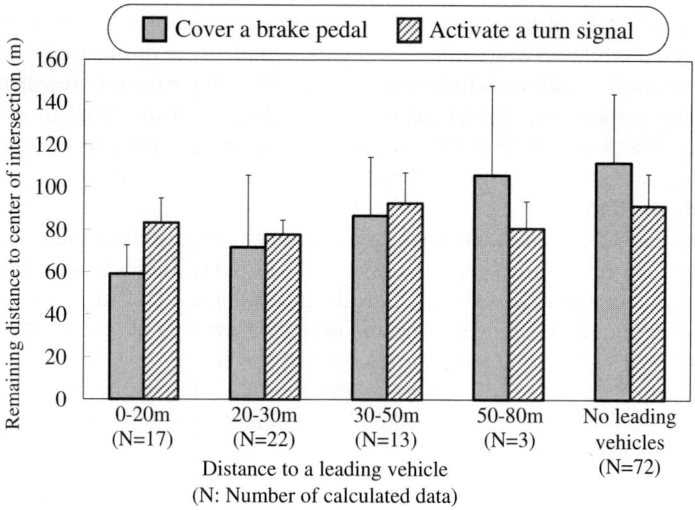

Fig. 3. Results of relationship between the onset location of each behavioral event and the relative distance to a leading vehicle. We have calculated the average remaining distance when each behavioral event occurred among the participants by classifying the relative distance to the leading vehicle and in the absence of a lead vehicle. The headway distance was categorized into four groups.

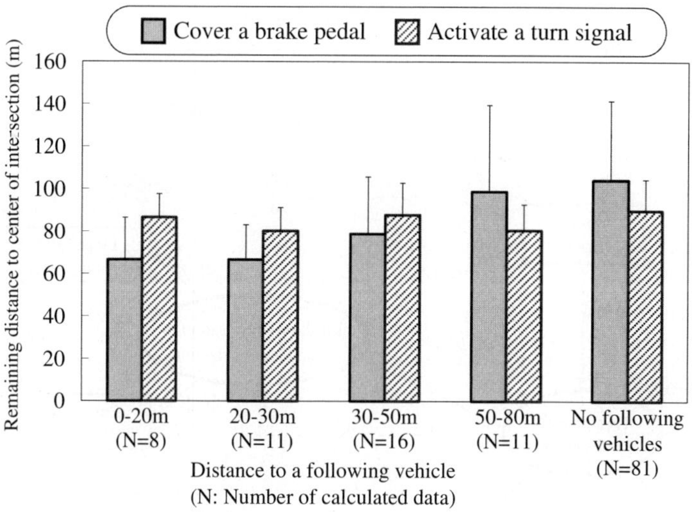

Fig. 4. Results of relationship between the onset location of each behavioral event and the relative distance to a following vehicle. We have calculated the average remaining distance when each behavioral event occurred among the participants, similar to the analyses based on the relative distance to a leading vehicle. The rear distance was categorized into the same four groups as the classification of the headway distance.

vehicle, including the absence of a following vehicle. The onset locations of covering the brake pedal while driving with a following vehicle were closer to the intersection compared to driving without a following vehicle. The closer the relative distance from the following vehicle, the shorter the remaining distance at the onset of covering the brake pedal. The results based on the relative distance to a following vehicle indicated similar tendencies to the relationship between the headway distance and the driver's decelerating operatoin.

The onset locations of activating the turn signal were almost stable in the five categories of the relation between the host and following vehicles. The experimental results in terms of the headway and rear distances suggest that the onset location of turn signal operation is almost constant and independent of the relative position between the host vehicle and a forward or following vehicle.

There were almost no differences in the relative velocity between the driver's vehicle and the following vehicle, which is similar to the relationship between the host and lead vehicles.

4 Modeling of Driver Preparatory Behavior Using Structural Equation Model

The observation results of the driver's preparatory behavior before making the right turn suggest that the vehicle velocity and the headway and rear distances influence the onset location of the foot movement to cover the brake pedal, and they have less influence on the onset location of the turn signal activation. We constructed a structural equation model using the measured data sets (vehicle velocity, relative distances to leading and following vehicles, and onset locations of covering the brake pedal and activating the turn signal) in order to estimate the observed relationships quantitatively.

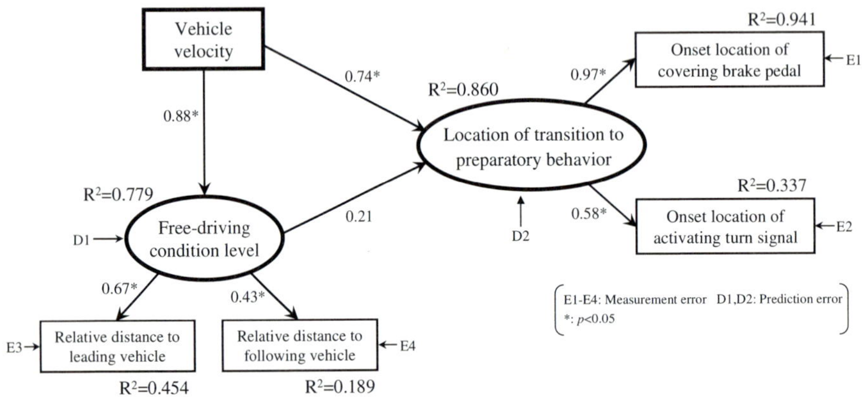

Fig. 5. Path diagram of the structural equation model and results of the estimation of path coefficients and factor loadings (standardized weights). The maximum likelihood method was used for the model estimation.

Figure 5 shows the structured model of driver preparatory behavior with estimated path coefficients and factor loadings. Two latent variables, the free-driving condition level and the location of transition to preparatory behavior, are input to the model specification. The free-driving condition level is defined using two indicator variables: the relative distance to the leading vehicle and the relative distance to the following vehicle, suggesting that shorter headway or rear distances mean close car-following conditions. The latter latent variable has two indicator variable: the onset location of covering the brake pedal and the onset location of activating the turn signal.

Table 1 presents the results of model fit indices. All model fit indices suggest an acceptable level of fit, supporting that the specified model depicts well the relationship among the driving speed, the free-driving condition level, and the location of transition to preparatory behavior.

Table 1. Goodness-of-fit indices for the proposed structured equation model. Non-significant chi-square value indicates little difference between the sample variance-covariance matrix and the reproduced variance-covariance matrix. GFI, AGFI and CFI more than 0.95 and RMSEA less than 0.05 reflect a good model fit [3].

χ^2 (df)	0.62 (3)
p value	0.892
GFI (Goodness-of-Fit Index)	0.998
AGFI (Adjusted Goodness-of-Fit Index)	0.990
CFI (Comparative Fit Index)	1.000
RMSEA (Root-mean-square error of approximation)	0.000

5 Implications for Route Guidance Presentation

5.1 Presentation Criteria of Route Guidance Information

The measurement of driver's preparatory behavior before making a right turn reveals that the remaining distances at the onset of the decelerating operation while driving without leading or following vehicles tend to be longer compared to the turn signal operation. On the other hand, the onset locations of the driver's decelerating behavior tend to be closer to the center of the target intersection compared to the turn signal activation when they drive with leading and following vehicles at close range, because the decelerating behavior is influenced by the relative position between the driver's vehicle and the forward or following vehicles.

It is effective to change the behavioral index based on the relative position to lead and following vehicles for the development of criteria of the presentation timing of route guidance information which helps drivers to begin preparation for making a right turn at a usual location. The information presentation criteria would be determined by measurement and accumulation of the driver's decelerating operation when driving without a leading or following vehicle and when approaching a target intersection under far car-following conditions. And the route guidance instruction would be provided for drivers at timing based on measurement and accumulation of the turn signal operation when driving with leading and/or following vehicles at close range. The voice instruction would be presented a few seconds earlier than the estimated criteria in order to accommodate the driver's response time to the provided information (see Fig. 6).

5.2 Detection of Unusual Driver Behavior Based on Prediction of Onset Location of Driver Preparatory Behavior

The specified structural equation model can predict the onset locations of right foot movement and turn signal activation based on the vehicle velocity and the headway and rear distances, as well as present the relationship among the driving speed, the traffic condition and the location of transition to preparation for making a right turn. The factor score for the free-driving condition level can be calculated by using the factor loadings for the relative distance to the lead vehicle and the relative distance to the following vehicle. The calculated factor score and the recorded vehicle velocity can predict the factor score for the location of transition to preparatory behavior, and the onset location of each preparatory maneuver can be inferred from the predicted factor score and each factor loading. Thus, the onset locations of covering the brake pedal and activating the turn signal can be predicted by using the measured variables and the estimated path coefficients and factor loadings of the structural equation model.

We compared the predicted onset locations and the observed values using the data sets which were not input to the model estimation. The prediction accuracy was the highest compared to predictions in which the vehicle velocity and the headway or rear distances are not taken into account or predictions using a single regression model with one independent variable, namely driving speed. The findings indicate that the specified structural equation model contributes to highly accurate prediction of when drivers move their right foot from the accelerator pedal to the brake pedal and activate the turn signal based on the driving speed and the traffic conditions while approaching the target intersection.

The prediction of the onset location of driver's preparatory maneuver can be applied to advanced functions of in-vehicle navigation systems. When the route guidance instruction is presented, the system linked with the sensors collecting driving speed and relative distances to the leading and following vehicles can predict the onset location of the driver's preparatory behavior using the specified structural equation model. If a driver does not begin to prepare to make a right turn after reaching the predicted onset location, the system can assess the driver's navigational error, i.e. the driver did not accurately identify the turning point, or did not notice the information provision, and then restate the route guidance or issue a warning. The detection method of the unusual behavior while approaching an intersection can be applied to measure a driver's fatigue and wakefulness level while driving and to reduce driver errors caused by temporary deterioration of the driver's internal states. Figure 6 shows the concept of the presentation criteria of route guidance information and the detection method of the driver's unusual behavior.

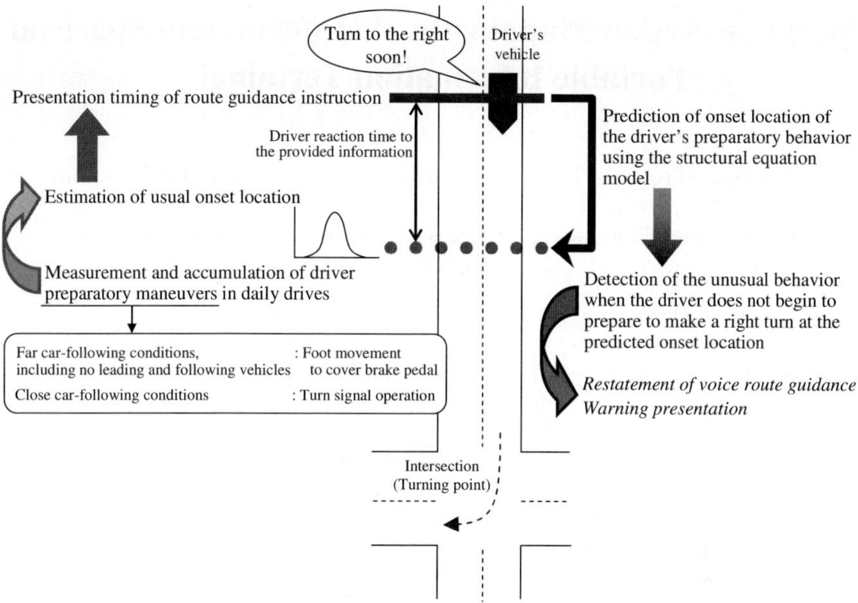

Fig. 6. Concept of the presentation criteria of route guidance instruction adapted to driver's usual preparatory behavior and the restatement method of the guidance based on the behavior prediction

References

1. Campbell, J.L., Carney, C., Kantowitz, B.H.: Human factors design guidelines for advanced traveler information systems (ATIS) and commercial vehicle operations (CVO). Technical Report No. FHWA-RD-98-057. U.S. Department of Transportation, Federal Highway Administration. (1998) Available from http://www.fhwa.gov/tfhrc/safety/pubs/atis/ch05/ch05.html
2. Akamatsu, M., Sakaguchi, Y., Okuwa, M.: Modeling of Driving Behavior When Approaching an Intersection Based on Measured Behavioral Data on an Actual Road. In: Proc. of the Human Factors and Ergonomics Society 47th Annual Meeting, pp. 1895–1899 (2003)
3. Schumacker, R.E., Lomax, R.G.: A Beginner's Guide to Structural Equation Modeling, pp. 79–122. Lawrence Erlbaum Associates, New Jersey (2004)

Operation-Action Mapping in 3D Information Space on Portable Information Terminal

Yu Shibuya, Hiromitsu Togeyama, Itaru Kuramoto, and Yoshihiro Tsujino

Kyoto Institute of Technology, Matsugasaki, Sakyo-ku, Kyoto 606-8585 Japan
shibuya@kit.ac.jp

Abstract. This paper focuses on finding the suitable mapping between the input action with the portable information terminal (PIT) in the real world and the operation in the 3D information space. From the experiments, following suitable mappings are found. They are changing the PIT position for changing the view position, changing the PIT orientation for changing the view orientation, and picking or grasping the object and moving or rotating hand for selecting the object and moving or rotating it.

Keywords: mobile interaction, portable information terminal, 3D information space, intuitive operation, human interface.

1 Introduction

In these days, many people carry their own portable information terminal (PIT), such as a PDA or a mobile phone. Some modern PITs have the power to process and display the 3D information space. The 3D information space is usable for various purposes, such as showing complex structures or representing some realistic objects. It should be useful to treat the 3D information space on the PIT but there are some restrictions because the PIT must be handheld. That is, it must be not so big and its output display size is limited. Furthermore, the number of input devices is limited too. On the other hand, it has an advantage that users can hand the PIT and easily change the orientation or position of it. Furthermore, the user can hold the PIT with one hand and use another hand freely. There are many possible input actions of the PIT but there have not been obvious description about the suitable mapping between the input action and the operation in the 3D information space. This paper focuses on finding the suitable mapping between the input action with the PIT in the real world and the operation in the information space.

2 Related Works

The Peephole system [1] relates physical movement of the PDA to logical movement in a virtual information space. The Rock 'n' Scroll input method handles scrolling, selection, and commanding an application based on tilting and gesturing with the

M.J. Smith, G. Salvendy (Eds.): Human Interface, Part II, HCII 2007, LNCS 4558, pp. 628–634, 2007.

device [2]. Its user experiments with a prototype system showed that users could quickly learn how to operate such devices. We have been developing intuitive navigation methods in the 3D information space using the PIT. In our previous study[3], a navigation method based on the "window metaphor" was proposed. With the method, the user can easily look around or navigate the 3D information space through the "virtual window", that is, the display of the PIT. For example, if users want to look at the left side of the window view, they can do it with moving their head to the right and looking into the window. If they come closer to the window, they can get the wider view of the outside.

3 Primitive Operations and Input Actions

In this paper, following five primitive operations in the 3D information space are treated. They are changing the view position, changing the view orientation, selecting the object, moving the object, and rotating the object. With these operations, users can explore the 3D information space, find their desired 3D information object, and browse the object. Other operations on object, such as annotation or modification, will be taken in future work.

In order to list the possible input actions, input devices of current ordinal PIT or mobile phone are taken into account. They are buttons, touch panels with stylus, gyro or acceleration sensors, and video cameras. A microphone is also usable but it might be used for voice conversation and is difficult to use in daily noisy environment, so it is not considered in this paper. With above input devices, possible input actions are pressing the button, tapping or dragging the stylus on the display, changing the position or orientation of the PIT, making the hand gesture, and changing the relative position of the user's face to the PIT. The hand gesture and face position can be detected by the image processing on the video captured image.

4 Experiment

4.1 Possible Combination

Experiments were made to find the suitable input action for each primitive operation. We firstly prepare the possible input actions for each operation in the 3D information space. They are shown in Table 1.

4.2 Prototype System for Experiment

The prototype system for experiment is shown in Fig. 1. It consists of a touch panel display, buttons, a motion tracker for tracking position and orientation of the PIT, and a data glove for detecting user's hand gesture. This prototype system is of no practical use. However, because we want to avoid the problems caused by the sensing accuracy, we decided to use above accurate sensors in the experiment.

Table 1. Possible input actions for each operation (x: candidate, -: non-candidate)

		Operation in the Information Space				
		Changing View Position	Changing View Orientation	Selecting Object	Moving Object	Rotating Object
Input Action	Pressing Button	x	x	x	x	x
	Tapping with Stylus	x	x	x	x	x
	Changing PIT Orientation	-	x	-	-	-
	Changing PIT Position	x	x	-	-	-
	Changing Orientation & Position of PIT	-	x	-	-	-
	Pointing Gesture with Finger	x	x	x	x	x
	Picking Gesture with Finger	x	x	x	x	x
	Grasping Gesture with Hand	x	x	x	x	x
	Opening Hand Gesture	x	x	x	x	x
	Changing Face Position	x	x	-	-	-

4.3 Participants and Procedure

Ten participants were recruited from our research laboratory. Five kinds of task were used for the experiment because there are five primitive operations as shown in Table 1.

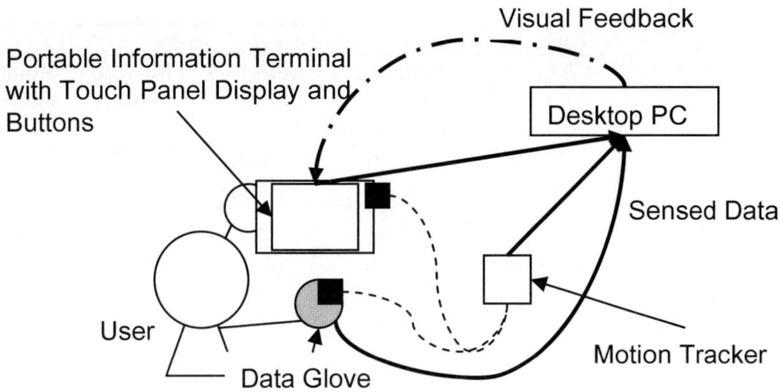

Fig. 1. System Configuration

Furthermore, as shown in Table 1, there were six to ten input actions for each operation respectively and a participant was asked to complete 36 tasks with assigned input actions.

4.4 Measures

Task completion time for each task was measured. Subjective evaluation was also made to find the intuitive operation-action mappings. There were following four questions for each task;

Q1 Was this input action intuitive and suitable for the operation? (1: definitely no – 7: definitely yes)
Q2 Please rate the easiness to learn. (1: very difficult – 7: very easy)
Q3 Please rate the fatigue of input action (1: very fatigue – 7: doesn't matter)
Q4 Was it fun to use? (1: dull – 7: fun)

4.5 Results

Task completion time of each task is shown in Table 2. In this table, with taking into account the subjective evaluation, the time of suitable input action for each operation is underlined. Detail results of subjective evaluation are shown in the appendix.

As shown in Table 2, suitable input actions for each operation are followings;

- changing the view position: changing PIT position, pointing gesture with finger, or grasping gesture with hand,
- changing the view orientation: changing PIT orientation, changing orientation and position of PIT, or pointing gesture with finger,
- selecting the object: picking gesture with finger or grasping gesture with hand,
- moving the object: picking gesture with finger or grasping gesture with hand,
- rotating the object: pointing gesture with finger, picking gesture with finger, or grasping gesture with hand.

Table 2. Task completion time ([sec]). Standard deviation is also shown in the parentheses.

		Task				
		Changing View Position	Changing View Orientation	Selecting Object	Moving Object	Rotating Object
Input Action	Pressing Button	32.68 (4.38)	18.70 (3.38)	36.13 (6.24)	36.46 (4.27)	12.78 (3.11)
	Tapping with Stylus	31.22 (6.09)	12.34 (4.19)	34.37 (4.72)	35.95 (3.20)	16.14 (3.83)
	Changing PIT Orientation	-	14.65 (1.82)	-	-	-
	Changing PIT Position	26.82 (3.39)	13.33 (1.38)	-	-	-
	Changing Orientation & Position of PIT	-	15.65 (2.08)	-	-	-
	Pointing Gesture with Finger	30.21 (7.94)	13.21 (1.40)	27.93 (6.97)	30.14 (5.33)	10.39 (2.62)
	Picking Gesture with Finger	31.04 (5.75)	13.51 (2.11)	25.16 (6.72)	29.51 (3.05)	9.63 (2.04)
	Grasping Gesture with Hand	28.13 (7.69)	13.32 (1.80)	23.68 (4.10)	31.68 (5.54)	9.65 (3.07)
	Opening Hand Gesture	32.91 (9.19)	13.69 (2.57)	27.01 (5.64)	31.80 (4.96)	9.63 (2.31)
	Changing Face Position	43.06 (6.72)	28.79 (3.63)	-	-	-

5 Discussion and Conclusion

In the above result, each primitive operation is mapped to some actions respectively. However, for practical operation in the 3D environment, multiple primitive operations are done concurrently. Furthermore, because users must select the object before move or rotate it, moving or rotating operation must follow the selecting operation. That is, the input action for selecting depends on the input action for moving or rotating. With taking into account of this concurrency and dependence, above suitable mapping between the operations in the 3D information space and the actions in the real world are followings;

- changing the view position: changing PIT position,
- changing the view orientation: changing PIT orientation,
- selecting and moving the object: picking or grasping the object and moving the hand,
- selecting and rotating the object: picking or grasping the object and rotating the hand.

In this paper, suitable operation-action mapping in the 3D information space on the portable information terminal is shown experimentally.

Acknowledgments. This research was supported by a grant-in-aid for scientific research (C) (18500096, 2006-2007) from Japan Society for the Promotion of Science (JSPS).

References

[1] Yee, K.: Peephole displays: pen interaction on spatially aware handheld computers. In: CHI '03: Proceedings of the conference on Human factors in computing systems, pp. 1–8 (2003)
[2] Bartlett, J.F.: Rock 'n' Scroll Is Here to Stay. IEEE Computer Graphics and Applications, pp. 40–45 (2000)
[3] Shibuya, Y., Narita, T., Yoshida, T., Kuramoto, I., Tsujino, Y.: An intuitive information space navigation method based on the window metaphor. Human-Computer Interaction: Theory and Practice (Part II). In: Proc. HCI International, vol. 2, pp. 278–282 (2003)

Appendix: Results of Subjective Evaluation

Followings are detail results of subjective evaluation in each task. The participants rated from 1 to 7 on each question. Here, "1" means bad and "7" means good.

Table 3. Subjective evaluation in changing view position task

		Q1	Q2	Q3	Q4
Input Action	Pressing Button	5.0	5.3	5.4	4.3
	Tapping with Stylus	4.8	5.2	5.0	4.4
	Changing PIT Position	5.7	4.9	3.5	5.8
	Pointing Gesture with Finger	5.3	4.2	3.5	5.2
	Picking Gesture with Finger	4.8	4.1	3.2	4.6
	Grasping Gesture with Hand	5.2	5.0	3.7	5.0
	Opening Hand Gesture	4.9	4.0	3.2	4.6
	Changing Face Position	4.1	2.2	2.1	3.7

Table 4. Subjective evaluation in changing view orientation task

		Q1	Q2	Q3	Q4
Input Action	Pressing Button	5.8	6.5	5.7	4.2
	Tapping with Stylus	5.5	6.3	5.9	4.8
	Changing PIT Orientation	5.4	5.9	4.2	4.9
	Changing PIT Position	5.8	5.9	4.6	4.9
	Changing Orientation & Position of PIT	6.1	5.9	4.4	5.3
	Pointing Gesture with Finger	6.2	6.1	4.4	5.7
	Picking Gesture with Finger	5.5	6.2	4.3	5.2
	Grasping Gesture with Hand	5.6	6.0	4.5	5.1
	Opening Hand Gesture	5.3	5.9	4.5	5.2
	Changing Face Position	3.6	1.5	2.0	2.5

Table 5. Subjective evaluation in selecting object task

		Q1	Q2	Q3	Q4
Input Action	Pressing Button	5.0	5.0	5.1	4.0
	Tapping with Stylus	5.1	6.1	5.6	4.4
	Pointing Gesture with Finger	4.9	4.9	3.7	5.3
	Picking Gesture with Finger	5.3	5.0	4.3	5.4
	Grasping Gesture with Hand	5.2	5.0	3.7	5.3
	Opening Hand Gesture	4.4	4.9	3.8	4.1

Table 6. Subjective evaluation in moving object task

		Q1	Q2	Q3	Q4
Input Action	Pressing Button	4.3	5.4	5.0	4.2
	Tapping with Stylus	4.7	5.3	5.6	5.1
	Pointing Gesture with Finger	5.1	5.2	4.5	5.6
	Picking Gesture with Finger	6.1	4.8	4.2	5.4
	Grasping Gesture with Hand	5.9	4.7	3.4	5.1
	Opening Hand Gesture	3.8	4.6	3.7	4.9

Table 7. Subjective evaluation in rotating object task

		Q1	Q2	Q3	Q4
Input Action	Pressing Button	5.0	5.5	5.6	4.7
	Tapping with Stylus	4.6	5.0	5.5	5.1
	Pointing Gesture with Finger	5.3	4.6	4.1	5.3
	Picking Gesture with Finger	5.4	4.4	4.0	5.5
	Grasping Gesture with Hand	5.4	4.2	4.3	5.1
	Opening Hand Gesture	4.8	4.6	4.4	5.1

Energy Efficient Route Discovery for Mobile HCI in Ad-Hoc Networks

Kwonseung Shin[1], Kwangjin Park[2], Min Young Chung[1], and Hyunseung Choo[1,*]

[1] School of Information and Communication Engineering, Sungkyunkwan University
440-746, Suwon, Korea +82-31-290-7145
{manics86,mychung,choo}@ece.skku.ac.kr
[2] Intelligent HCI Convergence Research Center
Sungkyunkwan University
440-746, Suwon, Korea
kjpark@disys.korea.ac.kr

Abstract. A mobile ad-hoc network [1] is a set of mobile nodes acting as routers in infrastructureless networking situations. In ad-hoc environments, to extend the lifetime of wireless mobile hosts, energy-efficient ad-hoc routing protocols must be designed. However, in conventional reactive ad-hoc routing, the route discovery operation is triggered whenever a source node has packets to send to a certain destination, but has no route information for the destination. Intermediate nodes, then, repeatedly broadcast the message until it is received by all nodes. This route discovery process could result in excessive drain of limited battery power and increase collisions in wireless transmission. In this paper, we propose an energy efficient route discovery scheme using the K-hop Pre Route Request (PRREQ) message which is only flooded within K hops from the source node. This reduces the number of nodes that participate in the route discovery process. Our empirical performance evaluation, which compares our proposed scheme to the conventional reactive adhoc routing schemes, demonstrates that proposed enhancement reduces energy consumed in the route discovery procedure by about 27% when each node initiates a new route at a rate of 0.05/sec.

Keywords: DSR, AODV, RREQ, RREP.

1 Introduction

In today's increasingly mobile computing world, people wish to be able to access a range of services at any time, at any place. However, the mobile computing environment is characterized by narrow network bandwidth and limited battery power. One of the greatest limitations is the finite power supplies inherent in mobile devices [2]. Accordingly,

* Corresponding author.

M.J. Smith, G. Salvendy (Eds.): Human Interface, Part II, HCII 2007, LNCS 4558, pp. 635–644, 2007.

numerous studies have focused on energy efficient wireless communications. Further, it has recently been recognized that networking techniques can also have a strong impact on the energy efficiency of such systems instead of the development of improved battery and low-power electronics [3]. Therefore, limited battery capacity of mobile host is one of the most challenging problems facing effective wireless communications.

An ad-hoc network is a set of wireless mobile hosts with the capability of cooperatively establishing communications independently. In ad-hoc environments, to extend the lifetime of wireless mobile ad-hoc hosts with limited battery capacity, an energy-efficient ad-hoc routing protocol must be designed. Ad-hoc routing protocols can be classified into two main categories, proactive [4-7] and reactive protocols [8-10]. In proactive routing protocols, node periodically exchanges routing information with its neighbors to consistently maintain valid routes on all destinations. Therefore, nodes are able to communicate with each other without the route discovery procedure. However, the exchange of control messages for maintaining all routes can consume large amounts of energy of nodes. In contrast, in reactive routing protocols, the operation of route discovery is triggered whenever a source node has packets and desires to send to a certain destination, but has no route information for the destination. Therefore, in general, reactive routing can be more energy efficient than proactive routing.

However, in traditional reactive routing protocols, when each node initiates a new route to a certain destination, it broadcasts the route request message to all nodes within its radio frequency coverage area. Then, all nodes within the area re-broadcast the message to their neighbor nodes. This process is repeated until the all nodes in a network receive the route request messages. This conventional route discovery scheme could lead to excessive drain of limited battery power of ad-hoc nodes, and increase collisions in wireless transmission. This energy is wasted in the overhead where the source node broadcasts route messages to an entire network, even when neighbor nodes are maintaining the route to the destination in its route cache, or the destination is close to the source node.

In this paper, an energy efficient route discovery using the K-hop Pre Route Request message which is flooded within K hops from an initiator, is proposed. This scheme reduces the number of broadcasting of route request messages in the route discovery process of reactive routing protocols. We demonstrate that the performance of the proposed scheme compared to the conventional reactive ad-hoc routing, is improved about 27% in terms of energy consumed in the route discovery process, and when each node initiates a new route at a rate of 0.05/sec. As the network traffic is increased, the relative efficiency of the proposed scheme becomes greater.

The rest of this paper is organized as follows. The section 2 presents traditional reactive routing protocols and policies of the route cache. The proposed energy efficient route discovery scheme is presented in section 3. In section 4, we evaluate the performance of our scheme, compared to conventional reactive routing schemes. Finally we conclude the paper in section 5.

2 Related Work

2.1 Reactive Routing Protocols for Ad-Hoc Networks

AODV. The Ad-hoc On Demand Distance Vector (AODV) routing protocol [10] uses a reactive approach for finding routes. Thus, a route is established for transmitting data, only when it is requested by the source node. When the source node wants to find a path to the destination node, it broadcasts a Route Request (RREQ) message.

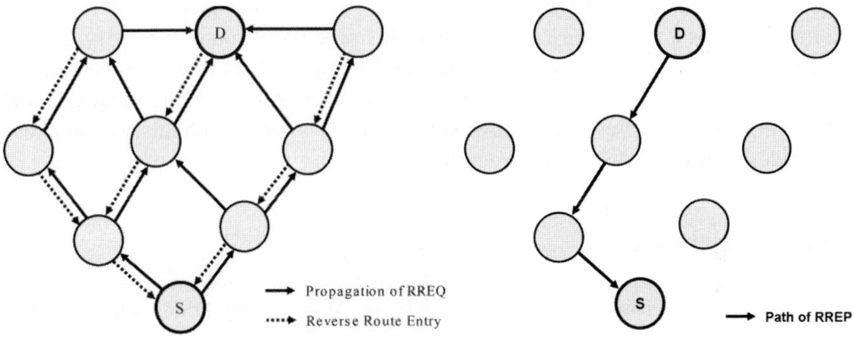

(a) Propagation of RREQ (b) Path of a RREP

Fig. 1. Path discovery mechanism of AODV

When an intermediate node receives a RREQ, it either forwards it on, or prepares a Route Reply(RREP) if it has a fresh route to the destination. The freshness of a route at the intermediate node is determined by comparing the destination sequence number in the RREQ. Intermediate nodes ignore the duplicate requests when the RREQ is received multiple times. The intermediate nodes set up a reverse path entry for the source node in its route table. In this way, the node knows where to forward a RREP to the source node, if a RREP is received later.

When the destination node eventually receives the RREQ, it creates a RREP and unicasts it toward the source node. When an intermediate node receives the RREP, it sets up a forward path entry to the destination in its route table. In this way, the forward path from source to destination node is established. Fig. 1 indicates the process of forwarding RREQ and RREP.

DSR. The Dynamic Source Routing (DSR) protocol [8] is based on a reactive approach to find routes by employing source routing. The source routing is a routing technique in which the sender of a packet determines the complete sequence of nodes through which to forward the packet. Therefore, each data packet carries the complete list of nodes through which the packet will pass, and the route cache maintains the information of routes which has the complete lists of nodes.

In DSR, when a source node sends a packet to destination node, the sender first checks its route cache for a source route to the destination. If a route is found, the sender uses this route to transmit the packet. If no route is found, the sender broadcasts a route request packet to discover a route to the destination node. Each route request packet contains the source address, the destination address, and a route record which contains the address of intermediate nodes that the route request packet has been forwarded. The nodes that receive a route request packet, if it is the target of the Route Discovery, or if it has a route to the target of the request in its own route cache, it returns a Route Reply to the initiator, giving a complete path from the source to destination. Otherwise, the intermediate nodes that receive the route request packet append its address to the route record in the route request packet, and re-broadcast the request. Additionally, when the route request packet has been received multiple times, the intermediate nodes discard the route request packet. The initiator of the route request that receives a route reply according to this process of route discovery, caches the route in its route cache for use in sending subsequent packets to the destination. In DSR, possible optimization such as using non-propagating route request and expanding ring search is presented. However, employing such mechanisms can cause increase of delay of route discovery process.

3 Proposed Scheme

In traditional reactive routing protocols, such as AODV and DSR, source nodes broadcast route request messages for route discovery. Then, all nodes in the ad hoc network participate in the route discovery process re-broadcasting the route request messages. Obviously, overhead is created, consuming energy when route messages are broadcasted to the entire network, even in the case neighbor nodes maintain the route to the destination in their route cache, or the destination is close to the source node. The frequent route discovery process causes not only the delay of route discovery and data transmission time, but also consumes the energy of nodes. In this section, an energy efficient route discovery scheme using the K-hop Pre Route Request (K-hop PRREQ) message is proposed. This reduces the number of broadcasting of route request messages in the route discovery process of reactive routing protocols.

3.1 Basic Mechanism

When the source node initiates a route to the destination node, it broadcasts K-hop PRREQ messages with the Time To Live (TTL) field set at K, while the conventional reactive routing protocols broadcast route request messages to the entire network. The intermediate nodes that receive the K-hop PRREQ, checks its route cache for a route to the destination. If a route is found, the intermediate nodes send back the RREP messages to the initiator of the K-hop PRREQ through the reverse path. If no route is found, the intermediate node re-broadcast K-hop PRREQ messages by decreasing the TTL value. Intermediate nodes discard the K-hop PRREQ when it receives the same K-hop PRREQ multiple times or the TTL value becomes 0.

The initiator that sends the PRREQ message waits a fixed time T, for receiving RREP messages from K-hop neighbor nodes. If the initiator receives several RREP messages in time T, it sends the data packet through the path that has the shortest hop count. If the initiator does not receive RREP messages for the K-hop PRREQ in time T, after sending the PRREQ, it attempts to discover a route to the destination using conventional route discovery protocol in which the source node broadcasts the route request message to the entire network.

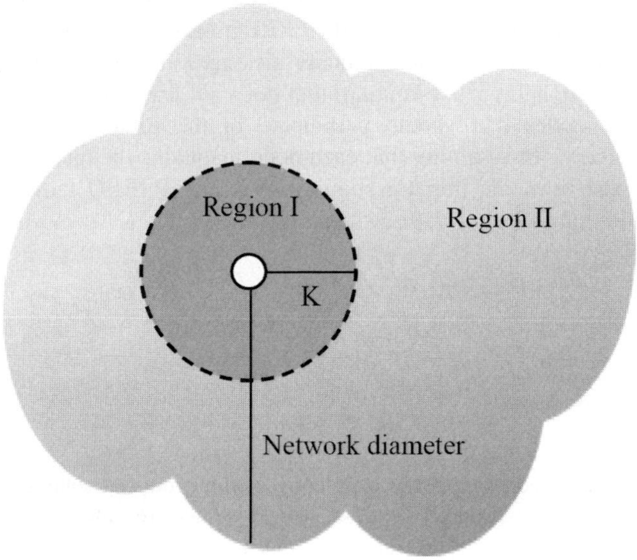

Fig. 2. Flood region of K-hop PRREQ in ad-hoc network

Let us illustrate with the example of Fig. 2. In the case that the source node successfully receive RREP messages for K-hop PRREQ in time T, as the K-hop PRREQ messages are flooded to nodes in the region I of the Fig. 2, the nodes in region II can save energy, which is required for participating in the route discovery process. Additionally, in the case that the hit ratio of the K-hop PRREQ is high, collisions in wireless transmission are decreased. Therefore, using k-hop PRREQ messages reduces not only the energy consuming process of route discovery but also delay of route discovery and data transmission time compared to conventional reactive routing protocols.

However, in the case that the initiator of the K-hop PRREQ does not receive RREP messages in time T, as it has to use the traditional route discovery process, the additional energy is needed to flood the K-hop PRREQ in region I of Fig. 2, compared conventional reactive routing. Moreover, it takes an additional T time to discover a route to a destination node, compared to traditional reactive routing protocols.

3.2 Enhancement

When there are many active routes in the ad-hoc networks, the hit ratio of K-hop PRREQ is increased, thereby raising energy efficiency. However, as mentioned in section 3.1, using the K-hop PRREQ message may cause the nodes consume more energy in some cases. Therefore, it is important to determine whether the initiator should use K-hop PRREQ message. In Fig. 2, let us assume that the nodes are distributed uniformly in an ad-hoc network, and D is the value of the default TTL of conventional reactive routing protocols. Then, when the PRREQ message is hit, the number of nodes that should participate in the broadcasting process of route discovery is proportional to πK^2. In the case that the PRREQ message is not hit, the number of nodes that should flood the route request messages is proportional to $\pi K^2 + \pi D^2$. Similarly, in traditional reactive routing, that does not use K-hop PRREQ messages, the number of nodes that should participate in the route discovery process is proportional to πD^2. In assuming that each node maintains the hit ratio of the K-hop PRREQ in fixed time, the initiator should use K-hop PRREQ messages under the following condition.

$$\pi K^2 p + (\pi K^2 + \pi D^2)(1-p) < \pi D^2 \tag{1}$$

That is,

$$\frac{K^2}{D^2} < p \tag{2}$$

As mentioned in section 3.1, the efficiency of the proposed scheme relies heavily on the hit ratio of K-hop PRREQ messages. [11] has presented an analysis of the effects of different design choices in caching strategies for reactive routing protocols in wireless ad hoc networks. As the proposed scheme uses route cache of neighbors within K hops, the hit ratio of K-hop PRREQ messages depends on caching strategy (i.e., choices of cache structure and cache replacement strategies based on the cache capacity and cache timeout) in each nodes. In our scheme, the proper caching strategy for the different network environments is applicable for enhancing the performance. In the environments that network traffic is concentrated on several nodes, for a example, caching policy that maintains the route which has been referenced from itself or neighbors frequently in certain time can improve the hit ratio of K-hop PRREQ compared to the policy which discard the route entries simply depending on the time when the route is obtained.

4 Performance Evaluation

To evaluate the efficiency of proposed scheme, we demonstrate the hit ratio of K-hop PRREQ and the number of nodes which participate in the route discovery process according to the fraction of active routes in network. In addition, we presented the result of energy consumed in the route discovery process as the number of flooding of route request messages. The proposed scheme is applied in AODV and DSR and implemented in C. 100 different networks are generated for 50 and 100 nodes, which are placed randomly with a density of 0.003 nodes/m^2. The transmission radius is set to 30m.

Random networks are the accepted model for explaining different kinds of networks. There are many algorithms and programs, however, the speed is usually the main goal, not the statistical properties. In the last decade the problem was researched by Waxman [12], Doar [13], Toh [14], Calvert et al. [15], and Kumar et al. [16]. They developed fast algorithms to generate random graphs with different properties, similar to real communication networks. However, none of them discussed the stochastic properties of generated random graphs. Rodionov and Choo [17] have formulated two major demands to the generators of random graph: attainability of all graphs with required properties and the uniformity of their distribution. If the second demand is sometimes difficult to prove theoretically, it is possible to check the distribution statistically. The random graph is similar to that used in real networks. The random networks used in this paper are generated by modifying Rodionov and Choo's mechanism to include the properties of ad-hoc networks.

Fig. 3 shows the results for the variation of hit ratio of K-hop PRREQ messages according to increasing the number of active routes in the network. In AODV, intermediate nodes do not maintain the complete list of nodes in the route cache. However, as DSR uses the source routing technique, the hit ratio of K-hop PPREQ in the proposed scheme applied in DSR is higher than that applied in AODV. In the certain range, when the value of K is 2, the hit ratio is much higher than when the value of K is 1. However, when the number of active route in the network is small or very large, the gap is relatively small.

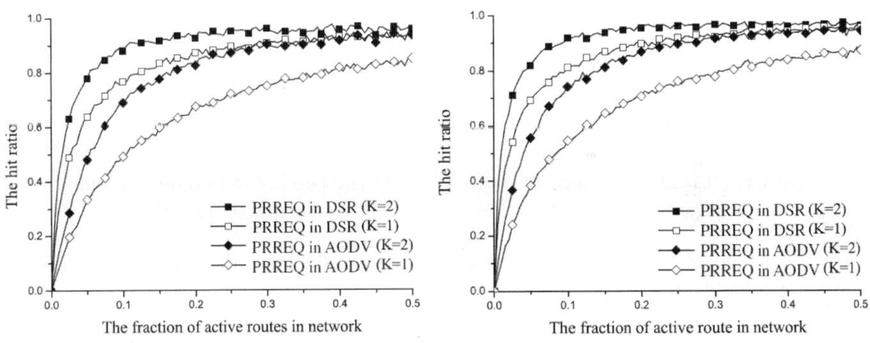

(a) Network with 50 nodes (b) network with 100 nodes

Fig. 3. The hit ratio of K-hop PRREQ according to the fraction of active routes in network

Fig. 4 presents the number of nodes that participate in the flooding of route request message when a node attempts to initiate a new route to the destination. When the number of active route is very small, using the K-hop PREEQ message is not efficient, considering the energy consumption of nodes in network. Using the 2-hop PRREQ message is more efficient than the 1-hop RREQ message only for a certain range.

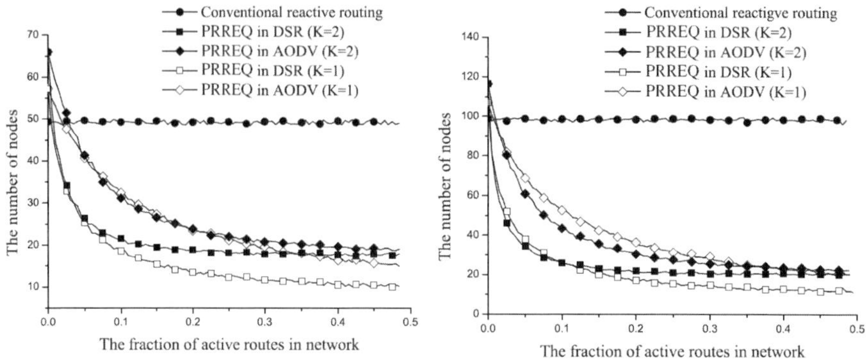

(a) Network with 50 nodes (b) network with 100 nodes

Fig. 4. The number of nodes which participate in the route discovery process according to the fraction of active in network

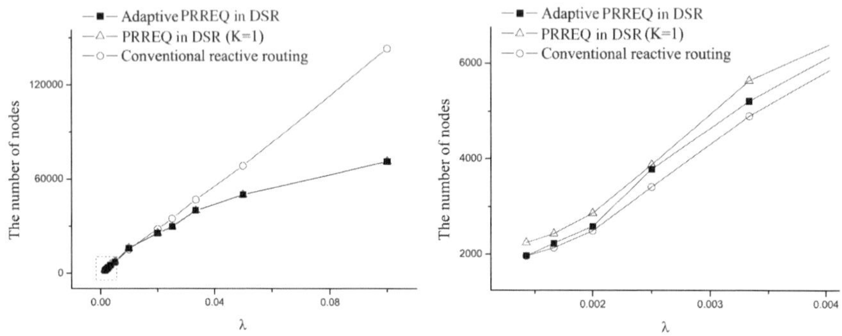

(a) The number of flooding of route request at (0.01, 0.1) (b) The number of flooding of route request at (0.01, 0.035)

Fig. 5. The number of flooding of route request messages in the network within 600 sec according to variation of route initiating rate at each node

Fig. 5 shows the number of flooding of route request messages in the network, over a period of 600 seconds. Each node initiate a new route at a rate of λ. It is assumed that event interval follows an exponential distribution. Fig. 5(b) represents the dotted rectangle area of Fig. 5(a). When network traffic is very low, the proposed scheme is inefficient compared with traditional reactive routing. The adaptive scheme described in section 3.2 is able to improve the efficiency of the proposed scheme using K-hop RREQ messages in fixed way. The simulation results represented in Fig. 5(a) demonstrate that the energy consumed in the route discovery process is reduced approximately 27% compared to conventional reactive routing when the each node initiate a new route at a rate of 0.05/sec. In addition, the network traffic is higher; and the efficiency of the proposed scheme is significantly increased.

5 Conclusion

In this paper, we have proposed an energy efficient route discovery scheme. In traditional reactive routing protocols, the route request messages are flooded to all nodes in the network when a source node initiates new route. This route discovery process causes an increase of energy consumption, as well as a delay in route discovery and data transmission.

The proposed scheme uses K-hop PRREQ message, which is only flooded within K hop neighbors from the source node, thereby reducing the number of nodes that participate in the route discovery process. Performance evaluation demonstrates the efficiency of the proposed scheme compared to that of conventional reactive routing protocols.

Acknowledgments

This research was supported by the MIC(Ministry of Information and Communication), Korea, under the ITRC(Information Technology Research Center) support program supervised by the IITA(Institute of Information Technology Assessment), IITA-2006-(C1090-0603-0046).

References

1. Corson, S., Macker, J.: Mobile Ad hoc Networking (MANET): Routing Protocol Performance Issues and Evaluation Considerations. Network Working Group, Request for Comments: 2501 (1999) http://www.ietf.org/rfc/rfc2501.txt
2. Jones, C.E., Sivalingam, K.M., Agrawal, P., Chen, J.C.: A Survey of Energy Efficient Network Protocols for Wireless Networks. Wireless Networks 7(4), 343–358 (2001)
3. Wieselthier, J.E., Nguyen, G.D., Ephremides, A.: Energy-Efficient Broadcast and Multicast Trees in Wireless Networks. Mobile Networks and Applications 7(6), 481–492 (2002)
4. Ogier, R., Templin, F., Lewis, M.: Topology broadcast based on reverse-path forwarding routing protocol (TBRF). Network Working Group, Request for Comments: 3684 (2004) http://www.ietf.org/rfc/rfc3684.txt
5. Chiang, C.-C.: Routing in clustered multihop mobile wireless networks with fading channel. In: Proceedings of IEEE SICON, pp. 197–211 (1997)
6. Clausen, E.T., Jacquet, E.P.: Optimised Link State Routing Protocol (OLSR). Network Working Group, Request for Comments: 3626 (2003) http://www.ietf.org/rfc/rfc3626.txt
7. Villasenor-Gonzalez, L., Ge, Y., Lamont, L.: HOLSR: a hierarchical proactive routing mechanism for mobile ad hoc networks. IEEE Communications Magazine 43(7), 118–125 (2005)
8. Johnson, D.B., Maltz, D.A., Hu, Y.-C.: The Dynamic Source Routing Protocol for Mobile Ad Hoc Networks (DSR). IETF MANET Working Group, INTERNET-DRAFT (2004) http://www.ietf.org/internet-drafts/draft-ietf-manet-dsr-10.txt
9. Ko, Y.-B., Vaidya, N.H.: Location-aided routing (LAR) in mobile ad hoc networks. In: Proceedings of the Fourth Annual ACM/IEEE International Conference on Mobile Computing and Networking (Mobicom'98) (1998)

10. Perkins, C., Belding-Royer, E., Das, S.: Ad hoc On-Demand Distance Vector (AODV) Routing. Network Working Group, Request for Comments: 3561 (2003) http://www.ietf.org/rfc/rfc3561.txt
11. Hu, Y.-C., Johnson, D.B.: Caching strategies in on-demand routing protocols for wireless ad hoc networks. MOBICOM 2000, pp. 231–242 (2000)
12. Waxman, B.M.: Routing of Multipoint Connections. IEEE JSAC 9, 1617–1622 (1993)
13. Doar, M.: A Better Mode for Generating Test Networks. In: IEEE Proc. GLOBECOM'96, pp. 86–93 (1996)
14. Toh, C.-K.: Performance Evaluation of Crossover Switch Discovery Algorithms for Wireless ATM LANs. In: IEEE Proc. INFOCOM'96, pp. 1380–1387 (1993)
15. Calvert, K.L., Doar, M., Doar, M.: Modelling Internet Topology. IEEE Communications Magazine, pp. 160–163 (June 1997)
16. Kumar, R., Raghavan, P., Rajagopalan, S., Sivakumar, D., Tomkins, A., Upfal, E.: Stochastic models for the Web graph. In: Proc. 41st
17. Rodionov, A.S., Hyunseung, C.: On Generating Random Network Structures: Connected Graphs. In: Kahng, H.-K., Goto, S. (eds.) ICOIN 2004. LNCS, vol. 3090, Springer, Heidelberg (2004)

Interaction Design and Implementation for Multimodal Mobile Semantic Web Interfaces

Daniel Sonntag

German Research Center for Artificial Intelligence
66123 Saarbrücken, Germany
sonntag@dfki.de

Abstract. Multimodal mobile interfaces to the Semantic Web offer access to complex knowledge and information structures. In SmartWeb we try to build multimodal interfaces to answer very specific closed and open domain questions by natural language dialogue and multimedia presentations. Advanced user interactions such as pointing gestures are also supported. We present the interaction design and its implementation for Semantic Web related knowledge structures, i.e., ontology instances and relations, and follow the principle of no presentation without representation for information content, its presentation, and interaction possibilities. In particular, we address the challenge to display interactive text and image results obtained from multiple homogeneous and heterogeneous information sources.

Keywords: Multimodal Dialogue Systems, Interaction Design.

1 Introduction

In dialogue systems dialogue managers are accountable for resolving linguistic phenomena such as anaphoric references and ellipses, handling topic and focus, keeping track of dialogue state and interaction, and planning next dialogue moves. In multimodal dialogue systems multimodal dialogue acts and multimodal interaction have to be taken into account, too. Not surprisingly interaction and presentation behaviour plays a major role in multimodal dialogue systems, since the desired intention is presented in different modalities, or the information content comprises of related multimedia artefacts. In SmartWeb [1] we try to build multimodal interfaces to the Semantic Web, to answer very specific closed and open domain questions by natural language dialogue and display multimedia presentations. Our primary goal in SmartWeb is to support multimodal interactions and multimedia presentations on portable mobile devices, such as a PDA, to answer natural language questions about football events, players, matches, and so on. The application scenario we report on in this contribution is a visit to a football game at the Football World Championship 2006. See figure 1 for an example question and the multimedia result.

M.J. Smith, G. Salvendy (Eds.): Human Interface, Part II, HCII 2007, LNCS 4558, pp. 645 – 654, 2007.

Fig. 1. On the left side: the speech recognition result of the user query *Who got a red card in the final ?* and its semantic interpretation. On the right side: the multimodal answer of the question answering process is a factoid answer snippet (synthesised) with additional supportive image evidence.

Question Answering (QA) in mobile domain means for us, for first, lack of computational power, for second, a small screen, and for third, very specific and interesting interaction possibilities such as pointing gestures on screen and the use of device buttons. In the scenario we envision the primary input modality is speech and the multimedia output is presented on the PDA screen, with additional speech synthesis of the most important information. In most cases, the multimedia results are text snippets, text documents, images, videos, and graphics. In order to support more advanced interaction possibilities, the user should be able to navigate through the semantically-represented answers, point to individual media items or other screen elements, and pose follow-up questions by speech. The following example illustrates some typical dialogue based interactions we support:

(1) *User:* "When was Germany world champion?"
(2) *System*: "In the following 4 years: 1954 (in Switzerland), 1974 (in Germany), 1990 (in Italy), 2003 (in USA)"
(3) *User*: "And Brazil?"
(4) *System:* "In the following 5 years: 1958 (in Sweden), 1962 (in Chile), 1970 (in Mexico), 1994 (in USA), 2002 (in Japan)" + [team picture, MPEG7-annotated]
(5) *User:* *Pointing gesture on player* Aldair + "How many goals did this player score?"

(6) *System:* "Aldair scored none in the championship 2002."
(7) *User:* "What can I do in my spare time on Saturday?"
(8) *System*: "Where?"
(9) *User:* "In Saarbruecken"
(10) *System: The cinema program, festivals, and concerts in Saarbruecken are listed.*

In order to allow these interaction possibilities, we developed a new dialogue system framework complementing other approaches to dialogue system architectures (e.g. [2–4]). Standardised interface descriptions (EMMA[1], SSML[2], RDF[3], OWL-S[4], WSDL[5], SOAP[6], MPEG7[7]) are used between the modules and components. Basically, we want to allow the user to send requests to various information services that are linked by a Semantic Web framework. The *dialogue server* [5] integrates the dialogue components, the speech interpretation component [6], the modality fusion and discourse component [7], the reaction and presentation component [8], and the natural language generation module [9] are the most important for interaction and presentation. The text is organised as follows: the core user interface, mainly its interaction and presentation part, is described in section 2. Consistent and homogeneous presentation for multiple answer streams is the result of ontological representation of common sense interaction and presentation knowledge patterns in interactive mobile domains. Compared to previous existing approaches to media representation and media allocation, we base all decision processes and presentation items on ontological structures as part of a discourse ontology, or upper-level ontologies, or domain ontologies. We conclude by describing the implementation process in short and the improvements we build in while implementing the interaction storyboard (section 3). Lack of computational power for task scheduling on mobile device processors remains an open problem for the synchronisation of speech and text output, and advanced graphics on the device. Content-planning is heavily influenced by previously selected media and modalities for presentation or interaction. We motivate the use of ontologies and Semantic Web data structures [10] for multimodal interaction design and implementation—spinning the Semantic Web for mobile dialogue system applications.

2 Core User Interface

The user interacts with the dialogue system by using a mobile PDA (T-Mobile's MDA pro) as core user interface; we explain the graphical screen interface in the following: In ontology-based interaction design for mobile devices [11], software development for the mobile context [12] has to be tailored toward ontology-driven

[1] http://www.w3.org/TR/EMMAreqs
[2] http://www.w3.org/TR/speech-synthesis
[3] http://www.w3.org/TR/rdf-primer/
[4] http://www.w3.org/Submission/OWL-S/
[5] http://www.w3.org/TR/wsdl
[6] http://www-w3.org/TR/soap
[7] http://www.chiariglione.org/mpeg/

discourse development for the mobile context. On the PDA screen (640 x 480 pixels) we defined three regions for the graphical user interface (figure 2), an upper region, a middle region, and a lower region. The title bar with status indications and corresponding icons are displayed in the upper region. Here we also display a direct feedback to the users input and the recognised words are presented with a direct editing function (cf. figure 4). We also display whether the microphone is on/off and the progress bar. In the middle region, we display the paraphrased question (cf. figure 3) in a semantic template-based or generated sentence-based form before the request is being sent to the Semantic Web access systems (see [13] for further information). In the lower section, we display result navigation buttons and a status bar. This means, for example, to display icons for result media types, a legend of symbols and colours when dynamic button allocation is used, or system control and system status information.

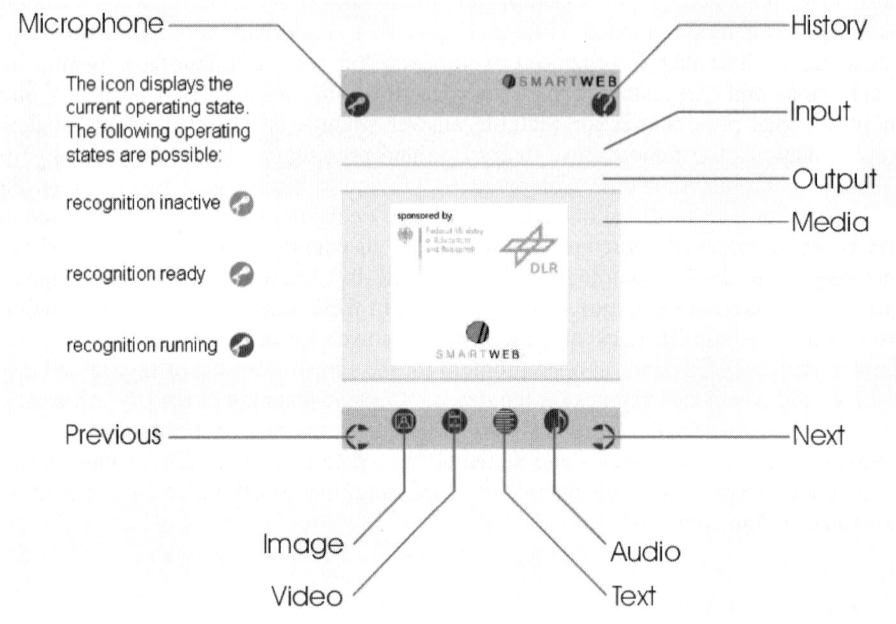

Fig. 2. Main GUI Widgets. Visit `http://smartweb.dfki.de/Intro_Demo/start.html` for an interactive demonstration.

Perceptual feedback allows the user to understand the current processing state of the system, in order to react accordingly. Although we do not use a life-like character like Smartakus [14], system activity and responsiveness can be expressed by liveliness of the presentation elements. Lively metaphors for system states that are used as perceptual states are (1) *Listening/Idle State*, (2) *Recording* (green and red icon), (3) *Understanding* by presenting a semantic query interpretation, (4) *Query Processing* (status bar), (5) *Presentation Planning*, and (6) *Presenting*.

The challenge we address is to give feedback on speech recognition and language understanding accuracy. We experimented with displaying best hypothesis confidences as probability values, or confidence scores in terms of discrete categorical classes (good, medium, bad) to indicate the status of the recognition process. The automatic speech recognition (ASR) component delivers these values as metadata. As informal user tests show, to simply present words with low probability in different colour works best as ASR feedback. Analogously, process metadata of language understanding accuracy is provided in terms of the semantic query interpretation we call semantic paraphrase. The dialogue management component [8] decides on-the-fly whether a verification dialogue is to be initiated, or a confirmation is needed, or the query is being sent without any confirmation. In this way, we try to obtain optimised dialogue prompts in specific data situations which is one of the main duties for dialogue managers. We give priorities to (1) users feedback from multimodal input parsing, (2) offer correction possibilities, (3) provide interface simplicity by progressive disclosure and a special link structure. We regard paraphrasing the natural language understanding output and displaying the proposed query as user feedback as the key element to the system's acceptability (figure 1 and 3). The language understanding module is used to analyse the input and to build up the internal semantic representation of the user utterance. This representation is in the same format as the Semantic Web knowledge bases. The textual semantic paraphrase is then generated from this query description. The ontological RDF/S structures resemble intentionally typed feature structures (TFS) [15] common in formal natural language processing. An example semantic query of the user enquiry: Show me the mascots of the Football World Cup is shown in figure 3.

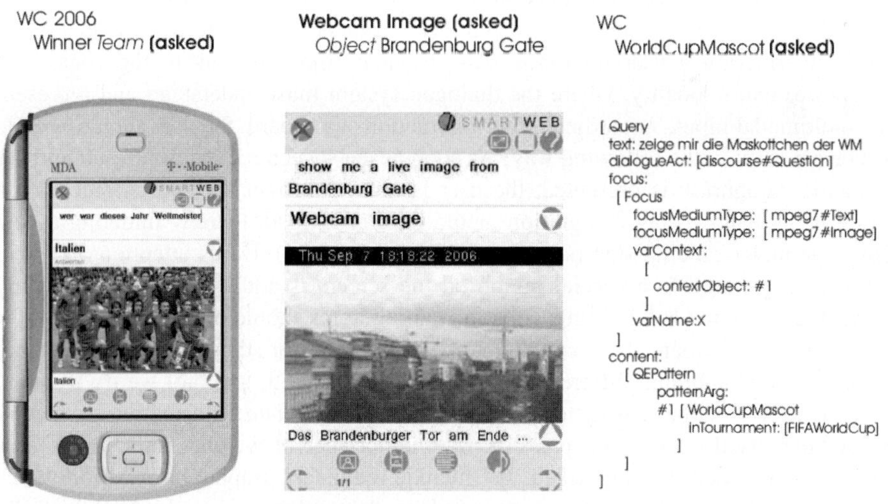

Fig. 3. Top Left: Textual semantic paraphrase as ontology concepts of winner team search and the MPEG7 image result shown on the PDA. Top Middle: Semantic Paraphrase of Webcam search and Web Service results . Top Right: Paraphrase of the world cup mascot search and the ontological query representation.

Fig. 4. On the left: User input corrections on *Werner (Who) was world champion in 1990 ?* On the right: Concept-icons present feedback of question understanding (a team instance is being asked) and answer presentation in a language-independent, conceptual way. The sun icon additionally complements a textual weather forecast result and conveys weather condition information.

2.1 Correction Possibilities and Focus Attention

One very important question concerning the user interaction model is how to correct invalid user input from automatic speech recognition errors[8], or from errors that occur while interpreting a user utterance. This becomes more serious in the context of composite multimodality, where the dialogue system must understand and represent the multimodal input. According to our interaction storyboard, the user should be able to correct input in the following way: As soon as the speech recognition result and the semantic paraphrase is presented, the user is able to correct the speech recognition result using handwriting recognition or the Qwerty keypad, thereby initiating a new language understanding step (see the left part of figure 4). Focus attention, to direct the user's attention to a special region on the screen, is addressed in the following way: According to [16], multiple attention calling items should be avoided, except for two situations: when they complement each other, or they express the same information. Combining different media as focus (text with graphic) we try to avoid that an user is either distracted or confused by redundant information—whereby redundancy itself is used to express the high importance of the understanding process of a concept which is being asked. We include redundant graphic information which presents easily-to-understand, language-independent information which additionally pleases the eye, at least by colouring a system response dominated by written

8 The accuracy of our recogniser started server-side is over 90%. Using out-of-vocabulary words, the accuracy, however, decreases drastically.

language. People do accept multimodal presentations which intersperse text and graphic, otherwise comic strips would not have become so popular. Figure 4 shows concept icons for system understanding feedback and presentation of results. In addition, we deal with layout as a rhetorical force, influencing the intentional and attentional state of the discourse participants [17]. As shown in figure 1, the reader is presented the multimodal paraphrase in the central display section, from upper left to lower right. Text is naturally presented from left to right. In addition to the textual speech recognition result, the textual part of the paraphrase is presented first. The concept icons are always placed in the lower right area, in order to exploit the rendering surface and to balance out the general text view from the left.

2.2 Navilink Structures

Having received a result list of multimodal items, we address the question, what kind of response is appropriate to be presented. Following the last example, the names of the mascots can be synthesised, and additional textual and image material can be presented on screen. The problem we address is how to display extensive text and image information on the small PDA screen. We implemented a hyperlink structure (*Navilinks*) to provide an interactive result summary and additional multimedia material through navigation. Navilink structures are generated automatically from the ontological descriptions (figure 5) obtained as results from the ontology servers. The constraints coded into the presentation ontology, augmented with result filtering and aggregation mechanisms, provide the automatic layouts and interaction possibilities shown in figure 5.

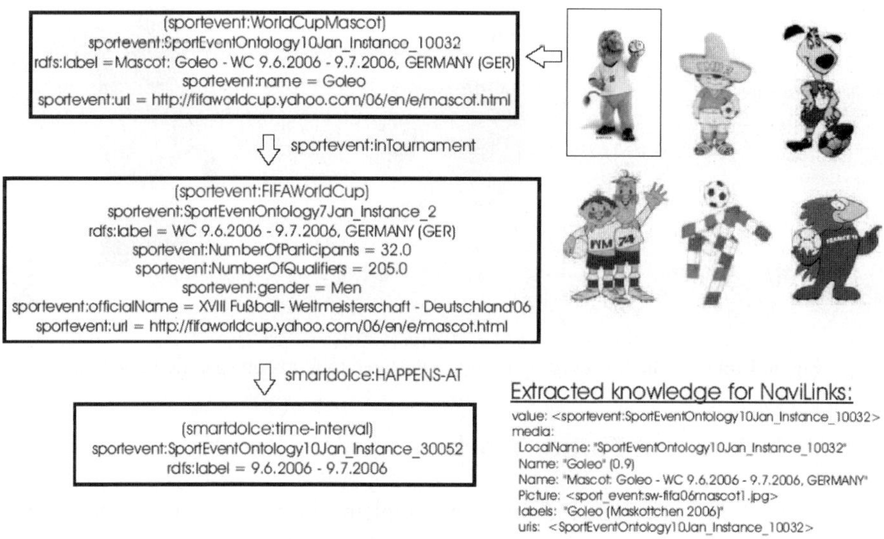

Fig. 5. On the left: the fragment of the ontology results for the 2006 World Cup mascot *Goleo*. The extracted knowledge for result presentation and link generation is shown on the right.

In the QA scenario a summary of the QA process should be displayed on the screen for fast visual inspection at first sight (*11 mascots*). Lateral navigation for displaying numerous parallel results must be applicable in this presentation situation, i.e., presenting the result of enumeration questions. Limiting the number of vertical layers of presentation elements (hierarchical links) by surface links such as *Help* and *Home* buttons which are active on each result page, is a common practice we adopted for our needs. The Navilink generator similar to hyperlink generation for desktops [18] takes the extracted knowledge for Navilinks and produces the level-1 hierarchical link structure shown in the middle region of the left screenshots in figure 6. The user is able to focus on the special result instance *Goleo* through (a), (b), and (c). The most suitable word within a text that contains enough content information as a Navilink headword is chosen by special natural language parsing and generation rules [9]. Pointing gesture on the screen allow users to activate Navilinks in red colour, or other textual material which is displayed on the screen for navigation and linguistic discourse reference. Pointing gestures can be used to complement language input as shown in the interaction example (utterance 5). GUI interaction is additionally leveraged by meta dialogue which the user utters as speech commands, such as accept, reject, going back to previous turns and question (history of system-user turns is available), or starting over the dialogue session, which deletes all active discourse referents.

Fig. 6. Links can be followed by pointing gestures to focus on specific data items

3 Conclusions and Future Work

We presented the interaction design and implementation for multimodal mobile Semantic Web interfaces. With an interaction ontology we provide a semantic-based data integration framework for symmetric multimodal question processing. We evaluated in our mobile application scenario that

- audio-repetition of user queries is useless, only implicit or non-intrusive feedback in textual form should be given.
- extra input correction modes with bigger fonts are valuable for handwriting recognition applications on small devices.
- editing the semantic paraphrase remains a challenge. The current implementation only allows to edit the ASR output.
- building integrated ontological representation is very time-consuming (about 30% of implementation effort) and experts are needed.
- co-ordination of short textual answer and its synthesis is impressive on PDAs if it works. Unfortunately, due to the PDA processor charge, the synthesis start is delayed for about 3 seconds for some questions, which remains an open problem. Synthesising augmented speech, where the synthesis is a complete sentence, instead of a short answer, compensates the effect.
- it is beneficial to follow the principle: no multimodal presentation is generated without representation. By rigorously following this design principle we can refer to all presentation elements at input processing as a side-effect of profound data structure design. Ontologies provide semantic-based data integration frameworks for multimedia.

An accompanying evaluation with real users of the whole SmartWeb-system is planned. Extensions are editing functions via concurrent pen and voice, to extend symmetric multimodality to symmetric multimodal query correction. More flexibility and robustness in speech recognition and understanding is much appreciated, since users formulate questions in various styles and phrase order, even in closed domains, such as football. Even closed-domain applications have to deal with many different linguistic surface forms. Our future investigations will explore more fine-grained co-ordination of multimodal presentations in mobile environments, and graph-like visualisation of ontological result structures (Semantic Navigation).

Acknowledgements. The research presented here is sponsored by the German Ministry of Research and Technology (BMBF) under grant 01IMD01A (SmartWeb). We thank our student assistants, the project partners, and Ralf Engel and Norbert Reithinger for valuable comments on earlier versions. The responsibility for this papers lies with the author.

References

1. Wahlster, W.: SmartWeb: Mobile Applications of the Semantic Web. In: Dadam, P., Reichert, M. (eds.) GI Jahrestagung 2004, pp. 26–27. Springer, Heidelberg (2004)
2. Herzog, G., Ndiaye, A., Merten, S., Kirchmann, H., Becker, T., Poller, P.: Largescale Software Integration for Spoken Language and Multimodal Dialog Systems.Natural Language Engineering 10, Special issue on Software Architecture for Language Engineering (2004)
3. Wahlster, W.: SmartKom: Symmetric Multimodality in an Adaptive and Reusable Dialogue Shell. In: Krahl, R., Günther, D. (eds.) Proceedings of the Human Computer Interaction Status Conference 2003, Berlin, Germany, DLR, pp. 47–62 (2003)

4. Wahlster, W. (ed.): SmartKom: Foundations of Multimodal Dialogue Systems. Springer, Berlin (2006)
5. Reithinger, N., Sonntag, D.: An integration framework for a mobile multimodal dialogue system accessing the Semantic Web. In: Proceedings of Interspeech'05, Lisboa, Portugal (2005)
6. Engel, R.: Robust and efficient semantic parsing of free word order languages in spoken dialogue systems. In: Proceedings of 9th Conference on Speech Communication and technology, Lisboa (2005)
7. Pfleger, N.: Fade - an integrated approach to multimodal fusion and discourse processing. In: Proceedings of the Dotoral Spotlight at ICMI 2005, Trento, Italy (2005)
8. Sonntag, D.: Towards combining finite-state, ontologies, and data driven approaches to dialogue management for multimodal question answering. In: Proceedings of the 5th Slovenian First International Language Technology Conference (IS-LTC 2006)
9. Engel, R.: Spin: A semantic parser for spoken dialog systems. In: Proceedings of the 5th Slovenian First International Language Technology Conference (IS-LTC 2006)
10. Fensel, D., Hendler, J.A., Lieberman, H., Wahlster, W.: Spinning the Semantic Web: Bringing the World Wide Web to Its Full Potential. MIT Press, Cambridge (2005)
11. Sonntag, D.: Towards interaction ontologies for mobile devices accessing the semantic web - pattern languages for open domain information providing multimodal dialogue systems. In: Proceedings of the workshop on Artificial Intelligence in Mobile Systems (AIMS). 2005 at MobileHCI, Salzburg (2005)
12. Longoria, R. (ed.): Designing software for the Mobile Context. A Practitioner's Guide. Springer, Heidelberg (2004)
13. Sonntag, D., Engel, R., Herzog, G., Pfalzgraf, A., Pfleger, N., Romanelli, M., Reithinger, N.: Smartweb handheld - multimodal interaction with ontological knowledge bases and semantic web services. In: Proceedings of the International Workshop on AI for Human Computing (AI4HC) in conjunction with IJCAI (2007)
14. Poller, P., Reithinger, N.: A state model for the realization of visual perceptive feedback in Smartkom. In: Proceedings of Interspeech 2004, pp. 265–268 (2004)
15. Carpenter, B.: The logic of typed feature structures (1992)
16. Arens, Y., Hovy, E., Vossers, M.: On the knowledge underlying multimedia presentations, pp. 280–306 (1993)
17. Wahlster, W.: Planning multimodal discourse. In: Proceedings of the 31st annual meeting on Association for Computational Linguistics, Morristown, NJ, USA, Association for Computational Linguistics, pp. 95–96 (1993)
18. Stock, O.: Human-computer interaction through natural language and hypermedia in Alfresco. SIGCHI Bulletin 28(3), 102–107 (1996)

Society of Mobile Interactions

Hiroshi Tamura[1] and Motoyuki Akamatsu[2]

[1] Tamura Institute for Human Interface
tamura@mobilergo.com
[2] National Institute of Advanced Industrial Science and Technology (AIST)

Abstract. Society of Mobile Interactions was established in April 2006. Short history to establishment and topics of interests are introduced. Mobile interactions suggest interactions among people instead of mobile technology. First the influences of mobile mail to life and communication of young and elderly. Some aspects to be studied are discussed. Some trials to adapt people to mobile interactions are explained. Hito navi (handheld navigation) and Kid's K-tie are discussed as the recent applications.

Keywords: car navigation, mobile phone, hand-held navi, safety, brain functions.

1 Introduction

Society of mobile interactions is newly established in Japan in April 2006, after about 8 years of Symposium activities in this field. This paper is to introduce brief history of mobile Interactions studies in Japan and emerging topics in this area. The first symposium was held Dec. 1998, in Kyoto. It was dawn of mobile telephone and car navigation in Japan. In Europe first mobile HCI conference was held in France 2001, and in 1998 a work shop on mobile HCI was held in Scotland.

The society is established for the studies of human interactions using information communication technology while some sites of interactions are in motion, or prepared to motion. The society is interdisciplinary, covering social, cultural, engineering, design, psychological and physiological sciences. The activities are participated by manufacturers, network operators, dealer, consumers, social scientists and public services.

Mobile phone and car navigation were small application area of computer technology, but they had given larger impacts to daily lives of people. Various aspects of studies are necessary about mobile interactions to now and to the future of society.

First the influences of mobile mail to life and communication of young and elderly are explained. Some aspects to be studied are discussed in chap.3. Some trials to adapt people to appropriate use of mobile interactions are explained. Hito navi (handheld navigation) and Kid's K-tie are discussed as the recent applications.

(Society of Mobile Interactions web: http://mobilergo.com)

M.J. Smith, G. Salvendy (Eds.): Human Interface, Part II, HCII 2007, LNCS 4558, pp. 655–663, 2007.

2 Ergonomics of Mobile Interactions

Car navigation systems, based on GPS signal and CDROM driven digital map, was already popular by 1998, when the first symposium was held in Kyoto. Public traffic information service VICS have started, and expected to provide congestion information to the subscribers. Mobile phone also was popular, but only 1 out of ten persons was using it. Telephone company regarded mobile phone might be necessary for top leaders of big company, but not normal citizens, because telephone and fax are available in most family and in office room, and public telephones are available most terminals and streets. Thus mobile phone charge for a call was expansive. But by this time some students began using mobile phone. House wives and business men followed. It was very convenient tool for a husband to call his wife from a train to pick up at railway terminal.

When text mail systems are introduced to mobile phones, it was only a few people who were convinced that normal phone users would like to use text mail on mobile phones. Soon mobile phone companies realized, mobile mail traffic increases around 23 o'clock at night. It was new ecological trend of the young, which might not be recommended before.

Now a day the major use of mobile phone is text mail, instead of voice talk. Not only the young but also the old are using kanji entering methods. Especially Japanese fonts seemed almost impossible to be operated on size limited display of mobile phones. Increased use of mobile text mail has evoked new problems regarding text entry on culture. One is that over use of text mail on mobile phone especially by young people, is degrading their hand writing capability. Issues of this type were also heard in the past. Over use of speech telephone degraded writing habit of the young. Comic books are poisonous for the young. It may be true in some cases, but not in other. Young people, who overused speech telephone, are now heavily using email on mobile phone.

Fig. 1. Symposium Mobile Interactions2007 indicating 8 years history

In 2001 symposium, we stopped using word "keitai denwa" (indicating pocket telephone) and adopted "k-tie (meaning simply Pocketable)" . The intention is to indicate that the mobile devices are no longer telephone. It is directing all round information tools. Soon mobile phone became accessible to internet and now air ticket, railway ticket and hotel are booked from K-tie.

Government has been pushing schools and universities to teach PC, internet and email, but not mobile communication. Students learn about PC to get necessary credits, but out of classroom, they need note PC which is heavy, have start up OS, and wait for network connection. To use K-tie, it is just open the top, and start push button. Communication starts at any place and time. What is necessary to teach is communication, but not PC. This is the great miss understanding school and social leaders. Many universities open official announcements of bulletin board. But recently students do not pay much attention to bulletin board. So it took more than 2 weeks until an announcement reach most of students. By mobile email it is a matter of 30 minutes to an hour. Time difference in communication is essential.

Until recently K-tie was evil for school. But this is simply because teachers and professors are ignorant of K-tie. Situations are quickly changing recently, because of increase of insecurity in Japan. People are realized the need to protect themselves by communication network, instead of guns.

3 Environment of Mobile Interactions

Mobile interactions might include information communication devices, actually in motion, prepared to motion, fixed base devices willing to contact with people or devices in motion as well as intact individuals. Wired and wireless network methods are included. Some professional people, like railway driver, airline attendants or long distance truck drivers use special network for their own needs. Professional interactions may not be included in the main interest of the society.

In most countries, permanent address, or living address is essential key of social identity. In the mobile age, some reliable network identity will be official key for personal identity. To meet somebody in the city, they tried to determine exact place and time together. Now a day, time and place are loosely determined, and the final time and place are fixed flexibly adjusted by mobile contact, depending on the traffic and previous experiences. Appointment with mobile device carrier is flexible, while one with non-mobile carrier, like kids or seniors is critical.

But it is not always assured to make contact with mobile device carriers. In some cases, the wireless wave is not strong enough in the shed of building, mobile contact will fail. If the user inattentively switched off the device, contact will not succeed. Even the daily mobile carrier occasionally leaves the device at home, or forgets battery charge. Although the conveniences have increased by mobile technology, the user has to take care of the devices and environment to which the user is moving in.

4 Operation of Mobile Devices

Many studies have been done about the CRT display and keyboard of fixed base stations. In many cases, the research result cannot be applied to mobile devices, because of limit of physical size of keys, display size, contrast and response time. Technical limit might be overcome in time, but the human characteristics may not change.

4.1 Text Entry

Entry speed for mobile devices need not be equivalent to office devices. Text entry speed of professional typist is very high. But entry speed is not so high for the normal office worker and sales men. Required speed to accomplish each work is different.

The required speed of text entry for mobile devices might be different by situations. It is important to know roughly the necessary entry speed for various business, social and living situations.

Mobile devices are used not only in front of desk. They enter text at various places, like railway terminal, shopping center, library, etc. Sometimes, they enter text while walking, standing, and lying on a sofa or in bed. Business men would enter meeting report standing just outside of meeting room. They might prefer a quick entry method. It is worth examining entry speed of standing and walking.

If they are walking on the way back, another problem occurs. The distraction from street traffic has to be taken care. More seriously, some people enter text while driving a car. Is it the role of technology to find methods for those users safely to enter text? If key entry is not safe, how about hand writing? Which is better? The author has experienced in Northern Europe, when a taxi driver received a mobile phone call while driving, he held mobile phone in his left hand, and he began to write down customer address and phone number on a paper by his right hand. He left his both hands from steering wheel for some time. It was fearful. Recently a driver, who is hands free set user, took memo by his right hand while driving on high way. He looked accustomed to do the action. He is a truck driver of his own, and he is tempted not to loose opportunity of receiving order from a customer.

4.2 Reading Environment

Reading environment is also different by place, and daylight. The display of mobile devices is almost impossible to read under strong sunshine. Font size suitable for mobile devices might different by age and visual conditions. How should we determine?

People dare to read mobile display while doing other work or actions. How reading performance change by the action the user is taking? Recently people read emails when they are walking down corridor, street, and even driving a car or riding a bicycle. There are problems in two sides. One is the reading performance, and the other is the effect to the safety of action. Distraction occurs not only in driving but also walking across the road. Should the reading, listening audio, watching TV be prohibited in the public road?

4.3 Practical Education

Many traffic accidents were related to user actions to take out the mobile phone from a bag or from pocket of jacket when it started to ring. Recently frequent users may keep the mobile phone on fixed places easy to access while driving. This sort of risk can be decreased by preparing in advance to start driving.

It is not recommended but some users enter short text while driving. Those people are normally conscious about safety. They think about mechanism of distraction and try to decrease most distractive part of action. They are in search of less distractive methods.

Recently text email is so popular, that many people read and write text mail even while driving in the car, although it is officially prohibited today. One professor did experiments to the students. First, he asked students if they respond email while driving a car. About 10% of students, answered yes. Most of them were skilled in quick text entry. Then the professor required those skilled students to enter a short sentence to their mobile phone, and send it to professor. In this first experiment, students responded smartly. Then professor proceeded to the second experiment. Professor required them to enter a sentence keeping the mobile phone under the top board of desk, so that the user student cannot see the display and key array of mobile phone. By the second experiment, most of the students, who are skilled users, failed to enter the sentence correctly and they were convinced themselves that it is not possible to enter text without glancing at the mobile phone.

Generally, doing two time critical jobs at once is risky and dangerous. People know it. There have been quite a lot of discussions about safety and sub jobs, since the introduction of mobile phone and car navigation. One extreme position is that the car driving alone is risky and time critical, so introduction of any sub task increases risk of danger which should be avoided. The other position is that human capability to avoid risks is limited. However below the limit, risky task can be overcome. The idea is good. But the limit maybe different by individual, and might change by time or situation. One way to overcome uncertainty of risk is practice. Many students train themselves to use text entry while driving or enter text in the small intervals while driving.

The professor recognized one, or two out of 100 students succeeded to enter the sentence without looking at the display or key array. They are capable of perfect touch typing. But touch typing was not the key to text entry while driving.

One student explained her strategy. She holds handset together with steering wheel, and wait until the car stop by the traffic signals. When the car is stopping she can enter some length of sentence, and as the car start moving she think of next sentences. We do not say the strategy is safe or not, but the strategy includes many suggestive ideas. And the strategy can apply only to some Japanese traffic where a lots of signals and slow moving traffics.

5 Immerging Applications

5.1 Hito Navi

An important trend in recent applications is hand held navigation. Hand held navigation, normally called Hito (human) Navi, was supposed to be a light weight navigator equipped with small size map display and GPS receiver. HITO navi actually is completely different in functions compared to Car Navi. Car Navi deals mainly high way and road, which are open to public. Hito navi now include small walk through, mostly public but some are private. People need to know shortest path, sometime path without stairway, or path less up and down, and sometime the path people can go through without wetting from the rain. The customer needs of Hito navi and Car navi are completely different, thus the development method and their revision are different. In short, the route qualification of Hito navi is based on human evaluation of map provider. The map design corresponding to the qualification is also interesting problem.

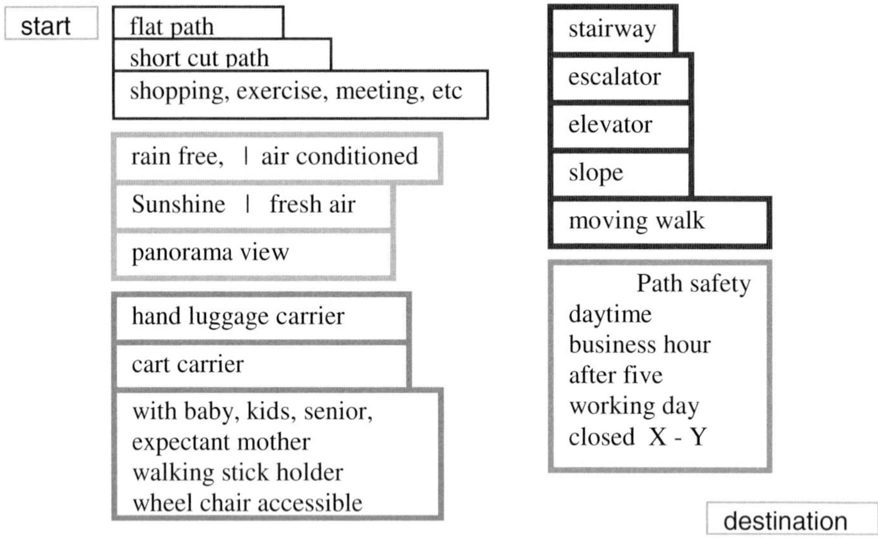

Fig. 2. Options to be considered for walking navigation

5.2 Kids' K-Tie

In the course of economic depression in the past 20 year, economy and safety of Japanese society changed drastically. Many house wives as well as the young are forced to work in non-regular partime jobs. Also students are working. To keep communication with other members of family, K-tie became the essential too for living. At the same time social unrest increased. Serious criminal like kidnap, killing on the way, swindle money using telephone, etc. To protect school children from suspicious strangers, local citizen try to exchange information with police using mobile phones. Information distributed from police is sometime too much, and too few. People response to stranger information is not easy to decide. Inhabitants need to to protect their security by themselves. School children disappeared on the way to school or back. Even at school children and teachers are killed and wounded. Systems to trace the children are proposed from various enterprises. Mobile Phone Operator Company is proposing a system, using GPS and antenna area.

Formerly mobile phone was an evil to the school. Parents and teachers were afraid of the children to connect to other children to much, and unfavorable agent might tempt children by mobile phone. School prohibited children to bring mobile device to school. Now schools gradually admit the need of mobile devices to use for daily lives of family. In case of emergency children have to contact with his relatives and in other case, school might need, help to pick up children. School assumes that some body of a family is at home. School believes that family can be reached by fixed station phone. But this is a big miss understanding. Most of a family members are away from home in the day time. It might not be appropriate to make a call to a parent on business by mobile phone, school might contact by a mail to have a call to the school.

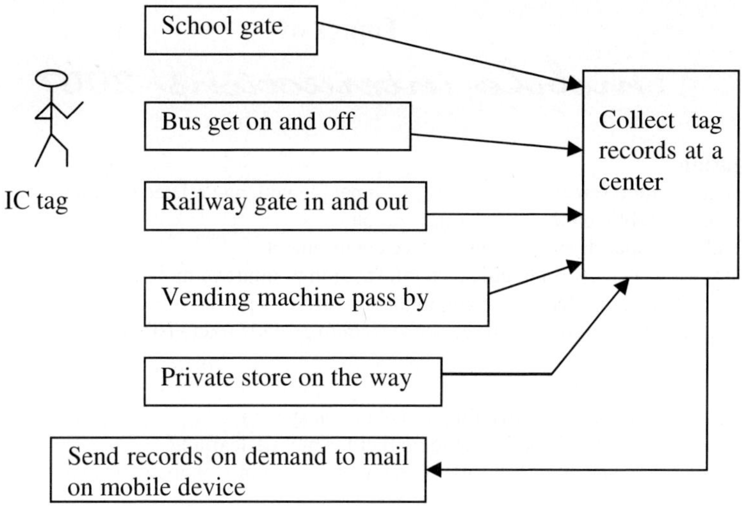

Fig. 3. Security system for school children

Recently IC tag application is in progress. By installing in a shoe or clothes of kid, whereabouts of the kid can be traced by IC tag sensors of various agents, distributed all around the city. This is a promising idea but the problems might be whether many different agent can get together to built unifies systems.

6 Safety and Security

Safety issue was seriously discussed with car navigation system when it was first introduced. Watching the navigation screen will distract attention from driving. There were some accidents related to car navigation watching and button operation. After 20 years, the response of car navigation is improved, and the users have understood the limit of computer search, and people do not expect beyond what they can do. It is important to differentiate what machine can do, and what user should do.

Electromagnetic interferences

Safety of pacemaker was emphasized in the past. After 20 years of experiences, hospitals permitted the use of mobile phones in the ward. It is good to be careful to new possibility, but they are also courageous to accept the evidence of experience. No serious problems ever reported thereafter.

Serious problem is education and public sectors. Schools still continue to teach children not to use mobile phone to avoid interferences to medical instruments. And in the train and bus, same announcements are repeated to the passengers. This is a false announcement. The public sector might think over prohibition is necessary. But over prohibition brings ill effect also. One case is like this. In a crowded train a young girl began to use mobile phone mail. Then an old men started to say, " You Stop using mobile phone. A pacemaker is in my heart, it is influenced by mobile phone." Stop using it. And get out of the train." Apparently the old man is not wearing a pacemaker.

Topics of
mobile interactions 2007

navigation

The navigation screen more accessible to people with age related cataract
Union of mobile phone and car navigation
Hands-free speech input in automotive environment
Study on in-vehicle route guidance information to improve mobility for foreigners
Effect of cellular telephone manipulation on driver's performance
Development of information service system to provide road visibility condition

visual

Readability of character size for car navigation system
A study of students' mobile phone market for mobile-learning
Method to evaluate the legibility of characters on mobile phone LCD and its application

input

Development of new input interface by quantitative usability evaluation based on user characteristic
On the actions of Japanese text-input on a mobile phone using a pressure mat
Design process of cellular phone based on Kansei engineering

learning

On broadcast of the great bonfire event 2006 with camera cell-phone
The function of mobile phone: a deprivation study
A model development on educational use of cellular phones at an elementary school
Taking leave to introduce inconvenient learning materials
Applications of non-grounding haptic interface for mobiles

medical

Telemedicine by movie transmission using TV mobile phone
Mental health support using mobile phone
Electronic patient's record and home monitoring system for high risk pregnant women
Remote patient monitoring system using Java enable mobile phones for doctors

Brain

Is the n-back task valid as the mental workload measure while driving?
Relationship between cognitive performance and change of oxy hemoglobin density in brain
Quantification of physiological change during a task of tracking and reaction by NIRS measurement
Activities of brain while operating the mobile phone
Variation of NIRS trajectories depending on the visual conditions of display
NIRS responses related to hand and finger actions

Fig. 4. Topics presented at symposium Mobile Interactions 2007 (May 10-11)

7 Conclusion

Generally higher attentions should be paid to operators in motion to avoid unpredictable risks. Risk warning are important for health and safety, but in many cases enough knowledge about the ways of using devices are not transferred to users and the designers do not know how the users use the devices. Health care specialists and risk psychologists are normally cautious. Careful checks to new interaction methods are essential and scientific investigation on interactions have to be further promoted. Societies pay attentions to the risk problems of mobile interactions, but points of discussion are more or less based on past experiences, which are some times appropriate and sometime not. Mobile interaction specialists have to show scientific basis to convince people in use and not in use.

The mission of the new born society is to support such social activities from academic side.

References

Tamura, H.: Human interface and ergonomics in the information society. In: Proceeding of the international ergonomics symposium 2000, pp.262–265, Ergonomics Society of Korea (2000)

Tamura, H.: Appropriate use of digital everyday appliances. In: 8th Symposium on Automated Systems based on Human Skill and Knowledge, Goteborg, SE (September 22-24, 2003)

Tamura, H.: For appropriate use of digital everyday appliances. In: IEA world conference, pp. 291–294, IEA (2003)

Hida, Y., Choui, M., Tamura, H.: Message transmission of public announcements by different media in university. In: Proc. of Human Interface symposium, pp. 991–994, Human Interface Society (In Japanese) (2004)

Towards an Optimal Information Architecture Model for Mobile Multimedia Devices

Timo-Pekka Viljamaa[1], Tuomas Vaittinen[1], and Akseli Anttila[2]

[1] Nokia Research Center, P.O. Box 407, 00045 Nokia Group, Finland
[2] Nokia Corporation, P.O. Box 407, 00045 Nokia Group, Finland
{Timo-Pekka.Viljamaa, Tuomas.Vaittinen, Akseli.Anttila}@Nokia.com

Abstract. In this paper we propose an optimal information architecture model for mobile multimedia devices. In our study we first reviewed the information architecture models of four currently popular mobile multimedia devices, Apple iPod Video, Nokia N80, Sony Ericsson W800i and Sony PlayStation Portable (PSP). Then we used Open card sorting method to find out what kind of mental models does people have for optimal information architecture model. Finally we compared these two study results together and created a design proposal.

Keywords: information architecture model, card sorting, cluster analysis, mobile multimedia device, mobile multimedia, user-centered design.

1 Introduction and Objective

Mobile multimedia devices, such as Apple iPods and Nokia Nseries multimedia phones, are becoming increasingly versatile convergence devices. The user can access a wide variety of different data entities, such as messages, calendar, music, photos, and even TV channels. Designer of such devices is faced with a problem of finding a user-acceptable model to combine the different features and data elements into one UI.

Managing digital documents in computers has been studied by many researchers in the past, and various solutions have been proposed [3]. Studies into document management in mobile multimedia devices however are few in number.

In the face of this increasing complexity and various implemented solutions we decided to conduct a study consisting from background and card sorting studies to determine the user expectations for data organization in mobile multimedia devices. Our main research interest was in finding out an information architecture model with access to all the possible types of data a device can hold from personal information items to various forms of multimedia. We also wanted to compare this model to the existing information architecture models of currently popular mobile multimedia devices.

First we conducted a background study, where four currently popular mobile convergence devices were reviewed to find out how different personal information and multimedia items are categorized in these devices.

M.J. Smith, G. Salvendy (Eds.): Human Interface, Part II, HCII 2007, LNCS 4558, pp. 664–673, 2007.

Next we used Open card sorting method to find out the Finnish and US participants' mental models for organizing and categorizing data. The cards contained both personal information and multimedia items, which participants were asked to sort and categorize in the most optimal way they could without any restrictions.

Finally we performed a cluster analysis on the card sorting data, compared the found information architecture to the background study findings, and also created a design proposal for optimal information architecture.

2 Background Study

Four currently popular mobile multimedia devices, Apple iPod Video, Nokia N80, Sony Ericsson W800i and Sony PSP, were selected to the background study. Each of the devices has a different main multimedia focus, but as an addition to that all of them offer pretty similar support for other multimedia features and personal information items. The information architecture model of each device for multimedia and personal information items was reviewed and the found main categories were also visualized as shown in Fig. 1. The devices had more functionality, but in this study we were only interested in items from these two categories.

Apple iPod Video (5th Generation)
- **Music**
  - Playlists, Artists, Albums, Songs, Podcasts, Genres, Composers, Audiobooks
- **Photos**
  - Photo library, Photo album 1, ..., Photo album N
- **Video**
  - Video playlists, Movies, Music videos, Video podcasts
- **Extras**
  - Games, Calendar, Notes

Nokia N80
- **Music player**
  - All tracks, Artists, Albums, Playlists, Genres, Composers
- **Gallery**
  - Images & video, Tracks, Sound clips, Streaming links, All files
- **My own**
  - Games, FM radio
- **Office**
  - Notes
- **Calendar**
- **Messaging**

Sony Ericsson W800i
- **File manager**
  - Pictures, Videos, Sounds, Games
- **Walkman**
  - Artists, Tracks, Playlists, Videos
- **Entertainment**
  - Games
- **FM radio**
- **Organizer**
  - Calendar, Notes
- **Messaging**

Sony PlayStation Portable (PSP)
- **Photo**
  - Photo album 1, ..., Photo album N
- **Music**
  - Music album 1, ..., Music album N
- **Video**
- **Game**

Fig. 1. Information architecture models of all reviewed devices

2.1 Background Study Results

The review of different devices showed that the prevalent method was to categorize the different multimedia and personal information items according to the personal information and media typology. The biggest differences were found in the subcategories under the quite similar main categories of each device.

Similarities were found especially in music where iPod's "Music" category and N80's "Music player" category had almost identical subcategories, namely "Playlists/Track lists", "Artists", "Albums", "Songs/All tracks", "Genres" and "Composers". As addition to this, iPod also had subcategories "Podcasts" and "Audiobooks". N80 on the other hand provided access to music also through "Gallery" category, but when accessed this way only songs and albums were shown without any sophisticated subcategories, like in "Music player" category. Similarly W800i provided access to music through two different top-level categories, "File manager" and "Walkman". Through "File manager" category the user could access "Sounds" subcategory containing both songs and folders, while "Walkman" category had similar, but fewer predefined subcategories as iPod and N80, namely "Artists", "Tracks" and "Playlists". PSP's "Music" category on the other hand only contained songs and folders without any predefined subcategories.

In all devices, "Photo" category contained both photos and photo albums. N80 was the only device, where photos and videos were stored under the same category, while in all other devices images and videos were separated to own categories. iPod had most subcategories for "Video" category, namely "Video playlists", "Movies", "Music videos" and "Video podcasts", while in other devices there wasn't any ready made subcategories for videos. Also, none of the reviewed devices made any distinction between the user's self-created videos and public/commercial videos. In W800i videos were accessible through two same categories as in music, "File manager" and "Walkman".

Only N80 and W800i contained FM radio, which in N80 was included to "My own" category together with "Games" subcategory, while in W800i it was placed on the top-level main category. iPod had "Games" subcategory under "Extras" category and PSP had own main category for games. In W800i games could be found from two separate categories, either through "File manager" or "Entertainment".

PSP was the only device that did not contain any personal information items, while all other devices contained at least calendar events and notes, N80 and W800i also messages. In iPod both calendar events and notes were stored as subcategories inside "Extras" category. N80 on the other hand had "Calendar" category for calendar events and notes stored under "Office" category, while in W800i calendar events and notes could be found under "Organizer" category. Both of them contained top-level "Messaging" category.

3 Card Sorting

3.1 Card Sorting Method

The card sorting method has been well established as a tool to find out the user's mental models for organizing and categorizing data [6, 10]. The method has been successfully adopted and used in different domains, such as World Wide Web design

[7, 9], decision support tools for the neonatal intensive care environment [2], requirements engineering [5] or even in sorting different wines to groups [4].

There are two different kinds of card sorting methods, Open and Closed, which can be used to understand how users categorize information in a given domain. Open card sorting method can be used for exploratory analysis of a domain, and for the generation of candidate structures, while Closed card sorting method can be used to comparatively test the suitability of several proposed structures and choose the most suitable one [8].

In our study, we chose to use Open card sorting method, because we wanted to see, what kind of categories people would create for multimedia and personal information items if there were not any predefined categories or restrictions. We were also interested to see how does the card sorting method work with applications and items stored in mobile multimedia devices.

3.2 Cluster Analysis

The model produced by participant in a card sorting exercise can be inspected visually and it will provide some understanding of users' mental models. Cluster analysis yields a richer understanding of the data through dendograms (hierarchical structures based on the user's cognitive representations of the domain). We decided to use IBM's EZSort analysis tool to perform cluster analysis on the data and display the results as a dendogram describing the strength of the perceived relationships between pairs of cards [1]. In addition to the quantitative material produced by the cluster analysis, we also conducted pre- and post-test interviews with participants to find richer data about their patterns of convergence device use, and the rationale behind their adopted organizational model.

3.3 Participants

The card sorting test was conducted in Helsinki, Finland and in Mountain View, US with total 27 participants. In Finland 18 people (6 female, 12 male) participated to the test, and in US we had 9 participants (2 female, 7 male). The main recruitment criteria were daily usage of media on the go, such as using a portable music player or a multimedia phone. Most of participants had a background in information technology or in arts and design, and had also different ethnic backgrounds. Age distribution was 24-43 (Finland 24-39, US 30-43) and mean 32 (Finland 29, US 38.5).

3.4 Cards

Before the test we created a set of 29 cards listing the data entities a user can access in a convergence device. These entities were known to be already available in high-end mobile phones and other mobile multimedia devices, and included both public and personal media, online and local content, communication and calendar data as shown in Table 1. The cards were then individually printed out in standard playing card size and enclosed in commercial protective plastic sheaths.

Table 1. Card sorting cards

Title	Description
Audio book	An audio book you saved
Audio note	Dictaphone recording you have made
Audio stream	Online service
Calendar entry	Time and place of event or meeting
E-mail	A message you received
MMS	Multimedia message you received and saved
Network folder	Online folder for sharing things
Newspaper	Online articles
Note	Text note you have created
Picture	Downloaded picture
Picture note	Photo taken as a reminder
Photo	You took with the phone camera
Podcast	Audio recording downloaded from the net
Radio channel	Tuning settings for a radio station
Radio reminder	Time and channel for a favorite radio program
Received image	A picture you got
Received video	A video clip someone sent you
Recorded music	You recorded with the phone
Ring tone	A tone you downloaded
Song	A music file (mp3)
Text message	You received and saved
TV channel	Tuning settings for a mobile TV station
TV recording	A recorded TV program
TV reminder	Time and channel for favorite program
URL	A web link you saved
Video clip	Downloaded video
Video recording	You made with the phone camera
Video stream	Online streaming service
Web notifier	A notification of change on a page

3.5 Test Set-Up

The test was conducted individually for each participant. Two facilitators were present, one taking notes and the other interacting with participant. The test was videotaped. Before the card sorting exercise, participants were asked about their current media usage in a semi-structured interview. The actual card sorting exercise was divided into three stages:

- First the shuffled deck of cards was given to participant who was asked to sort and cluster them according to any model that is logical in his/her mind. Participant was left in peace to sort the cards, but he or she was instructed to ask questions at any time.
- After participant had sorted the cards, he or she was asked to give a descriptive name to each cluster.
- Finally, participant was asked to describe why he or she decided to use this particular sorting strategy, and to elaborate on each cluster.

3.6 Card Sorting Results

In the interview, when participants were asked to describe their clusters, several different grouping schemes came up:

- Media type (e.g. text, radio, video, audio, images)
- Devices that participant associated the media (e.g. iPod, computer, phone)
- Origin (e.g. self-created, received from others, commercial)
- Content location (e.g. local, online, broadcasted)
- Time span (e.g. stuff that needs to be taking care of, things that come up any moment, things that wants to spend time on)

Typically, participants did not follow any of the grouping schemes literally, but combined ideas from several schemes.

The clusters created by 27 participants were entered into IBM's EZSort analysis tool and a dendogram was produced (the dendogram formulated as a table is shown in Table 2). The names describing the categories were created by us to describe the groups.

The dendogram showed a division into two main categories, which we named "Multimedia" and "Memories". "Multimedia" contained three subcategories "Web", "TV and Video" and "Audio". "Memories" on the other hand also contained three subcategories, "Calendar", "Messaging" and "Photos and Personal videos".

"Web" subcategory included clearly all web related content that wasn't related directly to audio, video or images. All public/commercial video content was placed in "TV and Video", which contained both live TV and video streams as well as TV recordings and downloaded video clips. However, it did not contain any photos and videos taken by the user as those were placed under "Photos and Personal videos" in "Memories". In a similar fashion, videos and photos received from others were considered separate from public/commercial multimedia content and placed also to the same subcategory. "Audio" on the other hand contained all audio related items from personal music and podcasts collection to FM radio and online audio feeds.

"Calendar" subcategory in "Memories" included items related to calendar events, multimedia reminders and notes, which was quite similar content than in current calendar applications both in mobile devices and computers. "Messaging" was also following relatively clearly the messaging model existing in current mobile devices as it contained all messaging related items. The only subcategory in "Memories" containing non-personal information items was the mentioned "Photos and Personal videos", which users placed in their clusters closer to personal information items than other multimedia items.

We wanted also to compare the clusters created by Finnish and US participants to see, what kind of an effect the double amount of Finnish participants had to the all participants' dendogram. The biggest difference was found from the location of "Photos and personal videos" subcategory. Finnish participants had placed it similarly than in all participants' dendogram to "Memories" with one small difference as the received photos and videos which were placed to "Messaging" subcategory. US participants on the other hand had clearly separated photos from videos. All self-created, received and downloaded images were placed together, while self-created and received videos were kept separated from photos (both under "Multimedia").

Table 2. Dendogram created from all participants' clusters in detail

Category	Subcategory	Folder	Content
Multimedia	Web	Network folder	Network folder
		Web pages	Web notifier (RSS), URL, Online newspaper
	TV and Video	Collection	TV recording, Downloaded video clip
		Online	TV channel, Video stream
	Audio	Online	Radio channel, Audio Stream
		Collection	Audio book, Podcast, Song, Ringtone, Recorded music
Memories	Calendar	Multimedia Reminders	Radio reminder, TV reminder
		Calendar entries	Calendar entry
		Notes	Picture note, Note, Audio note
	Messaging	-	Email, MMS, Text Message
	Photos and Personal videos	Received	Received video, Received image
		Albums	Video shot with camera, Downloaded picture, Photo taken with camera

4 Background Study and Card Sorting Results Comparison

Next, the dendogram created from all participants' clusters was compared to the information architecture models reviewed in the background study to see, how the information architecture models matched participants' mental models.

Similarly than in the background study results, the card sorting results quite clearly showed that the prevalent method was to categorize the different multimedia and personal information items according to personal information and media typology. Messaging and calendar items were included in own subcategories in both studies, and multimedia items had their own main and subcategories. The only difference to this was the inclusion of photos and personal videos closer to personal information items than public/commercial multimedia items as shown in the dendogram. This was drastically different compared to the reviewed information architecture models.

When comparing the personal information related subcategories, "Messaging" and "Calendar" were almost identical with the reviewed information architecture models. The only new items in "Calendar" were radio and TV reminders. Also notes were combined close to calendar items as in iPod, N80 and W800i currently.

With multimedia items there were more differences. In N80 and W800i, FM radio and music collection were clearly separated from each other. However, the dendogram showed clearly that participants saw closeness between them. Also, participants liked to have streamed audio close to other audio items.

In videos a clear difference could be seen. As stated in the background study results, none of the devices made any distinction between the self-created and public/commercial videos. However, the dendogram quite clearly showed that participants made this kind of distinction. And as an addition to that, participants also saw self-created photos and videos close together as already in N80, and place this new category closer to personal information items than multimedia items. With public/commercial videos there were similarities with music, as participants wanted both the local and the online content stored close to each others. Also, participants saw received items to be separated from self-created in "Photos and personal videos".

While doing the background study, we did not consider web content, such as web browser or web links, to be either multimedia or personal information and we left it out of the scope of the study. However, couple of web items was included to the card sorting study to see how participants would categorize them. The dendogram proposed web items to be included to "Multimedia", but this requires further research.

5 Design Proposal

Based on the background study and card sorting results we created a design proposal for optimal information architecture model in a mobile multimedia device as shown in Fig. 2. For the most part, it followed the dendogram.

The main categories of our design proposal were "Multimedia" and "Personal information". When compared to the dendogram, the biggest change was that "Photos and Personal videos" subcategory was moved to "Multimedia". We made this move, because participants' opinion was quite divided on this issue and we wanted to create a single entry point for all multimedia items. Other subcategories under "Multimedia" were "Audio", "TV and Video" and "Games".

As proposed by participants, both local and online items of the same multimedia type should be placed close together, but still clearly keep them separated from each other to different folders, namely "Collection" and "Online". The main reason for this was to clearly indicate, what content is always available and what might cause data traffic costs and delivery delays. Broadcasted content, such as Mobile TV and FM radio, was separated from other online content, because it usually has different pricing model and is only available in certain locations. Similarly audio and video podcasts were placed to own subcategories, because they contain both local and online aspects and for that reason do not fit to either subcategory. One more thing participants required was to separate your own content from received content. This was achieved by adding "Received items" folder to each collection/albums subcategory.

In "Audio", "Music collection" contained all local music items with typical artists, albums, songs, etc. categories, while "Online" had internet radio feeds and music shops. In "TV and Video" on the other hand, "Video collection" contained all local public/commercial video items, such as downloaded video clips and TV recordings, while "Online" had video streams and download services. In "Photos and Personal videos", "Albums" contained all photo albums including personal videos, while "Online" contained photo and personal video sharing services.

Multimedia	Personal Information
• Audio • Music collection, Podcasts, FM radio, Online • TV and Video • Video collection, Video podcasts, Mobile TV, Online • Photos and Personal Videos • Albums, Online • Games • Games collection, Online	• Calendar • Calendar events, Notes • Messaging • SMS/MMS, E-mail

Fig. 2. Design proposal for an optimal information architecture in a mobile multimedia device

Games was one multimedia type we had left out from the card sorting cards, but we saw that games would fit quite naturally to "Multimedia" as they are already included to current devices, like PSP. As mentioned in the previous chapter, "Web" subcategory requires further research and was now left out of this design proposal.

"Personal information", on the other hand, had the remaining subcategories "Calendar" and "Messaging" of "Memories" category of the dendogram. Instead of only calendar events, "Calendar" subcategory would also contain notes as already in calendar applications in computer world, such as in MS Outlook. The correct place for multimedia reminders remains an open question and needs further research, because it is both related to multimedia items and calendar reminders. "Messaging" subcategory would contain all messaging related items and types as already in current mobile multimedia devices.

6 Conclusions and Future Work

In our research, we first reviewed information architecture models of four currently popular mobile multimedia devices. Then we used Open card sorting method to see what kind of mental models participants had for an optimal information architecture model in mobile multimedia devices. The comparison between the results showed:

- Photos and personal videos were placed closer to personal information items than public/commercial multimedia items
- There was a clear distinction between self-created and public/commercial videos
- Both local and online content of same multimedia type should be combined under same subcategory, but still keep a clear distinction between these subcategories
- Received items should be separated from your own, but still placed close

Based on the study results we created a design proposal for optimal information architecture model that followed the dendogram, but it also had some differences:

- "Photos and Personal videos" subcategory was placed under "Multimedia" instead of "Personal information", to create a single entry point for all multimedia items
- Local, online, broadcasted and podcast items of same multimedia type were placed close to each other, but still clearly separated
- Gaming items were missing from the card sorting study, but "Games" was added to "Multimedia", because games are already included to mobile multimedia devices

- ”Web” subcategory was not included to ”Multimedia”, because the opinion of participants was quite divided on this issue. The card sorting results proposed it to be included to ”Multimedia”, but this requires further research.

We believe that the proposed information architecture model would make mobile multimedia device UIs clearer, more usable and closer to the users' mental models. The design proposal is going to be validated with a further user test by using a Flash Lite prototype implementation of a mobile media player.

Acknowledgments. We would like to thank Gilles Baudet and Petri Piippo for their participation to this research project.

References

1. Dong, J., Martin, S., Waldo, P.: A User Input and Analysis Tool for Information Architecture. In: CHI '01: Extended Abstracts on Human Factors in Computing Systems, pp. 23–24. ACM Press, New York (2001)
2. Ewing, G., Freer, Y., Logie, R., Hunter, J., McIntosh, N., Rudkin, S., Ferguson, L.: Role and Experience Determine Decision Support Interface Requirements in a Neonatal Intensive Care Environment. J. of Biomedical Informatics 4/5, 240–249 (2003)
3. Henderson, S.: Genre, Task, Topic and Time: Facets of Personal Digital Document Management. In: CHINZ'05: Proceedings of the 6th ACM SIGCHI New Zealand Chapter's International Conference on Computer-Human Interaction, pp. 75–82. ACM Press, New York (2005)
4. Hudson, W.: Playing Your Cards Right: Getting the Most from Card Sorting for Navigation Design. interactions 5, 56–58 (2005)
5. Maiden, N.A.M., Hare, M.: Problem Domain Categories in Requirements Engineering. Int. J. Hum.-Comput. Stud. 3, 281–304 (1998)
6. Nielsen, J.: Usability Engineering. Morgan Kaufmann Publishers, Washington (1994)
7. Nielsen, J., Sano, D.: Sun Web: User Interface Design for Sun Microsystem's Internal Web. Comput. Netw. ISDN Syst. 1(2), 179–188 (1995)
8. Sinha, R., Boutelle, J.: Rapid Information Architecture Prototyping. In: DIS '04: Proceedings of the 2004 Conference on Designing Interactive Systems, pp. 349–352. ACM Press, New York (2004)
9. Upchurch, L., Rugg, G., Kitchenham, B.: Using Card Sorts to Elicit Web Page Quality Attributes. IEEE Softw. 4, 84–89 (2001)
10. Vredenburg, K., Mao, J.-Y., Smith, P.W., Carey, T.: A Survey of User-Centered Design Practice. In: CHI '02: Proceedings of the SIGCHI Conference on Human Factors in Computing Systems, pp. 471–478. ACM Press, New York (2002)

Mach: A Content Generating Engine for Adaptive Multimedia Applications in the Mobile Environment

Chian Wang

Department of Information Management
National Changhua University of Education
Changhua, Taiwan
cwang@cc.ncue.edu.tw

Abstract. For adaptive content generation, it means that the application can recognize a user's preference and consequently retrieve and present the appropriate media objects stored in the databases. That is, the application presents only the media objects with the conditions evaluated to be true according to the user's demand or preference. The feature of adapting presentation contents can significantly improve browsing results through providing materials that are more suitable for each individual user. In the mobile environment, some more changing factors can affect a user's access to the contents generated by the presentation engine. These factors include location, available bandwidth, past visits, the currently closest media source, etc. In this paper, we propose an adaptive content generating engine called Mach for mobility-aware multimedia applications. Mach's goal is that a single content description file can be applied to compose various multimedia presentations. By following the statements of the description file and the user's preference, Mach can dynamically generate adaptive contents for each class of users. In this way, the content designer does not need to compose several description files for each class of users. Mach is based on the XML (eXtensible Markup Language) standard and thus its platform independence makes our work well-suited to the mobile environment.

Keywords: multimedia, adaptive presentations, content generation, XML, mobile applications.

1 Introduction

During the recent years, the rapid development of the Internet has broken the geographical and temporal barriers of information access. This is mainly driven by the integration of rich multimedia materials and advanced web technologies. Another factor that also contributes in this direction is the diversity in the population participating in web surfing activities. As a result, the presentation system needs to consider the diverse requirements of different types of users in order to deliver the

M.J. Smith, G. Salvendy (Eds.): Human Interface, Part II, HCII 2007, LNCS 4558, pp. 674–681, 2007.

most effective information. The typical changeable preferences include language, personal information, connection speed, presentation quality, etc. Considering these varieties, the application should incorporate a content generation engine to adapt the contents according to each user's preference. By adaptive content generation, it means that the presentation system can recognize a user's preference and consequently presents the appropriate media objects [2]. That is, when presenting information to a specific class of users, the system presents only the media objects with the conditions evaluated to be true according to the user's demands or preference. In the mobile environment, some other factors can affect a moving user's desired information. These factors include the user's location, the available bandwidth, his past visits, the currently closest media source, etc [1]. For example, the user may want to find a Japanese restaurant in a 10-minute walking distance. Applications that are capable of handling adaptive presentations can provide specific materials that are more attractive and interesting to the users.

In order to manage and integrate the presentation of media objects effectively and efficiently, every multimedia application needs a content generation mechanism [3]. The mechanism's job is to describe the interactive behaviors of a presentation and the spatial and temporal properties of the constituent media objects, e.g., where in the display window and when the media objects are to be presented. With these specifications, the application can retrieve the media objects from the corresponding databases and render the presentation accordingly. Though different multimedia applications usually have various structures and requirements, it's preferred to adopt a common description mechanism that does not need a specific presentation viewer for the sake of popularity.

Currently, HTML (Hyper Text Markup Language) [5] has been the most common presentation mechanism for web-based applications. It provides a simple and efficient way for content description. However, HTML lacks for some advanced capabilities when used in many applications. First, HTML can not handle adaptive presentations. Because with an HTML file, the browser can generate only a single presentation. If the designer wants to make minor changes in the presentation, e.g., adding Chinese subtitles to a video clip, another version of the description file is required. This makes the development of adaptive applications very inefficient because the designer must compose several versions of HTML files for different types of users. Second, HTML can not describe the temporal relationships of media objects in a presentation, e.g., playing an audio file when a specific video segment is over. For an adaptive application, an essential feature is that its mechanism can dynamically generate the appropriate presentation contents according to the user's requirements and preferences. It's also preferred that the designer only needs to compose a base content specification that can be applied to diverse types of users to generate the various contents automatically. From this point of view, HTML is not suitable for adaptive applications because it can generate only static presentations. Therefore, a more robust and dynamic content generation mechanism is required. The mechanism (i) should be simple but robust enough to specify the properties of the media objects in a presentation, (ii) is compatible with common web browsers, and (iii) can handle adaptive presentations.

Based on the above considerations, we proposed a content generation engine called Mach in this paper to improve the shortcomings of HTML when developing mobile adaptive applications. Our motivation for proposing Mach is to reduce the cost and time to develop adaptive multimedia applications. In the context of this paper, a multimedia presentation is considered adaptive if it is capable of identifying a user's preference and interpreting it to infer the user's requirements to present the most appropriate contents in order to dynamically facilitate the browsing process. Mach is a generic XML-based mechanism for dynamic content generation and it focuses on the assembly and contextualization of media objects that are intended for an individual user so that the "optimal" browsing result can be obtained. Mach can thus act as a front-end of multimedia applications to dynamically handle how the presentations are composed and their adaptive and mobile requirements.

2 System Architecture

To enhance web browsing effectiveness, this paper explores the concept of adaptability in the context of mobile applications. The intention is to provide a mechanism for specifying and handling the adaptive behaviors in the mobile environment. Many multimedia applications have to generate contents on request and present them to the user. The contents are usually available in pre-built media objects stored in the corresponding databases. Each generated content is described by using a content description file. In this paper, we will propose an mechanism called Mach for dynamic content generation and presentation based on the XML standard [4]. At first in this section, we will describe Mach's architecture that is depicted in Figure 1.

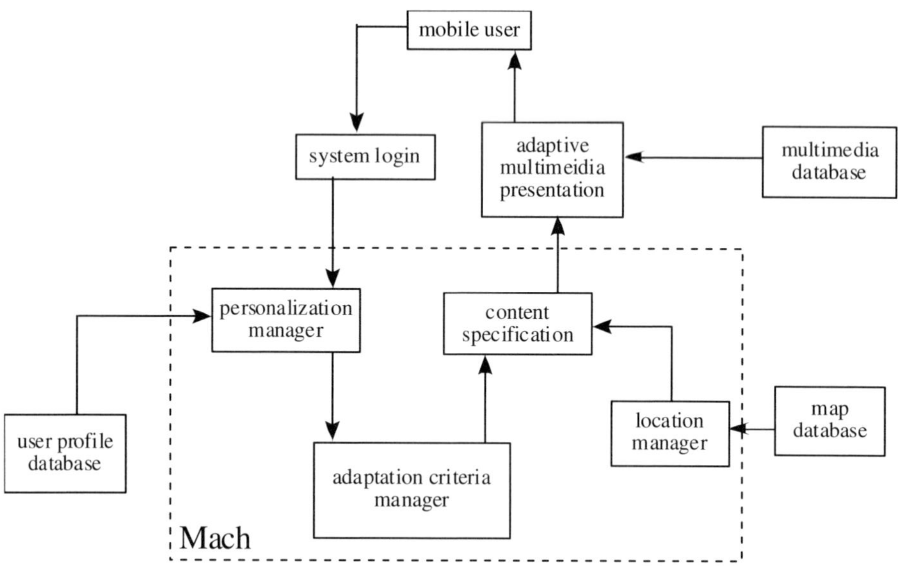

Fig. 1. System architecture

As shown in Figure 1, there are four major modules of Mach. Their functions are described as follows.

- personalization manager: to process user-related information stored in the user profile database as the basis to generate adaptive presentations.
- adaptation criteria manager: to determine the presentation layout and which media objects stored in the multimedia database are to be presented.
- location manager: responsible for retrieving the current location of the user with the map database and GPS information.
- content specification: After the presentation content is determined, this module generates the corresponding presentation specification file.

When a user wants to view a presentation, the browser retrieves the corresponding specification file generated by the content specification module. According to Figure 1, the specification file is generated by evaluating the adaptation criteria of the user. That is, the presentation contains only the media objects with the conditions evaluated to be true. For example, the browser presents appropriate speech and text according to the user's language preference, or it recommends only the restaurants that are less than 500m away from the user's current location. For each user, because the presentation content is specifically tailored for him, he can have more effective and satisfying results.

3 Content Specification

In this paper, we focus on describing Mach's multimedia content specification and presentation mechanism. As mentioned previously, the mechanism is based on XML so that no specific viewer is required to view the generated presentations. Any web browsers supporting XML can be used to view the presentations. The primary feature of the mechanism is that the actual presentation contents are generated dynamically based on the user's preferences and requirements. Also, the mechanism contains some control schemes to describe the temporal and spatial relationships of the media objects in the presentation. In this way, the presentation can be integrated and synchronous.

The core of Mach's presentation mechanism is the content specification language, called Adaptive Content Markup Language (ACML). ACML utilizes XML tags to describe the temporal, spatial, and adaptive properties of a multimedia presentation. Mach also contains the required Document Type Definition (DTD) of ACML to fully benefit from XML's flexibility and expandability. The tree structure of ACML tags is depicted in Figure 2. As shown in Figure 2, ACML consists of three major parts (i) The head block describes the spatial properties of the presentation, i.e., the layout of the presentation and how it looks like. (ii) The body block describes the temporal properties of the presentation, i.e., the relative timing sequence of the media objects. (iii) The adaptive block describes the adaptive behaviors.

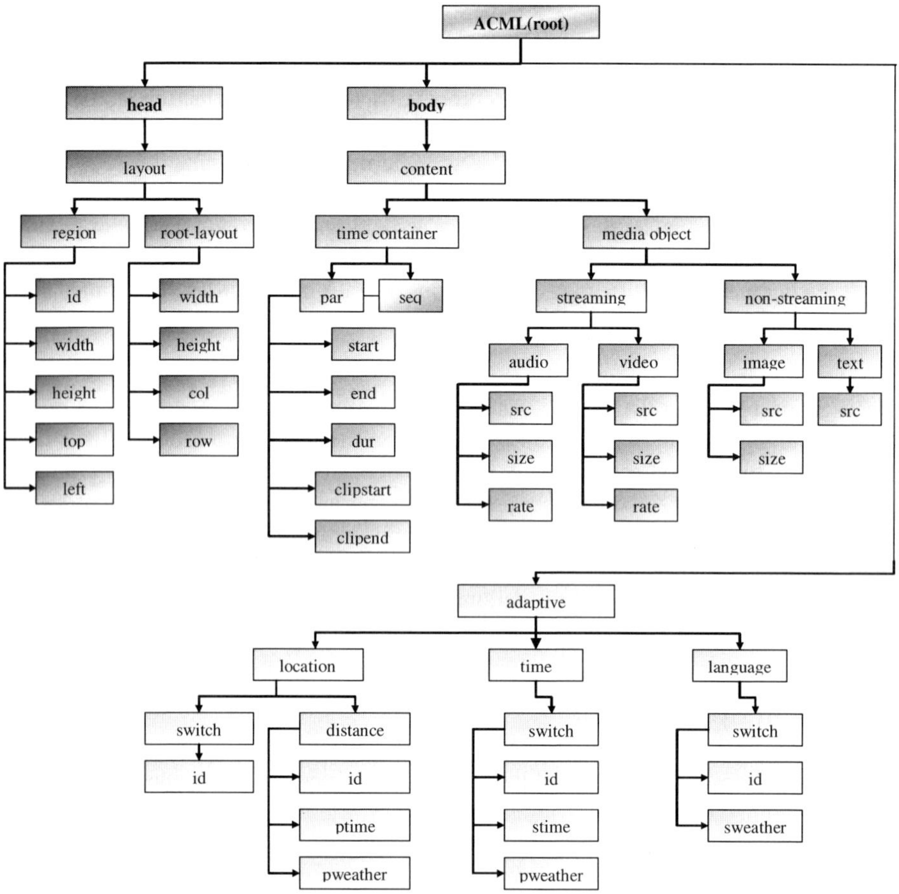

Fig. 2. ACML's tags

Currently at the prototype stage, Mach supports three types of classifying criteria: location, time, and language. Based on these criteria, each user can set his own preferences and then Mach generates the adaptive presentation contents accordingly. The three criteria are explained as follows.

(1) location: A basic characteristic of the mobile environment is that the user's location is changeable rather than fixed. Thus, it is reasonable to consider the user's location as the primary adaptive criterion. As depicted in Figure 1, Mach contains a location manager that is responsible for determining the user's location with the help of the map database and GPS information.

(2) time: In addition to the location, time is another factor affecting a mobile user's desired information. For example, he may want to go to different restaurants in the morning and in the afternoon. When generating the presentation contents, Mach considers the corresponding time to make the user more satisfied.

(3) language: With the Internet, web surfing becomes without any national or geographical boundaries. Because the users may come from all over the world, language has been a major consideration of web-based applications. To improve efficiency, the system should be able to dynamically adapt the language part of a presentation, e.g., subtitles and speech, instead of generating the whole presentation from scratch. Currently, most of the web browsers can automatically detect the preferred language of the user. With this feature, Mach can select the corresponding text and audio objects based on ACML's language adaptation tags in the content specification file.

4 System Development

To demonstrate the feasibility of Mach and ACML, a prototype adaptive application is developed. This section describes the system development of the application. Our goal is that Mach and ACML can be integrated into a complete mobile system to further improve business values.

Figure 3 depicts the components of the prototype application. There are three major modules in Figure 3: the XML parser, the synchronization processor, and the media content processor.

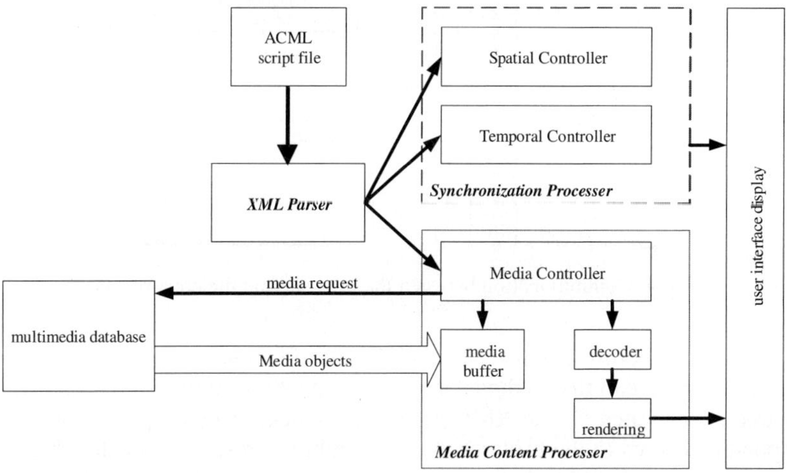

Fig. 3. System components

The presentation flow of the application is as follows. When the user wants to view a presentation, its specification file is passed to the XML parser for content generation. The XML parser parses the ACML tags in the specification file into a parse tree using the document object model (DOM) and keeps the corresponding attributes, including media type, start time and end time, and spatial location, of each media object in the tables. The synchronization processor includes the temporal controller and the spatial controller that allocate each media object according to its temporal and spatial attributes stored in the parse tables. After that, the media controller of the media content processor retrieves the corresponding media objects from the multimedia database. The media objects are first stored in the receiving buffer temporarily. When they are about to be presented, they are passed to the decoder and then the renderer. Finally, the user interface display plays them out under the control of the temporal controller and the spatial controller.

As to the communication among the components, Mach adopts Microsoft's ActiveX technology. ActiveX is based on COM (Computer Object Model) to provide programming interfaces that can handle the communication among web applications. With ActiveX, the components depicted in Figure 3 can send and receive messages via a common interface. In this way, the media objects and the related attributes can be exchanged and integrated more easily and efficiently. From the communication point of view, the relationship between the browser and Mach's components can be regarded as depicted in Figure 4.

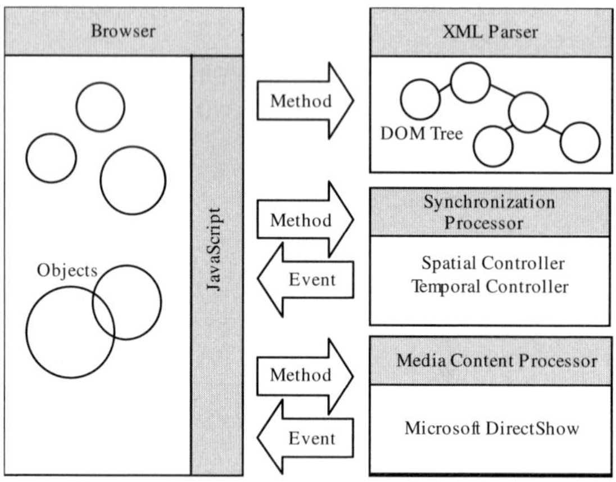

Fig. 4. Communication between the browser and the components

In Figure 4, the browser passes the ACML specification file to the XML parser by using JavaScript to call the methods of the corresponding ActiveX controls. After the DOM tree is generated by the XML parser, the synchronization processor sends the corresponding events to the browser via JavaScript to present the media objects. The media content processor handles the decoding and presentation of media objects by calling Microsoft DirectShow components in JavaScript. Additionally, because web browsers do not provide user interaction controls, e.g., pause, reverse, and fast forward/backward, as Microsoft Media Player and Real Network RealPlayer do, Mach uses JavaScript to handle user interactions. That is, the synchronization processor embeds JavaScript statements with the parsing result of the ACML specification file to provide the handling of user interactions to the web browser. With the embedded JavaScript statements, the web browser can control the presentation of each individual media object independently. In addition to VCR-like user interactions, the system also provides some other operations such as page-up and page-down to image and text objects.

5 Conclusions

As the source and background of web surfers vary a lot, it is preferred that the application can present the proper materials to different types of users. This is the basic concept of adaptive applications. Mach is a content generation engine for adaptive mobile applications. Compared with other similar works, the key feature of Mach is that it needs only one content specification file to generate the various presentations. This feature saves a lot of time and effort to develop adaptive applications. By browsing the dynamically adaptive materials and well-enabled user interactions, the users can have more better and effective browsing results.

References

1. Brunato, M., Battiti, R.: PILGRIM: A location broker and mobility-aware recommendation system. In: Proceedings of the 1st IEEE International Conference on Pervasive Computing and Communications, pp. 265–272 (2003)
2. Brusilovsky, P., Eklund, J., Schwarz, E.: Web-based education for all: a tool for developing adaptive courseware. Computer Networks and ISDN Systems 30(1), 291–300 (1998)
3. Zdun, U.: Dynamically generating web application fragments from page templates. In: Proceedings of Symposium of Applied Computing (2002)
4. Extensible Markup Language (XML), http://www.w3.org/XML/
5. Hyper Text Markup Language (HTML) Home Page, http://www.w3.org/MarkUp/

Part IV

Interacting with the World Wide Web and Electronic Services

e-Sports Live: e-Sports Relay Broadcasting on Demand

Woonhyuk Baek, Anjin Park, Jongin Kim, and Keechul Jung

HCI Lab., School of Media, College of Information Science, Soongsil University, 156-743, Seoul, S. Korea
bwhyuk@ssu.ac.kr, anjin@ssu.ac.kr, eyezzang01@nate.com, kcjung@ssu.ac.kr

Abstract. Electronic Sports (e-Sports) is computer and video games played as competitive sports over the Internet and a local area network, and are provided to users through a TV relay broadcasting system and the Internet due to tremendous popularity of the e-Sports recently. The main drawback of the previous e-Sports relay is to broadcast the e-Sports selected by the provider without viewers' intention, and thus the viewers can not watch desired information. Accordingly, this paper proposes a message-based semi-interactive relay broadcasting system called e-Sports Live over the Internet. The proposed system captures all regions of playing a game, and transfers the regions to a client's viewer. The client can watch desired information by selecting a slice of all regions. However, transferring continually all regions to the client over the Internet leads to a high-traffic due to high-capacity of data. To reduce the traffic, the system transfers all messages occurred in the game such as movements of characters instead of transferring continually all areas, based on resources repetitively used in game such as the whole map and game characters, and the resources are repetitively transferred to the client's viewer in an initial stage. Consequently, the proposed system can reduce the traffic over the Internet by transferring only message, and can perform a semi-interaction by watching desire information, as the client's viewer receives all areas occurred in the game based on resources repetitively used in the game and messages. Now, our system can not apply to TV relay broadcasting, as it does not have buffers for resources, but may be utilized for a variety of fields of interactive TV.

1 Introduction

Electronic Sports(e-Sports) is used as a general term used to describe computer and video games played as competitive sports over the Internet or a local area network [1]. Recently, in order to make the best use of e-Sports in entertainment industry, cable broadcasting companies[1] constructed a relay broadcasting system to broadcast the

[1] The e-Sports are serviced in Ongamenet [2] of OnMedia and MBCgame [3] of MBC which are the most popular cable relay broadcasting company in Korea, and the program has about 20% audience rating.

M.J. Smith, G. Salvendy (Eds.): Human Interface, Part II, HCII 2007, LNCS 4558, pp. 685–693, 2007.

e-Sports and has tendered the service for the e-Sports due to tremendous popularity of the e-Sports recently.

Previous relay broadcasting system did not make use of the notable characteristics of e-Sports, which means the system provides not information desired by the users, but only one-side information selected by the provider. To provide services for e-Sports desired by the user over the Internet[2], relay broadcasting system must continually send all areas. However, the system causes a heavy traffic if all areas are transferred to the client as images or videos because the whole areas are very wide. To reduce the traffic, several researchers proposed an effective video or image form for high-capacity of data [4,5], but they provide low-quality services, as the system compresses the video or image to reduce the size of data. In [6], almost all researchers for Internet-TV consider not the solution for low-quality services, but the solution of the traffic or application. Consequently, previous relay broadcasting system for e-Sports does not provide users with desired information, or causes the heavy traffic or low-quality information.

In this paper, we propose a message-based semi-interactive relay broadcasting system called e-Sports Live over the Internet. The proposed system can reduce the traffic by transferring only XML[3]-based text messages such as the movement of the game characters, based on information repetitively used in the game such as a whole map and game characters, instead of high capacity of multimedia data continually transferred to the client. Moreover, high-quality services are provided to the user using the uncompressed the information in an initial stage, and users can perform a semi-interaction, as they can watch the desired scene based on parts of the whole map selected by the user and variations of advance information occurred by the received messages corresponding to desired location. Consequently, the proposed broadcasting technique will be utilized for not only e-Sports but also a variety of fields of interactive TV due to previously mentioned characteristics which are the low capacity of message and the semi-interaction.

The remainder of this paper is organized as follows. We describe the e-Sports Live system and transmission techniques for e-Sports in detail in Section 2, and Section 3 describe the experimental results of e-Sports Live system applied to a sample of e-Sports. Finally, Section 4 summarizes our conclusions and presents possible future works.

2 e-Sports Live

Generally, the game repetitively uses identical resources such as the whole map and game characters. However, the previous relay system sequentially transfers the multimedia without re-using the resources, and transfers only scenes seen by the observer. Therefore, users can not watch the user's desired scenes or can miss a highlight when the game occurs frequently events in several regions. e-Sports Live transfers all regions of playing a game instead of scenes seen by the observer to be selectively displayed on a client's viewer. In a case of sending all regions as images to the client,

---

[2] In this paper, we do not consider the ordinary TV because the TV does not have buffer to save all regions.

[3] We use the XML data instead of meta-data for useful migration between several devices.

the relay system leads to a heavy traffic due to a high capacity of data. To reduce the traffic, our system previously sends game resources to client in an initial stage, and sends a message form instead of an image such as movements and/or variations of game characters. The game characters, the whole map, and so on are resources repetitively used in game, and these resources are sent to the client's viewers in an initial stage. The resources are also a high capacity of data because the whole map is a wide space and games have a lot of characters, but the system can reduce the traffic by sending only one resource per each game because same game uses same resources. Previous relay broadcast has two cases: a live TV broadcast and a filmed TV broadcast. In the live TV broadcast, the proposed system sends the resources before starting e-Sports if the resources are not saved in the buffer, and sends only messages while the e-Sports are playing. Moreover, all messages are saved in a memory to provide the e-Sports in the filmed TV broadcast. In the filmed TV broadcast, the proposed system sends the resources and saved messages. In two cases, the client's viewer displays all regions of playing e-Sports using received game resources and messages. In a case of StarCraft [11] as an example, resources repetitively used in the game are the whole map, game characters which are Marin, Medic, and so on, and buildings which are Command Center and so on, and are previously sent to the client in the initial stage. In playing a game, the messages are movements of characters, a creation and destruction of the buildings and so on, all messages are collected by the message grabber, and sent to the client or memory.

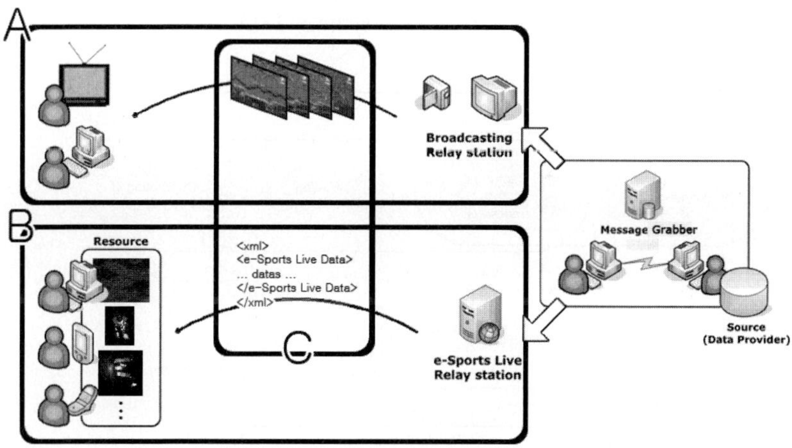

Fig. 1. Component diagram of the proposed system

Fig. 1 shows a component diagram to compare the proposed system with the previous relay broadcasting system. Symbol A in Fig. 1 indicates the previous relay technique using sequential images which mean a movie of e-Sports, and Symbol B indicates the relay technique of the proposed system. Our system transfers advance resources which means resources received in the initial stage to a client's program (Fig. 2) in advance, and the client's program receives only text-based XML data including messages such as the movement of the characters from the system to control the advance resources[4]. Symbol C indicates the messages transferred from the relay station. Previous system sends high capacity of data which is movie of e-Sports as the message, and therefore yields high traffic over the Internet or a local area network. Our system sends low capacity of data which is text-based XML data, and therefore user can watch the e-Sports in almost all devices in real-time by transferring quickly the messages from the system to client's program.

Fig. 2. Client's program on PDA

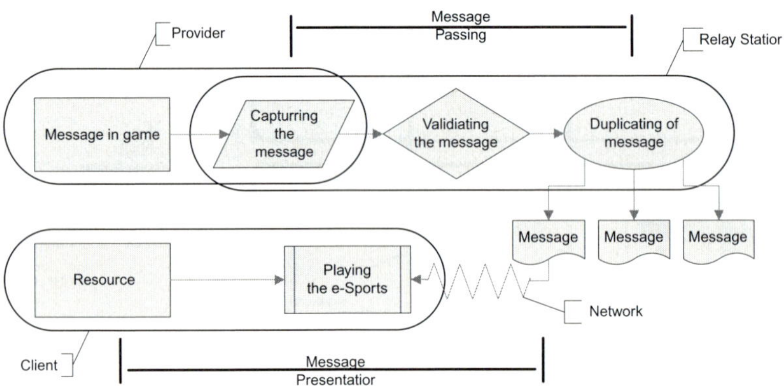

Fig. 3. Flow chart of the proposed system

Fig. 3 shows a flow chart of our system. All messages occurred in the game during gamer play the game are captured by the message grabber with the time, and the grabber validates the captured message by checking whether a part of game becomes different or not. Next, the system duplicate the message to send this message to all users,

[4] For example, the game characters move into the location corresponding to messages received from the relay system.

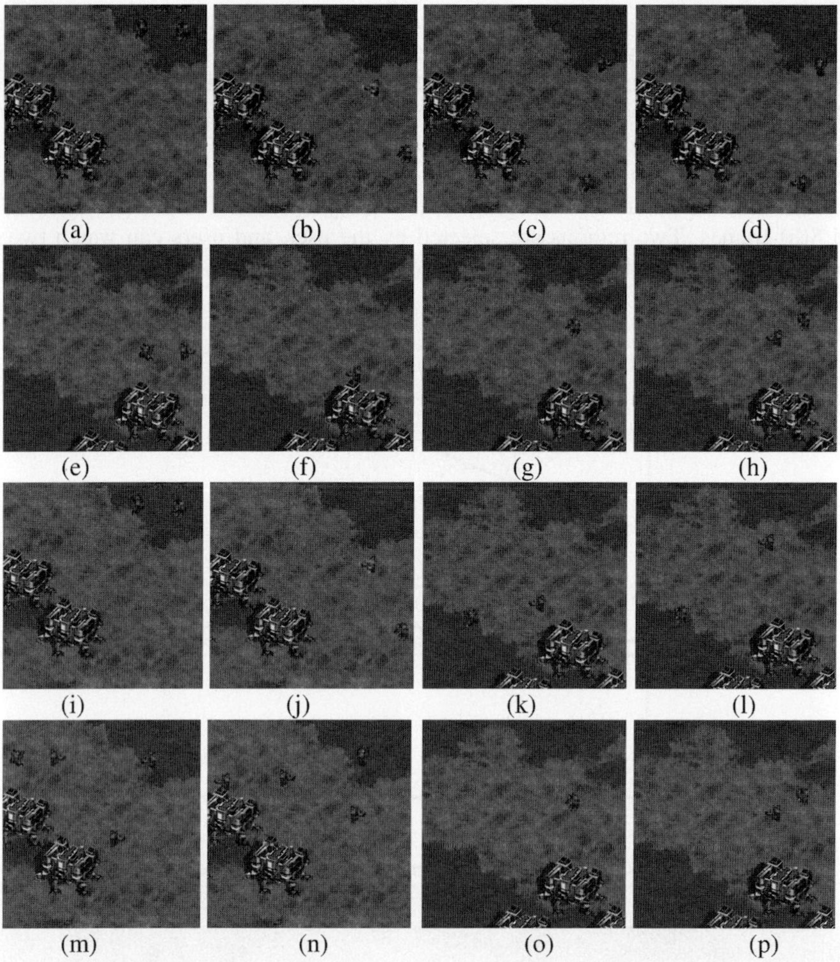

Fig. 4. e-Sports displayed on the PDA for evaluation of the interaction using a specific case

and the client's program play the e-Sports using both advance resources and messages transferred from the relay station.

3 Experimental Results

We implemented the proposed broadcasting system using a sample[5] of Starcraft which is the most popular e-Sports, and analyze difference between two systems which are previous relay broadcasting system using sequential image and e-Sport Live system using advance information and messages to verify the availability of our system. Fig. 4 shows e-Sports displayed on the PDA for evaluation of the interaction

[5] We directly made the sample of Startcraft because e-Sport Live system must control the resources of the game, but resources of the Startcraft are not open.

using a specific case that one player simultaneously attacks two bases of an opponent. In the previous system(Fig. 4(a-f)), the west region is displayed on the PDA from 1 frame(a) to 4 frame(d), and the east region is displayed on the PDA from 5 frame(e) to 8 frame(h). Two regions are selected by the observer of the provider, and a fight of the east region is half-over already, so the users can not watch half of the fight. In the previous system(Fig. 4(g-l)), the west regions is displayed on the PDA in 1(a), 2(b), 5(e) and 6(f) frames, and the east region is displayed on the PDA in 3(c), 4(d), 7(g) and 8(h) frames. Two regions are selected by the user, and users can watch two regions at the same time by selecting desire regions.

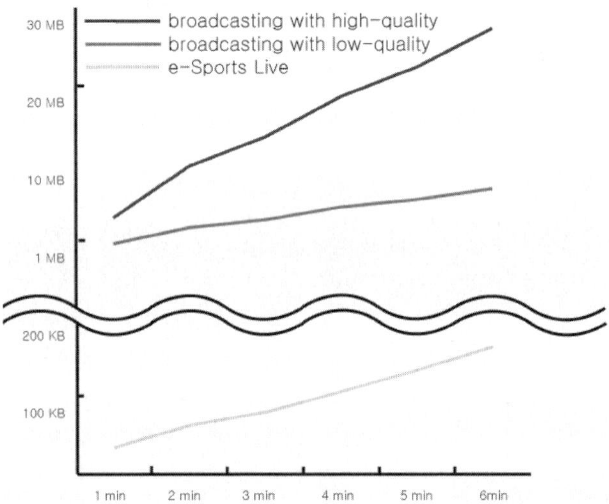

Fig. 5. Comparison of the size of transmission capacity between three systems

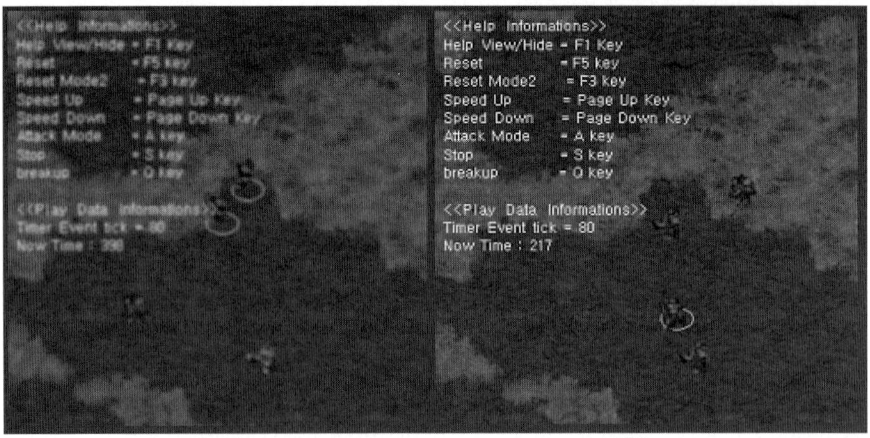

Fig. 6. Comparison of qualities of displayed scenes between two systems: (a) previous system, (b) e-Sports Live system

Fig. 5 shows differences of the transmission capacity between three systems over the Internet: transferring uncompressed high-quality images, transferring compressed low-quality images, and transferring XML-based messages, and X-axis of Fig. 5 is play time, and Y-axis is the transmission capacity. Although the proposed system using XML-based messages needs less capacity for messages than the previous relay system using compressed or uncompressed images as shown in Fig. 5, our system can perform semi-interaction by selecting directly desired scene by users, and can provides high-quality services (Fig. 6).

Table 1. Advance transmission capacity and time

	Previous Broadcasting System	e-Sports Live System using a sample game.
Capacity	None	5 MB
Time	None	40 sec.

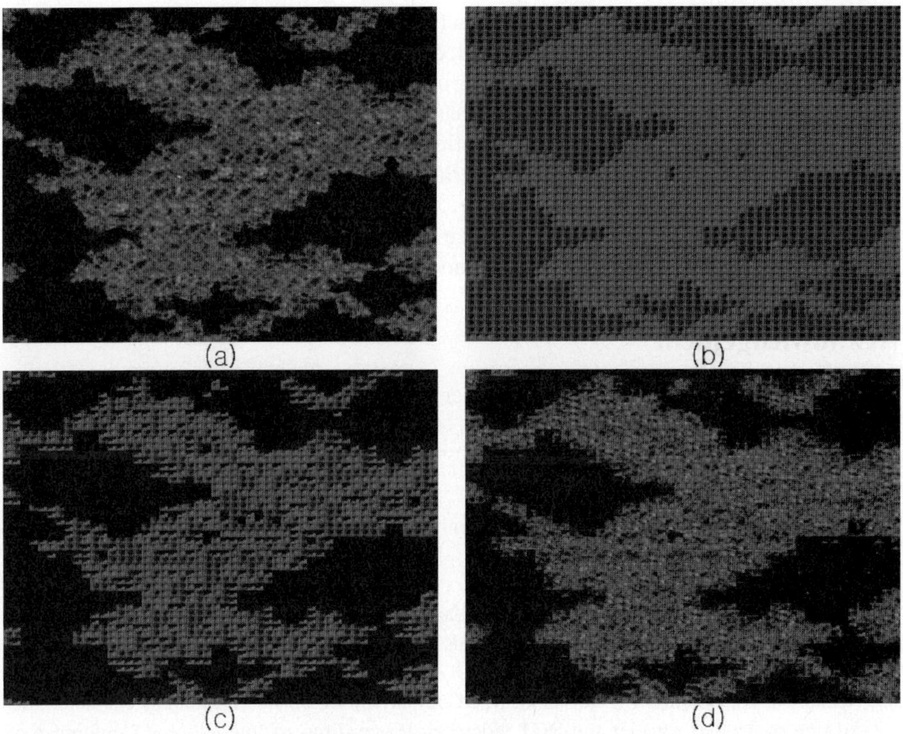

Fig. 7. Image displayed using code book in a specific case

When the system applies to a practical game instead of the sample game, it has a critical problem which needs high-capacitive advance information of the practical game. Therefore, the system needs much more times to play the e-Sports than the previous system if broadcasting time of the e-Sports is shorter than transmission time of advance information[6]. To solve the problem, from most well-defined image regions to less ones are sequentially sent to the client instead of sending all image regions. Well-defined image regions are found by k-Means algorithm[7], and the closer the image region is to means generated by k-Means, the more well-defined the image region becomes. Fig. 7 shows images sequentially displayed on a client display. Fig. 7(a) shows an original image, Fig. 7(b) shows an image based on the closest image regions to the means[8], Fig. 7(c) shows an image based on ten image regions close to the means, while Fig. 7(d) shows an image based on one hundred regions close to the means.

The second drawback of our system is closely associated with the game which will be broadcasted since the system must use the resources and engine of the game, and therefore development companies of the game must cooperate with service companies using our system.

4 Conclusion

This paper proposed a XML-based interactive system for e-Sports. Our system provides users with advance resources used continually in the game in advance, and broadcasts the e-Sports re-using the resources. Moreover, users selectively watch the desired scenes and can obtain more information in any time.

This e-Sports Live sample system satisfied efficiencies on a transmission, a quality of scene and demands of user, but has a critical problem which needs high-capacitive advance information of the practical game. As a future work we will investigate about trade-off between high-capacitive advance information and play time, and we will apply the general sports, such as baseball and football, based on e-Sports Live.

Acknowledgement

This is supported by Game Research Center.

References

1. http://en.wikipedia.org/wiki/Electronic_sports/
2. http://www.ongamenet.com/
3. http://www.mbcgame.co.kr/
4. Li, J., Yu, K., He, T., Lin, Y., Li, S., Zhang, Y.-Q.: Scalable Portrait Video for Mobile Video Communication. IEEE Transactions on Circuits and Systems for Video Technology 13(5), 376–384 (2003)
5. Kulapala, B., Seeling, P., Reisslein, M.: Comparison of traffic and quality characteristics of rate-controlled wavelet and DCT video. In: Proceedings of International Conference on Computer Communications and Networks, pp. 247–252 (2004)

---

6 For example, Table 1 shows the data size and time of the advance resources, and if the broadcasting time of the e-Sports is shorter then 40 seconds, our system needs much more times.

7 Image regions are expressed in 3-D that is R, G, and B to use the k-Means algorithm.

8 Fig. 7(b) consists of two image regions, as k is 2 in our implementation.

6. Kelly, E., Menard, R.: Internet TV: Guiding the Future of Broadcasting. International Broadcasting Convention, pp. 202–205 (September 1997)
7. Shim, H., Cho, Y., Chang, N.: Frame Buffer Compression Using a Limited-Size Code Book for Low-Power Display Systems. Embedded Systems for Real-Time Multimedia, pp.7–12 (September 2005)
8. Krishna, K., Ramakrishnan, K.R., Thathachar, M.A.L.: Vector quantization using genetic K-means algorithm for image compression Information. Communications and Signal Processing 3, 1585–1587 (1997)
9. Voinson, T., Guillemot, L., Moureaux, J.M.: Image compression using lattice vector quantization with code book shape adapted thresholding. Image Processing 2, II-641–II-644 (2002)
10. Bauer, C., Vinton, M.: Joint optimization of scale factors and Huffman code books for MPEG-4 AAC. In: IEEE Workshop on Multimedia Signal Processing, pp. 111–114 (October 2004)
11. http://www.blizzard.com/startcraft/

Privacy Requirements in Identity Management Solutions

Abhilasha Bhargav-Spantzel[1], Anna C. Squicciarini[1],
Matthew Young[2], and Elisa Bertino[1]

[1] Department of Computer Science, Purdue University
[2] Department of Industrial Technology, Purdue University
{bhargav, squiccia, mryoung, bertino}@purdue.edu

Abstract. In this paper we highlight the need for privacy of user data used in digital identity management systems. We investigate the issues from the individual, business, and government perspectives. We provide surveys related to the growing problem of *identity theft* and the *sociological* concerns of individuals with respect to the privacy of their identity data. We show the privacy concerns, especially with respect to *health* and *biometric* data, where the loss of privacy of that data may have serious consequences. Moreover, we also discuss how privacy concerns change according to the individual's disposition to provide the data. Voluntary disclosure of personal information is more acceptable to users than if information disclosure is involuntary, like in the case of surveillance. Finally, we highlight the shortcomings of current identity management systems with respect to the current privacy needs and motivate the need of hardened importance of privacy enabling functionalities in such systems.

Keywords: Identity Management, Security, Privacy, Social and Legal Concerns.

1 Introduction

Today a global information infrastructure connects remote parties worldwide through the use of large scale networks. Execution of activities in various domains, such as shopping, entertainment, business and scientific collaboration, and at various levels within those contexts, is increasingly based on the use of remote resources and services. The interaction between different remotely-located parties may be (and sometimes should be) based on little knowledge about each other. To support these rich experiences and collaborations, more convenient IT (Information Technology) infrastructures and systems are needed. We need, for example, for personal preferences and profiles of users to be readily available when shopping over the Web, without requiring the users to repeatedly enter them. In such a scenario, digital identity management (IdM) technology is fundamental in customizing user experience, underpinning accountability in business transactions, and in complying with regulatory controls.

Digital identity can be defined as the digital representation of the information known about a specific individual or organization. As such, it encompasses not only login names (often referred to as *nyms*), but many additional types of information,

M.J. Smith, G. Salvendy (Eds.): Human Interface, Part II, HCII 2007, LNCS 4558, pp. 694–702, 2007.

referred to as *identity attributes* or *identifiers*. The management of identity attributes raises a number of challenges, due to conflicting requirements. On the one hand, identity attributes need to be shared to speed up and facilitate authentication of users and access control. On the other hand, they need to be protected as they may convey sensitive information about an individual. In this paper we focus on the necessity to protect users' identity attributes and investigate related users' privacy issues. In the context of IdM, privacy is increasingly important due to the fact that the IdM systems deal with the sharing of user identity attributes with different entities. Improper and unsecured management of such attributes may lead to attacks and identity misuse as identity information can be exploited whenever authentication and authorization based on those identity attributes are required. One major fraud concerning digital identity is that of *identity theft* which is perpetrated by stealing digital attributes. Here, by identity theft we mean the act of impersonating others' identities by presenting stolen identifiers or other proofs of identities.

Malicious parties collect the sensitive identity attributes of (targeted) individuals and use them to impersonate these individuals or just to simply sell them. As such, attacks can be easily perpetrated when companies, often known as data brokers, collect very large amounts of data on consumers. Because those companies generally do not have direct relationships with the consumers, it is very difficult for consumers to control how their data is used. Therefore, there is a strong need to solve the problem of identity theft in the upcoming IdM solutions [3, 13, 19]. One way an IdM system can protect against identity theft is by preserving the privacy of its users. Privacy also relates to *sociological* concerns and acceptance of the shared identity information within an IdM system. Several Human Computer Interaction (HCI) studies [2] show that there is a strong relation between the users' privacy attitudes and their behavior in the context of e-transactions, i.e., their tendency to register for websites, provide incomplete data while registering, and provide inaccurate data. Privacy & American Business (P&AB) [18] classifies people as being privacy fundamentalists, pragmatists, or unconcerned. *Fundamentalists* "are passionate about what they [see] as business threats to their consumer privacy, and [favor] active government regulation of business and information practices." *Pragmatists*, who are a "middle group with balanced privacy attitudes," "ask what benefits they get as consumers in sharing their personal information to balance against risks to their privacy interests, and they usually favor a mixture of government and private solutions." *Unconcerned* individuals have "little to no concern about consumer privacy issues." Note that the concerns may also vary depending on the type of identity data. For example, most individuals consider health or biometric data more private than data such as product preferences and demographic data. Therefore, depending on the users of the IdM system, privacy requirements may vary. Moreover, the IdM system should ensure basic properties to hold cross domains because of the sharing of identity attributes and its ubiquitous usage.

Generally speaking the need for user control on his/her data is growing, but at the same time to ensure accountability and customized services the businesses do require identity information. In such cases, under what circumstances the identity data is provided by the user strongly relates to the privacy concerns associated with that information. For example, user providing his/her address while buying a book online is more acceptable to the user than a video of his/her activities be recorded in a workplace for surveillance purposes. Similarly, several government records are public in order to enable citizens to monitor their government and to ensure accountability in a democratic society. The challenge to policymakers is to balance the public's right to information with the individual's right to privacy. This is critical as in most countries every major change in life is recorded somewhere in a government document. This includes birth certificate, driver's license, marriage certificate and so on.

From the above discussion, we can thus see that it is important that IdM solutions accommodate the privacy needs and rules of all three constituents, the individual, the business and the government. Based on the privacy needs we identify a number of corresponding IdM system requirements. We highlight the privacy requirements and principles based on which IdM system can address the current gaps. The rest of the paper is organized as follows. In Section 2.2 the current social and legal concerns related to privacy in IdM systems are discussed. Identity theft is highlighted as a pressing concern related to the need for privacy. In Section 3 the types of identity data provisioning and privacy concerns related to them are elaborated. In Section 4 we discuss the shortcomings of current IdM systems with respect to the privacy requirements and then conclude in Section 5.

2 Surveys of Privacy Concerns

A classic reference to the meaning of privacy goes back to Aristotle's distinction between the public sphere of politics and political activity, the *polis*, and the private or domestic sphere of the family, the *oikos*, as two distinct spheres of life [20]. Ever since then, there have been several critiques concerning the exact meaning of privacy and several constitutional laws have been enacted in various countries stating the rights to privacy [7, 1]. In our current digital age in which an individual's information is extensively collected and distributed, the value of privacy is gaining significant importance. Individuals want to have control of their electronic identity information which can be misused if not properly secured. Managing identity attributes raises a number of challenges, and the identity information needs to be protected as it may convey sensitive information about an individual that can be target of attacks like identity theft. In what follows we first highlight data emphasizing the concerns related to the specific case of identity theft from the social and legal aspects. We then discuss some sociological concerns of users captured by various surveys. In particular, we show how the concern differs based on the type of identity data. For example, users show higher privacy concerns about their health and biometric data, versus their identity data indicating their demographic information.

2.1 Identity Theft

Usually, identity theft in the digital world occurs to obtain credit or to perform crimes, like accessing classified records without having the proper authorization. Americans are increasingly concerned about identity theft, and with good reason as it is a serious economic crime. In 2005, the Consumer Sentinel, an FTC complaint database, received over 685,000 consumer fraud and identity theft complaints. 37% of those complaints are related to identity theft [8]. As of July 2006, the U.S. Justice Department reported charging 432 individuals with aggravated identity theft for the current calendar year [13]. President George W. Bush called identity theft "one of the most harmful abuses of personal information" [11]. In fact, the average monetary loss per victim attributed to the crime of identity theft is more than the amount attributed to bank robbery [9]. Through attacks such as password cracking, pharming, phishing, and database attacks, malicious parties can collect sensitive identity attributes of (targeted) individuals and use them to impersonate these individuals or just to simply sell them. Even though solutions are available that mitigate such attacks [5], a comprehensive approach to the problem of identity theft cannot rely solely on these techniques and must be able to offer protection when these solutions fail.

Increased federal and state legislation regarding identity theft has brought a heightened awareness to identity theft in general and the special status of an individual's SSN as an identifier in particular. For instance, the Identity Theft and Assumption Deterrence Act of 1998 makes identity theft a federal crime [18 U.S.C. *x* 1028 (2003)]. The purpose of this statute is to criminalize the act of identity theft itself, before other crimes are committed. Many of the new laws enacted at the state level contain provisions addressing the circumstances under which SSN and other personally identifying information can be disclosed to third parties, confidential destruction of papers and electronic media containing SSN and personally identifying information of customers, and requirements for encryption of SSN and other sensitive personally identifying information held in electronically stored mediums.

2.2 Sociological Concerns

An individual typically possesses various types of identities, for example, workplace identity containing his/her financial information versus home identity containing details of family health records. Typically, information concerning health and biometric is of greater concern than other types of identity as is illustrated below. In 2000, a survey conducted by Privacy and American Business and ORC International [15] highlighted a paradox leading to the growing number of privacy 'pragmatists' in North America. The survey showed that as long as the information is not abused, most U.S. citizens support the free flow of information from public records. Citizens are concerned however, about their privacy in some instances when personal information is extracted from those records. The individuals surveyed believe it is less acceptable to allow private investigators or ordinary individuals to access such information. The 1,000 people surveyed in the report also said that they are against the government posting personally identifiable public information on the Internet unless specific safeguards are in place. These safeguards include the government requiring the consent of the individual before personal information is displayed on public record, and specifying a specific purpose for such information to be published.

In 2002, a survey by Gartner Inc. (NYSE: IT and ITB) [10], showed that 41% of U.S. citizens are opposed to the creation of a *national identification database* to identify citizens and visitors to the United States. Only 26% of U.S. citizens agreed that such a database be established. The survey showed overwhelming support for the use of national IDs at airports and as a means to gaining entry to the U.S., for example, and much less acceptance of national IDs as a means to controlling access to healthcare and banking services.

A survey carried out in 2005 [18] shown that consumers rate personal *health information* and financial information as the two most sensitive types of personal information. Individuals with chronic and especially genetically-based health conditions express sharp concerns about circulation and use of information about their health status in that they are afraid that if would result in denying them important consumer opportunities and benefits. Moreover consumers also express concerns about privacy and security in the current move to greater collection and use of electronic medical records. While 80% of online consumers go to health sites for information, they express high concerns about privacy and security of their surrounding. Because of their privacy concerns, many consumers using health information web sites do not share their personal data, and thus do not take full advantage of services and information from these sites. The 2001 Pew Internet and American Life Project survey found that 80% of respondents said that it is important for them to be able to access health information anonymously. Finally, consumers also express fears that their health information might be maliciously accessed to commit identity thefts as illustrated in Section 2.1.

In a 2006 survey, the perceptions of 391 individuals on issues relating to *biometric authentication* were analyzed [6]. In the recent years there has been an increased number of applications and biometric system deployments that affect individuals. One study projected revenues to grow from $2.1B in 2006 to $5.7B in 2010 [12], mainly due to incorporating biometrics in large-scale immigration programs such as the U.S. VISIT program (United States); CANPASS (Canada); Ben Gurion Airport (Israel); and Schipol Airport (Amsterdam). Biometric applications are widely employed by the banking sector, at ATM and advanced financial kiosks, for point-of-sale payment, and on personal consumer electronics (laptops, PCs, and tablets). Interestingly, from the survey 25% of the respondents had not heard of biometrics. Those who had heard of biometrics expressed several concerns, including cleanliness of the devices, safety (with respect to iris and retinal identification), applications for which their biometrics were going to be used, and who would have access to their biometric information. Results demonstrated overwhelming support for biometrics applications involving law enforcement and obtaining passports, while applications involving time and attendance tracking and access to public schools ranked lowest on the list. When asked whether the survey respondents thought that biometrics would enhance or threaten privacy, 26.3% were neutral, 34.8% were either concerned or very concerned, and 26% were either not concerned or hardly concerned. Age was significantly related to privacy concerns. Out of all of the respondents, those older than 35 thought that biometrics would threaten privacy, whereas those younger than 35 believed that biometrics would enhance privacy. Previous surveys on privacy concerns of biometrics [4] have also highlighted that people are concerned about privacy and the potential for misuse of information, especially by government agencies ([21]). According to [14], banking institutions are most trusted to manage identity credentials, perhaps because they are currently perceived as being first on a top-ten list as doing a good job of verifying identity. Since identity verification is a critical functionality of any IdM system, this observation also asserts the need of careful analysis of the IdM leading to better privacy assurance of the end users data.

3 Privacy Concerns While Providing User Identity Information

Problems like identity theft and other privacy concerns that we have discussed in Section 2 have pushed many users to request more control over their personally identifying information (PII). At the same time, businesses say their customers want customized products and services that are possible only through the collection, analysis and use of information about them. Thus, a tradeoff needs to be found. In the following two subsections we investigate two types of disclosures, the first in which the user voluntarily provides his/her identity information, and the second where such information is collected without an explicit approval or knowledge of the user. We show how the type of disclosure affects the privacy concerns for the information provided.

3.1 Acceptable and Voluntary PII Disclosure

Users often provide personal information without thinking about it or without knowing that the information can be used in ways they never intended to be used. This can happen in the course of common business activities such as filling out a product warranty card, sending in a magazine subscription, applying for a loan, or purchasing a product on the Internet. Users' name, mailing address, e-mail address, telephone number or other requested information can be used for filling lists that are the likely source of unsolicited offers. In general, if the user is obtaining a service in return of providing the information he/she tends to be more willing to disclose whatever information is required, without actually wondering whether or not the information is necessary for completing the transaction or if it's only for marketing.

Many companies operate under the assumption of implied consent regarding the use of personal information. This means that they very often decide how to use customer's data unless notified otherwise. On the contrary, users often have the opposite understanding, that is, that they will be informed of every action/use of their data from external retainers. Unfortunately, such perception does not often correspond to the PII management practices by such retainers.

Some companies inform consumers of their information practices through privacy policies. On the Internet, access to privacy policies (or privacy statements) is usually located at the bottom of a web sites home page (double click on the terms relating to privacy). Privacy policies are company specific statements that describe how personal information is used and with whom it might be shared or sold. Privacy policies are typically encoded using an XML based language, called P3P [17] which has been specifically conceived to describe servers' privacy practices with respect to data management and collection. P3P policies can be automatically matched against user privacy preferences, which are set up using internet browsers' features. Note however, that although many leading companies and popular web sites have privacy polices publicly available, only financial institutions (including brokerages and insurance companies) are currently required by law to have them.

3.2 Unknown PII Disclosure

Businesses may require strong surveillance to ensure accountability for business needs. Surveillance often implies disempowerment of the subject who has no control in the disclosure of his/her identity (and other) information. Businesses are however compelled through privacy legislation like HIPAA to be aware of the individuals privacy rights and preferences. When any form of surveillance is enforced for accountability purposes, users are not always aware of it. More often, if they are informed about how their personal information and actions are monitored, they do not accept and understand the ultimate goal of such invasive security precaution. Being monitored and under the "big eye" of a camera makes many individuals uncomfortable, who find these type of actions often needless and excessive. Other examples of these forms of monitoring for users exist when making queries in the web if the IP address is recorded for profiling or to hold abusive behavior accountable.

As discussed in Section 2.2, the willingness of individuals of disclosing their PII is definitely coupled with their understanding of the need and the benefits of releasing this information and/or data. In addition to the PII directly provided to companies, other sources of personal information include the Internet and the tons of discarded mail users throw in their recycling bins each year. Searching the Internet or mining the trash for personal information are techniques commonly used by attackers who may try to steal identity for personal gain.

4 Shortcomings of Identity Management Systems Related to Privacy

Several projects and initiatives are currently on-going to investigate digital identity management for federated systems [3, 13, 19]. A first limitation of most such approaches is that no information is provided about whether the strong and weak identifiers have been *verified* and the strength of this verification. We believe that if a party managing a protected resource is provided with such information for another requesting party, it is in a position to make a better judgment concerning whether to give access the requesting party access to the resource. Second, those approaches to improve acceptance by users, should allow the users be free to choose *any combination* of identity attributes for carrying on their online interactions. The identity attributes that service providers require may change in time, depending on the specific purpose of the interactions and on the providers security policies. Identity attributes may also have different privacy requirements. Another major drawback is that no specific techniques are provided to protect against *misuse of identity attributes*. Dishonest individuals can register fake attributes or impersonate other users of the federation. Moreover current approaches do not address how biometric and health data can be used together with other identity attributes without compromising the privacy of the user. In late 1980, the Organization for Economic Cooperation and Development (OECD) [16] issued a set of Guidelines concerning the privacy of personal records. These OECD guidelines are widely accepted and form the cornerstone of fair information practices and regulations designed to protect privacy of personal information around the world. These guidelines underpin most current international agreements, national laws, and self-regulatory policies. Current IdM systems do not satisfy several of the crucial OECD principles and would benefit with regards to privacy and security features by incorporating them.

5 Conclusions

In this paper we presented the privacy concerns in IdM systems. We also illustrated the fact that preserving privacy is more than just a technical dilemma. While some security techniques can be used to preserve privacy, the user him/herself may have some control on when and how to release his/her identity data. As such, the privacy laws, as highlighted in Section 2, also dictate the management of such data. Furthermore, unlike some aspects of security, privacy is reliant solely on the user's perceived level of privacy that is being upheld by other entity(s). While government metrics and regulations provide some level of understanding for the user about his/her privacy; the actual privacy provided by the system depends on the user perceptions and sociological concerns. Surveys of user perceptions and opinions are thus vital, and need to be conducted further and in greater frequency especially in the context of the emerging IdM systems.

References

1. Asia-Pacific Privacy Charter Initiative (2003) http://www.cyberlawcentre.org/appcc/
2. Brown, M., Muchira, R.: Investigating the relationship between internet privacy concerns and online purchase behavior. Journal of Electronic Commerce Research, 5(1) (2004)
3. Microsoft Windows CardSpace. http://cardspace.netfx3.com
4. Citizenship and Immigration Canada. Tracking public perceptions of biometrics. In: POLLARA Omnibus Survey (2003)
5. Dhamija, R., Tygar, J.D.: The battle against phishing: Dynamic security skins. In: SOUPS '05: Proceedings of the 2005 symposium on Usable privacy and security, pp. 77–88. ACM Press, New York (2005)
6. Elliot, S.J., Massie, S.A., Sutton, M.J.: The perception of biometric technology: A survey (2007)
7. Federal Trade Commission. United states federal trade commission privacy case summaries (2006)
8. Consumer Fraud and 2005 Identity Theft Complaint Data (January-December 2006) http://www.consumer.gov/sentinel/pubs/top10fraud2005.pdf
9. Identity Theft: A $2 Trillion Criminal Industry in 2005 FTC Fact Sheet, Aberdeen Group. http://www.ustreas.gov/o±ces/domestic-finance/financial-institution/cip/pdf/identity-theft-fact-sheet.pdf
10. Gartner Incorporation Gartner reports strong opposition to a US national identity program (2002)
11. National Consumer Protection Week Proclamation George W. Bush (2002) http://www.whitehouse.gov/news/releases/2002/ 02/200202 04-8.html
12. International Biometric Group. Biometrics market and industry report addresses biometric investment landscape through 2010 (In Press Release 2006)

13. Liberty Alliance Project. http://www.projectliberty.org
14. Moody, J.: Public perceptions of biometric devices: The effect of misinformation on acceptance and use. In: Informing Science and Information Technology (InSITE) (2004)
15. Mosquera, M.: Ranks of privacy 'pragmatists' are growing. TechWeb (July 2000)
16. Organization for Economic Cooperation and Development. http://www.oecd.org
17. P3P: The Platform for Privacy Preferences. http://www.w3.org/p3p/
18. Privacy & American Business. http://www.pandab.org/
19. Shibboleth, Internet2. http://shibboleth.internet2.edu
20. Stanford Encylopedia of Philosophy. http://plato.stanford.edu/entries/privacy/
21. TNS / TRUSTe Study. Consumer attitudes about biometrics in id documents (2005)

Collaboration Between People for Sustainability in the ICT Society

Gunilla Bradley

IT University, School of Information and Communication Technology
Royal Institute of Technology, Electrum 229, 16440 Stockholm, Sweden
Bradley@imit.kth.se

Abstract. At the present Net Work Period of the IT history deep changes are taken place in collaboration between people and human communication, its structure, quantity, and quality. A dominating steering factor for the design and structure of work life as well as private life is the convergence of three technologies, computer technology, tele technology and media technology (ICT). Telecommunication technology has come to play a more a more dominant role in this convergence, especially internet and web technology. Embedded (ubiquitous) computer technology is making the process invisible, and media technology converges within itself (multimedia or cross media). Well functioning organizational *and psychosocial communication* are an important prerequisite for successful industrial and social change in the ICT society. Managing and working in an organization organized as a network, involves communication between people, groups, units, other organisations, and various combinations of these entities. ICT applications together with deep knowledge and insights in organisational design and management (ODAM) are the keys to social change. The author describes her convergence theory on ICT and Psychosocial Life Environment with special emphasis on psychosocial communication and sustainability in the Net Era of the ICT society.

1 Sustainability

1.1 What Is Sustainability

The concept "Sustainability" was brought to the fore by The World Commission for Environment and Development (Brundtland commission, 1967) which was a follow up of the first UN World conference on Environment held in Stockholm, 1972. The so called "Agenda 21" stated four mutual independent dimensions of sustainability with a holistic approach (Agenda 21): Ecological, Social, Economic, and Cultural. It turned out to be various opinions in various countries and discussions started on issues like: who has the responsibility for sustainability, what actions should be taken, and what priorities should be set.

Aspects more directly connected to the e-work place and e-society were raised in a Swedish report (Swedish Ministry of Environment, 2003) "A strategy of sustainable development". Key areas in the sustainable development are covering a broad spectrum: Future physical environment, climate changes, *health and population,*

M.J. Smith, G. Salvendy (Eds.): Human Interface, Part II, HCII 2007, LNCS 4558, pp. 703–712, 2007.

social coherence, welfare and security, employment and learning in the E-society, sustainable economic growth and competition, regional development, sustainable social change. Italics above are used to point out the ones that could be related to the ICT society

There are some main approaches of research and policy to the cross disciplinary area of Sustainable Information Society.

1. The impact-based approach makes a substructure of the concept sustainable development with three dimensions or policy areas (environmental, social and economic sustainability). Regarding environmental sustainability, the most relevant ICT impacts are the optimization, substitution or induction of physical production or consumption processes. Regarding social sustainability, social inclusion, participation and wellbeing/health are viewed as the most important issues influenced by ICT development and use. Regarding economic sustainability, the long-term structural change towards a knowledge economy enabled by ICT infrastructures is the most important aspect discussed. New business models, so called social business, are being implemented and theories of market liberalism are challenged.
2. The Internet governance approach focuses the telecommunication part of the ICTs and internet technology. A central issue is the control and power over internet. Examples for the internet governance approach: Internet Society, ISOC, a non-profit organization dedicated to ensuring the open development, evolution and use of the Internet for the benefit of people throughout the world.
3. The Human Rights Approach asks the normative question about the use of ICT for a sustainable development. It explicitly mentions the fulfillment of the Human Rights as the fundamental goals for the information society. Examples for this approach are World IT Forum and World Summit on the Information Society (WSIS).

International Federation for Information Processing (IFIP) has founded a Working Group *ICT and Sustainable Development.* Community Informatics Network (CIRN) which is represented as a special track at the current HCII 2007 arranged an international symposium in 2003 on Community Informatics and Sustainability.

2 Convergence Model on ICT and the Psychosocial Life Environment

In the former HCII 2005 proceeding (Bradley 2005) the author described the convergence model, its concepts and their interrelations and the process of social change and ICT from a global perspective. In the current paper this model is taken further to encompass the concept of sustainability. Sustainability is a growing concern in the world mainly tied to climate changes and the green house effect. However the sustainability is also related to ICT as described in section 1.

The Network period in the evolution of computer technology is very much based on the convergence and integration of three main technologies; computer technology, tele technology and media technology. Telecommunication technology is playing a more and more dominant role in this convergence, especially internet and web technology. Embedded (ubiquitous) computer technology is making the process invisible, and media technologies converge within itself (multimedia and cross media). The convergence process is enforced all the time by smaller, cheaper, and more powerful components. ICT is being used in almost every activity and increasingly embedded in equipments and belongings (ubiquitous computing).

Converging circles are reflecting graphically the ongoing process, related to technology and the social processes at various levels of analysis. Both "convergence" and "interactions" are important features in the model. Convergence here means a move towards a common content. Interaction means that technology interacts with the social world with values and believes. There is also an ongoing interaction between the "clusters of circles". The thin double-directed arrows in figure 1 are representing interaction and the one-directed arrows are representing the main direction for the main movement and the process described in the circle model.

- Lower left of figure 1: A convergence is taking place between technology, economy, norms/values and labour market under the name of *Globalisation*. The geographical span is changing. At present our work life is based mainly on national and international trade which we will se softening in the future taking on a more global character. Electronic commerce and electronic market places are creating a strong change factor for the structure of work life. The geographical space in the future will be both global and beyond – including virtual reality (VR)
- Upper left: A convergence of computer technology, telecommunication technology and media is much evident, resulting in what is entitled *Information and Communication Technology (ICT)*. Knowledge is managed in new ways. In the 2000s there are multiple channels to choose from, word of mouth, writing, and audio visual, electronics. In the future we talk more about meta channels, e.g. meta media of virtual reality (VR) and controlled reality environments i.e. environments manipulated and managed in VR.
- Upper right: Work Environment, Home Environment and Public Environment are converging into a *Life Environment*, where the public issues tend to merge the private sphere of our homes. *A new emphasis* is put on certain dimensions in the psychosocial environment and new dimensions are appearing in the psychosocial environment. Openness for unforeseen implications is required.
- Lower right: Professional Role (Work Life), Private Role (Private Life) and Citizen's Role (Public Life) converge forming a *Life Role.*
- Middle of the figure: Effects on the Individual become more multi faceted and complex. This is valid for both psychological and the physical effects - *Effects on Humans.*

Four circles representing *Virtual Reality (VR)* are marked with dotted lines and are surrounding the set of converging circles. They could also be drawn as dotted circles that converge with the other circles in each of the four clusters/groups that represent ICT, Life Environment, Life role and Globalisation in figure 1. These circles reflect our participation in cyberspace on various levels. To the lower left part in figure we could talk about *Virtual Worlds* on the global level. Within the concept of ICT, the steps taken by applied *Embedded and ubiquitous technology* make technology more hidden to the individual and society as a whole (upper left). *Virtual Environments* in the upper right part of the figure are already a common concept. Finally we could talk about V*irtual Human Roles* (lower left), which could in more extreme forms could be another personality that people play e g avatars.

"Effects on Humans" is represented in the convergence model in figure 1 by a flower in the middle and the effects are multifaceted. The converging circles are forming a Life Role and special new life styles are created. Research focusing upon the individual is crucial that is research on how the use of ICT impact:

- identity and self perception
- social competence
- creativity
- integrity
- trust – security – privacy (the terms vary)
- addictiveness, dependency

The convergence model in figure 1 can also be applied to describe changes in academic disciplines such as Informatics. A change in focus regarding research and development is taken place. Analysis and design address increasingly both the work process and management connected to the sphere of production life and people's life environment. Not only professional roles but also our roles as citizens and private persons are going through a deep change. Community research in a broad sense comes to the fore, with respect to both physical and virtual communities. Analysis and design of ICT and societal systems both at local level and globally become important. There is also a requirement to involve new and additional actors at the deeper and broader integration of ICT in the society (children, elderly, and consumer organisations). Educational programs on Community Informatics and Social Informatics are appearing in many academics institutions.

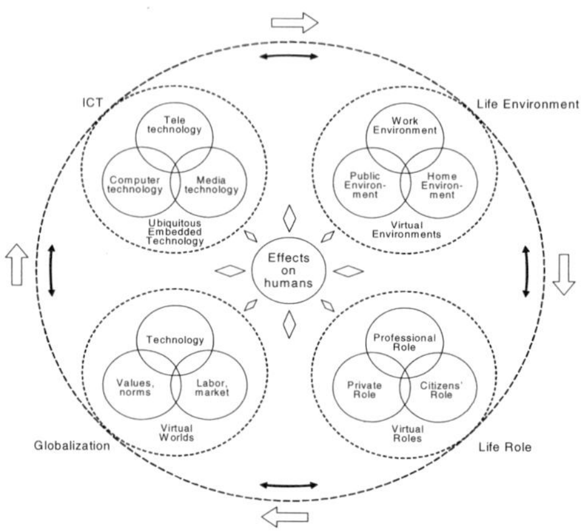

Fig. 1. Convergence Model on ICT and Psychosocial Life Environment (Source: Bradley, 2005, 2006)

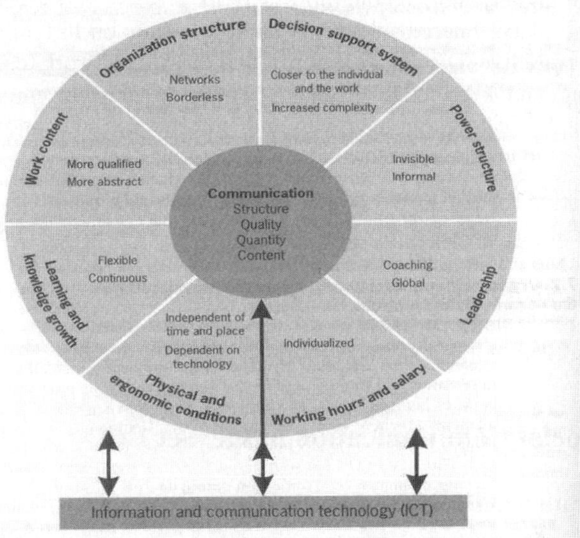

Fig. 2. Organizational Design and Management (ODAM) – Communication circle on Psychosocial Communication (Source: 2006)

International collaboration and communication in various forms, not at least at the widespread use of internet and web technology, has become an important component in *sustainability* in the workplace in the ICT-society. The message in the communication circle (figure 2) was initially (the authors publication from 1977, 1993) in short that technology *directly* effects communication because it is a tool and hence affects the structure, quality and quantity of communication, but also effects the communication between humans *indirectly* because of the changes in the phenomena which were described briefly in the segments of the circle. The technology (ICT) effects:

- Work content – Work tasks are becoming more qualified and need new and additional competences. There is an increase of abstract components in the work tasks.
- Organization structure – Networks organizations are developing and the borders of an organisation become harder to identify.
- Decision support system – Decisions are taken closer to the individual and the work place. Parallel to decentralization is also centralization. The complexity of decisions making is increasing both regarding content and process.
- Power structure – There is both a formal and an informal power structure. The informal power structure is invisible and in network organisations the power becomes increasingly invisible.
- Leadership – Global leadership is asked for and there is a trend towards a coaching model of leadership.
- Working hours and salary – A revision is going on regarding basic principles. An individualized salary system and individualized working hours are becoming more frequent.

- Physical and ergonomic conditions – Work is being performed independent of time and place. There is an increased dependency on ICT in work life as well as in private live. Distant work forms and distributed work teams are common.
- Learning and knowledge growth – There are new incentives for learning, learning by asking for and searching information and knowledge. Flexible learning and continuous learning are key words.

These objective factors in the organisational and work environment are effecting the communication between people directly and indirectly. The basic thinking reflected in the communication circle turned out to be applicable throughout the continuing technology phases and became fundamental at the introduction and use of internet and collaboration in distributed work environments. The later results were presented in a doctoral thesis in 2005 (Jansson 2005).

3 Psychosocial Communication in the Net Era

3.1 Collaboration in Distributed Environments

Psychosocial Communication was analyzed in Jansson and Bradley (2004). Jansson had performed a comprehensive field work which in 2005 resulted in a doctoral thesis in informatics and Bradley had in research projects during various periods of computerization studied psychosocial communication and ICT and reported theory and empirical findings in book chapters and articles. "Communication between people" was also identified as a main factor in Bradley's early research in psychosocial work environment (e g "Professional Roles and Life Environment"1979, in Swedish). We focused the analysis on the concept of sustainability in *collaborative network structures* based on the theoretical model - the Communication Circle and on the empirical experiences from team collaboration globally (Jansson & Bradley 2004).

Distributed working and learning environments exist not only in the multinational businesses but also in up starting companies world wide. Areas that become harder at a distance are building trust and reaching *sustainable relationships*, since distributed teams don't have the same awareness of each other and get fewer cues to "read" situations and handle teambuilding, motivational problems, and conflicts.

Learning as well as working in distributed environments are taking place in *virtual communities (web based communities)* and contribute to the creation and strengthening of these communities. When we now have integrated powerful technologies that strengthen the individual, various aggregations are important to focus on; organizations, communities, and societies. They are all formed by individuals and are forming the individuals; it is through these formations that relations become sustainable. The empirical results from the case studies above can be summarized as below (Jansson & Bradley 2004).

- Teams working in distributed projects go through the same stages and meet the same problems as any other team.
- Some problems become harder to overcome and some problems either disappear or decrease in magnitude. There are both new obstacles and new opportunities.

- It is possible to work at a distance without regular face-to-face meetings. It will always be different and the big danger to believe that it won't.
- When team members don't see each other daily, the base for collaboration changes. The different media that the team members use demand changes in behaviour. Media as communication tools also change the relation between the senses and change the perception of and relations to other people.
- At distance work people have to use technology for all communication, which means they have to be able to choose the right media for the right activity and handle sensitive topics through technology.
- Learning how to use it to reach the best results, learning how to feel comfortable in the new work environment, and learning how to be social and get to know each other through ICT are the main concerns at distance work - not the ICT per se.
- To reach a *sustainable communication and collaboration* it is vital that the team members get to know each other, have fun together and trust each other. It is important to have an organization that supports their work, to have sufficient ICT-support, and to have a team space. Learning about teambuilding, about conflict solving and learning communication skills are also necessary. Important is that somebody (often the coordinator or leader) prepares for problems that usually occur in the different stages of a team's lifecycle, is giving the participants enough guidance and support to keep them on track, and that somebody keeps the team motivated.
- At a distance it is *harder* to keep the motivation, build trust, and keep sustainability in relationships and interactions. Further it is harder to know each other's strengths and weaknesses, know other persons' competence, and read people.
- At a distance it *more important* to strive for and keep an open and direct communication, to be proactive, and to take initiate. Further it is more important to make sure to document the work and to actively work to maintain relationships.

3.2 Web Based Communities or Virtual Communities

There are ways in which Internet groups could meet parts of the strict definition of community. There are some identifiable circumstances which predispose a group to develop as a community, the more the group participants are mutually bound by necessity, the more like a community the social organization becomes. Groups tend to evolve towards communities when they engage in shared work. Shared commitment to a definable outcome creates bonds and increases interdependence, and multidimensionality. Through the use of technology and the development of virtual communities, a redefinition of the geographic communities occurs. By expanding the geographical communities global participation is increasing in managing real world events. There are new opportunities and prerequisites for both cities and rural areas to "colonize the world."

There are many different kinds of online communities or Web Based Communities (WBC) which are organised around such diverse online technologies as chat rooms, forums and multi user dungeons (MUD), and games. There are interrelated trends in the development of an online web-based *learning community* and *virtual collaboration spaces.* The first one is *integration of services.* An online meeting place will become more and more a portal that gives access to a wide number of services. The second one is *practicality of use.* The online meeting place is expected to be the networked desktop of the users at home, at the workplace, at seminars etc. The last trend is *personalisation.* The platform is expected to adapt to the preferences and competences of its members. The user just feels at home in an online community and this is feasible only when he/she is able to customize the spaces of the community that he/she participates in.

Community is a group of people who are related to some extent to each of the following minimal requirements:

- A common purpose: some form of reason for the co-existence of the members
- A common cultural context: a collective identity that includes values, beliefs, attitudes, behavioural norms, and accumulated experiences
- Co-location: all members of the community are sharing a common physical and/or virtual space
- Voluntary participation: the community chooses who participates in the community and members of the community choose to participate in the community
- Multiple, shifting and overlapping membership and participation: people can participate in more than one community at a time, they can join and leave the different communities to which they belong, and more communities may share purpose, location, cultural context or membership.

The strength and cohesiveness of the community can be enhanced by the following elements: common history, shared knowledge, common practices, common action, shared vision of the future and co-created future (*Bouras et al. 2005).* WBC could be considered as an *infrastructure* that coexists with the physical community in a temporary or permanent way. The perspective of *community sustainability* extends the notion of processes of sustainability as just an activity that keeps going. Definitions of sustainability are basically about the extension of human activities over a longer period of time based on the established social relationship among the participants. This description can be applied for the definition of community as well. On basis of the participants´ interests to learn about a common topic, to work towards a common goal, and to create sustainable relationships, sustainability in WBC is related to the broader context described in most of the circle sectors in the communication circle in figure 2 – both the technology and the context.

To achieve *a sustainable distributed work environment* it is important that people are aware of what variables affect the collaboration and communication. Preparation for and *planning* of the distributed project work becomes even more important at international team work. Training is needed for handling the communication problems as they occur in the actual work and in collaborative relations globally. Our model provides an aid to clarify what variables are important to consider at collaboration in distributed teams and from a *psychosocial and organizational perspective.*

4 Convergence Processes in Integration

To summarize there is a new focus in the ICT- age on *Societal and Humane Sustainability*. Key perspectives in this context are: system perspective, holism, human aspects, a bottom up approach, common good, and equality. The three circles in the lower part of figure 3 which are entitled Environmental sustainability, Economic Sustainability, and Social & Cultural Sustainability are main components of sustainability (and the sustainability is the converging part of the three circles). The Convergence Model on ICT and Psychosocial Life Environment in figure 1, which is repeated in the upper part of figure 3, is related to the three Sustainability aspects. Each aspects of sustainability can be traced back to the subsets of converging circles in figure 1. Hence the broader picture is illustrated above in figure 3, which describes how to perceive *sustainability* in distributed work environments as well as in the society as a whole. At defining sustainability, there may be additional concepts to be added such as connection between sustainability and social capital, time and virtual space, geographical and network sustainability, power relations in networking.

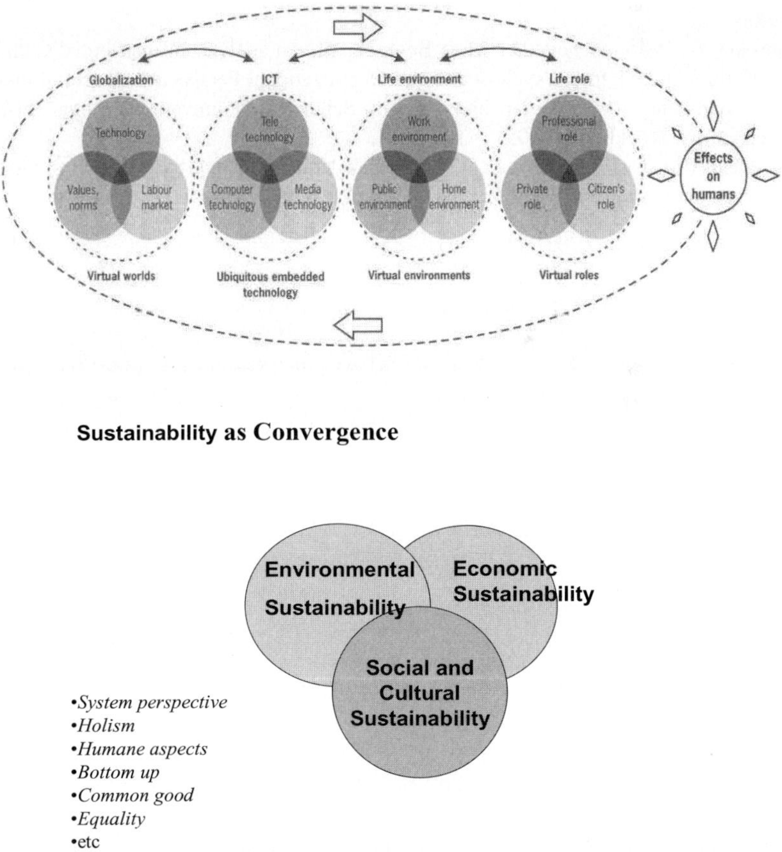

Fig. 3. Convergence processes in integration (Source: Bradley 2001, 2006)

References

1. Agenda 21. Rio Declaration on Environment and Development. In: Report of United Nations Conference on Environment and Development. United Nations: A/CONF.151/26/VOL 1
2. Bouras, C., et al.: A web – based virtual community. International Journal of Web. Based Communities 1(2), 127–139 (2005)
3. Bradley, G.: Humans on the Net - Information and Communication Technology (ICT), Work Organization and Human Beings. Stockholm: Prevent (2001)
4. Bradley, G.: Social and Community Informatics: Humans on the Net. London: Routledge (2006)
5. Bradley, G.: The Convergence Theory on Information and Communication Technology (ICT) and the Psychosocial Life Environment – The Connected Home. In: Smith, M., Salvendy, G. Proceedings of the HCI International 2005 conference, 22-27 July 2005, Las Vegas, USA. Mahwah, NJ: Lawrence Erlbaum Associates (available on CD) (2005)
6. Jansson, E., Bradley, G.: Sustainability in Collaborative Network Structures - with focus on the Psychosocial Work Environment in Distributed Teams. In: Proceedings at the CIRN conference on Sustainability and Community Technology: What does this Mean for Community Informatics, Prato, Italy, 29 September – 1 October 2004, vol. 1, pp. 271–283 (2004)
7. Jansson, E.: Working Together when Being Apart. An analysis of distributed Collaborative Work through ICT from Psychosocial and Organizational Perspective. Doctoral dissertation at Department of Computer and System Sciences, IT-university, Royal Institute of Technology, Stockholm (2005)
8. Swedish Ministry of Environment, En svensk strategi för hållbar utveckling (A Swedish Strategy for Sustainable Development. Report 2003/04:129 (2003)

The Impact of Verbal Stimuli in Motivating Consumer Response at the Point of Purchase Situation Online

Asle Fagerstrøm

The Norwegian School of Information Technology, Schweigaardsgt. 14, 0185 Oslo, Norway
asle.fagerstrom@nith.no

Abstract. This paper is a response to the lack of knowledge regarding actual online purchase behavior, and introduces behavior analysis as an alternative framework in studying consumers' purchase behavior. Motivation to confirm an order online can from the concept of motivating operation (MO) be analyzed as those antecedents in the environmental setting (included verbal stimuli) that; (1) have an effect on the consequences of responding, and (2) influence the responses (including purchase) related to those consequences. Using the functional analytic framework from behavior analysis, MO is identified as a likely predictor of consumer tendency to confirm their online orders.

Keywords: Online shopping, point of purchase, motivating operation, augmenting, functional analysis.

1 Introduction

When consumers go shopping on the Internet they usually have to go through a certain procedure; they will have to go to the web shop in the first place, search for a product, put the product in the shopping basket, and then finally confirm the order. It has been proved that many consumers leave the web shop in the check-out phase without confirming their order [e.g. 1]. Knowledge about the actual reason for this behavior, why this so often occurs and finally how to change this behavior, is of great interest to those who work with user interface design. By better explaining and predicting these unwanted sides of the purchase setting, companies could reduce this type of escape/avoidance behavior and consequently increase benefits and economic earnings from their web shop. Even small improvements in conversion rates can have a huge influence on sale. This paper seeks to identify those events that evoke (or abate) consumer response at the point of purchase situation online.

Monsuwé et.al. [2] define Internet shopping or online shopping as the use of online stores by consumers up to the transactional stage of purchasing and logistics. It is not unusual that research on online shopping end before the stage of actual purchase, like in the study of Monsuwé et.al. A meta-analysis shows that research on online shopping behavior can be placed in three categories [3]: (1) attitude toward online shopping, (2) intention to shop online and (3) actual online shopping behavior. Research on online shopping behavior has therefore explored the antecedents of intention to shop online, but not investigated the actual behavior online [3].

M.J. Smith, G. Salvendy (Eds.): Human Interface, Part II, HCII 2007, LNCS 4558, pp. 713–719, 2007.

Most of the studies on online shopping come from the disciplines of attitude research [3]. The concept attitude, understood within a cognitive framework, is also one of the most widely used foundations for understanding behavior in consumer research [4]. However, there has been an increased skepticism toward structural models like e.g. Theory of Reasoned Action, Theory of Planned Behavior and Technology Acceptance Model. The strongest critique has been that they do not take into consideration the full range of situational variables likely to influence the strength of the attitude-behavior relationship [5, 6]. As an answer to this, behavior analysis is suggested as an alternative, functional framework, to understand actual online purchase behavior and is the focus of this paper. Behavior analysis sees behavior as determined by the consumer learning history and situation, and thus includes one important predicting variable that is lacking in attitude theories [7].

This paper is structured as follows: In the first part the behavior analysis framework of motivation is presented together with a discussion regarding the impact of verbal stimuli in motivating consumers at the point of purchase setting online. Second, there will be a short discussion on methodological implication to researchers when using a functional approach as a foundation for studying online purchase behavior. Finally, concluding comments are given on the use of behavior analysis as a framework when studying consumers purchase behavior online.

2 Explaining Consumer Motivation to Purchase Online

In order to apply the experimental methods of science to the analysis of behavior, behavior analysts have broken down the stream of behavior into units called behavior responses and the changes in an individual's environment into arbitrary units called environmental stimuli. Behavior analysis describes the basic procedure as a three-term contingency involving stimulus, response and consequence. Within the three-term relationship, S^D is a *discriminative stimulus*, R is a *response*, and $S^{R/A}$ is a *reinforcing* or *aversive* stimulus [8]. The reinforcing consequences (S^R) are likely to increase the probability of repeated responses whereas the punishing consequences (S^A) have the contrary effect. This sequential process is illustrated as a chain of events as follows:

$$S^D - R - S^{R/A}.$$

A discriminative stimulus signals increased probability of reinforcement given certain responses. Discriminative stimuli are described as stimuli that set the occasion on which responses have consequences and are said to occasion responses [8]. A discriminative stimulus on a web shop can be the "Add to Shopping Cart" button which signals probability of reinforcement. When a person buys a product online his behavior is reinforced by acquiring the attributes of the product class and the consequences of consumption. Simultaneously his behavior is punished by the surrender of money, which is depriving him of opportunities to acquire other reinforcers. Therefore, the "Confirm order" button is another discriminative stimulus in the confirming order setting, which signals the aversive consequences as a result of confirming the online order. Purchase behavior within an online shopping context is accordingly the result of conflicting behaviors. These conflicting behaviors will be designated approach or avoidance. According to Alhadeff [9], the outcome of this conflict of behaviors – purchase or not purchase – is determined by the relative strength of reinforcers and aversive consequences in that specific setting.

2.1 Motivating Events at the Point of Purchase Situation

In recent years there has been a number of exciting new theoretical advances in the field of behavior analysis [e.g. see 10]. Among these is the concept of the *motivating operation* (MO)[1], a behavior analytic formulation of motivation developed by Michael [11-13]. The concept MO has made an important contribution to both basic and applied behavior analysis, inspiring new research and leading to innovative intervention strategies [see 14]. MO is defined as [12], an environmental event that first establishes (or abolishes) the reinforcing or punishing effect of another event and second, evokes (or abates) behaviors associated with that event. The first effect is related to the consequences of responding (*value-altering effect*) and the second is the effect of the responses related to those consequences (*behavior altering effect*).

Based on the different value-altering effects, Laraway et al. [15] distinguish between four MO subtypes: (1) Establishing operations related to reinforcement, (2) abolishing operations related to reinforcement, (3) establishing operations related to aversive consequenses, and (4) abolishing operations related to aversive consequences. From this definition we have four subtypes of MOs in the online purchase setting (fig. 1).

Behavior altering effect / Value-altering effect	Evocative effect on online purchase	Abative effect on online purchase
Establishing operations	REINFORCER ESTABLISHING EFFECT	AVERSIVE ESTABLISHING EFFECT
Abolishing operations	AVERSIVE ABOLISHING EFFECT	REINFORCER ABOLISHING EFFECT

Fig. 1. Four subtypes of MOs in the online purchase setting

Fig. 1 shows that MOs have two effects on consumer's purchase response online, the *evocative effect* and the *abative effect*. These two motivative effects on online purchase behavior will now be explained, and examples related to the impact of verbal stimuli on motivating consumer response at the point of purchase situation are given.

---

[1] To avoid unnecessary complication, the term motivating operations (MO) is in the present paper sometimes used to refer to events that were originally described by Michael as establishing operations.

2.2 Verbal Motivating Stimuli at the Point of Purchase Situation

Consumers' contingency-shaped behavior is behavior that has been shaped by its consequences. An example: a consumer learns to browse on the Internet through trial and success. However, what consumers do sometimes depends on what they are told to do. Consumers often follow recommendations from friends (word of mouth), online advertisements, customer product reviews, online instructions etc. Behavior which is mainly determined by verbal antecedents is defined as rule-governed behavior [8, 16]. Zettle and Hayes [16] argue that an understanding of the listener should lead to an understanding of the speaker. They define three main functional units of listeners' behavior: (1) *Pliance* (taken from the word compliance) is that the listener follows a verbal rule based on a history of socially mediated consequences for the correspondence between the rule and the rule-follower's behavior. (2) *Tracking* is rule-governed behavior under the control of a history of correspondence between the rule and "natural" social or non-social contingencies. Natural contingencies are those produced entirely by the exact form of the behavior in a particular setting. (3) *Augmenting* is defined as rule-governed behavior that alters the extent to which some event will function as a consequence.

There are two types of augmentals: (1) *motivative augmentals* which are rules that increase the value of an event that is already a functional consequence. A motivative augmental is simply an antecedent verbal stimulus that has an evocative or abative effect on consumer purchase behavior. Motivative augmentals in an online shopping setting can be the following message – "What about a holiday in Paris?" (2) *Formative augmentals* establish some new events as an important consequence. An example of a formal augmental is a message that says "booking flight ticket on the internet now gives you more frequent flyer points". It is obvious that augmentals have an important motivative function relate to consumer's purchase behavior online, and its evocative and abative effect on online purchase will therefore be more thoroughly explained.

The evocative effect. The evocative effect represents an increase in online purchase and is caused by reinforcer establishing effects and/or aversive abolishing effects. One operation that has an evocative effect on online purchase is the reinforcer establishing effect, which increases the reinforcing value of purchase and consequnences of consumption. An example related to the reinforcer establishing effect can be that a person goes to a web shop, finds a product, and puts the product in the basket and then goes to confirm the order. If the person sees that this particular product has been given a good rate by other customers it will probably have a reinforcing establishing effect on the consequence of confirming the order. Other events that may have reinforcer establishing effect on purchase behavior online could be scores in a product test (e.g., "this TV emerged as the test winner"), extended product use (e.g., "this mobile phone can also be used as an mp3 player"), sponsoring (e.g., "when purchasing the product online you support Red Cross and their humanitarian projects").

The other operation that has an evocative effect on online purchase behavior is the aversive abolishing effect, which decreases the aversive consequences of purchase and consumption. An example of aversive abolishing events on the web shop could be that the company offers account payment (e.g., "buy online now, and pay later"). Account payment will probably have an abolishing effect on the aversive consequences of purchasing. Another event that may also have abolishing effect on the aversive consequence of purchasing could be shipment (e.g., "buy online and get free freight").

The abative effect. The abative effect represents a reduction in online purchase and is caused by reinforcer abolishing effects and/or aversive establishing effects. The aversive establishing effect is the operation that has an abative effect on online purchase, and is defined as an operation that increases the aversive consequences of purchase and consumption. An example related to the aversive establishing effect can be that a person goes to a web shop, finds a product, and puts the product in the basket and then goes to confirm the order. If the person sees that the shipment costs for the product is high it will probably have an establishing effect on the aversive consequence of purchasing. Another example of events that may increase the aversive consequences of purchase and consumption could be sales tax (e.g., "when purchasing the product online 25 % tax must be added").

The other operation that has an abative effect on online purchase is the reinforcing abolishing effect, which decreases the reinforcing consequences of purchase and consumption. An example of reinforcing abolishing events on the web shop could be that the product has been bad rated in a product test. The bad product rating will probably evoke avoidance behavior related to the confirming order setting. Other events that may have a reinforcing abolishing effect on online purchase and consumption could be bad product rating by other customers.

3 Methodological Implication

In the introduction to this paper I distinguish between structural studies and functional analysis of online consumer behavior. In a structural study the relations between stimuli and responses are constant while varying critical properties of one or the other [8]. A functional analysis holds the stimuli and responses of interest constant while changing their relations. In the discussion above about MOs in the point of purchase setting online we hold the verbal stimuli and responses constant, but change their relations by looking at events that have establishing or abolishing effects on the reinforcing or aversive consequences of purchase behavior. Structure and function are complementary, not mutually exclusive. There is no reason why structural concerns should not enter into functional experiments, or vice versa [8]. Cognitivists prefer to summarize the organization of the consumer's behavior in terms of structures. Behavior analysts prefer to anchor their account of consumers purchase to the detailed functional relations between observable events.

Empirical evidence has been provided for the importance of MO in other areas. One challenge related to consumption is to what extent it is possible to do research on motivational operations in the complex world of contingencies and motivational variables. A solution to this challenge could be to accomplish research on MOs in an online purchase setting. By creating a simulated online store it would be possible in a PC lab to arrange variables and record responses in a way that is difficult to accomplish in a traditional shopping setting [see e.g. 17]. One experiment that could be accomplished in order to study the function of MO is to design different motivating events (e.g. product rating, special offers, payment, product delivery) in simulated online stores. Participants' retention at each stage of the shopping process could be recorded through clickstreaming as participants move through the site, and finally the conversion rate for a site could be recorded. The behavioral-analytic framework gives a unique possibility to study actual online consumer behavior in a sophisticated way which is difficult by more popular structural frameworks like cognitive attitude theories.

4 Conclusion

In the experimental analysis of consumer behavior, the behavior analytic paradigm has proven to be an insightful approach[18], and this paper has tried to show that it can also be an insightful way to explore verbal stimuli that have a motivative impact on consumers at the point of purchase situation. The behavior analytic paradigm describes the basic operant procedure as a three-term contingency involving pre-behavioral stimulus, response and consequence.

Consumer motivation to confirm order online can be analyzed from the concept of motivating operation, which are those antecedents in the environmental setting that influence consequences of responding. MOs have two effects on consumer's purchase response online, the evocative effect and the abative effect. The evocative effect is a result of an establishing reinforcing effect and/or abolishing aversive effect on the consequence of purchase and consumption. The abative effect is a result of abolishing reinforcing effect and/or establishing aversive effect on the consequence of purchase and consumption.

Cognitivists prefer to summarize the organization of the consumer's behavior in terms of structures. Behavior analysts prefer to anchor their account of consumers purchase to the detailed functional relations between observable events. Structure and function are complementary, not mutually exclusive. There is no reason why structural concerns should not enter into functional experiments, or vice versa.

References

1. Nielsen, J., Norman, D.A.: Web-Site Usability: Usability On The Web Isn't A Luxury. [Internet] [cited 2006 25.10.2006] (2000). Available from http://www.informationweek.com/773/web.htm
2. Monsuwé, T.P.y., Dellaert, B.G.C., Ruyter, K.d.: What drives consumers to shop online? A literature review. International Journal of Service Industry Management 15(1), 102–121 (2004)
3. Limayem, M., Cheung, C., Chan, G.: Online consumer behavior: what we know and what we need to know. In: European & Mediterannean Conference on Information Systems, Tunis Tunisia (2004)
4. Simonson, I., et al.: Consumer Research: In search for identity. Annual Review of Psychology 52, 249–275 (2001)

5. Olson, J.M., Zanna, M.P.: Attitudes and attitude change. Annual Review of Psychology 44, 117–154 (1993)
6. Brown, S.P., Stayman, D.M.: Antecedents and consequences of attitude toward the ad: a meta-analysis. Journal of Consumer Research 19, 34–51 (1992)
7. Foxall, G.R.: Consumer Behaviour Analysis: Critical Perspectives on Business and Management. Routledge 1416 (2002)
8. Catania, C.A.: Learning, 4th edn. Prentice Hall, Upper Saddle River, New Jersey (1998)
9. Alhadeff, D.A.: Microeconomics and Human Behavior: Toward a New Synthesis of Economics and Psychology. University of California Press, Berkeley (1982)
10. Roche, B.: New wave analysis. The. Psychologist 12(10), 498–499 (1999)
11. Michael, J.: Distinguishing between discriminative and motivational functions of stimuli. Journal of the experimental analysis of behavior 37(1), 149–155 (1982)
12. Michael, J.: Establishing operations. The Behavior Analyst 16, 191–206 (1993)
13. Michael, J.: Implications and refinements of the establishing operation concept. Jounal of Applied Behavior Analysis 33(4), 401–410 (2000)
14. Iwata, B.A., Smith, R.G., Michael, J.: Current research on the influence of establishing operations on behavior in applied settings. Journal of Applied Behavior Analysis (4), 411–418 (2000)
15. Laraway, S., et al.: Motivating operations and terms to describe them: some further refinements. Journal of Applied Behavior Analysis 36(3), 407–414 (2003)
16. Zettle, R.D., Hayes, S.C., Kendall, P.C.: Rule governed behavior: A potential theoretical framework for cognitive-behavior therapy. Advances in Cognitive-Behavioral Research and Therapy, vol. 1, pp. 73–118. Academic Press, New York (1982)
17. Smith, C.L., Hantula, D.A.: Pricing effects on foraging in a simulated Internet shopping mall. Journal of Economic Psychology 24, 653–674 (2003)
18. DiClemente, D.F., Hantula, D.A.: Applied behavioral economics and consumer choice. Journal of Economic Psychology 24, 589–602 (2003)

Measuring the Screen Complexity of Web Pages

Fongling Fu[1], Shao-Yuan Chiu[2], and Chiu Hung Su[1]

[1] Department of Management Information Systems, National Cheng-chi University, Taipei 11605, Taiwan
{flfu, 93356512}@nccu.edu.tw
[2] Department of Industrial and Manufacturing Systems Engineering, University of Missouri-Columbia, Columbia, MO 65201, USA
scdg5@mizzou.edu

Abstract. An increasing, degree of rich and dynamic content and abundant links are making Web pages visually cluttered. This paper presents a numerical tool to evaluate the screen complexity of a Web page using four critical measurements: size complexity, local density, grouping, and alignment. In the empirical study, we first translate the real screens from four first pages on Ebay auction web sites to serve as model screens that contain the structure of complexity without content. We subsequently compare the complexity values calculated from the model screens with the viewers' judgment from the real screens. The resemblance between the results indicates that the tool is useful.

Keywords: Screen complexity measurements, Web page design, Complexity on screen layout, GUI design.

1 Introduction

Both the function and interface of a system are important to making the system usable [7]. Previous empirical studies have proven that the screen complexity is important to the aesthetic perception of the graphic screen [3,4] as well as to the information searching time when utilizing the menu [2,6]. A growing amount of rich and dynamic content and abundant links are making Web pages visually cluttered. The complexity of screen layout on Web pages seems much more serious than the two aforementioned situations, and challenges the person who designed the graph menus of the Web pages. We need to understand what factors determine users' perception of the screen complexity of a Web page if we want to design the Web pages with less complexity. Therefore, the main purpose of the study is to propose a valid numerical tool to measure the screen complexity according to users' viewpoints when browsing a Web site. The tool is developed based on the four key elements of screen complexity. It is expected to help designers better understand how to decrease the complexity of Web pages.

M.J. Smith, G. Salvendy (Eds.): Human Interface, Part II, HCII 2007, LNCS 4558, pp. 720–729, 2007.

2 Complexity of Screen Layout

In GUI design, the key elements of complexity are size, local density, grouping and alignment, which are considered significant influences on the screen usability [2]. The aggregate complexity is the weighted average of these key elements' ratings, where the weight given is the element's subjective level of importance. We redefined the measurements of four complexities used by previous researchers in the following:

2.1 Measure of Size Complexity

Size complexity involved the categorization of elements into groups according to actual physical size and variation in those sizes. It was based partially on information order, and was calculated for each type of information so that the complexity did not increase significantly when different information appeared in different sizes (e.g. the icon of an advertisement is expected to be larger than that of a searching item on a catalogue). In terms of the magnitude size of the size complexity (CS), objects are related to size as follows:

$$\mathrm{CS} = 1 - \frac{\sum_{I}^{\mathrm{type}} (n_{size} - 1)}{n} \in [0,1] \tag{1}$$

where n_{size} is the number of sizes and n is the total number of objects.

2.2 Measure of Local Density

Local density is the extent to which the screen is filled with objects. Density is achieved by restricting screen density levels to the optimal percentage of 50% for graphic screens [4]. The method for calculating local density (LD) is as follows:

$$\mathrm{LD} = 1 - 2\left| 0.5 - \frac{\sum_{i}^{n} a_i}{a_{frame}} \right| \in [0,1] \tag{2}$$

where a_i and a_{frame} are the areas of object i and the frame, respectively; and n is the number of objects on the frame. Empirical studies proved that local density had an influence on users' searching time when using text menus both in English and in Chinese [1,2,6].

2.3 Measure of Group

The group measure reflects coherence: the degree to which all elements seem belong together and appear visually as one piece. Under the grouping technique, elements with similar functions or information are surrounded by a boundary by line, background color and/or space. Grouping makes it easier for the users to extract the information assigned to the group. Empirical study proved that the proper grouping results in shorter search times [6]. The method for calculating the grouping complexity is as follows:

$$\mathrm{CG} = \frac{g_i}{g} \in [0,1] \tag{3}$$

where g_i is the number of groups with clear boundary by line, background, color, or space and g is the total number of groups.

2.4 Measure of Alignment

In order to achieve simplicity on the screen, the smaller the number of alignment points, the better. Measuring the level of alignment for simplicity (AS) of a graphic screen involves counting the number of different rows and columns on the screen that are used as starting positions of objects, and which is calculated as follows:

$$\mathrm{AS} = \frac{3}{(n_{vap} + n_{hap} + n)} \in [0,1] \tag{4}$$

where n_{vap} and n_{hap} are the numbers of vertical and horizontal alignment points, respectively; and n is the number of objects on the frame [3]. The alignment had influence over the searching time on database menu [2,6].

3 Empirical Study

In Web site design, most people use the term "first page" when referring to the screen design of the interface, rather than the interface of entire screens. According to previous experimental studies, users had longer periods of fixation durations while on the first pages compared to the second pages [5], implying that the first page of a web site is more important than the other pages. To control the influence of the task effect, this study established complexity values for real screens on four "first pages" copied from auction web sites of Ebay USA, Hong Kong, Taiwan, and Korea. We first translated the real screens to serve as model screens. We then compared the viewers' judgment of the actual screens with the complexity values counted from our tool, and determined the level of consistency between the two.

3.1 Modeling the Screens

The study uses the model screen for further measurement of the complexity values. Ngo et al. [4] found that model screens can control content effects and facilitate interpretation of the data analysis in interface aesthetics. A sample of all screens of the first pages from Ebay auction sites – Ebay USA, Ebay Taiwan, Ebay Hong Kong, and Ebay Korea – are shown in Fig. 1 - 4. The four real screens (structure with content) and the corresponding models (structure without content) are displayed side by side.

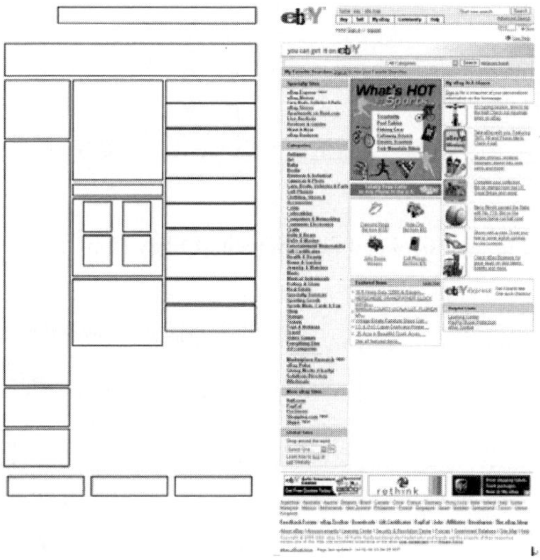

Fig. 1. Model and Real Screen of Ebay USA

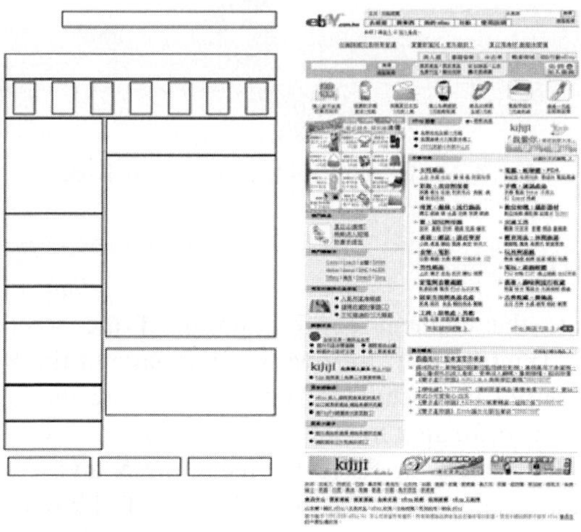

Fig. 2. Model and Real Screen of eBay Taiwan

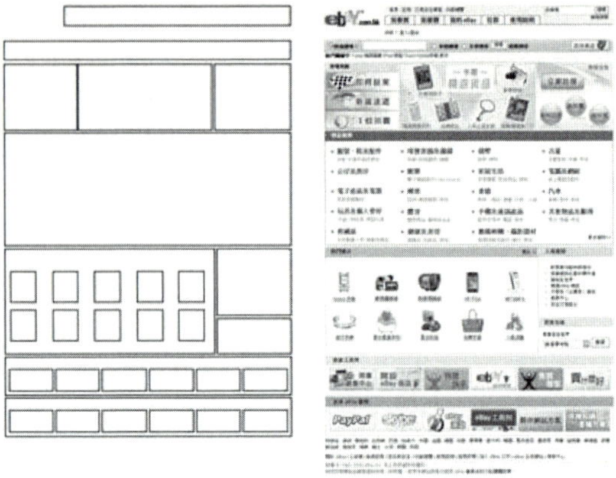

Fig. 3. Model and Real Screen of Ebay Hong Kong

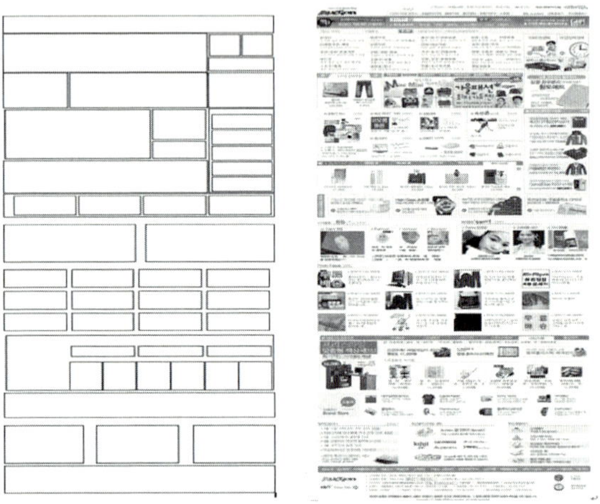

Fig. 4. Model and Real Screen of Ebay Korea

3.2 Layout Properties

Tables 1-4 presents the element configurations of the screens. We used the Windows function "Print Screen" to save the first page of the Web site as a BMP graph file and then used Photoshop software to measure the layout properties of objects. The axis of coordinates for the left top corner on the first page is set as (0, 0). The values of X and Y in Table 1-4 mean the X and Y axis of coordinates for the left top point of an object. The layout of Fig. 1 (Ebay USA) is 760 * 1661 pixels. The layout of Fig. 2 (Ebay Taiwan) is 760 * 1695 pixels. The layout of Fig. 3 (Ebay Hong Kong) is 760 * 1312 pixels. The layout of Fig. 4 (Ebay Korea) is 900 * 2446 pixels.

Table 1. Layout properties of Ebay USA

Object	X	Y	Width	Height
1 - 2	160/0	36/145	600/760	51/97
3 - 6	0	252/433/1162/1281	199	76/723/115/113
7 - 9	205	252/561/602	274	299/35/231
10 - 11	237/356	610	87	92
12 - 13	237/356	710	87	92
14 - 15	205/485	839/252	274/275	199/69
16 - 23	485	321/395/469/543/617/691/765/839	275	74
24	485	915	275	82
25 - 27	10/263/516	1423	233	59

Table 2. Layout properties of Ebay Taiwan

Object	X	Y	Width	Height
1 - 2	160/0	36/145	600/760	51/97
3 - 6	0	252/433/1161/1281	199	76/723/115/113
7 - 9	205	252/561/602	274	299/35/231
10 - 11	237/356	610	87	92
12 - 13	237/356	710	87	92
14	205	839	274	199
15	485	252	275	69
16 - 23	485	321/395/469/543/617/691/765/839	275	74

Table 3. Layout properties of Ebay Hong Kong

Object	X	Y	Width	Height
1 - 2	160/0	20/110	600/760	53/51
3 - 5	0/197/556	170	197/355/204	174
6 - 7	0	348/648	760/558	294/275
8 - 12	14/127/239/351/469	706	70	83
13 - 17	14/127/239/351/460	806	70	83
18 - 19	561	648/829	199	176/94
20	0	927	760	102
21 - 26	10/134/258/382/506/630	959	119	59
27	0	1034	760	102
28 - 33	10/134/258/382/506/630	1065	119	59

3.3 Complexity Values

Table 5-8 presents the complexity values according to our aforementioned formulae. The unit value of a_i and a_{frame} on Table 6 are in pixels. The Ebay Hong Kong Web

Table 4. Layout properties of Ebay Korea

Object	X	Y	Width	Height
1 - 2	0	59/146	900/678	87/193
3 - 4	683/791	151	105	108
5	680	254	220	68
6 - 8	0/220/679	339	215/459/2 21	173
9	4	520	485	242
10 - 11	494	520/675	178	149/87
12	0	770	678	161
13 - 18	680	516/544/622/699/ 776/853	220	415
19	0	933	900	115
20 - 23	32/248/463/679	944	209	95
24 - 25	0/471	1085	440	185
26 - 29	0/244/448/672	1305	209	91
30 - 33	0/244/448/672	1412	209	91
34 - 37	0/244/448/672	1519	209	91
38	6	1630	890	271
39 - 41	221/446/672	1683	215	54
42 - 47	216/328/440/552 /665/777	1761	112	140
48 - 49	0	1911/2072	900/264	124/188
50 - 51	305/626	2072	274	188
52	0	2270	900	137

Table 5. Size complexity for four Web pages

	USA	Taiwan	Hong Kong	Korea
n_{size}	15	13	11	24
n	27	23	33	52
CS	0.481	0.478	0.697	0.558

Table 6. Local density for four Web pages

	USA	Taiwan	Hong Kong	Korea
a_i	763595	977376	787764	1882704
a_{frame}	1262360	1288200	997120	2201400
LD	0.790	0.483	0.420	0.290

Table 7. Grouping complexity for four Web pages

	USA	Taiwan	Hong Kong	Korea
g_i	14	11	17	4
g	15	16	17	24
CG	0.933	0.688	1.000	0.167

Table 8. Alignment for four Web pages

	USA	Taiwan	Hong Kong	Korea
n_{vap}	9	13	16	27
n_{hap}	20	14	12	26
n	27	23	33	52
AS	0.054	0.060	0.049	0.029

page received better values for size complexity (Table 5) and grouping complexity (Table 7) while the Ebay USA page received a better value for local density (Table 6).

3.4 Comparison of Viewers' Judgments of Complexity Values

Each complexity value has its own weight which is determined by viewers' preferences. To simplify the problem, all weights in the study are set to equal one. A summary of the complexity values is shown in Table 9. According to the weighted average of all the complexity values, we consider the Ebay USA and Ebay Hong Kong Web pages to be better than the others; Ebay Taiwan is a little lower than the previous two while Ebay Korea is the ranks last.

From a comparison of individual complexity values among the screens, we found that Ebay USA (Fig. 1) has a relatively small number of groups (CG = 0.933) and the most space (LD = 0.790). Ebay Hong Kong (Fig. 2) has clear group boundary (CS = 1.0) and less size complexity (CS = 0.697). Ebay Korea (Fig. 4) consists of worst values for local density, grouping and alignment.

Table 9. Summary of complexity values for four Web pages

	USA	Taiwan	Hong Kong	Korea
CS	0.481	0.478	0.697	0.558
LD	0.790	0.483	0.420	0.290
CG	0.993	0.688	1.000	0.167
AS	0.054	0.060	0.049	0.029
Average	0.565	0.427	0.542	0.261

A survey study was conducted to test the robustness of the formulae for complexity using model screens. The subjects who judged the complexity of actual screens comprised of 53 undergraduate student volunteers from a university in Taiwan who each took the course "Introduction of software" and have completed a Web site design project. The questionnaire contained pictures of the Web pages of Fig. 1-4 and questions regarding the rating the complexity. For each picture, viewers assigned a numerical value between one and five, indicating their perception of the Web pages' degree of complexity. Values from one to five represented "very low," "low," "medium," "high," and "very high," respectively. The means and standard deviations of viewer's judgment are listed in Table 10.

Table 10. Statistics of the sample

	USA	Taiwan	Hong Kong	Korea
mean	3.49	3.43	2.59	4.26
S.D.	0.93	0.82	0.91	0.76

The means of viewers' judgments on complexity (Table 10) indicates that Ebay Korea is the worst, Ebay Taiwan and Ebay USA are similar, and Ebay Hong Kong is the best. The results of viewers' judgments of complexity almost resemble those obtained using the measures proposed by the above formulae, with the exception of Ebay USA. The complexity of Ebay Hong Kong (Fig. 3) was rated as 2.59 which indicated that its complexity is a little low to medium. The complexity of Ebay Taiwan was rated as 3.43 which indicated that its complexity is medium to a little high. Meanwhile the complexity of Ebay Korea was rated as 4.26, indicating that its complexity is high. These ratings are consistent with those obtained from the complexity values in Table 9. Based on the similar results generated, we conclude that the proposed formulae are useful.

However, the users' views of Ebay USA differ from the complexity value calculated using the formulae. We conjecture that there are two possible causes for the difference. One possibility is that the weight determination is actually a multi-dimensional optimisation problem [4] which assumes that the weights in the study are not reflective of the preference of the sample. If this is true, then from comparing the results of Table 9 and 10, we conjecture that the two measures – size complexity and grouping – should be given higher weights than the other two measures in the sample because both Ebay Hong Kong (Fig. 3) and Ebay Taiwan (Fig. 2) earn higher values in these two measures as compared to the other two figures. We also found that the Web page Ebay USA has the highest value of standard deviation (0.93), implying the high variance in judgment. The other possible cause behind the inconsistent results of Ebay USA may be that there still exists other measurements of complexity for Web pages, such as the number of items in an object, the ratio of width and height of objects, the number of colors, and so forth [7, 8].

4 Conclusions and Future Work

Having a way of objectively defining and measuring complexity could help us to more reliably design graphic menus for Web pages with better complexity. The purpose of this study is to propose determinants of and formulae for calculating the complexity of Web pages. Four measurements of complexity – size complexity, local density, grouping and alignment – were redefined for Web pages as formulae 1-4. In the empirical study, we suggest a method for translating a real screen (structure with content) of a Web page into a model screen (structure without content) and calculating the complexity values using the formulae. The results from the survey study on viewers' judgments of the real screens almost mirror the complexity values from the model screens, which confirms the usefulness of the tool measuring complexity. One inconsistent result which emerged from the real screen judgment of and aggregate complexity value for the Ebay USA site hints at the need for further studies on the method for calculating the preference weight of complexity values or other influencing variables on the screen complexity of Web pages.

References

1. Goonetilleke, R.S., Lau, W.C., Shih, H.M.: Visual Search Strategies and Eye Movements When Searching Chinese Character Screens. Int. J. Human-Computer Studies 57, 447–468 (2002)
2. Miyoshi, T., Murata, A.: A Method to Evaluate Properness of GUI Design Based on Complexity Indexes of Size, Local Density, Aliment, and Grouping. Man and Cybernetics. In: IEEE International Conference on, vol. 1, pp. 221–226 (2001)
3. Ngo, D.C.L, Teo, L.S., Byrne, J.G.: Formalising Guidelines for the Design of Screen Layouts. Display 21, 3–15 (2001)
4. Ngo, D.C.L, Teo, L.S., Byrne, J.G.: Modeling Interface Aesthetics. Information Science 152, 25–46 (2003)
5. Pan, B., Hembrooke, H.A., Gay, G.K., Granka, L.A., Feusner, M.K., Newman, J.K.: The Determinants of Web page Viewing Behavior: An Eye-Tracking Study. Access site: http://www.hci.cornell.edu/projects/pdfs%20of%20pubs/ETRA04.pdf
6. Parush, A., Nadir, R., Shtub, A.: Evaluating the Layout of Graphical User Interface Screens: Validation of a Numerical Computerlized Mode. International Journal of Human-Computer Interaction 10(4), 343–360 (1998)
7. Schneiderman, B., Plaidant, C.: Designing the User Interface: Strategies for Effective Human-Computer Interaction. 4th edn. Pearson Education, Inc (2005)
8. Woodward, R.M.: Proximity and Direction of Arrangement in Numeric Display. Human Factors 14, 337–343 (1972)

Reconciling Privacy Policies and Regulations: Ontological Semantics Perspective

Olga Krachina, Victor Raskin, and Katrina Triezenberg

CERIAS, Purdue University
686 Oval Drive
W. Lafayette, IN 47907-2086, USA
{okrachin, vraskin, kattriez}@purdue.edu

Abstract. How well the privacy policy follows a regulation is one of the current concerns of the user. Such a task can be accomplished by directly querying the policy statement with the regulation text. Automation of the process requires an expressive meaning-based framework for Natural Language Processing (NLP). This paper discusses the Ontological Semantics approach to the issue of verifying compliance and illustrates the potential of utilizing the framework in the domain of Privacy management for NLP-related tasks. As an example a section from BCBS and corresponding HIPAA regulations are used.

Keywords: Ontological Semantics approach, ontology, inference, Text-Meaning Representation, Privacy Policy, Natural Language Processing.

1 Introduction

Although formal language policies are being developed and deployed, it is unlikely that they will replace natural language privacy policies (PP) in the foreseeable future [1]. While the deployment of the privacy practices described in a particular policy takes place on the level of a formal language, certain tasks can and should be handled on a higher level, i.e. via a natural language processing (NLP) system. The motivation for the following division of labor between formal and natural language arises for two reasons. First, an intermediate step of natural-language to formal-language translation is required to work with the privacy policy on the level of a formal language; in the process, loss of potentially useful information is unavoidable if the formal language is not able to support the entire content. The notion of a formal language discussed in this context implies a domain-specific construct, thus placing apparent limitations on the range of the input to be processed. In such a case, an extensive research is necessary to account for all or most of the possible inputs for a specific domain.

The second reason is a logical consequence of the issue discussed above, namely, formal languages by design are not able to achieve the same level of expressiveness as a large-scale NLP system. If such an NLP system is designed to implement domain-independent mechanisms to process natural language input, it should comprise several extensive knowledge resources built and updated overtime.

M.J. Smith, G. Salvendy (Eds.): Human Interface, Part II, HCII 2007, LNCS 4558, pp. 730–739, 2007.

One of the central tasks in the field of privacy management is the issue of policy consistency with the outlined regulations. This paper will demonstrate how this issue is handled from the perspective of one specific NLP system.

2 Ontological Semantics in the PP Domain: Motivation

There is an apparent need in a comprehensive NLP system. Such system should be able to handle complex natural language structures; this implies disambiguation of the input not only on the lexicographic level but also on the semantic level, integrating the deduction mechanisms as well as those of induction. In other words, an NLP system has to employ mechanisms based on rather expressive semantics. One of the previously considered frameworks was P3P. There have been attempts at extending formal semantics of P3P and the underlying formal language APPEL [7].

Such attempts were directed towards creating a language that "avoids the APPEL's pitfalls but preserves the desirable functionalities in APPEL" [7]. The main objective of APPEL is to allow users to import preference rulesets created by other parties and to transport their own ruleset files between multiple user agents [7]. Among major drawbacks of APPEL is its syntax-based design, i.e. a new ruleset is generated for two P3P policies with the same meaning but different syntactic structure.

The new desiderata for an enhanced APPEL-like language included expressive power and semantic consistency as outlined in [7]. On the one hand, it appears reasonable to try and create a language exclusively for the purpose of analysis of privacy policies; on the other hand, it may be more efficient to extend a currently existing NLP system to the domain of PP. The framework proposed here for the task of handling problems relating to NLP is Ontological Semantics (Nirenburg and Raskin 2004 [9]). Its efficiency has been demonstrated in such applications as automatic translation, question-and-answering, information retrieval, text summarization, Internet search, and other NLP systems, largely due to the fact that it is not an *ad hoc* solution for a specific NLP problem but rather a comprehensive and systematic approach. Thus, in order to handle the domain of the privacy policy management within Ontological Semantics (OntoSem), the extension of existing resources may be the only requirement.

The rest of the paper is structured as follows: Ontological Semantics will be introduced, then a specific example will demonstrate mechanisms employed in evaluating compliance of an entity with a corresponding regulation, and finally, evaluation metrics will conclude the discussion.

2.1 OntoSem Framework

Text analysis in OntoSem relies on the results of several preprocessing modules performing lexicographic, morphological and syntactic analysis. The results of such a pre-semantic analysis, along with the support of the knowledge resources, contribute to the word-sense disambiguation and the establishment of semantic dependencies. Such dependencies are expressed in a basic Text-Meaning-Representation (TMR), which are further extended with more sophisticated dependencies, such as preconditions, effects, and complex events, or scripts, and also with modalities, time, aspect, and other parameters, if necessary to arrive at an exhaustive meaning representation of the input text. It is worth mentioning here that there is no one-to-one correspondence between a TMR and a sentence. It is often the case that a single TMR incorporates several syntactic units. TMR is the central construct of OntoSem as it underlies all text processing applications. It is an absolute knowledge representation module in a sense that TMR synonymy/paraphrase is not possible under OntoSem framework.

The OntoSem knowledge resources include the universal language-independent ontology (a large tangled hierarchy of concepts), language-specific lexicons, onomasticons (lexicons of proper names), and the Fact DataBase (FDB) containing instances of events occurred in the past processing of natural language texts, i.e. FDB is essentially a collection of TMRs. For further information on OntoSem see [8] and/or ontologicalsemantics.com.

2.2 More on the Goal of Compliance

As noted previously, the lack of a standard way to check compliance serves as one of the main motivating factors in finding a solution outside of the domain of formal languages for privacy policies, since their expressiveness is justifiably limited at the expense of other important factors such as the ease of use, etc [7]. Specifically, if the issue is approached from the perspective of natural language, its formulation can be reduced to the NLP question-and-answering problem. In other words, a PP notice can be queried with the regulation specifications.

One of the major hindrances in using an NLP system has been a necessity for "deep semantics" and inference capabilities and the ubiquitous "fear of semantics," due largely to the scarcity of adequately prepared semanticists. Those fears aside and the extended knowledge resources being available, this is still not an easily attainable goal, because such NLP system should be able to handle not only inductive and deductive reasoning but also reasoning under insufficient information or incomplete system resources. This research concerns itself with demonstrating how Ontological Semantics-based mechanisms handle inference problems.

3 Inference Processing with Ontological Semantics: Example

In Ontological Semantics, the TMR is the basic block for inference processing. Initially, the TMR corresponding to the query and the input text is constructed: TMRQ and TMRI, respectively. The Question-and-Answering mechanism is realized as a matching process of (see Section 3.1) between TMRQ and TMRI.

The following example is taken from HIPAA Section 164.522 on Rights to request privacy protection for protected health information and Section 164.528 on Accounting of disclosures[3]. This example not only illustrates (OntoSem) inference processing but also demonstrates the representation of a typical statement of the domain in Ontological Semantics.

Accounting of Disclosures:
164.522:
(iv) Individual Rights. The notice must contain a statement of the individual's rights with respect to protected health information and a brief description of how the individual may exercise these rights, as follows:

(E)[1] The right to receive an accounting of disclosures of protected health information as provided by §164.528

164.528:

An individual has a right to receive an accounting of disclosures of protected health information made by a covered entity in the six years prior to the date on which the accounting is requested, except for disclosures:

< …>

i. that occurred prior to the compliance date for the covered entity.

And the corresponding entry in BCBS PP [4] is as follows:

Right to Request an Accounting of Disclosures. You have a right to receive a list of certain instances in which we or our business associates disclosed your PHI for purposes other than our treatment, payment or health care operations and certain other activities. You are entitled to this accounting of disclosures for the six years prior to the date you make the request, but not for disclosures made before April 14, 2003.

Thus, relevant portions of TMRI are given in Listing (1), while Listing (2) is the selected portions of TMRQ. Indexing within FDB does not coincide with the indexing of the TMRI, since TMRI is not being committed to the FDB by default, i.e. the extension of FDB is controlled by the inference mechanism. Another thing to note is the proposition head REQUEST-INFORMATION which indicates the fact that the TMR is a representation of a query, i.e. TMRQ.

Listing 1. TMR of BCBS. Right to Request an Accounting of Disclosures.

```
TRANSFER-OBJECT-1
     BENEFICIARY                  HUMAN-1
     THEME                        LIST-1
LIST-1 (list of certain activities but not all)
     DESCRIBES                    INFORMATION-MANAGING-ACTIVITY-3
     THEME-OF                     TRANSFER-OBJECT-1
     SET                          TYPE(INFORMATION-MANAGING-ACTIVITY)
                          ELEMENTS(INFORMATION-MANAGING-ACTIVIITY-3)
                                  COMPLETE(NEG)

INFORMATION-MANAGING-ACTIVITY-3
     AGENT                        ORGANIZATION-2
                                  THIRD-PARTY
     HAS-EVENT-AS-PART            INFORM-1
     PURPOSE                      SET(PAY-1
                                  TREAT-ILLNESS
                                  SERVICE-EVENT) COMPLETE(NEG)
INFORM-1
     AGENT                        ORGANIZATION-1
     BENEFICIARY                  THIRD-PARTY
     THEME                        PHI
     PUBLIC-ATTRIBUTE
            time_1.value          PRIVATE
            time_1                < speech-act.time
            time_2.value          PUBLIC
            time_2                > = speech-act.time
```

[1] Skip sections A-D.

```
      TIME                                    |<16><04><2003>|
                                              < REQUEST-INFORMATION-1.TIME
               range                          ::years:: <= 6
               frequency                      default

  LEGAL-RIGHT
      THEME                                   TRANSFER-OBJECT-1
      AGENT                                   HUMAN-1
      BENEFICIARY                             HUMAN-1
```

A significant part of the process of verifying compliance lies in the construction of the query itself. There are two basic types of queries: 'wh'-type and 'yes-no'. The latter are more suitable for the task of determining compliance. This is due to the style of regulation documents: these texts as a rule indicate the obligative nature of existence of certain provisions or statements; such a structure, in turn, allows instantiation of REQUEST-INFORMATION field with THEME of main event in the TMRQ or relation as it turned out to be in this example (Listing 2).

Additionally, TMRQ contains a *LEVEL* slot, which indicates the desired depth of the resulting answer and takes on integer values. In other words, the mechanism accounts for cases where an exhaustive answer is not necessary. The motivation for this parameter stems from two reasons. On the one hand, the answer to a wh-query is a chain of dependencies (error estimation with a positive-valued response for yes-no queries) starting from the target variable itself and ending with the last relevant head of the frame, which in some cases is not necessary and is a waste of time and resources; on the other hand, introducing this parameter is an efficient step towards the goal of automation of query construction, i.e., instead of constructing a new query TMR from already existing TMRI, only an instantiation of REQUEST-INFOMRATION (and MODALITY in certain cases) fields with appropriate event/object focus are sufficient to formulate a query which is further constrained through the LEVEL parameter. In this example, the LEVEL parameter is set to its maximum. Furthermore, appropriate TMRQs can be constructed based on the filler (in case of a relation, a corresponding event or an object associated with it) of the slot 'scope' of the deontic modality [9], i.e., the modality that expresses obligation.

Listing 2

```
  MODALITY-1
      type                 DEONTIC
      value                1.0
      time                 > speech-act.time
      scope                DOCUMENT-1.TEXTUAL-RELATION
      attributed-to        ORGANIZATION-2

  DOCUMENT-1
      DESCRIBES            LEGAL-RIGHT-1
                           RULE-OF-CONDUCT
      AUTHORED-BY          ORGANIZATION-1
      LOCATION             WEB
      TEXTUAL-RELATION     SENTENCE

  SENTENCE
      DESCRIBES            LEGAL-RIGHT-1
      TEXTUAL-RELATION     DOCUMENT-1
```

```
AUTHORIZE
     AGENT                     ORGANIZATION-2
     THEME                     LEGAL-RIGHT-1
     BENEFICIARY               HUMAN-1

LEGAL-RIGHT-1
     AGENT                     HUMAN-1
     BENEFICIARY               HUMAN-1
     THEME                     INFORMATION-MANAGING-ACTIVITY
     CARDINALITY               PLURAL
     SET                       MEMBER-TYPE(LEGAL-RIGHT)
                               ELEMENTS ((LEGAL-RIGHT-2)) COMPLETE (NEG)
LEGAL-RIGHT-2
     THEME                     TRANSFER-OBJECT

TRANSFER-OBJECT
     THEME                     LIST-1
     AGENT                     ORGANIZATION-1
     BENEFICIARY               HUMAN-1

LIST-1
     DESCRIBES                 INFORM-1

INFORM-2
     AGENT                     ORGANIZATION-1
     THEME                     INFORMATION-1
     PUBLIC-ATTRIBUTE
            time_1.value       PRIVATE
            time_1             < speech-act.time
            time_2.value       PUBLIC
            time_2             > = speech-act.time
     BENEFICIARY               THIRD-PARTY
     TIME                      < OBEY.time
         range                 ::years:: <= 6
         frequency             default

OBEY
     AGENT                     HUMAN-1
     THEME                     RULE-OF-CONDUCT

RULE-OF-CONDUCT
     DESCRIBES                 INFORMATION-MANAGING-ACTIVITY-1

INFORMATION-MANAGING-ACTIVITY-1
     AGENT                     HUMAN-1
     THEME                     PHI

INFORM-1
     AGENT                     ORGANIZATION-1 (BCBS)
     THEME                     INFORMATION-1 (PHI)
     SET                       MEMBER-TYPE(INFORM)
     ELEMENTS                  (INFORM-2) COMPLETE (NEG)
     PUBLIC-ATTRIBUTE
            time_1.value       PRIVATE
            time_1             < speech-act.time
            time_2.value       PUBLIC
            time_2             > = speech-act.time
     BENEFICIARY               THIRD-PARTY
```

```
ORGANIZATION-1
     HAS-NAME                BCBS
     AGENT-OF                INFORM-1
     AUTHOR-OF               DOCUMENT-1

ORGANIZATION-2
     HAS-NAME                HIPAA
     AGENT-OF                AUTHORIZE
     AUTHORITY-ATTRIBUTE     1.0

REQUEST-INFORMATION
     THEME                   LEGAL-RIGHT-1
     LEVEL                   <ALL>
```

The query described in Listing 2 corresponds to the natural language statement: "Does the PP contain a statement of individual rights *as described in HIPAA regulation*". Typically, a 'yes-no' query involves an event, however, it is not limited to just events as long as it is formulated in a way compatible with the selectional restrictions on the filler of the THEME slot, i.e., the RANGE values of ontology-slot.

3.1 Inference Process: Brief Overview

The algorithm governing inference process in Ontological semantics is given in Listing 3 below: it is comprised of four basic procedures: FDB search, direct TMR matching, TMR expansion, and construction of PREMISE-SET. FDB search is invoked in case of wh-queries if a proper-name (object or event) is present. It is the least computationally expensive portion of the algorithm; a case statement is used to express the three possible outcomes of the search and each of them is handled appropriately--for a more detailed example see [5]. Direct TMR matching is a more general procedure triggered in case a proper-name is missing from FDB or in case of 'yes-no' queries where FDB search is not invoked, but rather matching of proposition in INFORMATION-REQUEST THEME of TMRQ and a proposition head in TMRI is performed. It is denoted as 'direct' because propositions are matched without utilizing any of the ontological resources. TMR expansion and PREMISE-SET are triggered in case of 'yes-no' queries, i.e., when the control is transferred to the "else" statement.

Listing 3

```
TARGET = THEME-OF REQUEST-INFORMATION
IF(((TARGET := OBJECT) || (TARGET := EVENT))
       && (HAS_NAME (non-empty)))
  INFO = SEARCH_FDB(TARGET)
   case_1: single entry
           RETURN INFO
   case_2: multiple entries
        do REFINE_VARIABLE(TARGET)^2
     case_3: no entry
        do DIRECT TMR MATCHING
```

[2] A discussion of FDB search-related specifics is omitted--for details see [5].

```
        APPEND_TO_FDB
        RESULT = TARGET
ELSE
     DIRECT_TMR_MATCHING (TARGET)
     DO TMR_EXPAND(PREMISE_SET)
     RESULT = EVALUATE(PREMISE_SET)
END IF
do ERROR_ESTIMATION
RESULT = RESULT + ERROR
RETURN (RESULT)
```

3.2 Inference Processing: Example

As the first step, the direct TMR matching procedure is invoked. It is assumed that, in the domain of PP, this particular procedure would be frequently used, especially in cases where PP aligns or is expected to align closely with the regulation due to the genre in which both texts are written. Once the THEME field of INFORMATION-REQUEST in TMRQ is matched to a proposition in TMRI, all its slots are being matched against those of the concept in TMRQ to the degree indicated in the LEVEL field. In other words, two dependency chains starting with THEME of REQUEST-INFORMATION are compared against each other and error estimates are attached in case of discrepancies or mismatches.

For the given example, mismatch occurs at a point of the LIST-1 DESCRIBE case role filler: INFORM-1 and INFORMATION-MANAGING-ACTIVITY-3 in TMRQ and TMRI, respectively. These mismatched items as well as the established dependencies are fed to the PREMISE-SET.

PREMISE-SET is a procedure that explores the OntoSem hierarchical structure; it updates two groups of relations in TMRQ: the case-roles and subsumption relations (see [5]). The system will attempt to establish a Most Common Intermediate Node (MCIN) by looking for a proposition in TMRI which results in the closest MCIN with the mismatched entry from the TMRQ. In this process, INFORM-1 will be found before MCIN is established, since INFORM appears as a proposition head and qualifies as a potential candidate for establishing MCIN. Its relation to the mismatched entry INFORMATION-MANAGING-ACTIVITY is established in the backtracking fashion.

At this point, PREMISE-SET contains the relation of the mismatched entries; further inconsistencies or discrepancies of the two TMRs are evaluated and passed into the result-formulating component in corresponding format.

The answer-formulating component collects error estimates as well as a reference dependency chain with respect to which the errors were estimated. The final answer is expressed as a natural language corresponding statement of the reference dependency chain with associated errors.

3.3 Error Estimation

Given the complexity of OntoSem text analysis, development of evaluation metrics is not an easy task: only preliminary efforts in this direction are outlined below.

The overall error assignment is based on the total number of tokens in the original query and the number of correctly inferred tokens [5]. Error estimation for each token proceeds differently depending on the type of mismatch; for cases of discrepancy in proposition heads, the *underspecification* error is calculated as the distance from the MCIN to the node in TMRI and normalized by the distance from MCIN to the corresponding target node in TMRQ. In case of *overspecification* in the TMRI node, i.e., a case where the target node in TMRQ is less restricted than the corresponding node in TMRI, error estimation is not considered.

The same exact technique cannot be applied in case of mismatched properties as it is not particularly informative, e.g. PURPOSE and EXPERINCER are both children of CASE-ROLE, however, semantically, their meaning is very different. In fact, there may not be meaningful error estimation when property mismatches occur. In such cases a relational or attribute error, E(r) or E(a), respectively, with the specific instance, is passed on to the answer-formulation component. The Relational or attribute error are not generally gradable, and the severity of such errors is determined by the user agent.

In the discussed example, the main discrepancy adding to the significance of alignment verification occurs in TIME field; in particular, there is an extra value associated with INFORM.TIME. It will be reflected in the result component. Additional information specified in TMRI that is not present in TMRQ has a restricting effect on the answer to the query and cannot be accounted for in a systematic way, however, it is reported as a part of the answer as E(+).

The error estimation for the example discussed in this paper is calculated below:

- total number of tokens to match: 9
- number of tokens successfully matched: 7

∴ probability of error = $1 - 7\times1/9 + 1/9\times E(r) + 1/9\times E(+)$, where E(r) is a relational error and E(+) is error due to extra information, which are set to constant value zero, hence the total error with respect to the reference dependency chain in Listing 4 is 2/9.

Listing 4

```
LEGAL-RIGHT-1 (SET (LEGAL-RIGHT-2))
LEGAL-RIGHT-2  (THEME  (TRANSFER-OBJECT))
TRANSFER-OBJECT (THEME (LIST))
LIST (DESCRIBES (INFORM-1))
INFORM-1 (SET (INFORM-2))
INFORM-2 (TIME (::YEARS = 6::))
INFORM-2 (THEME (PHI))
PHI (DESCRIBES (HUMAN))
THIRD-PARTY(BENEFICIARY-OF (INFORM-2, INFORM-1))
ORGANIZATION-1 (AGENT-OF (INFORM-1, INFORM-2))
```

E(+):

```
  INFORM-2.TIME                         |<16><04><2003>|
                                        < REQUEST-INFORMATION-1.TIME
              range                     ::years:: <= 6
              frequency                 default
```

E(r):

```
      LIST (DESCRIBES (INFORMATION-MANAGING-ACTIVITY))
      INFORMATION-MANAGING-ACTIVITY (HAS-EVENT-AS-PART (INFORM-1))
```

4 Conclusion and Future Research

Many problems that arise in the domain of PP management are due to the inherent ambiguity of natural language. Thus, it is expedient to look for solutions to such problems from the perspective of systems that incorporate adequate resources.

This paper focused on using the OntoSem framework in the domain of Privacy Policy management; specifically, it investigated the problem of establishing or checking PP compliance with regulations. The advantage of using OntoSem as an underlying framework is in the structure and interaction of its resources as well as in the fundamentals of knowledge representation. Given that the problem of policy compliance can be reduced to a question-and-answering problem, solution lies in the development and enhancement of NLP inference methods.

In order to accommodate the domain under OntoSem, current knowledge resources were extended: new lexical entries (500) as well as ontological concepts (50) were acquired (see [6]). FDB will be populated with the TMRs from domain with the processing of new texts. Even though, at the core of the inference processing, there lies a high-quality TMR, there still exists a degree of ambiguity, which is expressed in terms of error probabilities.

Future research will focus on exploring the application of OntoSem inference capabilities in the domain of Privacy Policies management--specifically, abstracting out a standard set of queries that are necessary to determine compliance based on a particular category of regulation document. Such a task would necessarily involve a development of a ranking system of the queries and will result in gradable outcomes. Implementation of this idea may require a development of a new framework particular query type. OntoSem-specific future research goals include the improvement and evaluation of automated TMR construction [cf. 8] and extension of current knowledge resources and further improvement and increasing automation of acquisition methodologies [10].

References

1. Anton, A., Baumer, D., Bertino, E., Dark, M., Li, N., Proctor, R., Rappa, M., Raskin, V., Vu, K., Yu, T.: CyberTrust Proposal, National Science Foundation (2004)
2. Blue Cross Blue Shield Privacy Policy of North Carolina
3. Health Insurance Portability and Accountability Act (1996)
4. KBAE. Knowledge Base Acquisition Editor
5. Krachina, O.: Ontology-Based Inference Methods. In: Proceedings of MCLC 2006, University of Illinois, Urbana (May 2006)
6. Krachina, O.: Role of Ontological Semantics in Handling Privacy Policies. Poster. 2005 CERIAS Annual Research Symposium, West Lafayette, IN. CERIAS TR-XX-2005 (2005)
7. Li, N., Yu, T., Anton, A.: A Semantics-Based Approach to Privacy Languages.To appear in Computer Science and Systems Engineering Journal
8. Nirenburg, S., Beale, S., McShane, M.: Evaluating the Performance of the OntoSem Semantic Analyzer (2004)
9. Nirenburg, S., Raskin, V.: Ontological Semantics. MIT Press, Cambridge (2004)
10. Triezenberg, K.E.: The Ontology of Emotion. An Unpublished Ph.D. Thesis, Linguistics, Purdue University (2006)

Evaluating Measurement Models for Web Purchasing Intention

Bing-Yi Lin[1], Ping-Ju Wu[2], and Chi-I Hsu[3]

[1,3] Department of Information Management, Kainan University, Taiwan, ROC
[2] Department of Information Management, Yuan Ze University, Taiwan, ROC
[1]m93221001@ms2.knu.edu.tw, [2]ceci.wu@gmail.com,
[3]imchsu@mail.knu.edu.tw

Abstract. This study is mainly to evaluate measurement models for web purchasing intention, which reflects to a tendency of attitudes and behaviors toward the online purchasing behaviors. Four dimensions of web purchasing intention are proposed based on a literature review, including (1) Information Provision: product-related information that e-retailing users receive from the Internet; (2) Alternative Evaluation: issues that affect an individual's decision making when purchasing online; (3) Customer Service: services that e-retailing websites offer for customers in the e-transactions and post-sale process; (4) Price: the final dealing prices of products on e-retailing websites. This study developed a questionnaire and delivered to consumers with web purchasing experience. The method of Structural Equation Modeling (SEM) is adopted to verify the internal quality of the proposed measurement.

1 Introduction

By providing a variety of products and services through the Internet, buyers and sellers can proceed with different transaction activities, such as searching for information, placing orders, making payments, delivering digitized products, or controlling physical logistics. Although many business practitioners have established their own industries on the Internet, many of them merely have the basic functions of e-catalogue provision and products distribution, with insufficient offers for a complete e-transaction process and post-purchasing customer services. This highlights the 'low entry threshold' and 'high ratio of being weeded out' nature of the web purchasing market.

Web purchasing intention can reflect the degree to which an individual intends to act such behavior during the web purchasing process, and further can lead to his/her actual purchasing behavior (Shim et al. 2001; Hsu and Chiu 2004). In order to survive in a highly competitive environment, it becomes very important for web purchasing practitioners to understand customers' purchasing intention and provide them with appropriate products and services.

This research tends to evaluate measurement models for web purchasing intention to understand the factors contributing to customers' web purchasing behaviors. Structural Equation Modeling (SEM) and LISREL 8.71 are used to analyze the data collected from questionnaires.

M.J. Smith, G. Salvendy (Eds.): Human Interface, Part II, HCII 2007, LNCS 4558, pp. 740–748, 2007.

2 Literature Review

Over the last decade, Internet commerce has been growing dramatically, and online shopping is becoming almost commonplace. When one purchases on the Web, a variety of factors and concerns may contribute to customers' intention to buy products online.

Due to the communication structures provided by the Internet, customers can become more empowered as they surf the net for the best prices, product information, the latest consumer innovations or special offers. The Internet serves as a channel allowing users to gather information; Bulletin boards and news groups provide the opportunity for consumers to exchange their information. Searching for and chatting with someone is one of the major uses of the Internet and E-shoppers can turn to each other for any kind of advices they need. E-consumers share the need for post purchase reassurance with their traditional counterparts and hence Internet virtual communities provide a forum for online shoppers to share their experiences and recommendations. Ramus and Nielsen's (2005) conducted a research investigating online grocery retailing and found out that although online shopping enabled customers to have a wide range of available information about stores and products, certain product categories such as food that is sensitive to shipping time and heavy set natural can limit the realistic range of stores and products to choose from. Lim and Palvia (2001) examined the impact of EDI (electronic data interchange) on customer service by arranging both the vendor and the customer to utilize EDI in their distribution operations. The results strongly supported the use of EDI to improve the following customer service components: order cycle time, product availability, distribution flexibility, distribution information, and distribution malfunction. However, an impact on post-sale product support was not significant from the data findings.

Other than the factors mentioned above to be involved in the decision-making process, it has been suggested that convenience is the main reason why consumers use the Internet for the purpose of purchasing (e.g., Jarvenpaa and Todd, 1997). Brown et al. (2003) conducted a research to investigate whether Internet shoppers are primarily motivated by convenience, and the empirical results showed that other factors such as product type, prior purchase, and gender are more likely to influence online customers' purchase intention.

Convenience is now giving way to price as the most important reason behind consumers making Internet purchases. Consumers' online purchasing experience may be enhanced by incentive programs such as saved shopping lists or personalized help, point- and incentive-based premium and gift programs and cumulative discounts or rebates based on purchase amount (Breitenbach and Van Doren 1998). Kim and Kim (2004) conducted a survey research by mailing to 303 adults who had a computer at home with the access to the Internet in the USA to investigate the perceived four factors of online shopping: transaction/cost; incentive programs, site design and interactivity. They found that the transaction/cost factor and the incentive programs factor were important predictors in determining the customer intention to purchase products via the Internet. Incentive programs also mediated the relationship between education level and online purchase intention.

Along with the attributes discussed above, other factors such as privacy concerns and customers' prior experiences with the Internet have been key reasons consumers do not make online purchases over the Internet. In addition, much of an individual's decision on making Internet purchases can be thought of in terms of trust. Many Internet users are concerned that people or companies they do not know will get information about them through the Internet. Concerns over trusting Web providers with personal information can be described in two aspects: concerns over trusting the Internet to be secure and trusting Internet transactions to be secure (Ratnasingham 1998). Internet security deals with general levels of security related to network, application, and system components of the Internet, while Internet transaction security deals with requirements for secure transactions over the Internet. Consumer participation in web commerce may be limited by customers' concerns about the risk to their sensitive information. Salisbury et al. (2001) created a scale applied in a causal model to measure perceived Web security and investigate the relative impact of beliefs about Web shopping on intent to purchase products using the Internet. Their findings indicated that security has greater influence on intent to purchase products using the Internet. Experience with the Internet is an essential consideration in making online purchases (Hoffman et al. 1999). Fox (2000) found out that novice Internet users are less likely to make a purchase online: 27 percent of users with less than six months experience using the Internet bought something online, compared to 60 percent of those with three or more years of Internet experience. Besides, new users worry more about online credit card theft (70 percent) than experienced Internet users (46 percent). George (2004) conducted a research using the theory of planned behavior (TPB) as the theoretical base to investigate if beliefs about privacy and Internet trustworthiness can help determine attitudes towards the Internet, which were thought to affect the intent to make Internet purchases. Taking Internet experience into account, TPB was suggested as a useful and powerful framework within which to investigate ideas about Internet purchasing.

Since security and privacy are two critical factors affecting consumers' intention to purchase online, it is important for online companies and retailers to improve e-transmission security and broader protection policies. Many of them are now more involved in providing customized strategies including return policies, interactivity and personalization on Web sites (Verton 2001; Detmer 2002). For example, the interactivity of the Web allows users to personalize and customize their experience through keyword search, dealer locator, comments and online ordering, and this experience has had a significant positive effect on the quality rating of Web sites (Ghose and Dou 1998).

3 Research Model

Based on related literatures reviewed, four constructs are established in this research; they are Information Provision, Alternative Evaluation, Customer Service and Price. Lin et al., (2003) viewed web purchasing as a serial information searching process and choices selecting evaluation, and the information and mechanisms provided on the websites for assisting customers to make choices and to evaluate will become influential to customers' purchasing intention. Hence, this research includes these two elements – Information Provision and Alternative Evaluation into the constructs established. Kalakota and Whinston, (1997) indicated that by including customer services into the main procedure of web purchasing can elevate customers' purchasing intention during the interactive process; thus, Customer Service is also included as one construct in this research. Covaleski, (1997) pointed out that since consumers can have all kinds of products information available online for comparison, companies tend to use those with the most competitive prices to attract consumers. Kannan and Kopalle, (2001) also mentioned that price is a key factor of web purchasing; price difference can significantly influence customers' purchasing behavior. Therefore, the element "Price" is also considered as one construct in our research.

Information Provision is defined in this research as "providing product related information for e-retailing users through the Internet". This construct is established based on related papers, such as searching for product information (INF1) (Ramus and Nielsen 2005), new-product information (INF2)(Lin, Fu et al. 2003), using the information service functions provided on the website (INF3)(Chang, Chen et al. 2005), and searching for specific products (INF4)(DeLone and McLean 2003).

Alternative Evaluation is defined as the factors affecting an individual's evaluation on alternative selections when purchasing products online. This construct is established based on customization (ALT1)(Kim and Kim 2004), convenience (ALT2)(Brown, Prope et al. 2003), and autonomy (ALT3)(Chang, Chen et al. 2005)(defined as no salesperson interrupting as what usually happens in a physical shop).

Customer Service is defined as the services provided by e-retailing websites for consumers during the whole transaction and post-purchasing process. This construct is established based on responses from retailing websites to the requests of customer problems (CUS1)(Yen, Chen et al. 2005), prompt post-sale service (CUS2)(Lim and Palvia 2001), and complete post-sale service (CUS3)(Lim and Palvia 2001).

Price is defined as the actual list price of products sold on e-retailing websites. This construct is established based on price on sales (PRI1)(Kim and Kim 2004), special offers for members (PRI2)(Yen, Chen et al. 2005), and discounts activity (PRI3)(Kim and Kim 2004).

As for the scale development, following the steps suggested by Davis (1989), we invited two online purchasing consumers, a website manager, and a graduate student from MIS doctoral program to categorize all measurement variables into four constructs according to their individual cognition, in order to ensure the construct validity and examine whether there is any ambiguous construct item. The categorized results are shown in Table 1, which indicate that the correct hit ratios of the four subjects are all higher than 84%, with an overall hit ratio of 91%. Since most of the items can be precisely classified into the estimated constructs, the measurement scales developed in this research for the four constructs have been recognized to have sufficient convergent validity and discriminate validity.

According to the measurement scales developed in this research, four models are proposed as shown in Figure 1~4 below. In Model 1 (M1), a single factor – Web Purchasing Intention is used to represent all 13 items; In Model 2 (M2), four independent factors are adopted; In Model 3 (M3), these four factors are made to be correlative to each other; Model 4 (M4) shows a structure with single second-order factor.

Table 1. Item Placement Ratios

	Actual Categories						
Target Category	INF	ALT	CUS	PRI	N/A	Total	% Hits
INF	14		2			16	87
ALT	1	10	1			12	84
CUS			12			12	100
PRI	1			11		12	92
Total Item Placement: 52			Hits: 47			Overall Hit Ratio: 91%	

4 Data Analysis

4.1 Data Analysis Process

The data analysis process in this research is to introduce the analysis method - Structural Equation Modeling (SEM) first, and then to explain the questionnaire development, descriptive statistics of samples, and the analysis results of normal distributed test, exploratory factor analysis (EFA), questionnaire reliability and SEM.

4.2 Data Collection and Analysis

This research collected data samples by questionnaire distribution aiming at users with online shopping experiences. Four hundred and fifty questionnaires were distributed to college students (including undergraduate and graduate), staffs and faculties with Internet shopping experiences from a university in northern Taiwan. Out of the total 450 questionnaires distributed, 412 valid samples were collected after deducting the samples with more than three items not answered. The effective response rate is 91.6%.

In the pretest, 100 out of 412 samples were selected randomly to conduct EFA. All the 13 items were used to run factor analysis by SPSS 13.0 for Windows; principal components were adopted to select those with eigenvalues bigger than 1, and varimax was used to turn the axle. The result of EFA supports the fitness of the factor structure proposed in this research.

The correlation matrix and reliability values for all constructs are shown in Table 2. The analysis result of reliability values shows that except for the construct of Customer Service, the values of Cronbach's α are all above the 0.7 level and thus satisfy the reliability requirement. In addition, composite reliability (CR) is used to measure the internal consistency of latent variables and the CR coefficients are all greater than 0.6 as suggested by Fornell and Larcker (1981), which means that the developed measurement items have significant CR on latent constructs.

4.3 The Measurement Model

Through CFA, SEM tests the validity of each questionnaire item by evaluating the convergent validity and discriminate validity in two parts: (1) whether all the regression coefficients are significant? (2) Whether the measurement model itself is sufficient to explain the data variations? The significance of coefficients can be decided by

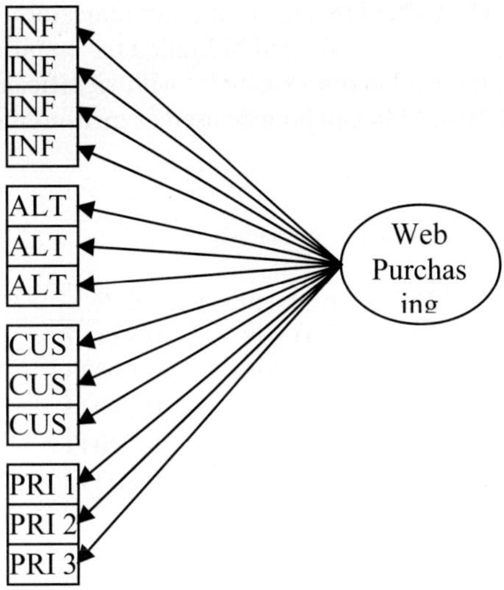

Fig. 1. M1: Single first-order factor

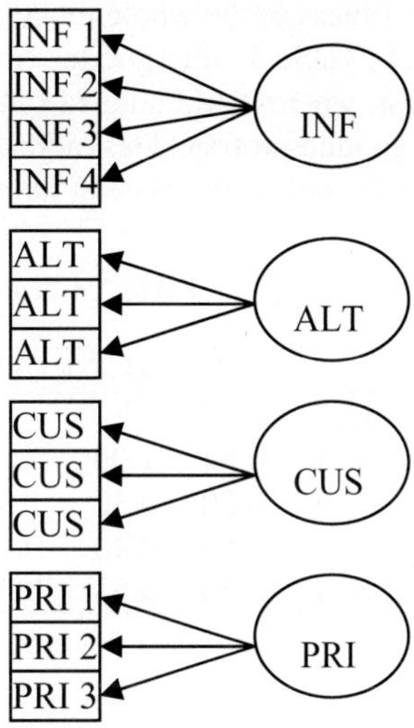

Fig. 2. M2: Four uncorrelated first-order factor

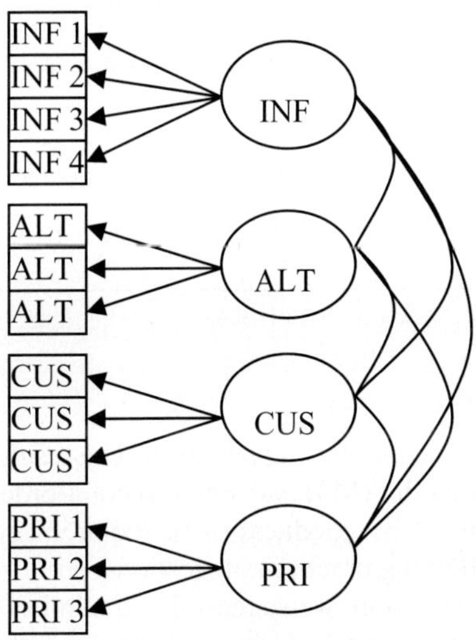

Fig. 3. M3: Four correlated first-order factor

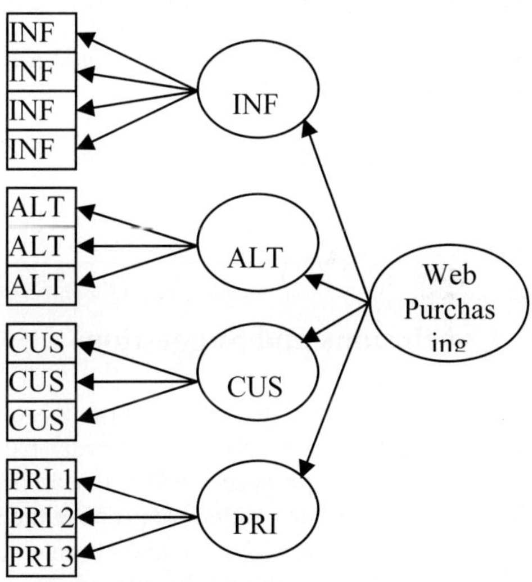

Fig. 4. M4: Single second-order factor

examining if the t-values are greater than 1.96 at the significant level of 0.05. The values of the factor loading between all latent variables and their first observable variable are fixed as 1 for the purpose of standardization. The overall fitness degree of the measurement model and real data can be assessed by Comparative Fit Index (CFI) (Joreskog and Sorbom 1993). The measurement model is accepted when the CFI value is greater than 0.95 (Bentler 1992; Bentler 1995).

The fitness of the whole model may also be judged by means of other indicators. As shown in Table 3, after goodness-of-fit index analysis, M1 and M2 failed the empirical data test, whereas M3 andM4 satisfied all the evaluation standards, with significantly better goodness-of-fit. This indicates that M3 and M4 can both be used as measurement models for online purchasing intention studies.

Table 2. Correlation Matrix & Reliability Analysis

	INF	ALT	CUS	PRI	LOY	Cronbach's α	C.R.
INF	1.00					0.808	0.83467
ALT	0.64	1.00				0.705	0.68073
CUS	0.20	0.32	1.00			0.653	0.68936
PRI	0.55	0.40	0.27	1.00		0.785	0.79043
LOY	0.50	0.53	0.38	0.49	1.00	0.832	0.80792

Table 3. The Result of Measurement Model

	χ2	df	χ2/df	CFI	NFI	NNFI	GFI	AGFI	PGFI	RMSEA	SRMR
Recommended Value			<3	>0.95	>0.90	>0.90	>0.80	>0.80	>0.50	<0.08	<0.08
M1(single first-order factor)	507.05	65	7.8007	0.82	0.79	0.78	0.80	0.72	0.57	0.148	0.10
M2(four uncorrelated first-order factor)	302.54	65	4.6544	0.90	0.88	0.88	0.87	0.82	0.62	0.108	0.19
M3(four correlated first-order factor)	100.03	59	1.69542	0.98	0.95	0.97	0.95	0.93	0.62	0.047	0.054
M4(single second-order factor)	107.08	61	1.75541	0.98	0.95	0.97	0.95	0.92	0.64	0.049	0.058

5 Conclusions and Suggestions

This research evaluate measurement models of online purchasing intention. The results indicate that the four correlated first-order factor model (M3) and single second-order factor model (M4) proposed in this research satisfy the goodness-of-fit requirement. Based on our research results, we propose the following practical suggestions:

As for information provision and alternative evaluation, we suggest that the links to product information should be placed on obvious spots of the website, and the searching mechanism should be improved so that customers can get what they want quickly and precisely. Moreover, web-mining techniques can be applied to establish a back-end mechanism to analyze the online behaviors of customers, and understand the correlation of products, in order to achieve Precision Marketing – to actively propose advertisements that can meet customers' preference. Nevertheless, the balance between customization and customer autonomy should be carefully dealt with; that is, customers' autonomy of purchasing online should not be invaded. For example, an option to allow users to activate/deactivate the pop-up advertisements and e-news subscription should be offered on the website.

In respect of customer service, a clear explanation of post-purchasing service procedures should be offered to protect customers' right. Meanwhile, an online instant communication service mechanism is also recommended. For example, through communication software services offered by Skype or VoIP can provide more diverse channels for customers to use. Besides, since the carriers of logistics companies are the first line to contact directly with customers, to build a good long-term cooperative relationship with logistics companies is also crucial to elevate the overall post-purchasing service level.

As for the price, along with discounted merchandise provided to loyal customers, double bonuses or e-coupons are also offered for customers to win through a series of funny promotion games. For example, a bonus with different points is given in a spin-the-wheel game and customers can only use the bonus they win to buy products on the same website. And lottery drawing games can keep the customers back by cashing the accumulated points they gained when buying products which can be deducted directly from a new order. Cross-industry cooperation is also a good choice; customers can have more options to use their e-coupons for purchasing again.

References

1. Bentler, P.M., Bonett, D.G.: Significance Tests and Goodness of Fit in the Analysis of Covariance Structures. Psychological Bulletion 88, 588–606 (1980)
2. Browne, M.W., Cudeck, R.: Alternative Ways of Assessing Model Fit. In: K.A. Bollen, Long, J.S. (eds.) Testing Structural Equation Models, pp. 136–162 (1993)
3. Chang, A.H., Chen, J.H., Chen, C.J.: Shopping Values, Motivations and Preference for E-commerce Marketing Strategies of Online Buying Behaviors. Fu. Jen. Management Review 12(1), 31–64 (2005)
4. Covaleski, J.: Panel: Internet Marketing may Reduce Prices. Best's Review / Property-Casualty Insurance Edition 97(12), 71–72 (1997)
5. Davis, F.D.: Perceived Usefulnes, Perceived Ease of Use, and User Acceptance of Information Technology. MIS Quarterly 13(3), 319–340 (1989)
6. DeLone, W.H., McLean, E.R.: The DeLone and McLean Model of Information Systems Success: A Ten-Year Update. Journal of Management Information Systems 19(4), 9–30 (2003)
7. Detmer, T.: Seeking the Complete Customer Experience: the Web as a Marketing Tool. CustomerInter@ctionSolutions 20(11), 44–46 (2002)
8. Fornell, C., Larcker, D.F.: Evaluating Structural Equation Models with Unobervables and Measurement Error. Journal of Marketing Research 18, 39–50 (1981)
9. Fox, S.: Trust and Privacy Online: why Americans Want to Rewrite the Rules. Pew Internet and American Life Project (2000), from http://www.pewinternet.org
10. George, J.F.: The Theory of Planned Behavior and Internet Purchasing. Internet Research 14(3), 198–212 (2004)
11. Ghose, S., Dou, W.: Interactive Functions and Their Impacts on the Appeal of Internet Presence Sites. Journal of Advertising Research 38(2), 29–42 (1998)
12. Hoffman, D.L., Novak, T.P., Peralta, M.: Building Consumer Trust Online. Communications of the ACM 42(4), 80–85 (1999)
13. Hsu, M.H., Chiu, C.M.: Internet Self-Efficacy and Electronic Service Acceptance. Decision Support Systems 38(3), 369–381 (2004)

14. Jarvenpaa, S.L., Todd, P.A.: Is there a future for retailing on the Internet. In: Peterson, R.A. (ed.) Electronic Marketing and the Consumer, pp. 139–154. Sage, Thousand Oaks (1997)
15. Joreskog, K., Sorbom, D.: LISREL 8: User's Reference Guide. Scientific Software International. Chicago (1993)
16. Kalakota, R., Whinston, A.B.: Electronic Commerce—A Manager's Guide. Addison-Wesley, London (1997)
17. Kannan, R.K., Kopalle, P.K.: Dynamic Pricing on the Internet Importance and Implications for Consumer Behavior. International Journal of Electronic Commerce 5(3), 63–83 (2001)
18. Kim, E.Y., Kim, Y.K.: Predicting Online Purchase Intentions for Clothing Products. European Journal of Marketing 38(7), 883–897 (2004)
19. Lim, D., Palvia, P.C.: EDI in Strategic Supply Chain: Impact on Cstomer Srvice. International Journal of Information Management 21(3), 193–211 (2001)
20. Lin, Y.L., Fu, F.L., Jou, Y.H., Lee, J.: The Comparison of Market Segment Variable on the Internet Marketing - Taking Web-Patronage as an Example. Sun Yat-Sen Management Review 11(4), 673–705 (2003)
21. Ramus, K., Nielsen, N.A.: Online Grocery Retailing: What do Consumers Think? Internet Research 15(3), 335–352 (2005)
22. Ratnasingham, P.: The Importance of Trust in Electronic Commerce. Internet Research 8(4), 313–321 (1998)
23. Salisbury, W.D., Pearson, R.A., W., P.A., W., M.D.: Perceived Security and World Wide Web Purchase Intention. Industrial Management & Data. Systems 101(4), 165–176 (2001)
24. Shim, S., Eastlick, M.A., Lotz, S.L., Warrington, P.: An Online Purchase Intention Model: The Role of Intention to Research. Journal of Retailing 77(3), 397–416 (2001)
25. Verton, D.: Churn. Computerworld 35(6), 50 (2001)
26. Yen, H.J., Chen, J.S., Lu, P.T., Sheu, T.G.: E-services and Customer Relational Benefits in the Context of Webs-Based Self-Service. Journal of Information Management 12(1), 1–28 (2005)

Web Application for Data Analysis by Means of Evidential Dominance

Zaw Aung Htwe Maung[1] and Yasufumi Kume[2]

[1,2] Programs in Mechanical Engineering, Interdisciplinary Graduate School of Science and Engineering, Kinki University
3-4-1 Kowakae Higashiosaka, Osaka 577-8502, Japan
bleu-ciel@s5.dion.ne.jp[1], kume@mech.kindai.ac.jp[2]

Abstract. This paper describes web application for data analysis by means of evidential dominance. Red Hat Linux 9, Apache, Tomcat, Java 2 Platform, and Eclipse are used to this system as tools. In conventional system, Java for Evidential Dominance is applied as Java servlet. Client accesses to the Web Server to analyse data. Java for evidential dominance is started and implemented on Web Server. The server processes and responses the result of analysis to the client. The information of data is not remained in server. This system not only responses the result of analysis to the client, but also save the information of data in server. Evidential dominance is applied to decide the priority relation of data which varies widely and have little amount of data. Evidential dominance not only is effective to focus on one phase a lot of alternatives, but also applies to decision making problem.

Keywords: evidential dominance, data analysis, eclipse web tools platform, decision making, apache, tomcat.

1 Introduction

The nature of much information available to decision makers is vague and imprecise. It is information for human managers in organizations or for process agents in a distributed computer environment [1]. Utility function could provide a mean of modelling value judgement of decision maker quantitatively. Scientific approach for value judgement has been discussed rigorously in the area of economics [2]. The probability density is allocated in the interval value in this research, and the data analysis by means of evidential dominance to decide the priority relation of interval value is described. Optimum alternatives are preferred from alternatives by comparing the expected utility of the evaluation item of each alternative. Optimum evidential dominance is decided by means of focus element (interval values) and basic probability assignments (weight) are generated by random number. Optimum evidential dominance is applied to the result data of conveyance of magnetic vertical conveyance supply machine, and the effectiveness of the evidential dominance is clarified.

M.J. Smith, G. Salvendy (Eds.): Human Interface, Part II, HCII 2007, LNCS 4558, pp. 749–758, 2007.

2 Web Application for Data Analysis

Web server construction and environment of web application development are described in this section. Web application for data analysis is shown in Fig. 1.

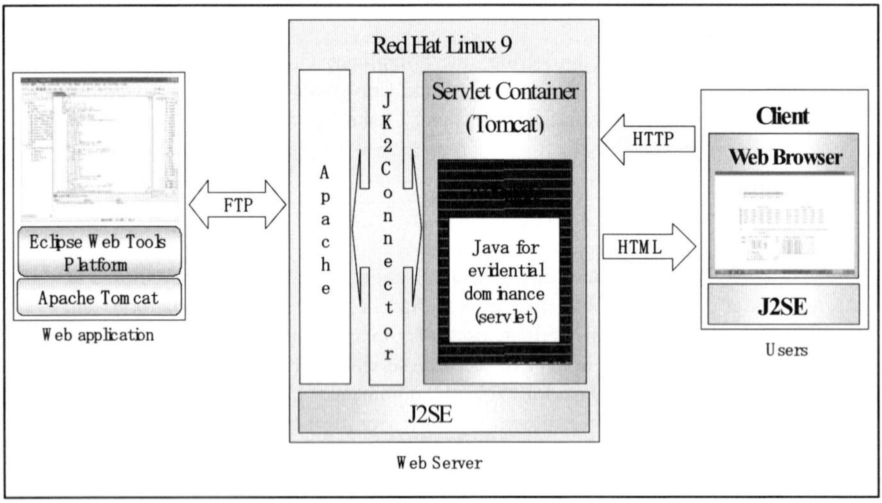

Fig. 1. Web application for data analysis

2.1 Construction of Web Server

Construction of web server is described. Red Hat Linux 9 is installed as a server operating system [3]. Apache software (Ver. httpd-2.0.59) was downloaded from the Apache Software Foundation; it was installed and configured in the server. Apache is a free software HTTP web server for Unix-like systems [4]. Java 2 Platform, Standard Edition (Ver. jdk1.5.0_09) was downloaded and installed in the server from Sun Developer Network. J2SE provides a complete environment for application development on server [5]. It is necessary to set an environment variable JAVA_HOME to the pathname of the directory. File of .bash_profile is opened by vi editor, and adds as follows:

```
PATH=/usr/local/jdk1.5.0_09/bin: ${PATH} :${ HOME}/bin
Export PATH
JAVA_HOME=/usr/local/jdk1.5.0_09;
export JAVA_HOME
```

File of .bash_profile is reflected by source command. Tomcat software (Ver. apache-tomcat-6.0.2) was downloaded and installed in the server from the Apache Software Foundation. Tomcat is operated as a plug-in of Apache to start the Java servlet on the server. Tomcat implements the servlet and Java Server Pages (JSP), and includes many additional features that make it a useful platform for developing web applications and web services [6].Web server specific connector is needed. If one application is written in Java using the Servlet API a connector is needed to route the requests from the web server to the Servlet Engine. Thus, tomcat connector is used in this system. It is plug-in that handles the communication between Tomcat and Apache. Tomcat connector (Ver. jakarta-tomcat-connectors-jk2-2.0.4) was installed to generate mod_jk2.so module. Mod_jk2.so module is the apache module. Httpd.conf file is necessary to edit for module reads in apache. Httpd.conf file which exists in /usr/local/apache2/conf/ adds as follows:

```
LoadModule jk2_module modules/mod_jk2.so
```

Server.xml file and workers2.properties file are necessary to edit and configure, if client accesses to Servlet (Ex.:kinki) of Servlet Container in web server. Workers2.properties file specifies the URL pattern forwarded from apache to tomcat. Workers2.properties file which exists in /usr/local/apache2/conf/ adds as follows:

```
[uri:/kinki/*]
info = For Kinki_Univ.
context = /kinki
debug = 0
```

Server.xml file specifies the mapping of URL which accepted by tomcat. Server.xml file which exists in /usr/local/tomcat/conf/ adds as follows:

```
   <Context path="/kinki" docBase="kinki"   debug="0" reloadable="true"
crossContext="true" trusted="fales">
   <Logger className="org.apache.catalina.logger.FileLogger" prefix="kinki_log."
suffix=".txt"   timestamp="true"/>
   </Context>
```

Client can be offered through Web browsers by using HTML generated dynamically by Java Server Pages (JSP) technology, or Java applets [7].

2.2 Environment of Web Application Development

Web application developer uses Eclipse software as tool for web application development. Eclipse is an open source community, whose projects are focused on building an open development platform comprised of extensible frameworks, tools and runtimes for building, deploying and managing software across the lifecycle. The tools and frameworks are primarily focused on building JEE, web services and web applications. Eclipse also provides support for other languages, such as C/C++, PHP, and others [8]. Eclipse software (Ver.SDK 3.0.1) was downloaded from the Eclipse Foundation; it was installed and configured in personal computer. Java 2 Platform, Standard Edition (Ver. J2sdk-1.4.2_08) was downloaded and installed in personal computer for java program. Tomcat software (Ver. apache-tomcat-5.0.28) was downloaded, installed and configured in personal computer. Java for server side development builds and run a J2EE web application using Eclipse Web Tools Platform and Apache Tomcat [9]. Web application developer generates the program of web application, and deploys to web server by means of FTP. The application is composed of a set of servlet that collaborate to provide the overall functionality of the application.

3 Evidential Dominance

The decision making using the evidential dominance in web application for data analysis is described. In the decision making problem under risk, probability distribution is ordered by maximisation of expected utility. Optimum alternative is decided. A cardinal utility function that expresses probability distribution of each alternative and preference of decision makers is necessary to identify in such as decision making. But, it is difficult to identify probability distribution and cardinal utility function in actual problem. The method is proposed that applies Dempster-Shafer theory of evidence to comparison of alternatives based on partial information of probability distribution. The importance of each alternative is expressed as basic probability assignment. The expected utility *E is* obtained as an interval [10]. The basic probability assignment *m* in universal set *X* is considered to be set of probability distribution [11]. The maximisation of expected utility is obtained as follows:

$$E = \max \sum_{x \in X} s(x) m(x), \; x_{ij} \in X \tag{1}$$

$$E = \max \left\{ \sum_{i=1}^{t} s(x_{i1}) m(x_{i1}), \sum_{i=1}^{t} s(x_{i2}) m(x_{i2}), .., \sum_{i=1}^{t} s(x_{in}) m(x_{in}) \right\} \tag{2}$$

$$\sum_{i=1}^{t} m(x_{ij}) = 1 (j = 1,2,...,n) \tag{3}$$

,

where $s(x)$ is interval values of alternatives. $m(x)$ is the basic probability assignment of a set x in X. The expected utility in evidential dominance is obtained as closed interval. Therefore, inequality relation "≥" of closed interval is necessary to define for concept that maximizes the expected utility [10]. L*R, LR, LC, RC, LW, RW, CW, L-R, and C-L evidential dominance can be defined on the basic of inequality relations of closed interval [12]. The inequality relations of four elements are the maximum value, the minimum value, the median value, and the range of closed interval A and B [13]. They are defined the as follows:

$A = [\, a_L, a_R \,], \; a_C = (\, a_L + a_R \,) / 2, \; a_W = a_R - a_L$
$B = [\, b_L, b_R \,], \; b_C = (\, b_L + b_R \,) / 2, \; b_W = b_R - b_L,$

where a_L and b_L are the minimum values of interval. a_R and b_R are the maximum values of interval. a_C and b_C are the median values of interval. a_W and b_W are the ranges of the interval. The intervals of alternatives are preferred by the necessary and sufficient condition for nine kinds of evidential dominance could be expressed as Table 1.

Table 1. The necessary and sufficient condition for nine kinds of evidential dominance

Evidential dominance		Necessary and sufficient condition	pereference	
Condition(1)	L*R	$①_L > ②_R$	alternative①≻	alternative②
Condition(2)	LR	$①_L > ②_L$ and $①_R > ②_R$	alternative①≻	alternative②
Condition(3)	LC	$①_L > ②_L$ and $①_C > ②_C$	alternative①≻	alternative②
Condition(4)	RC	$①_R > ②_R$ and $①_C > ②_C$	alternative①≻	alternative②
Condition(5)	LW	$①_L > ②_L$ and $①_W > ②_W$	alternative①≻	alternative②
Condition(6)	RW	$①_R > ②_R$ and $①_W < ②_W$	alternative①≻	alternative②
Condition(7)	CW	$①_C > ②_C$ and $①_W < ②_W$	alternative①≻	alternative②
Condition(8)	L-R	$①_L > ②_L$ or ($①_L = ②_L$ and $①_R > ②_R$)	alternative①≻	alternative②
Condition(9)	C-L	$①_C > ②_C$ or ($①_C = ②_C$ and $①_L > ②_L$)	alternative①≻	alternative②

where $①_L$ and $②_L$ are the minimum values of alternative. $①_R$ and $②_R$ are the maximum values of alternatives. $①_C$ and $②_C$ are the median values of alternatives. $①_W$ and $②_W$ are the ranges of alternatives. "≻" means the preference.

4 The Analysis Using Random Number

In evidential dominance, expected utility is obtained as closed interval and nine kinds of evidential dominances are proposed. In order to decide optimum evidential dominance, simulation is performed. Random numbers are used as basic probability assignment for each focus element (interval).

4.1 The Interval Values of Alternatives Using Random Number

When the number of basic probability assignment of each focus element is fixed at five items, the case of the number of alternatives are 7 items is case 1 and 15 items is case 2 .Basic probability assignment for first trial in case1 is shown in Table 2, focus element of alternatives (interval value) is shown in Table 3.The Lecture Notes in Computer Science volumes are sent to ISI for inclusion in their Science Citation Index Expanded.

Table 2. Basic probability assignments of case 1

Focus elements	A	B	C	D	E
Basic probability assignmei	0.21	0.09	0.31	0.18	0.21

The value of every time of focus element and focus element of alternatives (interval value) are changed by random number. Analysis is performed 250 times of trial. Alternatives are focused on one phase a lot of alternatives. Frequency distribution of alternatives focused one phase in case 1 is shown in Table 4.

Table 3. Focus element of alternatives (interval value) for case 1

Focus elements	A	B	C	D	E
Basic probability assignments	0.21	0.09	0.31	0.18	0.21
Alternative①	[1.40, 3.16]	[0.90, 3.06]	[1.62, 3.76]	[1.76, 2.62]	[1.24, 3.52]
Alternative②	[1.40, 3.62]	[1.72, 3.52]	[1.98, 3.78]	[1.40, 2.82]	[2.08, 3.54]
Alternative③	[2.32, 3.36]	[2.20, 3.62]	[1.28, 2.90]	[1.72, 3.16]	[1.24, 4.02]
Alternative④	[2.36, 3.92]	[1.28, 2.94]	[2.04, 3.80]	[2.36, 3.58]	[1.30, 3.00]
Alternative⑤	[1.76, 3.86]	[1.80, 3.76]	[2.38, 3.64]	[1.52, 3.74]	[1.56, 2.82]
Alternative⑥	[1.60, 4.24]	[1.38, 3.96]	[2.82, 3.92]	[0.80, 2.62]	[2.32, 3.94]
Alternative⑦	[1.86, 3.54]	[1.68, 3.52]	[2.54, 3.90]	[2.12, 2.94]	[1.76, 3.38]

Table 4. Frequency distribution of alternatives (case 1)

	Frequency distribution of alternatives focused on one phase							
Numbers of alternatives / Evidential	1	2	3	4	5	6	7	Times of failure for focus on one
L*R evidential dominance	0	0	0	0	0	0	250	0
L R evidential dominance	145	36	11	2	0	0	0	56
L C evidential dominance	72	83	59	26	10	0	0	0
R C evidential dominance	153	56	13	5	0	0	0	23
L W evidential dominance	37	16	7	3	0	0	0	187
R W evidential dominance	140	40	11	4	0	0	0	55
C W evidential dominance	71	77	60	30	12	0	0	0
L-R evidential dominance	70	81	59	29	11	0	0	0
C-L evidential dominance	53	68	66	37	23	3	0	0

Frequency distribution of alternatives focused one phase in case 2 are shown in Table 5. From Table 4 and Table 5, L R evidential dominance, RC evidential dominance and RW evidential dominance can be effectively focused on one phase in nine kinds of evidential dominance without increasing the number of alternatives.

4.2 The Value of Basic Probability Assignment for Focus Elements Using Random Number

When the number of alternatives is fixed as 7 items, the case of the number of basic probability assignment for each focus element in 6 items is case 3 and the case of the number of basic probability assignment for each focus element in 7 items is case 4. The value of every time of focus element and focus element of alternatives (interval value) are changed by random number. Analysis is performed 100 times of trial.

Table 5. Frequency distribution of alternatives (case 2)

Numbers of alternatives / Evidential Dominance	Frequency distribution of alternatives focused on one phase															
	1	2	3	4	5	6	7	8	9	10	11	12	13	14	15	Times of failure for focus on one
L*R evidential dominance	0	0	0	0	0	0	0	0	0	0	0	0	0	0	250	0
L R evidential dominance	123	44	7	1	0	0	0	0	0	0	0	0	0	0	0	75
L C evidential dominance	53	83	61	30	16	5	1	1	0	0	0	0	0	0	0	0
R C evidential dominance	117	67	20	7	0	0	0	0	0	0	0	0	0	0	0	39
L W evidential dominance	12	1	0	0	0	0	0	0	0	0	0	0	0	0	0	237
R W evidential dominance	113	50	9	3	0	0	0	0	0	0	0	0	0	0	0	75
C W evidential dominance	50	80	60	33	20	5	1	1	0	0	0	0	0	0	0	0
L-R evidential dominance	52	80	62	32	17	5	1	1	0	0	0	0	0	0	0	0
C-L evidential dominance	38	59	64	38	21	18	7	2	3	0	0	0	0	0	0	0

Alternatives are focused on one phase a lot of alternatives. L R evidential dominance, RC evidential dominance and RW evidential dominance can be effectively focused on one phase in nine kinds of evidential dominance without increasing the number of basic probability assignment of each focus element.

5 Application of Evidential Dominance to Conveyance Rate (Result Data) by Magnetic Vertical Supply Machine

Evidential dominance is applied to the result data of conveyance of magnetic vertical conveyance supply machine, and the effectiveness of the evidential dominance is clarified. As the method of experiment, numbers of five kinds of screw (Φ4×123L, Hexagonal M6×25L, M5×25L, M6×30L, M8×25L) are changed and put into a hopper of machine when each kind of screw is conveyed, and experiment. Numbers of conveyed screw are measured after conveyance after for 30 seconds. Times of experiment and measurement are carried out 30 times. It becomes easy to intertwine with the screw mutually by the length of screw, and varied widely in the conveyance

Table 6. Intervals of alternatives (each screw)

Numbers of screw / Kind of screw	600	700	800	900	1000
Φ4×123L	[1.90, 5.10]	[2.00, 4.80]	[2.10, 4.80]	[1.40, 4.20]	[1.80, 3.90]
Hexagonal M6×25L	[7.00, 8.20]	[6.30, 7.80]	[6.00, 7.30]	[6.20, 7.80]	[6.40, 7.40]
M5×25L	[11.30, 12.60]	[10.40, 12.60]	[9.70, 11.70]	[7.60, 9.40]	[8.20, 10.70]
M6×30L	[6.20, 8.50]	[6.30, 7.50]	[6.30, 7.40]	[6.30, 7.80]	[6.60, 8.10]
M8×25L	[5.00, 6.30]	[4.70, 6.30]	[4.50, 5.50]	[4.40, 5.30]	[4.70, 5.50]

rate. Minimum values and maximum values of intervals of alternatives (each screw) are obtained from result data of conveyance for each kind of screw by 30 times of experiments. Intervals of alternatives (each screw) are shown in Table 6.

Five focus elements (600 screws, 700 screws, 800 screws, 900 screws and 1000 screws) that want to be obtained as the probability. The total of the probability is one. The probability is represented as basic probability assignment $m(x)$ in equation (1). Basic probability assignment is generated by random number for simulation. The basic probability assignment of each focus element by random number is shown in Table 7.

Table 7. The basic probability assignment of each focus element by random number

times of simulation	basic probability assignment				
	600 screws	700 screws	800 screws	900 screws	1000 screws
1	0.27	0.04	0.05	0.59	0.05
2	0.2	0.1	0.48	0.08	0.14
3	0.03	0.04	0.22	0.3	0.41
4	0.09	0.2	0.48	0.08	0.15
5	0.65	0.11	0.13	0.09	0.02
6	0.15	0.07	0.23	0.19	0.36
7	0.16	0.06	0.36	0.29	0.13
*	*	*	*	*	*
*	*	*	*	*	*
94	0.24	0.28	0.06	0.37	0.05
95	0.1	0.42	0.3	0.15	0.03
96	0.6	0.06	0.01	0.02	0.31
97	0.38	0.09	0.1	0.06	0.37
98	0.16	0.1	0.05	0.44	0.25
99	0.26	0.31	0.11	0.13	0.19
100	0.12	0.12	0.37	0.17	0.22

Table 8. The result of focus on one phase by optimum evidential dominance with case of the basic probability assignment of each focus element is random number

evidential dominance	reference of alternative (kind of screw)
L R evidential dominance	M5×25L
R C evidential dominance	M5×25L
R W evidential dominance	M5×25L

This paper investigates how to prefer highly conveyance rate of screw by various basic probability assignments. The result of focus on one phase by optimum evidential dominance is shown in Table 8 when the basic probability assignment of each focus element is random number.

6 Discussion

LR evidential dominance prefers the alternative that is the larger value in minimum values and maximum values of interval of focus elements from necessary and sufficient condition (2) in Section 3. Alternatives are compared by expected utility or expected values when alternatives have various intervals. The expected values of maximum values of intervals for each alternative are obtained by equation (1). The expected value of minimum value of interval for focus elements of M5×25L screw is larger than other screws by Table 5. The expected values of maximum values of intervals for each alternative are obtained by equation (1). The expected value of maximum value of interval for focus elements of M5×25L screw is larger than other screws by Table 5. Therefore, M5×25L screw is focused on one phase by LR evidential dominance. In the result data of conveyance, M5×25L screw is highly rate of conveyance. Similarly, M5×25L screw is focused on one phase by RC evidential dominance and RW evidential dominance.

The program of Java for evidential dominance is composed in the web server. The server processes and responses the result of analysis to the client. Even if each client doesn't have program of analysis, the analysis became possible.

7 Conclusion

Numbers of focus elements and numbers of alternatives do not affect by means of result from simulation. LR evidential dominance, RC evidential dominance and RW evidential dominance are effective to focus in one phase from a lot of alternatives. The conveyance rate fluctuate by the length of screw and the weight of screw according to the result data of conveyance and the result of focus on one phase by optimum evidential dominance.

Acknowledgments. This paper was supported in part by Shimonishi Seisakusho Co., Ltd.

References

1. Danielson, M., Ekenberg, L.: A framework for analysing decisions under risk. European Journal of Operational Research 104, 474–484 (1998)
2. Tamura, H.: Behavioral models for complex decision analysis. European Journal of Operational Research 166, 655–665 (2005)
3. Red Hat Linux 9, Red Hat, Inc. (Retrieved December 1, 2004), from http://www.redhat.com/

4. Apache HTTP SERVER PROJECT, (Retrieved November 25, 2006), from http://httpd.apache.org/
5. J2SE, Sun Developer Network, (Retrieved November 25, 2006), from http://java.sun.com/javase/
6. The Apache Software Foundation, Apache Jakarta Tomcat, (Retrieved July 25, 2006) (1999-2006) from http://jakarta.apache.org/
7. Johnson, M., Stearns, B., Singh, I.: the Enterprise Team: Designing Enterprise Applications with the J2EETM Platform, 2nd edn. Sun Microsystems, Inc, pp. 7–128 (2002)
8. Eclipse, The Eclipse Foundation, (Retrieved December 18, 2006), from http://www.eclipse.org/
9. Shouda, T.: Introduction of Java Programming (Eclipse 3.2), Shuwasystem Co., Ltd (in Japanese), pp. 279–547 (2006)
10. Inuiguchi, M., Shirai, T., Sakawa, M.: Evidential Dominance for Decision Making in the Setting of Dempster-Shafer Theory. Trans. of the Society of Instrument and Control Engineers (in Japanese) 30(6), 720–728 (1994)
11. Inuiguchi, M., Inoue, Y., Kume, Y.: Ranking Indices in Dempater-Shafer Theory and Their Application to Decision Making Problems. Trans. of the Institute of Systems, Control and Information Engineers (in Japanese) 6(2), 71–81 (1993)
12. Shirai, T., Kume, Y., Toide, K.: Human Networking in Agile Manufacturing, Manufacturing Agility Hybrid Automation- II, Hong Kong, pp. 231–234 (1998)
13. Maung, Z.A.H., Liu, C.Y., Kume, Y.: Application of Evidential Dominance to Decision-making in e-Community. In: Proc. of The 5th ICMA 2004, pp. 393–398 (2004)

Usability of Electronic Medical Record System: An Application in Its Infancy with a Crying Need

Hal Miller-Jacobs and John Smelcer

Human Factors International

Abstract. There is almost universal agreement on the benefits of an Electronic Medical Records (EMR) System, yet successful implementations are few. While EMR systems focus on the important areas of functionality, interoperability and security, the area of usability has been overlooked. No electronic medical records system can be viable unless it addresses usability. We have had the opportunity to work with several systems and have employed field data gathering, expert reviews, usability testing and ScoreCarding to enhance the usability of EMR systems. Unless usability becomes more widespread and is given a higher priority, the benefits of an EMR system will remain elusive.

Keywords: Usability, Electronic Medical Records, Electronic Health Records, field data gathering, expert reviews, usability testing, ScoreCarding.

1 Introduction

There are very few areas in the health system on which there is almost total agreement. One of these areas is Electronic Medical Records (also referred to more generally as Electronic Health Records). Patients, physicians, and insurance companies see the potential of lowering costs, reducing medical errors and primarily improving patient care. In the State of the Union Address, January 23, 2007, President George Bush said, "We need to reduce costs and medical errors with better information technology" The US government is clearly behind the effort; it awarded contracts to four consortiums totaling $18.6 M for building EMR systems (Lohr, 2005).

To ensure that the process is done properly a certification commission has been established. The Certification Commission for Healthcare Information Technology (CCHIT) has the mission to accelerate the adoption of health information technology by creating an efficient, credible and sustainable product certification program (Certification Commission, 2005).

There have been some success stories, such as the Veteran's Health Administration, where there have been increases in patients served, decreases in cost per patient, and decreases in the number of staff per patient, resulting in an annual increase in productivity of nearly 6% (Evans et al 2006).

However, this success is not universal. At Cedars-Sinai Medical Center in Los Angeles, one of the most wired hospitals in the nation, physicians reported that it took much longer to use the new computer system, and there were limitations in their

M.J. Smith, G. Salvendy (Eds.): Human Interface, Part II, HCII 2007, LNCS 4558, pp. 759–765, 2007.

ability to make medical judgments. The implementation was finally cancelled after several hundred physicians refused to use the system (Connolly, 2005).

At Kaiser Permanente, users' attitude to implementation of an EMR indicated that the decision to adopt the system was flawed. The system reduced doctors' productivity, and the process fostered a climate of conflict between staff members that was resolved by withdrawal of the initial system (Scott et al, 2005).

Yet with all of these initiatives and funding at the outset of 2006, approximately 90% of physicians in the United States still use paper medical records in their practices (Provider's Edge, 2007). Why is that the case? One reason, for small medical groups, is cost. Another reason is that the technology is too complicated, even for tech-savvy physicians (Freudenheim, 2005).

The Electronic Medical Records industry has yet to discover usability. The Certification Commission for Healthcare Information Technology certifies EMR systems along three important criteria: functionality, interoperability and security. While all these are important, usability is absent! For an EMR system to be successful, usability must be considered.

Like any field where technology is relatively new, those advocating, developing, designing, purchasing and finally using EMR systems have put usability on the back burner; cost and functionality are the big drivers. But they soon come to realize that if the system is unusable it is less efficient than the old paper methods, and users abandon it.

Take the case of Mary Lightfoot, an American Indian, who appears at the clinic on the Indian reservation for a checkup. Her dedicated primary care physician asks when she had her last mammogram. Like many people she cannot remember. The doctor checks her record and there is no indication of when she had her last mammogram. When questioned, the doctor says: "what would you rather have me do, take care of patients, or spend time trying to figure out how to use this EMR system!" While Mary is fictitious, the case represents the state of affairs.

While Indian Health Service (IHS) is doing a great job taking care of their patients, the EMR system that has been implemented has a way to go. We have been working with IHS and others including the Mayo Clinic to improve the usability of their EMR systems.

2 Usability Techniques for EMR Systems

The techniques we have used to introduce and evaluate usability include:

- Field data gathering
- Expert Reviews
- Usability testing
- ScoreCarding

2.1 Field Data Gathering

Establishing the requirements for an EMR system was a critical part of the process. We worked with care providers to determine what exactly they do, with the aim of developing a system that facilitates rather than impedes their method of working.

This is particularly important when working in the medical community as physicians in general have little 'patience' for anything that is time consuming and does not work the way they would like it to work. In addition, there are others in the medical community who also use the system, such as support personnel, and their needs and perspectives must also be considered.

Working with an American Indian community and with medical personnel presented many challenges. Population diversity, multiple medical problems and issues of privacy had to be addressed when gathering data. There were multiple medical systems, such as a specialist's referral system for physicians, a behavioral health system for alcohol counselors and an integrated medical system for primary care providers. These systems have to be able to communicate and their interfaces must have commonality for the system to be usable.

Fortunately we found many physicians who were interested in improving the EMR system and were willing to have us shadow them during interactions with patients. In general most visits consisted of a preparatory phase wherein the physician reviewed his/her past notes and any other patient-relevant information. The examination and patient discussion phase then takes place, followed by the post-exam phase wherein the physician documented the encounter.

There were a variety of workstations for physicians to record the interaction. Figure 1, left, shows a wall mounted unit in the hallway, and on the right, a portable tablet computer that is used for recording the interaction and also used for electronically prescribing medications for the patient. Patients were asked permission if we could sit in on the interaction with the doctor to improve the computer system. Most agreed. In this way we were able to capture the entire flow of tasks for both primary care physicians as well as specialists.

Each venue had its challenges and each system had strengths and many weaknesses. To capture the needed data we had to be flexible and adapt our data gathering worksheets and techniques to various situations. In particular we had to be discreet when in the exam room with a physician and a patient as they discussed highly sensitive and complex medical issues.

2.2 Expert Reviews

As an output of the data gathering, we presented expert reviews of the various systems. Most of the medical personnel were pleased to hear the results although some were defensive. A few did not want a review of their system; we could guess why! We evaluated the systems along the usability dimensions of:

- Presentation
- Content
- Interaction
- Navigation

One of the biggest problems uncovered was lack of integration; that is the various systems within one facility were independent and did not 'talk' to one another. An integrated EMR system with consistent navigation and controls was desperately needed. While expert reviews were valuable, they were not nearly as important as usability testing.

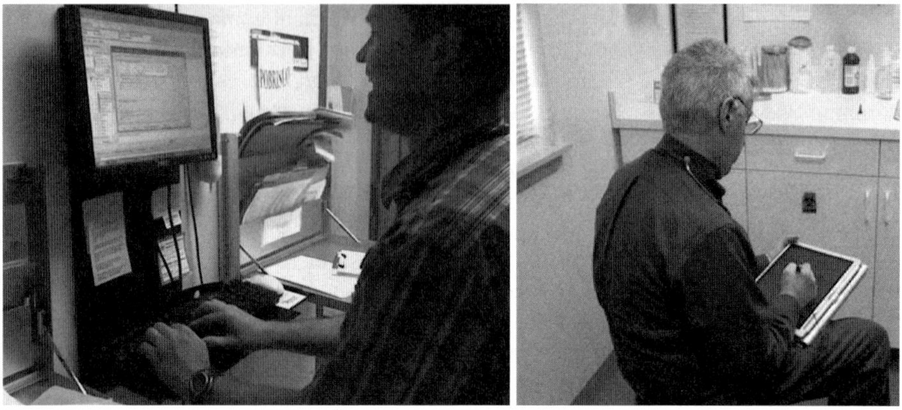

Fig. 1. Physicians entering post examination notes in hallway workstation (left) and via a tablet computer (right)

2.3 Usability Testing

The most effective technique we employed was usability testing. On most of the projects we worked on, after data collection and analysis was completed, paper prototypes were developed. These prototypes reflected how we thought physicians could best interact with the EMR system. They were then verified with physicians, medical staff personnel, and where appropriate with behavioral health professionals.

Rather than explain the new design, we simply presented the new prototype and asked the care provider to 'use it'. With this approach we were able to determine if the prototype matched the mental model of the care provider. When there was hesitation by the participant, we probed to uncover problem areas or misconceptions. Invariably the prototypes required revision and a rerun of the usability tests. While we were able to represent the functionality accurately, much of the revisions had to do with terminology. Figure 2 shows a screen with some of the terminology that had to be revised as a result of usability testing.

2.4 ScoreCarding

When selecting EMR systems, clinic and hospital administrators put primary emphasis on robust functionality. Vendors therefore typically emphasize the multitude of functionality in their EMRs. Forrester Research (Barrett, Holmes and McAulay, 2003) has identified fifteen categories of EMR functions. Similarly, CCHIT, which certifies EMRs according to functionality, interoperability, and security, has extensive checklists and scorecards. As a result scorecards for comparing EMRs according to the functions each provides have become very popular.

As usability professionals charged with evaluating multiple EMRs for usability, we found a usability scorecard to be very helpful. First, a usability scorecard encouraged a consistent method for evaluating multiple EMRs. Second, it allowed the evaluator

Prototype of Client Encounter form

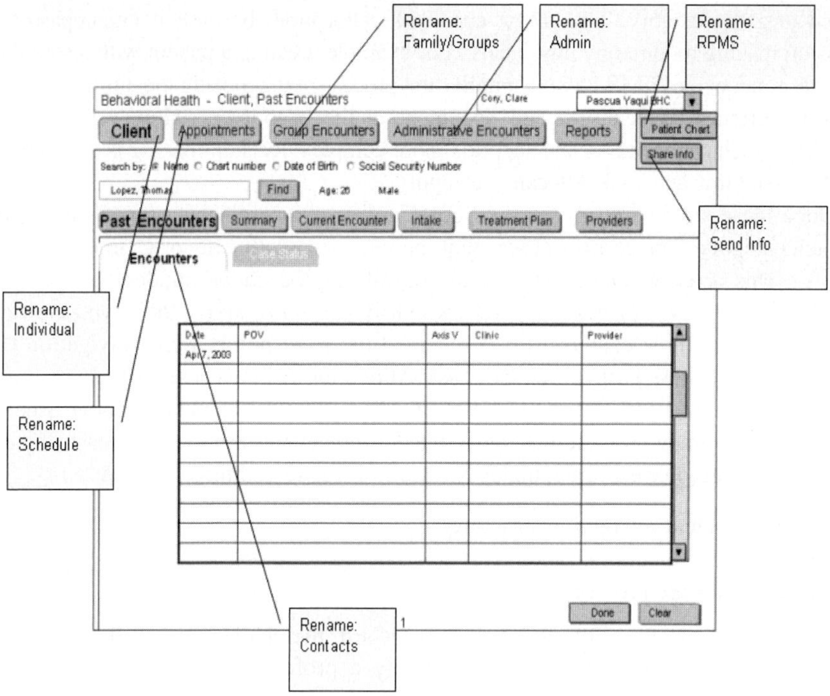

Fig. 2. A typical screen that accurately represented the functionality, but required extensive terminology changes as a result of the usability tests

to quantify the criteria used to evaluate usability. More specifically, it focused attention on key usability criteria, e.g. home page usability, navigation, interaction, presentation and task flow. Third, the format of the scorecard was very usable as it provided at-a-glance comparison across several EMRs and across several criteria. Finally, it provided a baseline for evaluating improvements to a single EMR.

With these scorecard benefits in mind, we introduce the Human Factors International (HFI) EMR Usability ScoreCard (Figure 3). At the highest level, the scorecard has six categories:

- Home page
- Navigation
- Windowing Elements
- Presentation
- Interaction
- Task Flow & Task Allocation

The first five categories, Home Page through Interaction, are drawn directly from prior HFI ScoreCards as used with clients such as Dell Computer and Neptune Software. We added a sixth category, Task Flow and Task Allocation, because it encompasses the decision making required by physicians. For example, treating a patient with a sore throat is fairly routine. An EMR should simplify the process of diagnosing the ailment, ordering labs, prescribing medications, and documenting the encounter. How well the EMR simplifies such routine tasks and supports more complex decision making is encompassed by the Task Flow and Task Allocation category.

Once these six categories for the ScoreCard were established, we assigned points to each category. The constraint was that the total across all categories must equal 100 points. This is a convenient anchor for simplifying the client's understanding of the resulting scores. Nearly everyone knows that if you score above 90%, you are doing very well. The three categories that received the most points were Navigation (20), Interaction (20) and Task Flow and Task Allocation (25). These reflect the criteria that are of greatest importance to hospital administrators, namely learning time and physician efficiency. For each category we crafted several criteria for measuring each category's score. For example, under Navigation we created three criteria:

- Consistent, clear navigation structure
- Sense of place on each screen
- Primary tasks are promoted

With this level of granularity to the ScoreCard, the challenging work of assigning scores began. Each EMR was evaluated by a professional usability analyst with

	EMR #1	EMR #2	EMR #3	EMR #4
Home Screen (15)	8	13	14	14
Navigation (20)	9	19	15	19
Windowing Elements (10)	5	7	6	7
Presentation (10)	5	7	8	9
Interaction (20)	10	16	16	17
Task Flow (25)	10	18	22	23
Total	**47**	**80**	**81**	**89**

Fig. 3. HFI EMR Usability ScoreCard

extensive EMR and medical experience. This required many days of on-site and off-site work. The on-site work included a briefing on the overall operation of the EMR, observing several physicians interacting with patients and the EMR, and a final briefing with technical and administrative staff on-site to answer any questions. The off-site work involved exhaustive review of notes and EMR screen shots taken during the on-site work.

The final EMR ScoreCard appears in Figure 3. We were somewhat surprised by our findings for the four EMRs under consideration. The lowest overall score was 47 points; the highest was 89 points. This represents a significant difference in overall usability. Our surprise was somewhat moderated by insights into our process. Part of the lowest score's problems came from familiarity. The old saying, "Familiarity breeds contempt" was certainly true here. The more we knew about EMR#1, the more we knew about its usability failings.

The ScoreCard fulfilled its objectives. It provided consistent evaluation of usability across several EMRs. Its measures of usability were clearly quantifiable with significant granularity to the analysis. Finally, it was quickly and easily understood by the client.

3 Summary

In summary, EMR systems are the wave of the future. Slowly, those developing and purchasing systems are just beginning to discover the value and importance of usability. We have the tools and techniques to make effective contributions to improve accuracy, efficiency, learnability, and ultimately to enhance medical services and the health of patients. Unless usability becomes one of the key factors in developing and implementing EMR systems, the benefits of such systems will continue to remain elusive.

References

1. Barrett, M.J., Holmes, B.J., McAulay, S.E.: Electronic Medical Records: A Buyer's Guide for Small Physician Practices. Forrester Research (2003) http://www.chcf.org/documents/ihealth/ForresterEMRBuyersGuideRevise.pdf
2. Certification Commission for Healthcare Information Technology (CCHIT), Overview of Mission (January 29, 2007) http://www.cchit.org/about/
3. Connolly, Ceci, Cedars-Sinai Doctors Cling to Pen & Paper, Washington Post, p. A01 (March 21, 2005)
4. Evans, D.C., Nichol, W.P., Perlin, J.B.: Effect of the Implementation of an Enterprise-Wide Electronic Health Record on Productivity in the Veterans Health Administration. Health Economics, Policy and Law. 1, 163–169 (2006)
5. Freudenheim, Milt, Doctors Collaborate to Find a Less Costly Way to Add Electronic Medical Records, New York Times, p. C4 (September 19, 2005)
6. Lohr, Steve, U.S. Awards Contracts to Help Automate Health Records, New York Times (November 11, 2005)
7. Provider's Edge (2007) http://www.providersedge.com/ehr_overview.htm
8. Scott, J.T., Rundall, T.G., Vogt, T.M., Hsu, J.: Kaiser Permanente's experience of implementing an electronic medical record: a qualitative study. BMJ 331, 1313–1316 (2005)

Usability of User Agents for Privacy-Preference Specification

Robert W. Proctor[1], Kim-Phuong L. Vu[2], and M. Athar Ali[1]

[1] Department of Psychological Sciences, Purdue University
703 Third St., W. Lafayette, IN 47907, USA
[2] Department of Psychology, California State University, Long Beach
1250 N. Bellflower Blvd. Long Beach, CA 90840, USA
proctor@psych.purdue.edu, kvu8@csulb.edu,
athar.ali@perficient.com

Abstract. The goal of this study was to determine (a) users' privacy concerns; (b) whether these privacy concerns can be checked by an existing Web-based privacy agent; and (c) whether users are able to easily specify their privacy preferences using this agent. Users were able to configure the agent correctly for about half of the desired privacy goals that could be checked by the agent. Of more significance, users thought that they had configured the agent successfully to achieve many privacy goals that cannot be accomplished with the version tested. We also examined alternative interface layouts to ascertain whether any of them allowed users to specify their preferences at a higher success rate than the current interface. We discuss implications of our findings for user agents designed to aid users' assessments as to whether a Web site's stated privacy practices are consistent with the users' preferences.

Keywords: privacy preferences; privacy policies; usability.

1 Introduction

In recent years there has been increased interest in protecting users' private information in Web transactions [1]. Surveys show that 75% of users express considerable concern about privacy issues [2, 3], especially when personally identifiable information (e.g., phone numbers and social security numbers) is used. Users are less concerned about providing information about their browsing patterns, and are quite willing to allow their personal information to be saved in order to reduce task-completion times for frequent transactions, as long as they can specify when this information will be released and to whom. Although prior surveys have documented that many users have concerns about privacy with respect to Web-based transactions, they have not provided information as to the specific concerns that users deem to be most important. The goal of our first study was to obtain such information.

Though many Web sites post privacy policies, these policies do not necessarily address consumer privacy concerns and are typically written at a reading level that is too difficult for the general user population [4]. Thus, many users do not even visit

M.J. Smith, G. Salvendy (Eds.): Human Interface, Part II, HCII 2007, LNCS 4558, pp. 766–776, 2007.

privacy policy Web pages when given the opportunity [5, 6]. Even if users visit the privacy-policy page, they may not understand the policy because it is often lengthy and may require as long as half an hour to find and analyze [7]. One solution to this problem is to employ a user agent that checks a site's privacy policy against the user's designated privacy preferences. Success of the user agent depends on the user being able to configure the agent properly. The second goal of the present study, therefore, was to evaluate whether participants can configure an existing user agent to check for a range of specified privacy preferences. We also examined whether alternative wordings for the privacy options in the interface would increase user success rates.

2 Study 1

The first study was a survey in which users rated whether they agreed or disagreed with each of 98 statements relating to privacy practices (or preferences). All of the statements were based on privacy practices extracted from existing privacy policies and used terminology common to many existing privacy policies.

2.1 Method

Participants. 32 students (20 males, 11 females, and one of unidentified gender) from an introductory Psychology class at Purdue University participated for experimental credits. All were experienced computer users.

Materials. Participants provided ratings regarding their concerns for the following categories of information, or subject types:

- Personally Identifiable Information (PII)
- Non-Personally Identifiable Information (Non-PII) / Cookies
- Financial Information FI (FI) / Credit Card Information (CCI)
- Health Information (HI) / Personal Health Information (PHI)
- Privacy Principles (e.g., certification seals)
- Passwords
- E-mail Addresses
- Privacy Preferences
- Information About Children

These subject types were identified by examining privacy policies and conducting content and goal-mining analyses on them [8]. Within each subject type, questions about privacy preferences were modeled along the following categories [9]: Collection (the amount and type of information collected; which organization is collecting the information), Personalization (customization for the individual user or group of users), Notice/Awareness (alerts given to users to let them know that information will be collected or that there is a change in the organization's privacy practices), Transfer (conditions under which information will be exchanged with other parties), Information Storage (where the information will be kept), and Access/Participation (who has permission to view the information). The questions were grouped by subject type. Four versions of the survey were assembled with different orders of the subject types in each. Within a subject type, the questions were kept in the same order in the four survey versions.

Procedure. Participants signed up for the survey and were tested individually or as part of a group of up to eight people, depending on the number who signed up for the time slot. Participants were first given an instruction sheet that defined key acronyms commonly found in privacy policies. They were told to make sure that they knew the meaning of each acronym and to refer back to the sheet as needed while answering the survey questions. They then received the survey, which consisted of 98 questions. For each question, each participant was to mark an answer of strongly disagree, disagree, neither agree nor disagree, agree, or strongly agree. For data analysis, these answers were transformed to a 5-point scale, with 5 = 'Strongly agree' and 1 = 'Strongly disagree'. Example statements are listed in Tables 1 and 2.

2.2 Results and Discussion

Table 1 shows the 25 statements with which the participants most strongly agreed. The topic with which the participants expressed the most concern was selling or sharing their personal information with other parties. Eight of the nine highest rated concerns involved this issue: Six of the statements were with respect to credit card and financial information, and the remaining two involved personally identifiable information and health information. The other question in the top nine involved the possibility of hackers gaining access to financial and credit card information.

The remaining statements in the top 25 included the following. Users noted that they did not want their e-mail address and username/password to be transferred to an acquiring company. Users also indicated that they wanted to have the option to receive electronic and printed copies of the privacy policies, to see privacy logos on Web pages, and the ability to edit their privacy preferences. They also indicated a desire for anonymous Web browsing.

Users expressed the least concern about the eight items in Table 2. These statements generally indicate that users are relatively unconcerned about cookies or non-personally identifiable information being used to customize their browsing experience, or about buying patterns being recorded or stored when their personal information is not identified. This lack of concern applies particularly to sites at which the user is voluntarily registering or purchasing products. Users also indicated relatively little concern with personally identifiable information being transferred to a company that bought out the company that originally collected the information. Although the participants did not seem very concerned about user profiling, this tendency may be a result of their not being aware of privacy threats associated with profiling and may be restricted to college-student Internet users.

3 Study 2

The privacy concerns identified in Study 1 were used to develop the tasks for setting privacy preferences in Study 2. This study was an experiment in which participants were instructed to try to set specific privacy preferences on Privacy Bird™, Beta 1.3. Privacy Bird is a user-agent tool that enables users to filter out, or be warned about, undesirable privacy practices stated in a site's privacy policy. We

Table 1. Rank order and mean rating (1 = Strongly Disagree and 5 = Strongly Agree) of the 25 statements with which users most strongly agreed. Asterisk indicates a task that can be configured in Privacy Bird.

Rank	Privacy Preference	Mean Rating
1*	I want the option of refusing to allow a company to share my CCI/FI with 3rd parties and affiliates.	4.66
2*	I mind when my CCI/FI is shared with a third party for promotions.	4.48
3*	I want the option of refusing to allow a company to share my PII with 3rd parties and affiliates.	4.38
4*	I mind when my email address is rented or sold.	4.34
5	I am concerned that hackers may be able access my PII.	4.24
6*	I mind when my PII is shared with a third party for promotions.	4.17
7	I want to see privacy logos on the privacy policy Web pages.	4.17
8*	I mind when my cookies/non-PI are rented or sold.	4.10
9*	I mind when my HI/PHI is shared with a third party for promotions.	4.10
10	I want the option to receive electronic/print privacy policy.	4.07
11	I want the option of having a manual way of editing privacy preferences and a machine readable option.	4.07
12	I mind when my email address is provided to an acquiring company.	4.03
13	I mind when my login/password is transferred to an acquiring company.	4.00
14	I mind that my CCI/FI is aggregated from third parties.	4.00
15	I want the option to restrict the company's employees from viewing my HI/PHI.	4.00
16	I want the option of refusing to allow a company to use cookies/non-PI for promotional purposes.	3.97
17	I mind that my HI/PHI is aggregated from third parties.	3.93
18	I am concerned that hackers can access my email messages.	3.93
19*	I mind that I am not allowed to update my CCI/FI.	3.83
20	I mind when my email account is used to monitor my purchase patterns.	3.79
21	I mind when my cookies/non-PI are transferred to an acquiring company.	3.72
22*	I mind when my CCI/FI is used to customize my browsing experience.	3.69
23*	I mind when my HI/PHI is used to contact me for health or drug promotions.	3.69
24	I mind that I cannot see my cookies/non-PI to verify their accuracy.	3.62
25	I want the option to share my privacy preferences with other Web sites.	3.62

chose Privacy Bird because it is readily available and allows users the option of customizing their privacy preferences. We developed tasks reflecting the top 25 rated privacy concerns in Study 1 (see Table 1), excluding statement 11, because Privacy Bird itself is a tool for setting privacy preferences that uses machine-readable code. We determined that Privacy Bird, if set properly, can warn users if a site's privacy policy does not address any of 10 of the privacy concerns identified in Study 1 (asterisk next to task number in Table 1), but it cannot do so for the remaining 14 concerns. It should be noted, though, that Privacy Bird can check for seven of the top 10 user concerns identified in Study 1.

Table 2. Mean rating (1 = Strongly Disagree and 5 = Strongly Agree) of the eight statements with which users most strongly disagreed

Privacy Preference	Mean Rating
I mind when my PII is used to customize my browsing experience.	3.38
I mind that my CCI/FI is collected when I purchase products/services.	3.31
I mind that my HI/PHI is recorded to my profile when I purchase products/services.	3.31
I mind that I am not allowed to update my HI/PHI.	3.31
I mind that my PII is collected when I purchase products/services.	3.24
I mind that my buying patterns are recorded to my profile.	3.17
I mind that my buying patterns are recorded to my cookies/non-PI.	3.17
I mind when cookies/non-PI are used to customize my browsing experience.	3.03

3.1 Method

Participants. 30 new students (9 male, 20 female, and one unidentified) from the same subject pool as in Study 1 participated. Ages ranged from 18 to 21, with mean age of 19 years.

Apparatus. Privacy Bird was used for subjects to set specific privacy preferences. The options available for privacy preferences in Privacy Bird can be viewed by selecting "My Preferences" in the main menu and "Privacy" in the sub-menu. The privacy configuration options on this screen are divided into four parts:

1. Health or Medical Information
2. Financial or Purchase Information
3. Personally Identifiable Information
4. Non-Personally Identifiable Information

The available privacy preference options are shown in Figure 1.

Procedure. Participants were initially required to complete a survey with questions about their Internet usage and experience with privacy tools. The survey asked their age, gender, ethnicity, previous computer experience (from 1 = none to 4 = very experienced), how often they access the Internet (1 = never to 4 = very often), knowledge of Internet privacy policies (1 = none to 4 = very knowledgeable), and experience with privacy tools (1 = none to 4 = very experienced).

For the main study, participants were asked to configure specified privacy preferences using Privacy Bird. Participants were to select appropriate options on the Privacy Bird configuration window accessible through "My Preferences" on the main menu and "Privacy" on the sub menu. The "Select Privacy Level" was set at "Custom" to allow participants to configure the tool.

Participants were provided with a list of 24 tasks. Each task was related to one of the privacy concerns identified in Study 1, and participants had to select options from those available on this screen to indicate the privacy setting needed in Privacy Bird to achieve each task. Because the tasks were developed from the concerns identified in Study 1, the task list consisted of a mixture of tasks that could be completed using the available options in Privacy Bird and tasks that could not be completed. Participants were asked to select the option in Privacy Bird that they felt would achieve the task goal, and they were allowed to select multiple options. After completion of each task, that participant was to say "done." If the participant thought that appropriate options were not available or that the task could not be completed, s/he was to say that the task could not be completed. In such cases, the participant moved on to the next task. The order of the tasks was different for each user.

The screen activities were videotaped during performance of the tasks. The researchers reviewed the tapes to determine the accuracy of participants' performance. A task was considered to be completed (correctly or incorrectly) when the participant said that the task was "done" or "cannot be done". Time taken to complete each task was recorded using the video counter when reviewing the tapes.

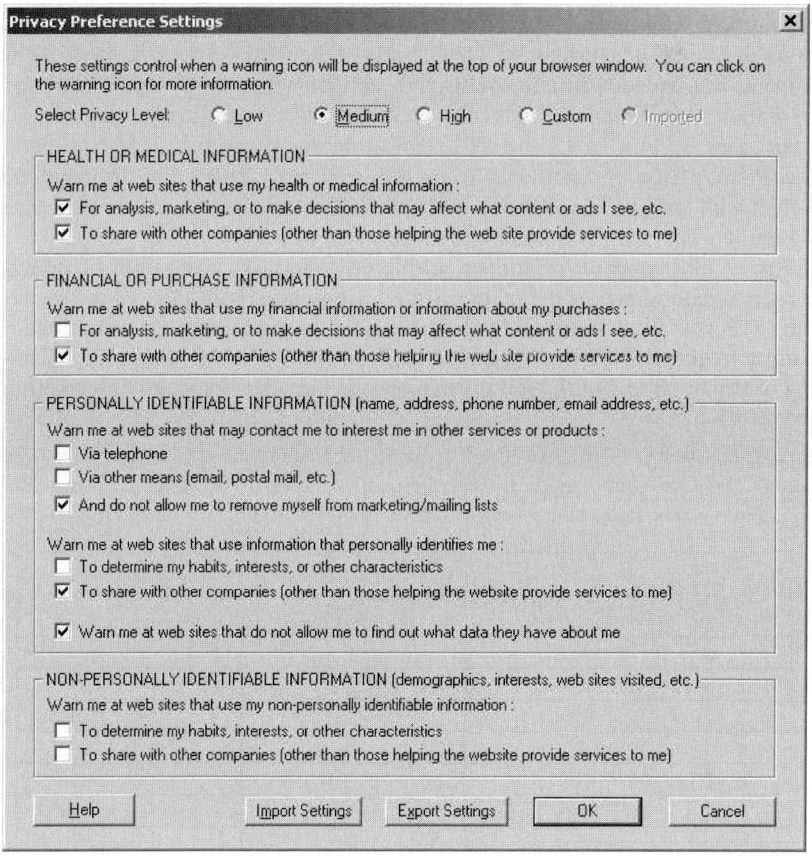

Fig. 1. The privacy preference options in Privacy Bird. Reprinted with permission.

3.2 Results and Discussion

Survey. All participants indicated that they had some previous computer experience, with 4 indicating that they were not very experienced, 20 fairly experienced, and 6 very experienced. All had access to the Internet, with 2 indicating that they accessed the Internet not very often, 7 often, and 21 very often. One participant indicated no knowledge of privacy policies, 18 that they were not very knowledgeable, and 11 that they were fairly knowledgeable. Three indicated that they had no experience with privacy tools, 22 very little experience, 4 a fair amount of experience, and 1 a lot of experience. In sum, the participants were relatively experienced with computers and using the Internet but not very knowledgeable about privacy policies and tools.

Performance with Privacy Bird. A configuration was considered correct if the participant selected the right settings for the tasks that could be configured, and said that the configuration cannot be set for the tasks that could not be configured. There was no difference in time to complete configurable ($M = 23.3$ s) and non-configurable ($M = 20.7$ s) tasks, $t(29) = 1.19, p > .24$. The percent correct was 66% for the 10 tasks that could be configured and 31% for the 14 tasks that could not be configured. For the tasks that could be configured, Table 3 shows the total number of errors for which participants responded "can't be configured" and for which they provided an incorrect configuration. The frequency of incorrect configurations was greater than that of non-configurations, $\chi^2(1) = 15.75, p < .001$, indicating that participants often thought they had set Privacy Bird appropriately when they had not. There was an interaction of this effect with task, $\chi^2(9) = 22.85, p < .01$, with 8 of the tasks showing more incorrect configurations than cannot-be-configured errors.

The three most popular erroneous settings for the tasks that could be configured were users selecting one or the other setting for Financial Information and the second setting for Personally Identifiable Information. These same settings were three of the four most frequently used for those tasks that could not be configured. The fifth setting of Personally Identifiable Information was the other frequently used setting.

Table 3. Total Number of Incorrect Configurations for Each of the 10 Tasks for which Privacy Bird could be Configured, and the Numbers of "Cannot be Configured" and Incorrect Configurations. *The task numbers correspond to the privacy concerns ranked in Table 1.

Task*	Total Incorrect	Cannot be Configured	Incorrect Configuration
1	9	1	8
2	9	2	7
3	8	0	8
4	14	5	9
6	10	3	7
8	9	3	6
9	8	0	8
19	22	6	16
22	6	3	3
23	17	12	5

4 Study 3: Usability Testing for Alternative Interfaces

Performance at setting Privacy Bird to address specific privacy concerns was not particularly good in Study 2. Participants' responses were incorrect for approximately 50% of the tasks they were asked to perform. Participants often thought that they had set Privacy Bird to accomplish goals that it could not and did not set it appropriately to accomplish goals that it could. Study 3 was designed to determine whether performance could be improved using alternative presentations for specifying privacy preferences on the Privacy Bird interface that used simple organizational or wording changes.

4.1 Method

Participants. 100 new students from the same subject pool as the previous studies participated.

Materials. Paper versions of four alternatives to the Privacy Bird interface were tested (see Figure 2), along with one that resembled the original Privacy Bird interface (Interface E). Interface A was similar to the Privacy Bird interface, with the main difference being that the names of all information types except non-personally identified information were prefixed with "my" to indicate that the information type is personal. For Interface B, the words "warn me" in the original interface were replaced with "DO NOT" to indicate an action verb. This interface's heading warns that the tool does not take any automatic action and only "warns". For Interface C, the options were categorized based on the action verb. Action verbs provided on the interface were USE, SHARE, CONTACT, and COLLECT. Within each action verb, options were categorized using the type of information such as financial or health. Again, personal information was prefixed with "my." For Interface D, the original options in the Privacy Bird interface were subdivided to provide more options. Thus, certain privacy preferences that were originally grouped together as one option in Privacy Bird were available as separate options. Under each category of information type, the sentence began with "warn me when…", but the options themselves did not have "warn me" written before them. Examples of the information types were provided, and some words were simplified from the original Privacy Bird interface.

Procedure. Participants were randomly assigned to receive one of the five paper versions of the Privacy Bird interface. Each of the interface versions had a list of 10 tasks to be completed in sequence with the interface sheet. The tasks were those from Study 2 that could be accomplished with appropriate settings of the interface. Participants were instructed to read each task in the list and select the options necessary to set the indicated privacy preference. The options were then to be noted alongside the task. If the participant felt that the options were not sufficient and that

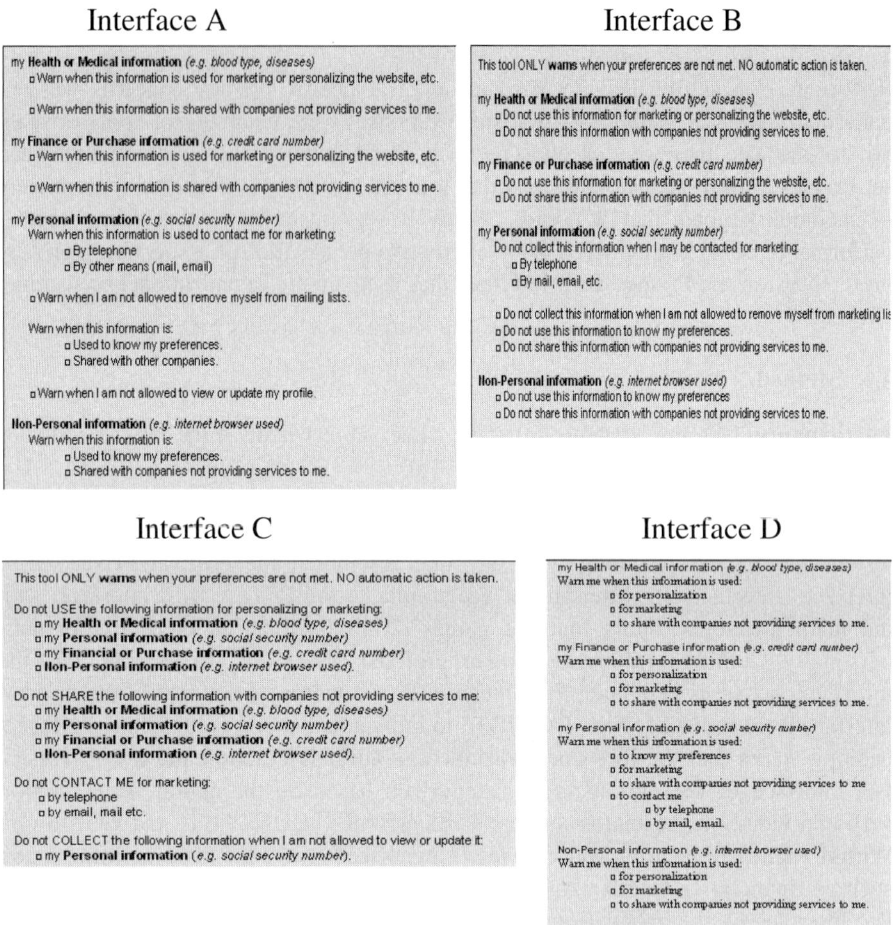

Fig. 2. The four modified versions of the Privacy Bird interface used in Study 3

the settings could not be completed, then s/he would write NA (for Not Applicable) alongside the task.

4.2 Results and Discussion

Across the five interfaces and 10 questions, the correct option was selected 54% of the time, including configurations for which additional options were also selected. Though the percent correct was highest for interface A (60%), lowest for interfaces E (the original Privacy Bird interface; 51%), B (52%), and D (52%), and intermediate for interface C (57%), an ANOVA indicated no significant difference, $F < 1.0$. Because interface A was the most similar to the Privacy Bird interface (interface E), it is likely that the higher percentage correct for it is due to sampling error. Regardless of whether there is indeed a small, real difference across the tested interfaces, it is clear that the relatively simple changes we made in organization and wording did not make much difference. Sizeable improvement in task performance apparently would require more than just surface changes to the Privacy Bird interface.

5 Conclusion

The privacy issue that users indicated to be of most concern in Study 1 was selling or sharing information with other parties. They also specified concern about hackers possibly gaining access to their financial information and with their e-mail addresses and username/passwords being transferred to an acquiring company. Users also indicated that they wanted a manual way of editing privacy preferences, which Privacy Bird is designed to do. Privacy Bird includes options that can address 7 of the 10 top privacy concerns endorsed by the users in Study 1. However, users showed evidence of confusion regarding exactly what they will be warned about with the various settings of Privacy Bird, both with the original interface and variations of it. For the privacy concerns that could be accommodated by Privacy Bird, users set the interface correctly only about 60% of the time, and they often thought that they had set Privacy Bird to provide protection that it could not provide. Thus, Privacy Bird provides a good start toward allowing users to determine whether Web sites adhere to their privacy preferences. However, because experienced computer users with little knowledge of online privacy issues show confusion about what can be accomplished with particular settings, it may be more effective users to rely on the default values (low, medium, or high privacy) rather than on custom settings of the user agent.

Acknowledgments

This research was supported in part by NSF ITR Cyber Trust grant #0430274.

References

1. Cranor, L.F., Garfinkel, F.: Security and usability: Designing secure systems that people can use. O'Reilly Media, Inc., Sebastopol, CA (2005)
2. Ackerman, M.S., Cranor, L.F., Reagle, J.: Privacy in E-commerce: Examining user scenarios and privacy preferences. In: E-commerce 99, pp. 1–8. ACM, New York (1999)
3. Spiekermann, S., Grossklags, J., Berendt, B.: E-privacy in 2nd generation E-commerce: Privacy preferences versus behavior. In: EC'01, pp. 38–47. ACM, New York (2001)
4. Proctor, R.W., Ali, M.A., Vu, K.-P.L.: Examining usability of Web privacy policies. International Journal of Human-Computer Interaction (in press)
5. Earp, J.B., Baumer, D.: Innovative Web use to learn about consumer behavior and online privacy. Communications of the ACM 46(4), 81–83 (2003)
6. Jensen, C., Potts, J.: Privacy policies as decision-making tools: An Evaluation of online privacy notices. In: CHI 2004, vol. 6, pp. 471–478 (2004)

7. Byers, S., Cranor, L.F., Kormann, D.: Automated analysis of P3P-enabled Web sites. In: ICES, Pittsburgh, PA, pp. 326–338 (2003)
8. Antón, A.I., Earp, J.B., He, Q., Stufflebeam, W., Bolchini, D., Jensen, C.: The lack of clarity in financial privacy policies and the need for standardization. IEEE Security and Privacy 2(2), 36–45 (2004)
9. Earp, J.B., Anton, A.I., Aiman-Smith, L., Stufflebeam, W.H.: Examining Internet privacy policies within the context of user privacy values. IEEE Transactions on Engineering Management 52, 227–237 (2005)

Achieving Usability Within E-Government Web Sites Illustrated by a Case Study Evaluation

Basil Soufi[1] and Martin Maguire[2]

[1] College of CS & IT, Abu Dhabi University, P.O.Box 59911, Abu Dhabi, UAE
[2] Ergonomics and Safety Research Institute, Loughborough University, Holywell Building, Holywell Way, Loughborough, leics, LE11 3UZ
Basil Soufi, basil.soufi@gmail.com, Martin Maguire, m.c.maguire@lboro.ac.uk

Abstract. Local government in the UK has made progress towards the delivery of services online. However individual local authorities do not necessarily have the experience or infrastructure to develop web sites that are usable and that can be maintained as content changes. This study used a three level model as a heuristic tool to evaluate a case study e-government web site. While the results were useful in proposing changes, barriers exist which slow down the process of making the changes. The paper proposes strategies that may be adopted to ease this process of web site development and change within local government environments and elsewhere.

Keywords: Web sites, heuristic evaluation, e-government, local government.

1 Introduction

Local government authorities have faced issues and challenges in meeting electronic delivery targets and in making their web services usable. These issues are highlighted via a 3-level model evaluation of a UK e-government web site. The study also provides the basis for a strategy to improve the usability of similar sites in the future.

Over the last 5 years there has been much development in the provision of electronic service delivery of both national and local government services in the UK. This has been driven by a central government agenda that is committed to making all services available on the Internet by the end of 2005 [2, 12]. To ensure successful take-up, these e-services need to be joined up in ways that make sense to the customer and be accessible to the widest range of users by being both accessible and usable. Although funding was allocated to these developments, the delivery of this strategy requires a great deal of work on back-end technology as well as the front end web sites. Making a Government organization's web site more usable has therefore to compete for resources with the other activities and projects undertaken to meet the overall e-government agenda. These include replacement of legacy systems, network infrastructure upgrades and integration with back office systems.

M.J. Smith, G. Salvendy (Eds.): Human Interface, Part II, HCII 2007, LNCS 4558, pp. 777–784, 2007.

The case study web site discussed in this paper was for a local authority area located in the south east of England. The site offered information to its citizens about the local area, council services, leisure and tourism, community living (e.g. schools and hospitals, events), support for business, etc.

Typically for a Government organization developing a web site for the first time, the initial development occurred in an ad-hoc manner. It was also left to the public relations department. The emphasis was to get content online to meet the target. The structure and presentation of this content and the usability of the web site were not given a great deal of attention.

The design strategy for the site is summarized below in comparison with a recommended user-centered design process (see Table 1 below).

Table 1. Comparison of user-centered design process with local authority approach

Recommended user-centered design process	Pragmatic process adopted by local authority
Plan to adopt a user-centered approach and allocate resources with management support.	Users considered but not formally.
Identify user needs with stakeholder analysis and user needs survey.	Documentation of user needs but second hand based on personal knowledge of design team.
Draw upon user interface and usability expertise to assist in developing design in an evolutionary way.	Proceed using 'common sense' approach of design team. Central government requirement for e-government with deadline encouraged some rushed developments.
Evaluate with human factors experts and users.	Lack of resources required web site to be launched without planned user feedback.
On-going reviews from user perspective to maintain web site.	Expert evaluation of site identified issues but lack of planned resources for further development has slowed down change.

2 Evaluation Case study

It was decided then to perform a usability evaluation of the current facility at the time. This was an independent study by one of the authors, employed by the government authority, with responsibility for information systems strategy. The evaluator went through the web site utilizing an understanding of the organization's business as well as usability expertise and knowledge of similar web sites. Usability heuristics were also used and the evaluation also considered the purpose of the web site, the tasks to be supported, and the needs of the end user as well as typical scenarios of users accessing the site. The evaluator was also guided by a 3-level model of user interface design [1, 11]. The levels can be described as (1) Required functions to meet user goals, (2) Suitable information architecture supporting navigation, and (3) Effective page layout, design and accessibility (See Fig. 1). These are explained below.

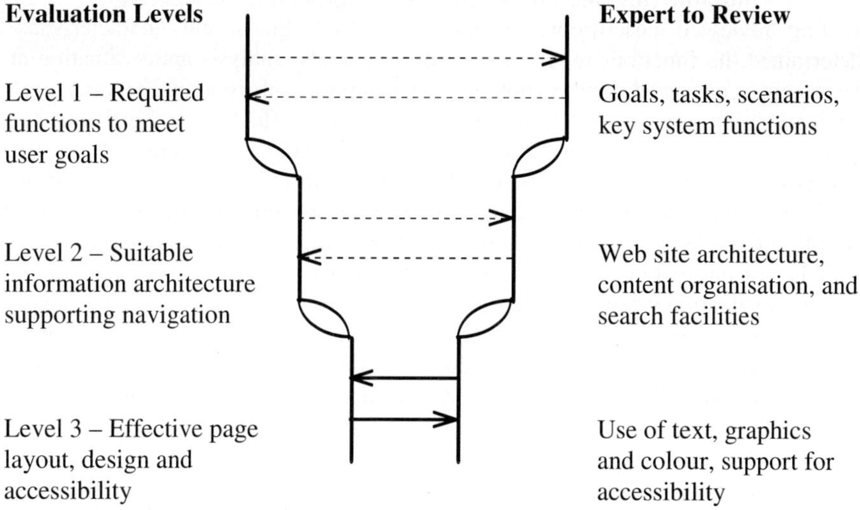

Fig. 1. Three level model used for evaluation

2.1 The Three Level Evaluation Model

Level 1 - Required Functions to Meet User Goals

A definition of usability according to Reddish [9] is that users can

- Find what they need,
- Understand what they find,
- Act appropriately on that understanding, and
- Do all that in the time and effort they think the task is worth

The web site must help the organization accomplish its business goals by helping users accomplish *their personal missions*. Users' goals will determine the functions that the web site will need to provide. Analysis and evaluation at this level will consider answers to the questions listed below and the scenarios developed based on user and task analysis [6].

- What are the main purposes of the web site?
- Who are the potential users?
- Why would they use the web site?
- What are their motivations and goals?
- What roles do they take?

These high level aspects were considered as part of the evaluation of the local government web site. Only by considering goals, tasks and scenarios could we ensure that the right content is provided. This is very important for such an information rich site serving a community of general public users.

Level 2 – Suitable Information Architecture Supporting Navigation

Having reviewed descriptions of the users, their goals and characteristics and determined the functions required to meet user goals, analysis and evaluation at this level are concerned with content organization, information architecture and navigation. In order for the web site to be learnable, effective, and efficient, the site structure must support user characteristics and cognitive abilities. Here we are concerned with site design in terms of the information architecture, content design, and navigation from the home page to other pages. Based on research into how people use the Web [7], web sites need to be skimmable and scannable to help users find the content that quickly answers their questions.

The evaluation examined the content for its relevance and appropriateness, the design of the home page, linking pages, and destination pages for effective navigation and content organization. Other web site features supporting navigation such as content map, search facilities, and A – Z directory lists were also evaluated.

Level 3 – Effective Page Layout, Design and Accessibility

The most important aspect of page layout and design is to design for the Web and avoid large chunks of text that are common in books and documents. Key questions to be asked are:

- Does the web site feature short sections with many headings or long scrollable pages?
- Does the web site feature effective use of space, bulleted lists and headings?
- Are the most important headings located in the right places on the web page?
- Do the visual cues enable appropriate chunking of the information contained on the web page?
- Does the text use an appropriate font, type size, and typeface?
- Does the color scheme enable adequate contrast between the foreground and the background?
- Is accessibility supported so that the web site can be easily viewed by people with impaired vision or by those using screen readers? If the site uses Flash™ content or PDF files, are they perceivable by people with visual impairment? Do the web page links make sense when read out of context as a list by a screen reader? Is there captioning for audio output for those people with impaired hearing.

Accessibility of web sites in the UK to people with disabilities is now a legal requirement upon web site developers [10]. Organizations are now realizing the social and financial benefits of making web sites accessible to people who are older or who have disabilities. Many local government sites include server side speech output facilities so that the many people without specialist screen reader technology can hear the content of the council's website easily.

2.2 Study Outcome

The evaluation revealed several usability problems and suggested areas of development. A summary of the evaluation findings for the three different levels is given in Table 2.

Table 2. Main Findings from the Evaluation Study

Evaluation Model	**Summary findings**
Level 1	Key information seeking and transactional services pertaining to properties and locations are not provided. Examples: • View the Council Tax rate for customer's property • Check account balance, make payment using secure web site • Search for planning applications approved for this location and neighboring locations
Level 2	Although the content is relevant and interesting the following problems were noted: • Poor search facility • No A – Z directory listing • Too many navigation options from the home page • Navigation from the home page, linking and destination pages can be improved • Clickable images contributed to inconsistency by introducing an additional and unnecessary navigation style
Level 3	• Simple design and good accessibility implemented (good points) • Text display was restricted to a central frame that caused unnecessary scrolling • Many web pages were based on print versions that should be redesigned for the web with bullets and short text

A report of the evaluation findings and recommendations was produced and presented to the government authority. The author's view was that based on the findings, there was a need for a radical redesign of the web site; one that is best undertaken by a specialist agency where a synthesis of good design and usability knowledge will lead to a greatly improved product.

For an organization with limited resources, and where no investment was planned in usability testing, the evaluation was extremely valuable in identifying problems and contributing to improvements, as the system is further developed. The most important finding of the evaluation was the need to provide information and transactions relevant to users' situations and locations. Steady progress has been made on this and to date all Level 1 issues have been either partially or fully addressed. With Levels 2 and 3, many improvements have been made to the web site subsequent to the initial revisions (See Fig. 2 for the new home page).

Several factors have however slowed down the progress of this development:

- Limited knowledge of user interface and usability issues within the team responsible for the development of the web site
- The redesign of the web site, though important, has to wait for its turn according to corporate prioritization of projects
- 'Who is in charge of, and makes decisions on the web site' is an organizational and political decision not necessarily based on existing skill sets.

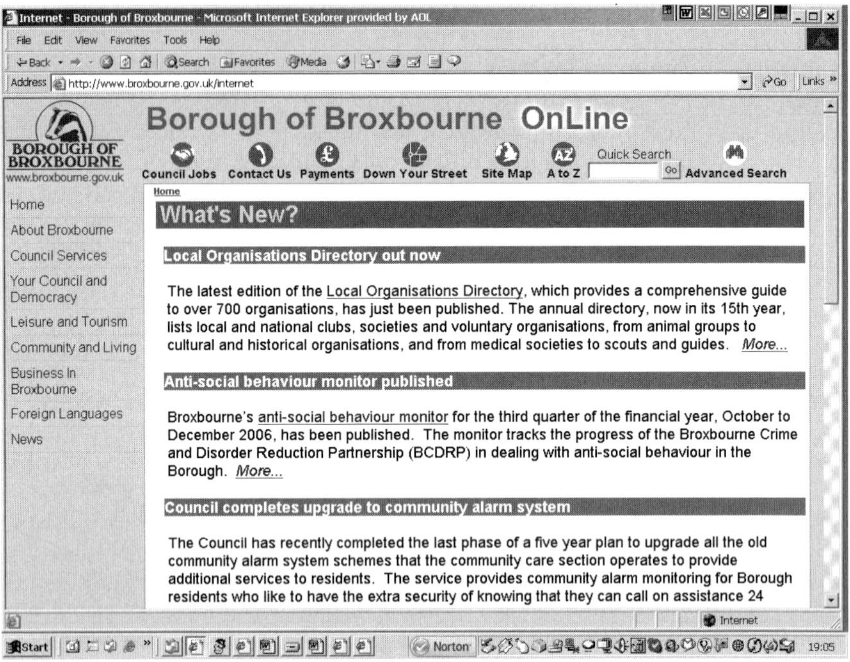

Fig. 2. The new home page for the Borough of Broxbourne

3 Discussion

The results of the local authority web site evaluation became a statement of 'what' needs to be done and 'why'. The course of action that was followed was that where it was easy to make the change, this was passed on to the development team to implement and this produced some improvements. Changes included navigational improvements, doing away with the initial awkward clickable images, and an A – Z list. What was needed, however, was more than just a tinkering with the existing web site (there was a need to provide certain types of information and support transactions) and this required the organization to answer many questions. When should this be tackled? How much should be spent? Who is going to be in charge of it? Should we do it in-house or outsource it?

The improvements that were made were also enabled by technology developments and in particular the use of a content management system (CMS) and through the integration of the web site with back end systems. The CMS could not only enable information to be updated quickly but also allowed templates to be set up where menu bars, logos and news areas could be placed in consistent areas throughout the site so users would learn where to find them easily. Without the CMS, the growth of the content could not be effectively managed. Without proper integration with back end systems, the key transactional and information services identified in Level 1 could not be provided.

Yet achieving this ideal within public sector projects and usability in e-government is not straightforward for the following reasons:

- In general, decisions on contracts and contractors take longer in the public sector following standards and procedures for supplier selection.
- Although conformance with accessibility standards has been prescribed by legislation, there were no clear and explicit requirements for usability testing, leading to inconsistent approaches amongst UK local authorities.
- The importance given to the availability of transactional services as a mark of maturity has encouraged authorities to work more on providing functional and transactional services on their web sites rather than focus on their end users.

This case study development of the web site can therefore be seen as an illustration of the need for a more workable strategy for e-government sites generally if they are to meet citizen needs in terms of functionality and usability. Initially a big effort was made to get content online. This served to meet the initial target and introduced the department and citizens to online information and information management. Following the usability evaluation, work has been done to identify user needs more fully and to fix the main usability problems. This can then be seen as a success. But what are the lessons to be learned?

By recognizing that many web sites will not get it right first time but will improve in a gradual evolutionary fashion, this is a recommended stance for a government department with many constraints on resources. Strategies may then be adopted to make the redevelopment work more effective, efficient, and planned. These are:

- Just produce a simple site to start with that attracts users (citizens), meets the needs of a limited set of user goals well, and opens the way to further development.
- Perform a usability review early on to address the main problems first (achieve quick wins) and refine the development plan.
- Plan out changes over time so that they can be made when skills and resources are available.
- Have regular expert and user reviews thereafter to see where the web site is failing and to steer it on course. This is best achieved through a user steering committee with representation from users and usability experts.

The experience of the Stockport Metropolitan Borough Council [5] who had to rebuild its web site is similar in some respects. This council needed a more robust, streamlined and "future proof" web site which was also more accessible. A key element of their new design was to purchase a content management system. The redesign was completed in 10 weeks and following launch there was an increase in visitors of 21%, and a 68% increase in returning visitors. The use of the CMS provided a base for relatively easy further development that suited the needs of the local council with limited resources.

Another way to address the scarcity of resources for development, adopted by Salford City Council [3], was to involve more members of Council staff in new content development. This was also facilitated by use of a content management system which staff were trained to use to make their contributions. Staff were encouraged to stretch their imaginations to produce information that is dynamic and interesting to the local citizens i.e. the users of Salford's web site. This approach is not only efficient in the use of resources but also created a buzz among members of staff for making contributions. This is reflected in the number of pages on the site rising from approximately 3,000 to 12,000.

4 Conclusions

This study has illustrated the challenges and complexities local government authorities face in developing usable web services. A three level model that supports heuristic evaluation has been used in a case study evaluation and subsequent redevelopment of a Government authority's web site. The three level model was shown to be extremely valuable in identifying problems and areas for improvements.

Together with the evaluation model, an evolutionary strategy for web site development with frequent expert reviews has been advocated. Such an approach is compatible with the limitations of a local government organization. At the same time, by showing gradual improvements in the site and increasing user appreciation of it, this will help drive usability work higher up the priority list.

References

1. Clarke, A.A.: A three level human-computer interface model. International Journal of Man. Machine Studies 24, 503–517 (1986)
2. DTLR: Towards a national strategy for local e-government, London (2002)
3. Fox, J.: Diversity adds value, p. 22 (October 1994) www.kablenet.com
4. Handy, C.: Inside Organizations: 21 Ideas for Managers, Penguin Books (1999)
5. Kirkham, A.: Rebuild and renew, Government Computing, p. 21 (October 1994) http://www.kablenet.com
6. McCracken, D., Rosalee, J W: User-centered web site development - A human computer interaction approach. Prentice Hall, New Jersey (2004)
7. Nielson, J.: Designing Web Usability - The Practice of Simplicity, New Riders, Indianapolis (2001)
8. Nielsen, J., Mack, R.L. (eds.): Usability inspection methods. Wiley, Chichester (1994)
9. Redish, J.: Organizing and writing content to support e-services. Hot Topics in Usability, London, UK (28th April – 2 May 2003)
10. Royal National Institute for the Blind: The legal case for accessible web sites (2006) URL: http://www.rnib.org.uk/xpedio/groups/public/documents/publicwebsite/public_legalcase.hcsp
11. Schneiderman, B., Plaisant, C.: Designing the User Interface, 4th edn. Addison Wesley, London (2005)
12. SOCITM (Society for Information Technology Management); Better Connected, Northampton, United Kingdom (2005)

A Conceptual Design for Children's WebOPAC Interface: Graphic Design Issues

Tengku Siti Meriam Tengku Wook[1] and Siti Salwa Salim[2]

[1] Faculty of Information Science and Technology
National University of Malaysia, 43600 Bangi Selangor
tsm@ftsm.ukm.my
[2] Faculty of Computer Science and Information Technology
University of Malaya, 50603 Kuala Lumpur
salwa@um.edu.my

Abstract. Usability testing on Ilmu (i.e. WebOPAC for children) showed that poor graphic design contributed the most to the usability problem. Three main issues of graphic design are layout, use of colors and human visual perception. A design approach is formulated in an effort to develop a conceptual design that specifies children's requirements and preferences on graphic issues. This involves two main activities which are identifying children's preferences on graphic design and generating design ideas using participatory design method. Four categories of requirements and preferences on graphic design are derived from children. Each category is demonstrated by design ideas. All these constitute towards the formulation of a conceptual design for children's WebOPAC interface and become the basis for redesigning Ilmu's interface.

Keywords: Graphic design, children's interface, conceptual design, participatory design and paper-based prototype.

1 Introduction

The most important issue in this study is the appearance or graphic design of the interface for children. Graphic design provides a communication mechanism between the user and the computer. The designing of graphic design plays an important role in arranging or placing information on children's interface of WebOPAC, which is one of Information Retrieval System (IRS). In Malaysia, it is known as Ilmu [1], a web-based system that enables students to search and browse the bibliography section of books in libraries.

Usability testing on Ilmu showed that poor graphic design contributed the most to the usability problem [1]. The testing involved 16 respondents using two techniques - observation and questionnaire. The parameters used were poor graphic design, lost navigation, less functionality, and display too much text derived from the studies on the IRS for children.

Investigation on the details of poor graphical design had led us to the identification of three issues which were layout, use of colors and human visual perception or aesthetics as illustrated by fishbone diagram in Figure 1 [2,3,4,5 and 6].

M.J. Smith, G. Salvendy (Eds.): Human Interface, Part II, HCII 2007, LNCS 4558, pp. 785–791, 2007.

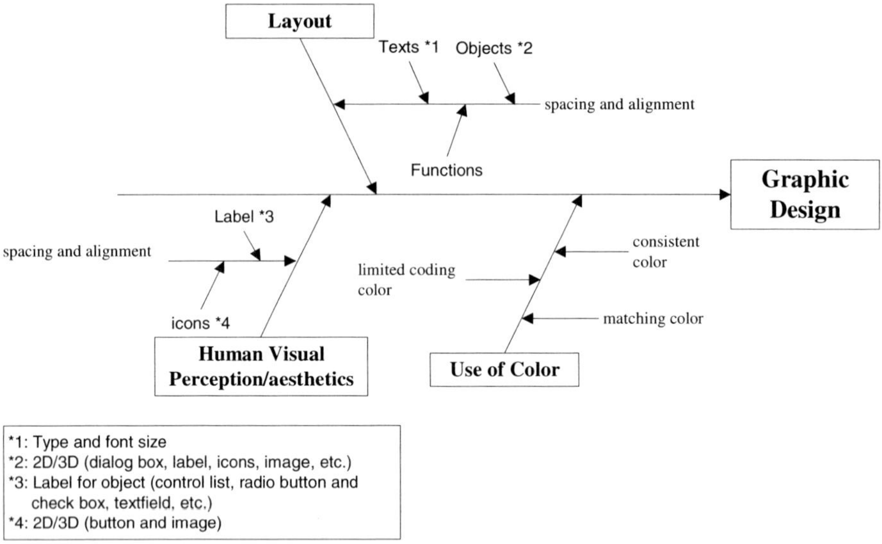

Fig. 1. Key issues in graphic design

Layout is a design of an interface appearance that involves screen and window. It is an important issue in graphic design because it refers to spatial usage. It shows the effect and arrangement of texts, objects (e.g. labels, dialog boxes, icons and menus) and functions on the interfaces which are connected through space. This connection ensures the symmetry and balancing in interface to organize and display information and objects appropriately [7].

The human visual perception or aesthetics is related to the layout issue. It consists of spacing and alignment of icons and labels in the interface. Human visual perception relies on aesthetical value of people which differs according to their age and requirements. However, it still has guidelines and principles to design a visually pleasing interface.

The use of color is very important in graphic design as it helps in changing user's feeling or perception for system interface [8]. The choice of colors must be consistent to avoid the user confused. Too much color effect will make the interface looks messy.

All of the above issues are important for the usability of graphic design on Ilmu's interface. Section 2 of this paper concerns the development of a conceptual design that specifies children's requirements and preferences on graphic design.

2 Design Approach

Figure 2 shows the design approach in developing the conceptual design for children's WebOPAC interface on graphic design issues. It involves 2 main phases which are:

a) Identifying children's preferences on graphic design.
b) Generating design ideas of the preferences identified using participatory design method. Design ideas are visualized through prototyping and evaluated. Based on the feedback from the evaluations, preferences are refined and alternative design is generated. Thus, specifications of conceptual design are emerged iteratively through the participation of the children.

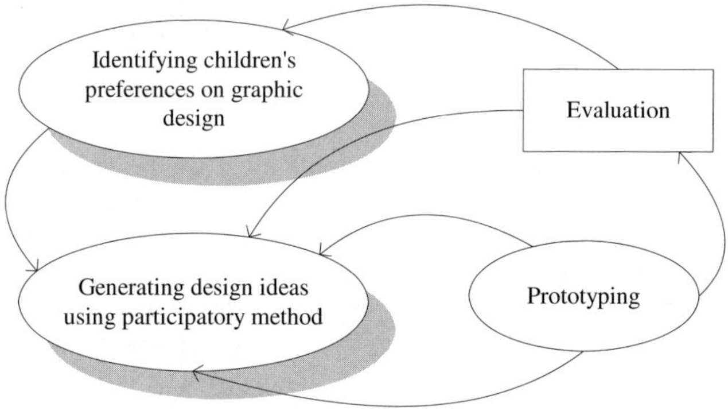

Fig. 2. Development of conceptual design

The two phases involved 8 students, aged from 7 to 11. They were briefed on the research objectives and their role as a design partner. The two phases took 2 months to complete from October to November 2006.

2.1 Identifying Children's Preferences on Graphic Design

The first phase was to recognize and identify children's interest on graphic design of WebOPAC interface. Students were given the following two samples of WebOPAC: Kid's Catalog Web [9] and KidsWeb [10]. The identifying process involved two activities. First, students were asked to search and browse books by keyword, location or subject. Next, is for them to express their preferences and expectations on searching and browsing of books focusing on the elements of layout, use of color and human visual perception. Prior to this, explanations on the three elements were given to the students by showing them the related examples. They recorded their preferences in a form given to them. Their preferences were gathered and a brainstorming session was conducted to clarify on any statements which were not cleared to the researchers.

Analysis and Findings. Table 1 shows a complete list of children's preferences on graphic design after eliminating all the redundant statements.

Table 1. Children's Preferences on graphic design

Elements of graphic	Children's Preferences	
Layout	▪ Use big button ▪ Use big font size ▪ Simple and less text ▪ Simple toolbar and menu	▪ View information in a simple way ▪ Use graphic and drawing to view information
Using color	▪ Color should be striking ▪ Blue is very interesting	▪ Variety of colors and striking ▪ Use clear background
Human visual perception	▪ Include game features ▪ Explore real library	▪ Use agent to search for books (e.g. Spiderman)

As shown in Table 1, students prefer a simple interface design with a bigger fonts and buttons. They enjoy searching and browsing books in an interface environment that uses attractive combination of colors. They are excited to have game features on the website. They are interested towards visual environment such as 3D library while exploring information. Besides that, they also have a preference for an agent who will help them finding information on books. In the next section, these preferences are explored and visualized using participatory design method.

2.2 Generating Design Ideas Using Participatory Method

Two participatory design sessions were conducted with children to generate design ideas and creativity of the preferences identified on graphic design.

Participatory Design Session (1). The participatory design (1) involved the PICTIVE activity where students brainstorm and visualize design ideas. Materials and stationeries to carry out the PICTIVE session such as pens, pencils, various sizes of Post-It notes, stickers and labels and a shared surface (i.e. drawing paper) were given to them. Sketches of their design ideas are the output for the following four task scenarios:

1. You are at the main interface of Ilmu's website. There are three search styles available: keyword, location and subject.
2. You are searching a book by keyword.
3. You are searching a book by subject.
4. You are at the subject menu page and searching for history book.

Figures 3(a) – 3(d) shows the four sketches produced for the four task scenarios. A walkthrough session was carried out with the students to gain an insight understanding of their design ideas produced.

Figure 3(a), the output of task scenario 1, shows a sketch of a train coaches carries a keyword, subject and location with rollover features. Figure 3(b), the output of task scenario 2, shows a superhero character that helps students searching and browsing information by title, author or abstract. Figure 3(c), the output of task scenario 3, shows a caterpillar character containing information of the available subjects such as

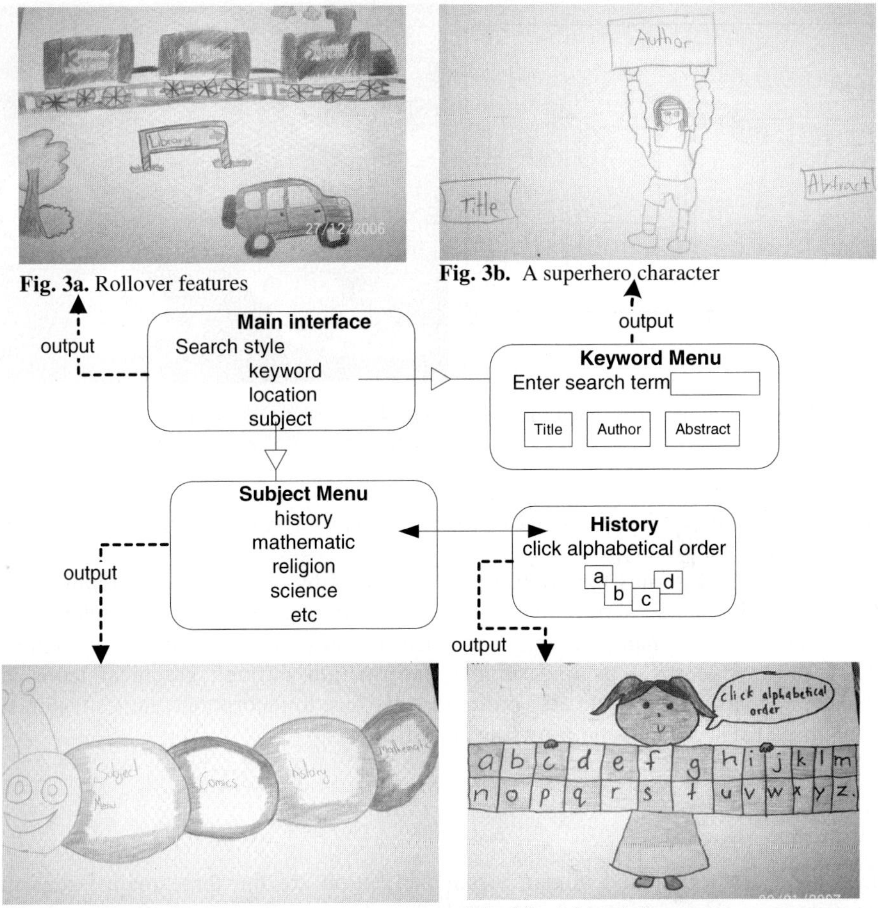

Fig. 3a. Rollover features

Fig. 3b. A superhero character

Fig. 3c. A character character

Fig. 3d. A friend character

Fig. 3. (a) – (d) Journal of design ideas for the four tasks scenarios

history, mathematic and science. Figure 3(d), the output of task scenario 4, shows a girl character that provides choices in alphabetical order of the available history books in the library when a student chooses a history subject.

Figures 3(a) – 3(d) visualize students' preferences on the concepts of agents in helping students locating books in the library; viewing information in graphics/ drawing, thus achieving a simple and less text page; and fun features as part of the interface on graphic design.

Participatory Design Session (2). The participatory design (2) started with a paper based prototype produced by the researchers as shown in figure 4(a) as an initial platform for students to work on the design ideas of browsing and searching

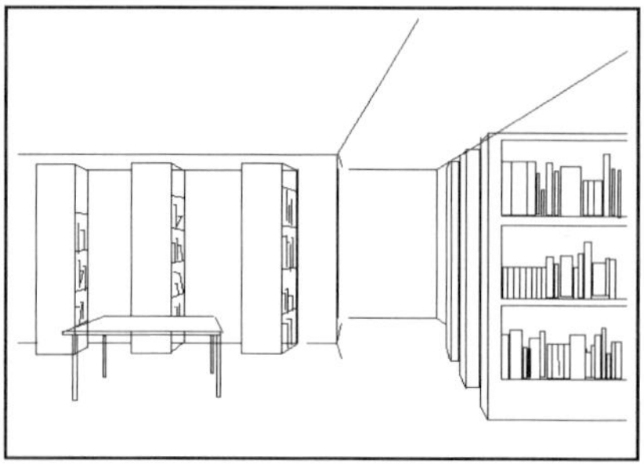

Fig. 4a. A real library in 3D environment

Fig. 4b. Design ideas from children on 3D environment

information in a real library in 3D environment. Figure 4(b) showed the design ideas evolved from the prototype given.

Figure 4(b) illustrates the environment preferred by children for searching books by subject. It clearly shows that students preferred an exciting and relaxing 3D environment rather than following sequence of steps as in the current WebOPAC. They illustrate easy steps in searching information through graphical icons and interface agents. They would also prefer the interface to incorporate games features in supporting the searching and browsing activities.

3 Conclusion

Layout, use of color and human visual perception are the three main issues that contribute to the usability of graphic design. Current Ilmu's interface encounters these issues of graphic design. The design approach formulated as described and illustrated in Section 2 concerns the development of a conceptual design that specifies children's requirements and preferences on graphic design. This is achieved through two main activities which are identifying children's preferences on graphic design and generating design ideas using a participatory design method. Figure 5 gives a schematic structure that conceptualizes the findings of this research which shows the:

- three main graphic design elements, which are the issues in the current Ilmu's interface
- the requirements and preferences on graphic design derived from children
- design ideas demonstrated by children in supporting their requirements and preferences

All these constitute towards the formulation of a conceptual design for children's WebOPAC interface and become the basis for redesigning Ilmu's interface.

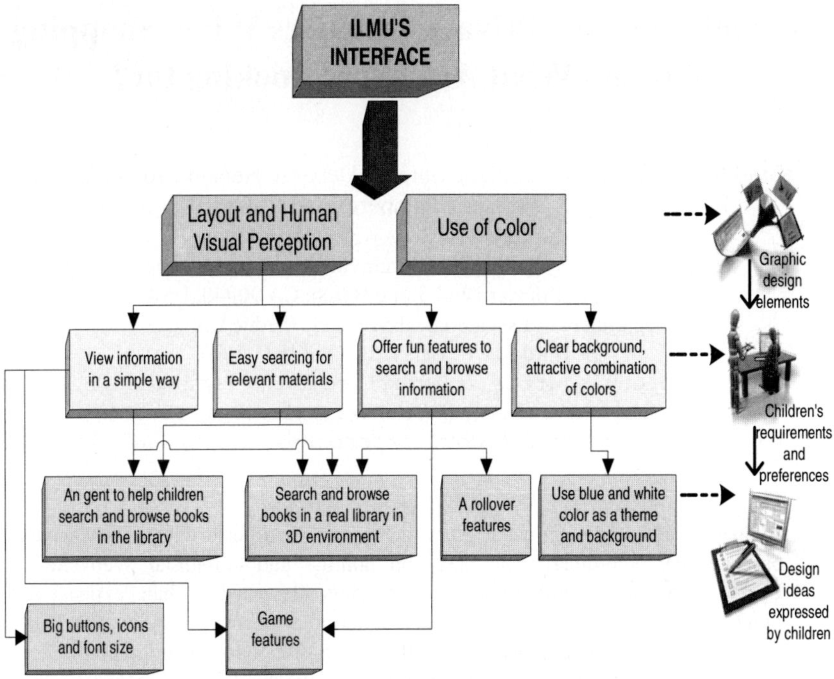

Fig. 5. Findings of research

References

1. Wook, T.S.M.T., Salim, S.S.: User Testing of Children's WebOPAC: A Malaysian Experience. In: The Seventh Asia-Pacific Conference on Computer Human Interaction. Taiwan (2006)
2. Dogget, A.M.: A Statistical Comparison of Three Root Cause Analysis Tools. Journal of Industrial Technology 20(2), 2–9 (2004)
3. Chaparro, B., Baker, J.R., Shaikh, A.D., Hull, S., Brady, L.: Reading Online Text: A Comparison of Four White Space Layouts. Usability News 6(2) (2004). Available at http://psychology.wichita.edu/surl/usabilitynews/62/whitespace.htm
4. Neilson, J.: Usability Engineering. Morgan Kaufmann Publishers, San Francisco (1993)
5. Van Barneveld, J.J.F.: User Interfaces for Personalized Information System. Telematica Instituut, 1–82 (2003)
6. Harst, G., Maijers, R.: Effective GI-ontwerp. Schoonhoven. The Netherlands (1999)
7. Vanderdonckt, J.: Visual Design Methods in Interactive Applications. In: Albers, M., Mazur, B. Content and Complexity: Information Design in Technical Communication, ch. 7, Part I, pp. 187–203, Lawrence Erlbaum Associates (2002)
8. Paal, K.: Graphic Design Using Color, EzineArticles (2004). Available at http://ezinearticles.com
9. Kid's Catalog Web, http://starship.snap.lib.ca.us/
10. KidsWeb, http://www.mln.lib.ma.us/ kids/
11. Muller, M.J.: Restrospective on a Year of Participatory Design using the PICTIVE Technique. Computer Human Interface, Bellcore, ACM (1992)

Examining User Privacy Practices While Shopping Online: What Are Users Looking for?

Kim-Phuong L. Vu[1], Fredrick P. Garcia[1], Deborah Nelson[1], John Sulaitis[1], Beth Creekmur[1], Vanessa Chambers[1], and Robert W. Proctor[2]

[1] Department of Psychology, California State University, Long Beach
1250 N. Bellflower Blvd. Long Beach, CA 90840, USA
[2] Department of Psychological Sciences, Purdue University
703 Third St., W. Lafayette, IN 47907, USA
kvu8@csulb.edu, garcia.fp@gmail.com, dnelson2@csulb.edu,
jsulaitis@gmail.com, beth.creekmur@gmail.com,
chambers.ve@gmail.com, proctor@psych.purdue.edu

Abstract. This study evaluated users' behaviors when performing inexpensive or expensive e-commerce purchases on familiar and unfamiliar Web sites. Users were more comfortable with making inexpensive than expensive purchases. They also felt more secure and that their privacy was better protected when shopping with a familiar Web site than an unfamiliar one, especially for expensive purchases. For inexpensive purchases, if the price was "right", participants were willing to purchase the product on unfamiliar Web sites. For expensive purchases, though, the reputation of the organization hosting the Web site was the most important factor. In both cases, privacy was a minor determinant for deciding whether to make a purchase from a Web site. Only 20% of the users regularly accessed the sites' privacy policies during their interactions with the sites. Moreover, less than half of the participants even looked at privacy-policy links during their interactions with the Web sites.

Keywords: e-commerce; privacy policies; usability; eye-tracking.

1 Introduction

Using Web-based systems to access and exchange information, and to perform online transactions, has become a daily activity for many users. Through these transactions, personal, financial, or health information is transmitted to various parties, who may or may not play necessary roles in the transaction. These secondary transmissions of personal information make the information more vulnerable to misuse. The potential for misuse of personal information has already been shown to raise concerns about privacy and security to users and law makers [1]. For example, the Health Insurance Portability and Accountability Act [2] established federal guidelines to protect privacy of personal health information.

For Web-based systems, users' privacy concerns are usually gathered through surveys and questionnaires that ask users about their worries regarding security and privacy practices [3, 4]. Although surveys and questionnaires can capture users' stated

M.J. Smith, G. Salvendy (Eds.): Human Interface, Part II, HCII 2007, LNCS 4558, pp. 792–801, 2007.

preferences, several studies have shown that users' actual practices may not live up to their stated concerns [5]. That is, participants may indicate that privacy assurance is extremely important to them, but when performing online transactions, these same participants may not access the Web site's privacy-policy page. Although evaluations of log files in previous studies [6] have shown that users do not visit a site's privacy-policy page often, it is unknown whether the lack of interest in a site's privacy policy was due to the site being familiar to users or other contextual factors. The use of more experimental, or quasi-experimental, methods allows researchers to manipulate variables that can influence user behaviors in specific situations.

The purpose of this study was to evaluate users' privacy concerns when performing e-commerce transactions through observing users' behaviors with respect to privacy-policy information while purchasing merchandise. Because ethical principles precluded us from asking users to enter their personal information while performing transactions, we gave them a scenario in which they were to help their grandmother make several purchases from a list of specified Web sites. While browsing and deciding whether to make a purchase on a particular Web site, each participant's eye-gaze patterns were captured to determine whether they were paying attention to the site's privacy features. We manipulated the type of purchase to be made by the participant (inexpensive or expensive) and used a range of Web sites that are well-known to users or unknown to them. We were particularly interested in whether users' decisions to view a site's privacy policy or attention to the privacy links were influenced by the type of purchase and/or the familiarity of the Web site.

2 Method

2.1 Participants

Ten participants were recruited from the Introductory Psychology subject pool at California State University, Long Beach. All participants received experiment participation credit for their Introductory Psychology course. The students ranged from 18-23 years of age, with the mean age being 20 years. The sample included 3 Caucasian, 3 Asian, and 4 Hispanic participants.

Users reported being relatively experienced with computers (M = 1.8, with 1 being very experienced and 4 not experienced), using the Internet very often (M = 1.4, with 1 being very often and 4 not often), and not very knowledgeable about privacy policy issues (M = 2.8, with 1 being very knowledgeable and 4 not knowledgeable). Half of the participants indicated that they had never read a privacy policy, and half indicated that they had read at least one privacy policy in the past. Of the 5 who indicated having read a privacy policy before, 4 stated that they read them regularly.

2.2 Apparatus

Each participant's eye gaze was captured using the Tobii 1750 eye tracker [7], a non-invasive eye tracker integrated into a 17" monitor. The study was conducted with a personal computer using Internet Explorer (version 6.0). The computer screen, eye-gaze, and keystroke information were captured and analyzed by Clear View (version 2) basic and Web recording software.

2.3 Materials

Participants were provided with a packet consisting of a cover story and two purchasing scenarios with associated follow-up questionnaires.

Cover Story. Participants were provided with the following cover story: *Your grandmother is interested in making several purchases online. However, she is unfamiliar with the Internet and trusts you with her information to make these purchases.*

Participants were also provided the following personal and financial information regarding their grandmother to use, if necessary, in their interaction with the Web site.

Mrs. Jane Merryweather
1500 Cherry Ave,
Long Beach, CA 90813
Phone: 562) 985-0107
Email: testingsite@yahoo.com
Password: online2research

Credit Card: Visa, #4209 8881 7463 5728, expires 01/08
Bank Info: Bank of America, account # 57483920, routing number 113324576
Driver's License: B2198844
Birthday: Dec. 1, 1933

Scenario 1. Book Purchasing Task: Participants were told, "Your grandmother would like to buy the book "Dear John" by Nicholas Sparks for your cousin's 17th birthday. She plans on spending between $10 and $20 for his present. You will determine whether you would make a purchase from each of the following sites and explain why/why not. You may explore as much of the site as you deem necessary to make your decision."

The following sites were given to users one at a time by the experimenter:

http://www.barnesandnoble.com/
http://www.bordersstores.com/index.jsp
http://www.amazon.com/
http://www.alibris.com/
http://www.abebooks.com/

Scenario 2. High-Definition Television (HDTV) Purchasing Task: Participants were told, "Your grandmother wants to purchase a 27 inch flat-screen HDTV for your parents as a Christmas gift. She can only afford to spend under $1000 for it. You will determine whether you would make a purchase from each of the following sites and explain why/why not. You may explore as much of the site as you deem necessary to make your decision."

The following sites were given to users one at a time by the experimenter:

http://www.bestbuy.com/
http://www.circuitcity.com/ccd/home.do
http://www.amazon.com/
http://www.ebay.com/
http://www.savinglots.com/
http://www.jr.com/

Follow-Up Questions. After visiting each Web site for each task scenario, participants were asked the following question: Would you purchase the book (or HDTV) from this site? Why? Participants were then asked to rate their:

- Experience level with the site (1 = no previous experience; 10 = extensive previous experience).
- Comfort level with the site (1 = not comfortable; 10 = very comfortable).
- Perception of the security provided by the site (1 = not secure; 10 = very secure)

- Perception of the privacy protection provided by the site (1 = no privacy protection; 10 = a lot of privacy protection).

2.4 Procedure

Each participant was observed individually in a single session lasting approximately 2 hours. Upon arrival to the experiment, the participants were given consent forms to read and sign. The consent forms indicated that they were participating in an experiment designed to gain an understanding of security and privacy concerns that people may have when using online services. Participants were also informed that their eye movements would be recorded using a non-invasive eye-tracker. All of the students were told that their participation was voluntary and that they could stop their participation at any time without penalty.

After signing the consent form, a participant was taken into a semi-anechoic chamber and seated in front of the eye-tracker/monitor. The experimenter then calibrated the eye-tracker to record the participant's eye gaze. The calibration process was repeated until it was successful. Participants were then asked to perform two e-commerce related tasks using the Web. One was to determine whether they would make a book purchase from a list of book-selling Web sites and the other was to determine whether they would make an HDTV purchase from a list of electronics-selling Web sites. Each task (see *Materials* section) was provided to the participants on a sheet of paper, to which they could refer at any time. The participants were also asked to think aloud while working on the task.

At the completion of each task, an experimenter asked the participants a set of follow-up questions associated with the scenario (see *Materials* section) and wrote down their answers. After the participants answered all of the questions, a short break was given. After completing both tasks, participants were asked whether they had ever read a privacy policy when making online purchases, and if so, whether they customarily read them. Participants were also asked whether they considered looking at the privacy policies for any of the sites during this study. Finally, the participants provided demographic information. All were thanked for their participation and debriefed prior to leaving the session.

3 Results and Discussion

3.1 Subjective Questions

Separate repeated-measures ANOVAs were performed on mean ratings (1-10) of participants' experience for the book and HDTV Web sites. For the book task, there was a significant effect of experience, $F(4,32) = 8.77$, $p < .001$: Participants were more experienced with the Amazon and Barnes & Noble sites than the other three sites. As a result, Amazon and Barnes & Noble were grouped as the familiar sites and the remaining sites as unfamiliar. For the HDTV task, there was a significant effect of experience, $F(5,40) = 10.35$, $p < .001$: Participants were most experienced with Amazon and eBay, familiar with Best Buy and Circuit City, and indicated no experience with JR.com and savinglots.com. As a result, the first four sites were grouped as the familiar sites and the remaining two sites as unfamiliar.

For the remaining scales (perceived comfort, perceived security, and perceived privacy protection), separate 2 Price (Expensive or Inexpensive) x 2 Familiarity (Familiar or Unfamiliar) ANOVAs were performed (see Table 1 for means in the book task and Table 2 for means in the HDTV task). One participant's data were not analyzed since s/he did not answer all of the questions for the HDTV task.

Perceived Comfort. There was a main effect of Price, $F(1,8) = 13.98$, $p < .007$. Participants were more comfortable making the inexpensive, book purchase ($M = 7.6$) than the expensive, HDTV ($M = 5.8$) purchase. There was also a significant main effect of Familiarity, $F(1,8) = 25.30$, $p < .001$, with participants feeling more comfortable with familiar ($M = 7.7$) than unfamiliar ($M = 5.7$) sites. The interaction between price of purchase and familiarity was not significant, $F < 1.0$.

Table 1. Mean rating levels of experience, comfort, security, and privacy protection for each of the five Web sites used in the book purchase scenario

	Web Site				
Rating	Barnes & Noble	Borders	Amazon	Abebooks	Alibris
Experience	4.89	2.00	6.44	1.67	1.00
Comfort	8.33	6.00	8.11	6.44	6.78
Security	8.89	7.89	8.33	7.11	7.11
Privacy Protection	8.11	7.56	8.00	7.22	7.00

Perceived Security. There was a main effect of Price, $F(1,8) = 14.00$, $p < .007$. Participants reported feeling more secure when making purchases from the book Web sites ($M = 8.2$) than from the HDTV sites ($M = 7.01$). There was also a significant main effect of Familiarity, $F(1,8) = 25.27$, $p < .001$ with participants perceiving that familiar ($M = 8.5$) sites provided more security than unfamiliar ($M = 6.8$) sites. The interaction between price of purchase and familiarity was marginal, $F(1,8) = 4.61$, $p = .064$. This marginal interaction reflected a trend of familiarity playing a bigger role in perceived security for expensive purchases.

Perceived Privacy Protection. There was a main effect of Price, $F(1,8) = 11.02$, $p < .012$. Participants reported that book Web sites ($M = 8.0$) provided more privacy protection than the HDTV sites ($M = 7.0$). There was also a significant main effect of Familiarity, $F(1,8) = 16.66$, $p < .004$, with participants perceiving that their privacy was protected more by familiar ($M = 8.3$) than unfamiliar ($M = 6.7$) sites. The interaction between price of purchase and familiarity was also significant, $F(1,8) = 18.33$, $p < .001$. Follow-up analyses indicated that when the purchase was inexpensive, there was no significant difference in the perceived privacy protection of familiar ($M = 8.4$) or unfamiliar ($M = 7.6$) Web sites, $F(1,8) = 4.12$, $p > .07$. However, for expensive purchases, participants felt that their privacy was more protected by familiar ($M = 8.2$) than unfamiliar ($M = 5.8$) sites, $F(1,8) = 30.75$, $p < .001$.

Table 2. Mean rating levels of experience, comfort, security, and privacy protection for each of the six Web sites used in the HDTV purchase scenario

	Web Site					
Rating	Best Buy	Circuit City	Amazon	eBay	JR	Savinglots
Experience	4.89	3.56	5.89	6.33	1.00	1.00
Comfort	8.67	7.22	7.22	4.67	5.33	4.11
Security	8.56	8.44	8.22	7.11	6.89	5.00
Privacy Protection	8.33	8.22	8.33	7.78	6.89	4.78

3.2 Decision to Purchase

Inexpensive, Book Purchase. A Chi-Square analysis was performed on the frequency with which users stated that they would purchase a book from each of the five Web sites. There was no significant difference in stated purchasing patterns for the different sites, χ^2 (4) = 2.70, p = .61, see Table 3.

Table 3. Number of participants who indicated "Yes," they would purchase a book or "No," they would not purchase a book from each of the five Web sites

	Web Site				
Purchase Book?	Barnes & Noble	Borders	Amazon	Abebooks	Alibris
Yes	8	6	9	7	7
No	2	4	1	3	3

For all Web sites, more users indicated "Yes", they would purchase a book than "No", they would not purchase a book. Price and company name were given as the most important factor in deciding whether to purchase a book from all sites. The site's usability and professional look also played a smaller role when participants decided to make a purchase from a particular site.

Only one user indicated that his/her purchasing decision was influenced by security and privacy concerns. Although half of the participants indicated that they normally read a company's privacy policy, only two accessed privacy policies for the Web sites during the present study. One participant visited the privacy policy for two of the sites (one familiar and one unfamiliar site) and the other for three of the sites (two unfamiliar and one familiar site).

Expensive, HDTV Purchase. A Chi-Square analysis was performed on the frequency with which users stated that they would purchase an HDTV from each of the six Web sites. There was a significant difference in stated purchasing patterns for the different sites, χ^2 (5) =17.05, $p < .004$, see Table 4.

Table 4. Number of participants who indicated "Yes," they would purchase an HDTV or "No," they would not purchase an HDTV from each of the six Web sites

	Web Site					
Purchase?	Best Buy	Circuit City	Amazon	eBay	JR	Savinglots
Yes	9	6	6	2	3	1
No	1	4	4	8	6	8

Users indicated that they would be more disposed to purchase the HDTV from Best Buy more than any other site. Price and familiarity with the company or site were the two main reasons why users were willing to purchase the HDTV from Best Buy, with usability and the professional look of the site also playing a smaller role. However, most users stated that they would not buy an HDTV from eBay or savinglots.com. When asked for reasons why they would not purchase the HDTV from those two sites, participants indicated that (a) prior experience (either bad or none) with the site, (b) the fact that the site was not an actual store (i.e., the seller could be an individual person and not a company), and (c) the lack of selection from the site, as being crucial reasons.

Across all of the sites, though, two users indicated that security/privacy concerns were reasons for why they would not make a purchase using the particular site. Furthermore, only two accessed the privacy policies (one familiar and one unfamiliar site) for the Web sites used in the present study (the same two participants who accessed the privacy policies from the book purchase task).

3.3 Eye-Gaze Data

Due to the small number of participants who looked at the privacy features of the Web sites while performing their online shopping task, only descriptive data regarding the eye gazes are reported.

Book Task. Participants' eye-gaze data were reviewed to determine whether they gazed on privacy links during their interaction with each of the five Web sites. Although all five sites had links to their privacy policy, none posted privacy logos on their homepages. Half of the participants ($N = 5$) looked at privacy links on four of the five sites, see Table 5, but only two participants actually visited the privacy policy pages for three of the five Web sites.

The location of the privacy link influenced whether participants would pay attention to it. Figure 1 shows the number of times that participants looked at the privacy policy links. There were five locations where users would look at the privacy policy links. Participants were more likely to look at the privacy link when it was placed above the fold or with the major navigation links. Sample gaze plots for the different navigation areas described above are shown in Figure 1, with the number of individual fixations for each location indicated in the text boxes.

HDTV Task. As with the book purchasing task, participants' eye-gaze data were reviewed to determine whether they gazed on privacy links or privacy/security logos during their interactions with each of the six Web sites. All of the sites had links to their privacy policies, and four of the six sites posted a security or privacy logo on their homepages, see Table 6.

Table 5. Number of participants who looked at privacy information, mean number of looks, mean duration of the time spent looking at privacy information, and where the privacy information was located on the page when it was attended to

	Web Site				
Description	Barnes & Noble	Borders	Amazon	Abebooks	Alibris
No. of Participants	1	4	1	1	2
\Mean Number of Looks	1	8	2	1	16
Mean Fixation Duration for Privacy Info (in ms)	777	1,717	379	1,695	2,594
Privacy Information Above the Fold when Looked at?	No	Yes	No	No	Sometimes
Location of Privacy Link	Bottom of Page	Left Side with Navigation	Bottom of Page	Bottom Navigation of Page	Bottom Navigation of Page; Link Provided with Check Out Fields

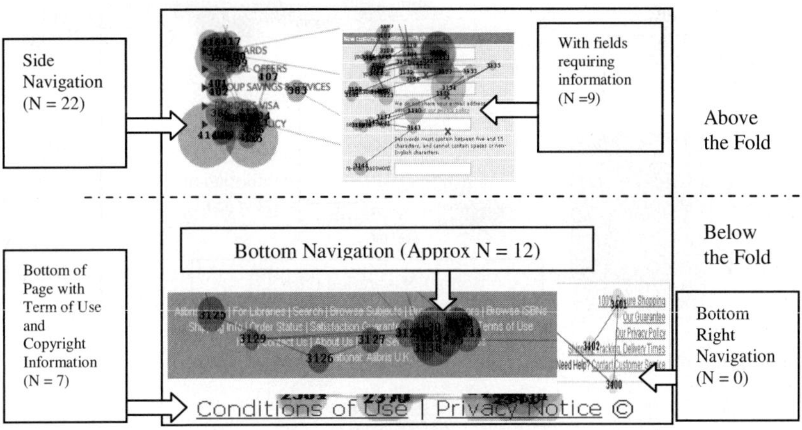

Fig. 1. Number of individual fixations on privacy information as a function of the location of privacy information, with sample gaze plots. The larger circles represent longer fixations.

Table 6. Number of participants who looked at privacy information, mean number of looks, mean duration of the time spent looking at privacy information, and where the privacy information was located on the page when it was attended to

	Web Site					
Eye Gaze	Best Buy	Circuit City	Amazon	eBay	JR	Savinglots
No. of Participants	1	4	1	1	2	1
Mean Number of Looks	8	8	2	4	6	4
Mean Fixation Duration for Privacy Info (in ms)	23,474	2,770	2,811	1,356	1,078	1,655
Privacy Information Above the Fold when Looked at?	Sometimes	No	No	No	Sometimes	Sometimes
Location of Privacy Link	Bottom Navigation of Page	Bottom Navigation of Page	Bottom of Page	Bottom of Page	Bottom Navigation of Page	Left Side Navigation as "Our Policies"
Security or Privacy Logo	VeriSign	CyberTrust	NA	NA	BBBOnline; VeriSign; HackerSafe	GeoTrust

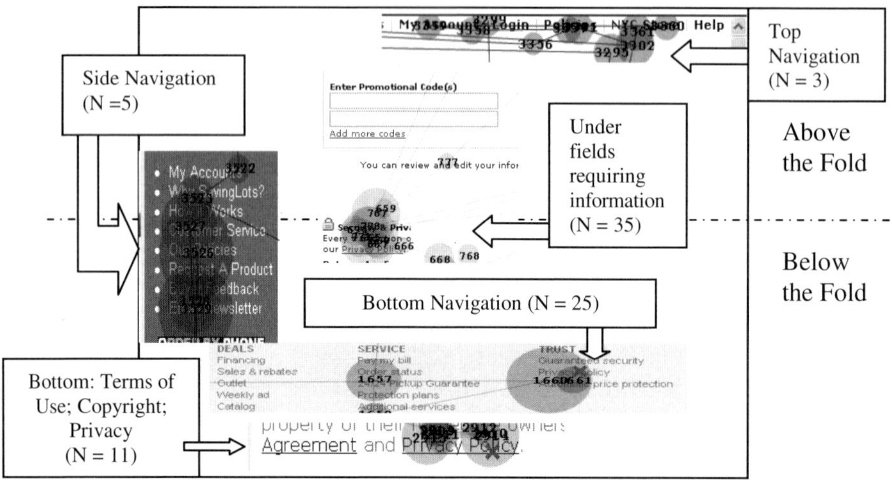

Fig. 2. Number of individual fixations on privacy information for HDTV sites as a function of the location of privacy information, with sample gaze plots. The larger circles represent longer fixations.

The same two participants who viewed the privacy policies for the book task visited the privacy policy pages for two of the six Web sites. Only one additional participant looked at privacy links during their Web interactions.

As with the book purchase task, the location of the privacy link influenced whether participants would pay attention to it, see Figure 2. Participants were more likely to look at the links to privacy information when the link was presented along with other navigation links or when the site provided the link underneath the fields where users were required to enter personal information.

3 Conclusion

Users indicated that they were more comfortable with making inexpensive than expensive purchases. They also felt more secure and that their privacy was protected when shopping with a familiar Web site than an unfamiliar one, especially for expensive purchases. Although financial information is transmitted to sites for purchases of any size, users were more willing to make a purchase on an unfamiliar Web site when it was inexpensive than when it was expensive. Price of the product, as well as the reputation or familiarity of company, emerged as the most important determinants of whether users will make a purchase from a Web site. Privacy concerns, however, played a smaller role. Several users indicated that they do not read the privacy policies because they either (a) trust well-known companies, (b) find the policies to be too long and confusing, or (c) do not care about who gets access to their information because it is not something they could control anyway.

Acknowledgments. This research was supported in part by NSF ITR Cyber Trust grant #0430274 and by an internal Psychology Department award to Kim-Phuong L. Vu. We thank Julie McMath for help with data collection.

References

1. Dark, M., McPherson, C.: Privacy and the public educator. Lecture Series in Computer Science (this volume)
2. Privacy Rights Clearinghouse.: HIPAA basics: Medical privacy in the electronic age. Privacy Rights Clearinghouse. (Retrieved October 22, 2006) from http://www.privacyrights.org/fs/fs8a-hipaa.htm
3. Ackerman, M.S., Cranor, L.F., Reagle, J.: Privacy in E-commerce: Examining user scenarios and privacy preferences. In: E-commerce 99, pp. 1–8. ACM, New York (1999)
4. Proctor, R.W., Ali, M.A., Vu, K.-P.L.: Examining usability of Web privacy policies. International Journal of Human-Computer Interaction (in press)
5. Spiekermann, S., Grossklags, J., Berendt, B.: E-privacy in 2nd generation E-commerce: Privacy preferences versus behavior. In: EC'01, pp. 38–47. ACM, New York (2001)
6. Jensen, C., Potts, J.: Privacy policies as decision-making tools: An Evaluation of online privacy notices. In: CHI 2004, vol. 6, pp. 471–478 (2004)
7. Tobii Technology: http://www.tobii.com/

How Users Read and Comprehend Privacy Policies

Kim-Phuong L. Vu[1], Vanessa Chambers[1], Fredrick P. Garcia[1], Beth Creekmur[1], John Sulaitis[1], Deborah Nelson[1], Russell Pierce[1], and Robert W. Proctor[2]

[1] Department of Psychology, California State University, Long Beach
1250 N. Bellflower Blvd. Long Beach, CA 90840, USA
[2] Department of Psychological Sciences, Purdue University
703 Third St., W. Lafayette, IN 47907, USA
kvu8@csulb.edu, chambers.ve@gmail.com, garcia.fp@gmail.com, beth.creekmur@gmail.com, jsulaitis@gmail.com, dnelson2@csulb.edu, russell.s.pierce@gmail.com, proctor@psych.purdue.edu

Abstract. This study examined participants' comprehension of Web privacy policies when the information from the policy had to be recalled from memory or when participants were able to view the policy while searching for the answers to specific questions. Eye-gaze data were analyzed to examine where users focus their attention when reading privacy policies and searching for information in them. Overall, participants showed poor comprehension of the information conveyed in the privacy policies even though they were written at the participants' level of education. When searching for information in the privacy policies, participants relied on the listing of individual sections provided at the beginning of a privacy policy, when available. When the listing of sections was not available, participants skimmed the entire policy, examining the headings and first few words of each paragraph to determine whether the sections or paragraphs were likely to contain the information for which they were searching.

Keywords: reading comprehension; privacy policies; usability; eye-tracking.

1 Introduction

Identity-theft costs consumers approximately $5 billion a year, and as many as 20% of identity-theft cases involve transactions performed over the Web [1]. The financial costs associated with identity theft, and emotional costs associated with invasion of privacy, have made the topic of privacy protection salient to users. Previous studies have shown that 75% of users express being concerned about privacy issues when conducting online transactions [2, 3]. To protect their privacy, online users need to be aware of how their information is being used, stored, and transferred by the Web sites with which the users perform transactions involving personal information.

Most Web sites that collect personal information have a privacy-policy page [4] that users can access to read the organization's stated privacy practices. Despite the concern that users express for privacy protection, log files have shown that users seldom access the privacy-policy pages of Web sites [5]. Users report this neglect is

M.J. Smith, G. Salvendy (Eds.): Human Interface, Part II, HCII 2007, LNCS 4558, pp. 802–811, 2007.

due to their belief that policies address the organizations' privacy concerns, not those of consumers, and that the policies are long and difficult to understand.

The goal of the present study was to examine how users read privacy policies, their understanding of those policies, and their strategies for finding specific types of information within the policies. Study 1 evaluated whether users are able to comprehend privacy policies when they read the entire policy and whether they focus on certain parts of a policy more than others. Study 2 evaluated how users go about locating a site's stated practices with regard to specific privacy concerns.

2 Study 1

In a prior study that we performed examining participants' online shopping behaviors [6], two of 10 participants visited the privacy-policy pages of several Web sites. These two subjects showed similar behaviors while reading the privacy policies: They read the information that was present "above the fold" and ignored or only skimmed the information "below the fold," see Figure 1. Thus, one reason why users may show poor comprehension of privacy policies is that they do not spend the time needed to read them carefully. However, it is also possible that poor comprehension results from the way privacy policies are worded.

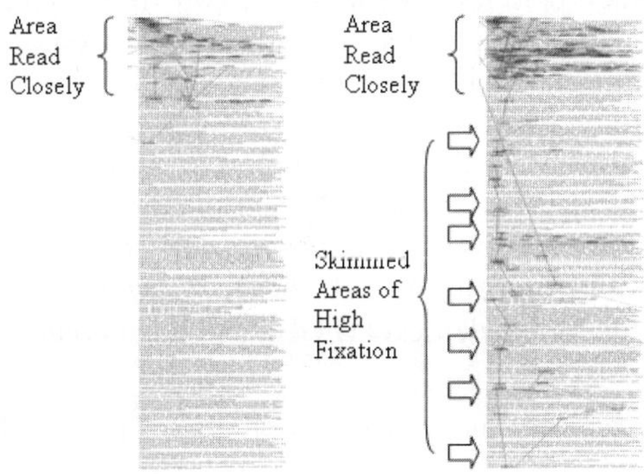

Fig. 1. Gaze plots of two participants who visited the privacy policy page while performing an online shopping task

The goal of Study 1 was to evaluate users' comprehension of Web privacy policies when they were asked to read them, in their entirety, at the participants' own leisure. To measure comprehension, we devised a list of 20 questions that reflect the top privacy concerns of users [7]. Because not all policies may address these concerns, we also evaluated whether the privacy policies examined met the information requirements of the end users.

To determine the types of Web sites most commonly used by college students for online shopping, we distributed a survey to 100 students. The survey showed that college students tend to purchase a range of items that vary in price, including books, electronic products, and clothing. We hypothesized that the amount of money spent for the online purchase and the length of the privacy policy would influence users' inclinations to read the policy. To assess this hypothesis, we used two privacy policies each from Web sites that sell books and electronic-products, since our survey showed that college students use sites from these categories. One policy from each category was short in length, and the other was long.

2.1 Method

Participants. Eight volunteers (*M* = 25 years, *SD* = 1.9 years) were recruited from California State University, Long Beach (6 graduate and 2 undergraduate students). All were paid $10/hr. The participants reported being experienced with computers (*M* = 3.3, with 1 being not experienced and 4 very experienced). They also reported that they used the Internet very often (*M* = 3.6, with 1 being not often and 4 very often), but that they were not very knowledgeable about privacy policy issues (*M* = 1.9, with 1 being not knowledgeable and 4 very knowledgeable).

Apparatus. The study was conducted with a personal computer using Internet Explorer (version 6.0). Each participant's eye gaze was captured using a Tobii 1750 eye tracker [8], which includes a non-invasive eye tracker integrated into a 17" monitor. The computer screen, eye-gaze, and keystroke information were captured and analyzed by Clear View 2 basic and Web recording software.

Materials. Four privacy policies (two long and two short) were selected from Web sites examined in the survey (two book sites and two electronics sites). To avoid any influence of prior experience with the organization on the participant's performance, the four policies were reproduced on a new Web page, with all identifiers removed. The format and organization of the privacy policy was maintained, but all hyperlinks were removed (i.e., formatted as underlined text) so that users could only access the information provided in the privacy policy page. The company's name was replaced by the generic names of Company A, B, C, or D. Table 1 shows the Flesch Reading Ease Score (FRES), Flesch Grade Level (FGL), and word count for each policy.

Table 1. Readability score and word count for each privacy policy. To compute FRES and FGL, the word count and number of sentences were obtained from MS Word and the syllable count was obtained from a ReadMe tool.

	Readability Score		**Word Count**
Web Site	FRES	FGL	
Book (Company A)	31.16	15.20	2,774
Book (Company B)	44.87	12.96	1,010
Electronic (Company C)	32.99	13.85	2,802
Electronic (Company D)	49.60	12.46	739

For each policy, participants were given the following questions to answer:

1. Does this site <u>give you the option</u> of being anonymous for browsing purposes?
2. Does this site <u>give you the option</u> of viewing the changes made to the privacy policy?
3. Does this site <u>use</u> privacy logo's on its privacy policy page?

4. Does this site rent or sell your credit card/financial information?
5. Does this site use your credit card/financial information to contact you for promotional purposes?
6. Does this site give you the option of refusing to allow them to use your credit card/financial information for promotional purposes?
7. Does this site share your credit card/financial information with *third parties or affiliates* for promotional purposes?
8. Does this site give you the option of refusing to allow them to share your credit card/financial information with *third parties and affiliates*?
9. Does this site allow all of its employees to view your credit card/financial information?
10. Does this site protect you from hackers accessing your credit card/financial information?
11. Does this site protect your password from hackers?
12. Does this site sell your personally identifiable information?
13. Does this site give you the option of refusing to allow the company to use your personally identifiable information for promotional purposes?
14. Does this site share your personally identifiable information with *third parties and affiliates* for promotional purposes?
15. Does this site give you the option of refusing to allow them to share your personally identifiable information with *third parties and affiliates*?
16. Does this site protect your personally identifiable information from hackers?
17. Does this site rent or sell your email address?
18. Does this site give you the option of refusing to allow the company to send emails to you for promotional purposes?
19. Does this site share your email address with *third parties and affiliates* for promotional purposes?
20. Does this site give you the option of refusing to allow them to share your email address with *third parties and affiliates*?

In addition, the following question was listed, "Would you purchase an item from this site based on their privacy policy? Why, or why not?"

Procedure. Each participant was tested individually in a single session lasting approximately 1 hour. Participants were given informed consent forms to read and sign upon arrival to the experiment. They were told that they were participating in an experiment designed to examine comprehension of Web privacy policies and that their eye movements would be recorded using a non-invasive eye-tracker while reading privacy policies online. Participants were reminded that they could stop their participation at any time without penalty.

After signing the consent form, a participant was taken into a semi-anechoic chamber and seated in front of a computer. The experimenter then calibrated the eye-tracker to record the participant's eye gaze. If the calibration process was not successful, the process was repeated until it was successful. To begin the experiment, the experimenter entered the URL for the privacy policy page.

Participants were asked to read the privacy policy in its entirety and informed that they would be asked some questions about the content of the policy afterwards. Immediately after the participant indicated that s/he had completed reading the privacy policy, s/he was given the list of questions. The participant answered "yes", "no", or "I don't know" to each question.

After reading and answering the questions for all four policies, participants answered a few questions regarding their concern for privacy and filled out a demographic questionnaire. All were thanked, paid for their participation, and debriefed prior to leaving the session.

2.2 Results and Discussion

Eye-Gaze Data. Participant's eye-gaze data indicated that they did read each policy in its entirety. The average time that participants spent reading the privacy policy was 9.88, 4.10, 9.04, and 2.79 minutes for Policy A, B, C, and D, respectively. This time was positively correlated to the word count for each of the policies ($r = .99$, $p < .007$).

The aggregate gaze data showed that, when reading privacy policies, participants paid particular attention to the first few sentences in the policy, the first few words of each paragraph in the policy and information in bold or underlined. Moreover, based on ten or more fixations, the percentage of headers read closely by participants was 30%, 14%, 7%, and 60% for policies A-D, respectively, and the percentage of paragraph or bulleted information read closely was 35%, 58%, 21%, and 57% for Policies A-D, respectively. Although the number of fixations on sections describing (a) the types of information gathered, (b) how information was shared with third parties and affiliates, and (c) security of stored information varied for the different policies, participants did spend more time looking at paragraphs containing these types of statements than other types of statements (see Table 2).

Table 2. Number of paragraphs with 10 or more fixations by type of information category

Information Category	Policy A (Book- Long)	Policy B (Book- Short)	Policy C (Electronics- Long)	Policy D (Electronics- Short)
Information gathered	5	3	2	1
Information shared with third parties & affiliates	5	5	3	2
Security of stored information	2	0	1	2
Number of paragraphs with intense fixations	12	7	10	4
Total paragraphs in the privacy policy	34	12	38	7

Analysis of the Privacy Policies. To determine whether participants were able to comprehend the information stated in the privacy policy, two arbiters went through each policy and answered each of the 20 questions independently. The arbiters referenced the privacy policy to determine their answers. They agreed on the answers for 13 questions for Policies A and B, 16 for Policy C, and 15 for Policy D. Because the policy was vague for the information addressed in the remaining 23 disputed questions, four additional arbiters went through the policies and answered them independently to come up with the final standard. If the policy does not address a specific concern, then the correct answer was judged as "don't know."

As shown in Table 3, the four privacy policies contained 40-65% of the privacy preferences that users indicated were most important [7]. However, the arbiters could not assess whether 25-45% of the privacy preferences were addressed in the policies due to the policies' vagueness on these issues.

Table 3. Number of questions (see materials section) reflecting users' privacy concerns that were followed, not followed or unclearly addressed by the policies

	Treatment of Users' Top Privacy Preferences in Privacy Policy			
Policy	**FOLLOWED**	**UNCLEAR**	**NOT FOLLOWED**	**TOTAL**
Policy A (Book- Long)	10	5	5	20
Policy B (Book- Short)	13	5	2	20
Policy C (Electronics- Long)	7	9	4	20
Policy D (Electronics- Short)	8	6	6	20

Performance Data. Separate 2 (Type of Company: Books or Electronics) x 2 (Policy length: Long or Short) analyses of variance (ANOVAs) were performed on percentage of correct answers for all questions, and for questions when the disputed items were excluded (see Table 4). Overall, comprehension was low (M = 50% when all questions were analyzed and 48% when disputed questions were removed).

Table 4. Percent correct for each privacy policy in Study 1 when all questions were analyzed and when disputed questions were excluded

Policy	**Including/Excluding Disputed Items**	**Percent Correct**
Policy A (Book- Long)	All Questions	42.50%
	Excluding Disputed Questions	40.38%
Policy B (Book- Short)	All Questions	38.75%
	Excluding Disputed Questions	39.42%
Policy C (Electronics - Long)	All Questions	49.38%
	Excluding Disputed Questions	51.56%
Policy D (Electronics - Short)	All Questions	55.63%
	Excluding Disputed Questions	57.50%

When all questions were analyzed, none of the effects reached statistical significance, $Fs(1,7) < 3.99$, $ps > .086$. However, when the disputed questions were removed, the effect of Type of Company was significant, $F(1,7) = 6.07$, $p < .043$. Comprehension was better for the electronic sites ($M = 55\%$) than the book sites

(M = 42%). Although the performance data may suggest that electronics sites have clearer privacy policies because they sell items at higher prices, the policy analyses showed that these policies were less clear with regard to certain privacy preferences expressed by users than the policies for the book sites.

All participants were graduate or undergraduate students, indicating that they were at the reading level for which the privacy policy was written. However, only half of the questions were answered correctly, based on the participants' reading of the privacy policy. Thus, the findings from this study are in agreement with previous studies [5, 6] indicating that privacy policies are not user-friendly. Although these policies were written at a level that should be appropriate for this group of users, they were written too vaguely for users to understand clearly how their personal information is being used and protected.

Follow-Up Questions. After answering all of the questions for each policy, participants were asked whether they would make a purchase from the company for which the privacy policy was written. Overall, only 60% of participants indicated, "Yes, they would make a purchase" from the sites, and this number did not differ significantly across the four sites (M = 50% for Company A and 62.5% for Companies B, C, and D), $\chi^2 < 1.0$. The main reason given by participants for not wanting to make purchases from the sites was the fact that, for the two long policies, their personal information would be shared with affiliates and third parties and, for the two short policies, the privacy policies were too brief.

Only two of the eight participants indicated that they had ever read or skimmed a privacy policy before this study. Several participants did not express being too concerned with privacy policies because they trusted the companies from which they purchase products.

3 Study 2

Participants in Study 1 showed poor comprehension of Web privacy policies when they read each policy with the general intent of understanding it in order to answer questions regarding the privacy practices of the organization immediately afterwards. The possibility exists that the participants could have shown better comprehension if they were reading the policies with the intent of looking for specific information within the policy. To test this possibility, participants in Study 2 were asked to answer privacy-related questions, one at a time, by searching the privacy policy to locate the answer. This method allowed us to evaluate whether participants are capable of comprehending what a policy states about specific, targeted privacy concerns. In addition, we were able to examine whether participants look at the sections in each privacy policy that address the specific question being answered and, if not, where in the policies they look to search for the answers.

3.1 Method

Participants. Ten undergraduate students (M = 22 years old, SD = 1.08 years) were recruited from California State University, Long Beach. All were paid $10/hr. The participants reported being experienced with computers (M = 3.30, with 1 being not experienced and 4 very experienced). They also reported that they used Internet very often ($M = 3.90$, with 1 being not often and 4 very often), but that they were not very knowledgeable about privacy policy issues ($M = 1.70$, with 1 being not knowledgeable and 4 very knowledgeable).

Apparatus and Materials. The apparatus was identical to that used in Study 1. For materials, only policies A and B were used because comprehension was poorest for those policies. Participants were asked to search the privacy policies to determine the answers to questions 1, 2, 3, 4, 5, 8, 16, and 18. These questions contain a mixture of four questions with clear answers and four questions with disputed or unclear answers from Study 1. The questions were presented to participants in random order.

3.2 Results and Discussion

Performance Data. Separate repeated-measures ANOVAs were performed on percent correct for all questions and when the questions with unclear answers were removed (see Table 5 for means). Although participants performed better on Policy A than Policy B, this difference was not significant for either dependent measure, $Fs(1,9) < 3.12$, $ps > .11$.

To compare comprehension of privacy policies when participants are able to reference the policy versus when they are recalling the information from memory, the percent correct scores on Policy A and B of the present study were compared with those for the same policies in Study 1. An ANOVA, with unequal *N*s, was performed on policy (A or B) with Study (1 or 2) as a between-subjects factor. The effect of Study was significant for both dependent measures (i.e., percent correct for all questions and when questions with unclear answers were removed), $Fs(1,16) > 118.7$, $ps < .001$. Participants performed better in the present study than in Study 1, which is not too surprising since they were able to view the policies while answering the questions. However, their comprehension scores in Study 2 were still relatively low, with more than 25% errors made to questions for which there were clear answers.

Table 5. Percent correct for each privacy policy in Study 2 when all questions were analyzed and when questions with unclear answers were excluded

Policy	Including/Excluding Disputed Items	Correct
Policy A (Book- Long)	All Questions	60.0%
	Excluding Questions with Unclear Answers	80%
Policy B (Book- Short)	All Questions	56.3%
	Excluding Questions with Unclear Answers	65%

Eye-Gaze Data. Participants' eye-gaze data were evaluated for each question of Policy A and B separately. For all questions on both policies, participants looked at the section(s) in the privacy policy that addressed the specific question for which they were trying to answer (see Table 6).

Although the number of irrelevant sections searched was similar for Policies A and B, the reason why was different. Policy A, which was the longer policy, has a listing of section headers at the beginning of the policy. Thus, participants were able to focus their attention on the relevant section for questions in which there was a clear answer (questions 1-3 and 18). For the questions with unclear answers, the gaze patterns show that participants were indeed confused and had to look at many areas of the privacy policy. For Policy B, which was the short policy, no listing of section headers was available. In this case, participants had to look at many areas of the policy to search for the section containing the critical information. Thus, the number of irrelevant areas focused on for Policy B was similar regardless of whether the information was clearly stated in the policy or not.

Table 6. Information regarding where participants' fixated most as they were reading Policies A and B in Study 2, as a function of the eight different questions. The * indicates that the answer to the question was unclear, as determined by the arbiters in Study 1.

	Most Visual Fixations on Area Containing Answer?		Number of Sections in the Policy with Intense Visual Fixation		Number of Irrelevant Sections with Intense Visual Fixation	
Quest	**Policy A**	**Policy B**	**Policy A**	**Policy B**	**Policy A**	**Policy B**
1	Yes	Yes	3	3	0	1
2	Yes	Yes	2	2	0	1
3	Yes	Yes	1	2	0	0
4*	Yes	Yes	5	4	2	1
5*	Yes	Yes	7	2	4	0
8*	Yes	Yes	4	4	2	3
16*	Yes	Yes	5	3	1	0
18	Yes	Yes	2	4	0	2

There were two reasons why participants had a difficult time finding the answers to questions 4, 5, 8, and 16. The first is that the information for which they were looking tended to be dispersed in several sections throughout a policy. The second and more critical reason was in the wording of the policy. The wording was either vague (e.g., the policy states, We "will not share personal information with third parties unless you have consented…or the disclosure is necessary…" but does not specify when that would be) or contained many conditional statements (e.g., "Other than as set out above, you will receive notice when information about you might go to third parties…").

Follow-Up Questions. After answering all of the questions for each policy, participants were asked whether they would make a purchase from the company for which the privacy policy was written. As in Study 1, only 60% of participants indicated, "Yes, they would make a purchase" from the sites (M = 40% for Company A and 80% for Company B). The main reasons given by participants for not wanting to make purchases from the sites were that their personal information would be shared with other organizations and the lack of clarity and/or organization of the privacy policy.

In general, participants expressed being concerned with privacy because they wanted to protect their personal information, with 60% indicating that they had read a privacy policy previously. However, most participants admitted that they do not necessarily follow good privacy practices, such as reading privacy policies of Web sites regularly, because (a) the policies are often long and boring (and many policies sound similar), (b) they tend to make purchases from companies they trust, and (c) they do not make enough online purchases to justify going through the trouble of reading the policies.

4 Conclusion

Participants' comprehension of information contained within privacy policies was found to be quite low. Comprehension was poor regardless of whether participants were asked to recall information after reading a policy or to search for information within a policy in response to specific questions. The fact that comprehension scores were still low when participants were able to search for the information in the policy suggests that important information in privacy policies is not stated in a way that can be easily understood by users. In fact, independent arbiters assigned to determine definitive answers to the same questions were unable to do so for 25 – 45% of the questions (depending on the policy). In Study 2, collected eye-gaze data on four of the unclear questions further suggest that participants were confused by the way the information was presented in the privacy policies. Participants searched many more areas of the privacy policies for answers to these questions than for answers to questions for which the policies provided clear answers. For a Web site's privacy policy to be of value to users, the host organization must ensure that the information desired by the users is provided in a form that they will be able to comprehend easily.

Acknowledgments. This research was supported in part by NSF ITR Cyber Trust grant #0430274 and by an internal Psychology Department award to Kim-Phuong L. Vu. We thank Mary Ngo for help testing participants.

References

1. Identity Theft Resource Center: http://www.idtheftcenter.org/alerts.shtml
2. Ackerman, M.S., Cranor, L.F., Reagle, J.: Privacy in E-commerce: Examining user scenarios and privacy preferences. In: E-commerce 99, pp. 1–8. ACM, New York (1999)
3. Spiekermann, S., Grossklags, J., Berendt, B.: E-privacy in 2nd generation E-commerce: Privacy preferences versus behavior. In: EC'01, pp. 38–47. ACM, New York (2001)
4. Adkinson, W.F., Eisenach, J.A., Lenard, T.M.: Privacy online: A report on the information practices and policies of commercial Web sites. The Progress and Freedom Foundation, Special Report. Washington, DC (2002)
5. Jensen, C., Potts, J.: Privacy policies as decision-making tools: An Evaluation of online privacy notices. In: CHI 2004, vol. 6, pp. 471–478 (2004)
6. Vu, K.-P.L., Garcia, F.P., Nelson, D., Sulaitis, J., Creekmur, B., Chambers, V., Proctor, R.W.: Examining user privacy practices while shopping online: What are users looking for? Lecture Note In Computer Science (this volume)
7. Proctor, R. W., Vu, K.-P. L., Ali, M. A.: Usability of user agents for privacy-preference specification. Lecture Notes In Computer Science (this volume)
8. Tobii Technology: http://www.tobii.com

Using Long Term Memory for Bookmark Management

Ming-Jen Wang[1], Kuen-Meau Chen[1], and Tee Koon Hau[2]

[1] National Center for High-Performance Computing, 7, R&D 6th Rd., Science Industrial Park, Hsinchu, Taiwan
[2] National Chiao Tung University, 1001 Ta Hsueh Road, Hsinchu, Taiwan
renny@nchc.org.tw, morris@nchc.org.tw,
achilles.cis92@nctu.edu.tw

Abstract. When using the bookmark or "Favorites" function included with most web browsers, it is difficult to retrieve a target bookmark if it is not properly categorized or tagged with meaningful keywords. For this project, we propose using a method of human-like long term memory (LTM) rather than directories or keywords for searching and managing bookmark data. This experimental model takes into account the user's browsing experience (i.e. the bookmark's added date, number of visits, and last date visited) to create and manage bookmarks. A pilot study of the model shows that the new system is significantly more efficient at retrieving information from large collections of bookmark data.

Keywords: Bookmark Management, Information Filtering, User Interface Design.

1 Introduction

Ever since the development of the Internet, the World Wide Web (WWW) has became an important platform for providing information. Because there is so much information on the Internet, it has become common task for many web users to find, record and keep track of important web pages. Most users use their browser's built-in bookmarks or "Favorites" function to record, organize and retrieve the web pages they've visited in the past. One survey [1] indicates that 58% of all web pages a user visits will be re-visited by the same user sometime in the future.

Currently, the bookmark function provided by most browsers relies on the user's categorizing and tagging abilities to make the information easily retrievable. Though this might be an easy task for experienced users, others may find it is difficult to categorize or tag every single website they visit. When it comes to retrieving dated information, even experienced users might find it a difficult and time-consuming task to use the current technology. This is because traditional bookmark management systems rely heavily on directory or keyword (tagging) mechanisms for labeling bookmarks. If the directory or keywords are not properly organized or, even worse, the directory or keywords are forgotten, the retrieval of the information can be a very daunting task. To solve this problem, we propose a new method which makes use of a human-like long term memory (LTM) instead of

M.J. Smith, G. Salvendy (Eds.): Human Interface, Part II, HCII 2007, LNCS 4558, pp. 812–820, 2007.

using short term memory (i.e. remembering directories and/or keywords) for managing bookmarks. This new method also utilizes visualized information to increase information retrieval efficiency.

2 Background

2.1 Related Works

Deficiencies of Traditional Bookmark Systems

In the past, several attempts have been made to improve upon the deficiencies found in most traditional browser bookmark systems. Ying Feng (2002) [2] created a system that automatically analyzed the semantic similarities of web pages and then grouped those web pages according to those similarities. The system improved the efficiency by auto-grouping the bookmarks. In another study, Data Mountain [3] used 3D space to help the user memorize and organize large amounts of information. PowerBookmarks [4] integrates auto bookmark categorizing, bookmark sharing, and a user-friendly interface to solve many of the problems associated with traditional bookmark systems. John Roberts (2005) [5] also improved bookmark metadata by integrating a new bookmark system that increased the usability of traditional bookmark systems.

In these studies, many researches proposed various solutions to the problems inherent in traditional bookmark systems. Unfortunately, all of the suggested solutions were still based on using directories and keywords to manage bookmarks. This mechanism requires the user to memorize directory or keywords when retrieving the information. This generally isn't a problem when the bookmark is accessed on a regular basis or if the bookmark collection is not too large but this is often not the case.

In this study, we propose a new method which uses a human-like LTM mechanism to enhance the efficiency of bookmark information retrieval. This new method reduces the amount of information (e.g. clutter) appearing on the screen and, as a result, decreases the amount of time required to scan through and retrieve the necessary information. The purpose of this paper is to propose a method by which to effectively locate the wanted bookmarks from a large number of bookmarks. This new method uses a human-like long-term memory mechanism to create an intuitive interaction for organizing bookmarks.

2.2 Long-Term Memory

Types of Long-term Memory

Long-term memory is memory stored as meaning that can last as little as 30 seconds or as long as decades [6]. Human memory is divided into short-term (STM) and LTM. The important characteristic of LTM and, consequently, the focus of this project, is on how long the memory "sticks" in the human brain. As an example, when asked to memorize a new telephone number, the telephone number would normally be stored in the STM of the individual being asked to memorize it. On the other hand, if the number is used and/or repeated regularly, it might not be forgotten for some time.

This process would require that the user repeat the number several time in order for it to be "transferred" into LTM.

This project asks the question, "Are there new methods by which to remember information other than by repetition?" The answer to this questions lies within a sub-category of LTM. LTM is typically divided into Declarative memory and Procedural memory. Declarative memory can be further divided into Episodic Memory and Semantic memory. Episodic memory is the recollection of events. It includes time, place and associated emotions [7]. Episodic Memory does not require the user to rehearse the event in order to memorize it. For example, if someone were to be having a conversation with a friend and in that conversation he tells his friend, "Last night I went to a good restaurant!" he would be recalling information stored in his Episodic memory. This type of memory is often what eye-witness testimony is compromised of and is especially susceptible to subsequent events like questioning, reading the newspaper, talking to others about an event, etc[8].

Currently, users must rely on the directory or keyword mechanisms to manage the ever-increasing amount of bookmark information they need to store. These two methods require the user to manually place bookmarks into a well-organized folder or tag it with the proper keywords. If the information is not stored in a well-organized directory or tagged with meaningful keywords, once some time has past, the information will most likely be removed from the user's STM and lost. He simply won't remember how he organized it. On the other hand, Episodic memory lasts longer and does not require the user to memorize information via repetition. For this project, we propose the use of this characteristic of Episodic Memory to manage large amounts of bookmark data and to improve upon retrieval efficiency.

Using Long-term Memory for Bookmark Management

It is often the case that a web page is visited only once or twice when a user adds it to his bookmark collection. For those bookmarks that are not used often, it is difficult to memorize their location and position in the bookmark directory. The only pieces of information the user has to help him locate the bookmarks are directories and keywords. Experienced users often name the directories and keywords with meaningful words which help them to recall where they placed the target bookmark.

In this study, we propose using LTM to help users organize and search large amounts of bookmark information. As an example of how the system works, let's say the user remembers that he adopted his pet dog three years ago and he remembers adding several pet-related web pages to his bookmarks during the period he was researching dogs. The user remembers that he has not visited those web sites for some time and that he visited them only a few times. This information alone is used in the development of our new bookmark system. In particular, the approximate date the bookmarks were added, the approximate number of times the bookmarks were visited and the most recent date the bookmarks were visited are all very useful.

The new bookmark system doesn't require the user to remember the exact date he added the bookmark, rather, he only needs to know "about" when he added it. All the information gathered is used to filter out the unnecessary information and highlight only the useful information that will guide him to his target bookmark. The more information the user is able to recall, the more accurate the search becomes. The new bookmark system can be used to increase the efficiency in searching bookmarks within a large number of bookmarks. From a management point of view, the new system can also be used to organized bookmarks according to personal life experiences.

3 Bookmark Management System

Many researchers have suggested various solutions to the difficulty inherent in managing a large number of bookmarks using traditional bookmark systems. All the previous solutions, though, use directories and keywords as the primary method of searching and managing bookmarks. With our new bookmark system, we take a new approach at organizing bookmark information which greatly improves on the efficiency of searching a large number of bookmarks.

3.1 Bookmark Metadata

The goal of this project is to improve bookmark metadata collection so as to provide users with a more meaningful and personal interaction with their bookmark management system. The new system will work in the background to automatically save and categorize the user's bookmark metadata. In fact, the user is normally not even aware of the background processes.

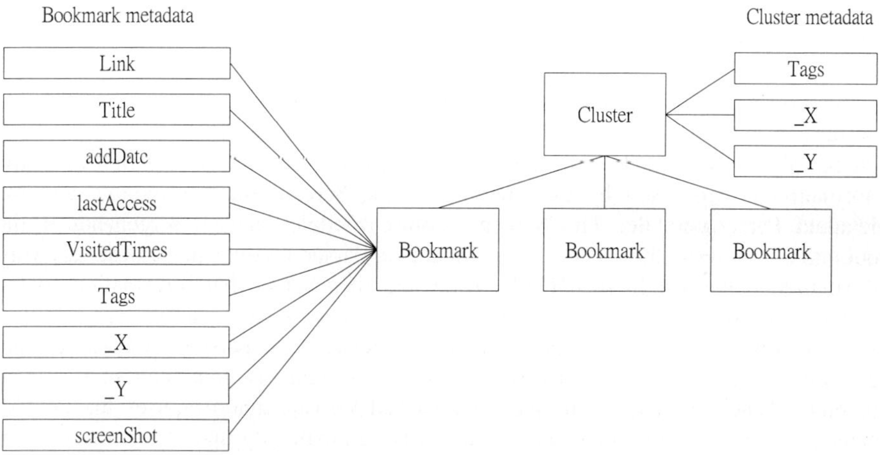

Fig. 1. Document Metadata, All the captured metadata will be saved as a XML format document on the local disk. The XML format is used to keep the cross platform capability of the bookmark information.

3.2 The System Model

The system model describes how the metadata is generated and updated. Using the interface, the user can actively manipulate the bookmark's 2D location and its parent cluster or he can passively change the metadata via his online browsing activities.

Using the 2D graphical user interface, the user can leverage the browsing parameters to dynamically change the information that appears on his screen, thus, avoiding wasted time reviewing irrelevant information.

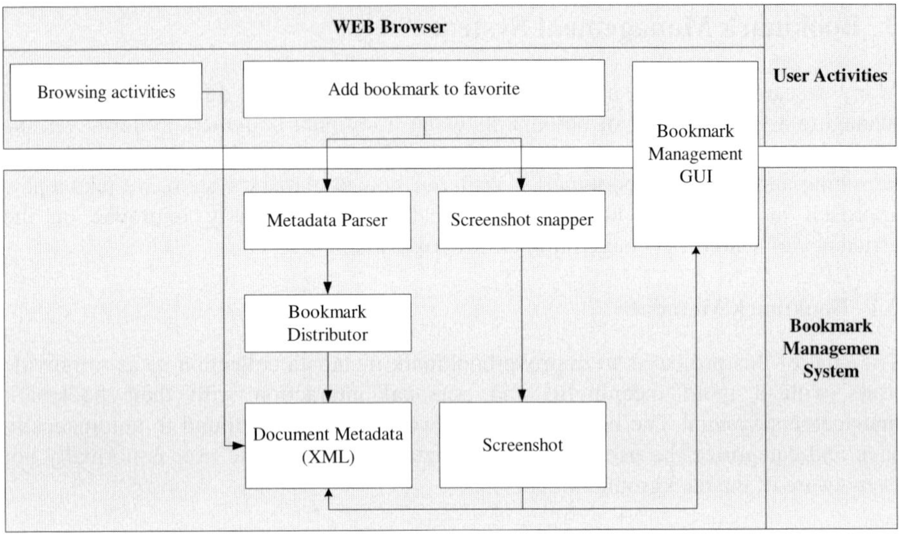

Fig. 2. The System Model

The user's activities such as browsing, adding bookmarks and using the 2D interface to modify the bookmark's metadata, are considered the "input" side of the system. When the website is added to the bookmark collection for the first time, the information is processed by two modules – the Screen Snapper module and the Metadata Parser module. The Screen Capture module takes a screenshot if the bookmark is new to the system. The Metadata Parser module parses the keyword information contained in the HTML code and then sends that information to the Bookmark Distributor. The Bookmark Distributor then suggests to the user where to place the new bookmark based on a comparison between the parsed keywords and the keywords in the collection. Finally, the XML metadata is generated in the local computer. When the user visits the bookmarked website again or uses the GUI to change its attributes, the document metadata is automatically updated to the system. Our system does not include all of the components illustrated in the Fig. 2. For this study, we implemented only the most essential system modules including the bookmark management GUI module and the Metadata module.

3.3 Visualization of Bookmarks and Folders

One of the problems associated with managing large amounts of information is dealing with the amount of time it takes to find the target information. Data Mountain [3] utilizes a semi-3D space which helps the user quickly locate the target information. It functions like human memory in that it helps the user to remember the location of everyday objects in real life. The problem with Data Mountain is that it uses screen shots as visual representations of the bookmarks. When there are hundreds of bookmarks appearing on the screen all at once, the user might find it difficult to locate his target information because of the visual "chaos." Our system utilizes the user's own browsing activities as a parameter for coding the bookmark's visual representation. This method allows large numbers of bookmarks to be shown on screen while still providing useful information about each individual bookmark.

Our system utilizes two parameters to represent each bookmark icon. The two parameters are 1) the last date the website was visited and 2) the total number of times the website was visited. In the first process, all the bookmarks are sorted and then divided into five groups according to the last date the website was visited. Each group is represented by a different icon opacity value, those being 100, 70, 50, 30, and 15 respectively. At the same time that the first process is taking place, the bookmarks are also being sorted by the second parameter - the total number of times the website was visited. This second parameter is represented by different color tones. The folder's visualization is based on the average value of the bookmarks contained within that folder.

Fig. 3. Bookmark Icon Visual Representation

3.4 Information Retrieval

The new bookmark system does not require the user to input precise information in order to view the search results. The filters include the date the bookmark was added, the last time it was visited, the relative total number of times visited and traditional keyword information. The system allows its user to combine any number of filters to help him effectively eliminate unwanted information and reduce the amount of time it takes to find the target information. As an example, for the user who has 200+ bookmarks to search, a forgotten directory or keyword could lead to an extremely time consuming ordeal. With our system, though, much of the irrelevant information is filtered out within seconds.

To further strengthen the connection between the user's bookmarks and his personal life experiences, he is able to add a "Memory Mark" to a time line which represents an important event. Memory Marks, can be used to reference when the website was added and/or when it was last visited. This information search model greatly improves the efficiency in searching a large amount of bookmark information.

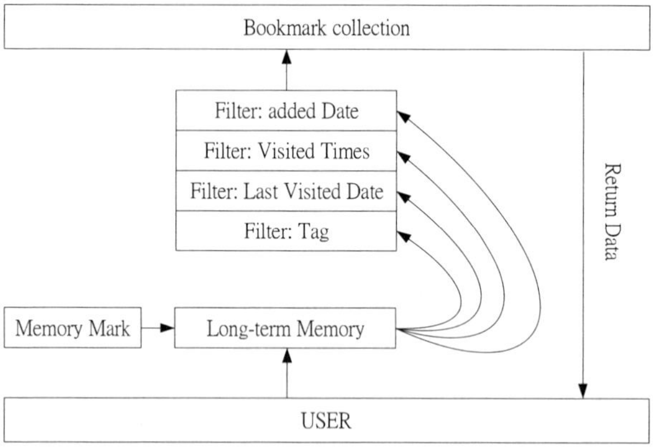

Fig. 4. Bookmark Searching Process

3.5 Graphical User Interface

To increase the usability of the new bookmark system, we incorporated a 2D graphical user interface (GUI). The user is able to use this GUI to get a rough idea of the number of bookmarks he has and how they are organized. The bookmarks are represented by small icons which are coded by the metadata. Another benefit of the 2D interface is that it allows the user to use space memory to organize his bookmarks.

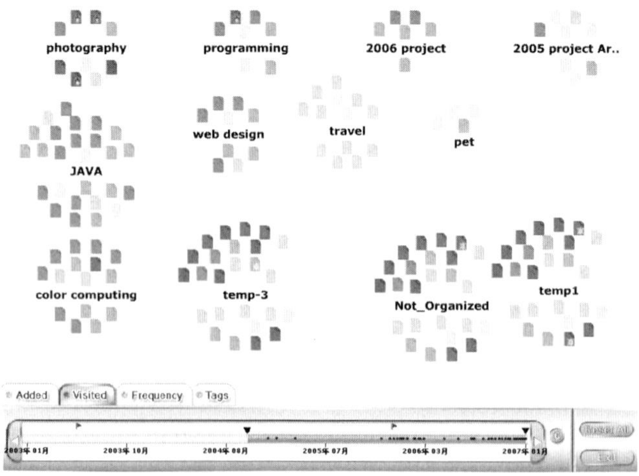

Fig. 5. System Screen Shot

By default, all the bookmarks belonging to the same cluster will gather around a main cluster name (Fig. 5). With the organized bookmark layout, it is very easy to distinguish one cluster from another. The user can also use the drag-and-drop technique to manipulate the position of the bookmarks and clusters.

The timeline contains many black dots which represent the bookmarks on the screen. This allows the user to understand the distribution of the bookmarks along the time line and use it as a reference for adjusting the time threshold. The little red flag above the time line is the Memory Mark which is used to strengthen the connection between time and personal life experiences.

4 Pilot Study and Results

We conducted a pilot study to compare traditional bookmark systems with our new system. In order to conduct neutral experiment, the two bookmark systems used the same bookmark collections and the same directory structure. In the experiment, the users were asked to look for a target bookmark within a specific directory. The users were asked to perform several different searches. They were tasked with searching 5, 10, 15, 25, 45, 65, and 190 bookmarks respectively.

All of the users in this experiment had prior experience with using traditional bookmarks and in using directories to organize bookmarks. There were five male and five female participants. They ranged in age from 25 to 35 years old. For the first experiment, the participants were asked to perform seven different tasks using traditional bookmark systems. The time needed to complete each task was recorded. Before the second experiment, there was a short tutorial on how to use the new bookmark system. The users were given some time to familiarize themselves with the new interface. After they had become familiar with how the new system works, they were asked to perform similar tasks they had performed in the first experiment.

The results of the experiment showed that when the number of bookmarks to be searched were few, the traditional bookmark system slightly outperformed the new bookmark system. Conversely, when the total number of bookmarks to be searched were many, the new bookmark system significantly outperformed the traditional bookmark system in that the users were able to locate the target information much more quickly.

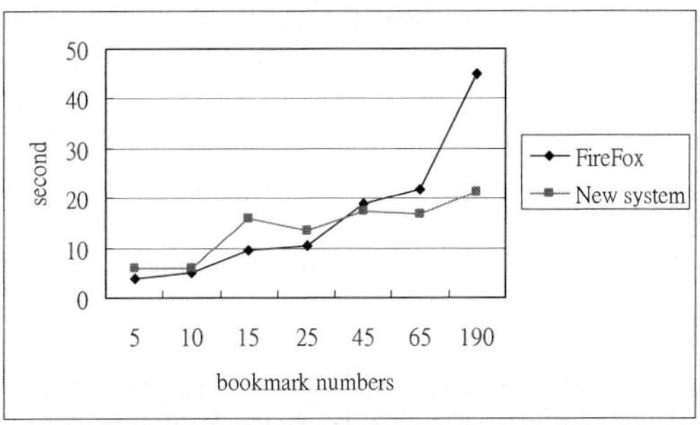

Fig. 6. Average Task Completion Time

5 Discussion and Future Development

In this study, we propose using a form of human-like LTM for managing bookmarks. The new system reduces the amount of information shown on-screen, thus, enhancing search performance. With the new system, users spend significantly less time scanning directories and bookmarks and are, therefore, able to find the target bookmarks more efficiently and quickly. With the new system, there is an increase in mouse movement due to the utilization of the 2D GUI. Also, the new system does require the user to spend additional time adjusting the filters. For these reasons, when searching a small number of bookmarks, the traditional bookmark system had better performance than the new system.

Fig. 6 illustrates the average amount of time it took users to find the target bookmarks using the traditional vs. the new bookmark systems. When using the new bookmark system, there is no direct relationship between the number of bookmarks needed to be searched and the time it took to find them, thus, the participants were able to locate the target bookmarks more quickly. In the pilot study, we also discovered that when a user is searching a large bookmark collection, he is much more likely to bypass the target information when using a traditional bookmark as opposed to using the new system.

In the future, the new bookmark management system's GUI will be further enhanced. Also, our bookmark management method will include document management. The traditional hierarchical directory structure and keyword management will be further incorporated into the user's personal life experience and LTM in order to enhance the overall performance on information search and management.

Acknowledgments. We would like to acknowledge and thank our colleagues at the NCHC for their help during the experimental stage of this research.

References

1. Tauscher, L., Greenberg, S.: Revisitation patterns in World Wide Web navigation. In: Proc. 1997 ACM CHI Conference, Atlanta, GA (March 1997)
2. Feng, Y., Borner, K.: Using Semantic Treemaps to Categorize and Visualize Bookmark Files. Visualization and Data Analysis 2002. In: Proceedings of SPIE, vol. 4665, pp. 218–227 (2002)
3. Robertson, G.: Data Mountain: Using Spatial Memory for Document Management. In: Proc. of the 11th annual ACM symposium on User interface software and technology, San Francisco, California (1998)
4. Li, W.-S., Vu, Q., Agrawal, D., Hara, Y., Takano, H.: PowerBookmarks: a system for personalizable Web information organization, sharing, and management. In: Computer Networks, vol. 31 (1999)
5. Roberts, J., Ruiz, J., Lank, E.: Making Favorites Useful. In: Proc. Of Human-Computer Interaction (2005)
6. (Retrived on December 21, 2006) http://en.wikipedia.org/wiki/Long-term_memory
7. Tulving, E.: Episodic and semantic memory. In: Tulving, E., Donaldson, W. (eds.) Organization of memory, pp. 381–403. Academic Press, New York (1972)
8. (Retrived on December 21, 2006) http://www.alleydog.com/glossary/definition.cfm?term=Episodic%20Memory

Part V

Business Management and Industrial Applications

Business Integration Using the Interdisciplinary Project Based Learning Model (IPBL)

Osama K. Alshara and Mohamed Ibrahim

Information Technology, Higher Colleges of Technology-ADWC, P.O. Box 41012, Abu Dhabi - UAE
osama.al-shara@hct.ac.ae, Mohamed.ibrahim@hct.ac.ae

Abstract. This paper recognizes the need for new approaches in IT based education. Our proposed model depends on two major concepts. The first is to believe in the learning approach not the teaching approach that is widely used. Learning delivery requires providers to alter their in-class approach and demands students to be more independent learners. The second concept is the interdisciplinary learning approach rather than the silo teaching model used currently. For Example, software engineers will not be allowed to work in isolation from the business people, the users, the client, etc. in the real world. Therefore, their college education must qualify them, as much as possible, to deal with the real world. This paper proposes an interdisciplinary project based learning model focusing on collaborative learning (CL) and Business Integration (BI) using the SEE [1] portal. This integral solution is designed to overcome the teaching and silo models used currently.

Keywords: Project based learning, integrated learning, interdisciplinary teams, interdisciplinary based learning.

1 Introduction

Due to the dynamic and innovative nature of the work place in today's business environment, educational institutes are required to provide graduates who fit in this dynamic environment. Calls have been initiated by different sources to reengineer the education delivery system to meet such a task [6], [22]. These calls have come at a time when advances in technology and in e-learning in particular will perhaps allow educators to meet this requirement.

Many studies [8],[10] show that the use of IT in the classroom improves student learning and the way they work [15],[21]. The implementation of technology in education requires innovation by teachers & the institute, and an organizational culture which supports such an implementation. Furthermore, technology must be used in an appropriate situation, be planned, monitored and evaluated against the goals of the educational initiative.

Advances in technology lend themselves to delivery methods such as independent learning, e-learning and project based learning. The author's experience in these methods triggered the desire for a more integrated approach to learning. Projects, especially real life ones, are multi-disciplinary by nature. For example, to work on an

M.J. Smith, G. Salvendy (Eds.): Human Interface, Part II, HCII 2007, LNCS 4558, pp. 823–833, 2007.

e-commerce website, the project team should be an interdisciplinary team of IT, Business and Media students in order to fully develop the business case and the website itself. The danger of developing a business application by IT people only is that the software developers run the risk of not achieving the client business goals and requirements fully.

Both students and faculty face problems of allocating enough projects for the students to work on. Moreover, there are major differences between the use of hypothetical case studies and projects when compared to real life projects [13], [7]. Local organizations are the best source for real life projects, which triggered the idea of establishing a relationship between the college and local organizations as the best source for these projects.

Allocating projects for students to work on is not an easy task let alone allocating real-life projects, which further enhances students learning. Real life projects are multi dimensional; in the sense that they require team based work. Moreover, team members represent different disciplines. Hence, educational institutes adapting the independent and project based learning require: 1) real life projects for the students to work on. 2) Team building mechanism. 3) Interdisciplinary students and teachers formation. 4) And strong support and participation from the local community. Hence, a mechanism has to be proposed and established to bring about an integral relationship between the higher education institutes and the local community.

Any one of the above requirements can be an obstacle of its own. This paper proposes an interdisciplinary project based learning model using the SEE [1] portal. This integral solution is designed to overcome these obstacles. See stands for **S**tandards, **E**nhancement, and **E**mpowerment, which depicts the students' life cycle at any educational institute. To join a higher education institute, a student must meet certain standards. The program of study is designed to enhance the students' knowledge etc, which should lead to an empowered student with a certificate in a particular field.

The SEE project [1] was proposed to serve the students' needs to implement real life projects as assessed course work tasks. This portal brings the students, instructors and local organizations together. Local organizations are required to register as members in the SEE portal before they can load their projects. Each project must include, among other information, description of the project, deadline date and contact person. Furthermore, organizations can browse through the students' innovative ideas/projects and ask to sponsor and implement a project at the organization.

Students would post their ideas/projects, browse posted projects by local organizations to find a suitable project to implement, bid for it and get the instructor's approval on the project. This online process would trigger direct communications between the students, instructor and the organization's representative to implement the project, which will be used by the organization.

Abu Dhabi Women's College http://adw.hct.ac.ae is part of the Higher Colleges of Technology (HCT) http://www.hct.ac.ae in the United Arab Emirates, and supports the independent and project-based learning approaches. Many initiatives have been taken system-wide by the HCT to enhance its educational delivery methods (e.g.. Laptop integration, e-learning strategy, Centers of Specialization). Senior management support for these early initiatives has further encouraged faculty and line management to generate new & innovative ways to utilize technology in the classroom.

2 Impact of Technology on Learning Methodologies

Organizations are increasingly paying attention to the concept of organizational learning in order to increase competitive advantage, innovation, and effectiveness. "Organizational learning is a competence that all organizations should develop" [3], and is defined as "the acquisition, application, and mastery, of new tools and methods that allow more rapid improvement of processes whose improvement is critical to the success of the organization" [19]. Educational organizations must learn how to employ and implement new technologies to improve the delivery of their learning methodologies.

Most organizations in today's work practices are constantly faced with environmental changes to which they must adapt. There is also increased pressure on these organizations to produce employees prepared for the knowledge-driven economy, and to serve as both a source and a producer of knowledge. Educational institutes must always pay attention to such environmental changes since they are the main human resource provider to the market.

Issues about the relationship between information technology and learning methodologies are becoming more critical as IT is becoming a focal point in organizations. The author asserts that in order to understand the role of IT in learning methodologies, educational institutions need to focus on the relationship between learning theories and understanding technology. Technology (especially information systems) can help achieve effective learning, hence improving students' and teachers' overall performance. Information systems can facilitate the learning process by supporting the processes of knowledge management and information distribution, sharing and communication [4]. These processes are very well represented in the educational concepts of:

- *Peer tutoring*: where students learn from each other directly;
- *Learning by example*: students learn from other students' work that was completed previously for the same or a similar course;
- *Independent learning*: this method requires students to access different resources such as peers, libraries, Internet, teachers etc. to acquire information and knowledge. Hence, the educational institute must facilitate access to such resources by its students;
- *E-learning*: promotes independent learning, both synchronous and asynchronous learning. Access to the technology allows for anytime, anywhere learning;
- *Project based learning*: Among other things it promotes team work, information sharing and communication skills [20], [7]
- *Market driven learning*: this is a new concept that we introduce here to emphasis the integral relationship needed between the local organizations (as the potential employer of graduates) and the educational institution that delivers the content [20].

- *Integrated learning*: based on the demand of today's job market, students must have well rounded knowledge about other fields to be able to work together in teams and communicate efficiently and effectively.
- *Interdisciplinary learning teams*: To be able to provide project based learning and integrated learning you need to form interdisciplinary learning teams of both students and teachers. It is through these teams that members learn the needs of the other disciplines and how to integrate their overall knowledge to form an integrated solution.

3 Interdisciplinary Team Based Learning

Multidisciplinary teams are not new in Education, and in fact have been used widely in the Health industry [5] and within large corporations such as Boeing, DEC and Chrysler [18]. The value of interdisciplinary teamwork is being recognized in many fields. In the educational setting, both students and teachers felt the interdisciplinary approach was a worthwhile and enlightening experience [7]. However, improvement and some fine-tuning of the different models are required.

"Utilizing interdisciplinary teams provides a more realistic development environment, but also comes with many challenges" [18] such as 1) project ownership and 2) team formation - who is the driving force behind this team formation?

3.1 Project Ownership

The problem of claimed ownership between participants from different disciplines affects the intra-team communication, consequently affecting the success rate of the overall experience. In [18] they considered the business professor as the real client and students as the developers. In our model, we use the outsourcing method, where a student **A** of any major would contract the project with a real industry-based client. In turn, this student **A** - being the primary contractor - would identify the help/skills that she requires to outsource by teaming up with students from other disciplines. This model ensures project ownership clearly lies with student **A**.

3.2 Interdisciplinary Team Formation

Most of the literature showed that teachers were the driving force in building the interdisciplinary teams. There appeared to be two primary methods of building interdisciplinary teams:

1. By synchronizing two interdisciplinary courses with 'obvious' synergies [18]. For example, the Business Dept course 'e-Business' has obvious synergies with the IT Dept. course 'Web Development I'; or
2. By one or more teachers delivering a course that is common across different program disciplines [16]. For example, an 'Introduction to Computing' course that is a core requirement in semester 1 of all programs.

The models that are illustrated by the literature show the formation of predefined interdisciplinary groups [16], [2], [12]. By this we mean that members of the groups were assigned by the teachers, based on their knowledge of the students' academic level, or conveniently grouped by virtue of the fact that the students were in the same class. To the contrary, our model supports non-predefined group formation. Depending on the amount of work needed for each task, student sub-teams can also be formed for each task. For example, if the development is a major task that requires more than one student to execute, an IT sub-team can be formed. Each student or sub-team then assumes a pivotal role of communicating the results from the sub-team back to the primary team. Thus, proper governance and effective communication are critical for project success.

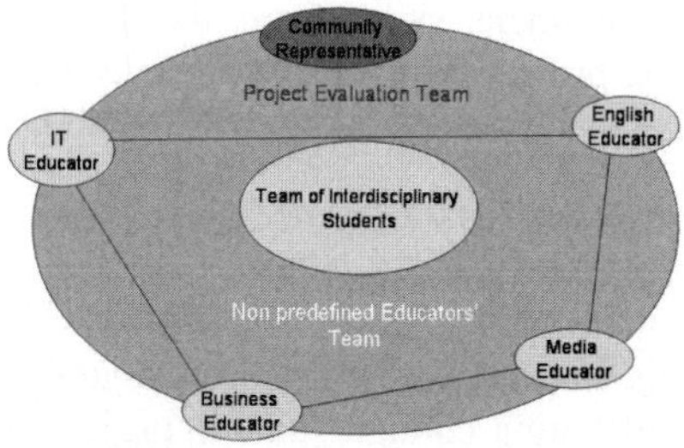

Fig. 1. Multi-interdisciplinary teams' formation

As depicted in Figure 1, our model allows for the formation of the following non-predefined teams:

1. Students' sub-team: students of the same major working on a major specific task
2. Students' team: students from different majors working on the same project
3. Educators' team: includes the teachers (from different disciplines) of the students' who worked on the project.
4. Project team/Community team: includes all parties involved in the development of the project: students, teachers and the project sponsor (usually from industry).

Linder [14] states that "Students in general like to work in teams to the contrary of teachers". Hence our challenge was to develop a model where the interdisciplinary team based learning would not be teacher driven, but rather student driven. Our model focuses on the students' teamwork regardless of the teachers' teamwork. It makes the students the center of the team, which in turn helps promote the formation of the educators' team. A non-predefined team of educators, that foster an environment where educators can formally or informally meet, leading to increased interdisciplinary cooperation would be the result of such formation. As a consequence, our model overcomes many of the obstacles faced when forming interdisciplinary educators' teams [14]. Having the presence of an outsider in the team, in the form of a project sponsor, also tends to bring the educators closer to present a unified image of their institute.

Students working within the SEE portal form their own groups as described earlier. Figure 2 further illustrates this concept by describing an e-Commerce project which is currently being undertaken. The skill-set needed to successfully implement this

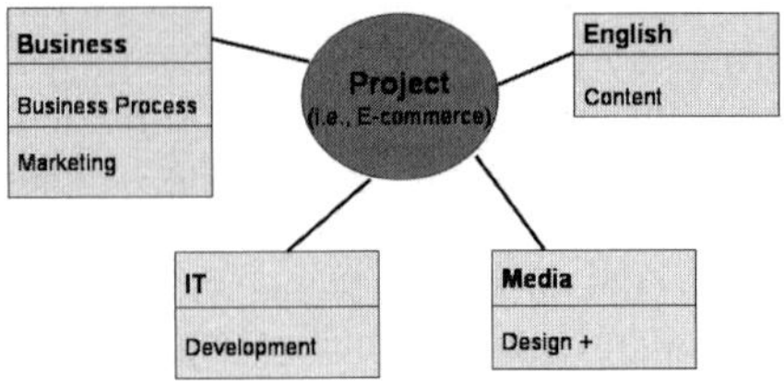

Fig. 2. Interdisciplinary project based learning format

project, include web development skills from the IT Department, business and marketing knowledge to be provided by the Business Department, graphic design by the Media Department, and content revision by the English Department.

4 Project Based Learning Using See

One of the major problems that instructors in our college face, especially in software development courses, is allocating enough suitable projects for students to work on. This is a major requirement when adopting independent and project based learning, which our college does in many advanced courses. So teaming up with the local community, as a rich source for such projects, works as a great idea that serves many objectives. Some of these objectives are real life project based learning, integration between the college and the local community, takes away some of the burden off of the teacher who constantly needs to come up with new projects, encourages students' innovation, etc.

We find the SEE portal to be a suitable virtual environment to implement the interdisciplinary teams. Most higher education institutes are structured, both physically and logically, in a secular manner. This means that colleges and even departments within the same college are not interweaved, rather are secular. Forming interdisciplinary teams in such environment is very difficult. Changing the structure and attitude of higher education institutions is an immense job if not impossible. Therefore, creating a virtual environment is a much easier approach.

5 SEE Structure

The SEE portal is a multi purpose portal that serves different requirements of any higher education institute. The main points that we are focusing on at this stage of the system are the following:

Real life projects, Learning by example, Recruitment, and *Events*. However, the recruitment part of the system will not be piloted at this stage due to other existing college plans for a recruitment portal.

SEE is a multi purpose portal that supports real life project based learning, learning by example, alon with other added value services. The OM (Organizational Memory) part of SEE employs the case based method. A literature review shows many OM and

OL (Organizational Learning) based systems that use this method such as HICAP [15], SEED [16] and BORE [17].

Project based learning (PBL): SEE is divided into three major components:

1. The Portal: This is the host of all the activities that take place between the actors. The portal houses the innovative ideas proposed by the students to be adopted by organizations, and projects proposed by the organizations to be booked and implemented by the students as shown in Figure 3.a.
2. Actors: are the students, instructors and the local organizations that interact and communicate through the portal.
3. The OM: it is the repository of all finished projects as shown in figure 3.b.

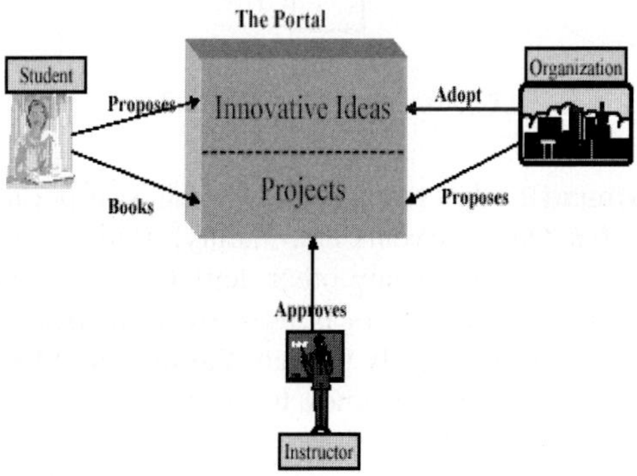

Fig. 3a. Project based learning

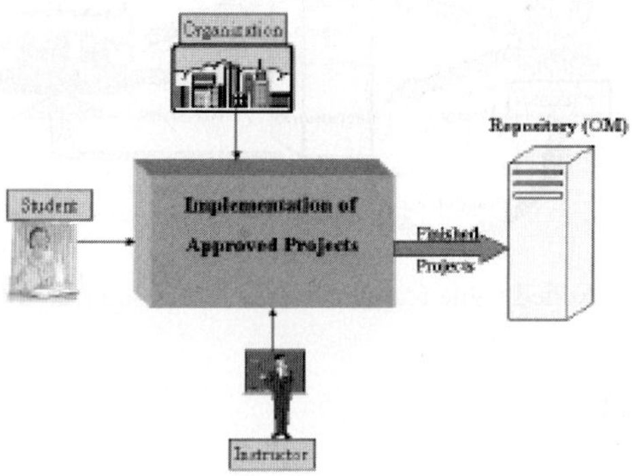

Fig. 3b. Project based learning with the OM

Learning by example: Another major issue that we face with our students is their need to view similar examples to their assignments. It is fairly easy to provide few examples to a specific problem or scenario. However, when dealing with different scenarios, an instructor does not have the time or, maybe, the resource to provide enough examples to each scenario.

It is a fact that students respond well to the learning by example method. Hence, the SEE portal builds a repository of examples to be used by both faculty and students in future work. Those examples are simply the finished projects worked on by previous students and stored in the OM. Teachers are very familiar with these stored projects and should be able to direct the current students to finished projects, see figure 4, as examples to be used in implementing the current tasks.

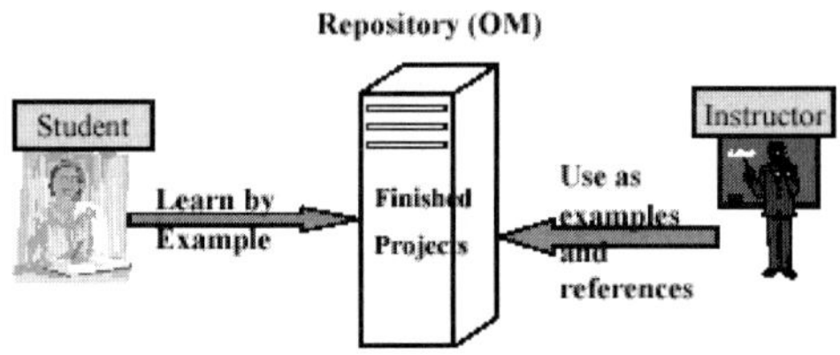

Fig. 4. Learning by example

Added value (recruitment): other focus issues for the SEE portal, as shown in figure 1, are the *Events* and *Job Offers*. Events are, mainly, used by the local organizations to list any conferences, seminars or any other activities that would be of interest to students as well as teachers. Job Offers is used by local organizations to post any vacancies, where the students can apply to them through the SEE portal directly. The students also use it to post their resumes to be viewed by the organizations for possible recruitment, as shown in figure 5.

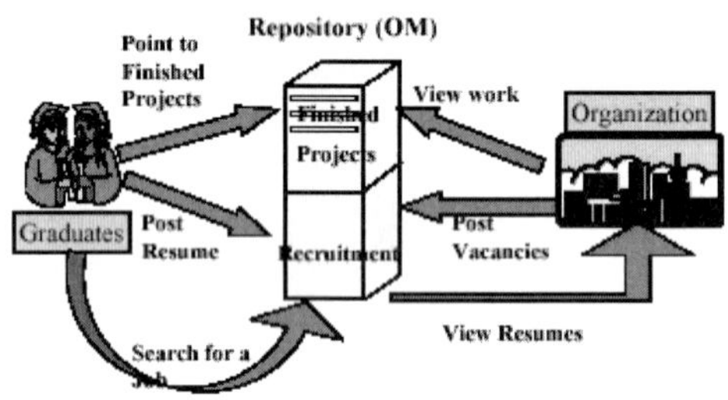

Fig. 5. Added value features of the SEE portal (recruitment)

6 SEE the TOOL

SEE is a web-based portal that was implemented using visual basic.net, ASP.net, HTML and Access as the back end database. However, we realize that Access would need to be replaced with a more powerful database such as SQL, Oracle, etc.

6.1 Users' Roles

The member student can post an innovative idea or search for a posted idea by a company. The student would go through the search results (figure 6) and read through the project details (figure 7) results of the search, book an idea and designate the course and teacher for approval. Once the student receives the approval from the teacher she can start working on the project. The same is true of a company that showed interest in a student's posted innovative idea.

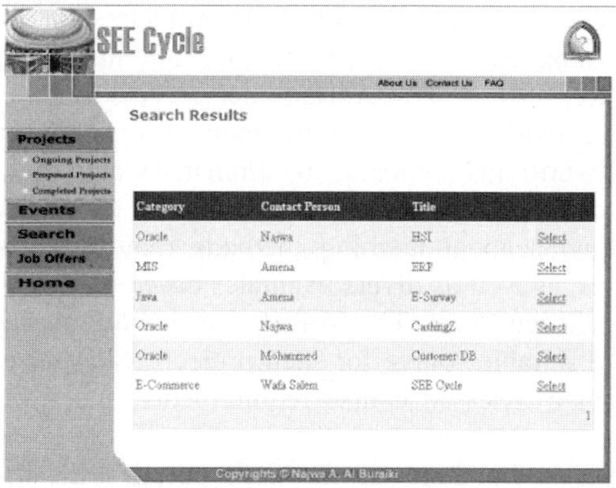

Fig. 6. The search results

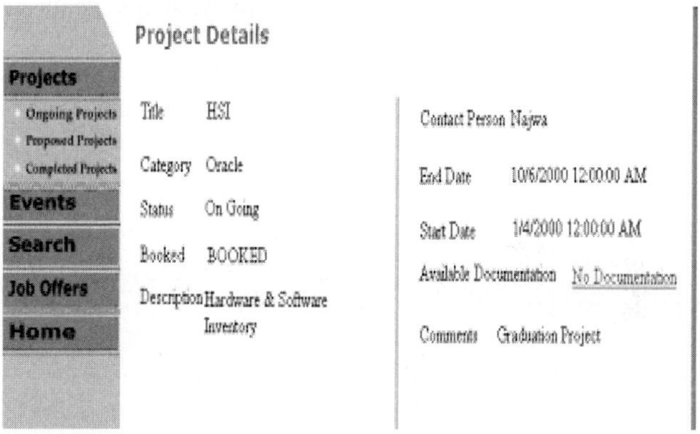

Fig. 7. Project details of the search results

Once the teacher logs into the portal, he gets a list of the booked projects by his students, which he needs to approve or disapprove the booking based on the course and assessment goals. The teacher also needs to be part of the communication between the student and the organization. On conclusion, the teacher assesses the student's work and rewards a grade.

A member organization posts project ideas for students to work on. The company's representative must work with the student and supervising teacher to assure the accurate completion of the project. Organizations also can book an innovative idea to be adopted by them. On conclusion of the project, the organization's representative needs to fill an evaluation form for the student.

7 Conclusion and Future Work

This paper sets out to espouse the benefits of using technology to enhance software engineering education. In particular, it advocates the adoption of integrated project based learning (IPBL) in an interdisciplinary environment, which very much describes the nature of software development projects. Our model illustrates the benefits of self-defining teams of both students and instructors. Our preliminary conclusions regarding IPBL and interdisciplinary teams is that together these two delivery strategies provide a rich environment for collaborative learning, the application of real-world (as opposed to simulated) activities and a chance for students to experience the positive & negative aspects of teamwork.

Project based and independent learning methods require teachers to provide the students with projects, as well as divers examples based on the given projects, which proved to be a very difficult task. The SEE system is built around the fact that local organizations are the suitable source for such projects. SE students work on real life projects as part of their assessed course requirements, and interdisciplinary project based learning.

More work is yet to be done on the SEE portal. SEE needs to be partially redesigned to allow for:

1. multi-team communication environments: This is to tailor for virtual meetings between the Educator's team, students sub-team, students' team and the project team.
2. the ability to break the project into identifiable tasks for outsourcing purposes: Since we established the concept of ownership and granted this ownership to a student, hence, this student needs to divide this project into identifiable (multi-disciplinary) tasks to be outsourced to other students (possibly from other disciplines).

References

1. Alshara, O.: SEE: A Project Based Learning Initiative To Integrate ADWC With The Local Community. In: (SETIT 2005) International Conference: Sciences of Electronic, Technologies of Information and Telecommunications, TUNISIA (2005)
2. Anewalt, K.: Utilizing Interdisciplinary Teams in Teaching E-commerce. Consortium for Computing Sciences in Colleges (CCSC) (2003)
3. Argyris, C.: On Organizational Learning, Blackwell Publishers (1999) http://www.monitor.com/binary-data/MONITOR_ARTICLES/object/94.pdf
4. Balasubramanian, V.: Organizational Learning and Information Systems, ISWorldNet (1995) http://www.e-papyrus.com/personal/orglrn.html
5. Carlton, B.: The Role of the Health Educator in Interdisciplinary Health Team Development: An Organizational Development Strategy. Health Education 15(6), 13–15 (1984)
6. Derntl, M., Renate, M.-P.R.: Engineering e-learning systems (ELS): Patterns for blended, Person-Centered learning: strategy, concepts, experiences, and evaluation. In: Proceedings of the, ACM symposium on Applied computing, pp. 916–923 (2004)

7. Fernandez, E., Williamson, D.M.: Software development: Using project-based learning to teach object oriented application development. In: Proceeding of the 4th conference on Information technology education (2003)
8. Fisher, D., Stolarchuk, E.: The effect of using laptop computers on achievement, attitude to science and classroom environment in science. In: Proceedings Western Australian Institute for Educational Research Forum (1998), Available from http://www.education.curtin.edu.au/waier/forums/1998/fisher.html
9. Flemming, U., Coyne, R., Snyder, J.: Case-based Design in the SEED System. American Society of Civil Engineers. In: The Proceedings of First Congress on Computing in Civil Engineering, Washington, D.C (1994) http://seed.edrc.cmu.edu/SD/asce-dist.html
10. Griffith, R.A., Gu, Y., Brown, D.G.: Assessment of the Impact of Ubiquitous Computing. The Annual Forum of the Association for Institutional Research (39th, Seattle, WA 1999)
11. Henninger, S.: Case-Based Knowledge Management Tools for Software Development. Journal of Automated Software Engineering 4(1), 319–340 (1992)
12. Johnson, P.M., Moffett, M.L., Pentland, T.: Lessons learned from VCommerce: A virtual environment for interdisciplinary learning about software entrepreneurship. Communications of the ACM 46, 12 (2003)
13. Levitt, B., March, J.: Organizational Learning. Annual Review of Sociology 14, 319–340 (1988)
14. Linder, A., Ibrahim, A.: Issues in technical communication education: Building cross-disciplinary teams in higher education institutions. In: Proceedings of IEEE professional communication society international professional communication conference and Proceedings of the 18th annual ACM international conference on Computer documentation: technology & teamwork (2000)
15. Lowther, D.L., Ross, S.M., Morrison, G.R.: Evaluation of a Laptop Program: Successes and Recommendations. In: Building on the Future. NECC 2001: National Educational Computing Conference Proceedings (22nd, Chicago, IL 2001)
16. Mosiman, S., Hiemcke, C.: Interdisciplinary Capstone Group Project: Designing Autonomous Race Vehicles. In: SIGCSE, Austin Tx, USA (2000)
17. Muñoz-Avila, H., Aha, D.W., Breslow, L.A., Nau, D.: HICAP: An Interactive Case-Based Planning Architecture and its Application to Noncombatant Evacuation Operations. In: AAAI/IAAI-99, AAAI Press, Stanford (1999)
18. Rogers, R.L., Stemkoski, M.J.: Reality-Based Learning and Interdisciplinary Teams: an Interactive Approach Integrating Accounting and Engineering Technology. Educational Resources Information Center (ERIC) (1995)
19. Schneiderman, A.M.: Measuring Organizational Learning (2000) http://www.schneiderman.com/The_Art_of_PM/measuring_learning/learning.htm
20. Tan, J., Phillips, J.: Challenges of real-world projects in team-based courses. The. Journal of Computing in Small Colleges 19, 2 (2003)
21. Tiu, F.S., Guglielmi, J.P., Walton, W.G.: Assessing a Technology Initiative: Lessons Learned While Integrating Technology into Teaching & Learning. In: Paper presented at the Annual International Forum of the Association for Institutional Research (42nd, Toronto, Canada 2002)
22. Vrasidas, C.: Engineering e-learning systems (ELS): Issues of pedagogy and design in e-learning systems. In: Proceedings of the 2004 ACM symposium on Applied computing (2004)

Business Insights Workbench – An Interactive Insights Discovery Solution

Amit Behal[1], Ying Chen[1], Cheryl Kieliszewski[1], Ana Lelescu[1], Bin He[1], Jie Cui[2], Jeffrey Kreulen[3], James Rhodes[1], and W. Scott Spangler[3]

[1] IBM Almaden Research Center, San Jose, California, USA
{abehal, yingchen, cher, lelescu, binhe, jjrhodes}@us.ibm.com
[2] IBM China Research Lab, Beijing, China
cuijie@cn.ibm.com
[3] IBM Almaden Research Center, San Jose, California, USA
{kreulen, spangles}@almaden.ibm.com

Abstract. Today's businesses increasingly rely on vast amount of information. Yet effective use of information is becoming more and more difficult. This paper describes a general purpose analytics solution, **"Business Insights Workbench" (BIW)**, which embeds two major classes of information analytics techniques and a unique set of visualizations to mine the available information and uncover critical business insights and enhance business performance. The two major classes of analytics technologies include the "taxonomy" analysis and the "relationship" analysis to facilitate understanding and detection of hidden concepts and patterns buried in the information respectively. The BIW technologies have been successfully applied in many application domains, e.g., Customer Relationship Management (CRM) for customer satisfaction analysis, Intellectual Property (IP) for patent portfolio analysis and licensee identification, and Healthcare Life-sciences (HCLS) for facilitating drug discovery by identifying the relationships among chemicals, DNA, proteins, drugs, and diseases. We show some BIW sample applications in this paper.

Keywords: Information Mining, Human Interaction, Visualization.

1 Introduction

As the amount of information available to business users is mounting, information mining technologies that discover hidden patterns and generate significant business insights become critical. Traditional approaches in On-Line Analytical Processing (OLAP) and Business Intelligence (BI) solutions [1, 2, 3, 4, 5], have started to address such challenges by mining structured data in relational databases (RDBMS) to find hidden associations, e.g., the relationship between the sales of a given product and buyer's gender, location of purchase, and time of purchase.

However, structured information accounts for only a small fraction of the total information size. The prominence of unstructured data such as emails, instant messages, and various forms of documents (e.g., Word, PowerPoint, PDF, WWW) demands for new information analytics approaches. In the past few years, a wide

M.J. Smith, G. Salvendy (Eds.): Human Interface, Part II, HCII 2007, LNCS 4558, pp. 834–843, 2007.

variety of text mining techniques for unstructured data have emerged, including smart information retrieval [6, 7], Natural Language Processing (NLP) to extract semantic entities out of text (also called "annotation") [8, 9, 10], clustering, classification, and taxonomy generation [11, 12].

Business Insights Workbench (BIW) goes beyond traditional OLAP/BI and text mining technologies by combining both structured and unstructured analytics in one platform. Such combination can significantly improve the quality of the insights discovered. For instance, structured data such as the product sale price, locations, and buyers would provide context to customer satisfaction surveys that may be in an unstructured text format. By analyzing the structured and unstructured data together using techniques such as taxonomy generation and co-occurrence analysis, one can gain insights in the relationship between price, location, and customer satisfaction.

Furthermore, unlike most of the existing OLAP/BI and text mining solutions that attempt to provide instantaneous analytical results to users with a few mouse clicks or in a fully automated fashion, BIW guides users through an interactive analytical process while incorporating users' domain knowledge whenever possible. Such an interactive platform is critical because typically data can be viewed from many different perspectives, depending on business needs. There is often no one fixed view of the data that would definitively lead to the most significant set of insights immediately upon interaction with the tools and the data. Often the insights are deduced and formulated by combining multiple informational perspectives derived from different analytical steps, while leveraging users' domain expertise wherever possible.

BIW incorporates a suite of tools to allow capturing of users' domain knowledge and utilizing them to guide and refine the analysis such as interactive editing of taxonomies, saving of intermediate analytics results, examining, and visualizing data and analytical results in different views throughout the analysis process.

To ensure simple interactions and precise presentations without lowering the quality of the insights, BIW embeds a graphic user interface (GUI) which anchors the text mining output on two major classes of insights that are generated by BIW, called "*taxonomies*" and "*relationships*" (see Sections 3 for details). Use of these classes then allow for nesting of information and grouping of relational views that allow the user to navigate through different data types (i.e., 1-dimensional to multi-dimensional) while staying within the context of the inquiry and affording the identification of interesting patterns to derive insights. Such an overall interaction approach is highly adaptive to different business environments and can significantly improve the quality of the insights, when compared to existing solutions.

In the reminder of the paper we discuss the overall BIW analytics processes in Section 2. Section 3 details the information analytics techniques and interactive visualization technologies employed in BIW. A patent licensing analysis is used as an example throughout the discussion to illustrate the approaches. Conclusions are provided in Section 4.

2 BIW Analytical Process

BIW guides users through a flexible analysis process by utilizing various analytical and visualization tools embedded in the system. The process typically contains an initial *exploratory phase*, in which a user queries, selects, and extracts information about an area of interest from an information warehouse, then an *understand phase* to help users to extract and understand the concepts embedded in the extracted document set, and finally, an *analyze phase* to identify and discover hidden patterns and relationships among the discovered concepts and other dimensions of data such as those represented by the structured fields. Fig. 1. shows the three-phase analytics approach BIW employs and the major functionalities available in each phase.

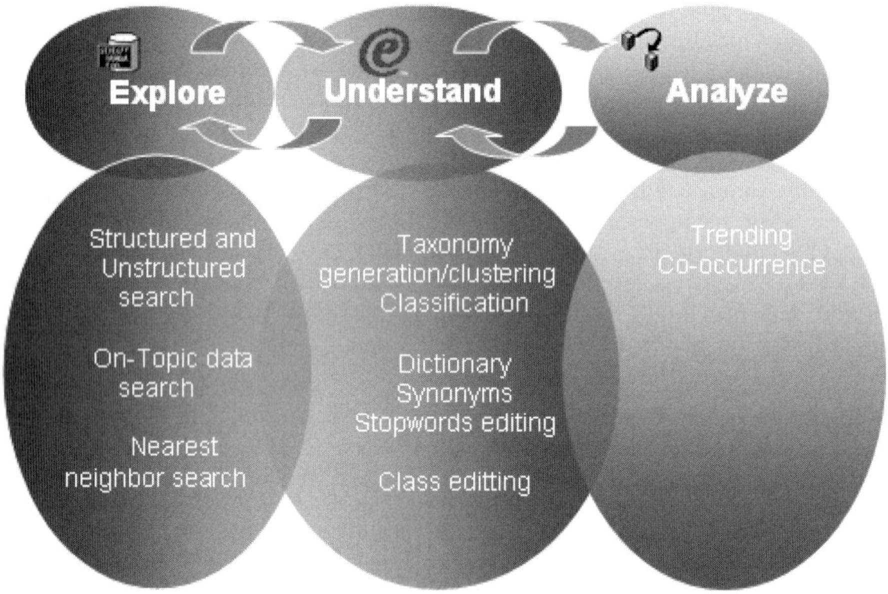

Fig. 1. BIW's three-phase analytical approach and the key functions in each phase

In a patent licensing analysis scenario, let us assume that a given company C wishes to understand the patent portfolio around a specific area, e.g., carbon nanotube, and derive Intellectual Property (IP) licensing strategies for such patents. To start such analysis, users can utilize BIW's "*explore*" tools to search and extract a set of seed patents related to carbon nanotube. For instance, users can extract patents based on structured fields or based on keyword/text searches from a patent warehouse. Such a patent warehouse can be built by acquiring or search patent information from agencies such as US Patent and Trademark Office (USPTO). Such

Table 1. Examples of structured and unstructured fields in USPTO patents

Structured Fields			
Inventors	Filing Date	Assignee	Citations
Unstructured Fields			
Patent title	Abstract	Claims	Text body

patent information typically contains both structured and unstructured fields as shown in Table 1 using some sample fields from the patent data.

The result of the "*explore*" step is a set of documents, called *seed documents*. In general, the seed documents can be produced as any combination of structured and unstructured field search query. The seed documents are meant to capture a sufficient amount of relevant information to allow meaningful subsequent analysis. Since the subsequent analytics steps will weed out irrelevant information as needed, the seed documents do not have to be extremely precise to begin with.

To aid the understanding of the extracted documents, an *understand phase* is carried-out. During the *understand phase,* a taxonomy generation technique is used to analyze the documents' text (unstructured data) and cluster the documents. These clusters represent naturally occurring concepts or themes revealed by the document contents. They also help users to categorize the documents. Without such taxonomy generation techniques, analysts must read each document in order to gain such insights. The technique is described in detail in Section 4. A document classification based on the pre-defined structured fields, such as filing date or inventors, can also help users to understand other dimensions of data. In BIW, naturally generated taxonomy based on clustering and classifications based on structured fields are generally referred to as *taxonomies*.

Table 2 shows a subset of the clusters from the naturally generated taxonomy for the carbon nanotube patents based on patent text content. Clearly, since the naturally generated taxonomy is based on the text content, it can reveal some unexpected categories/concepts, e.g., golf clubs, fuel cells, and diamonds. A deeper analysis shown later will explain the root cause of such results. It suffices to say that such concepts may not have been easily visible if they were not derived using a "content-driven" taxonomy generation approach.

Table 2. The carbon nanotube patent taxonomy with sample clusters

Golf clubs
Carbon layers
Composite materials
Electron emission source
Fabric
Field emission display
Fuel cells
Diamond

Taxonomies allow users to view information from different angles individually. To discover hidden patterns and relationships across multiple taxonomies, additional analytics techniques are needed. This is done in the *analyze phase*. In this phase, co-occurrence techniques are used to compare any two or more taxonomies, e.g., assignee's taxonomy against the naturally generated taxonomy, to identify what companies are highly associated with what kinds of patents. Furthermore, a trend analysis can associate the time dimension with the naturally generated taxonomy to understand the changes of the concepts over time.

In our carbon nanotube patent example, we performed the co-occurrence analysis between the carbon nanotube taxonomy as shown in Table 2 and the assignee taxonomy. The result showed that Callaway Golf is highly related to the Golf clubs category. A deeper look at the patents in such intersection showed that some very advanced carbon manufacturing technology has been used in golf shafts – a rather interesting and surprising insight.

Throughout the *"explore", "understand",* and *"analyze"* phases, users can zoom in and out between the global perspective created by the high level taxonomies and co-occurrence analysis as well as detailed perspectives created by drilling down into individual categories or clusters of documents. All intermediate analytical results can be saved for later refinement or reuse. Users can also edit the naturally generated taxonomy by splitting, merging clusters, or moving documents among clusters based on users' domain knowledge. Such a three-phase analytical approach helps the analyst to understand various aspects of an information set.

3 BIW Core Information Analytics Techniques

BIW contains two main classes of analytics capabilities: taxonomy generation and relationship analysis. We describe them in details below.

3.1 Taxonomy Generation

Taxonomy generation is typically done in the *understand phase* to examine information in different dimensions. The naturally generated taxonomy using clustering techniques based on document content can be especially interesting as unexpected categories may surface and bring new insights, even if the documents are already classified by some pre-defined structured fields, e.g., US Patent classes. This is because the structured fields may not accurately represent the information represented by the content or may not correctly reflect the specific business context. Machine-aided content analysis using text mining techniques will extract information naturally based on the content of the documents and thus reflecting categorization and concepts that are more relevant.

To generate a taxonomy based on document content, BIW's *taxonomy generation* approach first assigns each document in the seed document set a numeric vector that corresponds to its words, phrases, and structured information; then, classifies the documents into appropriate categories (also called *document clusters* or *classes*) using a clustering technology. Where, each document cluster represents a set of concepts common to the documents in that cluster such as those shown in Table 2 for carbon nanotube patents. BIW contains a suite of clustering algorithms that can be used under different conditions, such as K-means and intuitive clustering as described in [11] and [13] in details. We do not repeat in this paper.

BIW's taxonomy generation is interactive and allows the analyst to refine clusters and the taxonomy using various interactive editing tools. For example, a domain expert can edit the dictionary, stop-words, synonyms based on their domain knowledge. For instance, a company's product names could be included in the dictionary explicitly to avoid machine-based algorithms to filter them out. Users can also merge and split classes, or move documents from one class to another based on their expertise. BIW facilitates users with such operations by incorporating appropriate visualization tools and presenting various statistics along the way. [14] and [15] detail BIW's interactive taxonomy generation methodologies and tools.

Besides naturally generated taxonomy, other dimensions of information derived by structured fields can also be valuable, e.g., inventors, filing dates, and assignees fields in the patent warehouse. By categorizing documents based on those dimensions one can gain additional insights. Table 3 shows an assignee taxonomy with a subset of the assignees categories for the carbon nanotube patents. The classification indicates who might be the key owners of the carbon nanotube patents. With taxonomies created from both structured and unstructured fields the user begins to gain a more comprehensive understanding of what information is available and what the documents contain. Similar techniques can be applied in a finer granularity by selecting and refining individual classes in each of the taxonomies.

Table 3. The carbon nanotube patent assignee taxonomy (sample categories)

Aerospace Corp
Callaway Golf
3M Innovative Properties
Goodyear
Yokohama Rubber Co.
AMOCO
Canon

3.2 Relationship Analysis

Once the individual taxonomies are produced as described above, the next level of analysis is to understand various relationships among different dimensions and identify hidden associations, as typically carried out in the *analyze phase*. BIW contains two types of relationship analysis: *co-occurrence analysis* and *trend analysis*.

The *co-occurrence analysis* takes any two taxonomies produced in the *understand phase* and calculates affinity levels between the categories in the taxonomies. The result is a 2-dimensional co-occurrence matrix. Fig. 2 shows a portion of a co-occurrence matrix using the carbon nanotube naturally generated taxonomy and an industry taxonomy. Due to the space limitation, we do not show the full matrix, which is quite large. In this example, the industry taxonomy is generated by mapping the individual assignees of the patents to a pre-defined industry taxonomy, which includes aerospace, automotive, computers, etc., as shown in Fig. 2.

The values in the cells indicate the number of documents that occur at the intersection of an industry and a naturally generated category. In addition, each cell has an expected value that can be calculated based on the size of the category and the total number of documents that the category is included in. If the expected value is exceeded by the actual value in the cell, then the cell is highlighted. The highlighting indicates the degree of significance of the cell's value, which is calculated using a statistical test such as Chi-Square [16]. In other words, highlighted cells indicate a significant relationship between the two categories. Darker shading indicates a stronger relationship. For instance, automotive industry appears to be closely related to patents that are focused on "*carbon fibers*" and "*conductive materials*".

Very High Affinity = Moderate Affinity = Low Affinity = No Affinity =

	Size	Miscellaneous	aerospace	automotive	biotechnology-...	chemicals	computers
Miscellaneous	1862	1630 (19....	32 (16.93%)	38 (19.90%)	1 (20.00%)	52 (7.47%)	20 (14.29%)
batteries	138	88 (1.07%)	0 (0.00%)	6 (3.14%)	0 (0.00%)	10 (1.44%)	0 (0.00%)
carbon fiber bundles	70	45 (0.54%)	1 (0.53%)	0 (0.00%)	0 (0.00%)	19 (2.73%)	0 (0.00%)
carbon fibers	760	586 (7.09%)	22 (11.64%)	30 (15.71%)	0 (0.00%)	88 (12.64%)	16 (11.43%)
carbon layers	1181	973 (11.7...	18 (9.52%)	17 (8.90%)	1 (20.00%)	62 (8.91%)	20 (14.29%)
composite materials	478	396 (4.79%)	20 (10.58%)	17 (8.90%)	0 (0.00%)	16 (2.30%)	2 (1.43%)
compositions containing carbon fibers	395	299 (3.62%)	1 (0.53%)	6 (3.14%)	0 (0.00%)	60 (8.62%)	0 (0.00%)
conductive materials	227	182 (2.20%)	6 (3.17%)	12 (6.28%)	1 (20.00%)	4 (0.57%)	8 (5.71%)
conductor	322	270 (3.27%)	2 (1.06%)	6 (3.14%)	0 (0.00%)	6 (0.86%)	8 (5.71%)
crystals comprising nanotubes	20	19 (0.23%)	1 (0.53%)	0 (0.00%)	0 (0.00%)	0 (0.00%)	0 (0.00%)
diamond	329	261 (3.16%)	2 (1.06%)	0 (0.00%)	0 (0.00%)	4 (0.57%)	4 (2.86%)
electron emission source	184	138 (1.67%)	1 (0.53%)	2 (1.05%)	0 (0.00%)	1 (0.14%)	9 (6.43%)
fabric	164	129 (1.56%)	3 (1.59%)	2 (1.05%)	0 (0.00%)	24 (3.45%)	1 (0.71%)
field_emission	116	84 (1.02%)	0 (0.00%)	0 (0.00%)	0 (0.00%)	0 (0.00%)	0 (0.00%)
fuel cells	237	185 (2.24%)	11 (5.82%)	2 (1.05%)	0 (0.00%)	15 (2.16%)	2 (1.43%)
glass_fibers	317	280 (3.39%)	1 (0.53%)	4 (2.09%)	0 (0.00%)	17 (2.44%)	1 (0.71%)
golf clubs	138	88 (1.07%)	0 (0.00%)	0 (0.00%)	0 (0.00%)	7 (1.01%)	2 (1.43%)

Fig. 2. Portion of co-occurrence matrix for the industry and carbon nanotube taxonomies

Such techniques are extremely useful in analyzing relationships and deriving insights. In the carbon nanotube patent example, such analysis can help identify potential licensing targets for carbon nanotube patents. For instance, automotive companies might be interested in carbon nanotube patents in specific areas, such as "*carbon fiber*". Sports equipment companies might be interested in carbon nanotube patents since they can be related to certain sport equipments, such as golf shafts. When co-occurrence analysis is performed on the same taxonomy, such as between inventors and inventors, it can help understand the inventor relationship (e.g. based on which inventors co-author patents together), which can in turn be used to construct an inventors network.

A specialized co-occurrence analysis is the trend analysis. With trend analysis, the temporal information extracted from documents, such as patent filing dates or publication dates, can be overlaid on top of a taxonomy, a class in the taxonomy, or co-occurrence table cells to reveal trends over time. Fig. 3 shows an example of trends over the carbon nanotube patent taxonomy. There is a trend line shown for each category in the taxonomy.

The date shown in each trend line is the average date of the documents in this category. The X-axis for each of the mini-trend charts is not linear by time, but rather based on a chronological sorting of the data which is then "*quantized*" (into contiguous buckets). Each point on the mini-line graph then represents the number of mentions of that category in that bucket of time. The Y scale is normalized for each category. This helps to ensure that the specific trend for each category is well differentiated from the overall trend for the data as a whole. The categories that trend upwards indicate that the categories of concepts are more current or emerging in the marketplace. For instance, "*conductor*" and "*conductive materials*" are such categories. Downward trending shows a diminishing trend, e.g., "*manufacturing process*" and "*resin*" in Fig. 3. Trend analysis helps the analyst to further understand concepts with respect to time. In summary, such relationship-based analysis helps one to find hidden patterns and associations among different dimensions in data.

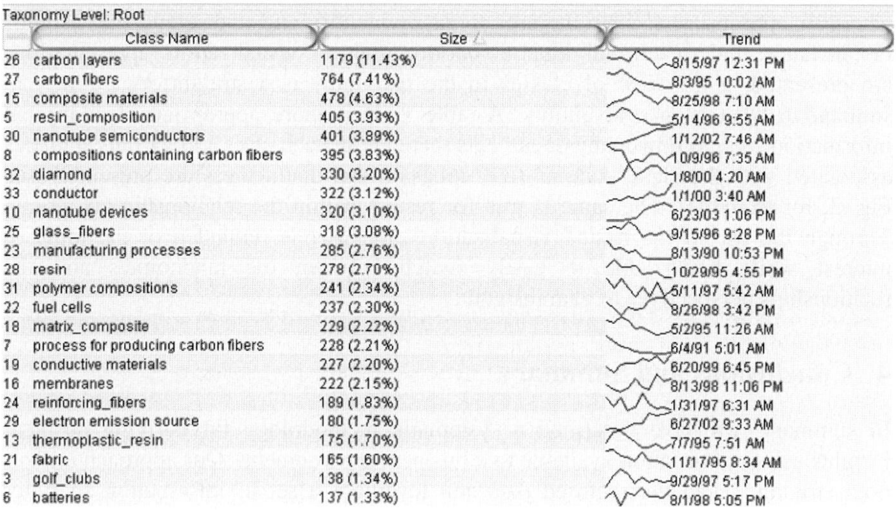

Taxonomy Level: Root

	Class Name	Size	Trend
26	carbon layers	1179 (11.43%)	8/15/97 12:31 PM
27	carbon fibers	764 (7.41%)	8/3/95 10:02 AM
15	composite materials	478 (4.63%)	8/25/98 7:10 AM
5	resin_composition	405 (3.93%)	5/14/96 9:55 AM
30	nanotube semiconductors	401 (3.89%)	1/12/02 7:46 AM
8	compositions containing carbon fibers	395 (3.83%)	10/9/96 7:35 AM
32	diamond	330 (3.20%)	1/8/00 4:20 AM
33	conductor	322 (3.12%)	1/1/00 3:40 AM
10	nanotube devices	320 (3.10%)	6/23/03 1:06 PM
25	glass_fibers	318 (3.08%)	9/15/96 9:28 PM
23	manufacturing processes	285 (2.76%)	8/13/90 10:53 PM
28	resin	278 (2.70%)	10/29/95 4:55 PM
31	polymer compositions	241 (2.34%)	5/11/97 5:42 AM
22	fuel cells	237 (2.30%)	8/26/98 3:42 PM
18	matrix_composite	229 (2.22%)	5/2/95 11:26 AM
7	process for producing carbon fibers	228 (2.21%)	6/4/91 5:01 AM
19	conductive materials	227 (2.20%)	6/20/99 6:45 PM
16	membranes	222 (2.15%)	8/13/98 11:06 PM
24	reinforcing_fibers	189 (1.83%)	1/31/97 6:31 AM
29	electron emission source	180 (1.75%)	6/27/02 9:33 AM
13	thermoplastic_resin	175 (1.70%)	7/7/95 7:51 AM
21	fabric	165 (1.60%)	11/17/95 8:34 AM
3	golf_clubs	138 (1.34%)	9/29/97 5:17 PM
6	batteries	137 (1.33%)	8/1/98 5:05 PM

Fig. 3. Portion of the trend chart for the carbon nanotube patent taxonomy

3.3 Visualization Techniques

Clearly, the BIW analysis process is not a linear process and users may carry out different kinds analysis throughout the course of discovery, i.e., taxonomy generation and relationship analysis. To allow users to interact with such a system in a flexible manner, we have designed a user interface that contains two main perspectives: *taxonomy* and *relationship*. The *taxonomy* perspective allows the user to create taxonomies and look through information via different taxonomies constructed by structured and unstructured data. The *relationship* perspective allows the user to generate and examine relationships in forms of trends and co-occurrence results. The user interface is always anchored on those two perspectives to help users focus on the end goals.

To ensure flexible user interactions and high quality insights, more detailed analysis information is provided and nested within the taxonomy perspective or the relationship perspective. For example, within the taxonomy perspective, multiple taxonomies can be generated and presented as nested elements. In our patent example, there may be a carbon nanotube natural taxonomy, an assignee taxonomy, and an inventor taxonomy as shown in Fig. 4. Similarly, nesting of information is also used for the relationship perspective, where multiple co-occurrence matrices or trends can be generated and presented.

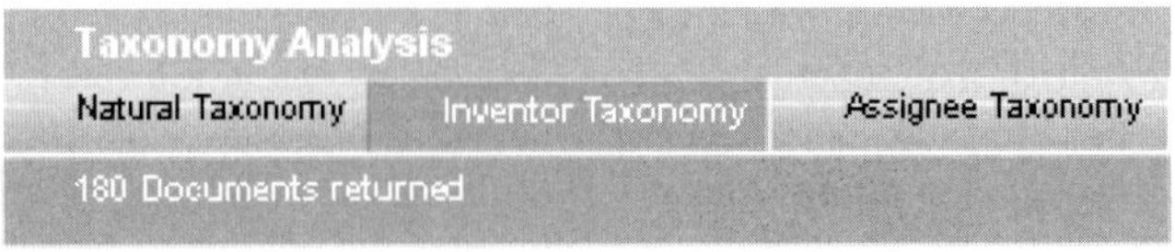

Fig. 4. Taxonomy nesting under Taxonomy perspective from BIW's user interface

Under each type of analysis different visualization representations are provided wherever appropriate to help the user to easily identify and understand the insights. For instance, under each individual taxonomy, multiple visualization representations are presented: a list view of classes in the taxonomy is a straightforward way to summarize what is in a taxonomy. A table view is more appropriate if additional information such as how many documents there are in each class and various statistics associated with the class, such as size, cohesion and distinctness are presented (see Fig. 3. for example). The same is true for results within the relationship taxonomy. Through the use of the nested visualizations, the user is reminded of the topic of interest while being able to easily switch between the taxonomies and the relationships perspectives to gain insight.

4 Conclusions and Summary

In summary, we have described a systematic methodology for deriving business insights using information available to a business environment. Our approach utilizes both structured and unstructured data and leverages a set of information analytics techniques to understand different aspects of information, including taxonomy generation, co-occurrence and trend analysis. An interactive, and yet focused and nested visualization user interface design allows the user to easily interact with the analytics tools embedded in BIW and identify insights intuitively without sacrificing the flexibility and the quality of insights.

Our approach is easily applicable to other environments, such as analyzing scientific articles, business related documents, and market information. The benefits of such analysis are multi-faceted to both individuals and organizations: individual researchers can identify key white-spaces to do research in; businesses can identify emerging markets and technology for growth opportunities. We have successfully applied the BIW technology in many different industry domains and for different purposes, including call center problem ticket analysis in Customer Relationship Management (CRM), drug discovery in HealthCare Life-Sciences, and patent portfolio analysis in Intellectual Property. In general, information is multi-faceted. Hence to discover insights buried in information, the flexibility of the analytical tool coupled with an appropriate user interface is critical. We have presented such a tool in this paper.

References

1. Srikant, R., Vu, Q., Agrawal, R.: Mining association rules with item constraints. In: Proceedings of the 3rd Int'l Conf. on Knowledge Discovery in Databases and Data Mining. Newport Beach, CA (1997)

2. Frawley, W., Piatetsky-Shapiro, G., Matheus, C.: Knowledge discovery in Databases: An overview. AI Magazine. Fall 1992, pp. 213–228 (1992)
3. Tan, P., Steinbach, M., Kumar, V.: Introduction to data mining. Addison-Wesley Publishing, London (2005)
4. Agrawal, R.: Data mining: Crossing the chasm. In: Keynote at the 5th ACM SIGKDD Int'l Conf. on Knowledge Discovery and Data Mining. San Diego, CA (1999)
5. Bayardo, R.J., Agrawal, R.: Mining the most interesting rules. In: Proc. Of the 5th CAN SIGKDD Int'l Conf. on Knowledge Discovery and Data Mining (1999)
6. Grossman, D., Frieder, O.: Information retrieval: Algorithms and heuristics, 2nd edn. Springer, Heidelberg (2006)
7. Baeza-Yates, R., Ribeiro-Beto, B.: Modern information retrieval. Addison-Wesley Publishing, London (1999)
8. Manning, C.D., Schutze, H.: Foundations of statistical natural language processing. MIT Press, Cambridge (1999)
9. Jackson, P., Moulinier, I.: Natural language processing for online applications: Text retrieval, extraction, and categorization. John Benjamins Publishing Co, Amsterdam (2002)
10. Gotz, T., Suhre, O.: Design and implementation of the UIMA common analysis system. IBM System Journal, vol. 43(3) (2004)
11. Modha, D., Spangler, S.: Feature weighting in K-Means clustering. Machine learning 52(3), 217–237 (2003)
12. Spangler, S., Kreulen, J., Lesser, J.: Generating and browsing multiple taxonomies over a document collection. Journal of Management Information Systems 19(4), 191–212 (2003)
13. Spangler, W.S., Kreulen, J.T., Newswanger, J.F.: Machines in the conversation: Detecting themes and trends in information communication streams. To appear in IBM Systems Journal (2006)
14. Kreulen, J., Spangler, W.S., Lesser, J.: MindMap: Utilizing multiple taxonomies and visualization to understand a document collection. HCCI (2002)
15. Spangler, S., Kreulen, J.: Interactive methods for taxonomy editing and validation. ACM CIKM (2002)
16. Press, W., et al.: Numerical Recipes in C, 2nd edn. pp. 620–623. Cambridge University Press, New York (1992)

A Decision Making Model for the Taiwanese Shipping Logistics Company in China to Select the Container Distribution Center Location

Chien-Chang Chou

Department of Shipping Technology, National Kaohsiung Marine University
Kaohsiung, Taiwan, Republic of China
chiench1@ms58.hinet.net

Abstract. The purpose of this paper is to propose a decision making model for the Taiwanese shipping logistics company in China to select the best container distribution center location. The representation of multiplication operation on fuzzy numbers is useful for the decision makers to solve the fuzzy multiple criteria decision making problems of container distribution center location selection. In the past, few papers discussed the representation of multiplication operation on multiple fuzzy numbers. Thus this paper first compute and obtain the representation of multiplication operation on multiple fuzzy numbers. Based on this representation, the decision maker can rank quickly the ordering of each alternative location and then select easily the best one. Finally, the representation of multiplication operation on multiple fuzzy numbers is applied to solve the fuzzy multiple criteria decision making problem of container distribution center location selection in China.

1 Introduction

In the recent years, a lot of Taiwanese business companies went to China to invest the shipping logistics industry, manufacturing industry, civil engineering industry, banking industry and so on. It is an important and difficult for the Taiwanese business company to select a best location for building the container logistics center or the manufacturing factory when beginning the investment planning. Thus, the purpose of this paper is to propose a decision making model for the Taiwanese shipping logistics company in China to select the best container distribution center location.

The selection of location is one of the most important decision issues for the decision makers of business companies or industrial organizations. Many precision-based methods for location selection have been developed. Dahlberg and May [7] utilized the simplex method to determine the optimal location of energy facilities. Tompkins and White [20] introduced a method that used the preference theory to assign weights to subjective factors by making all possible pairwise comparisons between factors. Spohrer and Kmak [18] proposed a weight factor analysis method to integrate the quantitative data and qualitative rating to choose a plant location from numerous alternatives. Stevenson [19] proposed a cost-volume analysis method to select the best plant location. Multiple criteria decision-making methods were

M.J. Smith, G. Salvendy (Eds.): Human Interface, Part II, HCII 2007, LNCS 4558, pp. 844–854, 2007.

provided to deal with the problem of ranking and selecting locations under multiple criteria [11, 17]. All the methods stated above are based on the concept of accurate measure and crisp evaluation.

Generally, the selection of a best location for business companies and industrial organizations from two or more alternatives locations on the basis of two or more factors is a multiple criteria decision-making problem. Under many situations, the values for the qualitative criteria are often imprecisely defined for the decision maker. It is not easy to precisely quantify the rating of each alternative location and the precision-based methods as stated above are not adequate to deal with the location selection problem [2, 13]. Since human judgments including preference are often vague and can not estimate his preference with an exact numerical value. A more realistic way may be to use linguistic terms to describe the desired value and important weight of criteria, e.g. "very low", "low", "fair", "high", "very high", etc [1, 24]. Due to this type of existing fuzziness in the location selection process, fuzzy set theory is an appropriate method for deal with uncertainty and the subjective evaluation data can be more adequately expressed in fuzzy linguistic variables [2, 10, 13, 21, 22, 24].

The concept of fuzzy sets, which was introduced by Zadeh [23], led to the definition of the fuzzy number and its implementation in fuzzy control and approximate reasoning problems. The basic arithmetic structures for fuzzy numbers were developed by Mizumoto and Tanaka [15], Nahmias [16], Dubois and Prade [9], Li [12], Ma et al. [14] and Delgado et al. [8]. The arithmetic operation was established either by the extension principle or by observing the fuzzy number as a collection of α-levels. Unfortunately, although there are many arithmetical operation approaches, none of these approaches focuses on the representation of multiplication operation on multiple fuzzy numbers. The representation of multiplication operation on multiple fuzzy numbers is useful for the decision makers to solve the fuzzy multiple criteria decision making problems of location selection. Chou [4] proposed the canonical representation of multiplication operation on two triangular fuzzy numbers by the Graded Multiple Integration Representation Method. Chou [5] further proposed the canonical representation of multiplication operation on two trapezoidal fuzzy numbers by the Inverse Function Arithmetic Representation Method and this representation is applied to a numerical example. Based on the Inverse Function Arithmetic Representation Method, Chou [6] further proposed the representation of multiplication operation on three fuzzy numbers and this representation is applied to solve the marine transshipment container port selection problems. In this paper, we will further propose the representation of multiplication operation on multiple triangular fuzzy numbers.

The rest of this paper is organized as follows. In section 2, we use the Inverse Function Arithmetic Representation Method for multiplication operation on multiple fuzzy numbers. Based on the Inverse Function Arithmetic Representation Method, it is easy to interpret clearly and compute quickly the canonical representation of multiplication operation with the membership functions of fuzzy numbers. Finally, The representation of multiplication operation on multiple fuzzy numbers proposed in this paper is applied to solve the fuzzy multiple criteria decision problem of container distribution center location selection of Taiwanese shipping logistics company in China in section 3 followed by conclusion in section 4.

2 The Representation of Multiplication on Multiple Fuzzy Numbers

In this section, we first introduce briefly the graded mean integration representation method for presenting the representation of one fuzzy number. Based on the graded mean integration representation method, we further propose the Inverse Function Arithmetic Representation Method for presenting the representation of multiplication operation on multiple fuzzy numbers.

2.1 The Graded Mean Integration Representation Method

Chen and Hsieh [3] proposed the graded mean integration representation method of fuzzy numbers based on the integral value of graded mean h-level of generalized fuzzy number. Here we describe the meaning as follows.

Suppose A=(c, a, b, d) is a fuzzy number. The membership function of A is

$$f_A(x)=\begin{cases} \dfrac{(x-c)}{(a-c)}, & c\le x\le a \\ 1, & a\le x\le b \\ \dfrac{(x-d)}{(b-d)}, & b\le x\le d \\ 0, & \text{otherwise} \end{cases}$$

Since $L_A(x)=\dfrac{(x-c)}{(a-c)}$, $c\le x\le a$, $R_A(x)=\dfrac{(x-d)}{(b-d)}$, $b\le x\le d$,

and $L^{-1}{}_A(h)=c+(a-c)h$, $0\le h\le 1$, $R^{-1}{}_A(h)=d+(b-d)h$, $0\le h\le 1$.

$L_A(x)$ and $R_A(x)$ are the function L and the function R of fuzzy number A, respectively. $L^{-1}{}_A(h)$ and $R^{-1}{}_A(h)$ are the inverse functions of function $L_A(x)$ and function $R_A(x)$ at h-level, respectively. Then the graded mean h-level value of the fuzzy number A is $h[L^{-1}{}_A(h)+R^{-1}{}_A(h)]/2$. The graded mean integration representation of A is

$$\begin{aligned} P(A) &= \int_0^1 \frac{h[L^{-1}(h)+R^{-1}(h)]}{2}\,dh \Big/ \int_0^1 h\,dh \\ &= \int_0^1 \frac{h[c+(a-c)h+d+(b-d)h]}{2}\,dh \Big/ \int_0^1 h\,dh \\ &= \frac{1}{6}(c+2a+2b+d) \end{aligned} \tag{1}$$

Formula (1) is equal to the formula proposed by Delgado, Vila, and Voxman et al. [8].

The triangular fuzzy number Y=(c, a, b) is a special case of generalized trapezoidal fuzzy number. The graded mean integration representation of triangular fuzzy number Y becomes

$$P(Y)=\frac{1}{6}(c+4a+b) \tag{2}$$

2.2 The Inverse Function Arithmetic Representation Method

Based on the above graded mean integration representation method, we further propose the Inverse Function Arithmetic Presentation Method for the multiplication operation on multiple fuzzy numbers as follows

Suppose the membership function of $A_1=(c_1, a_1, b_1)$ is

$$f_{A1}(x)=\begin{cases} \frac{(x-c_1)}{(a_1-c_1)}, & c_1 \le x \le a_1, \\ \frac{(x-b_1)}{(a_1-b_1)}, & a_1 \le x \le b_1, \\ 0, \text{otherwise} \end{cases}$$

Since $L_{A1}(x)=\frac{(x-c_1)}{(a_1-c_1)}, \quad c_1 \le x \le a_1, \quad R_{A1}(x)=\frac{(x-b_1)}{(a_1-b_1)}, \quad a_1 \le x \le b_1,$

and $L^{-1}{}_{A1}(h)=c_1+(a_1-c_1)h, \quad 0 \le h \le 1, \quad R^{-1}{}_{A1}(h)=b_1+(a_1-b_1)h, \quad 0 \le h \le 1.$

$L_{A1}(x)$ and $R_{A1}(x)$ are the function L and function R of fuzzy number A_1 respectively. $L^{-1}{}_{A1}(h)$ and $R^{-1}{}_{A1}(h)$ are the inverse functions of function $L_{A1}(x)$ and function $R_{A1}(x)$ at h-level, respectively.

Similarly, suppose the membership function of $A_2=(c_2, a_2, b_2)$ is

$$f_{A2}(x)=\begin{cases} \frac{(x-c_2)}{(a_2-c_2)}, & c_2 \le x \le a_2, \\ \frac{(x-b_2)}{(a_2-b_2)}, & a_2 \le x \le b_2, \\ 0, \text{otherwise} \end{cases}$$

Since $L_{A2}(x)=\frac{(x-c_2)}{(a_2-c_2)}, \quad c_2 \le x \le a_2, \quad R_{A2}(x)=\frac{(x-b_2)}{(a_2-b_2)}, \quad a_2 \le x \le b_2,$

and $L^{-1}{}_{A2}(h)=c_2+(a_2-c_2)h, \quad 0 \le h \le 1, \quad R^{-1}{}_{A2}(h)=b_2+(a_2-b_2)h, \quad 0 \le h \le 1.$

$L_{A2}(x)$ and $R_{A2}(x)$ are the function L and function R of fuzzy number A_2 respectively. $L^{-1}{}_{A2}(h)$ and $R^{-1}{}_{A2}(h)$ are the inverse functions of function $L_{A2}(x)$ and $R_{A2}(x)$ at h-level.

Similarly, suppose the membership function of $A_3=(c_3, a_3, b_3)$ is

$$f_{A3}(x)=\begin{cases} \frac{(x-c_3)}{(a_3-c_3)}, & c_3 \le x \le a_3, \\ \frac{(x-b_3)}{(a_3-b_3)}, & a_3 \le x \le b_3, \\ 0, \text{otherwise} \end{cases}$$

Since $L_{A3}(x)=\frac{(x-c_3)}{(a_3-c_3)}, \quad c_3 \le x \le a_3, \quad R_{A3}(x)=\frac{(x-b_3)}{(a_3-b_3)}, \quad a_3 \le x \le b_3,$

and $L^{-1}{}_{A3}(h)=c_3+(a_3-c_3)h, \quad 0 \le h \le 1, \quad R^{-1}{}_{A3}(h)=b_3+(a_3-b_3)h, \quad 0 \le h \le 1.$

$L_{A3}(x)$ and $R_{A3}(x)$ are the function L and function R of fuzzy number A_3 respectively. $L^{-1}{}_{A3}(h)$ and $R^{-1}{}_{A3}(h)$ are the inverse functions of function $L_{A3}(x)$ and $R_{A3}(x)$ at h-level.

Definition 1. Let $A_1=(c_1, a_1, b_1)$, $A_2=(c_2, a_2, b_2)$ and $A_3=(c_3, a_3, b_3)$ be all triangular fuzzy numbers. The representation of multiplication operation on multiple triangular fuzzy numbers at h-level is

$$A_{1(h)} \otimes A_{2(h)} \otimes A_{3(h)} = \frac{1}{8} h \times [L^{-1}_{A1}(h) \times L^{-1}_{A2}(h) \times L^{-1}_{A3}(h) + R^{-1}_{A1}(h) \times L^{-1}_{A2}(h) \times L^{-1}_{A3}(h)$$
$$+L^{-1}_{A1}(h)\times R^{-1}_{A2}(h)\times L^{-1}_{A3}(h)+L^{-1}_{A1}(h)\times L^{-1}_{A2}(h)\times R^{-1}_{A3}(h)$$
$$+L^{-1}_{A1}(h)\times R^{-1}_{A2}(h)\times R^{-1}_{A3}(h)+R^{-1}_{A1}(h)\times L^{-1}_{A2}(h)\times R^{-1}_{A3}(h)$$
$$+R^{-1}_{A1}(h)\times R^{-1}_{A2}(h)\times L^{-1}_{A3}(h) +R^{-1}_{A1}(h)\times R^{-1}_{A2}(h)\times R^{-1}_{A3}(h)]$$

Definition 2. Let $P(A_1 \otimes A_2 \otimes A_3)$ be the representation of $A_1 \otimes A_2 \otimes A_3$. And $P(A_1 \otimes A_2 \otimes A_3)$ is defined as follows.

$$P(A_1 \otimes A_2 \otimes A_3)$$
$$= \int_0^1\int_0^1\int_0^1 \frac{1}{8} [(h_{A1}L^{-1}_{A1}(h)\times h_{A2}L^{-1}_{A2}(h)\times h_{A3}L^{-1}_{A3}(h))+(h_{A1}R^{-1}_{A1}(h) \times h_{A2}L^{-1}_{A2}(h) \times h_{A3}L^{-1}_{A3}(h))$$
$$+(h_{A1}L^{-1}_{A1}(h) \times h_{A2}R^{-1}_{A2}(h) \times h_{A3}L^{-1}_{A3}(h))+(h_{A1}L^{-1}_{A1}(h) \times h_{A2}L^{-1}_{A2}(h) \times h_{A3}R^{-1}_{A3}(h))$$
$$+(h_{A1}L^{-1}_{A1}(h) \times h_{A2}R^{-1}_{A2}(h) \times h_{A3}R^{-1}_{A3}(h))+(h_{A1}R^{-1}_{A1}(h) \times h_{A2}L^{-1}_{A2}(h) \times h_{A3}R^{-1}_{A3}(h))$$
$$+(h_{A1}R^{-1}_{A1}(h) \times h_{A2}R^{-1}_{A2}(h) \times h_{A3}L^{-1}_{A3}(h))+(h_{A1}R^{-1}_{A1}(h) \times h_{A2}R^{-1}_{A2}(h) \times h_{A3}R^{-1}_{A3}(h))]$$
$$dh_{A1}dh_{A2}dh_{A3} / \left(\int_0^1 h_{A1}dh_{A1}\times \int_0^1 h_{A2}dh_{A2}\times \int_0^1 h_{A3}dh_{A3}\right) \quad (3)$$

By formula (3),

$$P(A_1 \otimes A_2 \otimes A_3)$$
$$= \int_0^1\int_0^1\int_0^1 \frac{1}{8} [(h_{A1}L^{-1}_{A1}(h)\times h_{A2}L^{-1}_{A2}(h)\times h_{A3}L^{-1}_{A3}(h))+(h_{A1}R^{-1}_{A1}(h) \times h_{A2}L^{-1}_{A2}(h) \times h_{A3}L^{-1}_{A3}(h))$$
$$+(h_{A1}L^{-1}_{A1}(h) \times h_{A2}R^{-1}_{A2}(h) \times h_{A3}L^{-1}_{A3}(h))+(h_{A1}L^{-1}_{A1}(h) \times h_{A2}L^{-1}_{A2}(h) \times h_{A3}R^{-1}_{A3}(h))$$
$$+(h_{A1}L^{-1}_{A1}(h) \times h_{A2}R^{-1}_{A2}(h) \times h_{A3}R^{-1}_{A3}(h))+(h_{A1}R^{-1}_{A1}(h) \times h_{A2}L^{-1}_{A2}(h) \times h_{A3}R^{-1}_{A3}(h))$$
$$+(h_{A1}R^{-1}_{A1}(h) \times h_{A2}R^{-1}_{A2}(h) \times h_{A3}L^{-1}_{A3}(h))+(h_{A1}R^{-1}_{A1}(h) \times h_{A2}R^{-1}_{A2}(h) \times h_{A3}R^{-1}_{A3}(h))]$$
$$dh_{A1}dh_{A2}dh_{A3} / \left(\int_0^1 h_{A1}dh_{A1}\times \int_0^1 h_{A2}dh_{A2}\times \int_0^1 h_{A3}dh_{A3}\right)$$
$$= \int_0^1\int_0^1\int_0^1 \frac{1}{8} \{h_{A1}[c_1+(a_1-c_1)h_{A1}]\times h_{A2}[c_2+(a_2-c_2)h_{A2}]\times h_{A3}[c_3+(a_3-c_3)h_{A3}]$$
$$+h_{A1}[b_1+(a_1-b_1)h_{A1}]\times h_{A2}[c_2+(a_2-c_2)h_{A2}]\times h_{A3}[c_3+(a_3-c_3)h_{A3}]$$
$$+h_{A1}[c_1+(a_1-c_1)h_{A1}]\times h_{A2}[b_2+(a_2-b_2)h_{A2}]\times h_{A3}[c_3+(a_3-c_3)h_{A3}]$$
$$+h_{A1}[c_1+(a_1-c_1)h_{A1}]\times h_{A2}[c_2+(a_2-c_2)h_{A2}]\times h_{A3}[b_3+(a_3-b_3)h_{A3}]$$
$$+h_{A1}[c_1+(a_1-c_1)h_{A1}]\times h_{A2}[b_2+(a_2-b_2)h_{A2}]\times h_{A3}[b_3+(a_3-b_3)h_{A3}]$$
$$+h_{A1}[b_1+(a_1-b_1)h_{A1}]\times h_{A2}[c_2+(a_2-c_2)h_{A2}]\times h_{A3}[b_3+(a_3-b_3)h_{A3}]$$
$$+h_{A1}[b_1+(a_1-b_1)h_{A1}]\times h_{A2}[b_2+(a_2-b_2)h_{A2}]\times h_{A3}[c_3+(a_3-c_3)h_{A3}]$$
$$+h_{A1}[b_1+(a_1-b_1)h_{A1}]\times h_{A2}[b_2+(a_2-b_2)h_{A2}]\times h_{A3}[b_3+(a_3-b_3)h_{A3}]\}$$
$$dh_{A1}dh_{A2}dh_{A3} / \left(\int_0^1 h_{A1}dh_{A1}\times \int_0^1 h_{A2}dh_{A2}\times \int_0^1 h_{A3}dh_{A3}\right)$$
$$= \int_0^1\int_0^1 \frac{1}{8} \{[\frac{1}{2}c_1h_{A1}^2+\frac{1}{3}(a_1-c_1)h_{A1}^3] \times h_{A2}[c_2+(a_2-c_2)h_{A2}] \times h_{A3}[c_3+(a_3-c_3)h_{A3}]$$

$$+[\frac{1}{2}b_1 h_{A1}^2+\frac{1}{3}(a_1-b_1)h_{A1}^3]\times h_{A2}[c_2+(a_2-c_2)h_{A2}]\times h_{A3}[c_3+(a_3-c_3)h_{A3}]$$

$$+[\frac{1}{2}c_1 h_{A1}^2+\frac{1}{3}(a_1-c_1)h_{A1}^3]\times h_{A2}[b_2+(a_2-b_2)h_{A2}]\times h_{A3}[c_3+(a_3-c_3)h_{A3}]$$

$$+[\frac{1}{2}c_1 h_{A1}^2+\frac{1}{3}(a_1-c_1)h_{A1}^3]\times h_{A2}[c_2+(a_2-c_2)h_{A2}]\times h_{A3}[b_3+(a_3-b_3)h_{A3}]$$

$$+[\frac{1}{2}c_1 h_{A1}^2+\frac{1}{3}(a_1-c_1)h_{A1}^3]\times h_{A2}[b_2+(a_2-b_2)h_{A2}]\times h_{A3}[b_3+(a_3-b_3)h_{A3}]$$

$$+[\frac{1}{2}b_1 h_{A1}^2+\frac{1}{3}(a_1-b_1)h_{A1}^3]\times h_{A2}[c_2+(a_2-c_2)h_{A2}]\times h_{A3}[b_3+(a_3-b_3)h_{A3}]$$

$$+[\frac{1}{2}b_1 h_{A1}^2+\frac{1}{3}(a_1-b_1)h_{A1}^3]\times h_{A2}[b_2+(a_2-b_2)h_{A2}]\times h_{A3}[c_3+(a_3-c_3)h_{A3}]$$

$$+[\frac{1}{2}b_1 h_{A1}^2+\frac{1}{3}(a_1-b_1)h_{A1}^3]\times h_{A2}[b_2+(a_2-b_2)h_{A2}]\times h_{A3}[b_3+(a_3-b_3)h_{A3}]\}\Big|_0^1 dh_{A2}dh_{A3}$$

$$/(\frac{1}{2}h_{A1}^2\Big|_0^1\times\int_0^1 h_{A2}dh_{A2}\times\int_0^1 h_{A3}dh_{A3})$$

$$=\int_0^1\int_0^1 \frac{1}{8}\{[\frac{1}{2}c_1+\frac{1}{3}(a_1-c_1)]\times h_{A2}[c_2+(a_2-c_2)h_{A2}]\times h_{A3}[c_3+(a_3-c_3)h_{A3}]$$

$$+[\frac{1}{2}b_1+\frac{1}{3}(a_1-b_1)]\times h_{A2}[c_2+(a_2-c_2)h_{A2}]\times h_{A3}[c_3+(a_3-c_3)h_{A3}]$$

$$+[\frac{1}{2}c_1+\frac{1}{3}(a_1-c_1)]\times h_{A2}[b_2+(a_2-b_2)h_{A2}]\times h_{A3}[c_3+(a_3-c_3)h_{A3}]$$

$$+[\frac{1}{2}c_1+\frac{1}{3}(a_1-c_1)]\times h_{A2}[c_2+(a_2-c_2)h_{A2}]\times h_{A3}[b_3+(a_3-b_3)h_{A3}]$$

$$+[\frac{1}{2}c_1+\frac{1}{3}(a_1-c_1)]\times h_{A2}[b_2+(a_2-b_2)h_{A2}]\times h_{A3}[b_3+(a_3-b_3)h_{A3}]$$

$$+[\frac{1}{2}b_1+\frac{1}{3}(a_1-b_1)]\times h_{A2}[c_2+(a_2-c_2)h_{A2}]\times h_{A3}[b_3+(a_3-b_3)h_{A3}]$$

$$+[\frac{1}{2}b_1+\frac{1}{3}(a_1-b_1)]\times h_{A2}[b_2+(a_2-b_2)h_{A2}]\times h_{A3}[c_3+(a_3-c_3)h_{A3}]$$

$$+[\frac{1}{2}b_1+\frac{1}{3}(a_1-b_1)]\times h_{A2}[b_2+(a_2-b_2)h_{A2}]\times h_{A3}[b_3+(a_3-b_3)h_{A3}]\}dh_{A2}dh_{A3}$$

$$/(\frac{1}{2}\times\int_0^1 h_{A2}dh_{A2}\times\int_0^1 h_{A3}dh_{A3})$$

$$=\int_0^1 \frac{1}{8}\{[\frac{1}{2}c_1+\frac{1}{3}(a_1-c_1)]\times[\frac{1}{2}c_2+\frac{1}{3}(a_2-c_2)]\times h_{A3}[c_3+(a_3-c_3)h_{A3}]$$

$$+[\frac{1}{2}b_1+\frac{1}{3}(a_1-b_1)]\times[\frac{1}{2}c_2+\frac{1}{3}(a_2-c_2)]\times h_{A3}[c_3+(a_3-c_3)h_{A3}]$$

$$+[\frac{1}{2}c_1+\frac{1}{3}(a_1-c_1)]\times[\frac{1}{2}b_2+\frac{1}{3}(a_2-b_2)]\times h_{A3}[c_3+(a_3-c_3)h_{A3}]$$

$$+[\frac{1}{2}c_1+\frac{1}{3}(a_1-c_1)]\times[\frac{1}{2}c_2+\frac{1}{3}(a_2-c_2)]\times h_{A3}[b_3+(a_3-b_3)h_{A3}]$$

$$+[\frac{1}{2}c_1+\frac{1}{3}(a_1-c_1)]\times[\frac{1}{2}b_2+\frac{1}{3}(a_2-b_2)]\times h_{A3}[b_3+(a_3-b_3)h_{A3}]$$

$$+[\frac{1}{2}b_1+\frac{1}{3}(a_1-b_1)]\times[\frac{1}{2}c_2+\frac{1}{3}(a_2-c_2)]\times h_{A3}[b_3+(a_3-b_3)h_{A3}]$$

$$+[\frac{1}{2}b_1+\frac{1}{3}(a_1-b_1)]\times[\frac{1}{2}b_2+\frac{1}{3}(a_2-b_2)]\times h_{A3}[c_3+(a_3-c_3)h_{A3}]$$

$$+[\frac{1}{2}b_1+\frac{1}{3}(a_1-b_1)]\times[\frac{1}{2}b_2+\frac{1}{3}(a_2-b_2)]\times h_{A3}[b_3+(a_3-b_3)h_{A3}]\}dh_{A3}/(\frac{1}{2}\times\frac{1}{2}\times\int_0^1 h_{A3}dh_{A3})$$

$$=\frac{1}{8}\{[\frac{1}{2}c_1+\frac{1}{3}(a_1-c_1)]\times[\frac{1}{2}c_2+\frac{1}{3}(a_2-c_2)]\times[\frac{1}{2}c_3+\frac{1}{3}(a_3-c_3)]$$

$$+[\frac{1}{2}b_1+\frac{1}{3}(a_1-b_1)]\times[\frac{1}{2}c_2+\frac{1}{3}(a_2-c_2)]\times[\frac{1}{2}c_3+\frac{1}{3}(a_3-c_3)]$$

$$+[\frac{1}{2}c_1+\frac{1}{3}(a_1-c_1)]\times[\frac{1}{2}b_2+\frac{1}{3}(a_2-b_2)]\times[\frac{1}{2}c_3+\frac{1}{3}(a_3-c_3)]$$

$$+[\frac{1}{2}c_1+\frac{1}{3}(a_1-c_1)]\times[\frac{1}{2}c_2+\frac{1}{3}(a_2-c_2)]\times[\frac{1}{2}b_3+\frac{1}{3}(a_3-b_3)]$$

$$+[\frac{1}{2}c_1+\frac{1}{3}(a_1-c_1)]\times[\frac{1}{2}b_2+\frac{1}{3}(a_2-b_2)]\times[\frac{1}{2}b_3+\frac{1}{3}(a_3-b_3)]$$

$$+[\frac{1}{2}b_1+\frac{1}{3}(a_1-b_1)]\times[\frac{1}{2}c_2+\frac{1}{3}(a_2-c_2)]\times[\frac{1}{2}b_3+\frac{1}{3}(a_3-b_3)]$$

$$+[\frac{1}{2}b_1+\frac{1}{3}(a_1-b_1)]\times[\frac{1}{2}b_2+\frac{1}{3}(a_2-b_2)]\times[\frac{1}{2}c_3+\frac{1}{3}(a_3-c_3)]$$

$$+[\frac{1}{2}b_1+\frac{1}{3}(a_1-b_1)]\times[\frac{1}{2}b_2+\frac{1}{3}(a_2-b_2)]\times[\frac{1}{2}b_3+\frac{1}{3}(a_3-b_3)]\}\ /(\frac{1}{2}\times\frac{1}{2}\times\frac{1}{2})$$

$$=\{(\frac{c_1+2a_1}{6})\times(\frac{c_2+2a_2}{6})\times(\frac{c_3+2a_3}{6})+(\frac{b_1+2a_1}{6})\times(\frac{c_2+2a_2}{6})\times(\frac{c_3+2a_3}{6})$$

$$+(\frac{c_1+2a_1}{6})\times(\frac{b_2+2a_2}{6})\times(\frac{c_3+2a_3}{6})+(\frac{c_1+2a_1}{6})\times(\frac{c_2+2a_2}{6})\times(\frac{b_3+2a_3}{6})$$

$$+(\frac{c_1+2a_1}{6})\times(\frac{b_2+2a_2}{6})\times(\frac{b_3+2a_3}{6})+(\frac{b_1+2a_1}{6})\times(\frac{c_2+2a_2}{6})\times(\frac{b_3+2a_3}{6})$$

$$+(\frac{b_1+2a_1}{6})\times(\frac{b_2+2a_2}{6})\times(\frac{c_3+2a_3}{6})+(\frac{b_1+2a_1}{6})\times(\frac{b_2+2a_2}{6})\times(\frac{b_3+2a_3}{6})\}$$

$$=(\frac{c_1+2a_1}{6})(\frac{c_2+2a_2}{6})[(\frac{c_3+2a_3}{6})+(\frac{b_3+2a_3}{6})]$$

$$+(\frac{b_1+2a_1}{6})(\frac{c_2+2a_2}{6})[(\frac{c_3+2a_3}{6})+(\frac{b_3+2a_3}{6})]$$

$$+(\frac{c_1+2a_1}{6})(\frac{b_2+2a_2}{6})[(\frac{c_3+2a_3}{6})+(\frac{b_3+2a_3}{6})]$$

$$+(\frac{b_1+2a_1}{6})(\frac{b_2+2a_2}{6})[(\frac{c_3+2a_3}{6})+(\frac{b_3+2a_3}{6})]$$

$$=[(\frac{c_1+2a_1}{6})+(\frac{b_1+2a_1}{6})]\times[(\frac{c_2+2a_2}{6})+(\frac{b_2+2a_2}{6})]\times[(\frac{c_3+2a_3}{6})+(\frac{b_3+2a_3}{6})]$$

$=\frac{1}{6}(c_1+4a_1+b_1) \times \frac{1}{6}(c_2+4a_2+b_2) \times \frac{1}{6}(c_3+4a_3+b_3)$

We have that $P(A_1 \otimes A_2 \otimes A_3)=\frac{1}{6}(c_1+4a_1+b_1) \times \frac{1}{6}(c_2+4a_2+b_2) \times \frac{1}{6}(c_3+4a_3+b_3)$ (4)

By formula (2), $P(A_1)=\frac{1}{6}(c_1+4a_1+b_1)$, $P(A_2)=\frac{1}{6}(c_2+4a_2+b_2)$, $P(A_3)=\frac{1}{6}(c_3+4a_3+b_3)$

$$P(A_1) \otimes P(A_2) \otimes P(A_3)=\frac{1}{6}(c_1+4a_1+b_1) \times \frac{1}{6}(c_2+4a_2+b_2) \times \frac{1}{6}(c_3+4a_3+b_3)$$

Thus, we have that $P(A_1 \otimes A_2 \otimes A_3)= P(A_1) \times P(A_2) \times P(A_3)$

3 Application to the Container Distribution Center Location Selection

The representation of multiplication operation on multiple triangular fuzzy numbers is applied to select the container distribution center location in China. This paper interviewed one Taiwanese shipping logistics company in China and then obtained the influential criteria, sub-criteria for container distribution center location selection. The influential criteria include the volume of containers aspect (C_1), cost aspect (C_2), land aspect (C_3) and government policies aspect (C_4). The sub-criteria of volume of containers aspect include the present volume of containers (C_{11}) and potential volume of containers in the future (C_{12}). The sub-criteria of cost aspect include the labor cost (C_{21}), transportation cost (C_{22}) and operation cost (C_{23}). The sub-criteria of land aspect include the availability of land (C_{31}), land cost (C_{32}), size of land (C_{33}) and distance from/to the city (C_{34}). The sub-criteria of government policies aspect include the efficiency of government department (C_{41}), co-operative relationship between the shipping logistics company and the government (C_{42}), tax break (C_{43}), other preferential treatment (C_{44}) and law on investment & investment restrictions (C_{45}). After preliminary screening, six candidate locations including the Shanghai (K_1), Qingdao (K_2), Ningbo (K_3), Tiangjin (K_4), Shenzhen (K_5) and Xiamen (K_6) remain for further evaluation. Besides, we also obtained the weights for criteria and sub-criteria from the Taiwanese shipping logistics company, including the weights W_1, W_2, W_3, W_4 for criteria C_1, C_2, C_3, C_4 respectively, and the weights W_{11}, W_{12}, W_{21}, W_{22}, W_{23}, W_{31}, W_{32}, W_{33}, W_{34}, W_{41}, W_{42}, W_{43}, W_{44}, W_{45} for sub-criteria C_{11}, C_{12}, C_{21}, C_{22}, C_{23}, C_{31}, C_{32}, C_{33}, C_{34}, C_{41}, C_{42}, C_{43}, C_{44}, C_{45} respectively. Similarly, we also obtained the performance (P_{ijk}) for candidate location K under each criteria i and sub-criteria j. The data of the weight for criteria and the weight for sub-criteria for the selection of container distribution center location are listed as follows.

Criteria Weights:
W_1=(0.3,0.35,0.4), W_2=(0.15,0.2,0.25), W_3=(0.15,0.2,0.25) and W_4=(0.2,0.25,0.3)
Sub-Criteria Weights:
W_{11}=(0.6,0.7,0.8),W_{12}=(0.2,0.3,0.4),W_{21}=(0.2,0.3,0.4),W_{22}=(0.3,0.4,0.5),W_{23}=(0.2,0.3,0.4) W_{31}=(0.2,0.3,0.4),W_{32}=(0.15,0.2,0.25),W_{33}=(0.15,02,025),W_{34}=(0.2,0.3,0.4),W_{41}=(0.15,0.2,0.25), W_{42}=(0.15,0.2,0.25),W_{43}=(0.15,0.2,0.25),W_{44}=(0.15,0.2,0.25) and W_{45}=(0.15,0.2,0.25).

Solution Process
Total Performance of candidate location K_1: TP_1

$$TP_1=\sum_{1}^{I}\sum_{1}^{J} W_i \otimes W_{ij} \otimes P_{ij1}$$

$=W_1 \otimes W_{11} \otimes P_{111}+ W_1 \otimes W_{12} \otimes P_{121}+ W_2 \otimes W_{21} \otimes P_{211} + W_2 \otimes W_{22} \otimes P_{221}+ W_2 \otimes W_{23} \otimes P_{231}+ W_3 \otimes W_{31} \otimes P_{311}+ W_3 \otimes W_{32} \otimes P_{321}+ W_3 \otimes W_{33} \otimes P_{331}+ W_3 \otimes W_{34} \otimes P_{341}+W_4 \otimes W_{41} \otimes P_{411}+ W_4 \otimes W_{42} \otimes P_{421}+ W_4 \otimes W_{43} \otimes P_{431}+ W_4 \otimes W_{44} \otimes P_{441}+ W_4 \otimes W_{45} \otimes P_{451}$
$=(0.3,0.35,0.4) \otimes (0.6,0.7,0.8) \otimes (10,10,10)+(0.3,0.35,0.4) \otimes (0.2,0.3,0.4) \otimes (7,8,9) +(0.15,0.2,0.25) \otimes (0.2,0.3,0.4) \otimes (6,7,8)+(0.15,0.2,0.25) \otimes (0.3,0.4,0.5) \otimes (7,8,9) +(0.15,0.2,0.25) \otimes (0.2,0.3,0.4) \otimes (6,7,8)+(0.15,0.2,0.25) \otimes (0.2,0.3,0.4) \otimes (8,9,10) +(0.15,0.2,0.25) \otimes (0.15,0.2,0.25) \otimes (6,7,8)+(0.15,0.2,0.25) \otimes (0.15,0.2,0.25) \otimes (7,8,9) +(0.15,0.2,0.25) \otimes (0.2,0.3,0.4) \otimes (10,10,10) +(0.2,0.25,0.3) \otimes (0.15,0.2,0.25) \otimes (8,9,10) +(0.2,0.25,0.3) \otimes (0.15,0.2,0.25) \otimes (7,8,9) +(0.2,0.25,0.3) \otimes (0.15,0.2,0.25) \otimes (7,8,9) +(0.2,0.25,0.3) \otimes (0.15,0.2,0.25) \otimes (7,8,9) +(0.2,0.25,0.3) \otimes (0.15,0.2,0.25) \otimes (7,8,9)$
$=8.56$

By the same solution process, we can obtain the total performance of candidate location $TP_2= 7.87$, $TP_3= 7.53$, $TP_4= 6.98$, $TP_5= 5.84$, $TP_6= 5.71$. The candidate locations Shanghai (K_1), Qingdao (K_2) and Ningbo (K_3) are selected as the container distribution center locations by the Taiwanese shipping logistics company in China.

4 Conclusion

This paper proposes the representation of multiplication operation on multiple fuzzy numbers and this representation is applied to the decision making model for Taiwanese shipping logistics company in China to select the container distribution center location under fuzzy multiple criteria decision making environment. Based on this representation, the decision maker of Taiwanese shipping logistics company in China can rank quickly the ordering of candidate location and then choose easily the best one. This representation of multiplication operation on multiple fuzzy numbers proposed in this paper not only can be applied to solve the location selection problem of container distribution center, but also can be applied to other fuzzy multiple criteria decision making problems.

On the other hand, we can see while selecting one container distribution center the Taiwanese shipping logistics company is much care about the present volume of containers, potential volume of containers in the future, labor cost, transportation cost, operation cost, availability of land, land cost, size of land, distance from/to the city, efficiency of government department, co-operative relationship between the shipping logistics company and the government, tax break, other preferential treatment, and law on investment & investment restrictions. Thus this paper suggests that the departments of government and the managers of ports should pay attention to the above 14 criteria to maintain the advantages and to improve the disadvantages.

References

1. Bellman, R.E., Zadeh, L.A.: Decision-making in a Fuzzy Environment. Management Science 17(4), 141–164 (1970)
2. Chen, C.T.: A Fuzzy Approach to Select the Location of the Distribution Center. Fuzzy Sets and Systems 118(1), 65–73 (2001)
3. Chen, S.H., Hsieh, C.H.: Graded Mean Integration Representation of Generalized Fuzzy Number. In: Proceedings of 6th Conference on Fuzzy Theory and Its Application (CD-ROM), Filename: 031.wdl, Chinese Fuzzy Systems Association, Taiwan, P.R. China, pp. 1–6 (1998)
4. Chou, C.C.: The Canonical Representation of Multiplication Operation on Triangular Fuzzy Numbers. Computers & Mathematics with Applications 45, 1601–1610 (2003)
5. Chou, C.C.: Representation of Multiplication Operation on Fuzzy Numbers and Application to Solving Fuzzy Multiple Criteria Decision Making Problems. In: Yang, Q., Webb, G. (eds.) PRICAI 2006. LNCS (LNAI), vol. 4099, pp. 161–169. Springer, Heidelberg (2006)
6. Chou, C.C.: A Fuzzy MCDM Method for Solving Marine Transshipment Container Port Selection Problems. Applied Mathematics and Computation 186, 435-444 (2007)
7. Dahlberg, M.D., May, J.H.: Linear Programming for Sitting of Energy Facilities. Journal of Energy Engineering, 5–14 (1980)
8. Delgado, M., Vila, M.A., Voxman, W.: On a Canonical Representation of Fuzzy Numbers. Fuzzy Sets and Systems 93, 125–135 (1998)
9. Dubois, D., Prade, H.: Operations on Fuzzy Numbers. Journal of Systems Sciences 9, 613–626 (1978)
10. Hsu, H.M., Chen, C.T.: Fuzzy Credibility Relation Method for Multiple Criteria Decision-making Problems. Information Sciences 96, 79–91 (1997)
11. Hwang, C.L., Yoon, K.: Multiple Attributes Decision Making Methods and Applications. Springer, Berlin, Heidelberg (1981)
12. Li, R.J.: Fuzzy Method in Group Decision Making. Computers and Mathematics with Applications 38, 91–101 (1999)
13. Liang, G.S., Wang, M.J.: A Fuzzy Multiple Criteria Decision-making Method for Facilities Site Selection. International Journal of Production Research 29(11), 2313–2330 (1991)
14. Ma, M., Friedman, M., Kandel, A.: A New Fuzzy Arithmetic. Fuzzy Sets and Systems 108, 83–90 (1999)
15. Mizumoto, M., Tanaka, K.: The Four Operations of Arithmetic on Fuzzy Numbers. Systems Comput. Controls 7(5), 73–81 (1976)
16. Nahmias, S.: Fuzzy Variables. Fuzzy Sets and Systems, 97–111 (1978)
17. Rietveld, P., Ouwersloot, H.: Ordinal Data in Multi-criteria Decision Making, a Stochastic Dominance Approach to Sitting Nuclear Power Plants. European Journal of Operation Research 56, 249–262 (1992)
18. Spohrer, G.A., Kmak, T.R.: Qualitative Analysis Used in Evaluating Alternative Plant Location Scenarios. Industrial Engineering 16, 52–56 (1984)
19. Stevenson, W. J.: Production/ Operation Management. Richard D. Irwin Inc. Illinois (1993)

20. Tompkins, J.A., White, J.A.: Facilities Planning. John Wiley & Sons Company, New York (1984)
21. Yong, D.: Plant Location Selection Based on Fuzzy TOPSIS. International Journal of Advanced Manufacturing Technology 28, 839–844 (2006)
22. Yong, D., Cheng, S.: Evaluating the Main Battle Tank Using Fuzzy Number Arithmetic Operations. Defence Science Journal 56, 251–257 (2006)
23. Zadeh, L.A.: Fuzzy Sets. Information and Control 8, 338–353 (1965)
24. Zadeh, L.A.: The Concept of a Linguistic Variable and Its Application to Approximate Reasoning. Information Sciences 8, 199–249 (1975)

A Study of Production / Transaction-Related Model Using Control Theory

Daisuke Doyo[1], Katsuhiro Sakamoto[1], and Katsuya Aoki[2]

[1] Aoyama Gakuin University 5-10-1 Fuchinobe, Sagamihara-shi, Kanagawa, 229-8558, Japan
{doyo, saksakamot}@ise.aoyama.ac.jp
[2] Mitsubishi Motors Corporation 1 Azanakashinkiri, Hashime-cho, Okazaki-shi, Aichi, 444-8501, Japan
katsuya1.aoki@mitsubishi-motors.com

Abstract. In recent years, needs for wide-ranging kaizen / improvements have arisen in relationships with suppliers and other transaction partners in response to rising demands including increases in profit and reductions in time. One of the technologies leading to kaizen / improvements is the application of servo-mechanism control to manage inventory ordering. It is thought that if this theory is applied to the overall supply chain, it may enable an optimization of the supply chain that fits the needs of modern society. This research proposes the prototype model of the supply chain model of the contemporary society applying the control theory, and the validity of that model is evaluated.

1 Introduction

In today's society, when an enterprise is confronted with demands to reduce inventories and shortening supply times, there are many factors that can not be controlled only within the enterprise and thus there are needs to consider related factors with other enterprises. One of the factors to be taken into consideration is "transaction," such as comprehensive Win-Win relationships that include suppliers and vendors. Applying servomechanism theory (control theory) to inventory management, H. A. Simon conducted research on the control of order quantity through a cycle that takes account of the difference between planned quantities and actual values and incorporates them into subsequent draft plans. The approach taken in this article is still one of the technologies in general use today in systems to keep control variables constant.

Nowadays, the transaction environment is such that an enterprise can not by itself exert enough control to yield results according to forecasts; thus there is a spreading approach of supply chain management that offers a scheme for overall optimization. In such an environment when an enterprise wants to appropriately manage its quantity of inventory and volume of production, it is crucial to aim at wide-ranging controls extending to transaction influenced strongly by relationships with other enterprises. It may be worth considering Simon's model, which is well-adapted to modern society,

M.J. Smith, G. Salvendy (Eds.): Human Interface, Part II, HCII 2007, LNCS 4558, pp. 855–862, 2007.

as a spreading approach to supply chain management. As such, it is an approach with a scope that is applicable to transaction with other enterprises, not just to production within one enterprise.

This research aims at practical application of a model that extends the scope of this control theory to cover production and transaction relationships. It also aims at evaluating the validity of the new model by first designing a prototype model and then looking both at the influence over the supplier provided by control on the manufacturer's side and at the influence over the manufacturer provided by control on the supplier's side.

2 Control Theory

The main distinctive characteristics of control theory a re its scheme of approximating the target values that have been set while providing feedback consisting of previously output result values to new input points. This makes it suitable for a loop structure such as a Plan-Do-See cycle. It is thought that if control theory is applied to production control so as to automatically control plant production rates and quantities of product and parts inventories, this would lead the way to control of numerical values such as planned production values, quantities of parts inventories, and the quantity of products produced. Doing so would thus result in fewer over- and under-runs of various numerical values. Moreover, this would enable a distinctive characteristic of control, namely, the equivalent conversion of a transfer functions. By taking advantage of this characteristic, it would be possible to alter the detailedness of models by part, with great detailedness in parts that are strongly affected and with only coarse detailedness in parts that are weakly affected. By taking advantage of this characteristic, it would thus be possible to derive results with the same precision without needing excessive calculations.

Moreover, it would be possible to treat control models as mathematical models. When mathematical models would be input into a computer, it would be possible to use them as such in simulations, predictions, and optimizations.

3 Model Design

3.1 SCM Model

The production of products is the focus manufacturers, and as such, it usually requires the most processes. Transaction takes place at external contact points, and these are the focus of model design because these are crucial points where matters are expected by both sides to proceed well. In addition, model design incorporates the occurrence of a number of problem factors. More specifically, problems may not be confined only to a single enterprise; problem factors that may arise among the enterprise's transaction partners may be expected to have a scope that could affect the enterprise itself. Such problems may include oversupplies of parts inventories due to problems on the production lines of the manufacturer, as well as decreases in order from suppliers and

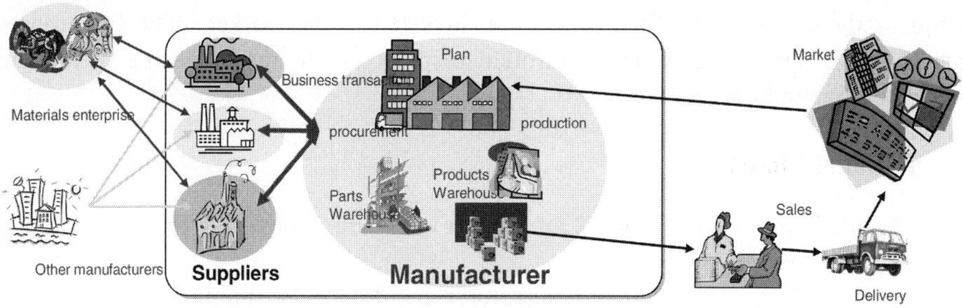

Fig. 1. Domain of the model

parts inventory shortages due to problems in the suppliers' product lines, and cases in which an enterprise can not place orders with a supplier because the supplier has already exceeded its production capacity with orders from other enterprises. This research will attempt to construct a simple supply chain centered on manufacturers and identify the flows of things and information by applying control theory.

3.2 Outline of Model

The scope of the domain of model design centers on transaction between manufacturers and suppliers from the viewpoint of the manufacturer. The transaction units of an enterprise consist of five divisions: product production, product warehousing, parts warehousing, parts procurement, and production planning. Only one manufacturer is modeled; however, elsewhere there are manufacturers with functions that affect only the production line operations of suppliers. In addition, product manufacturers and parts suppliers have together adopted the mini-max system with the market production method.

Figure 2 shows the manufacturer and suppliers in a format applying control theory. The inputs to the model are the volume of production that the market demands; the

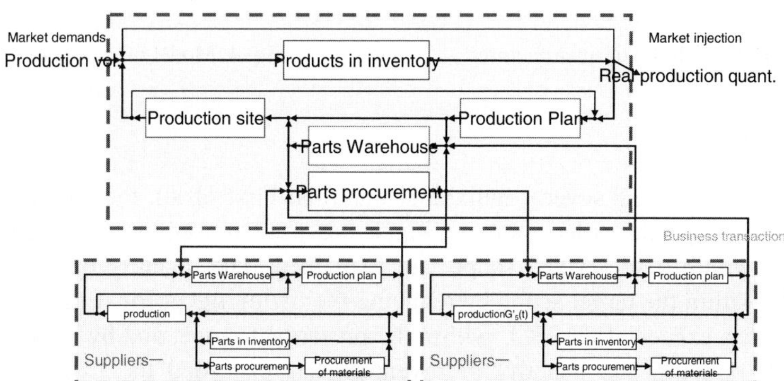

Fig. 2. Outline of model

output is the quantity of actual production invested in the market. The functions of manufacturer and suppliers are controlled respectively to minimize the differences between input and output formats.

3.3 Detail of Model

More detailed versions of Figure 2 are shown in Figure 3, covering the manufacturer, and Figure 4, covering the suppliers.

The information used when a manufacturer decides on the suppliers to place orders with consists of the order results (total quantity of orders), experience regarding postponements of the scheduled dates of delivery (total postponement time / total order quantity), production capacity of the suppliers (no. of units produced / time), the value of the supplier's product quality, and product prices. In addition, one more set of information included consists of the setting of indices reflecting the receptiveness of the supplier to orders and the depth of the relationship between manufacturer and suppliers. These information and data are weighted and orders are placed with suppliers that have high evaluations. It is possible to make a model that reflects the attitude of the suppliers quite well by weighting the receptiveness of the suppliers to orders. As for the process of determining order quantity, the resulting scheme channels orders from the manufacturer to high-priority suppliers, as reflected in the coordination between the manufacturer and the suppliers regarding the no. of units produced and the appointed date of delivery, etc.

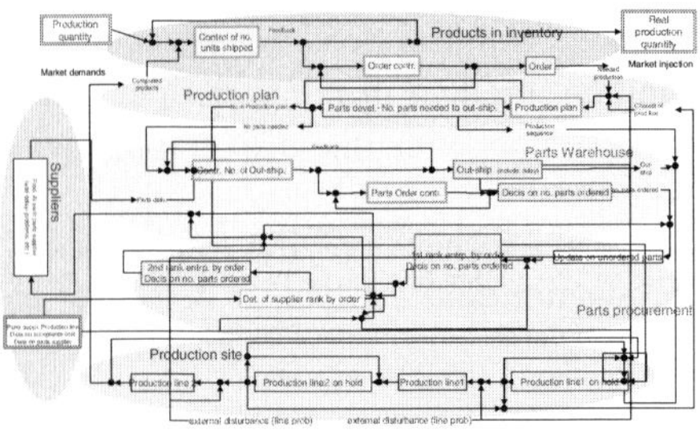

Fig. 3. Model of manufacturer

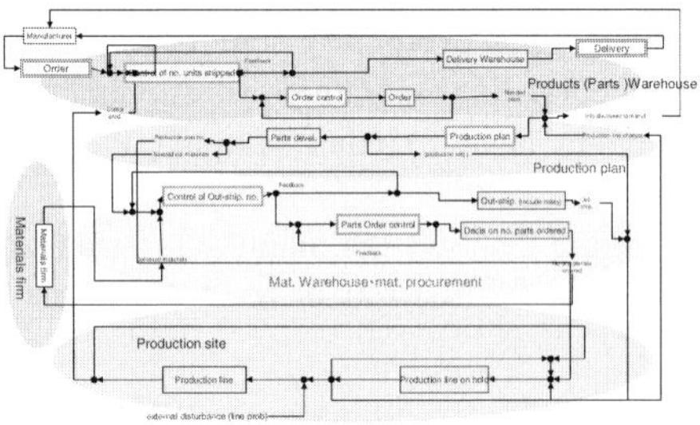

Fig. 4. Model of suppliers

3.4 Controls

It is worth mentioning several individual controls. First of all, the controls on the no. in inventory and order sources, which are used to stabilize the quantity of inventory within a prescribed scope, are explained. The quantity of inventory is stabilized within the target scope by reducing the ordering point if the quantity of inventory has exceeded the UCL within the prescribed scope, and by increasing the ordering point if the quantity of inventory has exceeded the LCL within the prescribed scope.

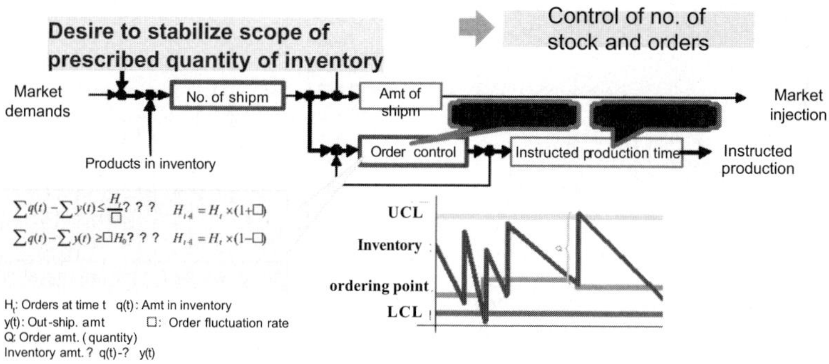

Fig. 5. Control of orders

Next, controls to alter production capacity so as not to exceed the no. in the production plan are paid attention.

In this case, when the numbers needed for production plans have continued to exceed a certain value, processing capacity is boosted by raising the value of basic production capacity. For example, if due to some accident the production line is halted and the number of unproduced units temporarily rises excessively, this situation is controlled by raising production capacity after the production line is restored and then returning the production value to the original value after stability has been restored.

4 Simulation Result

The model in situations of total operation is checked. In one environment the model was operated without any external disturbance in particular, and its production capacity control function was observed. This control function stabilizes production capacity by altering production capacity in accordance with input formulas. Under the formulas and environments that were input for the sake of the check, the model immediately ran out of inventory, assuming an initial production capacity of 50 units. Subsequently, fluctuations stopped and stability was restored when the value of production capacity was controlled by being raised to 70 units. As for the quantity of inventory, shortages of inventory stock did not take place after production capacity was adjusted. The results derived here can be used as reference materials for initial settings when establishing production plans.

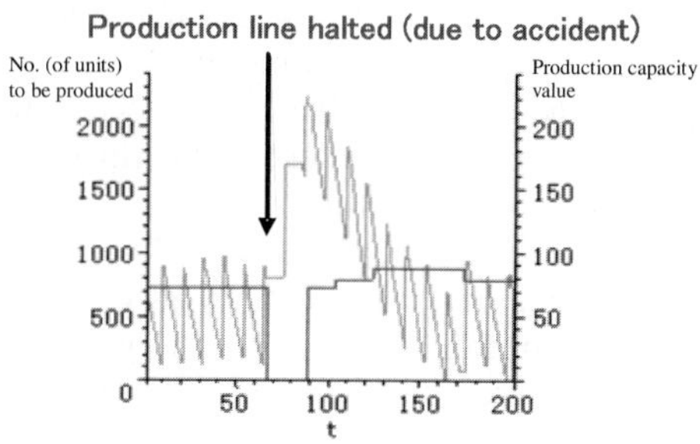

Fig. 6. Simulation of control of production capacity

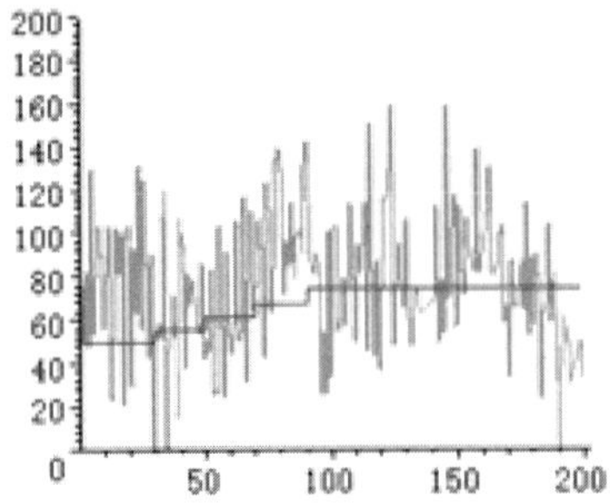

Fig. 7. Market requirements

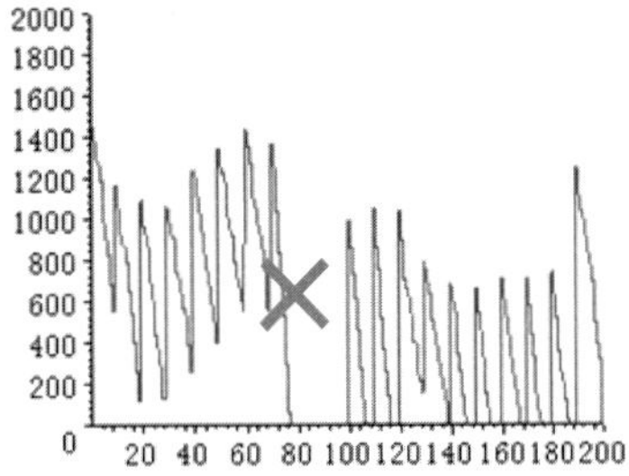

Fig. 8. Stock in inventory

Next, an accident was generated in the manufacturer's production line and the line was halted for a time. When the stoppage was 5 hours long, stability was restored immediately, even though the number of units in inventory dropped a little. Moreover, there were no effects on other sections besides the production area.

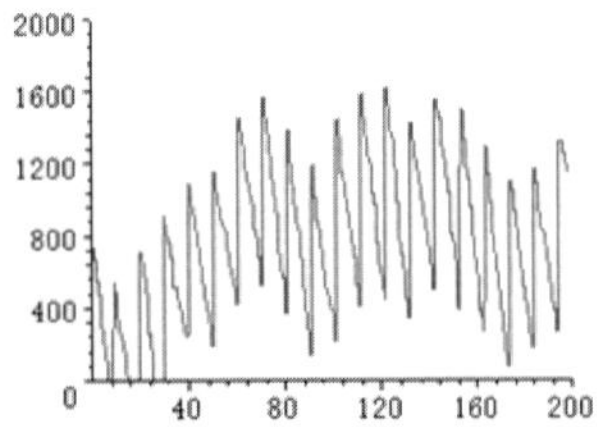

Fig. 9. Products in inventory in the case of a 5-hour stoppage

When the stoppage took ten hours, the number of stock in inventory was stabilized again, even though it took some time, because the manufacturer's production line was restarted before the parts warehouse on the manufacturer's side and the delivery warehouse on the supplier's side became filled up. In addition, the supplier's production line was not halted either.

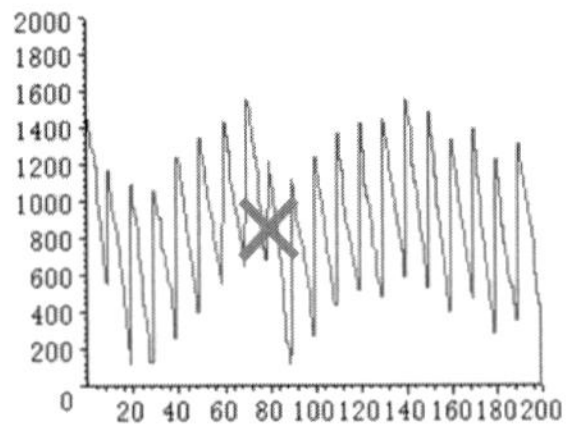

Fig. 10. Products in inventory in a 10-hour stoppage

In the case of a 20-hours stoppage, the supplier's production line stopped because the manufacturer's parts warehouse and the supplier's delivery warehouse became filled up. Because the production line of the supplier stopped and the supplier was faced with resulting increases in orders to be filled from other manufacturers, the quantity of parts supplied was not readily restored and the number of the inventory of finished goods was not stabilized either, even after the production line of the manufacturer in the model was restarted.

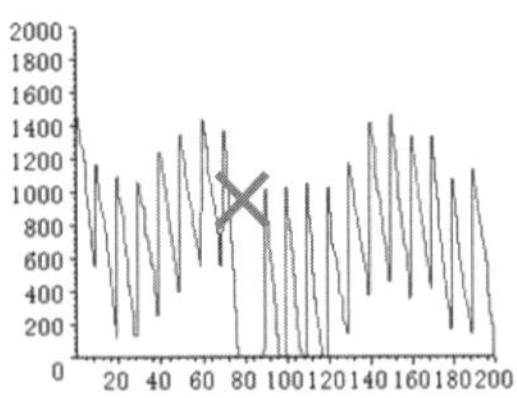

Fig. 11. Products in inventory in a 20-hour stoppage

According to these results, a stoppage has no effects in particular if the line can be restored within five hours. The line situation can also be restored to its pre-accident status in the case of a 10-hours stoppage. However, there is an effect on the supplier's side in the case of a 20-hours stoppage. Consequently, the results show potential for a considerable loss unless the line can be restarted within around 10 hours at the longest.

5 Conclusions

This research has proposed a production / transaction-related model applying control theory as a production control model and transaction model. The distinctive characteristic of this research is the modeling of transaction controls. However, a certain amount of decision-making must take place in some format in order to make control models applicable to actual transaction negotiations. Nonetheless, the control model in this research is only capable of accommodating decision-making by the model user in the stage of initial input. Consequently, some preconditions were set on the transaction model in this research. This is a format that allows determination of how well the numerical values approximate the objectives when the relationships with suppliers move ahead in patterns of varying conditions set by the controller.

The conditions were to hold steady the quality and prices of parts supplied by suppliers so as to allow the operating rate of the production line to be influenced as needed by transaction transactions. By doing so, the assessment of the controller may change depending on whether the situation allows an emphasis on price and quality or the situation demands an emphasis on speed above all. Such changes in assessment may also be chosen according to numerical assessments, including the occurrence of losses.

Thus, one of the differences between this and other simulation models is the fact that in this model all transaction-related decision-making may be performed numerically.

In addition, calculations to determine the final results are by composite expressions composed of mathematical expressions that express expressions according to the transaction unit. It may take some time to compose expressions; however, conventional simulations can assess whether the design of simulation models is applicable to operations in the same categories. Moreover, if the composite expressions are again disaggregated (returned to the status before composition of expressions) and different input formulas are substituted, it is possible to view the trends of value according to the individual composite expressions. And if inverse LaPlace transforms are applied after having composed feature expressions for each transaction unit subject to LaPlace transforms, this paves the way to have one composite formula representing multiple expressions. To carry out this task as a normal simulation, the task is carried out in two phases, with the results of Expression A inserted into the processing of Expression B. Then there is the possibility of instances requiring the redesign of models including both expressions.

As a result of the simulation by this research, it seems to be possible to extend the application of Herbert A. Simon's control of inventories and order quantity by servo-mechanism theory to cover production volume and transaction processes.

Moreover, it was possible to link this to a reduction of problem-resolution processes in the Plan-Do-See cycle through repetition of the PDS cycle, even though complete automation of all processes (which would require flawless results) was not achieved. Furthermore, this may be regarded as a model that allows automatic alteration (= automatic control) of the setting values to accommodate changes depending on status. When conventional simulation software is used, the user must learn how to operate it in accordance with the software's peculiar characteristics. However, mathematical software has a wide range of other applications and can be applied in other simulations. Thus it could reduce learning time.

Models using control theory were models that could acquire numerical values to stabilize the volume of production and quantity of inventory as well as critical points to restabilize the system against external disturbances.

They could also play a supporting role in plan proposals by using the data acquired to make production plans that take into account the supplier.

This has been a check of a prototype model. To feedback the scheme of this model into a real social setting will require the addition of model elements adapted to the scheme of the production worksite in order to enhance the model's precision.

6 Agenda for Future Research

In some views, control engineering itself may be said to already be in a period of maturity. However, it can not cope well enough with complex, nonlinear problems, only with problems that are easy to resolve.

The real world of society has vaguenesses and complexities that can not be measured physically at the stage of building mathematical models based on elimination and abstraction. It is not easy to approximate such matters nor to demonstrate the models' consistency with the real world. This has been an attempt to build a model incorporating elements of control engineering to cover part of non-linear supply chain management. However, it is only a first step toward complete modeling of all non-linear supply chains and corporate transaction and market trends.

It is better to think of this research in terms of the question "Is control theory useful as one technique to control supply chain management?" rather than in terms of the question "Does the scope of control theory also cover supply chain management?" It is not a research that applies the complete set of control theory. It is not applicable everywhere in the same format and requires altering format according to the conditions and usage requirements. Control theory may be regarded as one format of the alterations. Control theory for chemical, mechanical, and electrical applications is not directly applicable to supply chain management. However, "control of the supply chain" may well be established as one transaction model in the future.

References

1. Simon, H.A.: On the application of servomechanism theory in the study of production control. Econometrica 20, 247–268 (1952)
2. Najafi, H., Bennett, J.E.: Inventory-production optimization using optimal control theory techniques. In: Proc Annu Southeast Symp Syst Theory, vol.16, pp. 12–15, (1984)
3. Markus, S., Peter, V.: Supply system dynamics: Distributed control in supply chains and networks. Cybern Syst. 37(5), 375–415 (2006)
4. Ortega, M., Lin, L.: Control theory applications to the production-inventory problem: a review. Int. J Prod. Res. 42(11), 2303–2322 (2004)
5. Wiendahl, H.-P.: Backlog-Oriented Automatic Production Control. CIRP Ann. 50(1), 331–334 (2001)

Understanding a Sense of Place in Collaborative Environments

Simon Foley

School of Management, Royal Holloway, University of London Egham, Surrey, England
simon.foley@rhul.ac.uk

Abstract. The article explores how mobile, nomadic and distributed workers develop a sense of *place* to counter the lack of spatial determinants which increasingly characterize contemporary work and lifestyles. When developing collaborative technologies, designing for a sense of place becomes especially problematic when spatial realities and empirical coordinates are absent or no longer take precedence. What is lacking in the current knowledge is a coherent workplace approach that can offer an overview of placeness, delineating not only the core understandings of place but more importantly where these overlap and at times conflict. The intention of this article, therefore, is to present an approach that captures how key features of place as opposed to space, can be used as a motivation for the design of collaborative technologies. Most importantly, this approach must explain how the concept of place and its distinction from space can provide new understandings of the public, private and permanent places of work that help us to sustain productivity in the workplace. These at once discrete and overlapping concepts help to frame the key research questions: where is work? What constitutes a place of work? Who works there?

Keywords: place, space, affordance, knowledge work, collaborative technologies, permanence, productivity.

1 Introduction

When we say we are 'at work', where do we mean? Is it, for example, when we are in the office, at home, in the car or when travelling by train to a work related destination? Or is it simply whenever we happen to be immersed in task managing our workload through emails and diaries at our desk or laptops? In preparing this paper, I discussed these questions with a busy knowledge worker - a Sales Manager who spent his days travelling not only to and from clients, but also between regional offices and teams - mediating with others face to face and remotely through email, mobile phone and conferencing. When pressed to describe where he thought his work was actually located – where he experienced the feeling of being 'at work', after some thought, he replied "the second step down". He was referring to his staircase at home where he would sit at the end of the day with his diary, laptop and one or two files – 'far enough away' he explained, from the kids' TV in the front room and any tasks, chores and

M.J. Smith, G. Salvendy (Eds.): Human Interface, Part II, HCII 2007, LNCS 4558, pp. 863–872, 2007.

activities in the rest of the house. Most importantly from the day's demands and enquiries of colleagues, clients and line managers who might distract him from the essential as he saw it 'work to be done'.

Although anecdotal, this account illustrates some of the issues I shall discuss in this paper. Moreover it exemplifies how the ubiquity of contemporary work styles continues to challenge current workplace design initiatives. In particular, the transient nature of nomadic, remote and distributed working poses specific problems for knowledge workers in sustaining workplaces that are not only both flexible and productive, but that can also sustain some degree of identity and sense of ownership. In this article, I draw upon primary data gathered by Foley et al (2003) for the EU funded S.A.N.E (Sustainable Accommodation for the New Economy) project, EU, F5 IST-2000-25257 and related work by Lievonen, M and Rosenberg, D (2005). Researchers on this project studied knowledge workers in a variety of workplace settings in Europe and addressed the need for an appropriate framework that could accommodate how people might utilize their places of work in relation to the processes they undertake. Analysis of related findings in this study and ongoing research suggest that to design for the affordances of place as distinct from space might offer enhanced productivity for knowledge workers who need to retain a sense of 'belonging to' their work, their organisation and their place of work.

Although the literature is extensive on the need to distinguish between space and place in the context of workplace design it does not specifically address the notion of productive places or distinguish between types of place. (Harrison and Dourish, 1996: 67) introduce the notion of individuals possessing "a sense of place"; however, whilst they contend that where "space is the opportunity; place is the understood reality" it is unclear exactly how this notional reality is shared, discussed, negotiated with, concealed from, even communicated to others in workplace situations. Moreover, when applied to the design of collaborative technologies, designing for a sense of place becomes especially problematic when spatial realities and empirical coordinates are absent or no longer take precedence. What is lacking in the current knowledge is a coherent workplace approach that can offer an overview of placeness, delineating not only the core understandings of place but more importantly where these overlap and at times conflict.

The aim of this paper is therefore through interpretative analysis of primary data gathered from seven knowledge workers – project directors, managers and consultants in two organizations one based in Europe, the other in the UK – to outline the key elements of a workplace approach founded on notions of place as opposed to space. Moreover, in attempting to understand how respective informants conceive of a notion of place that is productive, my analysis addresses three key questions: Where is work? What constitutes a place of work? Who works there? My thesis is that a sense of work place can best be conceived of as a series of interrelated public, private and permanent affordances. Place has no intrinsic meaning as such, but it does have a purpose and this is individual, often negotiated and empowering. It is important not to conceive of place as zones or territories but as modes of workplace interaction and proximity - entered into or withdrawn for discrete purposes, and which incorporate a range of tasks, activities and social encounters. As Massey (1993: 66) writes, we should conceive of places as "articulated moments in networks of social relations and understandings" instead of "areas with boundaries around".

2 Background

Indeed, many writers concur in the literature with (Harrison and Dourish, 1996: 73) contention that "spaces are not places". They agree that whereas space can be located, measured and quantified, place is ephemeral and ubiquitous. For this reason it is helpful to consider that places and the ways in which individuals associate with and assign activities to them are significant as they most reflect the transience of contemporary work and life styles. Moreover, it is because of its ephemeral nature that we can identify and locate the notion of place by its absence in space – for example, as (Harrison and Dourish, 1996: 73) comment "we can all think of lifeless spaces in our buildings and cities; spaces that "don't work", that have no sense of place".

For Gibson too, places are also identified by omission – in particular, the absence of anything that might refer to fixed spatial coordinates such as names, labels or boundaries, for example "the habitat of an animal is made up of places" (Gibson, 1979: 34). Giddens (1991) refers to pre-modern societies where a notion of place was inextricably linked to fixed conceptions of time and space. In modern societies he contends that time has become separated from space and that it is this "empty dimension of time" that has "pulled [space] away from place" (Giddens, 1991: 16). Above all, places indicate appropriateness – a best fit between humans and their environment. As (Harrison and Dourish, 1996: 69) remark, we would be unlikely to refer to an individual's unsuitable behaviour as "out of space"; but it would most certainly be "out of place".

To discover which affordances of place provide the best fit between humans and environment it's important to separate the affordance of a place from its perceived use. As (Gaver, 1996: 113) makes clear "affordances are primarily facts about action and interaction, not perception". They don't create meaning; they simply allow it "to be understood in terms of the relations of humans and their environment" (Ibid). For example, returning to the account at the beginning of this article, the physical proximity of 'the second step down' to other rooms in the house has no inherent meaning. However, in terms of its functionality to our sales manager in affording a quiet place to do his work – it does. "It is an objective fact about a situation" (Ibid), linking and allowing the human to interact with his environment - so creating a sense of place.

The paradox of place is its ephemeral nature - its lack of distinct boundaries - its capacity to resist or invert any kind of label which seeks to identify where it is located. For example, it challenges our notions of private and public – terms we comfortably associate with space, such as home space or work space. Home we consider to be private and secluded – work to be public and visible. However, a sense of place manages to subvert and invert such fixed categorisations. As (Harrison and Dourish, 1996: 71) explain:

"A place-centric view emphasizes important distinctions: privacy is not the same as private events, nor is it a direct consequence of private places; and in contrast, being seen or even heard is not absolutely and inevitably public".

Places it appears can afford a person privacy without the necessity of being concealed in a space and afford a person publicity without the inevitability of being observable in a space. For example, it is possible to create one's own sense of private place working on a train whilst commuting to the office. The space may be public but the individual commuter becomes anonymous within that space by retreating to a private place of work. Similarly, it is possible to create one's own sense of public place working at home and away from the distractions of the office. The space may be private but the individual worker can be 'visible' and accessible to others through email, blogs, podcasts, phone, teleconferencing and video conferencing.

In this sense, such ICT inspired places might afford the anonymous but visible working commuter and the secluded but accessible home worker, the option and flexibility of permanence without commitment – to sustain or suspend identity within or without the environs of their respective organisations, without actually being there. Nonetheless, in the case of the home worker we must be careful here not to confuse "visual access" per se with a sense of place. Whilst there are "spatial features that the technology might provide" (Harrison and Dourish, 1996: 71) these do not afford "the place-oriented aspects of interaction which might arise there—formal and informal discussion, intimacy, a sense of ownership, and so forth" (Ibid).

In developing collaborative technologies that can complement a knowledge worker's place rather than space of work, the literature suggests that we should take account not only of the ephemeral context in which current work is undertaken but also employee resistance to cultures of organisational change that have motivated the transitory nature of contemporary work practices. Furthermore, we must consider that a simple polarisation of public and private places is insufficient and must incorporate into this place-centric framework a variable that conveys ownership and permanence. This need for a sense of permanence is well covered in the literature. (Ferguson et al., 1997) note

"The lack of personalization in workstations that must be used by multiple people conveys to occupants that they are part of a mobile and transient workforce. The portable partitions tell them that they may not be there tomorrow. Tomorrow there may be only empty space where a workstation stood today. The traditional private office was a metaphor for stability and permanence".

They cite numerous studies (Krohe, 1993), (Burgoon, 1982), (Brookes and Kaplan, 1972), 1972, (Sundstrom et al., 1980), (Boyce, 1974), (Justa and Golan, 1977), (Oldham and Brass, 1979) which document the transition from closed to open plan offices - where the removal of crucial indicators of permanence in the workplace correspond to an identifiable decrease in employee satisfaction. Whilst this paper does not contest these studies or presume the pre-eminence of a permanent over transient or closed over an open sense of place, it attempts to explore how knowledge workers might re-constitute markers of permanence in work settings founded on place rather than spatial determinants. The intention of this article, therefore, is to present an approach that captures how key features of place as opposed to space, can be used as a motivation for the design of collaborative technologies. Most importantly, this approach must explain how the concept of place and its distinction from space can provide new understandings of the public, private and permanent places of work that help us to sustain productivity in the workplace. These at once discrete and overlapping concepts help to frame the key research questions: where is work? What constitutes a place of work? Who works there? In contending that a sense of work place can best be conceived of as a series of interrelated public, private and permanent affordances, this article will attempt to demonstrate how individuals establish a sense of place by negotiating and interacting with these affordances to construct a best fit with their working environment.

3 Methods

The primary data for this article was collected as part of a wider study into the design of mobile workplaces for knowledge workers. The author together with fellow researchers conducted case studies over a one year period in UK and European organisations focusing on two teams of knowledge workers in their respective organisations. The seven knowledge workers who participated in the study were project directors, managers and communications consultants working in the design, engineering and construction industries. As the sample is statistically small it is not intended that the findings included here be generalised or applied to a wider population. However, the depth of data retrieved was sufficient to reveal theoretical insights that were of value to the project, in particular to this article and I suggest to future study. Over the year, informants were interviewed individually and within groups on two separate occasions and finally within focus groups to validate researchers' initial analyses, findings and reports. In the initial interviews they were asked to outline their specific roles, working procedures, modes of interaction and technology usage. In the second interviews, they were asked to review and discuss a series of new technology scenarios in an attempt to explore how such initiatives might enhance or detract from the exigencies of their own work practices. This aspect of the study was solely qualitative and data was gathered through interview and observation of working procedures in the respective field sites. The data for this article was analysed from related field notes and recorded transcripts of the interviews using qualitative software QSR N.U.D.I.S.T and NVivo.

4 Where Is Work?

It was clear that for many informants, regardless of its physical location, wherever work is experienced it is imperative that it conjures a feeling of permanence which can afford a sense of identity, ownership and belonging. Moreover, informants observed that hotdesking and the enhanced productivity that can ensue may suit some environments, but sometimes runs counter to existing working practices and procedures as well as cultural norms of the organisation. Indeed, as one informant remarked, some individuals go further in an attempt to preserve markers of personal space...

"Some people might feel comfortable sitting at any desk, but there are other people who do need their own space - it is like their little home; you can see people in our office who do actually come in and who do create a little home from home...".

Whilst such behaviour may be interpreted as resistance to changing practices and an inability to understand the wider organisational vision associated with distributed workplaces, it can also be an indicator of where individuals experience their place of work, where they feel themselves to be most productive and interactive. As such, their reluctance to de-personalise and remove specific individual markers from an open area should not always be discounted as irrational belligerence, but illustrative of the tension (subversion and inversion) between spaces that are designated as public but are then reconstituted as places that are private.

It was also felt that work occurs in mutually recognised locations where people can meet:

"It is interaction of people - if they are too peripatetic, then you can lose something in the structure... If everyone just sat in a different place - it is another barrier to actually going and dealing with that person face-to-face if you have to find where they are first".

However, whilst the lack of spatial coordinates in a place-centric view is problematic, it is perhaps the lack of these markers which identifies where work in many instances is actually undertaken. It could be argued that, the sharing of tacit knowledge is dependent not so much upon individuals inhabiting desks at fixed locations in their office building but in identifying the places where people can meet if they need to interact. Whilst the notion of being 'too peripatetic' is admittedly of concern to our informant above who relies on the constant exchange of tacit knowledge, he also makes clear it is the *interaction of people* that is significant. It is here, I would argue that knowledge is exchanged in a shared sense of place rather than in a space that delimits and specifies boundaries between what is perceived to be public, private and permanent. Whilst it follows that if we remove the "opportunity" which familiar spaces provide to exchange knowledge, we then, as one informant suggests, present a "barrier" to the "understood reality" of that exchange which it is claimed that place offers (Harrison and Dourish, 1996). However, if we concur with Giddens that place is simply space emptied of time, we have the opportunity to recreate the exchange of knowledge in ICT domains unhindered by the temporal determinants of location. Moreover, if as (Massey, 1993: 66) writes, we should conceive of places as "articulated moments in networks of social relations and understandings" instead of "areas with boundaries around", then it is important to identify the affordances such "articulated moments" offer if we are to accommodate distributed workforces that are unable to sustain the absolutes of spatial location. This paper contends that these 'moments… of understanding' occur in public, private and permanent places that respectively afford a sense of anonymity, seclusion and identity. A main finding of this study is that where people work is not necessarily determined by the characteristics of individual work spaces e.g. familiar office settings or furniture but in some instances by the intentions of those who meet in them. Whereas spatial markers may provide some degree of orientation as to where individuals are located at any one time, it is not at all conclusive that if these identifiers of permanence or ownership are removed, individuals are then unable to establish shared understandings of place. Some degree of balance is clearly required – between sustaining spatial opportunity and ensuring individuals share an understood reality of place where they can continue to work privately or publicly to meet and exchange knowledge with others. If the places where people work are established primarily through intention as opposed to setting we must identify distinct features of those places as well as the motivations for their formation.

5 What Constitutes a Place of Work?

Informants suggest that the underlying driver for the formation of place is enhanced productivity. The features we perhaps need to identify therefore are not spatial markers but the qualifiers of productivity that each place affords. For example one informant qualified the creative benefit of working from home as the facility "to move from room to room" or simply the ability to reflect by removing themselves from the view of others who may report to or work alongside them.

"I find work at home very productive. If you make your mind up and disciplined to get away out of office, you don't need to be in the office all the time, so that people can come and have access to you - you need time to go away and do that report..."

In such instances, qualifiers of productive merit apply predominantly to individual working. These can largely be measured against the degree of seclusion offered in constructing a sense of place that is private, say to work on a report as above or against the level of anonymity that a public sense of place affords, for example in a train carriage – a productive place for the creation of ideas.

"I commute quite a long distance every day from home, because I live about 60 miles away, so I spend quite a bit of time travelling to the office in the morning and travelling back in the evening.... I do actually do quite a lot of work on the train... and I think it is my most productive time actually - it's even more productive than sitting in the office. It is amazing the amount of new ideas that come to me just sitting on the train.".

Interestingly, there were no direct indicators in the informant responses as to what constitutes productivity in collaborative places. It could be argued that constructive phone conversations, meeting with others both face to face and online, informally and formally, are equally productive if well managed. Nonetheless, from my observations of informants in this study I would suggest that collaborative working is a facilitator to rather than qualifier of productive endeavour. Moreover, collaborative working should indeed foster productive relations in a shared sense of place. However, perhaps it is that the efficiency markers are less explicit here (and thus more difficult to qualify) than in individual working - an activity which, however creative, ultimately demands the production of some document or artefact that can act as a qualification of effort.

6 Who Works There?

If the places where people work are constituted primarily through intention as opposed to setting and the distinct features of those places are principally related to personal rather than collaborative work drivers, it identifies a curious paradox in that those who construct and inhabit a *shared* sense of place are individuals rather than groups or teams. This would concur with Felstead et al's aforementioned observation that worker type and occupation are clearer indications of the "social relations of production" (Felstead et al., 2001: 217).

Low discretion work that is "predictable, routine, standardised and rule-dominated" (Ibid) can be conducted within the fixed spatial collective settings of office, retail or factory workplaces. High discretion work, as instanced by the knowledge workers in this study, is "variable, complex and choice dominated" (Ibid) and demands more in terms of achieving an appropriate fit between human and environment. The fixed spatial settings are not always at their disposal so they must work harder to construct individualised *places* of work that are public, private or permanent, anonymous, secluded or owned, that reflect the variability, complexity and choice dominated aspects of their work, but which can be shared or understood by other high discretion workers. Indeed, informants identified areas of inappropriate fit between individuals and collaborative teams – designers meeting at a client site with the client present, but creating their own exclusive sense of place that seemed out of place to her own. A meeting that seemed productive to the designers yet to the client appeared wholly inefficient and detrimental to the work at hand:

"It can be quite challenging at times to have the client sitting so close to you, because we had an incident last week, where a few designers had a heated argument around the table, the client is sitting there watching them, and afterwards she said isn't this is a terrible thing? But we said, well, this is what the designers do and if they don't have arguments, they are not doing the job, basically. You know ... one person got up and kicked the chair, walked away and ... that is what the designers do, they have arguments during the projects, that is healthy and normal, but it can be quite oppressive if the client is sitting there watching you do it".

Informants also identified an inappropriate fit between their own understanding of permanent, owned organisational place and that of a work colleague who had returned from a prolonged period working at a client site:

"... there is a guy who works over at Project A permanently - and he's been there for nearly a year and a half. And you feel that they're not part of the group anymore - he's like a stranger when he comes into the office - you don't know really what he's doing there.

In effect, by not sustaining a shared sense of identity with work colleagues meant that as well as 'not belonging' at the client's place of work, he had become effectively anonymous within his own organisation. It appears that a sense of place is not only constructed individually between active participants but must be nurtured by the same if productive relations are to be sustained. Whilst individual endeavour continues to be the most productive it must be accommodated within an understood reality of place that is shared and owned by others. The informants in this study appeared to conduct a fair proportion of their work in public, private and permanent places that could afford a sense of anonymity, seclusion and ownership, respectively. These places did not constitute physical boundaries, territories, zones or other spatial terms but were qualified by the perceived markers of productivity which each place afforded. Moreover, these markers were predominantly individual as opposed to collective.

This does not mean such places are inhabited and sustained by isolates but shared by those who need to share an understanding of the variability, complexity and choice driven nature of their work.

7 Conclusion

The limited observations described above prompt me to suggest that the features researchers may need to identify are not the spatial but the efficiency markers that each place affords. Moreover, these qualifiers of productivity might allow us to conceive of place as "objective facts about a situation" (Gaver, 1996: 113) – a best fit between humans and environment. For example, when a corridor conversation between two colleagues becomes a secluded but creative opportunity to update each other on work in progress; an informal communal kitchen area at work encourages an opportunity to instil a sense of permanence and belonging though chat and banter, a train carriage provides an anonymous but productive moment to catch up on email. This concurs with Harrison and Dourish's view that places are distinct from spaces in that they cannot be described in terms of physical boundaries, territories, zones or other spatial terms (Gibson, 1979) but as a notional reality which is shared, discussed, negotiated with, concealed from, even communicated to others in workplace situations. This "understood reality" is qualified by the markers of productivity which each place affords.

Although this is essentially a scoping paper for further research, the policy implications discussed here are significant in that instead of seeking ICT solutions to accommodate fixed spatial limitations, researchers might focus on how individuals creatively work round the locational constraints to enhance their effectiveness. Moreover, when exploring how business users engage with contemporary work practices researchers must ensure that they don't simply observe the efficacy or otherwise of ICT systems in relation to the business processes and discrete user activities they are designed to map, but must also consider the intentions of individuals and how they use ICTs to improve their productivity at work. By understanding the places where people construct that work, the productive components that make up those places, we can develop an approach that is more suited to the types of worker who inhabit this ever-changing terrain. For example, when constructing ICT *spaces* for those who work in primarily fixed organisational locations, designers can transpose aspects of organisational process, describe particulars of task, and trace elements of activity. However, for nomadic, remote and distributed working, if we are to ensure an appropriate fit between these workers and their environment we require an approach that incorporates the notion of ICT *places* through which we can explore markers of productive merit, individual interpretation and creative intention.

References

1. Boyce, P.R.: Users' assessments of a landscaped office. Journal of Architectural Research, 3(3) (1974)
2. Brookes, M.J., Kaplan, A.: The office environment: Space planning and affective behaviour. Human Factors, 14 (1972)

3. Burgoon, J.K.: Privacy and communication. In: Burgoon, M. (ed.) Communication Yearbook, Sage, Beverly Hills, CA (1982)
4. Felstead, A., Jewson, N., Phizackea, A., Walters, S.: Working at Home: Statistical Evidence for Seven Key Hypotheses. Work, Employment & Society 15(2), 215–231 (2001)
5. Ferguson, S.D., Horan, H., Ferguson, A.M.: Indicators of Permanence in Workspace Features: Perceived Importance and Relationship to Workspace Satisfaction. Canadian Journal of Communication, 22(1) (1997)
6. Foley, S., Rosenberg, D., Crisp, M.-J., Kammas, S., Lievonen, M.: Human Environment Framework and Model, D22: Final report of Royal Holloway University of London contribution to the SANE (Sustainable Accommodation for the New Economy) (2003), EU, F5 IST-2000-25257
7. Gaver, W.: Situating Action: Affordances for interaction: The Social is Material for Design. Ecological Psychology 8(2), 111–129 (1996)
8. Gibson, J.J.: The ecological approach to visual perception, Dallas; London, Houghton Mifflin (1979)
9. Giddens, A.: Modernity and self-identity: self and society in the late modern age, Polity Press (1991)
10. Harrison, S., Dourish, P.: Re-Place-ing Space: The Roles of Place and Space in Collaborative Systems. In: Proceedings of the 1996 ACM Conference of Computer Supported Cooperative Work (1996)
11. Justa, F.C., Golan, M.B.: Office design: Is privacy still a problem? Journal of Architectural Research, 6(2) (1977)
12. Krohe, J.: What makes an office work? (1993)
13. Lievonen, M., Rosenberg, D.: Where Collaboration takes Place: Space and Place in the Study of Communication and Interaction in the Real-Life workplace (unpublished manuscript) (2005)
14. Massey, D.: Power-geometry and a progressive sense of place. In: Bird, J., Curtis, B., Putnam, T., Robertson, G. Tickner, L (eds.) Mapping the futures: local cultures, global change. London, Routledge (1993)
15. Nonaka, I.: The Knowledge Creating Company. Harvard Business Review, 96–104 (November-December 1991)
16. Nonaka, I., Takeuchi, H.: The knowledge creating company: How Japanese companies create the dynamics of innovation. Oxford University Press, New York (1995)
17. Oldham, G.R., Brass, D.J.: Employee reactions to an open-plan office: A naturally occurring quasi-experiment. Administrative Science Quarterly, 24 (1979)
18. Sundstrom, E., Burt, R.E., Kamp, D.: Privacy at work: Architectural correlates of job satisfaction and job performance. Academy of Management Journal, 23 (1980)

Development of an Affective Interface for Promoting Employees' Work Motivation in a Routine Work

Hidenori Fujino, Hirotake Ishii, and Hiroshi Shimoda

Graduate School of Energy Science, Kyoto University
Gokasyo, Uji, Kyoto, 611-0011, Japan
fujino@uji.energy.kyoto-u.ac.jp, hirotake@ieee.org,
shimoda@energy.kyoto-u.ac.jp

Abstract. A routine work requires few skills, so employees can be skilled easily. Therefore, the productivity and the quality of work greatly depend on employees' work motivation. However, as a routine work is often not interesting to almost all employees originally, a method for promoting and keeping employees' work motivation is required for keeping or increasing the productivity and the quality of work. Therefore, the objective of this study is to develop the method for promoting employees' work motivation in a routine work especially with an affective interface and applying the persuasion technique.

1 Introduction

A routine work, for example assembling products in factories, monitoring displays for controlling large plants and office work of inputting numerical data for accounting money and so on, requires few skills ordinary and employees can be skilled easily, so the productivity and the quality of work greatly depend on employees' work motivation. However, as a routine work is often not interesting to almost all employees originally and the significance of the work is often not clear and not easy to understand, employees' work motivation is decreased easily. Therefore method for promoting and keeping employees' work motivation in a routine work effectively is required for keeping or increasing the productivity and the quality of work.

In this study, the authors aim at promoting their understandings to the significance of the work by an affective interface as effective and efficient way for promoting and keeping employees' work motivation in a routine work. Especially the authors think "work motivation" of "attitude to the work", so an affective interface is developed with applying the force compliance which is one of persuasion techniques based on the theory of the cognitive dissonance.

2 Proposed Model for Employees' Promoting Work Motivation

2.1 Direction of Developing the Method

Motivation is generally related with three feels; the feel of autonomy or self determination to his behavior in the situation, the feel of competence like ability or

M.J. Smith, G. Salvendy (Eds.): Human Interface, Part II, HCII 2007, LNCS 4558, pp. 873–882, 2007.

pride to his behavior, and the feel of relatedness with the society through his behavior (Deci and Flaste, 1995;Gange and Deci, 2005). Especially in the context of work, the feel of autonomy and the feel of pride are connected with the feel of relatedness, so if employees could deeply understand the social significance of their work and feel the relatedness with the society strongly, their feels of autonomy and pride to their work would be improved. Since a monotonous work usually needs only a few special skills, they have already had the feels of ability. Therefore, even if the understandings of the social significance of their work are promoted, employees' work motivation would be increased or at least be kept. So the direction to develop the method for promoting employees' work motivation is toward developing the method for promoting the understandings of the social significance of their work.

2.2 Overview of the Process of Attitude Formation and Persuasion Technique

Since it can be thought that employees' work motivation is appearance of the employees' attitude to their work, the persuasion techniques are seemed to be applicable to promote the understandings of the social significance of their work.

Generally, the attitude consist of the cognitive component, the affective component, and the action component (Katz,1960).The cognitive component works at the first step of the attitude formation. This component manages the perception of the object and the information related with it. The affective component works at the second step of the attitude formation. This component manages the evaluation of the perceived object and information according to the value system of the person, for example "like" or "dislike", "positive" or "negative" and so on. The action component works at the last step of the attitude formation. This component manages the action and behavior to the object which are observed outside actually. When the person can behave autonomously to an object, his attitude to the object is expressed in his behavior in that situation. Therefore, in the case that his behavior affects the object positively, his attitude to the object is also positive and oppositely in the case that his behavior affects negatively, his attitude to the object is also negative.

If a new object and new information are inputted to the person, they are handled with these components sequentially like Fig.1 and the attitude to the object is formed. Formed attitude is memorized in the memory system of the person.

Many persuasion techniques have been proposed already. Those techniques are classified according to the component which each technique approaches. Typical persuasion techniques are shown in Fig.1.

2.3 The Methodology Applying the Theory of Cognitive Dissonance

For promoting employees' understandings of the social significance of their work, it is necessary not only to show the information of rational reason that their work is significant for the society, but also to support that the information would be evaluated more positively because the person who has had low motivation to their work already would not necessarily evaluate that information positively. That is, support to approach the affective component is required.

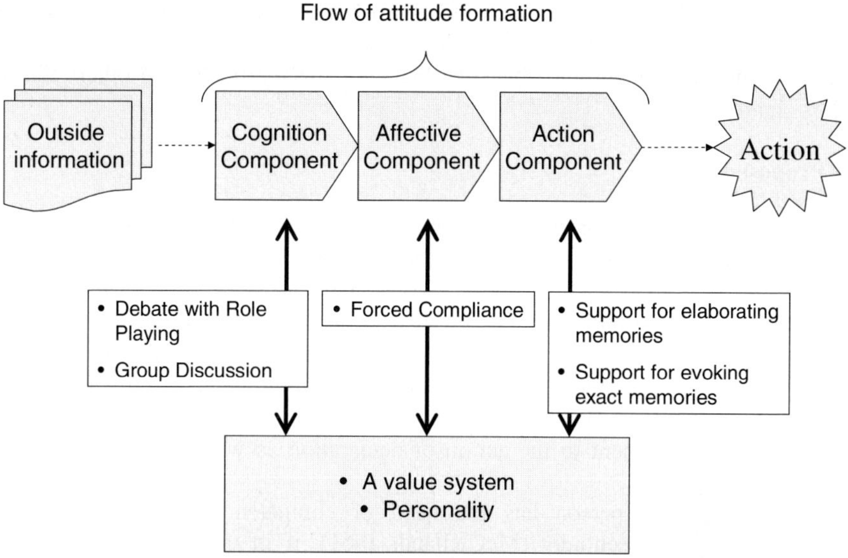

Fig. 1. The process of attitude formation and persuasion

In this study, therefore, the method to which "the force compliance" is applied, which is the persuasion technique based on the theory of cognitive dissonance (Festinger and Carlsmith, 1957). According to the theory of cognitive dissonance, when the person have the negative (or positive) attitude to an object and he behave to the object positively (or negatively), if he feel the strong stress of cognitive dissonance, he would change his own attitude from the negative to the positive (from the positive to the negative). Although having the negative (or positive) attitude to an object and behaving to the object positively (or negatively), if he feel only a little stress, for example because of large reward prepared for behaving so, his attitude would not be changed. For the prompting the strong stress of cognitive dissonance, it is necessary to make the person have the strong commitment to his own behavior.

The commitment is seemed to consist of three factors; the feel that the locus of the causality of his behavior is in himself, the publicness of his behavior, the impossibility of returning the state before his action (Kline and Peters, 1991). That is, when he thinks that he behave according to his own intention, when he thinks he would behave publicly, or when he thinks that the situation could not be returned to the state before his action be taken, he would have the strong commitment to his own behavior.

In conclusion, if it could be promoted to declare to others with rational reasons that his work is significant for the society for leading the feel of the impossibility of his behavior in the context of work for prompting his awareness of the publicness of his behavior and if the feel could be promoted that the locus of the causality of his behavior is in himself, he would be expected to evaluate the information positively that is the content of his own declaration and change his attitude to his work more positively, and to be more motivated to his work, because he would feel the strong stress of cognitive dissonance between his existing negative attitude and belief that his work is not significant and his own behavior that he declare to others that his work is significant.

2.4 Proposed Model

In this section, the model is proposed which is the basis for practicing the measure based on the methodology designed in the previous section more efficiently and effectively.

Generally, a person has the tendency of finding the personality to the artificial object when it would behave like human, even if he has known that the object is artifact. Therefore, it is expected that even when he would declare it to social agents like animation characters that his work is significant for the society, he would have such a strong commitment to the action of declaration as when he declare to other persons.

Moreover, since a person has the need of competence, in other word, the achievement motive essentially (McClelland, 1961), if in the situation of game, a person would be motivated to make his best to get the good result of the game. Therefore, it is also expected that if it is possible to prepare the situation of connecting

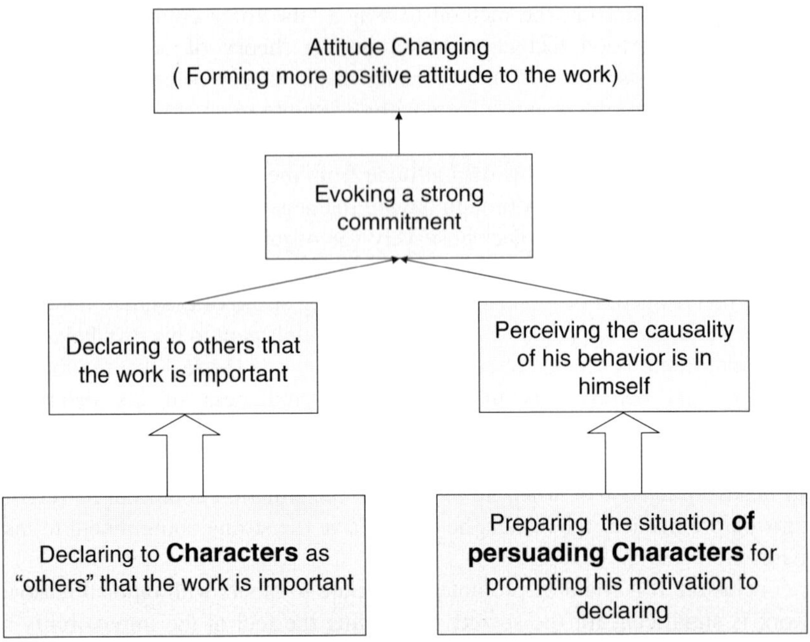

Fig. 2. Proposed model for promoting the understandings to the significance of the work

the action of declaration with the game, like persuasion, the person would be motivated to declaring that and feel the causality of his action of declaration is in himself.

The proposed model based on these considerations result is shown in fig.2.

In the next section, the experiment held for evaluating the validity of proposed model is explained.

3 Experiment for Evaluating the Validity of Proposed Model

3.1 Design of Experiment

3.1.1 Method for Evaluating the Validity of Proposed Model

The method for evaluating the validity of the model is firstly considered;

1. A simple and monotonous work is prepared and subjects engage in that work actually.
2. At a time during the work, subjects play the game based on the proposed model.
3. It is confirmed whether there is a difference measuring the attitude to the work before and after the game.

However, attitude changing is affected by not only playing the game but also other factors, for example, monotony of the work, experience to the work, and effect on elaboration of his memory about the significance of the work for considering the message whether he feel the strong stress of cognitive dissonance or not.

Therefore, since just the effect of the game to make subjects feel the strong stress of cognitive dissonance should be evaluated in this experiment, two following conditions is prepared and compared each other to confirm the effect of the game. First condition is the condition that subjects play the game based on the proposed model at the exact time during the work. This condition is called "Game Condition". Second condition is the condition that subjects input the message to tell the significance of the work to the prepared application which is very simple and in which any character is not displayed and any information about the result of persuasion is not returned to subject. This condition is called "Not Game Condition". After this, either playing the game in Game Condition or inputting the message in Not Game Condition is called "playing the game".

3.1.2 Questionnaire for Measuring the Attitude to the Work

The questionnaire prepared for measuring the attitude to the work is in table 1. All subjects are required to answer this questionnaire on a scale from 1 to 7 where 1 means "I don't think so at all" and 7 means "I think so strongly". Actually, because the Cronbach alpha of this questionnaire is 0.74 which is calculated from the result of all answers of all subjects in the experiment, it is thought that there is the adequate consistency of this questionnaire.

3.1.3 Prepared Works and Subjects

The work prepared for this experiment is to input the numerical dates which were written on papers by the hand and the dates consist of two kinds of dates. Subjects are

Table 1. Questionnaire for measuring the attitude to the work

Q.1	Engaging in this work, Do you think that you are useful for the society?
Q.2(inverse)	Engaging in this work, do you feel that you are rejected from the society?
Q.3(inverse)	Engaging in this work, do you feel a strong mental stress?
Q.4	Is this work enjoyable for you?
Q.5(inverse)	Do you feel that this work gets tired and boring easily?
Q.6	Are you proud to engage in this work?
Q.7	Do you feel that this work is significant for the society?
Q.8(inverse)	Do you think that you need not work hard because this work is not so important?
Q.9	Are you interested in this work?
Q.10(inverse)	Do you think that your good performance isn't related with someone's happiness even if you work hard?
Q.11(inverse)	Engaging in this work, do you get irritated?

explained with the backgrounds of these dates that each date were gathered in the quite different scene each other. All subjects are gathered in the condition of having experiences of engaging such a desk work by the temporary staffing agency.

Actually, subjects consisted of 15 women and 5 men. The oldest subject's age was 37 years old and the youngest subject's age was 21 years old. The average of all subjects' age was 30.5 and the S.D. was 4.6.

Considering the effect of learning the skill for the experimental tasks and of the order of the experimental tasks, subjects are grouped into four groups of 5 persons and subjects in each group engage in the experimental task according to the order shown by table 2.

Table 2. The order of experimental task of each group

	Group 1 (Subject1~5)	Group 2 (Subject6~10)	Group 3 (Subject11~15)	Group 4 (Subject16~20)
Work and Condition in the morning	Work A and Not Game	Work A and Game	Work B and Not Game	Work B and Game
Work and Condition in the afternoon	Work B and Game	Work B and Not Game	Work A and Game	Work B and Not Game

3.1.4 The Flow Chart of the Experiment

The time chart of the experiment is shown table.3. Although subjects should play the game for 40 minutes in each experimental condition according to this time chart, some subjects may finish the game in shorter. At that case, those subjects would be told to wait at his desk until next task after answering the questionnaire.

Table 3. Time table of the experiment

9:00 - 9:10	Briefly explaining the experiment
9:10 - 9:15	Explaining the work in the morning
9:15 - 9:20	1st answering questionnaires about the work in the morning
9:20 - 10:10	Engaging in the work in the morning
10:10 - 10:15	2nd answering questionnaires about the work in the morning
10:15 - 10:30	Engaging in the work in the morning
10:30 - 10:35	3rd answering questionnaires about the work in the morning
10:35 - 10:40	The break time
10:40 - 11:20	Playing the Game (Not Game)
11:20 - 11:25	4th answering questionnaires about the work in the morning
11:25 - 12:15	Engaging in the work in the morning
12:15 - 12:20	5th answering questionnaires about the work in the morning
12:20 - 12:35	Engaging in the work in the morning
12:35 - 12:40	6th answering questionnaires about the work in the morning
12:40 - 13:30	Lunch Time
13:30 - 13:40	Explaining the work in the afternoon
13:40 - 13:45	1st answering questionnaires about the work in the afternoon
13:45 - 14:35	Engaging in the work in the afternoon
14:35 - 14:40	2nd answering questionnaires about the work in the afternoon
14:40 - 14:55	Engaging in the work in the afternoon
14:55 - 15:00	3rd answering questionnaires about the work in the afternoon
15:00 - 15:05	The break time
15:05 - 15:45	Playing the Not Game (Game)
15:45 - 15:50	4th answering questionnaires about the work in the afternoon
15:50 - 16:40	Engaging in the work in the afternoon
16:40 - 16:45	5th answering questionnaires about the work in the afternoon
16:45 - 17:00	Engaging in the work in the afternoon
17:00 - 17:05	6th answering questionnaires about the work in the afternoon

The break time is prepared as shown, and if subjects want to go to the rest room for a few minutes, they can go freely even in any time.

3.2 Construction of the System for the Experiment

3.2.1 Game System

The rule of the Game is as follow; there are 10 characters to be persuaded in the game and a subject are required to make a message explaining the significance of the work which would be done at that time to persuade each character. Subjects are explained that each character would remain in the game window if the significance of the work would be adequately expressed in subject's message, and also explained that in the other hand the character would be back if not.

Actually, however, all characters accept any message. The reasons for setting so is as follow;

- The algorism of this setting is very simple algorism, so the game system is constructed very easily.

- When the case would be prepared that the persuasion be failed, the impossibility of returning the state before his action would be weakened compared with the case of succeeding the persuasion.
- When he would fail the persuasion many times, the subject would come to be less motivated to the game autonomously because the need of competence of him could not be satisfied through playing the game. Moreover, the many time failure might lead the subject's self persuasion which would change his attitude to more negative one.

If subjects could play the game repeatedly, it has a rational reason to prepare such a case for the game being more challenging. Actually in this experiment, however, subjects play only one time because of controlling all subjects' experimental conditions so it is necessary to be such a setting.

The one scene of the game system is shown in Fig.3. In this scene, the bird character is the character to be persuaded. The clerk character is also shown in this scene. This character is the organizer of this game and manages the proceedings of the game. For example, she calls in the character one by one to be persuaded and introduces each character's personality to the subject. The characters to be persuaded do not speak anything and they only response the subjects' message. All the messages from the game system is spoken by the clerk character.

Game started, the subjects need to persuade the characters one by one with referring the personality given by the clerk as a hint and.

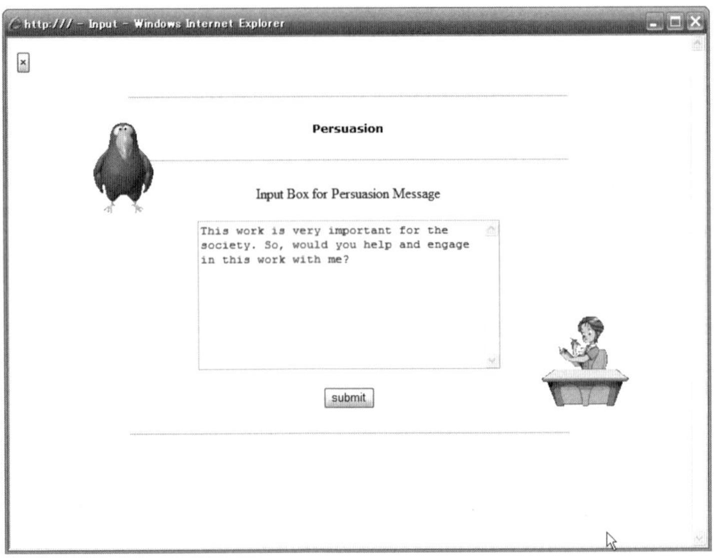

Fig. 3. Window of the Game system developed for the experiment

3.2.2 Not Game System

One scene of the system for the Not Game Condition, called Not Game system, is shown in Fig.4. A subject is needed to make the sentence of explaining the significance of the work which would be done at that time with referring the question sentence expressed in the window (following with Q.1 is the sentence in Fig.4) and input it to the textbox. It differs from the game that the character agent is not displayed at all, and that there is no feedback for the inputted sentence.

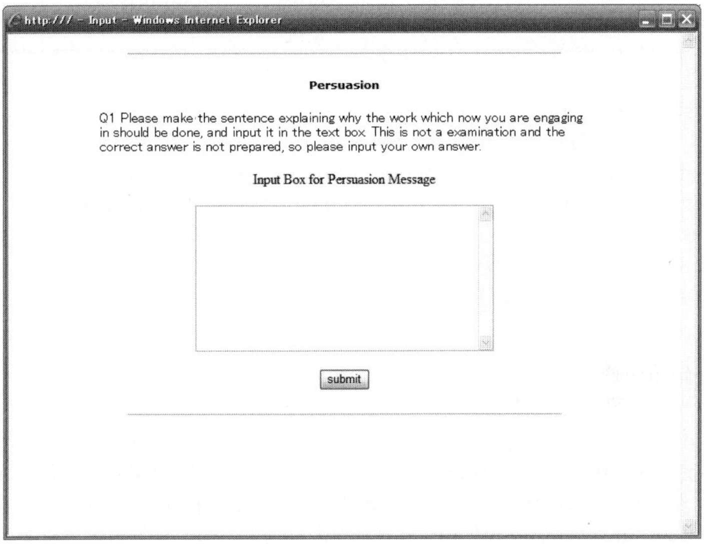

Fig. 4. Window of the Not Game system developed for the experiment

There are 10 questions in this system and each question is same meaning as the hint sentence given by the clerk character of the game as the character's personality. Actually it was confirmed by two university student that each the question sentence and each the hint sentence are same meanings.

3.3 Result

The result of the experiment is in table 4. The averages are calculating by 15 subjects' differences of the score of the questionnaire before and after playing the Game.

5 subjects of 20 subjects were discarded for the following reasons;

- 3 subjects (2 subjects in Group 1 and 1 subject in Group 3) understood about the purpose of the experiment by mistak. They understood that the questionnaire is prepared for measuring not the attitude to "the work" which was done at that time, but the attitude to the "experiment".
- 1 subject in Group 3 forgot to answer the questionnaire at the time of 2nd answering questionnaires about the work in the morning.
- The results of both conditions of 1 subject in Group 4 were seemed to be outlier by the Thompson's test for finding outliers, ($t = 0.005 < 0.01$ in Game Condition, and $t=0.007 < 0.01$ in Not Game Condition).

Table 4. Average of all subjects' differences of the score of questionnaire before and after Game or Not Game (N = 15)

Experimental Condition	Average	Variance	F test	T test(One-tail)
Game	1.93	3.77	0.32 (>0.05)	0.03 (<0.05)
Not Game	-1.13	4.96		

Obviously the average of the Game condition is significantly larger than the average of the Not Game condition. For this result, the Game system based on the proposed model is thought to promote subjects' attitude change effectively. This means that proposed model which consists of declaring to characters and preparing the situation of persuading characters is seemed to be valid as the model for promoting the understandings of significance of the work and promoting or at least keeping employees' work motivation.

4 Conclusion

In this study, for the purpose of promoting employees' work motivation in a routine work which, the model for promoting the understandings of the significance of the work was proposed which is applied the theory of cognitive dissonance. Moreover, for evaluating the validity of the proposed model, the subjects experiment was conducted by the game system constructed based on the proposed model. As the result of the experiment, the subjects' attitude to the monotonous work in Game Condition was significantly positive compared with in Not Game Condition. Therefore, the validity of the proposed model was confirmed.

References

Deci, E.L., Flaste, R.: Why We Do What We Do: The dynamics of personal autonomy. G. P. Putnam Son's, New York (1995)

Gagne, M., Deci, E.L.: Self-determination theory and work motivation. Journal of Organizational Behavior 26, 331–362 (2005)

Katz, D.: The functional approach to the study of attitudes. The Public Opinion Quarterly 24(2), 163–204 (1960)

Festinger, L., Carlsmith, J.M.: Cognitive consequences of forced compliance. Journal of Abnormal and Social Psychology 58(2), 203–210 (1959)

Kline, C.J., Peters, L.H.: Behavioral commitment and tenure of new employees: A replication and extension. Academy of Management Journal 34, 194–204 (1991)

McClelland, D.C.: The achieving society. Princeton, NJ: Van Nostrand Reinhold (1961)

Ecological Interface to Enhance User Performance in Adjusting Computer-Controlled Multihead Weigher

Yukio Horiguchi[1], Ryoji Asakura[1,*], Tetsuo Sawaragi[1], Yutaka Tamai[2], Kazufumi Naito[2], Nobuki Hashiguchi[2], and Hiroe Konishi[2]

[1] Graduate School of Engineering, Kyoto University, Japan
[2] ISHIDA Co. Ltd., Japan
horiguchi@me.kyoto-u.ac.jp, asakura1117@hotmail.com, sawaragi@me.kyoto-u.ac.jp, tamai@ishida.co.jp, naito@ishida.co.jp, hasiguci@ishida.co.jp, hiroe@ishida.co.jp

Abstract. This paper presents a case study on developing a smart user interface for supporting the adjustment works on automated weighing machines of computer-controlled multihead weighers. Based upon the theoretical and practical framework of Vicente and Rasmussen's Ecological Interface Design (EID), we clarified the functional structure of the work domain in terms of the means-end relations, and visualized it on the screen displays to encourage the human operator's "direct perception" of the meanings or values of his practicable actions to those automated processes. Comparative experiments using test subjects with a variety of skill levels confirmed the effectiveness of the redesigned user interface that can facilitate unskilled operators appropriately evaluating and effectively responding to their immediate work situations, and that will take the place of the experts' knowledge and insights on the works as one of the distributed resources for cognition.

Keywords: Ecological Interface Design, Work Domain Analysis, factory automation, computer-controlled multihead weigher, user interface.

1 Introduction

A computer-controlled multihead weigher (Fig. 1) is an automated weighing machine comprised of a plurality of balances or *heads* to measure weights of article batches, some of which would be selected and combined at every measurement cycle to produce a merchandise product within specified weight range (i.e., a "combination weighing" machine). Vibrating feeders constitute its article conveyance system that allows the machine to handle a variety of products ranging from fragile food products to detergent, pharmaceuticals and metal parts, while its rapid combination calculation system delivers higher speed and increased efficiency of weighing (up to 180 measurements per minute [1]). On the other hand, there are a number of control parameters in this automation such as the amplitude and duration of vibration of individual conveyers, and human operators have to adjust them adequately to changing

* He is currently with Hitachi, Ltd.

M.J. Smith, G. Salvendy (Eds.): Human Interface, Part II, HCII 2007, LNCS 4558, pp. 883–892, 2007.

measurement conditions for the machine to go on functioning. Adjustment of the conveyance system is required when the weighers get started running with new sorts of articles or when they tend to produce more no-good products. As to this human work, significantly greater accuracy and efficiency of automatic weighing do require much higher level of skill and knowledge of the operators in conditioning the systems. Dependence of the automation performance on the individuals' ability should be dissolved by some improvements in this work domain because flexible automation involves minimized effects of such variable human factors on production management. We seek a resolution of this problem in redesign of its user interface.

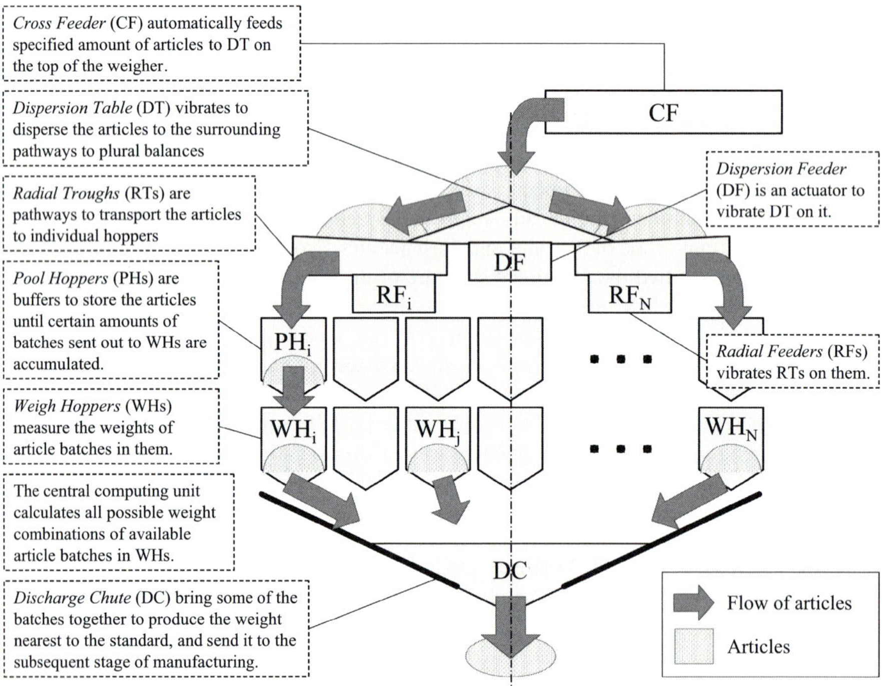

Fig. 1. Physical components of the Ishida multihead weigher [1] through which divided article batches are combined to produce a large amount of merchandise products within specified weight range with higher speed and increased efficiency

Development of a new user interface for the multihead weighers goes on the theoretical and practical framework of Vicente and Rasmussen's Ecological Interface Design (EID) [4-6]. Our former study [2] analyzed the skills of weigher experts compared with novice and intermediate operators, and it has clarified the experts' sufficient knowledge and deep insight to specify the actual state of the machine, both of which are closely associated with their available operations. This result suggests the experts have acquired the semantics of their work domain of the multihead weighers from their experiences, and thus direct our attention to the methodology of

EID. This paper presents our practical study on developing a "smart" user interface for supporting the adjustment works on the computer-controlled multihead weighers.

2 Developing Ecological Interface for Conditioning Automation of Multihead Weigher

EID is a theoretical and practical HCI design framework, which aims at embodiment of a functionally *transparent* mediator, i.e., "ecological" interface, between a human operator(s) and a complex mechanical system [4-6]. What is emphasized there is "direct perception" [7] of the *semantics* of work domains (which state the system is in now, and how much valuable possible actions are for the system's goal) as well as "direct manipulation" [7] of those displayed elements. "Smart" interfaces ask for no mental arithmetic of human operators. Our former study confirmed the experts have acquired the semantics of their work domain (i.e., of the multihead weigher) from their experiences, and thus direct our attention to the methodology of EID.

2.1 Analysis of the Work Domain Structure

As the first step in applying EID to our design problem, the inherent functional structure of the target work domain should be clarified in the form of connections among functional elements that describe their means-end relationships. *Work Domain Analysis* sort out those functional connections into hierarchy so that concrete elements are to be specified as the means to make abstract functions effective and to compose the whole system's behavior in the level of functional purpose [3] eventually. This analysis can give designers a useful guide for organizing the display of all relevant information variables to be "externalized" in accordance with their means-end relationships.

After a lot of discussion with an expert operator, we have gotten a *work domain model* of the multihead weigher system as shown in Fig. 2. This model describes the purpose of this system from two different perspectives: the efficiency and the accuracy of the combination weighing. The former perspective represents maximization of the productivity of the weighing process (i.e., elimination of defectives below the standard on product weight) while the latter pursues improved precision of the measured weights to meet the standard. In order to realize "accurate combination weighing," the worker should pay attention to the number of alternatives to a combination weight because much more candidates assure higher probability of products close to the target weight. The connection between "accurate combination weighing" and "adequate pop size for combination" represents this aspect of the domain knowledge. On the other hand, effective combination weighing demands the consideration of the mass flow in this process as well as the number of alternatives because it must be controlled adequately to achieve higher throughput of the weighing machine. The two connections from "effective combination weighing" represent these relationships. All other linkages in the work domain model derive from the means-end relationships of the same kind.

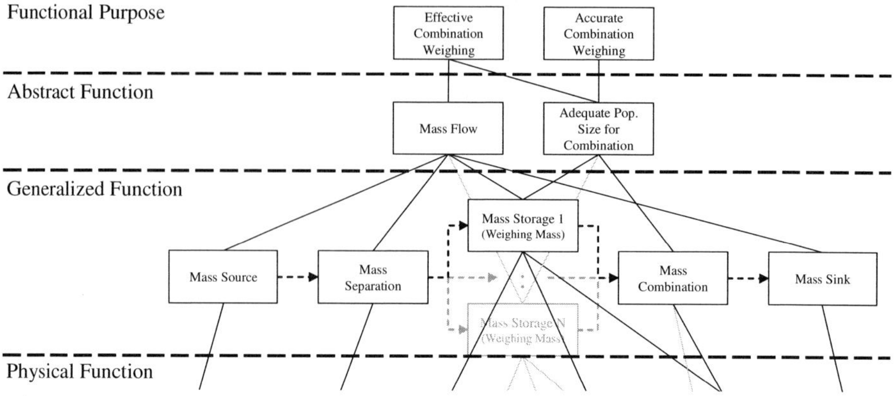

Fig. 2. Part of the work domain model for the multihead weigher derived from discussion with an expert operator

The work domain model shown in Fig. 2 was utilized to extract the basic requirements for the new version of the user interface. Table 1 is a partial list of variables extracted from this model by asking how we can measure those individual functional elements. They are to be organized in the interface display in a fashion consistent with the task demands.

Table 1. Information requirements derived from the work domain model. Variables in bold font are manipulatable by human operators.

Level	*Variables*
Functional Purpose	· Operation rate
	· Accuracy of product weight
Abstract Function	· # of heads selected for combination weight
	· Variation of batch weight among heads
Generalized Function	· Weights of batches
	· Selection of heads
	etc.
Physical Function	· **DF's vibration (amplitude & duration)**
	· **RFs' vibrations (amplitude & duration of each)**
	etc.
Physical Form	· Layout of physical components
	etc.

2.2 Development of Ecological Interface

Based on the former analysis on the work domain, we developed a new user interface for conditioning the automated weighing machine. Fig. 3 shows the two different interfaces for the control panel of the multihead weigher: the conventional version (a) and the newly developed, or ecological, version of the user interface (b). Fig. 4(a) highlights their difference, i.e., the redesigned part of the interface.

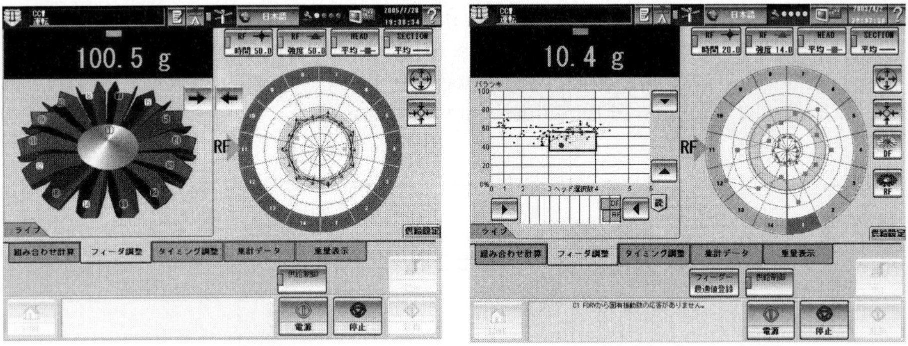

(a) Conventional UI (b) Ecological UI

Fig. 3. Two different user interfaces for the multihead weigher: the conventional version (a) and the (newly developed) ecological version (b)

This ecological interface makes visible a couple of intermediate variables that connects several operational variables of physical components to the resultant performance of the weighing machine. They are the (average) number of heads selected for combination weight and the variation of batch weight among heads. The discussions and analysis on the work domain revealed these parameters are of importance in skillful adjustments of the automation while not available for the operators in the user interface so far. As shown in Fig. 4(a), they are used to represent the state of the weighing process in the two dimensional chart (○,A) with the target area indicating the range where high measurement performance can be expected. This representation transforms the operators' task into a more concrete work of guiding the dot of the machine state to the inside of the target frame. It is, however, still a difficult work because there is no guidance for the operators to take the system state for the better in the chart using controls they can manipulate. Some intuitive ways to act on those two variables are necessary. For this purpose, we prepared new two pairs of button controls of ○,B and ○,C, which approximately serve as the horizontal and vertical controls of the system, respectively. Fig. 4(b) accounts for the relations that the functional design of the pairwise control ○,B depends on, indicating that the operators must manipulate the vibration parameters of DF and RFs in some way to regulate the average number of heads selected for combination weight. The function of the control B is tuned so as to balance the number of heads selected for combination weight by means of modifying DF's and RFs' vibration parameters. Fig. 4(c) presents the ground for the design of the pairwise control ○,C. By means of modifying RFs' amplitudes of their vibrations, the control ○,C tries to balance the batch weights among all the heads. These controls, so to speak, provide users with media for *quasi*-direct manipulation.

On the other hand, the weigher system has much delay in responding to manipulations of any vibration parameters. This attribute makes it difficult for the operators to recognize the effects of their operations onto the automated weighing process, which might depress the usability of this system seriously. On this issue, a pseudo indicator was introduced for providing a guesstimate of the future state transition in response to the current user operations. Fig. 5 zooms in the state chart ○,A in Fig. 4(a) and illustrates this function as the mark ○,D, which also presents the target range (a thick-lined rectangle in the center of the chart) for and the transition history (points connected with gray lines) of the weighing process. The operators hereby can approximate the direction of the future state transition.

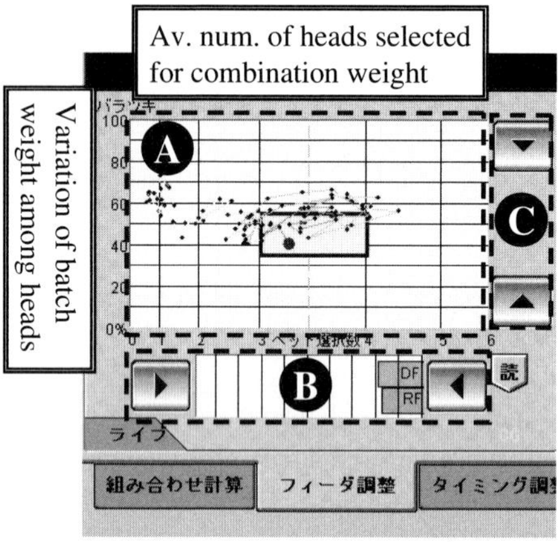

(a) Modified part in Ecological UI

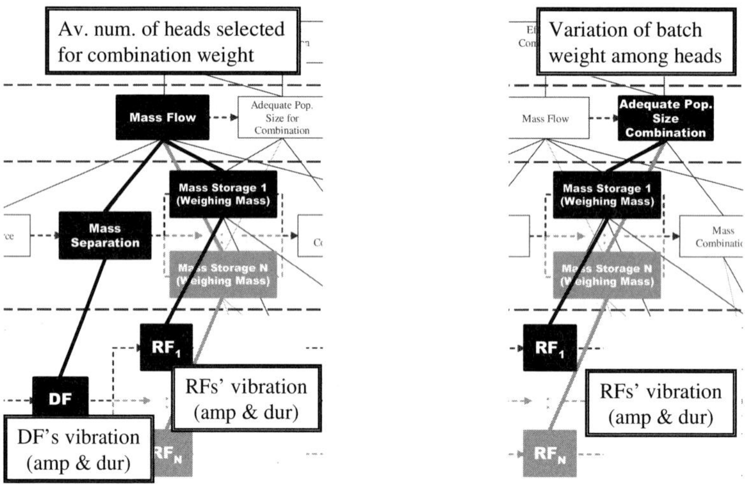

(b) (c)

Fig. 4. The modified part of the display in the ecological control panel is extracted in (a), and the two pairwise controls newly introduced as for the controls ○,B and ○,C are explained in terms of the means-end relationships (b) and (c) in the work domain model, respectively

All of the graphical items newly designed in the ecological interface were summarized as above. The other parts of the display are still common to the current or conventional user interface.

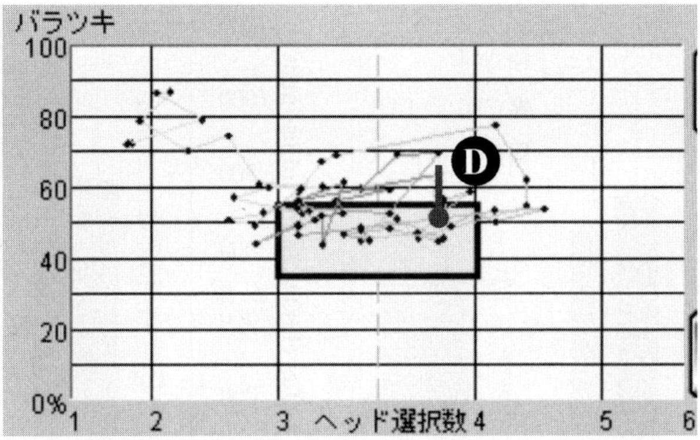

Fig. 5. The state chart representing the current state of the weighing process with the target range and its transition history. Where, a pseudo indicator ○,D provides a guesstimate of the future state transition in response to the current user operations.

3 Performance Evaluation of New User Interface

In order to evaluate the user interface newly developed, we conducted a comparative experiment of twelve test subjects [A to L] with a variety of skill levels using the two versions of user interfaces. The test operators divided into the two groups of subjects to balance the effect of order of trials. The first group of subjects (n = 6) performed one trial using the conventional UI and then another using the ecological UI (therefore labeled as "CUI → EUI") while the second group (n = 6) did in reverse order (labeled as "EUI → CUI"). The same experimental task was used in the both UI conditions, in which subjects were required to make adjustments of the control parameters of the automated weighing to the articles to be measured without any information except through the control panel. Their work performances were evaluated in terms of the accuracy of their conditioned weighers' performance and the amount of time required for their adjustment works, as well as their subjective workload ratings.

Fig. 6 compares the two user interfaces in the former two perspectives for evaluation, proving the successful conclusion of the new user interface redesigned for performance improvements. Fig. 6(a), on the one hand, plots mean values of product weights the weigher produced during 100 measurement cycles after conditioned (the target weight of products was 10[g]) where a whisker bar shows the plus one standard deviation for each data set. This result confirms the ecological interface can enable human operators to adjust the machine for more precise and accurate weighing than the conventional. On the other hand, Fig. 6(b) compares the two in terms of the average number of measurement cycles taken for adjustment, indicating the overall tendency that the subject operators finished their work using the ecological interface much faster than using the conventional. The both plots show smaller standard deviations, or reduced variations among individuals, in the ecological UI condition, which prove the dissolved dependence of the automation performance on one human factor, i.e. individuals' skill level.

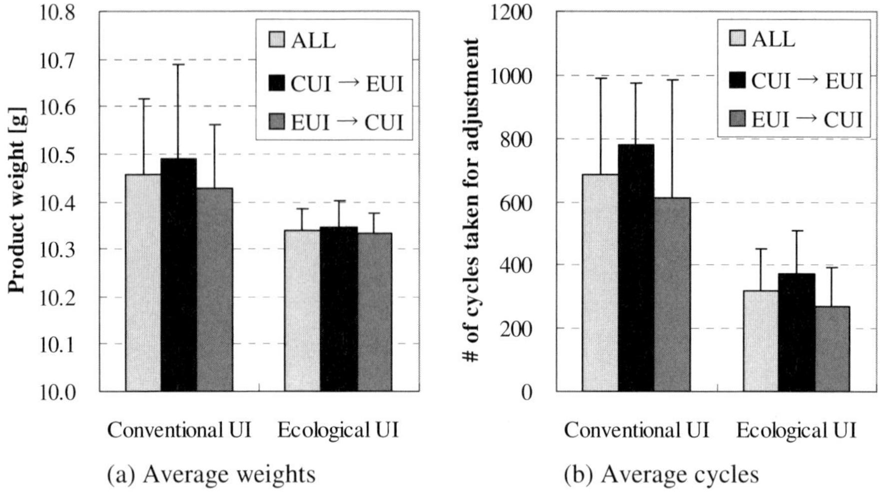

Fig. 6. Comparing the two different UIs in terms of the accuracy of the conditioned weighers' performance (a) and the time required for adjustment works (b)

Besides these objective performance measures of the subjects' conditioning the multihead weigher, we quantified the workload they felt on the adjustment works using NASA-TLX, which is one of the most popular assessment tools for subjective workload. Fig. 7 plots the results of all the subjects' works in terms of the measurement cycles required for adjustment and the weighted workload (WWL) score obtained from them. In the former work of this study [2], we have gotten a piece of knowledge that more skillful operators would rate the adjustment work through the conventional user interface as lower workload demands, and thus the WWL scores can be regarded as an index of individuals' skill. Fig. 7(a) shows this tendency again when you see the data in parts of the subjects A to E and the subjects F to L since the former group of subjects used a real machine in the experiment while the latter used a computer simulator of the weigher. The number of cycles required for adjustment increases in proportion to the WWL score. Fig. 7(b) does not show such a tendency, and confirms the alleviation of workload from the conventional adjustment work, especially for the operators unskilled.

All the results of this experiment validate the effectiveness of our ecological interface developed for the work domain of the multihead weighers. It can fairly improve the work performances of unskilled operators, thereby reducing deviations from the standard of product weight no matter who conditions the automation behavior. It can facilitate unskilled operators adequately evaluating and effectively responding to their immediate work situations. On the contrary, some of the operators who have sufficient skill with the conventional user interface put a low value on the ecological one because the latter changes the adjustment work into different one where they are required to guide a dot to the inside of the target frame. As weighings by the multihead weigher come from decisions based on combinational computing,

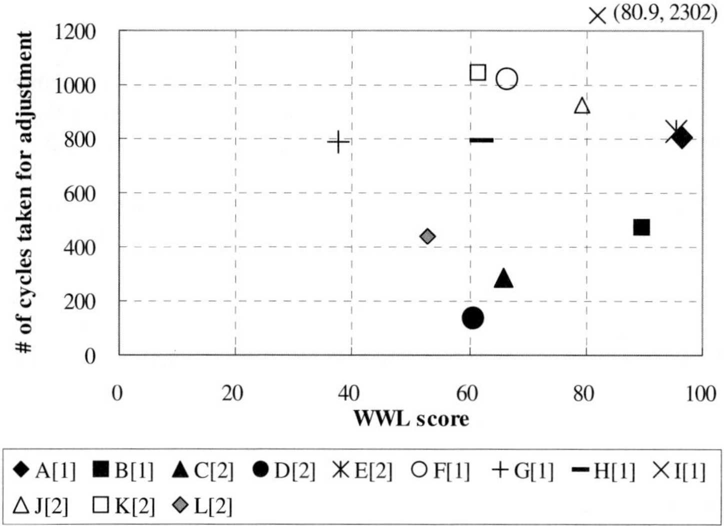

(a) Conventional UI

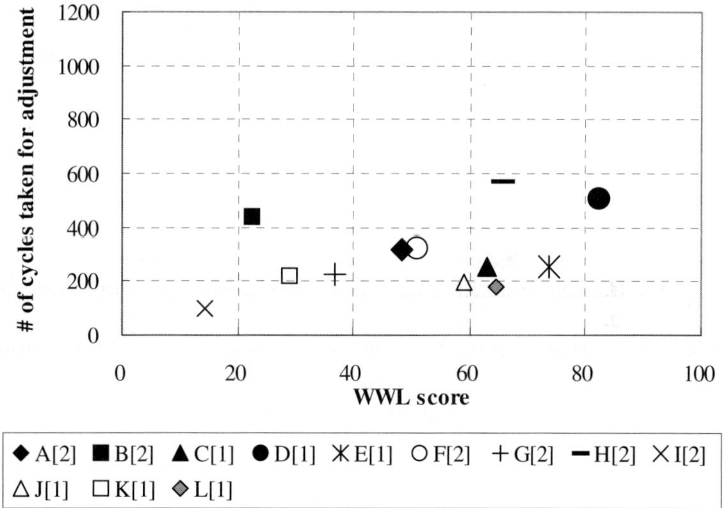

(b) Ecological UI

Fig. 7. Results of all the subjects' works plotted in terms of the measurement cycles required for adjustment and the WWL score, separated into the two UI conditions.

the dot in the state chart would often make "leaping" behaviors that might frustrate them. The current version of our ecological interface has no information resources to give them a reason for those behaviors and their mechanisms. This is the most important point we have to deal with in our future work.

4 Conclusion

This paper presented our study and development of a smart user interface for supporting the adjustment works on automated weighing machines of the computer-controlled multihead weighers. Based upon EID framework, we clarified the inherent functional structure of this work domain in terms of the means-end relations, and visualized part of it on the screen displays to encourage the human operator's "direct" perception of the meanings or values of his practicable actions to those automated processes. Comparative experiments using test subjects with a variety of skill levels confirmed the effectiveness of the redesigned user interface that can facilitate the unskilled operators appropriately evaluating and effectively responding to their immediate work situations. This externalization of core relational information will take the place of the experts' knowledge and insights on the works as one of the distributed resources for cognition.

References

1. http://www.ishidajapan.com/
2. Asakura, R., Horiguchi, Y., Sawaragi, T., Tamai, Y., Naito, K., Hashiguchi, N., Konishi, H.: Operational Skill Analysis of Computer-Controlled Checkweighing System for Improving User Interface toward Enhanced Performances. In: Proceedings of Human Interface Symposium 2005 (in Japanese), pp. 799–802 (2005)
3. Rasmussen, J.: Information Processing and Human-Machine Interaction: An Approach to Cognitive Engineering. Elsevier, Amsterdam (1986)
4. Vicente, K.J., Rasmussen, J.: The Ecology of Human-Machine Systems II: Mediating Direct Perception. Complex Work Domain: Ecological Psychology 2(2), 207–249 (1990)
5. Vicente, K.J., Rasmussen, J.: Ecological Interface Design: Theoretical Foundations. IEEE Transactions on Systems, Man. and Cybernetics 22(4), 589–606 (1992)
6. Burns, C.M., Hajdukiewicz, J.R.: Ecological Interface Design. CRC Press, Boca Raton (2004)
7. Shneiderman, B.: Designing the User Interface: Strategies for Effective Human-Computer Interaction, 3rd edn. Addison Wesley Longman, Inc. (1998)
8. Gibson, J.J.: The Ecological Approach to Visual Perception, Houghton Mifflin (1979)

Case-Based Snow Clearance Directive Support System for Novice Directors

Yoshinori Ikeda[1] and Yoshio Nakatani[2]

[1] Graduate School of Science and Engineering,
Ritsumeikan University, Kusatsu, Shiga, Japan
rs002025@se.ritsumei.ac.jp
[2] College of Information Science and Engineering,
Ritsumeikan University, Kusatsu, Shiga, Japan
nakatani@is.ritsumei.ac.jp

Abstract. Experts have issued directives for snow clearance in snowy regions.These directives are implemented in accordance with the experience of the director in charge, and snowplow drivers are reliant on directors to carry out their tasks efficiently. However, many experienced directors will be retiring when they reach retirement age in the next few years and their successors will be faced with difficult circumstances because of a lack of training and experience. In order to provide support to such neophyte directors, we propose a case-based directive support system, which reuses past directive cases. Each such case stores data on weather, snow, number of snowplows available, the time taken to complete a snow clearance task and an evaluation of the tasks carried out by the director. When a neophyte director specifies current task conditions, the system searches and presents cases similar to the current conditions. When the director selects the most similar case, the system estimates the time required to complete the snow clearance task, based on the snow and road conditions. The director can also manage the progress of snow clearance tasks with the system.

Keywords: case-based reasoning, artificial intelligence, snow clearance.

1 Introduction

Every winter, 60% of Japan's cities are covered in snow and snow removal becomes central to the maintenance of everyday life. Many studies have attempted to solve these problems by, for example, implementing efficient snow removal techniques or road heating systems, however, human experts have directed these tasks. Until recently, directing these tasks has been unproblematic but now many experienced directors are approaching mandatory retirement age. Their successors will be faced with difficult, unfamiliar situations but will be required to maintain quality standards, despite a lack of training. New directors cannot permit service levels to decline, whatever the weather conditions, because citizens have high expectations of them and their ability to provide quality snow clearance services. In order to provide support to neophyte directors, we propose a case-based directive support system that reuses past cases to derive a new snowplow deployment plan.

M.J. Smith, G. Salvendy (Eds.): Human Interface, Part II, HCII 2007, LNCS 4558, pp. 893–902, 2007.

The system's database records past data on the amount of snow; the task area; the number of snowplows required and available; time and distance; and evaluation of the operations. When a novice director uses the system, it automatically searches for similar past cases and adjusts the available data for the current snowfall situation. The system then suggests a solution in the form of a snow clearance plan and provides the director with visual feedback of snow removal progress. We have developed a prototype of the system for the Windows PC platform. At the end of this paper we have enumerated advice received from experts who have evaluated the system and our intended future work on the system.

2 Tasks and Their problems

We interviewed Sapporo city government officials to analyze their current situation regarding snow removal tasks and to determine the system's specifications. The following is a summary of the analysis performed.

(a) Sapporo city has 39 snowplow centers and 10 construction technology centers that manage snow removal centers. Snowplow directors control the snow removal centers and manage the snow clearance centers, snow removal planning, management of snow removal contractors and claim processing. Snowplow directors and snowplow drivers belong to the same private-sector construction companies and all snow removal operations (snow observation, plowing, instructions, among others) are outsourced to them. City government officers merely organize the budget because the city government officers are reassigned every four years, making it difficult for them to accumulate sufficient operational experience.
(b) Snow removal tasks commence when more than 10 cm of snow falls and operations take place at night to avert traffic congestion.
(c) Snowplows are slow moving vehicles so work must be completed outside of peak traffic times (7:00 to 8:30 am and 5:30 to 7:00 pm.)
(d) There are many weather forecasting sensors around Sapporo that provide prediction data about estimated snow depth supplemented by information from the Meteorological Agency. This provides predictions such as there is a 30% chance of a 10 cm snowfall in the next three hours.

We have listed four snow removal systems as shown in the organizational chart in Fig 1. Many local governments seem to use similar systems, but without weather forecasting sensors that are unique to Sapporo because it is the largest city in Japan that experiences heavy snowfall. In our research, we could not find other municipalities with a better system.

It is impossible for city officers to perform snow removal tasks so they have no choice but to outsource the work to construction companies. When snow falls: (1) the director drafts a snow removal plan; (2) snowplow drivers are alerted; (3) dispatched device; and (4) if necessary, the director repeatedly updates the snow removal plan until the task has been completed. Basically, snow removal is deficit work for construction companies. Companies previously covered their deficits by performing public construction in summer but public construction projects have recently declined

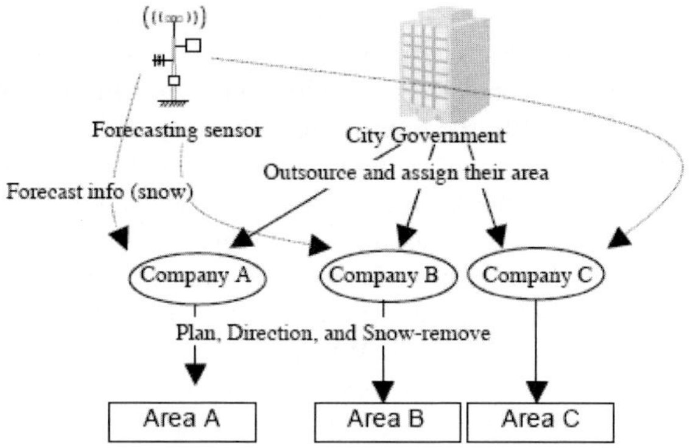

Fig. 1. Snow removal organizational chart in Sapporo

in number due to a shortage of local government funding. This has mitigated the bargaining power of local government with construction companies because the threat of withholding public construction orders is no longer effective and a number of construction companies have started rejecting snow removal work. Therefore, to maintain service levels, it seems that construction companies will have to work more efficiently, despite the problems of a rapidly aging and shrinking industry.

According to reports from the Cabinet Office, more than 80% of construction companies have less than 10 employees. Additionally, their employee turnover ratio is higher than in any other industry. This shows that their business bases are very weak and it is difficult to train new employees to replace old, experienced directors. The current experts are of the baby boomer generation and they are facing mandatory retirement within the next three years. In general, construction companies are facing some difficult maneuvering in the near future, with regard to organizational reformation.

During research interviews, we discovered some of the issues that the people involved in snow removal face.

a) Construction companies find it onerous to keep large fleets of snowplows so they have to efficiently use a limited number of snowplows.
b) Many expert directors will retire at the official retirement age limit within a few years and their expertise must be passed on to their successors. It is likely that snow removal officers will be changed every year because they are selected by means of a bidding process. There is a possibility that each new director's management tactics will differ from those of his/her predecessor.
c) Directors use wireless communications to communicate with snowplow drivers and gathering snowfall data. They also have to manage task progress by identifying the current locations of snowplows via GPS. For these reasons, whenever there is a heavy snowfall, the director is overwhelmed by a flood of calls.

For the issues raised in a) above, it appears impractical to persist with the previous tactic of maintaining a large fleet of snowplows in the face of a declining construction industry. Therefore, better performance with fewer snowplows and drivers is required and several studies have been conducted on this approach. Potential solutions suggested in these studies are, for example, one-operator snowplows (current snowplows require two operators) and retaining lane assist systems.

For the issues raised in b) above, every industry is subject to the problems of aging and generation shift, especially the construction industry. It has been speculated that snowplow drivers are comparatively young, whereas directors are close to retirement age so there will be a major change in directors in the near future. Every effort should be made to minimize the damage caused by many directors retiring in a short space of time, even if the construction industry's turnover ratio is high. These problems are exacerbated by a decline in the number of workers in the industry and growth in the areas requiring snow removal services. When the city experiences a heavy snowfall and all snow clearance tasks cannot be completed before rush hour, highways have to be prioritized for clearance work, bearing in mind that school routes have to be attended to as well. In short, the work must be done effectively under time pressure. Chaos may ensue if an inexperienced director is faced with the complex task of switching plowing areas or requesting backup from other areas. Moreover, it is foreseeable that snow removal decisions will change every year because officers, being subject to appointment by bidding, will change annually. In addition, construction workers are employed seasonally, seldom remaining with the same company for a long period of time. This affects efficiency because snowplow drivers do not have time to become acquainted with any particular area. Despite these problems, snow clearance work must be efficiently and quickly carried out.

For the issues raised in c) above, directors must deal with a communications overload during periods of heavy snowfall and information gathering is at risk of breaking down. There are no GPS locating systems or cameras fitted to the snowplows in Sapporo, so headquarters is obliged to communicate with snowplow drivers to ascertain the actual conditions on the ground. All communications are voice-based. The city hopes to motivate workers (directors and snowplow drivers) to dedicate themselves to their tasks by fitting GPSes and several other devices to snowplows in the near future, which automatically report snowplow conditions and locations.

We have identified three problems to solve, apart from the issues raised in c) for which many studies have suggested potential solutions. However, there have been few studies aimed at helping directors. We have focused especially on the issues raised in b) and propose a fast generation-shift methodology by using case-based reasoning.

We believe that it is effective to provide neophyte directors with a suggested plan of action that is based on references to past, similar cases. Consequently, we have developed a prototype suggestion system that has a database of past snow clearing operations. A neophyte director faced with fresh snowfall can consult the system for guidance.

3 Solution: Case-Based Reasoning

Case-based reasoning (CBR) is a process of solving new problems based on the solutions applied to similar, past problems. In the early 1980s a lawyer, Roger Schank, and his students at Yale University developed a CBR system to develop trial strategy from legal precedents and judge-made law.

Case-based reasoning has been formalized, for purposes of computer reasoning, as a four-step process: 1. retrieve; 2. reuse; 3. revise; and 4. retain (Fig 2.) Therefore, CBR has flexible, analogy-searching features. If no existing case perfectly matches a new problem, CBR is capable of making suggestions to the user. Other artificial intelligence methods, for example, rule-based reasoning and model-based reasoning, are less flexible because these are based on confirmed rules and exact matching with past cases. CBR stores past and current solutions in a database, together with the solutions and processes employed, as resources for solving similar problems in the future. CBR improves in accuracy as each new scenario and solution is stored in the database, making it easier to provide new suggestion patterns based on real cases.

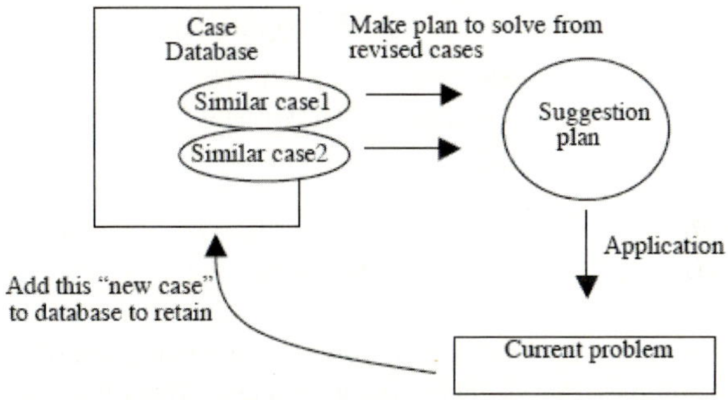

Fig. 2. Model of Case-Based Reasoning

Rule-based reasoning makes it difficult to retrieve similar cases when faced with an unknown situation. To address this problem, we have to use another reasoning method. The problem to be solved is to suggest a snow removal plan to a novice director by using past cases. Given this scenario, it is difficult to implement rulebased reasoning because of its heuristic rules and lack of basic scientific and physical evidence. In these circumstances, we have tried to provide neophyte directors, who must manage snow clearing operations, with a system that applies case-based reasoning.

4 Reasoning Algorithm

In order to complete snow removal tasks before rush hour, a director should prepare a snowplow operating schedule in advance. Therefore, it appears that predicting snowplow working hours would be an effective strategy. There is enough time to prepare for snow removal, such as to call to snowplow drivers, because no work commences before 10 cm of snow have fallen. Sufficient time helps in formulating a flexible response to rapidly changing climatic conditions.

Fig. 3 shows an outline of the system. By checking current snowfall against past snowfalls in the database, the system suggests that the director should advance the schedule to finish the task on time.

Sapporo weather bulletins contain many data items but to simplify the system, we have used the amount of snow that has fallen; the task area; the numbers of

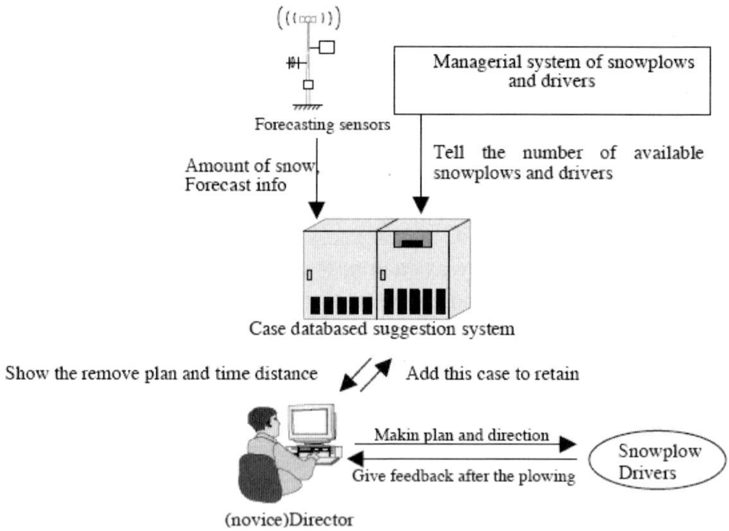

Fig. 3. System outline

snowplows available; the time/distance involved; and evaluation of the operations, as relevant items from past cases. The system rejects irrelevant cases from targeted information by performing evaluations and this will help to manage special circumstances that were hard to quantify at the time. The amount of accumulated snow is the primary factor in snow removal, so the system searches for similar accumulated amounts of snow cases first. Next, it searches for similar weather conditions and the number of snowplows available from the first search obtained and predicts likely solutions by comparing the search results with the current snowfall.

The following shows the specific reasoning method. Chart 1 shows the format of the case database.

Table 1. Format of the Case Database

Date:	Year/Month/Day (3 types)
Weather:	Sunny/Cloudy/Snowy (3 types)
Amount of snowfall:	Accumulated snow depth when snowplows started their task (Unit: cm)
Time/distance:	Time/distance for the task (Unit: minute)
Evaluations of operations:	Director evaluates his/her own work on a scale of 1 to 5 (bad) 1 - 5 (good)
Comments:	Director's comments. E.g., special circumstances

Basically, time/distance is inversely related to the number of snowplows available. Therefore, the prediction of time/distance (T) is such that:

$$r = \text{time/distance in the past case } (t) \times \text{number of snowplows available } (c) \tag{1}$$

$$\text{prediction of time/distance } (T) = \frac{r}{\text{Numbers of snowplows } (w)} \tag{2}$$

First, the system searches its database for similar weather conditions with snowfall within 5 cm of the current snowfall. We have used 5 cm for our own convenience and hope to narrow this number when the system goes into official use. The reason we have limited the search to the same weather conditions is to take into consideration that weather conditions may change during operations, and we have classified only three types of weather. In any event, the system will be capable of a flexible response when the database contains a sufficient number of cases. The matching of past scenarios to the present has to be based upon similarities and not precise matches because it is very rare for cases to match exactly. To accomplish this we have introduced a weighted reasoning element, k into the system as follows:

$$k = \frac{Cs - |Cs - Ps|}{Cs} \times \frac{E}{5} \tag{3}$$

k: Weight
Cs: Current snow depth
Ps: Snow depth in past case
E: 5 step evaluation of the case

Consequently, the maximum value of k is 1. The closer the match, the closer K will approach 1. We obtained time/distance by using equations (1,) (3) and (4.)

$$T = \frac{\sum_n (k_i \times t_i \cdot c_i)}{\sum_n k_i} \times \frac{1}{w} \tag{4}$$

K:	Weight
t:	Time/distance in the past case
C:	Number of snowplows in the past case
W:	Available snowplows
N:	Numbers of cases used for reasoning

The assumption is that the accuracy of the reasoning element will be enhanced when the number of cases (n) rise, as$k$ + k + ...+ k approaches n . In these circumstances the system will cope with a record snowfall. The system will then use the most similar case (the heaviest case recorded) and retain the data. Even if the number of stored cases further increases, the system's accuracy being reduced by this increase, the system will still be sustainable as a reference system.

5 System Implementation Plans

We have derived plans for a snow removal reference system from the perspectives of time/distance,weather condition, and the number of snowplows available. In virtually all cases the amount of the snow falls over an entire region but icy roads. Therefore, the snow removal procedure remains virtually static and the system suggests a removal plan for highway and parallel roads. However, when snow falls shortly before rush hour and the clearing task has to be completed before traffic congestion builds up, directors have to prioritize highways for snow clearance operations. That means that it is important to make an early decision to exclude less frequented streets from the clearance plan. We decided to focus on highways when the system predicts that currently deployed resources are insufficient to complete clearing all areas in time for rush hour traffic. The snowplow echelon can later time crunch as the area remaining to be cleared begins to shrink.

6 Prototype and Evaluation: System Outline

In order to support neophyte directors, we developed a prototype system of the snow removal directive support method described in this article. The system automatically collects the relevant data for conditions, such as depth of snow accumulated and temperature. Next, the system searches for cases with conditions that are similar to the current conditions and selects the most similar case. Then, the system estimates the time required to complete the task. After the snow removal work starts, the system acquires the current locations of snowplows via GPS and plots task progress on the display to show how long it will take to complete the task. Fig. 4 shows a screen image of the prototype system.

Fig. 4. System screenshot

We relegate the final task decisions to the director to avoid the frame problem in artificial intelligence as the system cannot understand all comments in the database and if the system provides all the answers, the director loses the opportunity to hone his/her own skills. When a director orders a snow removal operation, the database retains this information and the weather data for the new case. Next, the database retains the time/distance involved, the director's evaluation and the director's comments. While the snowplows are operating, the snowplow locations, roads that have already been cleared and those that have not, are shown on the system map, to make assessment of the situation easier. In addition, the system displays roads that have not yet been plowed in a lighter color during highway operations.

7 Evaluation and Future Work

We conducted interviews with employees of the Sapporo city government snow removal section and asked them to evaluate the system. The results reveal the advantages and disadvantages of the system.

(a) There has been little research on the issue of supporting neophyte directors. Casebased approaches to support directors are few in number and this system is expected to be helpful to them.
(b) The system should take additional conditions such as clouds and the quality of snow into consideration.
(c) Cities are constantly changing. New roads are constructed and the temperature changes from year to year. The system, however, does not take these factors into consideration and relies on past case scenarios, which occurred before the city changed.

The evaluation in (c) above is one of the principle problems with case-based systems. Of course, past cases can provide at least outline plans, which can be of some help to directors. This is the primary advantage of using a case-based approach: it can provide suggestions even if the historical cases do not match the current situation. The staff interviewed also observed that the reasoning order is quite simple to predict. The system first searches for a similar amount of snowfall and then performs a fivestep evaluation, but employees suggested that the system should use a specific temperature, weather conditions and the quality of snow as priority criteria. The initial reasoning method is limited to comparing current and past cases. Criteria such as the quality of snow and snow conditions will have a beneficial effect on the elements of snow weight and time/distance.

These tasks remain as future work. We designed the initial system to support novice snow removal directors. However, there are other problems in Sapporo, for example, a severe shortage of dump trucks and laborers. Our study may be designed to address future problems, but these are currently the primary tasks for Sapporo.

8 Conclusion

We propose a case-based directive support system for neophyte snow removal directors, based on an interview with the Sapporo city government. The basic idea is to support snow removal directors by searching for cases similar to the prevailing snow conditions. The system can be helpful by providing suggestions based on past cases, even if the current condition is not exactly the same.

Future tasks include taking subsidiary conditions such as clouds and the quality of snow into account, in addition to the depth of snow in judging similarity among cases and also to incrementally improve similarity metrics, based on the evaluation of tasks by directors.

References

1. Cabinet office, Government of Japan: Report of local economy (2001)
2. T. Yuki, T. Inoue, H. Okada: Control of maintenance for extraordinarily heavy snowfalls (2005)
3. M. Maruyama: Suggestion for a method that of planning for contract snowremoval (2005)
4. Kristian, J.: Hammond: Case-Based Planning. Academic Press, San Diego (1988)

The Relationship Between Working Conditions and Musculoskeletal/Ergonomic Disorders in a Manufacturing Facility

Dennis R. Jones

Department of Technology
College of Engineering and Engineering Technology
Northern Illinois University, DeKalb, Illinois, USA
jones@ceet.niu.edu

Abstract. This research investigates the relationship between working conditions and musculoskeletal disorders in an assembly facility. I believe that the physical and psychosocial aspects of work have a significant influence on the individual worker's health and well-being. The work organization at which I evaluated the employee's health and well-being was at a large manufacturing assembly facility. This research is based upon the Balance Theory Model of Smith & Carayon-Sainfort [5, 6].The overall aim of this research is to improve the long term health and well being of worker's in a manufacturing assembly facility. The overall purpose of this research is to try to identify the stressful working condition and attempt to control it. By trying to control the various musculoskeletal disorders in the work environment, you should also attempt to reduce or eliminate the level of stress, and to try to reduce or eliminate the level of strain.

Keywords: Human Factors, Ergonomics, Musculoskeletal Disorders, Psychosocial, Balance Theory Model, Applied Field Research, Manufacturing, Assembly, Production Processes.

1 Introduction

In the year 2005; 1.03 million people report taking time off work to recover from musculoskeletal pain due to overexertion or repetitive motion [1]. Workman's compensation costs due to musculoskeletal pain or discomfort exceed $20 billion annually. In reality in order to estimate the actual costs related to musculoskeletal pain and discomfort, you must also estimate the indirect costs. Therefore workman's compensation costs due to musculoskeletal pain or discomfort are estimated at over $50 billion annually [2].

I have done research on the relationship between working conditions and musculoskeletal disorders in an assembly facility. I believe that the physical and psychosocial aspects of work have a significant influence on the individual worker's health and well-being. The work organization at which I evaluated the employee's health and well-being was at a large manufacturing assembly facility. This research is based

M.J. Smith, G. Salvendy (Eds.): Human Interface, Part II, HCII 2007, LNCS 4558, pp. 903–909, 2007.

upon the Balance Theory Model of Smith & Carayon-Sainfort [4, 5]. This model specifies that the working conditions and other factors outside of work, can create a stress on the individual. This stress can have physiological and psychological consequences. If the stress exceeds the individual's capacity, the stress can produce a negative effect on the individual which could result in a strain. This is a bad fit between the individual resources and the work demands. If the stressful exposure continues for a prolonged time period, then this can result in serious musculoskeletal disorders.

The Balance Theory Model is as follows:

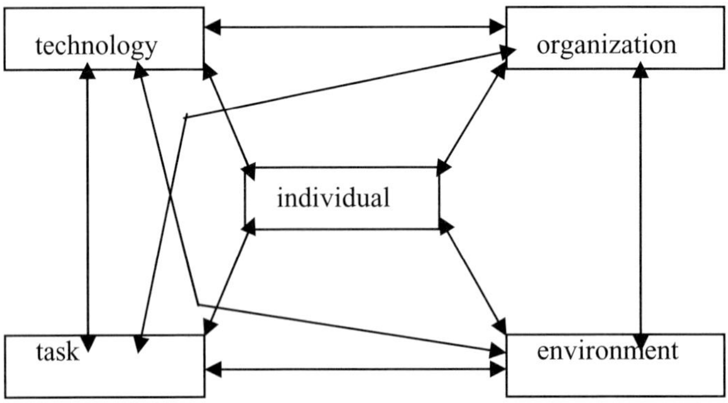

Fig. 1. The Balance Theory Model [4]

The Balance Theory Model is a system view concept for the various elements work. The Balance Theory Model shows the stress that working conditions can exert on the individual. These five elements of the Balance Theory Model all interact to define how work is to be performed. The individual is in the center of the Balance Theory Model. The individual has physical attributes and characteristics, previous experiences and knowledge, individual attitude and personality, and learned behaviors from which to draw from in order to cope with the working condition stress. The elements of task, technology, organization, and environment all influence the job content of work, the physical effort required of the individual, and the level of stress placed upon the individual [4].

The overall purpose of my research is to try to identify the stressful working condition and attempt to control them. Therefore the purpose of an intervention is to try to control the various musculoskeletal disorders in the work environment. By trying to control the various musculoskeletal disorders in the work environment, you should also attempt to reduce or eliminate the level of stress, and to try to reduce or eliminate the level of strain. There are a wide variety of things that can be attempted in the work environment in order to try to reduce or eliminate the existence of musculoskeletal disorders. These include engineering redesign, work method changes, administrative control, worker training, work hardening, and management organizational work rules to reduce exposures [6].

2 The Research Questions

Based on my objective to do research on the relationship over time between working conditions and musculoskeletal disorders in a manufacturing assembly facility, the following research hypotheses were explored:

H1: Self reports of physical aspects of work influence worker's musculoskeletal pain or discomfort. It is anticipated that the relationship will be stable over time.

H2: Self reports of psychosocial aspects of work influence worker's musculoskeletal pain or discomfort. It is anticipated that the relationship will be stable over time.

3 The Significance of This Research

The overall aim of this research is to improve the long term health and well being of worker's in a manufacturing assembly facility. The overall purpose of this research is to try to identify the stressful working condition and attempt to control it.

One of the contributions of this research is to determine the relationship between working conditions and musculoskeletal disorders in a manufacturing assembly facility. Generally the literature on this topic tends to be more heavily concentrated on the office worker, or service provider type of worker. Therefore the area of workers in a manufacturing assembly facility needs further investigation. Generally, the combination of both physical and psychosocial aspects of work influencing worker's musculoskeletal pain or discomfort, also needs further investigation.

A second contribution is to improve the research techniques and methodology which has been utilized in some previous scientific studies which attempted to control musculoskeletal disorders. Specifically the scientific evidence on interventions to control musculoskeletal disorders has some shortcomings. The lack of sound research methods and the lack of sound research designs are a significant shortcoming in some of the existing literature. This position is supported by the National Academy of Sciences paper of Smith, Karsh & Moro [6]. Their paper investigated and reviewed the research on interventions to control musculoskeletal disorders.

A third contribution is that the results of this research can assist industrial engineers in designing the jobs of workers in a manufacturing assembly facility. The incorporation of both physical and psychosocial aspects on worker job design should reduce the worker's musculoskeletal disorders, and therefore improve the long term health and well being of workers in a manufacturing assembly facility.

4 Literature Review - Summary

In summary, the National Research Council [3] concluded that there is theoretical evidence and some empirical evidence that links psychosocial factors and musculoskeletal discomfort. Stress appears to be a mediating variable that contributes to the development of musculoskeletal disorders. There also exists some theoretical evidence and some empirical evidence that links biomechanical factors and musculoskeletal discomfort.

In conclusion, more research studies need to be done in order to provide an answer to the existence of the link between biomechanical factors and psychosocial factors, and musculoskeletal disorders. Substantial research literature needs to be developed before a definitive answer could be given; yet the link seems very plausible.

5 Methodology – Study Design

The study was performed by using a multiple paged questionnaire to survey the three different assembly methods at the manufacturing assembly facility. Data was collected several times over a period of time; therefore a longitudinal study was conducted. These three different assembly methods were exhibited and displayed by three distinct individual manufacturing assembly lines. One of these manufacturing assembly lines consisted of the employee subjects of interest (i.e., the study group) which was called the study group. The other two assembly lines served as the control groups for the study group; these two areas were called the control group 1 and the control group 2.

The study group employee's assembly method consisted of utilizing an overhead powered conveyor system in order to assemble the product.

The control group 1 employee's assembly method consisted of utilizing automatic guided vehicles (AGVs) in order to assemble the product.

The control group 2 employee's assembly method consisted of utilizing a gravity roller conveyor (i.e., a manual push line) in order to assemble the product.

6 Data Collection

Data was collected from the three manufacturing assembly line area employee groups by utilizing a multiple page questionnaire survey. The multiple page questionnaire surveys were designed in order to obtain the information of desired pertaining to the research questions asked. The multiple page questionnaire surveys were given to both the study group and the control groups.

The multiple page questionnaire survey which was given to the study group and control groups was organized as follows:

A. Job Information (8 questions),
B. Characteristics of Work Environment (41 questions),
C. Quality of work Life (5 questions),
D. Health Information (23 questions),
E. Ergonomics and Physical Environment (28 questions),
F. Performance (11 questions),
G. Demographics (3 questions),
H. Implementation (3-5 questions) – this section was only given to the study group.

The multiple page questionnaire surveys for the study group and control groups were slightly different. The study group was asked additional questions in order to obtain information regarding their initial response to the new assembly method and the new assembly method implementation. Obviously the control group assembly methods had already existed for several years and therefore implementation questions pertaining to them were inappropriate and unnecessary; since we are only interested in the implementation of the study group assembly method.

7 Multivariate Analysis of Variance - MANONA

I concluded from the results from the multivariate analysis of variance (MANOVA) that musculoskeletal discomfort is primarily influenced by psychosocial factors (such as anxiety, and uncertainty), and physical demands. As when psychosocial factors (such as anxiety, and uncertainty), and physical demands increased; musculoskeletal discomfort also increased. It could be implied that musculoskeletal discomfort and psychosocial factors (such as anxiety, and uncertainty), and physical demands are linearly related.

Therefore, when the physical demands are increased, the stress on the individual is increased, and the ergonomic musculoskeletal discomfort is also increased.

Also, when the anxiety and uncertainty are increased, the stress on the individual is increased, and the psychosocial factor discomfort (i.e., negative psychosocial factors) is also increased.

So, an increase in physical demands, results in increased ergonomic musculoskeletal discomfort. And, an increase in anxiety and uncertainty, also results in increased negative psychosocial factors.

8 Discussion

Hypothesis 1 – Results

H1: Self reports of physical aspects of work influence worker's musculoskeletal pain or discomfort. It is anticipated that the relationship will be stable over time.

Hypothesis 1 is supported.

Hypothesis 2 – Results

H2: Self reports of psychosocial aspects of work influence worker's musculoskele tal pain or discomfort. It is anticipated that the relationship will be stable over time.

Hypothesis 2 is supported.

The control groups appear to be stable over time in terms of musculoskeletal discomfort, neck/shoulder/back discomfort, hand/arm discomfort, and leg discomfort. The control group 2 (i.e., the manual push line) consistently exhibits higher self reports of musculoskeletal discomfort, neck/shoulder/back discomfort, hand/arm discomfort, and leg discomfort than the control group 1 (i.e., the AGV line), over the three rounds of data collection. This is to be expected since the human factors, ergonomics, and working conditions of the control group 2 is significantly worse than the control group 1.

The study group (i.e., the overhead powered line) exhibited lower self reports of musculoskeletal discomfort, neck/shoulder/back discomfort, hand/arm discomfort, and leg discomfort than the control groups. This is to be expected since the human factors, ergonomics, and working conditions of the study group was significantly better than the control groups.

The control groups appear to be stable over time in terms of the physical aspects of work. The control group 2 consistently exhibits higher self reports of physical demands, repetitive motions, and loading on individual than the control group 1, over the three rounds of data collection. This is to be expected since the human factors, ergonomics, and working conditions on the control group 2 is significantly worse than the control group 1.

The study group exhibited lower self reports of physical demands, and repetitive motions than the control groups. This is to be expected since the human factors, ergonomics, and working conditions of the study group was significantly better than the control groups.

The control groups appear to be stable over time in some of the psychosocial aspects of work. The control group 2 exhibited lower self reports of task control, job control, resource control, and decision control than the control group 1, over the three rounds of data collection. This is to be expected since the human factors, ergonomics, and working conditions of the control group 2 is significantly worse than the control group 1.

The study group appears to be stable over time in some of the psychosocial aspects of work. The study group exhibited greater self reports of task control, job control, resource control, and decision control than the control groups. This is to be expected since the human factors, ergonomics, working conditions, and employee empowerment of the study group was significantly better than the control groups.

9 Summary/Conclusion

In summary, there is theoretical evidence and some empirical evidence that links psychosocial factors and musculoskeletal discomfort. Stress appears to be a mediating variable that contributes to the development of musculoskeletal disorders.

There also exists some theoretical evidence and some empirical evidence that links biomechanical factors and musculoskeletal discomfort.

In conclusion, more research studies need to be done in order to provide an answer to the existence of the link between biomechanical factors and psychosocial factors, and musculoskeletal disorders. Substantial research literature needs to be developed before a definitive answer can be given; yet the link seems very plausible.

References

1. Bureau of Labor Statistics, United States Department of Labor (2001). Occupational Injuries and Illnesses. Lost-Worktime Injuries: Characteristics and Resulting Time Away from Work (2005)
2. National Academy of Sciences, Musculoskeletal Disorders and the Workplace: Low Back and Upper Extremities. In: National Research Council, National Academy of Sciences, National Academy Press, Washington, D.C. (2001)

3. National Research Council, Work-Related Musculoskeletal Disorders: Report, Workshop Summary, and Workshop Paper. In: Steering Committee for the Workshop on Work-Related Musculoskeletal Injuries: The Research Base. National Academy Press, Washington, D.C. (1999)
4. Smith, M.J., Carayon-Sainfort, P.: A balance theory of job design for stress reduction. International Journal of Industrial Ergonomics 4, 67–79 (1989)
5. Smith, M.J., Carayon, P.: New technology, automation, and work organization: stress problems and improved technology implementation strategies. The International Journal of Human Factors in Manufacturing 5, 99–116 (1995)
6. Smith, M.J., Karsh, B., Moro, F.B.P.: A review of research on interventions to control musculoskeletal disorders. In: Work-Related Musculoskeletal Disorders: Report, Workshop Summary, and Workshop Papers, National Research Council, National Academy of Sciences, pp. 200–229. National Academy Press, Washington, D.C (1999)

A Method for Generating Plans for Retail Store Improvements Using Text Mining and Conjoint Analysis

Takumi Kaneko[1], Yuichiro Nakamura[1], Michiko Anse[1], Tsutomu Tabe[1], and Yumiko Taguchi[2]

[1] Graduate School of Science and Engineering, Aoyama Gakuin University
5-10-1, Fuchinobe, Sagamihara, Kanagawa, 229-0006, Japan
c5605046@cc.aoyama.ac.jp, c5605111@cc.aoyama.ac.jp, anse@ise.aoyama.ac.jp, tabe@ise.aoyama.ac.jp
[2] Department of Business Administration and Communication, Shohoku College
428, Nurumizu, Atsugi, Kanagawa, 243-8501, Japan
taguchi@shohoku.ac.jp

Abstract. Sales at retail stores in Japan have been declining for various reasons. One important factor has been a steady diversification in the lifestyles and needs of customers. If a retail store is to sustain itself and continue developing, it must search for the latent demands of customers and adapt itself to accommodate those demands. A support system for store improvement capable of viewing specific improvement plans will prove useful as a tool for improving stores on a sustainable basis. In this research we develop a method for generating effective store-improvement plans to accommodate actual customer demands. The method employs both Text Mining and Conjoint Analysis techniques. We also demonstrate a sequence and a test run of a prototype.

Keywords: Customer-satisfaction measurement, Text Mining, Definition Method, Conjoint Analysis, Store-improvement plan.

1 Introduction

Sales at retail stores in Japan have been declining. The declines are attributable not only to the slump in the national economy, but also intensifying competition with other stores and a diversification of the needs and lifestyles of customers. For a retail store to sustain itself and continue developing, it must continue to search for the latent demands of customers and adapt itself to accommodate them in a huge and fluctuating market. The demands of individual customers are vague and divergent. A method that enables us to understand the latent demands of customers has yet to be established.

In an earlier study, researchers developed a Support System for Improving Stores based on customer-satisfaction measurement (B. Ives, 1983) (Y. Taguchi 2004). When they attempted to elucidate the demands of customers using a 5-point questionnaire (Semantic Differential), their method showed clues for improvement. It proved difficult, however, to connect the improvements with specific actions or to develop a specific plan for improvement. Thus, the method developed through this

M.J. Smith, G. Salvendy (Eds.): Human Interface, Part II, HCII 2007, LNCS 4558, pp. 910–917, 2007.

earlier research turned out to an imperfect as a strategy for learning store improvements tailored to customer demands.

The purpose of the current research is to demonstrate the sequence applied in a new method of generating store-improvement plans tailored to customer demands.

2 Approach

2.1 Combining Text Mining with Conjoint Analysis

Text Mining is a technique to search for hidden information, trends, and correlations within inventory databases and to discover new factoids (A. Hearst, 1999). With Text Mining, we can measure qualitative data from unstructured text as determinate data and thereby grasp relevant issues for the store based on the real thinking of the customers. The technique is useful only for extracting clues on the issues and trends, however. It cannot elucidate how effective it will be for the store when the issues are solved. Therefore, it is difficult to show specific improvement plans based solely on this data.

Conjoint Analysis is a technique to provide a quantitative measure of the relative importance of one attribute as opposed to another when asking about preference (D. Aaker, 1990). With Conjoint Analysis, we can measure the preference of each participant for each plan and ask each participant to estimate the effectiveness of each plan. If an issue for store improvement (i.e., Selection, Price and Quality) is set as an Attribute for Conjoint Analysis and a specific content of Attribute(i.e., Handles Alcohol, A lot of perishable foods) is set as an Level for Conjoint Analysis, the virtual plan formulated will become a candidate as a store-improvement plan. As we attempt to estimate each virtual plan, Conjoint Analysis lets us measure the frequency of effect for each plan when the plan is executed. When we take this approach, the plan output will be the best for improving the store. The method for making the Attribute and the Level has yet to be established.

Therefore, to combine Text Mining with Conjoint Analysis, and to take advantage of the benefits of each (① Text Mining, "To measure qualitative data as determinate data and grasp the real demands of customers"; ② Conjoint Analysis, "To measure the frequency of the effect of each plan when each plan is executed, and then show an effective improvement plan"), we develop the sequence for store improvement and examine the method.

2.2 Generating the Improvement Plan

This support prototype system consists of two parts and uses two types of customer satisfaction questionnaire to confirm the behavior of the prototype system. The first part of the survey uses a free form questionnaire to grasp the real demand (see Fig. 1), and this is analyzed by Text Mining. Next, the Attribute and Level for the second part of the survey are distilled so that the issues can be ranked. The result of the first survey is used with the second part of the questionnaire to compare the combination of Attributes and Levels, then the data of the second questionnaire is analyzed by Conjoint Analysis. Lastly, the effective improvement plan is shown.

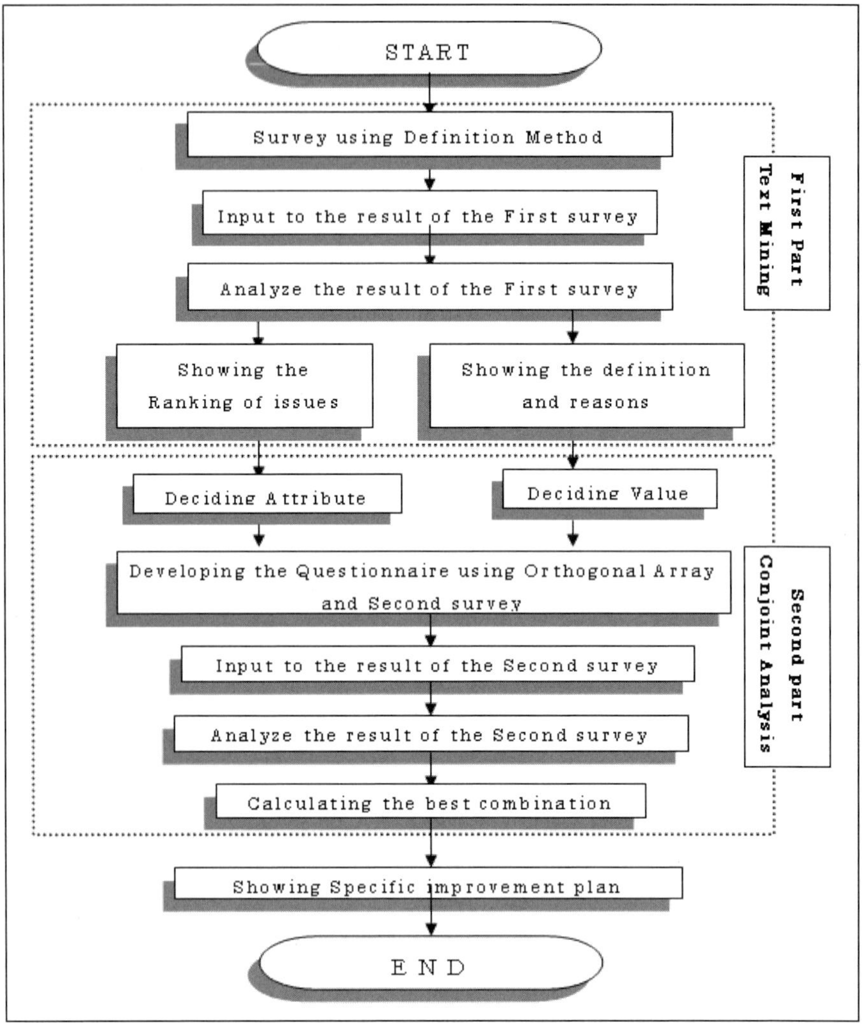

Fig. 1. Sequence leading to a specific improvement plan

3 Support System

3.1 Questionnaire of Definition Method

In the first survey, it is important to grasp the issues that combine the Attribute and Level, that is, the alternative improvement plan making use of the real demands of the customers. In the first survey we use a free-description type of questionnaire for Text Mining. The so-called Definition Method (T. Hayashi, 2001). The Definition Method has the questioner describe the store in a set format. The method is an efficient tool for deciding the Attributes for Conjoint Analysis, as it allows us to calculate the text data, measure the essential propositions without losing the real description of the questioner, and grasp information efficiently. TOTO LTD., a manufacturer of ceramic equipment, uses a Definition Method questionnaire to collect data in its research and development works (Chino, 2005). With this method, however, it can be difficult to grasp the real demand for combining the Level for Conjoint Analysis based merely on the description of the definition. Thus, we need to add a question comprising the reason for the description of the definition next to the definition, in order to grasp the effective customer demand for deciding the Level. As shown in Fig. 2, this clarifies the causation between the definition and the reasons, that is, what the customers focus on and what customers really need. The customer demands are measured by the gap between reality and the ideal, thus the survey asks about the reality of the store relative to competitors, and about an ideal store. Once the survey answers are calculated, compared, and analyzed by Text Mining, the issues to be improved are grasped.

①What do you think of the store?

Concrete Image (Definition)	The reason of describing the image
(*Fresh*)	(*There are always fresh vegetables*)
(*Useful*)	(*Open 24 hours*)
(*Cheap*)	(*Fish is very cheap*)
( )	( )
( )	( )

Fig. 2. The sample of first questionnaire

3.2 Process of Cleansing

To grasp the issues clearly and then smoothly apply the analysis to the second step, we have to analyze the result of the Definition Method by the same approach we would use to analyze determinate data. The definition described in the first survey is corrected and arranged into a common form (T. Nasukawa, 2001). The correction and arrangement of this common form collectively referred to as the "Process of Cleansing," and the sets of arranged words are referred to as "the Categories." To calculate the number of categories, the general trends and features are grasped.

There generally must be rules imposed on the Process of Cleansing, as the analytical results would diverge too widely from person to person without rules. Chart 1 below shows the formulation of a Classification Chart from a characteristic diagram, and how the chart is actually used. We can use the char if the definitions described for it are classifiable, whereas entirely new categories must be added for any definition which turns out to be unclassifiable (a no-category definition). If the questioner uses a negative word such as "expensive" to describe his or her own store or a competitor's store, the word is added as a descriptor for an ideal store, as the Category data is analyzed by the number of classified categories in the research.

Chart 1. Classification Chart from characteristic diagram

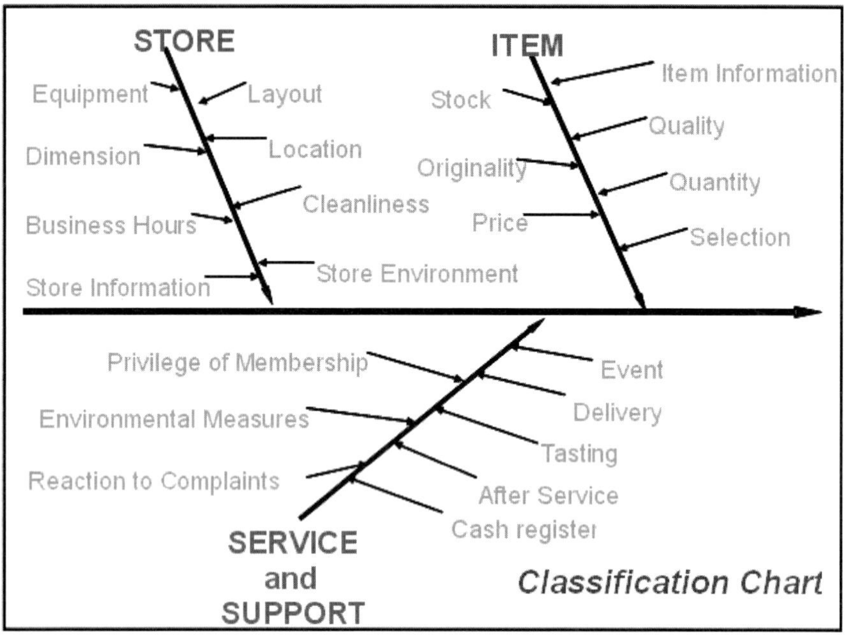

3.3 Developing the Ranking of Issues and Deciding Attributes

The classification of category is analyzed and the Ranking of issues is developed. The Ranking of issues calculates the frequency of each category described for each store (Own store, Competitor, Ideal store), measures the correspondence of the frequency of the category described for the customer's own store and for the competitor with the frequency of the same category for an ideal store, and then ranks the from the unfulfilled category. The frequency of each category is the rate at which each number of categories can be divided by the total number of categories (%). The Ranking of issues is helpful to evaluate the Attributes.

The interface is a method developed automatically by Excel VBA, to fill in the questionnaire results and support the Process of Cleansing and the Ranking (see Fig. 3). The interface is helpful for filling in the results without making mistakes, as it contains categories that have been already been entered. The person at the keyboard merely selects the category then the data is fed into the system automatically.

3.4 Deciding the Level

The Attributes for Conjoint Analysis are decided by the Ranking of the issues developed by Text Mining and the Level for Conjoint Analysis, as decided by the reasons given in the first part of the questionnaire.

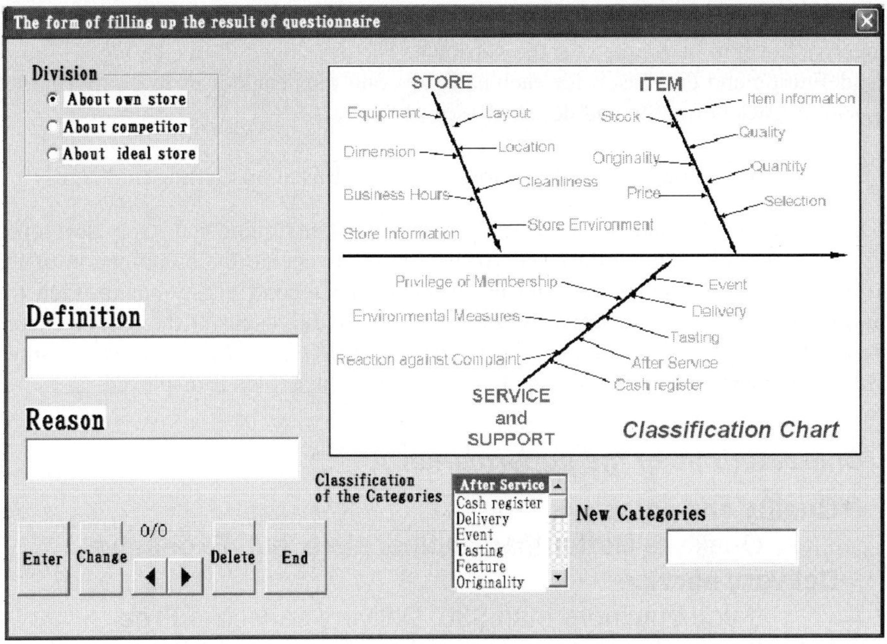

Fig. 3. Form for filling in the questionnaire results

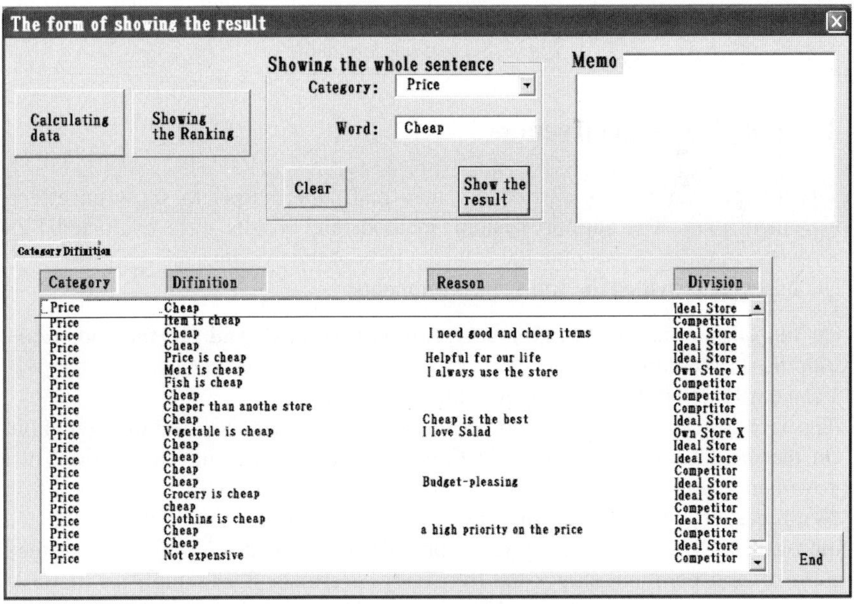

Fig. 4. Table of Definitions and Reasons

A great deal of text data has to be collected accurately in order to decide the effective Level to be adapted for the Attribute. The developed form (see Fig. 4) shows the definition and the reason for each category, and thus enables us to easily find the reason for each definition and decide the Level.

3.5 Second questionnaire and Conjoint Analysis Based on Orthogonal Array

In the second survey and Conjoint Analysis, we apply an Orthogonal Array developed by SPSS Conjoint. The Orthogonal Array is used to decide the combination of the Attribute and Level compared and evaluated in the second questionnaire, then the questionnaire itself is developed. As shown in Fig. 5, the results of the second survey are analyzed by SPSS Conjoint. The frequency of effect for each plan is measured when each plan is executed, and an effective improvement plan is displayed.

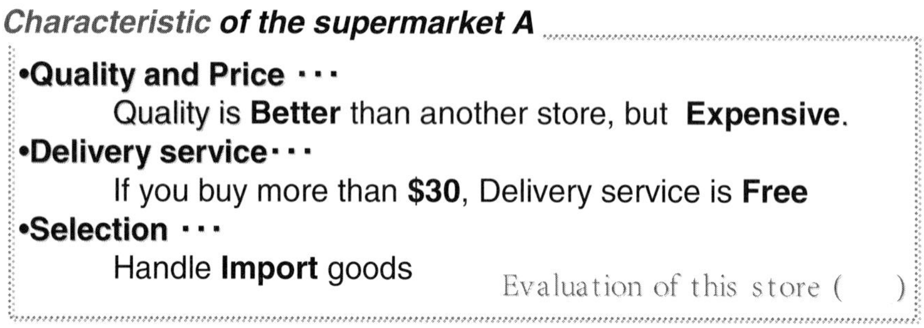
Characteristic of the supermarket A

•**Quality and Price** ・・・
Quality is **Better** than another store, but **Expensive.**
•**Delivery service**・・・
If you buy more than **$30**, Delivery service is **Free**
•**Selection** ・・・
Handle **Import** goods
Evaluation of this store ( )

Fig. 5. Second questionnaire

4 Examine the Effectiveness

This prototype system was carried out in a real supermarket to show an effective improvement plan. The support system, method, and results were evaluated by the store owner, in order to examine the effectiveness of the system.

The manager provided the following comments.

1) On the questionnaire used for the Definition Method, and the method used for developing the Ranking of issues.

"I can easily grasp the issues by looking at the Ranking"

"The Text Mining raised new and interesting opinions that I hadn't thought about."

2) On the display of the Table of Definitions and reasons, and on the flow of the improvement plan.

"With earlier support systems my employees had to think about the actual approaches. This support system, on the other hand, outputs an effective improvement plan and shows the frequency of effects. This simplifies the task of improvement and lets me clearly show my employees the reasons for the improvements we make. It helps us act with greater confidence."

3) On the method of combining Text Mining and Conjoint Analysis to show the store-improvement plan

"It raises surprising and unexpected issues and shows an effective improvement plan. If we follow this plan, our improvement activities will differ from those of other stores."

These comments from the store owner attest to the utility of the method developed as a tool for store improvement.

5 Results

In conclusion, we have demonstrated a sequence in which "the issues arising from the actual demands of customers are elucidated using Text Mining and an effective improvement plan is output with indications of the frequency of influence using Conjoint Analysis." Further, we have conducted a pilot study of a constructed prototype. The results of this research confirm that our method outputs effective store-improvement plans tailored to actual customer demands.

Three important tasks lie before us:

1) The prototype system should be capable of supporting improvement plans not only for one's own store, but also other stores. This is vital for grasping the demands of customers more comprehensively.
2) The territory of the store survey in this study was the whole store. The territory should be narrowed to enhance the efficiency of the store improvements.
3) The survey and analysis should be stratified and a consistent system should be constructed.

References

1. Aaker, D.A, Day, G.S: Marketing Research, 4th edn. Wiley, Chichester (1990)
2. Ives, B., Olson, M., Baroudi, J.: The measurement of user information satisfaction. Communications of the ACM 26(10), 785–793 (1983)
3. Luce, R.D., Turkey, J.W.: Simultaneous Conjoint Measurement: A New Type of Fundamental Measurement. Journal of Mathematical Psychology 1, 1–27 (1964)
4. Marti, A.: Hearst: Untangling Text Data Mining. In: ACL'99, pp.3–10 (1999)
5. Chino, M.: How to utilize the customer's voice in the business, Marketing Research with Text Mining, Kodansha Scientific, pp. 58–80 (2005)
6. Chiba, S., Iwamoto, T., Okamoto, S.: A Study on the Conjoint Analysis. Journal of Tokyo University of Information Science 1(2), 137–154 (1997)
7. Okamoto, S.: Conjoint Analysis Marketing research by SPSS Nakanishiya LTD (1999)
8. Hayashi, T.: Approach to Text Mining by Excel, Omusha,Ltd (2002)
9. Hayashi, T.: Customer Value measurement of Wine. In: Practice of Preference-based Design, Kaibundo, pp.81–100 (2001)
10. Nasukawa, T., Kawano, H., Arimura, H.: Base Technology for Text Mining. Journal of JSAI 16(2), 201–211 (2001)
11. Taguchi, Y., Tabe, T.: Trial Development of a Task Support System for Improving Supermarket Branches with the Use of Perceptual Maps. Journal of Japan

Coping with Information Input Overload: User Interface Concepts for Industrial Process Control

Tobias Komischke[1] and Luis E. Herrera[2]

[1] Siemens Corporate Research, User Experience,
Princeton, NJ 08540, USA
tobias.komischke@siemens.com
[2] Georgia Institute of Technology,
School of Industrial and Systems Engineering,
Atlanta, GA 30332, USA
lherrera@isye.gatech.edu

Abstract. Process control in control rooms is currently in a state of technological flux driven by recent developments in information technology. The continual integration of new control devices carries with it the risk of overburdening the human operators, with a major source of danger here being information input overload. This article addresses this phenomenon and the implications for user-centered process control.

1 Introduction

The internet search engine Google needs 0.35 seconds (as of July 20, 2006) to retrieve 4.3 million hits for the search string "Information Overload". This highlights two things: First, the search on the internet for this term ironically produces information overload because, without further filtering or other kinds of processing, the number of search hits cannot be handled. Second, the volume of hits points to the relative importance of the topic.

Although the phrase "information overload" conjures up an immediate intuitive understanding, the term "information" is difficult to define formally. The biologist Gregory Bateson thinks of it as "a difference that makes a difference" [1]. Claude Shannon, a mathematician, avoids a direct definition; however, his influential Information Theory [2] states that any arbitrary information can be expressed in terms of the unit that he introduced: the "bit" (short for "binary digit"). A bit is defined as the amount of information needed to halve a receiver's prior uncertainty, irrespective of the amount of prior uncertainty. Both approaches have in common that they are based on the notion that information generates a gain in knowledge or understanding.

Hall (1998) summed up the problem of information input overload in a single sentence: "*Too much information arriving too fast, much of which is irrelevant or of questionable quality, can harm both employees and their companies.*"

Possible causes for information input overload include (Hitz & Turoff, 1985; Infield, 1996; Hall, 1998; Marsters, 1998):

- A lack of appropriate information mangement techniques
- The amount of available and newly generated information

M.J. Smith, G. Salvendy (Eds.): Human Interface, Part II, HCII 2007, LNCS 4558, pp. 918–928, 2007.

- The limited capabilities of computer hardware and -software
- Limitations of the human cognitive capabilities

In the nineteen sixties, Miller (1960, 1962, and 1964) published studies of the effects of information input overload on cells, organs, individuals, groups and social institutions. The results show that it is possible to cope for a time with increasing quantities of information, but that after a certain point it is no longer possible for further processing resources to be made available.

Miller identified the following seven strategies used by individuals, groups and social institutions for coping with the flow of information (see Table 1):

Table 1. Seven strategies for coping with information input overload (after Miller, 1960, 1962 and 1964)

Strategy	Description
Omission	Temporary and arbitrary non-processing of information
Error	Incorrect processing
Queuing	Delaying response during high load in hope of catching up later
Filtering	Neglecting to process certain categories of information while processing others
Approximation	Being less precise for the sake of speed
Multiple channels	Distributing processing if possible
Escape	Escape from the task

Miller also investigated the frequency with which the various strategies were used. The most frequently used strategies are *filtering* and *omitting*. The difference between these two strategies is that *filtering* involves some preprocessing of data, whereas *omission* tends to be unconscious and unsystematic. The most seldom-used strategy is *error*, which is not surprising, as the consequences of this strategy cannot be justified: errors are easily discovered and the person responsible easily identified. Also, the consequences for a person's work/organization can be very negative. The maxim here, therefore, seems to be: "Better do nothing at all than make a mistake".

What can be done to reduce the incidence of information input overload? It is striking that in most of the literature, the responsibility is placed with the individual, who is said to require an appropriate qualification. Thus, Infield (1996) makes the following recommendations:

- Try never to handle a piece of paper more than once: Read it through once and then either archive it or throw it away.
- Try to read in a more disciplined way: By concentrating, focus on those subject areas that are directly relevant to the organizational goals and objectives.

Hall (1998) bases his recommendations on the need for information management skills which allow people to achieve the following:

- To be able to handle the flood of information to which they personally are exposed
- To become confident in their handling of information
- To acquire evaluation techniques for more targeted use of information sources

According to Wiggins (1983), information management skills can help employees to:

- Be able to initiate successful searches
- Distinguish between what is relevant and what is not
- Use relevant information appropriately and correctly, distributing it only to those who need it.

Hiltz & Turoff (1985) are also of the view that people should develop screening skills and strategies for coping with the quantity of information to which they are exposed. Results of longitudinal studies carried out by the authors of people working with electronic information systems show that although there is no doubt that information input overload is a serious and widespread problem, it is not a paralyzing one for the users of such systems. Rather, these people learn to cope with information input overload once they have gained sufficient experience of working with a system. Hiltz & Turoff (1985) do not believe that in the context of electronic information systems it is either possible or desirable for automatic filters to exist which could decide which information is useful and which is not. As justification of this thesis, they cite the popularity of flea markets, which shows that items which are worthless to some can be of great value to others. Another reason they hold this view is that a document, for example, can increase in value after a certain period of time, and they conclude from this that software cannot be the ultimate cure for information input overload.

Trappl (1995) looks at both conventional and newly-developed concepts. A conventional concept is that in texts, the most information is always found in the first sentence of a paragraph. Thus, here too, the responsibility for coping with information input overload is placed with the individual. New concepts mentioned are autonomous software agents, i.e. application packages with a preset goal, able to execute their mission even under changing requirements and conditions. These agents are able to evaluate information, to the extent that they do not even present content which is of no interest. For example, they might prioritize e-mails on the basis of both the sender and the content. This concept relieves the individual of their responsibility, since software agents are autonomous and capable of learning.

For a more comprehensive summary on causes and symptoms as well as general countermeasures refer to Mengis & Eppler (2004).

2 Current Challenges in Industrial Process Control

Modern industry is to a large degree dependent upon automated technical processes. They assist in the production of high-quality products with the minimal use of time, energy and raw materials (Charwat, 1992). Dynamic technical processes, because of their time behavior, have constantly changing process and system variables (Johannsen, 1993). The nature of the technical processes determine to a large degree how often operators, who supervise and control the processes at the plant floor or from a control room, intervene in running processes.

Today's process control systems have graphical user interfaces displaying an enormous amount of data – more than two million process variables and up to several thousand process images can be displayed. As human processing of these masses of data is not possible in real time, the difficult challenge for automation engineering is to ensure that the process control system makes the data required by the operator ready for use. Another challenge is presented by progress in information technology, which opens up the possibility of even more optimized process control and, with it, improved plant outputs. As human operators, with their limited cognitive skills, are responsible for the smooth running of processes, it is obvious that new concepts are needed for integrating modern information technology into existing automation structures. Owing to the demands of real-time affordance, the problems surrounding IIO are all the more acute. The consideration of IIO when designing human-machine interfaces has until now tended to be implicit, however. Today's process control systems therefore contain many features intended to suppress IIO, but these have tended to be integrated unsystematically, and not as the result of deliberate consideration of the problem (Hollnagel, Bye & Hoffmann, 2000).

Hollnagel, Bye & Hoffmann (2000) consider information input overload and Miller's strategies in the context of modeling operator behavior. At the same time, they point out that for more complete modeling, information input underload should also be taken into account. Some of the strategies (they mention *queuing* and *filtering,* in particular) could be anticipated in the design of process control systems and supported by integrating special user interface features. In their opinion, however, the strategies of *multiple channels* and *escape* should not be taken into account, because they are unrealistic. They argue that the *multiple channels* strategy, for example, is not practicable in acute situations, even where intelligent systems (expert systems) are used. And yet it is precisely this strategy which is undoubtedly employed, for example in power plants, and which can be supported in a targeted manner by the right design concepts. This project therefore puts the case for taking existing knowledge on information input overload as well as scenarios into account when developing these concepts.

3 Micro-scenarios, Consequences and First User Interface Design Implications

In the following, possible scenarios for each of Miller's seven strategies (1960, 1962 and 1964) are outlined, although other scenarios could of course apply. Scenarios are narrative descriptions of what a user may do and experience during his or her work (Carroll, 1997). They can be fragmentary, incomplete and informal, yet they should allow deriving implications for system design from them. Since – due to their volume – the following scenarios are reduced to a maximum they are called micro-scenarios. They were created in a workshop with usability engineers experienced in the area of industrial automation. These micro-scenarios also serve to highlight possible negative consequences of their application, the assumption being that these consequences could include lost time and production and premature wear of technical components. The consequences were derived from the micro-scenarios in the expert workshop as well.

Finally, as a last step of the workshop, implications were inferred by the experts. They are too abstract to be translated into a user interface concept directly, but they build the first user interface requirements.

For lack of space only one example can be presented here (for a more comprehensive information, refer to Komischke, 2006).

Table 2. Exemplary table showing micro-scenarios, consequences and user interface design implications

Coping Strategy: Error	
Micro-Scenarios	• The operator uses an inefficient strategy for overcoming plant faults. • The operator enters incorrect input into the process control system. These errors can be the result of insufficient understanding, typos or input resulting from following the wrong strategy. • Although the operator is working with the correct process data, s/he transmits this data incorrectly to other user classes, such as superiors or service personnel. • Although the operator is working with the correct process data, s/he sends this data to the wrong people. • The causes of errors are wrongly guessed at, instead of being derived from the process data.
Consequences	• Time is lost • Production output is impaired • Production quality is impaired • Staff, system and environment may be put at risk
User interface design implications	• As human errors cannot be avoided and it is hardly possible to anticipate all errors in advance at the design stage, it must be prevented as much as possible that operator errors result in serious consequences (see above) • The system should check entries for their plausibility. On the one hand this implies to intercept incorrect entries. Yet going further the system could compare current input sequences, in real time, with sequences recorded as resulting in a fault. If the system detects that the operator is again adopting the same incorrect strategy, it may warn him or her of the possible consequences and suggest alternatives. This error database could also produce error statistics from which tailor-made training courses could be devised that would improve process control skills in a targeted way. • In order to reduce the incidence of errors, the system should provide context-specific data, i.e. data geared to the particular process situation. Examples include action options and their prerequisites, as well as the names of colleagues with certain skills who can be called in to help solve problems.

It is possible for any of the strategies to have negative consequences. This however does not mean that the aim for the system design should be to prevent the application of each strategy as far as possible. This only applies to *Escape* and *Error*. *Multiple Channels*, for example, can also have very positive consequences, for example augmenting the operator's own skills through the positive transfer of knowledge during joint fault clearance operations.

4 Concepts and Prototypes

Sheridan's (2000,2002) concepts of different types and levels of human interaction with automation systems provided another dimension for the classification of the micro-scenarios. This dimension provides a set of useful strategies and concepts to enhance the design of human interaction with the automated system.

Miller's coping strategies, Komischke's micro-scenarios and Sheridan's supervisory control levels served in the development of a two-dimensional framework that (i) identifies the strategies human operators typically follow to cope with IIO, and (ii) implements a series of design strategies based on different levels of automation (Figure 1). A prototype was developed that integrates the design strategies into a simulated process control system. The prototype attempts to mitigate the negative consequences identified by the micro-scenarios and entails the implementation of varying automation levels.

Coping Strategies	Micro-Scenarios	Consequences	Automation Level Strategies				
			Automate	Enforce	Warn	Remind	Delegate (collaboration)
Omission							
Error							
Queueing							
Filtering							
Approximation							
Multiple channels							
Escape							

Fig. 1. Framework for the analysis methodology

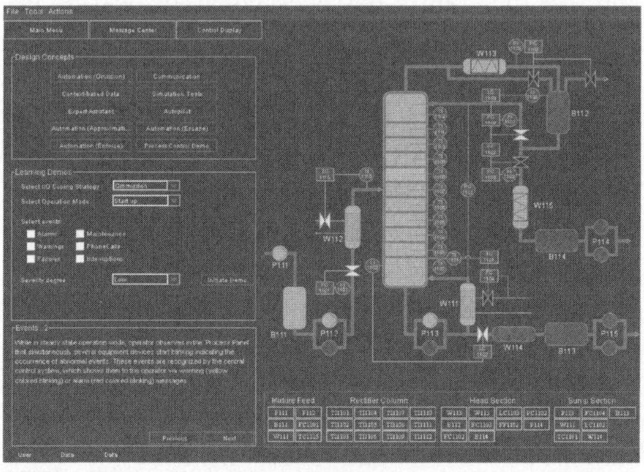

Fig. 2. Prototype showing the plant visualization together with controls to access different scenarios and design concepts

The prototype is a computational implementation of some of the design concepts developed in the conceptual framework discussed in the previous section. It is based on a simulation of a distillation column (Wittenberg, 2000). The prototype implements a set of eight scenarios that include some novel ideas to assist operators in situations of IIO. Subject matter experts in development and product management can use the prototype to get a hands-on experience of the novel concepts. They can follow the scenarios by interacting with the interface (Figure 2).

The prototype was written in Java using a variety of open source libraries. The social network structure and computational implementation is based on the research dissertation of one of the authors (Herrera, in press). It also extends the *Prefuse* information visualization toolkit. The network and profile data used are hypothetical, and the data are backed using a *MySQL* database server.

In the following, we present two of the scenarios that are covered by the prototype and that map design solutions to Miller's coping strategies.

Scenario 1: Multiple automation levels.

In this scenario, we present the concept of "Multiple automation levels" and its relation to the "Omission" coping strategy. We included various automation levels to assist operators in situations where—owing to the heavy workload—they omit the execution of key activities. The system is intelligent in that it tracks the operator's required responses during defined plant conditions. Once the system detects that an operator does not react, it starts to shift the control of the technical process gradually from the operator to the automation system, following pre-defined escalation steps.

For example, in one scenario, the control system issues an alarm message at a time when the operator is busy adjusting some process parameters in an attempt to fine-tune the technical process. The operator omits a response to the alarm, prompting the system to show a reminder asking her/him to execute the required activities; otherwise, the occurrence of certain negative consequences (e.g., impaired productivity, impaired quality, plant downtime) must be expected. If the system still does not detect a reaction from the operator, the control system warns her/him that doing nothing will result in the loss of her/his control, which will be taken over by the automation system. The final step in the escalation procedure is that the system issues a message indicating that the control has been taken away from the operator and has been given to the automation system, which will now autonomously stabilize the process. During this intervention, the operator cannot make inputs that could influence the automation system's problem handling, but the system still keeps her/him in the loop by showing what routines are carried out automatically. Once the process is stable, the task allocation can be shifted back; so the operator is granted control again.

Scenario 2: Context-based personnel localization.

In the following, we present a novel communication concept and its relationship to the "Multiple channels" strategy. The concept we propose to enhance operators' performance in a situation with IIO entails automatically finding appropriate personnel of whom to request help or to whom to delegate tasks. In the face of overwhelming situations, an operator can request the system to locate personnel based on context, who can assist her/him with the problem at hand. In this scenario, we used the concept of social networks to capture multiple interactions between personnel for the multiple dimensions that problems may possess. The proposed network features functionality that enhances the operator's capacity to explore staff in her/his search of appropriate personnel.

In order to model the process plant personnel as a social network, it was necessary to create a hypothetical hierarchical structure of the plant staff. Different categories of data were considered to enrich the power of the social network. These categories are part of the personnel profile. The visualization of the network includes features that enhance exploration and navigation, in addition to search tools (Figure 3).

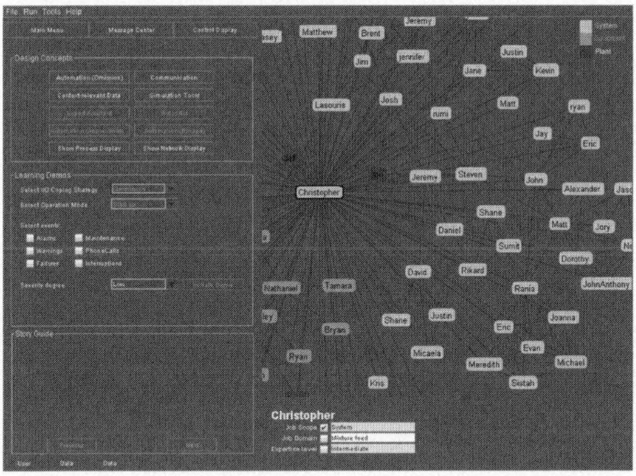

Fig. 3. Prototype showing the social network for the process plant staff. This view enables the user to visualize connections between operators and infrastructure they are assigned to: (i) system, (ii) equipment, (iii) plant.

Scenario 3: Context-based data location

The third scenario represents a variation of the previous scenario, only this one proposes a network of data –relevant elements. These elements (data) are presented on a per-need basis. With regard to Miller's strategies, this scenario relates to the *Filtering* strategy in that it provides the user with novel tools to search for relevant documents that are needed to solve critical situations. The specifics of this scenario are as follows: a hypothetical operator dealing with an overwhelming situation requires some sets of specific documents that are stored or are available somewhere in the system. The novel design concept enhances the operator's skills to locate those documents or data for a specific situation. A subset of documents and relevant data were modeled as a network structure. All documents were evaluated for their relevance towards different typical problems. As the operator is in need of some specific document / data, s/he can specify the characteristics of the problems at hand and the network representation retrieve a graphical network of relevant documents. Additional tools are provided to go deeper in the search of documents.

Scenario 4: Simulator tool

This scenario proposes a simulator tool that enhances Miller's *Approximate* strategy. It proposes the use of an alternative environment in which operators can safely test effects of a variety of actions in the system performance or other plants-level performance measures.

Scenario 5: Expert system tool

The fifth scenario provides a combination of scenarios 3 and 4. In this scenario, the operator receives direct assistance from the computer system to solve problems as if he received the assistance from a human expert. The expert system provides direct access to relevant data as well as hypothesis testing during a problem-solving process.

Scenario 6: Autopilot tool

The autopilot tool goes beyond the scope of the expert system in that it takes complete control of the system during defined periods. Most beneficial are times of high workload. Conventionally, the system may detect this condition by infering from the number of alarms and warning that are issued (system-centric inference). More advanced systems may also reason from physiological indications (human-centric inference, augmented cognition). In reality these situations are to a large extent characterically non-deterministic, therefore a more realistic application of the autopilot may be execution of monitoring and normal tune-up activities. Another feature of the autopilot is that it adjusts the alarms and warning settings so that during autopilot run the alarm and warning thresholds are more sensitive than during manual or semi-automatic control.

Scenario 7: Automation - approximation tool

To accommodate both Miller's *Error* and *Approximation* strategies, this story presents functionality to assist operators during the execution of tasks. It supports in carrying out simple and repetitive tasks such auto completion of data entering, auto-correction, and auto checking for the validity of data entries.

Scenario 8: Automation - escape tool

The last story is the combination of scenarios 1 and 6. The system enters the automation – escape mode when it detects that the operator is omitting the execution of critical tasks. By taking over the control while concurrently providing targeted information about what it is doing, the tool is geared towards Miller's *Escape* strategy. We assume that providing consistent situation awareness in these situation will prevent the operator to physically abandon the workplace, but rather motivates him/her to re-engage with the process control system once s/he is beyond the workload peak.

5 Conclusions

The project presented in this paper marks an important advancement in showcasing how advocating human capabilities in system design can solve real-world problems in process control. The user interface concepts the prototype is based on systematically integrate information input overload issues and consider the occurrence of Miller's coping strategies in the context of supervisory control. It could be shown that this approach resulted in novel user interface solutions that are not part of today's systems in use. Furthermore, the concepts include solutions that cope also with those strategies - *Multiple Channels* and *Escape* - that Hollnagel, Bye & Hoffmann (2000) viewed as unrealistic and negligible. The transformation of concepts into a tangible prototype makes it easy to discuss the proposed solutions with subject matter experts both in the academia and the industry. We believe that the results are generally transferable to domains other than industrial process control. Owing to shared characteristics in user tasks (supervisory control) and underlying processes (dynamic processes, real-time affordance, mass data), an application to traffic control, building management, network operations, and airport security may be appropriate. This again expands the target audience and emphasizes both the relevance and the applicability of this research.

References

1. Bateson, G.: Steps to an Ecology of Mind. Ballantine: New York (1972)
2. Carroll, J.M.: Scenario-Based Design. In: Helander, M.G., Landauer, T.K., Prabhu, P.V. (eds.) Handbook of Human-Computer Interaction, 2nd edn., Elsevier, Amsterdam (1997)
3. Charwat, H.J.: Lexikon der Mensch-Maschine-Kommunikation (Lexicon of human machine communication). R. Oldenbourg, München (1992)
4. Hall, H.: Networked information: dealing with overload. In: Proceedings of Information 1997, Strathclyde Business School, November 4th 1997. Library Association CIG Scotland, Paisley, 1998, pp. 37–44 (1998)
5. Herrera, L.: Design of decision support system for pro-duction planning and control in make-to-order manu facturing companies. Department of Industrial and System Engineering. Georgia Institute of Technology, Atlanta. PhD-thesis proposal (in press)
6. Hiltz, S.R., Turoff, M.: Structuring computer mediated communication systems to avoid information overload. Communications of the ACM 28(7), 680–689 (1985)
7. Infield, N.: Too much information? Library Manager 14, 25 (1996)
8. Johannsen, G.: Mensch-Maschine-Systeme (Man-Machine-Systems). Springer, Berlin (1993)
9. Komischke, T.: Micro-Scenarios as the Foundation for Preventing Information Input Overload in Supervisory Control. Automated Systems Based on Human Skills and Knowledge. In: Proceedings of the 9th IFAC Symposium (May 22-24, 2006) in Nancy, France (2006)
10. Marsters, C.: Push on the intranet. Managing Information 5(4), 28–29 (1998)
11. Mengis, J., Eppler, M.J.: The Concept of Information Overload: A Review of Literature from Organization Science, Accounting, Marketing, MIS, and Related Disciplines. The. Information Society 20(5), 325–344 (2004)
12. Miller, J.G.: Information input overload and psychopathology. American Journal of Psychiatry 116, 695–704 (1960)
13. Miller, J.G.: Information input overload. In: Yovits, M. C., Jacobi, G. T., Goldstein, G. D., (eds.) Proceedings of Self-Organizing Systems 1962. Washington, D.C.: Spartan Books, pp. 61–78 (1962)
14. Miller, J.G.: Coping with administrators' information overload. In: Boucher, R. M., Powers, L. (ed.) Report of the First Institute on Medical School Administration. Evanston: Association of American Medical Colleges, pp. 47–54 (1964)
15. Shannon, C.E.: A Mathematical Theory of Communication. Bell System Technical Journal 27, pp. 379–423, 623–656 (July & October, 1948)
16. Sheridan, T., Parasuraman, R., et al.: A model for types and levels of human interaction with automation. IEEE Transactions on Systems Science and Cybernetics A 30, 286–297 (2000)
17. Sheridan, T.B.: Humans and Automation. John Wiley & Sons, New York (2002)

18. Trappl, R.: Information Input Overload und die elektronische Lorelei. Interview in SUPERVISOR 1, 10–11 (1995)
19. Wiggins, R.E.: Managing information. In: Lock, D. Farrow, N. (eds.) The Gower Handbook of Management. Gower: Aldershot, p. 85 (1983)
20. Wittenberg, C.: Virtuelle Prozessvisualisierung am Beispiel eines Verfahrenstechnischen Prozesses (Virtual Process Visualization Illustrated Using Process Engineering). Fortschritt-Berichte VDI, Reihe 22. VDI Verlag, Düsseldorf, Germany (2000)

Impact of E-Commerce Environment on Selection of Sales Methods Considering Delivery Lead Time of Goods

Etsuko Kusukawa, Shinji Masui, and Ikuo Arizono

Department of Electrical Engineering and Information Systems, Osaka Prefecture University, 1-1 Gakuen-cho Nakaku Sakai Osaka, 577-0827, Japan
{kusukawa,arizono}@eis.osakafu-u.ac.jp

Abstract. In this study, we discuss a brick-and-mortar sales (the existing sales method through the sales channel 1) and an online sales under an e-commerce environment (the sales channel 2). We focus on each differences of goods return ratios and delivery lead-times to a customer through both sales channels for a selling period. Concretely, we consider a parallel sales method that the sales channel 2 is newly incorporated into the sales channel 1. In the parallel sales method, a manufacturer incurs returns processing costs and fixed system operational costs for both the sales channels. For an appropriate selection ratio to the sales channel 2, the additional system operational costs in the parallel sale method could be retrieved, and it could lead to improving the profit to the manufacturer. Utility of the parallel sales method is investigated from each difference of goods return rates and delivery lead-times between both the sales channels by comparing the expected profit of the manufacturer adopting the existing sales method with that adopting the parallel sales method.

Keywords: Sales method, e-commerce, inventory control, business model.

1 Introduction

From the widespread development of an e-commerce environment in recent years, the share of online sales through the Internet by a manufacturer / a retailer is increasing in total sales of the area of industrial/commercial business[1][2]. Under a situation, as an alternative sales method of a brick-and-mortar sales method through the sales channel (the existing sales method through the sales channel 1), an online sales under the e-commerce environment (the sales channel 2) has been developing rapidly in sales on orders from customers. On the sales channel 2, the manufacturer can not only receive an order of a goods directly from the customer on the primary market, but also delivery the goods directly to the customer without any intermediary of the traders. For the sales channel 2, it could lead to reduce the delivery lead-time of the goods to the customer. When a goods with the use-by date like beverage goods or periodical publications and so on are sold, the reduction of the delivery lead-time is an important problem for the manufacturer. However, some goods delivered might look different from pictures/descriptions on the online catalog. Thus, the sales channel 2 tends to

M.J. Smith, G. Salvendy (Eds.): Human Interface, Part II, HCII 2007, LNCS 4558, pp. 929–938, 2007.

have a much higher return rate than the sales channel 1 [1][3]. Under a situation, after the delivery of the goods to the customer, when the returns from the customer are considered, it could be assumed that the return rate of the goods sold through the sales channel 2 is generally higher than that through the sales channel 1, so the returns processing cost of the sales channel 2 might be cost more.

In this study, we discuss the impact of the e-commerce environment on the selection ratios of the sales channels 1 and 2 from the profit side of the manufacturer. Concretely, we focus on each differences of the goods return rates and the delivery lead-times to a customer through both the sales channels for a selling period, and then we consider a parallel sales method that the sales channel 2 is newly incorporated into the sales channel 1. In the parallel sales method, the manufacturer incurs the returns processing costs and the fixed system operational costs for both the sales channels. In an appropriate selection ratio to the sales channel 2, the additional system operational costs for the parallel sale method could be retrieved, and it could lead to improving the profit to the manufacturer. Utility of the parallel sales method is investigated from each difference of the goods return rates and the delivery lead-times between both the sales channels by comparing the expected profit of the manufacturer adopting the existing sales method with that adopting the parallel sales method.

2 Notation

T : the selling period of the goods corresponding to the use-by date
y_1: the initial goods inventory of the manufacturer adopting the existing sales method
y_2: the initial goods inventory of the manufacturer adopting the parallel sales method which the sales channel 2 is newly incorporated into the sales channel 1
$\pi_1(y_1)$: the expected profit of the manufacturer adopting the existing sales method
$\pi_2(y_2)$: the expected profit of the manufacturer adopting the parallel sales method
m: the production cost per unit of goods
P: the sales price per unit of goods
y_s: the disposal price on the salvage market per unit of surplus goods inventory including the returned goods
G: the shortage penalty cost per unit of goods
m_{ri} (i=1, 2): the return processing cost (including the collection cost, the inspection cost and the recovery cost) per unit of returned goods of the manufacturer adopting the sales channel i (i=1: the brick-and-mortar sales, i=2: the online sales)
h: the inventory holding cost per unit of goods
r_i (i=1, 2): the goods return ratio when the sales channel i is adopted ($0 \le r_i \le 1$)
L_i (i=1, 2): the delivery lead-time of goods when the sales channel i is adopted
k_i (i=1, 2): the ratio between the delivery lead-time L_i of the goods to the customer when the sales channel i is adopted and the selling period T, referred to the serviceable goods ratio, ($0 \le k_i \le 1$)
C_i (i=1, 2): the fixed system operational cost when the sales channel i is adopted
g_i (i=1, 2): the selecting ratio of the sales channel i by the customer ($g_1 + g_2 = 1$)

3 Model Assumptions

(1) It is assumed that an intended goods is a single type and a single period goods. The goods can be sold within a use-by date T (a selling period T) as like beverage goods, periodical publications and so on.

(2) It is assumed that the total customer demand of the goods for the selling period T is a stochastic variable x, and follows the normal distribution with mean μ and variance σ^2 . The probability density function (pdf) is $f(\cdot)$, and the cumulative distribution function (cdf) is $F(\cdot)$.

(3) As shown in Figures 1 and 2, the following two logistics models are considered: a single manufacturer receives an uncertain order (demand) of goods from the customer on the primary market during the selling period T by either a brick-and-mortar sales method (the existing sales method) or an parallel sales method which an online sales channel (the sales channel 2) is newly incorporated into the brick-and-mortar sales channel (the sales channel 1). Under the uncertain customer demand of the goods, the manufacturer finishes producing an anticipated order quantity of the goods, Q, from the customer before the beginning of the selling period T corresponding to the expiration date T for sale of the goods. Therefore, at the beginning of the selling period, T=0, the initial goods inventory of the manufacturer is equal to Q. The goods can be sold to satisfy the customer demand during the selling period T. For each sales method, both the production cost and the sales price per unit of goods are same. The optimal decision for the initial goods inventory is to be made by the manufacturer before the beginning of the selling period to maximize the expected profit of the manufacturer.

(4) The delivery lead-time of the sales channel 2, L_2, with the direct transaction between the manufacturer and the customer is shorter than that of the sales channel 1, L_1, with some intermediary of the traders for the transaction between the manufacturer and the customer. It is assumed that the customer demand per unit of time during the selling period T is constant. After the delivery lead-time L_i (i=1, 2) for each sales channel i (i=1, 2), the customer can receive the goods ordered. In this case, the serviceable goods ratio k_i (i=1, 2) to satisfy the customer demand of goods during the selling period T can be estimated as k_i=1- L_i / T. Due to the goods with the use-by date, after the end of the selling period T corresponding to the expiration date T, it is unable to sale the goods to the customer on the primary market.

(5) After the delivery of the goods to the customer, at the specific returned rates, r_1, r_2, for the sales channels 1 and 2, it is assumed that significant parts of the goods are returned to the manufacturer from the customer. During the selling period T, all returned goods are acceptable once only to the manufacture. The return rate of the goods for the sales channel 2, r_2, is higher than that for the sales channel 1, r_1, due to the difference between the actual goods and the pictures / the descriptions of the goods on the online catalog. After the selling period T, all the returned goods are inspected, and are recovered if the returns are damaged.

(6) After the selling period T (the expiration date T for sale of the goods), the manufacturer incurs the inventory holding cost for the surplus goods inventory,

meanwhile, the manufacturer incurs the shortage penalty cost for the unsatisfied customer demand. The inventory holding cost from the beginning of the selling period T to the end of the period is estimated roughly as the following average value: (the initial goods inventory + the end-of-period goods inventory) / 2.

(7) Regardless of any sales methods, all recovered returns and the surplus goods inventory that remains unsold at the end of the selling period T can be sold on the salvage market at a lower price as the disposal price.

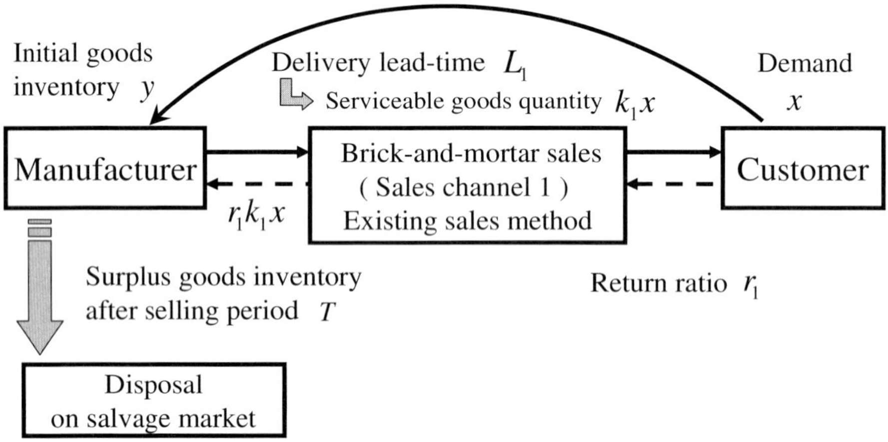

Fig. 1. A logistics model adopting the existing sales method with procedure of returned goods from the customer on the salvage market

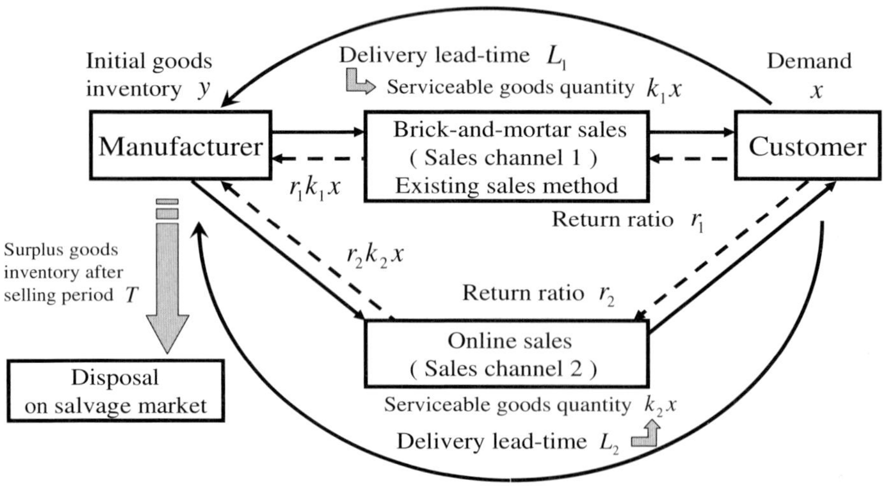

Fig. 2. A logistics model adopting the parallel sales with procedure of returned goods from the customer on the salvage market

4 Model Descriptions

4.1 Logistics Model Adopting the Existing Sales Method

Figure 1 shows a logistics model adopting the existing sales method with procedure of the returned goods from the customer on the salvage market. When the total customer demand of the goods for the selling period T to the manufacturer through the sales channel 1 is x, the goods can be delivered to the customer after the delivery lead-time L_1. In this case, it is assumed that the serviceable goods quantity to the customer for the selling period T can be estimated as k_1x. Therefore, the expect profit of the manufacturer $\pi_1(y_1)$ adopting the existing sales method can be obtained from the sales of the goods on the primary market, the disposal sales of the returned goods and the surplus goods inventory on the salvage market, the production cost of the goods, the inventory holding cost of the goods, the shortage penalty cost of the goods, the return processing cost and the fixed system operational cost as

$$\begin{aligned}\pi_1(y_1) &= -C_1 + \int_0^{y_1/k_1} Pk_1(1-r_1)xf(x)dx + \int_{y_1/k_1}^{\infty} P(1-r_1)y_1 f(x)dx \\ &+ \int_0^{y_1/k_1} P_S\{y_1 - k_1(1-r_1)x\}f(x)dx + \int_{y_1/k_1}^{\infty} P_S r_1 y_1 f(x)dx \\ &- \int_0^{y_1/k_1} my_1 f(x)dx - \int_{y_1/k_1}^{\infty} my_1 f(x)dx - \int_{y_1/k_1}^{\infty} G\{k_1x - y_1\}f(x)dx \\ &- \int_0^{y_1/k_1} m_{r1}r_1k_1xf(x)dx - \int_{y_1/k_1}^{\infty} m_{r1}r_1y_1f(x)dx \\ &- \int_0^{y_1/k_1} h\left[\frac{1}{2}\{2y_1 - (1-r_1)k_1x\}\right]f(x)dx - \int_{y_1/k_1}^{\infty} h\left\{\frac{1}{2}(1+r_1)y_1\right\}f(x)dx .\end{aligned} \tag{1}$$

From $\partial^2\pi_1(y_1)/\partial^2 y_1 < 0$ in equation (1), the expected profit of the manufacturer $\pi_1(y_1)$ is a convex function. In order to maximize the expected profit of the manufacturer $\pi_1(y_1)$, by differentiating equation (1) on y_1 and solving $\partial\pi_1(y_1)/\partial y_1 = 0$, the optimal initial goods inventory y_1^* of the manufacturer can be obtained as the following solution of the news boy problem:

$$y_1^* = k_1F^{-1}\left(\frac{-P(1-r_1) - G + m + \frac{h}{2}(1+r_1) - P_s r_1 + m_{r1}r_1}{(P_s - P - \frac{h}{2})(1-r_1) + m_{r1}r_1 - G}\right). \tag{2}$$

4.2 Parallel Sales Method Which Consists of Sales Methods 1 and 2

Figure 2 shows a logistics model adopting the parallel sales method with procedure of returned goods from the customer on the salvage market. When the parallel sales method is adopted, the delivery lead-time L_1 for the sales channel 1 and the shorter delivery lead-time L_2 for the sales channel 2 ($<L_1$) are simultaneously considered. In this case, the serviceable goods ratio for each sales channel can be estimated as k_1, k_2 ($> k_1$), respectively. Also, in the parallel sales method, the selection ratio to the sales channel i (i=1,2) from the customer, g_i, is considered. Based on k_i and g_i for each sales channel i, when the total customer demand of the goods is x, the serviceable goods quantity to the customer for the selling period T can be estimated as $(k_1g_1+ k_2g_2)x$. Therefore, the expect profit of the manufacturer $\pi_2(y_2)$ adopting the parallel sales method which consists of the sales channel 1 and the sales channel 2 can be obtained as summation of the expected profit of the manufacturer anticipated from the sales channel 1 and that from the sales channel 2. Concretely, based on the expected profit of the manufacturer adopting the existing sales method in equation (1), the expect profit of the manufacturer $\pi_2(y_2)$ adopting the parallel sales method can be obtained from the serviceable goods ratio for each sales channel, k_1, k_2, the selection ratio of each sales channel from the customer, g_1, g_2, the return ratio of the goods for each sales channel of r_1, r_2, the return processing costs for each sales channel, m_{r1}, m_{r2}, and the fixed system operational cost for each sales channel, C_1, C_2 as follows:

$$\begin{aligned}
\pi_2(y_2) &= -C_1 - C_2 + \int_0^{y_2/(k_1g_1+k_2g_2)} P\{(1-r_1)k_1g_1+(1-r_2)k_2g_2\}xf(x)dx \\
&+ \int_{y_2/(k_1g_1+k_2g_2)}^{\infty} P\{(1-r_1)k_1g_1+(1-r_2)k_2g_2\}\frac{y_2}{k_1g_1+k_2g_2}f(x)dx \\
&- \int_{y_2/(k_1g_1+k_2g_2)}^{\infty} G\{(g_1k_{f1}+g_2k_{f2})x-y_2\}f(x)dx \\
&- \int_0^{y_2/(k_1g_1+k_2g_2)} my_2f(x)dx - \int_{\frac{y_2}{(g_1k_{f1}+g_2k_{f2})}}^{\infty} my_2f(x)dx \\
&- \int_0^{y_2/(k_1g_1+k_2g_2)} h\left[\frac{1}{2}\{2y_2-\{(1-r_1)k_1g_1+(1-r_2)k_2g_2\}x\}\right]f(x)dx \\
&+ \int_0^{y_2/(k_1g_1+k_2g_2)} P_S\{y_2-(g_1k_1+g_2k_2)x+(r_1g_1k_1+r_2g_2k_2)x\}f(x)dx \\
&+ \int_{y_2/(k_1g_1+k_2g_2)}^{\infty} P_S(r_1k_1g_1+r_2k_2g_2)\frac{y_2}{k_1g_1+k_2g_2}f(x)dx \\
&- \int_{y_2/(k_1g_1+k_2g_2)}^{\infty} h\left[(r_1k_1g_1+r_2k_2g_2)+\frac{1}{2}\{(1-r_1)k_1g_1+(1-r_2)k_2g_2\}\right]y_2f(x)dx \\
&- \int_0^{y_2/(k_1g_1+k_2g_2)} (m_{r1}r_1k_1g_1+m_{r2}r_2k_2g_2)xf(x)dx \\
&- \int_{y_2/(k_1g_1+k_2g_2)}^{\infty} \frac{(m_{r1}r_1k_1g_1+m_{r2}r_2k_2g_2)}{k_1g_1+k_2g_2}y_2f(x)dx\,.
\end{aligned} \tag{3}$$

From $\partial^2\pi_2(y_2)/\partial^2 y_2 < 0$ in equation (3), the expected profit of the manufacturer $\pi_2(y_2)$ is a convex function. In order to maximize the expected profit of the manufacturer $\pi_2(y_2)$, by differentiating equation (3) on y_2 and solving $\partial\pi_2(y_2)/\partial y_2 = 0$, the optimal initial goods inventory y_2^* of the manufacturer can be obtained as the following solution of the news boy problem:

$$y_2^* = (k_1g_1+k_2g_2)F^{-1}\left(\frac{-P(1-R)-G+m+\frac{h}{2}(1+R)-P_sR+M}{(P_s-P+\frac{h}{2})(1-R)+m_rR-G}\right), \tag{4}$$

where, $R = \dfrac{r_1k_1g_1+r_2k_2g_2}{k_1g_1+k_2g_2}$, $M = \dfrac{m_{r1}r_1k_1g_1+m_{r2}r_2k_2g_2}{k_1g_1+k_2g_2}$.

5 Numerical Experiments

In this section, utility of the parallel sales method which the sales channel 2 is incorporated into the sales channel 1 is investigated from the aspect of the expected profit of the manufacturer. Concretely, by varying the values of the return rates of the goods for the sales channels 1 and 2, r_1 and r_2, the delivery lead-times of the goods to the customer through sales channels 1 and 2, L_1 and L_2, and the selection ratios to the sales channels 1 and 2 from the customer, g_1 and g_2, the optimal initial goods inventory and the expected profit of the manufacturer by adopting the parallel sales method are compared with those by adopting the existing sales method. From the numerical experiments, the optimal selection ratio to the sales channel 2 from the customer can be obtained to maximize the expected profit of the manufacturer. The system parameter values used in the numerical experiments are given as follows: T=50, P=10, P_S=6, G=10, m=7, h=1. The customer demand obeys the normal distribution with population mean μ =1000 and population variance σ^2=300^2. From the model assumption (4) in section 3, the serviceable goods ratio k_i for the sales channel i (i=1, 2) to satisfy the customer demand of goods can be estimated as k_i=1-L_i / T , based on the delivery lead-time L_i and the selling period T. The system parameters when the sales channel 1 is operated by the manufacturer are given as L_1=10, r_1=0.1, m_{r1}=1, C_1=50. The system parameters when the sales channel 2 is operated by the manufacturer are given as L_1=2-5, r_2=0.15, m_{r2}=1.5, C_2=150.

Table 1 shows the results of the optimal initial goods inventory and the expected profit of the manufacturer adopting the existing sales method and those adopting the parallel sales method. From Table 1, for r_1=0.1, r_2=0.15, L_1=10 and L_2=2, the expected profit of the manufacturer adopting the parallel sales method is higher than that adopting the existing sales method when the selection ratio to the sales channel 2 from the customer, g_2 , is expected to be from 0.3 to 0.6. This is the reason why the optimal initial goods inventory of the manufacturer adopting the parallel sales method is gradually larger as the selection ratio to the sales channel 2, g_2, is getting larger because the delivery lead-time L_2 of the sale channel 2 is shorter than the delivery lead-time L_1 of the sale channel 1. However, when g_2 is expected to be more than 0.7, the optimal initial goods inventory and the expected profit of the manufacturer adopting the parallel sales method are lower than those adopting the existing sales method. This is caused by the increase of the return processing cost for the parallel sales method when g_2>0.7 because r_2 and m_{r2} for the sales channel 2 are higher than r_1 and m_{r1} for the sales channel 1. Also, when not only g_2 is expected to be less than 0.3, but also the delivery lead-time of the sales channel 2 is more than 6, the expected profit of the manufacturer adopting the existing sales method are higher than that adopting the parallel sales method. This is reason why when g_2<0.3, increase of the system cost on returns of the manufacturer anticipated by adopting the parallel sales method is larger than that of the goods sales of the manufacturer expected in the case.

Table 2 shows the optimal initial goods inventory and the expected profit of the manufacturer adopting each sales method for the selection ratios g_1 and g_2 to the sales channels 1 and 2 from the customer when the delivery lead-time of the sales channel 2, L_2, is in the range between 4 and 6. Table 2 is used to investigate the minimal value of the delivery lead-time L_2 of the sales channel 2 that the expected profit of the

Table 1. Optimal initial goods inventory and the expected profit of the manufacturer adopting each sales method

Sales method	Selection ratio g_1	g_2	L_1 / L_2	10 / 2	10 / 4	10 / 6	10 / 8
Sales method 1 ($r_1 = 0.1$)	1.0	0.0	Optimal quantity	1151	1151	1151	1151
			Expected profit	1565	1565	1565	1565
Parallel sales which consists of sales channels 1 and 2 ($r_1 = 0.1$, $r_2 = 0.15$)	0.9	0.1	Optimal quantity	1172	1168	1159	1143
			Expected profit	1439	1437	1431	1423
	0.8	0.2	Optimal quantity	1210	1190	1173	1167
			Expected profit	1514	1472	1460	1438
	0.7	0.3	Optimal quantity	1257	1213	1208	1180
			Expected profit	1571	1512	1495	1455
	0.6	0.4	Optimal quantity	1310	1286	1232	1191
			Expected profit	1639	1599	1528	1472
	0.5	0.5	Optimal quantity	1330	1298	1249	1200
			Expected profit	1671	1622	1553	1484
	0.4	0.6	Optimal quantity	1258	1243	1219	1205
			Expected profit	1587	1544	1528	1491
	0.3	0.7	Optimal quantity	1237	1221	1164	1158
			Expected profit	1534	1524	1510	1480
	0.2	0.8	Optimal quantity	1181	1173	1166	1141
			Expected profit	1439	1398	1497	1476
	0.1	0.9	Optimal quantity	1147	1124	1116	1109
			Expected profit	1411	1403	1187	1166

Table 2. Optimal initial goods inventory of and the expected profit of the manufacturer adopting each sales method for difference of the delivery lead-times

Delivery Lead-time L_1	L_2	Selecting ratio g_1 / g_2	Sales method 1: 1.0 / 0.0	Parallel sales: 0.7 / 0.3	0.6 / 0.4	0.5 / 0.5	0.4 / 0.6
10	5.0	Optimal quantity	1151	1211	1262	1271	1230
		Expected profit	1565	1504	1565	1595	1534
10	5.2	Optimal quantity	1151	1210	1258	1268	1227
		Expected profit	1565	1502	1559	1587	1532
10	5.4	Optimal quantity	1151	1209	1252	1263	1224
		Expected profit	1565	1499	1552	1579	1532
10	5.6	Optimal quantity	1151	1208	1245	1256	1221
		Expected profit	1565	1497	1545	1568	1531
10	5.8	Optimal quantity	1151	1208	1239	1252	1220
		Expected profit	1565	1496	1534	1560	1530
10	6.0	Optimal quantity	1151	1208	1232	1249	1219
		Expected profit	1565	1495	1528	1553	1528

manufacturer adopting the parallel sales method is higher than that adopting the existing sales method. From Table 2, for $g_1 = g_2 = 0.5$ and $L_1 = 10$, when the delivery lead-time L_2 of the sales channel 2 can be reduced to less than 5.6, increase of the

Table 3. Optimal initial goods inventory of and expected profit of the manufacturer adopting each sales method for difference of the return rate of each sales channel

Sales method	g_1	g_2	L_1 / L_2	10 / 2	10 / 4	10 / 6	10 / 8
Sales method 1 ($r_1 = 0.1$)	1.0	0.0	Optimal quantity	1151	1151	1151	1151
			Expected profit	1565	1565	1565	1565
Parallel sales which consists of sales methods 1 and 2 ($r_1 = 0.1$, $r_2 = 0.125$)	0.9	0.1	Optimal quantity	1174	1170	1163	1145
			Expected profit	1456	1440	1435	1427
	0.8	0.2	Optimal quantity	1216	1196	1180	1173
			Expected profit	1570	1556	1467	1454
	0.7	0.3	Optimal quantity	1257	1223	1223	1196
			Expected profit	1595	1576	1506	1467
	0.6	0.4	Optimal quantity	1310	1298	1242	1224
			Expected profit	1668	1604	1543	1517
	0.5	0.5	Optimal quantity	1330	1312	1257	1200
			Expected profit	1703	1678	1573	1484
	0.4	0.6	Optimal quantity	1258	1247	1234	1205
			Expected profit	1630	1623	1569	1491
	0.3	0.7	Optimal quantity	1248	1231	1202	1158
			Expected profit	1610	1578	1552	1480
	0.2	0.8	Optimal quantity	1202	1189	1179	1145
			Expected profit	1572	1543	1497	1465
	0.1	0.9	Optimal quantity	1185	1124	1119	1117
			Expected profit	1532	1497	1464	1432

goods sales of the manufacturer expected by adopting the parallel sales method can be higher than that of the system cost on returns of the manufacturer anticipated in the case. It could lead to the improvement of the expected profit of the manufacturer. When the delivery lead-time L_2 of the sales channel 2 is longer than 5.6, it is caused by the reduction of the expected profit of the manufacturer due of increase of the system cost on returns of the manufacturer by adopting the parallel sales method.

Table 3 shows the optimal initial goods inventory of and the expected profit of the manufacturer adopting each sales method for the selection ratios g_1 and g_2 to the sales channels 1 and 2 from the customer when the return ratio r_2 of the sales channel 2 is reduced from 0.15 to 0.125. From the comparisons of Tables 1 and 3, it is verified that the range of the selection ratio to the sales channel 2 from the customer that the expected profit of the manufacturer adopting the parallel sales method is higher than that adopting the existing sales method can be improved from $0.3 \le g_2 \le 0.6$ when r_2 =0.15 to $0.2 \le g_2 \le 0.8$ when r_2 =0.125. This is reason why the return processing cost of the manufacturer adopting the parallel sales method can be reduced because the return ratio r_2 is reduced from 0.15 to 0.125. Next, we investigate the cost behaviors of the return processing costs of m_{r1} and m_{r2} for the sales channels 1 and 2. When r_1= r_2, m_{r1}=m_{r2}, k_1=k_2 and g_1=g_2=1/2 , it is evident that the terms on the return processing cost in the expected profit of the manufacturer in equation (3) can be reduced as

$$-r_1 k_1 m_{r1} \int_0^{2y_2/(k_1+k_2)} x f(x)dx - y_2 r_1 g_1 m_{r1} \int_{2y_2/(k_1+k_2)}^{\infty} f(x)dx.$$

Result of the investigation implies that it would be necessary to reduce the values of the return ratio r_2 and the return processing cost m_{r2} for the sales channel 2 to closer values of the return ratio r_1 and the return processing cost m_{r1} for the sales method 1 as much as possible in order to improve the range of the selection ratio to the sales channel 2 from the customer that the expected profit of the manufacturer adopting the parallel sales method is higher than that adopting the existing sales method.

6 Conclusions

In this study, we addressed the sales of a goods with the use-by date like beverage goods or periodical publications and so on. We discussed the impact of an e-commerce environment on the selection ratios to a brick-and-mortar sales channel (the sales channel 1) and an online sales channel (the sales channel 2) from the customer. Concretely, we focused on the return rates of the goods and the difference between the delivery lead-times of the goods to a customer through the sales channels 1 and 2 for a selling period, and then we considered a parallel sales method that the sales channel 2 was newly incorporated into the sales channel 1. By adopting the parallel sales method, the delivery lead-time of the goods to the customer could be reduced, however, the manufacturer would incur the returns processing costs and the fixed system operational costs for both the sales channels 1 and 2. From numerical experiments, the optimal selection ratio to the sales channel 2 from the customer was investigated to obtain the higher expected profit of the manufacturer adopting the parallel sales method than that adopting the existing sales method.

References

1. Choi, T., Li, D., Yan, H.: Optimal Returns Policy for Supply Chain with E-Marketplace. International Journal of Production Economics 88, 205–227 (2004)
2. Vlachos, D., Dekker, R.: Return Handling Options and Order Quantities for Single Period Products. European Journal of Operational Research 151, 38–52 (2003)
3. Gentry, C.R.: Reducing the Cost of Returns. Chain Store Age. 75, 124–126 (1999)
4. Meyer, H.: Many Happy Returns. The Journal of Business Strategy 20, 27–31 (1999)

Development of an Illumination Control Method to Improve Office Productivity

Fumiaki Obayashi[1], Misa Kawauchi[1], Masaaki Terano[1], Kazuhiro Tomita[2], Yoko Hattori[2], Hiroshi Shimoda[2], Hirotake Ishii[2], and Hidekazu Yoshikawa[2]

[1] Matsushita Electric Works, Ltd., 1048 Kadoma, Kadoma, Osaka, Japan
{obayashi,kawauchi.misa,terano}@mail.mew.co.jp
[2] Graduate School of Energy Science, Kyoto University Gokashou, Uji, Kyoto, Japan
{t-kazuhiro,h-yoko,shimoda,hirotake,yosikawa}
@energy.kyoto-u.ac.jp

Abstract. A lighting control method has been proposed which adjusts the circadian rhythms of office workers and eliminates a decrease in the arousal level in the afternoon by exposing a high intensity of light in the morning in order to improve their productivity. As a result, it was found that the task performance of the CPTOP test under 2,100 lx was significantly higher than that under 700 lx and more than a 9% improvement of task performance was achieved. Furthermore, a demonstration experiment has been conducted in order to confirm the effectiveness of the proposed lighting control method in actual office rooms. As a result, it was confirmed that the performance of their accounting work was over 10% higher than under normal conditions.

1 Introduction

A lot of people are spending considerable time within a day in the office and in public places, ex. libraries, and the approach on health and the comfort there is very important from the aspect of total QOL (Quality of Life). Health and the comfort of residents, however, it is generally indefinite cost-effectiveness, and this becomes a barrier to introduce equipment for residents. Therefore, the appeal power from which such "Health and Comfort for residents" technology is widely introduced into the office and public places are required. Then, the study where the index on the end user side (Health and Comfort) is caught by the index on the management side of clarification of cost-effectiveness again thus is being promoted.

So, productivity of office workers has become a new evaluation viewpoint of the office environment. Increase of office productivity by indoor environmental improvement produces great economic advantages. A trial calculation [1] shows the economic value of office productivity being $58,508/year in case productivity has improved by 3% in a 10,000-square foot office building. Office productivity is a new index to produce a better indoor environment for office workers, and to increase building value. A method which can evaluate office work productivity objectively and quantitatively has been developed [2].

M.J. Smith, G. Salvendy (Eds.): Human Interface, Part II, HCII 2007, LNCS 4558, pp. 939–947, 2007.

In this study, therefore, the lighting control method for the office has been developed and a basic experiment has been conducted to evaluate how much office work productivity is improved by controlling lighting conditions in the office space.

2 Office Productivity Index and Evaluation Method

Various definitions and interpretations of office productivity have been suggested. In this study, Office Productivity is defined as office working efficiency.

In order to calculate office productivity of a specified occupation category, the results of the performances of the elemental abilities are given added weight to each performance based on the deduced weight. The concrete productivity index can be calculated by equation (1).

$$\textbf{Office Productivity} = \sum_{i=1,\, j=1}^{I,J} (P_{ij} \times W_{ij}) \qquad \ldots(1)$$

P_{ij} :Score of Performance Test of Elementary Ability
W_{ij} :Weight of Elementary Ability
i :Occupation
j :Elementary Ability

This index can be calculated by the results of performance tests and weights of elementary abilities for a specific occupation and a specific office environment.

A performance test system, CPTOP (Cognitive Performance Tests for Office Productivity) has been developed to evaluate the productivity of office workers objectively and quantitatively based on the above productivity evaluation conception [2]. CPTOP is a web-based performance test system and it can be used by only PCs and Web browsers connected to the Internet as shown in Figure 1.

The CPTOP consists of 11 different task tests which correspond to 11 elementary abilities. By changing the configuration of the test set and weights, productivity in various office situations and environments can be evaluated. The results of the performance tests are dependent on individual abilities, the statistical calculation among plural subjects or comparison of the same subject before and after changing the environment is necessary to evaluate the office environment precisely. Note that this test is to evaluate productivity in a specific office environment, and NOT to evaluate the abilities of individual workers.

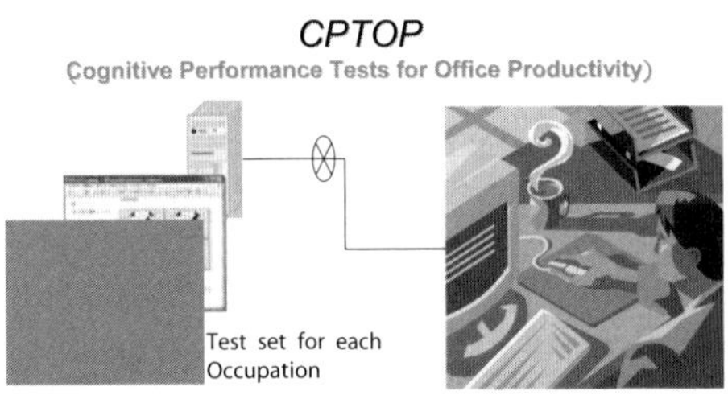

Fig. 1. Configuration of CPTOP

In order to evaluate any improvements of the office environment by using this method, productivity index before and after improvement is compared.

3 Productivity and Office Lighting

3.1 Design of Lighting Environment

Present office lighting is designed from the viewpoints of eye sight and stress reduction for the optic nerve, and it is considered as the minimum requirement for the office environment. In order to make the appropriate effect for productivity by lighting conditions, a lighting control method has been designed.

From the viewpoint of adjusting the human circadian rhythm, it is necessary to be exposed to thousands of lux lights in the morning for a few hours. It is known that exposure to a high intensity of illumination adjusts the human circadian rhythm and the arousal level can be increased as the intensity soars up to 10,000 lx [3]. In addition, if the circadian rhythm is synchronized with life rhythm and in keeping with a high arousal level in the day time, a high quality of sleep can be obtained. On the contrary, a high quality of sleep at the night causes higher arousal levels in the day time. This means that high illumination light keeps high arousal levels in the office. However from the standpoint of productivity in the long term, it is not appropriate because keeping high arousal levels all day may cause workers' fatigue. Therefore, the high illumination light should be applied only to regulate biological rhythm and improve the arousal level.

When aiming at adjusting biological rhythm, improvement of arousal level and reduction of long term fatigue at the same time, the lighting control method as shown in Figure 2 was proposed. It compensates for a lack of light by giving high illumination light in the morning and adjusts the biological rhythm by resetting the biological clock. In the early afternoon, high illumination light is increases to reduce sleepiness after lunch. After that the illumination decreases to 700 lx, which is the illumination of a normal office. During lunch time, the illumination is decreased to 500 lx to reduce energy consumption and promotes taking a nap to reduce sleepiness after lunch.

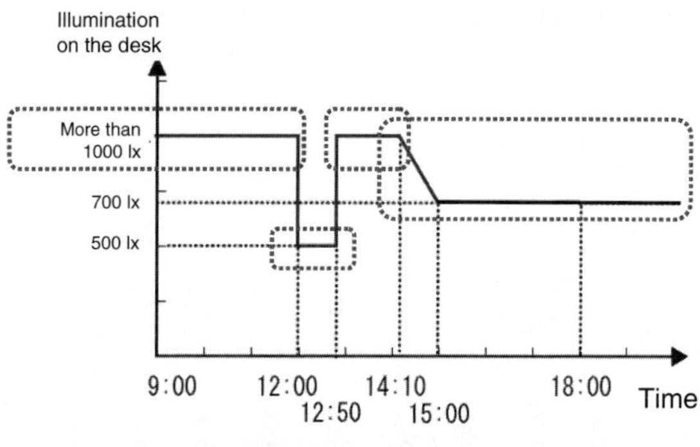

Fig. 2. Proposed lighting control method

It is expected that the lighting control method can reduce long term fatigue by adjusting biological rhythm and improve the work performance in the short term by keeping high arousal level and preventing from sleepiness after lunch.

3.2 Productivity Evaluation of Proposed Lighting Control Method

In order to evaluate effectiveness of the proposed lighting control method from the viewpoint of productivity, a subjective experiment was conducted to quantitatively evaluate it by using CPTOP. At the same time, a proper lighting control method was studied by comparing different maximum illumination of the lighting control pattern.

3.2.1 Experimental Method

Based on the control pattern as shown in Figure 2, two conditions in which the maximum illuminations are 1,400 lx (condition 1) and 2,100 lx (condition 2) were prepared. In addition, normal condition in which the illumination is fixed to 700 lx (normal condition) was also prepared.

The subjects of the experiment were 6 people with experience in office work. Their average age was 27.7 (from 22 to 36) and they were accustomed to using a PC. The experiment was conducted in an experiment room where the lighting condition can be controlled. Figure 3 shows the top view of the experimental room, while Figure 4 shows a view of the experiment. The lighting equipment were 12 illumination panels and 8 ceiling lights, which can give 500 to 3,000 lx illumination. Figure 5 shows the actual illumination measured on the surface of the desk by an illumination meter.

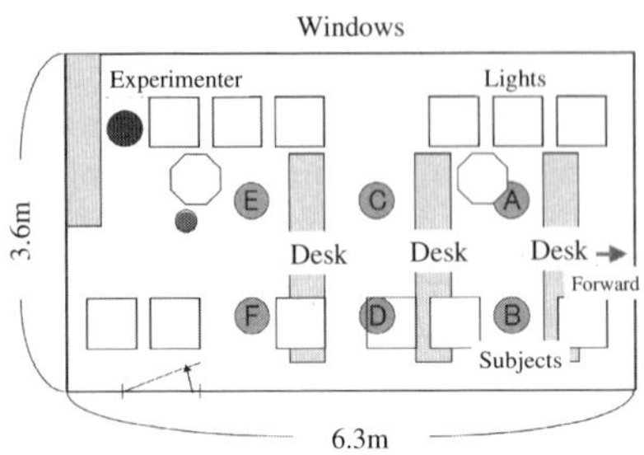

Fig. 3. Top view of experimental room

Fig. 4. A scene of experiment

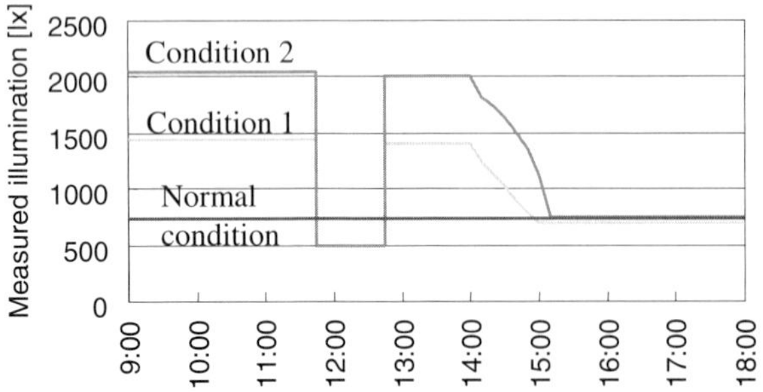

Fig. 5. Measured illumination

In order to evaluate the effectiveness of the lighting control method, CPTOP was conducted as a quantitative evaluation. In addition, a questionnaire was also conducted to evaluate subjective fatigue by using an examination sheet for subjective symptoms provided by the Japan Society for Occupational Health [4].

The experiment was conducted for 14 days along with the schedule as shown in Figure 6. And the experimental procedure of a day is shown in Figure 7. One day is divided into three parts morning, early afternoon and late afternoon. In each part, the questionnaire was conducted first, and simulated office tasks were given, then the CPTOP was conducted. The time for CPTOP was controlled to approximately 60 minute time lots. In each condition except for the first day and the fourth day, simulated office tasks were given instead of CPTOP as just a task workload.

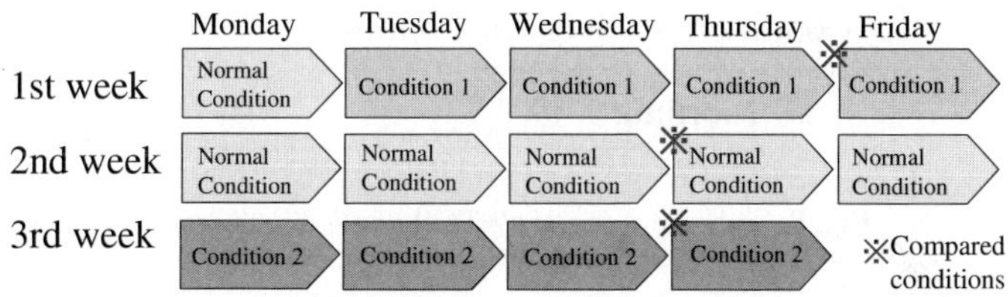

Fig. 6. Schedule of experiment

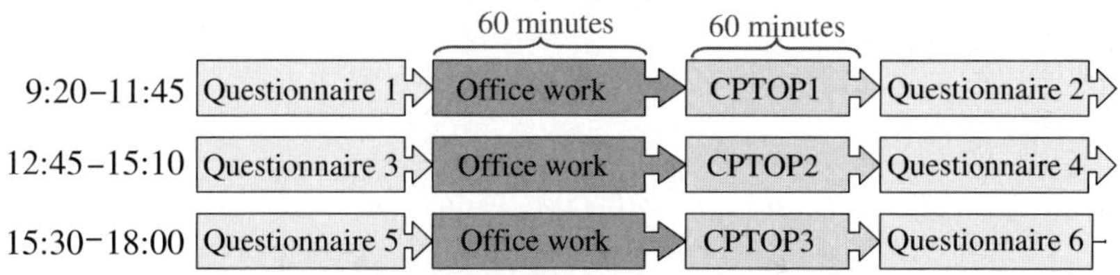

Fig. 7. Experimental Procedure

3.2.2 Result

Figure 8 shows the average score of CPTOP of all the subjects. Here, the scores were standardized based on the average score of all the experimental terms for each subject, and they were calculated as weights of the 11 task tests which were assumed to be equal.

As a result of CPTOP, the scores of condition 2 are significantly higher than those of normal conditions. Among condition 2, the scores of the third day and the fourth day are higher than that of the first day ($p<.05$ and $p<.01$, respectively). This means that the task performance was improved by the high illumination light. On the other hand, there is no difference between condition 1 and normal conditions. This means that the illumination of 1,400 lx is not bright enough to improve the performance.

Figure 9 shows the results of the questionnaire. As shown in Figure 12, total score of subjective fatigue in three days of condition 2 except for the third day which is significantly lower than those in normal conditions. It is presumed that illumination condition may affect subjective fatigue.

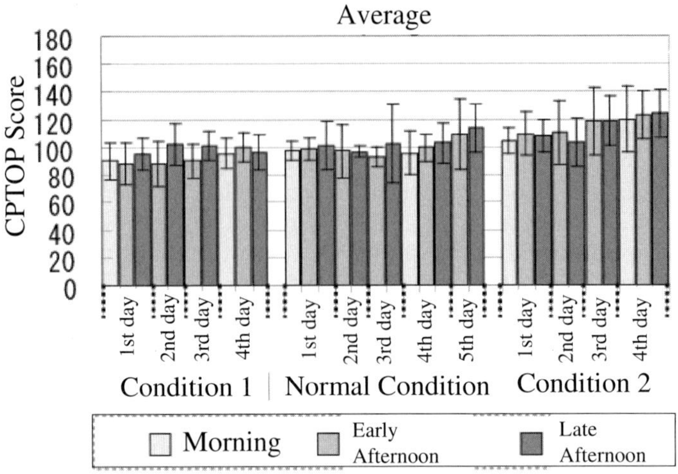

Fig. 8. Average score of CPTOP of all the subjects

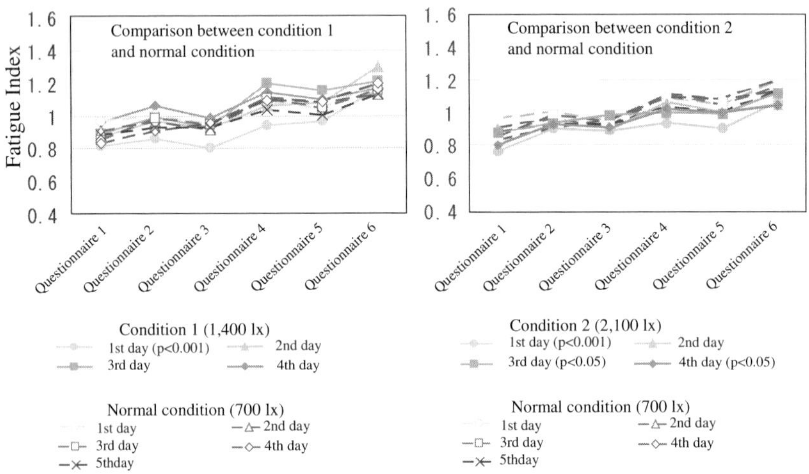

Fig. 9. Results of questionnaire for subjective fatigue

3.2.3 Evaluation of Productivity

The purpose of this experiment is the quantitative evaluation of productivity by the proposed lighting control method. The score of CPTOP in the fourth day of condition 2 and normal condition were compared statistically. As a result, the score of condition 2 is 9 points higher than that of normal condition ($p<.001$). The score of CPTOP is standardized to 100, therefore a 9% improvement of task performance can be expected when controlling lighting along with Figure 2 with 2,100 lx for four days.

4 Office Demonstration Experiment of Productivity Lighting System

In order to confirm the effectiveness of the proposed lighting control method, a demonstration experiment in an actual office has been conducted.

4.1 Experimental Method

Based on the control pattern as shown in Figure 2, one condition in which the maximum illumination is 3,500 lx (condition 3) was prepared. Also, normal conditions in which the illumination is fixed to 700 lx (normal condition) was prepared.

The subjects of the experiment were 6 people who worked in the accounting department of a company. The experiment was conducted in their office room where the additional light fitting was set up. Figure 10 shows the top view of the desks where subjects use, while Figure 11 shows a view of the office. The light equipments can give 300 to 4,000 lx illumination. Figure 12 shows the actual illumination measured on the surface of each desk by an illumination meter.

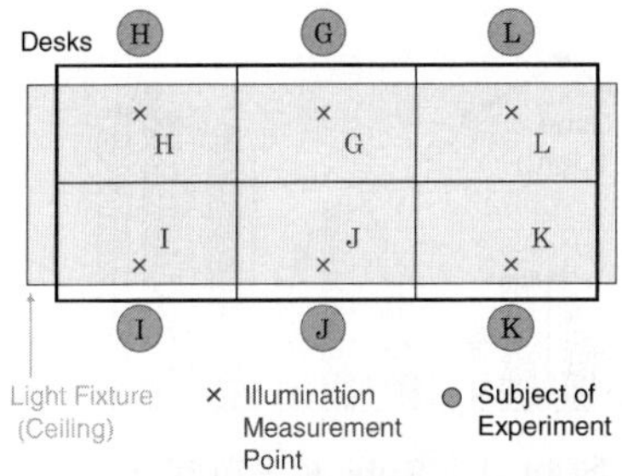

Fig. 10. Top view of experimental place

Fig. 11. A scene of office

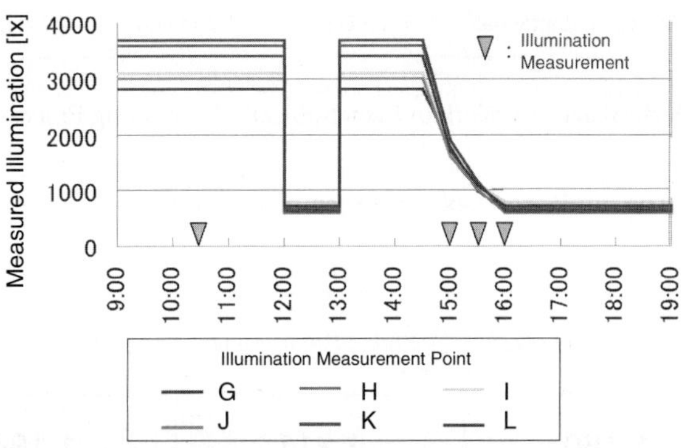

Fig. 12. Measured illumination

In order to evaluate the effectiveness of the lighting control method, the office workers quantity of work is measured. The subjects of the experiment elected one preferred hour in each day. The number of sheets of the slip processing in the accounting processing work is measured. The number of sheets a minute is assumed to be an amount of real work. The accounting processing works by extracting necessary data from the slip of paper, then recording the electronic data, and finally checking the correspondence of the slip. The experiment was conducted from late November to early January. The subjects of the experiment selected days which were an average amount of work for this period, and the measurement was conducted. The lighting condition 3 is conducted in one week in mid-December. About three weeks are made for the experiment period of a standard condition before and after. Each subject selected three days as measurement days from each period.

4.2 Evaluation

Figure 13 shows the performance of work in the each subject's accounting process work. For four of six subjects, the amount of work under the lighting condition 3 is higher than the amount of work under the standard conditions which were conducted before and after the lighting condition 3.

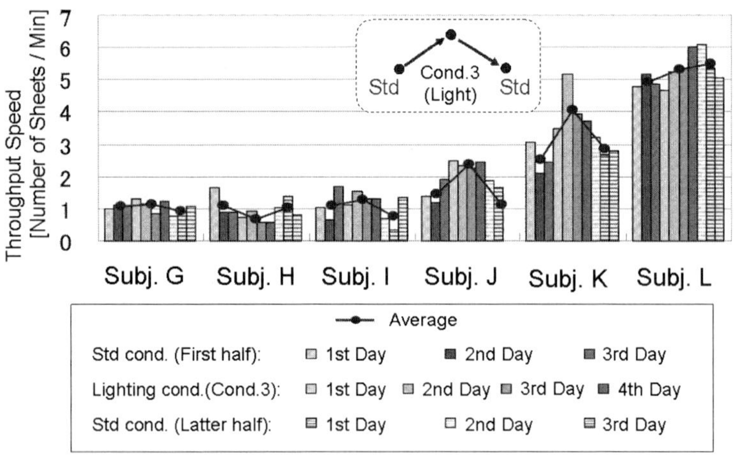

Fig. 13. Performance of Work in Each Subject's Accounting Processing Work

Table 1. Comparison of Performance between Lighting Control Condition and Standard Condition

Lighting / Std (First half) (Cond. 3)	Lighting / Std (Latter half) (Cond. 3)	Lighting / Std (Average) (Cond. 3)
1.180	1.211	1.184

As a result, it is shown that their performances have improved with the lighting system. About a ten percent or over improvement of work performance can be expected when controlling lighting along with Figure 2.

5 Conclusions

In this study, the lighting control method has been proposed which improves productivity of office workers and adjusts their circadian rhythms. A subjective experiment was conducted to evaluate a proposed lighting control method which can adjust circadian rhythm of office workers by exposing them to high illumination light. In the experiment, three lighting conditions were prepared which were normal condition (700 lx fixed), condition 1 (maximum illumination was 1,400 lx and control pattern was along with Figure 2) and condition 2 (2,100 lx as well). Under these lighting conditions, CPTOP and questionnaires for subjective fatigue were conducted. As a result, it was found that the CPTOP score under the condition 2 was 9% higher than that under the normal condition, and subjective fatigue under condition 2 was significantly lower than that of under normal conditions.

Furthermore, a demonstration experiment has been conducted in order to confirm the effectiveness of the proposed lighting control method in actual office rooms. As a result, it was confirmed that the performance of their accounting work was over 10% higher than that of under normal conditions.

In the future, we will confirm the effectiveness of the proposed lighting control method in various actual office rooms.

References

[1] Lomonaco, G., et al.: Environmental Satisfaction, Personal Control and Positive Correlation to Increased Productivity, Johnson Controls (1997)

[2] Obayashi, F., Kawauchi, M., Terano, M., Ito, K., Shimoda, H., Yoshikawa, H.: Objective Measurement Method for Productivity and Investigation on Environmental Factor. In: Proceedings of Human Computer Interaction International 2005, CD-ROM (2005)

[3] NEMA, Lighting and Human Performance: A Review, Lighting Equipment Division of National Electrical Manufactures Association and the Lighting Research Institute (1989)

[4] Japan Society of Occupational Health, Handbook of Industrial Fatigue, Roudou Chousakai (in Japanese) (1988)

The Inspiring Store: Decision Support System for Shopping Based on Individual Interests

Akihiro Ogino, Sae-ueng Somkiat, and Toshikazu Kato

Faculty of Science and Engineering,
Chuo University, 1-13-27 Kasuga Bunkyo-ku Tokyo, Japan
{ogino,somkiat,kato}@indsys.chuo-u.ac.jp

Abstract. This paper proposes a ubiquitous environment store, which is composed of cameras, sensors and displays, for supporting decision concerning commodities based on each customer's shopping interest. This store suggests that three analyzing views can be utilized to analyse relations between customers' interests and their behaviours towards commodities. This store collects data on the behaviour of each customer, who has agreed to be observed, and makes individual models, by analyzing collected data. This store also provides commodity information, which is decided by the individual's model, to each customer through a display, which is in front of this customer. Comparing this system to a non-system through a purchase experiment of clothes shows this effectiveness of this system.

1 Introduction

Recently, customers have become interested in a larger variety of commodities than ever before. They seek a diverse range of commodities when shopping. However, it is sometimes difficult for customers to find commodities of interest because there is such a vast range of commodities in the modern world. Therefore, customers now need a system to help them find interesting commodities.

In the business field, there is a growing need for finding customer interests in commodities [1], [2]. For example, many companies use the POS (point of sale) system, which accumulates the buying history of customers, to detect their interest in commodities. However, the POS cannot account for interest in commodities, which customers did not buy although the customer has shown interest in them, for example, by picking them up and studying them, before they reach the electronic cash register. There are also recommendation systems on the Internet, which support customer's decision concerning commodities using the purchase histories of customers on a certain site [3]. However, these computer systems have drawbacks in that recommendations sometimes don't match a customer's individual interest in commodities because the systems use a purchase history of other customers. In other words, current systems are too generalized, looking at the interest of large groups rather than individual.

M.J. Smith, G. Salvendy (Eds.): Human Interface, Part II, HCII 2007, LNCS 4558, pp. 948–954, 2007.

For recommending appropriate commodities to individual customers, computer systems need to know which commodities they are interested without asking for too much time and effort from customers. This paper attempts to remedy these problems by focusing on the individuals.

This paper proposes a personalized shopping store, which interprets individual interest in commodities and attempts to directly recommend commodities, in which customers may have an interest, to each customer. This paper calls it the inspiring store. The inspiring store observes each customer individually observing what the customer touches and views or where the customer moves using ubiquitous devices, such as cameras with infrared rays and RFID (radio frequency identification) Readers. This inspiring store organizes observed data and makes individual models from the observed data to detect each customer's interest in commodities. This store predicts which commodities suitable for the individual, by using their own model and recommends personalized information concerning commodities through displays for advertisement.

2 The Inspiring Store

Customers with an unclear idea about what they want often have trouble deciding which commodities to buy. Therefore, they sometimes select unsuitable commodities or buy almost the same commodity, which they had before.

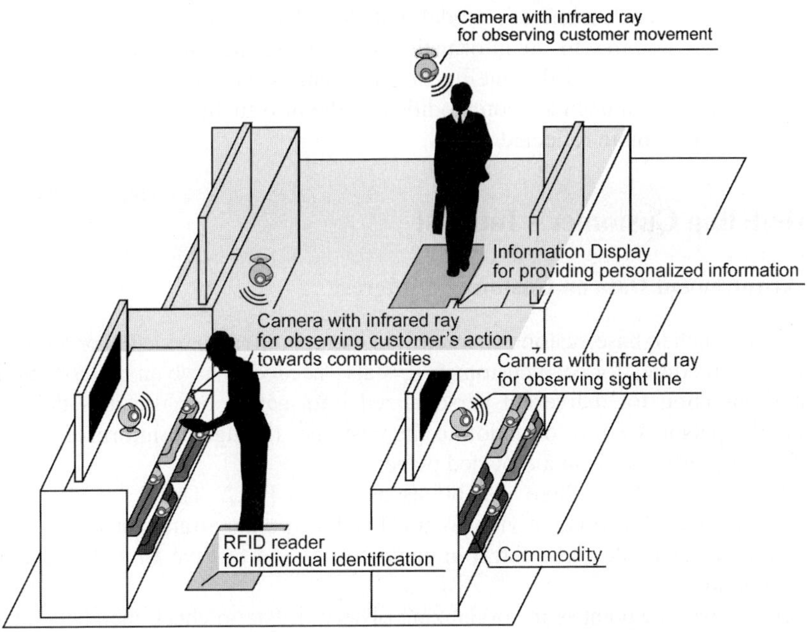

Fig. 1. The Inspiring Store: A store with a purchase support system that provides personalized information concerning commodities in which each customer has interests

If customers could directly and immediately get information concerning commodities suitable for them through a system in a store, such as information devices and advertisement displays, they will be happy because they could achieve their aims and receive the commodities they want.

During this study a prototype of The Inspiring Store was built based on the previously described motivation, as shown Fig.1 Customers in this store agreed to have their purchase observed and are given member's card, which were RFID cards. If customers didn't have member's cards, this store couldn't authenticate the identity of each customer.

This inspiring store is 30 square meters and has 6 intelligent shelves, which are composed of cameras with infrared ray (Panasonic Network Camera BL-C31) and a RFID reader (Omron V720). Each camera observes a designated commodity and observes customers' actions towards commodities by checking image differences between a previous image and a current image. The RFID reader identifies members and the member's position. The intelligent shelves also have an information display, which shows information using a web browser, for providing personalized information concerning commodities to each member. This store additionally has 6 cameras with infrared rays on the ceiling for observing the store's environment, such as the number of unknown customer in this store.

This inspiring store makes member's models, which simulate each member's decision concerning commodities based on their own criteria, by analyzing observed data on the relation between an individual's interest in a commodity and individual actions towards the commodity. The store selects commodities, in which they will have interest, based on each member's model and then delivers personalized information concerning commodities to an information display on the intelligent shelf when a customer comes in front of the intelligent shelf. This store collects data on customer reaction towards recommended commodities on the information display and then uses this data to make a more reflected model.

3 Modeling Customer's Interest

3.1 Acquirement Data on Customer's Interest

Criteria on which to base customer's interest in commodities is needed in order to select suitable commodities for each customer. It is also needed to have an understanding of the store situation for delivering personalized information because it is difficult to provide the personalized information to the customer through an information display directly if the customer is in a crowded place.

This study introduces three viewpoints as shown Fig.2. There are Microscopic, Macroscopic and Mezzoscopic viewpoints. The viewpoints arrange observed data, to make individual's models of interest in commodities and to have an understanding of store's situation.

Microscopic viewpoint is for organizing observed data to check members' identity and their actions towards commodities in the store. This study uses the Microscopic data to detect relations between a member's interest in commodities and the member's action towards the commodities. By analyzing each member's relations, this store can

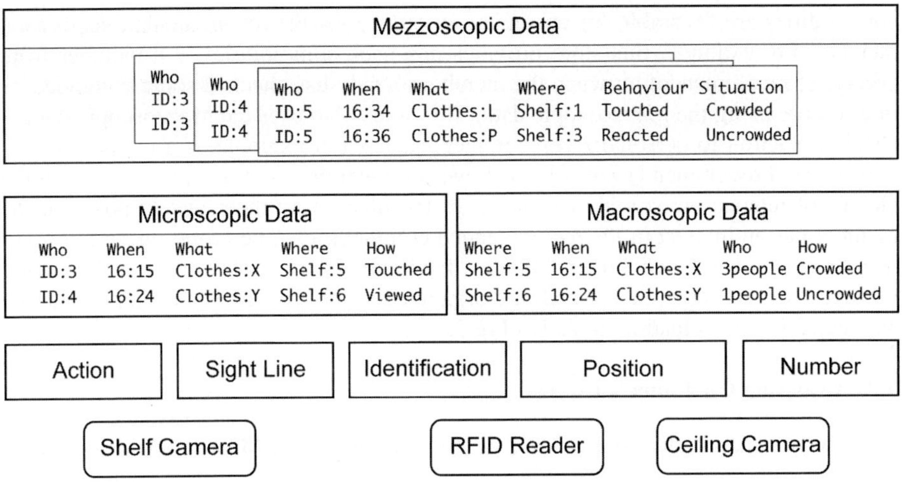

Fig. 2. Micro, Macro and Mezzoscopic Viewpoint: they are viewpoints that detect each customer's interest for making individual's models and to grasp store's situation for providing personalized information to each customer

understand common features of commodities, which each member has shown an interest in and then has acted towards. This study organized the Microscopic data using three types of observed data along with the time it was observed. One type of data was sensed identity from an RFID card, which belonged to each member. Other types of data were a customer's action and sightline captured on cameras, when customers handle or look at commodities. This study manages Microscopic data by 4W1H (Who, When, Where, What and How) as shown Fig.2 For example, Microscopic data of ID: 3 in Fig. 2 show that the member of ID: 3 touched the clothes of ID: X at 16:15 in the shelf of ID: Y.

Macroscopic viewpoint is for understanding the environment of the store, such as each customer's position. This study uses the Macroscopic data to decide when is the best time to provide personalized information or the contents of information provided to each member. For example, if there are many customers in front of a intelligent shelf, it's not the best time to show personalized information to one customer. Therefore, for example, this store decides that it is better to provide common information in which most customers will have interest or it would be better to send personalized information to each customer's mobile phone. This study organizes the macroscopic data by two types of observed data with time. One is data on sensed position of each member by their RFID card and captured position of unknown customer by ceiling cameras. Another is data on captured rough numbers of customer by ceiling cameras. This study also manages the Macroscopic data by 4W1H as shown Fig.2. For example, the Macroscopic data of shelf: 5 in Fig.2 show that there are 3 people in front of clothes of ID: X on the shelf of ID: 5 and the data also show a crowded situation for providing personalized information.

Mezzoscopic viewpoint is for detecting a current member's interest in commodities by analyzing the order, in which the commodities were touched, and the common traits of commodities. This study can use this Mezzoscopic data to decide which commodities are favorable for customers in order to make recommendations to each member. For example, this store provides un-check commodities to a member from criteria of recommendation when the member already has checked some commodities in this store using the Mezzoscopic data. This study also uses the Mezzoscopic data to check a reaction to personalized recommendations. For example, if a member sees a personalized recommendation, which shows a commodity with the place in which the clothes of interest are stored, and moves to the place from their current position, the member has an interest in the recommended commodities. Therefore, this study saves data on react towards recommendation as "having reacted", as shown Fig.2. This study manages Mezzoscopic data, which are organized by Micro and Macroscopic data, by 4W, behavior and situation as shown Fig.2.

3.2 Modeling Customer's Interest

This study makes individual models by combining Micro, Macro and Mezzoscopic data.

First, this study analyzes Microscopic data using the discriminant analysis and detects tendency of commodities' traits, which they like and act towards, individually. This study uses actions as the objective variable and uses quantified traits of commodities as the explanatory variable. This study determines commodities in which each customer will have interest and have action, by a discriminant function of each customer.

4 Experiment

4.1 Experiment's Scenario

This study evaluated each customer's satisfaction concerning personalized recommendations using this inspiring store. This experiment provided a task to each examinee: Each examinee entered the store, after determining a trait of interest from clothes, such as a color, type and material, and tried to find clothes that suited their decided trait from about 60 items of clothing. For example, an examinee selected the color blue as their criteria of trait, and tried to find blue clothes in the store during experiment.

The evaluation experiment was conducted by using 18 examinees acting as member's customer in this inspiring store and purchasing items of clothing twice as follows.

1. No use of personalized recommendation based on this inspiring store method.
2. Use of personalized recommendation based on each examinee's model made during the first experiment.

The identity of each examinee was registered in a member's database before starting the experiment and each examinee has the RFID card with their identity. The examinees with a RFID card came to the experimental clothes store independently and bought clothes, which they were interested in. The examinee can feel free to move in the store. Each time of experiment used different clothes set and the implementation interval of this experiment was 3 days.

In the first experiment, this store accumulated Macro, Micro and Mezzoscopic data of each examinee from observed data, which are their actions towards clothes. Then, the store made a model of each examinee by analyzing common feature of the clothes in which they showed interest though their actions.

In the second experiment, this store gave personalized recommendation of clothing commodities through a display and accumulated data on examinee's reaction to the recommended information.

Table 1. Number of examinees' response, Reaction results and Satisfaction

Examinee	Number of Checked	Reacted	Not Reacted	Ratio of reaction	Satisfaction
A	6	5	1	0.83	A
B	5	4	1	0.80	A
C	4	3	1	0.75	B
D	7	5	2	0.71	B
E	9	6	3	0.67	B
F	6	4	2	0.67	B
G	6	4	2	0.67	B
H	8	5	3	0.63	B
I	5	3	2	0.60	C
J	5	3	2	0.60	B
K	12	7	5	0.58	B
L	8	4	4	0.50	C
M	4	2	2	0.50	C
N	2	1	1	0.50	C
O	5	2	3	0.40	C
P	4	1	3	0.25	C
Q	6	1	5	0.17	C
R	1	0	1	0.00	C

The model of each examinee, which was made during the first experiment, decided the recommended information about clothes. The store displayed the personalized recommendation on clothes using the display in front of the examinee, when it observed identity or presence of a member.

4.2 Result

This study defined that the model was considered a success if the examinee reacts to the recommendation of the clothes and moves to the place on which clothes are put. Table.1 shows number of their responses to recommended information, the reaction results of each examinee and satisfaction of the recommendation. Therefore, they have checked many clothes concerning a target feature.

Examinees of about 56% have satisfaction concerning personalized recommendation by shown Table.1. This study could consider that examinees of A and B satisfied personalized recommendations because they react in a positive way towards the recommendation and showed satisfaction grade of A. This study could also consider that examinees of A and B satisfied personalized recommendations because they react in a positive way towards the recommendation and showed satisfaction grade of A.

5 Conclusions

This paper proposed the inspiring store with ubiquitous computing for supporting commodity decision based on each customer's interest in commodity in their shopping. This paper also describes new ideas, which are an acquisition and modeling method of customer's Interest. This study introduced three viewpoints, Macroscopic, Microscopic and Mezzoscopic, in this acquisition method and shows a mechanism used it for organizing data on relation between customer's interest in commodities and their action towards commodities.

References

1. Underhill, P.: Why We Buy,The Science of Shopping, TouchstoneBook, 1999 (2000)
2. Dennis, C., Marsland, D., Cockett, T., Hlupic, V.: Market segmentationand customer knowledge for shopping center. In: 25th int. Conf.Information Technology Interfaces ITI 2003, pp. 417–424 (June 16-19, 2003)
3. Linden, G., Smith, B., York, J.: Amazon.com recommendations: item-to-item collaborative filtering. Internet Computing, IEEE 7(1), 76–80 (2003)

A More Comprehensive Approach to Enhancing Business Process Efficiency

Seung-Hyun Rhee[1], Nam Wook Cho[2], and Hyerim Bae[3,*]

[1] Department of Industrial Engineering, Seoul National University, 151-742, Seoul, Republic of Korea
shyun.rhee@gmail.com

[2] Department of Industrial and Systems Engineering, Seoul National University of Technology, 139-743, Seoul, Republic of Korea
nwcho@snut.ac.kr

[3] Department of Industrial Engineering, Pusan National University, 635-709, Busan, Republic of Korea
hrbae@pusan.ac.kr

Abstract. Whereas Business Process Management (BPM) systematically guides employee participation in business processes, there has been little support, use or development of user-friendly functions to improve the efficiency of those processes. To enhance business process efficiency, it is necessary to provide automatic rational task allocation and work-item importance prioritization, so that task performers no longer need to be concerned with process performance. In the context of BPM, two different perspectives, the Process Engine Perspective (PEP) and the Task Performer Perspective (TPP), are considered. Accordingly, we developed a comprehensive method that considers those two perspectives, in combination rather than separately. We carried out simulation experiments to show the combinational effect of the two phases.

Keywords: Business Process Management, User-Oriented Support, Process Efficiency, Theory of Constraints, Dispatching Rule.

1 Introduction

Companies have come to perceive the importance of the process perspective in managing intra-organizational resources and inter-organizational relationships [3][6][10][11]. The advent of BPM (Business Process Management) is enabling them to employ their processes as the main backbone of the management of their resources and, through the linking of heterogeneous processes, to enter into partnerships involving Business Process Outsourcing and M&A [2][6][14].

---

* Corresponding author.

M.J. Smith, G. Salvendy (Eds.): Human Interface, Part II, HCII 2007, LNCS 4558, pp. 955–964, 2007.

Most notably, BPM enables even employees, as well as top executives, to easily trace the progress status of their business processes [15]. However, there has been little support, use or development of user-oriented functions in improving the efficiency of those processes. As explained in this paper, we devised a methodology for providing human-oriented features in the context of BPM. Such innovations ultimately lead to the improvement of process efficiency.

Process execution efficiency has been dealt with according to two different perspectives: one is task allocation to participants, and the other, the priorities of the assigned tasks. However, it should be noted that in order to operate business processes efficiently, those two perspectives must be considered together. This paper proposes a comprehensive methodology of process efficiency that combines the two approaches.

The paper is organized as follows. Section 2 explains process efficiency in the context of BPM. Section 3 discusses previous research on process efficiency. Sections 4 and 5 describe, from the BPM-engine- and user perspective, respectively, methodologies for efficient process execution. Section 6 provides an integrated methodology for business process efficiency. Section 7 offers the results of a simulation study to validate our overall concept.

2 Process Efficiency

As BPM functions become more effective and the number of processes managed by BPM increases, Business Process (BP) efficiency issues are beginning to attract more attention [1][17][19]. BP efficiency can be considered from two different perspectives: the Process Engine Perspective (PEP) and the Task Performer Perspective (TPP).

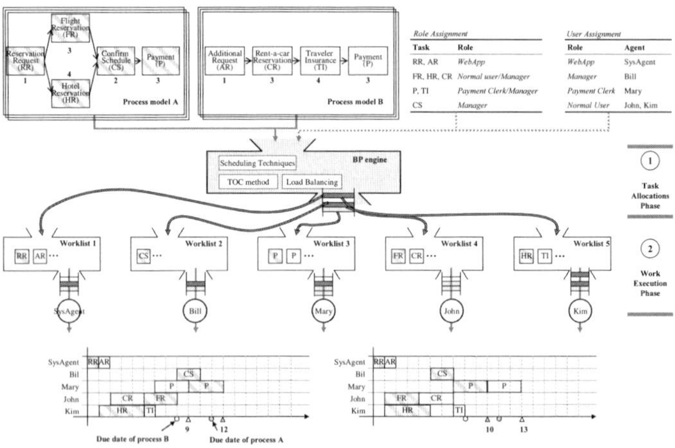

Fig. 1. Overall BP management procedure using BPM

Figure 1 illustrates the overall procedures of BP execution in a BPM system, showing the two perspectives. First, the PEP is the task allocation of the BPM system engine, a core function of BPM. After process models are prepared, the engine starts a process instance based on those models [2]. During process execution, the engine assigns tasks to proper users, using delicate rules [17], and these rules have a great influence on the efficiency of the processes [9]. For example, the FR task, which is assigned to John, can also be assigned to Bill. Once assigned, the FR task can start as early as time point 1, and thereby overall process efficiency, which is shown at the bottom-left of the chart, can be improved.

Second, BP efficiency can vary depending on how a user executes tasks assigned to him. In dealing with the assigned tasks, the priority defined by a user can influence the efficiency of the whole process. The two charts at the bottom of Fig. 1 show the two results from applying two different priority rules in executing tasks on worklists. As can be seen, the user's priority assignment affects overall efficiency by affecting the process completion time.

With respect to BP efficiency, another important issue is the means by which efficiency is measured [2]. A proper strategy has to be considered according to the kind of efficiency desired. The left result in Fig. 1 is more efficient in process completion time. Alternatively, if due-date is the important measure, the right result is better.

3 Related Work

A lot of research has aimed at improving BP efficiency through the use of scheduling technologies, or of frameworks specifically designed for smooth process routing. Baggio *et al.* [1] has provided a method to minimize the number of late jobs in a workflow by applying scheduling techniques. Chang *et al.* [4] proposed the identification and analysis of critical paths to manage time and resources within a workflow process. Zhao *et al.* [19] introduced a concept to predict the turnaround time of a time-driven process and to allocate the expected processing time to each activity in the process. In Kumar's study [12], a general framework was devised for efficient workflow management.

More specific research on task distribution and priority setting has been conducted as well. Ha *et al.* [9] developed a process execution rule that enables the balancing of the workload of agents, each of whom has a worklist. Rhee *et al.* [16] developed a TOC (Theory of Constraints) -based process execution method that contributes to the enhancement of efficiency by controlling task allocation to an overloaded participant and synchronizing process instance release with his/her work pace. Kumar *et al.* [13] dealt with the problem of the trade-off between observing a deadline and offering work items to an overloaded participant at run-time, and proposed a systematic approach to dynamically creating a balance between quality and performance issues in workflow systems.

Regarding priority setting, Eder *et al.* [7] employed a personal schedule that provides information on future work. They showed that, with this information, both the turn-around time and time-constraint violation rate can be decreased. Rhee *et al.* [17] calculated the slack time of tasks in workflow processes. Slack time guides task performers in processing urgent tasks first, which eventually improves BP efficiency. Whereas the previous research has addressed either the PEP or the TPP, this research comprehends both perspectives.

4 Process Engine Perspective (PEP)

As for the PEP, we adopted the TOC [8]-based process execution method developed by Rhee [16]. The TOC method is discussed in detail in the following sections.

4.1 Drum

We introduce BP-Drum, Buffer, and Rope (BP-DBR), modified from the original DBR in the TOC concept. Our method emphasizes the situation in which a specific resource is relatively limited and, therefore, can determine the performance of the entire system. We define a Constraint Capacity Resource (CCR) as a task performer who has the largest workload on his worklist.

First, the Drum is defined as the pace of the CCR. In order to calculate the Drum, two different cases are taken into account. The first case supposes that a CCR user (u_{CCR}) belongs to a workgroup with k ($k \geq 0$) users, and the workgroup processes a single task, a_i, within an expected processing time of ET_i. Since one agent can process $1/ET_i$ tasks per unit time, the Drum can be

$$\text{Drum } (d) = \frac{k}{ET_i} \text{ (task/time).} \tag{1}$$

In the second case, u_{CCR} can deal with multiple tasks in the respective processing times. To deal with this type of case, the concept of unit task was introduced. For example, suppose that a task a_1 has the smallest process time of 1, and that a task a_2's processing time is 2. Then, processing of a_2 corresponds to processing of two unit tasks. Therefore, for a BP-CCP responsible for J tasks, a unit task of a_i can be determined as

$$\text{Unit number of task } a_i\,(UN_i) = \frac{ET_i}{ET_{\min}}. \tag{2}$$

ET_i: Expected Time of a_i, $ET_{min} = \min\{ET_i : j=1,2,\ldots,J\}$

Let P_i denote the probability that a_i be assigned to u_{CCR}; therefore, the Drum can be

$$\text{Drum } (d) = \sum_{i=1}^{J} P_i \cdot \frac{1}{ET_i} \cdot UN_i \text{ (task/time).} \tag{3}$$

4.2 Rope and Buffer

The Rope is defined as a communication tool that controls the speed of input according to the Drum. A BPM engine plays a role as the Rope by controlling the release of a process instance into a system. The BPM engine synchronizes the process instance release rate (λ) with the Drum.

$$\lambda = d \text{ (task/time)} \tag{4}$$

Figure 2 illustrates the function of the Rope. The Rope in the BPM engine controls the entrance of instances so that the process release rate, which can be the instance arrival rate, equals the pace of the CCR.

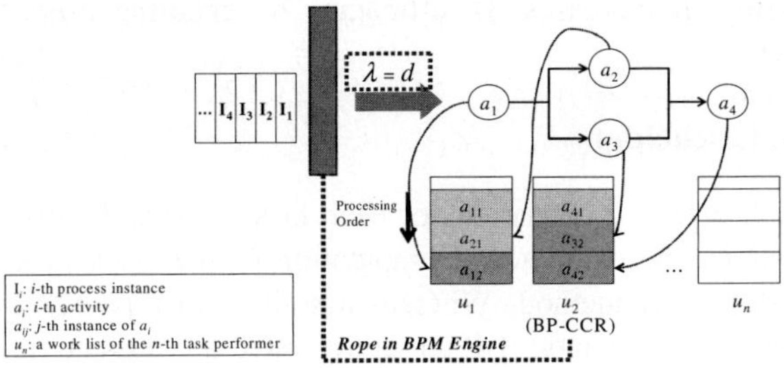

Fig. 2. The Function of the Rope

In the TOC, the Buffer protects the CCR from work starvation, guaranteeing continuous operation of the entire process. In order to achieve maximum utilization in a TOC environment, a system requires a sufficient amount of Buffer, whereas it also tries to minimize inventories and operating cost. In this paper, we suggest a simulation method to determine the size of the Buffer. An overall BP-DBR procedure is presented in Fig. 3.

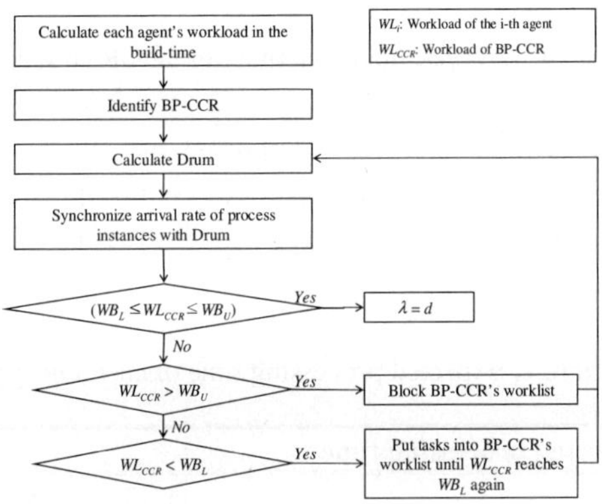

Fig. 3. BP-DBR Procedure

The application of the BP-DBR involves the assignment of tasks to agents through the Drum, Buffer and Rope, with special attention devoted to the BP-CCR. The maximum and minimum workload bounds (*WB*) are introduced in order to control business processes: if a Buffer size is greater than the maximum workload bound (WB_U), tasks are no longer assigned to the BP-CCR or are assigned to an alternative agent, if any, until the Buffer size falls below the WB_U. The minimum workload bound (WB_L) is also defined to maintain the minimum amount of Buffer.

5 Task Performer Perspective (TPP)

This chapter discusses BP efficiency from the TPP. In the present study, Rhee's method [17], was employed to prioritize tasks assigned to a user, based on the

PERT/CPM method. It increases BP efficiency by enabling users to deal with an urgent task earlier.

5.1 Slack Time Calculation

In order to furnish urgent information for each task, a critical path should be found and slack time should be computed. Computation of the slack time of each task is based on the PERT/CPM method. Whereas a split in PERT/CPM networks always entails use of the AND semantic, BP structures have an alternative path (that is, an OR block is included). In a BP model, the AND parallel structure can be handled in exactly the same way as the PERT/CPM method. However, for the OR structure, the multi-path is consolidated as a single path by a representative activity shown in Fig. 4.

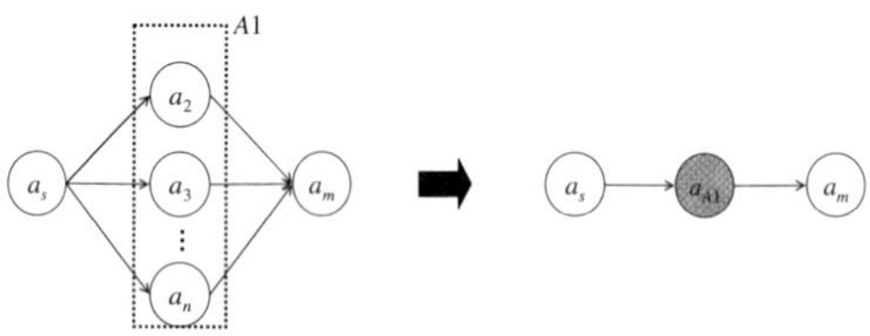

Fig. 4. A representative activity of an OR structure

Prior to finding the critical path and computing the slack time of tasks in an OR structure, the expectation time of each path is required. The expectation time is calculated for three representative types of OR structure; those calculations are summarized in Table 1.

Table 1. Expected processing time of each OR type

Type	Expected processing time	P_i (λ_i)
NOR	$\sum_{i=1}^{n}\prod_{j=1,j\neq i}^{n}(\frac{\lambda_i}{\lambda_i+\lambda_j})ET_i$	λ_i: The service rate with which the distribution of the actual processing time of the i-th path follows an exponential distribution
POR	$\begin{cases} P_1 \cdot ET_1 & , n=1 \\ P_1 \cdot ET_1 + \sum_{i=2}^{n}[\{\prod_{j=1}^{i-1}(1-P_j)\}P_i \cdot \{\sum_{k=1}^{i} ET_k\}] & , n \geq 2 \end{cases}$	**P_i:** The probability that, when the i-th path, r_i, is determined to be executed, the path is successfully completed
COR	$\sum_{i=1}^{n}(P_i \cdot ET_i)$	**P_i:** The probability that r_i is chosen for execution

With the representative activity of an OR structure, a slack time can be calculated for the structure, and each user prioritizes tasks according to this information. He first carries out an urgent task with the least slack time. We call this rule the Least Slack Time (LST) rule. More details on this rule can be found in [17].

6 Combination of the Two Perspectives

The two-phase perspectives, comprising the PEP and the TPP, have different target areas in improving process efficiency. Whereas the BP-DBR focuses on the engine's function of allocating and distributing tasks, the LST concentrates on fulfilling the assigned tasks. Although each method has been proved to be effective in each phase, the effect of combining the two phases has not yet been considered. The two-phase perspective covers an entire procedure from process launching to execution. We call the combined method *Comprehensive Rule for Process Efficiency* (CR4PE).

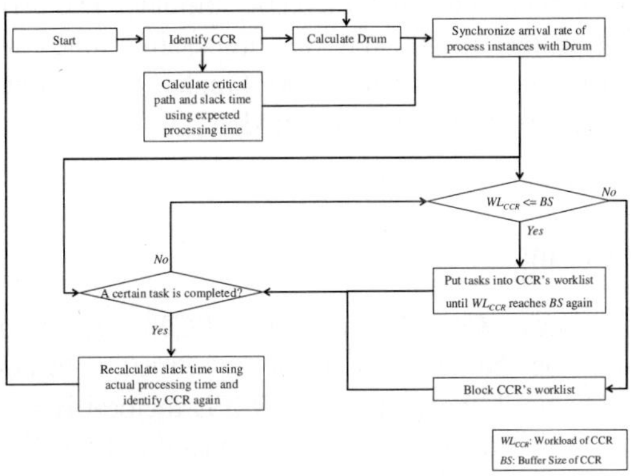

Fig. 5. Comprehensive Rule for Process Efficiency

The overall procedures of the method are illustrated in Fig. 5. Before a process is started, the CCR, the critical path and the slack time of each activity are identified. The Drum is also calculated. Once a process instance is launched, the Rope starts controlling the release of instances. Buffer management is also executed considering the Buffer size of the CCR. The Buffer is controlled with regard to the CCR's workload. In the course of executing a process instance, the completion of a certain task can result in changes to the CCR, the Drum, and the slack time. Whenever a certain task is finished, those data are recalculated in real time.

7 Simulation Experiments

7.1 Simulation Model and Experimental Environment

The BP-DBR is known to be effective when a heavy workload is concentrated on a few activities [16]. By contrast, the LST rule is effective when there is a heavy workload for many activities in parallel relations [17]. For simulation experiments, we used the process model shown in Fig. 6, which model includes features of both perspectives. As illustrated in the Figure, each box represents an activity, and the value inside a box represents an expected time. Participants are marked above each activity. A colored box is an activity with the heaviest workload and the activity's user, u_{25}, is a CCR.

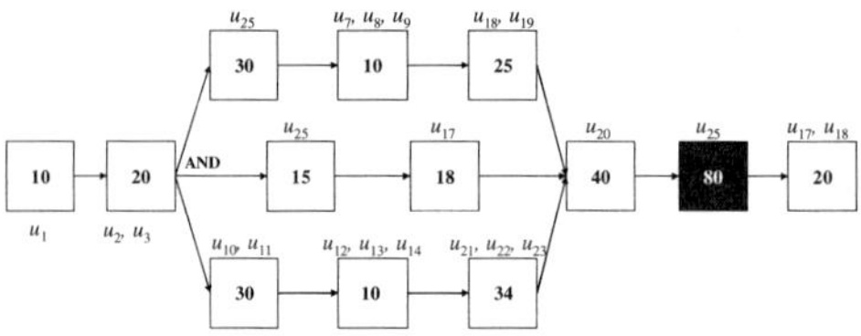

Fig. 6. Simulation Model

The simulation model was implemented with Arena 6.0. Each experiment was warmed up for 500 minutes and run for 10,000 minutes. The arrival rate of process instances with a parameter of 40, and the processing time were assumed to be exponential. The Buffer size was set to 1,800. Finally, thirty simulation experiments were carried out for each the following four models.

· I: Model with **FIFO Rule** · II: with **TOC** · III: with **LST Rule** · IV: with **CR4PE**

7.2 Experimental Results

We measured the Process Completion Time (PCT), the Number of Complete Instances (NCI) and the Number of Ongoing Process Instances (NOCI). The simulation results in Table 2 show that the CR4PE method is the most effective for the PCT and the NCI, which means that the combination of the TOC and the LST was more effective than the respective applications. However, the synergy effect was relatively less effective for the NCI, mainly because the TOC controls a number of instances by using the Drum and the Rope. As can be seen in Table 2, the NOCI of Models II and IV are much smaller than those of Models I and III, which means that in Model IV, a smaller number of instances are assigned to each task performer, and the effectiveness of the LST is relatively limited. We can still state, however, that Model IV is better than Model III, since in Model IV, each transaction can be completed in a shorter time.

Table 2. Summary of BP efficiency results

Model	PCT	NCI	NOCI
Model I	4223.30	58	108.27
Model II	2995.95	77	28.04
Model III	3780.85	80	96.66
Model IV	2747.46	81	26.66

7.3 Influence of Buffer Size

As discussed in Section 7.2, the effectiveness of the LST can be limited by the number of job instances on participants' worklists. To further study the effect of the Buffer, additional simulations were conducted with respect to different Buffer sizes.

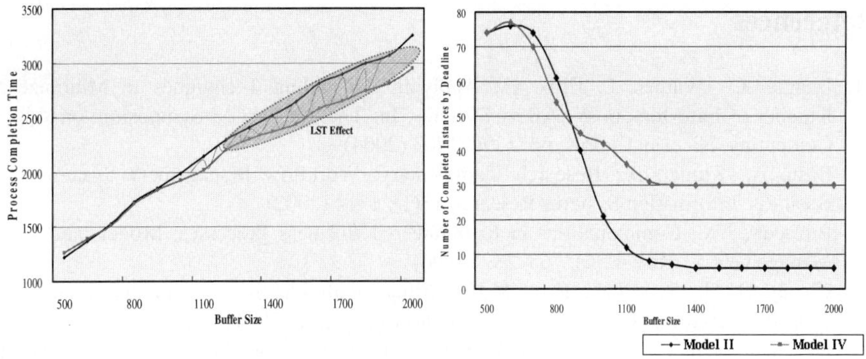

Fig. 7. Influence of Buffer Size

Figure 7 shows that, for PCT and NCI by Deadline, Model IV performs better than Model II as the Buffer size increases. The difference between the two models originates from the effectiveness of the LST Rule. Since a CCR has a greater number of job instances assigned with a larger Buffer size, the LST plays a more important role. The LST would be ineffective if a task performer had few jobs on his worklist. Therefore, the synergy effect becomes greater as the size of the Buffer increases.

8 Conclusions

Having perceived the absence of user-oriented support for BP efficiency, we devised a more comprehensive approach to enhancing it. Two different perspectives on BP efficiency, the PEP and the TPP, were identified. Based on our previous research in a single perspective, we combined both perspectives into an integrated model. According to the simulation results, the new methodology shows better performance than utilization of an individual perspective.

Our research has raised issues for future study. First, more process model patterns can be considered in our CR4PE. Second, whereas this research adopted one rule for each perspective, it would be interesting to analyze various rules for each perspective. Finally, if the economic values of process efficiency improvement are measured [5], we can acquire insight into the extent of the benefits our integrated methodology can guarantee compared with single-handed administration.

Acknowledgments. This work was supported by the Seoul R&BD Program. It was also supported by a Korea Research Foundation Grant funded by the Korean Government (MOEHRD) and the Regional Research Universities Program/Research Center for Logistics Information Technology.

References

1. Baggio, G., Wainer, J., Ellis, C.: Applying Scheduling Techniques to Minimize the Number of Late Jobs in Workflow Systems. In: The 2004 ACM Symposium on Applied Computing, Nicosia, Cyprus, pp. 1396–1403 (2004)
2. Basu, A., Kumar, A.: Research Commentary: Workflow Management Systems in e-Business. Information Systems Research 13(1), 1–14 (2002)
3. Buzacott, J.A.: Commonalities in Reengineered Business Processes: Models and Issues. Management Science 42(5), 768–782 (1996)
4. Chang, D.–H., Son, J.H., Kim, M.H.: Critical Path Identification in the Context of a Workflow. Information and Software Engineering 44(7), 405–417 (2002)
5. Davamanirajan, P., Kauffman, R.J., Kriebel, C.H., Mukhopadhyay, T.: System Design, Process Performance, and Economic Outcomes: An Empirical Study of Letter of Credit Systems Integration in International Banking. Journal of MIS 23(2), 65–90 (2006)
6. Davenport, T.: The New Industrial Engineering: IT and Business Process Redesign. Sloan Management Review, 11–27 (1990)
7. Eder, J., Pichler, H., Gruber, W., Ninaus, M.: Personal Schedules for Workflow Systems. In: van der Aalst, W.M.P., ter Hofstede, A.H.M., Weske, M. (eds.) BPM 2003. LNCS, vol. 2678, pp. 216–231. Springer, Heidelberg (2003)
8. Goldratt, E.M.: The Goal. North River Press, New York (1992)
9. Ha, B.–H., Bae, J., Park, Y.–T., Kang, S.–H.: Development of process execution rules for workload balancing on agents. Data. & Knowledge Engineering 56(1), 64–84 (2006)
10. Hammer, M.: The Agenda: What Every Business Must Do to Dominate the Decade, Crown Business, New York (2001)
11. Kaplan, S., Sawhney, M.: E-Hubs: The New B2B Marketplaces. Harvard Business Review 78(3), 97–103 (2000)
12. Kumar, A., Zhao, L.: Dynamic Routing and Operational Controls in a Workflow Management System. Management Science 45(2), 253–272 (1999)
13. Kumar, A., van der Aalst, W.M.P., Verbeek, H.M.W.: Dynamic Work Distribution in Workflow Management Systems: How to balance quality and performance? Journal of MIS 18(3), 157–193 (2001)
14. March, S., Hevner, A., Ram, S.: Research Commentary: An Agenda for Information Technology Research in Heterogeneous and Distributed Environments. Information Systems Research 11(4), 327–341 (2000)
15. McCoy, D.: The convergence of BPM and BAM. Gartner Research Note (2004) SPA-20-6074
16. Rhee, S.–H., Bae, H., Seo, Y.: Efficient Workflow Management through the introduction of TOC concepts. In: The 8th Annual International Conference on Industrial Engineering Theory, Applications and Practice, USA (2003)
17. Rhee, S.–H., Bae, H., Kim, Y.: A dispatching rule for efficient workflow. Concurrent Engineering – Research and Applications 12(4), 305–318 (2004)
18. Smith, H., Fingar, P.: Business Process Management – The Third Wave. Meghan-Kiffer Press, Florida (2003)
19. Zhao, J.L., Stohr, E.A.: Temporal workflow management in a claim handling system. SIGSOFT: Software Engineering Notes 24(2), 187–195 (1999)

Development of Productivity Evaluation Method to Improve Office Environment

Hiroshi Shimoda[1], Kyoko Ito[2], Yoko Hattori[1], Hirotake Ishii[1], Hidekazu Yoshikawa[1], Fumiaki Obayashi[3], and Masaaki Terano[3]

[1] Kyoto University, Gokasho, Uji, Kyoto, Japan
[2] Osaka University, Toyonaka, Osaka, Japan
[3] Matsushita Electric Work, Kadoma, Osaka, Japan
shimoda@energy.kyoto-u.ac.jp

Abstract. The authors have developed a performance test, CPTOP (Cognitive Performance Test of Productivity), which combines 11 basic task testss to evaluate intelligent abilities of office workers in order to quantitatively and objectively evaluate their productivity by controlling office environment. However, in the previous experiment, it was found that the results of CPTOP contain in-fluences of learning effect and variation of difficulties depending on each prob-lem set. In this study, therefore, subjective experiments have been conducted in order to compensate these influences and a correction method has been proposed.

1 Introduction

Energy saving is one of the countermeasures of increasing green house gas emission caused by increasing recent worldwide energy consumption. Although the energy saving in factories and industrial plants has been promoted, that in office buildings has not improved enough. In Japan, the government has promoted that the temperature of air-conditioning system in the summer should be 28 degree Celsius and office workers are recommended to wear casual style cloths. However, the drop in productivity of office workers caused by the energy saving may extend their working time and this may consume more energy[1].

On the other hand, recent studies have indicated that improvement of office environment may improve the work productivity. However, the method which evaluates office productivity objectively and quantitatively has not been established yet. If such method is developed, it is expected that it can be utilized for the design of office room and the evaluation of energy consumption based on office work productivity.

From this viewpoint, the purpose of this study is to develop a performance test which is a simulated task set of office work in order to evaluate work efficiency of office workers objectively and quantitatively.

M.J. Smith, G. Salvendy (Eds.): Human Interface, Part II, HCII 2007, LNCS 4558, pp. 965–972, 2007.

2 Productivity Evaluation Method

2.1 Conception of Office Productivity Evaluation

Although there are various studies and proposals for evaluation of office productivity index[2-5], there is no objective and quantitative evaluation method which reflects actual office work and useful evaluation method has been required.

In order to develop an appropriate evaluation method of productivity, it is necessary to analyze actual office work and pick up the factors of the productivity index, and then to configure suitable evaluation method and indices based on the analysis.

Since the office work has lots of atypical variations, the abilities which are necessary to execute the office work are first revealed, then the office productivity is measured by evaluating the performance how much the abilities can be applied for the work. In ad-dition, in order to conform it to the actual office work, the configuration of the abilities which are required for each occupation category is revealed and the total office pro-ductivity is evaluated based on the result of performance and configuration.

In order to evaluate the improvement of office environment by using this method, the productivity index before improvement is compared with that after the improvement.

2.2 Configuration of Productivity Indices

In order to reveal the abilities necessary for the office work, the authors defined essential abilities based on the human abilities by Fleishman[6]. Among the abilities, 21 cognitive abilities such as “Oral Comprehension” and “Memorization” were picked up as the essential abilities for the office work. In addition, interviews and questionnaires were conducted to the office workers who were general affairs, engineers, managers and so on. Based on the results, 11 elementary abilities are picked up as shown in Table 1 and weight of each abilities were deduced for each occupation category[7]. The example of the result is shown in Figure 1. In order to calculate a office productivity of specified occupation category, the results of the performances of the abilities are given added weight to each performance based on the deduced weight.

Table 1. 11 elementary abilities necessary to perform office work

Abilities		
1.Oral Comprehension	2.Written Comprehension	3.Oral expression
4.Written expression	5. Memorization	6. Mathematical Reasoning
7. Deductive Reasoning	8. Inductive Reasoning	9. Information Ordering Category
10. Perceptual Speed	11. Time Sharing	

Based on this conception, the concrete productivity index can be calculated by equation (1).

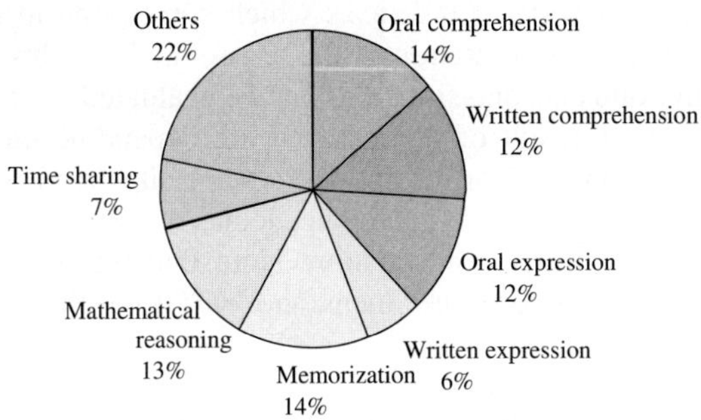

Fig. 1. An example of weights (clerical employee)

This index can be calculated by the results of performance tests and weights of elementary abilities for a specific occupation and a specific office environment.

$$Office\ Productivity = \sum_{i=1,j=1}^{I,J} W_{i,j} \cdot P_{i,j} \tag{1}$$

$W_{i,j}$: *Weights of elementary abilities*

$P_{i,j}$: *Scores of performance tests for elementary abilities*

i : *Occupation category*

j : *Elementary abilities*

2.3 Performance Test System, CPTOP

A performance test system, CPTOP (Cognitive Performance Tests for Office Productivity) has been developed to evaluate a productivity of office workers objectively and quantitatively based on the above productivity evaluation conception. CPTOP is a Web-based performance test system and it can be used by only PCs and Web browsers connected to the Internet as shown in Figure 2.

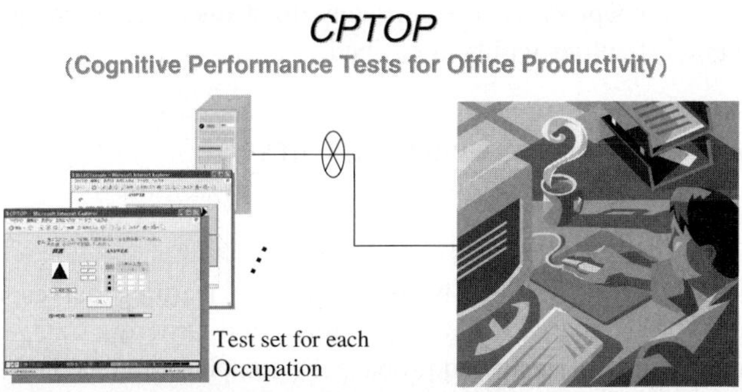

Fig. 2. Configuration of CPTOP

CPTOP consists of 11 sorts of task tests which correspond to the 11 elementary abilities. By changing the configuration of the test set and weights, the productivities in various office environments and situations can be evaluated.

Since the performance result of the task test are depending on each test such as throughput time and number of correct answers, the result score is standardized by dividing by the score average, then given the deduced weights to the standardized scores depending on the testee's occupation. Note that this test is to evaluate the productivity in a specific office environment, and NOT to evaluate the abilities of the individual workers

2.4 Problem of CPTOP and Policy of Its Improvement

Since the purpose of CPTOP is to evaluate office environment by testing how much cognitive abilities can be performed in different office environments, the results of the same person in the same environment should be always the same. However, there might be a learning effect when repeating the same test, and difference of difficulties of the problems except automatically generated problems. If there are such leaning effect and difference of problem difficulties, it is difficult to evaluate cognitive abilities exactly.

The tests in which the learning effect appears are "Memorization", "Inductive Reasoning", "Information Ordering Category", "Perceptual Speed" and "Time Sharing", which problems are automatically generated and difficulties of the problems are uniform. In this study, the authors assumed that the learning effect of each test is almost equal for anyone, and conducted a subject experiment to deduce the general learning curve for each test. And then, the learning effect will be removed by compensating each test result from the learning curve.

On the other hand, since the problems of "Oral Comprehension", "Written Comprehension", "Oral Expression", "Written Expression", "Mathematical Reasoning" and "Deductive Reasoning" cannot be generated automatically, they had been created in advance. Although a rule was laid down when creating the problem not to vary their difficulties, it is difficult to unify them in reality. In this study, therefore, the authors assumed that the difficulty of each problem is equal for anyone and conducted a subjective experiment to evaluate it. And then, the effect of difficulty variation will be removed by compensating each test result from the evaluated difficulties.

In the following chapters, two experiments to deduce the learning curve and to evaluate problem difficulties will be described.

3 Experiment to Deduce Learning Curve

3.1 Objective of Experiment

The objective of this experiment is to deduce the learning curves of five tests, "Memorization", "Inductive Reasoning", "Information Ordering Category", "Perceptual Speed" and "Time Sharing", and propose the compensation method.

3.2 Experimental Method

The experiment was conducted in an experimental room of Advanced Research Laboratory of Kyoto University. The room temperature was adjusted to 25 degree Celsius and the intensity of illumination on the desk was 1000 lux.

The subjects of the experiment were 15 persons (6 males and 9 females, age: 22-45, average age:30.9) who had experience to work with PC. They were explained the content and the purpose of the experiment and agreed to join it in advance.

In the experiment, the above five kinds of task tests of CPTOP were given to the subjects and the test results were measured. The experiment was conducted for three days for each subject. Since four sets of the task tests were given in each day, they performed 12 sets totally.

3.3 Result of Experiment

Figure 3 shows the average scores and their standard deviations of 12 sets in "Memorization". The horizontal axis indicates number of the set from 1 to 12, while the vertical axis indicates average scores and their standard deviations. From this graph, it was found that the scores are improved as the test set proceeded and the subjects learned the tests. Especially, the score improvement of "Memorization", "Deductive Reasoning" and "Time Sharing" tests were significant, and this indicated the necessity of com-pensation of learning effect.

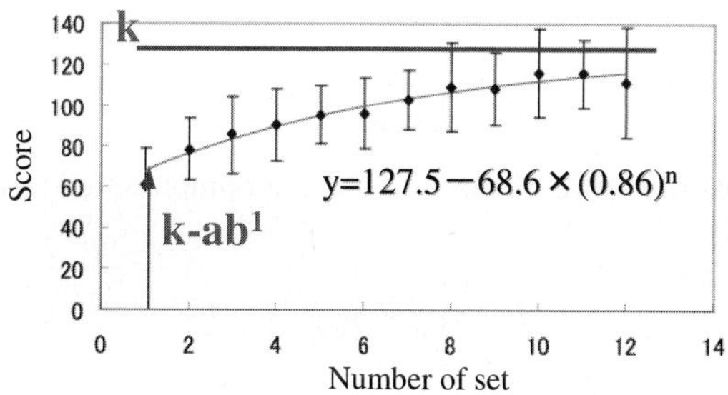

Fig. 3. Average scores and standard deviation of "Memorization"

Generally, it is assumed that the performance by the learning effect increases along with the experience and the performance will be saturated after infinite experience. Therefore, the following saturation model can be deduced. When supposing that the performance after n trials is Pn, the initial performance is P1, the saturated performance is Plim and improvement ratio after a trial is r, the performance after n trials can be described as equation (2).

$$P_n = P_{\lim} - (P_{\lim} - P_1) \cdot (1 - r)^{n-1} \tag{2}$$

This equation can be simplified as learning curve as equation (3).

$$y = k - ab^x \tag{3}$$

Table 2 shows the coefficients which are calculated by minimum square method for the results of all the subjects.

Table 2. Coefficients of learning curve

Task test	a	b	k
Memorization	68.6	0.86	127.5
Deductive Reasoning	56.8	0.81	116.5
Information Ordering	35.5	0.91	118.6
Category	32.1	0.92	118.0
Cognitive Speed	66.1	0.86	128.6
Time Sharing			

3.4 Compensation Method of Learning Effect

The conpensation method which removes the learning effect is proposed. In the equation (2), the saturated performance k is assumed as a standard score and the score is modified by the compensation ratio which is the saturated performance score divided by the performance of trial n. In other word, the conpensation ratio of trial n can be described as follows;

$$c_n = \frac{k}{k - ab^n} \tag{4}$$

When supposing that the score of trial n is Sn, the compensated task score, Tn will be described;

$$T_n = c_n \times S_n . \tag{5}$$

4 Difficulty Evaluation Experiment

4.1 Purpose of Experiment

The purpose of this experiment is to evaluate the difficulties of problems for the task performance tests in which the problems can not be generated automatically, and propose the compensation method.

4.2 Experimental Method

The experiment was conducted at an experimental room of the advanced research laboratory in Kyoto University. The temperature of the room was controlled to 25 degree Celsius and the illumination on the desk was 1,000 lux.

In the experiment, five task tests excluding "Oral Expression" were given to the subjects. The experimental term was 6 days and about 20 problems of the five task tests were given to the subjects for one day, and totally 110 to 120 problems were given. Before the experiment, 5 problems of each test had been given to be accustomed to solve the problems. Between the task tests, they took enough rest to remove the fatigue effect.

Four subjects (2 males and 2 females, age:26-35, average age:29.3) joins the experiment who had experience to work with PC.

4.3 Result of Experiment

Figure 4 shows the average scores of each problem of "Written Comprehension". The horizontal axis indicates problem number, while the vertical axis indicates the average scores. In this graph, the high average score means easy problem, while the low average score means difficult problem.

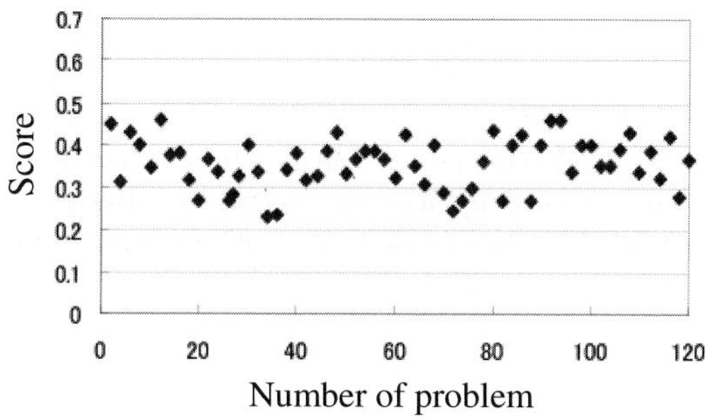

Fig. 4. Average score of each problem of "Written Comprehension"

4.4 Compensation Method for Variation of Difficulties

The difficulties of the problems can be described as the average score of all problems divided by the average score of each problem and the score of task test can be modified by the difficulty. In other word, the difficulty Di of problem i can be described using Tave which is the average score of each problem and Tall which is the average of all the problems as;

$$D_i = \frac{T_{all}}{T_{ave}} \tag{6}$$

When supposing that the score of the task test is Si, the compensated score Ti can be described as;

$$T_i = S_i \times D_i \tag{7}$$

5 Conclusion

In this study, the authors have proposed an office productivity evaluation method CPTOP and the compensation methods of its learning effect and variation of difficulty of each problem by conducting two subject experiments for deducing the learning curve and evaluating the difficulties of each problem respectively. As the results of the first experiment, the learning curves were deduced and the compensation method of the test scores of "Memorization", "Deductive Reasoning", "Information Ordcring Catc-gory", "Cognitive Speed" and "Time Sharing" was proposed by the learning curves. As the results of the second experiments, the difficulties of test problems were evaluated and the compensation method of test scores of "Oral Comprehension", "Written Comprehension", "Written Expression", "Mathematical Reasoning" and "Inductive Reasoning" was proposed by the evaluated difficulties.

In the future, the effectiveness of the proposed compensation methods will be confirmed by comparing CPTOP test score and the corresponding actual works in offices.

References

1. Lomonaco, G., Miller, D.: Environmental Satisfaction, Personal Control and the Positive Correlation to Increased Productivity, Johnson Controls Inc. (1997)
2. Sensharma, N.P., Woods, J.E.: An Extension of a Rational Model for Evaluation of Human Responses, Occupant Performance, and Productivity, Healthy Building 2000, Workshop 9 (2000)
3. Fisk, W.J., et al.: Worker Performance and Ventilation, Part 1,2. In: Proceedings of Indoor Air 2002, pp. 784–795 (2002)
4. Kroner, W.M., Stark-Martin, J.A.: Environmentally Resposive Workstation and Office-Worker Productivity. ASHRAE Transactions 100, 750–755 (1994)
5. Wargocki, P., et al.: Productivity is Affected by the Air Quality in Offices, Healty Building 2000, pp. 635–640 (2000)
6. Fleishman, E.A., Reilly, M.E.: Handbook of Human Abilities, pp.1–37. Consulting Psychologists Press (1992)
7. Obayashi, F., et al.: Objective Measurement Method for Productivity and Investigation on Environmental Factor. In: Proceedings of Human Computer Interaction International 2005, CD-ROM (2005)

Information on the Causal Relationship Between Store *kaizen* and Store Features That Attract Customers by Covariance Structural Analysis

Yumiko Taguchi[1], Yasushi Akiyama[2], and Tsutomu Tabe[3]

[1] Department of Business Administration and Communication, Shohoku College, 428 Nurumizu, Atsugi City, Kanagawa 243-8501 Japan
taguchi@shohoku.ac.jp
[2] Hewlett-Packard Japan, Ltd., System Integrator Sales, 7, Gobancho, Chiyoda-ku, Tokyo, 102-0076 Japan
yasushi.akiyama@hp.com
[3] Department of Industrial and Systems, Engineering College of Science and Engineering, Aoyama Gakuin University 5-10-1 Fuchinobe, Sagamihara City, Kanagawa 229-8558 Japan
tabe@ise.aoyama.ac.jp

Abstract. This research has two objectives: first, to develop a method to quantitatively clarify the causal relationship between store improvements (store *kaizen*) and store features that attract customers; second, to verify the effectiveness of the method developed. The results of this research confirm two points. (1) The causal relationship was quantitatively clarified through the use of a covariance structural analysis, and the degree of the relationship between the store features that attract customers and store kaizen could be understood. (2) The proposed methodology has proven to be useful in an actual store improvement activity. The owner of the store studied in the investigation commented, "It is very important to set up a hypothesis when seeking an improvement. This research provides a helpful framework for doing so."

Keywords: store features that attract customers, store *kaizen*, causal relationship, covariance structural analysis.

1 Introduction

The sales of small and medium-sized supermarket chains in Japan are hovering at low levels due to the changes of consumer needs and fiercer competition[1]. If the owner of a small and medium-sized supermarket chain in Tokyo wishes to make store improvements (referred to as "store *kaizen*" hereafter) he can take the following steps: 1) grasp the current problems, based on periodic surveys of customer satisfaction in his own stores and the stores of competitors; 2) consider management adjustments and

[1] According to a survey of commercial statistics by the Ministry of Economy, Trade and Industry, the growth rate of annual sales in "food stores" is decreasing (surveyed in 1994, 1997, 2002).

M.J. Smith, G. Salvendy (Eds.): Human Interface, Part II, HCII 2007, LNCS 4558, pp. 973–982, 2007.

improvements for his stores, and implement those that are feasible [1]. With this approach, however, it is difficult for the owner to determine the effects of *kaizen* on sales (improvements of the extracted problems). Thus, it is difficult for the owner to accurately grasp which problems should be improved.

Customers generally want to shop at stores with stable sales and abundant merchandise and services that the customers seek. In this research we think of these as stores with features that attract customers. To increase sales, it is generally thought that the owners must improve their stores (*kaizen*) by adding more store features that attract customers. In former research we developed several information systems targeting store *kaizen* [1][2]. These systems have drawbacks, however, as they fail to provide information concrete enough to allow the owners to: 1) grasp how the extracted problems influence the store features that attract customers, and 2) carry out the *kaizen* corresponding to the problems. If more concrete information on the causal relationship between store features that attract customers and the problems to be improved is provided to the owner, the owner can rapidly make improvements and maintain and enhance the features that attract customers in his store. Thus, we seek to develop a method capable of providing this type of information. This research aims to accomplish two tasks as means to facilitate the efforts of store owners to carry out store *kaizen* by maintaining and enhancing the store features that attract customers: first, to develop a method to clarify the causal relationship between store features that attract customers and store *kaizen*; second, to verify the effectiveness of the method developed.

2 Approach to Solve Problems

This chapter describes a basic strategy for grasping the causal relationship between the store features that attract customers, the problems to be improved, and the method to realize the above strategy.

2.1 Basic Strategy

This research focuses on the two following points.

- Quantifying and visualizing the causal relationship between store *kaizen* and the store features that attract customers

To carry out store *kaizen* as a means of effectively maintaining and enhancing the store features that attract customers, it is helpful to quantify and visualize the relationship between the store features that attract customers and store *kaizen*.

- Taking up the causal relationship which has a construct linked to "store features that attract customers."

This research proposes a method for clarifying the causal relationship, with mechanisms built in to consider concepts that cannot be directly measured by questionnaires.

2.2 Considering the Method

Regression analysis can be used as a representative method to quantify causal relationships. The method can capture a causal relationship between measured values, but not a causal relationship that includes abstract concepts that defy measurement. Factor analysis can explore common factors (referred to as "factors" hereafter) behind several variables, but cannot capture a causal relationship of factors and variables. Fortunately we can use a third alternative, covariance structural analysis [3], to compensate for the faults of regression and factor analyses in the handling of the same task. This research applies covariance structural analysis to clarify a causal relationship that includes abstract concepts.

3 Applying Covariance Structural Analysis

This chapter applies covariance structural analysis in order to clarify the causal relationship between store kaizen and store features that attract customers. We describe a tentative hypothesis, observed and latent variables, indexes to evaluate a model representing the hypothesis, and a method for developing such a model to apply covariance structural analysis.

3.1 Tentative Hypothesis

Prior research [1][2] has clarified that consumers base their impressions of shops on the merchandise, the counters, the services offered, etc. If store *kaizen* is performed based on the degree of customer satisfaction with each item, the store features that attract customers can be maintained and raised. Interviews of shop owners and large groups of housewives elicited comments such as the following: "The quality of the service, whether it's good or bad, forms my impression of the store"; "The cleanliness of the counter changes my impression of the quality of the merchandise." This informs us that the three factors of "service," "counter," and "merchandise," influence other factors. Conceivably, they may influence other factors that determine a store's success in enhancing store features that attract customers.

On this basis, we formed the following tentative hypothesis: "The services, the counter and the merchandise in the store are closely related to each other, hence they influence the store features that attract customers." After forming this tentative hypotheses, we verified it by the following methodology: 1) asking the owner to make improvements in their (store *kaizen*), 2) developing two models of covariance structural analysis, i.e., a model before store *kaizen* and a model after store *kaizen*, 3) comparing the two models.

3.2 Variables

Three types of variables, i.e., latent variables, observed variables, and error variables, are used to develop the models in the covariance structural analysis.

3.2.1 Latent Variables

There are two types of latent variables. The first is "store features that attract customers." The second is latent variables for the constructs if the constructs behind the observed variables can be found. The use of the latent variables allows us to arrange the observed variables and easily interpret the model.

3.2.2 Observed Variables

There are also two types of observed variables, as shown in table 1. First are variables for measuring the customers' evaluations of each factor targeted as store kaizen. Second are the variables for measuring the store features that attract customers. The factors include items i.e., merchandise, service, counters, and the total impression of the store, based on tentative hypotheses. Variables X1 to X12 are the degrees of customer satisfaction with three items, i.e., merchandise, service, and counters. Variable X13 is the degree of total customer satisfaction with the store.

Moreover, three variables are used to measure the store features that attract customers: 1) a variable indicating how frequently the customer visits the store, 2) a variable indicating how many times the customer has visited previously, and 3) a variable indicating the degree of the customer's intention to visit again, based on the definition of the store features that attract customers in Chapter 1. A questionnaire with a five-grade evaluation format was used to collect data in relation to these variables.

3.2.3 Error Variables

If the error variables have a large influence in the covariance structural analysis, the observed variable will not be useful for explaining the relation of the model. Therefore, we set up error variables for all of the observed variables.

3.3 Developing the Model

We verify the above by considering the alternative models before and after store *kaizen,* respectively. The flow of developing models has six steps.

- Step 1: collecting data on observed variables by conducting a customer survey in front of the store.
- Step 2: considering whether there are factors behind the observed variables (that is to say, constructs) in the customer evaluations of each factor that can be targeted as store *kaizen*, as shown in table 1, by conducing an exploratory factor analysis (hereafter described as "factor analysis") based on the response data in gained in step 1.
- Step 3: depicting one of the alternative models by a path diagram.
- Step 4: setting response data in the path diagram.
- Step 5: performing a covariance structural analysis.
- Step 6: repeating step 3 to step 5, as above, on other (alternative) models.

To consider the models effectively based on the flow, we use statistical software such as SPSS for factor analysis, and AMOS for covariance structural analysis.

Table 1. List of observed variables

Types	Variable : meaning
Customer evaluations of each factor which could be targeted as store *kaizen*	X1 : degree of satisfaction with the quality of vegetables X2 : degree of satisfaction with the assortment of vegetables X3 : degree of satisfaction with the quality of the fish X4 : degree of satisfaction with the assortment of the fish X5 : degree of satisfaction with the quality of the meat X6 : degree of satisfaction with the assortment of the meat X7 : degree of satisfaction with the quality of the delicatessen X8 : degree of satisfaction with the assortment of the delicatessen X9 : degree of satisfaction with the quality of other food X10 : degree of satisfaction with the assortment of other food X11 : degree of satisfaction with the staff's attitude toward customers X12 : degree of satisfaction with the cleanliness X13 : degree of customer satisfaction with the store overall
Store features that attract customers	X14 : how frequently customers visit the store X15 : how many times customers have visited the store X16 : the degree of customer intention to visit the store again

3.4 Evaluating the Models

One way to evaluate and select a final model from among a set of alternative models is to remove the models with a GFI (Goodness of Fit Index) of under 0.9 and adopt the model with the smallest AIC (Akaike's Information Criterion) from among those remaining. This research develops several alternative models by covariance structural analysis, then consider the models using GFI and AIC. We also consider the relationships between variables based on test statistics, then modify the models based on our consideration.

4 Developing a Model for a Real (Bricks-and-Mortar) Store

This chapter describes the development of two models, one before store *kaizen* and one after store *kaizen*, based on the targeted real store as mentioned in chapter 3, and examines each.

4.1 Model Before *Kaizen*

When developing several models of the store before *kaizen*, the three following models had GFIs of over 0.9.

– I. A model referenced based on the results of factor analysis, including latent variables.
– II. A model referenced based on advice from the store owner.
– III. A model encompassing paths directly drawing "store features that attract customers" and the "customer evaluations of each factor which could be targeted for store *kaizen*."

When executing covariance structural analysis based on the three models, the GFIs of all models are over 0.9. Thus, all models are considered to have good explanatory ability. The AICs of all models are shown in table 2. Model I has the smallest AIC, as shown in fig. 1. Therefore, we adopted model I as our model before *kaizen*.

Table 2. The values of AIC on alternative models before store *kaizen*

Model	I	II	III
AIC	93.13	96.58	106.50

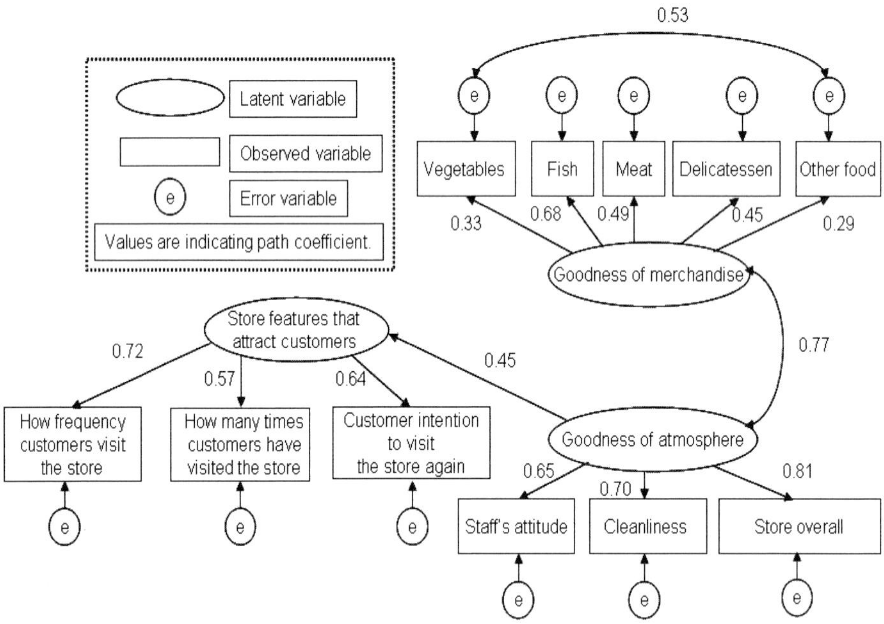

Fig. 1. Result of analysis in the model I

4.2 Model After *Kaizen*

Once we completed the store *kaizen* focused on merchandise, service, and counters, we developed alternative models of the store after *kaizen*. The six following models had GFIs of over 0.9.

- i. A model referenced based on the results of factor analysis, including latent variables.
- ii: A model which grouped the variables related to "quality" (i.e., X1) and the variables related to "assortment" (i.e., X2) together as one latent variable (i.e., goodness of vegetables) for each article (i.e., vegetables).
- iii: A model which grouped the variables related to "quality" (i.e., X1) and the variables related to "assortment" (i.e., X2) together as one latent variable (i.e., goodness of vegetables) for each article (i.e., vegetables) and then grouped them as one latent variable (i.e., goodness of merchandise).

- iv: A model referenced based on advice by the store owner based on model i above.
- v: referenced based on advice by the store owner based on model iii above.
- vi: A model encompassing paths directly drawing "store features that attract customers" and the "customer evaluations of each factor which could be targeted for store *kaizen*."

When executing covariance structural analysis based on the six models, the values of GFI in all models are over 0.9. Therefore, all models are considered to have good explanatory ability. The AICs of all models are shown in table 3. Model iii has the smallest AIC, as shown fig. 2. Therefore, we adopted model iii as the model after *kaizen*.

Table 3. AICs in alternative models after *kaizen*

Model	i	ii	iii	iv	v	vi
AIC	176.80	172.79	166.13	178.38	166.93	211.19

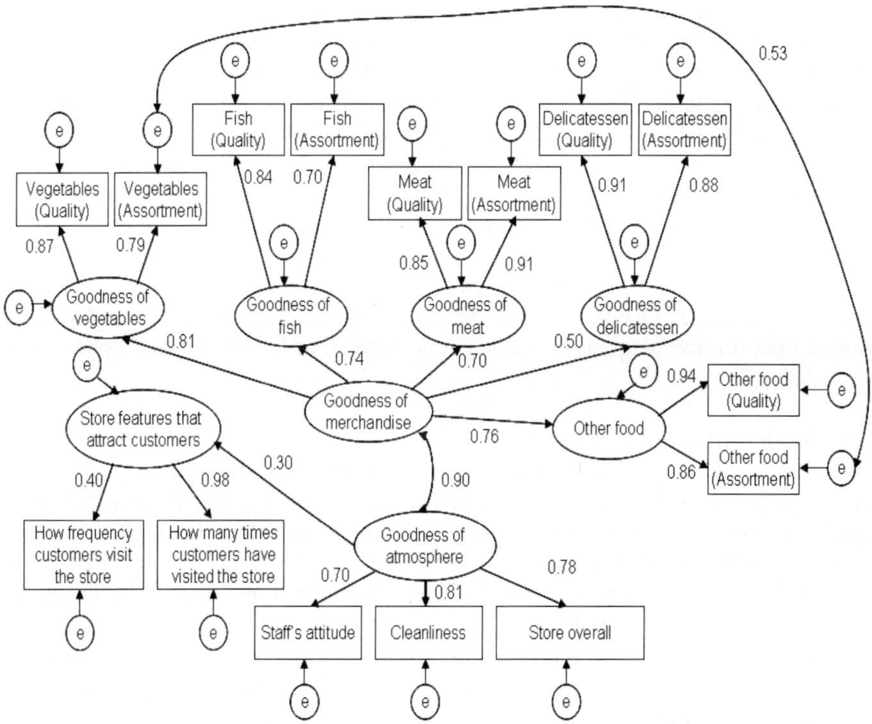

Fig. 2. Analytical results for model iii

4.3 Examining the Models

This section considers the relationship between the store features that attract customers and the store *kaizen* based on the results of analyses adopted in 4.1 and 4.2.

4.3.1 Model Before *Kaizen*

In the model before store *kaizen*, enhancement in "goodness of atmosphere" is ultimately linked to enhancement in "the store features that attract customers." Factors that can be targeted as store *kaizen*, such as merchandise, service, and counters, do not affect "the store features that attract customers." On the other hand, they can influence "the store features that attract customers" through "goodness of atmosphere." In addition, the features of this model can be understood based on three path coefficients (see fig. 1): 1) path "goodness of atmosphere" and "store features that attract customers," 2) path "goodness of merchandise" and "goodness of atmosphere," and 3) "goodness of merchandise" and "Fish."

In 1) mentioned above, an enhancement in "goodness of atmosphere" reflects enhancements in "the degree of satisfaction with the staff's attitude toward customers," "the degree of satisfaction with cleanliness," and "the degree of customer satisfaction with the store overall." By inference, therefore, these variables can be assumed to influence "the store features that attract customers." In 2) above, an enhancement in "goodness of merchandise" can be inferred to reflect an enhancement in the "goodness of atmosphere." Further, the "goodness of merchandise" influences_"the store features that attract customers" without directly influencing them, as consumers take it for granted that the merchandise at a supermarket should be good. Moreover, we can infer that a store's efforts take better care of its customers and improve in-store hygiene will enhance "the store features that attract customers" more than they enhance degree of satisfaction with the merchandise. In 3) above, "fish" is most robustly influenced by the "goodness of merchandise," as the targeted store was initially barraged by requests for fish and subsequently faced only a low demand for fish. Therefore, we can infer that an enhancement in the degree of satisfaction with fish is linked to "goodness of merchandise."

4.3.2 Model After Store Kaizen

In the model after store *kaizen*, enhancement in "goodness of atmosphere" is ultimately linked to enhancement in "the store features that attract customers." Factors that can be targeted as store *kaizen* do not directly influence "the store features that attract customers." On the other hand, they can influence "the store features that attract customers" through "goodness of atmosphere." In addition, the features of this model can be understood based on three path coefficients (see fig. 2): 1) path "goodness of atmosphere" and "the store features that attract customers," 2) path "goodness of merchandise" and "goodness of vegetables" and 3) "goodness of merchandise" and "goodness of delicatessen."

In 1) mentioned above, an enhancement in "goodness of atmosphere" reflects enhancements in "the degree of satisfaction with the staff's attitude toward customers," "the degree of satisfaction with the cleanliness of the store," and "the degree of customer satisfaction with the store overall." By inference, therefore, these variables can be assumed to influence "the store features that attract customers." Further, the "goodness of merchandise" influences "the store features that attract customers" without directly influencing them. In 2) above, the "goodness of vegetables" is most robustly influenced by the "goodness of merchandise." This could be explained by the soaring price of vegetables during the heavy snows at the end of 2005, when the customer survey was conducted. Vegetables were in short supply, hence we can infer that the degree of customer satisfaction dropped. In 3) above, the "goodness of delicatessen" is most weakly influenced by the "goodness of merchandise," due to relative abundance of New Year delicacies in the delicatessen at the end of the year, when the customer survey was executed. Further, we can infer that the store *kaizen* to "delicatessen" executed does not influence "the store features that attract customers."

5 Validation

In this chapter we examine whether the method to clarify the causal relationship between the store *kaizen* and the "store features that attract customers" is useful. To consider this, we ask the store owner two questions. First, we ask if the method developed clarifies the causal relationship between "store features that attract customers" and the factors which can be targeted as store *kaizen*. Second, we ask if the method is helpful for the store *kaizen* of a real store.

5.1 Method for Clarifying the Causal Relationship Between Store Features That Attract Customers and Factors Relative to Store *Kaizen*

When we asked the store owner to judge the result of the examinations based on the two analyses of the models (the models before and after *kaizen*) at the bricks-and-mortar store described in the last chapter, he responded as follows: "It was more interesting for me to generate new ideas for store *kaizen* with this method, compared with the methods used in your former research."

5.2 Helpfulness of the Method for the Store *Kaizen* of the Bricks-and-Mortar Store

When we asked the owner if the method was helpful for the store *kaizen* of the bricks-and-mortar store, he responded with two comments:

"It is very important to set up a hypothesis when seeking an improvement. This research provides a helpful framework for doing so."

"To enhance 'the store features that attract customers,' I can prioritize the *kaizen* by grasping the factors influence directly or indirectly 'the store features that attract customers'."

6 Conclusion

The results of this research confirm two points. First, the causal relationship was quantitatively clarified through the use of the covariance structural analysis and the degree of relationship between the store *kaizen* and the store features that attract customers. Second, the proposed methodology has been proved useful in an actual store *kaizen,* based on the comments of the owner of the store investigated. Therefore, the method proposed in this research is confirmed to be useful. The owner managing the store *kaizen* used the analytical results as a tool to form hypothesis for the store *kaizen*.

In the future it will be necessary to make good use of the proposed method by applying covariance structural analysis for store *kaizen* activities. To accomplish this we must focus on two new objectives. One is to produce a guideline on how stores owners should interpret path coefficients in their analytical results. The other is to help store owners anticipate the influences to be expected from specific store *kaizen* once executed. The store owner in this study suggested this.

References

1. Taguchi, Y., Tabe, T.: Developing of Decision Support Tool for Deciding Whether to Open a New Shop and How to Improve Existing Shops. In: Abridged Proceedings of the Ninth International Conference on Human-Computer Interaction (HCI' 2001), pp. 251–253 (2001)
2. Shibuya, T., Tabe, T., Taguchi, Y., Maki, N.: An Application of Fuzzy Integral to Measure the Degree of Consumer Satisfaction. In: The Proceedings of the 1st International Conference on Applied Ergonomics (ICAE '96), pp. 1032–1037 (1996)
3. Boollen, K.A.: Structural Equations with Latent Variables. Wiley, Chichester (1989)

Methodology for Constructing a Prototype Site for Finding Employment SPI Measures Using Mobile Phones

Shinsuke Wakita, Michiko Anse, and Tsutomu Tabe

Graduate School of Science and Engineering, Aoyama Gakuin University.
5-10-1, Fuchinobe, Sagamihara, Kanagawa, 229-0006, Japan
c5606138@cc.aoyama.ac.jp, anse@ise.aoyama.ac.jp,
tabe@ise.aoyama.ac.jp

Abstract. The written examination for job hunters in Japan is an important examination that cannot be skirted as the first step in the hunt for employment. However, there is no environment that can be studied using easy methods (mobile phones, etc.). In this research we clarify a methodology for constructing a prototype system for finding employment SPI measures. After constructing a prototype, we examine the effectiveness of the methodology. We use answer formats to achieve two purposes. (1) We clarified an answer format that can be efficiently studied through an experiment in 'no space' in order to produce a measure site that can be efficiently studied in little time. (2) We constructed a prototype for selecting answer formats in order to examine whether the prototype can be used to select the following two formats on a mobile phone: first, "the word, the meaning of the word, and an example of the word in use"; second, "the word, an explanation of the word, an example of the word in use, and the meaning of the word." (3) The test-takers studied the constructed prototype on mobile phones, then our group examined the effectiveness of the prototype based on the questionnaire results. The results indicate that the form can be efficiently studied. Thus, we can conclude that the prototype constructed can be used to select the following two answer formats: "the word, the meaning of the word, and an example of the word in use" and "the word, an explanation of the word, an example of the word in use, and the meaning of the word".

Keywords: e-learning, mobile phone, answer presentation form, problem display form.

1 Research Purposes

To construct an SPI (Synthetic Personality Inventory) examination support system for use on mobile phones; to clarify the methods for developing and progressing with the study and study content, to construct a prototype, to have test-takers use the prototype, and to examine the and effectiveness.

M.J. Smith, G. Salvendy (Eds.): Human Interface, Part II, HCII 2007, LNCS 4558, pp. 983–993, 2007.

2 Research Background

2.1 About the SPI Examination Support Using the Mobile Phone

After the written examination, many often reflect: "I would probably have solved it if I had reviewed a little more." The many company introductions during the job hunt leave little time to spare in a student's schedule]. At the end of the day, the job hunters return home exhausted. Even if they want to study, their fatigue prevents them. The SPI is a written examination no more difficult than the written portions of the high school entrance examination. Test takers can easily score well with little review. However, job seekers cannot proceed directly to the interview, by skipping this written examination (job recruitment process: applicant screening, the written examination, the interview, and the job offer). The written examination is often full of numbers, but it still can be easily passed by cramming in advance and listening to current events. Various books on employment SPI measures are now published, but most are too heavy for the students to lug around during their busy days. Our group therefore has speculated that the students could be more easily studied if students had a tool study tool suitable for use during the less busy hours of a student's life. In this research we propose methodology for constructing a prototype support system for mobile phones to help students prepare for the written examination. We focus on the answer format in this research, as the problem formats of SPI are almost uniformly multiple-choice (test takers choose from two or more answers). Specifically, we examine whether the amount of information placed in the answer format influences the efficiency of the student's study. This research is carried out in the following steps: (1) Give students a questionnaire to investigate the current status and needs of the written examination and the students' mobile phones. (2) Clarify the method for constructing the prototype support system for the written examination. (3) Construct the prototype based on the abovementioned result. (4) The effectiveness is examined.

2.2 About the SPI Examination

SPI stands for "Synthetic Personality Index," a measure of personality developed in 1974. When companies screen job applicants, many give aptitude tests. Typically this involves assessment of the SPI. The content of the written examination during the stages of applicant selection employment tests (applicant screening, written examination, interview) is shown in fig 1. SPI, the index referred to in (2), is the most widely used measure of aptitude.

One of the major stumbling blocks for students is the assessment of ability propriety during the aptitude test inspection. This assessment is divided into inspection I and inspection II. The former, Inspection I, is a set of 40 questions to be completed in 30 minutes for the examination of linguistic competence. Inspection II is a set 30 questions to be answered in 40 minutes for the examination of logical thinking and calculation abilities. The test takers must answer a relatively large number of problems within a relatively limited period. The problems themselves are not difficult: anyone with the scholastic attainments of a standard level junior high school education can solve them. A combination of the following factors does lead to mistakes, however: examine tension + short time allotted + many problems. Among the problems testing linguistic competence, test takers focus first on those they know. The most common questions are on idioms and proverbs made up of four Chinese characters. The test takers pay close attention to the answer format, usually a multiple-choice style format with four possible answers.

Written examination	(1)Common sense, Liberal learning ・Language study ability ・Expertise
	(2)Proper inspection (1)Potentiality examination (2)Character aptitude test

Fig. 1. Content of written examination

2.3 About Study by the Mobile Phone (Advantages and Disadvantages)

Two factors greatly influence study using mobile phones. The first is an environmental factor: the learner's place and situation. The other is a technical factor: the size of the screen and the transmission rate of the user's phone [D. Becking, 2005]. One advantage of mobile phones is freedom from operational problems: students are adept at using the devices, as almost all of the students use them in daily life[ANDO, 2006]. In our research we sent out a student questionnaire with questions on the use of the mobile phone for study. Students mentioned the following advantages: "It can be easily and readily used," "It's portable and convenient," "It doesn't become a burden to carry around (like luggage), and "I can use it on my free time." In a word, mobile phones can be used flexibly for study without constraints from the learner's location. The phone is "no place." Use is promoted in various places such as in coffee shops, restaurants, or during tedious waiting times, such as time spent at rendezvous spots or on bus journeys [T. Kopomaa, 2004]. One disadvantage of the mobile phone that detracts from its value in education is its reputation as a kind of toy for play [D. Morris, 1977]. Other disadvantages mentioned include the following: "The screen is small," "Books seem cheaper," "Only limited information is available," "The communication charge rises minute by minute," "It doesn't seem easy to use," "The charges make me anxious," "There is resistance to study with mobile phones," "Mobile phones make me tired." Regarding the communication charges, some questionnaire respondents indicated "The charges don't make me nervous, as the purpose is to study" and "My communication charges were covered by my charge plan" [ANDO, 2003]. These answers suggest that the mobile phone can be easily used, ideally via some kind of charge plan that exempts the user from paying.

3 Methodology of SPI Examination Support by Cellular Phone

We developed an effective format for presenting the answers, based on the results of our questionnaire and experiments.

3.1 Questionnaire Results

The questionnaire results dictated the following basic policy. (1) Learners use their mobile phones (i application program), a device at their disposal whenever they have free time. Mobile phones are often highly portable and are never a burden to carry (they can be carried more easily than a book). Our reason for choosing DoCoMo is presented in Table 1 below (there are many NTT DoCoMo users). (2) JAVA is used for the i application program. (3) It makes the tests for students searching for employment. This research focuses on "proverbs and four-character idioms" in problems from the SPI assessment for job seekers. For problems with synonyms and antonyms, etc., the only method for the test taker is to settle down and solve them. In this case, however, it makes a big difference if the test taker knows the correct answers or not. (4) The questions are arranged at random and the history function is activated to help the test takers can manage or correct their mistakes. Our experiment also clarified that test takers can efficiently memorize the answer format.

Table 1. Sep, 2006 Share comparison according to mobile

	pure increases	all users	share
DoCoMo	126,300	52,102,900	55.54%
KDDI	160,800	26,402,500	28.14%
SoftBank	23,400	15,307,000	16.32%
Sum	310,500	93,812,400	100.00%

3.2 About the Memory and the Forgetting Curve

What is problem format can be most efficiently memorized with a short time? The "forgetting curve" of Ebbinghaus illustrates the relation between time and a memory (Fig 2).

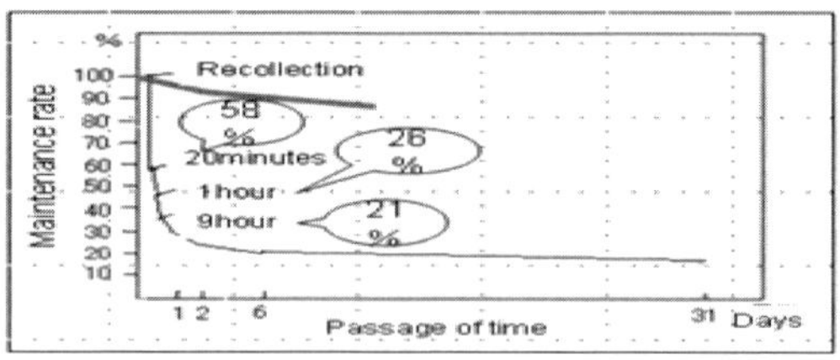

Fig. 2. Forgetting curve of Ebbinghaus (curve) and four character idiom (heavy line)

Ebbinghaus gave objective evidence of where the memory existed. At this time, a list of meaningless three-character strings is remembered from "Consonant, vowel, and consonant." As a test-taker, he measured how much information the test takers could recall after a fixed amount of time. The amount of maintenance (saving rate) is known to show a so-called negative acceleration curve: the memory will decrease rapidly just minutes after the information is studied, but thereafter the decline is not so sudden. (58% in 20 minutes, 26% in a day, 21% in 31 days) [John Henderson, 1999]. To this, 15 four-character idioms were divided into five titles and presented in three answer formats: “The word and the meaning of the word,” "the word, the meaning of the word, and an example of the word in use," and "the word, an explanation of the word, an example of the word in use, and the meaning of the word." In a similar experiment in which the meaning, the explanation, and the example are prepared for seven test-takers for complete memorization, the amount of maintenance fell by about 90% after a day and about 80% after one week average (Fig. 2). Nonetheless, the decrease was not as rapid as that observed by Ebbinghaus. This was because the four-character idiom has meaning and leaves a significant impression. Therefore, an unfamiliar word can be more effectively memorized if the material creates a significant (meaningful) impression using the meaning of the word, and explanation of the word, and an example of the word in use.

3.3 Experiment That Leads Effective Answer Form

The purpose of the experiment is to clarify whether to effect an alteration of the memory according to the answer form.

(1) Preliminary experiment 1 “Solving the problem in space.”
We conducted an experiment by having ten test-takers remember 15 four-character idioms. In this experiment, we presented "a four-character idiom and the meaning" in 1 to 5 titles, "a four-character idiom, its meaning, and an exercise" in 6 to 10 titles, and "a four-character idiom, its meaning, an explanation, and an exercise" in 11 to 15 titles. Test takers study them only once, and the number of words they remember immediately afterwards is tested. As a result, the correct answer rates were 28% for answer format 1, 53% for answer format 2, and 80% for answer format 3. Answer format 3, the format with the most detailed explanation provided, apparently left the strongest impression.

(2) Preliminary experiment 2 “Solving the problem on a PC.”
The mobile phone was displayed on a personal computer. We constructed a prototype for solving problems with 5 four-character idiom titles and 5 proverb titles (10 titles in total). Thirty test-takers were divided into three groups. All problem formats were multiple-choice with four options. The first group solved the problem using answer format 1, the second group solved the problem using answer format 2, and the third group solved the problem using answer format 3. The problems were to be solved in 2 or 3 days, one week, or two weeks in space. The acquisition levels will be shown later. (In the format shown here, the problem presents a set of ten words at the outset, then a list of 30 choices from which to choose the answers. The purpose is to determine the acquisition level for ten questions presented only once (for the first time.) The result is shown in Table 2 below: the answer rates were 56% for answer format 1, 72% for answer format 2, and 74% for answer format 3. No difference was seen between answer formats 2 and 3.

Table 2. Experiment 2

	1	2	3
At experiment	5.6	5.8	6.1
2days later	6	6.3	7.2
One week after	6.3	7.2	7.4

(3) Preliminary experiment 3 "Solving the problem with a mobile phone."
The content and problems studied were the same as those used in preliminary experiment 2. Three answer formats were downloaded to the mobile phone and studied by five test-takers, as shown in Table 3 below. The correct answer rates were 65% for format 1, 72% for answer format 2, and 73% for answer format 3. The answer rate did not change appreciably using either the PCs or mobile phones. However, some rest-takers indicated that the mobile phone is more convenient than the mobile phone in two ways: "It can be carried" and "I can use it anytime." From these three preliminary experiments, we can conclude that answer formats 2 and 3 are effective. Therefore, we adopt them.

Table 3. Experiment 3

	①	②	③
At experiment	5.8	5.7	6.2
2 days later	5.9	6.9	7
One week after	6.5	7.2	7.3

4 Construction of the Mobile Learning System

The purpose of constructing the prototype is to help learners study the SPI efficiently within little time. The content consists of (1) problem formats that can used effectively within little time; (2) the random ordering of the problems presented; (3) built-in answer formats that can be efficiently memorized (as clarified by the experiment); and (4) a history management function to manage problems handled erroneously by the test-taker.

4.1 The Development Environment

The i application program is the application service for the "i-mode" mobile standard introduced by NTT DoCoMo in January 2001. The execution environment runs on an installed version of Java (Java virtual machine), and the Java applications (scaled down version of the language) can be downloaded and executed. The system supports application of up to 30 Kbytes in size. For screen processing, the data-saving area for the local application "Scratchpad" stores up to 10 Kbytes of information.

4.2 Operation

(1) The system is designed for smooth cooperative function with the upper and lower buttons, as no scroll function is provided. (2) The prototype counts the number of characters in the displayed sentences and changes the line based on the appropriate number of characters, as no line-change function is provided. (3) On screens with two or more choices, the lines displayed on the screen move up and down with pushes of the upper and lower buttons, and the selected line is either highlighted or made to correspond to the figure key.

4.3 System Functions

Three system functions are built in: "solve the problem," "see history," and "see explanation." The user selects the desired function from the main menu screen. Once the function is selected, the user moves to the next screen by pushing the enter button on the cellular phone. The "solve the problem" function consists of three sub-functions: "select the style for answer presentation," "display the problem," and "check the answer and explain the correction." When the user chooses "Select the style for answer presentation," the system selects the currently highlighted items as the correct or incorrect answers are displayed. It does so by two methods: "display only the example" and "display the example and explanation" (Fig. 4). If the user chooses "Display the problem," the system displays the problem after the method for answer presentation is selected. The form for setting the questions is selected from two options: "Meaning" in the upper row and the "four answer candidates" displayed in the lower row. The answer is input from 1 to 4 with the figure key. The problem displayed is selected at random from the problems registered in the database. The "four-character idiom" offered as an answer candidate is also selected at random. When the user selects "check the answer and explain the correction," the system collates the input answer with the database and displays it according to "another correction" and "the method for presenting the selected answer." When the answer is corrected, the problem is removed from the problem list, and it displayed the answer. When the answer is incorrect, the problem is registered in the history management list as a "meaning" (Fig. 5). Through this method, the user can later study the problems that have incorrectly answered or left unanswered. Moreover, the system calculates the number of problems answered and the number of problems correctly answered. The "see history" function displays a list of incorrectly answered problems from earlier sessions ("Meaning") (Fig. 6). The "see explanation" function explains the function and the operation method to a person who uses the system for the first time. The flow chart below shows the entire flow of the system.

When the user pushes the enter button on the main menu screen, the system brings up the answer format selection screen. Initially it displays the message "the problem is solved" underlined in blue, and the blue underlining can be moved with by pressing the upper and lower buttons. If the user pushes enter button when "the history is seen" is underlined, the system brings up the history management screen. If the user pushes the enter button when "the explanation is displayed" is underlined, the system brings up the explanation screen.

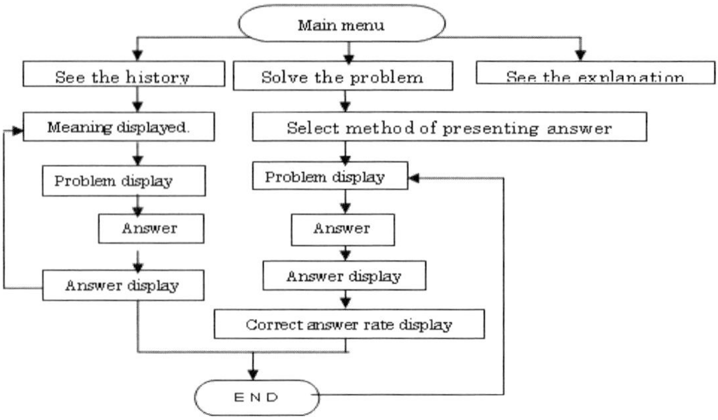

Fig. 3. Flow of system

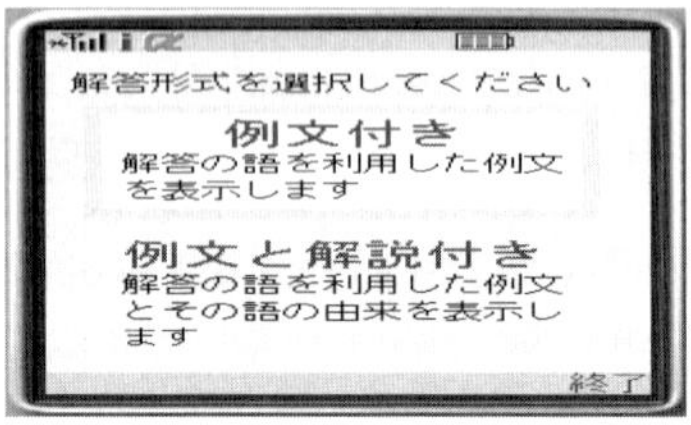

Fig. 4. Selection of method of presenting

The square on the answer format selection screen moves up and down with presses of the upper and lower buttons. The problem screen is displayed with a push of the enter button.

The flow up to the appearance of the problem screen generates random numbers first and then judges whether the problem is correctly answered. If the problem is correctly answered, random numbers are generated again. In the case of other problems (problem incorrectly answered or left unanswered), the array is called and the problem is displayed. Next, random numbers are generated in the answer group. If the same random numbers are generated, random numbers are generated again. If the same random numbers are not generated, choices are displayed. Moreover, as an answer format, choices of the answer correspond to the figure key. (The user goes to figure key 1 by selecting (1)

The flow chart for the answer screen (Figure 5) distinguishes whether or not the answer is correct. In the case of a correct answer, the system removed the problem from the problem list and displays the answer. In the case of an incorrect answer, the system registers the mistake in history management and displays the correct answer.

The line scrolls within the range when the user pushes the upper and lower buttons on this screen. 'Following' displayed in the lower right of the screen (corresponding to the 'Menu button' on the mobile phone). The system advances to the following

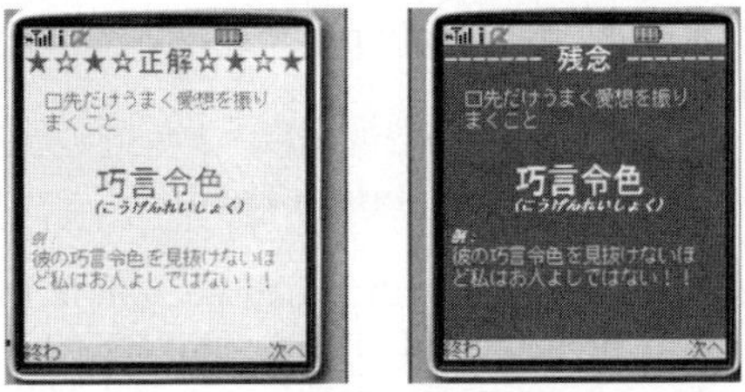

Fig. 5. Answer display

problem when the user pushes the 'Menu.' The 'Finish' message displayed at the bottom left of the screen corresponds to the 'i' button. The system advances to the correct answer rate display when the user pushes the 'i' button.

The meaning will scroll up and down the history management screen (Figure 6) when the user pushes the upper and lower buttons. If the user pushes the enter button, the system sets the problem with the meaning enclosed in the blue line. If the user pushes the 'i' button, the system returns to the main menu.

This is principally a screen for first-time users. The screen explains the method of operating the system and meaning of the system.

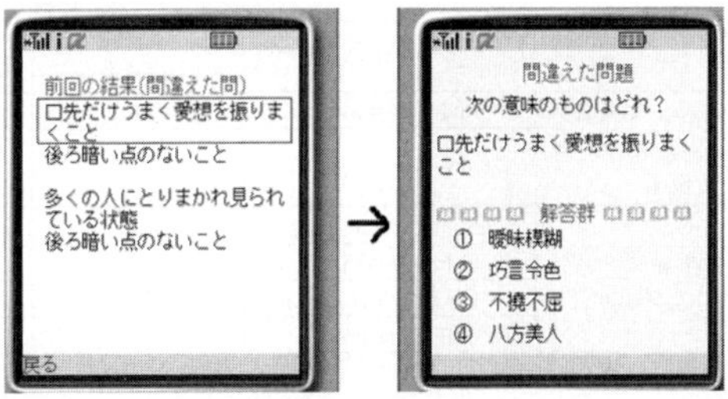

Fig. 6. History management screen

5 Experimental Results and Examination of the Effectiveness

Our purpose here was to examine whether the two answer formats obtained from the experiment were effective. The experiment was executed after a prototype registered on a website was downloaded to a cellular phone. In an experiment conducted over a 10-minute period, 10 test takers were asked to present two answers. After the study phase, the experiment proceeded in the following steps: "answer format," "problem format," "history management," and "operability of the keys." The results and considerations are described as follows.

5.1 About Operability

Over 90% of the test takers answered that the operability of present key arrangement was good, regardless of the method used to present the answer. With respect to the arrangement of the answer key, 20% of the test takers who studied using the

“Example” method for answer presentation chose the "2, 5, 8, 0" arrangement not used by the experiment.

5.2 About the Correct Answer Presentation Form

Over 30% people overall answered that they had been able to select the “necessity” about the method of answer presentation. The test takers who chose “only example” explained that “I want to answer many problems“ and “I don’t want to read the explanation sentences.” The test takers who chose “example and explanation” explained that “It was interesting to learn about the origin of the word” and “I expected it to help remember the etymology more easily.” There are clearly individual variations in the way people consider the methods used.

With respect to the impressions after study, test takers who chose “example” answered “four-character idiom shown in problem sentences with explanations of meaning,” while test takers who chose “example and explanation” answered “four-character idioms.” The reading of the explanation sentences apparently deepened the learners’ impressions of the words.

Table 4. Questionnaire of correct answer presentation form

Answer form		Unit(%) With explanation	Only the example
The answer form about being able to choose	Necessity	70	60
	Unnecessary	30	40
It remained in the impression.	Word of answer	80	50
	Meaning	30	60
	Explanation	40	×
	Example	10	10

5.3 Problem Display Form

With respect to the form of the problem-setting questions, all test takers answered that it was okay to newly set a problem that had been previously answered incorrectly. Moreover, the lowest number questions that answered in ten minutes were 20, and the highest number was 88. Individual variation was caused even when the same method was used for the answer presentation. 80% of test takers answered, “Good” about the number of problems being not set. This also suggests that it is effective not to set the number of problems.

5.4 About the History Management

History management is an effective option, as many people indicated that history management was "necessary.” Equal numbers of questionnaire respondents selected “answer the problem” and “display only the answer” in the history management. But the test takers who chose the “example” method of answer presentation tended to answer "display only the answer,” while the test takers who chose “example and explanation” tended to choose “answer the problem.”

6 Conclusion

Our examination of the effectiveness of the answer formats revealed that two effective types can be selected: "the word, the meaning of the word, and an example of the word in use" and "the word, the meaning of the word, an explanation of the word, and an example of the word in use." This clarifies the methodology for constructing the prototype site for supporting study of employment SPI measures using mobile phones.

Three tasks lie ahead. (1)The test takers suggested “if the answer is corrected, displayed the only example in the answer form and if the answer is incorrect, displayed the example and explanation in the answer form”, so there is a need to consider whether the answer format can be studied efficiently. (2) It will be useful to develop mechanisms to enable a beginner's class, intermediate class, and higher class to select problems appropriate for their abilities. (3) For managing the history, it will be good provide the capability to display the current score, the number of problems answered correctly before, the highest score, etc.

Acknowledgement

This researcher relied heavily on the efforts of Mrs. Yoshie Kawai and Mrs. Eriko Miyashita. We convey our gratitude.

References

1. Akinobu, A.: Introduction of Cellular Networking for Revitalization in University Classes. In: Human interface symposium thesis collection, pp77–80 (2003)
2. Akinobu, A.: Trends and Prospects of Utilizing Mobile Phones in Educational Fields, control and information, vol. 50(6), pp. 226–231, 20, 20060615(ISSN 09161600)
3. Becking, D.: How to match mobile learning resources with learners current needs: The didactic profiling? In: Proc. HCII2005 144 (2005)
4. Henderson, J.: Memory and Forgetting published, Routledge (1999) (ISBN 041518651x)
5. Morris, D., et al.: Man watching translated by Raija Mattila (1977). translated by Fujiita man watching, Shougakukan (1981)
6. Kopomaa, T.: The city in your pocket (2000), translated by Kawaura: The mobile phone changes the world, Kitaoji Bookstore, p. 52 (2004)

Production Information Management Interface for Operators in Production Process

Tomonori Yuasa and Yoshihisa Aizu

Department of Mechanical Systems Engineering, Muroran Institute of Technology
27-1, Mizumoto-cho, Muroran-shi, Hokkaido, 050-8585 Japan
{yuasa,aizu}@mmm.muroran-it.ac.jp

Abstract. The AMPAC [1, 2, 3] proposes a simple and unified format to describe parameters of the printing production process such as design, press and bookbinding. The database described by the format can be used commonly in any processes connected by the network production system, since the description accepts any parameters in any processes. A print production system assisted by computers should be very helpful to boot it after altering a parameter of the process, since preparation for starting up a printing machine requires abundant knowledge and wisdom. This paper gives details of the knowledge and wisdom stored in the AMPAC database used for booting the printing machine. Parameter data stored in the AMPAC database can easily be searched and found all of resetting targets in the printing machine when a parameter in the process is changed. This advantage comes from the fact that each parameter of the machine setting can be connected with the related parameters in this database.

Keywords: AMPAC, Database, Standard, Network, Printing, Workflow.

1 Introduction

For establishing an excellent assist system for initial booting, we have to analyze the booting process of expert operators on the machine and imitate it. In initial booting for a colour printing press, experts use various kinds of parameter data, such as a sectional image coverage area (image characteristics), image quality requirement, dynamic and static properties of used inks and papers. At the first stage of booting, experts decide setting points on the machine under given conditions. The related parameters between the setting points are essential in the second stage. Finally a state or value is given for each setting point on the machine. The first and second stages of booting process can be written as the knowledge of relation between parameters includes setting points which are also a kind of parameters. The final stage needs some algorithm describing a decision process, and this stage is a kind of skill or wise, so called "know-how".

An excellent expert requires a wide range area of parameters for getting his correct decision, so that the database should contain much kind of parameter data. The AMPAC database can store any kind of parameter data in a way the suitable for imitating of the expert decision process mentioned above. Furthermore, the AMPAC

M.J. Smith, G. Salvendy (Eds.): Human Interface, Part II, HCII 2007, LNCS 4558, pp. 994–1003, 2007.

database can easily be combined with the other AMPAC database that is generated separately, since every parameter has the same unified format.

It isn't effective for each company to construct an independent database in its own format for the above purpose, since it should be better to use more widely spread information. Furthermore, cooperation by using network between different companies belonging to different industries may become popular in the near future, even in printing production processes. The communication through a uniquely defined common database should be effectively used for the initial booting.

In this paper, the immediate future initial booting scheme of a printing process may be given in actual process.

2 Assurance of Uniqueness of Parameters Relating to Production by Using Hierarchical Structure

2.1 AMPAC Layered Structure Model

The AMPAC [1, 2, 3] consists of two parts. One defines a format for describing parameters classified by a layered structure model and a way of embedding the data into it. The set of parameters is used for controlling a production process after embedded the whole data in a design and workflow management process. If the complete data are embedded to the set of parameters, the initial booting may be performed. The sufficient sets of parameters seldom send to the operator on the machine, thus he needs to gather some of the control information about materials or machines for printing. These data may collect parameters from the database distributed site in different places, even in different companies. For keeping the simplicity and certainty of mutual understanding between senders and receivers, the unified definition of parameters is required. The AMPAC prepares the smart system to search the appropriate data and to collect them in order to consist of the set of sufficient parameters data for booting.

Although the focus of the AMPAC is settled on the printing and it's relating industries currently, the same concept can widely be adopted to other types of industries. It is a principle that parameter data and the related information in the AMPAC are open to the public. Therefore, the other industries can access freely to the distributed database and use them through the Internet or some other network lines. Any person is able to realize his own booting system proposed here and a flexible network production system.

The layered structure model used in the AMPAC classifies all the parameters of a production process into four categories. These are the information about design, manufacture, material / machine, and dictionary. Each of them is classified into more detail by using the layered structure of six stages (Table 1). Layers from the first to the third give conceptual classes that specify the object of data, and serve as a limited attachment for a parameter specified in the 4th layer. Parameters appeared in the 5th layer present relevance with parameters in the 4th layer. The 6th layer contains values to be embedded into parameters of the 4th layer with dimensions. This hierarchical structure ensures the clear and unified definition of each parameter. The contents of 5th and 6th layers play an important role for the embedded data and their relevant searching.

Table 1. Architecture model for classification of parameters in a printing system

	Dictionary	Design	Production process	Material / Machine
1st layer	0	- Specification - etc.	- Preprinting Process - Printing process - Finishing & converting process - etc.	- Media and material - Preprinting Machine - Printing Machine - etc.
2nd layer	0	**Specification for products** - Order and receipt spec. - Page layout - Spec. for print - etc.	**Kind of spec. for each process** - Image input - Edit and collection - Plate making - etc.	**Name of Material / Machine** - Paper - Web offset press - Drum scanner - etc.
3rd layer	0	**Element of spec** - Kind of products - Delivery information - etc.	**Elements of spec.** - Spec. for input - Spec. for output - etc.	**Functional classification** - Geometrical properties - Paper feeder - Stacker - etc.
4th layer	0:reference word (code)	**Parameter** - Products proper name - Name of client - Date of delivery - etc.	**Parameter** - Input file name - Magnification in Input - Angle of image - etc.	**Parameter** - Diameter of roll - Cutting position - Transfer guide position - etc.
5th layer		Relating parameter list to the parameter in 4th layer		
6th layer	Value	Values for the parameter or function name(pointer) for getting the value of the parameter		

2.2 AMPAC Data Format

Any parameter in the group of the AMPAC parameter is classified with the layer structure and is assigned a unique code number to ensure the uniqueness. In the previous version of the AMPAC parameter lists, English names are temporarily used as the parameter name, because complete sets of parameter groups should be arranged for assigning the code. The uniquely coded representation of parameters deduced from an arrangement of parameters by using the hierarchical structure prevents us from causing misunderstanding and mismatch when a designer or a technician wants to search or acquire the demanding parameter data from other sites belonging to an outer vender and a different field, or country. In expressing parameter data, a dimensional unit follows the parameter name to assure clear physical meaning.

The Database is composed of a set of parameter data each of which is written in a record containing knowledge. The record carries the dimensional unit of the target parameter assigned by the parameter name (or code), a set of the relevant parameters with the target parameter, and the numeric data to be embedded into it. The specification of a data description format for the record is shown in Fig. 1. This format for the record has the completely same format, and it can be adopted with any parameter for any production process and any properties of any machine and any material. Inversely saying, all of knowledge can be defined by the AMPAC format.

The type of embedded data to the target parameter can be selected from a numeric data, a set of numeric data, a state like "a file format", or functional representation. The set of relevant parameters, which is used to decide the state of the target parameter, can be written in the record. Those relevant parameters are used in two ways. One usage is an inverse searching of the related parameters for some application of the AMPAC database such as assistant of initial setting as shown below. Another usage is a variable of the functional algorithm for driving a state or a numeric value of the target parameter.

"Intelligence" or "Wisdom" grows in the AMPAC database in three steps. The first step is to give the group of the target parameters compose for expressing an object. In the AMPAC database, the first level of wisdom is given by the selection of combination of the parameter records. The second step is to put a set of related parameters into each record. The second level of wisdom is to know which parameter has a mutual relationship within the target system. The highest level of wisdom is to know the algorithm to estimate the numeric value or to decide the state of the target parameter.

AMPAC Ver.3.10; Uni Code; Japanese	Header part
manufacturer; [model identifier]; [information group]; parameter name [: reference :]; [identifier in parameter]; n1; [PR1 , PR2 , ... , PRn1]; [Q (name)]; DIM; [DIMN]; [DIML, DIMM, DIMT, DIMA, DIMK, DIMMOL, DIMCD]; [DIM10]; [n2]; [DATA1, DATA2, ... , DATAn2] Return code (0x0D)	
end of data	Footer part

Fig. 1. Data format for describing data contents in the database

3 Setting of Printing Machine for Booting

3.1 Principle of Assistant Algorithm for Boot-Up Machine

There are many points to set values on the equipment in printing and post printing processes. Some of these needs to reset when a new product is brought into the production process and some parameters (for example, paper thickness and width) are changed with this. An operator should alter the machine setting at the appropriate points with this change. The expression of wisdom in the AMPAC traces the operator's way of thinking and resetting actions, as shown in the previous section. In some cases, a designer designing a product specifies only a name of the product or type of his request for the used material. In this case, the operator can't receive the full information about actual physical parameters of the alternated target (for example, paper), otherwise only the product name (of paper). The AMPAC carries easily the specific data to the operator through the network, if the vendor has prepared the specification of materials in the AMPAC database previously.

Through his practice in long time, the operator usually studies how each setting point works and how the point should be reset to the appropriate value. The expert is trained to be so smart that he can select the appropriate setting point and can set the state or value according to his knowledge about the relation between change parameters and setting points.

In the AMPAC database, the setting points for a machine are listed in the 4th layer as the knowledge about the machine, for example, "roll width", "cutter work start position", and "transmission guide" etc., in the category of material/machine parameter. In a common dictionary of the AMPAC, the whole parameter set covering all kind of processes, facilities, materials and control parameters will be listed. The complete set of parameters describing an actual production process can be composed by selecting parameters from the common list. The set for specifying the instrument or material may be given from a manufacturer or vendor. Of course, a new parameter can be added for a specific set only by adding the record describing the parameter to the AMPAC database, if necessary. This is similar to the operator's way of study which to know the instrument he wants to operate. The AMPAC parameters and the assistant program for making the AMPAC files can be obtained from "http://www.d-ampac.com". The arrangement of parameter lists for the production system is the first step for establishing Wisdom by using the AMPAC.

Although classification from the 1st layer to the 4th layer is arranged to the target parameter, the 5th layer gives relevance of the target parameter with other parameters which are called "related parameters". Selection of the related parameter for each target parameter is a kind of wisdom or "know how". This is the second level of wisdom mentioned in section 2.2. Preparation for searching a group of the appropriate setting points is completed when the related parameters are put into each target parameter used in the system shown in Fig. 2.

Setting point (PR1, PR2, <u>PR3</u>)
Setting point (PR2, PR4)
Setting point (PR1, <u>PR3</u>, PR4)
Setting point (<u>PR3</u>, PR4, PR5, PR6)
.
.
.

Fig. 2. Target and related parameters and searching processes

In Fig. 2, the whole parameters for the system required to adjust the setting point are listed as "setting point 1", "setting point 2", and so on. The related parameters for each target parameter are given in the brackets such as PR1, PR2 and so on.

The complete set of these target parameter lists for a machine followed by related parameters are able to find out the setting point of it, when the change accompany with product replacement occurs. If a change in the value of PR3 occurs, the setting point required to be reset can be find out by searching the related parameters shown in Fig. 2. In this case, "setting point 1", "setting point 3" and "setting point 4" are selected as reset points, since these contain PR3 in their related parameter lists in the blanket. A simple and unified algorithm can perform this searching process for any kind of parameters that are encoded according with the AMPAC format, since any parameter in any process have the unified format in the AMPAC.

The highest level of usage of the AMPAC is to know the setting value or state for each point found out in the second level procedure mentioned just above. This strongly depends on the facilities of each process and it is difficult to treat in general. The appropriate type of algorithm to find out the state or values for each setting point should be selected for each point. The final derivation isn't given in this report, although the 6th layer in the AMPAC prepares the method of linkage to the function describing the procedure.

The principle of re-setting procedures stated above is summarized as follows:

I. System preparation

1. Arrangement of parameter list
   Determination of all items which demand adjustment and affect the process to including properties of material, environment and quality evaluation factor, etc.
2. Selection of the target parameter
   Selection of the target parameter for resetting
3. Embedding relevancy
   List-up the related parameters
4. Determination rule of adjustment value on printing machine (The 6th layer)
   A function formula or a correspondence procedure

II. Operation

5. Input changed parameters accompanied with a product alternation
   Procedure: automatic input or manual (operator) input
6. Searching setting point
   Procedure: automatic search by the proposed method in this paper
7. Presumption of setting states or values for all deduced targets in the process (6)
   Procedure: using information in the 6th layer in the AMPAC representation

An example of an application program for boot assisting of a printing machine is shown in the following section.

3.2 Selection of Change Parameters and Search of Adjusting Points

In this section, we show the example of the application program performing the operation of steps (5) and (6) stated in section 3.1. This application is created with JAVA language, to keep platform independence.

The initial change parameters are selected by the operator in this program, although it may be easy to carry out the automatic collection of the change parameters form the design stages. An operator's manual input is more convenient for checking the sequence of operation, because the system proposed here is a trial system. For input operation of the change parameters, the additional input program is made to assist an easy input of the parameter and to check a related parameter.

For demonstrating a simplified sample, supposing the case where the thickness of a paper is changed, then, "thickness of paper" should be selected as a change item. The operator chooses "Material" of the 1st layer to choose the item of the paper that belongs to the lower layer of the classification, as the entry shown in Fig. 3.

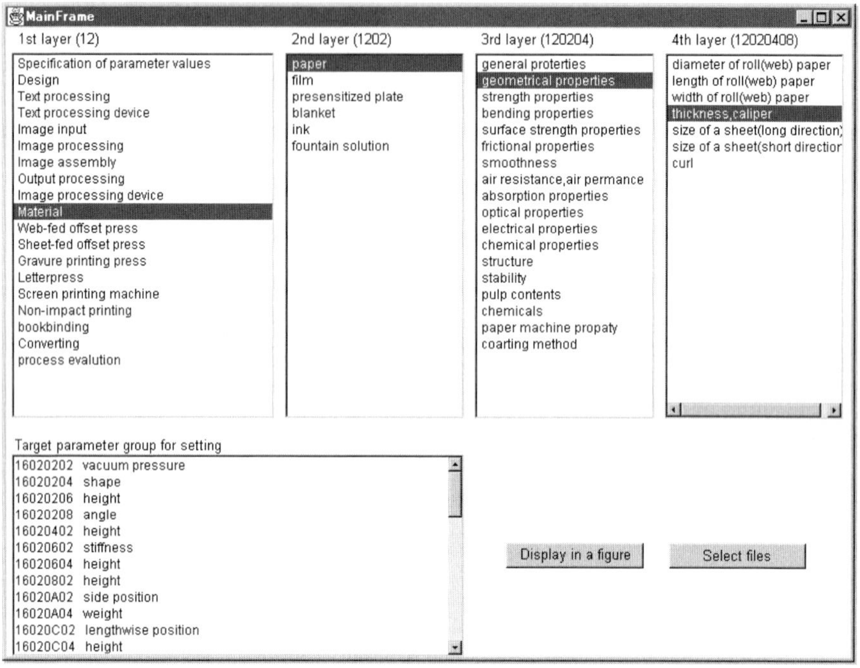

Fig. 3. The results up parameter selecting form

The lower layer items of "Material" automatically are displayed by marking it, and the 3rd layer can be displayed in the next column when "paper" is chosen. In the same way, "Geometrical properties" is chosen to reach "thickness, calliper". This is a physical-characteristic parameter finally required for changing. The parameter belongs to the 4th layer and inputs the value. In an actual operation, the parameter carries the value in the 6th layer.

When "thickness, calliper" in the 4th layer is chosen, the target parameter group for setting is searched and displayed in the lower part of the screen frame as shown in Fig.3. To find out the target parameter, the related parameters and all registered parameters representing the production system are searched. These target parameters are influenced by changed the "thickness, calliper" and need to be reset before the machine operation. The operator should reset all the states of these points referring the value appeared in the 6th layer of "thickness, calliper". Example of the sheet-fed offset printing system is shown in Fig. 4.

In the higher-grade smart system by using the AMPAC database, the states may be calculated by using the information embedded in 6th layer of each target parameter for setting.

3.3 Display of Printing Machines

The interactive display frames are prepared in this system, for promoting intuitive understanding. If the button of "Display in a figure" in the screen flame of Fig. 3 is chosen after displaying a setting target parameter list, the setting target points (parameters) are displayed with red marked point on the outline diagram of a printing system (Fig. 4).

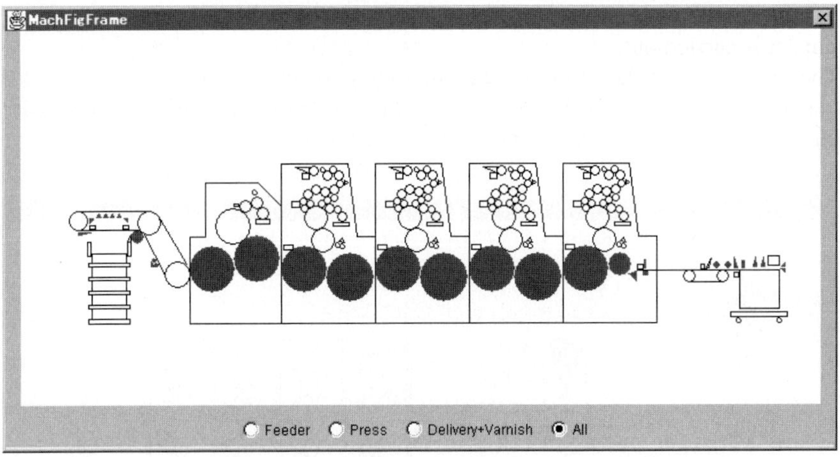

Fig. 4. The sheet-fed offset printing system (All)

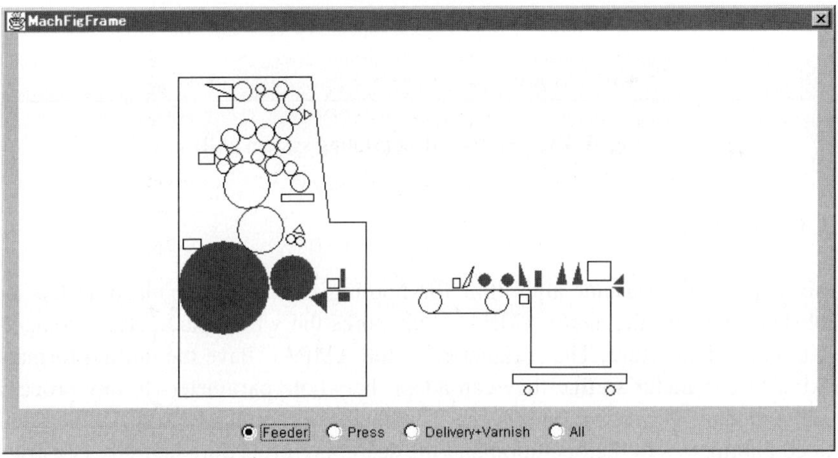

Fig. 5. The sheet-fed offset printing system (Feeder parts)

Moreover, it is possible to display an expanded frame of each portion by choosing one of the three buttons of "Feeder", "Press", and "Delivery + Varnish" to see it in more detail. The name of the setting point marked in red colour can be obtained by selecting the red marked point. An example of the resultant display is shown in Fig. 5, in which "Feeder" is chosen.

To simulate the other type of printing machine rather than the Sheet-fed offset printing machine, a "Select files" button may be pushed in Fig.3. Alternation of the machine type can be easily performed by choosing the data file which contains the element data of the machine. If a parameter of the 4th layer as an alternation target is chosen at this time in the same as the Sheet-fed machine, the setting points influenced by it are searched and updated automatically and then displayed. An example of the web-fed printing system is shown in Fig. 6, in which the same alternation of parameter is carried out.

The alternation of the printing machine may play an important role for workflow managements about a production process. In workflow managements, necessity to change a machine occurs for the sake of the schedule in process.

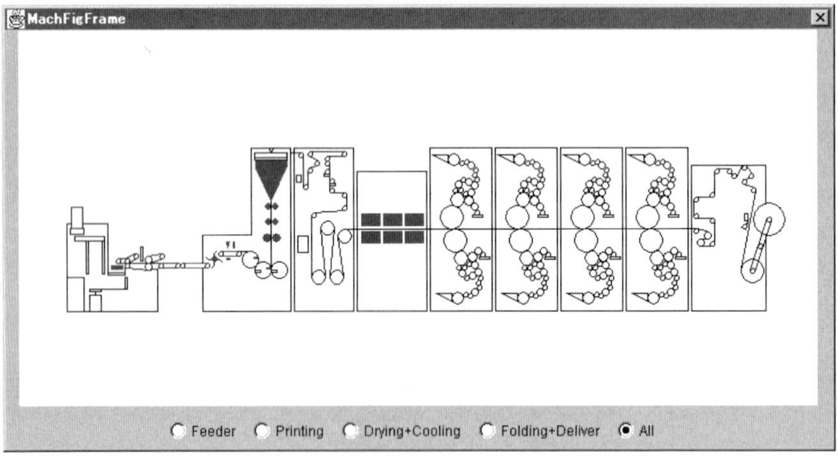

Fig. 6. The web-fed offset printing system (All)

4 Conclusion

In this paper, the assistant algorithm for booting of a printing machine has been established by using the smart AMPAC that stores the whole parameters arranged in the hierarchical structure. The parameters of the AMPAC have the unified format for any kind of parameter so that they can adapt the whole parameters to any process in the same manner of production process. The database can include so easy the relation between parameters in the operation and the searching procedure is easily constructed by using the AMPAC format. It is very important to search of the related parameters, since inference of resetting points is the same procedure of the inverse searching of the cross-related parameters. The booting system presented here shows that the format of the AMPAC works effectively for constructing the system using the wisdom or smartness such as the boot system having the complex parameter relationship.

Although the programmed booting system is a very simple because a database has not yet completed, the method can be adapted to a wide range of printing systems. The AMPAC database can be made in distributed sites at any place separately and can be combined easily. The common AMPAC parameter list can be used by any people in any country, although the AMPAC group of Japanese Industrial Standard (JIS) in "The Japanese Federation of Printing Industry" is mainly constructing the standards under cooperation with members of "The Architecture Model and Encoding for graphic arts parameter" project of IPTS in INSTAC and "Conformance Test of the AMPAC Coding to Design" members in Japanese Printing Machinery Manufacturers Association (JPMA) right now. The database is expanded, if more people join the AMPAC consortium to discuss parameters of the printing process. Many intelligent systems using the AMPAC for various kinds of production processes will come out in future.

References

1. Database architecture model and control parameter coding for process control and workflow, Part 1: Architecture model and database format. JIS X 9206-1:2000
2. Control parameter coding for process control and workflow. TR X 0092:2003
3. Database architecture model and control parameter coding for process control and workflow, Part 1: Architecture model and database format. ISO16044
4. Yuasa, T., Mishina, H.: A System Assisting in AMPAC Database File Creation. JSPST (in Japanese) 37(3), 96–102 (2000)
5. Yuasa, T., Fukumoto, H., Mishina, H.: Computer Aided Initial Adjustment of Printing Machine by using Databse AMPAC. JSPST (in Japanese) 37(6), 307–314 (2000)
6. Yuasa, T., Mishina, H.: Effective Use of Accumulated Printing Information with Database AMPAC. JSPST (in Japanese) 43(4), 279–287 (2005)

A Workflow Based Management System for Virtual Teams in Production Networks*

Bernhard Zimolong and Bjoern Krämer

Ruhr University Bochum
Work and Organizational Psychology
Bochum, 44780, Germany
Bernhard.Zimolong@rub.de

Abstract. The overall objective of this research project is the development and evaluation of a flexible and cooperative management system within SME based production networks for the ramp-up process. Inter-organizational processes of six previous ramp ups were analyzed and documented in flow charts. High-level executives and domain-experts of the production nets developed in moderated focus-groups standardized steps and phases of the ramp-up processes and agreed on one common reference workflow. The reference workflow is the basis for the further development of the inter-organizational management of virtual project teams supported by an IT-assisted workflow management system.

1 Production Ramp-Up as a Factor of Competition

A fast and stable production ramp-up is an essential factor of competition in dynamic market situations to gain financial advantages through the roll out of new or modified products to the market. Under the market condition of steady increasing product varieties, simultaneously reduced manufacturing depth, outsourcing of functions and strategic alliances with suppliers increasing costs of coordination and cooperation emerge (Lei & Hitt, 1995). During the ramp-up process, these are expressed through ramp-up delays, fluctuations in the ramp-up process and late cost-intensive changes.

As enterprises move from "pooled interdependence" of workflows to "sequential" and eventually to "reciprocal" interdependence, cause-effect relations become much more complicated and tenuous (Brass, 1985). The commonly used solutions in the automotive industry to manage complex production networks focus on a strict standardization of the process stages of the workflow in a stable work environment. Usually the Original Equipment Manufacturer (OEM) determines the type of standardization of the workflow, kind of production planning and scheduling, and

* Efficient ramp-up management within SME based customer-supplier-relationships (ELAN). Joint project of the Chair of Production Systems (LPS) and the Chair of Work and Organizational Psychology (A&O) of Ruhr-University Bochum. Funded by Ministry of Education and Research of the Federal Republic of Germany, project execution by Forschungszentrum Karlsruhe.

M.J. Smith, G. Salvendy (Eds.): Human Interface, Part II, HCII 2007, LNCS 4558, pp. 1004–1009, 2007.

quality control. In exchange of the requirements for suppliers to meet mandatory demands on production and quality procedures and schedules, the production network offers a stable work and market environment (Snell, 1992).

Under flexible product-market variation, i.e the degree of breadth and change in a firm's products or markets, particularly small and medium sized enterprises (SME) pursue multiple and at times conflicting goals. To deal with complexity and ambivalence in the market, SMEs develop adaptable workflows, and use more build-in flexibility in the operations planning and IT operating systems. Among other reasons, these resources support the ability for flexibility and innovation of SMEs to adapt to market changes (Nadler, Gerstein & Shaw, 1992).

Particularly employees of small and medium sized enterprises (SME) have a greater work autonomy, decision latitude and flexibility in task planning and scheduling due to a smaller division of labor. Though the OEM usually dominates production network by size, capital and higher division of labor, i.e. specialized functions, business relations with SMEs are based on a strategic partnership. Therefore, the management of production networks with SME partners needs to take into account adaptable workflows, flexible operations planning, less specialized functions and IT-systems diversity. On the other hand, net partners have to agree on a common reference workflow, satisfy scheduled dates of production planning and meet quality standards (Ring & van de Ven, 1992).

The overall objective of the research project is the development and evaluation of a flexible and cooperative management system within SME based production networks for the ramp-up process. The goals of this approach are: (1) to shorten ramp-up time, (2) to improve product quality and (3) to reduce costs.

The ramp-up phase is part of the product development process, which can be divided into product planning, construction, prototyping, initial batch, ramp-up and batch production (Baumgarten & Risse, 2001). The ramp-up traditionally begins with the Start of Production (SOP) and ends when the ridgeline for batch production has been met (see figure 1). To speed up the ramp-up time, the ramp-up process can be cut down, the start of production shifted forward and/or the ramp-up process stabilized. For each measure, former decisions in production development influence the shape of the ramp-up phase. Thus, the overall time and costs for the ramp up process depends on the entire previous product development process.

From an organizational perspective, production ramp-ups are temporary projects, which consist of interdependent tasks of strategic net partners. Employees with different background of disciplines and functions of various enterprises cooperate in a cross-organizational production network.

Planning and coordination of these virtual project teams is a challenging task. Implications of organizational behavior have to be foreseen and taken into account (Cascio, 1999). In comparison to an inter-organizational collaboration in ramp-ups in a stable work environment, standardization of the process stages of the net workflow is blurred and tenuous, coordination mechanisms are less distinctive, and tasks are accomplished with non-compatible IT-systems. As a rule, inter-organizational project responsibility is allocated onto many shoulders. In addition, implicit collaboration rules and common work standards are underdeveloped. Frequent face-to-face meetings usually establish them (Espinosa, Lerch und Kraut, 2004).

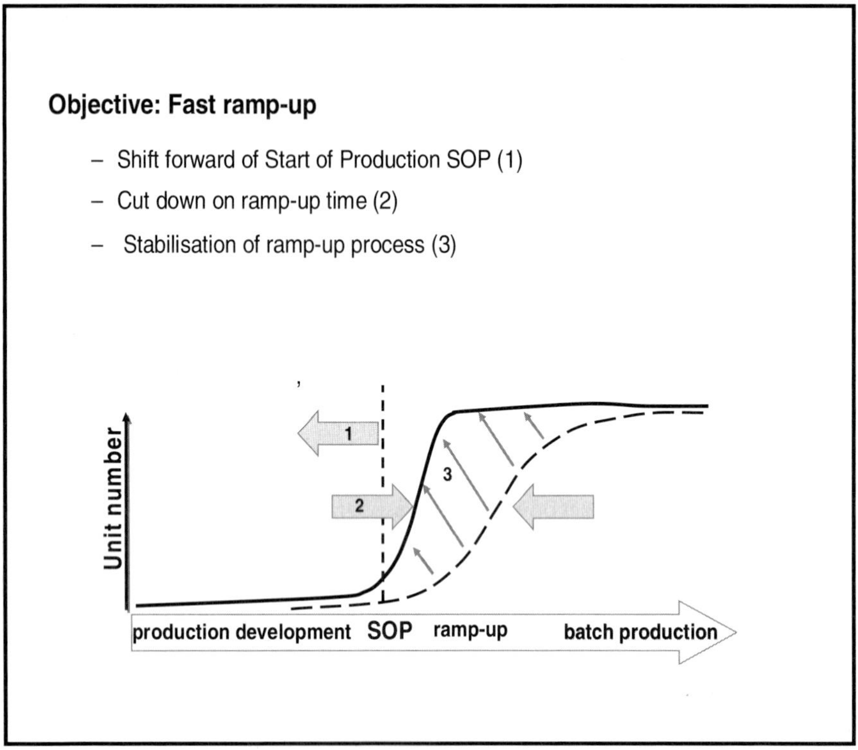

Fig. 1. Measures to speed-up production ramp-up processes

2 Concept of Solution

Important instruments to support work processes in virtual teams of autonomous net partners are standardization of interdependent workflows, control of reciprocal processes, supply of information and communication, support of semi-autonomous task planning and scheduling, immediate feedback on task completion, delays and quality standards. A preview tool for the effects of deviation from arranged schedules and standard processes allows the overall view on timely completion of ramp-up stages and tasks of the partners. Process controls and preview tools allow an evaluation of the progress or serve as a decisional base for interventions.

A determining factor for a fast ramp-up is a workable information and communication (I&C) management system. The system should allow a function- and task-based supply of information such as documents for order bookings, inventory management, statutory audits as well as the documentation and retrieval of personal and team communication. The suggested solution is the conception and development of a process oriented information and communication management. The goal is the timely and task specific provision of information for each step of the workflow process. The I&C management assists (1) the production of information, (2) the acquisition of information, (3) the exchange of information, (4) the storage of information, (5) the evaluation of information and (6) the use of information.

3 Method

The overall design approach follows the principles of socio-technical design (Cherns, 1987), and more specifically, the contextual design approach for costumer-centered systems proposed by Beyer & Holtzblatt (1998).

3.1 Participating Enterprises

The conceptual development, implementation and evaluation of the WFMS took place in cooperation with six enterprises of two customer-supplier nets. In one supply chain, the OEM produces high-priced washing machines and tumblers on a high batch size level. Technical improvements and innovations require frequent ramp-ups. Breadth and costs of ramp-ups vary on the complexity of technical and design changes. Total amount of the work force of the OEM is 15.000 with a sales volume of 2 billion Euros. The first tier is a medium sized company from electrical engineering industry with a workforce of 1.000 and a sales volume of 100 million Euros. The second tier is a specialist from plastics processing industry with 60 employees and a sales volume of 7 million Euros.

The OEM of the second production network manufactures customized compressors on a medium batch size level. OEM has 480 employees and a sales volume of 80 million Euros. The first tier is an enterprise from mechanical engineering with a workforce of 1.100 and a sales volume over 150 million Euros. The second tier is a specialist from electronics industry with 40 employees and a sales volume of 7 million Euro. All enterprises have globally distributed production locations.

3.2 Workflow Analysis

A useful method for the analysis of workflows is the graphical oriented and computer assisted method of an incident-driven process chain (Kuhn, 1999). High-level executives and domain-experts from each supply chain analyzed six previous ramp-ups. From the collected data, six process layouts were designed for the visualization of the ramp-ups. Process chain diagrams represent the inter-organizational workflows. In 14 sessions, participants discussed and completed the inter-organizational process chains and finally agreed on each of the workflows. Eventually these diagrams served as an access for the design of a common reference process in each of the production nets.

3.3 Focus Group Method

The reference workflow for each of the two production chains was developed and agreed on in four focus group workshops. Workshops lasted in each case half a day. The focus group method is a moderated group discussion with the help of a structured procedure that has been fixed in advance. In this study, the goal of the focus group method was to select and agree on the organizational workflow principles and the IT-support of the management system to be developed. During two additional meetings, all participants finally discussed the blueprints of the reference process and agreed on a common solution. The whole procedure took about six months.

4 Results and Discussion

The reference process starts with the phase of construction and ends with the beginning of batch production. To structure the ramp-up reference process stages for planning, directing and decision making was implemented on the operative level. This places special emphasis on a coordinated and standardized process organization while the intra-organizational structure of the net-partners remains unchanged. The phases, sub phases and tasks of the reference process are hierarchically arranged on four levels. The fundamental phases of product development are on the first level and describe the sequential dependencies. They are divided into six phases with the tasks (I) product planning, (II) construction, (III) prototyping, (IV) initial batch, (V) production ramp-up and (VI) batch production. Each of these phases is further divided into additional sub phases.

For the phase "initial batch", the following sub phases are located on the second level: (1) planning (2) coordination (3) resources and funding, (4) testing and approval (5) production (6) second testing and approval of initial batch. On the third level, tasks for the individual process units and net partners are defined. The sub phase "testing and approval of initial batch" includes e.g. the following tasks for the final equipment manufacturer: (1) control of the initial batch by a supplier, (2) company internal testing of the initial batch, (3) approval of the initial batch by the supplier, (4) company internal approval of initial batch and (5) delivery of initial batch to customer. The sequence of the work process determines the order of tasks.

To sum up, the reference process constitutes a generalized ramp-up process which can be adapted to different ramp-ups with specific complexity and breadth. The developed reference process describes the work flow of a ramp-up project on different levels of detail. In order to account for the differences in production ramp-ups, the phases, sub phases and tasks of the reference process can be adjusted to variable stages of complexity. This forms the preconditions of the development of an inter-organizational process organization and the base for the development of an IT-assisted work-flow-management system (Preuss et al.,2007). Currently IT engineers develop an IT system for the management of ramp-up workflows .The WFMS allocates functions and tasks, deals with and stores information, and provides information and communication tools to the users. It allows a function- and task-based supply of information such as documents for order bookings, inventory management, statutory audits as well as the documentation and retrieval of personal and team communication. Parallel to the technical development a thorough usability testing is performed (Konradt, Zimolong & Majonica, 1999).

References

1. Baumgarten, H., Risse, J.: Logistikbasiertes Management des Produktentstehungsprozesses. In: Hossner, R. (Hrsg.) Jahrbuch der Logistik 2001, S. 150–156. Düsseldorf: Verlagsgruppe Handelsblatt (2001)
2. Beyer, H., Holtzblatt, K.: Contextual design: Defining costumer-centered systems. Morgan Kaufmann, San Francisco (1998)

3. Brass, D.J.: Technology and the structuring of jobs: Employee satisfaction, performance and influence. Organizational Behavior & Human Decision Processes, 35, 216–240 (1985)
4. Espinosa, J.A., Lerch, F.J., Kraut, R.E.: Explicit vs. implicit coordination mechanisms and task dependencies: One size does not fit all. In: Salas, E., Fiore, S.M. (eds.) Team cognition: Understanding the factors that drive process and performance, pp. 107–129. Washington: APA (2004)
5. Cascio, W.E.: Virtual workplaces: Implications for organizational behavior. In: Cooper, C.L., Rousseau, D.M. (eds.) Trends in Organization Behavior: The Virtual Organization, vol. 6, pp. 1–14. Wiley & Sons, New York (1999)
6. Cherns, A.: Principles of socio-technical design revisited. Human Relations 40, 153–162 (1987)
7. Kuhn, A.: Prozeßkettenmanagement: Erfolgsbeispiele aus der Praxis. Dortmund: Verlag Praxiswissen (1999)
8. Konradt, U., Zimolong, B., Majonica, B.: User-centered software development: Methodology and usability issues. CRC Press, Boca Raton (1999)
9. Lei, D., Hitt, M.A.: Strategic restructuring and outsourcing: The effect of mergers and acquisitions and LBOs on building firm skills and capabilities. Journal of management 21, 835–858 (1995)
10. Martins, L.L., Gilson, L.L., Maynard, M.T.: Virtual Teams: What do we know and where do we go from here? Journal of Management 30(6), 805–835 (2004)
11. Preuss, S., Krämer, B., Steinborn, A., Zimolong, B.: Effizientes Anlaufmanagement innerhalb SME-basierter Kunden-Lieferanten-Beziehungen. Bochumer Berichte zur Angewandten Psychologie, Nr. 22, Ruhr Universität Bochum (2007) http://www.auo.psy.ruhr-uni-bochum.de/
12. Nadler, D.A., Gerstein, M.S., Shaw, R.B.: Organizational architecture: Designing for changing organizations. San Francisco: Jossey Bass (1992)
13. Ring, P.S., Van de Ven, A.H.: Structuring cooperative relationships between organizations. Strategic Management Journal 13, 493–498 (1992)
14. Snell, S.A.: Control theory in strategic human resource management: The mediating effect of administrative information. Academy of Management Journal 35(2), 292–327 (1992)

Part VI

Environment, Transportation and Safety

Driver Support System: Spatial Cognitive Ability and Its Application to Human Navigation

Kanako Ichikawa and Yoshio Nakatani

Ritsumeikan University
1-1, Noji-Higashi 1, Kusatsu, Shiga 525-0058, Japan
rs003027@se.ritsumei.ac.jp,
nakatani@is.ritsumei.ac.jp

Abstract. In this article, we examine individual spatial cognitive ability to improve the usefulness of a car navigation system. We developed a driver support system through which individuals can view an unfamiliar route before they actually drive the route. Using this system, we performed experiments to confirm the effectiveness of our approach, and we discuss the result of the experiment.

Keywords: spatial cognitive ability, navigation systems, rehearsal system.

1 Introduction

Intelligent Transportation Systems (ITS) offer a number of solutions for modern transportation problems. The advanced navigation system is one of the ITS effective solutions. Many car navigation tools, such as car navigation systems and the Vehicle Information and Communication System (VICS) in Japan, have been introduced, which have truly improved convenience in travel. However, these tools have not adequately addressed ease of use and visibility issues for individuals who have difficulty with orientation.

In this article, we propose a support system that enables automobile drivers to efficiently learn a route before actually driving it. This study examines human cognitive ability, particularly spatial cognitive ability. We also performed experiments to evaluate the effectiveness of drivers' way-finding ability while using the proposed system. We plan to evaluate the results of this study for application in human navigation systems in the near future.

2 Related Research

2.1 Cognition and Movement

Way-finding ability and orientation processes are integral to our daily lives. We applied Naiser's cognitive circulation model to human transfer behavior as a loop structure that consists of perception, cognition, environment, and action (Fig. 1).

M.J. Smith, G. Salvendy (Eds.): Human Interface, Part II, HCII 2007, LNCS 4558, pp. 1013–1019, 2007.

People collect geographic information and integrate that information as a mental map in their minds. The mental map is used to locate a destination while moving. The relation between the mental map and cognitive load during motion has been studied. For example, the thinking is that the complexity of an intersection and road width at a turn affect a person's perception of how difficult it will be to reach the destination [1]. Also, a study in the field of cognitive science identified some of the factors common to individuals who do not have a good sense of direction, such as problems with the ability to acquire adequate information from the environment and problems with mental rotation [2]. Way-finding behaviors depend on these individual cognitive abilities. Existing car navigation systems, however, are designed for the general population and are not designed with sufficient consideration for individuals who are weak in these cognitive abilities.

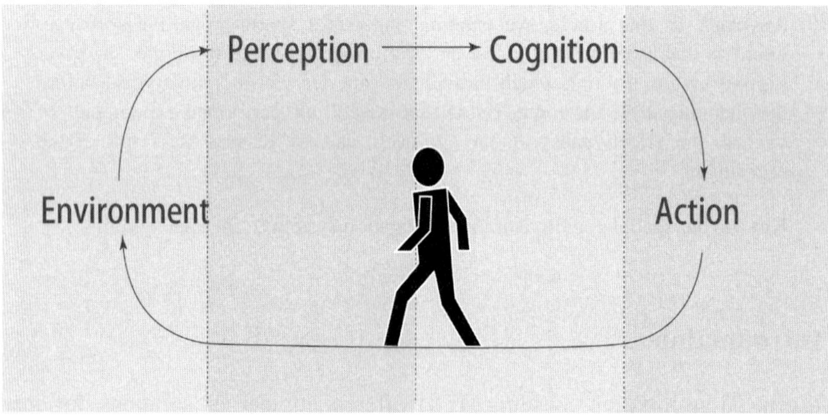

Fig. 1. Human transfer model

2.2 Learning and Movement

Some studies in the field of cognitive science have focused on the importance of mental rehearsal. Reports have shown that training with different media affects human navigation styles, and gaze targets in a mental rehearsal affect human navigation performance [3]. Studies have been conducted to discover the relation between individual mental rotation ability and navigation behavior [4]. These studies have proposed effective ways to conduct mental rehearsals. One such study suggests that people have a preference for the process by which they plan their route [5]. However, there is little argument about how people use what they have learned.

The distortion of a cognitive map is another reason why people easily lose their way. The finding is that perceived distance, which depends on how much information is acquired or searched for in learning a route, is responsible for the distortion [6]. On the basis of this finding, a learning method that allows the acquisition of information needed to learn a route for an intended destination, which promotes the acquisition of such information is important. Thus, an effective way to learn a route through rehearsal should be considered.

2.3 Car Navigation System

Car navigation systems have considered some aspects of these abilities, and especially have improved their functions in order to provide the effective display systems which support the drivers understand the information. For instance, some systems contrive ways to show a driver the result of the route search effectively as follows.

Car navigation systems have considered some aspects of cognitive abilities in their design and have made improvements in their display systems to make the information provided easier to understand. For example, the following features, which have improved user abilities so as to help them reach their destination, have been developed:

1. three-dimensional computer graphics, which provide realistic images of landmarks;
2. indication at the top of the map of the direction of travel (the top of the map always indicates north);
3. lines that connect the place of departure and the destination;
4. adjustability in the number of landmarks to be displayed;
5. change in the display to a bird's eye view when a car approaches an intersection; and
6. display of different types of maps in different windows.

Most of these functions are not developed based on the analysis of human cognitive features, but based on the proposals to develop a new business by using new technologies. Functions of car navigation systems remains insufficient for people who have problem with reading maps or are not good at sense of direction.

Most of these functions were not developed based on the analysis of human cognitive features, but rather to develop new business. Functions of car navigation systems remain inadequate for people who have problems reading maps or who have a poor sense of direction. Although many car navigation systems offer voice assistance, many users are uneasy about following these directions because they doubt their accuracy. Voice navigation prompts the driver with phrases, such as "x meters from here" and "soon come to a turn," which are often misleading and can increase the risk of traffic accidents.

3 Learning a Route by Rehearsal

3.1 Rehearsal

As we drive using the car navigation system, we need to learn the route at the same time. However, the mental load associated with multitasking makes driving difficult and dangerous. In fact, many researchers have cautioned against using car navigation systems. We previously mentioned that mental loads affect our moving behavior and have a relation to cognitive characteristics, which vary in individuals. Also, difficult to read geographic names and physical conditions are thought to affect driving tasks. We believe that driving routes should be learned in advance to reduce the cognitive load while driving.

The type of information needed by drivers is different depending on whether they are learning a route in advance or actually driving. In our study, we presented the subjects with three types of maps to determine which type of map is most suitable for learning a route in advance: a) *north-up* type map, which keeps the map top to the north; b) *heading-up* type map, which keeps the map top to the current direction; and c) *destination-up* type map, which keeps the map top to the destination. In addition to maps, we provided video images and text to support route learning.

3.2 Video Images and Maps

Most people occasionally use maps to identify a destination. Most maps, such as atlases, are the *north-up* type; therefore, this is the type of map that most of us are accustomed to. Nevertheless, many people have trouble confirming their current location when using a map. The global positioning system (GPS) has significantly improved location identification; however, many people still have some difficulty identifying a location when comparing a map with the surrounding environment. Information on maps provided as text or icons is often misleading. Video images and photos, which are more realistic, are thought to be more effective. Therefore, our system provides users with video images of the actual route in advance, which the user can recall while actually driving. This approach makes the drive easier.

3.3 Learning with Video Images

Learning a route in advance involves the recognition of landmarks, such as convenience stores and schools, along with learning the route. However, it is not always easy to find the names on landmarks while actually driving the vehicle, even when they are located directly in front of us. We all have preconceived images or schemas for things. Without access to actual photos or videos, we sometimes apply our own schema to the real world and fail to notice the real thing, the landmarks, if they do not match our schema. For example, if a school building is surrounded by a concrete wall, we will not notice the school. Another problem is that it takes longer to locate names in the real world than it does to find them on a map. Video images are effective because they provide drivers in advance with detailed real-world information related to landmarks, such as the color of a building, names of streets or buildings, and vegetation. Furthermore, advanced images of a landmark decrease the mental workload of drivers. The goal of our system is not to provide each landmark in detail, but to provide enough information about a landmark for drivers to feel confident that they are traveling the right route.

3.4 Approach to Car Navigation Systems

Our prototype system considers the abovementioned cognitive characteristics that are associated with the way-finding behaviors of drivers. A screenshot of the prototype system is provided in Figure 2.

Area “a” represents a *north-up* type map image of the entire route, which can be zoomed in or out and scrolled. The user can view images that can be seen from the car window by linking them with the location on the route. Explanatory text is displayed at area “e,” which identifies landmarks that may be helpful along the route. The text also indicates some characteristics of the route that may be confusing, such as road curves and distances. Users often read the text on the display very carefully [7]; therefore, the text should be brief to decrease the mental load.

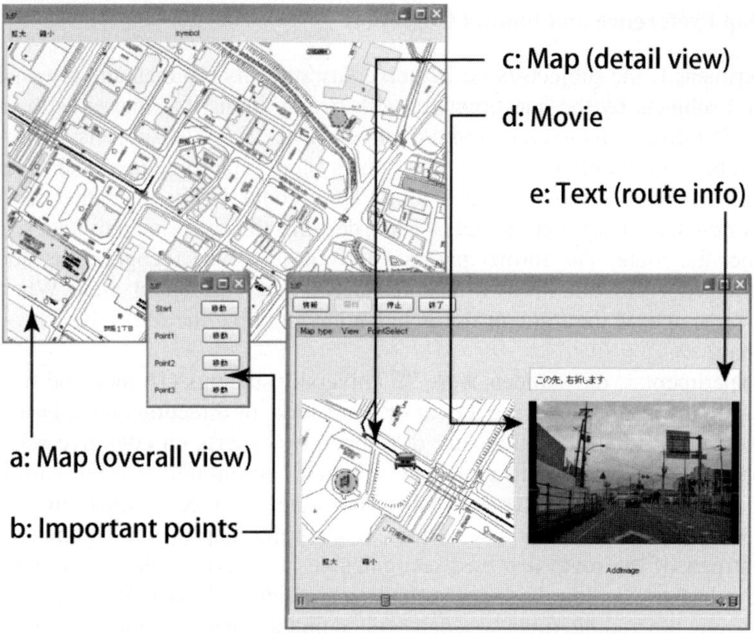

Fig. 2. An example of screen image of the system

Because of the aforementioned differences in cognitive abilities, the system provides three types of maps so that the user has a choice:

- *North-up* type map, which keeps the map top to the north
- *Heading-up* type map, which keeps the map top to the current direction
- *Destination-up* type map, which keeps the map top to the destination

Area "d" provides video images of geographic information, such as landmarks. These mental images are expected to make it easy for the driver to identify the actual images while driving. In situations where long distances are to be traveled, area "b" identifies important points along the route that may be helpful, such as a turn from a wider road to a narrower street. When a point is selected, the navigation starts. These functions will help the driver acquire and organize information and eliminate triggers of human error to avoid traffic accidents.

4 Experiments

Using this system, we performed two types of experiments to confirm the effectiveness of our approach. In Experiment 1, a map drawing task was used. In Experiment 2, the user navigated a vehicle on an actual road to confirm their recall of the route.

4.1 Map Preference and Spatial Cognition

In Experiment 1, the subjects were 22 university students (16 men and 6 women). We classified subjects by the questionnaire with twenty items that measures the level of bad sense of direction developed in cognitive science [8]. The subjects were required to remember an unfamiliar route by using the system. The route involved seven turns. The route video was three minutes long. They could freely scale the map on the computer display. They were asked to use a designated type of map to understand and remember the route. The former part of the route involved recognizable landmarks, while the latter did not involve any because of the residential area. Whether the subject remembered the route or not is judged by drawing the remembered route on a paper map.

In Experiment 1, the subjects were 22 university students (16 men and 6 women). We classified subjects as having a poor or good sense of direction on the basis of their responses to a 20-item questionnaire developed by experts on cognitive science [8]. The subjects were required to view and remember an unfamiliar route with the new system using a designated type of map. The route involved 7 turns, and the video lasted 3 minutes. The subjects were free to scale the map on the computer display. The first part of the route involved recognizable landmarks, whereas the last part of the route had no landmarks because it was a residential area. Whether or not the subjects remembered the route was determined on the basis of a drawing they made of the route on a paper map. The results were as follows:

1. Six of 11 subjects who were good at directions correctly drew the route, whereas only 3 of 11 subjects who were poor at directions correctly drew the route. This finding confirms that the sense of directions affects the task performance of remembering the route, even with this system.
2. The subjects generally reported that they referred to both the navigation map and the video image. The subjects who were not good at directions referred to them less and often referred to the whole route map and to the explanatory text.
3. The *north-up* type map was most effective for all subjects, regardless of their sense of direction. The *direction-up* type map was not helpful for any of the subjects.
4. The landmarks were very useful in helping the subjects remember the route. All subjects who failed to remember the route made mistakes in the latter part of the route; this part of the route had no landmarks.

4.2 Effectiveness in an Actual Stuation

In Experiment 2, we attempted to confirm the effectiveness of our approach in an actual driving situation. At the beginning of the experiment, we conducted a survey with the same questionnaire used in Experiment 1. Ten university students (7 men and 3 women) completed the questionnaire, and 4 subjects (1 who had a good sense of direction and 3 who had a poor sense of direction) were selected to participate in the experiment. The participants were required to learn the route with our system until they were able to perfectly trace it. They were then asked to navigate a driver to the destination on real roads while verbalizing the route as the driver followed their driving instructions. An experimenter rode with the navigator and driver, and recorded the process and utterances with a video recorder.

The results were as follows:

1. All the participants successfully navigated the driver to the destination.
2. The participants preferred to use the *north-up* type map in learning the route.
3. The participants often used demonstrative pronouns during the navigation.

These results led us to the following conclusions: a) the system is effective for actual route navigation, b) the *north-up* type map is better than the two other map types, and c) video images are useful for learning the route.

5 Future Work

We propose this rehearsal system as an advanced navigation system that takes into account the personal spatial cognitive abilities of drivers. Because drivers have different driving abilities, particularly different cognitive skills, cognitive ability must be considered in the design of navigation systems. In recent years, the number of older drivers and women drivers has increased rapidly. This trend requires improvements in the car design and support system for driver [9]. For the next step of this study, we plan to apply the current results to a human navigation system that helps pedestrians and tourists get around in unfamiliar places.

References

1. Matsuda, M., Sugiyama, H., Doi, M.: A Pedestrian Route Guidance System Based on Navigation Demand Model. IEICE Journal A (in Japanese) J87-A(1), 59–67 (2004)
2. Shingaki, N., Nojima, H.: Science of Person Who Has No Sense of Direction, Kodansha Ltd (in Japanese) (2001)
3. Kurano, F., Matsumoto, K., Ogata, M., Morimoto, K., Kurokawa, T.: Object Watched in Rehearsal of Street Navigation and Characteristics of Actual Navigation. In: Proceedings of Human Interface Symposium (in Japanese), pp. 611–616 (1999)
4. Ogata, M., Hamada, N., Morimoto, K., Kurokawa, T.: Relationship between Mental Rotation and Human Navigating Behavior. In: Proceedings of Human Interface Symposium (in Japanese), pp. 757–760 (2003)
5. Furukawa, H., Miyashita, Y., Endo, H., Inagaki, T.: A Study on Acquisition Process of Route Knowledge using Real-View Movie System. Journal of the Human Interface Society (in Japanese) 2(3), 211–216 (2000)
6. Milgram, S.: Introduction. In: Ittelson, W.H. (ed) Environment and cognition, Seminar Press, pp. 21–27 (1973)
7. Nakajima, Y.: Psychology of Picture -Base of Multimedia-, Saiensu-sha (in Japanese) (1996)
8. Takeuchi, Y.: Development, Individual Variety, Gender Difference and Environmental Factor of Spatial Perception, Kazama-shobo (in Japanese) (1998)
9. Kido, M.: Impact that Diversifying Mobility Needs Give to a Car. Journal-Society of Automotive Engineers of Japan (in Japanese), pp. 18–22 (2000)

A Development of Supporting Systems for Disaster Victims in the Initial Stage of Serious Earthquakes

Tamae Imafuku[1] and Sakae Yamamoto[2]

[1]Tokyo University of Science, Graduate school
[2]Tokyo University of Science

Abstract. The object of this research is to build supporting systems for disaster victims, especially for the elderly. These systems are considering as follows: support the people who belong to the specific area, work on the initial stage of disasters. By using these systems, the people in the area can be easy to refuge and to communicate each other. Therefore we suggest several ideas to build systems. And we focus on the system which concerns the elder's action to refuge and demands of the elderly in such situations. Then we aim at designing concrete functions based on these ideas.

Keywords: earthquakes, systems, the elderly, the initial stage, the characteristics of the area, the demand of the people in the specific area and the victims.

1 Introduction

1.1 To Support Disaster Victims

In Japan, it has said that big earthquakes will hit Tokyo. (Tokyo is a one of the largest cities in Japan.) About researches of earthquakes, there are generally two viewpoints. The one is a research about disaster prevention. The other is about decreasing damages in times of earthquakes. We think about the earthquake from the latter point of view. As the latter, we have to suggest decisions in consideration of anticipating the damages of the earthquakes. And it is necessary to reduce and cope with confusions, after they occur. The confusions will lead to a late rescue, secondary disaster, and secondary matter like mental illness.

To cope with confusions, we have to consider how to rescue and support the disaster-victims. And also we have to do that it is necessary to satisfy the demand of the victims.

About the former, first of all, we must support the disaster-wick: the elderly, the handicapped , the pregnant, the infancy and so on. As for the elderly, more than 60% of them died in *the Great Hanshin Awaji magunitude 7.3 earthquake in 1995.* In the case of *The Mid Niigata Prefecture Earthquake in 2004,* many elders failed to get out their houses or failed to run away to safety place, so they were isolated and were rescued late.

M.J. Smith, G. Salvendy (Eds.): Human Interface, Part II, HCII 2007, LNCS 4558, pp. 1020–1028, 2007.

The reasons that they were isolated are as follows. About 21% of people in Japan are the elderly. And in Tokyo, 25% of them lives alone or lives in couple. In large cities, people hardly know about neighbors, so they will be hardly able to identify, each other, their neighbors will be rescued and refuge. And it is difficult for the elderly to refuge by themselves because of the decline of strength. Therefore it will be the most important issues how to secure the safety of the disaster-wick, like the elderly.

About the latter, there were problems as shown in fig.1 in the case of *the Great Hanshin Awaji earthquake,* and *the Mid Niigata Prefecture Earthquake.* In the case of the latter, too many substances were offered, so staffs spent a lot of time in sorting out them. Eventually, some cities stopped receiving substances. Therefore it is necessary to cope with their demands quickly.

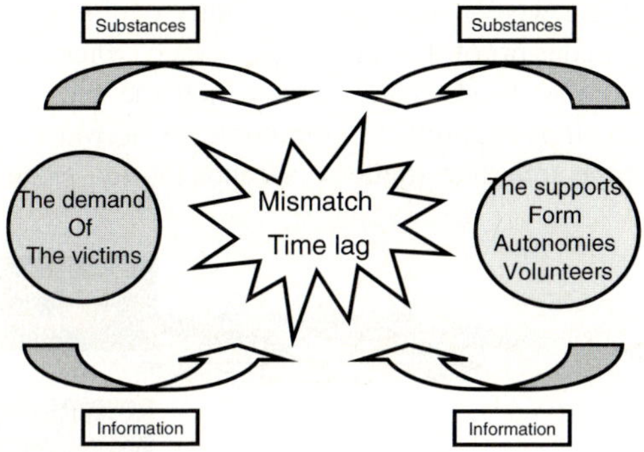

Fig. 1. The problem shown in past serious earthquakes

And it is required that autonomies offer the victims substances on the bases of them.

1.2 The Present Condition to Meet Earthquakes

Now in consideration of the supports of the victims, Japanese autonomies have prepared systems for big earthquakes and connections to private enterprises before earthquakes. However, on the initial stage, the term from just after to a few days later, autonomy's systems did not work well. There are mainly three reasons.

Taking *Odiya* city in *The Mid Niigata Prefecture Earthquake* as instance, firstly, the electricity infrastructure destroyed for about 3days. And also telephone did not work enough because of the regulation of the disaster-hit area of dispatch.

Secondly, existent devices and equipments did not work. Almost all power generators for disasters had wrecks in earthquakes shocks. Also autonomies did not enough prepare generators, so wireless radio for disaster did not work.

Finally, the demands of the victims were different following the characteristics of the areas. Also the supports are different by autonomies level: nation, county, city, footholds of the area. On the initial stage, it is likely to make more confusion. Also the supports from autonomies like nation or county levels were delay due to the damage of the roads. Therefore, it is required to support the victims within the area.

2 The Object of This Research

The object of this research is to build supporting systems for disaster victims, especially for the elderly. These systems are considering as follows: support the people who belong to the specific area, work on the initial stage of disasters. By using these systems, the people in the area can be easy to refuge and to communicate each other.

3 Characteristics of the Systems

After earthquakes occur, the victims will often gather and stay shelters. Then the demands of them are likely to be heard in shelters, first. Generally, the staffs who set up shelters are the people who involve in community of the area, like *town meeting*. Parking, parks, earthquake-proof buildings, the place where the victims gathered changed odd shelters. Also foods and waters which the victims need are saved in warehouses. Therefore, It is required to communicate between the units of the area and the institutions of the area, like shelters, warehouses and so on.

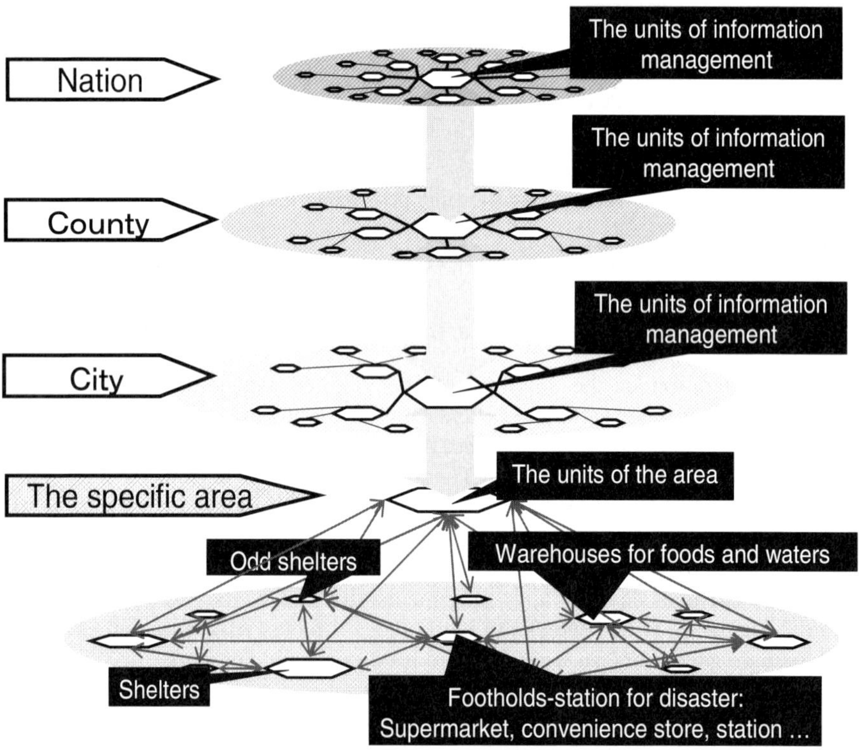

Fig. 2. Shows the image of the systems

And by using a website, we suggest communicating between units and institutions. If they prepare PC, robust networks, and power supply, then they can communicate, whenever and wherever they like. So they will be able to communicate more quickly and appropriately,

Therefore, we have to suggest building systems for supporting the victims. And we show what to do for building the system, as follows:

(1) To build robust network which can cope with troubles of the equipments.
(2) What information to insert on the websites
(3) What to do for supporting the elderly on the action to refuge

4 Ideas for Building Supporting Systems

4.1 Ideas for Building Network Which Can Cope with Trouble of the Devices

To cope with the demands and to communicate on the websites between units and institutions of the area, we suggest the wireless local area network (wireless LAN). This network has role of complementing another means for disaster. The merits of the wireless net work are as follows; easy to carry equipments, easy to operate and build equipments, cheaper than another equipments and so on. In Japan, it is remarked that the wireless local area network is worthy for the special circuit in times of disaster.

To build this network, we have to consider as follows;

1. Where to place the access points of the wireless LAN
2. How long does the access point and PC take to consume power supply
3. Prepare manuals to build the network and how to use (on the websites too)

About (1), we need to consider which equipments we use as standards and where to place access points. The distance to place access points are different between standard of the equipments and obstacles. As for obstacles, in times of earthquakes, many obstacles will occur, like collapse of the building and roads. Then we need to know where the obstacles are likely to occur in the area. Following the obstacles, we can expect to place access points. Generally, it is said that the access points are placed by straight lines and on tall buildings like the earthquake-proof buildings. So we need to research the place where tall buildings and open places area within the area. Then we set access points on these places, and we have to try whether the websites can be seen in the institutions like refuges of the area. Now we are researching them.

About (2), another unit of my university is under developing solar batteries, and we will be able to place them on the earthquake-proof buildings, refuges and so on. Also, we are carrying experiments on dry batteries. We are researching how many and how long, the batteries the access points will need in electricity consumption

4.2 Ideas About What to Insert on the Websites

Fig.3. shows an example of what to insert on the websites. After building the systems, we need information to operate and work them. The information should be on the bases of the system in the specific area.

About the information, the systems like figure3 are suggested. And we need to research what to insert from the viewpoints of the disaster-hit area and the area which the systems are focused on, as shown in figure4.

Then we need to consider what the demands of the victims are, what their troubles were and what to do. We also need to consider the demands of the people who belong to the area.

From our past researches, we showed that the demands are different between disaster victims and non-victims. The demands are different following the many conditions, and the characteristics of the area are likely to effect the victims the most. Also, the demands will change following time passage. We have to consider them too.

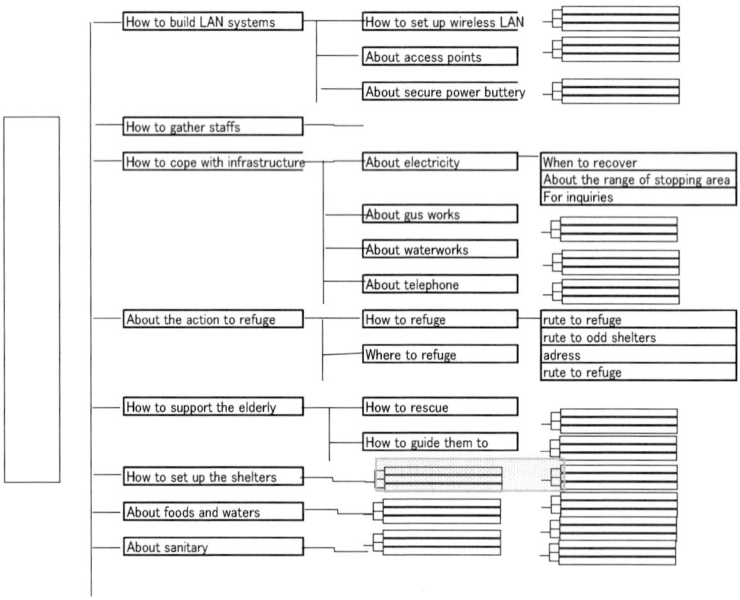

Fig. 3. example of the information of the systems

And, from these systems, we will focus on the system for the elderly which connects with the action to refuge. From our past researches, as shown in figure , the elderly are likely to be most worry about what they cannot judge what to do by themselves. And they want information about how to do just after earthquakes occur.

So we have to consider that there are varieties of the situations on the refuge as follows; where they were, injured or not, the damage of the area around them and so on. Also we need to consider that whether to know the refuges around them, how to go to the refuges. And the systems insert the data which shows their age, sex, address, families. Then, the staffs which stay around their elder's houses rescue them quickly. And also it is required to manage data on which refuges the elderly stay and so on.

And in the case of the past earthquakes, the place where the victims gathered naturally became odd shelters: parks, parking, earthquake-proof mansion, church, and so on. It is difficult for the elderly to go to refuges which are far from their places. And the elderly are likely to be tired and difficult to walk to shelters because of the damage of the road, too. So it is better that the elderly go and stay odd shelters than walk to assigned shelters. It is required to insert safe routs the elderly should take and ways to guide them.

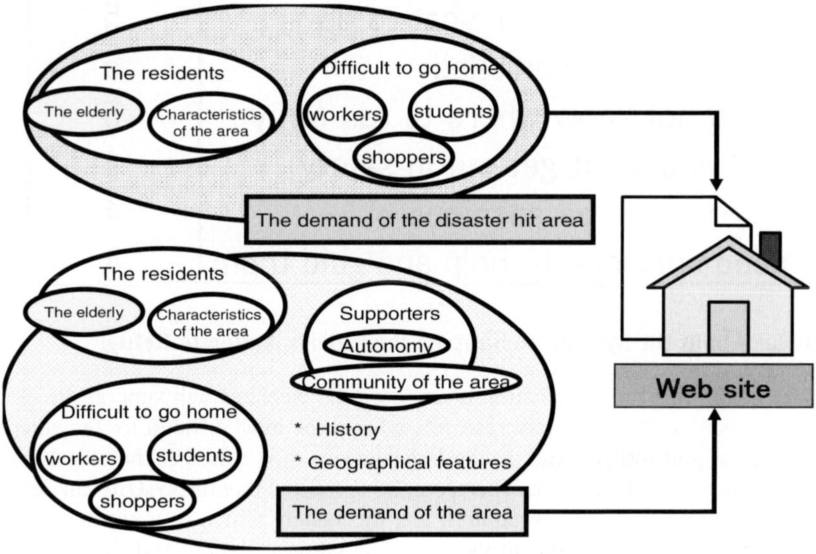

Fig. 4. What to insert on the website

To think what information to insert, we need to research the variety of situations which the elderly were set. For example, whereabouts following time passage, injured or not, the damage of the area around them and so on. In addition, we need the data about what they felt, what they did, what were your difficulties on refusing and what to do, how to go out their houses, and how to go to the shelters, when, where, and how to meet their family, and so on.

And we also have to research about the area which the system focuses on, from mainly two view points. The one is history and geography. The other is about the demands of the people who belongs to the area.

The former is a research as follows: the position of the footholds: refuges, odd shelters, where the danger points which are likely to destroy are, where are the place which are likely to be danger for the elderly on walking, where to refuge and what to do in times of the past disasters in Tokyo.

The latter is the demands of the people who belong to the area: what are the elderly worry about, what do the people, who have the elderly in their families, expect the elderly about the action to refuge.

Table 1. What were difficulties just after an earthquake has occur. Subject: the elderly who experienced the Mid Niigata Prefecture Earthquake

answer in free	number
did not command the situation	7
did not know what to do	5
panicked	5
did not know where to run away	5
could not get out in safety	4
be anxcious	4
need someone to help and guid them	4

4.3 Ideas About the Systems to Support the Elder's Action to Refuge

It is not enough that the information inserted on the websites will support the elderly in times of disaster. From the past research, what is the most difficult for the elderly is how to get out and refuge. And, in the largest cities, it is said that the elderly worry that there is no one to help them. Therefore, it is required that the system can decrease these conditions. And it is also required that the system which is connected with the action to refuge can communicate the elder's condition timely. And the elderly can tell their condition timely using something device.

As the idea of the system, we suggest the system as follows. The elderly have something device which is easy to carry. The device has some buttons which shows the conditions of the elderly. If the elderly push the button, the conditions of the elderly are inserted on the websites. And if the elderly need help, also if they are injured, then the staffs who stay the institutions around them can help them and cope with the situation more quickly after helping them. And if the staffs catch the information about their situations, the device will give the elderly something to feedback. Also we expect that number of the buttons is as few as possible. Because, in times of the earthquakes, the elder's need to judge the condition as fast as possible. And we expect that the device is small, like mobile pone, so the elderly can carry it if they go out.

Also as function of the device, we suggest that the staffs can tell the condition of the area with one-way. Just after an earthquake has occur, almost all of the people could not imagine what had happened and know what to do. So it is likely to be difficult to judge the condition of them and push the buttons on radios and TVs. Then the device turns on the switch automatically, and tell what has happened and what to do.

Therefore, to develop this system, we have to research the elderly in the disaster-damaged area. And it is required to collect information as follows: what they feel, what they do, when and how to refuge, and so on. Then, we need to see the validity of the system.

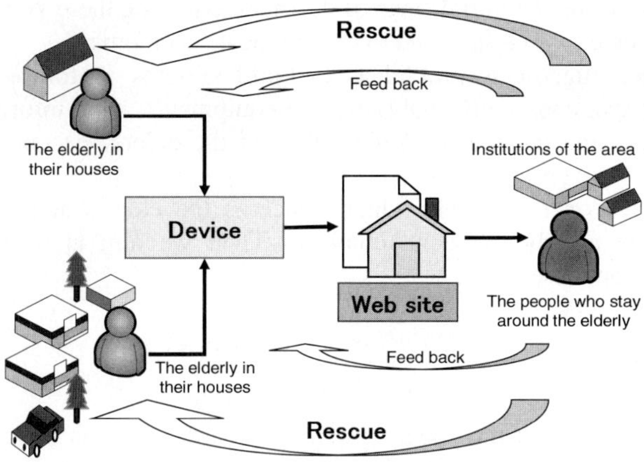

Fig. 5. Image of the system for supporting the action to refuge

However, we have to consider the costs and situation the elderly can not use the device because of the blowing out of the device and so on. Then it is considered to build systems as follows: simulator to guide the elderly to the safety place around their houses, supporting systems which urge the people who belong to collages, companies of the area and so on.

Fig. 6. Image of the function of the device

We have to clear what they want to do and what they want to be on action to refuge. Then, we have to build the system to approximate their demands more. And we have to reflect what the elderly require to the system as function and contents.

5 Conclusion

We suggest the supporting systems for disaster victims, especially for the elderly. These systems are considering as follows: support the people who belong to the specific area, work on the initial stage of disasters. By using these systems, the people in the area can be easy to refuge and to communicate each other.

Therefore we suggest several ideas to build systems as follows: build robust network which can cope with troubles of the equipments, what information to insert on the websites, and what to do for supporting the elderly except communicating information on the websites.

And we focus on the system which concerns the elder's action to refuge and demands of the elderly in such situations. Then we aim at designing concrete functions based on these ideas.

6 Application

We suggest that we built the social systems for earthquakes in consideration of the characteristics of the area and the demand of the victims. For that, it is considered how to secure the means of communicating information and how to insert. Now we are going to manage temporary websites focused on one local area.

Fig.7. shows the image of the system which we will aim at as the last stage. The institutions and community of the area can communicate information on the web. The victims can also get information through fm radio, electric bulletin board, and so on.

And in consideration of the rapid aging society, we have to consider the elderly's situation more hereafter.

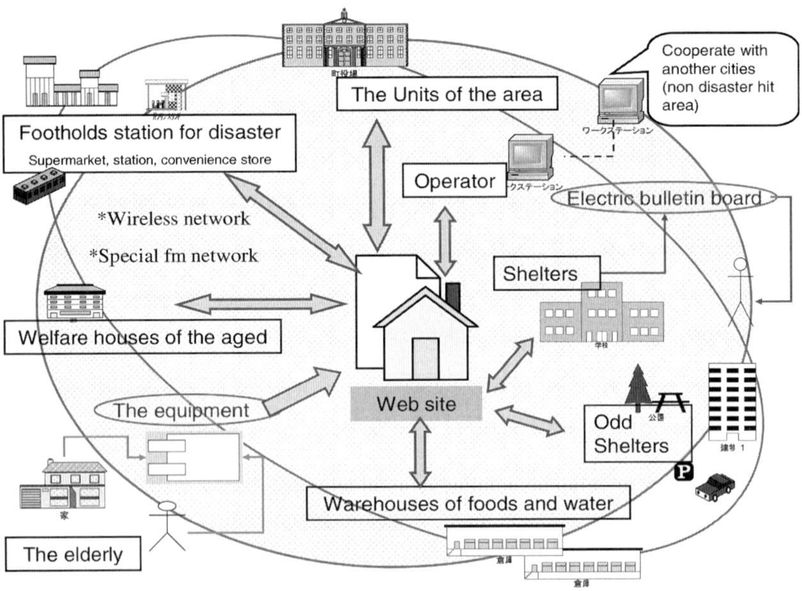

Fig. 7. Image of the system we will aim at

Web Services as a Solution for Maritime Port Information Interoperability

Pedro Teixeira Isaías[1] and Fernanda Maria Duarte Macedo[2]

[1] Universidade Aberta,
Rua Fernão Lopes, 9, 1-Esq,
Lisbon, Portugal
pisaias@univ-.ab.pt
[2] Administração do Porto de Lisboa,
EIDH, Doca de Alcântara Norte,Lisbon, Portugal
Lisbon, Portugal
fernanda.macedo@portodelisboa.pt

Abstract. Logistic integration is a essential condition for the development of maritime-port transportation. This paper proposes to demonstrate that this integration is dependant on a digital strategy. A conceptual architecture is proposed here. This architecture uses the emerging technology of web services as a good solution for enterprise application integration (EAI).

Keywords: Web Services, EAI, Maritime transportation, supply chain integration, interoperability.

1 Introduction

In order to pursue the goals of the maritime-harbour business, collaboration and integration enhancement is required [1, 2], aiming information reuse [3] throughout sharing flexible IT infrastructures. Supply-chain interoperability is becoming an essential requisite to accomplish [2, 4]. This business has many players with different roles, and goes over borders, and therefore, in order to characterize the system being studied and to find the right answers to the questions under research, it was chosen a triangulation approach, using different information sources.

An EAI system enables that organizations answer swiftly to the needs of change and innovation. With this goal, this paper proposes a Web Services technology architecture based in XML (eXtensible Markup Language), open norms and Internet technology that can boost the interoperability requirements of this industry.

The expression Web Services or services made available through the Web has associated to it an architecture, a set of norms, technology and business models [5]. According to [6], Web Services are like "plug and socket" universal XML based, simple and flexible like HTML, truly independent from the technological platform, being the default interfaces so that applications can talk to each other.

The Information Technologies (IT) architecture is the set of norms and policies that rule the use of IT and set the way to follow [7]. In this sense we talk about a Service

M.J. Smith, G. Salvendy (Eds.): Human Interface, Part II, HCII 2007, LNCS 4558, pp. 1029–1038, 2007.

Oriented Architecture (SOA) in which different applications are treated as components. Services can become gradually available, putting aside the traditional "architecture by application". This kind of architecture increased the negotiation power of the IT industry clients that demand better services and with lower prices [6].

2 Methodology Used

This work was based in a thorough literature review, on the triangulation of information from different European reports and from the application of the method of case studies to three specific Portuguese companies, key players in this specific industry.

The above methodology enabled the identification of the IT mission in this business and the proposal of broad application architecture to assure the management and control of the whole maritime-port net chain, promoting the co-operation and interoperability between the involved key players and enhancing an effective knowledge management.

The work started with a thorough literature review in the following domains: maritime-port logistics; IT and strategy; EAI and EDI/EDIFACT; and Web services as a solution for EAI.

Triangulation of information method has been used also. This resulted in a clear perspective and the possibility to scope the architecture proposed.

Case studies have been also produced to understand the requirements of key players. The studied node for the case studies was the Port of Lisbon. This was mainly because it was highly representative in terms of external commerce from the 13 Portuguese ports available [8]. The second reason is because port of Lisbon with port of Leixões together move more than 90 % of containers cargo in Portugal (IPTM . The last reason for choosing the Port of Lisbon was that this was the only Portuguese port since 2001 that had working the electronic integration of cargo manifestos, with shipping agents and shipping lines.

As mentioned before, three companies have been selected for this study. The selection process was based in the cargo volume, which has been considered as highly representing within the other companies, and also because they had on-going integration and collaborative processes with several partners, supported by IT. The contacted companies were: Navex, MacAndrews, and APL (Administração do Porto de Lisboa) – this is the port authority. In the first two, the interviewed people were top managers and in the latter, the coordinator for new information systems projects.

The interview was supported by an structured questionnaire. Before each person being interviewed, a brief explanation has been given of what was being analysed, how a port was seen from the author perspective, what was observed and what the goals were, using Fig. 1, using a mixed model of business in the Web, in which a port would be a Service Oriented Platform (SOP), supported by collaboration that will result from an electronic integration between all partners, based in two pillars: a new IT architecture and a continuous normalization process that will generate and effective information exchange and shared.

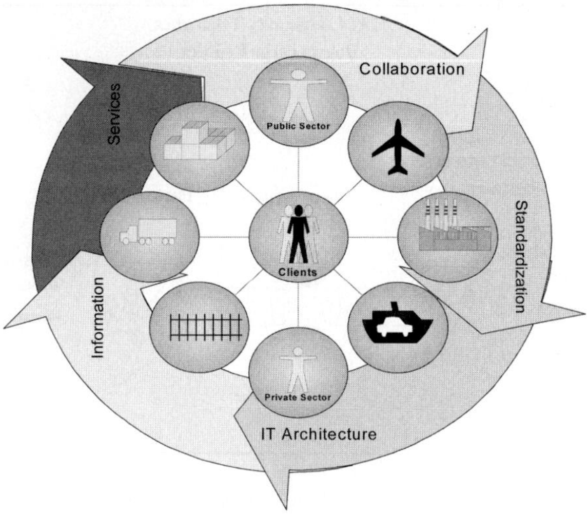

Fig. 1. Service Oriented Platform

One important conclusion from this study can be seen as the outline of a mission for the IT: to guarantee the management and control of all the maritime-harbour logistic chain, guaranteeing the co-operation and interoperability between the involved actors, enhancing an effective knowledge management.

Another important conclusion is that there is an integration from n to n, between partners acting on the same port and along the supply-chain: each port has a Port Community System leaded by the Port Authority, existing a weak co-operation between different ports in this way – co-operation in each port is informal and without a strong articulation between different actors in EU. According to the diagram presented in Fig. 2, between any two ports there is a unique physical flow of the ship and merchandises. Nevertheless, between the different actors present in the same two ports, numerous information pathways are set in each port and throughout the logistic chain between the two ports. Each partner in a port X send its information to its partner(s) in port Y making it arrive to the authorities present and their partners from hinterland. This is true even if port X and Y are both ports from the EU or even from the same country, like Portugal.

Other conclusions drawn from the information triangulation are:

- Guaranteeing the interoperability in the different modes and between the different actors;
- There are different EDI/EDIFACT implementations using different messages (e.g. Rotterdam is using IFTCMS and Lisbon and Spain are using IFCSUM but with different guide implementations) so each actor needs to implement data conversion and transformation depending on which partner is doing business;

Harmonization, standards, track and tracing of goods and vessels are needed.

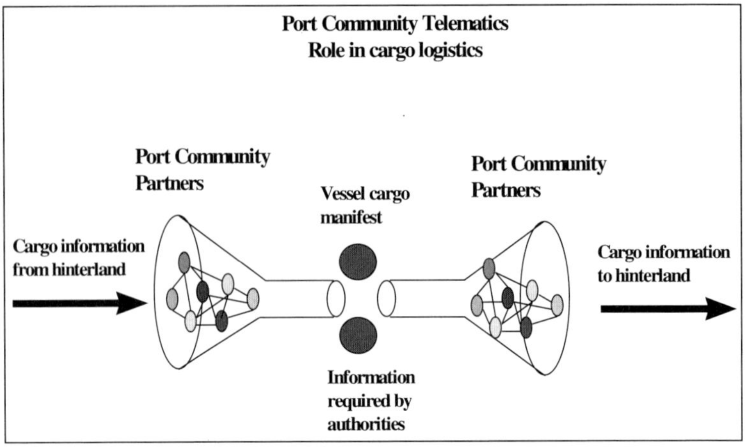

Fig. 2. PCT: The flow information (source: [9]: p.156)

3 Architecture and Implications

3.1 Proposed Architecture

The proposed architecture is presented in Fig. 3. It uses several components and details are provided.

XML is more and more often used in B2B integration. It makes available a vast number of norms to tag data structures and their content, organizing and managing text based information.

UDDI (Universal Description, Discovery and Integration) is also a service, more specifically a "finding" service. This is a web-distributed directory that makes it possible for businesses to know about each other, contributing for their own dissemination over the web. WSDL (Web Services Description Language) complements norm UDDI and it is a set of metadata that provide common semantics, in a standard way in XML language, regarding the description of interfaces, services and connection protocols of various services. A WDSL document contains all the information potential to interact with a Web Service but, in order to communicate that same information between the Web Service and the application, there is another norm in XML, the SOAP (Simple Object Access Protocol) that makes available a way to do it via HTTP.

This platform is developed in the Web Services architecture and supported in the UDDI registry. With this registry the process is made in a dynamic way being the key to keep the location independent of the Web Service, more specifically, the location concept of the SOP is relative since two or more ports can share the same SOP. Access to the UDDI registry is not made directly by the platform users but has the final goal of making the required pairing to the different users with different requirements about the services to which they access. As services are made available, there are documented in the WSDL and are registered in the UDDI registry. Each

SOP has a role of virtual Hub that will centralize different services in a distributed architecture, making available the maximization of collaboration in a port community and among different port communities.

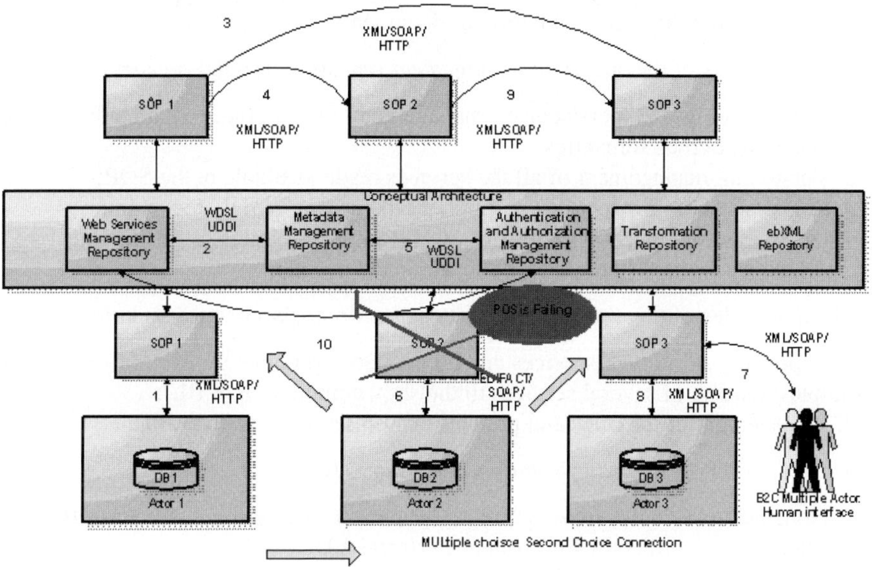

Fig. 3. SOA Architecture Proposal

Services can be implemented that through the outcome of several events can distribute the documentation between two or more SOP. For example, in relation to loaded merchandises in a port 1 (SOP 1) and that are to be unloaded in port 2 or 3, respectively, SOP 2 and 3, the information can flow from SOP 1 to SOP 2 (flow 4) and to SOP 3 (flow 3) and there be distributed to who requires it.

Communications can be held through Internet over SOAP/HTTP. The SOA is an IT architecture made of four big repositories that are functional services:

- Authentication and Authorization Management Repository;
- Metadata Management Repository (UDDI registry);
- Web Services Management Repository (Services Gateway);
- Transformation Repository.

Each of the aforementioned repositories is made of a set if aggregated services by roles. The first contact with the SOP is always made at the authentication and authorization management repository level.

The main roles of the *authentication and authorization management repository* are:

1. Authentication management that can be made through digital certificates. There is a directive that aims at creating an issuing entity in all the EU of digital certificates;

2. Access control to services only to duly authorized clients;

3. Confidentiality guarantee by the information management authorization – this specifies which information about a given document an authenticated actor is authorized to access and keep;

4. Management of the encryption and uncryption process, identifying if the partner, for the document to exchange, has previously agreed using an encryption service;

5. Guaranteeing the non-repudiation of documents.

The main roles of the *metadata management repository (UDDI registry)* are:

1. Guaranteeing the persistence, management and storage of metadata which contains all the agreed semantics;

2. Versioning management of all the services made available in the SOP;

3. Registration of new services, throughout the proper WSDL document;

4. the base of the external control of all the applicable rules of this architecture in the SOP, between the different actors, services and processes.

The main roles of the *web services management repository (services gateway)* are:

1. Repository of all web services and all available versions for each of the services (examples: request of several services to the ship; compilation of BL);

2. Management of the rules and policies of web services made available.

The main roles of the *transformation repository* are:

1. Nomenclature conversion – currently, each actor keeps these structures that may be as many as the partners with whom he performs EDI;

2. Translation of EDIFACT messages, given that, for each document there are several implementations, often using the same message;

3. Transformation of XML messages with different XDS or different versions of the same implementation;

4. Transformation of EDIFACT messages into XML and vice versa.

3.2 Implications

The interoperability and interconnectivity between different nodes, partners and processes throughout all the chain is a real need [2]. Therefore, as a result of the study main results, and also from the actors involved in this sector, there will necessarily be different systems, platforms, applications and database systems. This means that there is a real need to implement between the different systems from all partners, a set of middleware that makes it possible to have B2B between companies. The Web Services technology presents this characteristic [10]. Being that in this technology, the most important level is the norms and protocols level which facilitates the connectivity among applications and the electronic Exchange of messages, in which the applications present themselves as services [11]. For these reasons the Web Services technology is a guarantee of interoperability [12].

In this context, the results obtained in this work are adequate for the problem given: a strategy for the IT based in Web Services seems to be a good approach in the digital context in the maritime-port industry, enhancing the integration and co-operation throughout the chain, what can clearly play an important role in the creation of a competitive advantage [13], p.34.

In 3.1 a conceptual EAI model has been presented for the development of this very same strategy. This strategy may enhance and create the best conditions for generalization and interoperability among different heterogeneous existing systems [14]. Finally, the reuse concept that is associated to it [10], may contribute to an increased integration application-to-application (B2B). This last aspect may increase service transparency and efficiency because its inexistence may limit the efficiency of the transport operations as the companies Navex, APL and MacAndrews referred. The B2B integration is a constant challenge and supports the business strategy not only at the port level, but also widening itself to the whole logistic chain – this has also been referred by the interviewed companies and mentioned in European studies [9] and [15] and stated by several authors [16, 17, 1]. Therefore, this type of architecture enhances interoperability at two levels – IT and business, supported by:

- An increasing offer from the supply technologies industry that has been evolving in increasing interconnectivity and interoperability [18], based on open norms and the widely accepted language which is XML [6], to guarantee platform and programming languages' independence [12];
- The support to the entire existing technological legacy throughout the logistic chain in this industry [11]. The Web Services truly independent from the technological platform are considered the interfaces for the applications to talk to each other [6];
- A directory – the UDDI, "something that looks like the business yellow pages and also a technical specifications library" [19], will make it easy to manage this type of complex architectures, through the management and control of all the rules hereby implemented and also the control of the services versions from the IT viewpoint. From the business point of view, this may enable more actors to adhere and also a quicker adherence of existence actors.

4 Discussion and Conclusions

It is generally agreed that the Internet has been the phenomenon that has generated the first wave of normalization in the IT industry. The Web Services technology, based and inspired by Internet models, anticipates a new normalization wave in the IT Industry [20], based in open norms and in the XML commonly used language [6].

If the intermodality development, boosted by interoperability and interconnectivity between the different modes, partners and processes throughout all the value chain is a need [2], a IT strategy based in Web Services may be a good approximation in the digital context of maritime-port industry, enhancing integration and collaboration throughout the chain by guaranteeing the interoperability and interconnectivity between different systems.

Implementing the SOP Architecture will bring some advantages to the business:

1. Firstly, each partner in the supply-chain will need to develop process integration only with the SOP: SOP will be the systems boundaries system, the preferential B2B strategy. Today, between different actors, there is a n to n integration with different and complex solutions and architectures: a sort of islands of information. The SOP is a shared architecture, with which every actor in the maritime supply-chain would have to interact.

2. A unique physical flow from the ship and merchandises would correspond to a unique information flow between the two ports, between two SOP, with the inherent advantages to this simplification process of the circuit and control improvement. In this way there is a clear tendency to eliminate information corridors.

3. The reuse concept that is associated to it [10] may accelerate a greater application-to-application integration (B2B) throughout the chain and to a cost reduction, supported by the sharing of open norms, of the already developed and made available components in the SOP, and also the externalization of the transformation services: each transformation service will be developed only once and reused by all that need it.

4. The interoperability will succeed and will allow the reuse of information. The truly interoperable organizations are enabled to maximize the value and the potential of information reuse that they possess [3].

5. The existence of a metadata management repository may enable an harmonization, removing one of the barriers to have interoperability between partners [9] and making it possible to obtain useful information [21].

6. Guarantees the service availability 24x7 similarly to other network services (ex.: DNS or Domain Name Server) in Internet, are implemented today. In Fig. 3 and as an example, one of the other two SOP (1 or 3) could be the secondary SOP of SOP 2, in case this was not available.

7. The type of technology used in the SOP will even make it possible that partners with low application to application integration capacity, commonly use a set of utilities available to the final user which support XML [22], editing and manipulating XML files. This means that a partner, without sophisticated systems, but with Internet access, may capture documents in this format, and also manipulate, archive and afterwards, reuse the contained information.

SOP has some requirements and limitations that are now identified:

1. To implement this type of architecture in which all actors from various logistic chains converge to a single strategy it is necessary to abide to three requirements:

a) Strong leadership;

b) Great commitment from all the different players;

c) Architecture dissemination through training plans, workshops and by the SOP itself.

This first requirement would easily be eliminated if there was an actor in the business with these three tasks, like IATA or The International Air Transport Association [23] or ACI or the Airport Council International [24] for air transportation.

2. There is already a strong legate that should be carefully examined and that should be taken into consideration. All ports lead by the Port Authority already have a Port Community System (PCS). An alternative so that there is no abrupt disruption is that the SOP evolves from the current and already implemented PCS. This will make it possible that a balance between efficiency and innovation be met without putting at stake the IT architecture existing today in companies in this process of change [25].

3. There are also some drawbacks concerned with security inherent to Web Services technology yet to be solved.

4. As a consequence from the presented architecture, the norms hereby presented and related with Web Services technology have been approached in a generic way, namely regarding their role in a SOA architecture and their main characteristics. Its use, with the resulting advantages, will demand also a knowledge recycling process for all people that work in the IT industry, internal teams from companies and suppliers. If that does not happen, there may be the risk of developing Web Services that do not conform to norms, being these applications only available in Web environments, as well as many other applications that appeared with the "Dot-com" phenomenon, compromising the promise of this architecture: the interoperability guarantee and its reuse.

The main limitation of this work is that the SOP is a theoretical model. Given the importance of the themes being analysed, it would be important for some of the conclusions to reinforce its sustainability, by implementing it through a prototype. Other limitation of this work is connected with the fact that the SOP architecture was proposed at a macro level, mainly due to the fact that this is a complex business and within the level of the nature of this work, it could not be done in other way. This limitation could be minimised if two processes were analysed in detail.

In all this paper proposed an architecture for port information interoperability among port key players. It also describes the process taken to reach the architecture by describing the methodology used. The architecture is described in detail with its several components

The proposed architecture can be generalized to any port system in the world, since the proposed components can support and sustain the logistic integration effort.

Acknowledgments. The authors wish to thank the Port of Lisbon key players (i.e. Navex, MacAndrews, and APL (Administração do Porto de Lisboa)) that have participated in this study.

References

1. Lambert, D.L., Stock, J.R., Ellram, M.: Fundamentals of Logistics Management. McGraw-Hill, Singapura (1998)
2. EU, White Paper, European Transport Policy to 2010: Time to decide. Bruxells: Serviços das publicações das comunidades europeias (2001) (Available Accessed 01-08-2004) http://ec.europa.eu/comm/energy_transport/library/lb_texte_complet_en.pdf
3. Miller, P.: Interoperability. What is it and Why should I want it? (2000) (Available Accessed 15-02-2007) http://www.ariadne.ac.uk/issue24/interoperability/intro.html
4. Van Miert.: Group à Haut Niveau sur le réseau transeuropéen de transport, Report (June 27, 2003)
5. W3C, Web Services Architecture (February 11, 2004) (Available Accessed 15-02-2007) http://www.w3.org/TR/2004/NOTE-ws-arch-20040211/
6. Woods, D.: Enterprise Services Architecture, O'Reilly, Cambridge (2003)
7. Weill, P., Vitale, M.: What it infrastructure capabilities are needed to implement E-Business Models? MIS quarterly Executive 1(1), 17–34 (2002)
8. IPTM, Ports Statistics (2004) (Available Accessed 15-02-2007) http://www.imarpor.pt/main/anexos/Mov_Merc04.pdf

9. COST 330, TeleInformatics links between ports and their partners. Final report of the action, European Cooperation in the field of Scientific and Technical Research (1998) (Available Accessed 15-02-2007) http://www.cordis.lu/cost-transport/src/cost-330.htm
10. Karastoyanova, D., Buchmann, A.: Components, middleware and Web Services. In: Proceedings of the IADIS International Conference, Algarve, pp. 967–979 (November 5-8, 2003)
11. Hagel III, J., Brown, J.: Your Next IT Strategy. Harvard Business Review, pp. 105–113 (October 2001)
12. WS-I, Basic Profile Version 1.0. (April 16, 2004) (Available Accessed 15-02-2007) http://www.ws-i.org/Profiles/BasicProfile-1.0-2004-04-16.html
13. Porter, M.E.: Competitive Advantage, creating and sustaining superior performance, 1st edn. Free Press, New York (1985)
14. Karmarkar, A.: A Preview of WS-I Basic Profile 1.1. (2003) (Available Accessed 15- 02-2007) http://webservices.xml.com/pub/a/ws/2003/09/16/wsbp.html
15. EU, Study on Freight Integrators Final Report. European Commission: 16 Setembro (2003) (Available 15-02-2004) http://europa.eu.int/comm/transport/intermodality/highlights/doc/final_report_freight_integrators.pdfAccesses
16. Quaresma-Dias, J.: Integração Logística em Tempo de Globalização Que Estratégias e Políticas para Portugal (Cargo Edições), Cargo Edições, Lda, Camarate (2003)
17. ICC, Negotiations on Maritime Transport in the WTO General Agreement on Trade in Services GATS (2001) (Available Accessed 15-02-2007) http://www.iccwbo.org/home/statements_rules/statements/2001/negotiations_on_maritime_transport.asp
18. Carr, N.G.: Harvard Business Review: IT Doesn't Matter, pp.41–49 (May 2003)
19. Taft, D.: W3C, OASIS Recommended New Web Services Standards (2005) (Available Accessed 15-02-2007) http://www.eweek.com/article2/0,1759,1756174,00.asp
20. Daryl, P.: Tempo Bea: Web Services oriented. Número 11, p. 5 (2005)
21. Realise, European inventory of user requirements and evaluation of statistical methods (2004) (Available Accessed 15-02-2007) http://www.realise-sss.org/uploadfiles/Stat-D2-2.pdf
22. Microsoft, Office 2003—XML: What's it all about? (May 31, 2003) (Available Accessed 01-08-2004) http://office.microsoft.com/en-us/templates/TC011860771033.aspx?pid=CT102036971033
23. IATA, Cargo Services Conference Resolutions Manual, 25th edn. Montreal: IATA (October 2004)
24. IATA, IATA/ACI Guide for Interchange of Data & Shared Use of Equipment & Systems 1.0. (2005) (Available Accessed 15-02-2007) https://www.iataonline.com/Store/Products/Product+Detail.htm?cs_id=9543%2D01&cs_catalog=Publications
25. Prahalad, C., Krishnan, M.: The Dynamic Synchonization of Strategy and Information Technology. MIT Sloan Management Review 43, 24–33 (2002)

Walk Navigation System Using Photographs for People with Dementia

Hiroki Kaminoyama[1,2], Takashi Matsuo[2,3], Fumio Hattori[1,2], Kenji Susami[2], Noriaki Kuwahara[2], and Shinji Abe[2]

[1] Ritsumeikan University, College of Information and Science and Engineering, Hattori lab
1-1-1 Nojiurahigashi Kusatsu, Shiga, Japan
{rs012029@se, fhattori@is}.ritsumei.ac.jp
[2] Advanced Telecommunications Reaserch Institute International Intelligence Robotics lab
2-2-2 Hikaridai Keihanna Science City Kyoto 619-0288 Japan
{susami,kuwahara,sabe}@atr.jp
[3] Kinki University, College of Literature Nakatani lab
3-4-1 Kowakae, Higashi-Osaka City, Japan 577-8502

Abstract. Dementia occurs in every country of the world, especially in developed nations. Since such spatially and cognitive impaired people lack the ability to grasp positional relationship between landmarks, we have developed a walk navigation system using photographs of landmarks to benefit them and their families. Our system provides a photograph of the next target landmark to the mobile devices of cognitive impaired people. It can also identify points where cognitive impaired are likely to get lost. Our system can also support such tasks as taking tickets by photographs and animations.

Keywords: photograph, spatial cognitive ability, landmark, mobile device.

1 Introduction

In developed countries, the number of people with dementia (PWD) has been increasing year by year. This social problem occurs in developed nations where people desire to contribute to society and improve the quality of their lives. Such afflicted individuals often cause their families trouble. IT technology that supports their lives is becoming more important.

These people need support based on the particular details and circumstances of their lives. In particular we focus on the problem of going out and leaving the house. They often go to hospitals or nursing facilities with caretakers. But such excursions are burdensome and time-consuming, so navigation support is in great demand. Going out alone using a navigation system would ease the care burden on families.
Such navigation systems as satellite navigation (Global Positioning System), 3D maps, and electronic maps provide effective support. A human navigation system for PWD, whose short-term memory, judgment, and spatial cognitive ability are declining, has not been developed yet in Japan.

M.J. Smith, G. Salvendy (Eds.): Human Interface, Part II, HCII 2007, LNCS 4558, pp. 1039–1049, 2007.

Therefore, we suggest a human navigation system that uses photographs for early-stage dementia sufferers. In our system, users do not need to recognize their position and direction because they follow scenes displayed by a photograph. In addition, users do not have to memorize the course before leaving the house, and they can use this system in a mobile field environment. Furthermore, we believe that using arrows and effects can increase smooth navigation. In the case of task support, using animation helps them take a train or bus.

2 Design of Walk Navigation System

2.1 Requirements of Navigation System

Many problems exist when navigating cognitive impaired people. The following navigation requirements must be satisfied:

1) System doesn't require good spatial skills
   Based on recent studies, cognitive impaired people do not high spatial skills. In this paper, spatial skill denotes such cognitive ability as a sense of place, a grasp of approximate position, and recognizing a landmark during movement. The system should not expect users to organize space perception from maps or text instructions.
2) System doesn't rely on user memory
   Since cognitive impaired people have difficulty relying on their memory, they are prone to get lost even in familiar territory. Reading maps and looking at photographs before leaving the home is meaningless. The system must be used outdoors.
3) System does not depend on sentence direction
   PWD cannot understand text instructions. The system should direct them by a method that easily images a route.

2.2 System Overview

The Walk Navigation System shows photographs on a course at proper intervals with a mobile device. Users follow displayed scenes and reach a particular position, and then they press a button to advance to the next instruction and receive a photograph of a new landmark or a location to head toward. The system shows arrows at corners and effectively uses photographs to inform users of the course. The system also uses photographs in the task navigation scheme and exploits animation and texts to naturally advance to tasks. It avoids positioning technology due to unreliability in built-up environments and to maintain simple applications.

2.3 Walk Navigation System

Since users follow scenes displayed by photos, the system requires pictures taken at specific places to convey the course. We considered that users would understand the course by following photographed positions.

System shows a photograph at:

1) starting and end points;
2) corners to convey orientation;
3) such places as intersections to avoid confusion even if the user goes straight.

However only showing photos to create user recognition of a course is hard. We decided the following rules to convey a course by photographs:

1) When showing a photograph, the system tempts a user to slowly go forward by visual effect
2) Arrows clearly indicate course
3) Frames enclose landmarks

Fig. 1. Examples of Walk Navigation

2.4 Task Navigation System

In many cases, users have to walk and take public transportation. But since PWD have difficulty taking buses, trains, and other vehicles alone, we propose a task support system that uses animation, text, and photographs.

1) The system clearly displays a photograph of task objects. A photograph of an operational object helps users understand, for example, ticket machines and buttons on buses.
2) The system explains tasks. Photographs, animation, and text instruction allow users to visually imagine tasks.

3 Evaluation of Effects of Spatial Cognitive Ability

We experimented using the walk navigation system to check the correlation of spatial cognitive ability. We measured movement times and the number of times the device was referred to. Then we examined the relation between spatial skill and movement time.

Fig. 2. Photograph on left shows example of getting off a bus by pressing a button and right shows ticket machine

Our system was implemented in Flash and ran under a Windows Pocket PC on a HP iPAQ. This version does not use text instructions and has a forward/backward button at the display's bottom.

Fig. 3. Walk navigation system displayed on a mobile phone (left) and PDA (right)

3.1 Method

Observation
We compared the following three methods:

1) Our system
2) System without arrows and effects
3) Paper map (A4)

We measured the following by analyzing participant behavior:

1) Time
2) Number of times they referred to PDA/maps
3) Times referred they referred to PDA/maps
4) Number of times participants got lost

Participants

We recruited 30 healthy adult women/men aged 20 to 49 as participants who took a spatial skills test before the experiment. We grouped them into three navigation sets (our system, our system without arrows, and maps) and each group's score was equal. No participant knew about the PDA device and or our navigation system.

Course

Supposing a typical trip, we experimented in two areas. One is from station to bus stop, and the other is from bus stop to another bus stop. Each course was approximately a simple distance of 200 meters.

Procedure

We briefly explained PDA usage to those using it and then summarized the experiment for all participants. We did not show maps to participants until the experiment started. After receiving a map, participants were given starting points and destinations.

An observer accompanied participants during the experiment to record their behavior by digital video camera, walking behind them to avoid giving hints about the course. When a participant asked about the course, the observer answered, and observer counted it as a mistake. If participants got lost, observers hailed them and returned to a suitable position.

3.2 Results

Table 1 shows the observation results.

Table 1. Observation results: lost times means total number of errors; others are averaged

	Our system	System without arrows	Maps
Time (sec)	567.1	679.7	654.1
Reference number	66.5	75.6	35.1
Total reference time (sec)	233	237.2	173.9
Lost times	3	6	6

Effect of photographs

From the map results in which a correlation coefficient between movement time and spatial skill score is 0.6, a highly spatially skilled person can move faster than a lowly spatially skilled person. There is no correlation in walk navigation system (correlation coefficient is nearly 0) against a map. Using arrows and visual effects reduces the variance for movement time. By the above, the walk navigation system is effective for spatially impaired people.

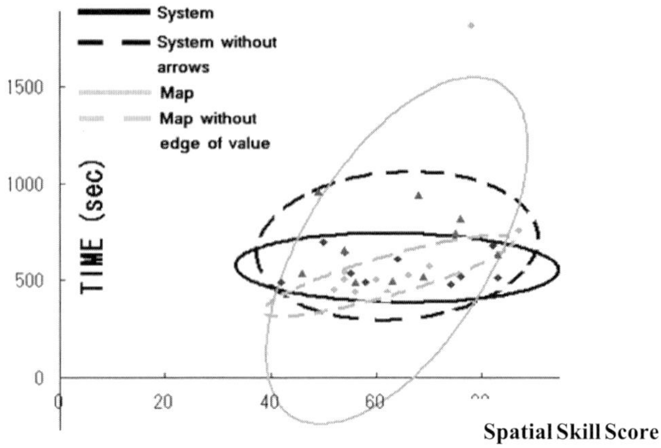

Fig. 4. Correlation between spatial skill and measured time. Graph shows 90% Probability ellipses for each method. High points of spatial skill mean low spatial cognitive ability.

Referring to device/maps
The number of times a PDA/map is referenced reflects the ease with which a course is understood. Results show map users did not read maps less than PDA users. Understanding a map is not easy due to the amount of confusing information, but if users can understand it, they do not need to refer to it often. Some participants actually only glanced at the map a few times, but a few participants repeatedly consulted the map. On the other hand, in our system, users tended to look at the device each time they turned at a corner and when operating it, and so on. This increased the number of references, but using arrows and effects can reduce a few seconds of referring.

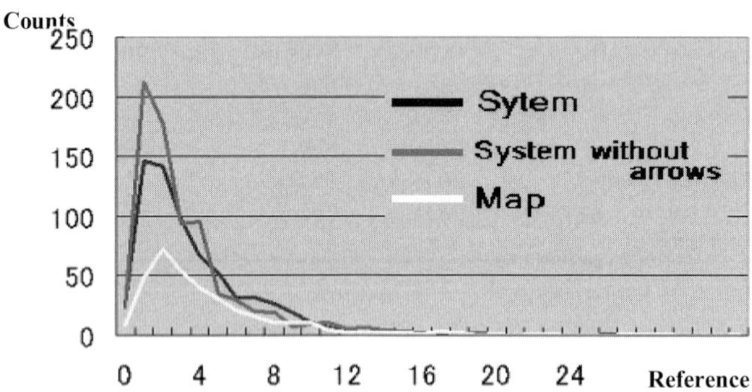

Fig. 5. Number of references participants made. Highest peaked line is system without arrows. Line in between two lines is our system. Lowest line is map.

3.3 Conclusion

We described an experiment of our walk navigation system concerning spatial cognitive ability in this section. We compared three methods that investigated a photograph's effectiveness for human navigation. The results show that the walk navigation system is effective for spatial cognitive impaired people, so PWD will also benefit. Moreover, it can teach a course using arrows and effects, but users have to constantly watch displays or scenes. Thus, users look at the device more than maps. This can be dangerous at crossroads and on stairs.

4 Evaluation of Memory and Landmarks for Navigation Aid

We suppose a situation where PWD use our Walk Navigation System in places they have visited before, not new places. In such places, landmarks become more important for remembering the course, so verifying our system as an alternative method for memorized landmarks is essential for our system. We examined the differences between movement by memory and movement by walk navigation system using difficult paths.

Specifications differences exist between the previous and current versions. First, our system was implemented in Flash and ran under a mobile phone on a W43H. Second, we explained each photo by adding such text instructions as "Go straight," or "Turn left here and go that way."

4.1 Method

Observations

We observed the following four points:

1) movement time from starting point to destination
2) mistake rate
3) landmark rate
4) our system rate

We asked participants for the decisive factor used for selecting the course at the 29 corners during the experiments. When they referred to photos and chose a course or looked at their mobile phone at the corner, we counted it as (4). When they used a building or something as a landmark, we counted it as (3).

Participants
All participants were university students who weren't very familiar with the place. We didn't tell them the experiment details not to remember the course. Ten proceeded to the destination by memory (Memory Group), and the other ten participants used our system (System Group). During the experiment, we tested participant spatial skill scores[9] by questionnaire.

Procedure
Participants made a round trip of the route, which included 29 corners and a distance that normally requires about 22 minutes on foot. On the outward trip, an observer accompanied participants to the starting point. After that, the spatial skills of the participants were tested, and the observer gave detailed explanations of the experiment. System Group participants received instructions for using the mobile

phones and moved to the destination. Memory Group moved to the destination relying on memory. After the experiment, the observer sought feedback of the walk navigation system and suggestions for refinements.

4.2 Results

Table 2 shows the observation results.

A 5% level of statistically significant difference was found in the rate of mistakes. Other data did not have it. There were individual differences for each data category.

Table 2. Observation results

	Spatial skill scores	Time	Mistake rates	Landmark rates	Our system rates	Both methods
Memory	67.2	24:01	27.50%	40.40%		
System	63.2	23:31	3.60%	26.40%	44.30%	1.20%

Interviews

The participants were asked to give feedback on the experiment. 80% believed that it was useful, but the others did not. The following comments are impressions of our system.

- It was easy to use and better than maps.
- Photographs should be replaced automatically.
- Some photos were too indistinct. There were also some differences between the photos and the actual scenes.
- I wish I knew the distance between each photo.

4.3 Study and Conclusion

Comparing the Memory and System Groups, there was a big difference in the residential areas. Pedestrians often get lost easily in residential areas due to a lack of landmarks. In short, the role photographs play in navigation is equally or more important than landmarks, and they are useful in areas of similar scenery like residential areas. Forgetful people have difficulty remembering landmarks, so our system is effective for people with declining memory.

From interviews and participant remarks, when participants made a mistake, they didn't look at the mobile phone. Such mistakes reflect the overconfidence of people who have high spatial skills scores. Of course, mistakes were made even by participants who checked their mobile phones. Such errors were caused by indistinct photographs or when our system didn't show photographs at proper intervals. Even without landmarks, users recognized the course by photographs; but if a photograph didn't have any characteristics at all, it was hopeless.

5 Task Navigation System Experiment

Persons suffering from dementia must accomplish many tasks when taking public transportation. We experimented to determine if task support could support such situations. All specifications are identical to Section 3.

5.1 Method

Observations
We compared the following two methods:

(1) our system (with task support)
(2) just showing a goal (without task support)
Participants were examined following task contents based on real life.
(a) In the bus:

i) get a numbered ticket
ii) push a button (buzzer) to get off
iii) insert coins and a numbered ticket

(b) Getting a ticket from a machine:

iv) insert coins
v) push the child button
vi) push the 120-yen button
vii) take a ticket and pass through the gate

Participants
Assuming they have similar cognitive ability levels as early-stage PWD, we recruited 20 boys and girls aged 8 to 9 and randomly divided them into two groups: with and without task support. None knew about either PDA or this navigation system.

Procedure
In this experiment, the children moved toward a destination with a PDA. They walked on the course and proceeded to the task part without interruption. An observer and parents followed them for protection and to record their actions.
Before the experiment, we explained how to use the PDA. We did not explain the task support system or the destination. After beginning the experiment, their parents and observers did not explain anything. When they asked a question, the observer answered it.

5.2 Results

Figure 6 shows the number of mistakes in this experiment. The task support group made six mistakes, but the without task group made 32 mistakes. The without task support groups made mistakes by asking passersby or their parents about the task. The task support group made several mistakes including forgetting their task during waiting a bus or failing to notice the task support system, but most understood their task.

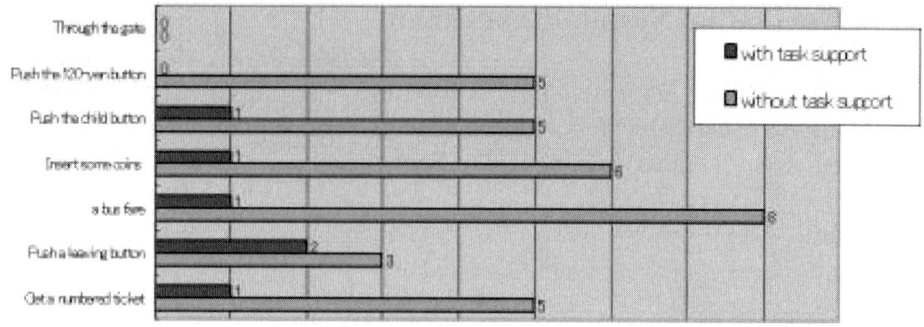

Fig. 6. Result of task support experiment; graph shows number of mistakes for each group

5.3 Summary

We described the task navigation system experiments in this section. Third grade children examined the task support navigation system. Most children with the task support system made a few mistakes in all parts, but children without task support made many mistakes. The task navigation system using animation and photographs is effective.

6 Conclusion

In this paper we described a human navigation system using photographs for early-stage PWD. Such individuals often go out with their caretaker to nursing facilities but they can't go alone because they have trouble understanding their location, direction, and distance from maps or memory. They need not only walk support but also task support for taking public transportation. We therefore developed Walk Navigation System, in which users do not have to be conscious of their position and direction because arrows and effects lead them to their destination. In walk navigation experiments with our system, spatially impaired people moved with large time reduction required to navigate routes as well as healthy people. Our system is effective in places without landmarks or for people with declining memories. However, uncharacteristic photos or places where scenes are similar such as intersections confuse users who easily get lost. In task support navigation, each task is carefully explained by animation, photographs, and text instruction. Our study has shown that the task support system is probably effective for children and PWD too.

7 Future Work

We have many problems based on the outcomes of experiments and interviews. In the future we will try the following:

1) system verification for people with dementia
2) improving the method of showing photos
   People with dementia get confused if they can't understand their direction when getting off a bus in complicated crossings. Text instruction is basically hopeless,

so we need more investigation of different methods for providing information from photographs.

3) improving usability
Comparing real scenes with photographs is demanding and dangerous in busy streets, on stairs, and so on. When they get lost, the system must re-orient them, but it can't do that yet.

4) develop an authoring tool
An authoring tool, which easily edits photographs taken by caretakers and volunteers to automatically create contents, must be developed.

Acknowledgements

We would like to thank all the people who participated in the experiments. Also we would like to thank the ATR Robotics and Communication Laboratories members for their detailed comments, suggestions, and constant support. This work was partially funded by The National Institute of Information and Communications Technology (NICT).

References

1. Goodman, J., Gray, P., Khammampad, K., Brewster, S.: Using Landmarks to Support Older People in Navigation. In: Mobile HCI2004 (2004)
2. Bosman, S., Groenendaal, B., Findlater, J.W., Visser, T., de Graaf, M., Markopoulos, P.: GentleGuide: An exploration of haptic output for indoors pedestrian guidance. http://www.industrialdesign.tue.nl/portfoliosAndPublications/publications/markopoulos/BossmanEtAl_GentleGuide.pdf
3. Bradley, N.A., Dunlop, M.D.: Understanding contextual interactions to design navigational context-aware applications, http://www.cis.strath.ac.uk/~mdd/research/publications/02bradleydunlopa.pdf
4. Liu, A.L., Hile, H., Kautz, H., Borriello, G., Brown, P.A., Harniss, M., Johnson, K.: Indoor Wayfinding: Developing a Functional Interface for Individual with Cognitive Impairments. In: Proceedings of Computers & Accessibility, ASSETS 2006, Portland, Oregon (October 2006)
5. Spatio-Temporal Association with Multiple Photographs (STAMP) http://home.csis.u-tokyo.ac.jp/~tanaka/stamp/stamp.html
6. Tanaka1, H., Arikawai, M., Shibasaki, R.: A Pseudo-3D Association of Photographs by using Spatial Hyperlinks, http://home.csis.u-tokyo.ac.jp/ ~tanaka/database_tod17.pdf
7. Miyake, S., Kusunoki, H., Mitani, T., Okabe, K., Torigoe, H., Yokota, K.: A tourist guidance system using the simple 3D functions, http://www.ieice.org/iss/de/DEWS/proc/2003/papers/4-P/4-P-02.pdf
8. Murakoshi, S.: Houkou onti no nazo ga wakaru hon (in Japanese) Shueis ha, p. 56 (2003)

Rapid and Precise Mobile Data Processing for Fire Brigades and Rescue Services (SAFeR/GÜTER/SHARE)

Rainer Koch, Rüdiger Harnasch, Bo-Sik Lee, and Jens Pottebaum

University Paderborn, Pohlweg 47-49,
D-33098 Paderborn, Germany
{r.koch,r.harnasch,lee,pottebaum}@cik.uni-paderborn.de

Abstract. The range of tasks for Search And Rescue (SAR) services is very comprehensive and requires extensive experience in addition to detailed knowledge. Actual and extended incident data, tactical information and technical knowledge are significant issues in this field. Within research projects information systems are designed to provide relevant information with the required/desired detail depth to the individual user groups. Mobile information processing instruments in combination with an internet based access to a modular knowledge database are predestined to improve the handling of emergencies as well as the data management for operation controls and fire-fighting forces. Emerging data transfer technologies like UMTS as well as WLAN are applied in order to access the database independent of the location. Semantic methods provide new possibilities to interlink information which accelerates the data preparation and retrieval. This paper describes the information and knowledge availability and adaptation.

1 Introduction

Search And Rescue (SAR) is a field of great potential demanding new initiatives in European security research [1]. The thematic field is characterised by well-established structures concerning the communication methods, the workflow of and the information and material support for the acting command levels.

SAR services span a big and comprehensive set of tasks. As they cover a lot of aspects this paper is focussed on fire brigades, exemplary the basic situation in Germany is analysed. Fire brigade and emergency medical service are organised on urban level and are under the same supervision and control. Their main task is fire fighting and danger averting at the front line. But directly coupled with that is the treatment of human beings. Another important aspect in this area is that not only professional, but also volunteer fire fighters are responsible for saving humans.

In an operation a lot of details are exchanged by the half-duplex radio transmission where use of a push-to-talk switch is required to communicate [2]. A single channel is used for transmit and receive. That means only one participant can speak while all others have to listen. The technique basically causes the problems in the area of emergency operations:

M.J. Smith, G. Salvendy (Eds.): Human Interface, Part II, HCII 2007, LNCS 4558, pp. 1050–1059, 2007.

- The distribution of information is undirected. The receiver has to be clearly mentioned and has to recognize that he is addressed. A direct call is not possible.
- The priority cannot be defined ('First come, first served'). All messages have the same talk priority.
- The ambient noise of an emergency restricts the communication. Sometimes one message has to be repeated several times. But misunderstands are still not impossible.
- The half-duplex radio transmission is specified for verbal communication. The capability to transfer data can be neglected.

Radio trunked services cannot solve all the problems mentioned above. Limited transfer rates reduce the data transmission possibilities to only small data volumes which are not enough to provide e. g. complex descriptions. The restrictions mentioned above cannot be solved by radio technologies which offer only low transfer rates. One solution can be found by wireless service provider or ad-hoc networks.

One of the main challenges is referred to the information availability. Even if high data rates are available information provider for emergency support are still missing. Even today a lot of information is paper based. This leads to the problems of actuality and availability on site. The research projects SAFeR and GUETER are focused on this problem in order to enhance the information support in order to assist the tactical considerations and actions. Tactical information like operational recommendations or technical details about complex objects can improve the course of an operation.

Another challenge within an emergency case is the data communication between the different command levels. Especially in disasters the management of different organisations and a high number of forces require an efficient data exchange. This leads to the need of precise and rapid data processing to realise real-time analysis, data control and specific distribution. The EU project SHARE is concerned with the data communication and computer aided distribution and semantic linkage.

2 Strategic and Process Supporting Information for Fire Brigades and Rescue Service (SAFeR)

The spectrum of different technology domains leads to various strategies with different options. The selection of the best-practice-method by combination of different strategic elements depends on wide spread of experience, which is one essential skill of the officer in charge. But in some cases special knowledge is needed but not necessarily available. On this basis the necessity for new data communication technologies and methods to provide relevant information on site is evident.

Within the publicly funded SAFeR[1] research project, mobile data processing facilities are used to provide general data and also extended information and/or technical knowledge to the task force in order to support and improve the processing of emergencies and rescue cases.

---

[1] Funded by the Federal Ministry of Education and Research (BMBF), 2002-2004.

Fig. 1. Information types with an exemplary representation by SAFeR

The application of a specialized information system for fire brigades and rescue services covers different domains to make the mentioned technical and strategic information available directly at the field.

The main focus of the SAFeR research project is the design of an information system that offers an "actual" support to all fire-fighting forces for the comprehensive spectrum of applications.

Hence the intention of the SAFeR project is the transformation of different knowledge areas into an intuitive and homogeneous system. Therefore different structuring, access and representation options were analyzed and combined. On the basis of various technical objects (for example motor vehicles, trains, agricultural machineries) an exemplary representation format was developed to represent the essential aspects in a compressed format in order to make the main facts fast and intuitively available (Fig. 1).

On the technical level mobile information processing instruments in combination with an internet based access to a modular knowledge database are used to improve the information availability within emergencies of fire brigades and rescue services [3]. Also the data providing for operation controls and fire-fighting forces are taken into account. The SAFeR-system supports the different end-users by customized graphical user interfaces and specific information and services to increase the usability.

Especially new data transfer technologies like UMTS (Universal Mobile Telecommunications Systems) and WLAN are applied in order to access the knowledgebase independent of time and location. Pocket-PCs, Smart-Phones as well as Notebooks and/or Tablet-PCs for the operations control on site are used as mobile clients to realize the information access in combination with the mentioned radio data services. The information are customized to the information demand respectively to the output device.

Depending on the current problem specific data with different levels of information-details are generated for the user. Forces which are engaged in rescue or fire fighting action need handy devices which do not affect their specific task. In this respect the information need depends on the specific function and is related to the size of the output device [5].

The core of the SAFeR-system is a stationary server, which provides recommendations for efficient tactics as well as special services like meteorological data, basic information for toxicities and hazardous substances, recommended special recourses/equipments with specific localization and handling. The recommendations are given by the different fire brigades with their specialists itself. The combination of internet technologies and a stationary server enables the SAFeR-system to give the fire forces the opportunity to publish their experiences and tactical recommendations via internet which can be useful in similar cases. In addition the fire brigades can describe their special equipment and offer the possibility to supply support by special resources.

The SAFeR system provides a specific information input service which can be used by the operation 'experts' to publish their individual experiences and recommendation for all kinds of emergency cases. The user can describe his descriptions in a graphical user interface. The system provides a list of existing progress reports. The user chooses a similar description as a sample. A copy of text passages which are editable can complement the new content and assures a uniform phrasing of similar contents.

Especially in the field of the rescue of injured persons alternative and complementary methods exist, their coordination and strategic use requires a wide knowledge spectrum. Technological innovations for example new materials and chemicals, changed structures of vehicles, automatic protective devices like Air-Bags, bring new demands to the fire brigades and rescue services. For the handling of the tasks that include the rescue of human lives as well as environmental protection and classical fire fighting an increasing number of devices and mission strategies are available - depending on kind and extent of the emergency. However, the training of the personnel, in particular of the volunteer members, up to now cannot be carried out in the extent the comprehensive missions and tasks require. From this a great requirement exists for technologies, to provide specialized/detailed information not only for operation control but especially for the fire-fighting forces on site.

The development of the system was done in a close co-operation with the end user. The end user has to get convinced of the advantages of a system. It is a recommendation to integrate the future user to achieve a high acceptance. Otherwise the user is lost and will not describe any requirements or will take part in the validation. Within the SAFeR project different prototypes are built to capture the needs and requirements precisely. In the beginning of the research power point presentations and static html mock-ups with 'faked' user interfaces and functions are built to discuss ideas and evaluate first visual approaches. On this basis the users gave detailed feedback of the realization status in early phases of the project. This was used to improve the system step by step and prevents expansive development failures and ensures the project schedule.

3 Strategic and Process Supporting Information for Fire Brigade for Accidents with Hazardous Materials (GÜTER)

Emergency cases with toxicities and hazardous substances often cause extreme human and environmental damages. Especially accidents in the field of transportation services are characterized by special circumstances and challenges which lead to the need of various information to decrease the complexity of the emergency case.

The intention of the research project GÜTER[2] is the improvement of an automatic alarm system in case of accidents and data communication on site [4]. One problem in transportation is the load itself: The arrangement of different toxicities e.g. on a truck is sometimes done with neglection of the chemical reaction in case of leakage which could occur as a result of an accident. As a solution for prevention the loading process has to be supported by giving the operator (e. g. warehouseman) specific advices and warnings according to a valid arrangement of chemicals. This can prevent the amount of damage in an emergency case.

The information of the chemical load is an important advice for the fire brigades for planning the tactics and specialized resources (equipment and personnel) [7]. The reaction time can be decreased by an automatic release of an alert. Actually there often is a significant time delay before the incident is reported and fire brigades are getting informed. In some cases the position of the incident is described inaccurate so that a lot of time is lost by search. Another problem is caused by missing load documents on site which causes additional loss of time. The precise identification of the chemical mix is often impossible. The spectrum of chemical compounds is widespread so that a fast and reliable determination is not possible.

The solution of the mentioned problems depends on the availability of all necessary data for decision making within an incident. To lower the described difficulties new methods for prevention and protection by improved data communication are researched. The aim of GÜTER is the combination of RFID- and GPS-Technologies with mobile information management to broadcast the necessary information. On this basis the command control is getting informed quicker, the fire fighters and rescue services have precise data for planning the measures as well as a monitoring and control of the load is possible [8].

In this area a lot of problems exist in all phases of transportation. The loading of a truck depends nowadays mainly on the optimization of the route. This means a high risk of a combination of goods, which are separately harmless but in case of a chemical reaction very dangerous. In this phase the project prevents a set of chemicals which is in case of an accident reactive or dangerous by mixture. The usage of RFID can support a safe load. Logistic operators get information for the loading process. A RFID reader realizes the specifications of the cargo. The system analyses the load and gives instructions to the logistic operator, which chemical is permitted or not.

During the transportation the route is controlled by the system. If there are extraordinary breaks or differences between the planned route and the actual position, the system gives at first an alarm for the logistic center. In case of an emergency case the fire brigade gets alarmed. The system provides the position and the loading data. On this basis the fire brigade can plan the resources and special equipment.

[2] Funded by the Federal Ministry of Education and Research (BMBF), 2005-2007.

During the approach of the fire and rescue forces the GÜTER system provides specific routing information for navigation and advices for the initial actions. The information about the load contains the specific danger and recommendation for counter measures. On this basis the fire forces can plan their proceeding and after care precisely.

The end-users can access the needed information very quick e. g. for getting specific maps and using meteorological services. The technical base is similar to the SAFeR-System but realized as a push-system, so that the end-users are getting informed automatically. In addition RFID-technologies are used to control and register the load of e.g. a truck. These data is stored in a server which can be accessed within an incident by the command control. GPS data services are implemented to control the transportation and to detect irregularities in order to alarm and route the fire brigades automatically.

The GÜTER system is advancement in a special area. By its design following a push system approach it informs all necessary organizations in all phases of transportation by specific data. By contrast the SAFeR is realized as a pull system which makes information available but the user has to take the initiative.

4 Mobile Support for Rescue Forces, Integrating Multiple Modes of Interaction (SHARE)

The acronym SHARE represents a project which is funded by the European Commission. Its goal is to develop an integrative communication platform for command levels in disaster management situations. It tries a compromise between innovation and utilization of established and working structure. Methods and instruments currently used are applied as far as possible and reasonable but enhanced by introducing new technology [6]. The final aim is an integrated system representing a full end-to-end solution with a valuable mobile work environment that is intelligent, robust and easy-to-use. It will connect all relevant operation data and make it easily accessible to specified user groups. In detail SHARE approaches

- a multimodal user interface,
- bi-directional communication using new communication technologies,
- robust speech and image processing under extreme conditions,
- interactive digital maps with linked multimedia information and
- dynamic resource management

In addition the processes of structuring required information using situation-dependent ontologies and semantic indexing of multimedia data are implemented. In this context the operation control needs support in coordinating all deployed formations, units and vehicles. The project allocates a significant amount of effort to involve the end users and to build up a system which can be integrated to the current workflow giving great benefit without changing the established and proved boundary parameters. Intensive co-operation is especially conducted for Requirements Engineering and System Evaluation.

The acquisition of end user requirements is an iterative and difficult process. The end users normally can not imagine which technical and conceptional options exist, how applications can be implemented, to what extent benefits can be achieved and to describe their needs precisely. In this context rapid prototyping methods can fasten the collaboration process between end users, researchers and developers to express detailed requirements and present solutions by visualizations of elementary functions. For the SHARE project clear boundaries for the system have to be set: a) The application is limited to the command level subset of firefighting forces. b) Because of current communication channel limitations the system is developed to support SAR forces on-site, the connection to the backend operation control center is not affected. c) The project focuses on firefighting and emergency medical services. Police and rescue organizations are not dealt with.

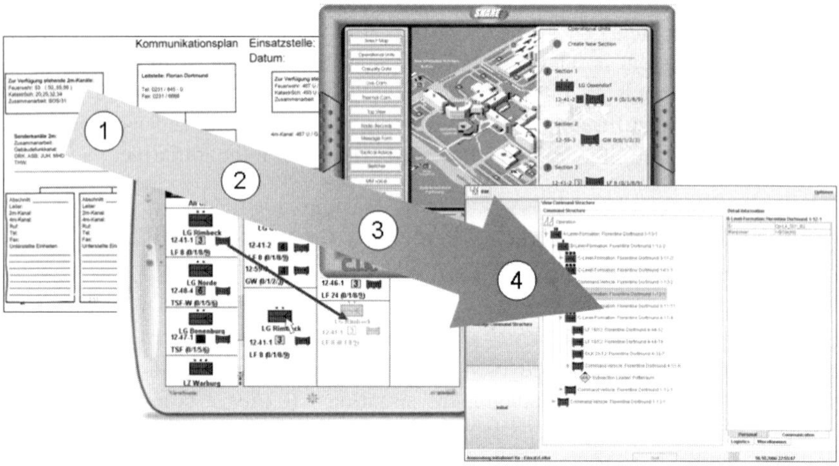

Fig. 2. Iterative phases of rapid prototyping approach

The limitation to on-site information on the one hand and to command levels as users on the other hand implies the different need for information support concerning the role of users. Therefore every action in the system underlies certain constraints related to the adopted role. In this context three main application fields were identified. Essential for operation management is a stable and reliable communication infrastructure. This includes speech communication (currently conducted by analogue radio) as well as paper-based forms. The latter is needed in areas where noise reduction is important, e.g. the command post. A third channel can be established by the introduction of video processing. SHARE includes the transmission of video streams from lower, operational command levels to higher instances; in addition thermal videos recorded by firefighters are analyzed on server-side and distributed with extended information to all responsible commanders. For the information management a smart map was implemented. It presents the information of formerly used maps and plans, extended by new features like 3D models. Additionally it integrates operation management capabilities and thereby gets interactive. The correlation of map annotations and resource hierarchy is realized by the application of

Semantic Web and Reasoning technologies. The innovation in all of these application fields is highly based on the knowledge base in the back-end.

The requirements engineering process was initialized by joint training sessions of firefighters and scientists resp. developers. With regard to Fig.2 the following phases were conducted: 1) All design issues were based on the currently used and thereby proved instruments (maps, plans, forms, structural diagrams, paper-based tools, etc.). 2) To get a requirements skeleton a Flash demonstrator was rapidly built showing representative characteristics of the GUI with basic functionality. 3) The production of the demonstrator was iterated on comment basis. Thereby the efficiency, profundity as well as the user motivation could be enhanced. 4) After the first project year a first integrated prototype based on draft end user requirements was presented.

An important effect of this iterative process is a very clear understanding about user needs and recommendations. The project started with a clear focus on the map application as the core component. Within the development steps it became obvious that at the critical stage of a starting operation the management of the command and communication structure is the main task for command levels. Therefore SHARE was extended by an Interactive Resource Management (IRM) tool. This enables the structural configuration of resources and their deployment based on guidelines and directives. Resources are defined as instances in an ontology underlying logical restrictions modeled by Description Logics (see [9] for more details). To avoid inconsistencies within this structure as well as between this structure and correlating annotations on top of the map every action is checked by reasoning procedures. IRM as well as text message, voice and video communication tools, map and retrieval applications are implemented following a service oriented approach. All services rely on the common knowledge base which can be addressed by a specific Ontology Data Service (see [10]). The ODS is implemented

This architecture was object to a first evaluation period already. The evaluation is splitted into a short phase (rapid evaluation/feedback), and an intensive phase with field trials at the end of the project (detailed validation). For the first set of tests a combination of two methods was selected to be most efficient: Expert oriented Heuristic Evaluation and user oriented Thinking Aloud. Nielsen's Heuristic evaluation [11] involves having a set of usability experts identify usability problems without carrying out a user test. When grading the interface, the expert checks it for compliance with a set of pre-approved usability heuristics. For think aloud testing task scenarios are provided to later users of the system. The test persons are requested to solve the task at the system and to speak out loud during their work what they think and do during the entire scenario

When selecting Nielsen's method the critical aspects emphasized by [12] mainly focusing on the surroundings of investigations and the comparisons drawn were considered. The advantage is that one will nearly every time get some clear problems and possible solutions. In our evaluation we found a rate of 56% of problems mentioned both by Heuristic Evaluation and User Testing. Having found 121 usability issues both methods found around 75% of these problems. Thereby and by the fact that the nature of the found problems was partly complementary we conclude that the combination of these methods was an effective decision.

As an additional outcome the need the gap between the adaptation to well-known structure and training needs became visible. The intensity of training measures was decreased to include the intuitiveness of the user interface to the evaluation objectives. In the final evaluation it has to be explored to what extend this affected the overall results. Besides these positive aspects we have to mention some points which have to be dealt in another session: Qualitative aspects must be complemented by quantitative metrics for to conclude on the effectiveness and efficiency of the users. With regard to [12] the final evaluation must reach for a statement with respect to a "total" amount of issues. Therefore additional tests are needed, using another evaluation method with other participants (to avoid effects between the different measures), but in the same setting. This setting must incorporate real surroundings. The artificial environment in an educational building of the fire brigade has to be taken into account when concluding on the current results.

The goal of SHARE is to generate a functional prototype having full functionality emphasizing potential of innovative technologies in crisis management environments. Because of this prototypical status the formal documentation of end user requirements, the evaluation of the first prototype, the documentation of the final prototype and another final and comprehensive evaluation period are given high priority.

5 Conclusions

The research projects SHARE, SAFeR and GÜTER are designed as information systems that offer an "actual" support to all fire-fighting forces for a comprehensive spectrum of applications. Mobile information processing instruments in combination with an internet based access to a modular knowledge database are used to improve the handling of emergencies as well as the data management for operation controls and fire-fighting forces.

Lessons leanrt brought out from these projects with respect to the integration of end users are:

- Involve the user at an early stage to gather all requirements
- Prepare interviews which can be easily analysed
- Give them feedback of the proceeding
- Use images or mock-up (like flash animations) to collect precise information and wishes

The situation especially of an emergency case leads to the need of a reliable and intuitive information support. Therefore the user interfaces have to provide an easy access, e.g. by specific retrieval functions or push methods. Time is a serious problem in this application. A proven technology or system in this application area is a good sample for industrial software.

From the technical point of view the actual radio technology (half-duplex) used in Germany is related only to speech communication. Data transfer is not possible. But this problem cannot be solved by trunked radio technologies (e.g. TETRA) which is the future technology used in the security area. Bandwidth issues will raise needs in establishing alternative networks to enable the application of innovative technologies in well established and proven workflows.

References

1. Koch, R., Barth, U., Tetzner, D., Aschenbrenner, D.: Managerial and technical aspects of tunnel safety regarding normal and emergency mode. At: International Conference Tunnel safety and ventilation – New developments in tunnel safety, Graz (2002). In: Pischinger, R. (ed.): VKM-THD Mitteilungen der technischen Universität Graz, Heft/ vol. 80, pp. 155–162 (2002)
2. Koch, R., Harnasch, R., Lee, B.-S.: Strategic and process supporting information for fire department and rescue service (SAFeR). Jahresbericht 2003 Deutsches Zentrum für Luft- und Raumfahrt (Englische Ausgabe), Bonn (2003)
3. Koch, R., Harnasch, R., Lee, B.-S., Schulz, A.: SAFeR – Strategische und Ablaufunterstützende Einsatzinformationen für Feuerwehr und Rettungsdienst. vfdb-Jahresfachtagung, Baden-Baden Mai 2003, Fachsitzung 5, Tagungsband S. pp. 305–316 (2003)
4. Finger, M., Lee, B.-S., Harnasch, R.: Hilfe für Feuerwehr –Integrierte Kommunikationssysteme im Forschungsprojekt SAFeR. In: Gefährliche Ladung, Ausgabe 1/2005, K.O. Storck & Co. Verlag und Druckerei GmbH, Hamburg, pp. 28–30 (2005)
5. Koch, R., Harnasch, R., Lee, B.-S.: Gezieltes Informationsmanagement zur Führungsunterstützung im Einsatz. In: Brandschutz – Zeitschrift für das gesamte Feuerwehrwesen, für Rettungsdienst und Umweltschutz, Ausgabe 2/2005, Verlag W. Kohlhammer GmbH, Stuttgart, pp. 88–93 (2005)
6. Koch, R., Löffler, J., Lee, B.-S., Harnasch, R., Köhler, J.: Datenfunkgestütztes multimodales Informationsmanagement für die Einsatzleitung bei Großschadensereignissen. Sicherheitskommunikation für mobile Hilfsorganisationen, Alcatel SEL Stiftung für Kommunikationsforschung, Berlin, Stiftungsreihe (2005) ISSN 0932-156x
7. Koch, R., Lee, B.-S., Harnasch, R., Finger, M.: Können Telematiksysteme den Feuerwehreinsatz bei Gefahrgutunfällen optimieren? In: Tagungsband vfdb-Jahresfachtagung, Münster, pp. 307–326 (September 25-28, 2005)
8. Koch, R., Harnasch, R., Lee, B.-S.: Mobilfunkgestützte Informationsbereitstellung zur Führungsunterstützung von Feuerwehr und Rettungsdienst. Hochschulkolleg E-Government, Alcatel SEL Stiftung für Kommunikationsforschung, Berlin, Stiftungsreihe 70 (2005) ISSN 0932-156x
9. Pottebaum, J., Konstantopoulos, S., Koch, R., Paliouras, G.: SaR Resource Management based on Description Logics. In: Klann, M., Löffler, J. (eds.) Proceedings of Mobile Response 2007 (to appear in spring 2007)
10. Konstantopoulos, S., Paliouras, G., Chatzinotas, S.: SHARE-ODS: An Ontology Data Service for Search and Rescue Operations. In: Antoniou, G., Potamias, G., Spyropoulos, C., Plexousakis, D. (eds.) SETN 2006. LNCS (LNAI), vol. 3955, pp. 525–528. Springer, Heidelberg (2006)
11. Nielsen, J.: Usability Engineering. AP Professional, New York (1993)
12. Wayne, D., Gray, M.C.S.: Damaged Merchandise? A Review of Experiments That Compare Usability Evaluation Methods. Human-Computer Interaction 13, 203–261 (1998)

Health Risk Estimation of Dichlorobenzene Exposure Workers by Using Computer Models

Pei-Shan Liu[1], Ren-Ming Yang [1], Yin-Mei Chiung [2], and Chung-Yong Liu[3]

[1] Department of Microbiology, Soochow University, Taipei, Taiwan
psliu@scu.edu.tw
[2] Institute of Occupational Safety and Health, Institute, Taipei, Taiwan
[3] Industrial Technology Research Institute, Taiwan
liu@itri-tokyo.jp

Abstract. 1,4-Dichlorobenzene (DCB), widely used as a moth repellent and a space deodorant, is one kind of chlorobenzene that is classified by the International Agency for Research on Cancer among chemicals possibly carcinogenic to humans on the basis of a sufficient evidence for carcinogenicity to rodents. DCB has been found to cause renal tubular-cell adenocarcinomas in rats and hepatocellular carcinomas in mice but without their clastogenic and genotoxic effects. In this study, we investigated the health risk of DCB workers from three steps. Firstly, we investigated the cytotoxicity of 1,4-dichlorobenzene and its metabolite 2,5-dichlorophenol in human lymphoma Jurkat cells, human colon adenocarcinoma cells and human hepatocellular carcinoma cells. The LD50 of DCB were 1.34 mM, 5 mM and 2 mM in Human lymphoma Jurkat cells, human colon adenocarcinoma cells and human hepatocellular carcinoma cells respectively. DCB was found to cause apoptosis itself and reduced the staurosporine-induced apoptosis in all three cells. Secondly, works in four factories were investigated to examine their blood cells. Two items, their apoptosis rate and their anti-apoptosis rate, were investigated on the work's blood. In an investigation on Dec, 2004, we found a significant higher apoptosis rate of exposure workers comparing with non-exposure workers. The blood cells of workers were significantly resistant to the staurosporine-induced apoptosis. In the other three investigations in 2005, we found similar phenomenon that the blood cells of exposure workers had higher apoptosis rate and higher resistant rate to staurosporine-induced apoptosis. Thirdly, the DCB content in the urine and blood of workers were analyzed. Based on the low to high level of DCB content, the workers were grouped into 10 degrees. The correlation of the level of DCB and apoptosis rate and anti apoptosis rate was analyzed by Pearson Correlation Program. The apoptosis rate and anti-apoptosis rate of worker's blood were both significant negatively correlated with the DCB content in their urine and air. Based on this computer program analysis, we suggest that the exposure of DCB induced significant changes of the blood cell of exposure workers and need to be concerned.

Keywords: 1,4-dichlorobenzene, apoptosis, anti-apoptosis, staurosporine, Pearson correlation analysis.

M.J. Smith, G. Salvendy (Eds.): Human Interface, Part II, HCII 2007, LNCS 4558, pp. 1060–1066, 2007.

1 Introduction

1,4-*para*-DCB (*p*DCB) has been widely used as a deodorant and a moth repellent, and as an intermediate for dyestuff, fungicides, pharmaceuticals and industrial chemicals[ATSAR, 1993]. Based on the literature, the exposure of *p*DCB will cause some damage to workers, including the irritation of eye and air way when the environmental dosage above 50 ppm; serious pain will occur when the dosage above 160 ppm [Bomhard et al., 1988; ACGIH, 1986]. In animal studies, DCB could induce hepatocarcinogenicity including hepatocellular carcinomas, hepatoblastomas [NTP, 1987; Giavini et al., 1986;]. The reports also mention kidney damages caused by *p*DCB [Umemura et al., 2000; Robbiano et al., 1999]. In vitro studies mentioned *p*DCB enhanced liver cell proliferation [Eldridge et al., 1990; 1992] and suppression of apoptosis induced by transforming growth factor β1[Nei et al., 1998]. When *p*DCB entered blood circulation into liver, it will be metabolized to 2,5-dichlorophenol (DCP). In this study, the cytotoxicity of *p*DCB and DCP on various human tissue cells were investigated firstly. Workers Investigation is the 2nd part of this study, we used computer program to distinguish the phenomenon related to the dosage of *p*DCB and DCP.

2 Materials and Methods

2.1 Chemicals

*p*DCB, *p*DCP, digitonin, EGTA, thapsigargin, ionomycin, α-estradiol and β-estradiol were all obtained from Sigma Chemical Co. NaCl, KCl, and other salts were obtained from Merck Co. Fura-2 acetoxymethyl ester was obtained from Molecular Probes Co.

2.2 Cell Culture

All the cell lines obtained from ATCC (American Type Culture Collection). Human lung adenocarcinomas H1355 cells and human leukemia Jurkat cells were cultured in RPMI medium with 10 % fetal calf serum. Human hepatocellular HepG2 cells and human HeLa cells were cultured in a MEM medium included 10% fetal calf serum. Subculture was carried every 4-5 days. The confluent cells were collected to process the cytotoxicity assay.

2.3 Cytotoxicity Assay

The cultured cells were collected in Phosphate-buffered saline (NaCl,137mM ; KCl,3mM ; Na_2HPO_4,10mM ; KH_2PO_4,2mM) . The cells were treated with *p*DCB or DCP at various concentrations for 30 min. After washing the cells with saline for 3 times, the cells were stained with 0.4% trypan blue [Eibl et al., 2003]. The dead cells will be blue and the cells alive will be clear. The viability will be calculated as (total cells – blue cells)/ total cells. The apoptosis experiments were carried by using caspase 3-FITC (PhiphiLux-G_1D_2) apoptosis Kit [Packard et a., 1996]. The mitochondria membrane potential analysis were carried by ApoAlertTM Mitochondrial Membrane Sensor Kit (BD cat. PT3308-2) [Green, 1998].

2.2 Analysis by Computer Program

All the data were analyzed by SPSS. Paired t test is used to analyze the difference between groups. The correlation between the DCB content (air or urine) and cytotoxicity of workers was analyzed by Birariate Correlation with Pearson correlation coefficient.

3 Results and Discussion

The entrance of p-DCB into human body can be through inhalation, skin contact and it will be metabolized in liver after its entrance. Therefore, we used the cytotoxicity of pDCB in human lung adenocaricinoma H1355 cells, human hepatoma HepG2 cells, human leukemia Jurkat cells and human HeLa cells. In human lung adenocaricinoma H1355 cells, pDCB caused some damage and its LD50 was 1.02 mM. While its LD 50 was 1.42mM in human HeLa cells and was 1.02mM in human hepatoma HepG2 cells.

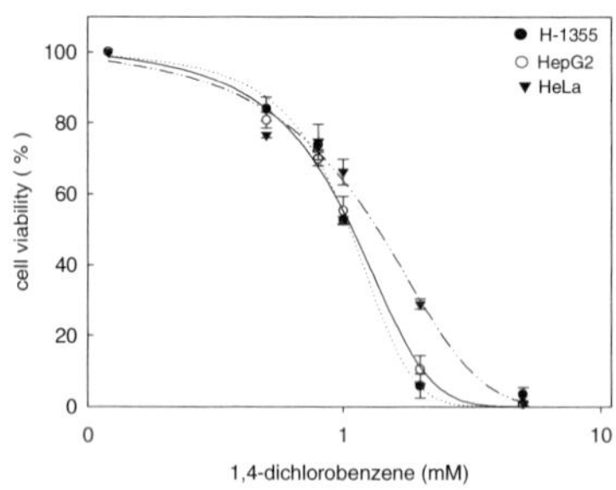

Fig. 1. Cytotoxicity of 1,4-dichlorobenzene in human lung adenocaricinoma H1355 cells, human hepatoma HepG2 cells, human HeLa cells

The cells were treated with 1,4-dichlorobenzene at various concentrations for 30 min at room temperature and the cell viability were observed under microscopy after the cells were stained by trypan blue.

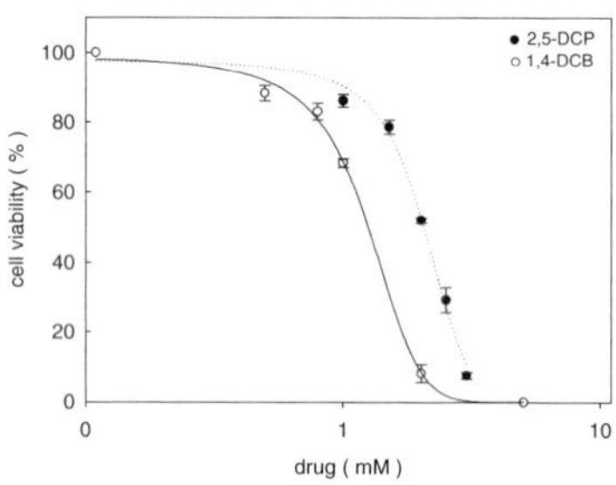

Fig. 2. Cytotoxicity comparison between 1,4-dichlorobenzene and its metabolite 2,5-dichlorophenol in human leukemia Jurkat cells

The cells were treated with 1,4-dichlorobenzene or 2,5-dichlorophenol at various concentrations for 30 min at room temperature and the cell viability were observed under microscopy after the cells were stained by trypan blue.

The cytotoxicity of pDCB and DCP was investigated in human leukemia Jurkat cells. The viability rate was 83.02 ± 2.45% (n=4) when the cells were incubated with 800 μM pDCB for 30 min. In the case of DCP, the viability rate will be will 86.13 ± 1.81%(n=6). The LD50 of pDCB and DCP was 1.18 and 2.06 mM, respectively. The acute toxicity of pDCB was higher than DCP. This data suggest DCB metabolizes into DCP will lower its acute toxicity.

The trypan blue staining is based on the selection of cellular leakage so its staining can not distinguish the apoptotic cells. The caspase 3 activity and PI staining were combined to analyze the cellular damage caused by pDCB. Figure 3 shows 500 μM pDCB caused an increase of cellular apoptosis at the concentration above 500 μM. Staurosporine caused a large increase of cellular apoptosis to represent as a positive control. Apoptosis have two main pathways: intrinsic pathway through mitochondria damage and extrinsic pathway through membrane Fas receptor. We firstly investigate the intrinsic pathway to analyze mitochondria membrane potential. We found that pDCB caused some cells loss mitochondria membrane potential. 500 μM pDCB caused 30 % cells loss mitochondria membrane potential (Fig 4). This cell survival rate is similar as the apoptosis rate induced by 500 μM pDCB analyzed by apoptosis kit that 500 μM pDCB induced about 30% apoptosis rate. Thus, we suggest pDCB caused cell damage, mainly apoptosis through mitochondria pathway. Some literature mentioned the apoptosis suppression by pDCB [Neil et al., 1998], we investigate the effects of pDCB on the apoptosis induced by staurosporine. The data showed pDCB not only induced apoptosis by itself but also suppressed apoptosis induced by staurosporine (data not shown).

In the second part, we investigated on workers in pDCB factory. The apoptosis rate and the influence on staurosporine-induced apoptosis in workers and non-workers were investigated. In non-workers shown in figure 5 panel A, staurosporine significantly elevated the level of necrosis (A1), late apoptosis (A2), and early apoptosis (A4) and suppressed the survival rate (A3). That illustrates human normal lymphocytes were sensitive to staurosporine to lead apoptosis. In workers, staurosporine only elevated the level of necrosis (Fig, 5, panel B, A1) and caused no further late and early apoptosis. The survival rate were similar in the lymphocytes treated with or without staurosporine. In another word, the lymphocyte of worker in pDCB factory will suppress the staurosporine-induced apoptosis or resistant to staurosporine. In order to eliminate the individual difference, the lymphocyte sample of individual person apoptosis rate in the absence and presence of staurosporine were carried (Figure 6). The purpose of this rating was to find the difference between workers and non-workers. We predict that staurosporine will lead apoptosis so the rate will exceed 1 in late (A2) and early apoptosis (A4). In non-workers, the rate will as higher than 3. The ratio will exceed 6 in the case of total apoptosis. In workers, the rate will be significant lower than non-workers. This data show that workers have lost the sensitivity to staurosporine.

The correlation between pDCB content and sensitivity to staurosporine was analyzed. In workers, urine pDCB content was negatively correlated with the staurosporine sensitivity (-0.488). pDCB will metabolized into DCP. Urine DCP content was negatively correlated with the staurosporine sensitivity (-0.461) in workers. However, this negative correlation was not found in non-workers (0.194 for DCB and 0.235 for DCP.

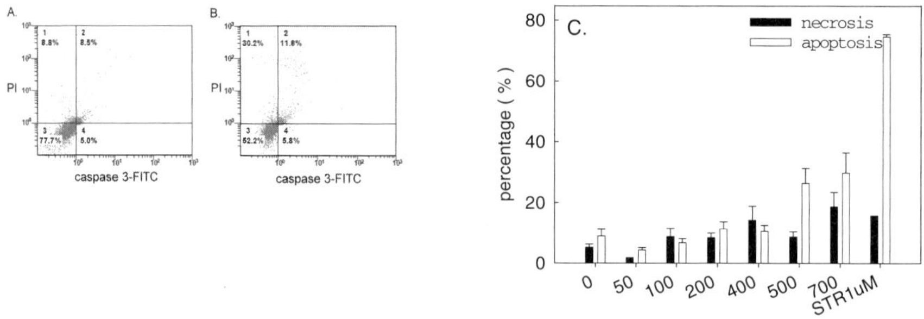

The cells were treated with drugs for 24 hr and stained with PI and caspase 3-FITC. Panel A: 1% DMSO; Panel B: 700 μM 1,4-dichlorobenzene; Panel C shows the apoptosis and necrosis percentage in the cells treated with 1,4-dichlorobenzene at various concentration and 1 μM staurosporine.

Fig. 3. The effects of 1,4-dichlorobenzene on apoptosis in human Jurkat cells

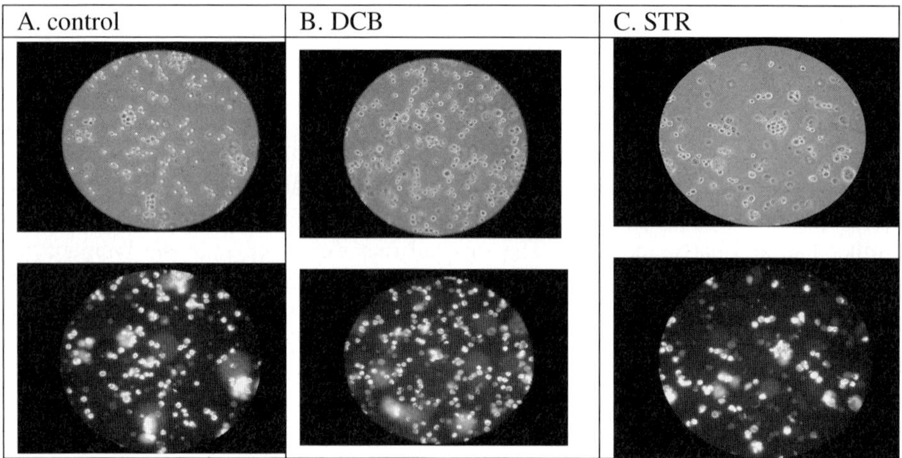

Fig. 4. The effects of 1,4-dichlorobenzene on mitochondria membrane potential in human HeLa cells

The cells were treated with 500 μM 1,4-dicholorobenzene or 2 μM staurosporine for 12 hr. After stained with Membrane Sensor, the cells were observed under fluorescent microscopy. The cells shown as orange were with normal mitochondria potential and the green cells were lost mitochondria potential.

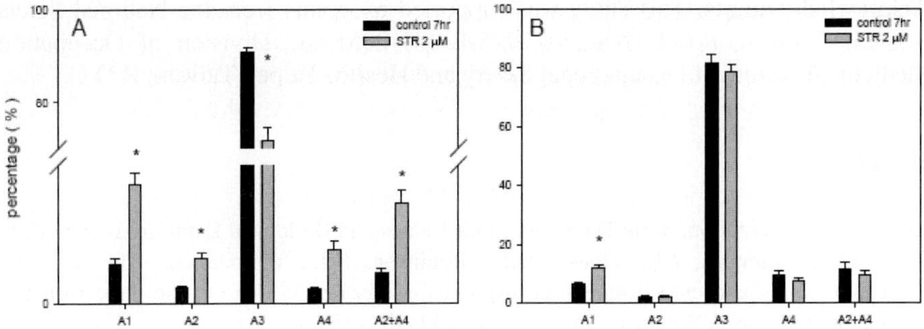

Fig. 5. The cytotoxicity of lymphocyte from workers and non-workers in DCB factory. The lymphocytes were isolated from the blood of donors and incubated with or without 2 M staurosporine for 7 hr, then, the cells were stained with PI/caspase 3 substrate and analyzed by flow cytometry. The data were analyzed by student t test. *: p< 0.05.

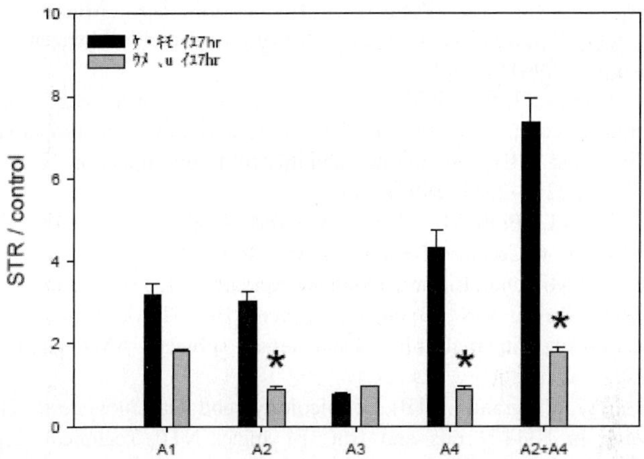

Fig. 6. The lymphocyte cytotoxicity difference between workers and non-workers in DCB factory. The data in figure 5 are used to gain the ratio between STR treatment and control. The data were analyzed by student t test. *: p< 0.05.

4 Conclusion

The cytotoxicity of pDCB was evidenced in this study. pDCB only slightly caused apoptosis in Human lymphoma Jurkat cells, human colon adenocarcinoma cells and human hepatocellular carcinoma cells. Moreover, the sensitivity to staurosporine was lost in the lymphocytes of workers in pDCB factory. Since staurosporine is an apoptosis causing agent, pDCB might have cause the cells to resist stress-induced apoptosis that might increase the risk of cancer.

Acknowledgements. This study was supported by grants from the National Science Council, Taiwan, R.O.C.(NSC-95-2745-B-031-005) and Division of Occupational Medicine, Institute of Occupational Safety and Health, Taipei, Taiwan, R.O.C.

References

ACGIH, Documentation of the Threshold Limit Values and Biological Exposure Indices (2001)

Allis, J.W., Simmons, J.E., House, D.E., Robinson, B.L., Berman, E.: The differential hepatotoxicity and cytochromeP450 responses of Fischer-344 rats to the three isomers of dichlorobenzene. J. Biochem. Toxicol. 7(4), 257–264 (1992)

ATSDR, Toxicological Profile for 1,4-Dichlorobenzene (1993)

Bomhard, E., Luckhaus, G., Voigt, W.H., Loeser, E.: Induction of light hydrocarbon nephropathy by p-dichlorobenzene. Arch Toxicol. 61, 433–439 (1988)

Eibl, K.H., Banas, B., Schoenfeld, C.L., May, C.A., Neubauer, A.S., Priglinger, S., Kampik, A., Welge-Lussen, U.: Alkylphosphocholines inhibit proliferation of human retinal pigment epithelial cells. Invest. Ophthalmol. Vis. Sci. 44, 3556–3561 (2003)

Eldridge, S.R., Goldsworth, T.L., Popp, J.A., Butterworth, B.E.: Mitogenic stimulation of hepatocellular proliferation in rodents following 1,4-dichlorobenzene administration. Carcinogenesis 13, 409–415 (1992)

Eldridge, S.R., Tilbury, L.F., Goldsworthy, T.L., Butterworth, B.E.: Measurement of chemically induced cell proliferation in rodent liver and kidney: a comparison of 5-bromo-2'-dexoyuridine and [3H] thymidine administered by injection or osmotic pump. Carcinogenesis 11, 2245–2251 (1990)

Giavini, E., Broccia, M.L., Prati, M., Vismara, C.: Teratologic evaluation of p-dichlorobenzene in the rat. Bull Environ. Contam. Toxicol. 37, 164–168 (1986)

Green, D.R., Reed, J.: Mitochondria and apoptosis. Science 281, 1309–1311 (1998)

Hollingsworth, R.L., Rowe, V.K., Oyen, F., Spencer, HC.: Toxicity of aradichlorobenzene: determinations on experimental animals and human subjects. AMA Arch. Ind. Health 14, 138–147 (1956)

National Toxicology Program (NTP), Toxicology and carcinogenesis studies of 1,4-dichlorobenzene in F344/N rats and B6C3F1 mice. NTP Technical Report 319, NIH Publication no. 87-2575 (1987)

James á Anthony, N.H., Soames á Ruth, R., Roberts, A.: Suppression of hepatocyte apoptosis and induction of DNA synthesis by the rat and mouse hepatocarcinogen diethylhexylphlathate (DEHP) and the mouse hepatocarcinogens 1,4-dichlorobenzene (DCB). Arch. Toxicol. 72, 784-790 (1998)

Packard, B.Z., Toptygin, D.D., Komoriya, A., Brand, L.: Profluorescenct protease substrates: intramolcular dimers described by the exciton model. In: Proc. Natl. Acad. Sci (USA) vol. 93, pp. 11640–11645 (1996)

Robbiano, L., Carrozzino, R., Puglia, C.P., Corbu, C., Brambilla, G.: Correlation between induction of DNA fragmentation and micronuclei formation in kidney cells from rats and humans and tissue-specific carcinogenic activity. Toxicol. Appl. Pharmacol. 161(2), 153–159 (1999)

Umemura, T., Kodama, Y., Kurokawa, Y., Williams, G.M.: Lack of oxidative DNA damage or initiation of carcinogenesis in the kidneys of male F344 rats given subchronic exposure to p-dichlorobenzene (pDCB) at a carcinogenic dose. Arch. Toxicol. 74(1), 54–59 (2000)

Economic Assessment of Industrial Accidents Caused by Abnormal Behaviors

Hunszu Liu

Industrial Engineering and Management, Ming Hsin University of Science and Technology, No. 1 Hsin-Hsing Road, Hsin-Fong, Hsinchu, Taiwan
hliu@must.edu.tw

Abstract. The economic assessment of industrial accidents caused by abnormal behaviors can not only improve the risk assessment results but also serve as a crucial criterion for prioritizing the correction plans. Three stages of works are designed to derive the loss figures from assessing the standard operation procedure. The objective of stage one is to identify potential abnormal behaviors through analysis of activities. The objective of stage 2 is to identify cost types initiated by different accident losses. The objective of stage 3 is to establish the accident cost bank through analyzing the historical accident data. Two industrial cases, including one semiconductor and one steel company, were studies to demonstrate the feasibility of this approach in both high-tech and traditional industry. This study proposes a framework to develop abnormal behavior identification and cost estimation model.

Keywords: abnormal behaviors, risk assessment, safety economics.

1 Introduction

Abnormal behaviors, defined as deviations from standard operation procedures, can create not only the quality problems, but also the interruption of production activities. Empirical studies have shown that abnormal behaviors are one of the major causes for industrial accidents. Although, industry tries to implement the safety observation and training programs to limit the probabilities and magnitudes of the abnormal behaviors, these types of unexpected acts continuously occur and create threats for line managers to maintain stable daily operations. As a result, operators and managers must constantly monitor abnormal behaviors and perform corresponding measurements based on the risk assessment results. This continuous process can easily be forgot or ignored by line managers if the economic assessment data were not properly provided. Moreover, the economic assessment data can not only improve the risk awareness of line managers but also serve as a crucial criterion for prioritizing the correction plans.

The economic assessment data can be derived from loss figures of past accidents. Unfortunately, these figures are currently not directly available due to the lack of collection mechanism and historical data. Different approaches are adopted by

M.J. Smith, G. Salvendy (Eds.): Human Interface, Part II, HCII 2007, LNCS 4558, pp. 1067–1071, 2007.

industry or insurance company to collect the loss figures. These approaches focus mainly on the estimation of total losses. However, they provide little information for identifying the causes. Often the accident investigation techniques are needed. The relationships between losses and abnormal behaviors are not established. It is likely for managers and operators to ignore the importance of controlling the abnormal behaviors. This study tries to develop an approach to assess industrial accidents economically, thus increasing the awareness of controlling abnormal behaviors.

2 Background Information

Various losses estimation methods are developed to help industry to assess the economic impacts of accidents. One approach calculates the total losses through estimating the direct cost and indirect cost. But the scopes of direct costs and indirect costs have not been formally established or accepted by practitioners or researchers. The loss figures calculated through this type of estimation approach are too rough to represent the cost of accidents. To solve this problem, industry use insured and uninsured costs to estimate the loss. The insured costs are easy to obtain. But the uninsured costs, which usually are the major loss of accidents, are hard to find.

Another approach is developing the cost typology. Various costs classification schemes were developed to calculate the costs of ergonomics solutions or safety and health accidents. Riel and Imbeau (1996) proposed a cost typology which comprises three different categories: insurance-related costs, work-related costs, and perturbation costs [1]. Each of these categories is further subdivided into cost components which can be used to develop detail safety and health cost typology. In recent year, there are increasing concerns on how to measure these costs. Unfortunately, there are still no agreements on how and what to measure yet. Different distinctions, such as economic vs. non-economic costs, fixed vs. variable costs, direct vs. indirect costs, and internal vs. external costs, were proposed to identify which costs can financially motivate the industrial managers to improve their working conditions. Dorman (2000) believed that, based on economic theory, industrial managers are more likely concerning of the economic, variable, direct, and internal costs [2].

The scope of cost estimation and the level of study play key factors in selecting cost typology. Cost estimation performed at the company level is primarily used by managers to assess the needs of implementing accident control measurements. Therefore, the scope of cost estimation and typology should satisfy the needs of industrial managers.

Literature reviews indicate that there are abundant of scientific literatures in studying safety and health and ergonomics costs. Anderson (1992) developed guidelines of economic evaluation of ergonomic solutions for practitioner, where the Net Present Value (NPV), Internal Rate or Return (IRR), Annuity Method (AM), and the Payback Period Method (PPM) were illustrated to examine the profitability of ergonomic solutions [3]. Alexander (1994) classified seventeen justification techniques of funds for ergonomics, safety and health into seven strategies [4].

The fundamental requirement of these approaches is that the accounting system can provide the necessary information. Unfortunately, no accounting model specially tailored to ergonomics has been designed [1], [5]. The information about different costs and their origins are not provided. As a result, incomplete cost information can not be used as a decision base for management and control purpose.

In order to solve the complex cost allocation and tracking problems, the Activity based costing (ABC) approach is introduced to measuring ergonomics, safety and health costs. In an ABC calculation, all manufacturing activities are specified, and the corresponding costs are estimated. These activities are grouped according to their background and connected to the product through cost drives that form an allocation basis for the different groups of activities. The cost drive is a measurable unit forming a link between activities and products. Riel and Imbeau combined the ABC and the hidden cost approaches to analyze the behavior of quantifiable and irreducible components of health and safety costs. The ABC provides a unifying framework for all quantifiable costs and the hidden cost approach is adopted to estimate irreducible costs.

Most studies of ergonomic, safety and health costs focus on the economic justification of project improvements or investments. The exception is that Dahlen and Wernersson (1995) tried to visualize the human factors, such as absenteeism and labor turnover, as costs in an accounting system. Only a few studies of costs modeling aim at the economic evaluation of impacts caused by human errors.

Previous discussions indicate that traditional accounting system can not directly provide loss information for assessing the impacts of accidents. This obstacles force industrial managers justify the proper accident and injury prevention process through experiences. As a result, the efficiency and effectiveness of the system safety barriers are in doubt and the smooth operation of manufacturing activities are insecure.

3 Methodology

To economic assess industrial accidents caused by abnormal behaviors; three important issues have to be considered. The first issue is how to identify abnormal behavior? In this study, abnormal behavior is defined as the deviations from normal operation. The abnormal behavior can be identified through safety observations. The standard operation procedure is used as the criteria to identify abnormal behaviors. Human Error Assessment System for manufacturing working environment has been developed through the combination of human errors diagnostic (HED) and human errors criticality analysis (HECA) methods. This system can help industry to develop its own human errors assessment model based on evaluation of daily plant operations. This system provides the decision-makers with the capability to assess the critical human errors and their impacts in manufacturing environment so that accidents can be avoided. The types of accident losses caused by each abnormal behavior are identified and its corresponding possibilities and magnitudes data are estimated by groups of senior workers including safety and health personnel and managers.

The second issue is to identify cost types initiated by different accident losses. At first, the cost typologies are summarized through reviewing related research works. The definitions and classifications of cost types are given by groups of safety experts, including safety and health practitioners based on industrial customs. These cost types are explained to factory managers and supervisors before they perform the criticality analysis of their contributions to each loss type through the Analytic Hierarchy Process (AHP). The Expert Choice software is used to conduct the pair-comparison between each cost type and identify the critical cost types for each type of accident.

The third issue is to establish the accident cost bank through analyzing the historical accident data. Group of managers and supervisors were asked to estimate the loss of every critical cost type for each accident after reviewing corresponding record and related information. These numbers were averaged to form the basic loss figures for each type of accident. To economically assess the accidents caused by abnormal behaviors the assessor first observing the operator and record the abnormal behaviors through work sampling schedule. Then the corresponding losses are look up from the accident cost bank and summation.

4 Discussions

Two industrial cases, including one semiconductor and one steel company, were studies to demonstrate the feasibility of this approach in both high-tech and traditional industry. To establish the accident cost bank for the semiconductor company, totally four thousands two hundred and sixteen accident cases were studied. Five hundred and four accident cases for local steel company were studied. These accidents were grouped into similar categories and the corresponding critical cost types were identified and respective costs are estimated based on the judgments of safety and health personnel.

In order to help the industrial manager to justify how to properly allocate limited company resources to change or maintain the current operation status, the mechanism of estimating the losses of abnormal behavior is required. Historical data usually can not provide enough information for managers to estimate the costs. Therefore, further refinery works are needed. Future works will focus on analyzing more accident cases and improve the reliability of cost bank.

Acknowledgments

The authors would like to thank the National Science Council for supporting this project (NSC#95-2221-E-159-005).

References

1. Riel, P.F., Imbeau, D.: Justifying investments in industrial ergonomics. International Journal of Industrial Ergonomics 18, 349–361 (1996)
2. Dorman, P.: Three preliminary papers on the economics of occupational safety and health, Safe Work, International Labor Organization, Geneva (2000)
3. Anderson, E.: Economic evaluation of ergonomic solutions Part I-Guidelines for the practitioner. International Journal of Industrial Ergonomics 10, 161–171 (1992)

4. Alexander, D.C.: The economics of ergonomics. In: Proceedings of the Human Factor and Ergonomics Society 38th Annual Meetings (1994)
5. Dahlen, P.G., Wernersson, S.: Human factors in the economic control of industry. International Journal of Industrial Economics 15, 215–221 (1995)

A Real-Time GPU-Based Wall Detection Algorithm for Mapping and Navigation in Indoor Environments

Hadi Moradi[1], Eun Kwon[2], Dae Neung Sohn[2], and JungHyun Han[2]

[1] Department of Computer Science, University of Southern California
LA, CA 90089, USA
moradi@usc.edu
[2] College of Information and Communications, Korea University
Seoul, 136-701, Korea
{argent1104,soonde93,jhan}@korea.ac.kr

Abstract. In robotic applications, there is a growing trend for developing human-like real-time interaction capabilities. A good example can be found in Simultaneous localization and mapping technique, where a robot or an autonomous vehicle builds up a map within an unknown environment while at the same time keeping track of its current position. Especially in indoor environments, wall detection is often a critical part of SLAM: it plays a key role in scene interpretation and 3D workspace modeling. Further, it also reduces the size of the map. This paper presents an effective and real-time approach for detecting walls in indoor environment using GPU (graphics processing unit). The experimental results show the feasibility of using GPU as a coprocessor in robotic applications.

1 Introduction

Wall detection is a serious issue for intelligent systems working in indoor environments. For instance, wall detection is a part of 3D environment modeling, and is needed in many applications such as Simultaneous Localization and Mapping (SLAM) [1]. In principle, wall detection is needed for two purposes: 1) Feature registration: having a map of an area, registration of the sensory data corresponding to a wall in the map allows a robot to localize itself. 2) Feature detection/extraction: having the vast sensory data, detection of the wall features in the environment allows better scene interpretation, and reduces the size of data needed to be stored.

Moradi *et al.* [2] proposed an intuitive method to detect walls. The core idea is based on the fact that the objects and environment behind an opaque wall are not visible. Consequently, any hypothetical wall in the environment can be verified using this simple fact. This article presents how to accelerate the wall detection algorithm of [2] using GPU (Graphics Processing Unit), which is a dedicated graphics rendering device for PC, workstation, or game console.

M.J. Smith, G. Salvendy (Eds.): Human Interface, Part II, HCII 2007, LNCS 4558, pp. 1072–1077, 2007.

2 Background

The experiment uses a robotic manipulator, which is equipped with a *stereo camera* in an eye-on-hand configuration. The camera is used for modeling the environment or workspace. Fig. 1-(a) shows an example of the captured data in the form of 3D *point clouds*, and Fig. 1-(b) shows the recognized edges. The recognized edges are vertically extruded to construct *hypothetical walls*, as shown in Fig. 2. The major strength of the hypothetical wall generation method is that it works in environments that include texture-less walls. Furthermore, using the edges in the ceiling, which are very small in number, leads to fast processing time. Finally, the ceiling features suffer less from feature obstruction since the ceiling area is fairly visible.

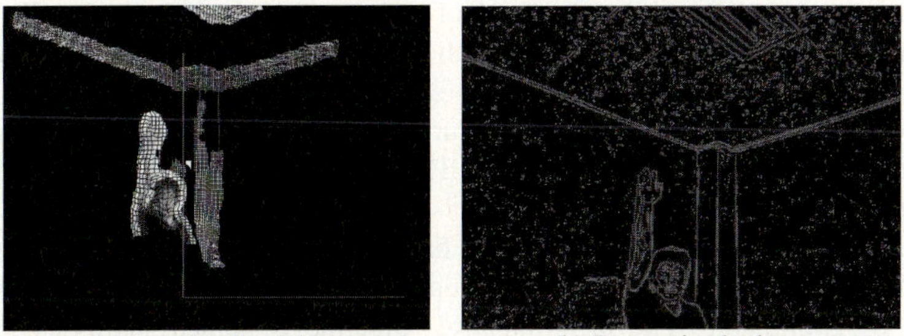

(a) Point clouds of the scene (b) Recognized edges

Fig. 1. Point clouds and edge detection (from [2])

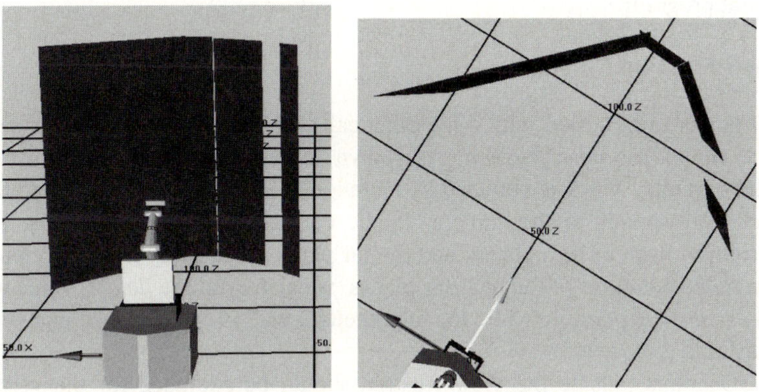

(a) Front view (b) Top view

Fig. 2. Hypothetical walls (from [2])

The walls are named hypothetical in the sense that they should be verified through geometric reasoning of the environment. The verification process uses GPU. Modern GPUs are very efficient at manipulating and displaying 3D objects, and their highly parallel structure makes them more effective than typical CPUs for a range of complex graphics algorithms. With the increasing programmability of commodity

GPUs, these chips are currently capable of performing more than the specific graphics computations for which they were designed. They are now capable coprocessors, and their high speed makes them useful for a variety of general applications[3-6], leading to the name of *GPGPU* (General-Purpose computation on GPU).

3 Wall Verification

If there are points behind a hypothetical wall, the wall should be rejected because no point behind a real wall is visible. However, the number of points is often too huge to be tested in real time. The first step in the wall verification method uses well-known *view frustum culling* technique to reduce the size of the 3D point set. Each of the remaining points is tested if it is in front of or behind the hypothetical wall. The test requires simple *geometric reasoning*, which can be implemented very efficiently with GPU.

All of today's commodity GPUs structure their graphics computation in a similar organization called the *graphics pipeline*. The pipeline is divided into several stages. In most of the contemporary GPUs, it is partitioned into *vertex processing* and *fragment processing* stages. Recently, we have witnessed the GPU revolution where the hardware capability has been exponentially growing and further the hardware has become programmable. The *programmable* pipeline is named *shader*. The shader program is a set of GPU instructions, and is partitioned into *vertex program* and *fragment program*. The first step of wall verification is automatically done between the *vertex processing* and *fragment processing* stages, and the second step is done by a fragment program.

3.1 View Frustum Culling

The 3D points to be tested with a hypothetical wall are restricted to be those within the *view volume* or *view frustum*. As shown in Fig. 3-(a), the view frustum is a truncated pyramid, which is bounded by rectangular *near* and *far planes*, and the peak of which corresponds to the camera. In the current experiment, the near plane is placed right in front of the camera, and the far plane is 5m away from the camera[1]. In addition to the positions of the near/far planes, we also need the *field of view*, which is defined as small as possible while the hypothetical wall is completely contained in the view frustum.

The 3D points outside the view frustum should be excluded for the verification process. It is called view frustum culling, and is automatically implemented within GPU. More specifically, between vertex processing and fragment processing, all primitives (such as 3D points/vertices) outside the view frustum are discarded. The result is shown in Fig. 3-(b).

---

[1] The error in stereo vision has been well studied, and it is shown that the major error happens in the depth or z measurement. For instance, in our experiments, the x and y errors are less than 1cm at 5m distance while the z or depth error is often around 15cm. Therefore, placing the far plane beyond 5m is not a reasonable decision.

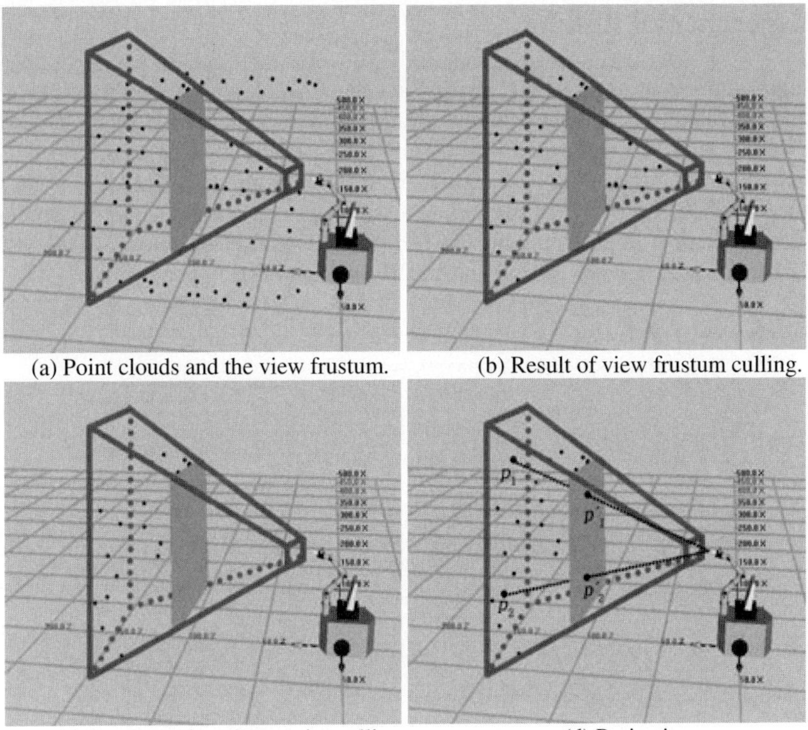

(a) Point clouds and the view frustum. (b) Result of view frustum culling.

(c) Preprocessing: front point culling. (d) Projection.

Fig. 3. Processing of 3D point clouds

3.2 Geometric Reasoning

As a preprocessing stage of the geometric reasoning step, the 3D points in front of the hypothetical wall are discarded, as shown in Fig. 3-(c). If the hypothetical wall is perpendicular to the viewing direction, it is set to the front face in defining the view frustum. Then, such a preprocessing is not needed. However the hypothetical wall is not necessarily perpendicular to the viewing direction.

As discussed in Section 3.1, the major error in stereo vision happens in the depth or z measurement. Consequently, a 3D point near the ceiling edge is not reliable in contradicting the hypothetical wall. We also have noise in the 3D point clouds. In contrast, a 3D point far away from the wall may be taken as reliable. When the hypothetical wall is not perpendicular to the viewing direction, an easy way to determine if a 3D point is reliable or not is to project the point to the hypothetical plane and measure the relative height of the projected point. In Fig. 3-(d), the height of the projected point p'_1 is higher than that of p'_2. When the ratio of the reliable points is above the threshold, the hypothetical wall is rejected.

The geometric reasoning is simple, easy to implement as a fragment program, and pretty fast. Further, the reasoning is accurate because it is 32-bit floating point calculation.

4 Experimental Results

The algorithms presented in this paper have been implemented in OpenGL and its shading language GLSL. Experiments are done on 3 platforms: i) Intel Pentium 3.2 GHz CPU + NVIDIA GeForce FX 5600 GPU (of 100M vertex processing per second and 1600M pixel processing per second), ii) Intel Pentium 3.2 GHz CPU + NVIDIA GeForce 6600 GPU (of 375M vertex processing per second and 4B pixel processing per second), and iii) Intel Core2duo 2.4 GHz CPU + NVIDIA GeForce 7800 GPU (of 1.1B vertex processing per second and 13.2B pixel processing per second).

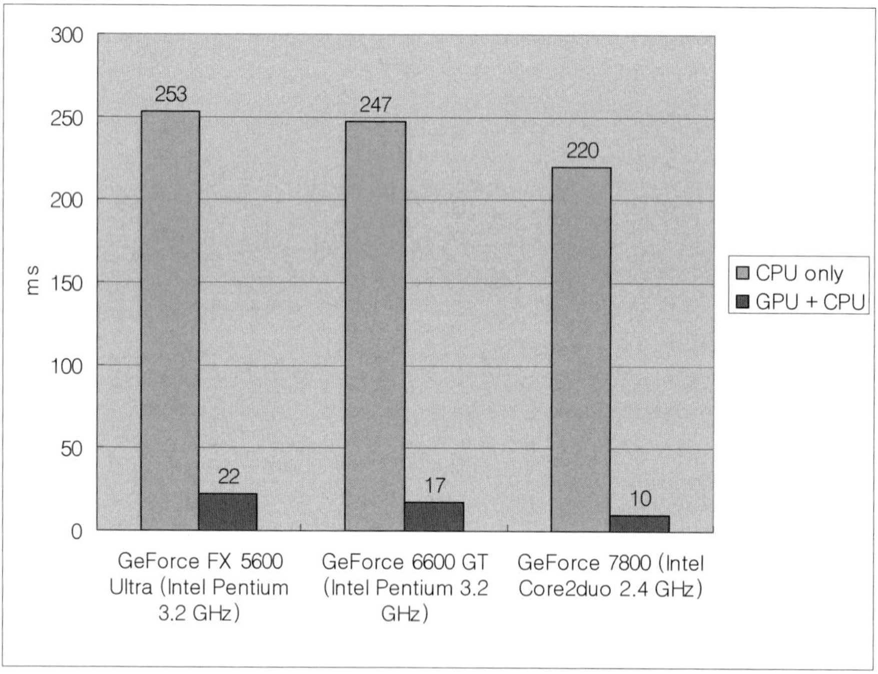

Fig. 4. Performance Comparison

The captured scene consists of 40,032 3D points and 120 hypothetical walls. Fig. 4 shows the algorithm performances on the 3 platforms. The running time is compared with that of the original algorithm of [2], which is run by CPU only. (The CPU of each platform is used.) The obvious performance gain demonstrates why GPU is taken as a useful coprocessor especially for 3D applications.

5 Conclusion

This paper presents an effective and fast GPU-based method for wall detection. The experiments show that GPU is a useful coprocessor for 3D applications. Future work will include, but not be limited to, handling windows, doors, tilted walls and curved corners. Furthermore, it is important to study the effects of newly-sensed data on an already detected wall. If the wall is long, and the sensed data is gathered in a specific area, then the wall may have to be split, rather than completely rejected.

Acknowledgements. This work was supported by grant No. R01-2006-000-11297-0 from the Basic Re-search Program of the Korea Science and Engineering Foundation. This research was also supported by MIC, Korea under ITRC IITA-2005-(C1090-0501-0019).

References

1. Leonard, J.J., Durrant-Whyte, H.F.: Simultaneous Map Building and Localization for an Autonomous Mobile Robot. In: Proc. IEEE/RSJ Int. Workshop on Intelligent Robots and Systems, pp. 1442–1447 (May 1991)
2. Moradi, H., Choi, J., Kim, E., Lee, S.: A Real-Time Wall Detection Method for Indoor Environments. In: Proc. IEEE/RSJ International Conference on Intelligent Robots and Systems (October 2006)
3. Owens, J.D., Luebke, D., Govindaraju, N., Harris, M., Krüger, J., Lefohn, A.E., Purcell, T.J.: A Survey of General-Purpose Computation on Graphics Hardware. Eurographics (2005)
4. Kass, M., Lefohn, A., Owens, J.: Interactive Depth of Field Using Simulated Diffusion on the GPU. Technical Report no.06-01, Pixar Animation Studios (January 2006)
5. Sud, A., Otaduy, M., Manocha, D.: Fast 3D Distance Field Computation Using Graphics Hardware. In: Proc. Eurographics (2004)
6. Govindaraju, N., Gray, J., Kumar, R., Manocha, D.: High Performance Graphics Coprocessor Sorting for Large Database Management. In: Proc. ACM SIGMOD (2006)

Development of Cooperative Building Controller for Energy Saving and Comfortable Environment

Yoshifumi Murakami[1], Masaaki Terano[1], Fumiaki Obayashi[1], and Mutuo Honma[2]

[1] New Product Technologies Development Department, Matsushita Electric Works, Ltd.
1048 Kadoma, Osaka 571-8686, Japan
{y_murakami,terano,obayashi}@mail.mew.co.jp
[2] Nikken Sekkei, Ltd., 2-18-3 Iidabashi, Chiyoda-ku Tokyo 102-8117, Japan
honmam@nikken.co.jp

Abstract. The purpose of automatic controllers for building services (air conditioning system, lighting and blinds system) is to save energy in the whole building while keeping occupants' comfort. Nevertheless, most of automatic controllers don't optimally operate from the viewpoint of total energy efficiency and occupants' satisfactions. The main reason is that each controller operates, neglecting states of other controllers and therefore, control point of each controller always doesn't correspond to best control point in the whole building system. In this study, we propose an innovative controller that communicates with other controllers to control building equipments optimally from the viewpoint of total energy efficiency and occupants' satisfactions. The developed controller functions as an agent in the control system, which can voluntarily collect available information from other controllers and decide set points to enhance total efficiency in cooperation with one another. Therefore, this controller is called as "cooperative controller". Several types of cooperative controllers have been developed and experimentally applied to an actual building located in Japan (Sogo et. al. 2002, Honma et. al. 2002, Terano et. al. 2002). In this paper, cooperative control using daylight, major example in series of cooperative control system, is reported. Daylight is useful to reduce energy consumption of lighting, however, it has not been widely used. The main reason is that as incoming daylight has complex effects on thermal and visual environment, it is difficult to optimally control amount of daylight. To solve this problem, we propose cooperative control system that can control amount of daylight effectively to optimize total energy efficiency and occupants' comfort. A series of numerical experiments was conducted to confirm effect of cooperative control system on improvement of total energy efficiency and indoor environment. The results showed that this cooperative control system could save energy use rate of almost 10% compared with no cooperative control system. In addition, in order to use in practice, cooperative controller is designed to apply an actual building easily and keep stability even in the other controllers' troubles.

Keywords: Energy saving, Cooperative control, Venetian blind.

M.J. Smith, G. Salvendy (Eds.): Human Interface, Part II, HCII 2007, LNCS 4558, pp. 1078–1087, 2007.

1 Introduction

The purpose of automatic controllers for building services (air conditioning system, lighting and blinds) is to save energy in the whole building while keeping occupants' comfort. Nevertheless, most of automatic controllers don't optimally operate from the viewpoint of total energy efficiency and occupants' satisfactions. The main reason is that each controller operates, neglecting states of other controllers and therefore, control point of each controller always doesn't correspond to best control point in the whole building. In this study, we propose an innovative controller to optimally control building equipments from the viewpoint of total efficiency of whole building system, communicating with other controllers. The developed controller functions as an agent in the control system, which can voluntarily collect available information from other controllers and decide set points to enhance total efficiency in cooperation with one another. Therefore, this controller is called as "cooperative controller", and several types of cooperative controllers have been developed and experimentally applied to an actual building located in Japan.

2 Cooperative Control System

2.1 Overview of Cooperative Control System

Fig.1 shows structure of cooperative control system that is composed of distributed controllers of air conditioning, lighting, and blind system. These controllers are integrated with communications based on open network, BACnet^{TM} and LON(Ashrae 1995).

a) "Blind control system"
Venetian blinds are controlled to enhance total energy efficiency and improve indoor environment near the window, using information of lighting, air conditioning system, outside weather and solar position.

b) "Air conditioning control system"
Air conditioning equipments are controlled to adjust setting temperature according to thermal comfort index calculated from information of lighting, air conditioning, and outside weather.

c) "Lighting control system"
Lighting equipments are controlled to adjust task illuminance to setting value according to detective-sensor of daylight on the ceiling.

2.2 Control Modules Inside Cooperative Controllers

Cooperative controllers can operate for improvement of total energy efficiency and indoor environment. However, if control algorithm is designed particularly to actual building layout, it doesn't work well in unexpected case such as layout renewal or adding new equipments. In this paper, algorithm in the cooperative controller is composed of several functional modules. The purpose of the proposed architecture is to adapt these logics to owners' wishes by changing combinations of these modules and to keep control behavior stable by separating unfunctional modules in other

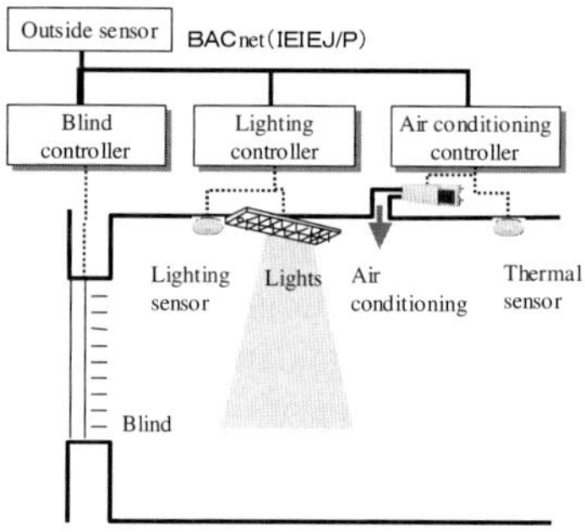

Fig. 1. Overview of Building equipment system

controllers' troubles(Fig.2). Controller C_i($i = 1,\ldots,N$) is organized of functional module J_m($m = 1,\ldots,M$), and modules are connected based on the relevant strength S_{ij} $_j$($j = 1,\ldots,M$, $j \neq i$). Module J_m includes software library and operates based on the information from neighbor controllers. Strength S_{ij} indicates relevant strength from Module J_i to J_j. Module J_i connects to the module which has the biggest value of S_{ij}. When an environmental change such as other controller's trouble happened, functional modules are re-organized voluntarily. For example, when troubles happen between controller C_2, functional modules are re-organized without connecting module J_2 which is the most relevant module to the controller C_2.

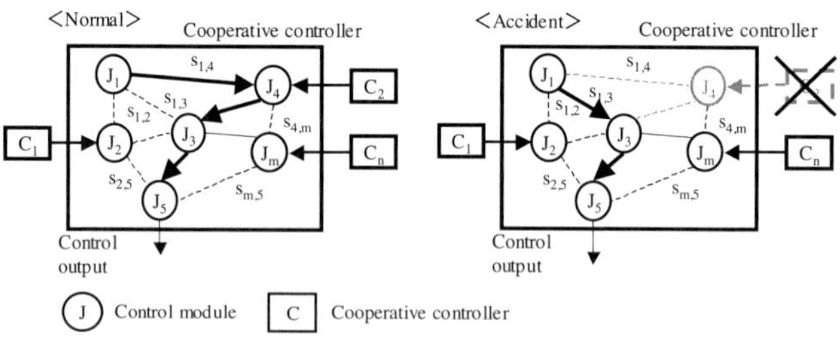

Fig. 2. Control modules at normal and accident time

2.3 Cooperative Control Algorithm

The aim of cooperative controllers is to optimize control state of energy consumption and occupants' satisfactions in the whole building. To achieve this aim, cooperative controllers are required to operate in cooperation with one another and help enhancing total efficiency in the whole building system. However, it is difficult to decide optimal control state at one optimization process without control server which can collect all type of information of building equipments.

Therefore, cooperative controllers predict some influences of set points on total energy efficiency and occupants' comfort and then, decide optimal set point c_i. These controllers repeat this process in every step and gradually lead to optimal state of whole building system.

In fact, each ccontroller can collect available information of other controllers' states $c_k(k=1,..,N, k\neq i)$ and calculate set point c_i using the following formula (1)-(3). Cooperative controllers calculate total energy consumption according to every solution candidate using formula (2) and find optimal set point to minimize total energy consumption. Formula (2) is used to predict effect of set point c_i on energy consumption of building equipments. Solution candidates are decided from formula (3) which predict comfort zone according to comfort indices such as PMV (Fanger et. al. 1970) and PGSV (Iwata et. al. 1992,1996).

$$\min . E(c_i) = \sum_{l=1}^{N} E_l(c_i) \tag{1}$$

sub.to.

$$E_l(c_i) = F_l(c_1 \cdots c_N, s_1 \cdots s_P) \tag{2}$$

$$Comf_{\min} \leq Comf(c_1 \cdots c_N, s_1 \cdots s_P) \leq Comf_{\max} \tag{3}$$

where

E_l is the energy consumption of $equipment_l$ (l=1,..,N) controlled by controller C_i;

F_l is the predicting function of energy consumption of $equipment_l$(l=1,..,N);

s_p is various type of information such as weather condition;

Comf is the predicting function of comfort;

$Comf_{min}$ is the minimum value of comfort zone according to thermal or visual comfort index;

$Comf_{max}$ is the maximum value of comfort zone according to thermal or visual comfort index.

2.4 Blind Cooperative Control

There are some problems of building control using daylight for energy-saving as daylight has complex influences on indoor comfort and energy consumption of air conditioning and lighting system. For example, daylight increases energy consumption of air conditioning system, while reducing energy consumption of lighting system in a cooled office. Also, glare from bright window in the use of daylight irritates office workers.

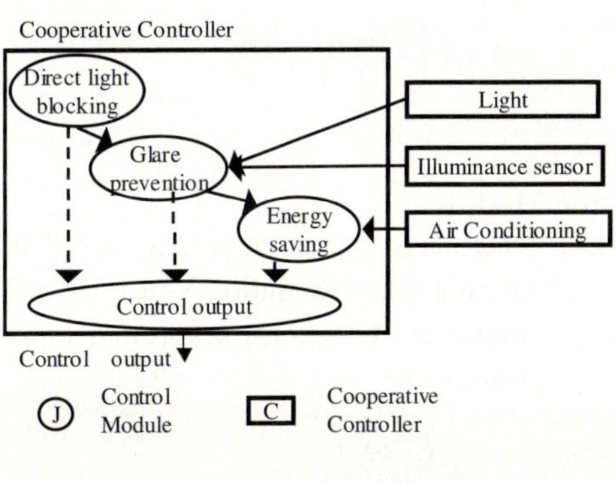

Fig. 3. Outline of cooperative control of blinds

Therefore, blind control system is designed not only to block direct sunlight, but also prevent glare from windows using visual comfort index PGSV and adjust amount of daylight to optimize balances between energy consumption of air conditioning and lighting system. In fact, blind cooperative controllers are composed of direct sunlight blocking module, glare prevention module, and energy saving module as shown in Fig3.These modules are hierarchically organized and therefore, lower module calculates slat angle using upper modules' result.

To calculate slat angle φ, the following formula (4)-(8) is proposed:

The optimal angle is decided to minimize total energy consumption of air conditioning and lighting system calculated from formula (5) and (6).In addition, formula (7) gives assurance of visual comfort according to PGSV(Predicted Glare Sensation Vote) as shown in 2.4.2 while formula (8) blocking direct sunlight.

$$\min .E(\phi) = P_Q + P_L \tag{4}$$

$$sub.to.\ P_L = Light(\phi, E_{lux}) \tag{5}$$

$$P_Q = \varepsilon * Hvac(\phi, E_{lux}, \theta_{out}, \theta_c) \tag{6}$$

$$PGSV(\phi, E_{lux}, c_L) \leq PGSV_{set} \tag{7}$$

$$\phi_{cut} \leq \phi \tag{8}$$

where

ε is the ratio of air conditioner's thermal load to energy transferring;
φis the slat angle(°) of blind;
$PGSV_{set}$ is the setting PGSVvalue whose scale is shown in Table 1;
θ_o is the outside temperature;
θ_c is the setting temperature of air conditioning system;
Light is the predicting function for consumption of lighting system;
Hvac is the predicting function for consumption of air conditioning system.

2.4.1 Direct Light Blocking Module

Direct light blocking module calculates slat angleφ(°) to block direct sunlight into a room using solar altitudes h(rad), solar azimuth A(rad) and azimuth of window A_0(rad).

$$\phi_{cut} = 90 - \tan^{-1}\left(\frac{\tan(h)}{\cos(A - A_0)}\right) * \frac{180}{\pi} \qquad 0 \leq \phi_{cut} \leq 90 \tag{9}$$

2.4.2 Glare Prevention Module

Glare prevention module calculates slat angleφ(°) to avoid discomfort glare from windows using PGSV（Predicted Glare Sensation Vote）. PGSV indicates predicting discomfort glare using comparison between background luminance L_b(cd/m2) and light source luminance L (cd/m2).

$$PGSV = \log \frac{L^{3.2}\omega^{-0.64}}{L_b^{0.61-0.79\log\omega}} - 8.2 \tag{10}$$

where

ω is azimuth(sr) to show window location, L is light source luminance, which is calculated from luminance on the slat and in the sky;

L_b is background luminance, which is luminance of lighting against the wall side, calculated using lighting output c_L and reflective characteristic.

Table 1. PGSV scale

PGSV value	Glare sensation
0	Just perceptible
1	Just acceptable
2	Just uncomfortable
3	Just intolerable

2.4.3 Energy Saving Module

Energy saving module calculates and adjusts slat angleφ to minimize total energy consumption of lighting and air conditioning system. Energy consumption of lighting and air conditioning system are calculated using vertical illuminance $E_{lux,}$ thermal status, layout or materials of equipments near the window. Light and Hvac, which predicts energy consumption of lighting and air conditioning, are indicated as followed.

Predicting function of lighting energy **Light :**

Lighting energy consumption P_L is calculated from adjusted lighting ratio and lighting capability pL. Lighting ratio are adjusted from task illuminance E_{desk}(lux) and setting illuminance E_{set}(lux). Task illuminance E_{desk} is calculated using slat angleφ, daylight factor(Shukuya et. al. 1982), and vertical illuminanceE_{lux}

$$Light(\phi, E_{lux}) = \frac{E_{set} - E_{desk}(\phi, E_{lux})}{E_{set}} * p_L \tag{11}$$

Predicting function of air conditioning system **Hvac:**

Air conditioning energy consumption P_Q is calculated using energy efficiencyεwhich is converted from thermal load quantity. Thermal load quantity is composed of transient load involving thermal transmission heat through window system Q_w(kW), transmission heat through window Q_t(kW), heat of lighting Q_L(kW), and transient heat load caused by change of air conditioning temperature Q_c.

$$Hvac(\phi, E_{lux}, \theta_{out}, \theta_c) = Q_w + Q_L + Q_C + Q_t \tag{12}$$

$$= K*(\theta_e - \theta_c) + \tau * \kappa * E_{lux} + P_L - M\frac{\Delta\theta_c}{\Delta t} \tag{13}$$

$$\theta_e = \theta_o + \rho * \kappa * E_{lux} \tag{14}$$

where

θe is the outside temperature including solar load elements;

ρ is the index for conversion from solar heat to temperature;

k is the index for conversion from illuminance to solar heat load;

K is the ratio of total thermal transmission through window system;

M is the thermal capacity in the air(kJ/(K·m^3)).

2.4.4 Control Output Module

Control output module gives restriction to slat angleφ_p(°) using restriction valueφ_ε(°) to keep control behaviour stable against change of weather condition.

2.5 Cooperative Control of Air Conditioning System

Cooperative control of air conditioning system adjusts setting temperature according to thermal comfort index which can evaluate influences of daylight on thermal comfort. Thermal comfort module, only module in the cooperative controller of air conditioning system, predict thermal comfort using PMV(Predicted Mean Vote), which is widely used as thermal comfort index.

2.5.1 Thermal Comfort Module

Thermal comfort module adjusts setting temperature to PMV, which is calculated from radical temperature affected by amount of daylight, temperature, and humidity. Daylight through blinds is calculated from slat angle$\varphi$and vertical illuminance E_{lux}, and radical temperature is calculated from thermal transmission of window system.

PMV predicts average occupants' comfort, whose scales is shown in Table2 and is calculated from the following equation (15):

$$PMV = (0.352 * \exp(-0.036M) + 0.028) * L \tag{15}$$

where

L is the human body's thermal load, calculated using comfort heat equation [5];
M is the amount of clothing or metabolism M (W/m2).

Table 2. PMV scale

PMV	Thermal sensation
3	Too hot
2	Hot
1	A little hot
0	None of hot and cold
-1	A little cold
-2	Cold
-3	Too cold

3 Simulation

3.1 Evaluation of Cooperative Control Algorithm Using Simulation

The evaluation is to confirm control behaviour of cooperative control system and effect of cooperative control on energy-saving using simulator to predict energy consumption, indoor comfort.

3.1.1 Simulation Conditions

Floor plan of testing office is shown in Fig.4. The office is located in OSAKA(in Japan). An air conditioning system is installed in the center of ceiling, and four lighting equipments are installed on the ceiling. Also, west-facing window has a white automatic blind.

An air conditioning system adjusts air temperature using information collected from thermal sensor, while lighting system adjusts task illuminance. Lighting equipment's capacity is 256(W), and COP (Coefficient Of Performance), which is efficiency for converting heat load to energy consumption of air conditioning system, is 2.0. Occupant is working with personal computer which generated 100(W) for heat. Airflow was 0.1(m/s), and humidity was 40%.

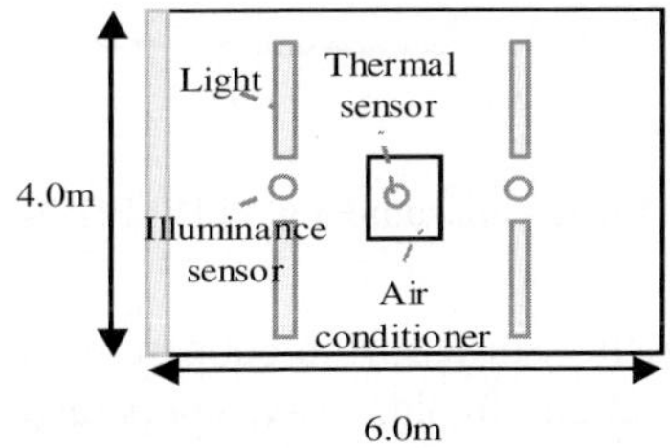

Fig. 4. system layout in test office

Table 3. Calculation condition

	Control type	Status
Case1	No-cooperative control	Normal
Case2	Cooperative control	Normal
Case3	Cooperative control	Lighting data accident at 15:45

Condition of calculation is shown in Table3. In case 1, building controllers basically operate with no cooperative control, while in case 2 with cooperative control. In case 3, building controllers shift from cooperative to no cooperative control because of connection trouble of outside illuminance sensor. We conducted numerical experiment during 15:00-16:30 on August 4th.

As for weather conditions, outside temperature is supposed to 30℃, and vertical illuminance is changed to 60,10,20(klx) in every 30minutes. Regarding control parameters of case 1, setting temperature is set up as 23.5℃ and additional angle is 40° to keep occupants' comfort in even undesirable environment. On the other hand, in case 2, setting value of PMV is 0.5 and PGSV is 2.0.

3.2 Results

Fig 5 shows blind slat angle and setting temperature of air conditioning system in case 1-3, and Fig 6 shows total energy consumption of air conditioning and lighting, PMV and PGSV.

Case1 shows that as blind system was controlled only for blocking direct sunlight, blind operation occasionally brought unnecessary energy consumption compared to case 2. For example, slat angle kept same in spite of vertical illuminance changed after 15:30. As a result, as PGSV and PMV changed according to vertical illuminance within comfort zone, total energy consumption was higher than case2. On the other hand, case 2 shows that blind and air conditioning systems were controlled in cooperation. Therefore, PMV and PGSV were kept to setting value and then, total energy of air conditioning and lighting system was efficiently consumed. Case3

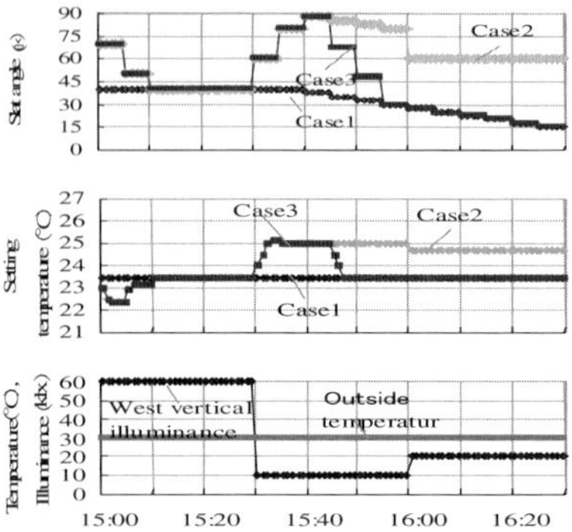

Fig. 5. Slat angle and setting temperature

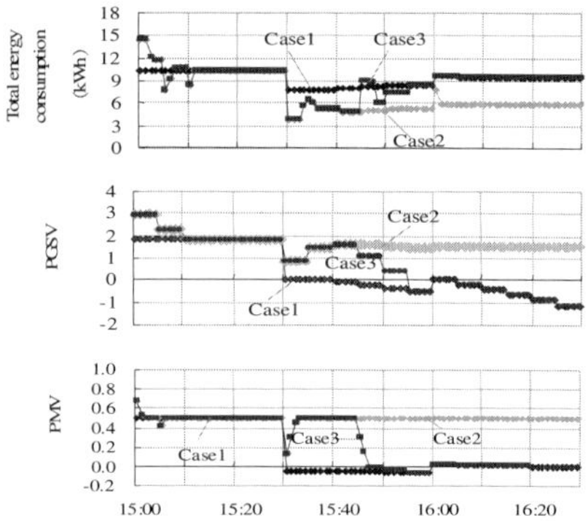

Fig. 6. Energy consumption, PMV, PGSV

shows that after the trouble at 15:45, operation smoothly shifted from cooperative to no cooperative control to keep stable control operation. In this way, cooperative control could provide comfort and energy efficiency when connection trouble happened.

In addition, simulation during cooling term of March-November is also shown in Fig.7, 8 and Table.4. Fig.7 shows total appearance frequency during cooling term of PGSV, and Fig.8 shows that of PMV. As shown in case 1, both of PMV and PGSV randomly appeared within comfort zone, on the other hand, in case 2, PGSV and PMV appeared almost in setting value(PGSV2.0 and PMV0.5).

Also, table.4 shows total energy consumption of air conditioning and lighting system during cooling term for case 1 and 2. Case 2 shows 9.9% of energy saving compared to case 1. In this way, cooperative control could provide energy efficiency while keeping occupants' comfort.

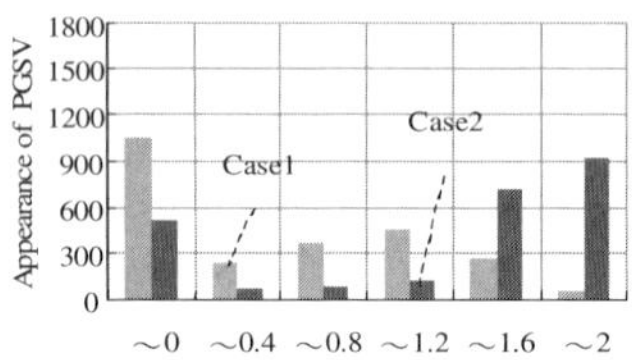

Fig. 7. PGSV appearance

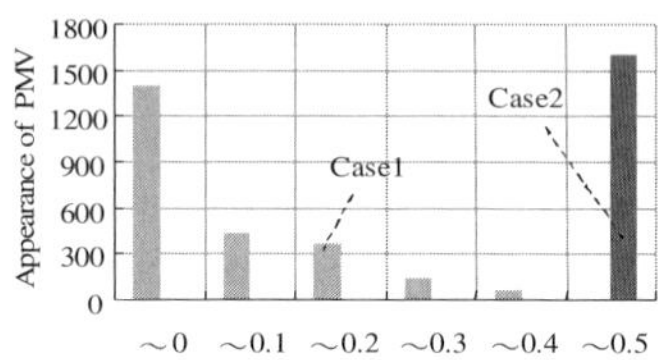

Fig. 8. PMV appearance

Table 4. Energy consumption of air conditioning and lighting system

	Air conditioning system(MJ)	Lighting system(MJ)	Ratio of energy saving system
Case1	2513	1805	-
Case2	2121	1771	9.9%

4 Conclusions

Cooperative controller to optimize total efficiency in the building system was proposed to enhance the total energy efficiency and building system, communicating with other controllers.

The numerical experiment using simulation confirmed that this control system could reduce energy consumption while keeping indoor comfort. In the future, we should attempt to apply to various type of actual building and evaluate effect of this control system.

References

Ashrae, ANSI/ASHRAE Standard, pp. 135–1995 BACnet (1995)

Honma, M.: Evaluation towards cooperative control for energy saving of air conditioning system, integration section lecture meeting and thesis collection (in Japanese), pp. 321–322 (2002)

Iwata, Experimental study on discomfort glare caused by windows. Journal of Archit. Plann. Environ. Engng., AIJ 432 (1992)

Iwata, Experimental study on discomfort glare caused by windows part 2. Journal of Archit. Plann. Environ. Engng., AIJ 439 (1992)

Iwata, Experimental study on discomfort glare caused by windows part 3. J. Archit. Plann. Environ. Eng., AIJ 489, 17–25 (1996)

Fanger, P.O.: Thermal Comfort. Danish Technical Press, Copenhagen (1970)

Shukuya, M.: Analysis on lighting using daylight and energy evaluation (in Japanese). J. Archit. Environ., UDC628.921 (1979)

Shukuya, M.: Theory on the determination of the work plane illuminance by daylight including direct sunlight through windows with reflective louvers or venetian blinds (in Japanese). J. Archit. Environ., UDC628.02 (1982)

Sogo, T.: Cooperative control for energy conservation, environment and comfort, integration section lecture meeting and thesis collection (in Japanese), pp. 323–324 (2002)

Terano, M.: Analysis on current situation for optimization of building management integration section lecture meeting and thesis collection (in Japanese), pp. 325–326 (2002)

Human-Centered Development of Advanced Driver Assistance Systems

Günther Nirschl

Fraunhofer Institute for Transportation Systems and Infrastructure Controls Engineering (IVI),
Zeunerstraße 38, 01069 Dresden, Germany
guenther.nirschl@ivi.fraunhofer.de

Abstract. A methodological approach is presented for the human-centered development of Advanced Driver Assistance Systems (ADAS). A procedure has been developed for the objective evaluation of potential impacts of ADAS usage on traffic safety. The procedure is based on the observation and analysis of the emergence resp. avoidance of driving errors (inadequate speed, insufficient safe distance, etc.) in conjunction with the usage of ADAS. Driving errors are assumed to occur by the coaction of functional ADAS features, the situational context and (psychological) behavioral backgrounds. The applicability of the approach is examplified with a simulated ADAS.

Keywords: Advanced driver assistance systems, traffic safety, driving errors, risk evaluation, driver workload.

1 Introduction

Advanced Driver Assistance Systems (ADAS) are auxiliary (primarily electronical) components in vehicles to support the driver in certain driving situations. They are dedicated to increase driving comfort as well as car safety and more generally road safety. Examples of ADAS are:

- Parking assistant, which detects and measures parking spaces and then steers the car automatically in the parking spot. With the current assistant versions accelerating and braking are still the driver's tasks.
- Adaptive Cruise Control (ACC) which uses either a radar or laser setup to allow the vehicle to automatically slow down when approaching another vehicle and accelerate again to preset speed when traffic allows.
- Lane departure warning (LDW) which warns the driver when the vehicle begins to move out of its lane (unless a turn signal is on in that direction).
- Lane change assistant (LCA) which warns the driver in case of an imminent collision when initiating a lane change manoeuvre. The LCA is typically released when the indicator is activated (contrary to the LDW which is deactivated in that case).

Since ADAS always work with a human driver co-existing, the overall vehicle performance will depend on, not how well the ADAS works by itself, but rather its

M.J. Smith, G. Salvendy (Eds.): Human Interface, Part II, HCII 2007, LNCS 4558, pp. 1088–1097, 2007.

interaction with the human driver. The evaluation of ADAS, therefore, needs to be conducted with human (driver) in the loop.

In particular it is important to design the human-machine interface of ADAS so that the systems provide actual support and do not pose additional stress or distraction on the driver. In this contribution a methodological approach is presented to assess the impact of ADAS design on traffic safety in an objective manner. The applicability of the approach is examplified with a simulated ADAS ("Stop&Go-ACC").

2 Evaluation of the Impact of ADAS on Traffic Safety

2.1 Interrelation Between ADAS, Driving Error and Accident Occurence

Up to now, the relationship between ADAS usage and road safety has been examined only in an incomplete way. A (human-centered) approach to analyse these interrelations consists in regarding changes in the appearance of driving errors in consequence of the application of ADAS. This approach has been applied already from time to time, e.g. by means of trained human observers (mainly driving instructors). What was missing up to now was an objective standardised method which follows the underlying principles, but (ideally) goes without a human observer.

A method which is geared to this objective was developed within the German research program INVENT [1]. The approach starts with a model for accident occurrence (see Fig. 1), which was already introduced by Reichart [2]. Whereas Reichart concentrated on the right part of Fig. 1 and worked out a fault tree analysis by means of probability calculations, the INVENT procedure is focusing on the occurrence of the driving error and asks for the ADAS characteristics, the context of the situation and the behavioral background leading to the occurrence or prevention of the error.

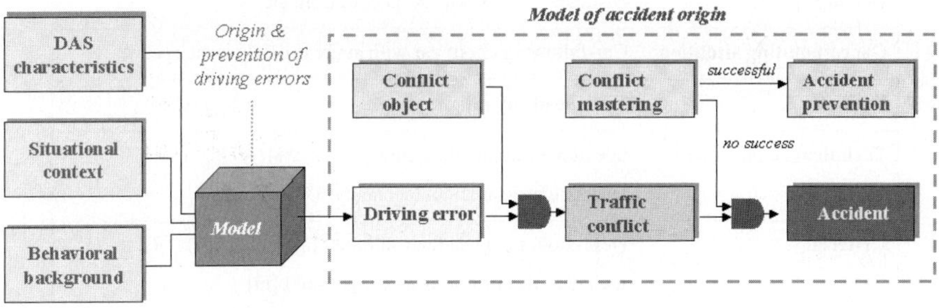

Fig. 1. Generic accident occurence model

It is assumed that the ADAS usage changes the character of the driving task. Depending on the current driving situation, this affects the behavioral background, i.e. the mental or visual workload, the situation awareness and the mental model. These changes result in measurable differences of driver characteristics and the driving behavior.

2.2 Evaluation Procedure and Criteria

2.2.1 Specification and Measurement of Driving Errors

Based on the concept of driving errors an evaluation procedure was developed for the objective assessment of the impact of ADAS usage on traffic safety. The basic principle of the evaluation procedure consists in specifying the set of potential (and measurable) driving errors for a given ADAS in the context of the traffic situations, in which the errors could appear. The error specifications are compiled in a generic error data base, structured into the following error categories:

- Speed too high
- Distance too low
- Braking too late/too weak
- Lane keeping/lane change error
- Illegal/risky overtaking
- Illegal/risky crossing or turning
- Omission/error with blinking
- Insufficient monitoring of traffic

The error specifications at first include the measurable indicators together with the technical requirements for the measurement, followed by the proper evaluation criteria and threshold values. When applicable, multiple grades of error severity are introduced. As an example, the specification for the driving error "distance too low with respect to front car" is depicted in Table 1.

Table 1. Example of driving error specification

Driving error:	Distance too low with respect to front car
Corresponding situation:	Car-following (front car with approximately same speed)
Indicator:	Time headway [s]
Technical requirements:	Speed measurement (accuracy +/- 2.5%) Distance measurement (accuracy +/- 2m)
Criteria:	Definition: time headway at time t [s]: $T_h(t) = d(t) / v(t)$ with d(t): distance to front car at time t [m] and v(t): speed of own car at time t [m/s]

Time headway / Duration of error	> 1.7s	1.2s – 1.7s	0.7s – 1.1s	< 0.7s
< 5s	---	---	---	A
5s – 30s	---	---	A	B
> 30s	---	A	B	C

Effect on traffic safety: ---: none A: low B: significant C: serious

In the further course of the evaluation procedure the general list of driving errors is restricted to a subset which includes only the situations relevant to the assessment of the specific ADAS under consideration. Thus, the restricted error list comprises the driving errors the occurrence of which has to be checked in driving simulator and/or field trials.

2.2.2 Specification and Measurement of Behavioral Background Parameters

According to the concept of the evaluation procedure, the measurement of driving errors is complemented by the recording of parameters which provide hints with regard to the behavioral background of the impact of ADAS usage on error occurrence resp. prevention. These indicators include:

- Mental workload to assess the available cognitive resources of the driver; to be measured by steering entropy or RSME (Rating Scale Mental Effort) questionnaire
- Visual workload to asses the amount of visual information the driver has to detect; to be measured by reaction time on peripheral detection task or eyelid closure frequency
- Situational awareness to assess the driver's understanding of the overall traffic situation; to be measured by reaction time on visual cues, e.g. road signs
- Mental model to assess the driver's understanding of the ADAS functionality; to be measured by questionnaires or recorded data of system usage.

2.2.3 Test Plan

The principal test plan for the experimental trials is specified as follows:

- Test cohort of 30 subjects with driving experience of at least 5 years and/or 50.000km, equally distributed from age groups < 31, 31-59, > 59 years.
- Test protocol with driving task under both conditions (with/without ADAS) in random order.
- Independent variables: driving with/without ADAS, driving situation, driver characteristics.
- Dependent variables: driving errors, behavioral background parameters.

The overall evaluation is based on statistical comparison of frequency and duration of the specified error categories.

3 Application of the Evaluation Approach with Simulated ADAS

In order to verify the evaluation procedure, criteria and threshold values introduced in Chapter 3, a prototypical version of an ADAS was implemented in the Fraunhofer IVI driving simulator. The ADAS prototype selected was a "Stop&Go-ACC", which is a normal Adaptive Cruise Control (ACC) system extended by specific features, which allow to support the driver in case of congestion situations by overtaking the steady stop and driveaway activities.

3.1 Fraunhofer IVI Driving Simulator

The Fraunhofer IVI driving simulator (see Fig. 2) is based on a series-production vehicle in which the driver finds all functions, operating controls and displays as in a real vehicle. A force-feedback steering wheel imparts a realistic driving feel to the driver. Various sensors at the steering wheel, at the pedals and at the transmission selector lever continuously gather data about the vehicle guidance interactions. The simulated vehicle environment is projected to the fore by 3 DLP projectors onto a spherical screen with a sight area of 180x50 degrees. The sight backwards is simulated by transmitting 2 channels of the 3D-model to 2 TFT displays, which are mounted at the positions of the left and of the middle rearview mirrors.

The simulator features specific audio components (sound pc, audio mixer), which allow to generate realistic environmental noise, to utilise the onboard audio systems and to communicate between supervisor and subjects during test trials.

The behavior of the driver can be logged extensively. To record the driver's visual behavior an eyetracking system (FaceLab) is installed, which operates contact-free on the basis of video cameras and optical image analysis.

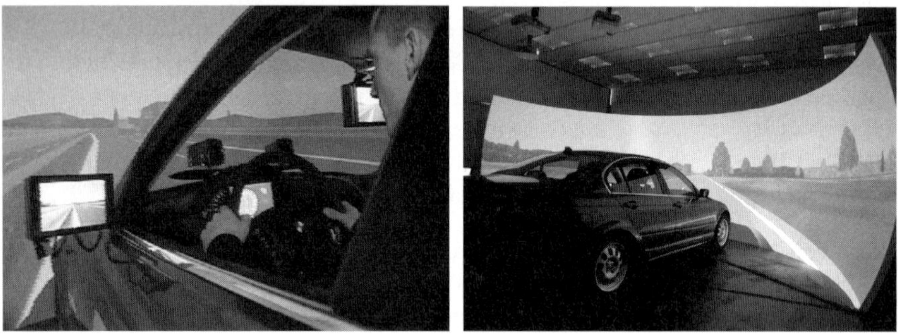

Fig. 2. Fraunhofer IVI Driving Simulator

3.2 Implementation of a "Stop&Go-ACC" in the Driving Simulator

Adaptive Cruise Control (ACC) systems are now actively being developed and introduced into the consumer market by vehicle manufacturers. The primary functional purpose of these systems is to maintain a safe distance between the ACC-equipped vehicle and its immediately preceding vehicle, when one is present.

In order to provide comfort and convenience to drivers, the ACC vehicles use many of the controls associated with conventional cruise control. In addition, the driver is usually provided with the option to select a desired time gap, which lies between maximum and minimum values determined by the designer of the ACC system. Using this control arrangement, each driver may select a free-driving speed (the "set speed") and a headway gap parameter, which is used by the system in determining the desired range. In this manner, each individual driver can obtain operating conditions that are favorable to that individual's driving preferences.

3.2.1 Functional Specification of the "Stop&Go-ACC"

The specification of the ADAS "Stop&Go-ACC", which was selected to be implemented as a prototype in the driving simulator, was developed and provided by a co-operating sub-project within the INVENT programme [3]. The "Stop&Go-ACC" is in principal a normal ACC (Adaptive Cruise Control) system, dedicated to support the driver in headway control when driving at a higher speed (> 50km/h). The ACC system is extended for the lower speed range (< 50km/h) by a specific functionality, which allows to disburden the driver by overtaking the control in stop&go congestion situations (i.e. to stop and driveaway again and again automatically).

The basic control loop of the ACC system is outlined in Figure 3. It is necessary to control the own velocity ($v_{vehicle}$) as well as the headway to a vehicle in front (dx). As a simplified approach a hierarchical controller is applied with a cascadic structure.

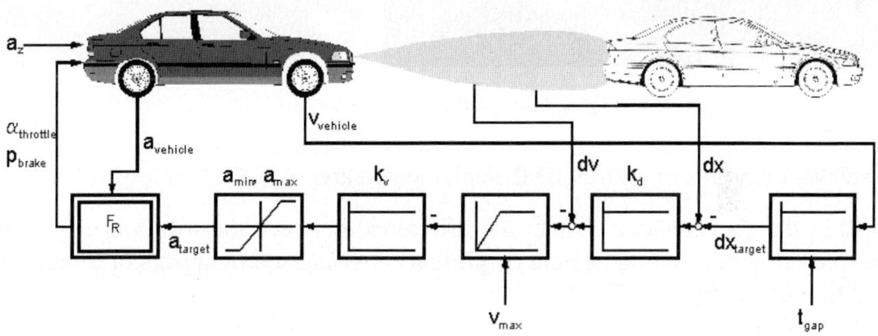

Fig. 3. Basic control loop of the ACC system

The dependent variable of the controller for the simulation is the nominal acceleration (a_z). If a relevant object is located in the own lane within the range of the sensor, the nominal acceleration is calculated as follows:

$$a_z = k_v * (k_d * (dx - dx_{target}) + dv) \quad (1)$$

with $kv = 1$ [1/s]
$kd = 0,2$ [1/s]
and dx = distance from ego vehicle to relevant object

$$d_v = v_{object} - v_{vehicle} \quad (2)$$

$$dx_{target} = t_{gap} * v_{vehicle} + d_{min} \quad (3)$$

with $d_{min} = 3$ [m]
$t_{gap} = 1,8$ [s]

If no relevant object is detected, the nominal acceleration a_z is calculated from the difference between a target speed (v_{target}) adjusted by the driver and the current own velocity ($v_{vehicle}$):

$$a_z = kv * (v_{target} - v_{vehicle}) \quad (4)$$

3.2.2 HMI of the "Stop&Go-ACC"

The HMI of the "Stop&Go-ACC" has been realised in the driving simulator as depicted in Figure 4. To be able to present the specific ACC speedometer and state symbols according to the given specification, an additional TFT display was installed into the cockpit of the simulator vehicle in front of the proper vehicle speedometer.

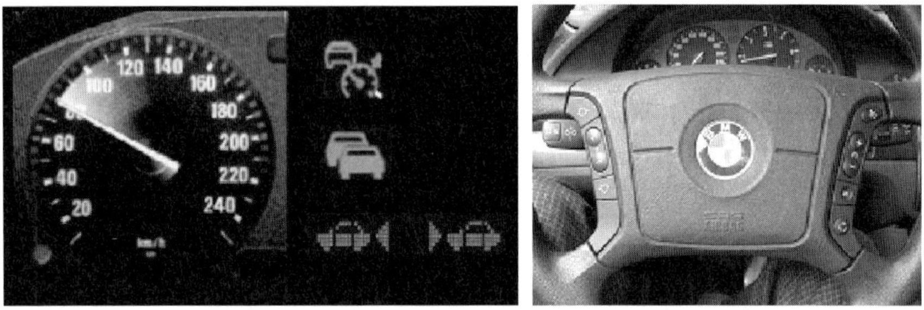

Fig. 4. HMI of the "Stop&Go-ACC"

The relevant features of ADAS HMI implemented are:

- Within the speedometer chaplet the difference between the own velocity and the velocity of a preceeding vehicle (if present) is visualised by means of a segment of lighting up diodes;
- Adjacent to the speedometer at the right there are 3 symbols which indicate (from top to down): ACC activated, preceding vehicle detected, target headway adjusted (1 out of 3 adjustments: 0,8s, 1,0s, 1,3s);
- The enter keys for the "Stop&Go-ACC" are integrated within the multifunction keys of the steering wheel

3.3 Experimental Tests

3.3.1 Test Set-Up

According to the evaluation procedure introduced in chapter 3 experimental tests were set-up to assess the impact of the usage of the "Stop&Go-ACC" on traffic safety. Because this type of ADAS is destined primarily to be applied on highways the selection of traffic situations was restricted to situations on highways like

- free driving,
- following preceding car,
- approaching preceding car,
- lane change / overtaking,
- stop&go driving.

32 subjects participated in the simulator tests.

3.3.2 Exemplary Results

Figure 5 shows for 4 exemplary driving error indicators the time slices within the 3 pre-defined endangering categories (low, considerable, serious endangering).

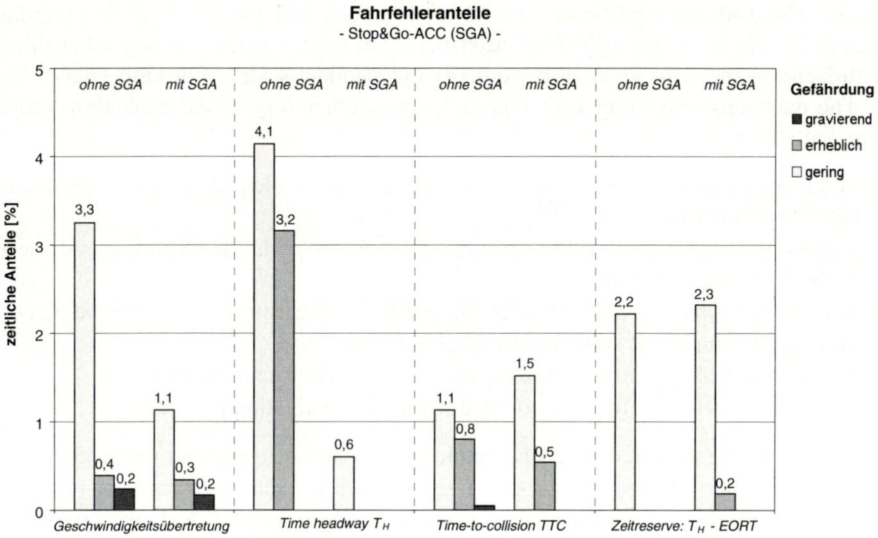

Fig. 5. Selected results (driving errors)

To improve the facility of inspection the category "no endangering" was omitted in each case.

The following remarkable statements can be derived from the analysis of the results:

- The share of the driving error "speed too high" in conjunction with "low endangering" is 3 times higher when driving without the "Stop&Go-ACC" than with the system.
- The time shares of the driving error "time headway too low" in conjunction with low or considerable endangering are clearly higher when driving with the "Stop&Go-ACC" than without it.
- With regard to the driving error indicator "Time-to-collision" (TTC), which is of relevance particularly when approaching with higher differential velocities to a preceding vehicle, there could not be detected remarkable variations when driving with the "Stop&Go-ACC" vs. without it.
- As a dependent parameter with regard to the visual distraction the "time reserve" t_{Res} was calculated as the difference between the time headway t_H and the measured "Eyes-off-the-road-time" (EORT). With regard to the glance behavior the results show now remarkable variations when driving with the "Stop&Go-ACC" vs. driving without the system.Discussion and conclusions

4 Discussion and Conclusions

The evaluation method presented in this contribution describes an objective and standardised procedure in the assessment of ADAS impacts on traffic safety. In a consecutive study [4] this approach was taken up and validated by means of tests in real traffic. The evaluation procedure was refined in particular to that effect that the data recorded in the trials are aggregated statistically in several stages, among other things with factor analysis, to seven dimensionally independent scales of driving safety.

The questions remaining with regard to the applicability of the evaluation procedure include:

- Is the comparison of driving with and without ADAS adequate to assess its overall safety applications?
- Can the selection of relevant situations be done for each ADAS in a specific way without loosing generality?
- Which variation is needed or tolerable for the realization of the experiments concerning the number of subjects, group effects, etc.?
- What is the best way to measure the behavioral background values?
- How can long-term effects be covered within the evaluation procedure?

The focus on the occurrence and prevention of driving errors offers the large advantage to overcome the societal luck and statistical issue that accidents are relatively rare events. On the other hand, the proposed evaluation method deals with the emergence of driving errors only. But, the conflict mastering strategy after the occurrence of a driving error may be affected by the ADAS usage as well. In this context, probability calculations as given by Reichart [2] can be of additional value. Again, the behavioral background of mental models as well as situation awareness play an important role in the conflict mastering strategy.

Accident data analysis are additional important means to detect deficiencies of all drivers or of specific groups [5], to make predictions of the safety benefits [6], or to analyze the safety impact by retrospective comparison of accident data with and without ADAS [7]. These techniques can be adopted supplementary to the evaluation procedure presented in this contribution, which already by itself achieves the goal to be an objective method applicable to all kind of ADAS at an early stage of development.

References

1. Nirschl, G.: Böttcher, St., Schlag, B., Weller, G.: Verfahren zur Bewertung der Verkehrssicherheit von Fahrerassistenzsystemen durch objektive Erfassung von Fahrfehlerrisiken. In: 21. Intern. VDI/VW Gemeinschaftstagung - Integrierte Sicherheit und Fahrerassistenzsysteme. VDI-Berichte 1864, VDI-Verlag, Düsseldorf, pp. 397–420 (2004)
2. Reichart, G.: Menschliche Zuverlässigkeit beim Führen von Kraftfahrzeugen. Fortschrittsberichte VDI, Reihe 22, Nr. 7, VDI-Verlag, Düsseldorf (2001) ISBN 3-18-300722-3
3. Audi, A.G., Man Nutzfahrzeuge, A.G., Adam Opel, A.G., Robert Bosch, G.: Siemens VDO Automotive AG: Simulationsspezifikation für die konzeptionelle Bewertung des Stauassistenten VRA-V. Entwurf Version 0.4 (24-01-2003)
4. Glaser, W., Waschulewski, H., Schmid, D.: I-TSA - A standardized procedure to assess the safety impact of driver-information and driver-assistance systems. In: Fahrer im 21. Jahrhundert - Der Mensch als Fahrer und seine Interaktion mit dem Fahrzeug. Internationaler Kongress des VDI Wissensforums IWB GmbH, Braunschweig, pp. 22–23 (November 2005), VDI-Berichte 1919. VDI-Verlag, Düsseldorf, 17–40 (2005)

5. Frank, P., Reichelt, W.: Fahrerassistenzsysteme im Entwicklungsprozess. In: Jürgensohn, T., Timpe, K.-P. (eds.) Kraftfahrzeugführung, Springer, Heidelberg (2001)
6. Busch, S., Stanzel, M., Zobel, R.: Prediction of the safety benefit of active safety systems. In: VDI conference Telematik im Kraftfahrzeug. Ottobrunn (November 2002)
7. Gwehenberger, J., Kiebach, H.: Sicherheitspotezial ausgewählter Fahrerassistenzsysteme. In: TÜV conference Aktive Sicherheit durch Fahrerassistenzsysteme. Technical University Munich (March 2004)

Disposition of Information Entities and Adequate Level of Information Presentation in an In-Car Augmented Reality Navigation System

Kyung S. Park[1], Il Haeng Cho[1], Gi Beom Hong[1], Tek-Jin Nam[2], Jinyung Park[2], Seong Ik Cho[3], and In-hak Joo[3]

[1] Department of Industrial Engineering, Korea Advanced Institute of Science & Technology
[2] Department of Industrial Design, Korea Advanced Institute of Science & Technology
[3] Telematics & USN Research Division, Electronics and Telecommunications Research Institute

{kspark, illy, aquaqaz, tjnam, vanilla0}@kaist.ac.kr
{chosi, ihjoo}@etri.re.kr

Abstract. As Augmented Reality Navigation systems receive increasing attention as the next generation navigation systems, the importance of the layout of the elements and searching for the appropriate level of information cannot be overstated. Due to this, a series of experiments of usability tests had been designed. According to the experiments, the results for the screen arrangement showed that the MBN screen displayed on the left and the ARN screen on the right was preferred the most while the results for changing lanes showed that displaying information directly on the road and highlighting the path was the best method. As for switching direction, using the upper icons and displaying information directly on the road was preferred. The adequate amount of information was found out to be an average of 3 POI with 20pt font sizes for ARN displays and 10 POI with 18pt font sizes for MBN displays. Lastly, the Eye-tracking experiments showed that the ARN screen was viewed more often with a ratio of 7:3.

Keywords: Augmented Reality, vehicle navigation, disposition of information, adequate information level, usability, user interface, AHP, Taguchi.

1 Introduction

With the advent of in-car navigation systems, drivers are exposed to various ranges of information that is primarily to aid driver's tasks to reach their destination. However, superfluous and obscurely perceived information may act as disturbance and cause unsafe acts of the driver which could lead to some serious accidents. Therefore, it is critical to distinguish which information to provide and decide how and where to display for the development of safe and efficient in-car navigation system.

In this study, we investigated the disposition issue of information entities and the adequate level of presenting them in an in-car Augmented Reality navigation (ARN)

M.J. Smith, G. Salvendy (Eds.): Human Interface, Part II, HCII 2007, LNCS 4558, pp. 1098–1108, 2007.

system, which brings on the issue of arranging even more complicated and sophisticated information entities than the conventional map-based navigation (MBN) system. An ARN system is a futuristic, yet feasible, concept of next generation in-car navigation system that presents driving guidance on real images of the immediate road ahead in real-time situation. In such a system, a camera is installed under the roof of a car, which enables to stream down the real-time images to a in-car display device after being processed through an image processing device. Then, the image is incorporated with a GPS, Gyro-sensor module, and a computer to generate accurate and necessary navigation information entities to be displayed as an overlay on top of the images of the forthcoming road [1].

Although various researches related to the technical aspects and embodiment of ARN system are known to be undertaken in some companies and R&D centers [1, 2, 3], few are published in respect to the user interface aspect or disposition of information entities in similar systems, since they are at very early stage of the development.

Providing a sufficient prediction for the next task of the driver is the key issue of a navigation system [4]. The ARN system has its distinctive advantage as it can visually point out with a direction indicator on the exact next road to take. By seeing the actual image of the road, drivers do not have to go through a mental mapping process, somewhat demanding to some users, trying to match the road indicated on the map to the real-world road as they did with the conventional MBN system [2].

However, even in the ARN system, it is required to present MBN screen abreast with the ARN screen due to the limitations of the visual field of the camera which only allows foreseeing the immediate road situation. Here, the issues of disposition of information entities and adequate level of information presentation are arisen. It is important to decide how to partition the display with the two types of the screen, and moreover, deciding which and how much information to present, and where to present them are also key factors to assure a good usability of the navigation system.

2 General Experiment Contents

This study involves four experiments of usability tests which proceeded one after the other, covering a six-month period. The four experiments are as follows; 1) finding the most preferred arrangement of ARN and MBN display, 2) deciding disposition of information entities, particularly focused on changing lanes and switching directions, 3) figuring out the appropriate level of information presentation, and 4) seeing where and for how long users put their eyesight in regards to using the navigation system.

Since the development of technical aspects of the ARN system still at a premature stage, the experiments were carried out without a fully working ARN system. To overcome the obstacles, navigation video clips of alternatives were produced according to experiment designs. All the experiments were conducted in a car simulator environment to provide a resemblance of real situation as shown in Fig 1. A large projector screen was set up in front of the car simulator to provide the images of forthcoming road, and a 7-inch navigation display device was installed on the right hand side of the handle, playing the alternative video clips prepared for each experiments.

Fig. 1. Car simulator with a large projector screen and 7-inch navigation display device

Experiment subjects were given sufficient time to practice driving in the simulator and were told to control the steering wheel and the accelerator as if he/she were actually driving during the experiment. After being shown each alternative video clip in a random balanced order, the subjects completed corresponding questionnaires.

2.1 Experiment Subjects

The experiments were carried out with 24 participants. All subjects had at least 2 years of driving experience. We selected the subjects having subgroups of navigation experienced users (at least for 1 year) / inexperienced users, male / female, young (20~34) / middle & old (37~55) aged, trying to represent potential users of the ARN system. In addition to that, their eyesight scores were recorded for further analysis.

3 ARN and MBN Display Arrangement Experiment

To find out the preference of arrangement of the display with ARN and MBN, experts in HCI evaluated possible alternatives, and extracted 6 alternatives to be tested with an experiment (Fig. 2). Short navigation video clips were produced with the alternatives applied to the partitions of the screen.

3.1 Experiment Design

The Analytic Hierarchy Process (AHP) was used to assess the preference levels. As assessment criteria *convenience of display operation, confusion, field of vision, and aversion of change* were chosen. After going through all six arrangements, each subject was asked to complete a questionnaire.

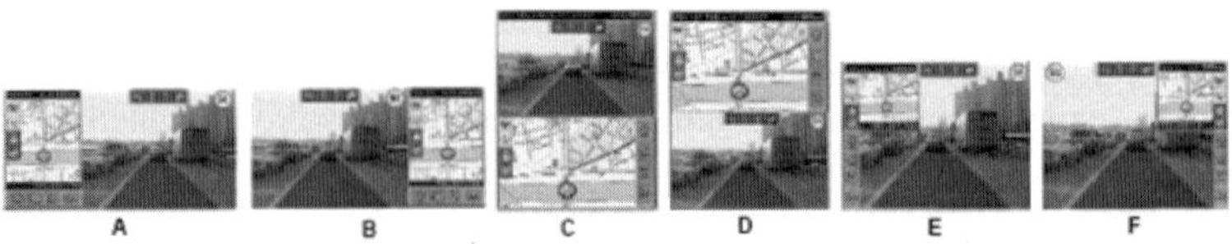

Fig. 2. Six types of ARN and MBN display arrangements

3.2 Results and Analysis

The preference levels were analyzed with 22 subjects' data out of a total of 24 experiment subjects. Two subjects were omitted from the analysis because their data had Consistency Ratio (CR) larger than 0.2 Normally, according to Saaty, a CR should be less than 0.1 [5]. However, Jensen et al. [6] asserted that a CR slightly above 0.1 may not have any significant influence on the level of consistency. Table 1 shows the results of the preference levels of certain groups for each alternative. High preference levels (> 0.15) are shaded gray.

Table 1. Total preference levels and group preference levels for each alternative

Alter-native	Total Preference	Male	Female	Young Aged	Middle & Old Aged	Experien-ced User	Inexperien-ced User
A	0.1998	0.1819	0.2190	0.1878	0.2087	0.1975	0.2088
B	0.1828	0.1942	0.1401	0.1520	0.2072	0.1906	0.1372
C	0.1538	0.1317	0.1399	0.1697	0.1153	0.1919	0.0861
D	0.1196	0.1169	0.1242	0.1511	0.1021	0.1127	0.1202
E	0.1178	0.1130	0.1204	0.1212	0.1134	0.0992	0.1491
F	0.0958	0.1056	0.0831	0.0735	0.1120	0.0841	0.1149

Among all the alternatives, Alternative A has the highest total preference level. Alternative A's arrangement is very similar to current navigation systems; thus, we could say that the subjects were already comfortable with the design. Moreover, for Alternative A, the proximity between the driver and the navigation device made it a preferable option because the icons operated by the user are located on the left side of the map display and the navigation device itself is located on the right-hand side of the driver (if the handle is located on the left as in Korea).

However, the difference in preference between each group suggests that, in due time, the preferred display arrangement will change depending on the target market of ARN systems. Alternative B has quite a high preference level, almost as high as Alternative A, and Alternative C was preferred among the young age-group and experienced users. This shows that it is advisable to design navigation devices so that users can switch it left, right, upwards, and downwards and also have a pivot function so that Alternatives A, B, and C are all possible with one device, as depicted in Fig 3.

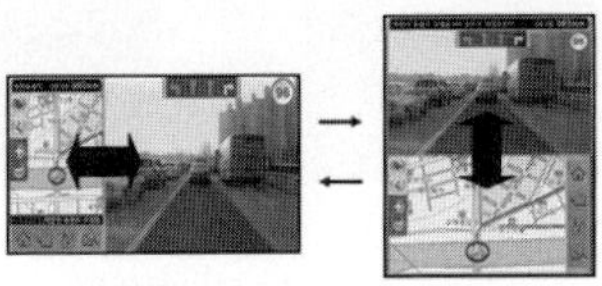

Fig. 3. Recommended alternative: A navigation system with switch and pivot functions

4 Information Entities Arrangement Experiment

By studying navigation systems that are out in the market and examining comments from the Focus Group Interview conducted previously, 23 information entities were enumerated and categorized depending on the priority and type.

As a heuristic evaluation, a program using Macromedia Flash was created, allowing us to change the size and ratio of the information entities, and to arrange them liberally on the display. Eight usability engineers produced their own design layouts of the entities, and the outcome was reviewed with a thorough discussion. As a result, commonly placed and indispensable information entities for efficient navigation guidance were selected and were arranged on a navigation screen. Information entities with a high frequency count were considered minimal information that was mandatory to display. These included *primary prediction information, lane information, map display, current direction, and the speed limit.* It was decided that further experiments were needed for deciding the position and method of presenting the primary prediction information which is comprised of lane and direction changing notifications, as they were considered to be the most important elements for the navigation functionality.

4.1 Experiment Design

To figure out the preference of the information display on changing lanes, three video clips were made: using an icon on the lower right (A), using the upper traffic lane panel (B), and highlighting the route to be taken on the road itself (C).

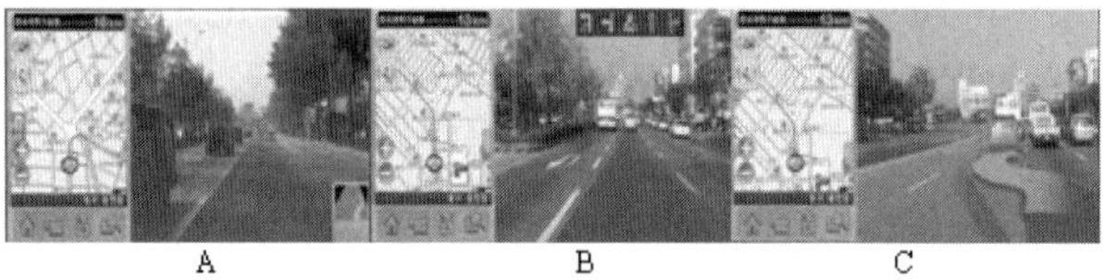

Fig. 4. Still cuts of the three different alternatives on changing lanes

To find out the preference of the information display on switching directions, four video clips were made: using an icon on the lower right (A), using an icon on the upper side (B), displaying it on the traffic light (C), and highlighting it on the road (D).

Fig. 5. Still cuts on the four alternatives for switching directions

We decided to omit the AHP method during this experiment after deciding that it would be a burden on the experiment subjects. Instead we created an easier questionnaire using a Thurstone's paired comparison method [7].

All subjective opinions and the paired comparison results were recorded through the questionnaires. For the lane-changing experiment, the handle signal time the experiment subjects actually changed lanes after the lane-changing information was displayed, were recorded. This was by using a custom-developed event capture program called e-Cap.

4.2 Results and Analysis

The results of the experiment showed that the consistency index was all 1.0, which meant that all 24 experiment subjects answered the questionnaire consistently. Fig. 6 is a graph of the questionnaire results on Thurstone's paired comparison method. The graph has been transformed so that the relative minimum preference level is zero and so, the higher the number, the higher the preference level.

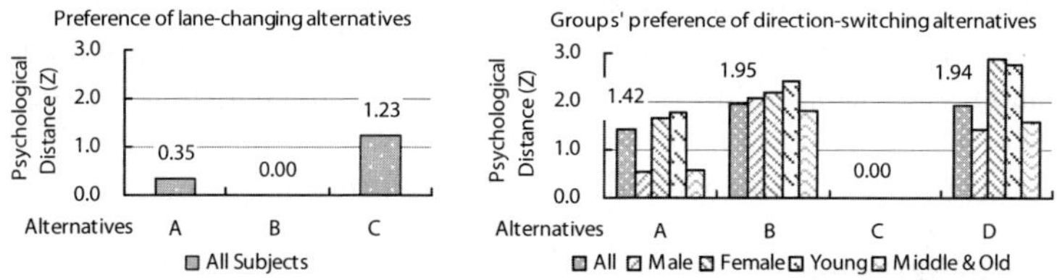

Fig. 6. Graphs of lane-changing and direction-switching alternatives' preference

For the lane-changing experiment, Alternative C was the most preferred while on the other hand, Alternative B was the least preferred. This result was the same throughout with no dependencies on sex, age, or whether experienced or not. In particular, the subjects commented on how easy it was to understand Alternative C because of the wide field of vision and intuitiveness of the design.

As for the direction-switching experiment, Alternatives B and D had close preference levels while Alternative C was preferred the least. The subjects remarked that Alternatives B and D had high recognizability while Alternative C was confusing and interfered with the view. There were slight differences in the preference levels among the groups with more men than women and more young than middle-aged and old-aged groups preferring Alternative D over Alternative B.

Finally, after analyzing the response time, we found that, for all experiment subjects, Alternative C had the shortest response time with 7.9949 seconds followed by Alternatives A and B, each with 11.9981 and 12.8776 seconds respectively. In other words, displaying the lane-switching information directly on the road and highlighting it caused the subjects to change lanes faster than the other alternatives. This also matches the subjective responses for Alternative C that stated it was intuitive, clear, and accurate. Looking at Fig. 7 we can see that the faster response time came from the inexperienced navigation users than experienced users, and middle & old aged rather than the young aged. We could interpret this as a result of inexperienced users who are not comfortable with navigation devices and middle & old aged subjects who have relatively low instant-response times, preparing for real situations ahead of time.

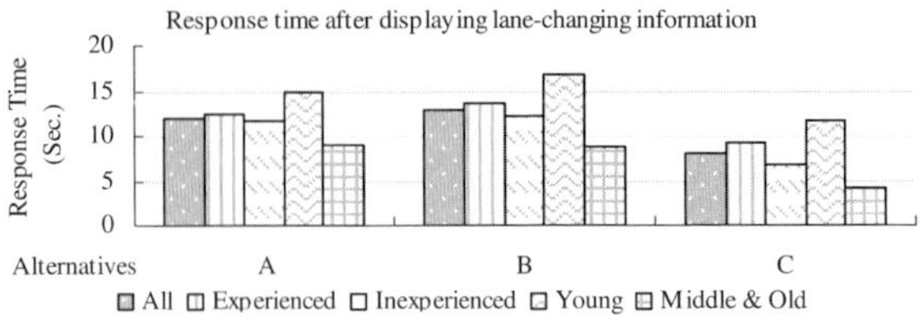

Fig. 7. Graph of each group's response time after lane-switching information was displayed

5 Adequate Information Level Experiment

Among the overlaid information presented on the ARN screen, landmarks had been examined with a Taguchi Method to find the adequate level of the presentation. Since the landmarks, such as buildings and road names, so-called POI (Point of Interest), occupy relatively large portion of the screen compared to the other fixed icons, they have a significant influence on overall amount of information.

5.1 Experiment Design

A Taguchi Experiment allows obtaining the optimal level of controlled parameters that has significant influence on the overall output with a reduced number of experiment iterations [8]. We used $L_9(3^4)$ Orthogonal Array, having three controlled parameters as A) the average number of POIs in ARN screen, B) average number of POIs in MBN screen, and C) respective text sizes in ARN & MBN screen [9].

Having applied the results from the prior experiments, 9 movie clips were produced according to the Taguchi's $L_9(3^4)$ Orthogonal Array design. In this experiment, subjects were asked to recall the names of POIs they had seen on each clip, and their scores later were used as performance measures. To minimize the learning effect, every clip was produced with evenly distributed POIs, but with different names. Questionnaires about the preference in overall consideration of safety, efficiency, intuitive perception, and visual field assurance, had also been answered.

Table 2. Controlled Parameters in Taguchi $L_9(3^4)$ design

	A) Average # of POIs in ARN	B) Average # of POIs in MBN	C) Text Size (ARN/MBN)
Level 1	1 (±1)	6 (±2)	16pt / 14pt
Level 2	3 (±1)	10 (±2)	18pt / 16pt
Level 3	5 (±1)	14 (±2)	20pt / 18pt

Since the experiment considered both preference and performance measure as the response variable, a procedure to combine the two measures was required. With the equations (1) and (2) below, the measures were scaled having values between 0 and 1.

$$Scaled\ Preference = \frac{(Preference - \frac{Min + Max}{2})}{Max - Min} + 0.5 \tag{1}$$

$$ScaledPerformance = \frac{Performance}{5} \tag{2}$$

After a comprehensive discussion with HCI experts, the magnitude of contribution of the two measures to the response variable had been decided as a case of a Multi-characteristics parameter [10].

$$Y = Preference^{0.7} \times Performance^{0.3} \tag{3}$$

5.2 Results and Analysis

The ANOVA in Table 3 indicates that the parameters A and C have significant influence, but not for B when α=0.05.

Table 3. Analysis of variance for means

Source	Sum of Squares	Degree of Freedom	Mean Sum of Squares	F_0=V/V(e)	p
A) # of POIs in ARN	0.126914	2	0.063457	74.24	0.013
B) # of POIs in MBN	0.028134	2	0.014067	16.46	0.057
C) Text Size	0.034811	2	0.017405	20.36	0.047
Residual Error	0.001710	2	0.000855		
Total	0.191568	8			

The optimal levels of the parameters (Fig. 8) are A) Level 2, C) Level 3, i.e. to have 3 POIs in average in ARN screen and 20pt text size in ARN and 18pt in MBN screen. Since the parameter B, the average number of POIs in MBN, turned out to be not significant, its level can be liberally decided in the needs of economical efficiency, convenience of technical implementation, or design issues. However, the p value of B is still comparatively small that it could be considered significant if we set the α generously to 0.1, suggesting us to have in average 10 POIs in MBN screen.

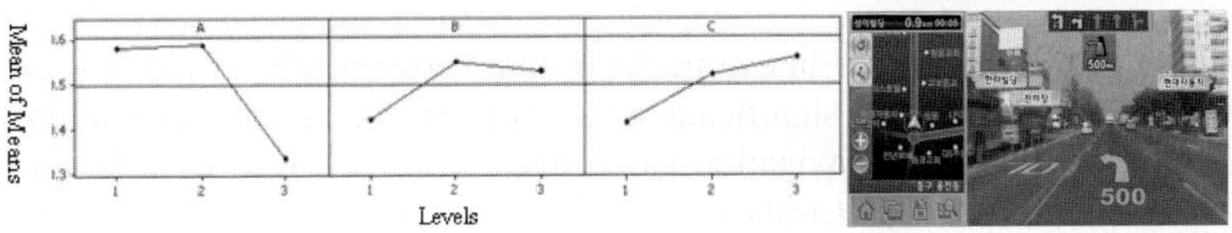

Fig. 8. Main effects plot for means (*left*), An example of navigation screen with the optimal level of information presentation (A: 2, B: 2, C: 3 Levels) (*right*)

A Further analysis with the subgroups of subjects had been carried out, deriving the result in Table 4.

Table 4. Analysis of main effects by subgroups. Significant parameters are shaded gray.

	A	B	C		A	B	C
Experienced Users	1	2	3	Inexperienced Users	2	2	3
Young Aged	2	3	3	Middle & Old Aged	1	2	3
Female	1	2	2	Male	2	2	3
Eyesight < 1.0	1	2	3	Eyesight ≥1.0	2	3	3

The results with significant parameters are rather similar to the results with all subjects, i.e. A:2 B:2 C:3. However, the experienced users and the ones with eyesight score less than 1.0 preferred less POIs in ARN screen, whereas the young aged group and the ones with eyesight score more than 1.0 preferred more POIs in MBN screen. Examining the comments of subjects recorded after the experiment, the former two groups pointed out the complexity of screen when there are too many POIs in ARN screen, which hinders perception ability. Correspondingly, the latter two groups' preference is thought to be due to their relatively higher ability in perceiving information in a shorter period of time. Although more research is called for to have an in-depth study on the cognitive cause of the difference, here we illuminated the existence of the difference of preference. These results could be put to practical use by developing a customizable user interface that sets the parameters according to the personal characteristics.

6 Eye Tracking Experiment

An eye tracking camera had been utilized to identify a preferring screen to which subjects gazed mostly to acquire necessary navigation information.

6.1 Experiment Design

With the results of prior experiments applied, two driving scenario navigation clips were produced; one with urban driving and the other with highway. The experiment was carried out using a faceLab 4.3 eye tracking camera, and subjects were carefully instructed to simulate their eye gaze as if they were on the actual road.

6.2 Results and Analysis

Although a certain difference in eye tracking data between urban roads and highway driving was anticipated, no significant differences in the rate of navigation gaze / overall gaze, MBN gaze / navigation gaze, ARN gaze / navigation gaze, and dwell time, were observed through t-tests.

Table 5. Average of eye tracking gaze data in urban roads and highway driving situation

	Navi./Overall	MBN/Navi.	ARN/Navi.	Dwell Time (sec.)
Urban roads	14.94%	32.26%	67.73%	0.5
Highway	14.10%	28.54%	71.46%	0.42
Average	14.52%	30.40%	69.60%	0.46

However, the gaze rate of ARN/navigation and MBN/navigation clearly showed a significant difference (with $p<0.0001$ in paired t-test), revealing that the ARN was viewed more often approximately with a ratio of 7:3. Combining with the subjects' remarks that ARN screen is perceived more intuitively than the MBN, the results of eye tracking experiment form an objective evidence proving that ARN holds a dominant position than the MBN. However, further research in cognitive engineering is necessary to clarify this assertion as one could interpret this phenomenon due to more mental workload perceiving information from ARN.

Additionally, the average of dwell time, which is the time of gaze dwelling on each glance at the navigation display, had been analyzed. This value (0.46 sec.) could be utilized as a cycle time or frequency of dynamic information to be implemented, for example, a direction indicator with flowing design attributes or flickering signals.

7 Conclusion

This study focused on usability issues of user interface in Augmented Reality navigation systems, putting aside the technical aspects of them. With the optimal alternatives found through the series of usability evaluation experiments done thus far, a basic design of the in-car Augmented Reality navigation system was developed. Through the additional analyzing done on different subgroups, it was verified that the usability could be increased by customizing the user interface depending on personal characteristics.

Because this study was conducted at the premature stage of the development of Augmented Reality navigation systems, there had been much limitations in conducting experiments, even though they were prerequisites for the initial design. Hence, when an operational AR navigation prototype is available, additional experiments in a *real driving situation* must be carried on to ensure the safety issues, as well as the usability.

Nonetheless, it is expected that the findings of our study in relation to the information disposition and the adequate level of information presentation, will form the basis of development of user interface in an in-car Augmented Reality navigation system. Furthermore, the research data and results could be used as the groundwork for future research, such as designing additional functions or creating new interface designs.

References

1. Hu, Z., Uchimura, K.: Real-time data fusion on tracking camera pose for direct visual guidance. In: IEEE Intelligent Vehicles Symposium, Proceedings, pp. 842–847 (2004)
2. Narzt, W., Pomberger, G., Ferscha, A., Kolb, D., Müeller, R., Wieghardt, J., et al.: Augmented reality navigation systems. Universal Access in the Information Society 4(3), 177–187 (2006)
3. Burnett, G.: A road-based evaluation of a head-up display for presenting navigation information. In: Int. Conference on Human-Computer Interaction, pp. 180–184 (2003)
4. Tonnis, M., Sandor, C., Klinker, G., Lange, C., Bubb, H.: Experimental evaluation of an augmented reality visualization for directing a car driver's attention. In: Proceedings. Int. Symposium on Mixed and Augmented Reality, pp. 56–59. Vienna, Austria (October 5-8, 2005)

5. Saaty, T.L.: The Analytic Hierarchy Process. McGraw-Hill, New York (1980)
6. Jensen, R.E., Spencer, R.W.: Matrix scaling of subjective probabilities of economic forecasts. Economics Letters 20, 221–225 (1986)
7. Thurstone, L.L.: The Measurement of Values. The University of Chicago Press, Chicago (1959)
8. Peace, G.S.: Taguchi Methods. Addison-Wesley, London (1993)
9. Taguchi, G., Konishi, S.: Orthogonal Arrays and Linear Graphs. ASI press, Dearborn, MI (1987)
10. Lee, P.H., Yum, B.J.: Multi-characteristics parameter design: A desirability function approach based on process capability indices. International Journal of Reliability, Quality and Safety Engineering 10(4), 445–461 (2003)

Validation of Critical Parameters for Predictive Evaluation of Notification System in Avionics Interfaces

Miguel A. Sánchez-Puebla, Ignacio Aedo, and Paloma Díaz

University Carlos III of Madrid. Computing Science Dpt., Avda. de la Universidad, 30
28911 Leganés (Madrid) Spain
{MiguelAngel.Sanchez-Puebla, Ignacio.Aedo, Mpaloma.Diaz}@uc3m.es

Abstract. The evaluation of notification systems is rare in critical systems. The only available experiences are based on empirical data on costly simulated environments and the subjective follow-up of some guidebooks probed by the experience. In case of absence of performance requirements, critical parameters can provide quantitative measures that may be used to determine the overall usability of interfaces that continuously present precious information (*id est.* Notification Systems) in Avionics (NSA) and should be validated before use to improve the confidence level in the measurement unit. In this paper, we introduce a set of critical parameters that could be used for usability evaluation of NSA's interfaces. A process for a formal validation is described and results of the applicability of these parameters to the real NSA interface of A400M's power plant are presented.

Keywords: Human Factors, Critical Parameters, Notification Systems, Predictive Evaluation, Evaluation/methodology.

1 Introduction

The user of a computer interface needs information, as any human being whatever his/her culture or activity. Different usage situations, expectations, and error consequences govern the growing breed of applications and devices being introduced to support multitasking information demands. Referred to as notification systems, these interfaces are generally conceived as a means to access valuable information in an efficient and effective manner without introducing unwanted interruption to a primary task.

On the other hand, a system will be considered as critical when a sole failure due to the system, the user, or the way the user executes the tasks that the system is designed for, could produce serious damage to the system itself, in its surrounds or even tragic damage to the user.

Evaluation of this type of interfaces should provide direct measurement units to determine their usability and to compare two interfaces objectively. However, even though computer users are extensively used for notification systems, their evaluation poorly developed on interfaces for systems in general [12] and rare in critical systems.

M.J. Smith, G. Salvendy (Eds.): Human Interface, Part II, HCII 2007, LNCS 4558, pp. 1109–1118, 2007.

For NSAs, the only available experiences are based on empirical data on very costly simulated environments and the follow-up of some guidebooks [5], [22].

Currently, the performance measurements on NSA interfaces do not refer to specific requirements prior to design. It is notable that there is an absence of this kind of requirements in a NSA's interface development life cycle. In such a case, a set of critical parameters could be introduced in order to define measurement units [14].

Newman introduced the concept of critical parameters for HCI as a mechanism to enable significant modeling and execution of usability evaluations that would allow systems to become gradually better [16]. These figures of merit, when defined and adopted, would help interface designers and evaluators to recognize the broader intentions of the technology, shifting focus away from interface specific details to qualities that could be directly measured, compared to benchmarks, and reengineered to better serve the user's purpose. According to [16], critical parameters have three essential characteristics: their satisfaction is critical to the success of the system, they are persistent across successive systems, and must be manipulable by designers.

The problem lies in the lack of identified critical parameters in the design of NSAs.

Following there are the main deficiencies of actual usability evaluation methods (heuristic evaluation, cognitive walkthroughs, empirical studies and inspections) applied to the interfaces of NSAs that can be mitigated by a predictive evaluation based on critical parameters:

- Costly evaluations because of the need of the real interface, expert evaluators and real users,
- Subjective evaluation due to the lack of cognitive theories as basis, and highly dependent on users and evaluators [7],
- Inability to conduct systematic and repeatable evaluations [13],
- Limited capacity to provide quantitative data [9].

In this paper, we introduce a set of critical parameters that could be used for predictive usability evaluation of NSA's interfaces. Next, in section 0, a formal validation process is also described. This is followed in section 0 by an example of application of this set to the interface of the power plant for the A400M. Lastly, in section 0, some conclusions and future works are made.

2 A Set of Critical Parameters for NSA

In [12], the authors consider interruption, reaction and comprehension as critical parameters for notification systems. Unfortunately, these parameters do not cover some key aspect of on-board critical systems as the workload due to multiple tasks and the situational context.

The identified key critical parameter for interfaces of NSAs may be stated as the cost of the notification. That is, the ratio between the information transmitted to the user, and the information lost because of the interruption inherent to the notification. However, this approach implies the availability in advance of the interface for the primary control system in opposition to benefits brought by a predictive and decoupled evaluation. Instead of the cost of the notification, other parameters may be used as an indirect measure of the usability. To illustrate the difference, we can make an analogy with the number of aircrafts using a runway: the track of the tyres due to the landings can be used to measure the number of aircrafts using the runway if it would not be possible to make a direct count.

The following set of critical parameters is then proposed for predictive evaluation:

(1) Information complexity and overload, (2) Multitask and workload, (3) Modes and multiple options, and (4) Context awareness and abnormal scenarios.

2.1 Information Complexity and Overload (ICO)

At any time the user needs to maintain a correct mental model of the system in a trade-off with a minimal amount of required information. As this increases, tasks become more difficult as secondary information can detract cognitive resource from the interpretation of information necessary for the primary task. This minimal and satisfactory set of information together with the right reception of the notification are the leading objectives of each NSA.

The evaluation of this parameter accounts for the amount of information with the well-known Equation 1:

$$H_{ave} = \sum_{i=1}^{N} P_i \left[\log_2 \left(\frac{1}{P_i} \right) \right]. \tag{1}$$

Where H_{ave} is the number of bits/chunks included in a notification, P_i is the probability for the event i to occur, and N is the total amount of possible events. The probability of occurrence of an abnormal or emergency event is sound analyzed and limited on avionics design per the classification stated in [2].Thus upper limit of probability ranges between 10^{-3} to 10^{-9}, corresponding to non- safety effects on the aircraft to events leading to catastrophic results.

This number of bits/chunks is then weighed-up according to the number of different orthogonal and correlated dimensions; if the same notification is sent through different channels (e.g. auditory, visual, hectic…) then the number of transmitted bits/chunks corresponds to the notification with the minimal number. Redundancy is required on critical systems to ensure that important notifications are transmitted and received by the pilots. In this way if one engine will be shutdown then this event will be notified on the displays but always with the same shape, size and on the same location. In addition, it will be notified through the loudspeaker (a repetitive chime) and a red light will start flashing; six different means of notification on two different orthogonal and correlated dimensions (i.e. auditory and visual).

The criterion for a successful evaluation is based on the magic number 7±2 (more than 3 bits implies likely information overload) and a redundancy level higher or equal to 100%, which corresponds to a notification sent, at least, for two orthogonal and correlated dimensions. The relationship between the entropy H and the magic number has also been addressed in other contexts like chess [18] and air traffic [25].

2.2 Multitask and Workload (MTW)

In [4], the workload in pilots is defined as an integration of physical and mental efforts perceived by the pilots when performing a specific task in flight. This definition implies a subjective perception stressed if there is a lack of acknowledge on the users performing the task. Human are very adaptable but they adapt at varying rates whith varying degrees of mental and physical efforts. Thus, what some pilots might find acceptable, others would reject as being unusable and unsafe.

Predictive evaluation activities should determine if the user is overloaded (qualitative data) and which degree (quantitative data) when a notification arrives whilst the *locus* [19] of attention of user was on the control task.

Qualitative data is provided based on the decision taken for the user overload; for all tasks T_i derived from any control C and notification N actions[1], the sum of all (n) allocated resources plus the required resources to attend the notification (R_i) shall be equal or less that the maximum capacity of the users MC as showed on Equation 2, related with the decision criteria for user overload. If this sum is bigger than the user's capacity s/he becomes overloaded and the NSA's interface must be redesigned or just discarded.

$$\sum_{1}^{n} R_i \leq MC \quad \forall T_i \in \{C\} \cup \{N\}. \tag{2}$$

This interaction is modeled in a similar way to the diode behavior in line with the model proposed by Waard in [23]. This behavior is also weight-up for each profile of users (i.e. depending of theirs cognitive resources and experience in similar interfaces) and the scenario/context of the interaction.

2.3 Modes and Multiple Options (MMO)

Interfaces designed with multiple options for the accomplishment of the same task may shift the *locus* of user attention from the task related with the NSA to the selection of the method for the task accomplishment [19]. The ideal situation presumes that the interface of the NSA will be evaluated by using GOMS without paying attention to the 'S' corresponding to the selection rules because there is only one way to complete successfully the task and then the user does not apply any cognitive resource choosing the selection rules. Multiple options are identified on a NSA's interface through the detection of keys combinations which include special keys (CTL, ALT, ESC).

The evaluation related with MMO could support the analysis activities by providing qualitative data showing whether each notification can be effectively handled with a unique sequence of tasks.

2.4 Context Awareness and Abnormal Scenarios (CAAS)

The occurrence of any hazardous event (like engine shutdown or engine fire) can lead to critical and degraded situations where the user is under stressful conditions. In this type of situation the higher the level of attention paid by the pilots the more intensive shall be the necessary stimulus for changing the *locus* of attention of the users [24]. In these scenarios with an increasing level of stress, the user becomes more and more

[1] All alerting systems generally perform four functions: monitoring, situation assessment, attention-getting and problem resolution [21].

concentrated on a few features of the environment with no attention at all to the others [25]. It is very improbable that the users will pay attention to any further notification provided by the NSA; as a result of this, additional notifications can add confusion and stress making the situation more difficult and/or the user can miss those notifications.

Garrison in [6] describes an accident on December of 1972, where 101 people died due to the aircraft crash in the Everglades. The three persons of the crewmembers were absorbed in a failure of the landing gear on the approach phase; even though there were continuous warnings (both aural and visual), they did not realize the low aircraft altitude and they could not avoid crashing into the mountains.

- The predictive evaluation has to cover all possible changes of scenarios and contexts in a dynamic way. In addition, the predictive evaluation has to verify that the less important notifications will not disturb the delivery of the most important ones by starving the resources of the user and consequently there is no loss of notifications whatever their notification priority level.

In consequence, the predictive evaluation related with CAAS parameter should support the analysis activities of the evaluation by examination of the following:

- All notifications have a priority, notifications, ranged 0 for information or recognition situations with non-safety effects up to 4 for alerts signaling catastrophic consequences either on the crew or in the aircraft.
- There is no loss of any notification; the highest priority notifications are delivered to the pilot before a predefined timeout occurs. Quantitative data accounting the number of transmitted, pending and lost notifications are reported.

3 Validation of Critical Parameters

The above set of parameter has been chosen taking into account the cognitive constraints of the user in a divided attention environment with abnormal or emergency scenarios according to the main cognitive theories (Adaptative Control of Though ACT-R [1], Interacting Cognitive Subsystems [1138], Executive-Process/Interactive Control [1138], and the 4-dimensions shared resources theory [1138]).

Because these critical parameters are mainly founded on HCI theory it seems to be possible a formal demonstration of their relationship with usability based on the comprehensive metrics validation criteria [20] that has six validity criteria: association, consistency, discriminative power, tracking, predictability and repeatability.

Principle of validation imposes a statistical relationship between each critical parameter i and the usability, with confidence level α (typically ≤ 0.05) and with threshold value β_i (≥ 0.7).

When parametric validation is unavailable or unfeasible, the principle of validation allows also nonparametric statistical methods to support validation for the consistency, discriminative power and tracking validity criteria. This could be applicable whether the relationship between usability and a critical parameter is collectively accepted, like the use of the Index 35 stock price average as leading indicator of the health of the Spanish economy. There are numerous references (ICO:

[3], [17], [24]; MTW: [25], [24], [4]; MMO [19], [17]; CAAS [19],[24], [25]) for the substantiation of these relations between usability and each of the proposed critical parameter.

Therefore it is possible a validation less demanding than those that apply to parametrical methods. However, this should not lead to the conclusion than nonparametric methods are less rigorous than parametric methods.

The main application of critical parameters as a metric of usability validated with nonparametric statistical methods can be derived from the ability to classify as being "less usability" or "higher usability".

4 Study Case – A400m Power Plant Notification System

The application example is the interface of the NSA for the power plant of A400M, a modern aircraft (currently in the final development phase) used for military transport. The interface of this NSA is composed of (see Fig. 1):

- Two redundant 6.17"x 8.22" usable display area (one per user) with video capability through optics fiber and bitmap capability through the Aircraft Full DupleX network (AFDX),
- Dedicated lights for caution and emergency advise for visual notification and
- Two loudspeakers for auditory notifications.

4.1 Method

The approach to usability evaluation for the design of Human-Computer Interface of notification systems for critical design in avionics is proposed in [12] in several steps: (1) to establish the set of critical parameters based on psychological models, (2) to carry out a predictive usability evaluation based on cognitive theories and the quantification of these parameters, and (3) to apply in an incremental way the design guidelines based on a scenario strategy.

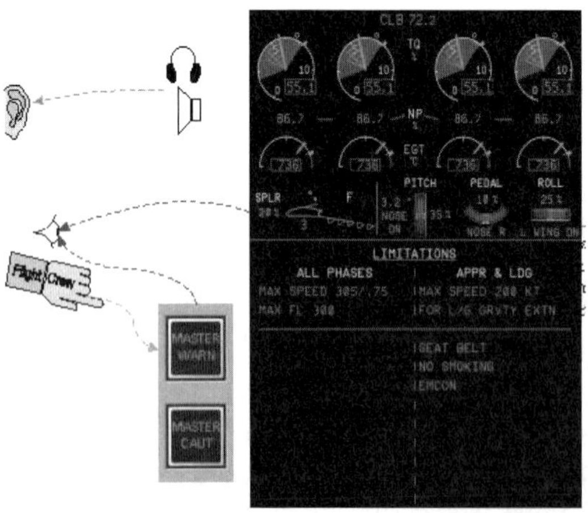

Fig. 1. The interface of NSA for the A400M power plant includes auditory and redundant visual notifications

According to this approach, the first step has been to define the set of critical parameters. The set of proposed critical parameters has been described in section 1136 and the selected scenarios for the evaluation are: (1) Normal operation, (2) Landing in combat, (3) Bad weather conditions and (4) System failures.

As part of the evaluation, a predictive model based on this set of critical parameters has been built and applied to the interface on Fig. 1. The results provided by the mentioned model have been checked against the earliest user testing data (provided in [8]) in a high level prototype with four expert pilots from different countries.

4.2 Results and Discussion

See below a précis of the expected values from the predictive model related with the set of parameters when applied to the notification system of Fig. 1:

ICO

```
Message : POWER_RATING_VALUE
Redundancy :  100.00 %
Component with 3.30 bits/chuncks of information. It
shall be reduced  0.30 bits
```

The previous results provide quantitative data to support analysis and critique activities of usability evaluation permitting to the evaluator/designer to decide, based on objective data, whether the notification can still carry more information or whether the interface should add more redundancies or else the notification should be split into lighter ones.

MTW

```
User COPILOT : expert and acquainted.
Estimated workload in BASIC context:   59.89%
Estimated workload in LANDING_IN_COMBAT context:
119.79%
Estimated workload in BAD_WEATHER_CONDITIONS context:
119.79%
Estimated workload in FAILURE_CONDITIONS context:
89.84%
```

Quantitative data permit the evaluator/designer selection of alternative design,s which implies less workload on the pilots.

MMO

```
DIGITAL_READ component includes modes to be eliminated.
```

Modes and multiple options are undesirable features for the interfaces of NSAs because they require dedicating some resources to the mode selection [1138]. Predictive detection of these features can be achieved in the very early stages of the design and simplifying (by removal of the additional options and modes) the design.

CAAS

```
Notification i-th : NUMERIC_VALUE_OF_LIMITATION_MODE
Next notification: CAUTION_WARNING
Sent notifications  1
Pending interruptions   2
Pending notifications  0
```

The main aim of any NSA is to ensure the delivery of the right notification just in time to the users without any loss of notifications. The evaluation of this aim includes a real-time random mechanism, which produces events with a random level of priority covering almost all of the possible cases. Thus, the designers know in the very early stages of the development if there are some bottlenecks in the NSA notification generation or in the event transmission dispatcher.

5 Conclusions and Future Works

We consider that predictive evaluation of NSAs can be successfully assessed on critical parameters in two cases:

1. When there are non-specific requirements previous to the interface design of the NSA.
2. When evaluation shall be conducted sufficiently early in the life cycle to permit usability to be assessed, controlled and predicted. We also note that because of the cost of solving errors grows rapidly with the life cycle, it is advantageous to have early indication of usability.

Analytical models of usability can be built for the interface of NSAs against these parameters and they can be established, as predictive evaluation methods, which can assess how better is the interface at doing the job intended.

As another consequence, value target for critical parameters can be set at the outset of the design as usability goals evaluated with these predictive methods, or even critical parameters can be used in writing requirements and specifications for NSAs.

We believe this approach holds promise for self-usability evaluation on NSAs, and are pursuing to refine the predictive models and increase the usability features covered to spread its use to any notification system. This proposal requires additional empirical studies in order to determine standard user parameters in cases where these parameters have not been previously defined and validated. Since few instances of user testing evaluation with NSAs exist today it is difficult to generalize the results obtained (i.e. four users is not statistically representative).

The present work represents the first cycle of an iterative process through which we seek to unify HCI theories and critical parameters in predictive models; validate and tune them with experimental data coming from user testing evaluations or extrapolate the results when user testing is not feasible (e.g. evaluate the behavior of the NSA in a scenario with total flame-out in all engines). We have enabled additional opportunities for further predictive models through this approach.

Opposite to general understanding, predictive usability evaluation for interfaces on NSAs may be seen as the basis and not a complement or a substitute. The approach described permits to designers the evaluation of their own interface without requiring expert evaluators. Its intensive use may lead to an eXtreme eValuation (XV) process supervising each decision design with objective data and used as decision criteria for next steps on the life cycle. Quantitative (for ICO, MTW and CAAS) data can be seen as a maturity measurement of the design as well: the number of iterations can stop when the quantitative set of values for critical parameter is reached. For MTW and MMO qualitative data an inspection evaluation can be conducted and constraints violations can be perceived as potential usability problems. The designer can alter the interface design or resolve violations, and ideally improve the interface as well from the very early stages of the design.

Critical parameters (ICO, MTW, MMO and CAAS) can be considered as dimensions to unambiguously specify an interface in the space of design defined by the set of critical parameters for NSAs. A formal characterization of this sort of interfaces will be considered in the next iteration of development of this evaluation framework.

Acknowledgments. This work is part of the project ARCE++ (TSI2004-03394) funded by MCT (Spain).

References

1. ACT-R06 (2005) http://act.psy.cmu.edu
2. ARINC. RTCA /DO-178B, Software Considerations in Airborne Systems and Equipment Certification, Special Committee 190, RTCA, Inc. (1992)
3. Baecker, R., Booth, K., Jovicic, S., McGrenere, J., Moore, G.: Reducing the Gap Between What Users Know and What They Need to Know. ACM Conference on Universal Usability. In: Proceedings on the 2000 conference on Universal Usability (2000)
4. Harper, C.y.: The use of Pilot Ratings in the Evaluation of Aircraft Handling Quality (1969)
5. Doernich, M.C., Ververs, P.M., Good, M.D.: Design and evaluation of an integrated avionics alerting system. Digital Avionic Systems. DASC. In: The 20th Conference, vol. 1, pp. 14–18, 5D1/1 – 5D1/12 (October 2001)
6. Peter, G.: Flying, vol. 121(12), p. 112 (1994)
7. Hart, S.G., Staveland, L.E.: Development of NASA-TLX (Task Load Index): Results of empirical and theoretical research. In: P.A. Hancock & N. Elsevier Science Publishers, B.V.: North-Holland (1988)
8. Colle, H.A., Reid, G.B.: Estimating a Mental Workload Redline in a Simulated Air-to-Ground Combat Mission. International Journal of Aviation Psychology 15(4), 303–319 (2005)
9. Ivory, M.Y., Hearst, M.A.: The State of the Art in Automating Usability Evaluation of User Interfaces. ACM Computing Surveys 33(4), 470–516 (2001)
10. Kieras, D.E.: EPIC Architecture Principles of Operation (2004)
11. Landauer, T.K.: Relations between cognitive psychology and computer system design. In: Carroll, J.M. (ed.) Interfacing Though: Cognitive aspects of Human-Computer Interaction, pp. 1–25. MIT Press, Cambridge (1987)
12. McCrickard, D.S., Chewar, C.M., Somerwell Jacob, P.N., Ali, N.: A model for notification systems evaluation – Addressing user goals for multitasking activity. In: Proc. TOCHI 2003, vol. 10(4), pp. 312–338. ACM Press, New York (2003)

13. Molich, R., Bevan, N., Butler, S., Curson, I., Kindlund, E., Kirakowski, J., Miller, D.: Comparative evaluation of usability tests. In: Proceedings of the UPA Conference, Washington, DC, pp. 189–200. Usability Professionals' Association, Chicago, IL (June 1998)
14. Newell, A.: Unified Theories of Cognition: The 1987 William James Lectures. Harvard University Press, Cambridge (1990)
15. Newman, W.M.: Better or just different? On the benefits of designing interactive systems in terms of critical parameters. In: Proceedings of the Conference on Designing Interactive Systems (DIS '97), pp. 239–245 (1997)
16. Newman, W.M., Taylor, A.S., Dance, C.R., Taylor, S.A.: Performance targets, models and innovation in interactive system design. In: Proc DIS'00, pp. 381–387 (2000)
17. Nielsen, J.: Usability Engineering. Academic Press, San Diego (1993)
18. Nievergelt, J.: Information Content of Chess Positions. SIGART Newsletter 62 (April 1977)
19. Raskin, J. (ed.): The Human Interface. In: New directions for designing Interactive Systems, Addison-Wesley, London (2000)
20. Schneindewing, N.F.: Methodology For Validating Software Metrics. IEEE Transactions on Software Engineering, 18(5) (1992)
21. Song, L., Kuchar, J.K.: Describing, Predicting, and Mitigating Dissonance Between Alerting Systems. In: 4th Int. Workshop on Human Error, Safety, and System Development (2001)
22. Swenson, H.N., Zelenka, R.E, Dearing, M.G., Hardy, G.H., Clark, R., Davis, T., Amatrudo, G., Zirkler, A.: Design and Flight Evaluation of an Integrated Navigation and Near-Terrain Helicopter Guidance System for Nighttime and Weather Operations. National Aeronautics and Space Administration (1994)
23. Waard, D.: The Measurement of Drivers' Mental Workload. The traffic Research Center VSC, University of Groningen 125 (1996)
24. Wickens, C.D., Hollands, J.G.: Engineering Psychology and Human Performance, 3rd edn. Prentice-Hall, Englewood Cliffs (2000)
25. Wickens, C.D.: Multiple resources and performance prediction. Theoretical Issues in Ergonomic Science 3(2), 159–177 (2002)

Development of an Augmented Vision Video Panorama Human-Machine Interface for Remote Airport Tower Operation

M. Schmidt, M. Rudolph, B. Werther, C. Möhlenbrink, and N. Fürstenau

German Aerospace Center, Institute of Flight Guidance, Lilienthalplatz 7, 38022 Braunschweig, Germany
{markus.schmidt, michael.rudolph, bernd.werther, christoph.moehlenbrink, norbert.fuerstenau}@dlr.de

Abstract. In this paper the development of a 180° high resolution video panorama system and results of initial field test at Braunschweig research airport are described. The system serves as main HMI for remote surface movement management of small airports or of movement areas not directly visible for the controller. It provides the framework for video-see-through augmented vision by integration of traffic and weather data and it allows for panorama replay. Preliminary evaluation of initial field tests quantify the visual resolution as compared to the real far view.

Keywords: Airport tower, traffic control, remote operation, video panorama, augmented vision, work analysis, cognitive modelling, field tests.

1 Introduction

Remote Tower Operation (RTO) describes the goal of surface movement management of small airports from a remotely located control center without direct far view to the airport surface. It is the first step on the way to the Tower Operations Center (TOC) for multiple airports and the Virtual Tower (ViTo) for large airports as long term goal [1]. Because small airfields often lack any advanced electronic surveillance system a high resolution augmented vision video panorama as a potential low cost system is proposed to replace the direct far view out of the tower windows as main component of the Human Machine Interface (HMI) [2].

A number of tower work analyses performed during the recent years found visual surveillance to be the most important activity of tower and apron controllers for creating their situational awareness, despite the availability of electronic surveillance [3][4]. In the tower environment of large airports the permanent refocusing between far view and displays contribute to the workload and increases head-down time which may both be reduced by a high resolution panorama display with distance to the operator comparable to radar and flight data displays. Consequently it is assumed that under the guideline of human centered automation, the reconstruction of the direct far view for small airports with a control tower, however without electronic surveillance,

M.J. Smith, G. Salvendy (Eds.): Human Interface, Part II, HCII 2007, LNCS 4558, pp. 1119–1128, 2007.

will improve the transition process to a towerless work environment and make it acceptable to the remotely located RTO controller. The detailed design of the RTO-HMI is supported by a structured tower work analysis [5][6]. The corresponding database is used as input into the simulation of different aspects of the tower controllers decision processes using cognitive and traffic process models realized with colored Petri nets [7]. The simulations in turn support the extraction of detailed expert knowledge during interviews with the controllers and they provide input for the human interface design.

Within the DLR project RapTOr (Remote Airport Tower Operation Research) an RTO experimental system is realized at the Braunschweig research airport [2] with an augmented vision video panorama as core of the RTO controller HMI. Information from real time image processing and electronic surveillance sensors like multilateration or GPS will be integrated into the digital videopanorama for realizing video see-through augmented tower vision (ATV). ATV has been proposed by several authors before, however aiming at augmenting the real far view by means of optical see through head mounted displays, e.g.[8]. Recently initial ATV demonstrations with superimposed information in the real tower environment have been performed by using a head-up holographic backprojection display [2].

In section 2 the tower work analysis and development of model based simulation are outlined which support the RTO HMI design. Section 3 describes the augmented vision video panorama system as basis of the experimental RTO HMI. The concept for video panorama validation experiments and initial results of field trials are described in section 4. Section 5 provides a conclusion and outlook.

2 Work Analysis and Model Supported RTO Workplace Design

The design and development of the experimental RTO HMI and the new Remote Controller working environment is supported by a formal cognitive work and task analysis (CWA) [5] by means of structured interviews of domain experts (controllers) from medium sized and small airports [2][6]. The formalised results serve as input data of a Formal Airport Control Model (FAirControl) for the simulation of the controller decision making processes at the tower work positions. In [9] it is shown how the results of a CWA on a medium size airport are transferred into an executable human machine model, based on Colored Petri Nets (CPN) for simulating the controllers work processes in relation to the airport processes. The executable model supports the identification of controllers' strategies in task organization and pursuance of goals. The formal model serves for evaluation of different variants of work organization, supports the design of a new working environment, and monitoring of psychological parameters, e.g. uncovering of reduced situational awareness.

Following Cacciabue [10] the human machine model is separated into submodels for the (1) human (controller), (2) interaction, and (3) machine (process) (Fig. 1). The interaction model defines the controller-process interactions and includes sub networks for description of information resources, such as radio communication and visual perception of the traffic situation. Consequently the human model(s) and machine model(s) can work independently from each other for certain time periods. The state of the airport process model determines the type and content of visual and electronic surface traffic information (e.g. usage of taxiways, landing clearance) which can be acquired and communicated by the controller. The controller model (human model) is implemented as a Formal Cognitive Resource (FCR) Model [11] and serves for the description of controller behaviour in the tower work environment. As most important feature this model considers the motivated character of human work as related to the limitations of cognitive resources [12]. The graphically represented formal work process model as depicted in Fig. 1 provides a valuable support for the communication between domain experts and system developers.

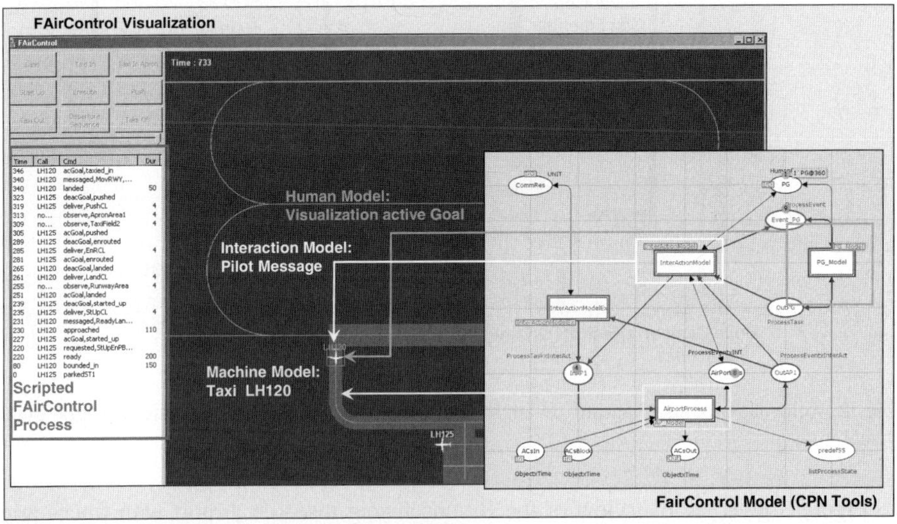

Fig. 1. FAirControl Visualization: CPN Model for simulation of interaction between the three model types (human, interaction, process; right side) and graphical visualization of the controlled work process on a simplified airport microworld. Currently pursued goal of the human model highlighted by a blue frame (orange arrow). By changing the colour of the call sign (here: LH120) the communication with the pilot is illustrated (white arrow).

3 Experimental RTO System

Motivated by the above mentioned relevance of visual information for tower work processes, a high resolution video panorama system has been set up at Brauschweig research airport as experimental environment for investigation of different aspects of the RTO HMI and development of a demonstrator [2]. A block diagram of the augmented vision video panorama system is depicted in Figure 2. The sensor component consists of four high resolution (1600 x 1200 pixels) high dynamic range (14 bit/pixel) CCD cameras ($P_{1,2,3,4}$) covering the Braunschweig airport within 180° and a remotely controlled pan-tilt zoom camera (P_5: PTZ), 400 m south of the runway which extends in E-W direction.

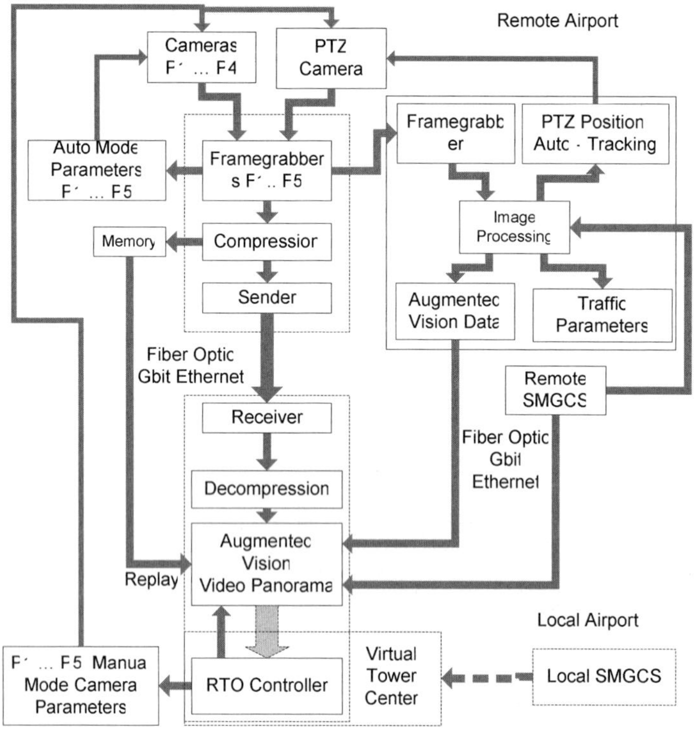

Fig. 2. Schematic block diagram of augmented vision video panorama system. Light blue: visual input.

Figure 3 gives an overview of the Braunschweig research airport with fiber-optic datalink connecting multilateration sensor containers with the main control center, and indicating camera position and viewing sectors. The cameras are positioned ca. 20 m above the airport surface, horizontally aligned on top of a building at the southern boundary of the airport with 100 m distance to Braunschweig tower, with horizontal alignment. The vertical aperture angle of about 20° (half angle with respect to the horizontal line of sight) allows for a closest surveillance distance of about 60 m. An optimistic estimate of the theoretically expected object resolution may be obtained by elementary optics and the given data of the electrooptical camera parameters using $G / B = (g/f - 1) \approx g / f$, with f = focal length = 12.5 mm, g = object distance, G = object size, B = image size. With a CCD pixel size of 7.5 µm (+ 0.5 µm gap) the vertical object size at g = 1 km distance corresponding to 1 Pixel is G / B = 0.6 m / 1 Pixel vertical, or ca. 2 arcmin angular resolution, and 1 m / 1 Pixel along the line of sight.

The observable resolution at the HMI, i.e. wide angle rear projection or monitor system, is reduced due to imperfect optics of the camera, the dynamic (illumination dependent) image compression, and resolution of the display system. The optimistic resolution value of about 2" (two times the value of the human eye) may be approached with decreasing camera aperture, which is of course possible only under good light conditions. This prediction is tested with known static objects on the airfield (see section 4). For realization of the panorama only 1424x1066 Pixels of each camera (50° viewing angle) are actually used in order to match the 180° panorama angle. For each camera the signals with 25 frames/s are split into two outputs. One feeds the data compression for transmission to the remote RTO HMI, while the other drives the simultaneous real time image processing running on a parallel workstation.

Fig. 3. Braunschweig research airport with with 1.6 km runway extending E-W, fiber optic data link (thin lines) connecting sensor containers. Circle with radiating lines indicate camera position and sectors respectively, viewing north (Photo: DLR).

A GBit ethernet switch feeds the images from the five sensors into a single mode fiber optic data link which transfers the typically 100 MBit/s data of the panorama system and PTZ over a distance of 450 m to the Advanced Control Center Simulator (ACCES). A second GBit ethernet switch splits the incoming data into five output channels for decompression with one PC per camera. Each camera is remotely controlled with respect to aperture and γ correction. The PTZ camera is controlled with respect to azimuth, vertical angle and zoom (23-fold, focal width 3.6 mm – 82.8 mm, corresponding to 54° - 2.5° visual angle).

The panorama visualization is realized in two different versions: a tiled wide angle backprojection system with one row of a high resolution SXGA projectors (1280 x 1024 Pixels) [2] and a display based system with four high resolution LCD-monitors (UXGA, 1600x1200 Pixels). Both options are realized with a PC cluster and a central workstation for display control and interaction. Figure 4 depicts the live video panorama of the monitor system, with the remotely controlled PTZ-camera displayed on a separate monitor. Interaction of the operator with the panorama system (cameras, weather station, microphone) is performed via pen touch-input display for modifying lens aperture, exposure time, γ correction of cameras and PTZ control.

For PTZ positioning the target can be defined manually or by automatic movement detection. A rectangular contour is positioned at any location of the panorama, defining the target area to be enlarged. With the tracking mode turned on the square moves coherently with the corresponding object. An algorithm for real time movement detection is running on a separate parallel processor of the image compression PC of each camera. An overall latency time between image acquisition and panorama visualization of 230 ms – 270 ms was measured by means of a special shuttered laser arrangement.

Fig. 4. Video panorama display system with additional PTZ display above (Photo:DLR)

The five recording PC's with the compression software at the camera position allow for storing panorama and zoom data (roughly 40 GByte of data per hour) and provide the possibility of complete panorama replay. Presently this feature is used for the augmented vision HMI development and validation experiments (see section 4).

Within the video panorama real-time aircraft position information is integrated as obtained from the multilateration system at the (local) Braunschweig airport via the aircraft (a/c) transponder (see Figure 4). Under reduced visibility this Augmented Tower Vision (ATV) feature allows for localizing the a/c near the correct position because the transponder code, a/c label and numerical information are integrated near the nominal a/c image location in real time. Contours of the movement areas are superimposed on the reconstructed panorama for guiding the operators attention during darkness or bad weather conditions to those areas where moving vehicles are expected.

One important advantage of the so called video see-through augmented vision technique using the digital video panorama is the easy integration of augmented vision features. This characteristic avoids the problem of (computational) delay between real scene and augmented information of the optical see-through technology as realized with the head–up and head mounted techniques (e.g. [8]). Initial laboratory experiments and theoretical investigations with superimposed information on the far view addressed the human performance such as response time and head down time reduction by using transparent displays for reducing the number of monitors [13][14], and the problem of spontaneous cognitive switching due to ambiguous stimuli [15].

4 Initial Field Trials

The main question to be answered refers to the comparability of the video panorama with the real view out of the tower windows. With the known size and distances of static objects on the airfield it is possible to evaluate the practically achieved video panorama resolution as compared to the optimistic estimate of 0.6 m / Pixel at 1 km given in section 3. We may take the red-white multilateration sensor-containers as reference objects (see Fig.3, height and width G = 2 m). The nearest containers as captured by the NE and E-looking camera $P_{3,4}$ are located at distances $g_E = 400.8$ m and $g_{NE} = 588$ m (dotted circle) respectively. With the lens equation of section 3 we obtain 7.8 and 5.3 pixels of the camera chip covered by the container image in the vertical direction. Evaluation of single video camera frames (cameras P3, P4) reveals 8-9 and 5-6 pixels, depending on the selected intensity threshold. The corresponding vertical display image size is 2.4 mm (ca. 9 Pixels) and 1.6 mm (6 Pixels) respectively. The size measured on the displays is 3 mm and 2.5 mm respectively. The red-white container coloring is resolved in both cases, however, as expected, significantly reduced as compared to the real view.

For initial steps towards validation of the system a flight-test plan was set up for experts and non-experts to evaluate identical scenarios under real view and video panorama conditions. Flight tests of two hour duration each, with the DLR DO-228 (D-CODE) test aircraft were designed with successions of approach, touch-and-go (or low approach) and takeoff. On December 13 2006 the first out of four planned 2-hour trials were performed. Five subjects (2 controllers of the Braunschweig Tower (S_1,S_2), and 3 non-experts (S_3, S_4, S_5, members of the human factors department)) observed the flyby from a position near the panorama camera system and monitored times of 11 characteristic events e_1 – e_{11}: a/c out of sight, low / steep dept. angle, take-off, touchdown, approach main / grass runway, landing gear down / up, steep approach, first sighting. The measurements were performed with time synchronized camera and notebook computers using a specially designed data input software. Significant time drifts of the individual notebooks were corrected for by comparing with the P_1-camera time as reference before and after the 2-hour experiment. Pilots received the flight plan for up to 16 approaches (with 11 realized). One out of the 11 recorded GPS trajectories is shown in Fig.5. The distance between the runway and approach turning points is 4 km and 14 km respectively. Flights were performed under VFR conditions with lower cloud boundary at 600 m. Each flyby was characterized by 6 parameters, with parameter values statistically mixed: 1. approaching main (concrete) or grass runway; 2. approach angle normal or high; 3. landing gear out: early, normal, late; 4. low level crossing of airport or touch and go; 5. touch down point early or late; 6. departure angle normal, low angle, steep angle.

While pilots had a detailed plan to follow for the sequence of approaches with different parameter values, the subjects only knew about the different possibilities (e.g. approach grass or main runway) within the approaches. They had to activate the corresponding field of their input display and set a time mark via tablet PC ($S_{1,2}$)or notebook/keyboard ($S_{3,4,5}$) at the time of their observation of one out of 11 possible events ($e_1 - e_{11}$) during each of the D-CODE approaches / flybys (e.g. a/c visible = first sighting of aircraft, recognized by the head light in the present experiment). Also all approaches of additional a/c were monitored. Experts and non-experts were briefed separately before the first experiment, with both groups filling separate questionaires. After the first 2-hour test raw data from all subjects and for all approaches under real view conditions were collected into a single data file. Evaluation of the different approach, touch-and-go, and departure conditions (altogether 14 approaches with 11 D-CODE and 3 other aircraft) yields the inter-subject time measurement scattering with mean and standard deviation (stdev) of the sample and standard errors (sterr) of mean for the n = 5 subjects.

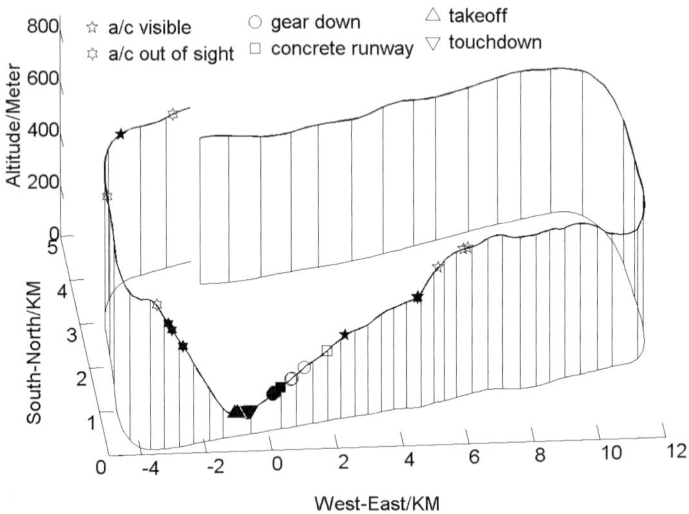

Fig. 5. GPS trajectory no. 4 out of 11 test flights of 13/12/06 (clockwise direction). Open / filled symbols represent event observation under real view / video panorama conditions.

Typical unbiased estimates of sample stdev for event e_{11} (first sighting during approach) are between 2 s and 25 s (sterr = 1 – 15 s). Comparing approach recognition time with low stdev with the GPS track yields first sighting of a/c (headlight) at distance 9 km. The minimum sterr of e.g. 1 s for e_{11} and 0.2 s for e_5 (touchdown) presumably represent the optimum observation conditions for all subjects (all n = 5 attending first sighting direction during expected apearance time).

Detailed information on the difference between real view and video panorama are obtained by repeating the experiments with the video panorama replay after a week or more in order for the subjects to no longer remember the different flight conditions. We state the hypothesis that due to lower resolution and contrast of the videopanorama (ca. 2 arc min) as compared to the real view (see section 3), distant events of approaching /departing a/c (like first / last sighting of a/c) should receive an earlier / later time mark under real view as compared to video observation. Correspondingly within-subject evaluations of the direct viewing and video panorama replay observations yields time differences t(real view, e_i) – t(video, e_i) < 0 and > 0 for approaching (app) and departing (dpt) a/c respectively. All three non-experts $S_{3,4,5}$ repeated the experiments with the videopanorama replay within 1/2007. In table 1 the results of four of the 11 possible observation types are evaluated, for the individual S_i with overall means and uncertainties. All displayed events exhibit reproducible and consistant pos.(dpt.) and neg.(app.) delays between real view and video panorama conditions. For example the significant negative delays measured as overall mean for e_8 (landing gear visible, -10 ± 2 s) and e_{11} (first sighting, -23 ± 4 s) show these events to be observable only 0.5 and 2 – 3 km respectively closer to the airport (a/c speed 100 and 200 knots respectively), as compared to the real view conditions (e.g. e_{11}(real view): a/c lights recognized at ca. 9 km).

Table 1. Time differences real view – video panorama of observation times for subjects $S_{3,4,5}$ for events e_1 (a/c out of sight), e_5 (tochdown), e_8 (landing gear visible), e_{11} (first sighting)

	t(real view) – t(video): mean(st.error of mean; st.deviation; sample size n)			
	S_3	S_4	S_5	$S_3 + S_4 + S_5$
e_1	+25(11;35;11)	+33(10;25;6)	+13(5;17;10)	+15(5;28;27)
e_5	+0.4(0.1;0.2;6)	+0.5(0.7;1.8;6)	+1.0(0.6;1.4;6)	+0.6(0.3;1.3;18)
e_8	-13(3;6;4)	-9(2;6;9)	-10(5;7;2)	-10(2;6;15)
e_{11}	-29(8;29;13)	-13(5;19;14)	-28(7;24;13)	-23(4;25;40)

5 Conclusion

Basic elements of DLR's experimental Remote Tower Operation (RTO) system at the Braunschweig Research Airport are described and initial field test results reported. The motivation for design of a high resolution augmented vision video panorama as basic RTO HMI is highlighted, based on work and task analyses. Important advantages as compared to the current work situation, such as zoom with tracking function, video-see-through augmented tower vision (ATV) for improving low visibility conditions, reduced head-down time, and panorama replay are presented. Preliminary quantitative evaluation of initial field tests for comparing real view and video panorama observation quantifies the expected reduced video resolution as compared to the real view condition, which however, may be compensated by the mentioned advantages. A RTO HMI demonstrator is presently under construction which will be integrated into the DLR tower simulator environment, allowing for shadow mode operation as well as simulation of different work scenarios, e.g. simultaneous control of two airports. Detailed evaluation of simulator and work model based RTO simulations and additional field tests will provide design guidelines for the prototype development.

Acknowledgements. We are indebted to Detlef Schulz-Rückert, Holger Uhlmann and Dieter Bensch of German Air Traffic Control (DFS) for supporting the work and task analysis by providing domain expert knowledge. The field tests would not have been possible without the cooperation of the Braunschweig tower controllers, Bernd Purwin and Uwe Mrozek, the DLR test pilot Jens Heider and flight engineer Andreas Grüttemann. Contributions of Monika Mittendorf to data evaluation are acknowledged. The project is funded by the DLR program directorate under contract no. 2393900 and supervised by Dr. Brigitte Brunner.

References

[1] Fürstenau, N., Rudolph, M., Schmidt, M., Werther, B.: Virtual Tower. In: Wettbewerb der Visionen 2001 – 2004, DLR-report, pp. 16–21 (2004)

[2] Schmidt, M., Rudolph, M., Werther, B., Fürstenau, N.: Remote Airport Tower Operation with Augmented Vision Video Panorama HMI. In: Proc. 2nd Int Conf. Res. in Air Transportation ICRAT, Belgrade, pp. 221–230 (2006)

[3] Tavanti, M.: Control Tower Operations: A Literature Review of Task Analysis Studies. EEC Note 05 (2006)

[4] Pinska, E.: An Investigation of the Head-up Time at Tower and Ground Control Positions. In: Proc. 5th Eurocontrol Innovative Research Workshop, pp. 81–86 (2006)

[5] Vicente, K.J.: Cognitive Work Analysis. Lawrence Erlbaum Associates, Mahwah (1999)

[6] Werther, B., Uhlmann, H.: Ansatz zur modellbasierten Entwicklung eines Lotsenarbeitsplatzes. In: Zustandserkennung und Systemgestaltung, Fortschritt Berichte VDI, vol. 22, pp. 291–294 (2005)

[7] Werther, B.: Kognitive Modellierung mit farbigen Petrinetzen zur Analyse menschlichen Verhaltens. PhD Dissertation, DLR-Inst. of Flight Guidance (2006)

[8] Ellis, S.: Towards determination of visual requirements for augmented reality displays and virtual environments for the airport tower. In: Proc. NATO workshop on Virtual Media for the Military, West Point /N.Y., HFM-121/RTG 042 HFM-136, 31-1-31-9 (2006)

[9] Werther, B.: Colored Petri net based modeling of airport control processes. In: Proc. Int. Conf. Comput. Intelligence for Modelling, Control & Automation (CIMCA) Sydney (2006)

[10] Cacciabue, P.C.: Modelling and simulation of human behaviour in System control. Springer, London (1998)

[11] Werther, B., Schnieder, E.: Formal Cognitive Resource Model: Modeling of human behavior in complex work environments. In: Proc. Int. Conf. Computational Intelligence for Modelling, Control & Automation (CIMCA), Wien, pp. 606–611 (2005)

[12] Dörner, D.: Eine Systemtheorie der Motivation. In: Khul, J., Heckhausen, H. (eds.) Enzyklopädie der Psychologie, Hogrefe, Göttingen, pp. 329–357 (1996)

[13] Peterson, S., Pinska, E.: Human Performance with simulated Collimation in Transparent Projection Screens. In: Proc. 2nd Int. Conf. Res. in Air Transportation, Belgrade, pp. 231–237 (2006)

[14] Fürstenau, N., Rudolph, M., Schmidt, M., Lorenz, B., Albrecht, T.: On the use of transparent rear projection screens to reduce head–down time in the air–traffic control tower. In: Proc. Human Performance, Situation Awareness and Automation Technology (HAPSA II), pp. 195–200. Lawrence Erlbaum Publishers, Mahwah (2004)

[15] Fürstenau, N.: Modelling and Simulation of spontaneous perception switching with ambiguous visual stimuli in augmented vision systems. In: André, E., Dybkjær, L., Minker, W., Neumann, H., Weber, M. (eds.) PIT 2006. LNCS (LNAI), vol. 4021, pp. 20–31. Springer, Heidelberg (2006)

TDARS, a Fusion Based AR System for Machine Readable Travel Documents

Yu Wu, Ling Xue, Chao Li, and Zhang Xiong

School of Computer Science and Technology, Beihang University
No. 37 Xue Yuan Road, Haidian District, Beijing, P.R. China
{wuyu, xueling}@cse.buaa.edu.cn, {licc, xiongz}@buaa.edu.cn

Abstract. A fuzzy set-based AR (Automatic Recognition) system named TDARS for machine readable travel documents (MRTDs) is designed and implemented for speeding up customs clearance. The system consists of three parts: text feature extraction, facial feature extraction and identity matching. Text feature extraction takes charge of locating machine readable zone (MRZ) and extracts text features automatically. Facial feature extraction focuses on locating the front face photo (FFP) in the MRTD and extracting the facial features of the MRTD possessor. Identity matching compares text and facial features with feature data in the Terrorist Database (TDB) respectively and fuses the matching results of text and facial features in the decision-making level using a fuzzy set based algorithm. Experimental results show that both the correct acceptance and refusal rates of the TDARS are over 90%, which evidently exceed those of the existing recognition systems that extract text features solely.

Keywords: information fusion; fuzzy integration; character recognition; face recognition.

1 Introduction

With the rapid development of international cooperation, customs clearance has become one of the crucial things in the management of floating population. The work of customs clearance consists of collecting basic information of travel documents and identity checking for terrorists and criminals. However, the traditional way for custom clearance collects and matches the information of travel documents by manpower, which is both inefficient and consuming in human resource. The usage of MRTDs makes it possible to automatically collect and match the information of document possessors. Machine readable travel documents (MRTDs) [1, 2, 3] indicate official documents used in the international travel including passport, visa, ID-card etc, which meet the international standard for machine readable travel documents (ICAO-9303) recommended by the International Civil Aviation Organization (ICAO).

In this paper, an automatic recognition system named TDARS is designed. Generally, the information contained in MRTDs includes the text information such as the possessors' nationalities and the possessors' names, and the facial feature

M.J. Smith, G. Salvendy (Eds.): Human Interface, Part II, HCII 2007, LNCS 4558, pp. 1129–1138, 2007.

information, such as the front face photos of the possessors. Both the text and the facial feature information can contribute to the automatic identity verification of the possessors. Thus, TDARS extracts both text and facial features from MRTDs to make full use of the feature information provided by MRTDs. To check up for terrorists, TDARS matches respectively the text and facial features with feature data stored in the TDB, which is built in advance. Then TDARS uses a fuzzy set based algorithm to fuse the matching results of the above two features to fulfill the requirement of identity matching. If the fusion result turns out to be larger than a given threshold T (its value should be in the interval (0, 1]), the document possessor will be deemed as a terrorist and vice versa. Through the above approaches, TDARS increase the correct identification rate and the processing speed, which is meaningful to custom clearance.

The remainder of this paper is organized as follows. In Section 2, we propose the architecture of TDARS, which consists of three parts, text feature extraction, facial feature extraction and identity matching. Section 3 presents the text feature extraction of MRTDs in details. Section 4 introduces the facial feature information extraction from MRTDs. Section 5 describes the matching of text and facial features with data stored in the TDB respectively. In addition, a fuzzy set based algorithm for the identity matching is also discussed in section 5. In section 6, we design an experiment to test the performance of TDARS and give the evaluation results of TDARS. Finally, we conclude with a discussion of TDARS and future work in section 7.

2 The Architecture of TDARS

In order to extract both text features and facial features of MRTDs and fulfill the requirements of customs clearance, the TDARS is composed of three modules: text feature extraction, facial feature extraction and identity matching. The architecture of TDARS is shown in Figure 1.

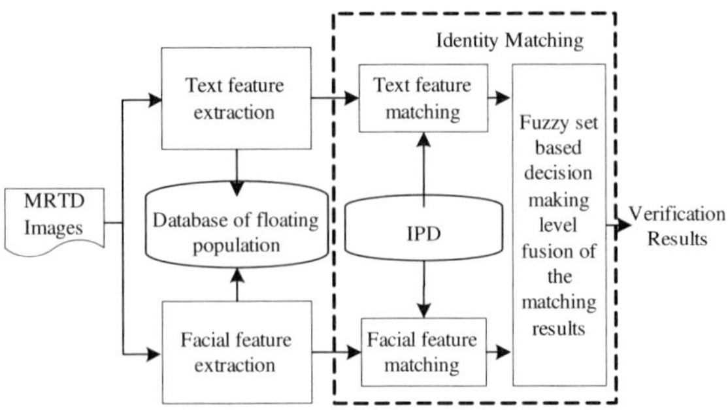

Fig. 1. TDARS architecture

2.1 Text Feature Extraction

The main task of text feature extraction can be divided into two steps: the location of Machine readable zone (MRZ) and the extraction of effective characters. In Figure 2, the label "MRZ" indicates the region on the bottom of the MRTD containing basic

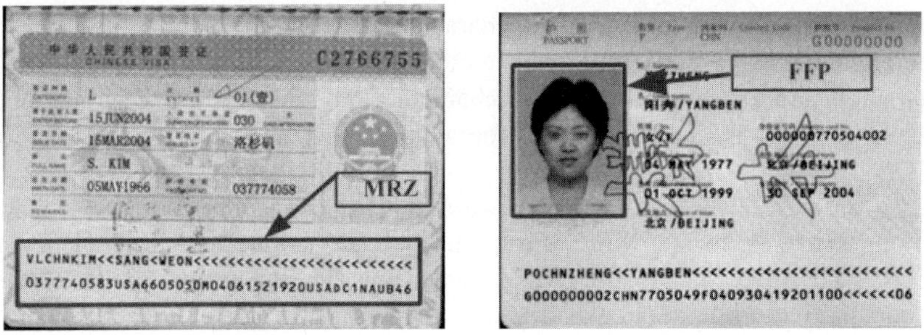

Fig. 2. Machine readable zone (MRZ) and front face photo (FFP)

information of the possessor. When text features are extracted, the system saves the corresponding information into the database of floating population to accomplish collecting basic information in the work of customs clearance.

2.2 Facial Feature Extraction

The module of facial feature extraction takes charge of locating the front face photo in MRTD using face detection algorithms and extracting the facial features. The extracted facial features will be also saved into the database of floating population. Thus the facial features of possessors can be used in the module of identity matching to verify the possessors' identities.

2.3 Identity Matching

The module of identity matching is designed to satisfy the requirement of terrorist checkup. In other words, it makes a judgment on whether the possessor is a terrorist or not. First, the extracted text and facial features is matched with the feature data stored in the TDB respectively. Then the matching results of the two kinds of features are fused in the decision-making level using a fuzzy based algorithm. Finally, it gives warnings if the value of the similarity between the features of the possessor and those of any terrorist in TDB is higher than a given threshold T.

3 Text Feature Extraction

As mentioned in section 2.1, text feature extraction includes two subparts: MRZ location and character recognition. The former locates the MRZ and the latter takes charge of the pre-processing of MRZ images and the recognition of the characters.

3.1 MRZ Location

Through the analysis of MRTDs, we can come to the 3 conclusions about the characteristics of the MRZ:(1) The MRZ is located on the bottom of the MRTD. (2) All the characters in MRZ are in black color and printed in the OCR-B font, which has a fixed size and style. Strokes of characters in MRZ are continuous. Thus each character can be represented as a connected component. (3) MRZ contains two rows of characters and each row contains 44 characters. The distance between neighboring characters in a row is small and fixed. The width of MRZ is larger than 3/4 that of the MRTD.

Based on the above characteristics, we adopt the connected component analysis-based approach to locate MRZ. First, detect each connected component in the MRTDs, and then locate the MRZ according to the analysis of connected components. According to (1), we scan only the bottom 1/3 area of the MRTD images to speed up the procedure of locating MRZ.

3.2 Text Feature Extraction

During the course of samplings, the MRTD may be placed somewhat slantwise to the camera so that the MRZ located in section 3.1 may also be slantwise. However, OCR algorithms can recognize only the binary image of a single upright character generally. Hence in this procedure, we first apply slant correction, binarization and character segmentation on the MRZ and then exert an optical character recognition (OCR) algorithm on the MRZ to get the text feature in the MRTD.

Because each character in the MRZ belongs to the character set consisting of numbers 0~9, capital characters A~Z and the sign of "<", after the above procedures we can obtain a similarity vector for each character in the MRZ, by calculating the similarities between the character and the characters in the character set. Thus for every MRZ, we can get a series of similarity vectors that are expressed by $VectorSeries = \{ SimVector_n \mid 1 \le n \le N\}$, where N is the number of characters in the MRZ. $SimVector_n$ is the similarity vector of the nth character in the MRZ, which can be expressed as $SimVector = \{Sim_m \mid 1 \le m \le M\}$, where M is the number of characters in the character set, and Sim_m is the similarity between the character in the MRZ and the mth character in the character set.

4 Facial Feature Extraction

The facial feature extraction is another part of the TDARS. It aims at extracting facial features as a complement to the text features. This part consists of two sub-parts: FFP detection and facial feature extraction. The former takes charge of locating the FFP in the MRTD and the latter focuses on extracting facial features from the photo.

4.1 Adaboost-Based FFP Detection

The Adaboost-based algorithm for front face detection is proposed by Viola and RainerLienhart [5] and improved by Rainer and Jochen [6]. Compared with other algorithms [7], the Adaboost-based algorithm achieves almost the highest correct recognition rate while keep the processing speed relatively high. Therefore, the sub-part of the FFP detection uses a front face detection framework based on the

Adaboost-based algorithm to implement the FFP location. The detection framework is composed of two modules: offline training and online detection.

Offline training. The goal of the offline training module is to get a FFP detector using training samples. Here the FFP detector is composed of a series of cascade classifiers. The training procedure includes 3 steps: Facial features selection, weak classifier training and combining the weak classifiers to form the detector.

Online detection. To locate the FFP on the MRTD, the online detection module provides a candidate window to move across the image and use the trained detector to check the region in the window to verify whether it is the FFP or not. If the candidate region passed the verifications of all the weak classifiers in the detector, the region is considered to be the FFP and vice versa.

4.2 EHMM-Based Face Feature Extraction

To extract facial features from the MRTDs for further identity verification, we use an EHMM based algorithm to extract facial features from the gained FFP region.

Embedded Hidden Markov Model (EHMM). The existing methods [8] for facial feature extraction can be divided into the following three categories: geometric feature-based methods, template-based methods, and Hidden Markov Model (HMM)-based methods [9]. The HMM approach exploits the fact that the most significant facial features of a front face image occur in a natural order, from top to bottom. This makes the HMM a robust algorithm for extracting the facial features. EHMM [10] is extended from HMM by allowing each state in a HMM to be extended as another HMM. In this way, the one-dimension HMM is extended to 2-dimensions. Thus the EHMM is more suitable for modeling 2D data such as face images than the 1D HMM.

Face feature extraction. To implement the facial feature extraction of the FFPs, this module builds an EHMM as indicated in Figure 3 according the characteristics of human faces. The EHMM is composed of 5 super states which are assigned to forehead, eyes, nose, and mouth and chin separately. This topology enables the super states to represent the macroscopical features of human faces while the sub-states embedded in the super states represent the local feature of the human face.

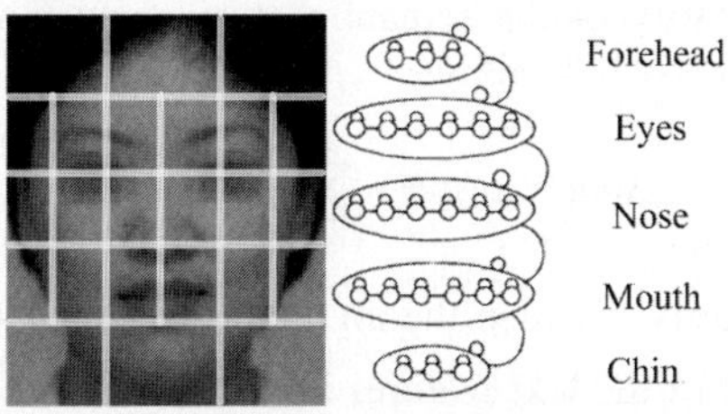

Fig. 3. EHMM of FFP

The details of the face feature extraction are given as follows:

Step 1: Applies a series of pretreatment procedures on the FFPs, including the scale normalization, gray scale normalization and background noises elimination, to implement the facial feature extraction and matching under the same conditions.

Step 2: Build the EHMMs for the FFPs. To achieve a high accuracy of representation, we set the number of super states to be 5, and set the number of embedded states in the super states as (3, 6, 6, 6, 3) [10].

5 Fusion Based Identity Matching

To verify the identity of the document processors in the work of customs clearance, we designed a fusion based identity matching to verify whether a document possessor is a terrorist or not.

5.1 Fusion System

In the system, we use following steps to fuse the matching results of text and facial features: First match respectively the extracted text features and facial features with the feature data in the TDB. Then Fuse the matching results of the features using a fuzzy set-based fusion algorithm. Finally give a binary decision, “accept” or “refuse”, according to the decision criteria.

5.2 Feature Matching

Text feature matching. Similarity between the text feature of the MRTD and the text feature data in the TDB can be calculated as follows:

Step 1: Divide the characters in the MRZ obtained in Section 3.2 into several semantic clusters. According to the ICAO-930, each cluster is a character string having a definite meaning, such as possessor’s name, nationality and etc. Moreover, these clusters are in a definite order [1, 2, 3]. Thus we can obtain the semantic clusters in order.

Step 2: Calculate similarities between the semantic clusters and the corresponding text features in the TDB. The similarity between the semantic cluster and the corresponding text feature is easy to get by looking up the *VectorSeries*. Repeat the looking up until all the similarities are obtained in order. Finally we sum up the values, obtain the similarity of the semantic cluster and the corresponding feature. The formula can be expressed as follows:

$$SimOfCluster = \sum_{i=1}^{N} Sim_i \tag{1}$$

where Sim_i is the similarity between the i*th* character in the semantic cluster and the corresponding character of the text features in TDB, N is the number of characters in the semantic cluster.

Step 3: Calculate the similarity of the text feature of the MRTD and the text feature data in the TDB. We use a weight based method to calculate the similarity. The method assigns higher weights to more important semantic clusters, while assigns relatively lower weights to less important semantic clusters. The similarity is calculated by:

$$\mu = \frac{\sum_{i=1}^{M} \alpha_i SimOfCluster_i}{\sum_{i=1}^{M} \alpha_i} \tag{2}$$

where α_i represents the weight of the ith semantic cluster, $SimOfCluster_i$ is the similarity between the ith semantic cluster and the corresponding text features in the TDB, which is obtained in step2, and M is the number of clusters. As there are a number of samples in the TDB, we can generate a "text feature channel" between the MRTD and the TDB, which can be presented as:

$$\Psi_1 = \{(o_1, \omega_{1,1}), (o_1, \omega_{1,2}), ..., (o_1, \omega_{1,i}), ..., (o_1, \omega_{1,p})\} \tag{3}$$

where $\omega_{1,i} = \mu_i$ is the similarity of text features with the ith sample o_i in the TDB.

Facial feature matching. In section 4.2.2, we built an EHMM for each FFP on the MRTD. During the course of the facial feature matching, we calculate the similarity between the EHMM of the FFP and the EHMMs stored in the TDB using the doubly embedded Viterbi recognizer algorithm [10]. Similar with text feature, we also can generate a "facial feature channel" between the MRTD and the TDB, which can be presented as:

$$\Psi_2 = \{(o_1, \omega_{2,1}), (o_1, \omega_{2,2}), ..., (o_1, \omega_{2,i}), ..., (o_1, \omega_{2,p})\} \tag{4}$$

where $\omega_{2,i}$ is the similarity between the facial features of the MRTD and the ith sample o_i in the TDB.

5.3 Fuzzy Set Based Fusion Algorithm

Theoretically, the identity matching results using the text features should be identical with that using the facial features. However, from the aspect of engineering, the identity matching result using the text and the facial features may be different due to a variety of factors, such as background noises and etc. Therefore we consider using the fuzzy set instead of the common set to fuse the identity matching results.

The theory of fuzzy set [11, 12] deems that every element can belong to a set in a certain degree and belong to many sets in different degrees as well. It provides a stretchable description of the large amount of existing concepts whose meaning are certain but not exact, which is considered as the most prorate way to solve AI problems, such as face recognition [13].

In section 5.2, we have got the "text features channel" and "facial feature channel" respectively. In this section, we fuse the value in these two channels using the fuzzy set. The fuzzy integration function is expressed as follows:

$$S(\omega_{1,i},\omega_{2,i}) = (\omega_{1,i} \cdot \omega_{2,i})^{1/2} \quad \text{(F1)}$$

Or

$$S(\omega_{1,i},\omega_{2,i}) = \frac{\omega_{1,i} + \omega_{2,i}}{2} \quad \text{(F2)}$$

If $\max S(\omega_{1,i},\omega_{2,i}) > T$ then the MRTD possessor is thought to be the *gth* terrorist in the TDB, where $g = \arg_i \max S(\omega_{1,i},\omega_{2,i})$, otherwise the possessor is not a terrorist.

6 Experiment Analysis

To verify the validity of the algorithm above, we design the following experiment. The number of samples used in the experiment is 300 groups of images. The images in one group include the passport image, the visa image and several front facial photos of one person. First, to construct the TDB, we select 50 groups of images randomly from the sample space, extract the text and facial features from them and store the features into the database. Generally, the number of FFPs of one person is limited. Hence, to simulate the real condition, the number of front facial photos in the group of images is limited to 6.

6.1 Identity Matching with Single Feature

Identity matching solely by text features. To test the text features extraction and matching method described in section 3 and 5.2.1, we apply these methods on the visa images of the 300 persons and implement the identity matching of these persons. As a result, the correct acceptance and refusal rate of the identity matching are 87.3% and 89.7% respectively.

Identity matching solely by facial features. Similarly, to test the facial features extraction and matching described in section 4 and section 5.2.2, we apply these methods on the passport images of the 300 persons and implement the identity matching of these persons. As a result, the correct acceptance and refusal rate of the identity matching are 82.2% and 83.9% respectively.

6.2 Identity Matching by Fusing Features

To test the fuzzy set-based fusion algorithm described in section 5.3, we apply the method on the images of the 300 persons and implement the identity matching of these persons. The experimental result is shown in Figure 4. Both the rates of correct acceptance and refusal are all remarkably increased. The correct acceptance and refusal rates using F1 in section 5.3 are 90.6% and 92.8%, and the correct acceptance and refusal rates using F2 are 90.5% and 93.1% respectively. Since the correct refusal rate is more important for customs clearance, the final fusion algorithm adopts F2. Experiments show that the fuzzy set based fusion algorithm is effective to promote

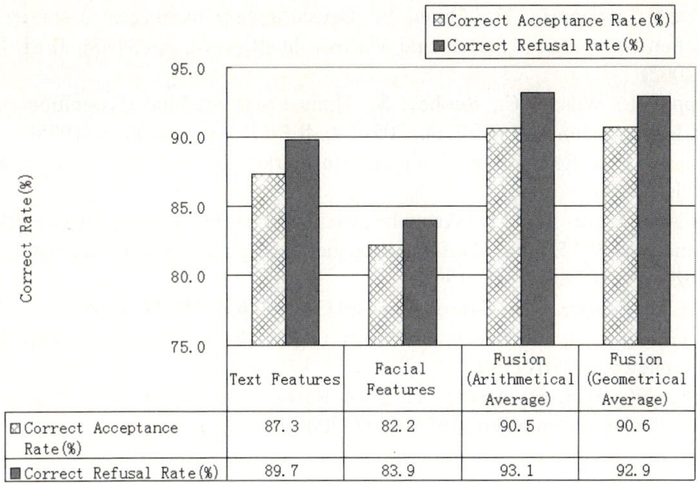

Fig. 3. Correct rate of each method

the correct acceptance and refusal rate of the recognition system for MRTDs, since it combines features from different channels.

7 Conclusion

We present a fusion-based automatic recognition system named TDARS for MRTDs. It extracts both text and facial features from the MRTDs and fuses them to fulfill the requirements of customs clearance. The experimental results show that the TDARS can effectively enhance the correct recognition rate of the recognition system, which make great senses to customs clearance.

References

1. Identification cards – Machine readable travel documents – Part 1: Machine readable passport (1997) ISO/IEC 7501-1
2. Identification cards – Machine readable travel documents – Part 2: Machine readable visa (1997) ISO/IEC 7501-2
3. Identification cards – Machine readable travel documents – Part 3: Machine readable official travel documents (1997) ISO/IEC 7501-3
4. Yu, B., Jain, A.K.: A Generic System for Form Dropout. In: IEEE Transactions on Pattern Analysis and Machine Intelligence, pp. 1127–1134. IEEE Press, New York (1996)
5. Viola, P., Jones, M.: Rapid object detection using a boosted cascade of simple features. In: Proc. of the IEEE Computer Vision and Pattern Recognition, pp. 511–518. IEEE Press, New York (2001)
6. Rainer, L., Jochen, M.: An Extended Set of Haar-like Features for Rapid Object Detection. In: Proceedings of the 2002 IEEE International Conference on Image Processing, pp. 900–903. IEEE Press, New York (2002)

7. Yang, M.H., Kriegman, D., Ahuja, N.: Detecting face in images: a survey. In: IEEE Transactions on Pattern Analysis and Machine Intelligence, pp. 34–58. IEEE Press, New York (2002)
8. Chellappa, R., Wilson, C., Sirohey, S.: Human and machine recognition of faces: A survey. In: Proceedings of IEEE, pp. 705–740. IEEE Press, New York (1995)
9. Samaria, F.: Face Recognition Using Hidden Markov Models, PhD thesis, University of Cambridge (1994)
10. Nefian, A.V., Hayes III, M.H.: An embedded HMM-based approach for face detection and recognition. In: IEEE Inter. Conf. on Acoustics, Speech and Signal Processing, pp. 3553–3556. IEEE Press, New York (1999)
11. Zadeh, L.A.: Fuzzy sets. In: Information and Control, pp. 338–383 (1965)
12. Klirm, G., Yuan, B.: Fuzzy sets and Fuzzy Logic: Theory and Applications. Englewood Cliffs, Englewood Cliffs (1995)
13. Yunhong, W., Wei, F., Tieniu, T.: Face Recognition Based on Information Fusion. Chinese Journal of Computers, 1657–1663 (2005)

AR-Navi: An In-Vehicle Navigation System Using Video-Based Augmented Reality Technology

Yoshihisa Yamaguchi[1], Takashi Nakagawa[1], Kengo Akaho[2], Mitsushi Honda[2], Hirokazu Kato[2], and Shogo Nishida[2]

[1] Advanced Technology R&D Center, Mitsubishi Electric Corporation,
8-1-1, Tsukaguchi-honnmachi, Amagasaki, Hyogo, 661-8661, Japan
Yamaguchi.Yoshihisa@dp.MitsubishiElectric.co.jp,
Nakagawa.Takashi@dy.MitsubishiElectric.co.jp
[2] Department of Systems Innovation, Graduate School of Engineering Science,
Osaka University,
1-3, Machikaneyama, Toyonaka, Osaka, 560-8531, Japan
{akaho, honda}@nishilab.sys.es.osaka-u.ac.jp
{kato, nishida}@sys.es.osaka-u.ac.jp

Abstract. In this paper we propose AR-Navi, an in-vehicle navigation system that uses video-based augmented reality technology. AR-Navi includes an in-vehicle camera, which captures real-time video, and displays the guidance which overlays the video display. To solve several problems that arise when real-time video is used, we propose registration adjustment based on recognizing crosswalks and filtering images based on evaluated values. We have built and tested an AR-Navi prototype that uses the proposed methods and confirmed that the system can achieve registration accuracy for driver route guidance at intersections.

Keywords: Route guidance, In-vehicle navigation systems, Augmented reality.

1 Introduction

In recent years, over 24 million in-vehicle navigation systems have been produced in Japan [1] as such systems have rapidly become popular. The basic functions of these systems, such as map indication and route planning, are sufficiently powerful for current needs, so manufacturers are now seeking to improve in-vehicle navigation systems with regard to their user interface design. For example, new functions are being provided, such as displays showing a city landscape using 3D computer graphics (CG) or images combining aerial photos and existing maps.

Unfortunately, the usefulness of these functions is limited because of the following problems.

- 3D CG data and aerial photos are obtained in the past, so discrepancies might exist between the data and the present state.
- Existing data is created by sampling the real world to some limited degree of accuracy. This means data is simplified and thinned. For example, 3D CG data is

M.J. Smith, G. Salvendy (Eds.): Human Interface, Part II, HCII 2007, LNCS 4558, pp. 1139–1147, 2007.

created according to the polygon number and images are created according to the resolution. Moreover, some data is deformed. That is, the data might not truly represent what a driver actually sees.

Because of such problems, a driver sometimes makes wrong turns or become confused, despite an in-vehicle system's correct route guidance, because the driver finds it difficult to make the correct correspondence between the real world and the displayed view.

In this paper, we propose AR-Navi, a system that uses video-based augmented reality technology. AR-Navi includes an in-vehicle camera, which captures real-time video, and displays the guidance which overlays the video display. Because the video is almost the same as what the driver sees, AR-Navi provides easily understandable guidance.

We give an overview of AR-Navi and the problems that limit existing systems in Section 2. We then propose solutions to these problems in Sections 3 and 4. We also describe an AR-Navi prototype and its performance.

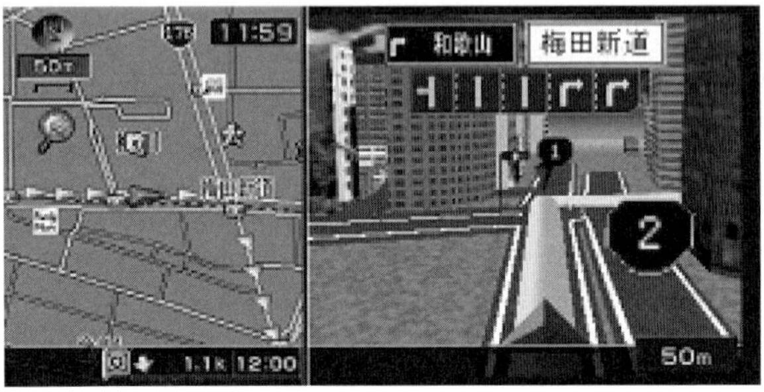

Fig. 1. An example of route guidance using 3D CG

2 AR-Navi

AR-Navi is an in-vehicle navigation system that shows driver guidance overlaid on a video image captured from the driver's front view. Figure 2 shows a screen image of AR-Navi. A turn direction and turn location are shown by CG in this figure.

Other research has described in-vehicle navigation systems which use augmented reality technology. Sawada [2] describes the extraction of road geometry information and a way to draw CG roads on real-time video. Hu [3] describes the extraction of future points along a road and a way to adjust the location between the guidance drawn on the roads and real-time video. These studies have demonstrated the feasibility of high-precision alignment between real-time video and CG with image analysis. The image analysis technology used in these systems, however, has some limitations. First, the time complexity of the technology is high. Second, the technology is not robust regarding environmental conditions, such as rain, darkness, and backlighting. Thus, these systems are not used in current in-vehicle navigation systems. Our goal is to develop a commercial system which needs only a camera as additional equipment and is robust with respect to such environmental conditions.

Fig. 2. An AR-Navi image

2.1 Problems

We had to achieve the following to create AR-Navi in a way that overcame the problems affecting earlier systems:

- Suitable registration: A drivers feels uncomfortable when they encounter location gaps between the real-time video and the CG image.
- Information control: Too much guidance can lead to a driver not paying attention to traffic conditions.
- Elimination of inappropriate images: Captured images are sometimes unfit for overlapping guidance because the camera signal is blurred.
  In the following sections, we explain how we solved these problems.

2.1.1 Suitable Registration

When AR-Navi overlays information on a video image, it adjusts the positions of the information and the image based on the location and direction of the vehicle. If the location and direction contain errors, gaps will appear in the overlaid image.

Almost all current in-vehicle navigation systems calculate their own location and direction based on two steps [4]. First, the systems estimate their own location and direction from a combination of global positioning system (GPS) data and autonomous navigation data, such as that from a gyro and speed sensor. Second, they use a map-matching method which compares the result of the first step with road patterns. The systems correct the vehicle location if the driving record changes significantly (such as when the vehicle makes a turn) or there is a difference between the driving record and the road patterns. Immediately after a vehicle turn, the location and direction may show high accuracy but huge errors sometimes occur if the vehicle goes straight for a long time without a sufficiently strong GPS signal. We divide the gap between the location and the direction into three categories, and consider the resolution as follows.

- Back-and-forth gap: The longer a vehicle travels in a straight direction, the larger the back-and-forth gap becomes. Since a large back-and-forth gap can cause a driver to confuse a wrong intersection with the correct one, it is necessary to reduce this gap. (Problem 1)

- Right-and-left gap: Because map matching is used, the calculated location rarely goes off road. Thus, the right-and-left gap generally does not exceed the road width. If the vehicle does not leave the road, AR-Navi will not provide incorrect guidance so this gap is not a practical issue.
- Rotation gap: Current in-vehicle navigation systems estimate their own direction based on gyros. A rotation gap easily accumulates, though, because gyros integrate the change of the angle. Moreover, a measurement delay occurs that misaligns the guidance and video when the vehicle direction changes abruptly (such as when the vehicle makes a U-turn). It is therefore necessary to adjust the rotation gap. (Problem 2)

2.1.2 Information Control

Various safety guidelines state that in-vehicle navigation systems should provide required guidance quickly and effectively without interfering with driving [5]. For example, "The visual information to be displayed shall be sufficiently small in volume to enable the driver to comprehend it in a short time or shall be presented in portions for the driver to scan them in two or more steps."

If AR-Navi overlays CG guidance onto real-time video as a general AR application and provides the driver with overlaid guidance, the driver might need more time to recognize the guidance because he or she might be distracted by the interactive CG guidance and the background movie, which change in real time. Thus, we had to find a way to provide guidance without interfering with vehicle operation. (Problem 3)

2.1.3 Elimination of Inappropriate Images

Because of the camera settings and environmental conditions, captured images are sometimes unfit to be used for overlapped guidance due to problems such as

- Inappropriate brightness: Failure to control the lens aperture when, for example, moving from a shadowed area to a sunny area (Fig. 3(a)).
- Blur: Image blurring due to camera movement, especially right-and-left movement; for example, while turning at an intersection or while changing lanes (Fig. 3(b)).
- Inappropriate composition: The center of the target is located outside or on the edge of the screen (Fig. 3(c)).
- Limited visibility: The driver cannot see ahead because a large vehicle blocks his view (Fig. 3(d)).

If AR-Navi overlays guidance on images which correspond to any of the above conditions and provide driver with overlaid guidance, the driver will find it difficult to understand the real world information provided, so using real-time video has no merit in these situations. (Problem 4)

2.2 Proposed Methods

In this paper, we propose two techniques: (1) registration adjustment based on recognizing crosswalks, and (2) image filtering based on evaluated values. The first technique solves problem 1, and the second solves problems 2 to 4.

(a) Inappropriate brightness (b) Blur

(c) Inappropriate composition (d) Limited visibility

Fig. 3. Inappropriate images

3 Registration Adjustment Based on Recognizing Crosswalks

To decrease the back-and-front gap, we propose adjusting the current location by detecting intersections, which can be recognized using image analysis, that correspond with intersections on the map. The following conditions must be met to use such an algorithm in current in-vehicle navigation systems.

- Low time complexity: To allow its application in current in-vehicle navigation systems.
- High robustness: Robustness with regard to environmental conditions, such as weather and time, is necessary.

To satisfy the above conditions, this method finds crosswalks by applying image analysis to one line set on a captured image and located a certain distance ahead of the driver's vehicle. This line is called the scan line.

The back-and-front gap can then be calculated by comparing each intersection found with a corresponding intersection shown on a map, because it is assumed that an intersection lies between a pair of crosswalks within a limited distance.

This method has the following merits.

- It has low time complexity because the image analysis is aimed at a narrow area; i.e., the scan line.
- If the scan line is set near enough, other vehicles will rarely interfere with it while the vehicle is in motion, so the image analysis will be stable.

- If environmental conditions are bad, such as during rain and at night, the scan line is lit by the vehicle's own headlights.

This method is applied through the following steps:

1. Judge whether a crosswalk is present.
2. Store the position where a vehicle enters and leaves an intersection based on the crosswalks.
3. Judge whether a pair of crosswalks is appropriate.
4. If the crosswalk pair is appropriate, calculate the midway point between the location of entering an intersection and the location of leaving an intersection.
5. Adjust the back-and-front gap using the difference between the calculated midway point and the corresponding intersection on a map.

3.1 Judging Whether a Crosswalk Is Present

To find crosswalks, a brightness emphasis filter is applied to the brightness of each pixel in the scan line. Figure 4 shows the brightness emphasis filter. It is easy to get a highly accurate binarized result because the filter turns the raw brightness into the brightness shown in Fig. 5.

This method judges whether a crosswalk exists on the scan line by checking the number of white runs and black runs. If the length of each run is almost the same as the width of the white line of a crosswalk on the image and the white and black runs alternate, this method judges that an intersection has been found.

-1	0	0	0	0	2	0	0	0	0	-1

Fig. 4. Brightness emphasis filter

Fig. 5. Result of applying a crosswalk empathies filter

3.2 Judging Whether a Crosswalk Pair Is Appropriate

As explained in Section 3.1, to locate an intersection it is necessary to find two crosswalks. If two crosswalks are found, however, there is a possibility of image analysis failure. For this reason, this method does not adjust the location if two crosswalks do not constitute an appropriate pair; for example, if a crosswalk is too narrow or the distance between two crosswalks is too short.

4 Image Filtering Based on Evaluated Values

Image filtering is necessary to provide effective information and estimate whether images are inappropriate. Our approach to this is to clip still images from a real-time video stream. Thus, the driver is not distracted by a movie. However, if still images are clipped at only certain time intervals, the clipped images can be inappropriate. Moreover, even if clipped images are not inappropriate, there is a possibility that other clipped images would be more appropriate. Thus, in our system still images are clipped at all times, and the optimal image is selected from all clipped images according to certain evaluated values.

In this paper, image filtering consists of three steps:

1. Detection of inappropriate images
2. Selection of optimal images
3. Image updating

Thus, this process solves problems 2 to 4 stated above. In particular, problem 2 is reduced by step 1 because images including an excessive rotation gap are not selected. Problem 3 is solved through the clipping of still images and step 4. Problem 4 is solved by step 2, except under limited visibility.

4.1 Detection of Inappropriate Images

This step uses the following conditions to detect inappropriate images. The problem of limited visibility has not been solved in this prototype because the time complexity is increased due to the image analysis to locate vehicles. Other conditions are considered in this process.

4.2 Selection of Optimal Images

After the detection of inappropriate images, optimal images are selected based on evaluated values. We define an optimal image as one having the target intersection in the center, and this image is clipped when the distance between the vehicle and the intersection is reasonably short. Because two conditions are considered, the evaluated value is set as

$$E = k * p + (1 - k) * | D - D_{best} |$$

where k is constant ($0 <= k <= 1$), D is the current distance between the vehicle location and the target intersection, D_{best} is the ideal distance between the vehicle location and the intersection on the route guidance, and p is the coordinate distance between the intersection center and the image center.

The image having the lowest value of E is treated as the optimal image. In this research, we found that a suitable value of k is 0.4 and D_{best} is 30 m.

4.3 Image Updating

The updating frequency is important for safe driving. If the updating frequency is too short, the driver might be distracted by frequently changing images. Conversely, if the updating frequency is too long, the image will differ considerably from the real world. After careful consideration of the movement distance and elapsed time, we decided to have updating occur when the movement distance exceeded 25 m or the elapsed time exceeded 2.0 seconds.

5 Discussion

5.1 Prototype

We have implemented AR-Navi based on our in-vehicle navigation system, using a 320 * 240 pixel lens with a 6-mm focal length and 47° angle of view. When the user's vehicle is far from a target intersection, real-time video offers few benefits because the intersection is out of sight. Thus, the prototype shows the conventional full CG guidance when the vehicle is far from an intersection, and then shows guidance using the real-time video stream as the vehicle nears the intersection. Figure 6 is a screenshot from AR-Navi showing how AR-Navi provides guidance by using a triangular prism set on the intersection center and an indicator arrow.

Fig. 6. An AR-Navi screenshot

5.2 Discussion

We drove through an urban area to evaluate the guidance image provided by the prototype. AR-Navi provided turn guidance at 12 intersections, and the driver clearly recognized the target for all 12 intersections. The CG guidance was correctly overlaid on the intersection center on 9 of the 12 intersections. On the three other intersections, there was a short distance between the intersection center and the guidance, but these errors did not present a problem in actual use. We confirmed that this system achieved registration accuracy for turn-by-turn guidance by using the proposed methods: registration adjustment based on the recognition of crosswalks and image filtering based on the best-shot-selection process.

The target intersection was sometimes hidden by other vehicles, but since the system drew an indicator arrow in the part of the image corresponding to the sky (Fig. 6), these vehicles did not interfere with the overlaid guidance. We think the current display is practical for drivers in most situations, but we plan to further evaluate the display design and improve it.

During the test drive, the vehicle location error was sometimes increased by the registration adjustment. This happened when the system incorrectly matched an intersection of the map with one calculated by recognizing crosswalks. It will be necessary to solve this problem to improve the registration adjustment algorithm.

6 Conclusion

AR-Navi, a newly developed in-vehicle navigation system, shows driver guidance overlaid on a video image captured from the driver's front view. To solve several problems that arise when real-time video is used in such a system, we propose registration adjustment based on recognizing crosswalks and filtering images based on evaluated values. We have built and tested an AR-Navi prototype that uses the proposed methods and confirmed that the system can achieve registration accuracy for driver route guidance at intersections.

References

1. Japanese Ministry of Land, Infrastructure and Transport, Road Bureau ITS Homepage (2007) http://www.mlit.go.jp/road/ITS/
2. Sawano, H., Okada, M.: A Study on An Augmented Reality-based Rendering Method for Car Navigation Utilizing Real-time Video. In: FCV2004 - Tenth Korea-Japan Joint Workshop on Frontiers of Computer Vision, Fukuoka Japan (February 2004)
3. Zhencheng, H., Keiichi, U.: Solution of Camera Registration Problem Via 3D-2D Parameterized Model Matching for On-Road Navigation. International Journal of Image and Graphics 4(1), 3–20 (2004)
4. Quddus, M.A., Ochieng, W.Y., Zhao, L., Noland, R.B.: A General Map Matching Algorithm for Transport Telematics Applications. The GPS Solutions Journal 7(3), 157–167 (2003)
5. Japan Automobile Manufactures Association: Guideline for In-vehicle Display Systems Virsion 3.0

A Study on Analysis Support System of Energy and Environmental System for Sustainable Development Based on MFM and GIS

Qi Zhang, Hidekazu Yoshikawa, Hirotake Ishii, and Hiroshi Shimoda

Graduate School of Energy Science, Kyoto University
Gokasho, Uji, Kyoto, Japan, 611-0011
ZhangQi@Uji.Energy.Kyoto-u.ac.jp

Abstract. An analysis support system has been developed of various energy and environmental system for the sustainable development. The proposed support system can model, and simulate the flows of matter, energy and information with tens or hundreds of elementary processes in the target energy and environmental system. It is based on the combination of Multilevel Flow Model (MFM) and Geographic Information System (GIS) and it can account various indices for sustainability. The values of the evaluation indices can be utilized to not only test the feasibility of one scenario but also carry out the inter-comparison of various optional scenarios. As a case study, introduction of hydrogen production system in a local town is investigated by the proposed analysis support system.

Keywords: MFM, GIS, Analysis Support System, Sustainability Development, Energy and Environmental System Evaluation.

1 Introduction

The sustainability of the energy and environment is a vital issue for the survival of entire humankind. Achieving sustainable development of energy and environment on a large scales from the world, a country, even to a local town requires the judicious use of various energy technologies[1]. In order to plan the suitable usage of energy technologies and to monitor the environmental consequences caused by energy technologies, an analysis support system is proposed in our study based on MFM and GIS. The MFM is a kind of semantic graphic modeling method not only to describe the hierarchical structure of objects from goal to function and to component, but also to represent the ''internal process'' of physical behavior by flow structure of mass, energy and information by using a series of standard symbols[3]. Traditionally, the MFM has been mainly applied in the process control areas ranging from signal validation, fault monitoring, fault diagnosis to procedure generation and human modeling, etc[4] [5][6] [7]. While in our study, the applications of MFM are extended to analyze and evaluate various energy and environmental systems from the aspects of sustainable development.

M.J. Smith, G. Salvendy (Eds.): Human Interface, Part II, HCII 2007, LNCS 4558, pp. 1148–1157, 2007.

Various energy and environmental systems have the same morphological character as that of the process plants because they have the flow networks of matter, energy and information in the process system, although the issues and application methods are rather different from the process plant. In this respect, one feature of energy and environmental system is the usage of geographical information. It should be also introduced in the study of energy and environmental systems.

The GIS is a computerized system designed to efficiently capture, store, update, manipulate, analyze, and display all forms of geographically referenced information which affect the energy and environment problems(especially various renewable energies) [10][11][12][13][14].

The combination of MFM and GIS can provide a comprehensive and easy-understanding interface to carry out the analysis and evaluation of energy and environmental system from many aspects of sustainable development.

2 Basic Issues for Evaluation of Sustainable Development

2.1 The Sustainability Energy and Environmental System

These days, "sustainable development" has become a popular keyword for the future of human civilization or national/international developments. It has been defined as "development that meets the needs of the present without compromising the ability of future generations to meet their own needs" [2]. A set of evaluation indicators called Energy Indicators (EI) have been proposed by IAEA for evaluating sustainable development at a national policy level from the three aspects of social, economical and environmental factors[1]. Although IAEA's EIs are rather national level, it is always necessary to set and think in mind such kinds of Evaluation Indicators (EIs) for the evaluation of energy and environmental systems of any level. The sustainable energy and environmental system such as metropolitan waste recycle system, natural gas delivery system, HTGR(High Temperature Gas Reactor) cogeneration system even the hydrogen economy society in future, are all hierarchical and complicated network system which is constituted of the various processes such as generation, conversion, transportation, distribution and consumption of various matters and energy in various forms. And there exist various interactions between the individual elemental processes. Those interactions are considered as constraints to the individual processes. The MFM seems to be an effective tool to represent the various processes and the constraints between them [8][9].

2.2 Aspects of Analysis and Evaluation

Generally speaking, when we consider the nature of sustainable energy and environmental system, there are two aspects of analysis and evaluation in common. The one is the concern of the basic technical feasibility of the scenario that the individual system would intend, and the other, the inter-comparison among various alternative scenarios from many aspects such as economic, environmental, energy efficiency even the exergy efficiency.

3 Prototype System Development

On the basis of the above-mentioned concept, an integrated design and simulation environment called as MFM&GIS Studio has been under development for the total analysis and evaluation of various energy and environmental systems.

3.1 Multilevel Flow Model -MFM

The basic functions, relationships and the connection rules, the computation rules based on connection are shown in Fig.1 and Table 1. In the computation rules columns, the "Fin", "Fout" and "Fs" mean the values of the matter or energy inflow, outflow and storage represented by one MFM function. Basically, there are five functions: "Source", "Sink", "Transport", "Storage" and "Balance" .The Balance connects more inflows and more outflows without storage, and the sum of the total inflow equals to that of total outflow.

In addition to those basic functions, the "Relation" is used to represent all the of MFM functions to realize the "Goal". Before constructing the Goal relations, the functions are grouped into "Networks", each of which describes a flow of matter, energy, or mixture of them. There are two relations in MFM: the "Achieve" relations between networks and goals which indicate that all functions in the network must work as designed to achieve a certain goal; the "Condition" relations between goals and functions which indicate that the function is available only after the goal is achieved.

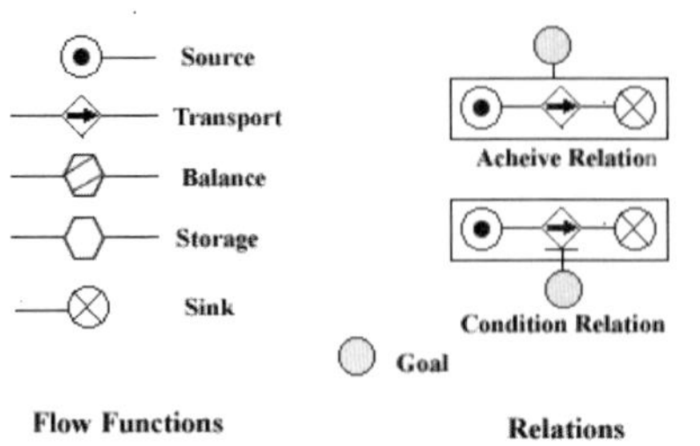

Fig. 1. The basic MFM functions and relations

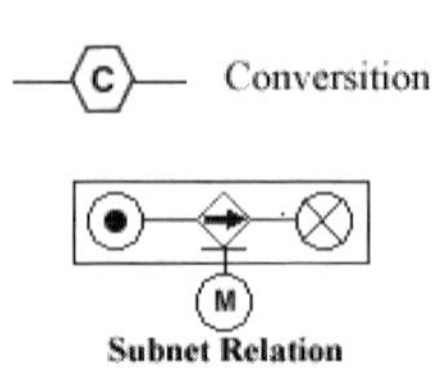

Fig. 2. New function and relation

Table 1. Basic connection rules and calculation relation

Type	Connection (Input/Output)	Computation rules
Source	Transport(0/1)	Fin=0,Fs=0
Sink	Transport(1/0)	Fout=0,Fs=0
Transport	Source, Balance, Storage and Sink(1/1)	Fin=Fout, Fs=0
Storage	Transport(1/1)	ΔFs=Fin-Fout
Balance	Transport (Multi/Multi)	Σ Fin= Σ Fout, Fs=0
Conversion	Transport (Multi/Multi)	Σ Fin= Σ Fout, Fs=0

By the definition of the authors' study, in order to express the matter flows and energy flows clearly, all the basic function of MFM only can be used to express one kind of matter or energy. However, when the conversions like chemical reactions happen, even though the total mass and energy are conservation before and after the reaction, the types of matter or the energy may change. So a new function called "Conversion" is introduced to express this situation as shown in Fig.2. The function-"Conversion" may have many inflows and many outflows of matter and energy at the same time while the total mass and energy are kept unchanged after the conversion process.

There is another problem in principle: that is the problem of where the boundary is. For the example of the MFM describing the hydrogen production from electrolysis of water, we can define the electricity is the boundary when you do not care about where the electricity comes from. But you want to know where the electricity comes from (the solar, the wind, hydro or even the conversional fossil fuels) because different power generation has different cost and environmental consequence. In this situation, the boundary of the system should be extended to the power generation.

So a new relationship "SubNet" is introduced to describe one MFM function by a sub network in detail, and carry out the account from the aspect of economic, environmental, energy even exergy. The "SubNet" function enables the users to express the flow processes in energy and environmental system by MFM in any degree of detail.

For ca. 20 years, MFM has been successfully applied for various process control issues based on reasoning algorithm [3][4][6]. However, to analyze the energy and environmental system, various types of computing would be needed on the basis of computation rules and the mathematic formula defined by users between the two or more MFM functions. As a simple example, Figs. 3 and 4 show MFM representation and the computation rules of an energy and environmental system which describes the import of LNG from overseas and distribution to local "LNG Satellite". In Fig.4, the computation rules based on the MFM connections can be understood by referring to Table.1, and the mathematic formula $y_t = f(x_{t-1})$ between the outflow of "Overseas LNG" and the inflow of the "LNG Satellite" is defined as $y_t = x_{t-1}$. In this simple example, if you take "t" as month, the import from overseas of current month equals to the consumption of the "LNG Satellite" of the previous month. (when computing from inverse direction, the "inverse function" of the formula should be used.) The result of this example is displayed in Table.2 where the initial data at "T0" is inputted, and at the first month-"T1", the value of the outflow of "Overseas LNG" and the value of the inflow of the "LNG Satellite" are inputted, the values of other flows at "T1" can be calculated by using the basic computation rules of MFM. And at the second month-"T2", the value of the outflow of the "Overseas LNG" equals to the value of the inflow of the "LNG Satellite" of "T1", and if you know the value of the inflow of the "LNG Satellite" of "T2", all of the value of other flows at "T2" can be calculated.

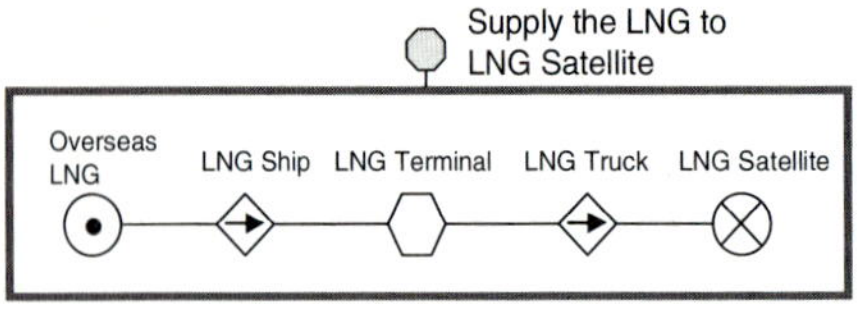

Fig. 3. A simple example for energy and environmental system analysis by MFM

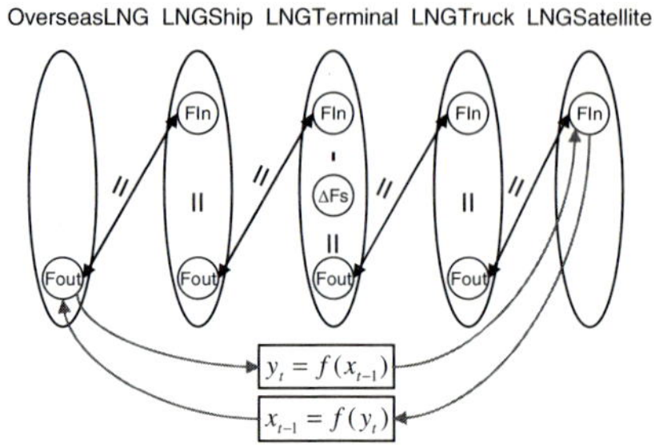

Fig. 4. Computation rules for the example system

Table 2. Computation result of the example system

	Overseas LNG	LNG Ship		LNG Terminal			LNG Truck		LNG Satellite
	Fout	Fin	Fout	Fin	Fs	Fout	Fin	Fout	Fin
T0	0	0	0	0	0.1*	0	0	0	0
T1	10*	10	10	10	0.3	9.8	9.8	9.8	9.8*
T2	9.8	9.8	9.8	9.8	0.2	9.9	9.9	9.9	9.9*
…									

*: the initial data inputted by the users

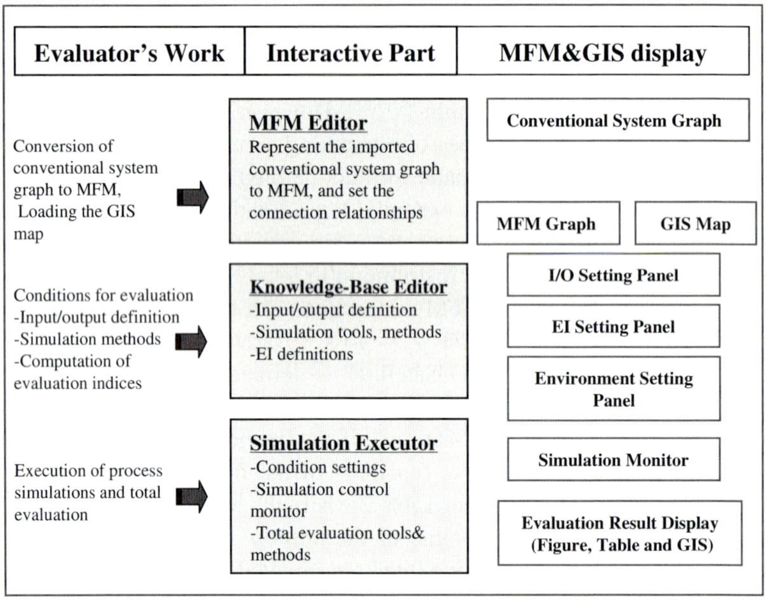

Fig. 5. System Design of MFM&GIS Studio

3.2 Geographical Information System -GIS

The GIS is a computer-based platform capable of handling spatial data that represent real world features such as urban settlements, roads, overhead lines, land type terrain, water features, wind distribution, solar intension, etc, in the form of digital maps and attributed geo-relational databases [10][11][12][13][14]. Modern GIS platforms are equipped with efficient tools able to handle enormous volumes of not only the common attribute data but also spatial data. The GIS has frequently been applied to resources and environment impact assessment, site identification as well as optimum path choice. In the developed MFM&GIS Studio, the GIS is also used to supply the necessary evaluation information and display the evaluation results visually and intuitively.

3.3 System Design

This system is developed by using the Microsoft Visual Studio .NET 2005 with an embeddable GIS component – MapObject2.3. And some necessary ArcGIS9.1 analysis tools (data management tool, spatial analysis tool, network analysis tool and 3D analysis tool and so on) are also used to preprocess the GIS data. As shown in Fig.5, the key components of this system are three interactive parts-"MFM Editor", "Knowledge-base Editor" and "Simulation Executor".

4 Evaluation Processes

Both Fig.5 and the process flow chart in Fig.6 show how these three parts work and what the user should do by the proposed support system in sequence. First, by using the "MFM Editor" the users can load conventional graph of the target energy and environmental system to be analyzed. Then the loaded conventional graph will be converted into the MFM graph based on the basic connection rules of MFM in terms of Goals-Means and Whole-Parts relationships, definition of connection relationships between the constructed MFM. In order to ensure the correctness of the constructed MFM, automatic checking will be carried out until the model become the correct ones. After that, the user defines the input/output items, the target map is loaded, and the necessary analysis is carried out to provide the spatial information for the input setting. After the I/O setting, the Flow Value (FV) computation is carried out. The "FV" not only includes the value of the inflow, outflow and storage at every MFM functions, but also includes the value of the other economic and environmental parameters such as capital cost, feedstock cost, CO_2 emission, etc. When the "FV" is calculated, the parameter statistics is conducted based on the FV result. The statistic can provide the total production, total cost, total CO_2 emission, total energy consumption, etc of the target system. After that, the users should define the other knowledge-base items such as Evaluation Indices (EI), the computing method of the EI, etc, by using the knowledge-base editor. Lastly execute the computation of EI by using the simulation executor, and the result of EI can be displayed intuitively by table or graph even on the GIS.

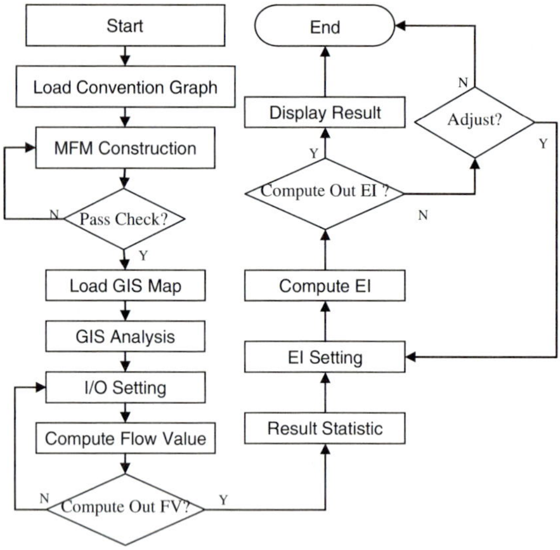

Fig. 6. Evaluation process of in the proposed analysis support system

5 Case Study

As a case study, the analysis system is applied to investigate the cost and the environmental impact of onsite water electrolysis hydrogen production system in the local city in Japan. The conventional system graph is converted into MFM representation based on the connection rules in "MFM Editor" as shown in Fig.7. The

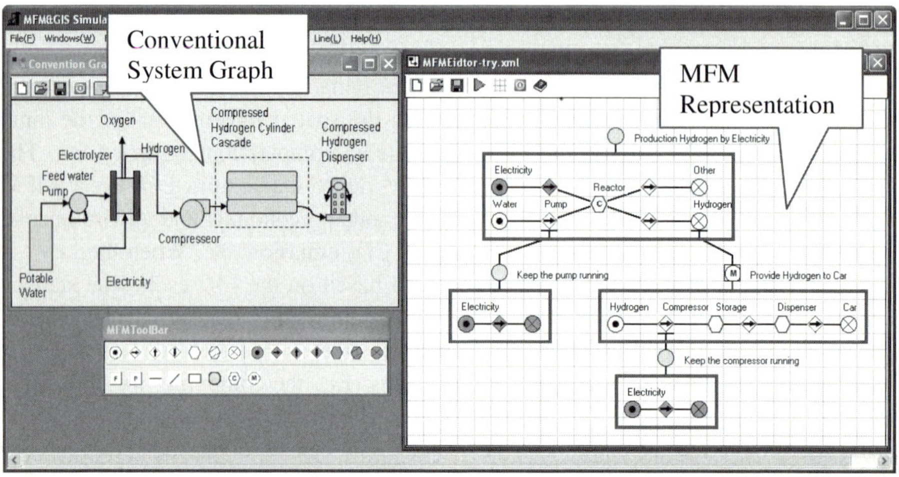

Fig. 7. Convert the conventional system graph to MFM in "MFM Editor"

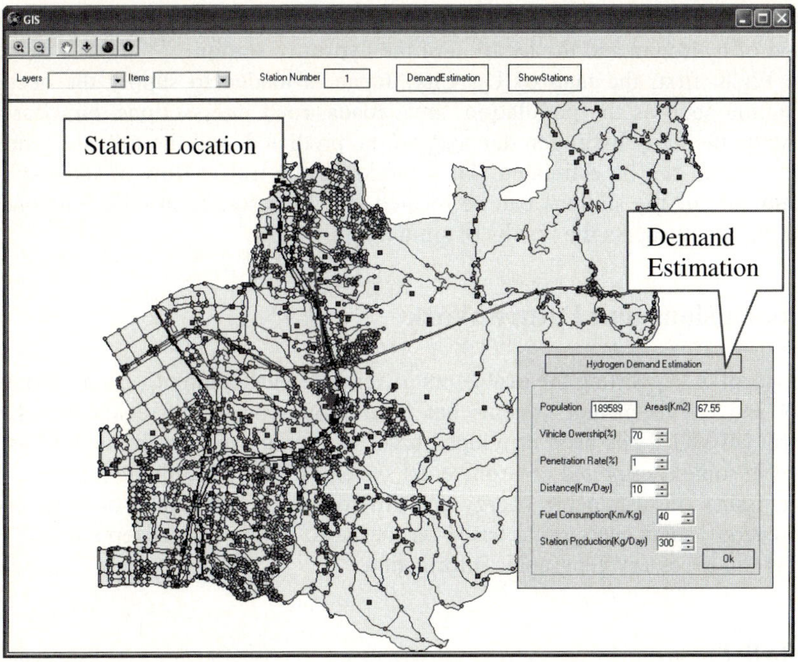

Fig. 8. Estimate hydrogen demand and locate optimum hydrogen station by GIS

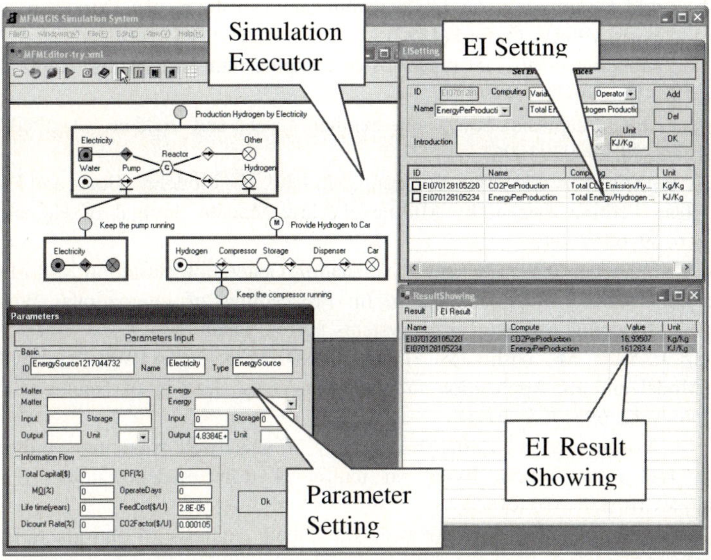

Fig. 9. Analysis result computation in "Simulation Executor"

GIS analysis is carried out to provide the necessary parameter values such as last demand of hydrogen and the location of the hydrogen station.[15] [16] [17] [18] As shown in the Fig.8, first, the map of Uji City, Japan is loaded to supply the necessary information such as the population, area, roads, road intersections, etc. Based on these data, the daily hydrogen demand can be predicted, and the necessary number of the stations can be estimated. Then based on the traffic flow of the roads, the optimum site of the stations can be located (the blue square in GIS component in Fig.8). Figure .9 shows the results of simulation.

6 Conclusions and Future Work

The concept of supporting the evaluators of versatile evaluation of various sustainable energy and environmental systems has been proposed. A new evaluation support platform MFMGIS has been developed based on the above concept, and it has been applied in some energy and environmental system. It will be further expanded for the various issues of sustainable energy and environment system such as wind energy, hydro energy, solar energy, etc based on the basic functional concepts of the target systems and necessary geographic information.

References

1. IAEA(International Atomic Energy Agency): Energy Indicators for Sustainable Development Guidelines and Methodologies. Vienna, Austria (2005)
2. WCED(World Commission on Environment and Development): Our Common Future. Oxford University Press, Oxford (1987)
3. Lind, M.: Modeling goals and functions of complex industrial plants. Applied Artificial Intelligence 18, 259–283 (1994)
4. Larsson, J.E.: Diagnosis based on explicit means-end models. Applied Artificial Intelligence 80, 29–93 (1996)
5. Zhou, Y.P., Yoshikawa, H., Wu, W., Yang, M., Ishii, H.: Modeling Goals and Functions of Micro Gas Turbine System by Multilevel Flow Models. Journal of Human Interface Society 6, 59–68 (2004)
6. Gofuku, A., Zheng, Y.Y.: Causality Explanation Generation Based on Multi-Level Flow Model for Operator Support System. In: Proceedings of International Workshop on Functional Modeling of Engineering Systems, Kyoto, Japan (2005)
7. OuYang, J., Yang, M., Yoshikawa, H., Zhou, Y.P.: Modeling of PWR Plant by Multilevel Flow Model and its Application in Fault Diagnosis. Journal of Nuclear Science and Technology 42, 695–705 (2005)
8. Liu, J.Q., Yoshikawa, H., Zhou, Y.P.: Study of visualized simulation and analysis of nuclear fuel cycle system based on multilevel flow model. Nuclear Science and Techniques 16, 358–370 (2005)
9. Zhou, Y.P., Yoshitake, J., Yoshikawa, H.: Trend Analysis of Introduction Effect of Deregulated Natural Gas Market in Japan By Multilevel Flow Models. In: The 8th Seminar of JSPS-MOE Core University Program on Urban Environment, Shanghai, P.R. China (2004)
10. Rodmana, L.C., Meentemeyer, R.K.: A geographic analysis of wind turbine placement in Northern California. Energy Policy 34, 2137–2149 (2006)

11. Baban, S.M.J., Parry, T.: Developing and applying a GIS-assisted approach to locating wind farms in the UK. Renewable Energy 24, 59–71 (2001)
12. Rylatt, M., Gadsden, S., Lomas, K.: Gis-based decision support for solar energy planning in urban environments. Computers, Environment and Urban Systems 25, 579–603 (2001)
13. Kaijuka, E.: GIS and rural electricity planning in Uganda. Journal of Cleaner Production 15, 203–217 (2007)
14. Yue, D.C., Yang, G.G.L.: Decision support system for exploiting local renewable energy sources: A case study of the Chigu area of southwestern Taiwan. Energy Policy 35, 383–394 (2007)
15. Huang, B., Liu, N., Chandramouli, A.: GIS supported Ant algorithm for the linear feature covering problem with distance constraints. Decision Support Systems 42, 1063–1075 (2006)
16. Committee of the handbook hydrogen and fuel cell, The handbook of hydrogen and fuel cell (in Japanese), Japan (2006)
17. Shayegan, S., Hart, D., Pearson, P., Joffe, D.: Analysis of the cost of hydrogen infrastructure for buses in London. Journal of Power Sources 157, 862–874 (2006)
18. Yanga, C., Ogdena, J.: Determining the lowest-cost hydrogen delivery mode. International Journal of Hydrogen Energy 32, 268–286 (2007)

Author Index

Printing: Mercedes-Druck, Berlin
Binding: Stein+Lehmann, Berlin

Lecture Notes in Computer Science

For information about Vols. 1–4508

please contact your bookseller or Springer

Vol. 4611: J. Indulska, J. Ma, L.T. Yang, T. Ungerer, J. Cao (Eds.), Ubiquitous Intelligence and Computing. XXIII, 1257 pages. 2007.

Vol. 4610: B. Xiao, L.T. Yang, J. Ma, C. Muller-Schloer, Y. Hua (Eds.), Autonomic and Trusted Computing. XVIII, 571 pages. 2007.

Vol. 4605: D. Papadias, D. Zhang, G. Kollios (Eds.), Advances in Spatial and Temporal Databases. X, 479 pages. 2007.

Vol. 4602: S. Barker, G.-J. Ahn (Eds.), Data and Applications Security XXI. X, 291 pages. 2007.

Vol. 4600: H. Comon-Lundh, C. Kirchner, H. Kirchner, Rewriting, Computation and Proof. XVI, 273 pages. 2007.

Vol. 4598: G. Lin (Ed.), Computing and Combinatorics. XII, 570 pages. 2007.

Vol. 4597: P. Perner (Ed.), Advances in Data Mining. XI, 353 pages. 2007. (Sublibrary LNAI).

Vol. 4596: L. Arge, C. Cachin, T. Jurdzinski, A. Tarlecki (Eds.), Automata, Languages and Programming. XVII, 953 pages. 2007.

Vol. 4595: D. Bošnački, S. Edelkamp (Eds.), Model Checking Software. X, 285 pages. 2007.

Vol. 4594: R. Bellazzi, A. Abu-Hanna, J. Hunter (Eds.), Artificial Intelligence in Medicine. XVI, 509 pages. 2007. (Sublibrary LNAI).

Vol. 4592: Z. Kedad, N. Lammari, E. Métais, F. Meziane, Y. Rezgui (Eds.), Natural Language Processing and Information Systems. XIV, 442 pages. 2007.

Vol. 4591: J. Davies, J. Gibbons (Eds.), Integrated Formal Methods. IX, 660 pages. 2007.

Vol. 4590: W. Damm, H. Hermanns (Eds.), Computer Aided Verification. XV, 562 pages. 2007.

Vol. 4589: J. Münch, P. Abrahamsson (Eds.), Product-Focused Software Process Improvement. XII, 414 pages. 2007.

Vol. 4588: T. Harju, J. Karhumäki, A. Lepistö (Eds.), Developments in Language Theory. XI, 423 pages. 2007.

Vol. 4587: R. Cooper, J. Kennedy (Eds.), Data Management. XIII, 259 pages. 2007.

Vol. 4586: J. Pieprzyk, H. Ghodosi, E. Dawson (Eds.), Information Security and Privacy. XIV, 476 pages. 2007.

Vol. 4585: M. Kryszkiewicz, J.F. Peters, H. Rybinski, A. Skowron (Eds.), Rough Sets and Intelligent Systems Paradigms. XIX, 836 pages. 2007. (Sublibrary LNAI).

Vol. 4584: N. Karssemeijer, B. Lelieveldt (Eds.), Information Processing in Medical Imaging. XX, 777 pages. 2007.

Vol. 4583: S.R. Della Rocca (Ed.), Typed Lambda Calculi and Applications. X, 397 pages. 2007.

Vol. 4582: J. Lopez, P. Samarati, J.L. Ferrer (Eds.), Public Key Infrastructure. XI, 375 pages. 2007.

Vol. 4581: A. Petrenko, M. Veanes, J. Tretmans, W. Grieskamp (Eds.), Testing of Software and Communicating Systems. XII, 379 pages. 2007.

Vol. 4580: B. Ma, K. Zhang (Eds.), Combinatorial Pattern Matching. XII, 366 pages. 2007.

Vol. 4578: F. Masulli, S. Mitra, G. Pasi (Eds.), Applications of Fuzzy Sets Theory. XVIII, 693 pages. 2007. (Sublibrary LNAI).

Vol. 4577: N. Sebe, Y. Liu, Y. Zhuang (Eds.), Multimedia Content Analysis and Mining. XIII, 513 pages. 2007.

Vol. 4576: D. Leivant, R. de Queiroz (Eds.), Logic, Language, Information and Computation. X, 363 pages. 2007.

Vol. 4575: T. Takagi, T. Okamoto, E. Okamoto, T. Okamoto (Eds.), Pairing-Based Cryptography. XI, 408 pages. 2007.

Vol. 4574: J. Derrick, J. Vain (Eds.), Formal Techniques for Networked and Distributed Systems – FORTE 2007. XI, 375 pages. 2007.

Vol. 4573: M. Kauers, M. Kerber, R. Miner, W. Windsteiger (Eds.), Towards Mechanized Mathematical Assistants. XIII, 407 pages. 2007. (Sublibrary LNAI).

Vol. 4572: F. Stajano, C. Meadows, S. Capkun, T. Moore (Eds.), Security and Privacy in Ad-hoc and Sensor Networks. X, 247 pages. 2007.

Vol. 4570: H.G. Okuno, M. Ali (Eds.), New Trends in Applied Artificial Intelligence. XXI, 1194 pages. 2007. (Sublibrary LNAI).

Vol. 4569: A. Butz, B. Fisher, A. Krüger, P. Olivier, S. Owada (Eds.), Smart Graphics. IX, 237 pages. 2007.

Vol. 4566: M.J. Dainoff (Ed.), Ergonomics and Health Aspects of Work with Computers. XVIII, 390 pages. 2007.

Vol. 4565: D.D. Schmorrow, L.M. Reeves (Eds.), Foundations of Augmented Cognition. XIX, 450 pages. 2007. (Sublibrary LNAI).

Vol. 4564: D. Schuler (Ed.), Online Communities and Social Computing. XVII, 520 pages. 2007.

Vol. 4563: R. Shumaker (Ed.), Virtual Reality. XXII, 762 pages. 2007.

Vol. 4562: D. Harris (Ed.), Engineering Psychology and Cognitive Ergonomics. XXIII, 879 pages. 2007. (Sublibrary LNAI).

Vol. 4561: V.G. Duffy (Ed.), Digital Human Modeling. XXIII, 1068 pages. 2007.

Vol. 4560: N. Aykin (Ed.), Usability and Internationalization, Part II. XVIII, 576 pages. 2007.

Vol. 4559: N. Aykin (Ed.), Usability and Internationalization, Part I. XVIII, 661 pages. 2007.

Vol. 4558: M.J. Smith, G. Salvendy (Eds.), Human Interface and the Management of Information, Part II. XXIII, 1162 pages. 2007.

Vol. 4557: M.J. Smith, G. Salvendy (Eds.), Human Interface and the Management of Information, Part I. XXII, 1030 pages. 2007.

Vol. 4553: J.A. Jacko (Ed.), Human-Computer Interaction, Part IV. XXIV, 1225 pages. 2007.

Vol. 4552: J.A. Jacko (Ed.), Human-Computer Interaction, Part III. XXI, 1038 pages. 2007.

Vol. 4551: J.A. Jacko (Ed.), Human-Computer Interaction, Part II. XXIII, 1253 pages. 2007.

Vol. 4550: J.A. Jacko (Ed.), Human-Computer Interaction, Part I. XXIII, 1240 pages. 2007.

Vol. 4549: J. Aspnes, C. Scheideler, A. Arora, S. Madden (Eds.), Distributed Computing in Sensor Systems. XIII, 417 pages. 2007.

Vol. 4548: N. Olivetti (Ed.), Automated Reasoning with Analytic Tableaux and Related Methods. X, 245 pages. 2007. (Sublibrary LNAI).

Vol. 4547: C. Carlet, B. Sunar (Eds.), Arithmetic of Finite Fields. XI, 355 pages. 2007.

Vol. 4546: J. Kleijn, A. Yakovlev (Eds.), Petri Nets and Other Models of Concurrency – ICATPN 2007. XI, 515 pages. 2007.

Vol. 4545: H. Anai, K. Horimoto, T. Kutsia (Eds.), Algebraic Biology. XIII, 379 pages. 2007.

Vol. 4544: S. Cohen-Boulakia, V. Tannen (Eds.), Data Integration in the Life Sciences. XI, 282 pages. 2007. (Sublibrary LNBI).

Vol. 4543: A.K. Bandara, M. Burgess (Eds.), Inter-Domain Management. XII, 237 pages. 2007.

Vol. 4542: P. Sawyer, B. Paech, P. Heymans (Eds.), Requirements Engineering: Foundation for Software Quality. IX, 384 pages. 2007.

Vol. 4541: T. Okadome, T. Yamazaki, M. Makhtari (Eds.), Pervasive Computing for Quality of Life Enhancement. IX, 248 pages. 2007.

Vol. 4539: N.H. Bshouty, C. Gentile (Eds.), Learning Theory. XII, 634 pages. 2007. (Sublibrary LNAI).

Vol. 4538: F. Escolano, M. Vento (Eds.), Graph-Based Representations in Pattern Recognition. XII, 416 pages. 2007.

Vol. 4537: K.C.-C. Chang, W. Wang, L. Chen, C.A. Ellis, C.-H. Hsu, A.C. Tsoi, H. Wang (Eds.), Advances in Web and Network Technologies, and Information Management. XXIII, 707 pages. 2007.

Vol. 4536: G. Concas, E. Damiani, M. Scotto, G. Succi (Eds.), Agile Processes in Software Engineering and Extreme Programming. XV, 276 pages. 2007.

Vol. 4534: I. Tomkos, F. Neri, J. Solé Pareta, X. Masip Bruin, S. Sánchez Lopez (Eds.), Optical Network Design and Modeling. XI, 460 pages. 2007.

Vol. 4533: F. Baader (Ed.), Term Rewriting and Applications. XII, 419 pages. 2007.

Vol. 4531: J. Indulska, K. Raymond (Eds.), Distributed Applications and Interoperable Systems. XI, 337 pages. 2007.

Vol. 4530: D.H. Akehurst, R. Vogel, R.F. Paige (Eds.), Model Driven Architecture- Foundations and Applications. X, 219 pages. 2007.

Vol. 4529: P. Melin, O. Castillo, L.T. Aguilar, J. Kacprzyk, W. Pedrycz (Eds.), Foundations of Fuzzy Logic and Soft Computing. XIX, 830 pages. 2007. (Sublibrary LNAI).

Vol. 4528: J. Mira, J.R. Álvarez (Eds.), Nature Inspired Problem-Solving Methods in Knowledge Engineering, Part II. XXII, 650 pages. 2007.

Vol. 4527: J. Mira, J.R. Álvarez (Eds.), Bio-inspired Modeling of Cognitive Tasks, Part I. XXII, 630 pages. 2007.

Vol. 4526: M. Malek, M. Reitenspieß, A. van Moorsel (Eds.), Service Availability. X, 155 pages. 2007.

Vol. 4525: C. Demetrescu (Ed.), Experimental Algorithms. XIII, 448 pages. 2007.

Vol. 4524: M. Marchiori, J.Z. Pan, C.d.S. Marie (Eds.), Web Reasoning and Rule Systems. XI, 382 pages. 2007.

Vol. 4523: Y.-H. Lee, H.-N. Kim, J. Kim, Y. Park, L.T. Yang, S.W. Kim (Eds.), Embedded Software and Systems. XIX, 829 pages. 2007.

Vol. 4522: B.K. Ersbøll, K.S. Pedersen (Eds.), Image Analysis. XVIII, 989 pages. 2007.

Vol. 4521: J. Katz, M. Yung (Eds.), Applied Cryptography and Network Security. XIII, 498 pages. 2007.

Vol. 4519: E. Franconi, M. Kifer, W. May (Eds.), The Semantic Web: Research and Applications. XVIII, 830 pages. 2007.

Vol. 4517: F. Boavida, E. Monteiro, S. Mascolo, Y. Koucheryavy (Eds.), Wired/Wireless Internet Communications. XIV, 382 pages. 2007.

Vol. 4516: L. Mason, T. Drwiega, J. Yan (Eds.), Managing Traffic Performance in Converged Networks. XXIII, 1191 pages. 2007.

Vol. 4515: M. Naor (Ed.), Advances in Cryptology - EUROCRYPT 2007. XIII, 591 pages. 2007.

Vol. 4514: S.N. Artemov, A. Nerode (Eds.), Logical Foundations of Computer Science. XI, 513 pages. 2007.

Vol. 4513: M. Fischetti, D.P. Williamson (Eds.), Integer Programming and Combinatorial Optimization. IX, 500 pages. 2007.

Vol. 4511: C. Conati, K. McCoy, G. Paliouras (Eds.), User Modeling 2007. XVI, 487 pages. 2007. (Sublibrary LNAI).

Vol. 4510: P. Van Hentenryck, L. Wolsey (Eds.), Integration of AI and OR Techniques in Constraint Programming for Combinatorial Optimization Problems. X, 391 pages. 2007.

Vol. 4509: Z. Kobti, D. Wu (Eds.), Advances in Artificial Intelligence. XII, 552 pages. 2007. (Sublibrary LNAI).